COLLEGE MATHEMATICS
A Graphing Calculator Approach

COLLEGE MATHEMATICS
A Graphing Calculator Approach

Ruric Wheeler
Samford University

Karla Neal
Louisiana State University

Roseanne Hofmann
Montgomery County Community College

JOHN WILEY & SONS, INC.
New York   Chichester   Brisbane   Toronto   Singapore

ACQUISITIONS EDITOR   Ruth Baruth
DEVELOPMENT EDITOR   Madalyn Stone
MARKETING MANAGER   Carl White
SENIOR PRODUCTION EDITOR   Cathy Ronda
DESIGN   A Good Thing, Inc.
MANUFACTURING MANAGER   Mark Cirilla
INDEXER   Julie S. Palmeter
ILLUSTRATION EDITOR   Sigmund Malinowski
ELECTRONIC ILLUSTRATION   Radiant Illustration & Design

This book was set in Times Roman by Progressive Information Technologies and printed and bound by R. R. Donnelley/Crawfordville.
The cover was printed by Leigh Press

Library of Congress Cataloging in Publication Data:
Wheeler, Ruric E., 1923–
    College mathematics: a graphing calculator approach / Ruric Wheeler, Karla Neal, Roseanne Hofmann.
        p.   cm.
    Includes bibliographical references (p.   –   ).
    ISBN 0-471-05720-7 (cloth : alk. paper)
        1. Mathematics—Data processing.   2. Graphic calculators.
    I. Neal, Karla.   II. Hofmann, Roseanne Salvatore.   III. Title.
    QA76.95.W47   1996
    510—dc20   95-41657
                                                                        CIP
Printed in the United States of America

10   9   8   7   6   5   4   3   2   1

PREFACE

Our primary goal in writing this book has been to create a book that is easy to read and comprehend by the average student, and yet also contains quality mathematics for students majoring in business management, social sciences, and life sciences. To accomplish this seemingly impossible goal, we have made extensive use of a calculator approach.

Most students will use throughout their lives the knowledge they gain to solve problems by using a calculator. We are assuming that every student using this book will have access to an inexpensive graphing calculator, for both class activities and homework.

The graphing calculator is revolutionizing the teaching and learning of mathematics. The time that used to be spent on pencil-and-paper computations can now be spent learning concepts and applications. Many people regard the graphing calculator as a tool, and it is true that it saves a substantial amount of time in performing matrix operations, drawing graphs, and other such operations. However, the graphing calculator is much more than a tool.

USE OF GRAPHING CALCULATOR

The graphing calculator can assist a teacher in using an investigative, exploratory approach to the teaching of mathematics. In this book, we use the graphing calculator in four ways.

1. **To investigate or explore new ideas, which we then validate by algebraic procedures.**
2. **To develop or define concepts algebraically, which we then support with numerical or graphical procedures on the calculator.**
3. **To mix calculating and algebraic procedures in solving problems.**
4. **To work a problem numerically or graphically.**

The teacher can choose to emphasize or not to emphasize any of these four uses of the graphing calculator.

We have attempted to introduce material independent of the specific calculator to be used. However, introductions to the use of the TI-82, the Sharp EL-9200, and the Casio 7700 GB are found in Appendix A. Whenever it is necessary to list calculator steps, these steps will be given for the TI-82.

DISTINGUISHING FEATURES

In addition to emphasizing a calculator approach, how else does this book differ from the typical book on this subject? The answer is summarized as follows.

Problem Solving   The ability to analyze problems and translate them into mathematical language is an important and, for the average student, a difficult skill. This book begins with a section on problem-solving and encourages the student to use problem solving procedures throughout. Thought processes and algorithmic procedures are introduced to improve problem solving.

Computer Use   Although no specific reference is given to computer use, this book can be used with as much emphasis on the computer as desired. In most places in this book the words ''graphing calculator'' can be replaced with the word ''computer,'' and the given statement will still be valid; that is, the computer can be used to carry out the same functions as the graphing calculator.

Ample Review   The beginning of the book contains ample review of the basic topics of algebra and the use of graphing calculators for students not adequately prepared.

Future Usability   Students often ask, ''Why should I study mathematics?'' In this book, we attempt to answer this question in two ways. First, every section of this book contains numerous applications, classified as ''Business and Economics'' or ''Social and Life Sciences.'' Then, CPA, CMA, and actuarial exam questions are scattered throughout the exercise sets to indicate that knowledge of the material is important in professional exams. These are denoted by the word ''EXAM.''

STUDENT-BASED APPROACH

First and foremost, this book is student oriented. A distinct effort is made to base each new concept on the student's prior experience or prior knowledge from the textbook. The book has a more intuitive than formal approach. To further help the student, the following features and teaching aids are included.

Gradual Development   This book begins with very easy material, giving the student the opportunity to develop mathematical maturity, problem solving procedures, and calculator skills. Once the student has developed this maturity, the material gradually increases in difficulty. This increase is so gradual that most students never realize it is happening. At the end of the course, students will be solving problems as complicated as those found in any textbook for this subject.

Practice Problems   Practice problems are scattered throughout each section as a check for understanding. That is, immediately after a topic is discussed, a simple problem is given that uses the material of the discussion. These practice problems allow students to evaluate for themselves whether they have understood the content before studying new material.

End-of-Chapter Tests   At the end of each chapter is a short test on the chapter material to assist students in evaluating their comprehension. All the answers to these problems are given in the back of the book.

Notes   Throughout the text, you will find the term ''NOTE,'' directed to the student and used to draw attention to an unusual idea or subject.

Exercises   The exercise sets contain more than 5000 problems that are arranged within each exercise set in order of difficulty from easy to medium to challenging. The problems are usually arranged in matching pairs with the answers to odd-numbered problems in the back of the book.

SUPPLEMENTS

Teacher's Manual to Accompany *College Mathematics: A Graphing Calculator Approach*
0471-13646-8

Testbank to Accompany *College Mathematics: A Graphing Calculator Approach*
0471-13648-4

Computerized Testbank IBM 3.5 to Accompany *College Mathematics: A Graphing Calculator Approach*
0471-13645-X

Computerized Testbank MAC to Accompany *College Mathematics: A Graphing Calculator Approach*
0471-13644-1

Student Solutions Manual to Accompany *College Mathematics: A Graphing Calculator Approach*
0471-13647-6

ACKNOWLEDGMENTS

We would like to thank the following individuals who made valuable contributions to this book:

Phil Beckman, Black Hawk College
Beth Borel, University of Southeastern Louisiana
Lou Hoelzle, Bucks County Community College
Stefan Hui, San Diego State University
Gary S. Itzkowitz, Rowan College of New Jersey
Rose Marie Kinik, Sacred Heart University
Anne Landry, Dutchess Community College
Frank A. Michells, Middle Tennessee State University
August J. Zarcone, College of DuPage

The production of this book has been a team effort, involving valuable contributions from many people, including Wiley staff members Ruth Baruth, Mathematics Editor, Madalyn Stone, Senior Developmental Editor, Cathy Ronda, Senior Production Editor, Sigmund Malinowski, Illustration Editor.

We are greatly indebted to our secretaries and student assistants, especially Jill Bailey.

We encourage all teachers to send us suggestions as they use this book. In many ways we, too, are exploring the use of the graphing calculator to enhance our methods of teaching.

RURIC WHEELER
KARLA NEAL
ROSEANNE HOFMANN

CONTENTS

LEARNING GUIDE FOR STUDENTS

Before you start reading, take a moment to look over the next few pages, which provide an overview of some of the book's built-in learning devices. Becoming familiar with these unique features can make your march through the material a lot easier.

Clarity

The most important characteristic of this book is clarity. This book is truly student oriented. It achieves a rare combination of outstanding exposition and sound mathematics. Note the specific instructions for the reduction of an augmented matrix.

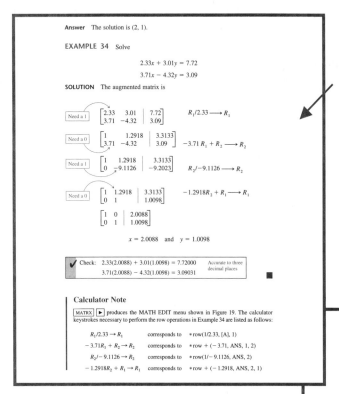

Answer The solution is (2, 1).

EXAMPLE 34 Solve

$$2.33x + 3.01y = 7.72$$
$$3.71x - 4.32y = 3.09$$

SOLUTION The augmented matrix is

$$\begin{bmatrix} 2.33 & 3.01 & | & 7.72 \\ 3.71 & -4.32 & | & 3.09 \end{bmatrix} \qquad R_1/2.33 \longrightarrow R_1$$

$$\begin{bmatrix} 1 & 1.2918 & | & 3.3133 \\ 3.71 & -4.32 & | & 3.09 \end{bmatrix} \qquad -3.71\,R_1 + R_2 \longrightarrow R_2$$

$$\begin{bmatrix} 1 & 1.2918 & | & 3.3133 \\ 0 & -9.1126 & | & -9.2023 \end{bmatrix} \qquad R_2/-9.1126 \longrightarrow R_2$$

$$\begin{bmatrix} 1 & 1.2918 & | & 3.3133 \\ 0 & 1 & | & 1.0098 \end{bmatrix} \qquad -1.2918R_2 + R_1 \longrightarrow R_1$$

$$\begin{bmatrix} 1 & 0 & | & 2.0088 \\ 0 & 1 & | & 1.0098 \end{bmatrix}$$

$$x = 2.0088 \quad \text{and} \quad y = 1.0098$$

✓ Check: $2.33(2.0088) + 3.01(1.0098) = 7.72000$ Accurate to three decimal places
$3.71(2.0088) - 4.32(1.0098) = 3.09031$ ∎

Calculator Note

[MATRX] [▶] produces the MATH EDIT menu shown in Figure 19. The calculator keystrokes necessary to perform the row operations in Example 34 are listed as follows:

$R_1/2.33 \to R_1$	corresponds to	*row(1/2.33, [A], 1)
$-3.71R_1 + R_2 \to R_2$	corresponds to	*row + (-3.71, ANS, 1, 2)
$R_2/-9.1126 \to R_2$	corresponds to	*row(1/-9.1126, ANS, 2)
$-1.2918R_2 + R_1 \to R_1$	corresponds to	*row + (-1.2918, ANS, 2, 1)

6. Since all elements in the bottom row are nonnegative, this matrix gives the maximum solution. Let $r = 0$ and $s = 0$, giving

$$1 \cdot x = 110$$
$$1 \cdot y = 60$$
$$1 \cdot P = 11,680$$

The maximum value is $P = \$11,680$ when $x = 110$, $y = 60$. ∎

EXAMPLE 17 Use a calculator to maximize

$$P = 100x + 150y$$

subject to

$$x, y \geq 0$$
$$2.11x + 3.02y \leq 13.28$$
$$0.06x + 4.31y \leq 13.05$$

SOLUTION The simplex tableau is

x	y	r	s	P	
2.11	3.02	1	0	0	13.28
0.06	4.31	0	1	0	13.05
-100	-150	0	0	1	0

Verify that 4.31 (second row, second column) is the pivot element. Then use

*row(1/4.31, [A], 2)
*row + (-3.02, ANS, 2, 1)
*row + (150, ANS, 2, 3)

As a second step verify that the element in the first row and first column is the pivot element. Now use

*row(1/2.067958, ANS, 1)
*row + (-.01392111, ANS, 1, 2)
*row + (97.911833, ANS, 1, 3)

getting

Calculator Exposition

Detailed steps for calculator computations are provided for students. Following these steps the student learns to use the calculator. This enables the teacher to spend more time on the mathematics of a given subject.

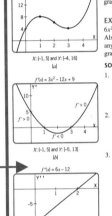

$F(x) = \int_0^x t^2\,dt$
$F(x)$

Calculator Note

You should be aware that

$$F(x) = \int_0^x t^2\,dt$$

is really the antiderivative of x^2, namely $x^3/3$. On your calculator, evaluate $F(x)$ at several points. Then, on paper, plot the points and connect them to make a graph as in Figure 33(a). Compare your graph with the graph of $y = x^3/3$ in Figure 33(b). Do these two graphs appear to be the same? We have plotted the points $F(-3) = -9$, $F(-2) = -\frac{8}{3}$, $F(-1) = -\frac{1}{3}$, $F(0) = 0$, $F(1) = \frac{1}{3}$, $F(2) = \frac{8}{3}$, $F(3) = 9$.

Practice Problem 2 Use the technique described in the preceding Calculator Note to plot the graph of

$$F(x) = \int_0^x t^3\,dt$$

Then plot the graph of $y = x^4/4$. Are they the same graph?

ANSWER After plotting $F(x)$ for $x = -2, -1, 0, 1, 2$ to Figure 34(a) and comparing this to the calculator graph of 34(b), the graphs appear to be the same.

Calculator Support of Mathematical Developments

Sometimes a calculator picture is just what the student needs to comprehend a mathematical exposition.

$f(x) = x^3 - 6x^2 + 9x + 4$

X: [-1, 5] and Y: [-4, 16]
(a)

$f'(x) = 3x^2 - 12x + 9$

$f' > 0$
$f' > 0$
$f' < 0$

X: [-1, 5] and Y: [-5, 13]
(b)

$f''(x) = 6x - 12$

X: [-1, 3] and Y: [-15, 5]
(c)

NOTE: Where $f''(x) = 0$ or where it does not exist a *possible* inflection point occurs. However, if $f''(x)$ does not change sign, there is no inflection point. The graph of the second derivative will help you see what is occurring at the point.

EXAMPLE 18 Determine where the graph of the function $f(x) = x^3 - 6x^2 + 9x + 4$ is concave up, is concave down, and has points of inflection. Also, determine where f is increasing, where it is decreasing, and the values of any relative extrema. Use critical values, possible inflection points, and the graphs of f, f', and f'' to find the information.

SOLUTION

1. Find the first derivative:

    $$f'(x) = 3x^2 - 12x + 9 = 3(x^2 - 4x + 3) = 3(x - 3)(x - 1)$$

    From this we see that the critical values are $x = 1$ and $x = 3$.
2. Find the second derivative:

    $$f''(x) = 6x - 12 = 6(x - 2)$$

    From this we determine that the only possible inflection point is at $x = 2$.
3. Using the information found in (1) and (2), we look at the graphs of the function and the derivatives in Figure 28 to obtain the following:

    Increasing on $(-\infty, 1) \cup (3, \infty)$ where $f' > 0$.
    Decreasing on $(1, 3)$ where $f' < 0$.
    Relative maximum at $(1, 8)$ where $f' = 0$ and changes from positive to negative.
    Relative minimum at $(3, 4)$ where $f' = 0$ and changes from negative to positive.
    Concave down on $(-\infty, 2)$ where $f'' < 0$.
    Concave up on $(2, \infty)$ where $f'' > 0$.
    Inflection point at $(2, 6)$ where $f'' = 0$ and concavity changes. ∎

Practice Problem 3 Draw the graph of $f(x) = x^3 - 12x^2$. Find all extrema and points of inflection.

ANSWER The graph is shown in Figure 29.

$f'(x) = 3x^2 - 24x = 3x(x - 8)$. Relative maximum at $(0, 0)$ and relative minimum at $(8, -256)$.
$f''(x) = 6x - 24$. Inflection point at $(4, -128)$.

Combination of Algebra and the Graphing Calculator

The mix algebraic procedures and calculator use helps improve your understanding.

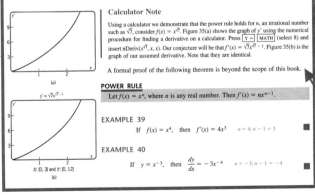

Calculator Note

Using a calculator we demonstrate that the power rule holds for n, an irrational number such as $\sqrt{7}$, consider $f(x) = x^{\sqrt{7}}$. Figure 35(a) shows the graph of y' using the numerical procedure for finding a derivative on a calculator. Press [Y =] [MATH] (select 8) and insert nDeriv($x^{\sqrt{7}}, x, x$). Our conjecture will be that $f'(x) = \sqrt{7}x^{\sqrt{7}-1}$. Figure 35($b$) is the graph of our assumed derivative. Note that they are identical.

A formal proof of the following theorem is beyond the scope of this book.

POWER RULE

Let $f(x) = x^n$, where n is any real number. Then $f'(x) = nx^{n-1}$.

EXAMPLE 39
  If $f(x) = x^4$, then $f'(x) = 4x^3$ $n = 4; n - 1 = 3$ ∎

EXAMPLE 40
  If $y = x^{-3}$, then $\dfrac{dy}{dx} = -3x^{-4}$ $n = -3; n - 1 = -4$ ∎

Investigative Mathematics

New ideas you can explore and investigate even when proofs are too complicated to present.

Future Usability

Students are interested in the mathematics they will use in the future. The exercise sets contain problems from professional examinations, denoted by *Exam*. Application problems, many of them taken from advanced textbooks in business, economics, and the social and life sciences, are abundant in this book.

Applications

An unusually large number of applications of the mathematics of a section to real-world problems illustrates the importance of a section.

18. *Exam.*   What is the inverse of the following matrix?

$$\begin{bmatrix} 1 & 1 & 1 \\ 0 & 1 & 1 \\ 0 & 0 & 1 \end{bmatrix}$$

(a) $\begin{bmatrix} 1 & 0 & 0 \\ 1 & 1 & 0 \\ 1 & 1 & 1 \end{bmatrix}$ (b) $\begin{bmatrix} -1 & 0 & 0 \\ -1 & 1 & 0 \\ 1 & -1 & 1 \end{bmatrix}$

(c) $\begin{bmatrix} 1 & -1 & 1 \\ 0 & 1 & -1 \\ 0 & 0 & 1 \end{bmatrix}$ (d) $\begin{bmatrix} 1 & -1 & 0 \\ 0 & 1 & -1 \\ 0 & 0 & 1 \end{bmatrix}$

(e) $\begin{bmatrix} 1 & -1 & 0 \\ 0 & 1 & 1 \\ 0 & 0 & 1 \end{bmatrix}$

Applications (Business and Economics)

19. *Investments.*   An investment club has $100,000 invested in bonds. Type A bonds pay 8% interest, whereas type B bonds pay 10%. How much money is invested in each type if the club receives $9360 in interest?

20. *Break-Even Analysis.*   A country club has been studying its cost of operation and has found the monthly cost to be $C = \$6000 + \$15x$, where x represents the number of members. A service company states that by remodeling, it can provide a similar service for a monthly cost of $C = \$10,500 + \$10x$. Use the inverse of a matrix to find the break-even point, and find the number of members necessary to undertake the remodeling.

21. *Manufacture.*   A company makes two kinds of tires. Type I requires 2 hours on machine A and 1 hour on machine B. Type II requires 3 hours on A and 2 on B. How many tires can be manufactured in a week if machine A is in use 40 hours and machine B 25 hours?

Applications (Social and Life Sciences)

22. *Herbicides.*   Three herbicides are available to kill weeds, grass, and vines. One gallon of one herbicide contains 1 unit of chemical A, which kills weeds, and units of chemical B, which kills grass. One gallon the second herbicide contains 2 units of chemical and 3 units of chemical C, which kills vines. One gallon of the third herbicide contains 3 units of chemical A and 4 units of chemical C. It is desired to spread 10 units of chemical A, 16 units of chemical B, and 20 units of chemical C. How many gallons of each herbicide should be purchased?

23. *Diet.*   A dietician wants to combine two foods as the main part of a meal to get 2600 calories and 70 units of vitamin C. Each ounce of food I contains 200 calories and 10 units of vitamin C. Each ounce of food II contains 400 calories and 5 units of vitamin C. How many ounces of each food would be required to obtain the desired number of calories and the desired units of vitamin C?

24. *Polling.*   A company has been engaged to make 4850 phone polls and 2650 home visits. The company has two teams of pollsters. Team I can make 60 phone polls and 40 home polls a day, whereas team II can make 70 phone polls and 30 home polls each day. How many days should each team be scheduled in order to complete the engagement?

Review Exercises

25. Discuss the solutions of the following.

(a) $\left[\begin{array}{cc|c} 1 & 0 & 3 \\ 0 & 0 & 2 \end{array}\right]$ (b) $\left[\begin{array}{cc|c} 1 & 0 & 0 \\ 0 & 1 & 1 \end{array}\right]$

(c) $\left[\begin{array}{cc|c} 1 & 0 & 3 \\ 0 & 0 & 0 \end{array}\right]$ (d) $\left[\begin{array}{ccc|c} 1 & 0 & 0 & 2 \\ 0 & 1 & 0 & 3 \\ 0 & 0 & 0 & 0 \end{array}\right]$

(e) $\left[\begin{array}{ccc|c} 1 & 0 & 0 & 4 \\ 0 & 1 & 0 & 2 \\ 0 & 0 & 0 & 3 \end{array}\right]$ (f) $\left[\begin{array}{ccc|c} 1 & 0 & 0 & 1 \\ 0 & 1 & 0 & 0 \\ 0 & 0 & 1 & 0 \end{array}\right]$

26. Determine whether or not the following systems have common solutions, and solve those that have solutions.

(a) $x + y + z = 5$
$2x - 3y + 2z = 4$
$4x - y + 4z = 10$

(b) $x - y - z = -3$
$3x - 4y - 2z = -11$
$5x - 6y - 4z = -17$

Problem Solving Strategies

Polya's suggestions for solving problems, introduced in Chapter 0, are helpful throughout the book. The concept of estimating is a recommended procedure for all calculator computations.

SOLUTION   Let's estimate that we make 8 small-car axles (at a profit of $10 · 8) and 6 large-car axles (at a profit of $15 · 6) for a total profit of

$$P = 10 \cdot 8 + 15 \cdot 6$$
$$= \$170 \qquad \text{See Table 1}$$

TABLE 1

Description	Small Axles		Large Axles
Number of each (guess)	8		6
Profit from guess	$10 \cdot 8$	+	$15 \cdot 6 = \$170$
Number of each (variable)	x		y
Profit	$10x$	+	$15y$
Limitation on number of hours			
Hours needed (machine A)	$1x$ hours	+	$2y$ hours ≤ 24
Hours needed (machine B)	$1.5x$ hours	+	$1y$ hours ≤ 24

We now replace our estimate of 8 small-car axles with x small-car axles and our estimate of 6 large-car axles with y large-car axles. The profit-objective function becomes

$$P = \$10x + \$15y$$

Each small-car axle is on machine A for 1 hour, so x small-car axles would utilize machine A for x hours. Each large-car axle uses machine A for 2 hours, or y large-car axles use machine A for $2y$ hours.

$$x + 2y \leq 24 \qquad \text{Machine A operates up to 24 hours each day.}$$

Likewise,

$$1.5x + 1y \leq 24 \qquad \begin{array}{l}\text{Each small axle needs 1.5 hours and each} \\ \text{large axle needs 1 hour on machine B.}\end{array}$$

$$x \geq 0$$
$$y \geq 0 \qquad \text{Cannot have a negative number of axles.}$$

The feasible region is plotted in Figure 13, and the corner points are (0, 0), (0, 12), (12, 6), (16, 0). At the corner points, the values of the objective function $P = 10x + 15y$ are

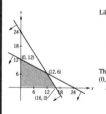

$$(0, 0): \quad P = 10(0) + 15(0) = 0$$
$$(0, 12): \quad P = 10(0) + 15(12) = \$180$$
$$(16, 0): \quad P = 10(16) + 15(0) = \$160$$
$$(12, 6): \quad P = 10(12) + 15(6) = \$210$$

A matrix with the same number of rows and columns is called a **square matrix**. The following are square matrices:

$$\begin{bmatrix} 2 & 1 \\ 3 & -3 \end{bmatrix}_{2 \times 2} \quad \text{and} \quad \begin{bmatrix} 3 & 7 & 2 \\ 4 & -1 & 6 \\ 0 & 2 & 1 \end{bmatrix}_{3 \times 3}$$

Likewise, a matrix consisting of a single row of elements is called a **row matrix** or a **vector**. Similarly, a matrix consisting of a single column is called a **column matrix** or a **vector**.

EXAMPLE 4   The matrix $\mathbf{A} = [1 \quad 3 \quad 7]$ is called a row matrix or vector with elements $a_{11} = 1$, $a_{12} = 3$, and $a_{13} = 7$. The matrix

$$\mathbf{B} = \begin{bmatrix} 7 \\ -1 \\ 4 \end{bmatrix}$$

is called a column matrix or vector with elements $b_{11} = 7$, $b_{21} = -1$, and $b_{31} = 4$. ∎

Calculator Note

To enter the matrix

$$\begin{bmatrix} 4.1 & 7.8 & 5.6 \\ 3.2 & 5.1 & -4.3 \\ -2.1 & 2.5 & 3.7 \end{bmatrix}$$

on a calculator first press ⬚MATRX ▶ ▶ to display the MATRX EDIT menu. Choose a storage position for the matrix [Figure 1(a)]. We choose 1: [A] and press ⬚ENTER. As shown in Figure 1(b), we insert the size of our matrix, 3 × 3, and input the elements of the matrix. To return to the home screen press ⬚2nd ⬚QUIT. To see the matrix displayed press ⬚MATRX, select matrix [A], and press ⬚ENTER. The matrix is displayed in Figure 1(c).

Just as there are times when we need to know whether two equations are equal, there are times when we need to know whether two matrices are equal.

DEFINITION: EQUAL MATRICES

Two matrices are **equal matrices** if and only if they have the same size (i.e., the same number of rows and columns) and if all corresponding elements are equal.

Calculator Notes

Found throughout the book are suggestions or instructions on using a calculator for a particular problem.

In general, a matrix can have m rows and n columns and be written as

Third
column

$$\begin{bmatrix} a_{11} & a_{12} & a_{13} & \cdots & a_{1n} \\ a_{21} & a_{22} & a_{23} & \cdots & a_{2n} \\ \vdots & \vdots & \vdots & & \vdots \\ a_{m1} & a_{m2} & a_{m3} & \cdots & a_{mn} \end{bmatrix} \quad \text{Second row}$$

The size of this matrix is $m \times n$, where m represents the number of rows and n the number of columns. The objects in the matrix are called **elements** of the matrix. Usually, the elements are numbers. Notice that the first part of the subscript notation for each element gives the row in which the element lies, and the second part gives the column location. For example, a_{23} is the element in the second row and third column. The general $m \times n$ matrix is sometimes written as $[a_{ij}]$, where a_{ij} represents each element.

Color Utilization

Color is used for instructional purposes throughout the book.

Special Helps

Special helps (set apart by a second color) are given to assist students through critical stages of a problem. These aids provide additional explanation of steps in the solution of a problem.

Visual Aids

Students will find an unusually large number of diagrams to assist in understanding an explanation.

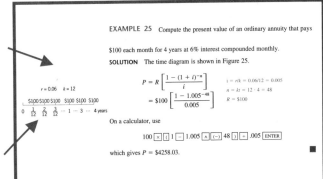

EXAMPLE 25   Compute the present value of an ordinary annuity that pays $100 each month for 4 years at 6% interest compounded monthly.

SOLUTION   The time diagram is shown in Figure 25.

$r = 0.06$ $k = 12$

$100 \; 100 \; 100 \quad 100 \; 100 \; 100$

$0 \quad \frac{1}{12} \quad \frac{2}{12} \quad \frac{3}{12} \quad \cdots \quad 1 \quad \cdots \quad 3 \quad \cdots \quad 4 \text{ years}$

$$P = R\left[\frac{1 - (1 + i)^{-n}}{i}\right]$$

$$= \$100\left[\frac{1 - 1.005^{-48}}{0.005}\right]$$

$i = r/k = 0.06/12 = 0.005$
$n = kt = 12 \cdot 4 = 48$
$R = \$100$

On a calculator, use

$100 \; ⬚× \; ⬚(\; 1 \; ⬚- \; 1.005 \; ⬚^ \; ⬚(-) \; 48 \; ⬚) \; ⬚÷ \; .005 \; ⬚ENTER$

which gives $P = \$4258.03$. ∎

Overviews

Each section begins with an overview, which enables the authors to make the material more personal and less formal in approach. The overviews summarize the goals of a section and relate the section to previous studied material or real-world experiences.

10.1 FINDING DERIVATIVES OF EXPONENTIAL FUNCTIONS

OVERVIEW   This is an important section because exponential functions are used extensively in applications. Growth and decay models are based on exponentials. Exponential functions are used by economists to study the growth rate of the money supply. They are needed by businesspeople to study the rate of change in sales, by biologists to study the rate of growth of organisms, and by social scientists to study the rate of population growth. In this section we

* Find the derivative of exponential functions
* Use the chain rule with exponential functions
* Study graphing techniques
* Introduce applications of exponential functions

Summary

At the end of each section is a summary of key ideas or formulas in the section.

SUMMARY

The derivative of a function f, at a number x in its domain, is defined to be

$$f'(x) = \lim_{h \to 0} \frac{f(x + h) - f(x)}{h}$$

Since the difference quotient

$$\frac{[f(x + h) - f(x)]}{h}$$

is the slope of the line joining $(x, f(x))$ and $((x + h), f(x + h))$, the derivative may be interpreted as the slope of a tangent line to the curve. Economists often refer to the derivative of a function as the marginal value of that function.

Using the Calculator as a Tool

The most common use of the graphical calculator is as a tool. Throughout much of the book a graphical picture of the problem is most helpful to students.

We can now complete the problem.

$$\int_0^4 x^2 \sqrt{1 + 2x}\, dx = \frac{x^2(1 + 2x)^{3/2}}{3} \Big|_0^4 - \frac{2x(1 + 2x)^{5/2}}{15} \Big|_0^4$$
$$+ \int_0^4 \frac{(1 + 2x)^{5/2}}{15} \cdot 2\, dx$$
$$= \left[\frac{x^2(1 + 2x)^{3/2}}{3} - \frac{2x(1 + 2x)^{5/2}}{15} + \frac{2(1 + 2x)^{7/2}}{105} \right]\Big|_0^4$$
$$= \left[\frac{16(9)^{3/2}}{3} - \frac{8(9)^{5/2}}{15} + \frac{2(9)^{7/2}}{105} \right] - \left[\frac{2}{105} \right]$$
$$= 144 - \frac{648}{5} + \frac{1458}{35} - \frac{2}{105}$$
$$= \frac{5884}{105} \approx 56.04$$

✓ Calculator Check: fnInt($x \wedge 2 \sqrt{}$ (1 + 2x), x, 0, 4) ≈ 56.04

Draw the graph of $y = x^2\sqrt{1 + 2x}$ with your graphing calculator and shade the area being found (see Figure 2). ■

$y = x^2\sqrt{1 + 2x}$

X: [0, 4] and Y: [0, 50]

Common Errors

Throughout the book a word of warning is given to the student about common errors made on the material being studied.

COMMON ERROR   Many times students will differentiate like this:

$$y(x) = (3x^2 + 4x)^5, \quad \text{so} \quad y'(x) = 5(3x^2 + 4x)^4$$

What is wrong with this? The answer is $5(3x^2 + 4x)^4(6x + 4)$.

CHAPTER 0

Linear Models and an Introduction to the Graphing Calculator

Application of mathematical concepts has become increasingly important in business, economics, the social sciences, the natural sciences, industry, and service. The ability to read technical material, use critical thinking skills and current technology, and provide solutions to problems is expected in most professions. The mathematical techniques presented in this book will help you meet the real needs of business and industry. In this chapter, we emphasize the use of linear models. To assist us in forming models to represent real situations, we introduce a number of problem-solving techniques for converting verbal statements into mathematical notation. We discuss how a picture of the linear mathematical model can be shown as a graph on a coordinate system.

Most of the mathematical concepts in this chapter are reinforced with applications involving a graphing calculator. These applications along with Appendix A provide the background needed for understanding the material in this book.

0.1 FOUNDATIONS: SETS AND THE REAL NUMBER SYSTEM

OVERVIEW   We begin this section with a review of the basic concepts and notation of sets. Additional topics on set theory will be reviewed in Chapter 5 on probability. Numbers are used every day, but we seldom think about what kind of numbers they are. In this section we discuss real numbers (the kind we use in this book) and introduce the number line to aid your understanding of the concept of real numbers. Properties of the four fundamental operations (addition, subtraction, multiplication, and division) are presented to help in solving problems throughout this book. A quick review of these operations is presented for integers and rational numbers. In this section we consider the following topics:

- Characteristics of sets
- Characteristics of real numbers; the number line
- Properties of the basic operations on real numbers

Review of Set Notation

We describe a set as follows.

DEFINITION: SET

A **set** is a collection of objects or symbols possessing a property that enables one to determine whether a given object is in the collection.

A set is *well defined* if we can determine whether an object is in a given set. For example, the set of all people who have been president of the United States is a well-defined set. However, the set of all people who are tall is not well defined, since the term tall is subjective. If an object is in a set, we say that it is an **element** of the set. We can say an object is *an element of* or *is a member of* a set if it is in the set.

DEFINITION: ELEMENT OF A SET

$x \in A$ means that x is an element of set A. $x \notin A$ means that x is not an element of set A.

It is often possible to specify a set by listing its members within braces. This method of describing a set is called the **tabulation method** (sometimes called the **roster method**). The set of counting numbers less than 10 can be written as $A = \{1, 2, 3, 4, 5, 6, 7, 8, 9\}$; furthermore, we can say that $4 \in A$, but $11 \notin A$. A set remains the same regardless of the order of its elements. For example, $\{1, 2, 3\}$ is the same set as $\{2, 1, 3\}$, $\{3, 2, 1\}$, $\{3, 1, 2\}$, $\{1, 3, 2\}$, or $\{2, 3, 1\}$.

DEFINITION: EQUAL SETS

Two sets are **equal sets** if and only if they contain exactly the same elements. $A = B$ if and only if A and B have exactly the same elements.

EXAMPLE 1   Glenda, Cathie, and Marcia are the only counselors in the Admissions Office. They constitute the set $A = \{$Glenda, Cathie, Marcia$\}$. Thus we can say Cathie $\in A$, but Linda $\notin A$.   ■

Sometimes sets have so many elements that it is tedious, difficult, or even impossible to tabulate them. Sets like this can be indicated by a descriptive statement or a rule. The following sets are well specified by a descriptive statement without tabulating the members: the counting numbers less than 10, the even numbers less than 1000, the past presidents of the United States, and the football teams in Pennsylvania.

The difficulty in tabulating sets can be minimized by using **set-builder notation**, which encloses within braces a letter or symbol representing an element of the set followed by a qualifying description of the element. For example, let A represent the set of counting numbers less than 10; then

$$A = \{n \mid n \text{ is a counting number less than } 10\}$$

This notation is read "the set of all elements n such that n is a counting number less than 10." Notice that the vertical line is read "such that."

Frequently, three dots (called an ellipsis) are used to indicate the omission of terms. The set of even counting numbers less than 100 may be written as $\{2, 4, 6, \ldots, 98\}$. This notation saves time in tabulating elements of large sets, but it can be ambiguous unless the set has been specified completely by another description. For example, $\{2, 4, \ldots, 16\}$ could be written as $\{2, 4, 8, 16\}$ or $\{2, 4, 6, 8, 10, 12, 14, 16\}$.

An ellipsis is also used to indicate that a sequence of elements continues indefinitely. For example, consider the set of natural, or counting, numbers $\{1, 2, 3, 4, 5, \ldots\}$. The set of natural numbers is an example of an **infinite set**, described informally as one that contains an unlimited number of elements. In contrast, a **finite set** contains either no elements or a natural number of elements that can be specified by a natural number.

DEFINITION: EMPTY SET

A set that contains no elements is called the **empty set** or **null set** and is denoted by either $\varnothing$ or $\{\ \}$.

COMMON ERROR   Do not use $\varnothing$ as 0. Note that $\varnothing = \{\ \}$ but $\varnothing \neq \{\varnothing\}$. By placing $\varnothing$ within braces, the set is no longer empty.

One relationship between two sets such as $A = \{1, 3, 5, 7\}$ and $B = \{1, 2, 3, 4, 5, 6, 7, 8\}$ is described by the term subset.

DEFINITION: SUBSET

Set A is a **subset** of set B, denoted by $A \subseteq B$, if and only if each element of A is an element of B.

For example, if $P = \{1, 4, 7\}$ and $Q = \{4, 7, 1\}$, then $P \subseteq Q$ because each element of P is an element of Q. Similarly, $Q \subseteq P$ and $P \subseteq P$ because every set is a subset of itself.

Since all dogs are animals, the set of dogs is a subset of the set of animals. Moreover, the set of dogs is a proper subset of the set of animals, since there are animals that are not dogs.

DEFINITION: PROPER SUBSET

Set A is a **proper subset** of set B, denoted by $A \subset B$, if and only if each element of A is an element of B and there is at least one element in B that is not an element of A.

The proper subsets of $\{a, b, c\}$ can be listed as $\{a\}$, $\{b\}$, $\{c\}$, $\{a, b\}$, $\{a, c\}$, $\{b, c\}$, and $\emptyset$. That is, $\{a, c\} \subset \{a, b, c\}$. Likewise, $\emptyset \subset \{a, b, c\}$. *The null set is a subset of any set.* The sets listed are all the proper subsets of $\{a, b, c\}$. Moreover, $\{a, b, c\}$ is a subset but is not a proper subset of itself.

If a discussion is limited to a fixed set of objects and if all elements to be discussed are contained in this set, then this overall set is called the **universal set**, or simply the **universe**. Venn diagrams, such as the one shown in Figure 1, are very useful in representing the relationship between a set and the universe. The universal set can be regarded as the region bounded by the rectangle, and the set under consideration as the region bounded by the circle (or some other closed figure within the rectangle). $x \in A$ means that x is a point in the circular region.

In Figure 2, A is a proper subset of B; that is, $A \subset B$. The region outside set A and inside the universe U represents the complement of A.

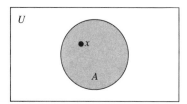

Figure 1

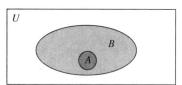

Figure 2

DEFINITION: COMPLEMENT OF A SET

The **complement** of set A, denoted by A', is the set of all elements in the universe that are not in set A.

For example, if the universe consists of all mutual funds, and if A consists of all mutual funds that have interest rates exceeding 8%, then the set of all mutual funds bearing interest of 8% or less is the complement of A. In Figure 3, the shaded region outside the circle but inside the rectangle represents A', whereas U consists of everything inside the rectangle: both A and A'.

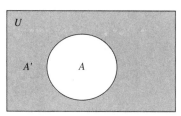

Figure 3

EXAMPLE 2 If the universe $U = \{a, b, c, d\}$ and $A = \{b, c\}$, find A'.

SOLUTION A' consists of the elements that are in U but not in A; thus, $A' = \{a, d\}$. ∎

Practice Problem 1 Given: $U = \{1, 2, 3, 4, 5\}$, $A = \{1, 3, 5\}$, $B = \{2, 3, 4\}$. Answer the following questions.

(a) Is $A \subset B$?   (b) Is $A \subset U$?
(c) Is $B \subset U$?   (d) Find A'.
(e) Find B'.

ANSWER

(a)   No   (b) Yes
(c)   Yes   (d) {2, 4}
(e)   {1, 5}

Characteristics of Real Numbers

The numbers that we will be using in this book are called **real numbers**. Certain subsets of real numbers, such as {1, 2, 3, . . . }, are called the **counting numbers** or **natural numbers**. These numbers along with 0 make up the **whole numbers**. The **integers** consist of {. . . , − 4, − 3, − 2, − 1, 0, 1, 2, 3, 4, . . . }. **Rational numbers** are numbers that can be expressed as quotients of integers, such as $\frac{3}{4}$ and $-\frac{7}{1}$. Some rational numbers can be simplified to integers (such as $\frac{8}{4} = 2$), and some rational numbers can be expressed as fractions (such as $\frac{3}{7}$). When a rational number is written as a decimal, it is either terminating, such as 2.13 (which is $\frac{213}{100}$), or it is repeating, such as 2.333. . . (which is $2\frac{1}{3}$). **Irrational numbers** are usually classified as numbers that are not rational; that is, they cannot be expressed as a quotient of two integers. Figure 4 illustrates the relationship between these sets of numbers. Note that a number can belong to more than one set of numbers. For example, − 2 is a negative integer, an integer, a rational number, and a real number.

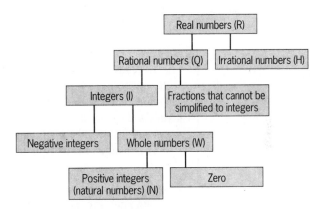

Figure 4

To represent real numbers, a **real number line** can be constructed in the following manner. Mark an arbitrarily chosen point as 0, called the **origin**. Then measure equal segments to the right to determine points labeled 1, 2, 3, 4, . . . and to the left to determine points labeled − 1, − 2, − 3, − 4, . . . , as illustrated in Figure 5. By this process, you set up a one-to-one correspondence between a subset of points on the line and the integers. The positive integers extend to the right of zero, and the negative integers extend to the left of zero. Points representing fractions such as $2\frac{1}{2}$ and points representing irrational numbers such as $-\sqrt{10}$ are also shown in Figure 5. Actually, there is a one-to-one correspondence between the real numbers and the points on the real number line (to each real number there corresponds one and only one point on the line, and vice versa). The

Figure 5

real number corresponding to a point on the line is called the **coordinate** of the point, and the point is called the **graph** of a number.

Basic Operations on Real Numbers

The basic mathematical operations on real numbers are addition, subtraction, multiplication, and division. Certain properties of these operations are important. With a calculator, add 42 + 33. Then clear and add 33 + 42. Note that you get the same answer, 75. You have demonstrated that the order in which numbers are added does not affect the sum. The fact that $a + b = b + a$ is called the **commutative property of addition**. Again with a calculator, multiply two numbers in one order and then reverse the order. Did you get the same answer? You have illustrated the **commutative property of multiplication**, $ab = ba$.

With a calculator, add 41 + 27. Then add 33 to this sum: (41 + 27) + 33. Can you find the sum of 41 + (27 + 33)? That is, to 41 add the sum of 27 + 33 (see Figure 6). Did you get the same answer each time? You have illustrated that when adding more than two numbers you may arrange the numbers in any order for addition. In general, we state the **associative property of addition** as $a + (b + c) = (a + b) + c$. Similarly, the **associative property of multiplication** is stated as $a(bc) = (ab)c$.

Again with a calculator, verify that 53(42 + 7) has the same value as $53 \cdot 42 + 53 \cdot 7$. Hence, $53(42 + 7) = 53 \cdot 42 + 53 \cdot 7$. In general, $a(b + c) = ab + ac$ and $(b + c)a = ba + ca$, which is the **distributive property of multiplication over addition**.

Zero is called the **additive identity** because $a + 0 = 0 + a = a$. Likewise, 1 is the **multiplicative identity** because $1 \cdot a = a \cdot 1 = a$. With a calculator, we can show that $42 + (-42) = 0$ (see Figure 7). These two numbers are **additive inverses**. In general, a and $-a$ are additive inverses. Likewise, a and $1/a$ (where $a \neq 0$) are **multiplicative inverses** because $a(1/a) = 1$.

We summarize the basic properties of numbers as follows.

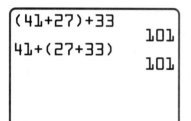

Figure 6

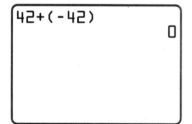

Figure 7

PROPERTIES OF NUMBERS

$a + b = b + a$	Commutative property of addition
$a + (b + c) = (a + b) + c$	Associative property of addition
$ab = ba$	Commutative property of multiplication
$a(bc) = (ab)c$	Associative property of multiplication
$a(b + c) = ab + ac$	Distributive property of multiplication over addition
$0 + a = a + 0 = a$	Additive identity
$1 \cdot a = a \cdot 1 = a$	Multiplicative identity
$a + (-a) = 0$	Additive inverse
$a\left(\dfrac{1}{a}\right) = 1 \quad (a \neq 0)$	Multiplicative inverse

Practice Problem 2

Find $4(3 + 2)$ in two ways to demonstrate the distributive property of multiplication over addition.

ANSWER

$$4(3 + 2) = 4(5) = 20$$

$$4(3) + 4(2) = 12 + 8 = 20$$

$$\text{Thus, } 4(3 + 2) = 4(3) + 4(2)$$

Calculator Note

Did you have trouble getting -42 into your calculator? On a graphing calculator, use the minus key $\boxed{(-)}$, not the $\boxed{-}$ key, which is the binary operation of subtraction. (See Figure 7.)

There are other operations in arithmetic besides addition, subtraction, multiplication, and division. Absolute value is one of them. Unlike the four basic operations $(+, -, \cdot, \div)$, which are binary, absolute value operates on only one element. The **absolute value** of a number x, denoted by $|x|$, is always positive or 0.

DEFINITION: ABSOLUTE VALUE

$$|x| = \begin{cases} x, & \text{if } x \text{ is greater than or equal to } 0 \\ -x, & \text{if } x \text{ is less than } 0 \end{cases}$$

This definition states that the absolute value of a nonnegative number is that number itself, and the absolute value of a negative number is its opposite (making it positive).

EXAMPLE 3

$$|-5| = -(-5) = 5$$

$$|15| = 15$$ ■

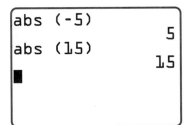

Figure 8

Calculator Note

Figure 8 illustrates absolute value using a calculator. To enter $|-5|$, press $\boxed{\text{2nd}}$ $\boxed{\text{ABS}}$ $\boxed{(-)}$ 5 $\boxed{\text{ENTER}}$. Note that you do not enter the absolute value bars and that the absolute value bars are not shown in the window.

The operations of addition and multiplication on the set of real numbers are performed by using the procedures for signed numbers at the top of page 8.

COMMON ERROR   The term $-x$ is read as the "additive inverse of x" or the "opposite of x"; it should not be read as "negative x" because x can be either a positive or negative number; that is, $-x$ is positive if x is negative. For instance, if $x = -5$, then $-x = 5$.

EXAMPLE 4   Evaluate $-(2x - 4x - 3)$ when $x = 1$.

SOLUTION $-(2 \cdot 1 - 4 \cdot 1 - 3) = -(2 - 4 - 3) = -(-5) = 5$ ■

DEFINITION: ADDITION AND MULTIPLICATION OF SIGNED NUMBERS

a) To add two numbers with the same sign, add their absolute values and keep their common sign.

$$-2 + -5 = -(2 + 5) = -7$$

b) To add two numbers with unlike signs, find the difference in their absolute values. Use the sign of the number whose absolute value is the largest.

$$-2 + 5 = (5 - 2) = 3 \quad \text{and} \quad -7 + 4 = -(7 - 4) = -3$$

c) The product (or quotient) of two numbers with unlike signs is negative.

$$-3 \cdot 2 = -6 \quad \text{and} \quad 4 \div -2 = -2$$

d) The product (or quotient) of two numbers with like signs is positive.

$$-4 \cdot -3 = 12 \quad \text{and} \quad -20 \div -5 = 4$$

NOTE: Always do multiplications and divisions in the order in which they occur in the problem, before additions and subtractions. This is the order in which most calculators perform the operations. However, be aware that parentheses alter that order.

EXAMPLE 5

(a) $|-3| = 3$ and $|17| = 17$
(b) $-(-6) = 6$
(c) $5 + (-9) = -4$ 9 is greater than 5.
(d) $-3 + 5 = 2$ 5 is greater than 3.
(e) $-5 + (-6) = -(5 + 6) = -11$ Like signs
(f) $(-3) \cdot (5) = -15$ Unlike signs
(g) $(-4) \cdot (-8) = 32$ Like signs

Practice Problem 3 Find $|2 - 5| - |5 - 4|$.

ANSWER   2

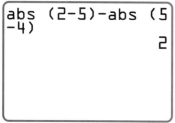

Figure 9

Calculator Note

When using a calculator for computing absolute values, parentheses are sometimes required. Figure 9 shows Practice Problem 3 done on a calculator with the following steps: $\boxed{\text{2nd}}\ \boxed{\text{ABS}}\ \boxed{(}\ 2\ \boxed{-}\ 5\ \boxed{)}\ \boxed{-}\ \boxed{\text{2nd}}\ \boxed{\text{ABS}}\ \boxed{(}\ 5\ \boxed{-}\ 4\ \boxed{)}$.

Properties of rational numbers expressed as fractions are summarized as follows.

PROPERTIES OF FRACTIONS

(a) $\dfrac{cp}{cq} = \dfrac{p}{q}$ $(c \neq 0, q \neq 0)$

(b) $\dfrac{p}{q} + \dfrac{r}{q} = \dfrac{(p + r)}{q}$ $(q \neq 0)$

(c) $\dfrac{p}{q} + \dfrac{r}{s} = \dfrac{ps}{qs} + \dfrac{qr}{qs} = \dfrac{(ps + qr)}{qs}$ $(q \neq 0, s \neq 0)$

(d) $\dfrac{p}{q} \cdot \dfrac{r}{s} = \dfrac{pr}{qs}$ $(q \neq 0, s \neq 0)$

For mixed fractions, recall that $2\frac{1}{3}$ means $2 + \frac{1}{3}$; thus, $2\frac{1}{3} = \frac{6}{3} + \frac{1}{3} = \frac{7}{3}$.

EXAMPLE 6

(a) $\dfrac{-10}{15} = \dfrac{-2 \cdot 5}{3 \cdot 5} = \dfrac{-2}{3}$

(b) $\dfrac{2}{7} + \dfrac{3}{7} = \dfrac{5}{7}$

(c) $\dfrac{2}{5} + \dfrac{3}{7} = \dfrac{2 \cdot 7}{5 \cdot 7} + \dfrac{5 \cdot 3}{5 \cdot 7} = \dfrac{14 + 15}{35} = \dfrac{29}{35}$

(d) $\dfrac{-2}{7} \cdot \dfrac{-5}{6} = \dfrac{-2(-5)}{7 \cdot 6} = \dfrac{10}{42} = \dfrac{5}{21}$ ∎

Many of the numbers in this chapter are expressed in decimal form. Use a calculator to review the rules for placing decimal points in the addition, subtraction, multiplication, and division of numbers in decimal form. (See Exercise Set 0.1, Exercises 31–33.)

Subtraction and division are defined in terms of addition and multiplication in the following manner.

DEFINITION: SUBTRACTION AND DIVISION

(a) Subtraction: $p - q = p + (-q)$

(b) Division: $p \div q = p \cdot \dfrac{1}{q} = \dfrac{p}{q}$ $(q \neq 0)$

(c) $\dfrac{p}{q} \div \dfrac{r}{s} = \dfrac{p}{q} \cdot \dfrac{s}{r}$ $(q, s, r \neq 0)$

EXAMPLE 7

(a) $3 - (-5) = 3 + [-(-5)] = 3 + 5 = 8$

(b) $-17 - 4 = -17 + (-4) = -21$

(c) $-3 \div 6 = -3 \cdot \left(\dfrac{1}{6}\right) = -\dfrac{3}{6} = -\dfrac{1}{2}$

(d) $-\dfrac{2}{3} \div \dfrac{5}{7} = \left(-\dfrac{2}{3}\right)\left(\dfrac{7}{5}\right) = -\dfrac{14}{15}$ ∎

Practice Problem 4 Simplify $-0.1\left(\dfrac{0.3}{2} - \dfrac{0.75}{3}\right)$ as much as possible.

ANSWER 0.01

Division is sometimes defined as the inverse of multiplication. That is, $a/b = c$ if and only if $a = bc$. For example, $15/5 = 3$ because $15 = (5)(3)$. This definition helps us to understand division involving 0.

$\dfrac{0}{6}$ is 0 because $(6)(0) = 0$.

$\dfrac{0}{0}$ is indeterminant because 0 times any number is 0.

$\dfrac{6}{0}$ is undefined because there is no number x such that $0(x) = 6$.

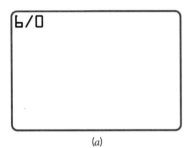

(a)

(b)

Figure 10

Calculator Note

On a graphing calculator, 0/0 and 6/0 result in an error (see Figure 10).

EXAMPLE 8 (Depreciation) Many items that we own do not have the same value today that they had a year ago. For example, an automobile may not be worth as much today as it was worth 2 years ago. In business terms, we say the value has **depreciated**. One method of depreciation, called the sum-of-digits method, provides an excellent application of the use of fractions. Suppose that an asset has a life of 6 years and a value of $1000. To find the amount of depreciation using the sum-of-digits method, we add the natural numbers representing each of the 6 years:

$$1 + 2 + 3 + 4 + 5 + 6 = 21$$

Then the depreciation for the first year is determined as 6/21 of $1000; for the second year, 5/21 of $1000; for the third year, 4/21 of $1000, until finally the depreciation for the sixth year is 1/21 of $1000.

Year	Fraction · Value	Allowable Depreciation
1	$\dfrac{6}{21} \cdot \$1000$	$285.71
2	$\dfrac{5}{21} \cdot \$1000$	$238.10
3	$\dfrac{4}{21} \cdot \$1000$	$190.48
4	$\dfrac{3}{21} \cdot \$1000$	$142.86

5	$\dfrac{2}{21} \cdot \$1000$	$95.24
6	$\dfrac{1}{21} \cdot \$1000$	$\dfrac{\$47.62}{\$1000.01 \approx \$1000.00}$

■

SUMMARY

In this section we reviewed the basic concepts and notation of sets and the characteristics of real numbers, including the representation of real numbers on a real number line. The properties of the four basic mathematical operations (addition, subtraction, multiplication, and division) were also illustrated by examples both with and without a graphing calculator.

Exercise Set 0.1

1. Let A be the set of all counting numbers less than 16. Which of the following statements are true and which are false?
   (a) $11 \in A$ (b) $81 \in A$
   (c) $\{1, 6, 21\} \subseteq A$
   (d) $\{1, 2, 3, \ldots, 15\} \subseteq A$
   (e) $14 \in A$ (f) $0 \in A$

2. List within braces the members of the sets below.
   (a) The counting numbers less than or equal to 16
   (b) The set of even counting numbers
   (c) The set of women presidents of the United States

3. Given $U = \{2, 4, 6, 8, 10\}$ and $A = \{2, 6, 10\}$.
   (a) Find A'.
   (b) Is $A \subset U$?
   (c) Is $A' \subset A$?

4. Which of the following sets are well defined?
   (a) The set of great baseball players
   (b) The set of beautiful horses
   (c) The set of students in this class
   (d) The set of counting numbers smaller than a million

5. Classify the following numbers as natural numbers (N), integers (I), rational numbers (Q), irrational numbers (H), and/or real numbers (R). (For example, -5 is I, Q, and R.)
   (a) $\sqrt{3}$ (b) $\dfrac{1}{2}$
   (c) 0.125 (d) 0
   (e) 0.17 (f) π
   (g) $\sqrt[3]{3}$ (h) $-1/\pi$
   (i) $0.343434\ldots$ (j) $0.10110111\ldots$

6. State whether each statement is true (T) or false (F).
   (a) Every integer is a rational number.
   (b) Every integer is a natural number.
   (c) Every rational number is a natural number.
   (d) No natural numbers are rational numbers.
   (e) No irrational numbers are natural numbers.

7. Perform the indicated operations using a calculator.
   (a) $\dfrac{13.1 \div (0.12 \cdot -4.7) - 3.621}{13.1 - 7.6(-0.17) - 16.3}$
   (b) $\dfrac{-7.2(-4.3) - 0.01 \div -2.5}{4.61 - 2.73 \cdot 0.1}$

8. Simplify the following expressions. (Remember that multiplication and division must be performed before addition and subtraction.)
   (a) $12 \div 4 - 8 \div 2$
   (b) $(-75 \div 3)5$
   (c) $\dfrac{-9 + (-4)(-3) \div 6}{2 \cdot 2 - (-3)}$
   (d) $(-9 + 2 \cdot 2 \cdot 3) \cdot (-3)$
   (e) $56 \div 7(8)$

Rewrite the expressions below without absolute value notation.

9. $|-11|$ 10. $|x|$

11. $|-(-11)|$ 12. $|-x|$

13. $|3 - 5 \cdot 2|$ 14. $-|-5| - |-2| - (-14)$

Perform the following operations.

15. $\left(-\dfrac{4}{3}\right)\left(\dfrac{1}{4}-\dfrac{2}{5}\right)$

16. $\dfrac{3}{ab}+\dfrac{5}{a}$

17. $-8\left(\dfrac{2}{3}-4\right)$

18. $-11\dfrac{8}{9}-3\dfrac{5}{9}$

19. $\dfrac{6-5}{4-4}$

20. $\dfrac{17}{-3-(2-5)}$

21. $-\left(\dfrac{3}{8}-\dfrac{7}{8}\right)$

22. $\dfrac{4}{3}+\left(\dfrac{7}{9}\right)\cdot\left(-\dfrac{3}{5}\right)$

23. $\dfrac{2}{3}\cdot\dfrac{0}{1}\cdot\dfrac{13}{2}$

24. $-\dfrac{4}{3}\cdot\dfrac{1}{3}\cdot-\dfrac{1}{2}$

25. $\left(-\dfrac{3}{2}+\dfrac{17}{2}\right)\left(-\dfrac{2}{5}+\dfrac{3}{8}\right)$

26. $\left(\dfrac{4}{3}\div-\dfrac{7}{8}\right)\left(-\dfrac{1}{2}-\dfrac{3}{4}\right)$

27. $6xy+(-4xy)$

28. $3\dfrac{1}{2}+\left(-7\dfrac{1}{4}\right)$

29. $-(-3+4)$

30. $-(7-10)$

31. Use a calculator to perform the following additions and subtractions. Note carefully what happens in each step. Then formulate a rule for placing decimal points in additions and subtractions.
    (a) $14.613 - 0.0001 + 2.6$
    (b) $8.0001 + 126.3 - 4.216$
    (c) $0.0021 - 4.67 + 360.1$
    (d) State your rule.

32. Use a calculator to perform the following multiplications. Note carefully what happens. Then formulate a rule for placing decimal points in multiplications.
    (a) $(0.021)(3.6)$
    (b) $(42)(3.006)$
    (c) $(6.2)(0.21)(3.76)$
    (d) State your rule.
    (e) Does your rule work for $(0.5)(0.02)$ on your calculator?

33. Use a calculator to perform the following divisions. Note carefully what happens. Then formulate a rule for placing decimal points in divisions.

(a) $14 \div 0.002$
(b) $16.2 \div 8.1$
(c) $0.0024 \div 0.012$
(d) State your rule.
(e) Does your rule work for all cases on the calculator?

34. If a universe consists of all counting numbers less than 20, and $A = \{1, 3, 5, 9, 11, 13\}$, find the complement A' and draw a Venn diagram showing both A and A'.

35. For universe $\{A, B, C, D, E, F, G\}$ find H' if H is $\{B, C, D, E\}$. Show both H' and H on a Venn diagram.

Applications (Business and Economics)

36. *Discounts.*   Compute the current selling price for an article with a list price of $50.00 and with a trade discount of 40%. [Discount = 0.40 × (List price); Current price = List price − Discount].

37. *Depreciation.*   Use the sum-of-digits method to find the yearly depreciation of an automobile valued at $15,000 with a life of 10 years.

38. *Depreciation.*   Use the sum-of-digits method to find the yearly depreciation of an office computer valued at $5000 with a life of 8 years.

Applications (Social and Life Sciences)

39. *Muscle Efficiency.*   Sometimes algebraic expressions are used to describe reactions as they occur in time. For example,

$$E = \frac{1 - 0.2t}{2 + t}$$

has been used to describe muscle efficiency E during maximum contraction over the time t that the muscle is contracted. Find a value for E when $t = 0, 0.5, 4\frac{1}{2}$, and 5.

40. *Politics.*   Lawrence Soafer is running for city mayor. He expects to spend 60% of his campaign budget on TV advertisements. If he spends $36,000 on TV advertisements, what is his campaign budget?

0.2 SOLVING EQUATIONS AND INEQUALITIES

OVERVIEW   In this section we review the procedures used to solve equations and inequalities. Knowledge of this material is required in Chapter 2 on matrix theory and Chapter 3 on linear programming. Both equations and inequalities are used in problems such as the following. A dietitian wishes to produce a diet composed of two foods, each containing a given number of calories and a given number of grams of protein, fat, and carbohydrate per unit. The problem to be solved is to minimize the number of calories but have the mixture satisfy the given requirements for grams of protein, fat, and carbohydrate. In this section we consider the following topics:

- Solutions of linear equations
- Solutions of linear inequalities

Sentences that can be identified as either true or false are called **statements**. Each of the following is a statement:

$$8 + 2 = 2 + 8 \qquad \text{(a true statement of equality)}$$
$$7 - 3 = 8 - 4 \qquad \text{(a true statement of equality)}$$

Statements of equality are called **equations**. Statements of inequality are called **inequalities** and can be written using any of the following symbols:

$$\neq \qquad \text{is not equal to}$$

$$> \qquad \text{is greater than}$$

$$< \qquad \text{is less than}$$

EXAMPLE 9

(a) $6 - 4 < 5 - 2$ (a true statement of inequality)
(b) $5 + 4 > 2 + 1$ (a true statement of inequality)
(c) $7 + 3 < 2 + 4$ (a false statement of inequality) ■

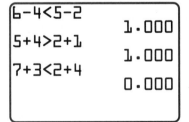

Figure 11

Calculator Note

An equation or an inequality can be tested on a calculator by using the TEST menu. To display the menu, press 2nd TEST . Then select for a given statement either $=$, $\neq$, $>$, $\geq$, $<$, or $\leq$. The calculator screen will show a 1 if a statement is true and a 0 if the statement is false. For example, to test whether $8 + 2$ equals $2 + 8$, press 8 + 2 2nd TEST (select $=$) 2 + 8 ENTER . A 1 will appear on the screen, indicating that the statement is true. For Example 9, to test whether $6 - 4$ is less than $5 - 2$, press 6 − 4 2nd TEST (select $<$) 5 − 2 ENTER . The screen in Figure 11 shows a 1, so $6 - 4 < 5 - 2$ is a true inequality.

Many sentences involving mathematical variables, or unknowns, cannot be classified as either true or false. These are not statements but rather are called **open sentences**. Two examples of open sentences are equations such as

$$x + 4 = 7$$

and inequalities such as

$$x + 1 < 10$$

Solutions of Linear Equations

The equation $2x - 1 = 13$ is of the first degree in x (i.e., the power of x is 1) and is called a **linear equation**. A linear equation in one variable is an equation that can be arranged in the form $ax + b = c$, where a, b, and c are constant ($a \neq 0$). If the number 8 is substituted for x in the linear equation $2x - 1 = 13$, the result is false, because $2(8) - 1 = 15$, not 13. If the number 7 is substituted for x in this linear equation, the result is true.

> ✓ Check: $2(7) - 1 = 13$
> $13 = 13$

Thus, 7 satisfies (or is a solution of) the equation $2x - 1 = 13$. In general, we have the following definition.

DEFINITION: SOLUTION OF AN EQUATION OR INEQUALITY

If an equation (or inequality) involves only one variable and there is a number that, if substituted for that variable, makes the equation (or inequality) a true statement, then that number is called a **solution** of the equation (or inequality).

Two equations are **equivalent equations** if and only if they have the same solutions. The following operations produce equivalent equations.

PROPERTIES THAT YIELD EQUIVALENT EQUATIONS

For any real numbers a, b, and c, if an equation $a = b$ is true, then

1. $a \pm c = b \pm c$ is true.
2. $ac = bc$ is true.
3. $a/c = b/c$ is true if $c \neq 0$.

EXAMPLE 10 Solve $3x + 5 = 11$.

SOLUTION To isolate the term involving x on a side by itself, we use properties 1 and 3.

$$(3x + 5) - 5 = 11 - 5 \qquad \text{Property 1}$$

$$3x = 6$$

$$(3x) \div 3 = 6 \div 3 \qquad \text{Property 3}$$

$$x = 2$$

The solution, $x = 2$, should be checked by substitution.

$$\text{Check:} \quad 3(2) + 5 = 11$$
$$6 + 5 = 11$$
$$11 = 11$$

Thus, the solution of $3x + 5 = 11$ is 2.   ■

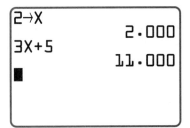

Figure 12

Calculator Note

To check the solution in Example 10 using a calculator, we first store $x = 2$. Press 2 $\boxed{\text{STO▶}}$ X $\boxed{\text{ENTER}}$ $3X$ $\boxed{+}$ 5 $\boxed{\text{ENTER}}$. In Figure 12 the output on the screen is 11, which equals the right side of the equation in Example 10.

EXAMPLE 11   Solve $\dfrac{x}{2} - 3 = 1$.

SOLUTION
$$\left(\frac{x}{2} - 3\right) + 3 = 1 + 3 \qquad \text{Property 1}$$

$$\frac{x}{2} = 4$$

$$2\left(\frac{x}{2}\right) = 2 \cdot 4 \qquad \text{Property 2}$$

$$x = 8$$ ■

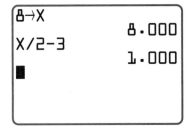

Figure 13

Calculator Note

To check Example 11 on a calculator, store $x = 8$ and then enter the left side of the equation. Press 8 $\boxed{\text{STO▶}}$ X $\boxed{\text{ENTER}}$ X $\boxed{\div}$ 2 $\boxed{-}$ 3 $\boxed{\text{ENTER}}$. In Figure 13 the value of the expression is 1, which equals the right side of the equation.

EXAMPLE 12   Solve $|x + 2| = 3$.

SOLUTION   The equation $|x + 2| = 3$ means

$$x + 2 = 3 \quad \text{or} \quad x + 2 = -3$$

We solve each equation:

$$x + 2 = 3 \qquad \text{Subtract 2}$$
$$x = 1$$
$$x + 2 = -3 \qquad \text{Subtract 2}$$
$$x = -5$$

> ✓ Check: Substitute -5 for x: $|-5+2| = |-3| = 3.$
> Substitute 1 for x: $|1+2| = |3| = 3.$

■

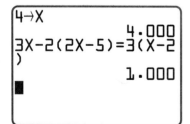

(a)

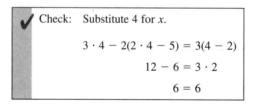

(b)

Figure 14

Calculator Note

To check the solutions in Example 12 using a calculator, store both $x = 1$ and $x = -5$ and evaluate the left side of the equation for each. Press 1 [STO▶] X [ENTER] [2nd] [ABS] [(] X [+] 2 [)] [ENTER]. Then press [(−)] 5 [STO▶] X [ENTER] [ABS] [(] X [+] 2 [)] [ENTER]. Since the value on the left side of the equation is 3 in both Figures 14(a) and 14(b), both 1 and -5 are solutions.

EXAMPLE 13   Solve $3x - 2(2x - 5) = 3(x - 2)$.

SOLUTION   This time we need to get all terms involving x on one side of the equation. First we simplify each side.

$$3x - 2(2x - 5) = 3(x - 2)$$
$$3x - 4x + 10 = 3x - 6$$
$$-x + 10 = 3x - 6$$
$$(-3x) + (-x) + 10 = (-3x) + 3x - 6 \qquad \text{Add } -3x \text{ to both sides.}$$
$$-4x + 10 = -6$$
$$-4x = -16 \qquad \text{Add } -10 \text{ to both sides.}$$
$$x = 4 \qquad \text{Divide both sides by } -4.$$

> ✓ Check: Substitute 4 for x.
> $$3 \cdot 4 - 2(2 \cdot 4 - 5) = 3(4 - 2)$$
> $$12 - 6 = 3 \cdot 2$$
> $$6 = 6$$

■

Calculator Note

To check Example 13 using the TEST menu, press 4 [STO▶] X [ENTER] 3X [−] 2 [(] 2X [−] 5 [)] [2nd] [TEST] (select =) 3 [(] X [−] 2 [)] [ENTER]. The 1 in Figure 15 indicates that the value at $x = 4$ on the left side of the equation equals the value on the right side.

Figure 15

Students are often asked to solve a formula for a particular variable in a physics, chemistry, biology, or economics course. These are called **literal equations** because the equations have letters, and the solution of the equation is given not as a number but in terms of other letters. The procedures for solving equations in the preceding examples may also be used for solving literal equations.

EXAMPLE 14   Solve $A = P + Prt$ for r.

SOLUTION $P + Prt = A$

$$Prt = A - P \qquad \text{Subtract } P.$$

$$\frac{Prt}{Pt} = \frac{A - P}{Pt} \qquad \text{Divide by } Pt.$$

$$r = \frac{A - P}{Pt}$$

■

EXAMPLE 15   Solve $y = mx + b$ for x $(m \neq 0)$.

SOLUTION $y - b = mx \qquad \text{Subtract } b.$

$$\frac{y - b}{m} = x \qquad \text{Divide by } m.$$

✓ Check: By substituting the expression for x, an identity results, indicating a valid answer.

$$y = m\left(\frac{y - b}{m}\right) + b$$

$$y = (y - b) + b$$

$$y = y$$

The solution for x is $x = (y - b)/m$, where $m \neq 0$.   ■

Practice Problem 1 Solve $A = ac + ab$ for a.

ANSWER $$a = \frac{A}{c + b}$$

The procedures used in solving the preceding equations can also be used for equations involving decimals.

EXAMPLE 16   Solve each equation.

(a) $3.126x = 12.87912$
(b) $x + 16.221 = 74.16$
(c) $16.1x - 14.2 = 13.3x + 84.3$

SOLUTION

(a) $$x = \frac{12.87912}{3.126} = 4.12$$

✓ Check: $3.126(4.12) = 12.87912$

(b) $x = 74.16 - 16.221 = 57.939$

> ✓ Check: $57.939 + 16.221 = 74.16$

(c) $(16.1 - 13.3)x = 84.3 + 14.2$

$$x = \frac{84.3 + 14.2}{16.1 - 13.3}$$

$$x = 35.179$$

> ✓ Check: Most calculators perform operations with eight or ten decimal-place accuracy. Round your answer to whatever accuracy you desire. However, remember that your check will be no more accurate than your rounded solution. Suppose that you use 35.179 as the solution.
>
> $$(16.1)(35.179) - 14.2 \approx (13.3)(35.179) + 84.3$$
>
> $$552.1819 \approx 552.1807$$

Practice Problem 2 Solve $\dfrac{x - 31.12}{17.6} = 0.05$.

ANSWER   32.0

Solutions of Linear Inequalities

The procedures used for solving inequalities are almost the same as those used for solving equations. However, let us first define the terms *less than* and *greater than*.

DEFINITION: LESS THAN

If a and b are any real numbers, then a is **less than** b, denoted by $a < b$, if and only if $b - a$ is positive, that is, if and only if there exists a positive number c such that $a + c = b$.

EXAMPLE 17   The integer $-7 < -5$ because $-5 - (-7) = 2$ (which is positive), that is, because $-7 + 2 = -5$ and 2 is positive.   ■

DEFINITION: GREATER THAN

If $a < b$, then $b > a$ (read "b is **greater than** a").

Sometimes the equality symbol is combined with inequalities to state a less-than-or-equal-to relationship. The statement $a \leq b$ means that "$a < b$ or $a = b$" or "a is not greater than b."

EXAMPLE 18

(a) $-4 \leq -4$ is true because -4 is equal to -4.

(b) $-3 \leq -2$ is true because $-3 < -2$. ◼

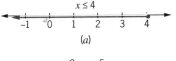

Figure 16

Calculator Note

To check the solutions of inequalities such as those in Example 18, we use the TEST menu. Press $\boxed{(-)}$ 4 $\boxed{\text{2nd}}$ $\boxed{\text{TEST}}$ (select ≤) $\boxed{(-)}$ 4 $\boxed{\text{ENTER}}$. The screen in Figure 16 shows a 1, so the inequality is true. Then press $\boxed{(-)}$ 3 $\boxed{\text{2nd}}$ $\boxed{\text{TEST}}$ (select ≤) $\boxed{(-)}$ 2 $\boxed{\text{ENTER}}$. The 1 on the screen indicates a true inequality.

The concepts of less than and greater than are easy to illustrate by points on a number line. For example, $x < 4$ is represented by all the points on a number line to the left of 4. The graph of $x < 4$ would show an open circle at 4 because $x = 4$ is not part of the solution set. The graph of $x \leq 4$ includes all of the preceding points and the point $x = 4$, so the graph in Figure 17(a) shows a closed circle at $x = 4$. Furthermore, $2 < x < 5$ (which means $2 < x$ and for the same x, $x < 5$) is represented by all points between 2 and 5 on a number line [see Figure 17(b)]. Note that the open circles indicate that 2 and 5 are not values of x.

Inequalities are often written in **interval notation**, as indicated in Table 1. Although the symbol ∞ (infinity) does not represent a real number, by (a, ∞) we mean an interval starting at a and extending indefinitely to the right. Similarly, $(-\infty, a)$ extends indefinitely to the left. A bracket indicates the value is included; a parenthesis means it is not. A bracket is never used with the symbol ∞.

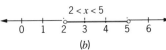

(a)

(b)

Figure 17

TABLE 1

Interval Notation	Inequality	Graph
$[a, b]$	$a \leq x \leq b$	●——————● a b
$[a, b)$	$a \leq x < b$	●——————○ a b
$(a, b]$	$a < x \leq b$	○——————● a b
$[a, \infty)$	$x \geq a$	●————————▶ a b
(a, ∞)	$x > a$	○————————▶ a b
$(-\infty, b]$	$x \leq b$	◀————————● a b
$(-\infty, b)$	$x < b$	◀————————○ a b

The following operations can be used to solve inequalities.

ADDITION AND MULTIPLICATION PROPERTIES OF INEQUALITIES

For any real numbers a, b, and c:

(a)   If $a < b$, then $a \pm c < b \pm c$.

(b)   If $a < b$ and $c > 0$, then $ac < bc$ and $\dfrac{a}{c} < \dfrac{b}{c}$.

(c)   If $a < b$ and $c < 0$, then $ac > bc$ and $\dfrac{a}{c} > \dfrac{b}{c}$.

EXAMPLE 19

(a)   For $-3 < 2$:

$$-3 + 5 < 2 + 5 \quad \text{and} \quad -3 - 4 < 2 - 4$$
$$2 < 7 \qquad\qquad\qquad -7 < -2$$

In words: You can add or subtract the same number to both sides of an inequality without changing the direction of the inequality sign.

(b)   For $4 < 6$:

$$4 \cdot 2 < 6 \cdot 2 \quad \text{and} \quad \frac{4}{2} < \frac{6}{2}$$
$$8 < 12 \qquad\qquad 2 < 3$$

In words: You can multiply or divide both sides of an equality by a positive number without changing the direction of the inequality sign.

(c)   For $4 < 6$:

$$-2(4) > -2(6) \quad \text{and} \quad \frac{4}{-2} > \frac{6}{-2}$$
$$-8 > -12 \qquad\qquad -2 > -3$$

In words: If you multiply or divide both sides of an inequality by a *negative* number, you *change the direction* of the inequality sign.   ■

Any real number substituted for a variable in an inequality that makes the inequality a true statement is a solution. All solutions comprise the **solution set**. Two inequalities that have the same solution set are **equivalent inequalities**. The operations listed above produce equivalent inequalities and thus can be used to solve inequalities.

EXAMPLE 20   Find the solution set of $x/3 - 2 < 4$.

SOLUTION
$$\frac{x}{3} - 2 < 4$$

$$\frac{x}{3} < 6 \qquad \text{Add 2 to both sides.}$$

$$x < 18 \qquad \text{Multiply both sides by 3.}$$

Consider all real numbers less than 18. By reversing our steps we discover that these real numbers satisfy $x/3 - 2 < 4$.

$$x < 18 \qquad \text{Given}$$

$$\frac{x}{3} < 6 \qquad \text{Divide both sides by 3.}$$

$$\frac{x}{3} - 2 < 4 \qquad \text{Subtract 2 from both sides.}$$

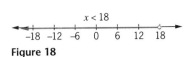

$x < 18$

Figure 18

The solution set for $x/3 - 2 < 4$ consists of all real numbers less than 18 or x in $(-\infty, 18)$. The graph of this solution set is shown in Figure 18.

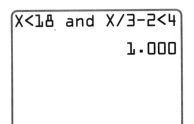

Figure 19

Calculator Note

To check the solutions of inequalities, we use the TEST LOGIC menu. For Example 20, press X $<$ 18 $2nd$ $TEST$ $\blacktriangleright$ (select "and") X $\div$ 3 $-$ 2 $<$ 4 $ENTER$. The 1 on the screen in Figure 19 indicates that the two inequalities are equivalent; that is, the solution given is correct. (**Hint:** Select $<$ using the TEST menu.)

EXAMPLE 21 Find the graph of the solution set of $-3x + 2 < 7$.

SOLUTION $-3x + 2 < 7$

$$-3x < 5 \qquad \text{Add } -2 \text{ to both sides.}$$

$$x > -\frac{5}{3} \qquad \begin{array}{l}\text{Divide both sides by } -3 \text{ and} \\ \text{reverse the inequality.}\end{array}$$

Figure 20

The solution set consists of all real numbers greater than $(-5/3)$ or x in $(-5/3, \infty)$. The graph of this solution set is shown in Figure 20.

Practice Problem 3 Solve $2x + 3 \geq 9$.

ANSWER The solution set consists of all real numbers greater than or equal to 3.

As with equations, inequalities can also involve absolute values. Let a be any positive number. Then $|x| < a$ if and only if $-a < x < a$; similarly, $|x| > a$ if and only if $x > a$ or $x < -a$. That is, $|x| < 6$ means that $-6 < x < 6$, and $|x| > 4$ means that either $x < -4$ or $x > 4$.

EXAMPLE 22 Graph the solution of $|2 - 6x| \geq 4$.

SOLUTION

$$2 - 6x \geq 4 \quad \text{or} \quad 2 - 6x \leq -4 \quad \text{Definition of absolute value}$$

$$-6x \geq 2 \qquad\qquad -6x \leq -6 \quad \text{Subtract 2 from both sides.}$$

$$x \leq -1/3 \qquad\qquad x \geq 1 \quad \text{Divide both sides by } -6 \text{ and reverse the inequality.}$$

Figure 21

The solution is x in either $(-\infty, -\frac{1}{3}]$ or $[1, \infty)$ and is graphed in Figure 21. ■

Practice Problem 4 Graph the solution set of $|3x - 2| \leq 7$ on a number line.

Figure 22

ANSWER   The graph is shown in Figure 22.

SUMMARY

In this section we showed the similarities between solving equations and solving inequalities. Both can be solved using algebra and a scientific calculator or a graphing calculator. For inequalities, remember that if you multiply or divide by a negative number, you must reverse the sense of the inequality.

Exercise Set 0.2

In Exercises 1–12, solve for the variable and verify your solution using a graphing calculator.

1. $2x - 7 = 3$
2. $1 - 2x = -5$
3. $4x - 7 = 5$
4. $17 - 5x = -3$
5. $4 - 2x = (8 + 3x) + 1$
6. $-5 - x = (3 + 2x) + 1$
7. $2x - (-5) = 6 - (-x)$
8. $4x + (-7) = 3x - (-1)$
9. $\dfrac{x}{5} - 3 = -2$
10. $-\dfrac{x}{7} - 2 = -3$
11. $\dfrac{x}{5} - \dfrac{1}{3} = \dfrac{x}{3} + \dfrac{1}{5}$
12. $\dfrac{x}{2} - \dfrac{1}{4} = \dfrac{3x}{4} + 1$

Solve each of the following inequalities and express the answer in interval notation.

13. $-x + (-4) < -7$
14. $-2x - 3 \geq 5$
15. $\dfrac{x}{3} + 2 < -5$
16. $-3x - 4 \geq -5$

17. $\dfrac{x}{3} + \dfrac{4}{6} < \dfrac{x}{2} - \dfrac{4}{15}$
18. $\dfrac{x}{7} + 3 \leq 4 - \dfrac{x}{3}$

Solve for the variable indicated.

19. $A = P + Prt; \quad r$
20. $I = A(1 - dt); \quad t$
21. $y = mx + b; \quad b$
22. $y = mx + b; \quad m$
23. $S = \dfrac{a}{1 - r}; \quad r$
24. $L = a + (n - 1)d; \quad d$

Solve the following equations by including a check.

25. $-[2x - (3 - x)] = 4x - 11$
26. $x^2 - 3 = 1 + (x + 1)(x - 2)$
27. $2 - x - x^2 = 1 - (x - 1)^2$
28. $\dfrac{3x + 4}{5} = \dfrac{7x + 6}{10}$
29. $\dfrac{x - 5}{4} = 1 + \dfrac{x - 9}{12}$
30. $\dfrac{x}{3} - 1 = \dfrac{2x - 3}{3}$
31. $|4 - x| = 1$
32. $|3 - x| = 4$
33. $|2x + 1| = 3$
34. $|3x - 2| = 4$

Solve and express the answer in interval notation; then draw the graph of the solution.

35. $-1 \le x + 5. \le 6$

36. $-4 \le \dfrac{3x + 2}{5} \le -2$

37. $5 \le \dfrac{2x + 1}{-3} < 8$

38. $-2 \le \dfrac{2x + 5}{2} < 0$

39. $|x - 2| < 7$

40. $|x - 5| < -1$

41. $|2x + 2| \ge 4$

42. $|2x - 4| \ge 6$

Solve for the variable indicated.

43. $A = P + Prt; \quad P$

44. $\dfrac{x}{a} + \dfrac{4}{b} = 1; \quad x$

45. $x^2y - 3x - 2z^3y = 1; \quad y$

46. $3x_1x_3 + x_1x_2 = x_4; \quad x_1$

Solve Exercises 47–50 using a calculator.

47. $6.310x - 8 < 1.60x - 0.011$

48. $-3.41(1 - 0.62x) > 4.071x - 6.3$

49. $7.61(x - 4.5) \le 3.01x - 3.7$

50. $0.01(x - 3.7) \ge 0.03x - 0.7$

51. $\dfrac{6x}{7} + \dfrac{5}{9} = \dfrac{8}{11}$

52. $\dfrac{4x}{6} + \dfrac{2}{3} = \dfrac{2x}{5} + \dfrac{8}{9}$

Review Exercises

53. Write the following in mathematical language.
    (a) Three times a number less 6 equals 21.
    (b) Eight less than 4 times a number equals 61.
    (c) A number less 42 is always less than 20.

54. According to the National Safety Council, about 60% of all automobile accident fatalities for a year occurred during daylight hours. If, during the year, 23,000 automobile accident fatalities occurred during daylight hours, how many such fatalities occurred that year?

0.3 SOLVING APPLICATION PROBLEMS WITH LINEAR EQUATIONS

OVERVIEW   It would be great if there were a magic formula that made all application problems easy. Unfortunately, such a formula does not exist. However, there are several procedures that we can follow to help us formulate linear equations to represent **mathematical models** of real-world phenomena. In Chapters 2, 3, and 6, matrices will serve as mathematical models for application problems. In Chapters 5, 6, and 7, probability functions will serve as models. In this section we consider the following topics:

- Polya's four steps in problem solving
- Suggestions for understanding a problem
- Forming linear models
- The strategy of using estimation

Polya's Four Steps in Problem Solving

In the 1950s, a very successful research mathematician named George Polya wrote a series of insightful articles and books on problem solving. He identified four steps that have characterized problem solving from the time of the ancient Greeks to the present day.

Polya's Four Steps in Problem Solving

1. Understand the problem.
2. Devise a plan.
3. Carry out the plan.
4. Look back; see if your solution makes sense.

In trying to understand a problem, we ordinarily ask a number of questions. Typically they include: What do we know? What is to be found?

EXAMPLE 23   On their way back to the university, Joy, Beth, and Dill took turns driving. Joy drove 50 miles more than Beth; Beth drove twice as far as Dill. Dill only drove 10 miles. How many miles is the trip back to the university? (List some questions you would ask in order to understand the problem.)

SOLUTION   Understanding the problem: Do you know how far anyone drove? (Yes, Dill drove 10 miles.) How many more miles did Joy drive than Beth? (50.) What is the relationship between the number of miles driven by Beth and the number driven by Dill? (Beth drove twice as far as Dill.) What are we trying to find? (The distance back to the university.)   ■

Although asking questions similar to the ones in the preceding example will usually give an understanding of a problem, some additional suggestions may be helpful.

HINT: Understanding the Problem

1. Read and reread the problem; look up words you don't know.
2. Identify what you are trying to find.
3. Strip the problem of irrelevant details.
4. Don't impose conditions that do not exist.

Once you have carefully summarized the information given in the problem, the suggestion ''Identify what you are trying to find'' enables you to find the answer easily for a simple problem such as the following.

EXAMPLE 24   A bank issues two sets of certificates of deposit. The interest rate for the one with the longer period of time is 7% and the other has an interest rate of 6%. If you decide to invest x dollars at 7%, how much do you invest at 6% if you have $50,000 to invest?

SOLUTION   We consider the following questions: What do we wish to find? (The amount invested at 6%.) Is this amount related to the $50,000? (Yes, it is part of the $50,000.) What is the other part of the $50,000? (The amount invested at 7%.) How much is invested at 7%? (x dollars.) If x dollars is invested at 7%, how much remains to be invested at 6%? ($\$50,000 - x$ remains to be invested.) ■

Linear Equations as Models

Problems that arise in business, economics, or the social and life sciences are usually too complicated to be defined precisely by a mathematical formula, but we are often able to formulate a model that satisfies the parameters of a problem fairly well. For example, in business, costs and prices are considered over a short period of time as part of manufacturing costs. (Of course, over a long period there are many economic variables: inflation, supply, demand, etc.) Usually the manufacturing costs consist of fixed costs and variable costs. **Fixed costs** are those that remain constant, independent of the number of units produced. Some examples of fixed costs include rent on buildings, insurance, and depreciation. **Variable costs** depend on the number of items produced, for example, labor, materials, and distribution. In the next example we write a linear equation as a model representing both fixed and variable costs.

EXAMPLE 25 (Cost Analysis) Write an expression for the total cost (C) if the fixed cost is $2000 and the variable cost is $1.20 per item produced. From this linear model, determine the cost of producing 1000 items.

SOLUTION Let x represent the number of items to be produced. The variable cost per item is $1.20. Thus, the variable cost of producing x items is $1.20x$ and the total cost is given by

$$C = \$2000 + \$1.20x$$

To find the cost of producing 1000 items, we substitute $x = 1000$ to obtain

$$C = \$2000 + \$1.20(1000)$$
$$= \$3200$$

■

In business, supply S and demand D are often expressed as linear equations in terms of the number of items x produced (or demanded). The second variable in the linear equation is price. That is, in a supply equation, price p_s often decreases as more items x are produced. Likewise, price p_d often increases as the demand for items increases. The value of x at which $p_s = p_d$ is called the **equilibrium point** or **point of equilibrium**.

EXAMPLE 26 (Supply and Demand) Consider a demand expressed as $D = p_d(x) = (400 - 5x)/2$ and a supply expressed as $S = p_s(x) = 5x/2$.

(a) Which has a larger price, p_d or p_s, when $x = 10$?
(b) When $x = 20$?

(c)   When $x = 50$?

(d)   When $x = 60$?

(e)   Find the equilibrium point for supply and demand.

SOLUTION

(a)   Demand: at $x = 10$, $p_d = 175$ and $p_s = 25$.

(b)   Demand: at $x = 20$, $p_d = 150$ and $p_s = 50$.

(c)   Supply: at $x = 50$, $p_d = 75$ and $p_s = 125$.

(d)   Supply: at $x = 60$, $p_d = 50$ and $p_s = 150$.

(e)   At the equilibrium point, $p_d = p_s$ or

$$\frac{400 - 5x}{2} = \frac{5x}{2}$$

$$400 - 5x = 5x$$

$$400 = 10x$$

$$x = 40$$

So $(40,100)$ is the point of equilibrium.

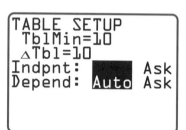

(a)

```
Y₁◼(400-5X)/2
Y₂◼5X/2
Y₃=◼
Y₄=
Y₅=
Y₆=
Y₇=
Y₈=
```

(a)

```
TABLE SETUP
 TblMin=10
 ΔTbl=10
Indpnt:  Auto  Ask
Depend:  Auto  Ask
```

(b)

X	Demand Y₁	Supply Y₂
10.000	175.00	25.000
20.000	150.00	50.000
30.000	125.00	75.000
40.000	100.00	100.00
50.000	75.000	125.00
60.000	50.000	150.00
70.000	25.000	175.00

X=40

(c)

```
WINDOW FORMAT
 Xmin=-3
 Xmax=93
 Xscl=10
 Ymin=-10
 Ymax=190
 Yscl=20
```

(d)

Figure 23

Calculator Note

For Example 26 we can use a calculator to display a table of values for demand and supply for given values of x [Figure 23(c)] and a graph showing the intersection of the two lines [Figure 23(e)]. Specific instructions for forming a table are found on page A-7 of Appendix A. You may wish to delay studying the material on page A-1 of Appendix A on drawing graphs until you complete the next section on graphing linear equations.

Building a Model

Polya's four steps for solving problems are also very helpful in forming models. In building models, however, several ideas for expanding on Polya's four steps are helpful.

Understand the Problem   Keep in mind all of the suggestions on understanding the problem. In particular, for complicated models be sure to use charts or tables and write down what you are trying to find.

Devise a Plan   In this book, most of our experience in forming models will involve translating a verbal problem into a mathematical model, usually an equation, inequality, system of equations or inequalities, or other mathematical concepts. A recommended step in working any problem using a calculator is to first **estimate the answer**. We make use of this essential step to assist in forming models.

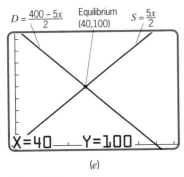

$D = \dfrac{400 - 5x}{2}$ Equilibrium $S = \dfrac{5x}{2}$
(40,100)

(e)

Figure 23

STRATEGY: Steps in Devising a Plan

1. Estimate an answer. Give your best estimate.
2. With a calculator, check your estimate to see how closely it fits the given relationships.
3. Note the mathematical operations involved in the check. Now replace your estimate with a variable or variables indicating the same mathematical operations as in step 2.

When step 3 above is completed, we usually have a mathematical expression to solve. This of course involves Polya's third step, "Carry out the plan," for which we will study procedures throughout this book. Finally, we certainly want to check to see if our answer or solution makes sense, Polya's fourth step.

EXAMPLE 27   Susan has invested $10,000, a portion at 8% and another portion at 10%. If her annual interest is $920, how much does she have invested at each rate?

SOLUTION   Estimate that Susan has $8000 invested at 10%, as noted in Table 2. Then she would have $10,000 − $8000 = $2000 invested at 8%. Her total interest for this estimate would be

$$(0.10)8000 + (0.08)2000 = ?$$

Using a calculator, we note that this is $960, which is more than the $920 given as interest.

TABLE 2

	Invested at 10%	Invested at 8%	Total Interest
Estimate	$8000	$10,000 − $8000	
Interest	(0.10)8000	(0.08)2000	$960 (too large)
Replace estimate with x	x	$10,000 − x$	
Interest	$(0.10)x$	$0.08(10,000 − x)$	$920

Now let's go back and replace our estimate of $8000 by x dollars. Note the wording, "Suppose she has x dollars invested at 10%." Then she would have $10,000 − x$ invested at 8%. Her total interest would be

$$(0.10)x + (0.08)(\$10,000 − x) = \$920$$

We now have a linear equation representing the verbal problem. Next, we solve the equation.

$$0.10x + (800 - 0.08x) = 920 \qquad \text{Subtract 800.}$$

$$0.02x = 120 \qquad \text{Divide by 0.02.}$$

$$x = 120 \div .02$$

$$x = 6000$$

Now let's look back to see if our answer makes sense.

$6000 invested at 10%:	$600
$10,000 − $6000 = $4000 invested at 8%:	$320
Total interest:	$920

Thus, our answer is correct.   ■

Note the pattern established in Example 27. First, estimate what you are trying to find and then work through the relationships given in the problem. Next, replace your estimate with x in the equations involving the estimate.

EXAMPLE 28   A will provides that an estate is to be divided among a wife and two children. The wife is to receive $10,000. From what remains, the wife is to receive twice as much as each child. If the estate is valued at $50,000, how much does each person receive?

SOLUTION   We begin by estimating that each child receives $5000. As a result, the wife receives $10,000 + 2($5000), as noted in Table 3. We see immediately that the answer of $30,000 is much less than the total estate of $50,000. Now we replace the estimate of $5000 by x dollars for the amount that each child receives.

TABLE 3

	Wife	Child 1	Child 2	Total
Estimate	$10,000 + 2($5000)	$5000	$5000	$30,000 (too small)
Replace estimate with x	10,000 + 2x	x	x	$50,000

$$\underset{\text{Wife}}{(\$10,000 + 2x)} + \underset{\text{Child 1}}{x} + \underset{\text{Child 2}}{x}$$

Thus we have the following linear equation, which we then solve.

$$(10,000 + 2x) + 2x = 50,000 \qquad \text{Combine like terms.}$$

$$10,000 + 4x = 50,000 \qquad \text{Subtract \$10,000.}$$

$$4x = 40,000 \qquad \text{Divide by 4.}$$

$$x = 10,000 \qquad \text{Each child's share}$$

$$10,000 + 2x = 10,000 + 2(10,000)$$

$$= 30,000 \qquad \text{Wife's share}$$

Thus the wife receives $30,000 and each child receives $10,000.

We now verify that the answer satisfies the facts. Each child receives $10,000 and the wife, $30,000. Does the wife receive $10,000 plus double what each child receives? (Yes.) Does the sum of the allotments equal $50,000? (Yes, because $30,000 + $10,000 + $10,000 = $50,000.)

EXAMPLE 29   A coin collector has $45 in quarters and dimes. How many coins of each kind does she have if the total number of her coins is 240?

SOLUTION   We begin by estimating the number of quarters and dimes, as noted in Table 4.

TABLE 4

	Quarters	Dimes	Total
Estimate	100	$240 - 100$	240
Value of estimate	$0.25(100)	$0.10(140)	$39.00 (too small)
Assignment of variables	x	$240 - x$	240
Value	$0.25x	$0.10(240 - x)$	$45.00

Next, we solve the equation. If we let x represent the number of quarters, then we have

$$0.25x + 0.10(240 - x) = 45.00$$

$$0.25x + 24 - 0.10x = 45.00$$

$$0.15x = 21.00$$

$$x = 140 \text{ quarters}$$

$$240 - 140 = 100 \text{ dimes}$$

We then check these results by substitution.

$$\$0.25(140) + \$0.10(100) = \$45.00$$

Practice Problem 1   Mark has invested $20,000, part at 10% and part at 8%. If his interest is $1760 per year, how much does he have invested at each rate? Estimate the amount invested at each rate and compute the interest to get a feel for the problem. Then set up an equation and solve it.

ANSWER $x = \$12,000$ (at 8%)
$\$20,000 - x = \$8000$ (at 10%)

Practice Problem 2 A stereo manufacturer has a fixed cost of $1650 and a variable cost of $105 for each item produced. Write the cost function. What is the cost of producing 200 items?

ANSWER $C = \$105x + \$1650$
$\$22,650$ for 200 items

SUMMARY

In this section we showed how to set up an algebraic model to solve application problems. Sometimes it is helpful to estimate an answer and perform the necessary operations to check the estimate; then substitute an unknown for the estimate in the operations. Remember Polya's four steps, especially the last one: Check your answer.

Exercise Set 0.3

Work Exercises 1–6 by first estimating an answer. With a calculator, determine whether your estimate is too large or too small; then substitute an unknown for your estimate. Verify that your answer is correct.

1. Carol is twice as old as Ed. Ed is 10 years younger than Carol. How old is Ed?

2. Tom has twice as many books as Joe. Together they have 75 books. How many does each boy have?

3. A rectangle has a length 3 inches less than 4 times the width. If its perimeter is 38 inches, what are the dimensions of the rectangle?

4. The difference between two integers is 21. The larger integer is equal to twice the smaller integer plus 20. What are the two integers?

5. Three times Ralph's weight added to 54 kg is equal to 300 kg. How much does Ralph weigh?

6. The longest side of a triangle is twice as long as the shortest side and 2 inches longer than the third side. If the perimeter of the triangle is 33 inches, what is the length of each side?

7. An automobile radiator contains 4 gallons of a solution that is 10% antifreeze and 90% water. How much solution must be drained and replaced with pure antifreeze to obtain a solution that is 25% antifreeze?

8. At weekend performances a theater sells 400 adult tickets and 520 children's tickets, which are $0.75 cheaper than adult tickets. If total ticket receipts are $1680, what is the cost of an adult ticket? A child's ticket?

9. **Exam.**   A company sells ties for $6 each. Variable costs are $2 per tie. Fixed costs per week are $37,500. How many ties must be sold per week to break even?
   (a) 9375   (b) 9740
   (c) 11,029   (d) 12,097

Applications (Business and Economics)

Express the following as mathematical statements or sentences.

10. **Manufacturing.**   A corporation makes valves and reducers. A valve requires 1 hour on machine A and 2 hours on machine B. A reducer requires 3 hours on machine A and 4 hours on machine B. Four reducers and x valves were produced last week. How many hours was machine A in operation? Machine B?

11. **Investments.**   An investment club has $27,000 to invest in bonds of two types, government securities and junk bonds. Government securities yield 7%, and junk bonds yield 10%. If x dollars are invested in government securities, what is the total yearly interest?

12. **Profit.**   At a local department store, a suit that costs $100 sells for x dollars. What is the profit? **Hint:** Profit = Selling Price − Cost.

13. **Revenue Equation.**   Suppose a manufacturer sells a product for $5 per unit. Write an expression for revenue R in terms of x units sold. **Hint:** Revenue = (Price per unit)(Number of units sold).

14. **Cost Equation.**   In the manufacturing of the x units in the previous exercise the fixed cost is $1000 plus a variable cost of $2.40 per unit for labor and materials. Write an expression for cost c.

15. **Supply Curve.**   A supply curve indicates the number of units of a good or service that will be offered for sale at different prices. Write an equation for the supply curve if price p is 20 plus the product of 2 and the number of units (x) for sale.

16. **Demand Curve.**   A demand curve indicates the total quantities that purchasers will buy at different prices. Write an equation for the demand curve if price p is 20 less the product of 2 and the number of units x people will buy.

17. **Profits.**   At a corporation this year's profits P will exceed last year's profits of $620,000. Express this mathematically.

18. **Sales Commissions.**   Each salesperson in Wargo Sales earned more than $20,000 in commissions c. Express a mathematical relationship between x, John Smith's earnings, and $20,000.

19. **Production.**   Refinery I produces 200 barrels of high-grade oil a day. Refinery II produces twice as much per day as refinery I. How much do both produce in x days?

Solve the following problems.

20. **Investments.**   A woman has an annual income of $6500 from two investments. She has $15,000 more invested at 10% than she has invested at 12%. How much does she have invested at each rate?

21. **Investments.**   A sum of $2000 is invested, part at 8% and the remainder at 10%. Find the amount invested at each rate if the yearly income from the two investments is $180.

22. **Mixture Problem.**   Carl is pricing different brands of coffee. Brand A sells for $0.25 a pound more than brand B, and brand C sells for $0.37 less than twice the price of brand B. Carl decides to taste the coffee before making a decision on which coffee is the best buy; so he buys 1 pound of each brand and spends $5.08. How much does a pound of each brand cost?

23. **Mixture Problem.**   A grocer mixes 40 pounds of $1.60-a-pound nuts with 60 pounds of $1.20-a-pound nuts. If he wants to receive at least the same amount of money as when he sold the nuts separately, how much should he charge for the mixture?

Applications (Social and Life Sciences)

24. **Diet.**   Food I contains 30 protein calories per unit and food II contains 20. Abdul eats x units of food I and y units of food II. How many protein calories does he get?

25. **Medicine.**   Medicine mixture I contains 2 milligrams per gram of drug X (for high blood pressure). A gram of medicine mixture II contains 4 milligrams. At the Morningside Hospital, g_1 grams of I and g_2 grams of II are administered to a patient. How much drug X does the patient receive? (**Hint:** Give the answer in milligrams.)

26. **Agriculture.**   A farmer has a 50-acre farm on which he plants two crops, wheat and corn. Wheat requires 2 days of labor per acre, and corn requires 3 days. Fertilizer and other costs amount to $40 for wheat and $30 for corn per acre. What is the labor requirement for w acres of wheat and c acres of corn? What is the cost of fertilizer and other costs for w acres of wheat and c acres of corn?

27. **Psychology.**   An intelligence quotient (IQ) is found by dividing the mental age x by the chronological age and then multiplying this result by 100. Write an expression for the IQ for a 10-year-old.

28. **Population Growth.**   A city with a population of 60,000 is increasing at a uniform rate of 4% of this year's population. What will the population be in t years?

29. **Wildlife Growth.**   The growth rate of fish in a lake is 10% per year of the current number of fish. If there are 20,000 fish in the lake, how many will there be in t years?

30. **Leisure Time.**   If you spend 65% of your leisure time reading, how much leisure time do you have if you spend 13 hours a week reading?

31. **Temperature.** $F° = \frac{9}{5}C° + 32°$ relates temperature in degrees Celsius to temperature in degrees Fahrenheit. Solve for $C°$ in terms of $F°$.

32. **SAT Scores.**   The average SAT verbal scores at a high school have been increasing since new graduation requirements were inaugurated in 1980. If y rep-

resents the average verbal score and t the number of years since 1980, then

$$y = 1.5t + 430$$

Solve for t in terms of y and then find when the school can expect a 450 verbal score. Will this actually happen? Why or why not?

33. **Protein.**   One serving of a cereal contains 4% of the recommended daily allowance of protein. If one serving contains 8 grams of protein, what is the recommended daily allowance for protein?

34. **Voting.**   In a presidential election, only 54% of the registered voters voted. If 42 million went to the polls, how many registered voters were there?

Review Exercises

35. Let $A = \{-2, -\frac{1}{2}, -\sqrt{2}, 0, 1/\sqrt{2}, \sqrt{2}, 2\}$. List the elements of A in the following sets.
    (a) Set of natural numbers
    (b) Set of integers
    (c) Set of rational numbers
    (d) Set of irrational numbers
    (e) Set of negative numbers
    (f) Set of real numbers

36. Solve the following equations and check your answer.
    (a) $6x - 5 = 14$ (b) $2x + 7 = x - 6$
    (c) $\dfrac{2x + 3}{5} = x + 1$ (d) $\dfrac{9x + 6}{4} = x - 1$

0.4   GRAPHING LINEAR EQUATIONS AND INEQUALITIES

OVERVIEW

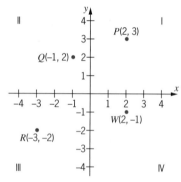

Figure 24

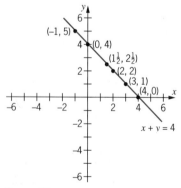

Figure 25

In this section we introduce the rectangular (or Cartesian) coordinate system and learn to draw graphs. Businesspeople, economists, and scientists must be proficient not only in interpreting graphs but also in drawing them. Graphs such as histograms and line graphs are important in statistics. Graphs of functions and of the derivatives of functions are important in differential calculus. Sometimes graphs are essential in finding area in integral calculus. In this section we consider the following topics:

- Graphing linear equations
- Graphing linear inequalities

We form a **rectangular (or Cartesian) coordinate system** with two perpendicular real number lines, as shown in Figure 24. The two lines are called **coordinate axes**. Traditionally, the horizontal line is called the **x-axis** and the vertical line is called the **y-axis**. The point of intersection of the two lines is called the **origin**. The plane in which the two axes lie is called the **coordinate plane**. The four parts into which the two axes divide the plane are called **quadrants**, which are labeled I, II, III, and IV, as shown in Figure 24.

To each ordered pair of numbers (x, y), there corresponds a point in the coordinate plane. For example, in Figure 24 the ordered pair of numbers $(2, 3)$, where $x = 2$ and $y = 3$, corresponds to the point P. Conversely, an ordered pair of real numbers (called coordinates) exists for each point in the coordinate plane. In Figure 24, W is identified by the coordinates $(2, -1)$, Q by $(-1, 2)$, and R by $(-3, -2)$.

Graphing Linear Equations

A linear equation in two variables is an equation that can be arranged in the form $ax + by = c$, where a, b, and c are real numbers and a and b are not both zero.

Note that x and y appear only to the first degree. Consider the linear equation $x + y = 4$. Let e and f be two real numbers. We say that the ordered pair (e, f) satisfies $x + y = 4$, or is a solution of $x + y = 4$, if the equation $x + y = 4$ becomes a true statement when e is substituted for x and f is substituted for y. To find a solution to this equation, we select a value for x and then solve the equation for y. For example, let $x = 2$ and substitute this value in $x + y = 4$. Then $2 + y = 4$ or $y = 2$. Thus $(2, 2)$ is a solution, and so are $(3, 1)$, $(1\frac{1}{2}, 2\frac{1}{2})$, $(-1, 5)$, $(4, 0)$, and $(0, 4)$. In fact, there are infinitely many solutions for this equation in the real number system.

Now that we have defined the coordinate axes, we can give a geometric description of all the solutions of $x + y = 4$. Each ordered pair (e, f) that is a solution corresponds to a unique point in the coordinate plane. If we plot all the points that are solutions, the resulting line is called the **graph** of the equation. Since the number of solutions is infinite, we cannot possibly plot every ordered pair that satisfies the equation. We can, however, get an idea of what the graph looks like by plotting several representative points. For example, we use six solutions of $x + y = 4$ to obtain the graph in Figure 25.

All six points in Figure 25 seem to lie on the same straight line. In fact, it can be shown that this straight line contains all points represented by ordered pairs satisfying $x + y = 4$ and each point on the line is represented by such a pair. The equation $x + y = 4$ is called a linear equation; recall that an equation of the form $ax + by = c$, where a, b, and c are real numbers and a and b are not both equal to zero, is a linear equation. The graph of an equation is the figure that is drawn by plotting the points whose coordinates (x, y) satisfy the equation.

Practice Problem 1 Using a calculator, draw the graph of $y = 4 - x$ and check your answer with the graph in Figure 25. (Use $-4.7 \le x \le 4.7$, $-6.2 \le y \le 6.2$.)

ANSWER   See the following calculator note.

Calculator Note

We review here the steps for drawing a graph as discussed on page A-1 of Appendix A. Press [WINDOW] to display the current WINDOW variable values. In Figure 26(b) note that we have selected the minimum value of x to be -4.7, the maximum value of x to be 4.7, the minimum value of y to be -6.2, and the maximum value of y to be 6.2. This gives the portion of the coordinate plane that we will see in the viewing window. Note also that we have selected the distance between tick marks to be 1. This is indicated by scl = 1 on the screen. Next press [Y_1 =] and insert in the first line of the window the equation $Y_1 = 4 - x$ [Figure 26(a)]. Press [GRAPH] to draw the graphs shown in parts (c) and (d) of Figure 26. Press [TRACE] and use either [▶] or [◀] to move the cursor from one plotted point to the next along the graph while displaying the cursor coordinates at the bottom of the screen. The [▶] key moves the cursor to the right and the [◀] key moves the cursor to the left along the graph (see page A-3 of Appendix A). Move the cursor until $x = 0$. Note that at the bottom of the screen in part (c) the y-intercept is at $y = 4$. In part (d) move the cursor until $y = 0$. Note that the x-intercept at the bottom of the screen is at $x = 4$.

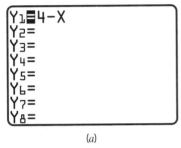

(a)

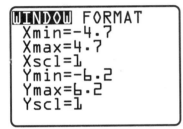

(b)

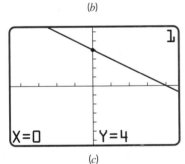

(c)

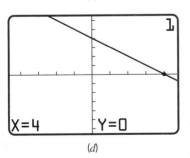

(d)

Figure 26

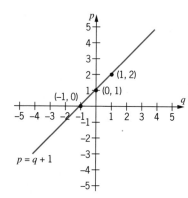

Figure 27

Note that the calculator graph in Figure 26(c) checks exactly with the graph in Figure 25.

EXAMPLE 30   Graph $p = q + 1$.

SOLUTION   In this example, we plot p along the vertical axis and q along the horizontal axis. The coordinates $(1, 2)$, $(0, 1)$, and $(-1, 0)$ are solutions of $p = q + 1$. If we represent these coordinates on a rectangular coordinate system, we see that a straight line contains all three points (Figure 27). Two of the points determine the line and the third point serves as a check.   ■

The x-intercept and y-intercept are the points (x, y) where the graph crosses the x-axis and the y-axis, respectively. However, in general practice we often designate the y-coordinate where the graph crosses the y-axis as the y-intercept and the x-coordinate where the graph crosses the x-axis as the x-intercept. To find the y-intercept, set $x = 0$ in the equation of a line and solve for y. To find the x-intercept, set $y = 0$ and solve for x. In Figure 25, the x-intercept is 4 and the y-intercept is 4. In Figure 27, the q-intercept is -1 and the p-intercept is 1.

EXAMPLE 31   Find the x-intercept and y-intercept of the graph of $2x - 3y = 6$, and use them to sketch the graph of the equation.

SOLUTION   To find the x-intercept, set $y = 0$. Then

$$2x - 3 \cdot 0 = 6$$

$$x = 3 \qquad x\text{-intercept}$$

To get the y-intercept, set $x = 0$. Then

$$2 \cdot 0 - 3y = 6$$

$$y = -2 \qquad y\text{-intercept}$$

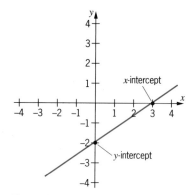

Figure 28

The graph of $2x - 3y = 6$ is shown in Figure 28.   ■

EXAMPLE 32   Graph $x = 4$.

SOLUTION   The x-intercept is $(4, 0)$. Since $x = 4$ for each point on the graph, the graph is parallel to the y-axis, as shown in Figure 29.   ■

Now let's look at one of the many applications of graphing linear equations. The intersection of the graph of a demand equation and the graph of a supply equation is called the equilibrium point for supply and demand (see the preceding section). The point of intersection (x, p) satisfies both the demand equation and the supply equation. At that point, supply is equal to demand.

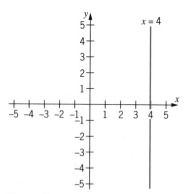

Figure 29

EXAMPLE 33   Suppose that the demand equation is

$$D = p_d = \frac{320 - 4x}{3}$$

and the supply equation is $S = p_s = 20x$. Find the point of equilibrium for these equations by graphing both the demand equation and the supply equation on the same coordinate system.

SOLUTION   Note that when $x = 2$, $p_d = 104$; and when $x = 5$, $p_d = 100$. We use points $(2, 104)$ and $(5, 100)$ to graph

$$p_d = \frac{320 - 4x}{3}$$

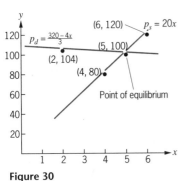

Figure 30

We use $x = 4$, $p_s = 80$ and $x = 6$, $p_s = 120$ to graph $p_s = 20x$, as shown in Figure 30. The intersection of the two graphs is at $x = 5$, $p = 100$. Thus, the equilibrium price is \$100. If the number of units is more than 5, the supply will exceed the demand. If the number of units is less than 5, the demand will exceed the supply.   ■

Practice Problem 2   Using a calculator, draw the graphs of $y = 20x$ and $y = (320 - 4x)/3$ and obtain the approximate point of intersection. Use $-2 < x < 7.4$ with a scale of 2 and $-4 < y < 128$ with a scale of 20. (Use the negative values for the viewing window even though negative numbers are not reasonable for the number of units. Otherwise the coordinate axes will not be in the window.)

ANSWER   The graphing procedure is shown in Figure 31. Instructions for finding the intersection of the two lines on a calculator are found on page A-3 of Appendix A.

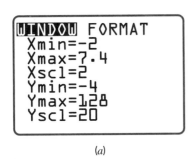

(a)

Figure 31

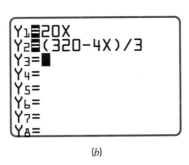

(b)

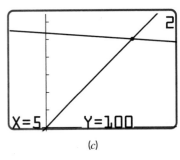

(c)

Graphing Linear Inequalities

As seen in the preceding example, graphing equations is useful in finding points of equilibrium. Graphing inequalities is equally important. As shown in Figure 32, a line divides the coordinate plane into two parts called **half-planes**. Just as the points on a line are described by the equation of the line, the points in a half-plane are described by an inequality. Consider the line $y - 2x - 1 = 0$

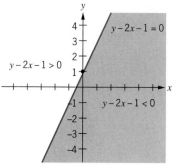

Figure 32

shown in Figure 32. If (x, y) is any point on the line, then it satisfies the equation $y - 2x - 1 = 0$. If (x, y) is any point above the line, then it satisfies $y - 2x - 1 > 0$. For example, $(-1, 2)$ is above the line, and $(-1, 2)$ satisfies the inequality $2 - 2(-1) - 1 = 3 > 0$. Similarly, $(0, 0)$ is in the solution set of $y - 2x - 1 < 0$, since $0 - 0 - 1 < 0$. This agrees with the fact that the half-plane below the line consists of all points that satisfy $y - 2x - 1 < 0$ (see Figure 32). An inequality using less than or greater than represents a half-plane, and the corresponding line ($<$ or $>$ replaced by $=$) represents the boundary of the half-plane. The following procedure is helpful when graphing inequalities.

HINT: Graphing Inequalities

1. Treat the inequality as an equation and plot points as if to graph the line.
2. If the relationship in the inequality is $<$ or $>$, draw the line with dashes, since the points on the line are not solutions of the inequality but rather are on the boundary. If the relationship is $\leq$ or $\geq$, draw a solid line, since the points on the line are solutions.
3. By testing a point on either side of the line in the inequality, determine which side of the line represents the set of solution points. (The origin is often a convenient point to check if it is not on the boundary line.)
4. Shade the side of the half-plane that was found to contain the solution points.

EXAMPLE 34 Graph $3x + 2y < 5$.

SOLUTION

1. Four points on the line represented by $3x + 2y = 5$ are given in Table 5.

TABLE 5

x	y
0	$\frac{5}{2}$
1	1
-1	4
3	-2

2. Sketch the line. Since no point on the line itself is a solution of $3x + 2y < 5$, draw the line with dashes.
3. Determine which side of the line consists of points (x, y) such that $3x + 2y < 5$. Select any point, say $(3, 3)$, that is to the right of the line. Does the inequality become a true statement if this point is substituted for the x and y values? Test it. Is $3(3) + 2(3)$ less than 5? The answer is no, since

$$3(3) + 2(3) = 9 + 6 = 15$$

$$15 \geq 5$$

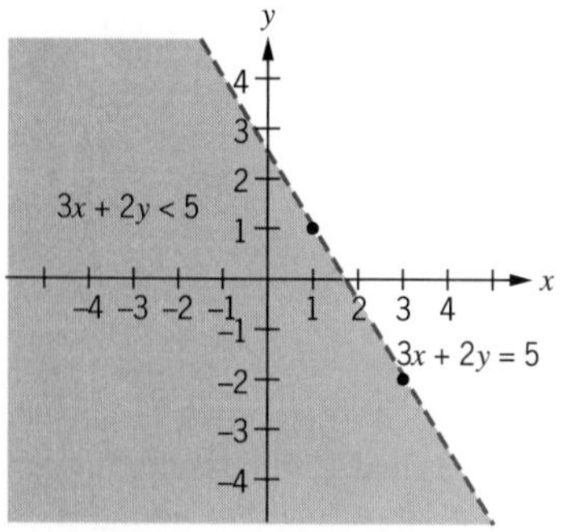

Figure 33

```
WINDOW FORMAT
 Xmin=-5
 Xmax=5
 Xscl=1
 Ymin=-3.3
 Ymax=3.3
 Yscl=.5
```

(a)

```
Y₁=(-3X+5)/2
Y₂=
Y₃=
Y₄=
Y₅=
Y₆=
Y₇=
Y₈=
```

(b)

```
ClrDraw
              Done
Shade(-3.3,
      (-3X+5)/2,2)
```

(c)

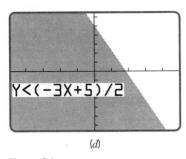

(d)

Figure 34

Therefore, points to the right of the line are not solutions of the inequality. Now test any point to the left of the line. Substitute the point $(1, -1)$ to see if it gives a true statement. Is $3(1) + 2(-1)$ less than 5?

$$3(1) + 2(-1) = 3 + (-2) = 1$$

$$1 < 5$$

This point is one solution of the inequality.

4. Shade the half-plane to the left side of the dashed line [containing $(1, -1)$], forming the graph shown in Figure 33.   ■

Calculator Note

As shown in Figure 34, a calculator can shade a function using the Shade command under the DRAW DRAW menu. To display this menu, press $\boxed{\text{2nd}}$ $\boxed{\text{DRAW}}$. As shown in Figure 34(a), use $X\text{min} = -5$, $X\text{max} = 5$, $Y\text{min} = -3.3$, $Y\text{max} = 3.3$, and $X\text{scl} = 1$ and $Y\text{scl} = 0.5$. We clear existing drawings from the screen by selecting 1: ClrDraw under the DRAW DRAW menu. Then select 7:Shade(. For this command we type within the parentheses the lower function and then the upper function bounding the region we wish to shade. We then type 2 and close the parentheses. For Example 34 rewrite $3x + 2y < 5$ as

$$y < \frac{-3x + 5}{2}$$

As shown in Figure 34(b), we place $Y_1 = (-3X + 5)/2$ as a line to bound one side of our shaded region. The other side is bounded by $Y\text{min} = -3.3$. Under 7:Shade(press $\boxed{(-)}$ 3.3, $\boxed{(}$ $\boxed{(-)}$ 3X $\boxed{+}$ 5 $\boxed{)}$ $\boxed{\div}$ 2, 2 $\boxed{)}$. The screen is shown in Figure 34(c). Press $\boxed{\text{ENTER}}$ to get the shaded region in Figure 34(d).

To get the region between $y > (-3x + 5)/2$ and $Y\text{max}$, use Shade$((-3X + 5)/2$, 3.1, 2) or Shade $(Y_1, Y\text{max}, 2)$.

EXAMPLE 35   Graph the solution set of $y \geq x + 1$.

SOLUTION

1. The ordered pairs in Table 6 help to determine the line representing $y = x + 1$.

TABLE 6

x	y
0	1
1	2
2	3
−1	0

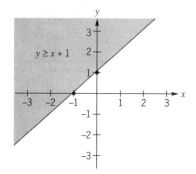

Figure 35

2. Since the relationship is $\geq$, the line $y = x + 1$ is sketched as a solid line in Figure 35.
3. Testing the points $(0, 4)$ and $(0, 0)$, we see that $4 > 0 + 1$ and $0 < 0 + 1$.
4. Shade the solution set shown in Figure 35.   ■

Calculator Note

To graph the inequality in Example 35, follow the instructions in the calculator note above. We set $Y_1 = X + 1$, enter under WINDOW the portion of the coordinate plane we will use, press [2nd] [DRAW] to obtain the DRAW DRAW menu, select item 1: ClrDraw to clear the screen, and select 7:Shade$(X + 1, 4.2, 2)$. Press [ENTER] to obtain Figure 36(d).

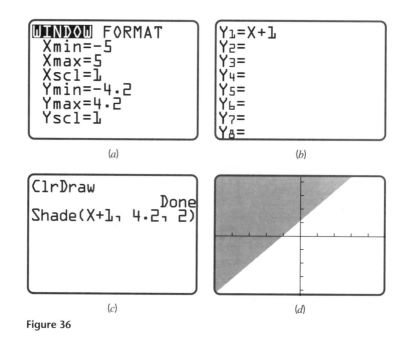

Figure 36

SUMMARY

In this section we discussed graphing linear equations and inequalities. Linear equations can be graphed by choosing an x value and finding the corresponding y value, and then choosing another value for x and finding the corresponding y value. They can also be graphed by finding the x- and y-intercepts. To graph a linear equation using a graphing calculator, we write the equation of the line as a function of y in terms of x and choose an appropriate viewing rectangle. The graph of a linear inequality is a region rather than a line.

Exercise Set 0.4

1. Graph the following points and name the quadrant where each is located. Verify on a graphing calculator.
   (a) $(2, 4)$ (b) $(-5, 6)$
   (c) $(-3, 2)$ (d) $(4, 1)$

2. Find y for each equation when $x = -1$, 0, and 1. Verify by using the List command on a graphing calculator. (**Hint:** For $y = -2x + 3$, input $Y_1 = -2\{-1, 0, 1\} + 3$, which outputs $\{5\ 3\ 1\}$.)
   (a) $y = -2x + 1$ (b) $2x + 5y = 3$
   (c) $\dfrac{x}{3} - 5y = -4$

3. Find five solutions for each of the following equations and inequalities. Verify using a graphing calculator.
   (a) $2x + 3y = 6$ (b) $y < x + 2$
   (c) $15 - 3y = 5x$ (d) $5x + 3y = 7$
   (e) $\left(\dfrac{3}{4}\right) x + y = 2y - 1$
   (f) $x + y + 3 > 0$

4. Plot five points (x, y) satisfying the expression $y = 3x + 1$. Draw a straight line through any two of these points and observe the results. Verify using a graphing calculator.

5. Construct a graph of the points satisfying each equation or inequality. Verify using a graphing calculator.
   (a) $y = 2x - 3$ (b) $y > x + 2$
   (c) $2x + 3y < 6$ (d) $5x + 3y = 7$
   (e) $2x + y = 1$ (f) $x + 1 = 0$
   (g) $y + 1 = 0$ (h) $y \le x - 1$
   (i) $y > x$ (j) $y = 2x$
   (k) $2s + t = 7$ (l) $x + y \le 9$

6. Find the x- and y-intercepts and graph each of the following equations. Verify using a graphing calculator.
   (a) $2x - 3y = -6$ (b) $\left(\dfrac{1}{2}\right) y = 2x + 4$
   (c) $x - 3y = 6$ (d) $5x - 3y = 15$

7. Graph $y = 2x - 3$ for $-1 \le x \le 4$. (**Hint:** The line extends from $x = -1$ to $x = 4$.)

8. Graph $y = 0.1x + 1.3$ for $-10 \le x \le 5$.

9. Graph $y = -2x + 4$ for $-2 \le x \le 4$.

10. Graph $y = 2x - 4$ for $-1 \le x \le 3$.

11. Graph $y \le 3x - 2$ for $-2 \le x \le 2$.

12. Give the equation of each of the axes in the coordinate plane.

13. On the same coordinate system graph $y = mx + 2$ for various values of m. (**Calculator hint:** Let $Y_1 = \{-3, 0, 3\}X + 2$.)

14. Graph $\dfrac{x}{2} + \dfrac{y}{2} \le 5$ for $x \ge 0$ and $y \ge 0$.

15. Draw a graph of the line that passes through the points $\{(1, -1),\ (0, -3),\ (2, 1),\ (3, 3),\ (\frac{3}{2}, 0),\ (-1, -5)\}$. Can you guess the equation of the line?

Applications (Business and Economics)

16. **Cost and Revenue.**   A revenue function is given as $R = 100x$. The cost function is defined by $C = 1000 + 80x$. Draw R and C on the same coordinate system and determine for what production there is or is not a profit. Verify using a graphing calculator. (**Hint:** Profit = Revenue − Cost.)

17. **Simple Interest.**   Draw a graph of the amount of money owed if $200 is borrowed at 8% simple interest for different periods. Use the equation $A = \$200(1 + 0.08t)$.

18. **Depreciation.**   Let x in the equation $y = -60x + 10{,}000$ represent months and y represent the dollar value of a machine less depreciation.
    (a) Prepare a table of values for $x = 0, 5, 10, 15, 20,$ $25, 30, 35, 40$.
    (b) Sketch the graph of the equation, letting x assume all real values greater than or equal to 0.

19. **Supply and Demand.**   Suppose the demand D (price per unit) for a certain item varies with the number of units x so that
    $$p_d = \frac{50 - 5x}{2}$$
    (a) What is the price when $x = 0$?
    (b) What is the price when $x = 4$?
    (c) What happens to the price when $x = 10$?
    (d) Graph the equation.

20. **Supply and Demand.**　Suppose the supply S (price per unit) for the item in Exercise 19 is given by

$$p_s = \frac{5x}{3}$$

(a) What is the price when $x = 3$? When $x = 9$?
(b) Graph this equation on the same coordinate axes that you used in Exercise 19.
(c) Find the equilibrium point.
(d) When is the supply greater than the demand?

21. **Supply and Demand.**　The demand D and supply S for a certain commodity are given by

$$D = p_d = -3x + 26 \quad \text{and} \quad S = p_s = 4x - 9$$

(a) Graph the equation involving D and the equation involving S on the same coordinate axes.
(b) Find the equilibrium point.
(c) When is the supply less than the demand?

Applications (Social and Life Sciences)

22. **Height and Weight Chart.**　A health club considers the following weights to be desirable for women who are either 60 or 72 inches tall.

Height x (in inches)	60	72
Weight y (in pounds)	100	146

Draw a graph of a straight line through these two points and then estimate from the line the desirable weight for a woman who is 63, 66, or 69 inches tall.

23. **Population.**　The population of a small city seems to be increasing linearly. In 1980, the population was 15,000. In 1986, it was 20,000. Graph the linear equation that represents population in terms of time. Estimate the population in 1990 and 1994. (**Hint:** Let $t = 0$ for 1980.)

Review Exercises

24. Solve each equation.

(a) $5 + \dfrac{x}{5} = \dfrac{7}{10}$

(b) $\dfrac{3x}{4} - \dfrac{5x - 1}{8} = \dfrac{1}{4}$

25. Solve each inequality.

(a) $\dfrac{y}{2} + 5 \geq 4y - 7$

(b) $\dfrac{x}{5} + 7 \leq \dfrac{x}{3} - 2$

(c) $\dfrac{x - 2}{-5} \leq 2$

(d) $-2(x + 1) > \dfrac{x}{4}$

26. Pam is holding $3.25 in her hand. If she has twice as many nickels as quarters and 1 more dime than quarters, how many coins does she hold? How many coins of each type does she have?

0.5　EXPONENTS AND RADICALS

OVERVIEW　Ten thousand dollars invested at 8% interest compounded annually for 20 years will grow to $10,000(1.08)^{20}$; this calculation involves the use of exponents. Exponents are sometimes classified as mathematical shorthand, but the properties of exponents must be learned before the shorthand can be used correctly. We use exponents in formulas that calculate compound interest, growth of bacterial cultures, the time for decay of radioactive material, and even the magnitude of earthquakes. We also study radicals in this section (square roots and higher roots) because radicals can be defined in terms of exponents. For example, in calculus we often use expressions involving radicals such as $y = \sqrt{x + 4}$, which can also be written as $y = (x + 4)^{1/2}$. In this section we consider the following topics:

- Properties of exponents
- Simplifying exponential expressions
- Properties of radicals
- Simplifying radicals

Properties of Exponents

When solving problems we often represent quantities by using symbols such as x^2, x^3, and x^4, which are read "x squared," "x cubed," and "x to the fourth power," respectively. These symbols express multiplication in a more condensed form and are defined as follows.

DEFINITION: EXPONENT

If x is any real number, variable, or algebraic expression, and if n is any nonnegative integer, then

$$x^n = \underbrace{x \cdot x \cdot x \cdot \cdots \cdot x}_{n \text{ factors}} \qquad (x^0 = 1, x \neq 0)$$

where n is the **exponent** and x is the **base**; x^n is read "the nth power of x" or "x to the nth power."

Let's investigate some of the properties of exponents. Since

$$x^3 = x \cdot x \cdot x \quad \text{and} \quad x^4 = x \cdot x \cdot x \cdot x$$

we have

$$(x^3)(x^4) = (x \cdot x \cdot x)(x \cdot x \cdot x \cdot x) = x^7 = x^{3+4}$$

That is, to multiply when the bases are the same, add the exponents. In general, if m and n are positive integers, then

$$(x^m)(x^n) = \underbrace{(x \cdot x \cdot x \cdot \cdots \cdot x)}_{m \text{ factors}} \underbrace{(x \cdot x \cdot x \cdot \cdots \cdot x)}_{n \text{ factors}} = x^{m+n}$$

EXAMPLE 36 Simplify $4^3 \cdot 4^2$.

SOLUTION Using the formula $(x^m)(x^n) = x^{m+n}$, the base x is 4, $m = 3$, and $n = 2$. Thus, $4^3 \cdot 4^2 = 4^{3+2} = 4^5$. We can evaluate this number using the exponential key $\boxed{\wedge}$ on a calculator. When we enter the keystrokes

$$4 \; \boxed{\wedge} \; 5 \; \boxed{\text{ENTER}}$$

We get the answer 1024. Also note that $4 \; \boxed{\wedge} \; 3 \; \boxed{\times} \; 4 \; \boxed{\wedge} \; 2 = 1024$. ■

Calculator Note

The graphing calculator is very versatile in calculating exponential expressions. By using parentheses, you can perform operations within an exponent. For example, to calculate $4^3 \cdot 4^2$, you can enter the calculation directly, or perform the addition within parentheses. For simple calculations, you will most likely simplify in your head first. Consequently, you could enter 4^5 much faster than you could enter $4^3 \cdot 4^2$ or 4^{3+2}. However, when you are faced with a much more complicated problem, such as $4^{\sqrt{2}+3-\pi}$, you can use parentheses after using the exponential key to perform the calculation very quickly. That is,

$$4 \boxed{\wedge} \boxed{(} \boxed{\text{2nd}} \boxed{\sqrt{}} 2 \boxed{+} 3 \boxed{-} \boxed{\text{2nd}} \boxed{\pi} \boxed{)} \approx 5.8371$$

Similar properties exist for the quotient of two expressions involving exponents.

$$\frac{x^5}{x^3} = \frac{x \cdot x \cdot x \cdot x \cdot x}{x \cdot x \cdot x} \qquad x \neq 0$$

$$= \frac{x \cdot x \cdot x}{x \cdot x \cdot x} \cdot x \cdot x$$

$$= 1 \cdot x \cdot x = x^2$$

This result could also have been obtained by subtracting the exponents:

$$\frac{x^5}{x^3} = x^{5-3} = x^2$$

Similarly,

$$\frac{x^4}{x^7} = \frac{x \cdot x \cdot x \cdot x}{x \cdot x \cdot x \cdot x \cdot x \cdot x \cdot x} = \frac{x \cdot x \cdot x \cdot x}{x \cdot x \cdot x \cdot x} \cdot \frac{1}{x \cdot x \cdot x} \qquad x \neq 0$$

$$= \frac{1}{x \cdot x \cdot x}$$

$$= \frac{1}{x^3} = \frac{1}{x^{7-4}}$$

In general, if m and n are positive integers and $x \neq 0$, then

$$\frac{x^m}{x^n} = \frac{\overbrace{x \cdot x \cdot x \cdots \cdot x}^{m \text{ factors}}}{\underbrace{x \cdot x \cdot x \cdots \cdot x}_{n \text{ factors}}} = \begin{cases} x^{m-n}, & \text{if } m \geq n \\ \dfrac{1}{x^{n-m}}, & \text{if } m < n \end{cases}$$

EXAMPLE 37   Simplify each expression.

(a) $\dfrac{(1.6)^5}{(1.6)^3}$ (b) $\dfrac{3}{3^5}$

SOLUTION

(a) $\dfrac{(1.6)^5}{(1.6)^3} = (1.6)^{5-3} = (1.6)^2$ or $1.6 \boxed{\wedge} 5 \boxed{\div} 1.6 \boxed{\wedge} 3 = 2.56$

(b) $\dfrac{3}{3^5} = \dfrac{1}{3^{5-1}} = \dfrac{1}{3^4} = \dfrac{1}{81}$ or $3 \boxed{\div} 3 \boxed{\wedge} 5 \boxed{\blacktriangleright \text{Frac}} = \dfrac{1}{81}$ ∎

Other useful properties involving exponents include the following.

$$(x^2)^3 = x^2 \cdot x^2 \cdot x^2 = (x \cdot x)(x \cdot x)(x \cdot x) = x^6 = x^{2 \cdot 3}$$

That is, to raise a power to a power, multiply the exponents. In general, if m and n are positive integers, then

$$(x^m)^n = x^{mn}$$

EXAMPLE 38

$(3^2)^4 = 3^{2 \cdot 4} = 3^8$ or $\boxed{(}\, 3 \boxed{\wedge} 2 \boxed{)}\boxed{\wedge} 4 = 6561$ and $3 \boxed{\wedge} 8 = 6561$ ∎

We also have

$$(xy)^3 = (xy)(xy)(xy) = x \cdot x \cdot x \cdot y \cdot y \cdot y = x^3 y^3$$

and

$$\left(\frac{x}{y}\right)^3 = \frac{x}{y} \cdot \frac{x}{y} \cdot \frac{x}{y} = \frac{x^3}{y^3}$$

In general, if m is a positive integer, then

$$(xy)^m = x^m y^m$$

and

$$\left(\frac{x}{y}\right)^m = \frac{x^m}{y^m}$$

For example,

$$(2 \cdot 3)^4 = 2^4 \cdot 3^4$$

The fact that $\dfrac{x^m}{x^m} = x^{m-m} = x^0$ ($x \neq 0$) provides a reason for defining x^0 to be

1. Any number (except 0) divided by itself is considered to be 1. Therefore, $\dfrac{x^m}{x^m} = 1$, and $\dfrac{x^m}{x^m} = x^0$. Thus, $x^0 = 1$.

Let's extend the multiplication of exponents to negative exponents, assuming that $x^m \cdot x^n = x^{m+n}$ holds for negative exponents.

$$
\begin{aligned}
x^n \cdot x^{-n} &= x^{n+(-n)} \\
&= x^{n-n} \\
&= x^0 \\
&= 1
\end{aligned}
$$

Now consider $x^n \cdot x^{-n} = 1$ and divide both sides by x^n, where $x \neq 0$, to obtain

$$\frac{x^n \cdot x^{-n}}{x^n} = \frac{1}{x^n}$$

$$x^{-n} = \frac{1}{x^n}$$

Therefore, when $x \neq 0$, x^{-n} is defined as

$$\frac{1}{x^n}$$

For example,

$$2^{-3} = \frac{1}{2^3} = \frac{1}{8}$$

Calculator Note

Can you evaluate 6^{-5}? On a calculator we can enter a negative exponent directly by using the $\boxed{(-)}$ key, not the subtraction key. However, the answer will be in decimal form, not in rational form. To display the answer in fraction form, we use $\boxed{\blacktriangleright \text{Frac}}$ from the MATH menu. Note that the calculator cannot display every number as a fraction; that is, there are limitations to the size of the denominator.

EXAMPLE 39 The properties of exponents are illustrated as follows. The problems involving numbers have been done both with and without a calculator. Note that the use of $\boxed{(-)}$ indicates the minus key, not the subtraction key.

(a) $11^3 \cdot 11^{-5} = 11^{3+(-5)} = 11^{-2} = \dfrac{1}{11^2} = \dfrac{1}{121}$ or

$11 \boxed{\wedge} 3 \boxed{\times} 11 \boxed{\wedge} \boxed{(-)} 5 \boxed{\blacktriangleright \text{Frac}} = 1/121$

(b) $\dfrac{7^4}{7^{-2}} = 7^{4 - (-2)} = 7^6 = 117{,}649$ or

$$7 \boxed{\wedge} 4 \boxed{\div} 7 \boxed{\wedge} \boxed{(-)} 2 = 117649$$

(c) $(2^3)^{-2} = 2^{3(-2)} = 2^{-6} = \dfrac{1}{2^6} = \dfrac{1}{64}$ or

$$\boxed{(} 2 \boxed{\wedge} 3 \boxed{)} \boxed{\wedge} \boxed{(-)} 2 = .015625$$

(d) $(3y)^2 = 3^2 y^2 = 9y^2$

(e) $\left(\dfrac{2}{y}\right)^2 = \dfrac{2^2}{y^2} = \dfrac{4}{y^2}$

(f) $8^0 = 1$ or $8 \boxed{\wedge} 0 = 1$

(g) $(3y)^{-2} = \dfrac{1}{(3y)^2} = \dfrac{1}{3^2 y^2} = \dfrac{1}{9y^2}$ ∎

The properties of exponents that we have discussed in this section are summarized as follows.

PROPERTIES OF EXPONENTS

For any integers m and n, where $x \neq 0$, $y \neq 0$ are any real numbers, variables, or algebraic expressions, each of the following is true.

(a) $x^m x^n = x^{m+n}$

(b) $\dfrac{x^m}{x^n} = x^{m-n}$

(c) $(x^m)^n = x^{mn}$

(d) $(xy)^m = x^m y^m$

(e) $\left(\dfrac{x}{y}\right)^m = \dfrac{x^m}{y^m}$

(f) $x^0 = 1$

(g) $x^{-n} = \dfrac{1}{x^n}$

Although stated for integers, these properties hold for any real numbers m and n.

COMMON ERROR   Students often confuse negative exponents with the sign of the number itself. The sign of the exponent does not affect the sign of the base.

Incorrect	Correct
$2^{-2} = -4$	$2^{-2} = \dfrac{1}{2^2} = \dfrac{1}{4}$

Another problem area is property (f) above. Even though it is a simple property, it can cause some confusion when the base is negative or there is a negative sign involved.

$$-8° = -(8)° = -1 \quad \text{but} \quad (-8)^0 = 1$$

Be careful when dealing with negative exponents.

Incorrect	Correct	Correct
$3y^{-2} = \dfrac{1}{(3y)^{-2}}$	$3y^{-2} = \dfrac{3}{y^2}$	$(3y)^{-2} = \dfrac{1}{9y^2}$

Practice Problem 1 Find the value of each of the following with and without a calculator.

(a) $\dfrac{3^{-2}}{3^{-4}}$ (b) $\dfrac{2^3 \cdot 2^5}{2^4}$ (c) $(3^{-2})^2$ (d) $(3x)^0$

ANSWER

(a) 9   (b) 16   (c) $\dfrac{1}{81}$ (d) 1

Practice Problem 2

(a) How is -3^2 different from $(-3)^2$? Evaluate these on a calculator as well.
(b) Simplify $-(3x)^2$ and $(-3x)^2$.

ANSWER

(a) The difference is demonstrated with parentheses:

$$-3^2 = -(3 \cdot 3) = -9 \quad \text{whereas} \quad (-3)^2 = (-3)(-3) = 9$$

(b) $-(3x)^2 = -(3x)(3x) = -9x^2$ but $(-3x)^2 = (-3x)(-3x) = 9x^2$.

Simplifying Exponential Expressions

Using the properties of exponents and the commutative and associative properties of multiplication, we can multiply and divide monomials (one-term expressions) and simplify, as demonstrated in the following examples.

EXAMPLE 40 Simplify each expression.

(a) $(2xy^3)(3x^3y)(-5y^2)$
(b) $(-14x^4y^3) \div (7xy^5)$

SOLUTION

(a) $(2xy^3)(3x^3y)(-5y^2) = 2 \cdot 3 \cdot (-5) \cdot x \cdot x^3 \cdot y^3 \cdot y \cdot y^2 = -30x^4y^6$

(b) $(-14x^4y^3) \div (7xy^5) = \dfrac{-14}{7} \cdot \dfrac{x^4}{x} \cdot \dfrac{y^3}{y^5} = \dfrac{-2x^3}{y^2}$

EXAMPLE 41   Simplify each expression. Your final answer should have only positive exponents.

(a) $\dfrac{x^5y^2}{xy^4} = \dfrac{x^5}{x} \cdot \dfrac{y^2}{y^4} = \dfrac{x^4}{1} \cdot \dfrac{1}{y^2} = \dfrac{x^4}{y^2}$

(b) $\dfrac{3x^2}{y^2} \cdot \dfrac{y^2}{9x^3} = \dfrac{3x^2y^2}{9x^3y^2} = \dfrac{3}{9} \cdot \dfrac{x^2}{x^3} \cdot \dfrac{y^2}{y^2} = \dfrac{1}{3} \cdot \dfrac{1}{x} \cdot 1 = \dfrac{1}{3x}$

(c) $\left(\dfrac{2x^{-6}}{y^{-2}}\right)^{-2} = \dfrac{(2x^{-6})^{-2}}{(y^{-2})^{-2}} = \dfrac{2^{-2}x^{12}}{y^4} = \dfrac{1}{2^2} \cdot \dfrac{x^{12}}{y^4} = \dfrac{x^{12}}{4y^4}$ ∎

Practice Problem 3   Perform the indicated operations and simplify.

(a) $\left(\dfrac{3x^2}{y^3}\right)\left(\dfrac{y^2}{2x^5}\right)$ (b) $\left(\dfrac{4x}{3y}\right)^{-1}\left(\dfrac{3x}{2y}\right)^2$

ANSWER

(a) $\dfrac{3}{2x^3y}$ (b) $\dfrac{27x}{16y}$

Properties of Radicals

As indicated earlier, the properties that we have developed for integral exponents hold when the exponents are any real numbers. For example,

$$3^{1/2} \cdot 3^{1/2} = 3^{1/2 + 1/2} = 3^1 = 3$$

The term $\sqrt{3}$ is defined to be a number that, when multiplied by itself, equals 3. That is,

$$\sqrt{3} \cdot \sqrt{3} = 3$$

and thus we have $\sqrt{3} = 3^{1/2}$. In general, $\sqrt{a} = a^{1/2}$ ($a \geq 0$). However, there is no real number x so that x^2 equals a negative real number. For example, there is no real number such that $x^2 = -16$; therefore, $\sqrt{-16}$ is not a real number. This discussion is summarized in the following definition.

DEFINITION AND PROPERTIES OF SQUARE ROOTS

(a) x is the square root of y if $x^2 = y$.
(b) $\sqrt{a}$ is the square root of a since $\sqrt{a} \cdot \sqrt{a} = a$ ($a \geq 0$).
(c) $\sqrt{a}$ can be written as $a^{1/2}$.
(d) $\sqrt{-a}$, where a is a positive number, is not a real number.
(e) $\sqrt{0} = 0$.

In the expression $\sqrt{a}$, the symbol $\sqrt{}$ is called a **radical** and a is called the **radicand**.

EXAMPLE 42   Evaluate, if possible. Check with a calculator.

(a) $\sqrt{16}$ (b) $-\sqrt{25}$ (c) $\sqrt{-9}$ (d) $\sqrt{0}$ (e) $\sqrt{3^{-2}}$

SOLUTION

(a) $\sqrt{16} = 4$ (b) $-\sqrt{25} = -5$ (c) $\sqrt{-9}$ is not a real number

(d) $\sqrt{0} = 0$ (e) $\sqrt{3^{-2}} = \sqrt{\frac{1}{9}} = \frac{1}{3}$ or $\boxed{\sqrt{}}\;\boxed{(}\;3\;\boxed{\wedge}\;\boxed{(-)}\,2\,\boxed{)}\;\boxed{\blacktriangleright \text{Frac}} = 1/3.$ ∎

Practice Problem 4 Evaluate

(a) $\sqrt{36}$ (b) $\sqrt{3^4}$ (c) $\sqrt{2^{-6}}$ (d) $\sqrt{-25}$

ANSWER

(a) 6   (b) $3^2 = 9$ (c) $\dfrac{1}{2^3} = \dfrac{1}{8}$ (d) Not a real number

Calculator Note

Some calculators calculate square roots of negative numbers and give the answer as an ordered pair. This is the complex form of the number. If you cannot suppress the complex mode on your calculator, be aware of this characteristic. On other calculators, if you attempt to calculate $\sqrt{-4}$, you get ERROR:DOMAIN, which means $\sqrt{}$ is not defined on the domain of negative numbers.

Let's now extend our discussion to the nth root of a number. First, we have

$$x^{1/3} \cdot x^{1/3} \cdot x^{1/3} = x^{1/3 + 1/3 + 1/3} = x^{3/3} = x$$

The real number raised to the third power to yield a number x is called the **cube root** of x and is denoted by $\sqrt[3]{x}$. Thus, $x^{1/3} = \sqrt[3]{x}$ and, in general,

$$x^{1/n} = \sqrt[n]{x}$$

if $\sqrt[n]{x}$ is defined. If n is even and x is negative, then $\sqrt[n]{x}$ does not exist over the set of real numbers. That is, $\sqrt{-4}$ and $\sqrt[4]{-16}$ are undefined over the real numbers. If n is odd and x is negative, then $x^{1/n} = \sqrt[n]{x}$ does exist and is a negative real number. That is,

$$\sqrt[3]{-27} = \sqrt[3]{(-3)^3} = -3 \quad \text{or} \quad \boxed{(-)}\,27\,\boxed{\wedge}\,\boxed{(}\,1\,\boxed{\div}\,3\,\boxed{)} = -3$$

EXAMPLE 43   The properties of radicals are illustrated as follows.

(a) $(32)^{1/5} = \sqrt[5]{32} = 2$ or $32\,\boxed{\wedge}\,\boxed{(}\,1\,\boxed{\div}\,5\,\boxed{)} = 2$

(b) $8^{1/3} = \sqrt[3]{8} = 2$ or $8\,\boxed{\wedge}\,\boxed{(}\,1\,\boxed{\div}\,3\,\boxed{)} = 2$

(c) $8^{2/3} = \sqrt[3]{8^2} = \sqrt[3]{64} = 4$ or $8\,\boxed{\wedge}\,\boxed{(}\,2\,\boxed{\div}\,3\,\boxed{)} = 4$

   Alternatively: $8^{2/3} = (\sqrt[3]{8})^2 = 2^2 = 4$ $\boxed{(}\,8\,\boxed{\wedge}\,\boxed{(}\,1\,\boxed{\div}\,3\,\boxed{)}\,\boxed{)}\,\boxed{\wedge}\,2 = 4$ ∎

We summarize the preceding definitions as follows.

DEFINITION AND PROPERTIES OF RADICALS

If x is any real number and m and n are any positive integers, then

(a) The nth root of x is $x^{1/n} = \sqrt[n]{x}$; x must be nonnegative if n is even.

(b) The cube root of x is given by

$$x^{1/3} = \sqrt[3]{x}$$

(c) $x^{m/n} = \sqrt[n]{x^m} = (\sqrt[n]{x})^m$; x must be nonnegative if n is even.

Note that property (b) is a special case of property (a).

NOTE: Since $(-2)^2 = 4$ and $2^2 = 4$, it seems that a square root of 4 can be either -2 or 2. However, for the answer to be unique, we restrict $\sqrt{x}$ to indicate only the positive or *principal* square root. Therefore, we write $\sqrt{4} = 2$. If we want the negative value, we use $-\sqrt{4} = -2$. Or, if we were solving an equation such as $y^2 = 4$ and wanted both values, we would write $y = \pm\sqrt{4} = \pm 2$. In general,

$$\sqrt[n]{a^n} = \begin{cases} |a|, & \text{if } n \text{ is even} \\ a, & \text{if } n \text{ is odd} \end{cases}$$

EXAMPLE 44   Simplify each expression.

(a) $(2x^{1/3})(3x^{1/2})$ (b) $\left(\dfrac{4x^{1/2}}{x^{-2}}\right)^{1/2}$

SOLUTION

(a) $(2x^{1/3})(3x^{1/2}) = 6x^{1/3 \,+\, 1/2}$ $\quad x^m \cdot x^n = x^{m-n}$

$\qquad\qquad\qquad\quad = 6x^{5/6}$

$\qquad\qquad\qquad\quad = 6\sqrt[6]{x^5}$

(b) $\left(\dfrac{4x^{1/2}}{x^{-2}}\right)^{1/2} = \dfrac{4^{1/2}x^{1/4}}{x^{-1}}$ $\quad \left(\dfrac{x}{y}\right)^m = \dfrac{x^m}{y^m}$ and $(x^m)^n = x^{mn}$

$\qquad\qquad\qquad\quad = 4^{1/2}x^{5/4}$ $\quad \dfrac{x^n}{x^m} = x^{n-m}$

$\qquad\qquad\qquad\quad = 2x \cdot x^{1/4}$

$\qquad\qquad\qquad\quad = 2x\sqrt[4]{x}$ ■

Practice Problem 5   Simplify each expression.

(a) $\left(\dfrac{2x^{1/2}}{y^{-1/2}}\right)^2$ (b) $\dfrac{\sqrt[3]{(-4)^3}}{\sqrt[3]{-8}}$ (c) $\left(\dfrac{3x^{1/6}y^{1/3}}{4y^{-2/3}}\right)^2$

ANSWER

(a) $4xy$ (b) 2   (c) $\dfrac{9x^{1/3}\,y^2}{16}$

The properties of exponents can be used to develop the following properties of radicals.

PROPERTIES OF RADICALS

If m and n are any positive integers and x and y are any real numbers for which the following exist, then

(a) $\sqrt[n]{x^n} = \begin{cases} |x|, & \text{if } n \text{ is even} \\ x, & \text{if } n \text{ is odd} \end{cases}$

(b) $\sqrt[n]{xy} = \sqrt[n]{x}\,\sqrt[n]{y}$

(c) $\sqrt[n]{\dfrac{x}{y}} = \dfrac{\sqrt[n]{x}}{\sqrt[n]{y}} \quad (y \neq 0)$

(d) $\sqrt[m]{\sqrt[n]{x}} = \sqrt[mn]{x}$

EXAMPLE 45   The properties of radicals are illustrated as follows.

(a) $\sqrt[3]{5^3} = 5$ or $(5 \boxed{\wedge} 3) \boxed{\wedge} (1 \boxed{\div} 3) = 5$

(b) $\sqrt[4]{y^4} = |y|$. Let $y = 3$ to check.

$$|3| = 3 \quad \text{and} \quad (3 \boxed{\wedge} 4) \boxed{\wedge} (1 \boxed{\div} 4) = 3$$

Also check by using $y = -3$.

(c) $\sqrt[4]{x^4 y^8} = \sqrt[4]{x^4} \cdot \sqrt[4]{y^8} = |x| y^2$. Let $x = 2$ and $y = 3$ to check.

$$2 \cdot 3^2 = 18 \quad \text{and}$$
$$((2 \boxed{\wedge} 4)(3 \boxed{\wedge} 8)) \boxed{\wedge} (1 \boxed{\div} 4) = 18$$

Also check by using $x = -2$ and $y = -3$.

(d) $\sqrt[3]{\dfrac{y^6}{x^3}} = \dfrac{\sqrt[3]{y^6}}{\sqrt[3]{x^3}} = \dfrac{y^2}{x}$. Let $x = 2$ and $y = 3$ to check.

$$((3 \boxed{\wedge} 6) \boxed{\div} (2 \boxed{\wedge} 3)) \boxed{\wedge} (1 \boxed{\div} 3) = 4.5$$

(e) $\sqrt[3]{\sqrt{64}} = \sqrt[6]{64} = 2$ or $(\boxed{\sqrt{}} 64) \boxed{\wedge} (1 \boxed{\div} 3) = 2$ ■

Simplifying Radicals   The preceding properties can be used to simplify radicals.

SIMPLEST RADICAL FORM

An expression containing square root radicals is said to be in **simplest radical form** if the following conditions are satisfied:

(a)   When the radicand is in factored form, no factor has an exponent greater than 1.

(b)   No radical appears in a denominator.

(c)   No rational expression is part of the radicand.

To simplify a square root radical, factor the terms in the radicand into factors whose exponents are multiples of 2 (whenever possible).

EXAMPLE 46 Change each radical to simplest form. Assume that all variables are nonnegative. In each part note that we select factors that are perfect squares, such as $25 = 5^2$.

(a) $\sqrt{4x^3} = \sqrt{2^2 \cdot x^2 \cdot x} = 2x\sqrt{x}$
(b) $\sqrt{75x^3} = \sqrt{25 \cdot 3 \cdot x^2 \cdot x} = \sqrt{5^2 \cdot x^2 \cdot 3 \cdot x} = 5x\sqrt{3x}$
(c) $\sqrt{100x^4y^3} = \sqrt{10^2 \cdot x^4 \cdot y^2 \cdot y} = 10x^2y\sqrt{y}$ ∎

COMMON ERROR

Expressions such as $\sqrt{2^3}$ and $1/\sqrt{2}$ are not in simplest radical form. Why not?

In general, an expression involving radicals with any index (root) is in simplest form when:

(a) No factor within the radicand has an exponent greater than or equal to the index.
(b) No radical appears in a denominator.
(c) No rational expression is part of the radicand.
(d) The index and the exponents in the radicand have no common factors.

Practice Problem 6 Simplify $\sqrt{72x^3}$. Assume that $x \geq 0$.

ANSWER $6x\sqrt{2x}$

EXAMPLE 47 Simplify each expression. Assume that all variables are nonnegative. As in Example 46, we select factors that will help us simplify the problem.

(a) $\sqrt[3]{54}$ (b) $\sqrt[4]{32x^5y^6}$ (c) $\sqrt[3]{-16y^5x^4}$ (d) $\sqrt[8]{x^4}$

SOLUTION

(a) $\sqrt[3]{54} = \sqrt[3]{27 \cdot 2} = \sqrt[3]{3^3 \cdot 2} = 3\sqrt[3]{2}$
(b) $\sqrt[4]{32x^5y^6} = \sqrt[4]{16 \cdot 2 \cdot x^4 \cdot x \cdot y^4 \cdot y^2} = 2xy\sqrt[4]{2xy^2}$
(c) $\sqrt[3]{-16y^5x^4} = \sqrt[3]{-8y^3x^3 \cdot 2y^2x} = -2yx\sqrt[3]{2y^2x}$
(d) $\sqrt[8]{x^4} = \sqrt{x}$ ∎

Practice Problem 7 Simplify $\sqrt[5]{96x^5y^7}$.

ANSWER $2xy\sqrt[5]{3y^2}$

When a radical expression contains a fraction or there is a radical in the denominator of an expression, we **rationalize the denominator** by eliminating the radical from the denominator. Note that $\sqrt{x}\,\sqrt{x} = \sqrt{x^2} = x\ (x \geq 0)$, since x^2 is a perfect square. Likewise, we see that $\sqrt[3]{x}\,\sqrt[3]{x^2} = \sqrt[3]{x^3} = x$, since x^3 is a perfect cube. A rational expression can be multiplied by 1 without changing the value of the expression. Therefore, we can multiply rational expressions by such terms as $\sqrt{2}/\sqrt{2}$ or $\sqrt[3]{x^2}/\sqrt[3]{x^2}$ without changing the actual value of the original expression. The term we select to multiply by is determined by the expression being rationalized.

EXAMPLE 48 Simplify each expression by rationalizing the denominator.

(a) $\dfrac{3}{\sqrt{2}} = \dfrac{3}{\sqrt{2}} \cdot \dfrac{\sqrt{2}}{\sqrt{2}} = \dfrac{3\sqrt{2}}{\sqrt{4}} = \dfrac{3\sqrt{2}}{2}$

Since $\sqrt{2}/\sqrt{2} = 1$, we have not changed the value of the original fraction.

(b) $\dfrac{7}{\sqrt[3]{3}} = \dfrac{7}{\sqrt[3]{3}} \cdot \dfrac{\sqrt[3]{3^2}}{\sqrt[3]{3^2}} = \dfrac{7\sqrt[3]{9}}{\sqrt[3]{3^3}} = \dfrac{7\sqrt[3]{9}}{3}$

Multiply by $\sqrt[3]{3^2}$ to get a perfect cube.

(c) $\dfrac{1}{\sqrt[4]{x}} = \dfrac{1}{\sqrt[4]{x}} \cdot \dfrac{\sqrt[4]{x^3}}{\sqrt[4]{x^3}} = \dfrac{\sqrt[4]{x^3}}{\sqrt[4]{x^4}} = \dfrac{\sqrt[4]{x^3}}{x}$ $\qquad (x > 0)$

(d) $\sqrt{\dfrac{x^2}{8}} = \dfrac{\sqrt{x^2}}{\sqrt{4 \cdot 2}} = \dfrac{x}{2\sqrt{2}} = \dfrac{x}{2\sqrt{2}} \cdot \dfrac{\sqrt{2}}{\sqrt{2}} = \dfrac{x\sqrt{2}}{2\sqrt{4}} = \dfrac{x\sqrt{2}}{4}$

(e) $\dfrac{5}{\sqrt{10}} = \dfrac{5}{\sqrt{10}} \cdot \dfrac{\sqrt{10}}{\sqrt{10}} = \dfrac{5\sqrt{10}}{\sqrt{100}} = \dfrac{5\sqrt{10}}{10} = \dfrac{\sqrt{10}}{2}$

✔ Calculator Check: By calculating the original expression and the rationalized expression, we can check our answers using a calculator. For example, in part (a) we have

$$3 \boxed{\div} \sqrt{2} \approx 2.121320344 \quad \text{and}$$

$$\boxed{(}\,3\sqrt{2}\,\boxed{)}\,\boxed{\div}\,2 \approx 2.121320344$$

When a problem involves a variable, substitute a value for the variable that is in the domain.

Practice Problem 8 Simplify each expression.

(a) $\dfrac{8}{\sqrt[3]{2}}$ $\qquad$ (b) $\dfrac{5}{\sqrt[3]{(5x)^2}}$

ANSWER

(a) $\dfrac{8}{\sqrt[3]{2}} \cdot \dfrac{\sqrt[3]{4}}{\sqrt[3]{4}} = 4\sqrt[3]{4}$ $\qquad$ (b) $\dfrac{5}{\sqrt[3]{(5x)^2}} \cdot \dfrac{\sqrt[3]{5x}}{\sqrt[3]{5x}} = \dfrac{\sqrt[3]{5x}}{x}$

We can use the distributive property of multiplication over addition to add and subtract expressions involving like radicals.

EXAMPLE 49   Simplify $6\sqrt{27} - 3\sqrt{12} + \sqrt{48}$.

SOLUTION $6\sqrt{27} - 3\sqrt{12} + \sqrt{48} = 6\sqrt{9 \cdot 3} - 3\sqrt{4 \cdot 3} + \sqrt{16 \cdot 3}$
$$= 6 \cdot 3\sqrt{3} - 3 \cdot 2\sqrt{3} + 4\sqrt{3}$$
$$= 18\sqrt{3} - 6\sqrt{3} + 4\sqrt{3}$$
$$= (18 - 6 + 4)\sqrt{3}$$
$$= 16\sqrt{3}$$

✔ Calculator Check: $6\sqrt{27}\ \boxed{-}\ 3\sqrt{12}\ \boxed{+}\ \sqrt{48} \approx 27.71281292$ and
$$16\sqrt{3} \approx 27.71281292$$

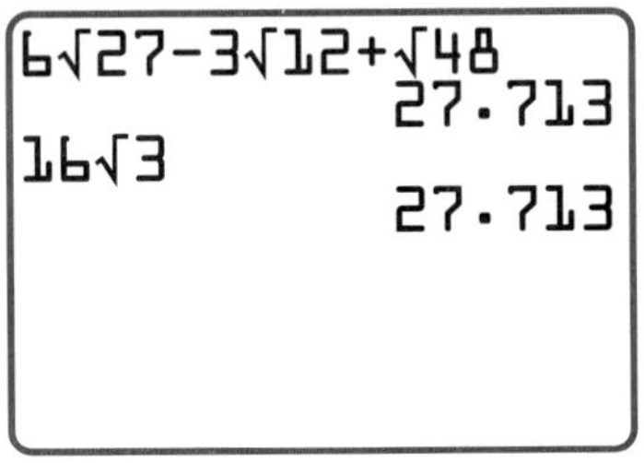

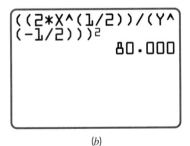

Figure 37

Practice Problem 9 Simplify $2\sqrt{48} - 3\sqrt{27}$.

ANSWER $2\sqrt{16 \cdot 3} - 3\sqrt{9 \cdot 3} = 8\sqrt{3} - 9\sqrt{3} = -\sqrt{3}$

Calculator Note

(a) We can use a calculator to determine if a numerical expression has been properly simplified. For instance, in Example 49 we stated that $6\sqrt{27} - 3\sqrt{12} + \sqrt{48}$ simplified to $16\sqrt{3}$. There are several ways that this can be checked. Perhaps the simplest way would be to subtract the term on the right of the equation from the expression on the left. If the result is zero or a number close to zero resulting from round-off, then the problem has been worked correctly. Another method is to calculate the value of the expression on the left and compare it to the value of the expression on the right, since both values can be seen on the screen at the same time (see Figure 37).

(b) For expressions involving variables, we can select any value for the variable for which the term is defined and then compare the numerical values (see Figure 38 for Practice Problem 10).

Practice Problem 10 Is

$$\left(\frac{2x^{1/2}}{y^{-1/2}}\right)^2 = 4xy \quad (x > 0, y > 0)$$

a true statement?

ANSWER   Yes, it is true.

SUMMARY

In this section we discussed the basic properties governing the use of exponents and radicals. Be sure to learn these rules, understand them, and use them. The rules of exponents apply for any real number as an exponent. It is important to be

Figure 38

able to reduce exponential expressions and to simplify radical expressions. We will be working with these expressions in later chapters and it is important to be proficient in simplifying all such expressions.

Exercise Set 0.5

Simplify each expression. The final answer should have only nonnegative exponents.

1. $3x^{-7}$

2. $4y^{-3}$

3. $\dfrac{5}{2x^{-4}}$

4. $\dfrac{4}{3y^{-5}}$

5. $2x^{-5}x^2$

6. $5y^{-9}y^7$

7. $\dfrac{x^{-5}}{x^{-2}}$

8. $\dfrac{z^{-7}}{z^{-3}}$

9. $(2x^{-3})^2$

10. $(y^3)^{-4}$

11. $(xy^2)^0$

12. $\left(\dfrac{x}{y}\right)^0$

13. $(-2x^{-2})^2$

14. $(-3x^{-3})^2$

15. $(-5x^{-1})^2$

Use the properties of exponents to simplify the following expressions. Use a calculator to check your answer.

16. $5^3 \cdot 5^{-4}$

17. $(7^2)^{-1}$

18. $4^2 \cdot 4^3 \cdot 4^{-1}$

19. $(3^{-2})^3$

20. $\left(\dfrac{2^2}{5^3}\right)^{-2}$

21. $5^0 \cdot 3^{-3}$

Write each of the following in exponential form.

22. $\sqrt[3]{x^4}$

23. $\sqrt{3y^2}$

24. $\sqrt[4]{x^2y^3}$

25. $\sqrt[3]{x^7y^4}$

26. $3\sqrt[5]{4x^2y}$

27. $4\sqrt[3]{x^3y^2}$

Find the indicated root, if it exists for real numbers. Check with a calculator.

28. $\sqrt[3]{-27}$

29. $\sqrt[4]{81}$

30. $\sqrt{-100}$

31. $\sqrt[5]{-32}$

32. $\sqrt{x^8y^4}$

33. $\sqrt[3]{x^9y^6}$

34. $\sqrt{9x^4y^2}$

35. $\sqrt[3]{\dfrac{125x^3}{8y^6}}$

36. $\sqrt[4]{-16}$

Evaluate each of the following, if it exists for real numbers. Check with a calculator.

37. $9^{1/2}$

38. $(-27)^{1/3}$

39. $\sqrt{16}$

40. $100^{-1/2}$

41. $(16^{-2})^{-1/2}$

42. $(-27)^{2/3}$

43. $(25^{-1/2})^{-1}$

44. $(2^{-7/4})^4$

45. $(-100)^{1/2}$

46. $-100^{1/2}$

Perform the indicated operations and leave each answer in simplest form with nonnegative exponents. Check with a calculator.

47. $(3x^3yz)(8xy^2z^4)$

48. $(x^2y^3)^4$

49. $\dfrac{10a^4b^3}{2a^5b^2}$

50. $\dfrac{2a^4}{b^3}$

51. $\dfrac{5x^3a^2z^4}{15x^4ay^2}$

52. $\left(\dfrac{x^2y^3}{x^4y}\right)^2$

Put each of the following in simplest form. Check by substituting values for the variables.

53. $\sqrt{8x^5}$

54. $\sqrt{9x^5y^4}$

55. $\sqrt[3]{2x^3y^4}$

56. $\sqrt{x} \cdot \sqrt{xy}$

Rationalize the denominator of each expression. Check with a calculator (substitute values for the variables).

57. $\dfrac{1}{\sqrt{2}}$

58. $\dfrac{2}{\sqrt{3x^3}}$

59. $\dfrac{4}{\sqrt[3]{2x}}$

60. $\dfrac{1}{\sqrt[3]{4}}$

In calculus, we often obtain expressions in the form cx^n, where c is a constant. Change each of the following to this form. Some exponents may be negative in your final answer. For example, $-3/x^2 = -3x^{-2}$.

61. $\dfrac{3}{x}$

62. $\dfrac{(2x)^3}{x^7}$

63. $\dfrac{3}{2x^{-1}}$

64. $2\sqrt{5y}$

65. $\dfrac{3}{2\sqrt[3]{x}}$

66. $\dfrac{9x}{2\sqrt[4]{3x}}$

Simplify each expression.

67. $\sqrt[3]{-125}$

68. $\sqrt[4]{81}$

69. $\sqrt{50}$

70. $\sqrt[3]{-343}$

71. $\sqrt[4]{1296}$

72. $\sqrt[3]{54}$

Simplify each of the following and express the result without negative exponents. (You may wish to verify your answers using a calculator.)

73. $\left(\dfrac{16x^{-2}y^3}{x^{-2}y^{-1}}\right)^{1/4}$

74. $\dfrac{(a^{-4}y^{-2})^{-1/2}}{(a^{-1}y^{-2})^2}$

75. $\left(\dfrac{25x^{-1}y^{1/2}}{16x^{-1/2}y}\right)^{1/2}$

76. $\dfrac{x^{-7}y^0z^{-1}}{x^2y^3z^2}$

77. $\left(\dfrac{4a^{-1}b^{-3}c}{2^0a^{-1}b^2c^{-2}}\right)^{-1}$

78. $3^{-3} \cdot \dfrac{9^{-3}}{9^{-5}}$

Simplify by reducing to positive exponents only, and then evaluate on a calculator.

79. $\dfrac{1.3^3}{1.3^{-2}}$

80. $\dfrac{2.07^{-3}}{2.07^{-5}}$

81. $\dfrac{1.001^{-3}}{1.001^{-2}}$

Put each of the following in simplest form. (You may wish to verify your answers using a calculator.)

82. $\sqrt{8x^3y^5}$

83. $\sqrt[3]{40a^7b^5}$

84. $\dfrac{6a^2}{\sqrt{3a}}$

85. $\dfrac{4y^2}{\sqrt{4y^3}}$

86. $\sqrt[3]{\dfrac{x}{2}}$

87. $\sqrt{\dfrac{4xy^5}{3x^2y}}$

Applications (Business and Economics)

88. **Sales Decay.**   A company finds that sales began to fall between November 1 and January 1 according to

$$S = \$1,000,000(2)^{-0.2t}$$

where t is the number of days after November 1. Find the sales volume on November 6, November 11, November 16, December 10, December 14, and December 22.

89. **Double-Declining Balance Depreciation.**   After k years, the value V of a machine that has a life of n years and an original cost of C dollars is given by

$$V = C\left(1 - \dfrac{2}{n}\right)^k$$

If a machine costs $10,000 and has a life of 10 years, find its value after

(a) 1 year   (b) 2 years

(c) 3 years   (d) 8 years

90. **Exponential growth.**   When compounded continuously, money grows according to the formula

$$A = P(2.718)^{rn}$$

where P is the money deposited and A is the value after n years at r percent per year (expressed as a decimal). Find the amount for $1000 invested at 8% per year for 20 years.

91. **Exponential Growth.**   Using the formula in Exercise 90, find the amount for $5000 invested at 6% per year for 10 years.

Applications (Social and Life Sciences)

92. **Decay.**   A decay model was found to be $P = 2^{-t}$, where t is the number of years of decay and P is the proportion left after t years of decay. Find P when $t = 4$, 6, and 10.

93. **Decay.**   The pressure in the aorta of a human adult can be approximated by $P = 100(2.718)^{-0.5t}$, where t is the number of seconds after the valves have closed. Find the aortic pressure when $t = 0$, 1, 2, 4, 6, 8, and 10. Notice how the pressure decays.

Review Exercises

94. Construct a graph for each of the following.

(a) $y < 3x + 2$ (b) $x = 2y - 7$

(c) $3x + 2y = 1$ (d) $3y + 2x \geq 1$

(e) $x + y < 5$ (f) $2x + y \leq 6$

0.6 OPERATIONS INVOLVING ALGEBRAIC EXPRESSIONS

OVERVIEW   In this section we review operations involving algebraic expressions that are used in the remainder of the book. The rules for combining algebraic expressions are based on the properties of numbers introduced in Section 0.1. When a number is written as the product of other numbers, each number in the product is called a **factor**. Similarly, when an algebraic expression is written as a product of other algebraic expressions, each algebraic expression in the product is called a factor. In this section, we are interested in writing algebraic expressions as products of factors. We consider the following topics:

- Adding, subtracting, and multiplying algebraic expressions
- Factoring algebraic expressions
- Simplifying rational expressions

Adding, Subtracting, and Multiplying Algebraic Expressions

Recall that to add algebraic expressions we add the coefficients of like terms (i.e., terms with the same variables having the same exponents). In the sum $3xy + 5xy$, we can use the distributive property of multiplication over addition to write $3xy + 5xy = (3 + 5)xy = 8xy$, which is the value we obtain if we simply add the coefficients.

EXAMPLE 50   Add $(5x^2 + 7) + 6x^2$.

SOLUTION   Using the associative and commutative properties of addition, we group like terms together:

$$(5x^2 + 6x^2) + 7 = 11x^2 + 7$$ ■

To combine terms involving a negative sign, such as $3x^4 - 2x^4$, we add the coefficients:

$$3x^4 - 2x^4 = 3x^4 + (-2)x^4 = [3 + (-2)]x^4 = x^4$$

EXAMPLE 51   Add $(3x^2 + 4xy + 7) + (6x^2 - xy - 4)$.

SOLUTION   Using the associative and commutative properties of addition, we have

$$(3x^2 + 6x^2) + (4xy - xy) + (7 - 4) = 9x^2 + 3xy + 3$$ ■

EXAMPLE 52   Using the distributive property of multiplication over addition, multiply and simplify $3xy(2x^2 - 3xy + y^2)$.

SOLUTION

$$3xy(2x^2 - 3xy + y^2) = 3xy(2x^2) + 3xy(-3xy) + 3xy(y^2)$$
$$= 6x^3y - 9x^2y^2 + 3xy^3$$

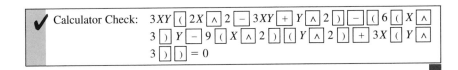

Practice Problem 1 Perform the operations, simplify, and then check with a calculator.

(a) $3x^2(4xy + y^2)$ (b) $3x^2 - (4x + 3 - x^2) + (6 - x^2)$

ANSWER

(a) $12x^3y + 3x^2y^2$ (b) $3x^2 - 4x + 3$

✓ Calculator Check: The original expression minus the answer equals zero in each part.

Algebraic expressions often involve symbols of grouping such as parentheses, brackets, and braces. These symbols always come in pairs and are often used to add clarity to a problem. When working a problem that has more than one set of grouping symbols, start with the innermost set and work your way to the outermost set.

EXAMPLE 53   Simplify $7x - [4x - (2x - 1)]$.

SOLUTION

$$7x - [4x - (2x - 1)] = 7x - [4x - 2x + 1]$$
$$= 7x - [2x + 1] = 7x - 2x - 1 = 5x - 1$$ ■

Calculator Note:

To check the solution in Example 53 using a calculator, store a value for x; next, evaluate the original expression and then the simplified version (see Figure 39).

```
3→X
           3.000
7X-(4X-(2X-1))
          14.000
5X-1
          14.000
```

Figure 39

Practice Problem 2 Simplify $x - [y - 2(x + y)]$. Check with a calculator.

ANSWER $3x + y$

✓ Calculator Check: $X \boxed{-} \boxed{(} Y \boxed{-} 2 \boxed{(} X \boxed{+} Y \boxed{)} \boxed{)} \boxed{-} \boxed{(} 3X \boxed{+} Y \boxed{)} = 0$

The distributive property of multiplication over addition is also the basis by which we multiply in the following example.

EXAMPLE 54   Multiply $(3x + 2)(6x + 1)$. Check with a calculator.

SOLUTION

$$(3x + 2)(6x + 1) = (3x + 2)(6x) + (3x + 2)(1)$$
$$= (3x)(6x) + 2(6x) + 3x(1) + 2(1)$$
$$= 18x^2 + 12x + 3x + 2$$
$$= 18x^2 + 15x + 2$$

✔ Calculator Check: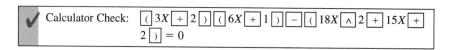

Recall that the multiplication of two binomials is often referred to as the FOIL method.

$$(3x + 2)(6x + 1) = \overbrace{(3x)(6x)}^{\text{First}} + \overbrace{(3x)(1)}^{\text{Outer}} + \overbrace{(2)(6x)}^{\text{Inner}} + \overbrace{(2)(1)}^{\text{Last}} = 18x^2 + 15x + 2$$

We can use a double application of the distributive property of multiplication over addition to verify the first five of the following properties involving products.

SPECIAL PRODUCTS

(a) $(a + b)(a - b) = a^2 - b^2$
(b) $(a + b)^2 = a^2 + 2ab + b^2$
(c) $(a - b)^2 = a^2 - 2ab + b^2$
(d) $(a + b)(a + c) = a^2 + a(b + c) + bc$
(e) $(a + b)(c + d) = ac + ad + bc + bd$
(f) $(a + b)^3 = a^3 + 3a^2b + 3ab^2 + b^3$
(g) $(a - b)^3 = a^3 - 3a^2b + 3ab^2 - b^3$

NOTE: For the remainder of this section, we will not show the calculator check for each problem, but you should continue to check each problem and exercise that you work.

EXAMPLE 55   Find the following products.

(a) $(3x^2 + 2y)(3x^2 - 2y)$ (b) $(3x^2 - 2y)^2$ (c) $(3x^2 + y)^3$

SOLUTION

(a)   Use the special product $(a + b)(a - b) = a^2 - b^2$.

$$(3x^2 + 2y)(3x^2 - 2y) = (3x^2)^2 - (2y)^2 = 9x^4 - 4y^2$$

(b)   Use the special product $(a - b)^2 = a^2 - 2ab + b^2$, where $a = 3x^2$ and $b = 2y$.

$$(3x^2 - 2y)^2 = (3x^2)^2 - 2(3x^2)(2y) + (2y)^2 = 9x^4 - 12x^2y + 4y^2$$

(c)   Use the special product $(a + b)^3 = a^3 + 3a^2b + 3ab^2 + b^3$, where $a = 3x^2$ and $b = y$.

$$(3x^2 + y)^3 = (3x^2)^3 + 3(3x^2)^2(y) + 3(3x^2)y^2 + y^3$$
$$= 27x^6 + 27x^4y + 9x^2y^2 + y^3 \qquad\blacksquare$$

The difference of two squares, $a^2 - b^2 = (a - b)(a + b)$, leads us to a useful method of rationalizing a denominator containing a sum or a difference. In the last section, we rationalized denominators with only one term. However, in calculus we often have to rationalize denominators or numerators such as $2 + \sqrt{3}$ or $x - \sqrt{x}$. To rationalize these, we multiply by a *conjugate* factor. The conjugate of $2 + \sqrt{3}$ is $2 - \sqrt{3}$ because $(2 + \sqrt{3})(2 - \sqrt{3}) = 4 - 3 = 1$ (no radicals).

EXAMPLE 56   Rationalize the denominator in the expression $\dfrac{6}{\sqrt{x} - 2}$.

SOLUTION   The conjugate factor of $\sqrt{x} - 2$ is $\sqrt{x} + 2$, so we multiply the numerator and the denominator by the conjugate.

$$\frac{6}{\sqrt{x} - 2} \cdot \frac{\sqrt{x} + 2}{\sqrt{x} + 2} = \frac{6\sqrt{x} + 12}{x - 4}$$

When checking this problem, remember not to substitute $x = 4$, which would make the denominator zero. $\blacksquare$

Practice Problem 3   Rationalize the denominator in the expression $\dfrac{4}{\sqrt{3} - 1}$.

ANSWER $\dfrac{4(\sqrt{3} + 1)}{2} = 2\sqrt{3} + 2$

Factoring

We now study the process of factoring polynomials. Factoring is often important in the process of solving equations or reducing rational expressions. For example, we factor 12 as a product of 3 and 4 or $12 = 3 \cdot 4$. We can also write $12 = 24 \cdot \frac{1}{2}$; however, $\frac{1}{2}$ is not an integer and we are not usually interested in factors other than integers. The easiest procedure for factoring polynomials (when factoring is possible) involves using the distributive property of multiplication over addition in reverse order.

EXAMPLE 57   Factor $6x + 3y$.

SOLUTION   Note that $6 = 3 \cdot 2$ and $3 = 3 \cdot 1$, so we factor out the 3.

3 is common factor

$$6x + 3y = \underline{3} \cdot 2x + \underline{3} \cdot y = \underline{3}(2x + y) \qquad\blacksquare$$

NOTE: Factoring can always be checked by reversing your work, that is, by performing the multiplication and seeing if you get the original product.

If there are several terms of a polynomial that have common factors, we can remove the common factors using the distributive property. Always be sure to remove the greatest common factor.

EXAMPLE 58   Factor $4x^2y + 8x^3y^3 + 12x^4y$.

SOLUTION

$$4x^2y + 8x^3y^3 + 12x^4y = \overbrace{4x^2y(1) + 4x^2y(2xy^2) + 4x^2y(3x^2)}^{4x^2y \text{ is common factor}}$$
$$= 4x^2y(1 + 2xy^2 + 3x^2)$$

Practice Problem 4 Factor each expression.

(a) $6x^3 - 8x$ (b) $(x + 2)(2x) + (x + 2)(7)$

ANSWER

(a) $2x(3x^2 - 4)$ (b) $(x + 2)(2x + 7)$

In Examples 57 and 58 the expressions are factored completely; that is, all the factors are removed that can be removed. Earlier in this section we listed several special products. We now use these in reverse order: from polynomial form to factored form. It is important to recognize the following special polynomials.

FACTORING SPECIAL POLYNOMIALS

(a)	Difference of two squares:	$a^2 - b^2 = (a + b)(a - b)$
(b)	Perfect square trinomial:	$a^2 + 2ab + b^2 = (a + b)^2$
(c)	Perfect square trinomial:	$a^2 - 2ab + b^2 = (a - b)^2$
(d)	Sum of two cubes:	$a^3 + b^3 = (a + b)(a^2 - ab + b^2)$
(e)	Difference of two cubes:	$a^3 - b^3 = (a - b)(a^2 + ab + b^2)$

The easiest of the special polynomials to recognize is the difference of two squares. Look for perfect square integers and exponents that are even.

EXAMPLE 59   Factor $4x^2 - 9y^4$.

SOLUTION   Note that $4x^2 = (2x)^2$ and $9y^4 = (3y^2)^2$. We let $a = 2x$ and $b = 3y^2$. Thus

$$4x^2 - 9y^4 = (2x)^2 - (3y^2)^2 = (2x + 3y^2)(2x - 3y^2) \qquad a^2 - b^2 = (a + b)(a - b)$$

It is a little more difficult to recognize and factor perfect square trinomials, but with a little practice it becomes easier to recognize them. When you want to determine whether a trinomial is a perfect square trinomial, the first step is to locate the two terms that are perfect squares. Then determine whether *two times the product of the square roots of these two terms* is the other term in the trinomial.

$$a^2 + 2ab + b^2 = (a + b)^2 \qquad a^2 - 2ab + b^2 = (a - b)^2$$

Same sign   Same sign

EXAMPLE 60   Factor $9y^4 + 12y^2x + 4x^2$.

SOLUTION   We see that $9y^4 = (3y^2)^2$, $4x^2 = (2x)^2$, and $12y^2x = 2 \cdot 3y^2 \cdot 2x$. Therefore, it is a perfect square trinomial and we factor as follows:

$$9y^4 + 12y^2x + 4x^2 = (3y^2)^2 + 2 \cdot 3y^2 \cdot 2x + (2x)^2 = (3y^2 + 2x)^2$$

$$a^2 + 2ab + b^2 = (a + b)^2$$

■

Practice Problem 5 Factor each expression.

(a) $25 - 64y^4$ (b) $16x^4 - 24x^2y + 9y^2$

ANSWER

(a) $(5 - 8y^2)(5 + 8y^2)$ (b) $(4x^2 - 3y)^2$

Now we turn our attention to factoring the sum or difference of two cubes. Pay careful attention to the signs when working with cubes.

EXAMPLE 61   Factor $27 - y^3$.

SOLUTION   Since $27 = 3^3$, we have the difference of two cubes and we apply the formula.

$$27 - y^3 = (3 - y)(3^2 + 3y + y^2) = (3 - y)(9 + 3y + y^2)$$

Signs differ

$$a^3 - b^3 = (a - b)(a^2 + ab + b^2)$$

■

Practice Problem 6 Factor $8x^3 + 125y^3$.

ANSWER $(2x + 5y)(4x^2 - 10xy + 25y^2)$

COMMON ERROR   Students often have trouble with cubes by confusing $a^3 + b^3$ and $(a + b)^3$.

Incorrect	**Correct**
$(a + b)^3 = a^3 + b^3$	$(a + b)^3 = a^3 + 3a^2b + 3ab^2 + b^3$
$a^3 - b^3 = (a - b)(a - b)(a - b)$	$a^3 - b^3 = (a - b)(a^2 + ab + b^2)$

All trinomials are not special trinomials; however, some trinomials will fit the formula $a^2 + a(b + c) + bc = (a + b)(a + c)$. A polynomial such as $x^2 + 7x + 10$ fits this formula because $x^2 + 7x + 10 = x^2 + x(5 + 2) + 5 \cdot 2 = (x + 5)(x + 2)$.

EXAMPLE 62   Factor $x^2 - 7x + 12$.

SOLUTION   In this polynomial 12 is not a perfect square (since $\sqrt{12}$ is not a rational number), so $x^2 - 7x + 12$ is not a perfect trinomial. We need to find two integers whose sum is -7 and whose product is 12. The two integers must be negative because the sum is negative and the product is positive. The negative integer factors of 12 are -1 and -12, -2 and -6, and -3 and -4. From this list we can see that -3 and -4 meet the requirements, so we now rewrite and factor according to the formula $a^2 + a(b + c) + bc = (a + b)(a + c)$.

$$
\begin{aligned}
x^2 - 7x + 12 &= x \cdot x - 7x + (-3)(-4) \\
&= \underbrace{x \cdot x}_{a^2} + \underbrace{(-3 + -4)}_{b + c}\, x + \underbrace{(-3)(-4)}_{b \cdot c} \qquad \begin{aligned} &a = x, b = -3, \\ &c = -4 \end{aligned} \\
&= (x - 3)(x - 4) \qquad\qquad\qquad \begin{aligned} &a^2 + a(b + c) + bc \\ &= (a + b)(a + c) \end{aligned}
\end{aligned}
$$

■

The special product $(a + b)(c + d) = ac + ad + bc + bd$ may occur with three terms or with four terms. For these we often have to use trial and error to find the right combination of factors. Note that when factoring a trinomial, we do not have a factor such as $(2x + 2)$ unless 2 is common to all terms. In that case, the 2 should be factored out before proceeding.

EXAMPLE 63   Factor $6x^2 + 7x + 2$.

SOLUTION   We need to find factors of $6x^2$ and 2 that give us a sum of $7x$ (middle term). We have

$$6x^2 = 3x \cdot 2x \quad \text{or} \quad 6x^2 = 6x \cdot x$$

and

$$2 = 1 \cdot 2 \quad \text{or} \quad 2 = (-1)(-2)$$

We know that we will not have the factor $(2x + 2)$ because 2 is not common to all terms. Therefore, by trial and error we find

$$6x^2 + 7x + 2 = (2x + 1)(3x + 2) \qquad (a + b)(a + c)$$

■

A four-term expression usually requires that we use *factoring by grouping*, which involves the use of a double application of the distributive property. Group two terms with a common factor or pattern or three terms with a common factor or pattern. Then try to factor each group. Sometimes it may be necessary to rearrange the terms for the grouping process. The following example illustrates the grouping technique.

EXAMPLE 64   Factor each expression.

(a) $3xy + 9x + 2y + 6$ (b) $x^2 - 4y^2 + x - 2y$.

SOLUTION

(a)   We group the first two terms and the last two terms and then determine whether there is a common factor in the two groups.

$$3xy + 9x + 2y + 6 = 3x\underline{(y + 3)} + 2\underline{(y + 3)} \qquad (y + 3) \text{ is common.}$$
$$= (3x + 2)(y + 3)$$

(b) $(x^2 - 4y^2) + (x - 2y) = (x + 2y)(x - 2y) + 1(x - 2y)$
$$= (x - 2y)(x + 2y + 1) \qquad \blacksquare$$

EXAMPLE 65   Factor $2x^2 + 2x + 1$.

SOLUTION   The factors of the first term, $2x^2$, are $2x$ and x. The factors of the last term, 1, are 1 and 1 or -1 and -1. Since the middle term $2x$ is positive, the product is positive and the sum is positive; thus we can use only 1 and 1.

$$(2x + 1)(x + 1) = 2x^2 + 3x + 1 \neq 2x^2 + 2x + 1$$

Thus, $2x^2 + 2x + 1$ cannot be factored with real coefficients. $\blacksquare$

Practice Problem 7   Factor the following expressions.

(a) $25x^2 - y^4$ (b) $36x^2 + 24x + 4$ (c) $x^2 + x - 6$
(d) $64 - 27y^3$

ANSWER

(a) $(5x - y^2)(5x + y^2)$ (b) $4(3x + 1)^2$ (c) $(x + 3)(x - 2)$
(d) $(4 - 3y)(16 + 12y + 9y^2)$

Simplifying Rational Expressions

In Section 0.1, we learned from a property of fractions that

$$\frac{ka}{kb} = \frac{a}{b} \quad (k, b \neq 0) \quad \text{and} \quad \frac{x(x - 2)}{3(x - 2)} = \frac{x}{3} \quad (x \neq 2)$$

Using factoring and this property, we can simplify rational expressions as demonstrated in the following examples.

EXAMPLE 66   Simplify the rational expression

$$\frac{4x^2 - 1}{4x^2 - 4x + 1}$$

SOLUTION

$$\frac{4x^2 - 1}{4x^2 - 4x + 1} = \frac{(2x - 1)(2x + 1)}{(2x - 1)(2x - 1)} \quad \text{Factor.}$$

$$= \frac{2x + 1}{2x - 1} \quad \text{Reduce.}$$

When checking this problem, do not set $x = \frac{1}{2}$. Why?

EXAMPLE 67   Multiply and simplify

$$\frac{x^2 - 5x + 6}{x + 2} \cdot \frac{x^2 - 4}{x - 3}$$

SOLUTION   We factor all terms that can be factored and then reduce the expression.

$$\frac{x^2 - 5x + 6}{x + 2} \cdot \frac{x^2 - 4}{x - 3} = \frac{(x - 3)(x - 2) \cdot (x - 2)(x + 2)}{(x + 2)(x - 3)}$$

$$= (x - 2)^2 \quad \text{Reduce.}$$

Practice Problem 8 Simplify

$$\frac{2x + 8}{12} \div \frac{3x + 12}{6} \qquad \text{Remember } \frac{a}{b} \div \frac{c}{d} = \frac{a}{b} \cdot \frac{d}{c} \quad (b, c, d \neq 0)$$

ANSWER $\dfrac{1}{3}$

Addition and subtraction of rational expressions follow the same procedures as in adding and subtracting real numbers. First, be sure to determine the **least common denominator**.

EXAMPLE 68   Add

$$\frac{x + 1}{x^2 - 9} + \frac{2}{x - 3}$$

What values of x cannot be used to check?

SOLUTION   To find the least common denominator, we factor all denominators.

$$\frac{x + 1}{x^2 - 9} + \frac{2}{x - 3} = \frac{x + 1}{(x - 3)(x + 3)} + \frac{2}{x - 3} \quad \text{Factor.}$$

$$= \frac{x + 1}{(x - 3)(x + 3)} + \frac{2(x + 3)}{(x - 3)(x + 3)} \quad \begin{array}{l}(x - 3)(x + 3) \text{ is the} \\ \text{least common denomina-} \\ \text{tor.}\end{array}$$

$$= \frac{x + 1 + 2x + 6}{(x - 3)(x + 3)} \qquad \text{Add numerators.}$$

$$= \frac{3x + 7}{(x - 3)(x + 3)} \qquad \text{Simplify.}$$

$$= \frac{3x + 7}{x^2 - 9}$$

The values $x = 3$ or $x = -3$ should not be used as values for x. Why?   ■

Calculator Note

A calculator can be used to check your work when combining rational expressions. Substitute values for the variables and determine whether your answer is correct. Since you are dealing with rational expressions, there may be a very slight difference in answers because of round-off.

Practice Problem 9 Use a calculator to determine whether the expressions are equivalent.

$$\frac{8}{6x + 3} + \frac{3}{10x + 5} \stackrel{?}{=} \frac{49}{15(2x + 1)}$$

ANSWER   By substituting any value for x that does not yield a zero denominator, you can evaluate each expression on your viewing screen. You will find that the expressions are equivalent. You can also use the TEST menu.

SUMMARY

The operations of addition, subtraction, factoring, multiplying and dividing algebraic expressions are basic and essential. Although a calculator can help you in this work, it is essential that you understand all the properties and formulas for special products and special polynomials that have been presented. Good algebra skills are essential for success in calculus.

Exercise Set 0.6

Write each polynomial in general form, as illustrated in Example 51. Check with a calculator.

1. $(3x^2 - 6x + 4) + (x^2 - 4x + 4) - (x^2 - 1)$
2. $(x^3 + 2x^2 - 4) + (2x^3 - x^2 + 2x) + (x - 1)$

Remove the grouping symbol and combine like terms. Check with a calculator.

3. $3x + [(x - 4z) - (4x - 3z)] - 2x$

4. $x + 4 - 2[2 - 3(x - y)]$
5. $4x - \{3x^2 - 2[x - 3(x^2 - x)] + 4\}$
6. $3(x - 2y) - (-x + 2y)$

Simplify each product. Check with a calculator.

7. $6x(x^2 + 2x)$
8. $3xy(x^2y - 4x)$
9. $(x + 2)(x - 3)$
10. $(x - 1)(x + 5)$
11. $(3x - 1)(x + 2)$
12. $(x + 3)(2x + 1)$
13. $4(2x - 1)(3x + 7)$
14. $-5(x - 5)(2x - 1)$

Perform the indicated operations and simplify. Check with a calculator.

15. $(2x + 3)(2x - 3)$ 16. $(x^2y - 0.2)^2$

17. $(4x + 0.3)^2$

18. $\left(2z^2 + \dfrac{1}{3}\right)\left(2z^2 - \dfrac{1}{3}\right)$

19. $(4x + 1)(3x - 2)$ 20. $(x^2 + 3)(3x^2 - 1)$

21. $(0.1 - 4x)(0.1 + 4x)$ 22. $(6x + 3)(5x + 4)$

Factor each of the following completely. Check with a calculator.

23. $x^2 - 5x$ 24. $3x^2 - 7x$

25. $3x^2 - 75$ 26. $4x^2 - 36$

27. $3x^2y - 12x^2$

28. $2x^3y^2 + 8x^2y + 16x^2y^2$

29. $x^2 + 10x + 25$ 30. $x^2 + 14x + 49$

31. $x^2 - 2x - 3$ 32. $x^2 - 2x - 15$

33. $2x^2 - 6x - 56$ 34. $x^2 + 2x - 24$

Reduce to lowest terms.

35. $\dfrac{8x + 12}{12x - 16}$ 36. $\dfrac{2x^2 + x}{4x^2 - 1}$

37. $\dfrac{3y^2 + y - 2}{y^2 + 3y + 2}$

Factor the following expressions completely. Check with a calculator.

38. $x^2 + 3x - 10$ 39. $x^2 + 7x + 10$

40. $xy + y - x - 1$

41. $xy^2 - 2y^2 + 2xy - 4y$

42. $x^3 - 2x^2 - 8x$ 43. $2x^2 + 6x + 4$

44. $8x^3 - 27y^3$ 45. $64x^3 + 125y^3$

46. $16 + 2y^3$ 47. $(x - 1)^3 - 8$

Perform the following operations and simplify. Check with a calculator.

48. $\dfrac{5x + 25}{10} \div \dfrac{6x + 30}{12}$

49. $\dfrac{3y - 6}{5y - 10} \div \dfrac{4y - 8}{3y + 6}$

50. $\dfrac{5}{3(x + 2)} - \dfrac{1}{6(x + 2)}$

51. $\dfrac{2}{3(x - 2)} + \dfrac{3}{4(x - 2)}$

52. $\dfrac{5}{x^2 - 2x - 3} + \dfrac{3}{x^2 - x - 6}$

53. $\dfrac{5}{y^2 - 3y - 10} + \dfrac{3}{y^2 - y - 20}$

54. $\dfrac{3x}{x^2 - 2x - 3} - \dfrac{5x}{x^2 - x - 6}$

55. $\dfrac{4x}{3x^2 + 7x - 6} - \dfrac{2x}{3x^2 - 14x + 8}$

56. $\dfrac{6x^2 - 5x - 6}{x^2 - 10x + 24} \div \dfrac{4x^2 - 9}{x^2 - 9x + 18}$

57. $\dfrac{6x^2 - 5x - 6}{6x^2 + 5x - 6} \div \dfrac{12x^2 - x - 6}{12x^2 + x - 6}$

Rationalize each denominator and simplify as much as possible. Check with a calculator.

58. $\dfrac{3}{\sqrt{2} - 1}$ 59. $\dfrac{1}{2\sqrt{2} - 2}$

60. $\dfrac{4}{a - 2\sqrt{x}}$ 61. $\dfrac{4}{\sqrt{a} - \sqrt{b}}$

62. $\dfrac{2}{\sqrt{x} - \sqrt{y}}$ 63. $\dfrac{3}{2\sqrt{x} - 4}$

Applications (Business and Economics)

64. **Profit and Cost Functions.** Recall that Profit = Revenue − Cost; that is, $P(x) = R(x) - C(x)$. Suppose that a company's revenue and cost of manufacturing x units are given by

$$R(x) = (175 - x)x \quad \text{and}$$
$$C(x) = 10{,}000 + 300x - 10 - 6x^2$$

State the profit as a simplified expression.

65. **Simple Interest.** The formula for simple interest is given by $I = Prt$. The amount owed, A, is given by $A = P + I$. Substitute for I and factor so that A is given as a product of P and a factor.

Applications (Social and Life Sciences)

66. **Blood Velocity.** The velocity v of blood x centimeters from the center of a given artery can be approxi-

mated by

$$v = 2 - 20{,}000x^2$$

where x is limited by the size of the artery. Write the velocity as the product of 20,000 and some factor.

67. **Response Function.**   An employee of United Appeal noted in a study that she could relate the number of responses y received per month to the number of invitations x to fund-raising social occasions by

$$y = x - 0.004(x - 1)(x + 200), \qquad 0 < x \le 200$$

Simplify this expression.

Review Exercises

Simplify each expression as much as possible. Check with a calculator by substituting a nonzero value for each unknown.

68. $\dfrac{(x^2y)^{-2}}{x^2y^{-1}}$

69. $\dfrac{(x^{-1}y^0)^{-2}}{x^{-2}}$

70. $\dfrac{30x^2y^3z^4}{5xyz^3}$

71. $\dfrac{4xy}{7z} \cdot \dfrac{15}{28x^2y}$

72. $\dfrac{9x^2}{16y^4} \div \dfrac{27x}{32y^2}$

Chapter Review

Important Terms

Absolute value
Additive identity
Additive inverse
Associative properties
Cartesian coordinate system
Commutative properties
Coordinate axes
Depreciation
Distributive property of
   multiplication over addition

Equation
Equilibrium point
Fixed costs
Graph
Greater than
Half-plane
Inequality
Integer
Intercepts
Inverse of addition
Inverse of multiplication

Irrational numbers
Less than
Linear equation
Multiplicative identity
Multiplicative inverse
Natural numbers
Origin
Point of equilibrium
Polya's four steps in
   problem solving
Proper subset

Radical
Rational numbers
Real numbers
Real number line
Set theory terminology
Suggestions for
   understanding a problem
Supply and demand
Variable costs
Whole numbers

Important Formulas

$-(-x) = x$

$|x| = \begin{cases} x, & \text{if } x \ge 0 \\ -x, & \text{if } x < 0 \end{cases}$

$x + (-y) = x - y \quad \text{if } |x| > |y|$
$\qquad\qquad = -(y - x) \quad \text{if } |x| < |y|$

$x + (-x) = 0$

$-x + (-y) = -(x + y)$

$\quad -x(y) = -xy \quad \text{(unlike signs)}$

$\quad -x(-y) = xy \quad \text{(like signs)}$

$\dfrac{cp}{cq} = \dfrac{p}{q} \quad (c \ne 0,\, q \ne 0)$

$\dfrac{p}{q} + \dfrac{r}{q} = \dfrac{p + r}{q} \quad (q \ne 0)$

$\dfrac{p}{q} + \dfrac{r}{s} = \dfrac{ps + qr}{qs} \quad (q,s \ne 0)$

$x^m x^n = x^{m+n}$

$x^{-m} = \dfrac{1}{x^m}$

$x^{m/n} = \sqrt[n]{x^m} = (\sqrt[n]{x})^m$

$\dfrac{x^m}{x^n} = x^{m-n} = \dfrac{1}{x^{n-m}}$

$x^0 = 1$

$(x^m)^n = x^{mn}$

$\sqrt[n]{x} = x^{1/n}$

$\dfrac{\sqrt[n]{x}}{\sqrt[n]{y}} = \sqrt[n]{\dfrac{x}{y}}$

$(xy)^m = x^m y^m$

$\sqrt[n]{x}\,\sqrt[n]{y} = \sqrt[n]{xy}$

$\sqrt[n]{x^n} = \begin{cases} x, & \text{for } n \text{ odd} \\ |x|, & \text{for } n \text{ even} \end{cases}$

$(a + b)^2 = a^2 + 2ab + b^2$

$(a - b)^2 = a^2 - 2ab + b^2$

$(a - b)(a + b) = a^2 - b^2$

$(a + b)(a + c) = a^2 + a(b + c) + bc$

$(a + b)(c + d) = ac + ad + bc + bd$

$(a + b)^3 = a^3 + 3a^2b + 3ab^2 + b^3$

$a^3 + b^3 = (a + b)(a^2 - ab + b^2)$

$a^3 - b^3 = (a - b)(a^2 + ab + b^2)$

Chapter Test

1. If $U = \{3, 4, 5, 6, 7, 8, 9, 10\}$ and $A = \{4, 5, 7, 8\}$, find the complement of A.

2. Simplify $\left(\dfrac{4}{3} + \dfrac{-7}{8}\right)\left(\dfrac{-1}{2} - \dfrac{3}{4}\right)$.

3. Solve for x: $-3(4 - x) = 5 - (x + 1)$.

4. Express Alice's age in mathematical terms, letting n be Tom's age now. Alice is 3 years younger than 5 times Tom's age 1 year from now.

Put each answer in simplest form.

5. $\dfrac{(x^4 y^{-3})^{-1}}{(y^{-2})^{-3}}$ 6. $\sqrt{\dfrac{18x^2}{5y}}$ 7. $\sqrt[3]{\dfrac{54x^5}{y}}$

Factor completely.

8. $9r^2 - 25s^2$ 9. $3k^2 - 8k - 35$ 10. $x^3 + 4x + x^2y + 4y$

11. Graph the line $2x + 3y = 12$ using a calculator. Indicate your viewing rectangle.

12. Solve $\left|\dfrac{2x + 4}{3}\right| = 5$.

13. Solve $S = \dfrac{a}{1 - r}$ for r.

14. Draw the graph of $x + 3y < 9$ with and without a graphing calculator.

15. Solve $-2x + 4 < 6$.

16. Solve $|2x + 1| < 5$ and use a calculator to verify.

17. Solve $|3x - 2| \geq 7$ and use a calculator to verify.

18. Graph $2x + 3y > 6$ using a calculator.

19. If you spend 20% of your monthly income on food, what is your monthly income if you spend $300 on food?

20. A woman has an annual income of $13,000 from two investments. She has $30,000 more invested at 10% than she has invested at 12%. How much does she have invested at each rate?

21. If the selling price for a product is $50, the variable costs are $30, and the fixed costs are $10,000, find the revenue function, the total cost function, and the profit function.

22. Given the following demand and supply functions, find the equilibrium point:

$$p_d(x) = -2x + 56 \quad \text{and} \quad 3p_s(x) = x + 34$$

Functions and Their Graphs

W̲e begin this chapter with a study of linear equations and discuss how to determine an equation of a line. Then we study one of the most important concepts in elementary mathematics, the **function**. After functions are defined, we consider several special functions. One objective of this book is to assist you in understanding certain characteristics of the graphs of functions and in interpreting those graphs. This ability is important to the economist and the businessperson as well as someone in the social or life sciences. It is essential that relationships among quantities in a business be expressed efficiently and precisely. Functions enable us to do this. A profit function can be used to analyze what change in profit occurs if production levels are changed. Furthermore, a function can tell a doctor what the level of concentration of a drug will be at a certain time after being administered.

In this chapter we study special cases of the polynomial function in one variable, that is,

$$f(x) = a_n x^n + a_{n-1} x^{n-1} + \cdots + a_1 x + a_0$$

where $a_n \neq 0$ and n is a positive integer. Later in this chapter, we study functions such as

$$f(x) = 3^x$$

You will learn to recognize the graphs of these functions and investigate applications in various fields of study.

1.1   SLOPES AND LINEAR EQUATIONS

OVERVIEW   The ability to develop an equation that accurately describes a real-life situation is an important skill. This section is important in your accumulation of strategies to solve application problems. If a company can make 20 air conditioners for $15,000 and 10 air conditioners for $7950, the points (20, 15,000) and (10, 7950) can be plotted in a plane. If we draw a line between these two points, then other points on that segment can be used to represent the number of units and the cost. Since we are dealing with air conditioners, only integer values are useful in this problem, and the line gives an overall view of the relationship between number of units and cost. If we want to know the cost of producing 15 units, we need to find the equation of the line. In this section we

- Develop equations of lines from data representing real-life situations
- Define the slope of a line
- Use slope to graph a line and to find an equation of the line
- Learn to find an equation of a line given two points on the line

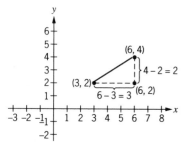

Figure 1

Any two points in a plane can be considered as endpoints of some line segment. Let (3, 2) and (6, 4) be endpoints of the line segment in Figure 1. If we now construct a line parallel to the x-axis through (3, 2) and a line parallel to the y-axis through (6, 4), the lines meet at the point (6, 2). The change in x as we move from (3, 2) to (6, 2) is $6 - 3 = 3$, and the change in y as we move from (6, 2) to (6, 4) is $4 - 2 = 2$. The ratio of the change in y to the change in x is $\frac{2}{3}$ (see Figure 1).

Practice Problem 1   For $P_1(1, 2)$ and $P_2(4, 4)$, find $y_2 - y_1$ (the change in y), $x_2 - x_1$ (the change in x), and the ratio of the change in y to the change in x.

ANSWER $y_2 - y_1 = 2$; $x_2 - x_1 = 3$, and the ratio of the change is $\frac{2}{3}$.

The preceding concepts are important in discussing the inclination of a line, which is measured by comparing the **rise** (the change in y: $y_2 - y_1$) to the **run** (the

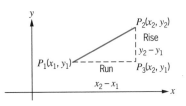

Figure 2

change in x: $x_2 - x_1$), as shown in Figure 2. This inclination, called the **slope**, is a useful characteristic of a line, telling us both the direction and the relative steepness of the line. In business, for example, this slope could represent the rate of change in profit.

DEFINITION: SLOPE OF A LINE SEGMENT

The ratio of the rise to the run of a line segment is called the **slope** of the line segment and is designated by the letter m. The slope of the line segment from $P_1(x_1, y_1)$ to $P_2(x_2, y_2)$ is

$$m = \frac{y_2 - y_1}{x_2 - x_1} \qquad (x_1 \neq x_2)$$

Since the ratio of the rise to the run on a line is always constant, the slope of a line segment is always the same no matter which two points on the line are selected to compute the slope. The slope of a line is defined as the slope of any of its line segments.

If P_1 is to the left of P_2, $x_2 - x_1$ will necessarily be positive, and the slope will be positive or negative as $y_2 - y_1$ is positive or negative. Consequently, positive slope indicates that a line rises from left to right; negative slope indicates that a line falls from left to right. For example, a slope of 2 means that y increases by 2 when x increases by 1. A slope of $-\frac{2}{3}$ means that y decreases by 2 when x increases by 3.

There is no restriction on which point is labeled P_1 and which is labeled P_2, since

$$\frac{y_2 - y_1}{x_2 - x_1} = \frac{-(y_1 - y_2)}{-(x_1 - x_2)} = \frac{y_1 - y_2}{x_1 - x_2}$$

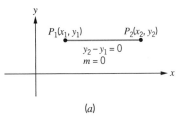

(a)

and the order in which the points are considered is immaterial in determining the slope. However,

$$\frac{y_2 - y_1}{x_2 - x_1} \neq \frac{y_2 - y_1}{x_1 - x_2}$$

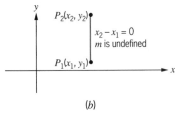

(b)

Figure 3

Slopes of Horizontal and Vertical Lines   Two points on the line $y = 3$ are $(1, 3)$ and $(2, 3)$. The slope is

$$m = \frac{3 - 3}{2 - 1} = 0$$

The slope of any horizontal line is 0 because the value of y does not change as x changes [see Figure 3(a)].

Two points on the line $x = 2$ are $(2, 1)$ and $(2, 4)$, but when we try to calculate the slope we find that the slope is undefined:

$$m = \frac{4 - 1}{2 - 2} = \frac{3}{0} \quad \text{(undefined)}$$

A vertical line has no slope; it is undefined because the value of x does not change and the denominator is always 0 [see Figure 3(b)].

EXAMPLE 1 Find the slope of the line shown in Figure 4.

SOLUTION

$$m = \frac{y_2 - y_1}{x_2 - x_1} = \frac{5 - 2}{2 - 1} = \frac{3}{1} = 3 \quad \text{or} \quad m = \frac{2 - 5}{1 - 2} = \frac{-3}{-1} = 3 \quad \blacksquare$$

Practice Problem 2 Find the slope of the line through points $(-1, 2)$ and $(3, -2)$.

ANSWER The slope is -1.

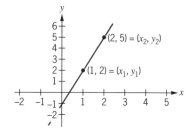

Figure 4

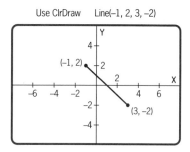

Figure 5

Calculator Note

Under the DRAW DRAW menu (obtained from $\boxed{\text{2nd}}$ $\boxed{\text{DRAW}}$), select 2:Line (to draw a line segment from $(-1, 2)$ to $(3, -2)$. Line $(-1, 2, 3, -2)$ gives the segment in Figure 5.

The relationship between two quantities in a business can often be expressed as a linear function in the form $ax + dy + c = 0$. For example, if y is the total cost to produce x units and the relationship is linear, the function may be linear and in this form. Since the slope of a line is important in analyzing a graph, let's look at slopes in relation to linear equations of the form $ax + dy + c = 0$, where a and d are not both zero. The graphs of equations such as $y = x + 1$, $x + y = 4$, and $2x - 3y = 6$ are straight lines. In fact, the graph of any linear equation with not more than two unknowns is a straight line. (If there are more than two unknowns, the graph is not a line.) Any line may be described by an equation of the form $ax + dy + c = 0$. Therefore, a given graph is a line if and only if it has an equation that can be written in the form $ax + dy + c = 0$, where a and d are not both zero.

A linear equation may be solved for y, if $d \neq 0$, to obtain

$$y = \frac{-a}{d}x + \frac{-c}{d}$$

By letting $m = -a/d$ and $b = -c/d$, the expression becomes

$$y = mx + b$$

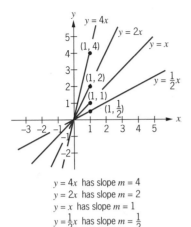

$y = 4x$ has slope $m = 4$
$y = 2x$ has slope $m = 2$
$y = x$ has slope $m = 1$
$y = \frac{1}{2}x$ has slope $m = \frac{1}{2}$

Figure 6

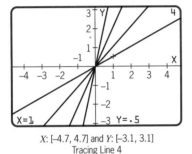

X: [-4.7, 4.7] and Y: [-3.1, 3.1]
Tracing Line 4

Figure 7

Consider the graph of $y = mx$ ($b = 0$ for this example) when m has values $\frac{1}{2}$, 1, 2, and 4 (see Figure 6). Notice the change in the lines as the value of m increases.

Practice Problem 3 Using a calculator, draw the lines $y = 4x$, $y = 2x$, $y = x$, and $y = \frac{1}{2}x$.

ANSWER We can enter each line as Y_1, Y_2, Y_3, Y_4. By pressing the Trace key, we can scroll through each of the lines. Notice the small number in the upper right-hand corner of the screen (see Figure 7). That number indicates which line we are tracing.

By studying the graphs in Figures 6 and 7, we can make some observations. Since (0, 0) and (1, $\frac{1}{2}$) are two points on the line $y = \frac{1}{2}x$, the slope can be evaluated by $(\frac{1}{2} - 0)/(1 - 0) = \frac{1}{2}$. We can also see from studying the two points that the value of y increases by $\frac{1}{2}x$. You may have already realized that in the form $y = mx + b$, the value of m is the slope of the line.

If (x_1, y_1) is a fixed point on a given nonvertical, straight line, and (x, y) is any point on the line, then the slope from (x_1, y_1) to (x, y) is

$$m = \frac{y - y_1}{x - x_1} \quad \text{or} \quad y - y_1 = m(x - x_1)$$

Since the coordinates x and y are variables denoting any point on the line, the equation $y - y_1 = m(x - x_1)$ represents the relationship between x and y. Therefore, the equation of the line with slope m passing through the fixed point (x_1, y_1) is $y - y_1 = m(x - x_1)$. A linear equation written in this form is said to be in **point-slope form**. If the slope and one point on the line are known, then the equation of the line can be obtained.

DEFINITION: POINT-SLOPE FORM

If a line has slope m and passes through the point (x_1, y_1), then the equation of the line in **point-slope form** is given by $y - y_1 = m(x - x_1)$.

EXAMPLE 2 Find an equation of the line through (2, 1) with a slope of 3.

SOLUTION

$$y - y_1 = m(x - x_1)$$
$$y - 1 = 3(x - 2) \qquad x_1 = 2, y_1 = 1, m = 3$$
$$y = 1 + 3x - 6$$
$$y = 3x - 5 \qquad \blacksquare$$

As a special case, the fixed point can be chosen as the point where the line crosses the y-axis (the y-intercept). The coordinates of this point are usually written as (0, b). Then the equation of the line becomes $y = mx + b$, which is the equation that we discussed earlier. The b in this equation is the value of y when $x = 0$, or the y-intercept.

DEFINITION: SLOPE-INTERCEPT FORM

If a line has slope m and y-intercept b, then the equation of the line in **slope-intercept form** is given by $y = mx + b$.

EXAMPLE 3   If the slope of a line is 3 and the y-intercept is 2, what is an equation of the line?

SOLUTION   Since $m = 3$ and $b = 2$, an equation is $y = mx + b$ or $y = 3x + 2$. ■

Practice Problem 4 Find an equation of the line that crosses the y-axis at $(0, -5)$ and has a slope of 2.

ANSWER   An equation is $y = 2x - 5$.

Suppose that we are given two points that are on a line. How can we find an equation of the line? Since we can find the slope with two points, we can then select one of the points and the slope to put into the point-slope form to find the equation.

EXAMPLE 4   Find an equation of the line that contains the points $(2, 3)$ and $(-1, 4)$.

SOLUTION   We can use the two points in either order to find the slope:

$$m = \frac{4 - 3}{-1 - 2} = -\frac{1}{3}$$

Now select either one of the points (but only one) and put that point and the slope into the point-slope form. Then simplify the resulting equation. Let's use the point $(2, 3)$. Note that we could also use the point $(-1, 4)$, which would yield the same simplified equation.

$$y - 3 = -\frac{1}{3}(x - 2) \qquad x_1 = 2,\, y_1 = 3,\, m = -1/3$$

$$y = -\frac{1}{3}x + \frac{11}{3}$$

The equation $y = -\frac{1}{3}x + \frac{11}{3}$ is the equation in slope-intercept form. Substitute the point $(-1, 4)$ to verify that the resulting equations are the same. ■

Practice Problem 5 Find an equation of the line through the points $(-1, 2)$ and $(3, -2)$.

ANSWER $y = -x + 1$

The graph of a linear equation can be a horizontal line, which we illustrate using a calculator.

Practice Problem 6   Draw the line $y = 3$ using a calculator.

ANSWER   We can enter the function $Y_1 = 3$ and use an appropriate viewing rectangle to get a graph such as that in Figure 8.

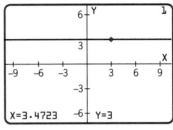

X: [–10.4, 10.4] and Y: [–7.4, 7.4]

Figure 8

The equation of a vertical line is of the form $x = h$, where h is a constant and the slope is undefined. The equation of the line shown in Figure 9 is $x = 3$. By selecting two points on the line, say $(3, 5)$ and $(3, 2)$, we can see that the slope is undefined. That is,

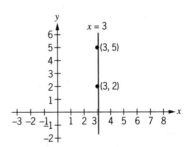

Figure 9

$$m = \frac{5 - 2}{3 - 3} = \frac{3}{0}$$

which does not exist.

Practice Problem 7   Find an equation of the vertical line through $(1, -2)$.

ANSWER $x = 1$

Calculator Note

A vertical line such as $x = 1$ can be drawn on a calculator using the DRAW DRAW menu (obtained from $\boxed{\text{2nd}}$ $\boxed{\text{DRAW}}$). Three different graphs are shown in Figure 10. For (a) select 4: Vertical and insert 1 then $\boxed{\text{ENTER}}$. For (b) select 2: Line (and insert the y values of the endpoints of a line segment (say $y = -2$ to $y = 2$). Use Line $(1, -2, 1, 2)$. For (c) follow the same procedure as in (b) except use Ymin and Ymax for the endpoints , or Line $(1, \text{Ymin}, 1, \text{Ymax})$.

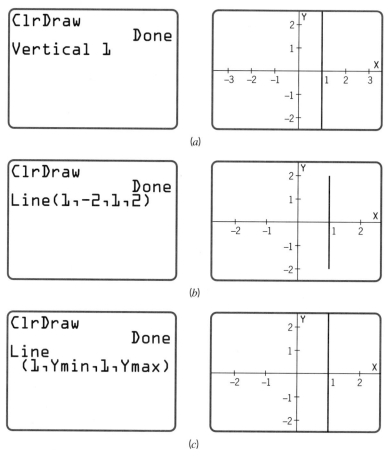

(a)

(b)

(c)

Figure 10

Parallel and Perpendicular Lines

It can be proved that if two nonvertical lines are **parallel**, then they have the same slope. Conversely, two lines with the same slope and different y-intercepts are parallel. Similarly, if two nonvertical lines are **perpendicular**, then the slope of one line is the negative reciprocal of the other; that is,

$$m_2 = \frac{-1}{m_1} \quad \text{or} \quad m_1 \cdot m_2 = -1$$

Conversely, if the slope of one line is the negative reciprocal of another, then the two lines are perpendicular.

EXAMPLE 5   Find an equation of the line passing through (2, 1) and perpendicular to the line given by $x + 2y = 4$.

SOLUTION   Putting $x + 2y = 4$ into slope-intercept form, we have $y = -\frac{1}{2}x + 2$. Therefore, the slope of this line is $-\frac{1}{2}$ and any line perpendicular to it will have a slope of 2, since $-\frac{1}{2} \cdot 2 = -1$. We are given the point (2, 1) and now we have the slope of 2. Putting this information into point-slope form, we have

$$y - 1 = 2(x - 2) \quad \text{or} \quad y = 2x - 3 \qquad ■$$

Practice Problem 8 Find the y-intercept of the line containing the points (2, 1) and (4, −2).

ANSWER   The y-intercept is 4.

Calculator Note

The equation of a line can be verified by putting the equation in Y_1, graphing the line, and then using the Trace command to check that the given points are on the line. Or, the Value command under the CALC menu can also be used.

SUMMARY

1. *Point-slope form:* If the slope is m and the line passes through (x_1, y_1), then

$$y - y_1 = m(x - x_1)$$

2. *Slope-intercept form:* If the slope is m and the y-intercept is $(0, b)$, then

$$y = mx + b$$

3. *General form:* $ax + dy + c = 0$.
4. *Horizontal line:* If the y-intercept is $(0, b)$ and the line has a slope of 0, then

$$y = b$$

5. *Vertical line:* If the x-intercept is $(h, 0)$ and the slope of the line is undefined, then

$$x = h$$

Exercise Set 1.1

Compute the slope or indicate that the slope is undefined for the line through each pair of points.

1. (3, 6), (4, 1)   2. (0, 1), (2, 3)
3. (−3, −5), (4, 2)   4. (7, −1), (−3, 1)

5. (0, 4), (4, 0)   6. (−1, −7), (−6, −5)
7. (4, 3), (4, −1)   8. (7, −1), (7, 4)
9. (3, 1), (7, 1)   10. (−1, 2), (7, 2)
11. Find an equation for and graph the line that has the following conditions. Verify using a graphing calculator.
    (a) Slope of 4; passes through the point (2, 3)

(b) Slope of -2; passes through the point $(4, -1)$

(c) Slope of $\frac{1}{2}$; passes through the point $(-1, 1)$

(d) Slope of $-7/2$; passes through the point $(3, 4)$

(e) Slope is undefined; passes through the point $(2, -4)$

(f) Slope is 0; passes through the point $(-3, -5)$

Find the slope and the y-intercept in each of the following linear equations. Graph each equation using a calculator.

12. $y = 3x + 2$

13. $y + 2x - 1 = 0$

14. $y = 3x - 1$

15. $2y = 10x$

16. $y = \dfrac{x - 4}{2}$

17. $x = \dfrac{y - 1}{3}$

18. $4x + 3y - 7 = 0$

19. $3x - 2y = 5$

20. Classify the following statements as either true or false.

(a) The slope of the y-axis is 0.

(b) The line segment joining (a, b) and (c, b) is horizontal.

(c) A line with a negative slope rises to the right.

(d) A line that is almost vertical has a slope close to 0.

Find an equation of each line through the given point with the given slope. Verify using a graphing calculator.

21. $(1, 3)$, $m = \dfrac{1}{2}$

22. $(0, 2)$, $m = 1$

23. $(-1, -2)$, $m = -\dfrac{1}{3}$

24. $(-3, 1)$, $m = 0$

Find an equation of the line through each of the following pairs of points. Verify using a graphing calculator.

25. $(1, 1)$, $(2, 5)$

26. $(-1, 1)$, $(2, 5)$

27. $(1, 3)$, $(1, -2)$

28. Find an equation of each line with the following characteristics and verify using a graphing calculator.

(a) The line contains the two points $(1, -3)$ and $(4, 5)$.

(b) The line has a slope of -3 and goes through the point $(7, 1)$.

(c) The line has a slope of 1 and goes through the point $(-7, 1)$.

(d) The line contains the two points $(0, 1)$ and $(4, 3)$.

(e) The line has a y-intercept of 4 and a slope of 5.

(f) The line has a y-intercept of 6 and a slope of -3.

29. Find an equation of the horizontal line through $(-4, -6)$.

30. Find an equation of the vertical line through $(-5, 4)$.

31. Write the equation of the x-axis.

32. Write the equation of the y-axis.

33. Suppose that the equation of a line is written in the form

$$\frac{x}{a} + \frac{y}{b} = 1$$

What is the x-intercept? The y-intercept?

34. Use the intercept form of the equation of a line (see Exercise 33) to find equations for lines with the following intercepts.

(a) $x = 2$ and $y = -3$ (b) $x = 3$ and $y = 5$

35. Find the y-intercept of the line that passes through the point $(3, -2)$ with a slope of 2.

Applications (Business and Economics)

36. An electric company charges a $6-per-month customer charge plus $0.07186 per kilowatt-hour used during the month. Write an equation that relates the monthly bill, in dollars, to the number of kilowatt-hours used. What would be the charge for 1500 kilowatt-hours?

37. Every Monday, a newsstand sells x copies of a weekly sports magazine for $2.50. The owner of the newsstand buys the magazines for $1.70 a copy, plus a delivery fee of $50.

(a) Write an equation that relates the profit, in dollars, to the number of copies sold; graph this equation.

(b) How many copies must be sold to make a profit?

(c) What will the profit be if 200 copies are sold?

1.2 FUNCTIONS

OVERVIEW

We study functional relationships in just about every chapter of this book: functional relationships between supply and demand, between cost and production, between the value of a piece of equipment and the age of the equipment, between IQ and accomplishment, and in many other applications. In this section, we introduce the mathematical concept of a *function*. We begin our discussion by considering a function as a rule associating one set with another. Numerous examples help to explain the meaning of a function. In this section we

- Define a function as a set of ordered pairs with certain characteristics
- Specify a function by an equation
- Introduce function notation

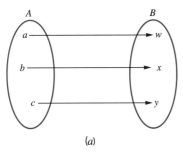

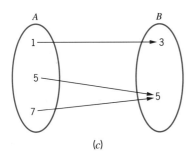

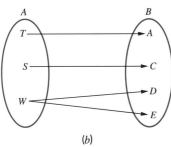

(a)

(b)

(c)

Figure 11

To introduce the concept of a function, we start with a correspondence between the elements in two sets. In Figures 11(*a*), 11(*b*), and 11(*c*), we have set up a correspondence between the elements of set *A* and the elements of set *B*. Each correspondence is a relation. A **relation** is a set of ordered pairs. In Figure 11(*a*), we have the relation defined by the set of ordered pairs $\{(a, w), (b, x), (c, y)\}$. The relations in Figures 11(*b*) and 11(*c*) can be similarly listed.

A function is a special type of relation; it is a rule that sets up a correspondence between a set *A* and a set *B* so that for every element of *A* there is a unique element of *B*. This way of thinking of a function can be demonstrated by what we call input–rule–output. Set *A* can be considered as the input, called the *domain*, and then from a rule, set *B*, called the *range*, can be obtained as the output.

EXAMPLE 6   As shown in Figure 12(*a*), we insert an input (the domain), operate with the rule, and obtain an output (the range). Suppose that we establish the rule to be "add 4 to the input." This rule can be expressed as $x + 4$ when the input is *x*. For example, when the input is 3, the rule operates to give

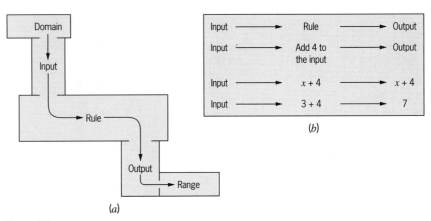

(a)

(b)

Figure 12

3 + 4 and the output is 7 [see Figure 12(*b*)]. This creates the ordered pair (3, 7). An infinite number of ordered pairs can be created because *x* can be any real number and we can add 4 to any real number. ◼

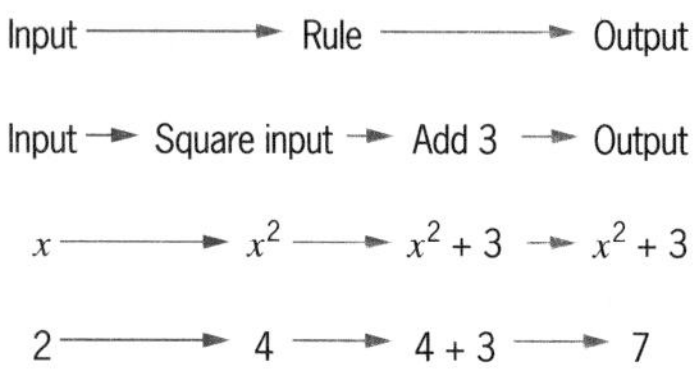

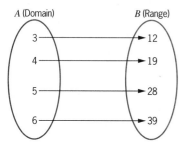

Figure 13

EXAMPLE 7   As a more complex example, consider the rule described in Figure 13. Note that this rule also sets up a correspondence. By examining the correspondence closely, we see that each element of the input is paired with exactly one element of output, creating a set of ordered pairs. ◼

Figure 14 shows some examples of the ordered pairs that are created by the function in Example 7.

From the preceding discussion we can say that a rule describes a function if it produces a correspondence between one set of elements (input set), called the domain, and a second set of elements (output set), called the range, in such a way that to each element in the domain there corresponds one and only one element in the range. Looking back at the rule in Figure 12 (*b*), we can see that the set of ordered pairs

$$\{. . . , (\tfrac{1}{2}, 4\tfrac{1}{2}), (1, 5), (2, 6), (3, 7), . . .\}$$

can be created. Also, from the rule in Figure 13, we can obtain the set

$$\{. . . , (\tfrac{1}{2}, 3\tfrac{1}{4}), (1, 4), (2, 7), (3, 12), ...\}$$

What is characteristic of both sets of ordered pairs? In each set, no two ordered pairs have the same first element. This is a defining characteristic of a function.

A (Domain)   B (Range)

3 ────▶ 12

4 ────▶ 19

5 ────▶ 28

6 ────▶ 39

Figure 14

DEFINITION: FUNCTION

A **function** is a set of ordered pairs with the property that no two ordered pairs have the same first element. The set of first elements constitutes the **domain** and the set of second elements constitutes the **range**.

EXAMPLE 8   Consider the following sets of ordered pairs (*x*, *y*), where *x* is an element of the domain {2, 3, 4} and *y* is an element of the range {3, 4, 5, 6}. Which of these relations are functions?

(a) $S = \{(2, 3), (3, 4), (4, 5)\}$
(b) $T = \{(3, 3), (3, 4)\}$
(c) $U = \{(2, 3), (3, 4), (4, 5), (2, 6)\}$
(d) $V = \{(2, 5), (3, 5), (4, 5)\}$

SOLUTION

Using the preceding definition, we see that *S* and *V* are functions, since the first element is not repeated. The sets *T* and *U* are relations but not functions be-

TABLE 1

x	1	3	4	6
y	4	6	12	18

TABLE 2

x	1	0	1	2
y	-2	3	4	5

cause in T the element 3 is paired with both 3 and 4 and in U the 2 is paired with both 3 and 6. ∎

NOTE: It is acceptable for the second element in the list of ordered pairs to be repeated. In fact, there are functions that contain only one element in the range.

Sometimes a set of ordered pairs is given in a table. In the two given tables, notice that Table 1 represents a function because for each x there corresponds only one y, and Table 2 does not represent a function because the two ordered pairs $(1, -2)$ and $(1, 4)$ have the same first element and different second elements.

Practice Problem 1 Does the relation $R = \{(2, 1),\ (3, 2),\ (4, 5),\ (4, 6),\ (5, 9)\}$ represent a function?

ANSWER The relation R does not represent a function because $(4, 5)$ and $(4, 6)$ have the same first element.

In Figure 12(b), the rule could be written as $y = x + 4$ and in Figure 13 as $y = x^2 + 3$. That is, the rule can be expressed in equation form. The equation representing a function assigns to each x in the domain a unique value y in the range. Often, the domain and range are restricted to a subset of the real numbers by either the equation itself or the nature of the function. The variable x in these equations is called the **independent variable** and the variable y the **dependent variable**. We can say that if an equation in two variables specifies exactly one value of the dependent variable for each value of the independent variable, then the equation represents a function.

If the domain of a function is not stated, we assume that it is the largest set of real numbers for which the rule or equation gives a real-valued function.

Calculator Note

When using a calculator to determine the domain and range of a function, it is essential that you graph enough of the function to observe all relevant behavior. You must have an algebraic understanding of the function to use the calculator properly. Using a calculator and your algebraic knowledge together can greatly facilitate your understanding of the function and its behavior.

EXAMPLE 9 Draw the graph of each function on a calculator (see page A-1 in Appendix A) and use the graph to determine the domain and range of each function.

(a) $y = x^2 + 2x + 1$ (b) $y = \dfrac{2}{x + 3}$

(c) $y = \sqrt{x + 2}$ (d) $y = \sqrt{9 - x^2}$

SOLUTION

(a) Since $y = x^2 + 2x + 1$ is a polynomial, it is defined for all values of x and the domain is the set of real numbers. The domain is $D = (-\infty, \infty)$ and the range is $R = [0, \infty)$. The graph is shown in Figure 15(a).

(b) When $x = -3$, $x + 3 = 0$ and $2/(x + 3)$ is undefined. However, for any other real number $2/(x + 3)$ is defined. Therefore, the domain is the set of all real numbers except -3. So $D = (-\infty, -3), (-3, \infty)$ and the range is $R = (-\infty, 0), (0, \infty)$. The fraction cannot equal zero because the numerator is a constant, so zero is not included in the range [see Figure 15(b)].

(c) For $y = \sqrt{x + 2}$ to be a real number, $x + 2$ must be nonnegative. Furthermore, $x + 2 \geq 0$ implies that $x \geq -2$. Therefore, the domain is $D = [-2, \infty)$ and the range is $R = [0, \infty)$, since y is always nonnegative. The graph is shown in Figure 15(c).

(d) Since $9 - x^2 \geq 0$ only over the interval $[-3, 3]$, we see in Figure 15(d) that the domain of the function is $D = [-3, 3]$ and the range is $R = [0, 3]$.

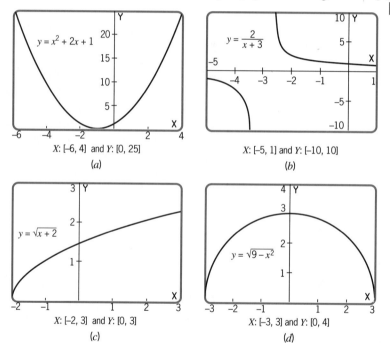

X: [-6, 4] and Y: [0, 25]

(a)

X: [-5, 1] and Y: [-10, 10]

(b)

X: [-2, 3] and Y: [0, 3]

(c)

X: [-3, 3] and Y: [0, 4]

(d)

Figure 15

Practice Problem 2 Find the domain and range of $y = \sqrt{x + 1}$.

ANSWER The domain is $D = [-1, \infty)$ and the range is $R = [0, \infty)$.

EXAMPLE 10 Is the relation $y^2 = 2 + x^2$ a function?

SOLUTION Since $y^2 = 2 + x^2$ implies that $y = \pm\sqrt{2 + x^2}$, the assignment of a real value to x will result in two different values of y (one positive and one negative), and hence the relation is not a function. By restricting y to $\sqrt{2 + x^2}$

or $-\sqrt{2 + x^2}$, we have a function. To graph this relation, you would have to graph $y_1 = \sqrt{2 + x^2}$ and $y_2 = -\sqrt{2 + x^2}$ or $y_1 = \sqrt{2 + x^2}$ and $y_2 = -y_1$. ■

Graphically, the fact that a function associates each element in its domain with one and only one element in the range implies that no two of the ordered pairs in a function correspond to points on the same vertical line. That is, if a vertical line cuts the graph at more than one point, then the graph does not represent a function. Figures 16(a) and 16(c) show the graphs of functions. Figure 16(b) shows a graph that is not a function, since a vertical line could be drawn to intersect the figure at more than one point.

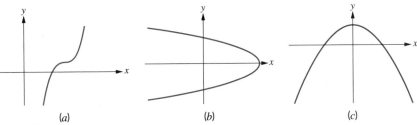

(a)   (b)   (c)

Figure 16

Vertical Line Test

The **vertical line test** can be applied as follows: If there is no vertical line that intersects the graph of an equation in more than one point, then the equation represents a function; if any vertical line passes through two or more points of the graph, the equation does not specify a function.

EXAMPLE 11   Which of the graphs in Figure 17 represent functions?

SOLUTION   The graphs in Figures 17(a) and 17(c) represent functions. The circle in Figure 17(b) is not a function. ■

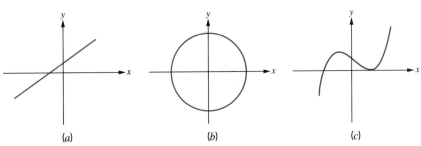

(a)   (b)   (c)

Figure 17

Practice Problem 3 Use a calculator to sketch the graphs of $x^2 - y = 3$ and $y^2 - 2x = 1$. (**Hint:** Use two equations for the second graph.) Do the graphs represent functions?

ANSWER The equation $x^2 - y = 3$ defines a function but $y^2 - 2x = 1$ does not define a function (see Figure 18).

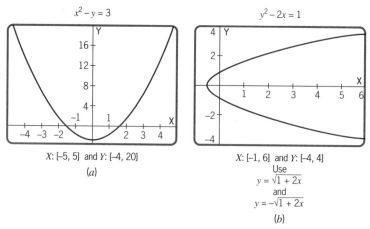

X: [–5, 5] and Y: [–4, 20]

(a)

X: [–1, 6] and Y: [–4, 4]
Use
$y = \sqrt{1 + 2x}$
and
$y = -\sqrt{1 + 2x}$

(b)

Figure 18

Practice Problem 4

Use a calculator to find the domain and range of each function.
(a) $y = \sqrt{x^2 + 2x - 3}$
(b) $y = |x^2 - 9|$

ANSWER

(a) Since a calculator graphs only real values, we can see in Figure 19(a) that the domain is $D = (-\infty, -3], [1, \infty)$ and the range is $R = [0, \infty)$.
(b) As shown in Figure 19(b), the domain is $D = (-\infty, \infty)$ and the range is $R = [0, \infty)$.

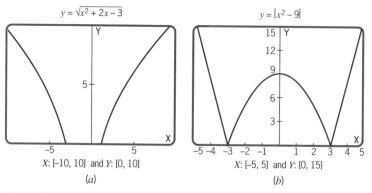

X: [–10, 10] and Y: [0, 10]

(a)

X: [–5, 5] and Y: [0, 15]

(b)

Figure 19

Function Notation

Often, we write a function using function notation. That is, we write $y = f(x)$, which is read "y equals f of x." It is important to remember that f is the name of the function and that $f(x)$ does not indicate multiplication. Any letter can be used to name a function; we often give a function a name relating to its nature. For example, a cost function could be $C(x)$. This notation indicates that with each x there is associated a unique C.

Suppose that we write $y = f(x) = x + 7$. When $x = 2$, we substitute 2 for x into the function to obtain

$$y = f(2) = 2 + 7 = 9$$

Likewise, $f(4) = 11$, and $f(6) = 13$.

EXAMPLE 12   For $y = f(x) = -x^2 + 2x + 1$ find each of the following:

(a) $f(1)$ (b) $f(0)$ (c) $f(-2)$ (d) $f(x + 1)$

SOLUTION

(a) $f(1) = -(1)^2 + 2(1) + 1 = 2$
(b) $f(0) = -(0)^2 + 2(0) + 1 = 1$
(c) $f(-2) = -(-2)^2 + 2(-2) + 1 = -4 - 4 + 1 = -7$
(d) $f(x + 1) = -(x + 1)^2 + 2(x + 1) + 1 = -(x^2 + 2x + 1) + 2x + 2 + 1$
$$= -x^2 - 2x - 1 + 2x + 3 = -x^2 + 2$$

Practice Problem 5   For $f(x) = 3x^2 - 2x + 1$, find each of the following:

(a) $f(-3)$ (b) $f(x + 1)$

ANSWER

(a) $f(-3) = 34$ (b) $f(x + 1) = 3x^2 + 4x + 2$

Calculator Note

Some calculators create tables that allow you to see the values of a variable and the function in a table. If your calculator has this feature, study several functions using this feature. See page A-7 of Appendix A.

COMMON ERROR   When evaluating a function such as $f(x) = -x^2 + 3x - 2$, students often include the negative sign as part of the squared term. Why is this not correct?

Correct	**Incorrect**
$f(4) = -4^2 + 3 \cdot 4 - 2 = -6$	$f(4) = (-4)^2 + 3 \cdot 4 - 2 = 26$

There is an important expression that we will use many times in calculus when we study the way functions look in their graphs and the way that they change. It is called the difference quotient and is given as

$$\frac{f(x + h) - f(x)}{h} \qquad (h \neq 0)$$

Finding the difference quotient for a function is important. It will be done often in calculus as important formulas are developed. Let's look at an example of how a difference quotient is found and simplified.

EXAMPLE 13   For the function $f(x) = x^2 - 2x$, find the difference quotient.

SOLUTION

$$\frac{f(x + h) - f(x)}{h} = \frac{\overbrace{[(x + h)^2 - 2(x + h)]}^{f(x+h)} - \overbrace{[x^2 - 2x]}^{f(x)}}{h}$$

$$= \frac{x^2 + 2xh + h^2 - 2x - 2h - x^2 + 2x}{h}$$

$$= \frac{2xh + h^2 - 2h}{h}$$

$$= \frac{h(2x + h - 2)}{h}$$

$$= 2x + h - 2 \qquad (h \neq 0) \qquad \blacksquare$$

Practice Problem 6   Find the difference quotient for $f(x) = 2x - 3$.

ANSWER   2

SUMMARY

Make certain that you understand the definition and concept of a function. It is an important concept and will be used throughout this book.

1. *Function:* A relation that creates a set of ordered pairs, each consisting of an independent variable and a dependent variable and for which each independent variable has a unique value of the dependent variable associated with it.
2. *Notation:* In the equation $y = f(x)$, f is the *name* of the function and its *value* at x is $f(x)$.
3. *Domain:* All values of the independent variable for which the function is defined over the real numbers.
4. *Range:* The set of all values of the dependent variables.

Exercise Set 1.2

1. Which of the following relations are functions?
   (a) $\{(1, 3), (3, 3), (5, 3)\}$
   (b) $\{(1, 3), (3, 3), (5, 7)\}$
   (c) $\{(1, 3), (3, 5), (5, 1)\}$
   (d) $\{(1, 1), (3, 3), (5, 5)\}$
   (e) $\{(3, 4), (5, 10), (6, 4), (7, 1)\}$
   (f) $\{(1, 5), (1, 6), (2, 5), (3, 10)\}$
   (g) $\{(3, 7), (7, 3), (8, 3)\}$

(h) $\{(4, 6), (5, 6)\}$

(i) $\{(5, 3), (5, 4)\}$

(j) $\{(5, 5), (6, 6)\}$

2. Which of the following tables define functions? (The variable x is the independent variable.)

(a)

x	2	2
y	4	1

(b)

x	1	3
y	-1	-1

(c)

x	1	1
y	1	2

(d)

x	0	0
y	0	1

(e)

x	1	2
y	1	1

(f)

x	0	2	3	4
y	2	4	7	2

(g)

x	-2	1	3	2	1
y	1	2	4	3	4

(h)

x	-1	2	4	4
y	3	4	6	5

3. State whether or not each correspondence represents a function.

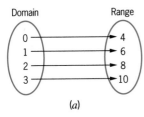

(a)

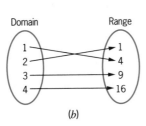

(b)

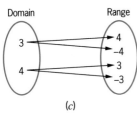

(c)

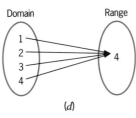

(d)

4. If $y = f(x) = x^3 - 2x$, find $f(-2)$, $f(1)$, and $f(0) - f(3)$.

5. If $y = f(x) = (x + 2)(x - 1)$, find $f(3)$, $f(2)$, and $f(-1) - f(0)/2$.

6. Which of the following graphs represent functions?

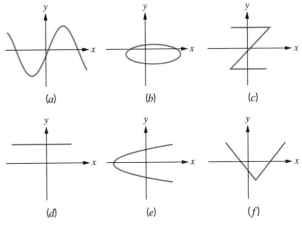

(a)   (b)   (c)

(d)   (e)   (f)

7. If $y = f(x) = x^3 - 2x$, find $f(-1)/f(1)$, $f(2) \cdot f(4)$, $f(w)$, $f(2z)$, $f(t + 2)$, and $f(3x - 1)$.

8. If $y = f(x) = (x + 2)(x - 1)$, find $f(-2)$, $f(3)$, $f(z)$, $f(3w)$, $f(t + 3)$, $f(2w - 1)$, and $f(4 + h)$.

Determine the domain of each function using a graphing calculator.

9. $y = \sqrt{x}$

10. $y = \dfrac{x^2 + 2}{x^2 - 1}$

11. $y = \dfrac{x}{x^2 - 4x - 5}$

12. $y = \dfrac{3}{x - 2}$

13. $y = \sqrt{3 - x^2}$

14. $y = \dfrac{1}{x}$

15. $y = \dfrac{3}{x(x - 3)}$

16. $y = \sqrt{9x^2 - 16}$

17. $y = \sqrt{x(x + 4)}$

18. $y = \sqrt{x^2 - 5x + 4}$

19. $y = \sqrt{\dfrac{x - 1}{x + 1}}$

20. $y = \sqrt{\dfrac{|x - 3|}{x - 2}}$

21. $y = \sqrt{(x - 1)(x + 2)(x - 3)}$

22. $y = \dfrac{x + 7}{x - 1}$

23. $y = x^2 + 6x + 4$

24. $y = \sqrt{x + 1}$

25. Find the expression

$$\dfrac{f(2 + h) - f(2)}{h}$$

for each function.

(a) $f(x) = 3x - 1$ (b) $f(x) = 2x + 4$
(c) $f(x) = 4x^2$ (d) $f(x) = x^2 - 3$
(e) $f(x) = \sqrt{x}$ (f) $f(x) = 1/x$

26. Find the difference quotient

$$\frac{f(x + h) - f(x)}{h}$$

for each of the functions in Exercise 25.

In Exercises 27–30, match each graph to the correct function listed below. Try to do this without using a graphing calculator. Then verify your answer using a graphing calculator.

(a) $y = |3x + 2|$ (b) $y = x^2 - 4x + 3$

(c) $y = \sqrt{2x + 1}$ (d) $y = \dfrac{x}{x + 1}$

27.   28.

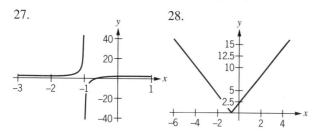

29.   30.

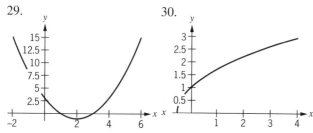

Applications (Business and Economics)

31. **Revenue.**   A parking lot charges by the hour as follows:

$$f(x) = 2 + 1.5(x - 1) \quad \text{for } x \geq 1$$

Find the charge for 1 hour, 2 hours, 10 hours, and 24 hours. Try to state the parking rate in words.

32. **Revenue Function.**   A travel agency books a flight to Europe for a group of college students. The fare in a 200-passenger airplane will be $400 per student plus $2.00 per student for each vacant seat.
   (a) Write the total revenue $R(x)$ as a function of empty seats, x.
   (b) What is the domain of this function?
   (c) Calculate $R(x)$ for 5, 10, 20, 40, and 100 empty seats.

33. **Demand Function.**   Suppose that the demand function (in price per unit) for a certain item is given by

$$p = D(x) = \frac{50 - 5x}{4}$$

when x units are demanded by the consumer at price $p = D(x)$.
   (a) What is $D(x)$ when $x = 0$?
   (b) What is $D(x)$ when $x = 4$?
   (c) What happens to $D(x)$ when $x = 10$?
   (d) Sketch the graph of $D(x)$.

34. **Supply Function.**   Suppose that the supply function for the item in Exercise 33 is given by

$$p = S(x) = \frac{5x}{6}$$

where $p = S(x)$ is the price per unit of an item at which the seller is willing to supply x units.
   (a) What is $S(x)$ when $x = 3$? When $x = 9$?
   (b) Sketch the graph of $S(x)$ on the same axis system that you used in Exercise 33.
   (c) Estimate the point of equilibrium from the intersection of the graphs.
   (d) Estimate the equilibrium price.
   (e) When is supply greater than demand?

35. **Demand and Supply Functions.**   The supply function $S(x)$ and the demand function $D(x)$ for a certain commodity in terms of units available, x, at price p, are

$$p = S(x) = \frac{400 - 5x}{2} \quad \text{and} \quad p = D(x) = \frac{5x}{2}$$

   (a) Graph $S(x)$ and $D(x)$ on the same axes.
   (b) Find the equilibrium point.
   (c) Find the equilibrium price.
   (d) When is the supply less than the demand?

Applications (Social and Life Sciences)

36. ***Bacteria Count***.   The number of bacteria in a culture x hours after an antibacterial treatment has been administered is given by

$$N(x) = 1000 - 150x, \qquad 0 \le x \le 6$$

Find $N(0)$, $N(2)$, $N(4)$, and $N(6)$.

37. ***Data***.   Write a function that shows the relationship between the gallons of gasoline that your car uses and the miles you drive. What effect would getting more miles per gallon have on the function?

Review Exercises

38. Write an equation of the line meeting the following criteria.

   (a) Has slope -3 and passes through the point $(-1, 2)$
   (b) Has no slope and passes through the point $(1, 1)$
   (c) Passes through the points $(1, -2)$ and $(3, 1)$
   (d) Passes through the point $(2, -4)$ and is parallel to the line having the equation $2x - 3y + 4 = 0$
   (e) Has x-intercept 4 and y-intercept -2

1.3 QUADRATIC FUNCTIONS

OVERVIEW

The equation $y = f(x)$, where $f(x)$ is a polynomial of degree n in the variable x, is called a polynomial function. A special case of the polynomial function when $n = 1$ (first-degree equation) has a graph that is always a line. In this section, we study another special case of a polynomial function by letting $n = 2$. This second-degree equation is called a quadratic function. In this section we

- Study the characteristics of graphs of quadratic functions
- Find the vertex and the axis of symmetry of graphs of quadratic functions
- Graph quadratic functions

All functions that can be written in the form

$$y = f(x) = ax^2 + bx + c \qquad (a \ne 0)$$

are called **quadratic functions**. The graphs of such functions are curves called **parabolas**. For example, the path a ball takes when thrown or hit in the air is parabolic in shape. Parabolic shapes are also found in such places as satellite dishes or spotlights. All parabolas are symmetric about a line called the **axis of symmetry**. In addition, all parabolas have a **vertex**, which is the point at which the parabola and the axis of symmetry intersect. This is shown in Figure 20.

We begin our discussion by examining the graph of the simplest quadratic function, $y = x^2$.

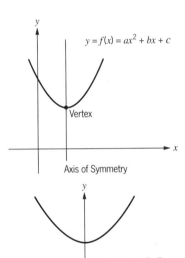

Note the axis of symmetry will be the y-axis for some quadratic functions.

Figure 20

EXAMPLE 14   Draw the graph of $y = x^2$ and then compare it to the graphs of $y = 3x^2$ and $y = \frac{1}{3}x^2$. Describe how changing the value of a (coefficient of x^2) alters the graph.

SOLUTION   Using a calculator, we can easily graph $y = x^2$ and $y = 3x^2$ on the same screen. We can then add the graph of $y = \frac{1}{3}x^2$ to the screen (see Figure 21). By examining these graphs, we see that as the coefficient a is changed, the apparent "width" of the parabola is changed. If $|a| > 1$, the parabola is not as wide as when $0 < |a| < 1$. The domain of all the functions is the real

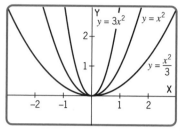

X: [–3, 3] and Y: [–1, 3]

Figure 21

numbers, so in actuality, they all have the same width. What we are referring to here is the appearance of the graph. Instead of "width," we could say that a affects the "rate of change" of the function.

In Figure 21, notice that all of the parabolas pass through the point $(0, 0)$ and are symmetric about the y-axis. Thus, $(0, 0)$ is the vertex and the y-axis is the axis of symmetry for these parabolas. [When the y-axis is the axis of symmetry, the vertex will be the point $(0, c)$.]

EXAMPLE 15   Use a graphing calculator to graph $y = x^2$, and $y = -x^2$; then graph $y = 2x^2$ and $y = -2x^2$. What effect does the change in a (coefficient of x^2) have on the graph?

SOLUTION   We say that the relationship between $y = x^2$ and $y = -x^2$ and between $y = 2x^2$ and $y = -2x^2$ is a reflection across the x-axis. We can see in Figure 22 that when $a < 0$ in $y = -x^2$ and $y = -2x^2$, the parabola is turned downward. The fact that $a = -2$ means that the parabola turns down and that it is not as wide as $y = -x^2$. Again we see that the vertex is $(0, 0)$ and the axis of symmetry is the y-axis for these parabolas.

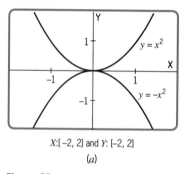

X:[–2, 2] and Y: [–2, 2]

(a)

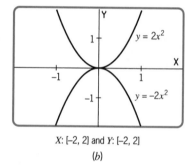

X: [–2, 2] and Y: [–2, 2]

(b)

Figure 22

We can summarize the characteristics of quadratic functions discussed in Examples 14 and 15 as follows.

CHARACTERISTICS OF $y = ax^2$

1. If $|a| > 1$, the graph of $y = ax^2$ is not as wide for a fixed x as the graph of $y = x^2$.
2. If $0 < |a| < 1$, the graph of $y = ax^2$ is wider than that of $y = x^2$.
3. If $a > 0$, the graph of the function has an upward direction.
4. If $a < 0$, the graph of the function has a downward direction.
5. The vertex of $y = ax^2$ is $(0, 0)$.
6. The axis of symmetry of $y = ax^2$ is the y-axis.

EXAMPLE 16   Draw the graphs of $y = x^2 + 1$ and $y = x^2 - 3$. Discuss the relation of these graphs to that of $y = x^2$.

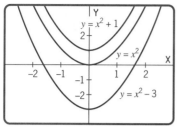

X: [–3, 3] and Y: [–4, 4]

Figure 23

SOLUTION In Figure 23, we can see that the graphs have been shifted vertically. A **vertical shift** or **translation** occurs when a constant is added to $y = ax^2$, giving $y = ax^2 + c$. If $c > 0$, the shift is upward. If $c < 0$, the shift is downward. The vertex of $y = x^2 + 1$ is $(0, 1)$ and the vertex of $y = x^2 - 3$ is $(0, -3)$. The y-axis is the axis of symmetry for both parabolas. ■

Vertical Shifts
1. If $c > 0$, to obtain the graph of $y = x^2 + c$, move the graph of $y = x^2$ **up** c units.
2. $y = x^2 - c$, move the graph of $y = x^2$ **down** c units.

Calculator Note

Experiment with a graphing calculator by entering different values for both a and c in $y = ax^2 + c$. Leave the graph of $y = x^2$ on the viewing rectangle to compare what is happening with each change in a and c.

Up to this point, we have looked only at quadratic functions with the vertex on the y-axis. However, all quadratic functions do not have graphs that are symmetric about the y-axis, as illustrated in the next example.

EXAMPLE 17 Draw the graphs of $y = (x - 3)^2$ and $y = -(x + 2)^2$ and compare these to the graph of $y = x^2$. What is the vertex and axis of symmetry for each function?

SOLUTION These graphs are examples of **horizontal shifts** or **translations**. The graph of the function $y = (x - 3)^2$ is the same as the graph $y = x^2$ shifted 3 units to the right. Its vertex is $(3, 0)$ and the line $x = 3$ is the axis of symmetry [see Figure 24(*a*)]. The graph of the function $y = -(x + 2)^2$ is the same as the graph $y = x^2$ shifted 2 units to the left and reflected across the x-axis. Its vertex is $(-2, 0)$ and the line $x = -2$ is the axis of symmetry [see Figure 24(*b*)]. ■

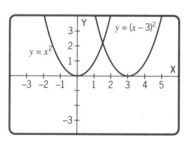

X: [–4, 6] and Y: [–4, 4]

(*a*)

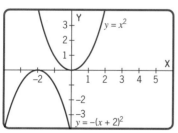

X: [–4, 6] and Y: [–4, 4]

(*b*)

Figure 24

Horizontal Shifts
1. For $c > 0$, to obtain the graph of $y = (x - c)^2$, move the graph of $y = x^2$ c units to the **right**.
2. $y = (x + c)^2$, move the graph of $y = x^2$ c units to the **left**.

Practice Problem 1 Draw the graph of $y = (x - 2)^2$, $y = (x + 1)^2$. How do these graphs compare to the graphs of $y = x^2 - 2$ and $y = x^2 + 1$? What is the vertex and axis of symmetry for each parabola?

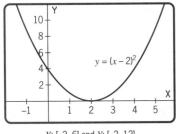

X: [-2, 6] and Y: [-2, 12]

(a)

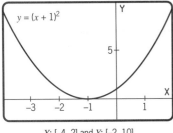

X: [-4, 2] and Y: [-2, 10]

(b)

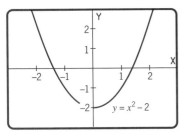

X: [-3, 3] and Y: [-3, 3]

(c)

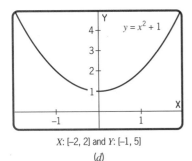

X: [-2, 2] and Y: [-1, 5]

(d)

Figure 25

ANSWER   The graph of $y = (x - 2)^2$ in Figure 25(a) shifts the parabola $y = x^2$ horizontally to the right 2 units. The vertex is (2, 0) and the axis of symmetry is $x = 2$. The graph of $y = (x + 1)^2$ Figure 25(b) is a horizontal shift of $y = x^2$ 1 unit to the left. The vertex is (−1, 0) and the axis of symmetry is $x = -1$. The graph of $y = x^2 - 2$ in Figure 25(c) is a vertical shift of $y = x^2$ two units downward. It has vertex (0, −2) and the axis of symmetry is the y-axis. The graph of $y = x^2 + 1$ in Figure 25(d) is a vertical shift of $y = x^2$ one unit upward. It has vertex (0, 1) and the axis of symmetry is the y-axis.

When graphing and analyzing a parabola, the intercepts and the vertex are the points we use most often. The x-intercept is obtained by setting $y = 0$ and the y-intercept is found by setting $x = 0$. We discuss the algebraic techniques for finding the x-intercepts in Section 1.4.

The vertex is not always obvious when looking at a parabola. There are several methods of finding the vertex. Look at the graphs shown in Figures 24 and 25. Can you see any correlation between the equation and the vertex? Notice that when there is only a vertical or horizontal shift, we can determine the vertex very quickly. Thus $y = x^2 + 1$ has its vertex at (0, 1), whereas $y = (x + 1)^2$ has its vertex at (−1, 0). Before we give a specific method for finding the vertex, let's examine some more graphs to see if we can find a pattern. Let's look at some graphs that have both a horizontal and vertical shift and are in what is called the **standard form of a quadratic**:

$$y = f(x) = a(x - h)^2 + k$$

EXAMPLE 18   Draw the graphs of $y = (x - 2)^2 - 3$ and $y = (x + 2)^2 + 3$. Can you determine the vertex of each parabola? How do the points of the vertex correspond to the equation of the function when compared to the standard form of a quadratic?

SOLUTION   We see in Figure 26(a) that the vertex of $y = (x - 2)^2 - 3$ is (2, −3) and that in Figure 26(b) the vertex of $y = (x + 2)^2 + 3$ is (−2, 3). These points correspond to the point (h, k) from the standard form given above. ∎

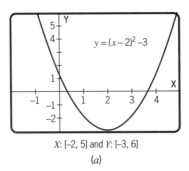

X: [-2, 5] and Y: [-3, 6]

(a)

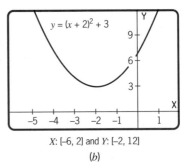

X: [-6, 2] and Y: [-2, 12]

(b)

Figure 26

Practice Problem 2 Experiment using a calculator with equations in the form $y = f(x) = a(x - h)^2 + k$. What do the changes in a, h, and k do to the graph of $y = x^2$?

ANSWER Note that a determines the direction and width of the parabola, and changes in (h, k) alter the vertex of the parabola.

We see that the form used in Example 18 and Practice Problem 2 is extremely useful because it helps us to quickly identify the vertex and the direction of the parabola.

DEFINITION: STANDARD FORM OF A QUADRATIC FUNCTION

A quadratic function $y = f(x) = ax^2 + bx + c$, is in **standard form** when it is written as

$$y = f(x) = a(x - h)^2 + k$$

and has as its vertex the point (h, k). Its axis of symmetry is the line $x = h$. If $a > 0$, the parabola turns upward and has a minimum value of k. If $a < 0$, the parabola turns downward and has a maximum value of k.

COMMON ERROR   Students often confuse the y-intercept $f(0) = c$ with the value of k in the standard form. If $h \neq 0$, then the y-intercept is not k.

It is important to note that not all quadratic functions in x have x-intercepts, but they always have a y-intercept. Since the domain of the function is all real numbers, x can take on the value 0 and there will be a y-intercept.

EXAMPLE 19   Find the axis of symmetry, the coordinates of the vertex, the maximum or minimum value of the function, and the y-intercept for the function

$$y = f(x) = x^2 - 4x + 3$$

SOLUTION   The first step is to write the function in standard form by obtaining the expression $(x - h)^2$. We do this by completing the square. (Later, we will learn another way to find the vertex.) Remember, to complete the square, find $(-4/2)^2$. When this is added and subtracted from the function, the net change is zero, but we are able to factor the perfect square trinomial that results.

$$y = [x^2 - 4x + (-2)^2] - (-2)^2 + 3 \qquad \frac{b}{2} = \frac{-4}{2} = -2$$
$$= (x - 2)^2 - 1$$

By comparing this to the standard form, we determine that the vertex is at $(2, -1)$. Since $a > 0$, there is a minimum value, which occurs at $x = 2$ and is $f(2) = -1$. The y-intercept is $f(0) = 3$. The graph is shown in Figure 27. ∎

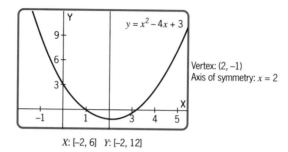

X: [-2, 6]   Y: [-2, 12]

Figure 27

Practice Problem 3 For $y = x^2 - 2x - 3$, find the equation for the axis of symmetry, the vertex, and the y-intercept. Does the function have a maximum value or a minimum value? If so, what is the value?

ANSWER The equation $x = 1$ is the axis of symmetry. The vertex is $(1, -4)$ and the y-intercept is -3. Since $a > 0$, the minimum value of the function is -4.

As stated earlier, we discuss the algebraic techniques for finding the x-intercepts in Section 1.4. For now, we can use a calculator to locate the values (or approximate values).

Practice Problem 4 Using the equation in Practice Problem 3, $y = x^2 - 2x - 3$, draw the graph and estimate the x-intercepts.

ANSWER The graph is shown in Figure 28. The x-intercepts are -1 and 3.

We can describe the coordinates of the vertex and the axis of symmetry for a quadratic function as follows.

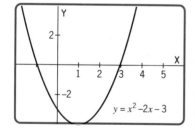

X: [-2, 6] and Y: [-4, 4]

Figure 28

Vertex and Symmetry

For the graph of the quadratic function $y = f(x) = ax^2 + bx + c$:

1. The vertex is $\left(\dfrac{-b}{2a}, f\left(\dfrac{-b}{2a}\right)\right) = \left(\dfrac{-b}{2a}, c - \dfrac{b^2}{4a}\right)$.

2. The axis of symmetry is $x = \dfrac{-b}{2a}$.

We verify these characteristics as follows:

$$f(x) = ax^2 + bx + c$$

$$= a\left(x^2 + \frac{b}{a}x\right) + c \qquad\qquad \text{Factor out } a.$$

$$= a\left(x^2 + \frac{b}{a}x + \frac{b^2}{4a^2}\right) + c - \frac{b^2}{4a} \qquad \text{Complete the square.}$$

$$= a\left(x + \frac{b}{2a}\right)^2 + \left(c - \frac{b^2}{4a}\right)$$

Comparing this result to $y = a(x - h)^2 + k$, we can see that the axis of symmetry is $x = -b/2a$. The x-coordinate of the vertex is also $-b/2a$ and the y-coordinate is

$$f\left(\frac{-b}{2a}\right) = c - \frac{b^2}{4a}$$

EXAMPLE 20   For $f(x) = 2x^2 - 12x + 19$, find the vertex, y-intercept, and the axis of symmetry and determine whether there are any x-intercepts. Write the function in standard form. What is the maximum or minimum value of f?

SOLUTION   Let's use the technique that we just developed to find the vertex. For $f(x) = 2x^2 - 12x + 19$, we have $a = 2$, $b = -12$, and $c = 19$. Since the x-coordinate of the vertex is $-b/2a$, we have

$$x = \frac{-(-12)}{2(2)} = \frac{12}{4} = 3$$

Since the x-coordinate of the vertex is 3, the y-coordinate is

$$f(3) = 2(3)^2 - 12(3) + 19 = 1$$

Therefore, the vertex is $(3, 1)$ and we can write f in standard form as

$$f(x) = 2x^2 - 12x + 19 = 2(x - 3)^2 + 1$$

The minimum value of f is 1, the axis of symmetry is $x = 3$, and the y-intercept is 19. We can see in Figure 29 that there are no x-intercepts.

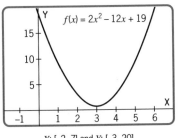

X: [–2, 7] and Y: [–3, 20]
Vertex: (3, 1)
Axis of symmetry: $x = 3$

Figure 29

Practice Problem 5   For $f(x) = -3x^2 + 6x - 5$, find the axis of symmetry, the vertex, y-intercept; put in standard form; and approximate any x-intercepts. What is the maximum or minimum value of f?

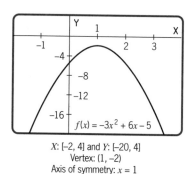

X: [-2, 4] and Y: [-20, 4]
Vertex: (1, -2)
Axis of symmetry: $x = 1$

Figure 30

ANSWER   As shown in Figure 30, the axis of symmetry is $x = 1$; the vertex is $(1, -2)$; and we write $f(x) = -3(x - 1)^2 - 2$. The maximum value of f is -2. The y-intercept is -5 and there are no x-intercepts.

SUMMARY

1. The graph of a quadratic equation in standard form, $f(x) = a(x - h)^2 + k$, has vertex (h, k) and axis of symmetry $x = h$.
2. If the quadratic is not written in standard form, the vertex of $f(x) = ax^2 + bx + c$ is

$$\left(\frac{-b}{2a}, f\left(\frac{-b}{2a} \right) \right)$$

with the axis of symmetry $x = -b/2a$. From this, we see that $h = -b/2a$ and $k = f(h)$.
3. The direction and width of a parabola is determined by the value of a. If $a > 0$, the parabola opens upward. If $a < 0$, the parabola opens downward.
4. The maximum value or minimum value of the function is k.

Exercise Set 1.3

Use a calculator to draw the three given functions in the same viewing rectangle and discuss why they are different.

1. $y = 4x^2, \quad y = 4(x + 1)^2, \quad y = 4(x - 1)^2$
2. $y = 2x^2, \quad y = 2(x - 1)^2, \quad y = 2(x + 1)^2$
3. $y = -x^2, \quad y = -(x - 3)^2, \quad y = -(x + 3)^2$
4. $y = -x^2, \quad y = -x^2 + 1, \quad y = -x^2 - 1$
5. $y = x^2 - 2, \quad y = (x - 3)^2, \quad y = (x + 3)^2 + 2$
6. $y = x^2 + 1, \quad y = 2x^2 + 1, \quad y = -2x^2 + 1$

Describe how the graph of each function can be obtained from the graph of $f(x) = x^2$ in terms of horizontal and vertical shifts, magnification, and reflection.

7. $f(x) = 2x^2$
8. $f(x) = -\frac{1}{3}x^2$
9. $f(x) = (x + 1)^2$
10. $f(x) = (x - 3)^2$
11. $f(x) = x^2 + 3$
12. $f(x) = -x^2 - 2$
13. $f(x) = 2(x - 1)^2 + 3$
14. $f(x) = -3(x + 2)^2 - 1$

For each function in Exercises 15–24:

(a) *Find the line of symmetry.*
(b) *Find the vertex.*
(c) *Find the y-intercept.*
(d) *Draw the graph.*
(e) *Approximate any x-intercepts.*
(f) *Determine whether the vertex is a maximum point or minimum point.*

15. $f(x) = x^2 + x - 6$
16. $f(x) = 2x^2 - 9x - 5$
17. $f(x) = -x^2 - 2x + 24$
18. $f(x) = -2x^2 + 7x - 3$
19. $f(x) = 3x^2 + 9x - 12$
20. $f(x) = 2x^2 + 6x - 30$
21. $f(x) = x^2 + 6x$
22. $f(x) = -2x^2 + 4x$
23. $f(x) = 3x^2 - 4$
24. $f(x) = -4x^2 + 2$

Review Exercises

Determine the domain and range of each function using graphs drawn with a calculator.

25. $f(x) = \sqrt{9x^2 - 16}$
26. $y = \dfrac{x + 3}{x - 1}$
27. $f(x) = \dfrac{|4 + x|}{x}$
28. $y = \sqrt{\dfrac{3 - x}{x + 2}}$

1.4   QUADRATIC EQUATIONS AND APPLICATIONS

OVERVIEW   A quadratic function can have zero, one, or two x-intercepts. The number of x-intercepts can be determined graphically; however, to find the exact value of these intercepts, algebraic techniques are used. These intercepts are called **zeros of the function**. Models involving quadratic functions are used in business, economics, and social and life sciences. In this section we

- Find the zeros of a quadratic function (solve quadratic equations)
- Locate a point of equilibrium for supply and demand functions
- Locate break-even points for cost and revenue functions

We use two procedures in this section for solving quadratic equations: factoring and the quadratic formula. If a quadratic equation $ax^2 + bx + c = 0 \, (a \neq 0)$ can be factored, the solutions can be attained easily. If the product of two numbers is zero, then one or both of the numbers must be zero. If, for example, $(x + 3)(x + 4) = 0$, then either $x + 3 = 0$ or $x + 4 = 0$.

COMMON ERROR   When solving an equation such as $(x + 3)(x + 7) = 10$, you cannot set each factor equal to 10 and solve. The equation must be written so that one side is zero.

EXAMPLE 21   Solve $2x^2 = 32$ by factoring.

SOLUTION

$$2x^2 - 32 = 0 \qquad \text{Rewrite the equation so one side is 0.}$$

$$2(x^2 - 16) = 0 \qquad \text{Factor out the common factor of 2.}$$

$$2(x - 4)(x + 4) = 0 \qquad \text{Factor as the difference of two squares.}$$

This equation is only true if $x - 4 = 0$ or $x + 4 = 0$. Hence

$$x = 4 \quad \text{or} \quad x = -4$$

The factor 2 is not considered because only factors that can attain a value of 0 will yield the solutions.

✓ Check: $2(-4)^2 = 2(16) = 32$ and $2(4)^2 = 2(16) = 32$

Thus, the solution set is $\{-4, 4\}$ (see Figure 31). ■

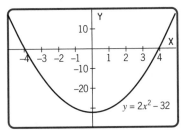

X: [−5, 5] and Y: [−40, 20]
x-intercepts: $x = -4$ and $x = 4$

Figure 31

EXAMPLE 22   Find the solution set of $x^2 + x - 12 = 0$.

SOLUTION   The equation $x^2 + x - 12 = 0$ is equivalent to $(x + 4)(x - 3) = 0$. This equation is only true if $x + 4 = 0$ or $x - 3 = 0$. If $x + 4 = 0$, then $x = -4$ and if $x - 3 = 0$, then $x = 3$.

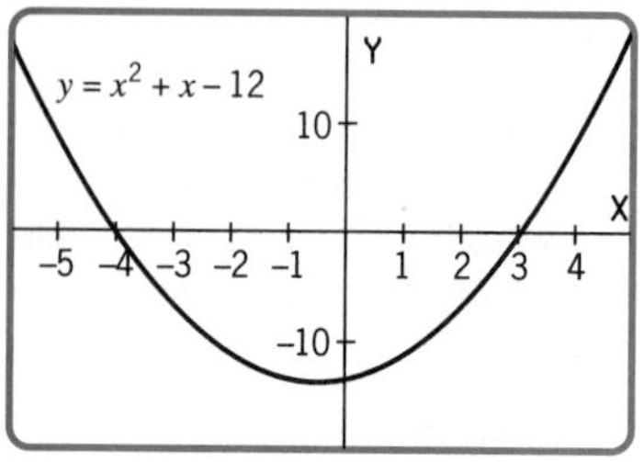

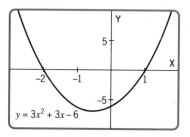

$y = x^2 + x - 12$

X: [-6, 5] and Y: [-20, 20]
x-intercepts: $x = -4$ and $x = 3$

Figure 32

$y = 3x^2 + 3x - 6$

X: [-3, 2] and Y: [-10, 10]
x-intercepts: $x = -2$ and $x = 1$

Figure 33

Check: $3^2 + 3 - 12 = 12 - 12 = 0$ and $(-4)^2 + (-4) - 12 = 16 - 16 = 0$

Thus, the solution set is $\{-4, 3\}$ (see Figure 32). ■

Practice Problem 1   Solve $3x^2 + 3x = 6$ and draw the graph using a calculator to reinforce your answer.

ANSWER   The solution set is $\{-2, 1\}$ (see Figure 33).

There are quadratics that do not factor easily and some that do not factor at all over the rational numbers. Also, some quadratic equations have no real solutions. In order to work effectively with these quadratics, let's return to the work we did in Section 1.3—completing the square of the quadratic $f(x) = ax^2 + bx + c$ to find the vertex. From this equation, we obtained

$$f(x) = a\left(x + \frac{b}{2a}\right)^2 + c - \frac{b^2}{4a}$$

which can also be written as

$$f(x) = a\left(x + \frac{b}{2a}\right)^2 - \left(\frac{b^2 - 4ac}{4a}\right)$$

Now let $f(x) = 0$ (for the equation $ax^2 + bx + c = 0$) and solve for x.

$$\left(x + \frac{b}{2a}\right)^2 = \frac{b^2 - 4ac}{4a^2} \qquad \text{Divide by } a \neq 0.$$

$$x + \frac{b}{2a} = \pm\sqrt{\frac{b^2 - 4ac}{4a^2}} \qquad \text{Take the square root.}$$

$$= \pm\frac{\sqrt{b^2 - 4ac}}{2a} \qquad \text{Simplify the denominator.}$$

$$x = \frac{-b \pm \sqrt{b^2 - 4ac}}{2a}$$

This result is called the **quadratic formula** and can be used to solve any quadratic equation.

THE QUADRATIC FORMULA

If a, b, and c are real numbers and $a \neq 0$, then the solutions of $ax^2 + bx + c = 0$, if they exist, are

$$x = \frac{-b + \sqrt{b^2 - 4ac}}{2a} \quad \text{or} \quad \frac{-b - \sqrt{b^2 - 4ac}}{2a}$$

You can program the quadratic formula into your graphing calculator and use it to get a decimal approximation of the roots, but you will need to use the quadratic formula without a calculator or program your calculator to give you the exact expression if a problem requires that the answer be in exact form (i.e., a rational number or a radical expression).

Calculator Note

You must be aware of the capabilities of your graphing calculator. As mentioned previously, some graphing calculators are always in complex mode and give complex answers as an ordered pair.

EXAMPLE 23 Solve $2x^2 + 2x = 12$.

SOLUTION In order to use the quadratic formula, we need to put the equation in the form $ax^2 + bx + c = 0$. Thus, we have $2x^2 + 2x - 12 = 0$, where $a = 2$, $b = 2$, and $c = -12$. Using the quadratic formula, the two solutions are

$$x = \frac{-2 + \sqrt{(2)^2 - 4(2)(-12)}}{2(2)} = \frac{-2 + \sqrt{4 + 96}}{4}$$

$$= \frac{-2 + \sqrt{100}}{4} = \frac{-2 + 10}{4} = 2$$

$$x = \frac{-2 - \sqrt{(2)^2 - 4(2)(-12)}}{2(2)} = \frac{-2 - \sqrt{4 + 96}}{4}$$

$$= \frac{-2 - \sqrt{100}}{4} = \frac{-2 - 10}{4} = -3$$

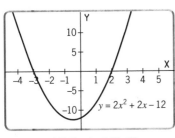

X: [−5, 6] and Y: [−15, 15]
x-intercepts: $x = -3$ and $x = 2$

Figure 34

Check: $2(2)^2 + 2(2) = 8 + 4 = 12$ and $2(-3)^2 + 2(-3) = 18 - 6 = 12$

The solution set is $\{-3, 2\}$ (see Figure 34).

EXAMPLE 24 Solve $x^2 - 6x + 25 = 0$.

SOLUTION We have $a = 1$, $b = -6$, and $c = 25$, so

$$x = \frac{6 \pm \sqrt{36 - 100}}{2} = \frac{6 \pm \sqrt{-64}}{2}$$

Since $\sqrt{-64}$ is not a real number, this equation has no real solution. The graph is shown in Figure 35. Notice that there are no x-intercepts; hence, there are no real solutions to the equation.

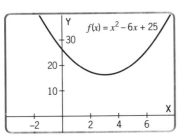

X: [−4, 8] and Y: [−5, 40]
No x-intercept: no real solution

Figure 35

Practice Problem 2 Solve $3x^2 - 0.1x - 12 = 0$ using a graphing calculator. Draw the graph and locate the x-intercepts using the CALC menu and selecting root.

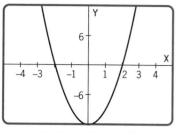

X: [-5, 5] and Y: [-12, 12]
The roots are approximately
-1.983402777 and 2.01673611

Figure 36

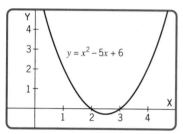

X: [-1, 5] and Y: [-1, 5]

Figure 37

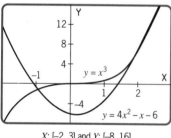

X: [-2, 3] and Y: [-8, 16]

(a)

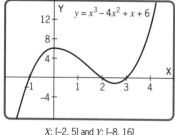

X: [-2, 5] and Y: [-8, 16]

(b)

Figure 38

ANSWER   Using the range X: $[-5, 5]$ and Y: $[-12, 12]$, we locate roots close to -2 and 2 (see Figure 36). You can change the viewing rectangle to get a better approximation of the roots. Try to locate the roots in this manner (see page A-3 of Appendix A). After this, you can have your graphing calculator state the intercepts. The solution set of approximate answers is $\{-1.983402777, 2.01673611\}$.

The material on quadratic equations can be easily extended to inequalities or higher degree equations by using a graphing calculator. For example, we can use a graphing calculator to solve inequalities and to solve cubics and higher-degree equations and inequalities.

EXAMPLE 25   Use a graphing calculator to solve the inequality $x^2 - 5x + 6 < 0$.

SOLUTION   Draw the graph and locate the x-intercepts. Then determine the intervals over which the graph is below the x-axis. We can see from Figure 37 that the solution interval is the interval $(2, 3)$.   ∎

Practice Problem 3   Solve the inequality $-2x^2 + 3x + 4 \leq 0$.

ANSWER $(-\infty, -0.85], [2.35, \infty)$

EXAMPLE 26   Solve the equation $x^3 = 4x^2 - x - 6$.

SOLUTION   If we graph the two equations $y_1 = x^3$ and $y_2 = 4x^2 - x - 6$, we can locate the points of intersection. However, as shown in Figure 38(a), the points of intersection are not clear and we must take precautions to find all the points of intersection. This problem can be approached in another way. We know that if $y_1 = y_2$, then $y_1 - y_2 = 0$; so we graph the equation $y = y_1 - y_2 = x^3 - 4x^2 + x + 6$ and then solve the equation for the zeros. From Figure 38(b) we see that the solutions are $x = -1$, $x = 2$, and $x = 3$.   ∎

Using algebra to find the point of intersection of the graphs of two quadratic functions, or of a linear function and a quadratic function, we can eliminate y and obtain a quadratic in x.

EXAMPLE 27   Find the intersection of $y = -x + 1$ and $y = x^2 - 4x + 3$.

SOLUTION   Set the two equations equal and simplify to a quadratic:

$$x^2 - 4x + 3 = -x + 1$$
$$x^2 - 3x + 2 = 0$$
$$(x - 2)(x - 1) = 0$$

Thus $x = 2$ and $x = 1$. Substituting into $y = -x + 1$, we get the points of intersection $(1, 0)$ and $(2, -1)$ (see Figure 39).   ∎

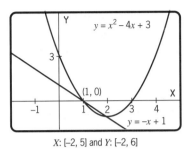

$y = x^2 - 4x + 3$

3

(1, 0)

$y = -x + 1$

X: [-2, 5] and Y: [-2, 6]

Figure 39

Supply and Demand Curves

In economics, a supply curve represents the price p that is necessary for manufacturers to produce x number of an item and is written as $p = S(x)$. A demand curve represents the price necessary for consumers to purchase x items produced and is written as $p = D(x)$. Since all values of x and p are nonnegative in the first quadrant, the parts of parabolas that are in the first quadrant can often be used for supply and demand curves. The point at which these curves are equal is called the equilibrium point or point of equilibrium, that is, where supply equals demand.

EXAMPLE 28   If the supply function for a commodity is given by $p = S(x) = 2x^2 + 30x$ and the demand function by $p = D(x) = -15x + 5000$, where x is the number of units of the commodity and p is the price, find the equilibrium point.

SOLUTION   At the point of equilibrium, we have

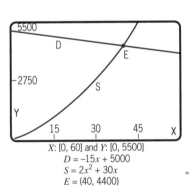

5500

D

E

-2750

S

Y

15   30   45   X

X: [0, 60] and Y: [0, 5500]
$D = -15x + 5000$
$S = 2x^2 + 30x$
$E = (40, 4400)$

Figure 40

$$D(x) = S(x) \qquad \text{Supply = Demand at equilibrium point}$$

$$-15x + 5000 = 2x^2 + 30x \qquad \text{Substitute for } S \text{ and } D.$$

$$2x^2 + 45x - 5000 = 0$$

$$(x - 40)(2x + 125) = 0 \qquad \text{Factor and solve.}$$

$$x = 40 \quad \text{and} \quad x = -\frac{125}{2} \qquad \begin{array}{l}\text{Eliminate } x = -125/2, \text{ since} \\ x \text{ is the number of} \\ \text{units of a commodity} \\ \text{and must be positive.}\end{array}$$

$$D(x) = -15(40) + 5000 = 4400 \qquad \text{Find the equilibrium point.}$$

The equilibrium quantity is 40 units and the equilibrium price is $4400, since $S(40) = D(40) = 4400$. The graphs of both the supply function and the demand function are shown in Figure 40. ■

Break-Even Points and Profit Maximization

Some costs tend to decrease sharply after a certain level of production. A quadratic function is often used to represent the total cost of producing a product. For example,

$$C(x) = 400 + 30x - 0.4x^2$$

might represent the cost of producing x units of a product. A revenue function is found by the following product: Number of units sold · Price per unit. Often, revenue functions are quadratic functions. When both are sketched on a coordinate system, the intersections of the two graphs should represent the **break-even** points—the points where the functions are equal. Algebraically, by setting $C(x) = R(x)$, we obtain a quadratic equation that we solve either by factoring or by the quadratic formula. Furthermore, when $R(x) > C(x)$, a profit is made, but when $R(x) < C(x)$, there is a loss.

EXAMPLE 29   The cost of producing x small portable vacuum cleaners is given by

$$C(x) = 400 + 8x + 0.1x^2$$

when the price per unit is

$$p = 32 - 0.1x$$

Find the break-even point, draw the graphs of both cost and revenue, and show areas of profit and loss.

SOLUTION   Since Revenue = Units sold · Price per unit, the revenue equation is calculated as $R(x) = x(32 - 0.1x) = 32x - 0.1x^2$. To find the break-even point, set

$$C(x) = R(x)$$

$$400 + 8x + 0.1x^2 = 32x - 0.1x^2$$

$$0.2x^2 - 24x + 400 = 0$$

$$x^2 - 120x + 2000 = 0 \qquad \text{Divide by 0.2.}$$

$$(x - 20)(x - 100) = 0 \qquad \text{Factor.}$$

$$x = 20 \quad \text{and} \quad x = 100 \qquad \begin{array}{l}\text{Therefore, the break-}\\\text{even points occur at}\\ x = 100 \text{ and } x = 20.\end{array}$$

Thus, at $x = 100$ we have

$$C = R = 32(100) - 0.1(100)^2$$
$$= \$2200$$

and at $x = 20$ we have

$$C = R = 32(20) - 0.1(20)^2$$
$$= \$600$$

Figure 41(a) shows the areas of profit and loss, and Figure 41(b) shows how the graphs look on a graphing calculator.   ■

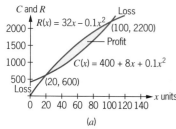

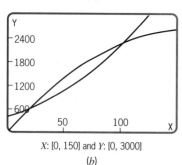

X: [0, 150] and Y: [0, 3000]

(b)

Figure 41

SUMMARY

1. The solutions of $y = ax^2 + bx + c = 0 \ (a \neq 0)$ are

$$x = \frac{-b \pm \sqrt{b^2 - 4ac}}{2a}$$

2. Revenue = Number of units sold · Price per unit
3. A break-even point occurs when Revenue = Cost.
4. Profit occurs when revenue is greater than cost.
5. The equilibrium point occurs when supply and demand are equal.

Exercise Set 1.4

Find only real solutions for the following exercises. If possible, solve by factoring. Verify by drawing a graph on a calculator.

1. $x^2 - 16 = 0$
2. $3x^2 - 27 = 0$
3. $2x^2 - 14 = 0$
4. $3x^2 - 15 = 0$
5. $x^2 - 5x = 0$
6. $3x^2 - 7x = 0$
7. $3x^2 - 75 = 0$
8. $4x^2 - 36 = 0$
9. $x^2 - 2x = 3$
10. $x^2 - 2x = 15$
11. $2x^2 - 6x = 56$
12. $x^2 + 2x = 24$
13. $x^2 + x = 2$
14. $x^2 - 5x - 10 = 4$
15. $x^2 = 6 - x$
16. $x^2 - 2x = 3$

Use the quadratic formula to solve each equation.

17. $x^2 + 8x - 9 = 0$
18. $x^2 - 4x = 4$
19. $x^2 + 11x = -30$
20. $2x^2 = 5x + 2$

In Exercises 21–23, determine the point(s) of intersection of the graphs of the functions. Verify your answers using a calculator.

21. $f(x) = x^2 - 4x + 7; \quad g(x) = x + 1$
22. $f(x) = -2x^2 + 5x + 3; \quad g(x) = -3x^2 + x$
23. $f(x) = 6x^2 + 2x - 2; \quad g(x) = 3x^2 + x$

Solve each of the following quadratic inequalities using a calculator to draw the graph.

24. $(x - 2)(x + 1) > 0$
25. $x^2 \geq 4$
26. $r^2 \leq 9$
27. $b^2 + 5b + 6 > 0$
28. $w^2 + 6 < 5w$
29. $6k^2 - k \leq 2$
30. $2y^2 \geq 5y + 3$

Find the zeros (x-intercepts) and discuss the sign of y between the zeros. Use a calculator.

31. $y = x(x + 2)(x - 2)$
32. $y = x^2(x - 4)$
33. $y = x(x + 1)^2$
34. $y = x(x + 1)(x^2 - 4)$
35. $y = x^2(x^2 - 9)$
36. $y = 3x^4 + x^3 - 2x^2$
37. $y = x^4 + x^3 - 2x^2$

Applications (Business and Economics)

38. **Maximum Profit.** Find the maximum profit for the following profit functions.
    (a) $P(x) = 4 + 6x - 3x^2$
    (b) $P(x) = 80x - 8x^2$

39. **Maximum Revenue.** Find the maximum revenue for the revenue function $R(x) = 386x - x^2$, where x is the number of units sold.

40. **Equilibrium Point.** Sketch the graph of the demand function $p = D(x) = (x - 6)^2$ for $x \geq 6$ and the supply function $p = S(x) = x^2 + x + 10$ on the same coordinate system and locate the equilibrium point. What is the price at this point?

41. **Equilibrium Point.** The demand function for a commodity is $p = D(x) = (x - 4)^2$ and the supply function is $p = S(x) = 2x$, where $x \geq 4$. Sketch the graphs of these two functions on a coordinate system and locate the equilibrium point. What is the price at this point?

42. **Profit and Loss Regions.** Suppose that a company has a fixed cost of $150 per day and a variable cost of $x^2 + x$. Further suppose that the revenue function is $R(x) = xp$ and the price per unit is given by $p = 57 - x$. Sketch the cost and revenue functions and locate the regions of profit and loss.

43. **Maximum Profit.**   In Exercise 42, where does the maximum profit occur? What is the maximum profit? (**Hint:** $P = R - C$.)

Applications (Social and Life Sciences)

44. **Blood Velocity.**   In the formula

$$V_r = V_m \left(1 - \frac{r^2}{R^2} \right)$$

V_r is the velocity of a blood corpuscle r units from the center of the artery, R is the constant radius of the artery, and V_m is the constant maximum velocity of the corpuscle. If $V_r = \frac{1}{2} V_m$, solve for r in terms of R and then draw the graph.

45. **Pollution.**   An environmental study is used to approximate the annual growth of carbon monoxide in the air (parts per million) as a quadratic function. The approximating function is $M(t) = 0.01t^2 + 0.05t + 1.6$, where M is parts per million and t is measured in minutes starting at 7:00 A.M. ($0 \leq t \leq 180$). Show this relationship with a graph.

46. **Population Growth.**   A small city is expecting a decline in growth for a few years and then a rapid growth. This trend has been approximated by $P(t) = 2t^2 - 100t + 10,000$ ($t \geq 0$), where $P(t)$ is the population and t is time in years from now. Show this growth trend with a graph.

47. **Laffer Curve.**   At the endpoints of the Laffer curve shown in the figure, there is no revenue (R) for given tax rates (r). Suppose that a Laffer curve is approximated by

$$R = \frac{r(50r - 5000)}{r - 110} \text{ for } 0 \leq r \leq 100$$

What tax rates provide no revenue? What rate provides maximum revenue?

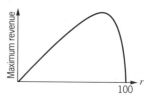

Review Exercises

48. Which of the following specify functions?
    (a) $y = 2x + 3$ (b) $3x^2 + 4y^2 = 25$
    (c)   (d)

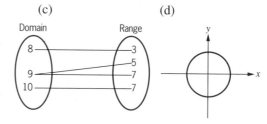

49. Find an equation of the line passing through the points $(-2, 1)$ and $(7, -4)$.

50. Determine the vertex and sketch the graph for $y = -2(x + 1)^2 - 3$.

1.5   EXPONENTIAL FUNCTIONS

OVERVIEW   We have been studying functions such as $f(x) = x^2$, where the base is the variable and the exponent is a constant; we now study functions such as $f(x) = 2^x$, where the base is constant and the exponent is a variable. These are called exponential functions. Although this type of function may be unfamiliar to you, such functions are used by economists to study the growth of money supply, by businesspeople to study declines in sales, by biologists to study the growth or decay of organisms in a laboratory culture, by social scientists to study population growth, and by archeologists to study the carbon dating of fossils. In this section we consider

- Exponential functions with any base $a > 0$ $(a \neq 1)$
- Exponential functions with base e
- Graphs of exponential functions
- Applications involving exponential functions

We know how to evaluate the expression 2^x when x assumes integral values. For example,

$$2^0 = 1 \qquad 2^1 = 2 \qquad 2^2 = 4$$

$$2^{-1} = \frac{1}{2} \qquad 2^{-2} = \frac{1}{2^2} = \frac{1}{4} \qquad 2^{-3} = \frac{1}{2^3} = \frac{1}{8}$$

However, in this section we allow x to be any real number. We call the function $f(x) = 2^x$ an exponential function with base 2.

DEFINITION: EXPONENTIAL FUNCTIONS

Let a be any positive number, where $a \neq 1$. The function

$$f(x) = a^x$$

is called the **exponential function f with base a**. The domain of the independent variable x is any real number.

There are two important things to note here: the domain of f is the set of real numbers and a must be positive, $a > 0$. If a were negative, then x could not be any real number. The function would not have meaning when $x = \frac{1}{2}$ and $a < 0$. For example, if $a = -6$, then $(-6)^{1/2} = \sqrt{-6}$ has no meaning over the set of real numbers. The base a cannot be 1 because $f(x) = 1^x = 1$ is a constant function and not an exponential function.

Exponential functions can be evaluated easily on a calculator. We have already evaluated expressions such as $16^{3/4} = 8$ and $16^2 = 256$. Now we look at irrational exponents, such as $a^{\sqrt{3}}$ or $4^{\pi+1}$. We will not go into a technical definition for terms like $a^{\sqrt{3}}$, but we will state that we can get as close as we like with closer and closer approximations for $\sqrt{3}$.

Calculator Note

Use $\boxed{\wedge}$ when evaluating values of an exponential function. However, be sure to use some care with irrational and negative exponents. By using parentheses around the exponent, you should be able to get the correct answers consistently.

EXAMPLE 30 Calculate each of the following. Round to three decimal places.

(a) $3^{\sqrt{2}}$ 　　　(b) 5^{π} 　　　(c) $7 \cdot 8^{2/5}$ 　　　(d) $4^{-3/7}$

SOLUTION

(a) $3^{\sqrt{2}} \approx 4.729$ (b) $5^{\pi} \approx 156.993$

(c) $7 \cdot 8^{2/5} \approx 16.082$ (d) $4^{-3/7} \approx 0.552$ ■

   In the last section, we looked at the graphs of quadratic functions. Now let's turn to graphs of exponential functions.

EXAMPLE 31   On a calculator draw the graphs of $y = 2^x$ and $y = 3^x$.

SOLUTION   The graphs are shown in Figure 42. Table 3 shows some values for both functions. Notice that as x increases, the functions also increase. Use the Trace key to move the cursor along the graph of each function and locate the values shown in Table 3. Note that you can create this table on a calculator. We can see that the y-intercept for both functions is 1, since $a^0 = 1$ and $a \neq 0$.   ■

TABLE 3

x	-2	-1	0	1	2
$y = 2^x$	$\frac{1}{4}$	$\frac{1}{2}$	1	2	4
$y = 3^x$	$\frac{1}{9}$	$\frac{1}{3}$	1	3	9

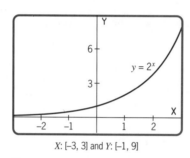
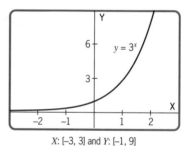

X: [-3, 3] and Y: [-1, 9]   X: [-3, 3] and Y: [-1, 9]

Figure 42

EXAMPLE 32   Using a calculator, draw the graphs of $y = 2^{-x}$ and $y = 3^{-x}$.

SOLUTION   The graphs are shown in Figure 43, and Table 4 shows some values for both functions. Notice that as the value of x increases, the value of the function decreases. Also, note that the y-intercept is 1, as was the case in Example 31. Use the Trace key to move the cursor along the graph of each function and locate the values shown in Table 4. Create this table with a calculator.   ■

TABLE 4

x	-2	-1	0	1	2
$y = 2^{-x}$	4	2	1	$\frac{1}{2}$	$\frac{1}{4}$
$y = 3^{-x}$	9	3	1	$\frac{1}{3}$	$\frac{1}{9}$

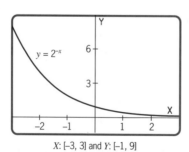
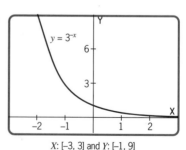

X: [-3, 3] and Y: [-1, 9]   X: [-3, 3] and Y: [-1, 9]

Figure 43

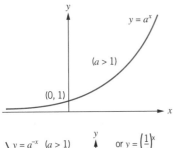

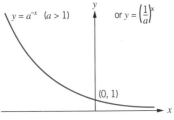

Figure 44

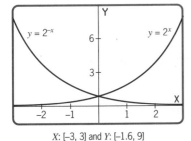

X: [–3, 3] and Y: [–1.6, 9]

Figure 45

Practice Problem 1 Using a calculator, draw the graphs of several functions of the form $y = a^x$: for example, $y = 4^x$, $y = 4^{-x}$, $y = (\frac{1}{4})^x$, and $y = (\frac{1}{4})^{-x}$. Remember that $a > 0$. What do you observe about these graphs? Try some other numbers as bases.

ANSWER Note that when $0 < a < 1$, the function decreases as x increases. When $a > 1$, the function increases as x increases. What happens when $a = 1$, $a = 0$, $a < 1$? Consider these cases.

The graphs in Figures 42 and 43 are typical exponential graphs. Now let's look at the properties of exponential functions (see Figure 44).

PROPERTIES OF EXPONENTIAL FUNCTIONS

Each of the following is true for the exponential function

$$y = f(x) = a^x \qquad (a > 0, a \neq 1, x \in \mathcal{R})$$

(a) The domain of the function is all real numbers.
(b) The graph lies above the x-axis for all x; thus, the range of the function is $(0, \infty)$.
(c) If $x = 0$, then $a^x = 1$ or $f(0) = 1$, so the y-intercept is at $(0, 1)$.
(d) There are no x-intercepts; the graph does not cross the x-axis.
(e) If $a > 1$, the function increases as x increases.
(f) If $0 < a < 1$, the function decreases as x increases.
(g) By the rules of exponents we can write $f(x) = a^{-x} = (1/a)^x$.

EXAMPLE 33 Draw the graphs of $y = 2^x$ and $y = 2^{-x}$ on the same coordinate axes.

SOLUTION The calculator graphs of these functions are shown in Figure 45. From the rules of exponents, we know that $2^{-x} = (\frac{1}{2})^x$. We can also see in Figure 45 that the graph of $y = 2^{-x}$ is the reflection of the graph of $y = 2^x$ across the y-axis. (The y-axis is the axis of symmetry for the two graphs.) ∎

Obviously, the graphs of exponential functions can be moved around just as those of quadratic functions. What functions produce such changes? For example, how do the graphs of $f(x) = 2^{x+1}$ and $f(x) = 2^x + 1$ compare to the graph of $f(x) = 2^x$? Let's look at this in the following example.

EXAMPLE 34 Draw the calculator graphs of $f(x) = 2^x$, $f(x) = 2^{x+1}$, and $f(x) = 2^x + 1$. What conclusions can you draw from these graphs?

SOLUTION In Figure 46(a) we can see that the graph of $f(x) = 2^{x+1}$ is a horizontal shift to the left one unit, just as we saw in the last section that $f(x) =$

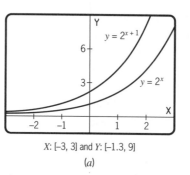

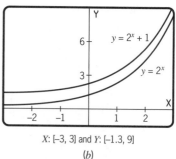

X: [-3, 3] and Y: [-1.3, 9]

(a)

X: [-3, 3] and Y: [-1.3, 9]

(b)

Figure 46

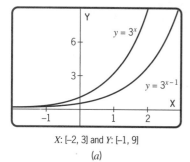

X: [-2, 3] and Y: [-1, 9]

(a)

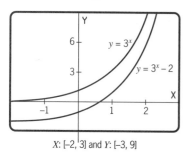

X: [-2, 3] and Y: [-3, 9]

(b)

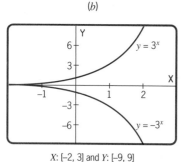

X: [-2, 3] and Y: [-9, 9]

(c)

Figure 47

$(x + 1)^2$ shifts the graph of $f(x) = x^2$ one unit to the left. Likewise, in Figure 46(b) we can see that $f(x) = 2^x + 1$ is a vertical shift upward 1 unit. ■

Practice Problem 2 Compare the graphs of $f(x) = 3^{x-1}$, $f(x) = 3^x - 2$, and $f(x) = -3^x$ with the graph of $f(x) = 3^x$. Check by drawing the calculator graphs.

ANSWER

$f(x) = 3^{x-1}$ is a horizontal shift 1 unit to the right in Figure 47(a).
$f(x) = 3^x - 2$ is a vertical shift 2 units down in Figure 47(b).
$f(x) = -3^x$ is a reflection across the x-axis in Figure 47(c).

Doubling Concept

Certain bacteria split at regular intervals. That is, where there is one, after a given amount of time there are two [see Table 5(a)].

TABLE 5

(a)

Time	0	1	2	3	4 . . .
Bacteria	1	2	4	8	16 . . .

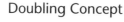

(b) After $t = t_0$

Time	0	d	$2d$	$3d$
Bacteria	$N(t_0)$	$2N(t_0)$	$4N(t_0)$	$8N(t_0)$

In general, after a given time $t = d$, the number of bacteria doubles. Then, if $N(t_0)$ is the number at time t_0, the number at time t is equal to

$$N(t) = N(t_0)2^{t/d} \qquad t \text{ is the time after } t_0; d \text{ is the time after } t_0 \text{ for the population to double.}$$

Note that when $t = d$, $N(t) = 2N(t_0)$; that is, the number has doubled [see Table 5(b)]. This concept applies to the growth of populations involving bacteria, animals, insects, or people. This same concept can be applied to the increase in value of a purchase.

EXAMPLE 35   An abstract painting was purchased in 1930 for $600. Its value has doubled every 10 years. What was its value in 1990?

SOLUTION

$$V(t) = V(t_0) \cdot 2^{t/d}$$
$$= 600 \cdot 2^{t/10} \qquad V(t_0) = 600 \text{ and } d = 10 \text{ years}$$

$$V(60) = 600 \cdot 2^{60/10} \qquad \text{From 1930 to 1990 is 60 years.}$$
$$= \$38,400$$

The value of the painting in 1990 is $38,400.   ■

A good example of exponential growth is compound interest. The formula for compound interest is

$$A = P(1 + i)^n$$

where P is the principal, i is the interest rate per compounding period, and n is the number of compounding periods.

EXAMPLE 36   If $1000 is invested at 7% interest compounded annually for 3 years, how much will be in the account?

SOLUTION   The amount will be

$$A = 1000(1 + 0.07)^3 = \$1225.04$$ ■

The Natural Exponent

A logical question to ask is, What happens to the amount A if the interest is compounded more and more frequently? Later in the book we will show that as n increases without bound, the value of $[1 + (1/n)]^n$ approaches an irrational number that is denoted by e. The function $f(x) = e^x$ is called the **natural exponential function**. This number occurs naturally in many areas of study. We use the approximation

$$e \approx 2.71828$$

We often see the natural exponential function in the form

$$f(x) = ce^{bx}$$

This function is one of the most useful exponential functions in mathematics, economics, and applications in the fields of life science and social science.

> ## Calculator Note
>
> Graphing calculators have an e^x key. It is most likely a 2nd function. Look above the $\boxed{\text{ln}}$ key and see if e^x is written there. Try to evaluate e^2 to see if you get ≈ 7.38906.

Practice Problem 3 Use a calculator to evaluate e^{-3}, $e^{2/3}$, and $e^{-5/6}$. Round to five decimal places.

ANSWER

$$e^{-3} \approx 0.049787$$

$$e^{2/3} \approx 1.94773$$

$$e^{-5/6} \approx 0.43460$$

The graph of $f(x) = e^x$ is shown in Figure 48(a). In Figure 48(b), the graph of $f(x) = e^x$ is shown with the graphs of $f(x) = 2^x$ and $f(x) = 3^x$.

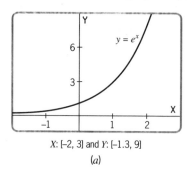
X: [-2, 3] and Y: [-1.3, 9]
(a)

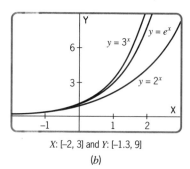
X: [-2, 3] and Y: [-1.3, 9]
(b)

Figure 48

Practice Problem 4 With a calculator, draw the graphs of several functions of the form $y = ce^{bx}$. Use such functions as $y = -2e^{3x}$ or $y = e^{x-1}$. What can you determine from the graphs?

ANSWER Regardless of the functions that are graphed, notice that the graph of $y = ce^{bx}$ behaves the same as the graphs of the exponential functions that we graphed at the beginning of this section.

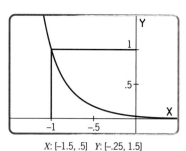
X: [-1.5, .5]   Y: [-.25, 1.5]
When $y = 1$, $x = -1$

Figure 49

Practice Problem 5 Use a calculator to locate the value of x for $y = 1$ on the graph of $y = e^{x^2 - 2x - 3}$ over the interval $[-1.5, .5]$.

ANSWER The graph is shown in Figure 49.

The natural exponential function for money compounded continuously is as follows.

DEFINITION: COMPOUNDED CONTINUOUSLY

The compound amount of an investment at the end of t years is given by

$$A = Pe^{rt}$$

where r is the annual rate (expressed as a decimal) **compounded continuously**, and P is the principal or money invested.

EXAMPLE 37   A person wishes to place \$10,000 in a savings account for $2\frac{1}{2}$ years. A savings and loan association advertises that the interest rate is 7% compounded continuously. To what amount will the \$10,000 accumulate?

SOLUTION   Since $A = Pe^{rt}$, we have that $P = 10,000$, $r = 0.07$, and $t = 2\frac{1}{2} = 2.5$. The \$10,000 will accumulate according to the equation

$$A = 10,000e^{0.07 \cdot 2.5} = \$11,912.46$$

Note that the answer is rounded to the nearest cent.   ■

Many problems that describe the growth of bacteria or the growth of a city's population, the decrease of sales activity, or the decay of radioactive materials can be formulated to use e as the base of the exponential function. We illustrate this concept with an example involving decay.

EXAMPLE 38   Suppose that the amount in grams of a radioactive material present at a time t is given by

$$A(t) = 100e^{-0.002t}$$

where t is measured in days and 100 grams is the original amount. Find the amount of radioactive material at the end of 80 days.

SOLUTION   The amount of radioactive material remaining after 80 days is

$$A(80) = 100e^{-0.002 \cdot 80} = 100e^{-0.16} = 85.2144 \text{ grams} \quad ■$$

Practice Problem 6   If you deposit \$10,000 today in a bank that pays interest at a rate of 8% compounded continuously, how much would you have at the end of 6 years?

ANSWER   \$16,160.74

Equations with Exponential Functions

Equations often contain exponential functions. Solving some exponential equations, such as $2^x = 5^{2-x}$, requires the use of topics that we will cover later;

however, if the bases are the same, the equations can be solved using only the properties that we have covered in this chapter. For a^x and a^y, where $a > 0$ and $a \neq 1$, if $a^x = a^y$, then $x = y$ (since a^x and a^y are one-to-one functions). Likewise, if $x = y$, then $a^x = a^y$.

EXAMPLE 39 Solve $\dfrac{1}{27} = 9^x$.

SOLUTION

$$\frac{1}{27} = 9^x$$

$$\frac{1}{3^3} = (3^2)^x \qquad \text{27 and 9 are powers of 3.}$$

$$3^{-3} = 3^{2x} \qquad \text{Write as } a^x = a^y.$$

$$-3 = 2x \qquad \text{Exponents are equal.}$$

$$x = -\frac{3}{2}$$

$\blacksquare$

SUMMARY

Exponential functions have wide applications in business to describe growth of money as well as decrease in sales. In science, exponential functions are used to describe the growth of bacteria and populations and the decay of radioactive materials.

1. An exponential function is given as $f(x) = a^x$, where $a > 0$ and $a \neq 1$.
2. The amount of an investment at the end of t years and rate r, if interest is compounded continuously, is given by $A = Pe^{rt}$.

Exercise Set 1.5

Evaluate each of the following using a calculator.

1. 3^{-6}
2. $5^{3/4}$
3. $1000(2.01)^{-3}$
4. $20^{\sqrt{3}}$
5. $e^{-1/3}$
6. $5e^{2/5}$
7. $-e^{1.10}$
8. $6^{-\pi+1}$
9. $7^{(1/2)+\sqrt{2}}$

Evaluate each function at $t = 0$, $t = 0.5$, and $t = -2$.

10. $y = 2^{-t}$
11. $y = \left(\dfrac{1}{2}\right)^t$
12. $y = 2^t - 2$
13. $y = 10^t - 1$

14. $y = 4(3)^{-2t}$
15. $y = 3(2)^{-t}$
16. $y = 3e^{-2t}$
17. $y = e^{-t} + e^t$

Draw the graphs of each exponential function.

18. $y = 4^x$
19. $y = 3^{-x}$
20. $y = 4e^x$
21. $y = 2^x + 1$
22. $y = 2^{x+1} - 2$
23. $y = -e^x$
24. $y = 3^{x-2} + 1$
25. $y = -2^{-x}$
26. $y = 4^{x+1} - 2$

Graph each exponential function. What is the range of each function?

27. $y = 4e^{-0.05x}$
28. $y = 3.2^{2x}$
29. $y = 5 - 3e^{2x}$
30. $y = 1 - e^{-x}$

31. $y = 6e^{0.02x}$ 32. $y = 3e^{-0.13x}$

33. Use a calculator to graph $y = 2^x$ and $y = 4^x$ on the same viewing screen. From this determine where $2^x < 4^x$.

34. Use a calculator to graph $y = 2^{-x}$ and $y = 4^{-x}$ on the same viewing screen. From this determine where $2^{-x} < 4^{-x}$.

Graph each of the following functions. Determine maximum or minimum values that may occur.

35. $f(x) = x^2 e^{-x}$ 36. $f(x) = x3^{2+x}$

37. $f(x) = e^{x^2 - 1}$

Solve each equation.

38. $5^{x^2} = 25^4$ 39. $3^{2x+1} = 27^{-x}$

40. $e^{x^2+1} = e^{2x}$ 41. $16^{2x} = 64^{x-3}$

Applications (Business and Economics)

42. **Depreciation.**   A machine is depreciated by a given amount each year. The value of the machine (usually designated by S) at the end of n years, when the machine depreciates r percent per year and C is the original cost is

$$S = C \left(1 - \frac{r}{100}\right)^n$$

Find the value at the end of 8 years of a machine costing $100,000 at an annual rate of depreciation of 12%.

43. **Sales Functions.**   The sales for a product are represented by the following exponential function, where t represents the number of years the product has been on the market:

$$S(t) = 600 - 200e^{-t}$$

Graphically represent sales as a function of time t. From the graph describe how sales change with time.

44. **Money Doubling.**   An antique automobile was purchased in 1970 for $4000. Its value has doubled every 8 years. What is its value in 1990?

45. **Demand Function.**   The demand function for a certain product is given as

$$p = D(x) = 4500 \left(1 - \frac{3}{3 + e^{-0.003x}}\right)$$

(a) What is the price p for a demand of $x = 100$ and $x = 400$?

(b) Draw the graph using a graphing calculator. What are the domain and range of the function?

46. **Inflation.**   If the annual rate of inflation averages 3% over the next 5 years, then the approximate cost C of goods or services during any year in that interval is given by

$$C(t) = P(1.03)^t$$

where t is the time in years and P is the present cost. If the price of a set of tires is presently $269, estimate the price in 5 years.

47. **Depreciation.**   After t years, the value of a car that originally cost $28,000 is given by

$$V(t) = 28,000(0.75)^t$$

Use a graph of the function to estimate the value of the car (a) 2 years from the date of purchase and (b) 5 years from the date of purchase. When will the car be worth less than half of its original cost?

Applications (Social and Life Sciences)

48. **Bacteria Culture.**   A certain bacteria splits every 10 hours. If a culture starts with 10,000 bacteria, what is the size of the culture after 24 hours?

49. **Bacteria Culture.**   A bacteria culture starts with 10,000 bacteria. After 3 hours the estimated count is 160,000. Find the period necessary for doubling the number of bacteria by estimating with a graph on your calculator. (**Hint:** Use X: [0, 1], Y: [0, 170,000].)

50. **Learning Curves.**   The graph of an equation in the form

$$y = k - ke^{-ct}$$

is called a learning curve. Learning increases rapidly with time and then levels out and tends to an upper limit. Suppose that $c = 0.1098$ and $y = 200$ when $t = 10$ minutes.

(a) Find k.

(b) Find y when $t = 5$ minutes.

(c) Find y when $t = 20$ minutes.

(d) Draw the graph of the function. At approximately what point does learning level out?

51. **Learning Curves.** A special learning curve is found to be represented by

$$R = 100(1 - e^{-ct})$$

where R is the number of responses and t is the time involved. Suppose that $c = 0.0695$.
(a) What is R when $t = 0$? Explain.
(b) Find R when $t = 1$.
(c) Draw the graph of the function. Approximately where does the curve level out?

52. **Bacteria Culture.** A certain type of bacteria increases according to the function

$$P(t) = 100e^{0.3124t}$$

where t is the time in hours. Find (a) $P(0)$, (b) $P(10)$, and (c) $P(24)$.

53. **Population Growth.** In 1993, the population of a small town is found to be growing according to the function

$$P(t) = 3200e^{0.0157t}$$

where t is the time in years and $t = 0$ corresponds to the year 1993. Use the function to approximate the population in (a) 1975 and (b) 2000.

Review Exercises

54. Solve each quadratic equation.
    (a) $9x^2 - 5 = 0$ (b) $7x^2 - 14 = 0$
    (c) $(y - 4)^2 = 9$ (d) $(k - 2)^2 = 16$

55. Solve each equation by factoring.
    (a) $x^2 - 3x - 4 = 0$ (b) $2x^2 - 5x + 3 = 6$

56. Solve each equation.
    (a) $3x^2 + x = 2$ (b) $4x^2 - 5x - 7 = 0$

Sketch the graph of each equation using a graphing calculator. Locate the vertex and all intercepts.

57. $y = -3x^2 + 4x$ 58. $y = x^2 - 7x + 10$
59. $y = 2x^2 - 4x + 3$

Chapter Review

Review the properties of slopes of lines, the solution of quadratic equations by factoring, the properties of exponential functions, and the quadratic formula

$$x = \frac{-b \pm \sqrt{b^2 - 4ac}}{2a}$$

Functions	Quadratic functions and graphs
Rule	Line of symmetry
Ordered pair	Upward or downward
Table	Point of maximum or minimum
Notation	Intercepts

Chapter Test

Find an equation of the line with the given conditions.

1. The line passes through the points $(2, -4)$ and $(7, 1)$.
2. The line passes through the point $(3, 1)$ and is perpendicular to the line having the equation $y = 2x + 3$.

Solve each equation algebraically and then verify your answers using a calculator.

3. $2x^2 - 14 = 0$ 4. $(x + 3)^2 - 5 = 0$ 5. $6x^2 + x = 1$

Solve each equation using a graphing calculator.

6. $1.26 = 2^x$ 7. $3x^2 - 11x = 10$ 8. $x^3 - 2x + 1 = 0$

9. If $f(x) = x^2 + 2x$, find
   (a) $f(2) - f(3)$
   (b) $\dfrac{f(x + h) - f(x)}{h}$

Find the domain of each function using a calculator if necessary.

10. $y = \dfrac{2x}{x^2 - 4x - 5}$ 11. $y = \sqrt{25 - x^2}$

12. $y = x^3 - 2x + 3$ 13. $y = 3^{x + 1}$

14. For $y = x^2 - 2x - 3$:
    (a) Find the axis of symmetry.
    (b) Does the curve turn upward or downward?
    (c) Find the point of maximum or minimum value.
    (d) Find the x- and y-intercepts.
    (e) Put the function in standard form for graphing.
    (f) Draw the graph by hand and then verify using a calculator.

15. Draw the graph of $y = 3^x + 1$ by hand and then verify using a calculator.

Determine whether each relation is a function.

16. $\{(2, 1), (3, 1), (4, 2), (5, 3)\}$ 17. $x^2 + y^2 = 36$

18. The profit function for a firm making widgets is $P(x) = 88x - x^2 - 1200$. Find the number of units at which maximum profit is achieved. Draw the graph using a calculator to verify your answer.

19. The demand equation for a certain product is given by $p = D(x) = 600 - 0.6e^{0.003x}$. Find the price p for $x = 100$.

CHAPTER 2

Linear Algebra

René Descartes was one of the first mathematicians to relate algebra and geometry. Descartes represented a point, which is a geometric concept, by means of an ordered pair of real numbers. The set of all ordered pairs of real numbers that satisfy a linear equation in two unknowns was then interpreted geometrically as being the graph of the set of all points on a line. When the graphs of two lines are plotted on the same coordinate system, the intersection of the two lines consists of all points common to both lines. The intersection may be found by geometric methods and also by algebraic methods. Associated with a system of equations is an array of numbers called a matrix. The theory of matrices was developed by an English mathematician, Arthur Cayley.

In this chapter we consider an important tool used by business executives, government agencies, social scientists, and life scientists. With the increased use of computers and large quantities of data, a convenient scheme is necessary for storing information in the computer. One such scheme is to arrange the data in the form of a matrix. We also illustrate the application of matrices to input–output analysis, developed by a well-known contemporary economist, Wassily Leontief.

2.1   GETTING ACQUAINTED WITH MATRICES

OVERVIEW   In this section we use matrices to represent systems of equations. We introduce the concept of matrices with an inventory matrix. An inventory is a list of all the items that a company has stored. Since this amount of data can be quite large, it is important to organize the data into a compact and easily read format. This is where a matrix can be most useful. We also illustrate why matrices provide a convenient and concise way to store data and why they are used so frequently in computer applications. In this section we consider

- The meaning of a matrix
- The size of a matrix
- Addition of matrices
- Scalar multiplication
- The negative of a matrix

A **matrix** is a rectangular array of objects (most often numbers). We use brackets to indicate that the numbers represented form a matrix.

EXAMPLE 1   The following rectangular arrays are matrices:

$$\begin{bmatrix} 2 & 1 & 4 \\ 3 & 2 & 7 \end{bmatrix}, \quad \begin{bmatrix} 2 & 1 \\ 3 & 2 \\ 4 & 5 \end{bmatrix}, \quad \begin{bmatrix} 2 & 3 \\ 1 & 5 \end{bmatrix}, \quad [10 \quad 3 \quad 1]$$

Let's look at how we can use matrices in a business situation. To keep track of inventory, a store uses an **inventory matrix** to store all of its inventory data.

EXAMPLE 2   A department store has three warehouses in which sports equipment is stored. There are 10 tennis rackets, 17 baseball bats, 11 footballs, and 5 pairs of handball gloves in warehouse A. In warehouse B, there are 12 tennis rackets, 11 baseball bats, 15 footballs, and 7 pairs of handball gloves. Warehouse C contains 20 tennis rackets, 12 baseball bats, 32 footballs, and 20 pairs of handball gloves. This information is shown in Table 1. Express the data given in Table 1 in matrix form.

TABLE 1

	Warehouse A	Warehouse B	Warehouse C
Tennis rackets	10	12	20
Baseball bats	17	11	12
Footballs	11	15	32
Pairs of handball gloves	5	7	20

SOLUTION

$$\begin{bmatrix} 10 & 12 & 20 \\ 17 & 11 & 12 \\ 11 & 15 & 32 \\ 5 & 7 & 20 \end{bmatrix}$$

Matrices are classified by their size (sometimes called dimension or order), that is, by the number of rows and columns they have, with the number of rows being given first. In Example 1, the first matrix has two rows and three columns, so it is a 2×3 (read "2 by 3") matrix. The second matrix has three rows and two columns, so it is 3×2. The third matrix is 2×2; the fourth matrix is 1×3; and in Example 2 the matrix is 4×3.

In general, a matrix can have m rows and n columns and be written as

$$\begin{matrix} & & \text{Third} \\ & & \text{column} \end{matrix}$$

$$\begin{bmatrix} a_{11} & a_{12} & a_{13} & \cdots & a_{1n} \\ a_{21} & a_{22} & a_{23} & \cdots & a_{2n} \\ \vdots & \vdots & \vdots & & \vdots \\ a_{m1} & a_{m2} & a_{m3} & \cdots & a_{mn} \end{bmatrix} \quad \text{Second row}$$

The size of this matrix is $m \times n$, where m represents the number of rows and n the number of columns. The objects in the matrix are called **elements** of the matrix. Usually, the elements are numbers. Notice that the first part of the subscript notation for each element gives the row in which the element lies, and the second part gives the column location. For example, a_{23} is the element in the second row and third column. The general $m \times n$ matrix is sometimes written as $[a_{ij}]$, where a_{ij} represents each element.

EXAMPLE 3 The matrix

$$\mathbf{A} = [a_{ij}] = \overbrace{\begin{bmatrix} 2 & 3 & 4 \\ 1 & 5 & 7 \end{bmatrix}}^{\text{3 columns}} \left.\vphantom{\begin{bmatrix} 2 \\ 1 \end{bmatrix}}\right\} \text{ 2 rows}$$

is a 2 × 3 matrix. Its elements are $a_{11} = 2$, $a_{12} = 3$, $a_{13} = 4$, $a_{21} = 1$, $a_{22} = 5$, and $a_{23} = 7$.   ■

Practice Problem 1   Determine the size of matrix **B** and find the element b_{32}.

$$\mathbf{B} = \begin{bmatrix} 1 & 4 \\ 3 & -1 \\ 2 & 7 \end{bmatrix}$$

ANSWER   The matrix is 3 × 2 and element $b_{32} = 7$.

A matrix with the same number of rows and columns is called a **square matrix**. The following are square matrices:

$$\underset{2 \times 2}{\begin{bmatrix} 2 & 1 \\ 3 & -3 \end{bmatrix}} \quad \text{and} \quad \underset{3 \times 3}{\begin{bmatrix} 3 & 7 & 2 \\ 4 & -1 & 6 \\ 0 & 2 & 1 \end{bmatrix}}$$

Likewise, a matrix consisting of a single row of elements is called a **row matrix** or a **vector**. Similarly, a matrix consisting of a single column is called a **column matrix** or a **vector**.

EXAMPLE 4   The matrix $\mathbf{A} = [1 \quad 3 \quad 7]$ is called a row matrix or vector with elements $a_{11} = 1$, $a_{12} = 3$, and $a_{13} = 7$. The matrix

$$\mathbf{B} = \begin{bmatrix} 7 \\ -1 \\ 4 \end{bmatrix}$$

is called a column matrix or vector with elements $b_{11} = 7$, $b_{21} = -1$, and $b_{31} = 4$.   ■

Calculator Note

To enter the matrix

$$\begin{bmatrix} 4.1 & 7.8 & 5.6 \\ 3.2 & 5.1 & -4.3 \\ -2.1 & 2.5 & 3.7 \end{bmatrix}$$

on a calculator first press $\boxed{\text{MATRX}}$ $\boxed{\blacktriangleright}$ $\boxed{\blacktriangleright}$ to display the MATRX EDIT menu. Choose a storage position for the matrix [Figure 1(a)]. We choose 1: [A] and press $\boxed{\text{ENTER}}$. As shown in Figure 1(b), we insert the size of our matrix, 3 × 3, and input the elements of the matrix. To return to the home screen press $\boxed{\text{2nd}}$ $\boxed{\text{QUIT}}$. To see the matrix displayed press $\boxed{\text{MATRX}}$, select matrix [A], and press $\boxed{\text{ENTER}}$. The matrix is displayed in Figure 1(c).

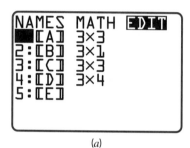

(a)

(b)

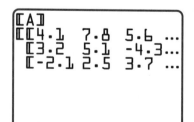

(c)

Figure 1

Just as there are times when we need to know whether two equations are equal, there are times when we need to know whether two matrices are equal.

DEFINITION: EQUAL MATRICES

Two matrices are **equal matrices** if and only if they have the same size (i.e., the same number of rows and columns) and if all corresponding elements are equal.

EXAMPLE 5

$$\begin{bmatrix} 3 & 2 & 1 \\ 5 & 0 & 2 \end{bmatrix} = \begin{bmatrix} \dfrac{-6}{-2} & \dfrac{14}{7} & \dfrac{1}{1} \\ \dfrac{10}{2} & \dfrac{0}{3} & 2 \end{bmatrix}$$

These two matrices are equal because they have the same size (i.e., they are both 2×3), and all corresponding elements are equal:

$$3 = \frac{-6}{-2}, \quad 2 = \frac{14}{7}, \quad 1 = \frac{1}{1}, \quad 5 = \frac{10}{2}, \quad 0 = \frac{0}{3}, \quad 2 = 2 \qquad \blacksquare$$

EXAMPLE 6

$$\begin{bmatrix} 3 & 2 & 1 \\ 5 & 0 & 2 \end{bmatrix} \neq \begin{bmatrix} 3 & 2 & 1 & 0 \\ 5 & 0 & 2 & 0 \end{bmatrix} \quad \text{and} \quad \begin{bmatrix} 2 & 1 \\ 3 & 5 \end{bmatrix} \neq \begin{bmatrix} 2 & 7 \\ 3 & 5 \end{bmatrix}$$

The first two matrices are not equal because they are not the same size; that is, the first matrix is 2×3 and the second is 2×4. The last two matrices are unequal because $a_{12} \neq b_{12}$ or $1 \neq 7$. $\qquad \blacksquare$

In the next chapter on linear programming, we will interchange the elements in the rows and columns of a matrix. If the rows and columns of a matrix **A** are interchanged, then a matrix, $\mathbf{A}^T$, is called the **transpose** of matrix **A**. For example,

$$\mathbf{A}^T = \begin{bmatrix} 2 & 1 \\ 3 & 2 \\ 1 & 4 \end{bmatrix} \quad \text{is the transpose of} \quad \mathbf{A} = \begin{bmatrix} 2 & 3 & 1 \\ 1 & 2 & 4 \end{bmatrix}$$

The first column of $\mathbf{A}^T$ is the first row of **A**. The second row of **A** is the second column of $\mathbf{A}^T$.

Practice Problem 2 If **A** is the matrix

$$\begin{bmatrix} 4.1 & 7.8 & 5.6 \\ 3.2 & 5.1 & -4.3 \\ -2.1 & 2.5 & 3.7 \end{bmatrix}$$

display $\mathbf{A}^T$ using a calculator.

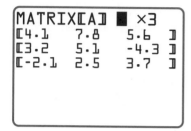

(a)

(b)

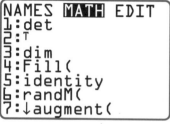

(c)

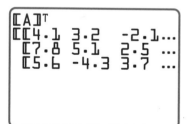

(d)

Figure 2

ANSWER   Enter matrix [A] as discussed in the Calculator Note on page 120. The matrix is displayed in Figure 2(*a*). Press MATRX (Select 1: [A]) ENTER. From the MATRX MATH menu (obtained by MATRX ▶) in Figure 2(*b*), choose 2: , as shown in Figure 2(*c*). Press ENTER and $\mathbf{A}^T$ is displayed as in Figure 2(*d*).

One of the main uses of matrices is to solve systems of linear equations. This requires that we have definitions and algebraic rules that will allow us to manipulate matrices. Let's begin our study of matrix algebra with addition. Two matrices can be added if they have the same size. We find the sum of two matrices by adding corresponding elements.

DEFINITION: SUM OF TWO MATRICES

The **sum** of two matrices **A** and **B** of the same size is the matrix with elements that are the sum of the corresponding elements of **A** and **B**; that is, the entry in the ith row and jth column is $a_{ij} + b_{ij}$.

For instance, the sum of

$$\begin{bmatrix} 1 & 0 \\ 2 & 3 \end{bmatrix} \quad \text{and} \quad \begin{bmatrix} -2 & 1 \\ 1 & 2 \end{bmatrix}$$

is

$$\begin{bmatrix} 1 & 0 \\ 2 & 3 \end{bmatrix} + \begin{bmatrix} -2 & 1 \\ 1 & 2 \end{bmatrix} = \begin{bmatrix} 1 + (-2) & 0 + 1 \\ 2 + 1 & 3 + 2 \end{bmatrix} = \begin{bmatrix} -1 & 1 \\ 3 & 5 \end{bmatrix}$$

Note that according to the preceding definition, matrices can be added only when they have the same size. For example,

$$\begin{bmatrix} 3 & -2 & 8 \\ 5 & -1 & 1 \\ 2 & 7 & 0 \end{bmatrix} + \begin{bmatrix} 3 & 4 \\ 9 & -5 \\ -6 & 2 \end{bmatrix}$$

cannot be added because one is 3×3 and one is 3×2.

Practice Problem 3   Using a calculator, add the matrices.

$$\begin{bmatrix} -1 & 2 & 3 \\ 1 & 5 & 1 \end{bmatrix} + \begin{bmatrix} 3 & 1 & 0 \\ 0 & 1 & -5 \end{bmatrix}$$

ANSWER   Insert matrices **A** and **B** as discussed in the Calculator Note on page 120. Display **A** and **B** in Figure 3(*a*) using MATRX (Select 1) ENTER MATRX (Select 2) ENTER. To add the two matrices press MATRX (Select [A]) + MATRX (select [B]) ENTER, getting the answer in (*b*).

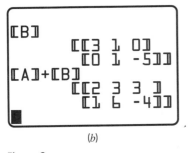

(a)

(b)

Figure 3

One matrix can be subtracted from another if they have the same size. To subtract matrix **B** from matrix **A**, obtaining **A** − **B**, subtract each entry of **B** from the corresponding entry of **A**; that is,

$$\mathbf{A} - \mathbf{B} = [a_{ij} - b_{ij}]$$

EXAMPLE 7

$$\begin{bmatrix} 2 & 6 & 7 \\ 5 & 4 & 1 \end{bmatrix} - \begin{bmatrix} 5 & 2 & 4 \\ 3 & -1 & 6 \end{bmatrix} = \begin{bmatrix} 2-5 & 6-2 & 7-4 \\ 5-3 & 4-(-1) & 1-6 \end{bmatrix}$$

$$= \begin{bmatrix} -3 & 4 & 3 \\ 2 & 5 & -5 \end{bmatrix}$$ ∎

Let's return now to the inventory problem involving a department store.

EXAMPLE 8   The following matrix indicates shipments made last week from the warehouses in Example 2.

Warehouses
A B C

$$\begin{matrix} \text{Rackets} \\ \text{Bats} \\ \text{Footballs} \\ \text{Gloves} \end{matrix} \begin{bmatrix} 2 & 1 & 3 \\ 4 & 0 & 2 \\ 5 & 1 & 1 \\ 1 & 0 & 0 \end{bmatrix}$$

For example, 4 baseball bats were shipped from warehouse A, and 3 tennis rackets were shipped from warehouse C. After this shipment, tabulate an inventory.

SOLUTION

$$\begin{bmatrix} 10 & 12 & 20 \\ 17 & 11 & 12 \\ 11 & 15 & 32 \\ 5 & 7 & 20 \end{bmatrix} - \begin{bmatrix} 2 & 1 & 3 \\ 4 & 0 & 2 \\ 5 & 1 & 1 \\ 1 & 0 & 0 \end{bmatrix} = \begin{bmatrix} 8 & 11 & 17 \\ 13 & 11 & 10 \\ 6 & 14 & 31 \\ 4 & 7 & 20 \end{bmatrix}$$

Inventory − Shipment = New inventory

Note that there are now 14 footballs in warehouse B. ∎

Practice Problem 4   Using a graphing calculator, find **A** + **B** for

$$\mathbf{A} = \begin{bmatrix} 4.1 & 7.8 & 5.6 \\ 3.2 & 5.1 & -4.3 \\ -2.1 & 2.5 & 3.7 \end{bmatrix} \quad \text{and} \quad \mathbf{B} = \begin{bmatrix} -3 & 2.1 & 5.6 \\ 4.3 & 0 & 7.1 \\ -2.1 & 3 & 0 \end{bmatrix}$$

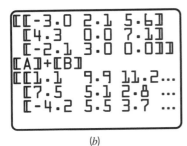

(a)

(b)

Figure 4

ANSWER   Follow the procedures of Practice Problem 3 to obtain the answer as shown in Figure 4.

A matrix with each element equal to 0 is called an additive identity or a **zero matrix**, also denoted by 0. For example,

$$0 = \begin{bmatrix} 0 & 0 & 0 \\ 0 & 0 & 0 \end{bmatrix}$$

is a 2×3 zero matrix. Note that

$$\begin{bmatrix} 2 & 3 & 1 \\ -1 & 0 & 5 \end{bmatrix} + \begin{bmatrix} 0 & 0 & 0 \\ 0 & 0 & 0 \end{bmatrix} = \begin{bmatrix} 2 & 3 & 1 \\ -1 & 0 & 5 \end{bmatrix}$$

The additive identity or zero matrix must have the same dimension as the matrix to which it is being added; for example,

$$\begin{bmatrix} 1 & 2 & 3 \\ -1 & 2 & 1 \\ -6 & 5 & -3 \end{bmatrix} + \begin{bmatrix} 0 & 0 & 0 \\ 0 & 0 & 0 \\ 0 & 0 & 0 \end{bmatrix} = \begin{bmatrix} 1 & 2 & 3 \\ -1 & 2 & 1 \\ -6 & 5 & -3 \end{bmatrix}$$

In this book we are interested in two kinds of mathematical products involving matrices: (1) the product of a real number and a matrix and (2) the product of two matrices. For the product of a real number and a matrix, the real number is multiplied by each element of the matrix. For example,

$$2\begin{bmatrix} 3 & 1 \\ 0 & 2 \end{bmatrix} = \begin{bmatrix} 2 \cdot 3 & 2 \cdot 1 \\ 2 \cdot 0 & 2 \cdot 2 \end{bmatrix} = \begin{bmatrix} 6 & 2 \\ 0 & 4 \end{bmatrix}$$

In working with matrices, a real number is called a **scalar**.

DEFINITION: SCALAR PRODUCT

The product of a real number c and a matrix $\mathbf{A} = [a_{ij}]$ is the matrix $[ca_{ij}]$. That is, $c[a_{ij}] = [ca_{ij}]$.

EXAMPLE 9

$$5\begin{bmatrix} 1 & -1 & 2 \\ 3 & 0 & 4 \end{bmatrix} = \begin{bmatrix} 5 \cdot 1 & 5 \cdot (-1) & 5 \cdot 2 \\ 5 \cdot 3 & 5 \cdot 0 & 5 \cdot 4 \end{bmatrix} = \begin{bmatrix} 5 & -5 & 10 \\ 15 & 0 & 20 \end{bmatrix}$$ ■

The following are properties of scalar multiplication.

PROPERTIES OF SCALAR MULTIPLICATION

If c and d are any real numbers, then
(a) $c(d\mathbf{A}) = (cd)\mathbf{A}$
(b) $c\mathbf{A} + d\mathbf{A} = (c + d)\mathbf{A}$
(c) $c(\mathbf{A} + \mathbf{B}) = c\mathbf{A} + c\mathbf{B}$

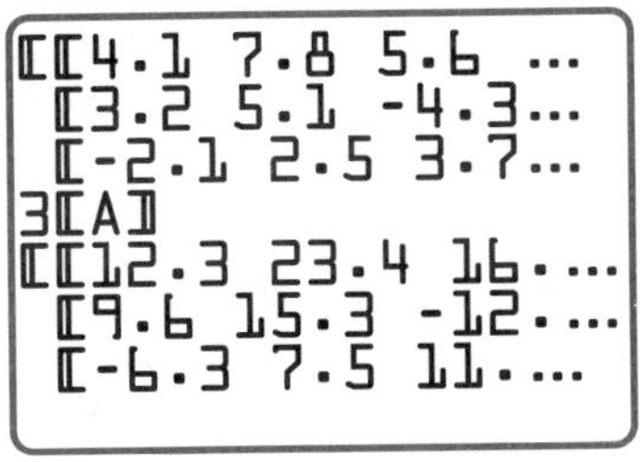

```
[[4.1  7.8  5.6  ...
 [3.2  5.1  -4.3...
 [-2.1 2.5  3.7...
3[A]
[[12.3 23.4 16....
 [9.6  15.3 -12....
 [-6.3 7.5  11....
```
(a)

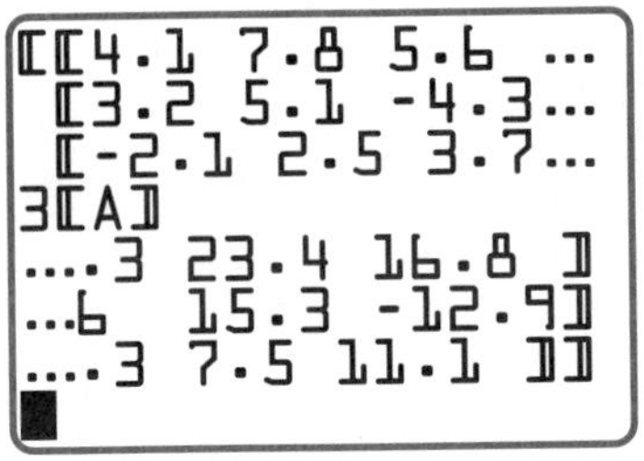

```
[[4.1  7.8  5.6  ...
 [3.2  5.1  -4.3...
 [-2.1 2.5  3.7...
3[A]
....3  23.4 16.8 ]
...6  15.3 -12.9]]
....3  7.5  11.1 ]]
■
```
(b)

Figure 5

Practice Problem 5 For matrix **A** in Practice Problem 2, display 3**A** on your calculator.

ANSWER   Store the matrix under [A] as discussed in the Calculator Note, on page 120. Use 3 [MATRX] (select [A]) [ENTER]. The answer is shown in Figure 5. To view the entire matrix, push the right arrow key to scroll to the right.

The preceding definition of the product of a scalar and a matrix can be used to find the negative of a matrix. One matrix is the negative of another if their sum is a zero matrix.

DEFINITION: NEGATIVE OF A MATRIX

The **negative of a matrix A** $= [a_{ij}]$, denoted by $-\mathbf{A}$, is the scaler product of -1 and **A**. That is, $-\mathbf{A} = [-a_{ij}]$.

EXAMPLE 10   Find the negative of $\begin{bmatrix} 2 & 3 & 1 \\ -1 & 0 & 2 \end{bmatrix}$.

SOLUTION   The negative of a matrix is found by multiplying each element by -1. Hence, the negative is

$$\begin{bmatrix} -2 & -3 & -1 \\ 1 & 0 & -2 \end{bmatrix}$$

To show that these matrices are negatives, note that

$$\begin{bmatrix} 2 & 3 & 1 \\ -1 & 0 & 2 \end{bmatrix} + \begin{bmatrix} -2 & -3 & -1 \\ 1 & 0 & -2 \end{bmatrix} = \begin{bmatrix} 0 & 0 & 0 \\ 0 & 0 & 0 \end{bmatrix}$$ ■

EXAMPLE 11

If $\mathbf{A} = \begin{bmatrix} 2 & -1 \\ 4 & -2 \\ -3 & 3 \end{bmatrix}$,   then $-\mathbf{A} = \begin{bmatrix} -2 & 1 \\ -4 & 2 \\ 3 & -3 \end{bmatrix}$

since

$$\mathbf{A} + (-\mathbf{A}) = \begin{bmatrix} 0 & 0 \\ 0 & 0 \\ 0 & 0 \end{bmatrix}$$ ■

Practice Problem 6 With your calculator, perform the following computation.

$$\begin{bmatrix} 2 & 3 \\ 5 & 1 \end{bmatrix} - 3\begin{bmatrix} 0 & 1 \\ 2 & 1 \end{bmatrix}$$

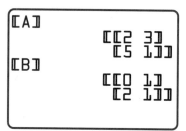

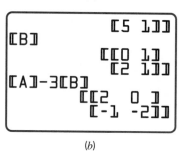

(a)

(b)

Figure 6

ANSWER   Store the two matrices under [A] and [B] as discussed in the Calculator Note, page 120. Use $\boxed{\text{MATRX}}$ (select [A]) $\boxed{-}$ 3 $\boxed{\text{MATRX}}$ (select [B]) $\boxed{\text{ENTER}}$. [A], [B] and [A] $-$ 3[B] are shown in Figure 6.

It is interesting to note the relationship between **B** subtracted from **A**, (**A** $-$ **B**), and the sum of **A** and the negative of **B**, **A** $+$ ($-$ **B**); A $-$ B $=$ A $+$ ($-$ B).

EXAMPLE 12

$$\begin{bmatrix} 1 & -1 & 2 \\ 3 & 1 & 4 \end{bmatrix} - \begin{bmatrix} -1 & 2 & 2 \\ 2 & -4 & 3 \end{bmatrix} = \begin{bmatrix} 1 & -1 & 2 \\ 3 & 1 & 4 \end{bmatrix}$$

$$+ \begin{bmatrix} 1 & -2 & -2 \\ -2 & 4 & -3 \end{bmatrix}$$

$$= \begin{bmatrix} 1+1 & -1-2 & 2-2 \\ 3-2 & 1+4 & 4-3 \end{bmatrix}$$

$$= \begin{bmatrix} 2 & -3 & 0 \\ 1 & 5 & 1 \end{bmatrix} \quad \blacksquare$$

Practice Problem 7 With a calculator, find the negative of the matrix

$$\begin{bmatrix} -1 & 1 & 2 \\ -2 & 4 & -5 \end{bmatrix}$$

ANSWER   Insert [C] after selecting 3: that under the MATRX EDIT menu. Then use $\boxed{(-)}$ $\boxed{\text{MATRX}}$ (select 3) $\boxed{\text{ENTER}}$ to obtain the display shown in Figure 7.

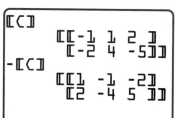

Figure 7

Graph Theory and Matrices

One of the many applications of matrix theory involves graph theory. Assigning broadcast frequencies, scheduling work assignments, and making production runs at a factory are just a few examples of the uses of graph theory. In this discussion we show how matrices are useful in representing graphs. In graph theory, a dot represents a **vertex**, and a line or a curve connecting two dots is called an **edge**. For example in Figure 8(a), A, B, and C are vertices, and an edge connects A and B, B and C, but not A and C. Sometimes graphs are used to represent relations between elements of a set. In this case, there could be a relation from a vertex to itself. Such an edge is called a **loop**. A loop appears at V_1 in Figure 8(b).

A representation of a graph can be put into a computer using a matrix. One way is to define the elements of the matrix as follows.

$$a_{ij} = \begin{cases} 1, & \text{if there is an edge joining } V_i \text{ and } V_j \\ 0, & \text{if there is no edge joining } V_i \text{ and } V_j \end{cases}$$

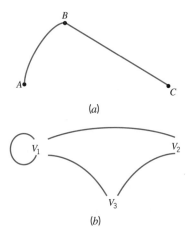

(a)

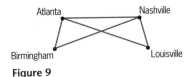

(b)

Figure 8

Matrices representing the graphs in Figure 8 are

$$
\begin{array}{c}
\\ A \\ B \\ C
\end{array}
\begin{array}{c}
A \quad B \quad C \\
\begin{bmatrix}
0 & 1 & 0 \\
1 & 0 & 1 \\
0 & 1 & 0
\end{bmatrix}
\end{array}
\quad \text{and} \quad
\begin{array}{c}
\\ V_1 \\ V_2 \\ V_3
\end{array}
\begin{array}{c}
V_1 \quad V_2 \quad V_3 \\
\begin{bmatrix}
1 & 1 & 1 \\
1 & 0 & 1 \\
1 & 1 & 0
\end{bmatrix}
\end{array}
$$

The graph in Figure 9 shows routes of direct air flights among four cities. The corresponding matrix is

$$
\begin{array}{c}
\\ \text{Atlanta} \\ \text{Nashville} \\ \text{Birmingham} \\ \text{Louisville}
\end{array}
\begin{array}{c}
\text{Atlanta} \quad \text{Nashville} \quad \text{Birmingham} \quad \text{Louisville} \\
\begin{bmatrix}
0 & 1 & 1 & 1 \\
1 & 0 & 1 & 1 \\
1 & 1 & 0 & 0 \\
1 & 1 & 0 & 0
\end{bmatrix}
\end{array}
$$

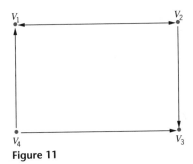

Figure 9

Note that there are flights between Birmingham and Nashville, so both a_{23} and a_{32} are equal to 1 in the matrix above. However, both a_{34} and a_{43} are 0. Why?

In Figure 8(*b*), an edge was used to represent a two-way or a symmetric relationship between two vertices. However, there are situations where relationships hold in only one direction (e.g., one-way streets in an urban downtown area). In these cases we use directions (say, from vertex V_3 to vertex V_2 in Figure 10). In a microcomputer, information flows from the input vertex to the memory vertex of a graph and from the memory vertex to the output vertex. However, information travels in both directions between the CPU and the memory.

The matrix for the graph in Figure 10 would be

$$
\begin{array}{c}
\\ V_1 \\ V_2 \\ V_3 \\ V_4
\end{array}
\begin{array}{c}
V_1 \quad V_2 \quad V_3 \quad V_4 \\
\begin{bmatrix}
0 & 1 & 0 & 0 \\
1 & 0 & 0 & 1 \\
0 & 1 & 0 & 0 \\
0 & 0 & 0 & 0
\end{bmatrix}
\end{array}
$$

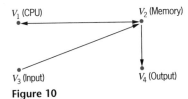

Figure 10

Check to see if the directed graph in Figure 11 and its matrix are in agreement.

$$
\begin{array}{c}
\\ V_1 \\ V_2 \\ V_3 \\ V_4
\end{array}
\begin{array}{c}
V_1 \quad V_2 \quad V_3 \quad V_4 \\
\begin{bmatrix}
0 & 1 & 0 & 0 \\
1 & 0 & 1 & 0 \\
0 & 0 & 0 & 0 \\
1 & 0 & 1 & 0
\end{bmatrix}
\end{array}
$$

Note that the sum of the numbers in the ith row equals the number of directed edges from vertex V_i. Likewise, the sum of the numbers in the jth column equals the number of directed edges to vertex V_j.

Figure 11

SUMMARY

In this section we introduced the basic operations on matrices. Using examples, we demonstrated how to add, subtract, and perform scalar multiplication. A graphing calculator relieves the tedium of computation, but make sure you understand the following terms:

1. Matrix
2. Equal matrices

3. Transposed matrix
4. Matrix algebra

Exercise Set 2.1

Determine the size of each matrix and find the element requested.

1. $A = \begin{bmatrix} 4 & 1 & 3 \\ 2 & -1 & 0 \end{bmatrix}$, a_{12}

2. $B = \begin{bmatrix} 2 & 1 \\ 3 & 4 \end{bmatrix}$, b_{22}

3. $C = \begin{bmatrix} 1 & 2 & 1 \\ 3 & 1 & 4 \\ 5 & 2 & 6 \end{bmatrix}$, c_{32}

4. $D = [2 \quad 1 \quad 7 \quad 4]$, d_{13}

Perform the indicated operations if possible.

5. $\begin{bmatrix} 1 & 3 & 1 \\ 6 & 1 & 4 \end{bmatrix} + [2 \quad 1 \quad 7]$

6. $\begin{bmatrix} 4 & 3 \\ -1 & 2 \end{bmatrix} - \begin{bmatrix} 2 & 1 \\ 1 & -1 \end{bmatrix}$

7. $6\begin{bmatrix} 3 & 4 & 5 \\ 1 & 2 & 3 \end{bmatrix}$

8. $-3\begin{bmatrix} 2 & 1 \\ 7 & -1 \\ -2 & 3 \end{bmatrix}$

Perform the following indicated operations when possible. If impossible, explain why.

9. $[1 \quad 3] + \begin{bmatrix} 4 & 2 & 3 \\ -1 & 0 & 2 \end{bmatrix}$

10. $\begin{bmatrix} 4 & 2 & 3 \\ -1 & 0 & 2 \end{bmatrix} + [1 \quad 3]$

11. $5\begin{bmatrix} 4 & 2 & 3 & 3 \\ -1 & 0 & 2 & -1 \end{bmatrix}$

12. $\begin{bmatrix} 3 & 2 \\ 4 & 1 \end{bmatrix} + \begin{bmatrix} 3 \\ 1 \end{bmatrix}$

13. $4\begin{bmatrix} 3 & 2 & 5 \\ 1 & 6 & 7 \\ 0 & 5 & -2 \end{bmatrix} - 3\begin{bmatrix} -2 & 5 & 1 \\ 1 & 2 & 3 \\ 3 & 2 & 1 \end{bmatrix}$

14. $2\begin{bmatrix} 2 & 1 \\ 3 & 5 \end{bmatrix} + 3\begin{bmatrix} 3 & 0 & 5 \\ 2 & 1 & 3 \\ 5 & 2 & 1 \end{bmatrix}$

Combine each into a single matrix.

15. $3\begin{bmatrix} 2 & 1 \\ 0 & 5 \end{bmatrix} - 4\begin{bmatrix} 2 & 1 \\ 0 & 3 \end{bmatrix}$

16. $2\begin{bmatrix} 2 & 1 & -1 & 3 \\ 4 & 2 & 0 & -1 \\ 0 & 0 & 2 & -1 \end{bmatrix}$
$-3\begin{bmatrix} 1 & 2 & -1 & 4 \\ 2 & 3 & 1 & 4 \\ -1 & -1 & -3 & -4 \end{bmatrix}$

17. $5\begin{bmatrix} 7 & 3 & 1 \\ 2 & 4 & -1 \end{bmatrix} - 4\begin{bmatrix} 0 & 5 & -1 \\ 3 & -2 & 4 \end{bmatrix}$

Find the transpose of each of the following matrices.

18. $\begin{bmatrix} 1 & -2 \\ 3 & 4 \end{bmatrix}$

19. $\begin{bmatrix} 2 & 3 & 4 & 5 \\ 1 & 0 & 5 & 2 \end{bmatrix}$

20. $\begin{bmatrix} 0 & 0 & 0 \\ 0 & 0 & 0 \end{bmatrix}$

Find x, y, and z (where given) in the following matrices.

21. $\begin{bmatrix} x \\ y \\ z \end{bmatrix} = 4\begin{bmatrix} -1 \\ 2 \\ 5 \end{bmatrix}$

22. $\begin{bmatrix} 3 \\ 4 \end{bmatrix} + \begin{bmatrix} x \\ y \end{bmatrix} = \begin{bmatrix} 7 \\ 10 \end{bmatrix}$

23. $\begin{bmatrix} 0 \\ 0 \\ 0 \end{bmatrix} = 3 \begin{bmatrix} x \\ y \\ z \end{bmatrix}$

24. $\begin{bmatrix} 6 \\ x \end{bmatrix} = \begin{bmatrix} x + y \\ -3 \end{bmatrix}$

25. If

$$\begin{bmatrix} 3 \\ 2 \end{bmatrix} + \mathbf{X} = \begin{bmatrix} -9 \\ 1 \end{bmatrix}$$

find **X**. **Hint**: Let $\mathbf{X} = \begin{bmatrix} x_1 \\ x_2 \end{bmatrix}$.

Given

$$\mathbf{A} = \begin{bmatrix} 0.02 & -1.7 & 2.3 \\ 1.7 & -2.7 & -1.1 \\ -3.4 & 5.1 & -4.3 \end{bmatrix}$$

and

$$\mathbf{B} = \begin{bmatrix} 14.1 & 5.7 & -8.6 \\ -2.3 & 15.2 & -10.1 \\ 17.2 & 16.1 & -4.3 \end{bmatrix}$$

find each of the following using a calculator.

26. $\mathbf{A}^\mathsf{T}$ 27. $5\mathbf{B}$

28. $\mathbf{A} + \mathbf{B}$ 29. $\mathbf{A} - \mathbf{B}$

30. $3\mathbf{A} - \mathbf{B}$ 31. $5\mathbf{A}^\mathsf{T} + 3\mathbf{B}$

Applications (Business and Economics)

32. **Investments.** An investment club purchases the following stocks.

Company	Number of Shares	Price per Share
IBM	150	61
Pfizer	200	63
American Home Products	300	56
Delta Airlines	100	45

(a) Form a row matrix showing the number of shares purchased.
(b) Form a column matrix showing the price per share of each stock.

33. **Inventory.** A carpet distributor ships whosesale from three warehouses in Atlanta (A), in Houston

(H), and in Chicago (C). For pattern W34 there are 100 rolls stored in (A), 50 in (H), and 200 in (C). For pattern G47, there are 70 in (C), 60 in (A), and 40 in (H). For pattern B71 there are 40 in (H), 24 in (C), and 13 in (A).

(a) Show this storage with an inventory matrix.
(b) Last week the following shipments were made: 10 W34, 5 G47, and 8 B71 from (A); 30 G47, 15 W34, and 10 B71 from (C); and 8 G47, 11 W34, and 16 B71 from (H). Show the shipments as a matrix.
(c) Compute a new inventory matrix after the shipments.
(d) After the shipments, the distributor decided to triple all inventories. Show this process as multiplication of a matrix by a scalar.

34. **Markov Matrices** (to be discussed in Chapter 6).   If knowledge of a current state of a situation aids in the prediction of the next state, the next state of the situation is often modeled by a Markov matrix. Two stores in a city, A and B, are competing for customers. The elements in the first row of the matrix represent the proportion of customers who remain with A and who change to B. A survey shows that last year store A lost 20% of its customers to B and kept 80% (first row of the matrix).

Next State

$$\begin{array}{cc} & \begin{array}{cc} A & B \end{array} \\ \begin{array}{c} \text{Present} \;\; A \\ \text{State} \;\;\;\; B \end{array} & \begin{bmatrix} 0.80 & 0.20 \\ 0.24 & 0.76 \end{bmatrix} \end{array}$$

The second row represents what happened to the customers of store B. Interpret the second row.

35. **Marketing.**   Suppose two newspapers, the *Star* and the *Times,* are competing for customers. During the last year it was found that the *Star* kept 60% of its customers and lost 40% to the *Times.* The *Times* kept 70% of its customers and lost 30% to the *Star.* Write a matrix for this survey (see Exercise 34).

Applications (Social and Life Sciences)

36. **Transportation.**   The following graph shows the routes of direct air flights among five cities. Represent this information as a matrix using the cities in the following order: New Orleans, Birming-

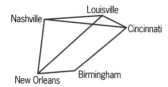

ham, Nashville, Louisville, and Cincinnati. Let 1 represent a flight and 0 no flight. Complete the matrix given below.

	New Orleans	Birming- ham	Nash- ville	Louis- ville	Cincin- nati
New Orleans	0	1	1	1	0
Birmingham					
Nashville					
Louisville					
Cincinnati					

37. Form matrices representing the following graphs.

(a)   (b)

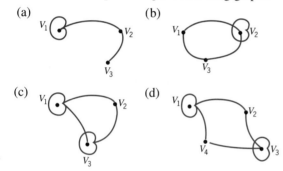

(c)   (d)

38. Draw graphs represented by the following matrices.

(a)
	V_1	V_2	V_3
V_1	0	1	0
V_2	1	0	1
V_3	0	1	1

(b)
	V_1	V_2	V_3
V_1	1	0	1
V_2	0	0	1
V_3	1	1	1

(c)
	V_1	V_2	V_3	V_4
V_1	0	1	0	0
V_2	1	0	1	1
V_3	0	1	0	0
V_4	0	1	0	0

39. **Communication.** Matrices are used in communication-network studies and to show sociological re-

lationships. A reception is indicated by 1 and no reception by 0.

		Possible Receivers			
Sender		1	2	3	4
1		0	1	0	1
2		1	0	0	1
3		1	1	0	0
4		1	1	1	0

Receivers 2 and 4 receive a message from sender 1, but 3 does not. Discuss for a row at a time the meanings of the 1's in this matrix.

40. Construct a directed graph for the following matrices.

(a)
$$\begin{bmatrix} 0 & 1 & 0 & 1 \\ 1 & 0 & 1 & 0 \\ 0 & 0 & 0 & 0 \\ 0 & 1 & 1 & 0 \end{bmatrix}$$

(b)
$$\begin{bmatrix} 0 & 1 & 1 & 1 \\ 1 & 0 & 1 & 1 \\ 1 & 1 & 0 & 1 \\ 0 & 1 & 0 & 0 \end{bmatrix}$$

41. Write a matrix for each of the following graphs.

(a)   (b)

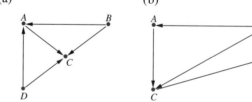

42. **Demographics.** Can you write a matrix showing the changes in three classifications of employment? The first row, representing changes from business to the three classifications, would be [0.75 0.15 0.10]. That is, 75% of those in business remain in business. Now complete the matrix.

From business (or industry) to business (or industry)	75%
From business (or industry) to unemployment	15
From business (or industry) to self-employment	10
From unemployment to business (or industry)	25
From unemployment to unemployment	60
From unemployment to self-employment	15

From self-employment to business (or industry)   5

From self-employment to unemployment   5

From self-employment to self-employment   90

43. *Genetics*. The characteristics of an inherited trait are given by the following matrix.

$$
\begin{array}{c}
 & & \text{Offspring} \\
 & & \begin{array}{ccc} R & H & D \end{array} \\
\text{Parents} & \begin{array}{c} R \\ H \\ D \end{array} & \left[\begin{array}{ccc} 0.75 & 0.25 & 0 \\ 0.25 & 0.50 & 0.25 \\ 0 & 0.75 & 0.25 \end{array}\right]
\end{array}
$$

Assume that an inherited trait is governed by a pair of genes of type G or g. Possible combinations are gg (recessive, denoted by *R*), gG (hybrid, *H*), and GG (dominant, *D*). In the first row of the matrix, 75% of the offspring of recessive parents are recessive, 25% are hybrid, and none is dominant. Interpret the remainder of the matrix.

2.2 MATRIX MULTIPLICATION

OVERVIEW   A manufacturer of ice cream determines that 1 gallon of vanilla ice cream uses 3 quarts of milk, 1 pint of cream, 2 cups of sugar, and 6 fluid ounces of vanilla extract. What will be the cost of producing 10,000 gallons of ice cream? To facilitate computer solutions to problems such as this, we write the information about the ingredients as a row matrix:

$$[3 \quad 1 \quad 2 \quad 6]$$

Suppose that milk costs $0.65 per quart, cream $0.50 per pint, sugar $0.12 per cup, and vanilla extract $0.14 per fluid ounce. We write the costs as a column matrix:

$$\begin{bmatrix} 0.65 \\ 0.50 \\ 0.12 \\ 0.14 \end{bmatrix}$$

Your common sense multiplication to get the cost

$$10,000[3(0.65) + 1(0.50) + 2(0.12) + 6(0.14)] = \$35,300$$

illustrates the dot product of two matrices.

In this section we consider the product of two matrices. Do not confuse matrix multiplication with the multiplication of a matrix by a constant, as we considered in the preceding section. Matrix multiplication is a bit more involved than addition of matrices. Furthermore, not all of the properties of multiplication apply to matrix multiplication. In this section we consider

- The dot product of two matrices
- The definition of matrix multiplication

- The arithmetic of matrix multiplication
- Applications involving matrix multiplication

We begin our study of matrix multiplication by considering the dot product of the row matrix $[1 \quad -2 \quad 3]$ and the column matrix

$$\begin{bmatrix} 2 \\ -1 \\ 1 \end{bmatrix}$$

The **dot product** of these two matrices is

$$[1 \quad -2 \quad 3] \cdot \begin{bmatrix} 2 \\ -1 \\ 1 \end{bmatrix} = 1(2) + (-2)(-1) + 3(1) = 7$$

The answer is a real number that is the sum of the products of corresponding entries (first element in a row times first element in a column plus second element in a row times second element in a column, etc.).

DEFINITION: DOT PRODUCT

The **dot product** of a $1 \times n$ row matrix and an $n \times 1$ column matrix is defined by

$$[a_1 \quad a_2 \quad \cdots \quad a_n] \cdot \begin{bmatrix} b_1 \\ b_2 \\ \cdot \\ \cdot \\ \cdot \\ b_n \end{bmatrix} = a_1 b_1 + a_2 b_2 + \cdots + a_n b_n \qquad \text{A real number}$$

The dot between the two matrices indicates the dot product and distinguishes this product from matrix multiplication, which we define later in this section.

EXAMPLE 13

$$[1 \quad 2 \quad 3] \cdot \begin{bmatrix} 2 \\ -1 \\ 1 \end{bmatrix} = 1(2) + 2(-1) + 3(1) = 3$$

$$\underset{1 \times 3}{} \qquad \underset{3 \times 1}{}$$

Same

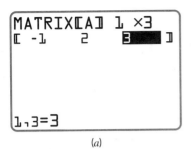

(a)

(b)

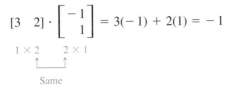

EXAMPLE 14

$$[3 \quad 2] \cdot \begin{bmatrix} -1 \\ 1 \end{bmatrix} = 3(-1) + 2(1) = -1$$

$1 \times 2 \qquad 2 \times 1$

Same

EXAMPLE 15

$$[2 \quad 3 \quad -1 \quad 4] \cdot \begin{bmatrix} 5 \\ 2 \\ 8 \end{bmatrix}$$

$1 \times 4 \qquad\qquad 3 \times 1$

Not the same

The dot product is not defined because the size of the two matrices is not compatible.

Practice Problem 1 Using a calculator, find the dot product of the two matrices.

$$[-1 \quad 2 \quad 3] \cdot \begin{bmatrix} 2 \\ 0 \\ -1 \end{bmatrix}$$

ANSWER First we review the procedure for storing row and column matrices in a calculator. Using the MATRX EDIT menu (obtained from MATRX ▶ ▶), store $[-1, 2, 3]$ under [A] by choosing 1: [A] and pressing ENTER. Input the 1×3 size and the elements of the matrix as seen in Figure 12(a). Use 2nd QUIT. Store the column matrix under [B] by choosing 2: [B]. Press ENTER and input the size, 3×1, and the elements of the matrix. See Figure 12(b). Then press MATRX (select [A]) × MATRX (select [B]) ENTER. The answer is -5 as seen in Figure 12(d).

Now let's use the ideas of the preceding examples to define the product of two matrices. The product of a 2×3 matrix and a 3×4 matrix is a 2×4 matrix; the element in the first row and second column, C_{12}, is the dot product of the first row of the 2×3 matrix and the second column of the 3×4 matrix. The element c_{23} is the dot product of the second row and third column.

(c)

(d)

Figure 12

DEFINITION: PRODUCT OF MATRICES

The **product** of an $m \times p$ matrix **A** and a $p \times n$ matrix **B** is the $m \times n$ matrix **AB** whose element in the ith row and the jth column is the dot product of the ith row of **A** and the jth column of **B**. The answer will be an $m \times n$ matrix.

Consider the multiplication of two matrices such as

$$A = \begin{bmatrix} 3 & 1 & 2 \\ -1 & 0 & 5 \end{bmatrix} \quad \text{and} \quad B = \begin{bmatrix} 1 & 0 \\ 4 & 6 \\ -2 & 3 \end{bmatrix}$$

Since A is 2×3 and B is 3×2, the answer will be 2×2. To obtain our product matrix, we find the element c_{11} in the first row and first column of the product matrix by finding the dot product of the first row of A and the first column of B. In this example,

$$c_{11} = \begin{bmatrix} 3 & 1 & 2 \end{bmatrix} \cdot \begin{bmatrix} 1 \\ 4 \\ -2 \end{bmatrix} = (3)(1) + (1)(4) + (2)(-2) = 3 + 4 - 4 = 3$$

This element goes in the first row, first column of the product matrix. If the product matrix is denoted by c_{ij}, then $c_{11} = 3$. Next, to find c_{12}, we take the dot product of the first row of A and the second column of B:

$$c_{12} = \begin{bmatrix} 3 & 1 & 2 \end{bmatrix} \cdot \begin{bmatrix} 0 \\ 6 \\ 3 \end{bmatrix} = 3(0) + 1(6) + 2(3) = 0 + 6 + 6 = 12$$

This element goes in the first row, second column of the product matrix. Then we take the dot product of the second row of A and the first column of B:

$$c_{21} = \begin{bmatrix} -1 & 0 & 5 \end{bmatrix} \cdot \begin{bmatrix} 1 \\ 4 \\ -2 \end{bmatrix} = (-1)(1) + 0(4) + 5(-2) = -1 + 0 - 10 = -11$$

This element goes in the second row, first column of the product matrix. Then we take the dot product of the second row of A and the second column of B:

$$c_{22} = \begin{bmatrix} -1 & 0 & 5 \end{bmatrix} \cdot \begin{bmatrix} 0 \\ 6 \\ 3 \end{bmatrix} = (-1)(0) + 0(6) + 5(3) = 0 + 0 + 15 = 15$$

This element goes in the second row, second column of the product matrix. Thus,

$$AB = \begin{bmatrix} 3 & 1 & 2 \\ -1 & 0 & 5 \end{bmatrix} \begin{bmatrix} 1 & 0 \\ 4 & 6 \\ -2 & 3 \end{bmatrix} = \begin{bmatrix} 3 & 12 \\ -11 & 15 \end{bmatrix}$$

Note that the product of two matrices is not defined unless the number of columns in the first matrix is equal to the number of rows in the second matrix. In other words, the middle two dimensions must be the same and the outer dimensions give the dimensions of the product. A 3×2 matrix multiplied by a 2×3 matrix gives a 3×3 matrix, whereas, if the order is reversed, the product of a 2×3 matrix and a 3×2 matrix gives a 2×2 matrix.

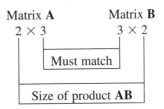

Thus, the order of multiplication is important when multiplying matrices.

The following example indicates that the commutative property of multiplication does not hold for all pairs of matrices. Many times, the product will not be defined if the order of multiplication is reversed. For example, a 2×3 matrix times a 3×5 matrix results in a 2×5 matrix, but a 2×5 matrix times a 3×5 matrix is not defined. However, when both products are defined, **AB** is often not the same as **BA**. In the example given earlier,

$$\begin{bmatrix} 3 & 1 & 2 \\ -1 & 0 & 5 \end{bmatrix} \begin{bmatrix} 1 & 0 \\ 4 & 6 \\ -2 & 3 \end{bmatrix} = \begin{bmatrix} 3 & 12 \\ -11 & 15 \end{bmatrix} \qquad \begin{aligned} c_{21} &= -1(1) + 0(4) + 5(-2) \\ &= -11 \end{aligned}$$

If the order is reversed, the product becomes

$$\begin{bmatrix} 1 & 0 \\ 4 & 6 \\ -2 & 3 \end{bmatrix} \begin{bmatrix} 3 & 1 & 2 \\ -1 & 0 & 5 \end{bmatrix} = \begin{bmatrix} 3 & 1 & 2 \\ 6 & 4 & 38 \\ -9 & -2 & 11 \end{bmatrix} \qquad \begin{aligned} c_{21} &= 4(3) + 6(-1) \\ &= 6 \end{aligned}$$

Hence, the commutative property of multiplication, $\mathbf{AB} = \mathbf{BA}$, does not hold for the product of two matrices.

Practice Problem 2 Find the following product:

$$\begin{bmatrix} 2 & 1 \\ -1 & 3 \end{bmatrix} \begin{bmatrix} 0 & 4 & 1 \\ 1 & 2 & 0 \end{bmatrix}$$

ANSWER

$$\begin{bmatrix} 1 & 10 & 2 \\ 3 & 2 & -1 \end{bmatrix}$$

Practice Problem 3 Find the product **AB** using a graphing calculator if

$$\mathbf{A} = \begin{bmatrix} 4.1 & 7.8 & 5.6 \\ 3.2 & 5.1 & -4.3 \\ -2.1 & 2.5 & 3.7 \end{bmatrix} \text{ and } \mathbf{B} = \begin{bmatrix} -3 & 2.1 & 5.6 \\ 4.3 & 0 & 7.1 \\ -2.1 & 3 & 0 \end{bmatrix}$$

ANSWER Following the procedures of Practice Problem 1 store [A] and [B] in the calculator, as seen in Figures 13(*a*) and (*b*). Then press ⌞MATRX⌟ (select [A]) ⌞MATRX⌟ (select [B]) ⌞ENTER⌟. The answer is shown in Figure 13(*c*), the beginning of the matrix, and (*d*), the end of the matrix.

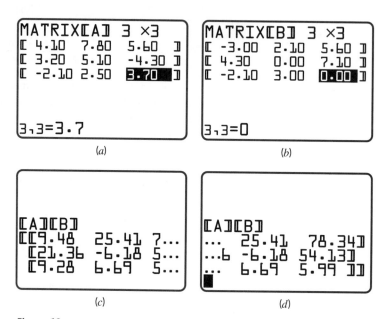

Figure 13

In much of the work with matrices in this book, we use square matrices; that is, we use matrices that have the same number of rows and columns. The product of square matrices of the same size always exists. In particular, we denote the products as follows:

$$\begin{array}{ll} (\mathbf{A})(\mathbf{A}) & \text{as } \mathbf{A}^2 \\ (\mathbf{A})(\mathbf{A})(\mathbf{A}) & \text{as } \mathbf{A}^3 \\ \cdot & \\ \cdot & \\ \cdot & \\ (\mathbf{A})(\mathbf{A}) \cdots (\mathbf{A}) & \text{as } \mathbf{A}^n \end{array}$$

n factors

For example, if

$$\mathbf{A} = \begin{bmatrix} 2 & 3 \\ -4 & 1 \end{bmatrix}, \quad \text{then} \quad \mathbf{A}^2 = \begin{bmatrix} 2 & 3 \\ -4 & 1 \end{bmatrix}\begin{bmatrix} 2 & 3 \\ -4 & 1 \end{bmatrix}$$

$$= \begin{bmatrix} -8 & 9 \\ -12 & -11 \end{bmatrix}$$

**COMMON
ERROR**

If $\mathbf{A} = \begin{bmatrix} 2 & 3 \\ -4 & 1 \end{bmatrix}$, then $\mathbf{A}^2 \neq \begin{bmatrix} 2^2 & 3^2 \\ (-4)^2 & 1^2 \end{bmatrix}$

Practice Problem 4 If

$$\mathbf{A} = \begin{bmatrix} 4.1 & 7.8 & 5.6 \\ 3.2 & 5.1 & -4.3 \\ -2.1 & 2.5 & 3.7 \end{bmatrix}$$

find $\mathbf{A}^2$ using a calculator.

ANSWER Store matrix [A]. Then press $\boxed{\text{MATRX}}$ (select [A]) $\boxed{x^2}$ $\boxed{\text{ENTER}}$. The answer is shown in Figure 14(*a*), the beginning of the matrix, and in Figure 14(*b*), the end of the matrix.

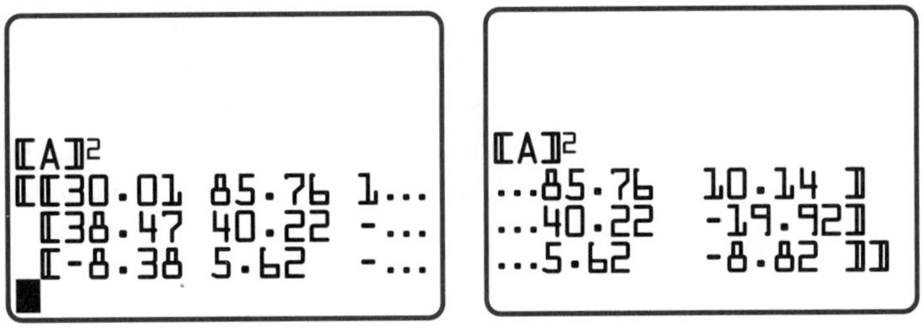

(*a*)   (*b*)

Figure 14

The identity matrix for multiplication is an important property of matrices. Just as 1 is the identity for multiplication of real numbers (i.e., $1 \cdot 6 = 6 \cdot 1$ or $1 \cdot x = x \cdot 1$), we define

$$\mathbf{I} = \begin{bmatrix} 1 & 0 \\ 0 & 1 \end{bmatrix}$$

as the identity matrix for 2×2 matrices. For example, $\mathbf{IA} = \mathbf{AI} = \mathbf{A}$; that is, if

$$\mathbf{A} = \begin{bmatrix} 3 & -4 \\ 2 & 7 \end{bmatrix}, \quad \text{then} \quad \begin{bmatrix} 1 & 0 \\ 0 & 1 \end{bmatrix}\begin{bmatrix} 3 & -4 \\ 2 & 7 \end{bmatrix} = \begin{bmatrix} 3 & -4 \\ 2 & 7 \end{bmatrix}$$

Likewise,

$$\begin{bmatrix} x & y \\ z & w \end{bmatrix} \begin{bmatrix} 1 & 0 \\ 0 & 1 \end{bmatrix} = \begin{bmatrix} x & y \\ z & w \end{bmatrix}$$

Note that the identity matrix is a square matrix. The matrix

$$\begin{bmatrix} 1 & 0 & 0 \\ 0 & 1 & 0 \\ 0 & 0 & 1 \end{bmatrix}$$

is the 3×3 identity matrix; in general, the $n \times n$ matrix

$$\begin{bmatrix} 1 & 0 & 0 & \cdots & 0 \\ 0 & 1 & 0 & \cdots & 0 \\ 0 & 0 & 1 & \cdots & 0 \\ & & & & \\ \cdot & \cdot & \cdot & & \cdot \\ \cdot & \cdot & \cdot & & \cdot \\ \cdot & \cdot & \cdot & \cdots & \cdot \\ 0 & 0 & 0 & \cdots & 1 \end{bmatrix}$$

is the $n \times n$ identity matrix.

Many applications involve the multiplication of matrices.

EXAMPLE 16   At an ice cream company a matrix lists the various quantities of milk (in quarts), cream (in pints), sugar (in cups), vanilla extract (in fluid ounces), and baking chocolate (in ounces) to make a gallon of different flavors of ice cream (only two of which are listed).

$$\begin{array}{c} \\ \text{Vanilla} \\ \text{Chocolate} \end{array} \begin{array}{ccccc} \text{M} & \text{C} & \text{S} & \text{VE} & \text{BC} \\ \begin{bmatrix} 3 & 1 & 2 & 6 & 0 \\ 3 & 1 & 2\frac{1}{2} & 1 & 5 \end{bmatrix} \end{array}$$

If milk costs \$0.65 per quart, cream \$0.50 per pint, sugar \$0.12 per cup, vanilla extract \$0.14 per fluid ounce, and chocolate \$0.20 per ounce, find the cost of 1 gallon of vanilla and 1 gallon of chocolate ice cream.

SOLUTION

$$\begin{bmatrix} 3 & 1 & 2 & 6 & 0 \\ 3 & 1 & 2\frac{1}{2} & 1 & 5 \end{bmatrix} \begin{bmatrix} 0.65 \\ 0.50 \\ 0.12 \\ 0.14 \\ 0.20 \end{bmatrix} = \begin{bmatrix} 3.53 \\ 3.89 \end{bmatrix}$$

Not including the cost of labor, vanilla ice cream costs \$3.53 per gallon and chocolate ice cream costs \$3.89 per gallon.   ■

Matrices can also be used to replace or represent a system of equations.

EXAMPLE 17 Write the following system of equations in matrix form.

$$\begin{array}{rrrrr} 3x_1 + 4x_2 - 2x_3 + 1x_4 &=& 5 \\ -x_1 + 3x_2 - 2x_4 &=& 7 \\ 2x_1 + 3x_2 + x_3 - 1x_4 &=& 0 \\ 1x_1 - 1x_2 - x_3 + 3x_4 &=& -2 \end{array}$$

SOLUTION This set of four equations in four unknowns can be replaced by the single matrix equation

$$\begin{bmatrix} 3 & 4 & -2 & 1 \\ -1 & 3 & 0 & -2 \\ 2 & 3 & 1 & -1 \\ 1 & -1 & -1 & 3 \end{bmatrix} \begin{bmatrix} x_1 \\ x_2 \\ x_3 \\ x_4 \end{bmatrix} = \begin{bmatrix} 5 \\ 7 \\ 0 \\ -2 \end{bmatrix}$$

as can be seen by multiplying the two matrices on the left to obtain

$$\begin{bmatrix} 3x_1 + 4x_2 - 2x_3 + 1x_4 \\ -x_1 + 3x_2 - 2x_4 \\ 2x_1 + 3x_2 + 1x_3 - 1x_4 \\ 1x_1 - 1x_2 - 1x_3 + 3x_4 \end{bmatrix} = \begin{bmatrix} 5 \\ 7 \\ 0 \\ -2 \end{bmatrix}$$

Since the two matrices are equal, the definition of equality demands that their corresponding elements be equal. By setting corresponding elements equal we obtain the original system of equations. If we denote the three matrices by **A**, **B**, and **X**,

$$\mathbf{A} = \begin{bmatrix} 3 & 4 & -2 & 1 \\ -1 & 3 & 0 & -2 \\ 2 & 3 & 1 & -1 \\ 1 & -1 & -1 & 3 \end{bmatrix}, \quad \mathbf{X} = \begin{bmatrix} x_1 \\ x_2 \\ x_3 \\ x_4 \end{bmatrix}, \quad \text{and} \quad \mathbf{B} = \begin{bmatrix} 5 \\ 7 \\ 0 \\ -2 \end{bmatrix}$$

then the given system becomes the simple matrix equation **AX** = **B**. ∎

The product of matrices is a short-hand procedure for showing the repetition of procedures in graph theory. For example, if **A** represents the scheduling of no-stop flights or direct air flights between cities, then $\mathbf{A}^2$ would represent a repetition of no-stop schedules or one-stop flights. Similarly, $\mathbf{A}^3$ would represent three repetitions of the same schedule or two-stop flights.

Let's return to the matrix that we discussed in Section 2.1 representing the routes of direct air flights among the following four cities.

We found the matrix representing the graph to be

$$
\begin{array}{c}
 \\
\text{Atlanta} \\
\text{Nashville} \\
\text{Birmingham} \\
\text{Louisville}
\end{array}
\begin{array}{cccc}
\text{Atlanta} & \text{Nashville} & \text{Birmingham} & \text{Louisville}
\end{array}
\left[
\begin{array}{cccc}
0 & 1 & 1 & 1 \\
1 & 0 & 1 & 1 \\
1 & 1 & 0 & 0 \\
1 & 1 & 0 & 0
\end{array}
\right]
$$

To find the number of one-stop flights we compute the product $\mathbf{A}^2$.

$$
\mathbf{A}^2 =
\begin{bmatrix}
0 & 1 & 1 & 1 \\
1 & 0 & 1 & 1 \\
1 & 1 & 0 & 0 \\
1 & 1 & 0 & 0
\end{bmatrix}
\begin{bmatrix}
0 & 1 & 1 & 1 \\
1 & 0 & 1 & 1 \\
1 & 1 & 0 & 0 \\
1 & 1 & 0 & 0
\end{bmatrix}
=
\begin{bmatrix}
3 & 2 & 1 & 1 \\
2 & 3 & 1 & 1 \\
1 & 1 & 2 & 2 \\
1 & 1 & 2 & 2
\end{bmatrix}
$$

The 2 in the first row of $\mathbf{A}^2$ indicates there are 2 one-stop flights from Atlanta to Nashville (by Birmingham and Louisville). The two 1's in the first row indicate that there is only 1 one-stop flight between Atlanta and Birmingham (through Nashville) and between Atlanta and Louisville (through Nashville).

EXAMPLE 18   For the preceding illustration, what does the 3 mean in the first row? Interpret the second row.

SOLUTION   The 3 means that you can go from Atlanta to Atlanta in 3 one-stop flights: Atlanta to Nashville to Atlanta; Atlanta to Birmingham to Atlanta; and Atlanta to Louisville to Atlanta. The second row lists the number of one-stop flights from Nashville to Atlanta, Nashville to Nashville, Nashville to Birmingham, and Nashville to Louisville.   ■

We conclude this section with an example involving the associative property of multiplication of matrices.

ASSOCIATIVE PROPERTY OF MULTIPLICATION OF MATRICES

If **A**, **B**, and **C** are any three matrices for which multiplication is defined, then

$$\mathbf{A}(\mathbf{BC}) = (\mathbf{AB})\mathbf{C}$$

EXAMPLE 19 Compute the following product of three matrices by grouping in two different ways:

$$\begin{bmatrix} 2 & 1 \\ -1 & 3 \end{bmatrix} \begin{bmatrix} 3 & 1 \\ 0 & 1 \end{bmatrix} \begin{bmatrix} 1 & 0 \\ 0 & 1 \end{bmatrix}$$

SOLUTION Multiply the first two matrices first:

$$\left(\begin{bmatrix} 2 & 1 \\ -1 & 3 \end{bmatrix} \begin{bmatrix} 3 & 1 \\ 0 & 1 \end{bmatrix} \right) \begin{bmatrix} 1 & 0 \\ 0 & 1 \end{bmatrix} = \begin{bmatrix} 6 & 3 \\ -3 & 2 \end{bmatrix} \begin{bmatrix} 1 & 0 \\ 0 & 1 \end{bmatrix}$$
$$= \begin{bmatrix} 6 & 3 \\ -3 & 2 \end{bmatrix}$$

Multiply the last two matrices first:

$$\begin{bmatrix} 2 & 1 \\ -1 & 3 \end{bmatrix} \left(\begin{bmatrix} 3 & 1 \\ 0 & 1 \end{bmatrix} \begin{bmatrix} 1 & 0 \\ 0 & 1 \end{bmatrix} \right) = \begin{bmatrix} 2 & 1 \\ -1 & 3 \end{bmatrix} \begin{bmatrix} 3 & 1 \\ 0 & 1 \end{bmatrix}$$
$$= \begin{bmatrix} 6 & 3 \\ -3 & 2 \end{bmatrix}$$

Since the two answers are equal, we have shown that the associative property of multiplication of matrices holds for these three matrices. ■

SUMMARY

Matrix operations have many similarities to algebra and arithmetic. The associative property of multiplication holds; that is, $(AB)C = A(BC)$. However, matrix multiplication is not commutative; that is, $AB \neq BA$. The identity matrix for multiplication is a square matrix with 1's along the main diagonal and 0's elsewhere. A matrix equation can be used to express a system of equations, that is, $AX = B$.

Exercise Set 2.2

Find the dot product of the following matrices if defined.

1. $[3 \quad -2 \quad 5] \cdot \begin{bmatrix} 3 \\ 4 \end{bmatrix}$

2. $[3 \quad -2 \quad 5] \cdot \begin{bmatrix} 1 \\ 0 \\ 4 \end{bmatrix}$

3. $[0 \quad 1 \quad 3] \cdot \begin{bmatrix} 2 \\ -1 \\ -1 \end{bmatrix}$

4. $[3 \quad -2 \quad -5] \cdot \begin{bmatrix} -1 \\ 0 \\ 4 \end{bmatrix}$

5. $[1 \quad 2] \cdot \begin{bmatrix} 1 \\ 2 \\ 3 \end{bmatrix}$

6. $[1, \quad 0 \quad 3] \cdot \begin{bmatrix} 1 \\ 2 \\ 3 \\ 1 \end{bmatrix}$

*Suppose that **A** is a 2 × 3 matrix, **B** is a 3 × 4 matrix, **C** is a 4 × 4 matrix, and **D** is a 4 × 3 matrix. Determine whether or not each of the following products is defined, and if it is defined, specify the size of the matrix answer.*

7. **AB**

8. **BA**

9. **AC**

10. **BC**

11. **CB**

12. **CD**

13. **B(CD)**

14. **A(BC)**

15. **(AB)D**

Compute the matrix answer if defined.

16. $[1 \quad 3] \begin{bmatrix} 2 & 1 & 6 \\ 1 & -1 & 0 \end{bmatrix}$

17. $\begin{bmatrix} 3 & 1 \\ -4 & 1 \\ -1 & 1 \end{bmatrix} \begin{bmatrix} 1 \\ 2 \end{bmatrix}$

18. $[1 \quad 3 \quad -1] \begin{bmatrix} 2 & 0 \\ 1 & 0 \\ 3 & 1 \end{bmatrix}$

Find each of the following products if possible.

19. $\begin{bmatrix} 3 & 5 & 1 & 2 \\ 2 & 4 & 0 & -1 \end{bmatrix} \begin{bmatrix} 1 & 1 & 3 \\ 2 & -1 & 5 \\ 3 & 0 & 2 \\ 4 & 2 & 1 \end{bmatrix}$

20. $\begin{bmatrix} 1 & 1 & 3 \\ 2 & 1 & 5 \\ 3 & 0 & 2 \\ 4 & 2 & 1 \end{bmatrix} \begin{bmatrix} 3 & 5 & 1 & 2 \\ 2 & 4 & 0 & -1 \end{bmatrix}$

21. $\begin{bmatrix} 3 & 2 & 1 \\ 1 & 0 & -1 \\ 2 & 1 & 1 \end{bmatrix} \begin{bmatrix} 1 & 5 & 2 \\ 0 & 2 & 1 \\ 0 & 0 & 5 \end{bmatrix}$

22. $\begin{bmatrix} 1 & 5 & 2 \\ 0 & 2 & 1 \\ 0 & 0 & 5 \end{bmatrix} \begin{bmatrix} 3 & 2 & 1 \\ 1 & 0 & -1 \\ 2 & 1 & 1 \end{bmatrix}$

23. $\begin{bmatrix} 2 & 0 & 0 \\ 0 & 1 & -1 \\ 7 & 0 & 0 \end{bmatrix} \begin{bmatrix} 0 & 0 & 0 \\ 3 & 3 & 3 \\ 3 & 3 & 3 \end{bmatrix}$

24. $\begin{bmatrix} 0 & 0 & 0 \\ 3 & 3 & 3 \\ 3 & 3 & 3 \end{bmatrix} \begin{bmatrix} 2 & 0 & 0 \\ 0 & 1 & -1 \\ 7 & 0 & 0 \end{bmatrix}$

25. $\begin{bmatrix} 1 & 4 \\ 2 & 5 \\ 3 & 6 \end{bmatrix} \begin{bmatrix} 3 & 1 & 5 \\ 2 & 0 & 3 \end{bmatrix}$

26. $\begin{bmatrix} 3 & -1 & 5 \\ 2 & 0 & 3 \end{bmatrix} \begin{bmatrix} 1 & 4 \\ 2 & 5 \\ 3 & 6 \end{bmatrix}$

27. $\begin{bmatrix} 1 & -2 \\ 2 & 0 \end{bmatrix}^2$

28. $\begin{bmatrix} 2 & 1 \\ 3 & 4 \end{bmatrix}^2$

Perform the indicated operations when possible. When impossible, explain why.

29. $\begin{bmatrix} 3 & 5 & 1 & 2 \\ 2 & 4 & 0 & -1 \end{bmatrix} \begin{bmatrix} 1 & 1 \\ 2 & -1 \\ 3 & 0 \\ 4 & 2 \end{bmatrix}$

30. $\begin{bmatrix} 1 & 1 & 3 \\ 2 & 1 & 4 \\ 3 & 1 & 5 \end{bmatrix} \begin{bmatrix} 3 & 5 & 2 \\ 6 & 1 & 4 \end{bmatrix}$

31. $\begin{bmatrix} 2 & 0 & 0 \\ 0 & 1 & -1 \\ 7 & 0 & 0 \end{bmatrix} \begin{bmatrix} 0 & 0 & 0 \\ 3 & 3 & 3 \\ -1 & -1 & -1 \end{bmatrix}$

32. $\begin{bmatrix} 3 & -1 & 5 \\ 2 & 0 & 3 \end{bmatrix} \begin{bmatrix} 2 & 3 \\ 1 & -1 \end{bmatrix}$

Write the following systems of equations as $\mathbf{AX} = \mathbf{B}$*, where*

$$\mathbf{X} = \begin{bmatrix} x \\ y \end{bmatrix}$$

A *is the matrix of coefficients, and* **B** *is the column matrix of constant terms.*

33. $4x + 3y = 7$
$6x - y = 10$

34. $3x - y = 7$
$5x + 2y = -1$

35. $y = 4x + 3$
$x = 2y - 7$

36. $y + x = 2$
$y = 4$

37. $x = 4$
$x + 2y = 7$

38. $x = 6$
$y = 4$

Follow the instructions for Exercises 33–38, where

$$\mathbf{X} = \begin{bmatrix} x \\ y \\ z \end{bmatrix}$$

39. $6x + 3y + 7z = 2$
    $5x + 4y + z = 5$
    $2x - y + z = 6$

40. $y + z = 6$
    $x = 7 - y$
    $x = z - 6$

41. $x = 4$
    $y = 7$
    $z = 9$

In Exercises 42–45 where

$$A = \begin{bmatrix} 0.07 & -0.01 & -0.08 \\ 0.16 & 0.31 & 0.32 \\ -0.11 & -0.42 & -0.57 \end{bmatrix}$$

and

$$B = \begin{bmatrix} 61 & 74 & -82 \\ 43 & -55 & 64 \\ -82 & 71 & -64 \end{bmatrix}$$

find each of the following.

42. **AB**   43. **BA**   44. A^2 45. B^2

Applications (Business and Economics)

46. *Stocks.*   In the previous exercise set we had the following table of the purchases of stock by an investment club. Using matrices, find the total cost of all stocks.

Company	Number of Shares	Price per Share
IBM	150	$61
Pfizer	200	63
American Home Products	300	56
Delta Airlines	100	45

47. *Investments.*   A bank has facilities in three cities. The amount in millions of dollars of money invested in various categories is given by

City	Bonds	Mortgages	Consumer Loans
City A	6	14	18
City B	3	10	12
City C	2	6	8

If the average return on bonds is 8%, on mortgages 10%, and on consumer loans 15%, determine earnings at each bank by using matrices.

48. *Markov Processes.*   In Exercise 34 of Exercise Set 2.1, suppose that, at the beginning of a year, store A has 60% of the customers and store B has 40%. This is indicated by the initial row matrix $X_0 = [0.60 \quad 0.40]$. If we multiply this matrix by the matrix **A**, we get $X_0A = X_1$, a matrix at the end of the period. For example,

$$[0.60 \quad 0.40] \begin{bmatrix} 0.80 & 0.20 \\ 0.24 & 0.76 \end{bmatrix} = [0.576 \quad 0.424]$$

Check this multiplication and then see if you can give an interpretation of the meaning of $[0.576 \quad 0.424]$.

Applications (Social and Life Sciences)

49. *Demographics.*   Find A^2 for Exercise 42 of Exercise Set 2.1, and interpret the elements of the product as the percentages after 2 periods of time.

50. *Genetics.*   Find A^2 for Exercise 43 of Exercise Set 2.1, and interpret the elements of the product as traits from parents to grandchildren.

51. *Transportation.*   The square of the matrix in Exercise 36 of Exercise Set 2.1 gives one-stop flights. A 2 in the product indicates 2 one-stop flights. What cities are connected with 2 one-stop flights?

52. *Communications.*   Interpret A^2 in Exercise 39 of Exercise Set 2.1 as a transmission from one station to another through an intermediate station. Describe the elements of A^2.

53. *Pollution.*   A chemical company is accused of polluting Bacon Creek by dumping industrial wastes from four manufacturing processes. Three pollutants are found in the creek. The following quantities of pollutants, expressed in milliliters, are found per 1000 liters of water for each process.

	Pollutants		
Process	1	2	3
A	6	1	4
B	3	2	2
C	7	4	1
D	8	10	3

Equipment is purchased to reduce the three pollutants as follows:

$$\begin{array}{c} \text{Pollutant 1} \\ \text{Pollutant 2} \\ \text{Pollutant 3} \end{array} \begin{bmatrix} 0.80 \\ 0.40 \\ 0.60 \end{bmatrix}$$

How many milliliters of pollutants still remain for each process?

Review Exercise

54. Find the following if possible.

(a) $2 \begin{bmatrix} 1 & 3 & -1 \\ 2 & -3 & 1 \end{bmatrix} - 3 \begin{bmatrix} 4 & 1 & -7 \\ -2 & 3 & -1 \end{bmatrix}$

(b) $4 \begin{bmatrix} 2 & 1 \\ 1 & 1 \end{bmatrix} - 3 \begin{bmatrix} 2 \\ 3 \end{bmatrix}$

2.3 SOLVING SYSTEMS OF EQUATIONS

OVERVIEW   Systems of equations are used extensively in economics and physical sciences. For example, a person may want to find the best way to invest money at two or more interest rates given certain conditions. A scientist may need to mix certain chemicals to reach a particular proportion. These are both problems that lend themselves to being solved using two or more equations. In this section we solve systems of equations

- Graphically
- Using addition and subtraction
- Using substitution

First let's consider a system of two equations in two unknowns such as

$$2x - 3y = 1$$
$$x + y = 3$$

The solution of any equation in two unknowns is the set of ordered pairs that satisfy the equation. A solution of a system of two equations is defined to be a common solution of the individual equations. Such solutions are called **simultaneous solutions** because each ordered pair satisfies both equations simultaneously. A solution of the system above is the ordered pair (2, 1). Check to see that this ordered pair satisfies both equations.

Our first procedure for obtaining such a solution is to solve the system graphically. The graph of each linear equation in two unknowns is a straight line. Suppose now that two such equations are graphed on the same coordinate system. The solution of the system of two linear equations in two unknowns is then given by the intersection of the two lines.

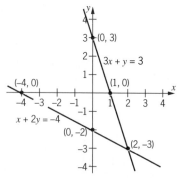

Figure 15

EXAMPLE 20   Solve the following system graphically:

$$3x + y = 3$$
$$x + 2y = -4$$

SOLUTION   The first equation is satisfied by infinitely many ordered pairs, three of which are $(0, 3)$, $(1, 0)$, and $(3, -6)$. Likewise, some ordered pairs that satisfy $x + 2y = -4$ are $(-4, 0)$, $(0, -2)$, and $(2, -3)$. The graphs of these two equations are shown in Figure 15. The intersection of the two lines in Figure 15 is the point $(2, -3)$. We can see whether this ordered pair is a solution of the system of equations by checking it in each equation. Substituting $x = 2$ and $y = -3$ in the first equation gives

Check: $3(2) + (-3) = 3$
$6 - \quad 3 = 3$

Substituting $x = 2$ and $y = -3$ in the second equation gives

Check: $2 + 2(-3) = -4$
$2 - \quad 6 = -4$

Hence, $(2, -3)$ is a solution of the system. ∎

In the next example we illustrate for a system of linear equations consisting of two equations in two unknowns that one and only one of the following is true:

1.   The two lists of solutions contain exactly one common ordered pair, called the solution of the system.
2.   The two lists of solutions are identical.
3.   The two lists of solutions contain no common ordered pairs.

Each of these possibilities is illustrated in the following example.

EXAMPLE 21   Find the solution set of each of the following systems of equations.

(a) $6x + 2y = 8$ (b) $3x + y = 3$ (c) $3x + y = 3$
$\quad\;\; 3x - 2y = 1$ $\quad\;\; 6x + 2y = 6$ $\quad\;\; 6x + 2y = 12$

SOLUTION

(a)   The solution is $(1, 1)$. See Figure 16(a).
(b)   There is an infinite number of solutions (all the points of a line, since all the solutions of one equation are solutions of the other). See Figure 16(b). Note that one equation can be obtained from the other by multiplying both sides of the equation by a constant; that is,

$$2(3x + y = 3) \quad \text{gives} \quad 6x + 2y = 6$$

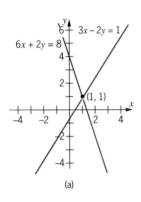
(a)

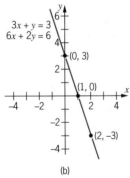
(b)

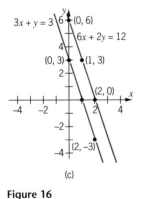
(c)

Figure 16

(c)   There are no common ordered pairs in (c) because the lines are parallel. See Figure 16(c).   ■

Geometrically, we are confronted with three possibilities for the straight-line graphs of equations in a system of two linear equations in two unknowns.

Intersection of Lines

One of these possibilities must occur for the graph of two lines in a plane:

1.   The two lines intersect at exactly one point.
2.   The two lines coincide.
3.   The two lines are parallel.

Although graphing the solution of a system of two linear equations with two variables gives an excellent picture of the relationship between the two variables, the method is time-consuming and may not be accurate if the numbers that compose the ordered pairs in the solution set are not integers. Furthermore, graphical techniques do not easily generalize to systems with more than two variables and/or more than two equations. Consequently, algebraic methods for solving the system are often more practical. We now consider how to obtain algebraically a second system of equations that is *equivalent* (has the same solution) to the given system. One such method, called the **addition–subtraction method**, is used extensively.

If we know that $A = B$ and $C = D$, it follows that $A + C = B + D$. This property can be extended to equations and will help you understand the idea behind the addition–subtraction method. If we add the two equations together, we eliminate the y variable.

$$\begin{array}{ll} A = B & 2x - 3y = 5 \\ \underline{C = D} & \underline{3x + 3y = 10} \\ A + C = B + D & 5x = 15 \end{array}$$

Now let's use this procedure to solve the following system of equations:

$$\begin{array}{ll} 4x + 2y = 8 & (1) \\ 3x - y = 1 & (2) \end{array}$$

Let's eliminate x. What multiplied by $3x$ and what multiplied by $4x$ will make the coefficients inverses? In this illustration we multiply both sides of the second equation by -4 and both sides of the first equation by 3:

$$12x + 6y = 24 \quad (1)$$

$$-12x + 4y = -4 \quad (2)$$

This system is equivalent to the preceding system; that is, it has the same solution. If we add equations (1) and (2) term by term, we obtain

$$10y = 20 \quad \text{or} \quad y = 2$$

Substituting $y = 2$ in equation (1) yields

$$4x + 2(2) = 8$$
$$4x \qquad = 4$$
$$x \qquad = 1$$

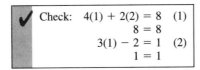

Check: $4(1) + 2(2) = 8$ (1)
                     $8 = 8$
        $3(1) - 2 = 1$ (2)
                 $1 = 1$

So the solution is the ordered pair $(1, 2)$.

EXAMPLE 22   Find the solution of the system

$$3x + \ y = \ \ 3 \quad (1)$$
$$x + 2y = -4 \quad (2)$$

SOLUTION   In order to make the coefficients of y additive inverses in the two equations, multiply each term of the first equation by -2 to obtain $-6x - 2y = -6$. Add to this equation the like terms of the second equation:

$$-6x - 2y = \ -6 \quad (1)$$
$$\underline{\ \ \ x + 2y = \ -4 \quad (2)}$$
$$-5x \qquad = -10$$
$$x = 2$$

Substituting $x = 2$ into the first equation gives $3(2) + y = 3$ or $y = -3$. The solution is $(2, -3)$, and we can check that this point lies on both lines by substituting its coordinates into both equations. Note that this is the same solution that we obtained by graphical procedures in Figure 15.   ■

Practice Problem 1 Solve the system of equations

$$3x + \ y = 5$$
$$x = 2y - 3$$

ANSWER   The solution is the ordered pair $(1, 2)$.

EXAMPLE 23   Find all the solutions of the system

$$3x + y = 3 \quad (1)$$
$$6x + 2y = 6 \quad (2)$$

SOLUTION   Multiply both sides of the first equation by -2 to obtain

$$-6x - 2y = -6 \quad (1)$$

Add this equation to the second equation to obtain $0 = 0$.
   If (x, y) satisfies

$$3x + y = 3 \quad (1)$$

then multiplying by 2 gives

$$2(3x + y) = 2 \cdot 3 \quad (1)$$

or

$$6x + 2y = 6 \quad (2)$$

Hence, any point that satisfies the first equation will satisfy the second equation. That is, the graphs of the two equations coincide [see Figure 16(b)].   ■

EXAMPLE 24   Find the solution of

$$3x + y = 3 \quad (1)$$
$$6x + 2y = 12 \quad (2)$$

SOLUTION   Multiplying the first equation by -2 gives

$$\begin{array}{r} -6x - 2y = -6 \quad (1) \\ \underline{6x + 2y = 12 \quad (2)} \\ 0 = 6 \end{array}$$

Adding equations (1) and (2) yields 0 on the left side of the equation and 6 on the right. Since $0 \neq 6$, no numbers x and y satisfy both equations. We see that the two lines are parallel in Figure 16(c).   ■

Practice Problem 2 Solve

$$3x - 2y = 4$$
$$4y - 1 = 6x$$

ANSWER   The solution is the empty set because the two lines are parallel.

EXAMPLE 25   Solve the following system of equations using a graphing calculator.

$$3x + 5y = 7$$

$$5x + 7y = 9$$

SOLUTION   Each equation can be solved for y in terms of x for input in the Y = menu. $Y_1 = (7 - 3x)/5$ and $Y_2 = (9 - 5x)/7$. See Figure 17(*b*). Use instructions on page A-1 of Appendix A to obtain the two graphs in Figure 17(*c*). By using the TRACE key, you can locate the point where the lines intersect [Figure 17(*c*)]. By using the TABLE key (see page 8 of Appendix A), you can see the values for Y_1 and Y_2 at $x = -1$ [Figure 17(*e*)]. The point of intersection is $(-1, 2)$. ∎

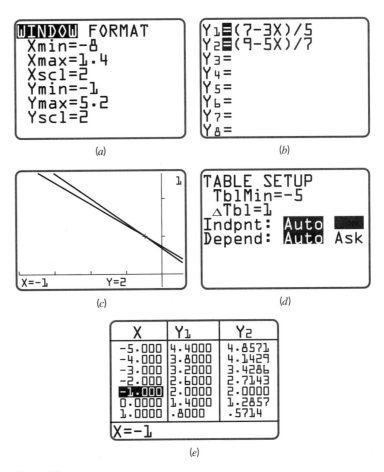

Figure 17

In calculus we often use a method of solving a system of equations called **substitution**. This procedure involves solving for one variable in terms of other variables in one equation and then substituting this value for a variable in the other equation.

EXAMPLE 26   Solve

$$3x + 2y = 7$$
$$4x - y = 2$$

using the substitution method.

SOLUTION   First we solve for y in the second equation:

$$y = 4x - 2$$

Then we substitute for this value of y in the first equation:

$$3x + 2(4x - 2) = 7$$
$$3x + 8x - 4 = 7$$
$$11x = 11$$
$$x = 1$$
$$y = 4(1) - 2 = 2$$

The solution is $(1, 2)$.

✔ Check: $3(1) + 2(2) = 7$
$$7 = 7$$
$$4(1) - 2 = 2$$
$$2 = 2$$

In the following examples we use systems of two linear equations in two variables to solve practical problems.

EXAMPLE 27   The total number of Democrats and Republicans in a community is 40,000. In a recent election, 60% of the Democrats and 40% of the Republicans voted. If only Democrats and Republicans voted and the total vote was 21,400, find the number of Democrats and the number of Republicans in the community.

SOLUTION   Let's apply the problem-solving techniques of Chapter 0 as a start toward the solution to this problem. First we estimate an answer and then check with a calculator in Table 2. Then we replace our estimates with variables x and y.

TABLE 2

	Democrats		Republicans		Total
Estimate	30,000	+	10,000	=	40,000
Number who voted	0.60(30,000)	+	0.40(10,000)	=	22,000
					(Too large; answer 21,400)
Assignment of variables	x	+	y	=	40,000
Number who voted	0.60x	+	0.40y	=	21,400

The equations that form a mathematical model for this problem are

$$x + y = 40{,}000$$

$$0.60x + 0.40y = 21{,}400$$

By the method of substitution,

$$x = 40{,}000 - y$$

$$0.60(40{,}000 - y) + 0.40y = 21{,}400$$

$$0.20y = 2600$$

$$y = 13{,}000$$

$$x = 40{,}000 - 13{,}000 = 27{,}000$$

Solving this system gives $x = 27{,}000$ Democrats and $y = 13{,}000$ Republicans in the community, and the solution is (27,000, 13,000). ■

Always check an application problem by seeing if the conditions of the stated problem are satisfied (not by substituting the values into the equations). To check this problem, note that the sum of 27,000 Democrats and 13,000 Republicans gives a total of 40,000. If 60% of the Democrats voted, the number of Democrats who voted was (0.60)(27,000) = 16,200. Similarly, if 40% of the Republicans voted, the number of Republicans who voted was (0.40)(13,000) = 5200. The total vote cast was 16,200 + 5200 = 21,400. Hence, the statements given in the problem are satisfied.

Break-Even Point   When an equation expressing total cost C in terms of x (the number of items produced) and an equation expressing revenue R in terms of x are graphed on the same coordinate system, then the break-even point is the intersection of the two lines. Also, the break-even point can be found by solving the system of linear equations by setting $C = R$; that is, the revenue and cost are equal.

EXAMPLE 28 A university is offering a special course in crafts for which tuition is $60 per student. The university has found that the cost of the course is $600 plus $20 for each student who registers for the course. How many students must take the course for the university to break even?

SOLUTION Let x = the number of students taking the course. Then the revenue equation becomes $R = 60x$, and the cost equation becomes $C = 20x + 600$. Setting these two equations equal to each other gives

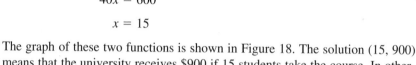

$$R = C \qquad \text{\$600 is the fixed cost.}$$

$$\$60x = \$20x + 600 \qquad \text{\$20x is the variable cost.}$$

$$40x = 600$$

$$x = 15$$

The graph of these two functions is shown in Figure 18. The solution (15, 900) means that the university receives $900 if 15 students take the course. In other words, $900 is the cost for 15 students. ∎

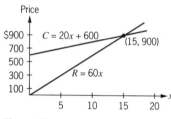

Price

Figure 18

SUMMARY

Systems of equations can be solved graphically, algebraically, or using a graphing calculator. When using algebra, we can use the addition–subtraction method or the substitution method. The possible graphs for two lines are two lines that intersect at exactly one point; two lines that coincide; or two lines that are parallel. In general, any system of two linear equations in two variables will have either one solution, no solution, or an infinite number of solutions.

Exercise Set 2.3

1. Solve the following systems graphically. Verify using a graphing calculator.

(a) $x + y = 5$
$\quad\ x - y = -1$

(b) $x + y = 1$
$\quad\ x - y = 5$

(c) $2x + y = 4$
$\quad 4x + 2y = 8$

(d) $2x + y = 4$
$\quad\ x - y = -1$

(e) $3x - y = 0$
$\quad\ x + y = 4$

(f) $3x - y = 0$
$\quad 6x - 2y = 2$

(g) $3x - 2y = 1$
$\quad 6x - 4y = 1$

(h) $2x - y = 3$
$\quad 4x - 2y = 6$

(i) $x - 2y = -3$
$\quad 2x + y = 4$

2. Solve the problems in Exercise 1 using the addition–subtraction method, and explain why no solution exists in some problems.

3. Solve the following systems by the addition–subtraction method.

(a) $3x + 2y = 1$
$\quad 5x + 3y = 4$

(b) $4x - 3y = 5$
$\quad 8x - 6y = 7$

(c) $2x + 3y = 9$
$\quad 3x + 5y = 11$

(d) $4x - 3y = 5$
$\quad 8x - 6y = 10$

(e) $5x - 2y = -4$
$\quad 3x + 5y = 10$

(f) $4x - 3y = 5$
$\quad 3x + 2y = 8$

4. Solve the following systems of equations:

(a) $\dfrac{5}{2}x - \dfrac{2}{3}y = -\dfrac{4}{3}$

$$\frac{3}{5}x + \frac{5}{2}y = 2$$

(b) $0.4x - 0.3y = 0.05$
$0.3x + 0.2y = 0.08$

(c) $0.01x - 0.2y = -0.1$
$0.02x + 0.3y = \quad 1.9$

(d) $\frac{2}{3}x + \frac{1}{4}y = \frac{1}{6}$

$-\frac{3}{4}x = \frac{y}{3} + \frac{5}{2}$

5. The sum of two test scores is 175. The difference between the two scores is 11. Find the scores.

6. Fifty coins totaling $7.40 are removed from a soft-drink machine. If the coins are all dimes and quarters, determine the number of each.

7. Use a calculator to solve the following systems.

(a) $17.05x - 3.24y = 22.63$
$3.21x - 4.56y = \quad 3.96$

(b) $\quad 35.7x + 103y = 104.759$
$-23.4x + \quad 37y = 118.412$

(c) $123x - 37y = 6890$
$47x - 31y = \quad 896$

(d) $137x - 41y = -10,688$
$105x + 13y = \quad -2,994$

8. **Exam**.   Given the following notations, what is the break-even sales level in units? (**Hint**: $C = FC + (VC)(x)$, and $R = (SP)(x)$, where $x =$ Sales level, $SP =$ Selling price per unit, $FC =$ Total fixed cost, and $VC =$ Variable cost per unit.)

(a) $\dfrac{SP}{FC \div VC}$ 　　　　(b) $\dfrac{FC}{VC \div SP}$

(c) $\dfrac{VC}{SP - FC}$ 　　　　(d) $\dfrac{FC}{SP - VC}$

Applications (Business and Economics)

9. **Mixture Problem**.   A candy-store proprietor wishes to mix candy that sells for $3 per pound with candy selling for $4 per pound to make a mixture to sell for $3.60 per pound. How many pounds of each kind of candy should be used to make 80 pounds of the mixture?

10. **Commission**.   A man is trying to decide between two positions. The first pays $225 per week plus 5% commission on gross sales. The second pays only 9% on gross sales. Graph the two pay functions and find where they are equivalent.

11. **Break-Even Point**.   A producer knows that she can sell as many items at $0.25 each as she can produce in a day. If her cost is $C = \$0.20x + \$70$, find her break-even point.

12. **Break-Even Point**.   A firm knows that it can sell as many items at $1.25 each as it can produce in a day. If the cost is $C = \$0.90x + \$105$, find the break-even point.

13. **Break-Even Point**.   If the firm in Exercise 12 can change the cost equation to $C = \$0.80x + \$120$, should the change be made? Explain.

Applications (Social and Life Sciences)

14. **Nutrition**.   A special diet requires 4 milligrams of iron and 52 grams of protein each day. A person decides to attain these requirements by drinking skim milk and eating fish. A glass of skim milk provides 0.2 milligrams of iron and 1 gram of protein. One-fourth pound of fish provides 8 milligrams of iron and 10 grams of protein. How many glasses of milk and how many pieces of fish ($\frac{1}{4}$ pound) are needed to attain the diet's requirements?

15. **Population**.   A town has a population of 1000. The number of men is 80 less than twice the number of women. Find the number of men and the number of women.

16. **Psychological Attraction and Repulsion**.   A psychologist has been studying reactions of attraction and repulsion by first feeding mice and then later giving them mild electric shocks from the same box. With this procedure the psychologist established the following functions where a represents attraction, r represents repulsion, and x represents the distance in centimeters of the mouse from the box.

$$a = -\frac{1}{4}x + 70$$

$$r = -\frac{4}{3}x + 200$$

Graph the attraction function and the repulsion function on the same coordinate axes, and find the distance where attraction equals repulsion. (**Hint**: Find the intersection point of the two lines.) Check your graphical result by setting $a = r$ and solving for x.

18. Multiply

$$\begin{bmatrix} -1 & 1 & 3 \\ 2 & 0 & -2 \end{bmatrix} \begin{bmatrix} 1 & 1 \\ 0 & 1 \\ 3 & 2 \end{bmatrix}$$

Review Exercises

17. $3 \begin{bmatrix} 2 & 3 \\ 1 & 1 \\ 5 & 7 \end{bmatrix} - \begin{bmatrix} 1 & 0 \\ 0 & 1 \\ 1 & 1 \end{bmatrix} = \begin{bmatrix} a & b \\ c & d \\ e & f \end{bmatrix}$

Find a, b, c, d, e, and f.

2.4 SOLVING SYSTEMS OF LINEAR EQUATIONS USING MATRICES

OVERVIEW   In practical applications, most systems of linear equations involve a large number of equations and unknowns. Usually these systems are solved using computers. In this section we introduce a procedure that can be used for a computer or calculator solution of a system of equations in any number of variables and any number of equations. Our first step involves forming an augmented matrix. We then perform permissible matrix operations to put the matrix in a form where the solution can be obtained. The step-by-step procedure we use is called the Gauss–Jordan elimination method. In this section we

- Find augmented matrices
- Perform appropriate row operations
- Determine the solution of a system of equations

An **augmented matrix** for a system of equations contains the coefficients of the unknowns as elements, in all except the last column. The elements are listed in the same order as they appear in the system of equations. A vertical line usually replaces the equal signs. The last column consists of the constant terms. For example, the augmented matrix for the system

$$2x + 3y = -5$$
$$x - 2y = 8$$

is

$$\begin{bmatrix} 2 & 3 & \bigm| & -5 \\ 1 & -2 & \bigm| & 8 \end{bmatrix}$$

EXAMPLE 29   Find the augmented matrix for the system

$$2x = 7 + 3y$$
$$4y = -4 - 3x$$

SOLUTION   First we write the two equations with the variables occurring in the same order.

$$2x - 3y = 7$$
$$3x + 4y = -4$$

The augmented matrix is

$$\left[\begin{array}{rr|r} 2 & -3 & 7 \\ 3 & 4 & -4 \end{array}\right]$$

■

Practice Problem 1 Write an augmented matrix for the following system of equations:

$$2x = 4 - 3y + z$$
$$2z - 3y = 4 - x$$
$$6 + 3y = 2x - 4z$$

ANSWER

$$\left[\begin{array}{rrr|r} 2 & 3 & -1 & 4 \\ 1 & -3 & 2 & 4 \\ 2 & -3 & -4 & 6 \end{array}\right]$$

EXAMPLE 30   Using x, y, and z, write a system of linear equations having the augmented matrix

$$\left[\begin{array}{rrr|r} 3 & -1 & 4 & 7 \\ 2 & 2 & -5 & 3 \\ 3 & 2 & 1 & 4 \end{array}\right]$$

SOLUTION   The system of equations is

$$3x - y + 4z = 7$$
$$2x + 2y - 5z = 3$$
$$3x + 2y + z = 4$$

■

Practice Problem 2 Write a system of equations from the augmented matrix

$$\begin{bmatrix} 1 & 3 & -1 & | & -2 \\ 4 & -1 & 3 & | & 0 \\ 3 & -2 & -4 & | & -1 \end{bmatrix}$$

ANSWER

$$x + 3y - z = -2$$
$$4x - y + 3z = 0$$
$$3x - 2y - 4z = -1$$

EXAMPLE 31 Using x and y, write a system of linear equations from the augmented matrix

$$\begin{bmatrix} 1 & 0 & | & 2 \\ 0 & 1 & | & 3 \end{bmatrix}$$

SOLUTION

$$1 \cdot x + 0 \cdot y = 2 \qquad x = 2$$

or

$$0 \cdot x + 1 \cdot y = 3 \qquad y = 3$$ ■

From the preceding example it seems that if there is a set of operations that we can perform on an augmented matrix that will reduce it to the form shown in Example 31, this may give a procedure for solving a system of equations.

Now we demonstrate three operations on rows of a matrix that enable us to put a matrix in the form of Example 31. One row operation is to interchange two rows. Another operation is to multiply any row by a nonzero constant. This operation enables us to get the 1's along the main diagonal in Example 31. To get the appropriate 0's (see Example 31), we multiply a row of the matrix by any constant and add (or subtract) the result to any other row. Each of these operations gives a new matrix representing a system of equations with the same solution as the original system. These operations are summarized as follows, where R_i represents the ith row.

Row Operations

1. Interchange two rows ($R_i \leftrightarrow R_j$, or interchange the ith row and the jth row).
2. Multiply (or divide) each element of a row by a nonzero constant ($cR_i \rightarrow R_i$, or replace the ith row by a constant times the ith row).
3. Replace any row by the sum (or difference) of that row and any other row times a constant (can be 1) or replace the jth row by a constant times the ith row added to the jth row ($(cR_i + R_j) \rightarrow R_j$).

1. *Interchange two rows.* For the matrix

$$\left[\begin{array}{cc|c} 2 & 1 & 3 \\ 3 & -1 & 2 \end{array}\right]$$

interchange the rows $(R_1 \leftrightarrow R_2)$:

$$\left[\begin{array}{cc|c} 3 & -1 & 2 \\ 2 & 1 & 3 \end{array}\right]$$

The matrix

$$\left[\begin{array}{cc|c} 2 & 1 & 3 \\ 3 & -1 & 2 \end{array}\right]$$

represents

$$2x + y = 3$$
$$3x - y = 2$$

while

$$\left[\begin{array}{cc|c} 3 & -1 & 2 \\ 2 & 1 & 3 \end{array}\right]$$

represents

$$3x - y = 2$$
$$2x + y = 3$$

You can verify that $x = 1, y = 1$ satisfies both systems of equations. Thus, interchanging rows does not affect the solution in this example.

2. *Multiply (or divide) each element of a row by a nonzero constant.* For the matrix

$$\left[\begin{array}{cc|c} 3 & 2 & 1 \\ 2 & -2 & 4 \end{array}\right]$$

we randomly select a constant (say 4) and multiply the first row by this constant. Then we show that the new system of equations and the old system have the same solutions. The new matrix is

$$\left[\begin{array}{cc|c} 12 & 8 & 4 \\ 2 & -2 & 4 \end{array}\right]$$

These matrices represent the systems of equations

$$
\begin{array}{ccc}
3x + 2y = 1 & & 12x + 8y = 4 \\
& \text{and} & \\
2x - 2y = 4 & & 2x - 2y = 4
\end{array}
$$

Verify that these two systems have the same solution, namely, $x = 1$, $y = -1$.

3. *Replace any row by the sum (or difference) of that row and any other row times a constant.* Using the matrix

$$
\left[
\begin{array}{cc|c}
4 & 1 & 5 \\
3 & 2 & 0
\end{array}
\right]
$$

let's replace the second row with the sum of the second row and 2 times the first row, or $2R_1 + R_2 \rightarrow R_2$. Replace 3 with $2 \cdot 4 + 3 = 11$, 2 with 4, and 0 with 10, getting

$$
\left[
\begin{array}{cc|c}
4 & 1 & 5 \\
11 & 4 & 10
\end{array}
\right]
$$

Show that $(2, -3)$ is a solution of both

$$
\begin{array}{ccc}
4x + y = 5 & & 4x + y = 5 \\
& \text{and} & \\
3x + 2y = 0 & & 11x + 4y = 10
\end{array}
$$

The row operations as defined seem to produce systems with the same solution. If, by these row operations, we can get an augmented matrix in the form

$$
\left[
\begin{array}{cc|c}
1 & 0 & a \\
0 & 1 & b
\end{array}
\right]
$$

we will have the solution

$$
x = a
$$
$$
y = b
$$

and this satisfies the original system of equations. The procedure for transforming an augmented matrix to this form is called Gauss–Jordan elimination after Carl Friedrich Gauss (1777–1855) and Camille Jordan (1838–1922).

EXAMPLE 32   Solve

$$2x + 3y = -5$$
$$x - 2y = 8$$

by augmented matrices.

SOLUTION   The augmented matrix is

$$\left[\begin{array}{cc|c} 2 & 3 & -5 \\ 1 & -2 & 8 \end{array}\right]$$

In the first two steps we will be working to obtain a matrix that looks like

$$\left[\begin{array}{cc|c} 1 & e & f \\ 0 & g & h \end{array}\right] \qquad e, f, g, \text{ and } h \text{ can be any real numbers.}$$

Get the 1 first and then the 0. In the last two steps, we will be working to obtain a matrix that looks like

$$\left[\begin{array}{cc|c} 1 & 0 & j \\ 0 & 1 & k \end{array}\right] \qquad j \text{ and } k \text{ can be any real numbers.}$$

We will again follow the procedure of getting in the second column the 1 first and then the 0.

NOTE: We could get a 1 in the upper left corner of this example by interchanging the first and second rows. We could also solve this small system more easily using something other than augmented matrices. However, on this small system we will practice procedures that will, in general, work on more complicated systems.

$$\boxed{\begin{array}{c}\text{Need a}\\\text{1 here}\end{array}} \quad \left[\begin{array}{cc|c} 2 & 3 & -5 \\ 1 & -2 & 8 \end{array}\right]$$

Divide each element of the first row by 2:

$$\frac{1}{2}R_1 \longrightarrow R_1$$

$$\boxed{\begin{array}{c}\text{Need a}\\\text{0 here}\end{array}} \quad \left[\begin{array}{cc|c} 1 & \frac{3}{2} & -\frac{5}{2} \\ 1 & -2 & 8 \end{array}\right]$$

Now multiply the first row by -1 and add to the second row:

$$-R_1 + R_2 \longrightarrow R_2$$

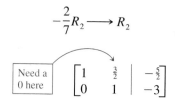

Need a
1 here

$$\begin{bmatrix} 1 & \frac{3}{2} & \Big| & -\frac{5}{2} \\ 0 & -\frac{7}{2} & \Big| & \frac{21}{2} \end{bmatrix}$$

Multiply the second row by $-\frac{2}{7}$:

$$-\frac{2}{7}R_2 \longrightarrow R_2$$

Need a
0 here

$$\begin{bmatrix} 1 & \frac{3}{2} & \Big| & -\frac{5}{2} \\ 0 & 1 & \Big| & -3 \end{bmatrix}$$

Multiply the second row by $-\frac{3}{2}$ and add to the first row:

$$-\frac{3}{2}R_2 + R_1 \rightarrow R_1$$

$$\begin{bmatrix} 1 & 0 & \Big| & 2 \\ 0 & 1 & \Big| & -3 \end{bmatrix}$$

Thus,

$$x + 0(y) = 2$$
$$0(x) + y = -3$$

The solution is $(2, -3)$.

Practice Problem 3 Write a system of equations from the augmented matrix

$$\begin{bmatrix} 1 & 0 & 0 & \Big| & 2 \\ 0 & 1 & 0 & \Big| & 3 \\ 0 & 0 & 1 & \Big| & 4 \end{bmatrix}$$

ANSWER $x = 2$

$y = 3$

$z = 4$

EXAMPLE 33 Solve

$$3x + 2y = 12$$
$$4x - 3y = -1$$

using augmented matrices.

SOLUTION

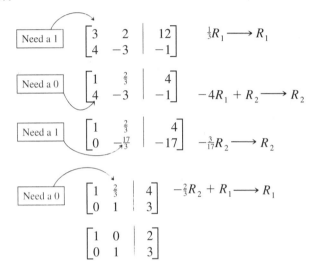

The solution is (2, 3). ■

Practice Problem 4 Using augmented matrices, solve the system of equations

$$x + 3y = 5$$
$$x + 5y = 7$$

Answer The solution is (2, 1).

EXAMPLE 34 Solve

$$2.33x + 3.01y = 7.72$$
$$3.71x - 4.32y = 3.09$$

SOLUTION The augmented matrix is

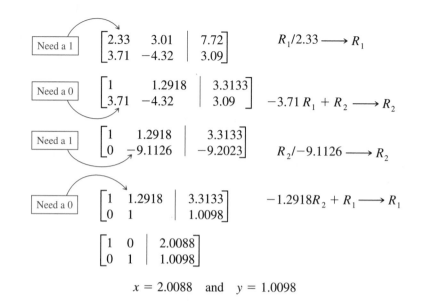

$$\left[\begin{array}{cc|c} 1 & 0 & 2.0088 \\ 0 & 1 & 1.0098 \end{array}\right]$$

$$x = 2.0088 \quad \text{and} \quad y = 1.0098$$

Check: $2.33(2.0088) + 3.01(1.0098) = 7.72000$ Accurate to three
         $3.71(2.0088) - 4.32(1.0098) = 3.09031$ decimal places

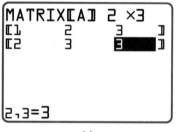

Figure 19

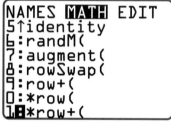

(a)

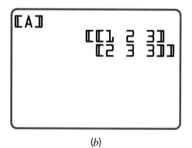

(b)

Figure 20

Calculator Note

[MATRX] [▶] produces the MATH EDIT menu shown in Figure 19. The calculator keystrokes necessary to perform the row operations in Example 34 are listed as follows:

$R_1/2.33 \to R_1$	corresponds to	*row(1/2.33, [A], 1)
$-3.71R_1 + R_2 \to R_2$	corresponds to	*row + (-3.71, ANS, 1, 2)
$R_2/-9.1126 \to R_2$	corresponds to	*row(1/-9.1126, ANS, 2)
$-1.2918R_2 + R_1 \to R_1$	corresponds to	*row + (-1.2918, ANS, 2, 1)

Practice Problem 5 With a graphing calculator, solve the system of equations

$$x + 2y = 3$$
$$2x + 3y = 3$$

Answer Store the 2×3 augmented matrix of the system under [A] as seen in Figures 20(a) and (b). Press [MATRX] [▶] Select *row + (−2, [MATRX] (Select [A]), 1, 2 [)] [ENTER]. The new matrix is seen in Figure 20(c). Next press [MATRX] [▶] Select *row (−1, [2nd] [ANS], 2 [)]. See Figure 20(d). Then press [MATRX] [▶] Select *row + (−2, [2nd] [ANS], 2, 1 [)]. From Figure 20(d) we read $x = -3$, $y = 3$.

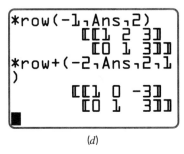

(c)

(d)

Figure 20

EXAMPLE 35   Show that the following system has no solution by using permissible row operations on the augmented matrix of the system

$$2x + y = 5$$
$$2x + y = 7$$

SOLUTION   The augmented matrix is

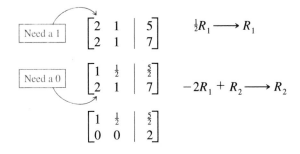

This is the augmented matrix for the system

$$x + \frac{1}{2}y = \frac{5}{2}$$
$$0 \cdot x + 0 \cdot y = 2$$

and because the second equation has no solution, the system has no solution. ■

EXAMPLE 36   Solve the system of equations

$$3x - 2y = 1$$
$$6x - 4y = 2$$

by using augmented matrices.

SOLUTION   The augmented matrix of the system can be written as

Need a 1 $\begin{bmatrix} 3 & -2 & | & 1 \\ 6 & -4 & | & 2 \end{bmatrix}$ $\frac{1}{3}R_1 \longrightarrow R_1$

Need a 0 $\begin{bmatrix} 1 & -\frac{2}{3} & | & \frac{1}{3} \\ 6 & -4 & | & 2 \end{bmatrix}$ $-6R_1 + R_2 \longrightarrow R_2$

$\begin{bmatrix} 1 & -\frac{2}{3} & | & \frac{1}{3} \\ 0 & 0 & | & 0 \end{bmatrix}$

The system of equations can be written as

$$x - \frac{2}{3}y = \frac{1}{3}$$

$$0 \cdot x + 0 \cdot y = 0$$

Since any pair of values that satisfies the first equation will satisfy the second, there are infinitely many solutions. For example, let $y = t$. Then

$$x = \frac{1}{3} + \frac{2}{3}t$$

If the equations were

$$x + ay = b$$

$$0 \cdot x + 0 \cdot y = 0$$

then, when $y = t$, $x = b - at$. ■

Summary of Types of Solutions

$$\left[\begin{array}{cc|c} 1 & 0 & a \\ 0 & 1 & b \end{array}\right] \qquad \left[\begin{array}{cc|c} 1 & a & b \\ 0 & 0 & 0 \end{array}\right] \qquad \left[\begin{array}{cc|c} 1 & a & b \\ 0 & 0 & c \end{array}\right]$$

Unique solution (a, b) | Infinitely many solutions $(b - at, t)$, where t is any real number | $c \neq 0$; no solutions

---

EXAMPLE 37 A toothpaste company manufactures two kinds of toothpaste in quantities measured in units equivalent to 1000 5-ounce tubes. Some of the toothpaste is sold to discount stores under several house brands at a net profit of \$20 per unit; toothpaste sold under the company's own brand nets \$30 profit per unit. If sales last month were 50,000 units and the profit was \$1,300,000, how many units were sold under the company's own brand and how many under another brand?

TABLE 3

	Number of Units		Other Brands		Total
Estimate	20,000	+	30,000	=	50,000
Profit	20,000(\$30)	+	30,000(\$20)	=	\$1,200,000 (too small)
Assignment of Variables	x	+	y	=	50,000
Profit	$30x$	+	$20y$	=	1,300,000

SOLUTION   As shown in Table 3, we use our problem-solving techniques to estimate that the company sold 20,000 units of its own brand. Then it sold $50,000 - 20,000 = 30,000$ units of other brands. After finding the profit for these estimates, we replace 20,000 and 30,000 with x and y.

The augmented matrix of the system is

$$\left[\begin{array}{cc|c} 1 & 1 & 50{,}000 \\ 30 & 20 & 1{,}300{,}000 \end{array}\right] \qquad -30R_1 + R_2 \rightarrow R_2$$

which has the following solution:

$$\left[\begin{array}{cc|c} 1 & 1 & 50{,}000 \\ 0 & -10 & -200{,}000 \end{array}\right] \qquad (-1/10)R_2 \rightarrow R_2$$

$$\left[\begin{array}{cc|c} 1 & 1 & 50{,}000 \\ 0 & 1 & 20{,}000 \end{array}\right] \qquad -R_2 + R_1 \rightarrow R_1$$

$$\left[\begin{array}{cc|c} 1 & 0 & 30{,}000 \\ 0 & 1 & 20{,}000 \end{array}\right]$$

Thus, $x = 30{,}000$ units and $y = 20{,}000$ units. The profit on 30,000 units of their own brand is \$900,000 and the profit on 20,000 units of other brands is \$400,000, making a total profit of \$1,300,000. ■

SUMMARY

In this section we demonstrated how to solve systems of equations using augmented matrices, instead of using algebra or the graphical approach discussed earlier. The procedure that we used is called Gauss–Jordan elimination. One advantage of this method is that it uses just three elementary row operations on a matrix. From the reduced matrix we can determine whether there is one solution, multiple solutions, or no solution. The Gauss–Jordan method can be programmed to be done on a computer. Very large systems of equations are usually done on a computer.

Exercise Set 2.4

Write the system of equations for each of the following augmented matrices.

1. $\left[\begin{array}{cc|c} 3 & 1 & 13 \\ 2 & -1 & 2 \end{array}\right]$ 2. $\left[\begin{array}{cc|c} 16 & -4 & 0 \\ 8 & 1 & 12 \end{array}\right]$

3. $\left[\begin{array}{cc|c} 1 & 0 & 4 \\ 0 & 1 & 6 \end{array}\right]$ 4. $\left[\begin{array}{cc|c} 1 & 0 & 2 \\ 0 & 1 & -1 \end{array}\right]$

Write the augmented matrix for each of the following systems.

5. $2x + 3y = 5$
$4x - y = 3$

6. $7u - 4w = 6$
$3u - w = 8$

7. $5x + 2y = 3$
$3y + x = -2$

8. $6x + 4 = 5y$
$2y + 3 = 7x$

9. $6 + 2y = 2x$
$x + 5 = 2y$

10. $2s - 1 = 7t$
$5 - t = s$

11. For the augmented matrix

$$\left[\begin{array}{cc|c} 3 & 2 & 5 \\ 7 & 4 & 4 \end{array}\right]$$

find the matrix with each of the following changes.

(a) $2R_2 \rightarrow R_2$ (b) $R_1 \leftrightarrow R_2$

(c) $3R_1 + R_2 \rightarrow R_2$ (d) $4R_1 \rightarrow R_1$

(e) $2R_2 + R_1 \rightarrow R_1$ (f) $-\frac{7}{3}R_1 + R_2 \rightarrow R_2$

12. Discuss the solutions of the following augmented matrices.

(a) $\begin{bmatrix} 1 & 0 & | & 2 \\ 0 & 1 & | & 3 \end{bmatrix}$ (b) $\begin{bmatrix} 1 & 2 & | & 3 \\ 0 & 0 & | & 0 \end{bmatrix}$

(c) $\begin{bmatrix} 1 & 4 & | & 7 \\ 0 & 0 & | & 4 \end{bmatrix}$

Solve each of the following systems of equations by using augmented matrices.

13. Exercise 1   14. Exercise 2   15. Exercise 5

16. Exercise 6   17. Exercise 7   18. Exercise 8

19. Exercise 9   20. Exercise 10

Discuss the solutions for the following systems of equations.

21. $4x + 3y = 7$
    $8x + 6y = 14$

22. $2x + 4y = 6$
    $3x + 6y = 9$

23. $2x - y = 3$
    $4x - 2y = 9$

24. $2x + 4y = 7$
    $3x + 6y = 11$

Use a calculator to solve the following systems of equations.

25. $1.1x - 2.5y = 7$
    $2.1x - 3.5y = 6$

26. $0.1x + 0.6y = 7$
    $1.1x + 1.6y = 4$

Applications (Business and Economics)

27. **Production Scheduling.** In a small furniture-man-ufacturing plant, 400 hours of labor are available for making tables and chairs and 107 hours are available for finishing (painting or staining). If it takes 8 hours to make a table and 5 hours to make a chair, and if it takes 2 hours to finish a table and 3 hours to finish a chair, how many of each can be manufactured using this schedule?

28. **Investments.** An investment club has $200,000 invested in bonds. Type A bonds pay 8% interest, and type B bonds pay 10%. How much money is invested in each type if the club receives $18,720 in interest?

29. **Mixture Problem.** A paper company uses both scrap paper and scrap cloth to make their paper. Their best paper requires 3 tons of cloth and 15 tons

of paper for each run, but their good paper requires 1 ton of cloth and 12 tons of paper for each run. How many runs of their best paper and how many runs of their good paper should be made if the company has 34 tons of scrap cloth and 261 tons of scrap paper on hand? Assume that the company wishes to use all its scrap paper and scrap cloth.

30. **Mixture Problem.** At a recent concert, tickets were $5 for adults and $3 for students. There were 2000 more adult tickets sold than student tickets. How many tickets of each kind were sold if the total receipts were $42,000?

31. **Break-Even Point.** A revenue equation is given by $R = 1.40x + 60$, where $x \geq 10$ is the number of items produced. The cost equation is $C = 0.95x + 105$. Find the break-even point.

Applications (Social and Life Sciences)

32. **Nutrition.** A dietician plans to combine food A and food B to make a meal containing 2000 calories, and 30 units of a combination of vitamins. Each ounce of food A contain 200 calories and 5 units of vitamins. Each ounce of food B contains 250 calories and 4 units of vitamins. How many ounces of each food should be in the meal?

33. **Diet.** A man is on a low-carbohydrate diet. He is planning a meal composed of two foods: food I with 7 grams of carbohydrate per unit and food II with 4 grams of carbohydrate per unit. To keep him from becoming discouraged with his diet, his doctor has insisted that he consume 500 calories at each meal, of which 210 calories must be protein. Both foods contain 100 calories per unit, but food I contains only 30 protein calories per unit, whereas food II contains 50 protein calories per unit. How many units of each food should he eat to consume the calories allowed?

Review Exercises

34. Determine whether or not the following systems have common solutions, and solve by graphing and by the addition–subtraction method.

(a) $3x + 2y = 1$
    $5x - 3y = 27$

(b) $3x - 7y = 2$
    $12x - 28y = 6$

(c) $11x - 3y = 7$
    $55x - 15y = 35$

(d) $2x + 3y = 3$
    $3x - y = 10$

2.5 SYSTEMS WITH THREE OR MORE VARIABLES OR THREE OR MORE EQUATIONS

OVERVIEW   Now that we know how to solve a system of two equations in two variables by using augmented matrices and have a better understanding of the theory of matrices, there is no reason why we should not extend this theory to three or more variables or systems with three or more equations. Note in this section that it is not necessary for a system of equations to have the same number of equations as variables. In fact, later in this section as we solve m equations with n unknowns, we consider three possibilities:

- $m = n$: The number of equations is the same as the number of unknowns.
- $m > n$: There are more equations than unknowns.
- $m < n$: There are more unknowns than equations.

Systems with a large number of equations and variables are very common today because of the accessibility of high-speed computers.

The solution of an equation in three variables, such as

$$x + 2y - 3z + 6 = 0$$

is an **ordered triplet** (x, y, z). For example, $(-3, 0, 1)$ is a solution because

$$1(-3) + 2(0) - 3(1) + 6 = 0$$
$$-3 - 3 + 6 = 0$$

The solutions for a system of three equations in three unknowns are the ordered triplets that satisfy the three equations.

The augmented matrix for three equations with three variables is very similar to that for two variables. It has one additional column and possibly additional rows.

EXAMPLE 38   Find the augmented matrix for the system

$$3x - y + 2z = 5$$
$$x \qquad - 4z = 3$$
$$y + 2z = 4$$

SOLUTION

$$\begin{bmatrix} 3 & -1 & 2 & 5 \\ 1 & 0 & -4 & 3 \\ 0 & 1 & 2 & 4 \end{bmatrix}$$

EXAMPLE 39   Write a system of equations for the augmented matrix

$$\begin{bmatrix} 1 & 0 & 0 & 3 \\ 0 & 1 & 0 & -1 \\ 0 & 0 & 1 & 2 \end{bmatrix}$$

SOLUTION

$$1 \cdot x + 0 \cdot y + 0 \cdot z = 3 \qquad x = 3$$
$$0 \cdot x + 1 \cdot y + 0 \cdot z = -1 \quad \text{or} \quad y = -1$$
$$0 \cdot x + 0 \cdot y + 1 \cdot z = 2 \qquad z = 2$$

■

From the preceding example we note that if, with a set of row operations, we can transform a given augmented matrix to one like the matrix in Example 39 (identity matrix on left and then a column of numbers), we will have a solution for the equations in three unknowns.

EXAMPLE 40   Solve the system of equations

$$x + 2y - 3z = -6$$
$$2x - y + z = -1$$
$$3x + 2y + z = 4$$

SOLUTION   The augmented matrix for this system is

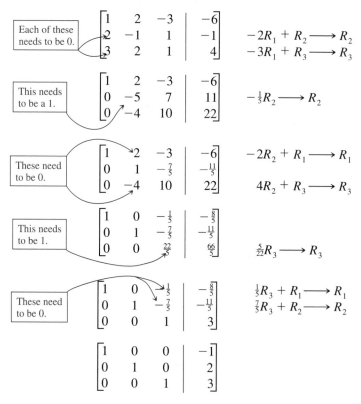

The solution is $(-1, 2, 3)$.

■

Although most of the examples we have worked involve relatively simple computations, in practice, most application problems involve matrices with several rows and columns and numbers with several digits. These problems are easily solved with a computer. Some can be solved with a calculator.

EXAMPLE 41   Using augmented matrices, solve

$$2x - y + z = -1$$
$$x + 2y - 3z = -7$$
$$3x + 2y + z = 7$$

SOLUTION   The augmented matrix is

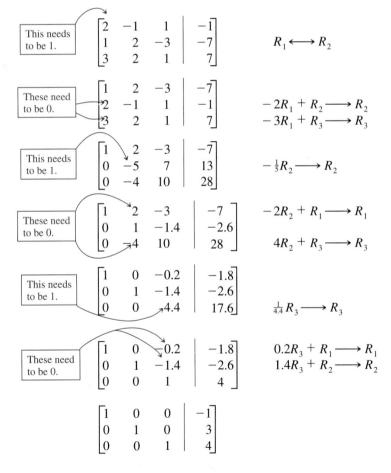

The solution is $x = -1$, $y = 3$, and $z = 4$.

Since linear equations in three variables represent planes in three-dimensional space, visualizing how three planes intersect can give you insight as to what type of solutions are possible with a system of planes (Table 4).

Sometimes it is not possible to reduce a matrix of coefficients to an identity matrix. Note that only one graph in Table 4 has a unique point on all three planes—(1). This case corresponds to reducing the left side of an augmented matrix to the identity. All other graphs correspond to cases where the matrix of coefficients does not reduce to the identity matrix.

TABLE 4

Description	Graph
1. The common intersection consists of a single point.	
2. The intersection is a line, and the system has infinitely many solutions.	
3. The three planes coincide, and the system has infinitely many solutions.	
4. The three planes are parallel and have no common intersection, and the solution set is the null set.	
5. Two planes are parallel, and thus the solution set is the null set.	
6. The planes intersect only in pairs of parallel lines, and the solution is the null set.	

To understand these six possibilities we need to discuss matrices in reduced form. The following matrices are reduced matrices:

$$\left[\begin{array}{cc|c} 1 & 0 & 4 \\ 0 & 1 & 2 \end{array}\right], \quad \left[\begin{array}{ccc|c} 1 & 0 & 0 & 5 \\ 0 & 1 & 0 & 2 \\ 0 & 0 & 1 & -1 \end{array}\right],$$

$$\left[\begin{array}{ccc|c} 1 & 0 & 3 & 2 \\ 0 & 1 & 2 & 4 \\ 0 & 0 & 0 & 1 \end{array}\right], \quad \left[\begin{array}{ccc|c} 1 & 0 & 3 & -2 \\ 0 & 1 & 1 & 4 \\ 0 & 0 & 0 & 0 \end{array}\right]$$

Check to see if they satisfy the given definition.

DEFINITION: REDUCED MATRIX

A matrix is in **reduced form** if:

1. The leftmost nonzero element in each row is 1.
2. The leftmost nonzero element in a row has all 0's above it and all 0's below it in its column.
3. The first nonzero element in each row is to the right of the first nonzero element in each row above it.
4. Rows containing all 0's are below the rows containing nonzero elements.

Suppose that the last equivalent matrix of a system is of the form

$$\left[\begin{array}{ccc|c} 1 & 0 & 2 & -3 \\ 0 & 1 & -2 & 4 \\ 0 & 0 & 0 & 0 \end{array}\right]$$

This matrix is in reduced form but does not give a unique solution:

$$1 \cdot x + 0 \cdot y + 2 \cdot z = -3$$

$$0 \cdot x + 1 \cdot y - 2 \cdot z = 4$$

$$0 \cdot x + 0 \cdot y + 0 \cdot z = 0$$

The value of z, which usually comes from the last equation, can be anything and the last equation is satisfied [Table 4: (2)]. Thus, we say the value of z is arbitrary; that is, we can assign any real number as the value of z. For example, let's assign $z = 3$. Then

$$1 \cdot x + 0 \cdot y + 6 = -3$$

$$0 \cdot x + 1 \cdot y - 6 = 4$$

$$0 \cdot x + 0 \cdot y + 0 \cdot 3 = 0$$

In the first equation $x = -9$, and in the second equation $y = 10$. One solution is $(-9, 10, 3)$.

Suppose that we let z be an arbitrary constant c. Then

$$\begin{array}{ll} y - 2c = 4 & y = 4 + 2c \\ \text{or} & \\ x + 2c = -3 & x = -3 - 2c \end{array}$$

Then x, y, and z are all expressed in terms of a constant c. The solution can be expressed as

$$(-3 - 2c, 4 + 2c, c)$$

for any real number c. When $c = 1$, the solution is $(-5, 6, 1)$. When $c = 2$, the solution is $(-7, 8, 2)$.

Practice Problem 1 Find the solution of the system of equations with variables x, y, and z for the augmented matrix

$$\begin{bmatrix} 1 & 0 & 2 & | & 4 \\ 0 & 1 & 3 & | & 5 \\ 0 & 0 & 0 & | & 0 \end{bmatrix}$$

ANSWER Let z equal the constant c. Then the solution can be written as $(4 - 2c, 5 - 3c, c)$.

EXAMPLE 42 Discuss the solution of the system of equations when the reduced augmented matrix is of the form

$$\begin{bmatrix} 1 & 0 & 0 & | & -4 \\ 0 & 1 & 2 & | & 2 \\ 0 & 0 & 0 & | & 1 \end{bmatrix}$$

SOLUTION The last equation is

$$0 \cdot x + 0 \cdot y + 0 \cdot z = 1$$

There is no triplet of numbers that will satisfy this equation, so no solution exists [Table 4: (4), (5), or (6)]. ■

When there are more variables than equations, the reduced matrix generally gives answers that must be expressed in terms of an arbitrary constant c. For example, the reduced matrix for the system

$$2x - y + z = 4$$
$$x + y + 2z = 5$$

is

$$\left[\begin{array}{cc|c} 1 & 0 & 1 \\ 0 & 1 & 1 \end{array}\begin{array}{c} 3 \\ 2 \end{array}\right]$$

So $x = 3 - z$ and $y = 2 - z$. Now let z be any constant c. Then $x = 3 - c$ and $y = 2 - c$. The solution is $(3 - c, 2 - c, c)$.

Practice Problem 2 Find the solution of the system of equations with augmented matrix

$$\left[\begin{array}{ccc|c} 1 & 0 & 2 & 3 \\ 0 & 1 & 3 & 2 \\ 0 & 0 & 0 & 1 \end{array}\right]$$

ANSWER No solution exists.

Practice Problem 3 Write a system of equations in terms of x, y, and z for the augmented matrix

$$\left[\begin{array}{ccc|c} 1 & 0 & 2 & 3 \\ 0 & 1 & 1 & 2 \end{array}\right]$$

and discuss values of x and y in the solution.

ANSWER

$$x + 2z = 3$$
$$y + z = 2$$

Assign z any value; say, $z = c$. Then x and y can be solved in terms of this constant. There are an infinite number of solutions of the form $(3 - 2c, 2 - c, c)$.

EXAMPLE 43 Solve the following system by operations on the corresponding augmented matrix:

$$2y + 3z - 7t = \quad 3$$
$$2x \qquad + z - 2t = -5$$
$$4x + y + z - 5t = -6$$
$$y + 2z - 4t = \quad 1$$

SOLUTION The augmented matrix is

$$\left[\begin{array}{cccc|c} 0 & 2 & 3 & -7 & 3 \\ 2 & 0 & 1 & -2 & -5 \\ 4 & 1 & 1 & -5 & -6 \\ 0 & 1 & 2 & -4 & 1 \end{array}\right]$$

The reduced matrix is

$$\left[\begin{array}{cccc|c} 1 & 0 & 0 & -\frac{1}{2} & -2 \\ 0 & 1 & 0 & -2 & 3 \\ 0 & 0 & 1 & -1 & -1 \\ 0 & 0 & 0 & 0 & 0 \end{array}\right]$$

Since the bottom row of the matrix consists entirely of 0's, the value of t is arbitrary. Let c be a number chosen in any fashion but left fixed for a moment; then let

$$t = c$$

Then, from the third equation,

$$z = c - 1$$

and from the second equation,

$$y = 2c + 3$$

and finally from the first equation,

$$x = \frac{1}{2}c - 2$$

For each value of c a solution is obtained, so there is an infinite number of solutions. Assigning values to c gives some of the solutions. For example, $c = 0$ gives, by substitution, $x = -2$, $y = 3$, $z = -1$, and $t = 0$, which can be expressed as $(-2, 3, -1, 0)$. Similarly, $c = 2$ gives $(-1, 7, 1, 2)$.   ■

SUMMARY

In this section we applied the Gauss–Jordan elimination method that we discussed in Section 2.4 to systems of equations with three or more variables. This method can be used for systems of equations that have the same number of equations as unknowns, more equations than unknowns, and fewer equations than unknowns. The row reduction options under the MATRX EDIT menu on a graphing calculator will help you avoid computational mistakes.

Exercise Set 2.5

Indicate whether each matrix is in reduced form.

1. $\begin{bmatrix} 1 & 0 & | & 3 \\ 0 & 1 & | & 0 \end{bmatrix}$

2. $\begin{bmatrix} 1 & 0 & | & 2 \\ 1 & 0 & | & 3 \end{bmatrix}$

3. $\begin{bmatrix} 1 & 0 & | & 3 \\ 0 & 1 & | & 2 \\ 0 & 0 & | & 3 \end{bmatrix}$

4. $\begin{bmatrix} 1 & 0 & 0 & | & 2 \\ 0 & 1 & 0 & | & 3 \\ 0 & 0 & 0 & | & 0 \end{bmatrix}$

5. $\begin{bmatrix} 1 & 0 & 2 & 3 & | & 7 \\ 0 & 1 & 1 & 4 & | & 8 \end{bmatrix}$

6. $\begin{bmatrix} 1 & 0 & 2 & | & 1 \\ 0 & 1 & 3 & | & 4 \end{bmatrix}$

7. $\begin{bmatrix} 1 & 0 & 0 & | & 2 \\ 0 & 1 & 0 & | & 3 \\ 0 & 1 & 1 & | & 1 \end{bmatrix}$

8. $\begin{bmatrix} 1 & 0 & | & 0 \\ 0 & 1 & | & 0 \\ 0 & 0 & | & 1 \end{bmatrix}$

The augmented matrices below are given in reduced form. Determine whether each system has a solution and find the solution or solutions if they exist. Use variables x, y, z, and possibly w.

9. $\begin{bmatrix} 1 & 0 & 0 & | & 3 \\ 0 & 1 & 0 & | & -1 \\ 0 & 0 & 1 & | & 4 \end{bmatrix}$

10. $\begin{bmatrix} 1 & 0 & 0 & | & 0 \\ 0 & 1 & 0 & | & 0 \\ 0 & 0 & 1 & | & 2 \end{bmatrix}$

11. $\begin{bmatrix} 1 & 0 & 2 & | & 3 \\ 0 & 1 & 1 & | & 1 \\ 0 & 0 & 0 & | & 0 \end{bmatrix}$

12. $\begin{bmatrix} 1 & 0 & 0 & | & 2 \\ 0 & 1 & 0 & | & -4 \\ 0 & 0 & 1 & | & 0 \end{bmatrix}$

13. $\begin{bmatrix} 1 & 0 & 0 & | & 6 \\ 0 & 1 & 0 & | & 1 \\ 0 & 0 & 0 & | & 4 \end{bmatrix}$

14. $\begin{bmatrix} 1 & 0 & 0 & | & 7 \\ 0 & 1 & 0 & | & 3 \\ 0 & 0 & 0 & | & 2 \end{bmatrix}$

15. $\begin{bmatrix} 1 & 0 & 0 & 0 & | & -2 \\ 0 & 1 & 0 & 0 & | & 4 \\ 0 & 0 & 1 & 0 & | & 6 \\ 0 & 0 & 0 & 1 & | & 1 \end{bmatrix}$

16. $\begin{bmatrix} 1 & 0 & 0 & 0 & | & 2 \\ 0 & 1 & 0 & 0 & | & 0 \\ 0 & 0 & 1 & 0 & | & 0 \\ 0 & 0 & 0 & 1 & | & 0 \end{bmatrix}$

17. $\begin{bmatrix} 1 & 0 & 0 & 0 & | & 3 \\ 0 & 1 & 0 & 0 & | & 4 \\ 0 & 0 & 1 & 0 & | & 5 \\ 0 & 0 & 0 & 0 & | & 2 \end{bmatrix}$

18. $\begin{bmatrix} 1 & 0 & 0 & 0 & | & 3 \\ 0 & 1 & 0 & 0 & | & 4 \\ 0 & 0 & 1 & 0 & | & 5 \\ 0 & 0 & 0 & 0 & | & 0 \end{bmatrix}$

Use augmented matrices to solve the following systems of three equations in three unknowns.

19. $\begin{aligned} x \quad\quad + z &= 1 \\ y - 2z &= 3 \\ 3x + y + z &= 6 \end{aligned}$

20. $\begin{aligned} x + 2y + z &= 8 \\ 2x - y + 3z &= 9 \\ x - 2y - z &= -6 \end{aligned}$

21. $\begin{aligned} z - 4y &= 3 \\ 2x - z &= 5 \\ x - 3y &= 1 \end{aligned}$

22. $\begin{aligned} x + 2y &= 3 \\ x - 2z &= 7 \\ 3y + z &= 9 \end{aligned}$

23. $\begin{aligned} 3x + y + z &= 3 \\ 5x + 2y - 3z &= 0 \\ x + 2y + 2z &= 1 \end{aligned}$

24. $\begin{aligned} 2x - y + z &= 3 \\ x + 2y - z &= 2 \\ 3x - y + z &= 4 \end{aligned}$

25. $\begin{aligned} x \quad\quad - z &= 1 \\ y + 2z &= 3 \\ 3x + y + z &= 6 \end{aligned}$

26. $\begin{aligned} x + 3y &= 10 \\ x - y - 4z &= -6 \\ 2x + 4y - 2z &= 12 \end{aligned}$

27. $\begin{aligned} x + 3y &= 10 \\ 2x + 6y &= 20 \\ 2x + 4y - 2z &= 12 \end{aligned}$

28. $\begin{aligned} x + 3y &= 10 \\ 2x + 6y &= 5 \\ 2x + 4y - 2z &= 12 \end{aligned}$

Solve the following systems in terms of an arbitrary parameter c. Let z = c.

29. $\begin{aligned} x - y + z &= 4 \\ 2x + y - 2z &= 6 \end{aligned}$

30. $\begin{aligned} 2x - y + 3z &= 5 \\ x + 2y - z &= 2 \end{aligned}$

31. Determine whether or not the following systems have solutions and solve those that have solutions.

(a) $\begin{aligned} x + y + z &= 5 \\ 2x - 3y + 2z &= 4 \\ 4x - y + 4z &= 10 \end{aligned}$

(b) $\begin{aligned} x - y - z &= -3 \\ 3x - 4y - 2z &= -11 \\ 5x - 6y - 4z &= -17 \end{aligned}$

(c) $\begin{aligned} x - 2y - 3z &= -3 \\ 7x - 14y - 21z &= -21 \\ 11x - 22y - 33z &= -33 \end{aligned}$

(d) $\begin{aligned} x + y + z &= 5 \\ 2x + 3y - 2z &= -19 \\ 3x + 2y + 3z &= 20 \end{aligned}$

Solve using Gauss-Jordan elimination.

32. $\begin{aligned} 4x_1 - 2x_2 &= 0 \\ 9x_1 + 6x_2 &= 2 \\ 3x_1 + 3x_2 &= -3 \end{aligned}$

33. $\begin{aligned} 4x_1 - 2x_2 &= 0 \\ 6x_1 + 4x_2 &= 6 \\ 3x_1 - 3x_2 &= -6 \end{aligned}$

34. $\begin{aligned} 2x_1 - 6x_2 &= 10 \\ x_1 + x_2 &= 3 \\ 2x_1 - 4x_2 &= 10 \end{aligned}$

35. $\begin{aligned} x_1 + x_2 &= 1 \\ 2x_1 + 4x_2 &= 6 \\ x_2 &= 2 \end{aligned}$

36. $x_1 - 3x_2 + 2x_3 = 2$
    $x_1 - 3x_2 + 4x_3 = 2$

37. $\quad x_1 + 2x_2 - x_3 = 1$
    $-3x_1 - 6x_2 + 3x_3 = -3$

38. **Exam.** How many solutions does the following system have?

$$x - y + z = 2$$
$$2x - y + 2z = 2$$
$$3x + 3y + 3z = 4$$

(a) 0
(b) 1
(c) 2
(d) 3
(e) ∞

39. Solve $\quad 4.2x + 6.3y + 8.4z = 10.5$
    $\quad 3.3x - 1.1y - 3.3z = 2.2$

40. Solve $\quad 0.3x + 1.2y + 5z = 1.5$
    $\quad 3.6x - 1.2y - 3z = 2.4$

Applications (Business and Economics)

41. **Manufacture.** A battery company manufactures four kinds of batteries, grades A, B, C, and D. Each battery must go through three process stations in manufacturing. Each A requires 10 minutes on station I, 5 minutes on station II, and 15 minutes on station III. Each B requires 15, 10, and 5 minutes on I, II, and III, respectively; each C requires 20, 15, and 5 minutes, respectively; and D requires 5, 10, and 15 minutes, respectively. Find the number of batteries of each kind that can be manufactured in a day if station I is available for 740 minutes, station II for 480 minutes, and station III for 660 minutes.

42. **Demand Curve.** The economist at a corporation has noted that the demand curve for a given product is in the form of a parabola, $p = ax^2 + bx + c$. The demand price for 2 units is $1; that is, $p = 1$ when $x = 2$; $p = 2$ when $x = 1$; and $p = \frac{1}{2}$ when $x = 3$. Substitute these values for x and p and solve the three linear equations for a, b, and c. Predict the demand price when x is 4.

Applications (Social and Life Sciences)

43. **Pollution.** The pollution count for Clanton on a particular day is 600. Assume that this pollution is produced by 3 industries: A, B, and C. Industry A contributes twice as much to the pollution count as industry B. It is known that the pollution count would

be 500 if the pollution count from industry A were reduced by 50%. Find the pollution count of industries A, B, and C, respectively.

44. **Traffic Flow.** The number of cars entering and leaving four intersections of one-way streets has been tabulated as shown in the diagram. The number

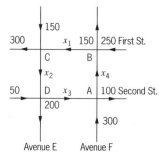

of cars entering an intersection must equal the number that leave. At B, $250 + x_4 = 150 + x_1$. Find the equations at C, D, and A. Set up and find a reduced augmented matrix. Note that you can solve for the three variables in terms of x_4. What is the smallest value that x_4 can attain to have meaning? What are the other variables when x_4 takes on this value?

45. **Diet.** A meal is planned around three foods to obtain 100% of carbohydrates, 100% of protein, and 100% of a given set of vitamins. Each small serving of food I has 20% of daily requirements of carbohydrates, 20% of protein, and 30% of vitamins. Each serving of food II contains 20% of required carbohydrates, 10% of protein, and no vitamins. Each serving of food III contains 20% of required carbohydrates, 50% of protein, and 20% of vitamins. How many servings of each would be necessary to get exact daily requirements?

Review Exercise

46. Solve the following systems of equations using graphs, the addition–subtraction method, and augmented matrices.

(a) $2x + y = 4$
    $3x - 2y = -1$

(b) $5x + 3y = -2$
    $4x - 3y = -7$

(c) $5x - 3y = 1$
    $4x + y = 11$

(d) $5x - 7y = -1$
    $2x - 3y = 0$

2.6   INVERSE OF A MATRIX AND ITS USE

OVERVIEW   In this section, we present a concept of matrix algebra that is not only useful as we discuss another method for solving simultaneous systems of linear equations, but is also essential in constructing and finding solutions of models of various application problems. For example, in the next section as we consider an economics model called the Leontief model, we need the inverse of a matrix. When the product of two square matrices is equal to the identity matrix, each matrix is said to be the inverse of the other. We study a procedure for finding the inverse of a square matrix, if the inverse exists. In this section we

- Define the inverse of a matrix
- Perform row operations to obtain the inverse
- Show applications involving the inverse of a matrix

Two 2×2 square matrices **A** and **B** are inverse matrices if

$$\mathbf{AB} = \mathbf{I}$$

EXAMPLE 44   The matrices

$$\mathbf{A} = \begin{bmatrix} 1 & -\frac{1}{2} \\ 0 & \frac{3}{2} \end{bmatrix} \quad \text{and} \quad \mathbf{B} = \begin{bmatrix} 1 & \frac{1}{3} \\ 0 & \frac{2}{3} \end{bmatrix}$$

are inverse matrices, since

$$\begin{bmatrix} 1 & -\frac{1}{2} \\ 0 & \frac{3}{2} \end{bmatrix} \begin{bmatrix} 1 & \frac{1}{3} \\ 0 & \frac{2}{3} \end{bmatrix} = \begin{bmatrix} 1 & 0 \\ 0 & 1 \end{bmatrix}$$

DEFINITION: INVERSE OF A MATRIX

If there exists a matrix $\mathbf{A}^{-1}$, such that $\mathbf{AA}^{-1} = \mathbf{A}^{-1}\mathbf{A} = \mathbf{I}$, then $\mathbf{A}^{-1}$ is called the **inverse** of the matrix **A**.

This definition states that for the inverse $\mathbf{A}^{-1}$ to exist, both $\mathbf{A}^{-1}\mathbf{A} = \mathbf{I}$ and $\mathbf{AA}^{-1} = \mathbf{I}$. However, for square matrices it can be proven that if $\mathbf{AA}^{-1} = \mathbf{I}$, then $\mathbf{A}^{-1}\mathbf{A} = \mathbf{I}$ and vice versa. Hence, to prove that a square matrix is the inverse of another square matrix, it is necessary to check only one of these conditions, not both.

Practice Problem 1   Show that $\mathbf{A}^{-1}\mathbf{A} = \mathbf{I}$ and $\mathbf{AA}^{-1} = \mathbf{I}$ for

$$\mathbf{A} = \begin{bmatrix} 1 & 3 \\ 2 & 4 \end{bmatrix} \quad \text{and} \quad \mathbf{A}^{-1} = \begin{bmatrix} -2 & \frac{3}{2} \\ 1 & -\frac{1}{2} \end{bmatrix}$$

ANSWER

$$\mathbf{A}^{-1}\mathbf{A} = \begin{bmatrix} -2 & \frac{3}{2} \\ 1 & -\frac{1}{2} \end{bmatrix} \begin{bmatrix} 1 & 3 \\ 2 & 4 \end{bmatrix} = \begin{bmatrix} 1 & 0 \\ 0 & 1 \end{bmatrix}$$

$$\mathbf{A}\mathbf{A}^{-1} = \begin{bmatrix} 1 & 3 \\ 2 & 4 \end{bmatrix} \begin{bmatrix} -2 & \frac{3}{2} \\ 1 & -\frac{1}{2} \end{bmatrix} = \begin{bmatrix} 1 & 0 \\ 0 & 1 \end{bmatrix}$$

It can be shown that not every matrix has an inverse. See if you can find the inverse of

$$\begin{bmatrix} -1 & 2 \\ -2 & 4 \end{bmatrix}$$

Let

$$\begin{bmatrix} a & c \\ b & d \end{bmatrix}$$

be such an inverse. Then

$$\begin{bmatrix} -1 & 2 \\ -2 & 4 \end{bmatrix} \begin{bmatrix} a & c \\ b & d \end{bmatrix} = \begin{bmatrix} 1 & 0 \\ 0 & 1 \end{bmatrix}$$

$$\begin{bmatrix} -a + 2b & -c + 2d \\ -2a + 4b & -2c + 4d \end{bmatrix} = \begin{bmatrix} 1 & 0 \\ 0 & 1 \end{bmatrix}$$

This is impossible. There are no numbers a and b such that

$$-a + 2b = 1$$

$$2(-a + 2b) = 0$$

That is, $-a + 2b$ cannot be both 0 and 1. Similarly, no numbers c and d exist to satisfy the matrix equation. Thus,

$$\begin{bmatrix} -1 & 2 \\ -2 & 4 \end{bmatrix}$$

has no inverse.

Practice Problem 2 Show that the inverse of

$$\begin{bmatrix} 3 & 2 & -2 \\ 1 & 2 & -1 \\ -1 & -1 & 1 \end{bmatrix} \quad \text{is} \quad \begin{bmatrix} 1 & 0 & 2 \\ 0 & 1 & 1 \\ 1 & 1 & 4 \end{bmatrix}$$

ANSWER

$$\begin{bmatrix} 3 & 2 & -2 \\ 1 & 2 & -1 \\ -1 & -1 & 1 \end{bmatrix} \begin{bmatrix} 1 & 0 & 2 \\ 0 & 1 & 1 \\ 1 & 1 & 4 \end{bmatrix} = \begin{bmatrix} 1 & 0 & 0 \\ 0 & 1 & 0 \\ 0 & 0 & 1 \end{bmatrix}$$

EXAMPLE 45 Find the inverse of the matrix

$$\begin{bmatrix} 2 & 3 \\ 1 & 2 \end{bmatrix}$$

SOLUTION To find the inverse of

$$\begin{bmatrix} 2 & 3 \\ 1 & 2 \end{bmatrix}$$

define the inverse to be

$$\begin{bmatrix} a & c \\ b & d \end{bmatrix}$$

$$\begin{bmatrix} 2 & 3 \\ 1 & 2 \end{bmatrix} \begin{bmatrix} a & c \\ b & d \end{bmatrix} = \begin{bmatrix} 1 & 0 \\ 0 & 1 \end{bmatrix} \qquad AA^{-1} = I$$

$$\begin{bmatrix} 2a + 3b & 2c + 3d \\ a + 2b & c + 2d \end{bmatrix} = \begin{bmatrix} 1 & 0 \\ 0 & 1 \end{bmatrix}$$

So

$$\begin{matrix} 2a + 3b = 1 \\ a + 2b = 0 \end{matrix} \quad \text{and} \quad \begin{matrix} 2c + 3d = 0 \\ c + 2d = 1 \end{matrix}$$

Solving these two systems of equations, we find $a = 2$, $b = -1$ and $c = -3$, $d = 2$. Thus, the inverse is

$$\begin{bmatrix} 2 & -3 \\ -1 & 2 \end{bmatrix}$$

✔ Check: $\begin{bmatrix} 2 & 3 \\ 1 & 2 \end{bmatrix} \begin{bmatrix} 2 & -3 \\ -1 & 2 \end{bmatrix} = \begin{bmatrix} 1 & 0 \\ 0 & 1 \end{bmatrix}$

In the preceding example the augmented matrices for the system of equations in a and b and the system in c and d are, respectively,

$$\begin{bmatrix} 2 & 3 & | & 1 \\ 1 & 2 & | & 0 \end{bmatrix} \quad \text{and} \quad \begin{bmatrix} 2 & 3 & | & 0 \\ 1 & 2 & | & 1 \end{bmatrix}$$

Since the coefficient matrices are the same, we can solve both systems at the same time using

$$\left[\begin{array}{cc|cc} 2 & 3 & 1 & 0 \\ 1 & 2 & 0 & 1 \end{array}\right]$$

Thus, the inverse of a matrix may be found by the following steps.

Steps for Finding an Inverse of a Matrix

1. Write the identity matrix (of same size as **A**) adjacent to the matrix **A** to form an augmented matrix:

$$\left[\begin{array}{ccc|ccc} a_{11} & a_{12} & a_{13} & 1 & 0 & 0 \\ a_{21} & a_{22} & a_{23} & 0 & 1 & 0 \\ a_{31} & a_{32} & a_{33} & 0 & 0 & 1 \end{array}\right] \qquad \mathbf{A} = \left[\begin{array}{ccc} a_{11} & a_{12} & a_{13} \\ a_{21} & a_{22} & a_{23} \\ a_{31} & a_{32} & a_{33} \end{array}\right]$$

2. Perform permissible row operations on this augmented matrix until the matrix that was **A** is reduced to the identity matrix:

$$\left[\begin{array}{ccc|ccc} 1 & 0 & 0 & b_{11} & b_{12} & b_{13} \\ 0 & 1 & 0 & b_{21} & b_{22} & b_{23} \\ 0 & 0 & 1 & b_{31} & b_{32} & b_{33} \end{array}\right]$$

3. The matrix now in the position of the original identity matrix is the inverse of **A**:

$$\mathbf{A}^{-1} = \left[\begin{array}{ccc} b_{11} & b_{12} & b_{13} \\ b_{21} & b_{22} & b_{23} \\ b_{31} & b_{32} & b_{33} \end{array}\right]$$

This procedure is illustrated by the following examples.

EXAMPLE 46   Find the inverse of

$$\left[\begin{array}{cc} 5 & -3 \\ 2 & -1 \end{array}\right]$$

SOLUTION   Place the 2×2 identity matrix next to this original matrix to form the new 2×4 augmented matrix

Need a 1 here
$$\left[\begin{array}{cc|cc} 5 & -3 & 1 & 0 \\ 2 & -1 & 0 & 1 \end{array}\right] \qquad \tfrac{1}{5}R_1 \longrightarrow R_1$$

Need a 0 here
$$\left[\begin{array}{cc|cc} 1 & -\tfrac{3}{5} & \tfrac{1}{5} & 0 \\ 2 & -1 & 0 & 1 \end{array}\right] \qquad -2R_1 + R_2 \longrightarrow R_2$$

Need a 1 here
$$\left[\begin{array}{cc|cc} 1 & -\tfrac{3}{5} & \tfrac{1}{5} & 0 \\ 0 & \tfrac{1}{5} & -\tfrac{2}{5} & 1 \end{array}\right] \qquad 5R_2 \longrightarrow R_2$$

$$\begin{array}{c}\text{Need a}\\\text{0 here}\end{array} \begin{bmatrix} 1 & -\frac{3}{5} \\ 0 & 1 \end{bmatrix} \begin{array}{|cc} \frac{1}{5} & 0 \\ -2 & 5 \end{array} \qquad \tfrac{3}{5}R_2 + R_1 \longrightarrow R_1$$

$$\begin{bmatrix} 1 & 0 \\ 0 & 1 \end{bmatrix} \begin{array}{|cc} -1 & 3 \\ -2 & 5 \end{array}$$

Thus, the inverse of

$$\begin{bmatrix} 5 & -3 \\ 2 & -1 \end{bmatrix} \quad \text{is} \quad \begin{bmatrix} -1 & 3 \\ -2 & 5 \end{bmatrix}$$

which can be proven by noting that

$$\begin{bmatrix} 5 & -3 \\ 2 & -1 \end{bmatrix}\begin{bmatrix} -1 & 3 \\ -2 & 5 \end{bmatrix} = \begin{bmatrix} 1 & 0 \\ 0 & 1 \end{bmatrix} \qquad \blacksquare$$

Practice Problem 3 Find the inverse of

$$\begin{bmatrix} 2 & 1 \\ 0 & -\frac{1}{2} \end{bmatrix}$$

ANSWER

$$\begin{bmatrix} \frac{1}{2} & 1 \\ 0 & -2 \end{bmatrix}$$

EXAMPLE 47   Find the inverse of

$$\begin{bmatrix} 1 & 2 & 1 \\ -1 & 1 & 1 \\ 2 & 3 & 1 \end{bmatrix}$$

SOLUTION   Write the third-order identity adjacent to the given matrix to give the new 3 × 6 matrix.

$$\begin{array}{c}\text{Need 0's}\\\text{here}\end{array} \begin{bmatrix} 1 & 2 & 1 \\ -1 & 1 & 1 \\ 2 & 3 & 1 \end{bmatrix} \begin{array}{|ccc} 1 & 0 & 0 \\ 0 & 1 & 0 \\ 0 & 0 & 1 \end{array} \quad \begin{array}{c} R_1 + R_2 \longrightarrow R_2 \\ -2R_1 + R_3 \longrightarrow R_3 \end{array}$$

$$\begin{array}{c}\text{Need a}\\\text{1 here}\end{array} \begin{bmatrix} 1 & 2 & 1 \\ 0 & 3 & 2 \\ 0 & -1 & -1 \end{bmatrix} \begin{array}{|ccc} 1 & 0 & 0 \\ 1 & 1 & 0 \\ -2 & 0 & 1 \end{array} \quad \tfrac{1}{3}R_2 \longrightarrow R_2$$

$$\begin{array}{c}\text{Need 0's}\\\text{here}\end{array} \begin{bmatrix} 1 & 2 & 1 \\ 0 & 1 & \frac{2}{3} \\ 0 & 1 & -1 \end{bmatrix} \begin{array}{|ccc} 1 & 0 & 0 \\ \frac{1}{3} & \frac{1}{3} & 0 \\ -2 & 0 & 1 \end{array} \quad \begin{array}{c} -2R_2 + R_1 \longrightarrow R_1 \\ \\ R_2 + R_3 \longrightarrow R_3 \end{array}$$

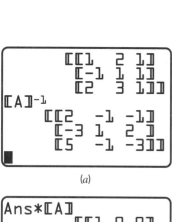

(a)

(b)

Figure 21

Need a 1 here

$$\begin{bmatrix} 1 & 0 & -\frac{1}{3} \\ 0 & 1 & \frac{2}{3} \\ 0 & 0 & -\frac{1}{3} \end{bmatrix} \begin{array}{|ccc} \frac{1}{3} & -\frac{2}{3} & 0 \\ \frac{1}{3} & \frac{1}{3} & 0 \\ -\frac{5}{3} & \frac{1}{3} & 1 \end{array} \qquad -3R_3 \longrightarrow R_3$$

Need 0's here

$$\begin{bmatrix} 1 & 0 & -\frac{1}{3} \\ 0 & 1 & \frac{2}{3} \\ 0 & 0 & 1 \end{bmatrix} \begin{array}{|ccc} \frac{1}{3} & -\frac{2}{3} & 0 \\ \frac{1}{3} & \frac{1}{3} & 0 \\ 5 & -1 & -3 \end{array} \qquad \begin{array}{l} \frac{1}{3}R_3 + R_1 \longrightarrow R_1 \\ -\frac{2}{3}R_3 + R_2 \longrightarrow R_2 \end{array}$$

$$\begin{bmatrix} 1 & 0 & 0 \\ 0 & 1 & 0 \\ 0 & 0 & 1 \end{bmatrix} \begin{array}{|ccc} 2 & -1 & -1 \\ -3 & 1 & 2 \\ 5 & -1 & -3 \end{array}$$

Hence, the inverse of

$$\begin{bmatrix} 1 & 2 & 1 \\ -1 & 1 & 1 \\ 2 & 3 & 1 \end{bmatrix} \text{ is } \begin{bmatrix} 2 & -1 & -1 \\ -3 & 1 & 2 \\ 5 & -1 & -3 \end{bmatrix}$$

This fact can be proven by multiplying:

$$\begin{bmatrix} 1 & 2 & 1 \\ -1 & 1 & 1 \\ 2 & 3 & 1 \end{bmatrix} \begin{bmatrix} 2 & -1 & -1 \\ -3 & 1 & 2 \\ 5 & -1 & -3 \end{bmatrix} = \begin{bmatrix} 1 & 0 & 0 \\ 0 & 1 & 0 \\ 0 & 0 & 1 \end{bmatrix}$$

As indicated earlier, many square matrices do not have inverses. The procedure for finding the inverse indicates this fact; that is, sometimes it is impossible to reduce the matrix **A** to the identity matrix by permissible row operations.

Calculator Note

To find the inverse of the matrix in Example 47 using a calculator, we use $\boxed{x^{-1}}$. Store the matrix in [A]. Then press $\boxed{\text{MATRX}}$ (select [A]) $\boxed{x^{-1}}$. [A]$^{-1}$ is shown in Figure 21(a). To prove it is in fact the inverse, multiply [A]$^{-1}$ [A] by pressing $\boxed{\text{2nd}}$ $\boxed{\text{ANS}}$ $\boxed{\times}$ $\boxed{\text{MATRX}}$ (select [A]) $\boxed{\text{ENTER}}$. The product of [A]$^{-1}$ [A] is equal to the 3 × 3 identity in Figure 21(b).

EXAMPLE 48   Show that the matrix

$$\begin{bmatrix} 1 & 2 \\ 3 & 6 \end{bmatrix}$$

does not have an inverse.

SOLUTION   If the procedure as outlined for determining the inverse is applied to this matrix, we have

$$\left[\begin{array}{cc|cc} 1 & 2 & 1 & 0 \\ 3 & 6 & 0 & 1 \end{array}\right] \quad -3R_1 + R_2 \longrightarrow R_2$$

Multiply the first row by -3 and add the result to the second row to obtain

$$\left[\begin{array}{cc|cc} 1 & 2 & 1 & 0 \\ 0 & 0 & -3 & 1 \end{array}\right]$$

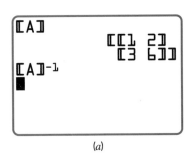

Since it is impossible to reduce the matrix

$$\left[\begin{array}{cc} 1 & 2 \\ 0 & 0 \end{array}\right] \quad \text{to} \quad \left[\begin{array}{cc} 1 & 0 \\ 0 & 1 \end{array}\right]$$

by permissible row operations, the procedure fails and the inverse does not exist. ■

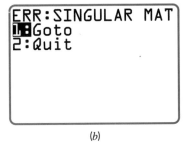

(a)

(b)

Figure 22

Calculator Note

What happens when a matrix that you entered into the calculator does not have an inverse? The calculator indicates that the matrix does not have an inverse as shown in Figure 22(b).

If the inverse of the coefficient matrix of a system of equations exists, it can be used to solve the system. For example, the system

$$5x - 3y = 29$$
$$2x - y = 11$$

can be written as

$$\left[\begin{array}{cc} 5 & -3 \\ 2 & -1 \end{array}\right] \left[\begin{array}{c} x \\ y \end{array}\right] = \left[\begin{array}{c} 29 \\ 11 \end{array}\right]$$

or $\mathbf{AX} = \mathbf{B}$, where $\mathbf{A}$ is the coefficient matrix

$$\left[\begin{array}{cc} 5 & -3 \\ 2 & -1 \end{array}\right]$$

and

$$\mathbf{X} = \left[\begin{array}{c} x \\ y \end{array}\right] \quad \text{and} \quad \mathbf{B} = \left[\begin{array}{c} 29 \\ 11 \end{array}\right]$$

The inverse of

$$\mathbf{A} = \left[\begin{array}{cc} 5 & -3 \\ 2 & -1 \end{array}\right]$$

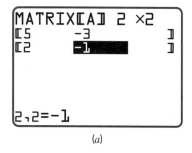

(a)

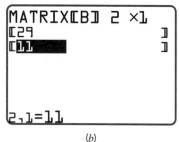

(b)

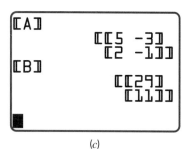

(c)

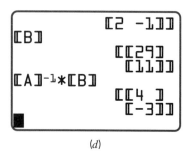

(d)

Figure 23

has been found to be

$$A^{-1} = \begin{bmatrix} -1 & 3 \\ -2 & 5 \end{bmatrix}$$

Multiplying both sides of the equation $AX = B$ on the left by A^{-1} yields

$$A^{-1}(AX) = A^{-1}B$$

$$(A^{-1}A)X = A^{-1}B \qquad A^{-1}A = I$$

$$IX = A^{-1}B \qquad IX = X$$

$$X = A^{-1}B$$

or

$$\begin{bmatrix} -1 & 3 \\ -2 & 5 \end{bmatrix} \begin{bmatrix} 5 & -3 \\ 2 & -1 \end{bmatrix} \begin{bmatrix} x \\ y \end{bmatrix} = \begin{bmatrix} -1 & 3 \\ -2 & 5 \end{bmatrix} \begin{bmatrix} 29 \\ 11 \end{bmatrix}$$

$$\begin{bmatrix} 1 & 0 \\ 0 & 1 \end{bmatrix} \begin{bmatrix} x \\ y \end{bmatrix} = \begin{bmatrix} 4 \\ -3 \end{bmatrix}$$

$$\begin{bmatrix} x \\ y \end{bmatrix} = \begin{bmatrix} 4 \\ -3 \end{bmatrix}$$

Hence, $x = 4$ and $y = -3$, and the solution is $(4, -3)$. This example suggests the following theorem.

SOLUTION BY INVERSE

If the coefficient matrix A of the system has an inverse A^{-1}, the solution X of the matrix equation $AX = B$ can be found as $X = A^{-1}B$.

Calculator Note

To find the solution using the inverse of the coefficient matrix, if it exists, input the coefficients in matrix A, the constants in matrix B, and then use the equation $A^{-1}B$ (see Figure 23).

Practice Problem 4 Find the solution of the system

$$2x + 3y = 7$$
$$3x + y = 7$$

by finding the inverse of the coefficient matrix.

ANSWER   The solution is $(2, 1)$.

EXAMPLE 49   Find the solution of the system

$$x + 2y + z = \quad 0$$
$$-x + \quad y + z = -4$$
$$2x + 3y + z = \quad 1$$

SOLUTION   In matrix notation this system becomes

$$\begin{bmatrix} 1 & 2 & 1 \\ -1 & 1 & 1 \\ 2 & 3 & 1 \end{bmatrix} \begin{bmatrix} x \\ y \\ z \end{bmatrix} = \begin{bmatrix} 0 \\ -4 \\ 1 \end{bmatrix} \qquad \mathbf{AX = B}$$

The inverse of

$$\mathbf{A} = \begin{bmatrix} 1 & 2 & 1 \\ -1 & 1 & 1 \\ 2 & 3 & 1 \end{bmatrix}$$

was shown to be

$$\mathbf{A}^{-1} = \begin{bmatrix} 2 & -1 & -1 \\ -3 & 1 & 2 \\ 5 & -1 & -3 \end{bmatrix}$$

Hence,

$$\begin{bmatrix} x \\ y \\ z \end{bmatrix} = \overset{\mathbf{A}^{-1}}{\begin{bmatrix} 2 & -1 & -1 \\ -3 & 1 & 2 \\ 5 & -1 & -3 \end{bmatrix}} \overset{\mathbf{B}}{\begin{bmatrix} 0 \\ -4 \\ 1 \end{bmatrix}} \quad \text{or} \quad \begin{bmatrix} x \\ y \\ z \end{bmatrix} = \begin{bmatrix} 3 \\ -2 \\ 1 \end{bmatrix} \qquad \mathbf{X = A^{-1}B}$$

Thus, the solution is $(3, -2, 1)$. ∎

Practice Problem 5 Solve

$$4.2x + 2.1y = 12.6$$
$$12.6x - 4.2y = 10.5$$

ANSWER

$$\begin{bmatrix} 4.2 & 2.1 \\ 12.6 & -4.2 \end{bmatrix} \begin{bmatrix} x \\ y \end{bmatrix} = \begin{bmatrix} 12.6 \\ 10.5 \end{bmatrix} \quad \text{and}$$

$$\begin{bmatrix} x \\ y \end{bmatrix} = \begin{bmatrix} 0.09524 & 0.04762 \\ 0.28571 & -0.09524 \end{bmatrix} \begin{bmatrix} 12.6 \\ 10.5 \end{bmatrix}$$

$$= \begin{bmatrix} 1.7 \\ 2.6 \end{bmatrix}$$

SUMMARY

In this section we introduced another way of solving a system of equations. The system must have the same number of equations and unknowns. The matrix equation $X = A^{-1}B$ is used to solve the system of equations. In this method, we must find the inverse of the coefficient matrix, if it exists, and then multiply it by the column matrix of constants.

Exercise Set 2.6

1. Show that the following matrices are inverses of each other.

(a) $\begin{bmatrix} 1 & 2 \\ 1 & 3 \end{bmatrix} \begin{bmatrix} 3 & -2 \\ -1 & 1 \end{bmatrix}$

(b) $\begin{bmatrix} 4 & -6 \\ 2 & 2 \end{bmatrix} \begin{bmatrix} \frac{1}{10} & \frac{3}{10} \\ -\frac{1}{10} & \frac{2}{10} \end{bmatrix}$

(c) $\begin{bmatrix} 5 & 7 \\ 3 & 4 \end{bmatrix} \begin{bmatrix} -4 & 7 \\ 3 & -5 \end{bmatrix}$

(d) $\begin{bmatrix} 2 & 0 & 2 \\ 4 & 2 & 0 \\ 2 & -2 & 2 \end{bmatrix} \begin{bmatrix} -\frac{1}{4} & \frac{1}{4} & \frac{1}{4} \\ \frac{1}{2} & 0 & -\frac{1}{2} \\ \frac{3}{4} & -\frac{1}{4} & -\frac{1}{4} \end{bmatrix}$

2. Find the inverses of the following matrices if they exist.

(a) $\begin{bmatrix} 5 & -3 \\ 2 & 3 \end{bmatrix}$ (b) $\begin{bmatrix} 3 & 0 \\ 2 & 0 \end{bmatrix}$

(c) $\begin{bmatrix} 5 & 19 \\ 1 & 4 \end{bmatrix}$ (d) $\begin{bmatrix} 3 & 3 \\ 3 & 3 \end{bmatrix}$

Find the systems of equations represented by each of the following.

3. $\begin{bmatrix} 2 & -1 \\ 3 & 4 \end{bmatrix} \begin{bmatrix} x \\ y \end{bmatrix} = \begin{bmatrix} 4 \\ 1 \end{bmatrix}$

4. $\begin{bmatrix} 1 & -1 \\ 2 & 1 \end{bmatrix} \begin{bmatrix} x_1 \\ x_2 \end{bmatrix} = \begin{bmatrix} -1 \\ 3 \end{bmatrix}$

5. $\begin{bmatrix} 1 & 3 & 2 \\ -1 & 2 & 1 \\ 0 & 1 & 2 \end{bmatrix} \begin{bmatrix} x \\ y \\ z \end{bmatrix} = \begin{bmatrix} 5 \\ 1 \\ 4 \end{bmatrix}$

6. $\begin{bmatrix} 1 & 0 & 2 \\ 0 & 2 & 5 \\ 1 & 0 & 6 \end{bmatrix} \begin{bmatrix} x_1 \\ x_2 \\ x_3 \end{bmatrix} = \begin{bmatrix} 5 \\ 6 \\ 8 \end{bmatrix}$

Write each of the following systems as a matrix equation in the form $AX = B$.

7. $5x + 3y = 7$
   $4x - y = 3$

8. $3x_1 + 5x_2 = 9$
   $8x_1 - 2x_2 = 3$

9. $3x + 4y + 5z = 9$
   $2x - y = 4$
   $7x + 2z = 5$

10. $2x_1 + 4x_2 = 7$
    $2x_2 + 5x_3 = 4$
    $5x_1 + 2x_3 = 5$

11. Solve the following systems by finding the inverse of the coefficient matrix and then using the theory of this section.

(a) $x + y = 2$
    $2x + 3y = 2$

(b) $x + y = 1$
    $3x + y = 7$

(c) $x - y = 1$
    $x + y = 5$

(d) $2x + y = 3$
    $x + 3y = 4$

12. Find the inverses of the following matrices if they exist.

(a) $\begin{bmatrix} 0 & 0 & 1 \\ 0 & 1 & 0 \\ 1 & 0 & 0 \end{bmatrix}$ (b) $\begin{bmatrix} 1 & 0 & 1 \\ 0 & 1 & 0 \\ 1 & 0 & 0 \end{bmatrix}$

(c) $\begin{bmatrix} 0 & 4 & 0 \\ 1 & 0 & 0 \\ 0 & -1 & 1 \end{bmatrix}$ (d) $\begin{bmatrix} 1 & -1 & 1 \\ 1 & 1 & 1 \\ 1 & 1 & 1 \end{bmatrix}$

(e) $\begin{bmatrix} 1 & -1 & 2 \\ 3 & 1 & 0 \\ 2 & 3 & 1 \end{bmatrix}$ (f) $\begin{bmatrix} 1 & 0 & 2 \\ 1 & 3 & 1 \\ 0 & -2 & 1 \end{bmatrix}$

13. Solve the following systems by finding the inverse of the coefficient matrix and then using the theory of this section.

(a) $x - y + 3z = 8$
    $2x + y + 2z = 6$
    $x + 2y + z = 0$

(b) $x + y + z = 5$
    $x - y + z = 7$
    $2x + y + z = 9$

(c) $-2x - 3y + z = 3$
    $x + 2y + z = 1$
    $-x - y + 3z = 6$

(d) $x + y + z = 2$
    $x - y + z = 6$
    $-x + y + z = -4$

14. Show that the following matrices do not have inverses.

(a) $\begin{bmatrix} 2 & 1 \\ 6 & 3 \end{bmatrix}$ (b) $\begin{bmatrix} 6 & -3 \\ 4 & -2 \end{bmatrix}$

(c) $\begin{bmatrix} 2 & 4 & -2 \\ 3 & -1 & 0 \\ 5 & 3 & -2 \end{bmatrix}$

Use the inverse key of your graphing calculator to solve the following systems of equations.

15.
$$2x_1 - x_2 + x_3 - 2x_4 = 3$$
$$3x_1 + x_2 - x_3 + 5x_4 = -8$$
$$4x_1 - 2x_2 + x_3 - x_4 = 4$$
$$6x_1 + 4x_2 - 5x_3 + x_4 = 4$$

16.
$$2x_1 + 4x_2 + x_3 - 2x_4 = 2$$
$$3x_1 + 5x_2 - x_3 - x_4 = -2$$
$$7x_1 + 3x_2 - 2x_3 + x_4 = 4$$
$$2x_1 + 7x_2 - 3x_3 + 3x_4 = 1$$

17. **Exam.** What is the multiplicative inverse of the following matrix?

$$\begin{bmatrix} 1 & 0 \\ 1 & 1 \end{bmatrix}$$

(a) $\begin{bmatrix} 1 & 0 \\ -1 & 1 \end{bmatrix}$ (b) $\begin{bmatrix} 1 & 0 \\ 1 & -1 \end{bmatrix}$

(c) $\begin{bmatrix} 0 & 1 \\ 1 & 1 \end{bmatrix}$ (d) $\begin{bmatrix} 1 & 1 \\ 1 & 0 \end{bmatrix}$ (e) $\begin{bmatrix} 1 & 1 \\ 0 & 1 \end{bmatrix}$

18. **Exam.** What is the inverse of the following matrix?

$$\begin{bmatrix} 1 & 1 & 1 \\ 0 & 1 & 1 \\ 0 & 0 & 1 \end{bmatrix}$$

(a) $\begin{bmatrix} 1 & 0 & 0 \\ 1 & 1 & 0 \\ 1 & 1 & 1 \end{bmatrix}$ (b) $\begin{bmatrix} -1 & 0 & 0 \\ -1 & 1 & 0 \\ 1 & -1 & 1 \end{bmatrix}$

(c) $\begin{bmatrix} 1 & -1 & 1 \\ 0 & 1 & -1 \\ 0 & 0 & 1 \end{bmatrix}$ (d) $\begin{bmatrix} 1 & -1 & 0 \\ 0 & 1 & -1 \\ 0 & 0 & 1 \end{bmatrix}$

(e) $\begin{bmatrix} 1 & -1 & 0 \\ 0 & 1 & 1 \\ 0 & 0 & 1 \end{bmatrix}$

Applications (Business and Economics)

19. **Investments.** An investment club has $100,000 invested in bonds. Type A bonds pay 8% interest, whereas type B bonds pay 10%. How much money is invested in each type if the club receives $9360 in interest?

20. **Break-Even Analysis.** A country club has been studying its cost of operation and has found the monthly cost to be $C = \$6000 + \$15x$, where x represents the number of members. A service company states that by remodeling, it can provide a similar service for a monthly cost of $C = \$10,500 + \$10x$. Use the inverse of a matrix to find the break-even point, and find the number of members necessary to undertake the remodeling.

21. **Manufacture.** A company makes two kinds of tires. Type I requires 2 hours on machine A and 1 hour on machine B. Type II requires 3 hours on A and 2 on B. How many tires can be manufactured in a week if machine A is in use 40 hours and machine B 25 hours?

Applications (Social and Life Sciences)

22. **Herbicides.** Three herbicides are available to kill weeds, grass, and vines. One gallon of one herbicide contains 1 unit of chemical A, which kills weeds, and 2 units of chemical B, which kills grass. One gallon of the second herbicide contains 2 units of chemical B and 3 units of chemical C, which kills vines. One gallon of the third herbicide contains 3 units of chemical A and 4 units of chemical C. It is desired to spread 10 units of chemical A, 16 units of chemical B, and 20 units of chemical C. How many gallons of each herbicide should be purchased?

23. **Diet.** A dietician wants to combine two foods as the main part of a meal to get 2600 calories and 70 units of vitamin C. Each ounce of food I contains 200 calories and 10 units of vitamin C. Each ounce of food II contains 400 calories and 5 units of vitamin C. How many ounces of each food would be required to obtain the desired number of calories and the desired units of vitamin C?

24. **Polling.** A company has been engaged to make 4850 phone polls and 2650 home visits. The company has two teams of pollsters. Team I can make 60 phone polls and 40 home polls a day, whereas team II can make 70 phone polls and 30 home polls each day. How many days should each team be scheduled in order to complete the engagement?

Review Exercises

25. Discuss the solutions of the following.

(a) $\left[\begin{array}{cc|c} 1 & 0 & 3 \\ 0 & 0 & 2 \end{array}\right]$ (b) $\left[\begin{array}{cc|c} 1 & 0 & 0 \\ 0 & 1 & 1 \end{array}\right]$

(c) $\begin{bmatrix} 1 & 0 & 3 \\ 0 & 0 & 0 \end{bmatrix}$

(d) $\begin{bmatrix} 1 & 0 & 0 & 2 \\ 0 & 1 & 0 & 3 \\ 0 & 0 & 0 & 0 \end{bmatrix}$

(e) $\begin{bmatrix} 1 & 0 & 0 & 4 \\ 0 & 1 & 0 & 2 \\ 0 & 0 & 0 & 3 \end{bmatrix}$

(f) $\begin{bmatrix} 1 & 0 & 0 & 1 \\ 0 & 1 & 0 & 0 \\ 0 & 0 & 1 & 0 \end{bmatrix}$

(b)
$$x - y - z = -3$$
$$3x - 4y - 2z = -11$$
$$5x - 6y - 4z = -17$$

(c)
$$x - 2y - 3z = -3$$
$$7x - 14y - 21z = -21$$
$$11x - 22y - 33z = -33$$

(d)
$$x + y + z = 5$$
$$2x + 3y - 2z = -19$$
$$3x + 2y + 3z = 20$$

26. Determine whether or not the following systems have common solutions, and solve those that have solutions.

(a)
$$x + y + z = 5$$
$$2x - 3y + 2z = 4$$
$$4x - y + 4z = 10$$

2.7   INPUT–OUTPUT ANALYSIS (Optional)

OVERVIEW   In recent years, matrix arithmetic has played a significant role in economic theories, especially in the branch of economics called **input–output analysis**. The first significant work in this field of economics was done by the famous economist Wassily Leontief, who was awarded the Nobel prize in economics in 1973 for his use of input–output analysis to study how much output must be produced by each segment of an economy in order to meet consumption and export demands. His study involved 500 sectors of the American economy interacting together. This analysis needed matrix calculations and, in particular, inverses of matrices. In this section we

- Define Leontief matrices
- Distinguish between open and closed systems
- Solve input–output problems

Suppose that we divide an economy into a number of industries—aluminum, steel, transportation, and so on. Each industry produces a certain output using raw materials (input). Of course, the input of some industries is the output of others (automobile manufacturers, for example, use steel as input). This interdependence among the industries is recorded in a matrix, called the **input–output matrix**. Consider the following simplified example involving services, manufacturing, and farming.

| | | Output | | |
		Services	Manufacturing	Farming
Input	Services	0.3	0.2	0.3
	Manufacturing	0.5	0.2	0.1
	Farming	0.1	0.3	0.1

The first column (read from top to bottom) indicates that the production of 1 unit of services consumes 0.3 unit of services, 0.5 unit of manufacturing, and 0.1

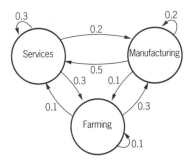

Figure 24

unit of farming. The second column indicates that the production of 1 unit of manufacturing consumes 0.2 unit of services, 0.2 unit of manufacturing, and 0.3 unit of farming. The third column indicates that the production of 1 unit of farming consumes 0.3 unit of services, 0.1 unit of manufacturing, and 0.1 unit of farming. This is further illustrated in Figure 24.

Given the internal demands for each other's outputs, the theory of input–output analysis attempts to establish conditions in an economy where the output will satisfy exactly each sector's demands as well as outside demands.

Suppose that there are n industries, each of which produces a single item. Each item can be used as an input to any industry. a_{12} is the fractional amount of item 1 needed by industry 2 to produce one unit of item 2. a_{23} is the fractional amount of item 2 needed by industry 3 to produce item 3. In general, a_{ij} is the fractional amount of the ith item required as input by industry j to produce one unit of the jth item as output. Written as a matrix, this ecomes

$$
\begin{array}{c}
\text{Output Item} \\
\begin{array}{cc}
& \begin{array}{ccccc} \text{Industry} & 1 & 2 & \cdots & n \end{array} \\
\begin{array}{c} \\ \\ \text{Input} \\ \text{Requirements} \\ \\ \\ \end{array}
\begin{array}{c} 1 \\ 2 \\ . \\ . \\ . \\ n \end{array}
& \left[
\begin{array}{ccccc}
a_{11} & a_{12} & \cdots & a_{1n} \\
a_{21} & a_{22} & \cdots & a_{2n} \\
. & . & . & . \\
. & . & . & . \\
. & . & . & . \\
a_{n1} & a_{n2} & \cdots & a_{nn}
\end{array}
\right]
\end{array}
\end{array}
$$

For example, in the preceding matrix, a_{53} can be thought of as the fractional part of items produced by industry 5 and used by industry 3 to produce one unit of industry 3.

The simplest input–output model is called a **closed system**, and it occurs when the output of each industry is used as input by the other industries. In a closed system the output exactly equals the input, since all outputs are used within the system. The sum of the elements in each column is 1. To introduce a closed system, consider the following, oversimplified example.

EXAMPLE 50   Consider a city that has only a farmer (to provide food) and a tailor (to provide clothes). Suppose that the farmer uses half of the food he grows. Assume that the tailor uses two-thirds of the clothes he produces. Write these requirements as a Leontief input–output matrix.

SOLUTION   The input–output matrix (called a Leontief matrix) for this closed system is

$$
\begin{array}{c}
\hspace{4cm}\text{Outputs} \\
\hspace{3cm}
\begin{array}{cc}
\text{Food} & \text{Clothes} \\
\text{produced} & \text{produced}
\end{array} \\
\begin{array}{cc}
\begin{array}{c} \\ \text{Input}\quad \text{Farmer} \\ \text{Requirements}\quad \text{Tailor} \end{array}
& \left[
\begin{array}{cc}
\frac{1}{2} & \frac{1}{3} \\
\frac{1}{2} & \frac{2}{3}
\end{array}
\right]
\end{array}
\end{array}
$$

This matrix indicates that in order for the farmer to produce food, he requires one-half of the food produced. If 1 unit of food is produced, $a_{11} = \frac{1}{2}$ and $a_{21} = 1 - \frac{1}{2} = \frac{1}{2}$. Similarly, in order for the tailor to produce clothes, he requires two-thirds of the clothes produced. If 1 unit of clothes is produced, $a_{22} = \frac{2}{3}$ and $a_{12} = 1 - \frac{2}{3} = \frac{1}{3}$. Let x_1 represent the number of units of food produced and x_2 represent the number of units of clothes produced. The farmer, who is responsible for x_1 units of food, needs $\frac{1}{2}x_1 + \frac{1}{3}x_2$ to produce x_1 units of food. The tailor needs $\frac{1}{2}x_1 + \frac{2}{3}x_2$ to produce x_2 units of clothes. These requirements can be represented by the equations

$$x_1 = \frac{1}{2}x_1 + \frac{1}{3}x_2$$
$$x_2 = \frac{1}{2}x_1 + \frac{2}{3}x_2$$

In matrix notation this system can be written as

$$\begin{bmatrix} x_1 \\ x_2 \end{bmatrix} = \begin{bmatrix} \frac{1}{2} & \frac{1}{3} \\ \frac{1}{2} & \frac{2}{3} \end{bmatrix} \begin{bmatrix} x_1 \\ x_2 \end{bmatrix}$$

■

Note in the preceding example that the sum of the elements in each column is 1. For a closed Leontief system, the sum of each column of the Leontief input–output matrix must equal 1 (unity) because all of the output is consumed as input.

The expression $a_{i1}x_1 + a_{i2}x_2 + \cdots + a_{in}x_n$ represents the total input requirements of industry i of all the items. For a closed Leontief system, the number x_i of output units of item i by industry i must equal the total input requirements by industry i of all the items. Hence,

$$x_1 = a_{11}x_1 + a_{12}x_2 + \cdots + a_{1n}x_n$$
$$x_2 = a_{21}x_1 + a_{22}x_2 + \cdots + a_{2n}x_n$$
$$\cdot$$
$$\cdot$$
$$\cdot$$
$$x_n = a_{n1}x_1 + a_{n2}x_2 + \cdots + a_{nn}x_n$$

In matrix notation this becomes $\mathbf{X} = \mathbf{AX}$, where

$$\mathbf{A} = \begin{bmatrix} a_{11} & a_{12} & \cdots & a_{1n} \\ a_{21} & a_{22} & \cdots & a_{2n} \\ \cdot & \cdot & \cdot & \cdot \\ \cdot & \cdot & \cdot & \cdot \\ \cdot & \cdot & \cdot & \cdot \\ a_{n1} & a_{n2} & \cdots & a_{nn} \end{bmatrix} \quad \text{and} \quad \mathbf{X} = \begin{bmatrix} x_1 \\ x_2 \\ \cdot \\ \cdot \\ \cdot \\ x_n \end{bmatrix}$$

Often the input and output of a Leontief system are expressed in terms of dollar value instead of items produced.

EXAMPLE 51   A builder moves to our city in Example 50 and now buys food from the farmer and clothes from the tailor. Suppose that the input–output matrix is

$$
\begin{array}{cccc}
 & \text{Farmer} & \text{Tailor} & \text{Builder} \\
\begin{array}{c}\text{Farmer}\\\text{Tailor}\\\text{Builder}\end{array} &
\left[\begin{array}{ccc}
0.4 & 0.2 & 0.050 \\
0.3 & 0.3 & 0.275 \\
0.3 & 0.5 & 0.675
\end{array}\right]
\end{array}
$$

The farmer, the tailor, and the builder use 0.3, 0.5, and 0.675 units, respectively, of the builder's production in order for each to produce 1 unit. Instead of an amount of production, let **X** represent the amount paid to each by the others.

SOLUTION   Let

$$
\mathbf{X} = \begin{bmatrix} x_1 \\ x_2 \\ x_3 \end{bmatrix}
$$

be a column matrix representing the dollar amount paid to each by the others (it also includes the dollar amount one individual would pay to himself; for example, the farmer pays himself for the food he consumes). Since the total amount paid out by each must equal the total amount received by each one,

$$
\begin{bmatrix} x_1 \\ x_2 \\ x_3 \end{bmatrix} =
\begin{bmatrix}
0.4 & 0.2 & 0.050 \\
0.3 & 0.3 & 0.275 \\
0.3 & 0.5 & 0.675
\end{bmatrix}
\begin{bmatrix} x_1 \\ x_2 \\ x_3 \end{bmatrix}
$$

$$
\begin{bmatrix} x_1 \\ x_2 \\ x_3 \end{bmatrix} =
\begin{bmatrix}
0.4x_1 + 0.2x_2 + 0.050x_3 \\
0.3x_1 + 0.3x_2 + 0.275x_3 \\
0.3x_1 + 0.5x_2 + 0.675x_3
\end{bmatrix}
$$

$$x_1 = 0.4x_1 + 0.2x_2 + 0.050x_3$$

$$x_2 = 0.3x_1 + 0.3x_2 + 0.275x_3$$

$$x_3 = 0.3x_1 + 0.5x_2 + 0.675x_3$$

or

$$-0.6x_1 + 0.2x_2 + 0.05x_3 = 0$$

$$0.3x_1 - 0.7x_2 + 0.275x_3 = 0$$

$$0.3x_1 + 0.5x_2 - 0.325x_3 = 0$$

The augmented matrix

$$
\left[\begin{array}{ccc|c}
-0.6 & 0.2 & 0.05 & 0 \\
0.3 & -0.7 & 0.275 & 0 \\
0.3 & 0.5 & -0.325 & 0
\end{array}\right]
$$

reduces to

$$\left[\begin{array}{ccc|c} 1 & 0 & -0.25 & 0 \\ 0 & 1 & -0.50 & 0 \\ 0 & 0 & 0 & 0 \end{array}\right]$$

Thus, $x_1 = 0.25x_3$, and $x_2 = 0.50x_3$. Choose $x_3 = \$8000$. Then $x_1 = \$2000$, and $x_2 = \$4000$. Since this is a closed system, no matter what we choose for x_3, x_1 and x_2 are determined. ∎

Practice Problem 1 An input–output matrix for a closed system is given by

$$\begin{array}{cc} & \begin{array}{ccc} A & B & C \end{array} \\ \begin{array}{c} A \\ B \\ C \end{array} & \left[\begin{array}{ccc} 0.3 & 0.2 & 0.1 \\ 0.1 & 0.4 & 0.2 \\ 0.6 & 0.4 & 0.7 \end{array}\right] \end{array}$$

Discuss the meaning of

(a) a_{11} (b) a_{31} (c) a_{12} (d) a_{23}

ANSWER

(a) We need 0.3 of a unit of A to produce 1 unit of A.
(b) We need 0.6 of a unit of C to produce 1 unit of A.
(c) We need 0.2 of a unit of A to produce 1 unit of B.
(d) We need 0.2 of a unit of B to produce 1 unit of C.

Now let's consider the main problem of input–output analysis. Given the internal demands for each industry's output, we want to determine outputs in order to meet an external (or outside) demand. For example, suppose Intel Corporation makes both computers and computer chips. The computer-chip division supplies the computer division and also other computer manufacturers. The computer division supplies the computer-chip division and sells computers on the open market.

In general, an **open Leontief system** consists of n production industries, as before, and a consumer section that demands items in addition to their use as inputs for production. Suppose that X and A are defined as before, and d_i represents the number of units of item i demanded by the consumer. Instead of

$$[a_{ij}][x_i] = [x_i] \quad \text{or} \quad AX = X \quad \text{Closed system}$$

we now have

$$[a_{ij}][x_i] + [d_i] = [x_i]$$

where $A = [a_{ij}]$ and $X = [x_i]$ are defined as before and d_i represents the number of units of item i demanded by the consumer. If $D = [d_i]$, then

$$AX + D = X$$

or

$$\mathbf{D} = \mathbf{X} - \mathbf{AX} \qquad \text{Subtract } \mathbf{AX} \text{ from both sides.}$$

$\mathbf{D} = \mathbf{X} - \mathbf{AX}$ represents the output available to the consumer sector of the economy. A very fundamental problem is whether or not the economy can satisfy the consumer demand. Another way of stating this problem is: Given a demand matrix $\mathbf{D}$ of the consumer sector, does there exist a matrix $\mathbf{X}$ with nonnegative components such that $\mathbf{X} - \mathbf{AX} = \mathbf{D}$? This question will be answered after the next example.

EXAMPLE 52   To illustrate an open Leontief system, let us return to our city and assume that the farmer and the tailor decide to sell some of the items they produce to a consumer. Suppose that the farmer requires one-third of the food and one-third of the clothes to produce 1 unit of the food. Suppose that the tailor requires one-half of the clothes and one-fourth of the food to produce 1 unit of the clothes. Write these requirements as an open Leontief input–output matrix.

SOLUTION

$$\begin{array}{cc} & \begin{array}{cc} \text{Food} & \text{Clothes} \\ \text{produced} & \text{produced} \end{array} \\ \begin{array}{c} \text{Input} \quad \text{Farmer} \\ \text{Requirements} \quad \text{Tailor} \end{array} & \begin{bmatrix} \frac{1}{3} & \frac{1}{4} \\ \frac{1}{3} & \frac{1}{2} \end{bmatrix} \end{array}$$

This matrix indicates that in order for the farmer to produce food, he requires one-third of the food produced and one-third of the clothes produced. Similarly, in order for the tailor to produce clothes, he requires one-fourth of the food produced and one-half of the clothes produced. Let x_1 represent the number of units of food produced and x_2 represent the number of units of clothes produced. Suppose that d_1 represents the units of food and d_2 the units of clothes required by the consumer. The requirements can be written as

$$\begin{bmatrix} \frac{1}{3} & \frac{1}{4} \\ \frac{1}{3} & \frac{1}{2} \end{bmatrix} \begin{bmatrix} x_1 \\ x_2 \end{bmatrix} + \begin{bmatrix} d_1 \\ d_2 \end{bmatrix} = \begin{bmatrix} x_1 \\ x_2 \end{bmatrix} \qquad \blacksquare$$

We are now ready to return to the fundamental question raised concerning the open Leontief system. Remember that the question was: Given a demand matrix $\mathbf{D}$ of the consumer sector, does there exist a matrix $\mathbf{X}$ such that

$$\mathbf{X} - \mathbf{AX} = \mathbf{D}$$

If $\mathbf{I}$ is the identity matrix, this equation becomes

$$\mathbf{IX} - \mathbf{AX} = \mathbf{D} \quad \text{or} \quad (\mathbf{I} - \mathbf{A})\mathbf{X} = \mathbf{D}$$

If $(I - A)^{-1}$ exists, then

$$X = (I - A)^{-1}D$$

Hence, if $(I - A)^{-1}D$ exists and has nonnegative components, it provides an affirmative answer to the fundamental question.

EXAMPLE 53 In Example 52, $AX + D = X$ was

$$\begin{bmatrix} \frac{1}{3} & \frac{1}{4} \\ \frac{1}{3} & \frac{1}{2} \end{bmatrix} \begin{bmatrix} x_1 \\ x_2 \end{bmatrix} + \begin{bmatrix} d_1 \\ d_2 \end{bmatrix} = \begin{bmatrix} x_1 \\ x_2 \end{bmatrix}$$

or $(I - A)X = D$, which can be written as

$$\left(\begin{bmatrix} 1 & 0 \\ 0 & 1 \end{bmatrix} - \begin{bmatrix} \frac{1}{3} & \frac{1}{4} \\ \frac{1}{3} & \frac{1}{2} \end{bmatrix} \right) \begin{bmatrix} x_1 \\ x_2 \end{bmatrix} = \begin{bmatrix} d_1 \\ d_2 \end{bmatrix}$$

$$\begin{bmatrix} \frac{2}{3} & -\frac{1}{4} \\ -\frac{1}{3} & \frac{1}{2} \end{bmatrix} \begin{bmatrix} x_1 \\ x_2 \end{bmatrix} = \begin{bmatrix} d_1 \\ d_2 \end{bmatrix}$$

$$\begin{bmatrix} x_1 \\ x_2 \end{bmatrix} = \begin{bmatrix} 2 & 1 \\ \frac{4}{3} & \frac{8}{3} \end{bmatrix} \begin{bmatrix} d_1 \\ d_2 \end{bmatrix}$$

where

$$\begin{bmatrix} 2 & 1 \\ \frac{4}{3} & \frac{8}{3} \end{bmatrix}$$

is the inverse of

$$\begin{bmatrix} \frac{2}{3} & -\frac{1}{4} \\ -\frac{1}{3} & \frac{1}{2} \end{bmatrix}$$

If the consumer requires 300 units of food and 500 units of clothes, then the number of units that must be produced is

$$\begin{bmatrix} x_1 \\ x_2 \end{bmatrix} = \begin{bmatrix} 2 & 1 \\ \frac{4}{3} & \frac{8}{3} \end{bmatrix} \begin{bmatrix} 300 \\ 500 \end{bmatrix} = \begin{bmatrix} 1100 \\ 1733.3 \end{bmatrix}$$

Therefore, 1100 units of food and 1733.3 units of clothes must be produced. ■

EXAMPLE 54 A three-industry open Leontief system has the following input-coefficient matrix

$$\begin{matrix} \text{Services} \\ \text{Manufacturing} \\ \text{Farming} \end{matrix} \begin{bmatrix} 0.3 & 0.2 & 0.3 \\ 0.5 & 0.2 & 0.1 \\ 0.1 & 0.3 & 0.1 \end{bmatrix}$$

If the demands from the consumer sector are for 21, 5, and 1 units, respectively, find the output needed to satisfy these demands.

SOLUTION

$$\mathbf{I - A} = \begin{bmatrix} 1 & 0 & 0 \\ 0 & 1 & 0 \\ 0 & 0 & 1 \end{bmatrix} - \begin{bmatrix} 0.3 & 0.2 & 0.3 \\ 0.5 & 0.2 & 0.1 \\ 0.1 & 0.3 & 0.1 \end{bmatrix}$$

$$\mathbf{I - A} = \begin{bmatrix} 0.7 & -0.2 & -0.3 \\ -0.5 & 0.8 & -0.1 \\ -0.1 & -0.3 & 0.9 \end{bmatrix}$$

$$(\mathbf{I - A})^{-1} = \begin{bmatrix} 2.14 & 0.84 & 0.81 \\ 1.43 & 1.86 & 0.68 \\ 0.71 & 0.71 & 1.43 \end{bmatrix}$$

(correct to two decimal places). Then

$$(\mathbf{I - A})^{-1}\mathbf{D} = \begin{bmatrix} 2.14 & 0.84 & 0.81 \\ 1.43 & 1.86 & 0.68 \\ 0.71 & 0.71 & 1.43 \end{bmatrix} \begin{bmatrix} 21 \\ 5 \\ 1 \end{bmatrix} = \begin{bmatrix} 49.95 \\ 40.01 \\ 19.89 \end{bmatrix}$$

Rounding the components off to the nearest whole number gives

$$\mathbf{X} = (\mathbf{I - A})^{-1}\mathbf{D} = \begin{bmatrix} 50 \\ 40 \\ 20 \end{bmatrix}$$

Hence, the outputs needed to satisfy the demands are 50 units of services, 40 units of manufacturing, and 20 units of farming. ∎

SUMMARY

In this section we learned to interpret Leontief matrices, distinguish between open and closed systems, and use Leontief models to solve input–output problems.

Exercise Set 2.7

1. Find one solution for each of the following systems.

(a) $\begin{bmatrix} 0.7 & 0.2 & 0.2 \\ 0.2 & 0.6 & 0.3 \\ 0.1 & 0.2 & 0.5 \end{bmatrix} \begin{bmatrix} x_1 \\ x_2 \\ x_3 \end{bmatrix} = \begin{bmatrix} x_1 \\ x_2 \\ x_3 \end{bmatrix}$

(b) $\begin{bmatrix} 0.3 & 0.4 & 0.5 \\ 0.2 & 0.3 & 0.2 \\ 0.5 & 0.3 & 0.3 \end{bmatrix} \begin{bmatrix} x_1 \\ x_2 \\ x_3 \end{bmatrix} = \begin{bmatrix} x_1 \\ x_2 \\ x_3 \end{bmatrix}$

2. In Example 50, if the farmer pays himself $10,000 for food, how much does the tailor pay for food?

3. For the given input–output matrix, determine the following.

	A	B	C
A	0.3	0.2	0.1
B	0.1	0.4	0.2
C	0.2	0.3	0.1

(a) The production of 1 unit of B requires how much from A?

(b) The production of 1 unit of C requires how much from B?

(c) If units are in millions of dollars, what is the value of C needed to produce $100,000,000 of B?

(d) Which sector, A, B, or C, consumes the greatest proportion of C?

(e) Which sector, A, B, or C, is least dependent on B?

4. In Example 50, if the farmer pays the tailor $2000 for clothes, how much does the tailor pay himself?

In Exercises 5–10, matrix A is an input–output matrix associated with an economy, and matrix D (in millions of dollars) is the demand matrix. In each exercise, find the final inputs of each industry so that the demands of both industry and the open sector are met.

5. $\mathbf{A} = \begin{bmatrix} 0.20 & 0.15 \\ 0.25 & 0.10 \end{bmatrix}$ and $\mathbf{D} = \begin{bmatrix} 12 \\ 20 \end{bmatrix}$

6. $\mathbf{A} = \begin{bmatrix} 0.30 & 0.50 \\ 0.20 & 0.40 \end{bmatrix}$ and $\mathbf{D} = \begin{bmatrix} 8 \\ 10 \end{bmatrix}$

7. $\mathbf{A} = \begin{bmatrix} \frac{3}{5} & \frac{2}{5} \\ \frac{1}{5} & \frac{3}{5} \end{bmatrix}$ and $\mathbf{D} = \begin{bmatrix} 10 \\ 6 \end{bmatrix}$

8. $\mathbf{A} = \begin{bmatrix} \frac{4}{10} & \frac{1}{10} \\ \frac{5}{10} & \frac{2}{10} \end{bmatrix}$ and $\mathbf{D} = \begin{bmatrix} 12 \\ 7 \end{bmatrix}$

9. $\mathbf{A} = \begin{bmatrix} 0.1 & 0.4 & 0.3 \\ 0.3 & 0.2 & 0.2 \\ 0.1 & 0.1 & 0.2 \end{bmatrix}$ and $\mathbf{D} = \begin{bmatrix} 12 \\ 10 \\ 8 \end{bmatrix}$

10. $\mathbf{A} = \begin{bmatrix} \frac{1}{5} & \frac{2}{5} & \frac{3}{10} \\ \frac{2}{5} & \frac{1}{10} & \frac{1}{5} \\ \frac{3}{10} & \frac{2}{5} & \frac{3}{5} \end{bmatrix}$ and $\mathbf{D} = \begin{bmatrix} 100 \\ 90 \\ 80 \end{bmatrix}$

11. Suppose that, in a three-industry open system, the Leontief input-coefficient matrix is

$$\mathbf{A} = \begin{bmatrix} 0.3 & 0.2 & 0.3 \\ 0.5 & 0.2 & 0.1 \\ 0.1 & 0.3 & 0.2 \end{bmatrix}$$

If the output matrix is

$$\mathbf{X} = \begin{bmatrix} 7 \\ 5 \\ 3 \end{bmatrix}$$

compute $\mathbf{X} - \mathbf{AX}$ to determine the output available for the consumer section.

12. The input–output matrix of a closed Leontief system is

$$\begin{array}{ccc} & \text{I} & \text{II} & \text{III} \end{array}$$
$$\mathbf{A} = \begin{array}{c} \text{I} \\ \text{II} \\ \text{III} \end{array} \begin{bmatrix} \frac{1}{3} & \frac{1}{4} & 0 \\ \frac{1}{3} & \frac{1}{2} & \frac{1}{2} \\ \frac{1}{3} & \frac{1}{4} & \frac{1}{2} \end{bmatrix}$$

Suppose that

$$\mathbf{X} = \begin{bmatrix} 3 \\ 8 \\ 6 \end{bmatrix}$$

Show that $\mathbf{X} = \mathbf{AX}$.

13. In a three-industry open Leontief system, industry 1 requires 0.3 of item I, 0.2 of item II, and 0.3 of item III. Industry 2 requires 0.5 of item I, 0.2 of item II, and 0.1 of item III. Industry 3 requires 0.1 of item I, 0.3 of item II, and 0.2 of item III. Write these requirements to produce one unit each as a Leontief input–output matrix $\mathbf{A}$. If the output is

$$\mathbf{X} = \begin{bmatrix} 10 \\ 7 \\ 5 \end{bmatrix}$$

compute $\mathbf{X} - \mathbf{AX} = \mathbf{D}$ to determine how much of the output is available for the consumer.

14. A closed Leontief system consists of three factories —1, 2, and 3—that produce items I, II, and III, respectively. Industry 1 requires $\frac{1}{3}$ of item I, $\frac{1}{4}$ of item II, and $\frac{5}{12}$ of item III. Industry 2 requires $\frac{1}{3}$ of item I, $\frac{1}{2}$ of item II, and $\frac{1}{6}$ of item III. Industry 3 requires $\frac{1}{4}$ of item I, $\frac{1}{4}$ of item II, and $\frac{1}{2}$ of item III. Write these requirements as a Leontief input–output matrix $\mathbf{A}$. If

$$\mathbf{X} = \begin{bmatrix} 10 \\ 11 \\ 12 \end{bmatrix}$$

show that $\mathbf{X} = \mathbf{AX}$.

15. A two-industry open Leontief system has the following input-coefficient matrix:

$$\begin{array}{c} \text{Services} \\ \text{Manufacturing} \end{array} \begin{bmatrix} 0.5 & 0.5 \\ 0.2 & 0.3 \end{bmatrix}$$

If the demands from the consumer section are for 14 units of service and 16 units of manufacturing, find the output needed to satisfy these demands.

16. A two-industry open Leontief system consisting of a farmer and a tailor has the following input–output matrix:

	Food produced	Clothes produced
Farmer	0.5	0.5
Tailor	0.2	0.3

Can this system satisfy a consumer demand for 16 units of food and 26 units of clothes?

17. The energy sector of an economy consists of electricity, oil, and coal. The production of 1 unit of electricity demands 0.05 unit of electricity, 0.2 unit of oil, and 0.6 unit of coal. The production of 1 unit of oil requires 0.1 unit of electricity, 0.4 unit of oil, and 0.001 unit of coal. The production of 1 unit of coal requires 0.2 unit of electricity, 0.6 unit of oil, and 0.002 unit of coal. What gross production is required to meet an external demand for 50 units of electricity, 60 units of oil, and 30 units of coal?

Review Exercises

18. Solve the following systems of equations by using augmented matrices.

(a) $\begin{aligned} x + y + 2z &= 4 \\ 3x - y + z &= 3 \\ 5x + 3y - 4z &= 4 \end{aligned}$

(b) $\begin{aligned} 3x - y - z &= 2 \\ x + 4y - 3z &= 11 \\ x - y + 2z &= 1 \end{aligned}$

19. Solve the systems of equations in Exercise 18 by using inverses of matrices.

20. Determine whether or not the following systems have solutions, and solve those that do.

(a) $\begin{aligned} x + y + z - 2t &= 2 \\ 2x + y + 6z + t &= 18 \\ 3x + 2y + 4z - 4t &= 11 \\ 4x + 3y + 8z - 3t &= 22 \end{aligned}$

(b) $\begin{aligned} x + y + z + t &= 2 \\ x - y + z - t &= 10 \\ 2x + y + z + t &= 6 \\ x + 2y + z - 2t &= 2 \end{aligned}$

21. Given the input–output matrix

$$\begin{bmatrix} 0.2 & 0.3 \\ 0.4 & 0.3 \end{bmatrix}$$

and the demand matrix

$$\begin{bmatrix} 1 \\ 5 \end{bmatrix}$$

find **X**.

22. Discuss the solutions of the following systems of equations.

(a) $\begin{aligned} 3x + 2y &= 5 \\ 4x - y &= 3 \\ x + y &= 2 \end{aligned}$

(b) $\begin{aligned} 2x - y &= 0 \\ 3x + 2y &= 8 \\ -x + 2y &= 0 \end{aligned}$

(c) $\begin{aligned} 5x + 7y + z &= 13 \\ 2x - y + z &= 2 \end{aligned}$

(d) $\begin{aligned} 6x + 5y + z &= 17 \\ x - 2y + z &= -2 \end{aligned}$

Chapter Review

Review and make certain you understand the following terms and concepts:

Matrices

Row matrix	Zero matrix	Column matrix
Identity matrix	Equal matrices	Transpose of a matrix

Matrix Operations

Sum	Product of a real	Dot Product
Negative of a matrix	number and matrix	Multiplicative inverse
	Product of matrices	A^n where **A** is a square matrix

System of Two Linear Equations

Graphical solutions   Addition and Subtraction   Break-even analysis
Algebraic solutions   Two parallel lines   Coincident lines
Two intersecting lines

System of Three or More Linear Equations

Geometric interpretation   More unknowns than equations
More equations than unknowns

Augmented Matrices

Permissible row operations   Gauss–Jordan elimination
Reduced matrix   Solutions of systems of equations

Inverse of a Matrix

Solution of systems of equations   Input–output analysis

Chapter Test

1. Solve by graphs.

$$3x - 2y = 4$$
$$3y = 2x - 1$$

2. Solve by algebraic manipulations.

$$3x - 2y = 4$$
$$3y = 2x - 1$$

3. $2\begin{bmatrix} 1 & 2 \\ 1 & 1 \\ 5 & 6 \end{bmatrix} - 2\begin{bmatrix} 2 & 3 \\ 0 & 1 \\ 1 & -1 \end{bmatrix} = \begin{bmatrix} a & b \\ c & d \\ e & f \end{bmatrix}$

   Find a, b, c, d, e, and f.

4. Multiple choice: An augmented matrix of a system of equations is

$$\begin{bmatrix} 1 & 0 & 0 & | & 2 \\ 0 & 1 & 0 & | & 3 \\ 0 & 0 & 0 & | & 4 \end{bmatrix}$$

   Classify the solution as
   (a) $x = 2$, $y = 3$, z can be anything
   (b) No solution
   (c) An infinite number of solutions
   (d) Both (a) and (c)
   (e) None of these

5. Multiply

$$\begin{bmatrix} -1 & 2 & 0 \\ 3 & 0 & -2 \end{bmatrix} \begin{bmatrix} 2 & 1 \\ 1 & 1 \\ 0 & 3 \end{bmatrix}$$

6. Find the inverse of

$$\begin{bmatrix} 3 & -2 \\ -1 & 1 \end{bmatrix}$$

7. Solve by augmented matrices.

$$3y = 2x + 3$$
$$x = y + 2$$

8. Solve the system in Problem 2 by using augmented matrices and a graphing calculator.

9. Solve the system in Problem 2 by inverses.

10. Solve by augmented matrices.

$$x + 3y + z = 5$$
$$4y + 2z = 6$$
$$3y - z = 2$$

11. Solve the system in Problem 7 by using inverses.

12. Change and then solve the system in Problem 10 by introducing the constants 7, 9, 11; that is,

$$x + 3y + z = 7$$
$$4y + 2z = 9$$
$$3y - z = 11$$

13. A man has $10,000 invested, some at 6%, some at 8%, and some at 10%. His interest on the investment at 10% is $540 more than the interest on the investment at 6%, and $360 more than at 8%. How much does he have invested at each rate?

14. In a three-industry open input–output system, Industry 1 requires 0.3 of item I, 0.2 of item II, and 0.3 of item III. Industry 2 requires 0.5 of item I, 0.2 of item II, and 0.1 of item III. Industry 3 requires 0.1 of item I, 0.3 of item II, and 0.2 of item III. Given that $X - AX = D$, find D, the output available to the customer when

$$X = \begin{bmatrix} 10 \\ 7 \\ 5 \end{bmatrix}$$

15. Find the equilibrium point for the following supply and demand functions:

$$3y = 5x - 1$$
$$y = -4x + 11$$

CHAPTER 3

Introduction to Linear Programming

We all like to get as much as possible in exchange for the least amount of effort or expenditure. Remember the last time you went shopping, you wanted to get the most for your money. And have you ever tried to satisfy your appetite with only 1000 calories a day? All of us have struggled with our daily schedule, trying to find enough time for classes, for study, for a part-time job, and still have the maximum amount of leisure time for club activities, movies, dates, and other activities.

Of course, the businessperson, the manufacturer, the planner, as well as the researcher, are interested in minimizing expenditures and maximizing benefits. The mathematical model we are about to study, linear programming, is a valuable tool.

This application of mathematics is rather young compared with other mathematical ideas. During World War II, the Allies faced daily problems concerning the transportation of troops, the movement of supplies, and the scheduling of operations. Linear programming models helped solve these problems. In fact, linear programming is one of the important developments in applied mathematics in the last half-century. Three men have been honored for work in this field.

Leonid Kantorovich, the Russian linear programming pioneer, started his work in linear programming in the late 1930s. He published the first paper in linear programming in 1939. However, his research remained unknown until the late 1950s. Meanwhile, several mathematicians and scientists in the United States were trying to solve the complex military problems of supply and deployment of both personnel and material. They constructed a mathematical model called a **linear program**. After construction of the model, a process that has come to be known as **linear programming** was developed to obtain a solution. Among those studying these problems were the mathematician George Dantzig, who received the National Medal of Science for his pioneering contributions, and the economist Tjalling Koopmans. Kantorovich and Koopmans in 1975 received the Nobel prize for their work in linear programming as applied to economics. Dantzig originated the technique for solving linear programs called the **simplex method**, which is introduced in this chapter.

3.1 SOLVING SYSTEMS OF INEQUALITIES GRAPHICALLY

OVERVIEW   Mathematical models often contain inequalities that express limitations. For example, the size of a school building limits the number of students who may attend the school; the amount of money a person has limits the amount he or she can invest; the quality of steel a company wishes to produce places conditions on the ingredients; and so on. In this section we

- Solve systems of inequalities graphically
- Formulate special definitions for concepts associated with a region of solutions for a system of inequalities
- Apply systems of inequalities

In Chapter 0 we studied the graphs of linear equations and linear inequalities. In Chapter 2 we studied the graphs of a system of linear equations. In this section, we study the graphs of a **system of inequalities**. The **solution region** for a system of inequalities in two unknowns consists of the ordered pairs that satisfy all the inequalities.

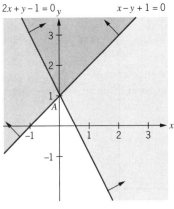

$2x + y - 1 = 0$ $x - y + 1 = 0$

Figure 1

EXAMPLE 1   Graph the solution region of the system

$$2x + y - 1 \geq 0$$
$$x - y + 1 \leq 0$$

SOLUTION   Solutions of linear inequalities were graphed in Chapter 0. In Figure 1 the lightly colored shading represents the solutions of $2x + y - 1 \geq 0$, and the colored shading represents the solutions of $x - y + 1 \leq 0$. These two solution regions are also indicated by two arrows perpendicular to the boundary of each inequality. The doubly shaded region represents the solution region, or solution set, of the system, consisting of all ordered pairs satisfying both inequalities.   ■

EXAMPLE 2   Solve the following linear system graphically:

$$x \geq 0$$
$$x \leq 4$$
$$y \geq 0$$
$$y \leq 3$$

SOLUTION   The graphs of the four inequalities are shown in Figure 2. The solution region associated with each line is indicated by arrows. The intersection of the four graphs (or the solutions common to the four graphs) is the solution region of the system and is shaded.   ■

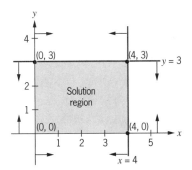

Figure 2

Certain points of a solution region are named and used in the remainder of this chapter.

DEFINITION: CORNER POINT

A **corner point** of a solution region is a point in the solution region that is the intersection of two boundary lines of the region.

Note in this definition that the intersection of two boundary lines may or may not be a corner point. The intersection must be in the solution region. In Example 1, $(0, 1)$ is a corner point. There are four corner points in Example 2: $(0, 0)$, $(0, 3)$, $(4, 0)$, and $(4, 3)$.

EXAMPLE 3   Use a calculator to graph the solution region and find the corner points of the system

$$2x + y \leq 3$$
$$-x + y \leq 0$$
$$x \geq 0$$
$$y \geq 0$$

SOLUTION We will use the shade function twice for this problem. For the first shading we input 3 to indicate the amount of shading, and we input 2 for the second shading to distinguish between the two solutions. We use $Y_1 = 3 - 2x$, $Y_2 = x$, and $Y_4 = -3.1$, the minimum value of y. Choose X: $[-4.7, 4.7]$ and Y: $[-3.1, 3.1]$. Press 2nd DRAW and choose 7: Shade (command). Then press 2nd Y-VARS and choose 1:Function. Choose Y_4 and input a comma. Then press 2nd Y-VARS and choose 1:Function. Choose Y_1. Then input a comma followed by 3 and close the parentheses. The shade command is Shade(Y_4, Y_1, 3). Press ENTER to draw the graph. To shade $Y_2 = x$, press 2nd ENTRY and repeat the process except change Y_1 to Y_2 and the amount of shading from 3 to 2. The shade command is Shade(Y_4, Y_2, 2). Since $x \geq 0$ and $Y \geq 0$, the solution region, double shaded in Figure 3, is only in the first quadrant. By using the TRACE key, we can find the corner points to be (1.5, 0), (0, 0), and (1, 1). ∎

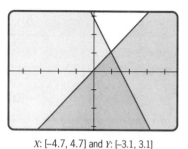

X: [-4.7, 4.7] and Y: [-3.1, 3.1]

Figure 3

Practice Problem 1 Find the corner points for the system

$$x + y \leq 3$$
$$3x - 2y \leq 4$$
$$x \geq 0$$
$$y \geq 0$$

ANSWER The corner points are (0, 0), (0, 3), (2, 1), $(\frac{4}{3}, 0)$.

Solution regions may be bounded or unbounded.

DEFINITION: BOUNDED SOLUTION REGION

A solution region of a system of linear inequalities is said to be **bounded** if one can draw a circle that will contain all points of the solution region. If the solution region cannot be contained in some circle, it is said to be **unbounded**.

The solution regions of Figures 2 and 3 are bounded regions; however, the solution region of Figure 1 is unbounded.

EXAMPLE 4 Find the corner points and graph the solution region of the system

$$2x + y \leq 3$$
$$-x + y \leq 0$$
$$x \leq 0$$

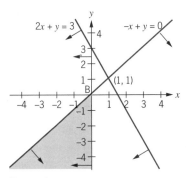

Figure 4

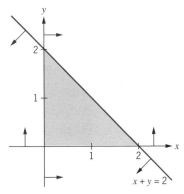

Figure 5

SOLUTION   The lines $2x + y = 3$, $-x + y = 0$, and $x = 0$ are graphed, and then the solution set of the points satisfying all of the inequalities is shaded in Figure 4. The only corner point is $B = (0, 0)$. The points $(0, 3)$ and $(1, 1)$ are intersections of boundary lines but are not corner points because they are not part of the solution region of the system. Note that this solution region is unbounded. ∎

Practice Problem 2   Find the solution region of

$$x + y \leq 2$$
$$x \geq 0$$
$$y \geq 0$$

and locate all corner points.

ANSWER   As shown in Figure 5, the corner points are $(0, 0)$, $(2, 0)$, and $(0, 2)$.

In the first of the two applications of a system of inequalities, we use the problem-solving ideas detailed in Chapter 0.

EXAMPLE 5   A supermarket has a supply of beef that has been advertised as virtually fat-free. Because this beef was too expensive to sell, the store wishes to mix this beef with fat to produce regular ground beef (60% beef and 40% fat) and extra-lean ground beef (75% beef and 25% fat). The store has up to 225 pounds of beef and 125 pounds of fat available to make ground beef. Write the system of inequalities that expresses these conditions, graph the solution region, and find the corner points.

SOLUTION   Suppose we estimate that the store wants 200 pounds of regular ground beef and 40 pounds of extra-lean beef. First, we determine whether we can get this estimate from 225 pounds of beef and 125 pounds of fat. The regular ground beef is 60% beef, so we need

$$0.60(200) \text{ pounds of beef}   \qquad \text{Our estimate}$$

The extra-lean beef is 75% beef, so we need

$$0.75(40) \text{ pounds of beef}   \qquad \text{Our estimate}$$

Since we have only 225 pounds of beef,

$$0.60(200) + 0.75(40) \leq 225   \qquad \text{Verify with a calculator that this inequality is satisfied.}$$

Regular ground beef is 40% fat. So

$$0.40(200) \text{ pounds of fat are needed}   \qquad \text{From our estimate of 200 pounds}$$

The extra-lean ground beef is 25% fat. So

$$0.25(40) \text{ pounds of fat are needed} \qquad \text{From our estimate of 40 pounds}$$

Since we have only 125 pounds of fat,

$$0.40(200) + 0.25(40) \leq 125$$

If both of the above inequalities are satisfied, we have guessed a solution. However, we want all solutions, so we replace 200 with x (representing the number of pounds of regular ground beef) and 40 with y (the number of pounds of extra-lean beef) to obtain

$$0.60x + 0.75y \leq 225 \qquad \text{Condition on beef}$$

In a similar manner,

$$0.40x + 0.25y \leq 125 \qquad \text{Condition on fat}$$

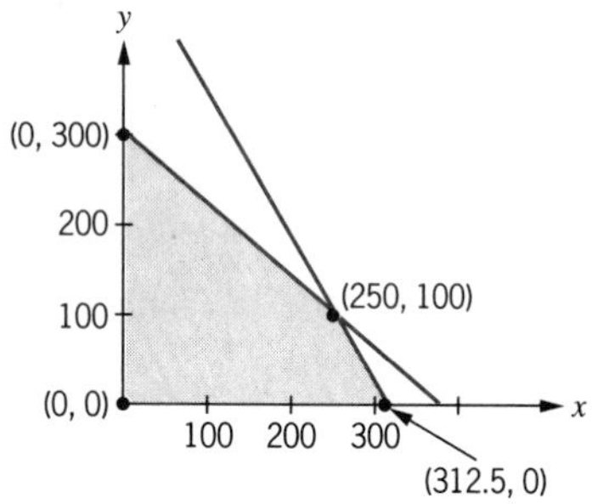

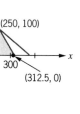

Since x and y represent numbers of pounds, we have $x \geq 0$ and $y \geq 0$. The solution region is shown in Figure 6. The corner points are $(0, 0)$, $(312.5, 0)$, $(250, 100)$, and $(0, 300)$.

Figure 6

EXAMPLE 6   A doctor has prescribed a diet in which the total number of calories must not exceed 1200. She insists that twice the number of protein calories added to the number of carbohydrate calories must equal or exceed 1600. Write these conditions as a system of inequalities and graph the possible solutions.

SOLUTION   Let x represent the number of protein calories and y represent the number of carbohydrate calories. The conditions are

$$x + y \leq 1200$$

$$2x + y \geq 1600$$

$$x \geq 0$$

$$y \geq 0$$

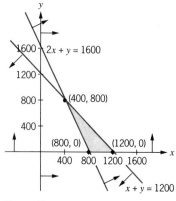

The solution region is shaded in Figure 7. Any point (x, y) in the solution region would satisfy the diet prescribed.

Figure 7

SUMMARY

In this section we learned to sketch the solution region of a system of inequalities. Two terms associated with the solution region are important in the remainder of this chapter: corner point and bounded solution region.

Exercise Set 3.1

Graph the solution regions, find the corner points, and classify the regions as bounded or unbounded.

1. $x \geq 0$
   $y \geq 0$

2. $x \leq 3$
   $y \leq 2$

3. $x \geq 0$
   $y \geq 3$

4. $x \leq 2$
   $y \geq 0$

5. $y \geq x$
   $x \geq 0$

6. $y \geq x$
   $y \geq 0$

7. $y \geq x$
   $x \geq 0$
   $y \leq 3$

8. $y \geq x$
   $y \geq 0$
   $y \leq 3$

9. $y \leq x$
   $x \geq 0$
   $y \leq 3$

10. $y \leq x$
    $y \geq 0$
    $y \leq 3$

11. $-x + y \geq 0$
    $x + y \geq 0$

12. $-x + y \geq 0$
    $2x + y \leq 3$

13. $y \geq x - 1$
    $y \geq -x + 1$
    $y \leq 1$

14. $y \geq x - 1$
    $y \geq -x + 1$
    $y \geq 1$

15. $\quad 2y \leq x + 6$
    $y + x \geq 2$
    $y - x \geq -4$
    $\quad y \geq 0$

16. $\quad 2y \leq x + 6$
    $y \geq -x + 2$
    $y - x \geq -4$
    $\quad y \geq 0$
    $\quad x \geq 0$

17. $\quad 2y \leq x + 6$
    $y + x \geq 2$
    $y - x \geq -4$
    $\quad -y \geq 0$

18. $x + y \geq 1$
    $x + y \leq 2$
    $x \geq 0$
    $y \geq 0$

19. $y \geq x - 1$
    $y \leq -x + 1$
    $y \leq 1$
    $x \geq 0$
    $y \geq 0$

20. $\quad x + \quad y \leq 10$
    $4x - \quad y \geq 8$
    $6x - 2y \leq 12$
    $\quad x \geq 0$
    $\quad y \geq 0$
    $\quad y \leq 3$

Applications (Business and Economics)

21. **Manufacturing.**   A manufacturer makes two products, valves and reducers. A valve requires 1 hour on machine A and 2 hours on machine B, and a reducer requires 2 hours on A and 2 hours on B. Let x be the number of valves produced in a day and y the number of reducers produced in a day. If each machine operates up to 8 hours a day, write the mathematical model for this manufacturer, graph the solution region, and find the corner points.

22. **Manufacturing.**   Rework Exercise 21 if the manufacturer decides to operate each of the machines up to 16 hours a day.

23. **Investment.**   An investor has up to $30,000 to invest in AA bonds or B bonds. The AA bonds yield 8%, and the B bonds yield 12%. If the investor wishes to invest at least two times as much in AA as in B bonds, write the mathematical model, graph the solution region, and find the corner points for the investor.

24. **Assignment.**   A coal company operates two mines with different production capacities. The following is the output in tons of coal per day from each mine.

Quality	Mine A	Mine B
High-grade	8	2
Medium-grade	3	2
Low-grade	4	10

For a week, the company needs at least 22 tons of high-grade, 15 tons of medium-grade, and 32 tons of low-grade coal. How many days should the company operate each mine in order to meet requirements? Shade the solution region and find the corner points.

Applications (Social and Life Sciences)

25. **Education.**   A classroom has space for up to 50 students. Let x represent the number of girls and y represent the number of boys. Graph the region that represents the possible number of boys and girls in a class and find the corner points. (Actually the solution region will consist of only the points with whole-number coordinates in the solution region.)

26. **Nutrition**. I decide to feed my pet dog a combination of two dog foods. Each serving of brand X contains 1 unit of protein, $\frac{1}{2}$ unit of carbohydrates, and 2 units of fat. Each serving of brand Y contains $\frac{1}{2}$ unit of protein, 1 unit of carbohydrates, and 3 units of fat. I feel that my dog should have at least 2 units of protein, 3 units of carbohydrates, and 2 units of fat each day. How many servings of each dog food will be necessary to meet minimum requirements? Find the corner points.

27. **Agriculture**. A farmer has a 50-acre farm on which he plans to plant two crops, wheat and corn. Wheat requires 2 days of labor per acre, and corn requires 3 days of labor per acre. The other costs amount to $40 per acre for wheat and $30 per acre for corn. The farmer has evaluated his assets and found that he has up to 150 days of labor available and up to $1800 capital. Let x represent the number of acres of wheat planted and y the number of acres of corn. Graph the solution region and find the corner points.

3.2 FINDING AN OPTIMAL SOLUTION

OVERVIEW   In the preceding section, we learned to locate a solution region. In this section we want to find the largest and the smallest value of a linear function over the solution region. This study presents the mathematical background for the graphical solution of linear programming problems in the next section. In this section we introduce the following terms associated with linear programming:

- Objective function
- Constraints
- Feasible region
- Optimal solution

We introduce the idea of **optimal value** by considering a simple selection process in taking a test.

EXAMPLE 7   On a test a multiple-choice question is worth 10 points and a true–false question 4 points. It takes, on the average, 5 minutes to work a multiple-choice problem and 3 minutes to work a true–false problem. The test is 18 minutes in length and you are not to answer more than 4 questions. If you feel certain that you can get the answer to each multiple-choice problem in the 5 minutes allowed and the answer to each true–false question in the 3 minutes allowed, how many of each type should you answer to get the best score?

SOLUTION   Let x be the number of multiple-choice questions that you answer and y the number of true–false questions. Since you cannot answer more than 4 questions,

$$x + y \leq 4$$

It takes $5x$ minutes to work x multiple-choice problems and $3y$ minutes to work

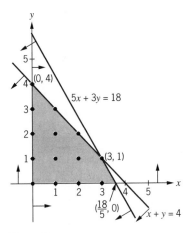

Figure 8

y true–false. You are limited to 18 minutes, so

$$5x + 3y \leq 18$$

In addition, since you can work only a positive number of problems (you cannot work -3 problems),

$$x \geq 0$$

$$y \geq 0$$

The solution region for this set of inequalities is shown as the shaded region in Figure 8. ∎

In linear programming such inequalities as the four in Example 7 are called **constraints**. Note in Figure 8 that the corner points are $(0, 0)$, $(\frac{18}{5}, 0)$, $(3, 1)$, and $(0, 4)$. The point of intersection of $x + y = 4$ and $5x + 3y = 18$ is $(3, 1)$. For this special problem only integral values have meaning, so the solution region consists of a finite set of points: $\{(0, 0), (0, 1), (0, 2), (0, 3), (0, 4), (1, 0), (1, 1), (1, 2), (1, 3), (2, 0), (2, 1), (2, 2), (3, 0),$ and $(3, 1)\}$. A multiple-choice problem is worth 10 points and a true–false question 4 points. If we answer x multiple-choice questions, we have $10x$ points. Likewise, y true–false questions produce $4y$ points. The total number of points we can attain is

$$P = 10x + 4y$$

We want to select x and y so that P is a maximum on our region of solutions. One way to work this problem is to evaluate P at each of our 14 points in the region of solutions.

$$P(0, 0) = 10 \cdot 0 + 4 \cdot 0 = 0 \qquad P(0, 1) = 10 \cdot 0 + 4 \cdot 1 = 4$$

$$P(0, 2) = 10 \cdot 0 + 4 \cdot 2 = 8 \qquad P(0, 3) = 10 \cdot 0 + 4 \cdot 3 = 12$$

$$P(0, 4) = 10 \cdot 0 + 4 \cdot 4 = 16 \qquad P(1, 0) = 10 \cdot 1 + 4 \cdot 0 = 10$$

$$P(1, 1) = 10 \cdot 1 + 4 \cdot 1 = 14 \qquad P(1, 2) = 10 \cdot 1 + 4 \cdot 2 = 18$$

$$P(1, 3) = 10 \cdot 1 + 4 \cdot 3 = 22 \qquad P(2, 0) = 10 \cdot 2 + 4 \cdot 0 = 20$$

$$P(2, 1) = 10 \cdot 2 + 4 \cdot 1 = 24 \qquad P(2, 2) = 10 \cdot 2 + 4 \cdot 2 = 28$$

$$P(3, 0) = 10 \cdot 3 + 4 \cdot 0 = 30 \qquad P(3, 1) = 10 \cdot 3 + 4 \cdot 1 = 34$$

The maximum value occurs at $(3, 1)$. Note that $(3, 1)$ is a corner point. We will see the significance of this later.

Two other terms are important in the remainder of this chapter. The solutions to the constraint inequalities are called **feasible solutions** (or feasible regions), and the linear function to be maximized or minimized is called the **objective function**.

EXAMPLE 8 Find the maximum and the minimum values of an objective function $C = 10x + 5y$ subject to the constraints

$$x + y \leq 12$$

$$2 \leq x \leq 6$$

$$3 \leq y \leq 8$$

SOLUTION The set of feasible solutions is shown in Figure 9. Note that this set is bounded by a five-sided polygon with corner points (2, 3), (2, 8), (4, 8), (6, 6), and (6, 3). To find the maximum and minimum values of $C = 10x + 5y$, subject to the constraints, let us assign values to C and plot the lines obtained. When C is assigned a particular value and the graph drawn, any point (x, y) in the set of feasible solutions that lies on this line would produce this same cost.

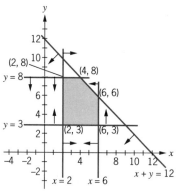

Figure 9

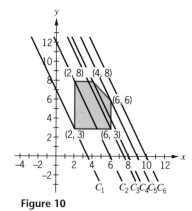

Figure 10

This line would be called a **constant-cost line**. As C takes on other values, we obtain a family of constant-cost lines (see Figure 10) that are parallel to each other, since they have the same slope. This cost function, in slope intercept form, is

$$y = -2x + C/5$$

which has a slope of -2 and a y-intercept of $C/5$. Note that for this cost function, as C increases the y-intercept increases, and the line moves up from the origin. Also note that the last line to contain any values of the feasible solution, C_5, is the line through the corner point (6, 6). C assumes a maximum value at this point.

$$C = 10 \cdot 6 + 5 \cdot 6$$
$$= \$90$$

C assumes a minimum value at (2, 3).

$$C = 10 \cdot 2 + 5 \cdot 3$$
$$= \$35$$

Finding a maximum or minimum value of the objective function over a feasible region is often called **optimizing the function**. The maximum or minimum value of the function is the **optimal solution**.

FUNDAMENTAL THEOREM OF LINEAR PROGRAMMING

Consider a linear function

$$P = ax + by + c$$

over a feasible region defined by linear constraints. If this function has an optimal solution, it will occur at one (or more) of the corner points.

EXAMPLE 9   Find the maximum and minimum values of the function

$$P = 4x + 3y$$

subject to the constraints

$$x \geq 0$$
$$y \geq 0$$
$$5x + 3y \leq 30$$
$$2x + 3y \leq 21$$

SOLUTION   The constraint inequalities are shown in Figure 11, and the feasible region of solutions is shaded. The corner points are (0, 0), (0, 7), (3, 5), and (6, 0). In addition, the objective function, $P = 4x + 3y$ or $y = (-4/3)x + P/3$, with slope $-4/3$, is drawn for several values of P. Note that the maximum value of P seems to occur at (3, 5) and the minimum value at (0, 0). This we verify by evaluating P at each corner point.

(0, 0): $P = 4(0) + 3(0) = 0$ Minimum

(0, 7): $P = 4(0) + 3(7) = 21$

(3, 5): $P = 4(3) + 3(5) = 27$ Maximum

(6, 0): $P = 4(6) + 3(0) = 24$

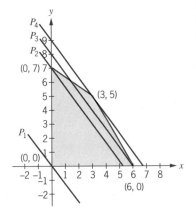

Figure 11

The maximum and minimum values occurred at corner points.

Practice Problem 1 Maximize

$$P = 10x + 15y$$

subject to

$$3x + 2y \leq 15$$
$$x + 3y \leq 12$$
$$x \geq 0$$
$$y \geq 0$$

ANSWER The maximum value of P is 75 at (3, 3).

If the objective function is defined on a region that is not bounded on each side by a line segment (i.e., the region is not a closed polygon), the objective function might not have a maximum or a minimum value on the solution region.

EXAMPLE 10 Find the maximum and minimum values of

$$f = 2x + 3y$$

subject to the constraints

$$x + y \geq 0$$
$$-2x + y \geq -4$$
$$y \geq 0$$

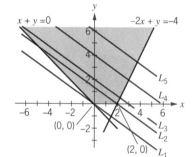

Figure 12

SOLUTION From a graph of the constraint inequalities, Figure 12 shows that the feasible region is not bounded. The only two corner points are (0, 0) and (2, 0). At the corner points

$$(0, 0): \quad f = 2(0) + 3(0) = 0 \quad \text{Minimum}$$
$$(2, 0): \quad f = 2(2) + 3(0) = 4$$

However, let f take on values 6, 12, and 18, and draw the corresponding lines.

$$L_3: \quad 6 = 2x + 3y$$
$$L_4: \quad 12 = 2x + 3y$$
$$L_5: \quad 18 = 2x + 3y$$

All these lines have points in the solution region. The maximum is not attained at a corner point, because 6, 12, 18, and many other values are larger than 4—the biggest value attained at a corner point.   ■

Existence of Maximum or Minimum

When a feasible region is bounded, maximum and minimum values of the objective function will always exist on the region. When a feasible region is unbounded, maximum or minimum values of the objective function may or may not exist.

Practice Problem 2 Find the maximum and minimum values of

$$P = 3x + 4y + 6$$

under the constraints

$$2x + 3y \leq 12$$
$$-x + 2y \leq 1$$
$$x \geq 0$$
$$y \geq 0$$

ANSWER   The maximum value of P is 24, and the minimum value is 6.

SUMMARY

The following important concepts were introduced in this section:

1. Objective function
2. Constraints
3. Feasible region
4. Optimal solutions

The fundamental theorem of linear programming states that if

$$P = ax + by + c$$

(over a feasible region defined by linear constraints) has an optimal solution, it will occur at one or more of the corner points.

Exercise Set 3.2

Draw the region of feasible solutions and find the indicated optimal values and the corner points at which they occur. If no solution exists, explain why.

1. Maximize

$$P = 25x + 35y$$

subject to

$$2x + 3y \leq 15$$
$$3x + y \leq 12$$
$$x \geq 0$$
$$y \geq 0$$

2. Maximize

$$P = 35x + 25y$$

subject to

$$2x + y \leq 7$$
$$3x + y \leq 8$$
$$x \geq 0$$
$$y \geq 0$$

3. Maximize

$$F = 25x + 35y$$

subject to

$$x + y \leq 7$$
$$2x + y \leq 9$$
$$3x + y \leq 12$$
$$x \geq 0$$
$$y \geq 0$$

4. Maximize and minimize

$$P = 6x + 2y + 3$$

subject to

$$x + y \geq 0$$
$$2x + 3y \leq 6$$
$$x \geq 0$$
$$y \geq 0$$

5. Maximize and minimize

$$F = 12x - 2y + 20$$

subject to

$$3x + 4y \leq 37$$
$$y \geq 2$$
$$y \leq 4$$
$$x \leq 3$$
$$x \geq 0$$

6. Maximize and minimize

$$F = 16x - 4y + 20$$

subject to

$$2x + 3y \leq 23$$
$$y \geq 1$$
$$y \leq 3$$
$$x \leq 4$$
$$x \geq 1$$

7. Maximize and minimize

$$C = 3x + 2y + 4$$

subject to

$$4x + 3y \geq 24$$
$$3x + 4y \geq 8$$
$$x \geq 0$$
$$y \geq 2$$

8. Maximize and minimize

$$P = 2x + 3y + 2$$

subject to

$$2x + y \geq 10$$
$$x + 2y \leq 8$$
$$x \geq 2$$
$$y \geq 0$$

9. Maximize and minimize

$$P = 25x + 35y$$

subject to

$$x + y \geq 2$$
$$2x + 3y \leq 15$$
$$3x + y \leq 12$$
$$x \geq 0$$
$$y \geq 0$$

10. Maximize and minimize

$$P = 35x + 25y$$

subject to

$$x + y \geq 2$$
$$2x + y \leq 7$$
$$3x + y \leq 8$$
$$x \geq 0$$
$$y \geq 0$$

11. Maximize and minimize

$$P = 25x + 35y$$

subject to

$$2 \leq x + y \leq 7$$
$$2x + y \leq 9$$
$$3x + y \leq 12$$
$$x \geq 0$$
$$y \geq 0$$

12. Maximize and minimize

$$P = 25x - 35y + 100$$

subject to

$$x + 3y \leq 21$$
$$2x + 3y \leq 24$$
$$2x + y \leq 16$$
$$x \geq 0$$
$$y \geq 0$$

13. Maximize and minimize

$$P = 50x + 20y$$

subject to

$$2x + 3y \geq 6$$
$$x + 3y \leq 21$$
$$2x + 3y \leq 24$$
$$2x + y \leq 16$$
$$x, y \geq 0$$

14. **Exam.**   A manufacturing company makes only two products, with the following two production constraints representing two machines and their maximum availability:

$$2x + 3y \leq 18$$
$$2x + y \leq 10$$

where x = the units of the first product, and y = the units of the second product. If the profit function is $Z = \$8x + \$4y$, the maximum possible profit is
(a) $40   (b) $42
(c) $36   (d) $48
(e) Some profit other than those given

Review Exercises

Sketch the solution set of the following systems of linear inequalities. Locate the corner points. Is the solution set bounded?

15. $x \geq 0$ 16. $x \geq 0$
         $y \geq 0$ $y \geq 0$
         $3x + 5y \leq 30$ $x + y \geq 5$
         $3x + 5y \leq 21$ $3x + 5y \geq 21$

3.3 GRAPHICAL SOLUTIONS TO LINEAR PROGRAMMING PROBLEMS

OVERVIEW   Many practical problems in business, economics, life sciences, and the social sciences involve relationships among variables that can be expressed linearly as inequalities. When this occurs, we can find a region of feasible solutions and can then find a maximum or minimum of some objective function. In this section we examine a large number of application problems called linear programming problems. To solve such problems, we

- Locate corner points
- Evaluate the objective function at each corner point

Let us now summarize some of the concepts introduced in the previous section. A **linear programming problem** is a problem for which we are to find a maximum or minimum value of a linear function in two variables $P = ax + by$ (in this section) or, in general, an n-variable linear function such as $P = c_1 x_1 + c_2 x_2 + \cdots + c_n x_n$ (in the last part of the chapter). The linear function we are to optimize is the objective function in n variables: $x_1, x_2, x_3, \ldots, x_n$, subject to constraints in the form of linear inequalities. In addition, each of the variables must be nonnegative, $x_i \geq 0$, where $i = 1, 2, \ldots, n$. The region of solutions satisfying both the constraint inequalities and the nonnegative requirements is called the **feasible region** of the problem. When the feasible region is bounded, the objective function has both a maximum value and a minimum value in the feasible region. When the feasible region is unbounded, maximum and minimum values of the objective function may not exist.

The following steps should be helpful as we formulate a mathematical model to represent a linear programming problem.

Solutions of Linear Programming Problems

1. Introduce the problem variables. (For difficult problems you may want to use the problem-solving technique of estimating values before assigning variables.)
2. Express the objective function (that which is to be optimized) in terms of the variables in the form

$$P = ax + by$$

3. Write the problem constraints using linear inequalities (and/or equations). For maximum-type problems, the inequalities are usually of the form

$$cx + dy \leq e$$

For minimum-type problems, the inequalities are usually of the form

$$cx + dy \geq e$$

4. The variables are to be nonnegative:

$$x \geq 0, \qquad y \geq 0$$

5. Find the graph, or the feasible solution region, of the system of constraints.
6. Locate the corner points by solving pairs of equations formed by adjoining constraints changed to equations.
7. Evaluate the objective function at each corner point to determine the maximum or minimum value.

We now follow these steps on both a maximum-type and a minimum-type linear programming problem.

EXAMPLE 11   The parts division of a manufacturing corporation shapes all axles to meet specifications on two machines: A and B. A small-car axle requires 1 hour on machine A and 1.5 hours on machine B. A large-car axle requires 2 hours on A and 1 hour on B. If each machine operates up to 24 hours a day and the manufacturer makes a $15 profit on large-car axles and a $10-per-axle profit on small-car axles, find the number of each of the axles that should be produced for maximum profit.

SOLUTION   Let's estimate that we make 8 small-car axles (at a profit of $10 · 8) and 6 large-car axles (at a profit of $15 · 6) for a total profit of

$$P = 10 \cdot 8 + 15 \cdot 6$$
$$= \$170 \qquad \text{See Table 1}$$

TABLE 1

Description	Small Axles		Large Axles
Number of each (guess)	8		6
Profit from guess	$10 \cdot 8$	+	$15 \cdot 6 = \$170$
Number of each (variable)	x		y
Profit	$10x$	+	$15y$
	Limitation on number of hours		
Hours needed (machine A)	$1x$ hours	+	$2y$ hours ≤ 24
Hours needed (machine B)	$1.5x$ hours	+	$1y$ hours ≤ 24

We now replace our estimate of 8 small-car axles with x small-car axles and our estimate of 6 large-car axles with y large-car axles. The profit-objective function becomes

$$P = \$10x + \$15y$$

Each small-car axle is on machine A for 1 hour, so x small-car axles would utilize machine A for x hours. Each large-car axle uses machine A for 2 hours, or y large-car axles use machine A for $2y$ hours.

$$x + 2y \leq 24 \qquad \text{Machine A operates up to 24 hours each day.}$$

Likewise,

$$1.5x + 1y \leq 24 \qquad \text{Each small axle needs 1.5 hours and each large axle needs 1 hour on machine B.}$$

$$x \geq 0$$

$$y \geq 0 \qquad \text{Cannot have a negative number of axles.}$$

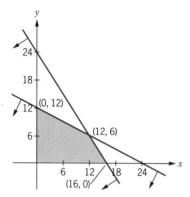

Figure 13

The feasible region is plotted in Figure 13, and the corner points are (0, 0), (0, 12), (12, 6), (16, 0).

At the corner points, the values of the objective function $P = 10x + 15y$ are

$$(0, 0):\quad P = 10(0) + 15(0) = 0$$
$$(0, 12):\quad P = 10(0) + 15(12) = \$180$$
$$(16, 0):\quad P = 10(16) + 15(0) = \$160$$
$$(12, 6):\quad P = 10(12) + 15(6) = \$210$$

The maximum profit in one day on the two machines is $210 obtained by producing 12 small axles and 6 large axles. Note that the minimum value is $0, which is what one would guess. ∎

Practice Problem 1 A sociologist wishes to maximize the time he can spend on his research project. He plans 14 morning sessions and 12 afternoon sessions for a week. The sociologist does not want more than 3 morning sessions and 2 afternoon sessions during 8 hours, nor more than 2 morning sessions and 4 afternoon sessions during 8 hours. How long should each session be in order for him to maximize the time spent on his project for the week?

ANSWER Let x be the length in hours of morning sessions and y the length in hours of the afternoon sessions.

$$3x + 2y \le 8$$
$$2x + 4y \le 8$$
$$x \ge 0, y \ge 0$$
$$T = 14x + 12y$$

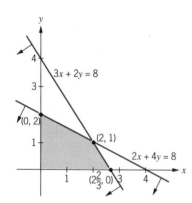

Figure 14

The feasible region of solutions and the corner points are shown in Figure 14.

$$(0, 2):\quad T = 14(0) + 12(2) = 24$$
$$(2, 1):\quad T = 14(2) + 12(1) = 40$$
$$(2\tfrac{2}{3}, 0):\quad T = 14(2\tfrac{2}{3}) + 12(0) = 37\tfrac{1}{3}$$

The morning sessions should be 2 hours long, and the afternoon sessions 1 hour long, to maximize the time spent on the project.

EXAMPLE 12 A hospital dietician is experimenting to get a prescribed amount of calcium and vitamin A and at the same time minimize cholesterol-producing units. Each ounce of food I produces 20 units of calcium, 10 units of vitamin A, and 6 cholesterol-producing units. Each ounce of food II produces

10 units of calcium, 20 units of vitamin A, and 4 cholesterol-producing units. If the minimum daily requirements are 300 units of calcium and 180 units of vitamin A, how many ounces of each food are needed to meet daily requirements and at the same time minimize cholesterol-producing units? What is the minimum number of cholesterol-producing units?

SOLUTION   Let's estimate 10 ounces of food I and 8 ounces of food II. The 10 ounces of food I would produce $6 \cdot 10$ cholesterol-producing units, and the 8 ounces of food II, $4 \cdot 8$ cholesterol-producing units. The number of cholesterol-producing units would be $6 \cdot 10 + 4 \cdot 8 = 92$. See Table 2.

TABLE 2

Description	Food I		Food II
Number of ounces (estimate)	10		8
Number of cholesterol-producing units (estimate)	$6 \cdot 10$	$+$	$4 \cdot 8 = 92$
Number of ounces (variable)	x		y
Number of cholesterol-producing units	$6x$	$+$	$4y = C$
Minimum requirement			
Calcium (units of)	$20x$	$+$	$10y \geq 300$
Vitamin A (units of)	$10x$	$+$	$20y \geq 180$

The 10 ounces would give 20(10) units of calcium and the 8 ounces would give 10(8) units of calcium for a total of

$$20(10) + 10(8) = 280 \quad \text{units of calcium}$$

and

$$10(10) + 20(8) = 260 \quad \text{units of vitamin A}$$

Now let's replace our guess of 10 ounces of food I by x ounces of food I and 8 ounces of food II by y ounces of food II. Then replace the 280 by a lower bound, 300 units of calcium, and the 260 by a lower bound, 180 units of vitamin A. That is,

$$20x + 10y \geq 300$$
$$10x + 20y \geq 180$$
$$x \geq 0$$
$$y \geq 0$$

and the objective function becomes

$$C = 6x + 4y \qquad \text{C is cholesterol-producing units.}$$

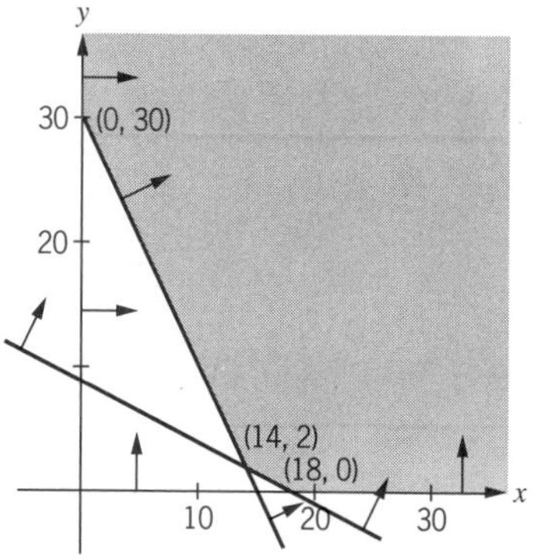

Figure 15

The feasible region of solutions and the corner points are found in Figure 15.

$$(0, 30): \quad C = 6(0) + 4(30) = 120$$

$$(14, 2): \quad C = 6(14) + 4(2) = 92 \qquad \text{Minimum}$$

$$(18, 0): \quad C = 6(18) + 4(0) = 108$$

The minimum number of cholesterol-producing units that meets the requirements for calcium and vitamin A is 92. Why is it reasonable that there is no maximum number of cholesterol-producing units? ∎

SUMMARY

The following steps are important in finding graphical solutions to linear programming problems:

1. Find the graph of the system of constraints.
2. Locate the corner points of the graph (feasible solution region).
3. Evaluate the objective function at each corner point to determine the maximum or minimum value.

Exercise Set 3.3

Applications (Business and Economics)

1. *Manufacturing*. A manufacturer makes two products, type I and type II. Type I requires 4 hours on machine A and 3 hours on machine B; type II requires 2 hours on A and 3 hours on B. If the machines operate up to 12 hours a day and the manufacturer makes a profit of $14 for each unit of type I and $13 for each unit of type II, find the number of units of each product that should be produced for maximum profit.
   (a) First understand the problem by completing the following table.
   (b) Translate to a linear programming problem by constructing the
      (i) objective function
      (ii) system of constraints
   (c) Graph the system of constraints, shade the feasible region, and find all corner points of this region.
   (d) Complete the solution of the problem.

Description	Type I	Type II
Number of each (estimate)		
Profit from (estimate)		=
Number of each (variable)		
Profit from (variable)		= P
		Maximum number of hours per day
Hours needed (machine A)		
Hours needed (machine B)		

2. *Manufacturing*. Redo Exercise 1 for profits of $80 per unit for type I and $200 per unit for type II.

3. *Minimum Cost*. An oil company requires 800, 1400, and 500 barrels of low-, medium-, and high-

grade oil, respectively. Refinery I produces 200, 300, and 100 barrels per day of low-, medium-, and high-grade oil, respectively; refinery II produces 100, 200, and 100 barrels per day of low-, medium-, and high-grade oil, respectively. If it costs $3500 per day to operate refinery I and $2000 per day to operate refinery II, how many days should each be operated to satisfy the requirements at minimum cost?

(a) First understand the problem by completing the following table.

(b) Translate to a linear programming problem by constructing the
   (i) objective function
   (ii) system of constraints

(c) Find the feasible region and all corners of this region.

(d) Complete the solution of this problem.

Description	Refinery 1	Refinery 2
Number of days (estimate)		
Cost of operation (from estimate)		=
Number of days (variable)		
Cost of operation (variable)		= C
		Minimum requirements
Low-grade oil		
Medium-grade oil		
High-grade oil		

4. *Assignment*. The sanitation department of a city has 100 garbage trucks and 250 employees. A full-strength collection team consists of 1 truck and 3 employees. A full-strength team collects 12 tons of garbage per day, and a partial-strength team with 2 employees collects only 6 tons. The city manager wishes to collect the maximum amount of garbage each day, but the numbers of operating trucks and available employees vary. How many full-strength teams and how many partial-strength teams should be formed in order to maximize the amount of garbage collected on a day when all trucks are operating and a maximum of 240 employees is available for

assignment? Write the mathematical model and graph the solution for the assignment problem of the city; then find the corner points and the answer.

5. *Transportation*. A company has two warehouses, A and B, and two stores, I and II. Warehouse A contains 100 tons of a product, and warehouse B contains 150 tons. Store I needs 50 tons of the product, and store II needs 75 tons. The shipping costs are

   A to I, $5 per ton
   B to I, $6 per ton

   A to II, $8 per ton
   B to II, $10 per ton

   Find shipping instructions that will satisfy the stores' needs at minimum shipping cost.

6. *Transportation*. Redo Exercise 5 with shipping costs of

   A to I, $8 per ton
   B to I, $6 per ton

   A to II, $5 per ton
   B to II, $10 per ton

7. *Investment*. An investor has up to $50,000 to invest in AA bonds or B bonds. The AA bonds yield 10%, and the B bonds yield 15%. If the investor wishes to invest at least 3 times as much in AA bonds as in B bonds, find the amount she should invest in each type of bond to maximize income.

8. *Mixture*. A shop sells a mixture of cashews and pecans for $3 a pound. $1000 has been allocated for buying cashews at $2.00 per pound and pecans at $1.50 per pound. The mixture must contain at least twice as many pecans as cashews (by weight), but no more than 3 times as many (by weight). How many pounds of each type should be ordered to maximize profit?

9. *Manufacturing*. A company manufactures two types of electric saws, M (medium size) and S (small). It takes 2 hours to assemble each small saw and 3 hours to assemble each medium-sized saw. The company can work up to 800 hours per week. However, the company can pack and ship only 300 saws per week. If the company sells the small saw

for $30 and the medium saw for $40, how many of each type should be produced in order to maximize revenue?

10. **Assembly Line.** A manufacturing company produces two products, A and B. Both products require time on three assemblies. The following table gives the time required in each assembly as well as the total time available in an assembly. If product A can be sold for a profit of $8 each and product B for a profit of $6 each, how many of each would produce a maximum profit? (**Hint:** Remember to convert the maximum time to minutes.)

Assembly	Product A (minutes)	Product B (minutes)	Maximum Time Available (hours)
1	10	8	8
2	12	4	24
3	4	10	12

Applications (Social and Life Sciences)

11. **Agriculture.** A vegetable farmer in Florida is to ship 1000 boxes of vegetables to New York City. He must ship at least 400 boxes of lettuce and at least 300 boxes of celery. His profits are $2.00 per box of lettuce and $2.50 per box of celery. How many boxes of each type of vegetable should he include in this shipment in order to maximize profit?

12. **Agriculture.** Work Exercise 11 if the constraint is added that the number of boxes of lettuce must be twice the number of boxes of celery.

13. **Education.** A university plans to use instructors and graduate assistants in a research project. The university needs at least 400 hours of labor spent in gathering data, with each instructor spending 8 hours and each graduate student spending 20 hours gathering data. After the data are gathered, at least 304 hours are required for processing. Each instructor can process data for 8 hours and each graduate assistant for 4 hours. If instructors cost $10 per hour and graduate assistants cost $7 per hour, find the number of each the university should use for minimum cost. Assume that the university uses the same instructors and graduate assistants to gather and process the data.

14. **Education.** Redo Exercise 13 for an instructor cost of $15 per hour, where the graduate assistant's cost remains at $7 per hour.

15. **Nutrition.** A diet company is planning to market a diet product composed of two foods. Food I contains 30 calories per unit, and each unit has 20 grams of protein, 20 grams of carbohydrate, and $\frac{7}{6}$ grams of fat. Food II contains 40 calories per unit, and each unit contains 10 grams of protein, 40 grams of carbohydrate, and $\frac{1}{2}$ gram of fat. How many units of each food should the product contain if the company wishes to minimize the number of calories, yet have the mixture satisfy the requirements of 60 grams of protein, 4 grams of fat, and 100 grams of carbohydrate?

16. **Nutrition.** A person on a low-carbohydrate diet plans to eat two foods. A unit of food I contains 5 grams of carbohydrate and 100 calories, of which 10 calories are protein. A unit of food II contains 6 grams of carbohydrate and 100 calories, of which 30 calories are protein. He wishes to minimize the number of grams of carbohydrate while eating at least 400 calories, at least 60 of which are provided by protein. Find the number of units that he should eat of each type of food.

17. **Nutrition (Animals).** A calf feed mixture is to be obtained from peanut hulls and soybean meal. The mixture must contain at least 100 pounds of protein, 40 pounds of fat, and 10 pounds of minerals. Each 100 pounds of peanut hulls contains 15 pounds of protein, 5 pounds of fat, and 1 pound of minerals. Each 100 pounds of soybean meal contains 50 pounds of protein, 10 pounds of fat, and 4 pounds of minerals. If each 100-pound bag of peanut hulls costs $8 and each 100-pound bag of soybean meal costs $12, how many bags of each must be purchased to minimize cost and satisfy minimum requirements?

18. **Medicine.** A pharmaceutical company is planning to make an antacid drug. There are two substances that seem to be effective as a mixture. Substance A contains 2 units of magnesia per ounce and 3 units of alumina per ounce. Substance B contains 3 units of magnesia per ounce and 4 units of alumina per ounce. The minimum requirement per pill is 6 units

of magnesia and 10 units of alumina. If an ounce of A costs $0.10 and an ounce of B costs $0.30, find the amount of each that should be used to minimize the cost of a pill.

19. **Exam.** A company manufactures and sells two products, A and B. Time requirements and selling prices for each product are given in the following table.

Products	A	B
Hours of labor for each item	1.0	1.5
Hours on a machine for each item	0.5	2.0
Selling price of each item	$27.50	$75.00

The company can work a total of 600,000 hours of labor and can use up to 200,000 hours on machines. The following questions relate to the use of linear programming to maximize gross sales, where x is the number of product A produced, and y is the number of product B produced. The objective function is

(a) $10.50x + 32.00y$
(b) $8.50x + 24.00y$
(c) $27.50x + 75.00y$
(d) $19.00x + 51.00y$
(e) $17.00x + 43.00y$

20. **Exam.** For the preceding problem, the constraint function for labor is
(a) $1x + 1.5y \le 200,000$
(b) $8x + 12y \le 600,000$
(c) $8x + 12y \le 200,000$

(d) $1x + 1.5y \le 4,800,000$
(e) $1x + 1.5y \le 600,000$

21. **Exam.** For Exercise 19, the constraint function for machine capacity is
(a) $6x + 24y \le 200,000$
(b) $0.5x + 2y \le 200,000$
(c) $(0.5 + 1)x + (2 + 1.5)y \le 200,000$
(d) $0.5x + 2y \le 600,000$
(e) $\dfrac{1x}{0.5} + \dfrac{1.5y}{2} \le 800,000$

Review Exercises

22. Find the maximum and minimum values, if they exist, of $P = 3x + 4y$ subject to the constraints

$$x \ge 0$$
$$y \ge 0$$
$$3x + 5y \le 30$$
$$3x + 5y \le 21$$

23. Find the maximum and minimum values, if they exist, of $C = 8x + 4y$ subject to the constraints

$$x \ge 0$$
$$y \ge 0$$
$$x + y \ge 5$$
$$3x + 5y \ge 21$$

3.4 A GRAPHICAL INTRODUCTION TO THE SIMPLEX METHOD

OVERVIEW   In the preceding sections we considered the geometric method of solving linear programming problems. Practically speaking, this procedure is useful only for problems involving two, or at most three, variables. However, our study of geometric procedures provides an excellent background for the study of the algebraic procedures that are needed for linear programming problems involving more than three variables. The number of variables in a linear programming problem often becomes quite large. For example, an investment company might consider 100 different sources for investments with the desire to maximize yield under given restrictions. In this section, we introduce an algebraic approach called the simplex method. This procedure makes use of our knowledge of row operations on matrices. Such a procedure is readily extended to any number of variables, and with computers linear programming problems involving hundreds, and even thousands, of variables can be considered. In this section we consider

- Slack variables
- Basic feasible solutions
- Pivot operations
- The simplex method

The **simplex method** (developed by George Dantzig in 1947) is a procedure for algebraically solving linear programming problems involving maximization. Let's return now to the manufacture of axles discussed in Example 11 of the preceding section. Maximize

$$P = 10x + 15y \qquad \text{Objective function}$$

subject to

$$x + 2y \leq 24 \qquad \text{Problem constraints}$$
$$1.5x + 1y \leq 24$$
$$x \geq 0 \qquad \text{Nonnegative constraints}$$
$$y \geq 0$$

where x and y are the number of small-car axles and the number of large-car axles, respectively, produced each day. Notice that all the problem constraints involve "less than or equal to" inequalities with positive constants on the right. In the next three sections we consider only maximization problems that satisfy this condition. Such a problem is sometimes called the **standard maximization problem**.

Standard Maximization Problem

A linear programming problem is said to be a **standard maximization problem** if it can be written in the following form:

Maximize

$$P = c_1 x_1 + c_2 x_2 + \cdots + c_n x_n \qquad \text{Objective function}$$

subject to

Standard Maximization Problem *(Continued)*

$$a_1x_1 + a_2x_2 + \cdots + a_nx_n \leq k_1$$
$$b_1x_1 + b_2x_2 + \cdots + b_nx_n \leq k_2$$
$$\vdots \qquad \vdots \qquad \qquad \vdots \qquad \vdots \qquad \text{Problem constraints}$$
$$d_1x_1 + d_2x_2 + \cdots + d_nx_n \leq k_i$$

$$(\text{all } k\text{'s} \geq 0)$$

$$x_1, x_2, \ldots, x_n \geq 0 \qquad \text{Nonnegative constraints}$$

To take advantage of matrix methods, we will first convert the constraint inequalities into equations. Recall that $a \leq b$ if and only if there is a nonnegative number s such that $a + s = b$. That is,

$2 < 5$ because there is an $s = 3$ such that $2 + s = 5$
$3 < 7$ because there is an $s = 4$ such that $3 + s = 7$

Note that s has a different value in each equation, so we use s as a variable.

DEFINITION: SLACK VARIABLE

A nonnegative variable that makes up the difference (or takes up the slack) between the left side and the right side of an inequality in the form $a < k$ to make it an equation is called a **slack variable**.

EXAMPLE 13   Convert the constraint inequalities

$$x + 2y \leq 24 \tag{1}$$
$$1.5x + 1y \leq 24$$

in the preceding example to equations using slack variables.

SOLUTION

$$x + 2y + r = 24 \qquad \text{Add slack variable } r \tag{2}$$
$$1.5x + y + s = 24 \qquad \text{Add slack variable } s$$

The constraint inequalities have been converted to equations; however, we have an enlarged set of nonnegative constraints

$$x \geq 0 \qquad r \geq 0$$
$$y \geq 0 \qquad s \geq 0$$

which we write in concise notation as

$$x, y, r, s \geq 0 \qquad \blacksquare$$

Although we omit the details, we can see that the system of inequalities (1) and the system of equations (2) are equivalent in the sense that if $x = a$ and $y = b$ is any solution of (1), then there are values of the slack variables so that $x = a$, $y = b$, $r = c$, and $s = d$ is a solution of (2). That is, $x = 2$, $y = 4$ satisfies (1) and $x = 2$, $y = 4$, $r = 14$, and $s = 17$ satisfies (2). Conversely, if $x = a$, $y = b$, $r = c$, and $s = d$ is a solution of (2), then $x = a$, $y = b$ is a solution of (1).

Next we write the objective function

$$P = 10x + 15y$$

in a form with all the variables on the left and 0 on the right.

$$-10x - 15y + P = 0$$

We then have the following system of equations:

$$x + 2y + r = 24$$
$$1.5x + y + s = 24$$
$$-10x - 15y + P = 0$$

Each solution of this system is a set of numbers

$$x, y, r, s, P$$

Now recall that for three equations in five unknowns, an infinite number of solutions may exist; by assigning values to two of the variables and solving for the other three, you can obtain a solution (provided that a solution exists). When we set two of the variables equal to 0 and solve for the other three, the answer is called a **basic solution**. In general, with m equations and n unknowns, where $n > m$, we set $n - m$ variables equal to zero and solve for the remaining m variables to get a basic solution.

DEFINITION: BASIC FEASIBLE SOLUTION

When a linear system consisting of m equations and n unknowns such that $n > m$ is associated with a linear programming problem, any basic solution of the system that has no negative values is called a **basic feasible solution**. The $n - m$ variables set equal to 0 in obtaining a basic solution are called **nonbasic variables**; the m variables for which we solve are called **basic variables**.

EXAMPLE 14   Discuss the basic feasible solutions for the following problem. Maximize

$$P = 4x + 3y$$

subject to

$$5x + 3y \leq 30$$

$$2x + 3y \leq 21$$

$$x, y \geq 0$$

SOLUTION   Using slack variables, the corresponding problem as a system of equations is

$$5x + 3y + r = 30$$

$$2x + 3y + s = 21$$

$$-4x - 3y + P = 0$$

$$x, y, r, s, P \geq 0$$

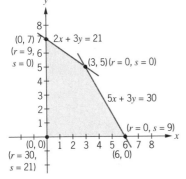

Figure 16

The five variables x, y, r, s, and P are all nonnegative and satisfy three equations. The feasible region for this problem is shown in Figure 16 in terms of the original variables x and y. We know that the optimum value of any linear function such as $P = 4x + 3y$ must be attained at one (or more) of the corners of the region.

Let's examine the solutions (x, y, r, s, P) of the three equations at the corner points. At the origin, $x = 0$, $y = 0$, so $r = 30$, $s = 21$, and $P = 0$; at $(0, 7)$, $x = 0$, so $s = 0$, $r = 9$, and $P = 21$; at $(3, 5)$, $s = 0$, $r = 0$, and $P = 27$; and at $(6, 0)$, $r = 0$, $s = 9$, and $P = 24$. These are basic feasible solutions of the system of equations.

Since the basic feasible solutions in the preceding example include all the corner points of the solution region, our discussion suggests, without proof, the following fundamental theorem.

BASIC FEASIBLE SOLUTION

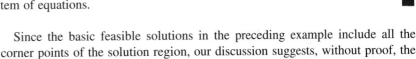

If a linear programming problem has an optimal solution (maximum or minimum), it must be at one of the basic feasible solutions.

Simplex Tableau

The simplex method consists of choosing a particular basic feasible solution as a starting point and then transforming from this to another basic feasible solution in such a way that the objective function gets closer and closer to being optimal. In the next section we will study the procedure to accomplish this result.

The transformation that moves from one basic feasible solution to another consists of row operations on a matrix called the **simplex tableau**. For the system of equations in Example 14

$$5x + 3y + r \qquad\quad = 30$$
$$2x + 3y \qquad + s \qquad = 21$$
$$-4x - 3y \qquad\quad + P = \quad 0$$

where $x \geq 0$, $y \geq 0$, $r \geq 0$, and $s \geq 0$, the simplex tableau is as follows:

$$
\begin{array}{ccccc}
x & y & r & s & P \\
\end{array}
$$

$$
\left[
\begin{array}{ccccc|c}
5 & 3 & 1 & 0 & 0 & 30 \\
2 & 3 & 0 & 1 & 0 & 21 \\
\hline
-4 & -3 & 0 & 0 & 1 & 0 \\
\end{array}
\right]
\tag{3}
$$

For tableau (3) set the variables x and y equal to zero. Note that after assigning $x = 0$ and $y = 0$, the coefficient matrix (under r, s, and P) in (3) is the identity matrix

$$
\begin{bmatrix}
1 & 0 & 0 \\
0 & 1 & 0 \\
0 & 0 & 1 \\
\end{bmatrix}
$$

Solving for r, s, and P, where $x = 0$ and $y = 0$, we have

$$5(0) + 3(0) + r = 30$$
$$2(0) + 3(0) + s = 21$$
$$-4(0) - 3(0) + P = \quad 0$$

We get $x = 0$, $y = 0$, $r = 30$, $s = 21$, and $P = 0$, which is the basic feasible solution corresponding to $(0, 0)$ in Figure 16.

By performing row operations on (3) we can place the identity matrix under three other variables, which will produce another basic feasible solution.

From Corner Point to Corner Point

We have already noted that a corner point corresponds in a natural way to a basic feasible solution. We now show how to move from one corner point (in the tableau) to a new corner point (represented by another tableau). By this procedure, a search can be made for the corner point that will optimize the objective function. To accomplish this goal, we use elimination procedures of matrix theory. Our goal is to obtain the identity matrix as the coefficient matrix of different variables. This will give a new corner point of the basic feasible solutions. For example, we perform row operations on tableau 3 to get a new tableau where the identity matrix is the coefficient matrix of the variables x, s, and P instead of r, s, and P.

$$\begin{array}{ccccc} x & y & r & s & P \\ \end{array}$$
$$\left[\begin{array}{ccccc|c} 5 & 3 & 1 & 0 & 0 & 30 \\ 2 & 3 & 0 & 1 & 0 & 21 \\ \hline -4 & -3 & 0 & 0 & 1 & 0 \end{array}\right] \tag{4}$$

To accomplish this goal, use the following three elementary row operations in order, where R_1 is the first row, R_2 is the second row, and R_3 is the third row.

$$\tfrac{1}{5}R_1 \rightarrow R_1 \qquad \text{Makes coefficient of } x \text{ become } 1$$

$$-2R_1 + R_2 \rightarrow R_2 \qquad \text{Makes coefficient of } x \text{ become } 0$$

$$4R_1 + R_3 \rightarrow R_3 \qquad \text{Makes coefficient of } x \text{ become } 0$$

We obtain the following tableau.

$$\begin{array}{ccccc} x & y & r & s & P \\ \end{array}$$
$$\left[\begin{array}{ccccc|c} 1 & \frac{3}{5} & \frac{1}{5} & 0 & 0 & 6 \\ 0 & \frac{9}{5} & -\frac{2}{5} & 1 & 0 & 9 \\ \hline 0 & -\frac{3}{5} & \frac{4}{5} & 0 & 1 & 24 \end{array}\right] \tag{5}$$

Note that the identity matrix

$$\begin{bmatrix} 1 & 0 & 0 \\ 0 & 1 & 0 \\ 0 & 0 & 1 \end{bmatrix}$$

is now the coefficient matrix of x, s, and P in tableau 5. If we set $y = 0$ and $r = 0$, we have from the constant column

$$x = 6, \quad s = 9, \quad \text{and} \quad P = 24$$

corresponding to the point $(6, 0)$ in Figure 16.

This method will be replaced in the next section by a procedure that locates a pivot element. The **pivot element** is the element that is made to equal 1, and all the other entries in the column containing the pivot element are made equal to 0.

DEFINITION: PIVOT OPERATIONS

Pivot operations are row operations (other than interchanges) performed on a matrix to make the pivot element 1 and to make all other entries in the pivot column 0.

EXAMPLE 15 Perform a pivot operation on the following simplex tableau with pivot element 3 (circled) to obtain a second matrix. Write the system of equations represented by the second matrix.

SOLUTION

$$
\begin{array}{c}
\boxed{\text{Need a 1}} \\
\boxed{\text{Need a 0}}
\end{array}
\quad
\begin{array}{ccccc}
x & y & r & s & P \\
\end{array}
$$

$$
\left[
\begin{array}{ccccc|c}
③ & 2 & 1 & 0 & 0 & 8 \\
2 & 4 & 0 & 1 & 0 & 8 \\
-14 & -12 & 0 & 0 & 1 & 0
\end{array}
\right]
\quad
\begin{array}{ll}
\frac{1}{3}R_1 \rightarrow R_1 & (1) \\
-2R_1 + R_2 \rightarrow R_2 & (2) \\
14R_1 + R_3 \rightarrow R_3 & (3)
\end{array}
$$

The pivot operation gives the following second matrix:

$$
\begin{array}{ccccc}
x & y & r & s & P \\
\end{array}
$$

$$
\left[
\begin{array}{ccccc|c}
1 & \frac{2}{3} & \frac{1}{3} & 0 & 0 & \frac{8}{3} \\
0 & \frac{8}{3} & -\frac{2}{3} & 1 & 0 & \frac{8}{3} \\
0 & -\frac{8}{3} & \frac{14}{3} & 0 & 1 & \frac{112}{3}
\end{array}
\right]
$$

This second matrix represents the system

$$
x + \frac{2}{3}y + \frac{1}{3}r = \frac{8}{3}
$$

$$
\frac{8}{3}y - \frac{2}{3}r + s = \frac{8}{3}
$$

$$
-\frac{8}{3}y + \frac{14}{3}r + P = \frac{112}{3}
$$

Practice Problem 1 Form a simplex tableau for the information given in the following problem. Maximize

$$
P = 7x + 3y
$$

subject to

$$
3x + 2y \le 4
$$

$$
5x + y \le 8
$$

$$
x, y \ge 0
$$

ANSWER

$$
\begin{array}{ccccc}
x & y & r & s & P \\
\end{array}
$$

$$
\left[
\begin{array}{ccccc|c}
3 & 2 & 1 & 0 & 0 & 4 \\
5 & 1 & 0 & 1 & 0 & 8 \\
-7 & -3 & 0 & 0 & 1 & 0
\end{array}
\right]
$$

Calculator Note

In Chapter 2 we learned to perform all the row operations on a calculator in order to change the pivot element of a matrix to 1 and to change all the elements in the pivot column to 0. Review these operations. The designated row operations listed to the right of the matrix in Example 15 are listed in a form that facilitates calculator computation. First we store the matrix in Example 15 in position [A], and then we perform the row operations (1), (2), and (3) in order.

Row operation (1), $\frac{1}{3}R_1 \to R_1$, corresponds on a calculator to $*$ row (1/3, [A], 1). This operation multiplies row 1 of [A] by 1/3:

$$[1 \quad 0.66667 \quad 0.33333 \quad 0 \quad 0 \quad 2.66667] \qquad \text{Rounded}$$

Row operation (2), $-2R_1 + R_2 \to R_2$, corresponds to $*$ row$+(-2, \text{ANS}, 1, 2)$, which multiplies row 1 of the new [A] by -2 and adds the result to row 2:

$$[0 \quad 2.66667 \quad -0.66667 \quad 1 \quad 0 \quad 2.66667]$$

Row operation (3), $14R_1 + R_3 \to R_3$, corresponds to $*$ row$+(14, \text{ANS}, 1, 3)$, which multiplies row 1 by 14 and adds the result to row 3:

$$[0 \quad -2.66667 \quad 4.66667 \quad 0 \quad 1 \quad 37.33333]$$

Note that the 3×6 matrix obtained is the same as that found in Example 15.

Practice Problem 2 Using a calculator, perform a pivot operation on the simplex tableau in Practice Problem 1 to get a second tableau if the pivot element is 3 in the first row and first column:

$$
\begin{array}{ccccc}
x & y & r & s & P \\
\end{array}
$$

$$
\left[
\begin{array}{ccccc|c}
③ & 2 & 1 & 0 & 0 & 4 \\
5 & 1 & 0 & 1 & 0 & 8 \\
\hline
-7 & -3 & 0 & 0 & 1 & 0 \\
\end{array}
\right]
$$

ANSWER

$$
\begin{array}{ccccc}
x & y & r & s & P \\
\end{array}
$$

$$
\left[
\begin{array}{ccccc|c}
1 & 0.6667 & 0.3333 & 0 & 0 & 1.3333 \\
0 & -2.3333 & -1.6667 & 1 & 0 & 1.3333 \\
\hline
0 & 1.6667 & 2.3333 & 0 & 1 & 9.3333 \\
\end{array}
\right]
$$

Practice Problem 3 Using the simplex tableau

$$
\begin{array}{ccccc}
x & y & r & s & P \\
\end{array}
$$

$$
\left[
\begin{array}{ccccc|c}
2 & 3 & 1 & 0 & 0 & 6 \\
4 & 2 & 0 & 1 & 0 & 2 \\
\hline
-3 & -5 & 0 & 0 & 1 & 0 \\
\end{array}
\right]
$$

(a) Write the equations represented by this tableau.

(b) Let x and y be zero. Find the basic feasible solution.

(c) List a set of three consecutive row operations that will change the basic variables to y, s, and P.

ANSWER

(a) $2x + 3y + r = 6$
$4x + 2y + s = 2$
$-3x - 5y + P = 0$

(b) Since x and y are nonbasic variables, set $x = 0$ and $y = 0$ in part (a).

$$0 + 0 + r = 6$$

$$0 + 0 + s = 2$$

$$0 + 0 + P = 0$$

So $r = 6$, $s = 2$, and $P = 0$.

(c) Now the basic variables are y, s, and P. The coefficients of these variables must be the corresponding components of the identity matrix. We need to find a set of row operations that will change the y column to

$$\begin{bmatrix} 1 \\ 0 \\ 0 \end{bmatrix}$$

Hence,

$$\tfrac{1}{3}R_1 \rightarrow R_1$$

$$-2R_1 + R_2 \rightarrow R_2$$

$$5R_1 + R_3 \rightarrow R_3$$

Where We Are Headed

In the next section we will use what we have learned in this section to develop a very powerful procedure to move quickly to the corner point that gives the optimal solution without checking all corner points. That is, each time we move to a corner point the procedure will be such that P increases in value.

SUMMARY

In this section we learned to

1. Work linear programming problems as standard maximization problems.
2. Write a set of constraints as a set of equalities using slack variables.
3. Obtain a set of basic feasible solutions.
4. Write a linear programming problem as a simplex tableau.
5. Perform pivot operations to make the pivot element 1 and all other elements in the pivot column 0.

Exercise Set 3.4

Write the equations represented by each of the following tableaus and find a basic feasible solution.

1.

$$
\begin{array}{ccccc}
x & y & r & s & P \\
\end{array}
$$

$$
\left[
\begin{array}{ccccc|c}
3 & 2 & 1 & 0 & 0 & 8 \\
2 & 4 & 0 & 1 & 0 & 8 \\
\hline
-14 & -12 & 0 & 0 & 1 & 0 \\
\end{array}
\right]
$$

2.

$$
\begin{array}{ccccc}
x & y & r & s & P \\
\end{array}
$$

$$
\left[
\begin{array}{ccccc|c}
2 & 3 & 1 & 0 & 0 & 2 \\
4 & 1 & 0 & 1 & 0 & 3 \\
\hline
5 & -7 & 0 & 0 & 1 & 4 \\
\end{array}
\right]
$$

3.

$$
\begin{array}{ccccc}
x & y & r & s & P \\
\end{array}
$$

$$
\left[
\begin{array}{ccccc|c}
2 & 0 & 4 & 1 & 0 & 7 \\
3 & 1 & 2 & 0 & 0 & 5 \\
\hline
4 & 0 & -3 & 0 & 1 & 12 \\
\end{array}
\right]
$$

4.

$$
\begin{array}{cccccc}
x & y & r & s & t & P \\
\end{array}
$$

$$
\left[
\begin{array}{cccccc|c}
2 & 0 & 1 & 0 & 4 & 0 & 5 \\
3 & 1 & 0 & 0 & 1 & 0 & 1 \\
4 & 0 & 0 & 1 & 2 & 0 & 6 \\
\hline
2 & 0 & 0 & 0 & -3 & 1 & 8 \\
\end{array}
\right]
$$

Form the simplex tableau for each of the following linear programming problems, where $x \geq 0$ and $y \geq 0$.

5. Maximize

$$P = 25x + 35y$$

subject to

$$2x + 3y \leq 15$$
$$3x + y \leq 12$$

6. Maximize

$$P = 35x + 25y$$

subject to

$$2x + y \leq 7$$
$$3x + y \leq 8$$

7. Maximize

$$P = 35x + 25y$$

subject to

$$2x + 3y \leq 15$$
$$3x + y \leq 12$$

8. Maximize

$$P = 25x - 35y$$

subject to

$$2x + 3y \leq 15$$
$$3x + y \leq 12$$

9. Maximize

$$P = 35x + 25y$$

subject to

$$x + y \leq 7$$
$$2x + y \leq 9$$
$$3x + y \leq 12$$

10. Maximize

$$P = 25x + 35y$$

subject to

$$x + y \leq 7$$
$$2x + y \leq 9$$
$$3x + y \leq 12$$

11. Maximize

$$P = 35x + 25y$$

subject to

$$x + y \leq 7$$
$$2x + y \leq 9$$
$$3x + y \leq 12$$

12. Maximize

$$P = 25x + 35y$$

subject to

$$x + 3y \leq 21$$
$$2x + 3y \leq 24$$
$$2x + y \leq 16$$

13. Maximize

$$P = 25x + 35y$$

subject to

$$2x + 3y \le 6$$
$$x + 3y \le 21$$
$$2x + 3y \le 24$$
$$2x + y \le 16$$

14. Maximize

$$P = 25x + 35y$$

subject to

$$x + y \le 2$$
$$x + y \le 7$$
$$2x + y \le 9$$
$$3x + y \le 12$$

For the following tableaus, perform the appropriate row operations to change the basic variables to those listed.

15. Tableau in Exercise 1; basic variables: y, s, P
16. Tableau in Exercise 2; basic variables: x, r, P
17. Tableau in Exercise 3; basic variables: r, s, P
18. Tableau in Exercise 4; basic variables: y, s, t, P

Using your calculator, find the second tableau for the following simplex tableaus, with the pivot elements circled. Write the system of equations represented by the second tableau.

19.
$$\begin{array}{ccccc} x & y & r & s & P \\ \end{array}$$
$$\left[\begin{array}{ccccc|c} 5 & 2 & 1 & 0 & 0 & 10 \\ ④ & 3 & 0 & 1 & 0 & 6 \\ \hline -10 & -5 & 0 & 0 & 1 & 0 \end{array}\right]$$

20.
$$\begin{array}{ccccc} x & y & r & s & P \\ \end{array}$$
$$\left[\begin{array}{ccccc|c} 5 & 2 & 1 & 0 & 0 & 10 \\ 4 & ③ & 0 & 1 & 0 & 6 \\ \hline -5 & -10 & 0 & 0 & 1 & 0 \end{array}\right]$$

21.
$$\begin{array}{ccccc} x & y & r & s & P \\ \end{array}$$
$$\left[\begin{array}{ccccc|c} 1 & \frac{2}{3} & \frac{1}{3} & 0 & 0 & \frac{8}{3} \\ 0 & ⑧\!\!/\!\!3 & -\frac{2}{3} & 1 & 0 & \frac{8}{3} \\ \hline 0 & -\frac{8}{3} & \frac{14}{3} & 0 & 1 & \frac{112}{3} \end{array}\right]$$

22.
$$\begin{array}{cccccc} x & y & r & s & t & P \\ \end{array}$$
$$\left[\begin{array}{cccccc|c} 1 & 2 & 3 & 0 & 0 & 0 & 5 \\ 0 & 5 & ⑥ & 1 & 0 & 0 & 7 \\ 0 & 4 & 2 & 0 & 1 & 0 & 9 \\ \hline 0 & -5 & -7 & 0 & 0 & 1 & 4 \end{array}\right]$$

23.
$$\begin{array}{cccccc} x & y & r & s & t & P \\ \end{array}$$
$$\left[\begin{array}{cccccc|c} 3 & 2 & 1 & 0 & 0 & 0 & 3 \\ 2 & 3 & 0 & 1 & 0 & 0 & 4 \\ 1 & ④ & 0 & 0 & 1 & 0 & 5 \\ \hline -4 & -6 & 0 & 0 & 0 & 1 & 0 \end{array}\right]$$

24.
$$\begin{array}{cccccc} x & y & r & s & t & P \\ \end{array}$$
$$\left[\begin{array}{cccccc|c} 1 & ⑤ & 1 & 0 & 0 & 0 & 3 \\ 2 & 6 & 0 & 1 & 0 & 0 & 4 \\ 3 & 7 & 0 & 0 & 1 & 0 & 5 \\ \hline 4 & -2 & 0 & 0 & 0 & 1 & 6 \end{array}\right]$$

Applications (Business and Economics)

For the applications problems, set up the appropriate linear programs and for maximization problems the simplex tableaus.

25. **Advertising.** An electronics firm has \$40,000 available for advertising. The following table gives the cost of an advertising package and the number of people in the advertising audience for each advertisement in two different media (numbers in thousands).

	Newspaper	TV
Cost	\$ 4	\$10
Audience	20	60

If the maximum number of newspaper advertisements is 8 and the maximum number of television advertisements is 2, how many advertisements of each type should be purchased to maximize advertising exposure?

26. **Investment.** An investor has \$200,000 to invest in bonds, municipal bond mutual funds, and money market funds. The average yields for these three are 8%, 10%, and 12%, respectively. For tax purposes the amount invested in bonds and money market funds must not exceed \$100,000. How much should be invested in each type of investment and what is the maximum yearly interest?

27. **Manufacturing.** A bicycle assembly plant consists of two shops. In shop A the bicycle frame is built, and in shop B wheels, transmission, and brakes are

added. Two bicycles are produced, a standard 3-speed and a deluxe 10-speed. The 3-speed bicycle sells for a profit of $10. It requires 1 hour of work in shop A and 2 hours in shop B. The 10-speed produces a $25 profit. It requires 2 hours in shop A and 3 hours in shop B. If shop A has 48 hours available each day and shop B has 84 hours available each day, how many of each type of bicycle should be produced to maximize profit?

Applications (Social and Life Sciences)

28. **Health Care**.   The government has mobilized to inoculate the student population against sleeping sickness. There are 200 doctors available and 450 nurses. An inoculation team can consist of either 1 doctor and 3 nurses (team A) or 1 doctor and 2 nurses (team B). On the average, team A can inoculate 180 students per hour, while team B can only inoculate 100 students per hour. How many teams of each type should be formed to maximize the number of inoculations per hour?

29. **Farming**.   A farmer has available 6000 acres of land to raise cattle, horses, and sheep. The farmer cannot graze more than three head of cattle, or five head of sheep, or one horse per acre on this land. In the winter (approximately 15 weeks) he will need three bales of hay per week for a horse, 2 bales per week for a cow, and 1 bale per week for a head of sheep. He has a total of 5000 bales of hay available. If he averages a profit of $20 per head for cattle, $15 per head for horses, and $10 per head for sheep, how many of each should he have in order to maximize profit?

3.5 THE SIMPLEX METHOD OF MAXIMIZATION

OVERVIEW   In the previous section we learned to form an initial simplex tableau. We suggested that the simplex method is a technique for moving from one basic feasible solution to another until the maximum of the objective function is obtained. In this section, we learn how the simplex technique is performed to obtain a final solution. In this section we learn

- How to select the pivot element
- How to work toward the next feasible solution
- When to look for the maximum value
- How to obtain a maximum value

Let's return now to a linear programming problem we have studied. Maximize

$$P = 14x + 12y \qquad \text{Objective function}$$

subject to

$$3x + 2y \le 8 \qquad \text{Problem constraints}$$
$$2x + 4y \le 8$$
$$x, y \ge 0 \qquad \text{Nonnegative constraints}$$

with corresponding tableau

$$
\begin{array}{ccccc}
x & y & r & s & P \\
\end{array}
$$

$$
\left[
\begin{array}{ccccc|c}
3 & 2 & 1 & 0 & 0 & 8 \\
2 & 4 & 0 & 1 & 0 & 8 \\
\hline
-14 & -12 & 0 & 0 & 1 & 0
\end{array}
\right]
$$

The row below the dashed line always represents the objective function. The numbers in this row are called **indicators**. They are used to select the variable that produces the largest increase in the value of the objective function. That is, in order to determine the pivot element, we first scan the indicator row and find the negative number with the largest absolute value. This yields the pivot column. For our example the first column is the pivot column because $|-14|$ is larger than $|-12|$. Now, how much can x be increased when $y = 0$ without causing r and s to become negative? For r and s to remain positive, $3x \le 8$ and $2x \le 8$ or $x \le 2\frac{2}{3}$ and $x \le 4$. That which satisfies both is $x \le 2\frac{2}{3}$. Note that this can be determined by dividing each positive element in the pivot column above the dashed line into the corresponding row element in the right column. From these calculations, choose the minimum value of the resulting quotients. This indicates the pivot row. For our example, the computations give

$$
\frac{8}{3} = 2\tfrac{2}{3} \quad \text{and} \quad \frac{8}{2} = 4
$$

The minimum quotient is $2\frac{2}{3}$; therefore, the first row is the pivot row. The element in the pivot column and pivot row is the pivot element. In our example, the 3, which has been circled, is the pivot element.

$$
\begin{array}{ccccc}
x & y & r & s & P \\
\end{array}
$$

Pivot row $\rightarrow$
$$
\left[
\begin{array}{ccccc|c}
③ & 2 & 1 & 0 & 0 & 8 \\
2 & 4 & 0 & 1 & 0 & 8 \\
\hline
-14 & -12 & 0 & 0 & 1 & 0
\end{array}
\right]
\quad
\begin{array}{l}
\frac{8}{3} = 2\tfrac{2}{3} \\[4pt]
\frac{8}{2} = 4
\end{array}
$$

$\uparrow$

Pivot column

Note that the pivot element is always positive and is never on the bottom row.

Procedure for Finding the Pivot Element

1. Choose as the **pivot column** the variable column that has a negative number in the bottom row that is the largest number in absolute value. If the negative number in the last row with the largest absolute value occurs in two or more variable columns, choose any one of these columns as the pivot column and proceed to the next step.

2. Divide each *positive* element above the dashed line of the pivot column into the right column entry in the same row of the positive element. If the pivot column has no positive elements above the dashed line, then there is no solution.

Procedure for Finding the Pivot Element *(Continued)*

3. Choose as the **pivot row** the row for which the quotient obtained in step 2 is smallest. If two or more quotients are the same, choose any one of these rows as the pivot row.
4. The **pivot element** is the element in both the pivot row and the pivot column and is denoted with a circle.

Now that we have located the pivot element in the example under discussion, our next step is to perform appropriate matrix row operations so that our pivot element will be 1 and the remaining elements in the pivot column will be 0.

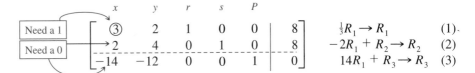

$$
\begin{array}{c}
\text{Need a 1} \\
\text{Need a 0}
\end{array}
\begin{array}{ccccc|c}
x & y & r & s & P & \\
③ & 2 & 1 & 0 & 0 & 8 \\
2 & 4 & 0 & 1 & 0 & 8 \\
-14 & -12 & 0 & 0 & 1 & 0
\end{array}
\qquad
\begin{array}{ll}
\tfrac{1}{3}R_1 \to R_1 & (1) \\
-2R_1 + R_2 \to R_2 & (2) \\
14R_1 + R_3 \to R_3 & (3)
\end{array}
$$

This procedure can be summarized as follows.

The second tableau is

$$
\begin{array}{ccccc|c}
x & y & r & s & P & \\
1 & \tfrac{2}{3} & \tfrac{1}{3} & 0 & 0 & \tfrac{8}{3} \\
0 & \tfrac{8}{3} & -\tfrac{2}{3} & 1 & 0 & \tfrac{8}{3} \\
0 & -\tfrac{8}{3} & \tfrac{14}{3} & 0 & 1 & \tfrac{112}{3}
\end{array}
$$

Now we take the second matrix as a new simplex tableau and obtain a third matrix from it by pivot operations. As we go through the procedure again, you should work it out on paper yourself, using this development only as a check. The *y* column is chosen as the pivot column because it is the variable column containing the negative number with the largest absolute value in the last row. Dividing $\tfrac{8}{3}$ by $\tfrac{8}{3}$ gives 1, while dividing $\tfrac{8}{3}$ by $\tfrac{2}{3}$ gives 4. Since 1 is the smaller value, the second row is chosen as the pivot row. Hence, the pivot element is $\tfrac{8}{3}$.

$$
\begin{array}{c}
\text{Need a 0} \\
\text{Need a 1}
\end{array}
\begin{array}{ccccc|c}
x & y & r & s & P & \\
1 & \tfrac{2}{3} & \tfrac{1}{3} & 0 & 0 & \tfrac{8}{3} \\
0 & ③ & -\tfrac{2}{3} & 1 & 0 & \tfrac{8}{3} \\
0 & -\tfrac{8}{3} & \tfrac{14}{3} & 0 & 1 & \tfrac{112}{3}
\end{array}
\qquad
\begin{array}{ll}
-\tfrac{2}{3}R_2 + R_1 \to R_1 & (2) \\
\tfrac{3}{8}R_2 \to R_2 & (1) \\
\tfrac{8}{3}R_2 + R_3 \to R_3 & (3)
\end{array}
$$

The third tableau is

$$
\begin{array}{ccccc|c}
x & y & r & s & P & \\
1 & 0 & \tfrac{1}{2} & -\tfrac{1}{4} & 0 & 2 \\
0 & 1 & -\tfrac{1}{4} & \tfrac{3}{8} & 0 & 1 \\
0 & 0 & 4 & 1 & 1 & 40
\end{array}
$$

P cannot be made larger by increasing any of the variables. This is seen by writing the equation represented by the last row of the matrix:

$$4r + s + P = 40$$

or

$$P = 40 - 4r - s$$

As r and s increase, P decreases. Since we cannot make P larger, we stop the process and read the answer.

$$x + \frac{1}{2}r - \frac{1}{4}s = 2$$

$$y - \frac{1}{4}r + \frac{3}{8}s = 1$$

$$4r + s + P = 40$$

When $r = s = 0$, this system gives

$$x = 2, \quad y = 1, \quad \text{and} \quad P = 40$$

The maximum value of P is 40 at the corner point (2, 1).

Calculator Note

A program to solve a simplex linear programming problem using a graphing calculator is given on page 12 of Appendix A. Some of the windows are shown for the preceding example.

Practice Problem 1 Find the pivot element for the simplex tableau

$$
\begin{bmatrix}
x & y & r & s & P & \\
3 & 1 & 1 & 0 & 0 & 6 \\
2 & 5 & 0 & 1 & 0 & 8 \\
-6 & -8 & 0 & 0 & 1 & 0
\end{bmatrix}
$$

ANSWER   The pivot element is 5.

Practice Problem 2 Maximize

$$P = 40x + 25y$$

subject to

$$x + 2y \leq 3$$

$$x + 3y \leq 5$$

$$2x + 3y \leq 7$$

$$4x + 2y \leq 9$$

$$x, y \geq 0$$

Set up a first tableau, circle the pivot element, and find a second tableau.

ANSWER

$$\begin{array}{ccccccc}
x & y & r & s & t & u & P \\
\end{array}$$

$$\left[\begin{array}{ccccccc|c}
1 & 2 & 1 & 0 & 0 & 0 & 0 & 3 \\
1 & 3 & 0 & 1 & 0 & 0 & 0 & 5 \\
2 & 3 & 0 & 0 & 1 & 0 & 0 & 7 \\
④ & 2 & 0 & 0 & 0 & 1 & 0 & 9 \\
\hline
-40 & -25 & 0 & 0 & 0 & 0 & 1 & 0
\end{array}\right] \quad \text{First tableau}$$

$$\begin{array}{ccccccc}
x & y & r & s & t & u & P \\
\end{array}$$

$$\left[\begin{array}{ccccccc|c}
0 & 1.5 & 1 & 0 & 0 & -0.25 & 0 & 0.75 \\
0 & 2.5 & 0 & 1 & 0 & -0.25 & 0 & 2.75 \\
0 & 2 & 0 & 0 & 1 & -0.5 & 0 & 2.5 \\
1 & 0.5 & 0 & 0 & 0 & 0.25 & 0 & 2.25 \\
\hline
0 & -5 & 0 & 0 & 0 & 10 & 1 & 90
\end{array}\right] \quad \text{Second tableau}$$

The key steps in the preceding procedure can be summarized as follows.

KEY STEPS: Simplex Method

If the problem constraints are of the form $\leq$ with nonnegative constants on the right, the key steps for maximization are as follows:

1. Introduce slack variables and write the appropriate equations.
2. Form the first tableau.
3. Determine the pivot element.
4. Perform pivot operations on the pivot column.
5. Repeat steps 3 and 4 until all indicators on the bottom row are nonnegative.
6. When this occurs we stop the process and read the optimal solution in the lower right corner (last row, last column).

EXAMPLE 16   Use the simplex method and a calculator to solve the following problem. Maximize

$$P = \$68x + \$70y$$

subject to

$$x \geq 0$$
$$y \geq 0$$
$$6x + 5y \leq 960$$
$$10x + 11y \leq 1760$$

SOLUTION

1. Introduce slack variables and write the appropriate equations.

$$6x + 5y + r = 960$$
$$10x + 11y + s = 1760$$

2. Form the first tableau.

$$
\begin{array}{ccccc}
x & y & r & s & P \\
\end{array}
$$

$$
\left[
\begin{array}{ccccc|c}
6 & 5 & 1 & 0 & 0 & 960 \\
10 & 11 & 0 & 1 & 0 & 1760 \\
\hline
-68 & -70 & 0 & 0 & 1 & 0 \\
\end{array}
\right]
$$

3. Determine the pivot element. The absolute value of -70 is larger than the absolute value of -68. The second column is the pivot column. Now divide the right column entry by the corresponding row entry of the pivot column.

$$\frac{960}{5} = 192, \qquad \frac{1760}{11} = 160$$

Since $160 < 192$, the second row is the pivot row. The pivot element will be the second-column, second-row entry.

4. Perform matrix operations on the pivot column to achieve the desired format using a graphing calculator.

$$
\begin{array}{ccccc}
x & y & r & s & P \\
\end{array}
$$

$$
\left[
\begin{array}{ccccc|c}
6 & 5 & 1 & 0 & 0 & 960 \\
10 & \boxed{11} & 0 & 1 & 0 & 1760 \\
\hline
-68 & -70 & 0 & 0 & 1 & 0 \\
\end{array}
\right]
\qquad
\begin{array}{ll}
-5R_2 + R_1 \rightarrow R_1 & (2) \\
\frac{1}{11}R_2 \rightarrow R_2 & (1) \\
70R_2 + R_3 \rightarrow R_3 & (3) \\
\end{array}
$$

Row operation (1), $\frac{1}{11}R_2 \to R_2$, is accomplished by *row $(\frac{1}{11}, [A], 2)$; row operation (2), $-5R_2 + R_1 \to R_1$, by *row $+ (-5, \text{ANS}, 2, 1)$; and row operation (3), $70R_2 + R_3 \to R_3$, by *row $+ (70, \text{ANS}, 2, 3)$ to obtain

$$
\begin{array}{ccccc}
x & y & r & s & P
\end{array}
$$
$$
\left[
\begin{array}{ccccc|c}
1.4545 & 0 & 1 & -0.4545 & 0 & 160 \\
0.9091 & 1 & 0 & 0.0909 & 0 & 160 \\
\hline
-4.3636 & 0 & 0 & 6.3636 & 1 & 11,200
\end{array}
\right]
$$

5. Since there is still a negative element in the first column of the bottom row, repeat the process of finding the next pivot element by using the first column as the pivot column. Since

$$160 \div 0.9091 \approx 176$$

$$160 \div 1.4545 \approx 110$$

1.4545 is chosen as the pivot element. Then $(1/1.4545)R_1 \to R_1$ is accomplished by *row$(1/1.4545, \text{ANS}, 1)$; $-0.9091\, R_1 + R_2 \to R_2$ is accomplished by *row $+ (-.9091, \text{ANS}, 1, 2)$; and $4.3636R_1 + R_3 \to R_3$ by *row $+ (4.3636, \text{ANS}, 1, 3)$ to obtain the third matrix:

$$
\begin{array}{ccccc}
x & y & r & s & P
\end{array}
$$
$$
\left[
\begin{array}{ccccc|c}
1 & 0 & 0.6875 & -0.3125 & 0 & 110 \\
0 & 1 & -0.625 & 0.3750 & 0 & 60 \\
\hline
0 & 0 & 3 & 5 & 1 & 11,680
\end{array}
\right]
$$

6. Since all elements in the bottom row are nonnegative, this matrix gives the maximum solution. Let $r = 0$ and $s = 0$, giving

$$1 \cdot x = 110$$

$$1 \cdot y = 60$$

$$1 \cdot P = 11,680$$

The maximum value is $P = \$11,680$ when $x = 110$, $y = 60$.   ■

EXAMPLE 17   Use a calculator to maximize

$$P = 100x + 150y$$

subject to

$$x, y \geq 0$$

$$2.11x + 3.02y \leq 13.28$$

$$0.06x + 4.31y \leq 13.05$$

SOLUTION   The simplex tableau is

$$
\begin{array}{ccccc}
x & y & r & s & P
\end{array}
$$

$$
\left[
\begin{array}{ccccc|c}
2.11 & 3.02 & 1 & 0 & 0 & 13.28 \\
0.06 & 4.31 & 0 & 1 & 0 & 13.05 \\
\hline
-100 & -150 & 0 & 0 & 1 & 0
\end{array}
\right]
$$

Verify that 4.31 (second row, second column) is the pivot element. Then use

$$*row(1/4.31, [A], 2)$$

$$*row + (-3.02, \text{ANS}, 2, 1)$$

$$*row + (150, \text{ANS}, 2, 3)$$

As a second step verify that the element in the first row and first column is the pivot element. Now use

$$*row(1/2.067958, \text{ANS}, 1)$$

$$*row + (-.01392111, \text{ANS}, 1, 2)$$

$$*row + (97.911833, \text{ANS}, 1, 3)$$

getting

$$
\left[
\begin{array}{ccccc|c}
1.00000 & 0 & 0.483568 & -0.33883 & 0 & 2 \\
0 & 1 & -0.00673 & 0.23673 & 0 & 3 \\
\hline
0.00001 & 0 & 47.3471 & 1.62685 & 1 & 650
\end{array}
\right]
$$

Setting r and s equal to zero yields

$$x = 2, \quad y = 3, \quad P = 650$$

SUMMARY

In this section we studied the following key steps in the simplex method of maximization:

1. Find the pivot element.
2. Perform pivot operations on the pivot column to
   (a) Make the pivot element 1.
   (b) Make all other elements in the pivot column 0.
3. Terminate the simplex process.
4. Read your linear programming answer.

Exercise Set 3.5

All answers for Exercises 1–12 are obtained from calculator solutions. Solve using the simplex method.

1. Maximize

$$P = 3x + 2y$$

   subject to

$$x \geq 0$$
$$y \geq 0$$
$$x + y \leq 4$$
$$3x + 2y \leq 9$$

2. Maximize

$$P = 2x + 3y$$

   subject to

$$x \geq 0$$
$$y \geq 0$$
$$x + y \leq 4$$
$$3x + 2y \leq 9$$

3. Maximize

$$P = 5x + 4y$$

   subject to

$$x \geq 0$$
$$y \geq 0$$
$$x + y \leq 4$$
$$3x + 2y \leq 9$$

4. Maximize

$$P = 4x + 3y$$

   subject to

$$x \geq 0$$
$$y \geq 0$$
$$x + y \leq 4$$
$$3x + 2y \leq 9$$

5. Maximize

$$P = 35x_1 + 25x_2$$

   subject to

$$x_1 \geq 0$$
$$x_2 \geq 0$$
$$2x_1 + x_2 \leq 7$$
$$3x_1 + x_2 \leq 8$$

6. Maximize

$$P = 35x_1 + 25x_2$$

   subject to

$$x_1 \geq 0$$
$$x_2 \geq 0$$
$$2x_1 + 3x_2 \leq 15$$
$$3x_1 + x_2 \leq 12$$

7. Maximize

$$P = x + 8y + 9z$$

   subject to

$$x \geq 0$$
$$y \geq 0$$
$$z \geq 0$$
$$5x - 2y - 3z \leq 0$$
$$-3x + y + z \leq 0$$
$$-5x + 3y + 4z \leq 200$$

8. Maximize

$$P = x - 3y + 2z$$

   subject to

$$x \geq 0$$
$$y \geq 0$$
$$z \geq 0$$
$$x + 6y + 3z \leq 6$$
$$x + 2y + 4z \leq 4$$
$$x - y + z \leq 3$$

9. Maximize

$$P = 2x + 9y + 5z$$

   subject to

$$x \geq 0$$
$$y \geq 0$$
$$z \geq 0$$
$$3x + 2y - 5z \leq 12$$
$$-x + 2y + 3z \leq 3$$
$$x + 3y - 2z \leq 2$$

10. Maximize

$$P = 2.36x + 4.23y$$

subject to

$$x \geq 0$$
$$y \geq 0$$
$$0.23x + 0.37y \leq 1.5$$
$$0.33x + 0.42y \leq 3.5$$

11. Maximize

$$P = 1.25x + 2.35y$$

subject to

$$x \geq 0$$
$$y \geq 0$$
$$0.17x + 0.38y \leq 33$$
$$0.35x + 0.23y \leq 13$$

12. Maximize

$$P = 2.36x + 4.23y$$

subject to

$$x \geq 0$$
$$y \geq 0$$
$$0.23x + 0.37y \leq 0.97$$
$$0.33x + 0.42y \leq 1.17$$

Use the simplex calculator program in Appendix A for Exercises 13 and 14.

13. Maximize

$$P = 40x + 60y + 90z + 70w$$

subject to

$$2x + 3y + 7z + 5w \leq 170$$
$$5x + 5y + 3z + 3w \leq 120$$
$$z + w \leq 20$$
$$x, y, z, w \geq 0$$

14. Maximize

$$P = 220x + 200y + 18z$$

subject to

$$x + 0.5y + 2z \leq 50$$
$$2x + 4y + 4z \leq 200$$
$$x + y + 0.5z \leq 50$$

$$x, y, z \geq 0$$

Applications (Business and Economics)

15. **Manufacturing (Resource Allocation).** A company is considering two products, type I and type II. Type I requires $\frac{1}{4}$ hour on a drill and $\frac{1}{8}$ hour on a lathe. Type II requires $\frac{1}{2}$ hour on a drill and $\frac{3}{4}$ hour on a lathe. The profit from type I is $50 per product and the profit from type II is $102 per product. If the machines are limited to 8 hours per day, how many of each product should be produced to maximize profit? What is the maximum daily profit?

16. **Shale–Bituminous Processors.**[1] Shale–bituminous processors use gasification and pressurization to make low-sulfur and high-sulfur crude oil. The two processes require different mixtures of coal and shale solids. A batch processed by gasification requires an input of 1 ton of coal and 2 tons of shale to yield 100 gallons of low-sulfur crude and 200 gallons of high-sulfur crude. Under pressurization, each batch requires 2 tons of coal and 1 ton of shale to provide 150 gallons of low-sulfur crude and 100 gallons of high-sulfur crude.

   The refinery manager wishes to determine how many batches should be converted to crude oil under each process to maximize total profits when supplying an order for exactly 10,000 gallons of low-sulfur and 5000 gallons of high-sulfur crudes. Available tonnages for filling this order are 100 tons of coal and 150 tons of shale. The costs of the solids per ton are $20 for coal and $25 for shale. The processor receives $0.50 per gallon for low-sulfur crude and $0.30 for high-sulfur crude.

17. **Manufacturing.** A factory manufactures two items, A and B. Sixty cubic feet of gas and two days

[1] Lawrence Lapin, *Quantitative Methods for Business Decisions,* 4th ed., Harcourt Brace Jovanovich, Orlando, Florida, 1988, p. 207.

of labor are required to fabricate item A. Item B requires 40 cubic feet of gas and 3 days of labor. Because of a severe winter, only 1480 cubic feet of gas are available to the factory each week. Sixty-six days of labor are available weekly. If item A produces a profit of $36 and item B yields a profit of $40, how many of each item should be manufactured weekly to maximize profit?

18. *Investments*. An investment club has at most $27,000 to invest in bonds of two types, good-quality and high-risk. Good-quality bonds average a yield of 7%, and high-risk yield 10%. The policy of the club requires that the amount invested in high-risk be not more than twice the amount invested in good-quality. How should the club invest its money to receive the maximum return subject to its investment policy?

19. *Investments*. Redo Exercise 18 assuming that a conservative member of the club gets the club policy changed so that the club must invest at least twice as much money in good-quality as it invests in high-risk bonds.

20. *Mixture Problem*. A nut company has at most 64 pounds of pecans and at most 132 pounds of peanuts that it wishes to mix. A package of mixture I contains 2 ounces of pecans and 11 ounces of peanuts, and mixture II contains 8 ounces of pecans and 12 ounces of peanuts per package. If a profit of $0.64 per package is obtained from mixture I and a profit of $0.70 per package is obtained from mixture II, how many packages of each mixture should be made to obtain the maximum profit?

21. *Advertising*. An electronics firm has $40,000 available for advertising. The following table gives the cost of an advertising package and the number of people in the advertising audience for each advertisement in two different media (numbers in thousands).

	Newspaper	TV
Cost	$ 4	$10
Audience	20	60

If the maximum number of newspaper advertisements is 8 and the maximum number of television advertisements is 2, how many advertisements of each type should be purchased to maximize advertising exposure? (Complete the solution of Exercise 25 in Exercise Set 3.4.)

22. *Assemblies*. A company makes four types of briefcases: the standard model, x_1; the large model, x_2; the deluxe model, x_3; and the economy model, x_4. Relative to time on machines it takes 4 times as much time for the deluxe model and 2 times as much time for the large model as it does for the standard and economy models. For the machines available the company determines that $2x_1 + x_2 + 4x_3 + x_4 < 120$. Under ideal conditions the company cannot produce more than 100 briefcases a day. Due to the special work involved, the company cannot produce more than 30 large and deluxe models per day. If the profit on the deluxe model is $12, on the extra large $11, on the standard $10, and on the economy $9, how many of each should the company produce each day for maximum profit?

23. *Production*. A company manufactures three kinds of electronic assemblies for automobiles. Each assembly requires labor, test labor, and resistors. Type I assembly requires 2 hours of labor, 1 hour of test labor, and 4 resistors. Type II requires 2, 1, and 10 respectively. Type III requires 4, 3, and 1 respectively. The profit on Type I is $80, on Type II is $90, and on Type III is $50. Find the number of each to produce for a maximum profit if the number of hours of labor cannot exceed 340 hours, the number of hours of test labor cannot exceed 850 hours, and the number of resistors must be less than 400?

Applications (Social and Life Sciences)

24. *Agriculture*. A farmer has available 200 acres on which to plant either wheat or corn or both. The planting costs (seed, fertilizer, etc.) are $150 per acre for corn and $60 per acre for wheat. The farmer has only $16,500 capital to spend on planting costs. If the anticipated income is $200 per acre for corn and $190 per acre for wheat, how many acres of each should be planted to maximize the income?

25. **Pollution.** Government regulations for pollution control have forced Langley Chemical to install a new process to help reduce pollution caused by production of a certain chemical. The old process releases 12 grams of pollutant A and 30 grams of pollutant B into the air for each liter of chemical produced. The new process releases 5 grams of pollutant A and 15 grams of pollutant B for each liter produced. The company makes a profit of $0.40 for each liter produced by the old process and only $0.18 for each liter produced by the new process. If the regulations do not allow more than 10,000 grams of pollutant A and no more than 27,000 grams of pollutant B each day, how many liters of the chemical should be produced by the old process and how many liters should be produced by the new process to maximize profit?

26. **Health Care.** The government has mobilized to inoculate the student population against sleeping sickness. There are 200 doctors available and 450 nurses. An inoculation team can consist of either 1 doctor and 3 nurses (team A) or 1 doctor and 2 nurses (team B). On the average, team A can inoculate 180 students per hour, while team B can only inoculate 100 students per hour. How many teams of each type should be formed to maximize the number of inoculations per hour? (Complete the solution of Exercise 28 in Exercise Set 3.4.)

27. **Farming.** A farmer has available 6000 acres of land to raise cattle, horses, and sheep. The farmer cannot graze more than three head of cattle, or five head of sheep, or one horse per acre on this land. In the winter (approximately 15 weeks) he will need three bales of hay per week for a horse, 2 bales per week for a cow, and 1 bale per week for a head of sheep. He has a total of 5000 bales of hay available. If he averages a profit of $20 per head for cattle, $15 per head for horses, and $10 per head for sheep, how many of each should he have in order to maximize profit? (Complete the solution of Exercise 29 in Exercise Set 3.4.)

Review Exercises

28. Set up the simplex tableau with the pivot element circled to maximize

$$P = 3x + 4y$$

subject to

$$x, y \geq 0$$
$$3x + 5y \leq 30$$
$$3x + 5y \leq 21$$

3.6 MINIMIZATION USING THE DUAL PROBLEM

OVERVIEW   In our use of the simplex procedure in the preceding sections, our constraints have satisfied conditions that we classified as standard maximum linear programming problems. In this section, we consider problems that do not satisfy these conditions. For example, a company that produces dog food mixes five different products to produce a dog food. The cost of the five products varies significantly. The company wishes to produce the dog food at minimum cost subject to stated constraints. Such problems require that we minimize the objective function.

Associated with each maximum problem is a minimum problem called its **dual**. Similarly, each minimum linear programming problem has a corresponding maximum problem, also called its dual. To solve a minimum problem, we simply use the simplex method to solve the dual maximum problem. In this section, we study

- Formation of the dual problem
- Solution of the minimization problem

As we just mentioned, there is associated with each linear programming problem a dual problem. For instance, the following are dual problems.

	Maximize	Minimize
Objective functions	$8x_1 + 6x_2 = P$	$2y_1 + 5y_2 = C$
	subject to	subject to
Constraint inequalities	$3x_1 + x_2 \leq 2$	$3y_1 + 2y_2 \geq 8$
	$2x_1 + 4x_2 \leq 5$	$y_1 + 4y_2 \geq 6$
	$x_1 \geq 0$	$y_1 \geq 0$
	$x_2 \geq 0$	$y_2 \geq 0$

The 8 and 6 in the maximum objective function, $8x_1 + 6x_2 = P$, are the constants for the constraint equations

$$3y_1 + 2y_2 \geq 8$$
$$y_1 + 4y_2 \geq 6$$

for the minimum problem; and the constants 2 and 5 for the constraint equations

$$3x_1 + x_2 \leq 2$$
$$2x_1 + 4x_2 \leq 5$$

for the maximum problem are the coefficients in the objective function $2y_1 + 5y_2 = C$ for the minimum problem. Also note that the column numbers in the system of constraints on the left become row numbers in the system of constraints on the right. That is, one matrix of coefficients is the transpose of the other matrix of coefficients.

Relationship: Dual Problems

1. (a) The dual of a maximum problem is a minimum problem.
   (b) The dual of a minimum problem is a maximum problem.
2. (a) The constraints in a maximum problem are given as $\leq$.
   (b) The constraints in a minimum problem are given as $\geq$.
3. (a) The coefficients of the variables in the objective function to be maximized are the constants in the $\geq$ constraints of the dual minimum problem.
   (b) The coefficients of the variables in the objective function to be minimized are the constants in the $\leq$ constraints of the dual maximum problem.
4. The matrix of coefficients of the constraint equations in the x's is the transpose of the matrix of the constraint equations in the y's.

Since we will be solving the minimum problem in terms of the maximum problem using the simplex method for the standard maximum problem, we should outline how to change a minimum problem to a maximum problem.

EXAMPLE 18   Minimize

$$C = 2y_1 + 5y_2$$

subject to

$$3y_1 + 2y_2 \geq 8$$

$$y_1 + 4y_2 \geq 6$$

$$y_1 \geq 0$$

$$y_2 \geq 0$$

SOLUTION

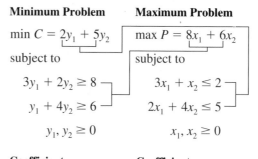

Minimum Problem   **Maximum Problem**

$\min C = 2y_1 + 5y_2$ $\max P = 8x_1 + 6x_2$

subject to   subject to

$\quad 3y_1 + 2y_2 \geq 8$ $\quad 3x_1 + x_2 \leq 2$

$\quad y_1 + 4y_2 \geq 6$ $\quad 2x_1 + 4x_2 \leq 5$

$\quad\quad y_1, y_2 \geq 0$ $\quad\quad x_1, x_2 \geq 0$

Coefficient   **Coefficient**
Matrix   **Matrix**

$$\begin{bmatrix} 3 & 2 \\ 1 & 4 \end{bmatrix}$$ $$\begin{bmatrix} 3 & 1 \\ 2 & 4 \end{bmatrix}$$

One matrix is the transpose
of the other.

We use the following steps to obtain the simplex tableau for this problem.

1. Write the coefficient matrix of the minimum problem and find the transpose. On a calculator use $[A]^T$.

$$\begin{bmatrix} 3 & 2 \\ 1 & 4 \end{bmatrix}^T = \begin{bmatrix} 3 & 1 \\ 2 & 4 \end{bmatrix}$$

2. For constant terms of the maximum simplex tableau use the coefficients of the objective function, $C = 2y_1 + 5y_2$, and for slack variables use the y_1 and y_2 of the minimum problem.

3. To get the last line of the maximization tableau, use the negatives of the constant terms in the minimization inequalities in y, namely -8 and -6 along with a P column. The last line represents $P - 8x_1 - 6x_2$.

The tableau for the given minimum problem is the tableau for the dual maximum problem.

$$
\begin{array}{ccccc}
x_1 & x_2 & y_1 & y_2 & P \\
\end{array}
$$

$$
\left[
\begin{array}{ccccc|c}
③ & 1 & 1 & 0 & 0 & 2 \\
2 & 4 & 0 & 1 & 0 & 5 \\
\hline
-8 & -6 & 0 & 0 & 1 & 0 \\
\end{array}
\right]
$$

We have replaced the slack variable headings of the columns by the variables in the dual problem. This replacement will enable us to give the solution to the dual problem when the final matrix is obtained.

Perform the row operations

$$\tfrac{1}{3}R_1 \to R_1 \quad (1)$$

$$-2R_1 + R_2 \to R_2 \quad (2)$$

$$8R_1 + R_3 \to R_3 \quad (3)$$

on the matrix to obtain

$$
\begin{array}{ccccc}
x_1 & x_2 & y_1 & y_2 & P \\
\end{array}
$$

$$
\left[
\begin{array}{ccccc|c}
1 & 0.3333 & 0.3333 & 0 & 0 & 0.6666 \\
0 & 3.3333 & -0.6666 & 1 & 0 & 3.6666 \\
\hline
0 & -3.3333 & 2.6666 & 0 & 1 & 5.3333 \\
\end{array}
\right]
\quad
\begin{array}{l}
-0.3333R_2 + R_1 \to R_1 \quad (2) \\
0.3R_2 \to R_2 \qquad\qquad\quad (1) \\
3.3333R_2 + R_3 \to R_3 \quad (3) \\
\end{array}
$$

The third matrix is

$$
\begin{array}{ccccc}
x_1 & x_2 & y_1 & y_2 & P \\
\end{array}
$$

$$
\left[
\begin{array}{ccccc|c}
1 & 0 & 0.4 & -0.1 & 0 & 0.3 \\
0 & 1 & -0.2 & 0.3 & 0 & 1.1 \\
\hline
0 & 0 & 2 & 1 & 1 & 9 \\
\end{array}
\right]
$$

Since all the elements of the last row are positive or zero, the third matrix gives the solution to the maximum problem:

$$x_1 = 0.3, \quad x_2 = 1.1, \quad \text{and} \quad P = 9$$

The solution to the minimum problem is obtained on the bottom row:

$$
\begin{array}{cc}
y_1 & y_2 \\
\end{array}
$$

$$[0 \quad 0 \quad 2 \quad 1 \quad 1 \mid 9]$$

The minimum value of C is 9 and this occurs at

$$y_1 = 2 \quad \text{and} \quad y_2 = 1$$

■

Solution: Minimum Problem

Given a minimum problem with nonnegative coefficients in the objective function.

1. Write all problem constraints as inequalities.
2. Form the dual maximum problem.
3. In the initial tableau, use the variables in the minimum problem as slack variables in the maximum problem.
4. Solve the maximum problem by the simplex method.
5. Read the solution for the minimum problem from the bottom row of the solution tableau for the maximum problem.

Note that if the maximum problem has no solution, then the minimum problem has no solution.

Practice Problem 1 Write the simplex tableau for the standard maximization problem in order to solve the following minimization problem. Minimize

$$C = 2y_1 + 5y_2$$

subject to

$$3y_1 + 2y_2 \geq 8$$
$$y_1 + 4y_2 \geq 6$$
$$2y_1 + y_2 \geq 4$$
$$y_1, y_2 \geq 0$$

ANSWER   Maximize

$$P = 8x_1 + 6x_2 + 4x_3$$

subject to

x_1	x_2	x_3	y_1	y_2	P	
3	1	2	1	0	0	2
2	4	1	0	1	0	5
-8	-6	-4	0	0	1	0

EXAMPLE 19   Minimize

$$C = 200y_1 + 500y_2$$

subject to

$$y_1, y_2 \geq 0$$

$$3.112y_1 + 2.013y_2 \geq 8.237$$

$$y_1 + 4.113y_2 \geq 6.113$$

SOLUTION   The tableau for the dual maximum problem is

$$
\begin{array}{ccccc}
x_1 & x_2 & y_1 & y_2 & P \\
\end{array}
$$

$$
\left[
\begin{array}{ccccc|c}
3.112 & 1 & 1 & 0 & 0 & 200 \\
2.013 & 4.113 & 0 & 1 & 0 & 500 \\
\hline
-8.237 & -6.113 & 0 & 0 & 1 & 0 \\
\end{array}
\right]
$$

The pivot element is 3.112. Perform the following row operations in order:

$$*\text{row}(1 \div 3.112, [B], 1)$$

$$*\text{row}+(-2.013, \text{ANS}, 1, 2)$$

$$*\text{row}+(8.237, \text{ANS}, 1, 3)$$

$$
\begin{array}{ccccc}
x_1 & x_2 & y_1 & y_2 & P \\
\end{array}
$$

$$
\left[
\begin{array}{ccccc|c}
1 & 0.3213 & 0.3213 & 0 & 0 & 64.2674 \\
0 & 3.4661 & -0.6469 & 1 & 0 & 370.6298 \\
\hline
0 & -3.4661 & 2.6469 & 0 & 1 & 529.3702 \\
\end{array}
\right]
$$

The new pivot element is 3.4661 in the second row, second column.

$$*\text{row}(1 \div 3.4661, \text{ANS}, 2)$$

$$*\text{row}+(-.3213, \text{ANS}, 2, 1)$$

$$*\text{row}+(3.4661, \text{ANS}, 2, 3)$$

$$
\begin{array}{ccccc}
x_1 & x_2 & y_1 & y_2 & P \\
\end{array}
$$

$$
\left[
\begin{array}{ccccc|c}
1 & 0 & 0.3813 & -0.0927 & 0 & 29.9107 \\
0 & 1 & -0.1866 & 0.2885 & 0 & 106.9299 \\
\hline
0 & 0 & 2 & 1 & 1 & 900 \\
\end{array}
\right]
$$

$$y_1 = 2, \quad y_2 = 1, \quad \text{and} \quad C = 900$$

Practice Problem 2 A simplex tableau for a minimization problem (obtained from the dual maximization problem) is

$$
\begin{array}{ccccccc}
\;\;x_1 & x_2 & x_3 & y_1 & y_2 & P & \\
\end{array}
$$

$$
\left[
\begin{array}{cccccc|c}
\frac{3}{4} & 1 & 0 & \frac{1}{2} & -\frac{1}{4} & 0 & \frac{225}{4} \\
-\frac{1}{12} & 0 & 1 & -\frac{1}{6} & \frac{1}{4} & 0 & \frac{325}{12} \\
\hline
15 & 0 & 0 & 50 & 15 & 1 & 16{,}625
\end{array}
\right]
$$

where $C = 250y_1 + 275y_2$. Find the minimum value of C and the values of y_1 and y_2 that produce a minimum.

ANSWER The minimum value of C is 16,625 at $y_1 = 50$, $y_2 = 15$.

SUMMARY

In this section we learned to write a minimum problem in terms of the dual maximum problem. After performing the simplex procedure for the maximum problem, the solution for the minimum problem is read from the bottom row of the solution tableau for the maximum problem.

Exercise Set 3.6

Find the dual problem that corresponds to each of the following linear programming problems.

1. Maximize

$$P = 25x_1 + 35x_2$$

   subject to

$$2x_1 + 3x_2 \le 15$$
$$3x_1 + x_2 \le 12$$
$$x_1 \ge 0$$
$$x_2 \ge 0$$

2. Maximize

$$P = 3x_1 + 2x_2$$

   subject to

$$x_1 + x_2 \le 4$$
$$3x_1 + 2x_2 \le 9$$
$$x_1 \ge 0$$
$$x_2 \ge 0$$

3. Minimize

$$4y_1 + 7y_2 = C$$

   subject to

$$y_1 + y_2 \ge 5$$
$$3y_1 + y_2 \ge 21$$
$$y_1 \ge 0$$
$$y_2 \ge 0$$

4. Minimize

$$4y_1 + 5y_2 = C$$

   subject to

$$y_1 + y_2 \ge 5$$
$$3y_1 + y_2 \ge 21$$
$$y_1 \ge 0$$
$$y_2 \ge 0$$

Write the linear programming minimization problems that are represented by the following simplex tableaus for maximization in Exercise Set 3.5.

5. Exercise 1
6. Exercise 2
7. Exercise 3
8. Exercise 4

Write the linear programming problems and duals that are represented by the following simplex tableaus.

9.

$$
\begin{array}{ccccc}
x_1 & x_2 & y_1 & y_2 & P \\
\left[\begin{array}{ccccc|c}
3 & 2 & 1 & 0 & 0 & 5 \\
1 & 3 & 0 & 1 & 0 & 8 \\
\hline
-7 & -4 & 0 & 0 & 1 & 0
\end{array}\right]
\end{array}
$$

10.

$$
\begin{array}{ccccc}
x_1 & x_2 & y_1 & y_2 & P \\
\left[\begin{array}{ccccc|c}
4 & 3 & 1 & 0 & 0 & 7 \\
2 & 4 & 0 & 1 & 0 & 9 \\
\hline
-5 & -8 & 0 & 0 & 1 & 0
\end{array}\right]
\end{array}
$$

11.

$$
\begin{array}{cccccc}
x_1 & x_2 & y_1 & y_2 & y_3 & P \\
\left[\begin{array}{cccccc|c}
2 & 3 & 1 & 0 & 0 & 0 & 4 \\
1 & 2 & 0 & 1 & 0 & 0 & 7 \\
3 & 1 & 0 & 0 & 1 & 0 & 6 \\
\hline
-6 & -8 & 0 & 0 & 0 & 1 & 0
\end{array}\right]
\end{array}
$$

12.

$$
\begin{array}{cccccc}
x_1 & x_2 & y_1 & y_2 & y_3 & P \\
\left[\begin{array}{cccccc|c}
1 & 2 & 1 & 0 & 0 & 0 & 5 \\
2 & 3 & 0 & 1 & 0 & 0 & 6 \\
3 & 5 & 0 & 0 & 1 & 0 & 8 \\
\hline
-7 & -4 & 0 & 0 & 0 & 1 & 0
\end{array}\right]
\end{array}
$$

Solve the following linear programming problems using the simplex method.

13. Minimize

$$4y_1 + 5y_2 = C$$

subject to

$$
\begin{aligned}
y_1 + y_2 &\geq 5 \\
3y_1 + 5y_2 &\geq 21 \\
y_1 &\geq 0 \\
y_2 &\geq 0
\end{aligned}
$$

14. Maximize

$$5y_1 + 4y_2 = C$$

subject to

$$
\begin{aligned}
y_1 + y_2 &\geq 5 \\
3y_1 + 5y_2 &\geq 21 \\
y_1 &\geq 0 \\
y_2 &\geq 0
\end{aligned}
$$

15. Minimize

$$4y_1 + 7y_2 = C$$

subject to

$$
\begin{aligned}
y_1 + y_2 &\geq 5 \\
3y_1 + y_2 &\geq 21 \\
y_1 &\geq 0 \\
y_2 &\geq 0
\end{aligned}
$$

16. Minimize

$$7y_1 + 4y_2 = C$$

subject to

$$
\begin{aligned}
y_1 + y_2 &\geq 5 \\
3y_1 + 5y_2 &\geq 21 \\
y_1 &\geq 0 \\
y_2 &\geq 0
\end{aligned}
$$

17. Minimize

$$25y_1 + 35y_2 = C$$

subject to

$$
\begin{aligned}
2y_1 + 3y_2 &\geq 15 \\
3y_1 + y_2 &\geq 12 \\
y_1 &\geq 0 \\
y_2 &\geq 0
\end{aligned}
$$

18. Minimize

$$35y_1 + 25y_2 = C$$

subject to

$$
\begin{aligned}
2y_1 + 3y_2 &\geq 15 \\
3y_1 + y_2 &\geq 12 \\
y_1 &\geq 0 \\
y_2 &\geq 0
\end{aligned}
$$

Minimize the following problems using a graphing calculator.

19. Minimize

$$25y_1 + 35y_2 = C$$

subject to

$$
\begin{aligned}
1.05y_1 + 0.5y_2 &\geq 2 \\
1.2y_1 + 3y_2 &\geq 15 \\
1.3y_1 + y_2 &\geq 12 \\
y_1 &\geq 0 \\
y_2 &\geq 0
\end{aligned}
$$

20. Minimize

$$25y_1 + 35y_2 = C$$

subject to

$$
\begin{aligned}
1.2y_1 + \ 3y_2 &\geq 6.2 \\
2.2y_1 + \ 3y_2 &\geq 15.1 \\
3.3y_1 + 1.2y_2 &\geq 12.5 \\
y_1 &\geq 0 \\
y_2 &\geq 0
\end{aligned}
$$

Use the simplex program found in Appendix A for Exercises 21 and 22.

21. Minimize

$$1000y_1 + 1000y_2 + 1000y_3 = C$$

subject to

$$20y_1 + 10y_2 + 1000y_3 \geq 220$$

$$10y_1 + 20y_2 + \ 20y_3 \geq 200$$

$$20y_1 + 20y_2 + \ 10y_3 \geq 180$$

$$y_1, y_2, y_3 \geq 0$$

22. Minimize

$$300y_1 + 300y_2 + 100y_3 = C$$

subject to

$$2y_1 + \ y_2 - 2y_3 \geq 10$$

$$y_1 + 2y_2 - 2y_3 \geq 20$$

$$2y_1 + 2y_2 + \ y_3 \geq 30$$

$$y_1, y_2, y_3 \geq 0$$

Applications (Business and Economics)

23. *Transportation.* A company has two warehouses, A and B, and two stores, I and II. Warehouse A contains 40 tons of a product and warehouse B contains 100 tons of the same product. Store I needs 50 tons of the product and store II needs 75 tons. The shipping costs are

A to I, $5 per ton
B to I, $6 per ton

A to II, $8 per ton
B to II, $10 per ton

Find the shipping instructions that will satisfy each store's need at a minimum shipping cost. Solve graphically.

24. *Mixture Problem.* A pharmaceutical company wishes to manufacture a vitamin mixture so that each bottle contains at least 40 units of vitamin A, 22 units of vitamin B, and 22 units of vitamin C. The mixture will be made from products R and S. Product R costs $0.10 per gram, and each gram contains 5 units of vitamin A, 2 units of vitamin B, and 4 units of vitamin C. Product S costs $0.15 per gram, and each gram contains 10 units of vitamin A, 6 units of vitamin B, and 1 unit of vitamin C. How many grams of each product should the company use in the mixture in order to minimize costs?

25. *Production.* A pharmaceutical firm produces two strengths of a newly developed drug. (Drug A is of greater strength than B.) Each costs $1000 per gallon to produce. It has been determined that at least 30 gallons of A and at least 20 gallons of B must be produced within the next month. Drug A requires 1 pound of a perishable raw material per gallon, and drug B requires 2 pounds. The current inventory of this perishable raw material is 80 pounds, and this much perishable raw material must be used within the month. Find the number of gallons of each to produce in order to minimize total cost.

Applications (Social and Life Sciences)

26. *Nutrition.* A man is on a special diet, which requires a minimum number of grams of protein and carbohydrates. He must have at least 40 grams of protein and 43 grams of carbohydrates each meal. For lunch he will eat food A, which costs $2.00 per unit, and food B, which costs $1.70 per unit. A unit of A contains 8 grams of protein and 6 grams of carbohydrates. A unit of B contains 5 grams of protein and 7 grams of carbohydrates. What is the cheapest meal he can eat (with foods A and B) that contains the minimum requirements?

27. *Nutrition.* A man is on a low-carbohydrate diet. He is planning a meal composed of two foods: food I with 7 grams of carbohydrate per unit and food II with 4 grams of carbohydrate per unit. In order to keep him from becoming discouraged with his diet,

his doctor has insisted that he consume at least 500 calories at each meal, of which at least 210 calories must be protein. Each food contains 100 calories per unit, but food I contains only 30 protein calories per unit, whereas food II contains 50 protein calories per unit. How many units of each food should the man eat to minimize the amount of carbohydrate?

28. *Medicine.*   A doctor has decided her patient needs at least 14 milligrams of drug I and at least 16 milligrams of drug II each day. These drugs are to be obtained by taking medicine I and medicine II. Both medicine I and medicine II contain the undesirable drug X. One gram of medicine I contains 3 milligrams of drug I, 1 milligram of drug II, and 3 milligrams of drug X. One gram of medicine II contains 2 milligrams of drug I, 2 milligrams of drug II, and 2 milligrams of drug X. How many grams of medicine I and medicine II should the doctor prescribe each day if she wishes to minimize the amount of drug X?

29. *Nutrition.*   A person on a low-carbohydrate diet plans to eat two foods. A unit of food I contains 5 grams of carbohydrate and 100 calories, of which 10 calories are protein. A unit of food II contains 6 grams of carbohydrate and 100 calories, of which 30 calories are protein. The individual wishes to minimize the number of grams of carbohydrate while eating at least 400 calories, of which at least 60 are protein calories. Find the number of units that the individual should eat of each type of food.

30. *Nutrition.*   Suppose that 16, 20 and 8 units of proteins, carbohydrates, and fats, respectively, are the minimum weekly requirements per person for a diet. Foods in category A contain (on the average) 3, 6, and 1 units of proteins, carbohydrates, and fats per pound, respectively. Foods in category B contain 2,

3, and 2 units, respectively. The average cost of category A is $1.50 per pound and category B is $1.00 per pound. How many pounds of food in category A and in category B should be purchased to minimize cost?

31. *Nutrition.*   The "miracle" diet of Eaters Anonymous requires that each lunch contain at least 26 units of protein and 24 units of nondigestible fiber. In addition, lunch is to consist of salmon patties and a special natural bread. Each salmon patty has 8 units of protein, 4 units of fiber, and 120 calories. Each slice of bread has 1 unit of protein, 6 units of fiber, and 45 calories. How many salmon patties and bread slices must be eaten to minimize calories and still satisfy the requirements of the diet?

32. *Nutrition.*   Two breakfast cereals, A and C, supply varying amounts of vitamin B and iron: cereal A supplies 0.15 mg/oz of Vitamin B and 1.67 mg/oz of iron; cereal C supplies 0.10 mg/oz of Vitamin B and 3.33 mg/oz of iron. The daily minimum requirements are 0.36 mg/day and 6.0 mg/day, respectively. If cereal A costs $0.14 per ounce and cereal C costs $0.20 per ounce, how can we supply our minimum daily requirements of Vitamin B and iron at minimum cost by eating cereal?

Review Exercises

33. Using graphing techniques, find the maximum value of $P = 3x + 4y$ subject to

$$x, y, \geq 0$$

$$3x + 5y \leq 30$$

$$3x + 5y \leq 21$$

3.7 MIXED-CONSTRAINT LINEAR PROGRAMMING (Optional)

OVERVIEW   The standard form of a maximization problem involves constraints written as

$$a_1x_1 + a_2x_2 + \cdots + a_nx_n \leq b \qquad (b \geq 0)$$

with objective function

$$P = k_1x_1 + k_2x_2 + \cdots + k_nx_n$$

The standard form of a minimization problem involves constraints

$$c_1 y_1 + c_2 y_2 + \cdots + c_n y_n \geq d \qquad (d \geq 0)$$

with objective function

$$C = h_1 y_1 + h_2 y_2 + \cdots + h_n y_n$$

However, not all linear programming problems fit exactly these two problems. The constraint equations for a maximization problem involve $\leq$, whereas the constraint equations for a minimization problem involve $\geq$. Note that both occur in the following problem: A mining company operates two mines, A and B. Each mine produces high-grade ore and low-grade ore. Mine A produces 2 tons of high-grade ore and 0.5 ton of low-grade ore per day. Mine B produces 1 ton of each grade of ore per day. The company has contracted to provide a local smelter a minimum of 14 tons of high-grade ore per week and a maximum of 6 tons of low-grade ore per week. Determine the minimum cost to meet this contract if it costs $4000 per day to operate mine A and $3000 per day to operate mine B (Exercise 21). In this section we consider problems involving both $\leq$ and $\geq$. We solve such problems, called mixed-constraint problems, using

- Geometric procedures
- Simplex procedures

Consider the following example involving both $\leq$ and $\geq$ in the constraint equations.

EXAMPLE 20   Maximize

$$P = 4x_1 + x_2$$

subject to

$$x_1 + x_2 \leq 8$$

$$-x_1 + x_2 \geq 4$$

$$x_1, x_2 \geq 0$$

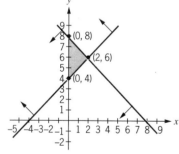

Figure 17

SOLUTION   Geometrically the feasible solution and corner points are shown in Figure 17. The maximum value of the function occurs at $(2, 6)$.

$$\text{Maximum} \quad P = 4(2) + 6 = 14$$

Let's solve the preceding example using the simplex procedure. In typical fashion we write the first equation as

$$x_1 + x_2 + s_1 = 8$$

where s_1 is our usual slack variable. We now introduce a second nonnegative variable r_2 and subtract it from the left side of the second inequality to get

$$-x_1 + x_2 - r_2 = 4$$

This variable is not a slack variable because we add slack variables to get equality. The variable r_2 is called a **surplus variable** because it is the amount or surplus by which the left side of the equation is more than 4. Just like slack variables, surplus variables are always nonnegative.

This linear programming problem can now be expressed as

$$x_1 + x_2 + s_1 = 8$$
$$-x_1 + x_2 - r_2 = 4$$
$$-4x_1 - x_2 + P = 0$$
$$x_1, x_2, s_1, r_2 \geq 0$$

First, we check to see if this system satisfies our requirements for a solution. Let's look for a possible basic feasible solution when $x_1 = 0$ and $x_2 = 0$. The solution is

$$x_1 = 0, \quad x_2 = 0, \quad s_1 = 8, \quad r_2 = -4, \quad \text{and} \quad P = 0$$

This does not satisfy the requirement, $r_2 \geq 0$. Therefore, the simplex procedure will not work on this set of equations.

To overcome this difficulty we introduce another nonnegative variable a_3 such that

$$-x_1 + x_2 - r_2 + a_3 = 4$$

The a_3 is often called an **artificial variable** because it has no real meaning relative to the original set of constraints. It will have a value that allows r_2 to remain positive. Since a_3 is now a variable in our constraint equations, it must be a variable in our objective function, which we write as

$$P = 4x_1 + x_2 - Ka_3 \qquad K \text{ considered as large and } \geq 0$$

We have now modified our system to the following:

$$x_1 + x_2 + s_1 \qquad\qquad\qquad = 8$$
$$-x_1 + x_2 \qquad - r_2 + a_3 \qquad = 4$$
$$-4x_1 - x_2 \qquad\qquad + Ka_3 + P = 0$$
$$x_1, x_2, s_1, r_2, a_3 \geq 0$$

Now setting $x_1 = 0$, $x_2 = 0$, and $a_3 = 0$, we see that r_2 is still negative. To resolve this difficulty we eliminate K under a_3 in the last line of the following tableau:

$$
\begin{array}{cccccc}
x_1 & x_2 & s_1 & r_2 & a_3 & P \\
\end{array}
$$

$$
\left[
\begin{array}{cccccc|c}
1 & 1 & 1 & 0 & 0 & 0 & 8 \\
-1 & 1 & 0 & -1 & 1 & 0 & 4 \\
\hline
-4 & -1 & 0 & 0 & K & 1 & 0
\end{array}
\right] \quad -KR_2 + R_3 \rightarrow R_3
$$

$$
\begin{array}{cccccc}
x_1 & x_2 & s_1 & r_2 & a_3 & P \\
\end{array}
$$

$$
\left[
\begin{array}{cccccc|c}
1 & 1 & 1 & 0 & 0 & 0 & 8 \\
-1 & 1 & 0 & -1 & 1 & 0 & 4 \\
\hline
-4+K & -1-K & 0 & K & 0 & 1 & -4K
\end{array}
\right]
\begin{array}{l}
-R_2 + R_1 \rightarrow R_1 \quad\quad (1) \\
\\
(1+K)R_2 + R_3 \rightarrow R_3 \quad (2)
\end{array}
$$

Now if x_1, x_2, and r_2 are set equal to 0, we have $x_1 = 0$, $x_2 = 0$, $s_1 = 8$, $r_2 = 0$, $a_3 = 4$, and $P = -4K$. This system satisfies our requirements for a solution, so we can now apply the simplex procedures to this problem. Since K is positive and large, $-1 - K$ is more negative than $-4 + K$, and the second column is the pivot column. Since $4 < 8$, the second row is the pivot row. The next tableau is given as

$$
\begin{array}{cccccc}
x_1 & x_2 & s_1 & r_2 & a_3 & P \\
\end{array}
$$

$$
\left[
\begin{array}{cccccc|c}
2 & 0 & 1 & 1 & -1 & 0 & 4 \\
-1 & 1 & 0 & -1 & 1 & 0 & 4 \\
\hline
-5 & 0 & 0 & -1 & 1+K & 1 & 4
\end{array}
\right]
\begin{array}{l}
\tfrac{1}{2}R_1 \rightarrow R_1 \quad\quad (1) \\
R_1 + R_2 \rightarrow R_2 \quad (2) \\
5R_1 + R_3 \rightarrow R_3 \quad (3)
\end{array}
$$

The pivot element is now 2 in the first row and first column.

$$
\begin{array}{cccccc}
x_1 & x_2 & s_1 & r_2 & a_3 & P \\
\end{array}
$$

$$
\left[
\begin{array}{cccccc|c}
1 & 0 & \tfrac{1}{2} & \tfrac{1}{2} & -\tfrac{1}{2} & 0 & 2 \\
0 & 1 & \tfrac{1}{2} & -\tfrac{1}{2} & \tfrac{1}{2} & 0 & 6 \\
\hline
0 & 0 & \tfrac{5}{2} & \tfrac{3}{2} & -\tfrac{3}{2}+K & 1 & 14
\end{array}
\right]
$$

Let K be a large positive number; then all elements in the last row are positive. The solution to our problem is $x_1 = 2$, $x_2 = 6$, $s_1 = 0$, $r_2 = 0$, $a_3 = 0$, and $P = 14$.

Check this solution with the solution obtained graphically in Example 20. Note that $P = 14$ is the maximum value obtained at $x_1 = 2$ and $x_2 = 6$.

As this example illustrates, if $a_3 = 0$ in the optimal solution of the modified problem, then deleting a_3 produces the same optimal solution in the original problem. What happens if $a_3 \neq 0$? It can be shown that the original problem has no optimal solution.

Sometimes one of the inequalities in a system of constraints is replaced by an equation. Usually when this occurs, we can solve the system by using the equation without a slack variable. If this does not yield a solution, we can try introducing surplus and artificial variables.

EXAMPLE 21   Maximize

$$P = 2x_1 + 3x_2$$

subject to

$$x_1 + x_2 \le 4$$
$$x_1 + 2x_2 = 6$$
$$x_1, x_2 \ge 0$$

SOLUTION   The system of equations can be written as

$$x_1 + x_2 + s_1 = 4$$
$$x_1 + 2x_2 - r_2 + a_3 = 6$$
$$-2x_1 - 3x_2 + Ka_3 + P = 0$$

This system can be represented by the following tableau:

$$
\begin{array}{cccccc|c}
x_1 & x_2 & s_1 & r_2 & a_3 & P & \\
\left[\begin{array}{cccccc|c}
1 & 1 & 1 & 0 & 0 & 0 & 4 \\
1 & 2 & 0 & -1 & 1 & 0 & 6 \\
-2 & -3 & 0 & 0 & K & 1 & 0
\end{array}\right] & & & & & &
\end{array}
\qquad -KR_2 + R_3 \rightarrow R_3
$$

In the preceding tableau, we eliminated K under a_3 from the last line. This we accomplish as follows:

$$
\begin{array}{cccccc|c}
x_1 & x_2 & s_1 & r_2 & a_3 & P & \\
\left[\begin{array}{cccccc|c}
1 & 1 & 1 & 0 & 0 & 0 & 4 \\
1 & 2 & 0 & -1 & 1 & 0 & 6 \\
-2-K & -3-2K & 0 & K & 0 & 1 & -6K
\end{array}\right]
\end{array}
\begin{array}{ll}
-R_2 + R_1 \rightarrow R_1 & (2) \\
\frac{1}{2}R_2 \rightarrow R_2 & (1) \\
(3+2K)R_2 + R_3 \rightarrow R_3 & (3)
\end{array}
$$

The 2 in the second row, second column is the pivot element.

$$
\begin{array}{cccccc|c}
x_1 & x_2 & s_1 & r_2 & a_3 & P & \\
\left[\begin{array}{cccccc|c}
\frac{1}{2} & 0 & 1 & \frac{1}{2} & -\frac{1}{2} & 0 & 1 \\
\frac{1}{2} & 1 & 0 & -\frac{1}{2} & \frac{1}{2} & 0 & 3 \\
-\frac{1}{2} & 0 & 0 & -\frac{3}{2} & \frac{3}{2}+K & 1 & 9
\end{array}\right]
\end{array}
\begin{array}{ll}
2R_1 \rightarrow R_1 & (1) \\
-\frac{1}{2}R_1 + R_2 \rightarrow R_2 & (2) \\
\frac{1}{2}R_1 + R_3 \rightarrow R_3 & (3)
\end{array}
$$

Using $\frac{1}{2}$ in the first row, first column as the pivot element, we get

$$
\begin{array}{cccccc}
x_1 & x_2 & s_1 & r_2 & a_3 & P \\
\end{array}
$$

$$
\left[
\begin{array}{cccccc|c}
1 & 0 & 2 & 1 & -1 & 0 & 2 \\
0 & 1 & -1 & -1 & 1 & 0 & 2 \\
\hline
0 & 0 & 1 & -1 & 1+K & 1 & 10
\end{array}
\right]
$$

If $s_1 = 0$, $r_2 = 0$, and $a_3 = 0$, then $x_1 = 2$, $x_2 = 2$, and $P = 10$. When $x_1 = 2$ and $x_2 = 2$, the function takes on its maximum value of 10. In the last step, pick the fourth column as the pivot column and see if you get the same answer. Both r_2 and a_3 must be 0. ∎

Other types of maximization problems involve negative constants on the right of one or more constraint equations. If this occurs, multiply both sides of the inequality by -1 to obtain a constraint inequality with a nonnegative constant. (Don't forget to reverse the inequality.)

Possible Steps for Mixed Problem Constraints

1. If any inequality has a negative constant on the right, multiply both sides by -1.
2. Introduce a slack variable for each $\leq$ inequality.
3. Introduce a surplus variable and an artificial variable for each $\geq$ inequality.
4. Use each equality as given or introduce surplus and artificial variables with each equality.
5. For each artificial variable, a_i introduced with constraints, put $-Ka_i$ in the objective function.
6. Form the preliminary tableau for the equations formed in the first five steps.
7. Use row operations to eliminate the K's under artificial variables in the bottom row.
8. Solve the problem by using the simplex technique on the tableau given in step 7.
9. If all artificial variables are 0, you should have a solution of the original problem.

Mixed-constraint minimization problems are converted to dual problems and the preceding techniques are used to find the optimum solution. However, in converting to the dual problem, the mixed-constraint difficulty sometimes disappears, as seen in the following example.

EXAMPLE 22 Minimize

$$
C = 2y_1 + 5y_2 + 6y_3 + 1
$$

subject to

$$-3y_1 + 2y_2 - 4y_3 \leq 1$$
$$y_1 + 2y_2 + 2y_3 \geq 2$$
$$2y_1 + 3y_2 + y_3 \geq 5$$
$$y_1, y_2, y_3 \geq 0$$

SOLUTION Notice that the first inequality constraint is stated as less than or equal to instead of greater than or equal to as required in order to set up an initial matrix. To change the sense of the first inequality, multiply each term by -1. The problem then becomes: Minimize

$$C = 2y_1 + 5y_2 + 6y_3 + 1$$

subject to

$$3y_1 - 2y_2 + 4y_3 \geq -1$$
$$y_1 + 2y_2 + 2y_3 \geq 2$$
$$2y_1 + 3y_2 + y_3 \geq 5$$
$$y_1, y_2, y_3 \geq 0$$

The transpose of

$$\begin{bmatrix} 3 & -2 & 4 \\ 1 & 2 & 2 \\ 2 & 3 & 1 \end{bmatrix}$$

is

$$\begin{bmatrix} 3 & 1 & 2 \\ -2 & 2 & 3 \\ 4 & 2 & 1 \end{bmatrix}$$

The simplex tableau for this problem is

x_1	x_2	x_3	y_1	y_2	y_3	P		
3	1	②	1	0	0	0	2	$\frac{1}{2}R_1 \rightarrow R_1$ (1)
-2	2	3	0	1	0	0	5	$-3R_1 + R_2 \rightarrow R_2$ (2)
4	2	1	0	0	1	0	6	$-R_1 + R_3 \rightarrow R_3$ (3)
1	-2	-5	0	0	0	1	1	$5R_1 + R_4 \rightarrow R_4$ (4)

The second matrix is

$$
\begin{array}{ccccccc}
x_1 & x_2 & x_3 & y_1 & y_2 & y_3 & P \\
\end{array}
$$

$$
\left[
\begin{array}{ccccccc|c}
\frac{3}{2} & \frac{1}{2} & 1 & \frac{1}{2} & 0 & 0 & 0 & 1 \\
-\frac{13}{2} & \frac{1}{2} & 0 & -\frac{3}{2} & 1 & 0 & 0 & 2 \\
-\frac{5}{2} & \frac{3}{2} & 0 & -\frac{1}{2} & 0 & 1 & 0 & 5 \\
\hline
\frac{17}{2} & \frac{1}{2} & 0 & \frac{5}{2} & 0 & 0 & 1 & 6 \\
\end{array}
\right]
$$

Since all the elements of the last row are nonnegative, the second matrix is the final matrix and gives the solution $C = 6$, when $y_1 = \frac{5}{2}$, $y_2 = 0$, and $y_3 = 0$. ■

Note that the suggestions on page 260 do not enable you to solve all mixed-constraint linear programming problems. Do not use the method of multiplying an inequality by -1 in a maximum problem if this will result in negative values in the last column of the simplex tableau. Remember that we assumed that all of the elements of the last column were nonnegative with the possible exception of the element in the last row and last column. A special starting procedure is required for problems in which some of the constants in the last column other than the constant in the last row and last column are negative.

SUMMARY

In this section we considered mixed-constraint problems, that is, constraint inequalities involving both $\geq$ and $\leq$. Many times such problems can be solved by inserting surplus variables and artificial variables.

Exercise Set 3.7

Solve the following mixed-constraint problems graphically.

1. Maximize

$$P = 8x_1 + 10x_2$$

subject to

$$8x_1 + 5x_2 \leq 40$$
$$4x_1 + 3x_2 \geq 12$$
$$x_1, x_2 \geq 0$$

2. Maximize

$$P = 10x_1 + 8x_2$$

subject to

$$8x_1 + 5x_2 \leq 40$$
$$6x_1 + 2x_2 = 36$$
$$x_1, x_2 \geq 0$$

3. Minimize

$$C = 4x_1 + 10x_2$$

subject to

$$8x_1 + 5x_2 \leq 40$$
$$4x_1 + 3x_2 \geq 12$$
$$x_1, x_2 \geq 0$$

4. Maximize

$$P = 5x_1 + 4x_2$$

subject to

$$x_1 + 2x_2 \leq 4$$
$$-x_1 + x_2 \geq -6$$
$$x_1, x_2 \geq 0$$

5. Minimize

$$C = 3x_1 + 8x_2$$

subject to

$$3x_1 + x_2 \leq 9$$
$$x_1 + 3x_2 \geq 6$$
$$x_1, x_2 \geq 0$$

6. Minimize

$$C = 4x_1 + 3x_2$$

subject to

$$3x_1 + 4x_2 \leq 24$$
$$5x_1 + 3x_2 \geq 15$$
$$x_1, x_2 \geq 0$$

Find the initial simplex tableau in terms of surplus and artificial variables as if you were solving the following problems. Don't forget to eliminate K for the last row.

7. Maximize

$$P = 8x_1 + 4x_2$$

subject to

$$x_1 + 2x_2 \leq 8$$
$$x_1 + x_2 \geq 2$$
$$x_1, x_2 \geq 0$$

8. Maximize

$$P = 10x_1 + 8x_2$$

subject to

$$x_1 + 3x_2 \leq 12$$
$$x_1 + x_2 = 8$$
$$x_1, x_2 \geq 0$$

9. Maximize

$$P = 5x_1 + 7x_2$$

subject to

$$-x_1 + 2x_2 \geq 4$$
$$x_1 + x_2 \leq 6$$
$$x_1, x_2 \geq 0$$

10. Maximize

$$P = 5x_1 + 3x_2$$

subject to

$$x_1 - 2x_2 \leq 4$$
$$x_1 + x_2 \geq 6$$
$$x_1, x_2 \geq 0$$

11. Maximize

$$P = 4x_1 + 6x_2$$

subject to

$$-x_1 + x_2 \geq -2$$
$$x_1 + x_2 \geq 4$$
$$x_1, x_2 \geq 0$$

12. Maximize

$$P = 10x_1 + 12x_2$$

subject to

$$-x_1 + 2x_2 \geq -4$$
$$x_1 + x_2 \geq 6$$
$$x_1, x_2 \geq 0$$

13. Minimize

$$C = 2x_1 + 3x_2$$

subject to

$$x_1 + x_2 \leq 8$$
$$3x_1 + 2x_2 \geq 6$$
$$x_1, x_2 \geq 0$$

14. Minimize

$$C = 5x_1 + 7x_2$$

subject to

$$-x_1 + x_2 \geq -2$$
$$x_1 + x_2 \leq 4$$
$$x_1, x_2 \geq 0$$

Solve the following problems using the simplex techniques involving surplus and artificial variables.

15. Exercise 1

16. Exercise 2

17. Exercise 3

18. Exercise 4

19. Exercise 5

20. Exercise 6

Applications (Business and Economics)

21. **Resource Allocation.** A mining company operates two mines, A and B. Each mine produces high-grade ore and low-grade ore. Mine A produces 2 tons of high-grade ore and 0.5 ton of low-grade ore per day. Mine B produces 1 ton of each grade of ore per day. The company has contracted to provide a local smelter a minimum of 14 tons of high-grade ore per week and a maximum of 6 tons of low-grade ore per week. Determine the minimum cost to meet this contract if it costs $4000 per day to operate mine A and $3000 a day to operate mine B.

Applications (Social and Life Sciences)

22. **Nutrition.** A dietician of a hospital has classified foods in categories A, B, and C. In order to include sufficient protein, at least 15% of a meal should come from category A. To regulate the amount of carbohydrates eaten, at least 20%, but not more than 50%, should be in category B. To provide suitable variety, the amount of category C should not exceed the sum of A and B. The average costs of A, B, and C per pound are $1.10, $0.90, and $0.80, respectively. Find the amount of foods A, B, and C in order to cook a meal of 200 pounds at minimum cost.

23. **Nutrition (Plants).** A fertilizer company has available 10 tons of nitrate and 8 tons of phosphate to produce two types of fertilizer. A bag of fertilizer A will have 20 pounds of nitrate and 10 pounds of phosphate, and a bag of fertilizer B will have 10 pounds of each. If there is a profit of $0.20 on each 50-pound bag of A and a profit of $0.15 on each 50-pound bag of B, how many of each would be mixed for maximum profit if all of the phosphate must be used in this mixing endeavor?

Review Exercises

24. Write the dual minimum problem to finding the maximum value of $P = 3x + 4y$ subject to

$$x, y \geq 0$$

$$3x + 5y \leq 30$$

$$3x + 5y \leq 21$$

25. A chicken-feed manufacturer mixes dried algae and crushed eggshells together and sells the mixture under the trade name "Start-Up." Each bag of Start-Up must contain at least 52 units of vitamins and 52 units of minerals. Each kilogram of dried algae costs $0.65 and contains 5 units of vitamins and 2 units of minerals. Each kilogram of crushed eggshells costs $0.28 and contains 1 unit of vitamins and 3 units of minerals. How much of each ingredient should be in each bag in order to minimize cost?

26. Write the dual maximum problem to finding the minimum value of $C = 8x + 4y$ subject to

$$x, y \geq 0$$

$$x + y \geq 5$$

$$3x + 5y \geq 21$$

27. Use simplex procedures to solve the minimum problem in Exercise 26.

Chapter Review

Make sure you understand each of these key ideas.

Systems of Linear Inequalities

Graph of a linear inequality   Unbounded region
Solution region   Corner point
Bounded region

Geometric Linear Programming

Linear programming problem   Problem constraints
Objective function   Constant-profit line

Simplex Method of Linear Programming

Slack variables   Pivot operation
Basic solution   Simplex tableau
Basic feasible solution   Indicators
Basic variables   Optimal solution
Nonbasic variables   Solution of maximization problem

Chapter Test

1. For

$$2x \leq 8 - 3y$$
$$4x + y \leq 6$$
$$2y \leq 8 - 6x$$
$$x, y \geq 0$$

    graph the solution region, find the corner points, and classify the solution region as bounded or unbounded.

2. Find the maximum and minimum of

$$P = 10x + 20y$$

    under the constraints of Exercise 1.

3. Set up an initial tableau to maximize $P = 10x + 20y$ under the constraints of Exercise 1.

4. Locate the pivot element in Exercise 3.

5. Determine the basic variables and the basic feasible solution of the simplex tableau

$$\begin{array}{ccccccc} x & y & z & r & s & t & P \end{array}$$
$$\left[\begin{array}{ccccccc|c} 0 & 0 & 0 & 2 & 1 & 1 & 0 & 12 \\ 1 & 0 & 0 & 3 & 0 & 0 & 0 & 21 \\ 0 & 6 & 1 & 4 & 2 & 0 & 0 & 28 \\ 0 & -1 & 0 & -5 & 3 & 0 & 1 & 50 \end{array}\right]$$

6. Find the next tableau in the simplex solution for Exercise 5 using 4 as the pivot element. Does this tableau give a solution?

7. A simplex tableau is given as

$$\begin{array}{ccccc} x & y & r & s & P \end{array}$$
$$\left[\begin{array}{ccccc|c} 0 & 1 & \frac{2}{3} & -\frac{1}{3} & 0 & 6 \\ 1 & 0 & -\frac{1}{3} & \frac{2}{3} & 0 & 18 \\ 0 & 0 & 4 & 1 & 1 & 54 \end{array}\right]$$

Is the simplex solution complete? Why? What is the maximum value of P and where does it occur?

8. Minimize

$$C = 5y_1 + 2y_2$$

subject to

$$3y_1 + 2y_2 \geq 8$$

$$y_1 + 4y_2 \geq 6$$

$$y_1 + y_2 \geq 4$$

Set up the simplex tableau of the dual problem. Label the columns.

9. A simplex tableau of a standard maximization problem is given as

$$\begin{array}{ccccc} x_1 & x_2 & y_1 & y_2 & P \end{array}$$
$$\left[\begin{array}{ccccc|c} 1 & 0 & \frac{4}{5} & -\frac{2}{5} & 0 & \frac{3}{20} \\ 0 & 1 & -\frac{3}{5} & \frac{3}{5} & 0 & \frac{12}{20} \\ 0 & 0 & 4 & 2 & 1 & 16 \end{array}\right]$$

Find the solution for the dual minimum problem.

10. An electronics store has $10,000 per month available for advertising. Newspaper ads cost $800 each and can occur a maximum of 8 times a month. Radio ads cost $300 each and can occur a maximum of 20 times a month. Each newspaper ad reaches 50,000 prospects and each radio ad reaches 30,000 people. How many of each ad should the store purchase in order to maximize advertisement exposure?

CHAPTER **4**

Mathematics of Finance

I n this chapter we study common topics of finance. Nearly everyone can use the concepts of this chapter in some way in everyday activities. Few people pay cash for all of their purchases, and financing a car or a home has become common practice. When items are financed, the total amount paid exceeds the price of the purchase; much of this difference is interest. When we save equal amounts of money each month, our savings account grows not only because of our deposits but also because of the interest we earn. The manager of a business must understand the importance of interest growth because the cost of borrowing money affects many business decisions.

In this chapter we introduce the following concepts: simple interest, compound interest, geometric progressions, annuities, amortization tables, sinking funds, perpetuities, and equations of value.

Although there are tables that perform certain financial calculations, we prefer to use a calculator or a computer because even the most complete set of tables is limited, and with a calculator you can handle a much wider range of problems.

4.1   SOME COMPARISONS OF INTEREST RATES

OVERVIEW   A sixth grader can deposit her $100 in a savings account paying 6% per year compounded quarterly, or she can deposit her savings in a local bank paying 6% simple interest. If at the end of a year her statement indicates that she has $106 in her account, did she place her money in the savings account or in the local bank?

The phrase *compounded quarterly* indicates that this problem involves compound interest. What is the difference between compound and simple interest? In this section we learn the difference and we use both concepts to solve a variety of problems. In this section we study

- Terms associated with simple interest
- A simple-interest formula relating amount and principal
- The concept of compound interest
- A compound-interest formula relating amount and principal
- Time diagrams

The cost of borrowing money is called **interest**. **Simple interest** is computed as a constant percentage of the money borrowed for a specific time, usually a single year or less, and is paid at the end of the specified time.

The sum borrowed is called the **principal**, P, or sometimes it is called the **present value**; r denotes the rate of interest, usually expressed in percentage per year; and t is time expressed in years or fractions of years. By definition, simple interest, I, equals principal multiplied by the interest rate multiplied by the time in years.

$$I = Prt$$

EXAMPLE 1   Four hundred dollars is borrowed at 8% simple interest for 3 years. What is the interest?

SOLUTION

$$I = Prt$$

$$I = (\$400)(0.08)(3) = \$96$$

8% as a decimal is 0.08

At the end of the term of the loan, the borrower must pay not only the sum that was originally borrowed, the principal, but also interest. The amount due at the end of the term is given by

$$\text{Amount} = \text{Principal} + \text{Interest}$$
$$A = P + I = P + Prt$$

Using the distributive property, we can write this formula as

$$A = P(1 + rt)$$

DEFINITION: SIMPLE INTEREST

$$I = Prt$$
$$A = P + Prt$$
$$= P(1 + rt)$$

where P = principal or present value, r = annual simple-interest rate expressed as a decimal, t = time in number of years, and A = amount or future value.

EXAMPLE 2   A loan of $1000 is made for 6 months at a simple-interest rate of 8%. How much does the borrower owe at the end of 6 months?

SOLUTION   In the formula replace P, r, and t with their values.

$$P = \$1000$$
$$r = 0.08$$
$$t = \frac{1}{2} \qquad \text{Convert 6 months to years; } \tfrac{6}{12} = \tfrac{1}{2} \text{ year.}$$
$$A = P(1 + rt) = \$1000\left[1 + (0.08)\left(\frac{1}{2}\right)\right]$$
$$= \$1040$$

The borrower owes $1040.

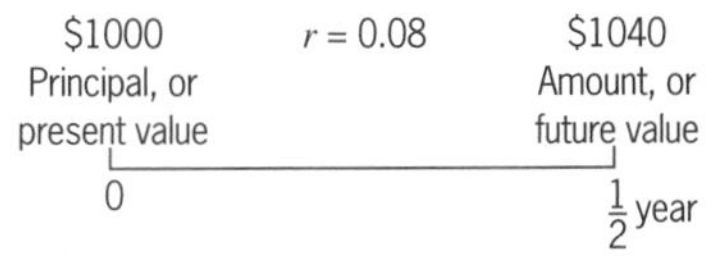

The relationship between principal, or present value, and amount, or future value, for Example 2 is shown in the time diagram of Figure 1.

$1000
Principal, or
present value

$r = 0.08$

$1040
Amount, or
future value

0

$\frac{1}{2}$ year

Figure 1

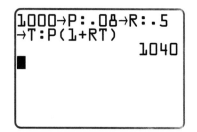

Figure 2

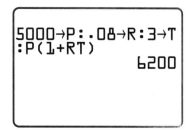

Figure 3

Calculator Note

When we use a formula repeatedly, it is helpful to store the formula in a calculator. We will now store the solution of Example 2 and recall the formula for Practice Problem 1. To store the variables press 1000 $\boxed{\text{STO} \blacktriangleright}$ $\boxed{\text{ALPHA}}$ $\boxed{\text{P}}$ $\boxed{\text{2nd}}$ $\boxed{:}$ .08 $\boxed{\text{STO} \blacktriangleright}$ $\boxed{\text{ALPHA}}$ $\boxed{\text{R}}$ $\boxed{\text{2nd}}$ $\boxed{:}$ .5 $\boxed{\text{STO} \blacktriangleright}$ $\boxed{\text{ALPHA}}$ $\boxed{\text{T}}$ $\boxed{\text{2nd}}$ $\boxed{:}$ $\boxed{\text{ALPHA}}$ $\boxed{\text{P}}$ $\boxed{(}$ 1 $\boxed{+}$ $\boxed{\text{ALPHA}}$ $\boxed{\text{R}}$ $\boxed{\text{ALPHA}}$ $\boxed{\text{T}}$ $\boxed{)}$ $\boxed{\text{ENTER}}$. Figure 2 shows not only what has been stored but also the answer 1040.

Practice Problem 1

A college student borrows $5000 to complete his senior year in college. If the bank charges 8% simple interest, how much will Steve owe in 3 years?

ANSWER

$$A = P(1 + rt), \text{ or } A = 5000(1 + 0.08 \cdot 3) = \$6200.$$

Calculator Note

If we are using a formula repeatedly, we can press $\boxed{\text{2nd}}$ $\boxed{\text{ENTRY}}$ to display a previous problem. We work Practice Problem 1 from Example 2 by replacing 1000 by 5000, and .5 by 3. The answer, as shown in Figure 3, is 6200.

To find the present value of an amount at simple-interest rate r, replace A, r, and t with the given values and solve for P.

EXAMPLE 3 Compute the present value of $1000 due in 3 months with interest at 12% annually.

SOLUTION

$$\$1000 = P\left[1 + (0.12)\left(\frac{3}{12}\right)\right] \qquad \begin{aligned} A &= \$1000 \\ r &= 0.12 \\ t &= \tfrac{3}{12} \end{aligned}$$

Using a calculator, press

1000 $\boxed{\div}$ $\boxed{(}$ 1 $\boxed{+}$ .12 $\boxed{\times}$ 3 $\boxed{\div}$ 12 $\boxed{)}$ $\boxed{\text{ENTER}}$

which gives $P = \$970.87$. ■

In this chapter, we round all final answers to two decimal places (nearest cent). Hence, the present value is $970.87. The time diagram for Example 3 is shown in Figure 4.

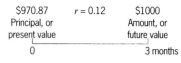

$970.87	$r = 0.12$	$1000
Principal, or present value		Amount, or future value
0		3 months

Figure 4

Practice Problem 2 In 5 years, you will need $10,000 as a down payment on a condominium. How much must you deposit today if a savings and loan will pay you 8% simple interest?

ANSWER $A = P(1 + rt)$,   or $10,000 = P(1 + 0.08 \cdot 5)$,   or
$10,000 = P(1.40)$; $P = \$7142.86$.

EXAMPLE 4

One thousand dollars is borrowed for 2 years. At the end of that time, $1120 is repaid. What percent of simple interest was charged?

SOLUTION

$$A = P(1 + rt)$$

$A = \$1120$
$P = \$1000$
$t = 2$ years

$$\$1120 = \$1000(1 + 2r)$$

$$\$1120 = \$1000 + \$2000r$$

$$r = 0.06 = 6\% \text{ simple interest}$$ ∎

$1000
Principal, or
present value

$1120
Amount, or
future value

0   2 years

Figure 5

Let's work the preceding example another way after examining the time diagram (Figure 5). Note that the difference between $1120 and $1000 is the interest.

$$I = \$1120 - \$1000 = \$120$$

$$I = Prt$$

$$\$120 = \$1000(r)(2)$$

$$r = 0.06 = 6\% \text{ simple-interest rate}$$

Practice Problem 3 You borrow $1000 from a bank that requires you to pay $1240 in 2 years. What simple-interest rate is the bank charging?

ANSWER $I = 1240 - 1000 = 240$. Then $I = Prt$, or $240 = 1000(r)(2)$, or $240 = 2000r$. $r = 0.12 = 12\%$.

Simple interest is not always as simple as it appears. Consider the story of Jill. Jill anticipated an IRS tax refund of $300 in early May. Being short on cash, on January 20 she decided to borrow $300 until May 10. Her banker agreed to give her the loan at 7.5% simple interest. First the banker calculated the number of days she would have the money.

Days remaining in January	$31 - 19 = 12$
Days in February	28
Days in March	31
Days in April	30
Days in May	10
	111

Note that January 20 is included as a day. She was comfortable with this computation, but she was surprised when the banker computed her interest:

$$I = Prt$$

$$I = (\$300)(0.075)\left(\frac{111}{360}\right)$$

$$I = \$6.94$$

The banker pointed out that the fraction $\frac{111}{360}$ indicated the portion of the year that she would have the money. Bankers often compute interest on the basis of a 360-day year. In the days prior to widespread use of calculators, this convention made computations easier. Today most bankers use a 365-day year, but some still use a 360-day year.

COMMON ERROR

Suppose you must pay $6000 at the end of 1 year. If the bank charges 6% simple interest, how much did you borrow? Answer: Using $P = 6000(1 + 0.06 \cdot 1)$ is incorrect because it is an accumulated amount and not a present value. The correct answer is $P = 6000 \div (1 + 0.06 \cdot 1)$.

Compound Interest

We have learned that simple interest I is found by using the formula $I = Prt$, where P represents the principal, r the rate, and t the time. When interest is computed by this formula, the principal always remains the same. If the interest is added to the principal at the end of each interest period, so that the principal is increased, the interest is said to be **compounded**. The sum of the original principal and all the interest is called the **compound amount**, and the difference between the compound amount and the original principal is the **compound interest**. A comparison of simple and compound interest is given in the following example.

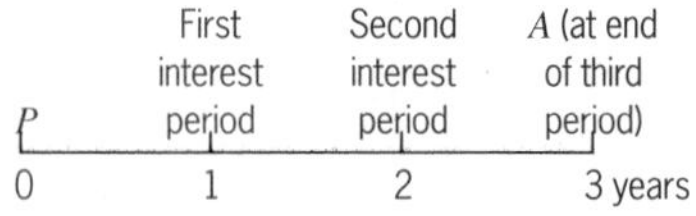

	First interest period	Second interest period	A (at end of third period)
P			
0	1	2	3 years

Figure 6

EXAMPLE 5   Find the simple interest on $1000 for 3 years at 6%. Then find the compound interest on $1000 for 3 years at 6%, compounded annually.

SOLUTION   First we summarize the problem on a time diagram (see Figure 6). To emphasize the difference between simple and compound interest, we will compute both year by year. At the end of the first year, the interest for both will be

$$I = \$1000(0.06) = \$60$$

For compound interest, this amount is added to the principal, and the new principal becomes $1000 + $60 = $1060. Interest for the second year is

$$I = \$1060(0.06) = \$63.60$$

Now, how do you get the principal for the third year?

TABLE 1

	Simple Interest	Compound Interest
For the first year	$I = \$1000(0.06) = \ \$60$	$I = \$1000.00(0.06) = \ \$60$
For the second year	$I = \$1000(0.06) = \ \$60$	$I = \$1060.00(0.06) = \ \$63.60$
For the third year	$I = \$1000(0.06) = \ \underline{\$60}$	$I = \$1123.60(0.06) = \ \underline{\$67.42}$
	$\$180$	$\$191.02$

From Table 1 we see that the simple interest for 3 years totals $180, whereas the compound interest totals $191.02. The compound amount is $1191.02. Notice that the principal changes each year when interest is compounded, but the principal remains the same when simple interest is used. ■

For the preceding example, let's compute the compound amount at the end of each year in another way. At the end of the first year, the compound amount is

$$\$1000 + \$1000(0.06) = \$1000(1 + 0.06) = \$1000(1.06)$$

During the second year, the principal is $1000(1.06). At the end of the second year the compound amount is

$$\$1000(1.06)(1 + 0.06) = \$1000(1.06)(1.06)$$
$$= \$1000(1.06)^2$$

TABLE 2

Period	Principal	+ Interest	= Amount (at end of period)
1	P	$+ \ Pi$	$= P(1 + i)$
2	$P(1 + i)$	$+ \ P(1 + i)i$	$= P(1 + i)^2$
3	$P(1 + i)^2$	$+ \ P(1 + i)^2 i$	$= P(1 + i)^3$
$\vdots$	$\vdots$	$\vdots$	$\vdots$
n	$P(1 + i)^{n-1}$	$+ \ P(1 + i)^{n-1} i$	$= P(1 + i)^n$

This pattern extends easily to the most general case. Suppose that P dollars are deposited at an interest rate of i per period for n periods. Table 2 shows the amount that has accrued at the end of that time. Table 2 suggests that the compound amount can be found by multiplying the principal by $(1 + i)^n$, where i is the interest rate per period [or $i = r/k$, where r is annual (nominal or quoted) rate and k is the number of interest periods per year] and n is the number of interest periods. For example, for a rate quoted as 12% compounded monthly, $r = 0.12$, $k = 12$, and $i = r/k = 0.12/12 = 0.01$.

DEFINITION: COMPOUND INTEREST

$$A = P\left(1 + \frac{r}{k}\right)^{kt} = P(1 + i)^n$$

where $A =$ compound amount after n periods, $P =$ principal invested (present value), $r =$ annual rate (sometimes called the nominal rate), $k =$ number of compound-interest periods per year, $t =$ time in years, $i = r/k$, and $n = kt$.

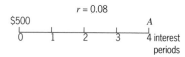

Figure 7

EXAMPLE 6

Find the compound amount that would result from investing $500 at 8% interest compounded annually for 4 years.

SOLUTION We first examine the time diagram in Figure 7. Substituting $P = 500, $r = 0.08$, and $k = 1$ in the formula $A = P[1 + (r/k)]^{kt}$ gives

$$A = \$500(1 + 0.08)^4 = 500(1.08)^4$$

Using a calculator, press

$$500 \boxed{\times} 1.08 \boxed{\wedge} 4 \boxed{\text{ENTER}}$$

which gives $A = \$680.24$. ∎

Tables in Appendix B are provided for those who need practice in working problems using a table. The value of $(1 + i)^n$ for various values of i and n are found in Table 1 of Appendix B.

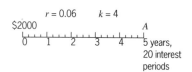

Figure 8

EXAMPLE 7 What is the interest obtained from an investment of $2000 at 6% compounded quarterly for 5 years?

SOLUTION The time diagram is shown in Figure 8. We have $r = 0.06$, $k = 4$, and $t = 5$.

$$A = P\left(1 + \frac{r}{k}\right)^{kt}$$
$$= \$2000(1 + 0.015)^{20} \qquad i = 0.06/4 = 0.015$$
$$= \$2693.71 \qquad\qquad n = 4 \cdot 5 = 20$$

The interest is

$$\$2693.71 - 2000 = \$693.71$$

Once we have established the relationship $A = P[1 + (r/k)]^{kt}$, we can use this formula to change the direction of our thinking. Consider the question, "How much principal must we invest now at 8% per year compounded quarterly in order to have the \$6000 we need to buy a used car in 4 years?" When the question is asked in this way, the principal for which we are searching is called the **present value**. Thus, we are looking for the present value of \$6000 due in 4 years at 8% per year compounded quarterly (Figure 9).

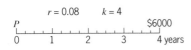

$r = 0.08$ $k = 4$

P |———|———|———|———| \$6000
0   1   2   3   4 years

Figure 9

$$A = P\left(1 + \frac{r}{k}\right)^{kt}$$

$$\$6000 = P(1 + 0.02)^{16}$$

$$P = \frac{\$6000}{(1.02)^{16}}$$

$A = \$6000$
$r = 0.08$
$k = 4$
$\dfrac{r}{k} = 0.02$
$n = kt = (4)(4) = 16$

With a calculator use $6000 \;\boxed{\div}\; 1.02 \;\boxed{\wedge}\; 16 \;\boxed{\text{ENTER}}$ to get \$4370.67. Thus, the present value of \$6000 due in 4 years at 8% compounded quarterly is \$4370.67.

Some students like to enter the numbers in the calculator all at one time. Since $P = A/(1 + r/k)^{kt}$,

$$P = 6000 \;\boxed{\div}\;\boxed{(}\;\boxed{(}\; 1 \;\boxed{+}\; .08 \;\boxed{\div}\; 4 \;\boxed{)}\;\boxed{\wedge}\;\boxed{(}\; 4 \;\boxed{\times}\; 4 \;\boxed{)}\;\boxed{\text{ENTER}}$$
$$= \$4370.67$$

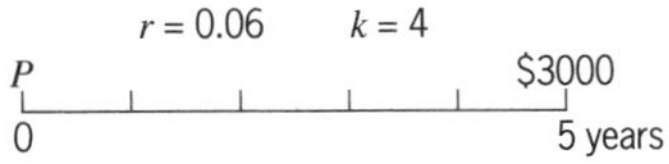

$r = 0.06$ $k = 4$

P |———|———|———|———| \$3000
0   5 years

Figure 10

EXAMPLE 8   How much money should be deposited in a savings and loan association paying 6% compounded quarterly in order to have \$3000 in 5 years?

SOLUTION   The time diagram is shown in Figure 10.

$$A = P\left(1 + \frac{r}{k}\right)^{kt}$$

$$\$3000 = P(1 + 0.015)^{20}$$

$$P = \frac{\$3000}{(1.015)^{20}}$$

$i = 0.06/4 = 0.015$
$n = kt = (4)(5) = 20$

Using a calculator, we find

$$P = \$2227.41$$

Thus, \$2227.41 must be deposited now in a savings and loan association in order to have \$3000 in 5 years.

COMMON ERROR   A college student borrows $10,000 at 6% compounded semiannually. At the end of 5 years he does not owe $A = 10{,}000 \cdot (1.06)^5$. This answer is incorrect because interest is compounded two times a year. The answer is $A = \$10{,}000\,(1.03)^{10}$.

EXAMPLE 9   How long will it take a dollar to double at 8% compounded semiannually?

SOLUTION   Suppose that we start with $P = \$1$. This amount doubles or $A = \$2$. r is 0.08 and $k = 2$ (semiannually). Substitute these values in the formula

$$A = P\left(1 + \frac{r}{k}\right)^{kt}$$

$$2 = (1.04)^n$$

Use the $\boxed{\wedge}$ key of your graphing calculator and the very useful mathematical tool of ''guessing and then checking'' to get an approximate answer. Try any number for n. Let's try a 5.

$1.04 \boxed{\wedge} 5 = 1.217$	Much too small
$1.04 \boxed{\wedge} 10 = 1.480$	Still too small
$1.04 \boxed{\wedge} 20 = 2.191$	A bit too large
$1.04 \boxed{\wedge} 17 = 1.948$	
$1.04 \boxed{\wedge} 18 = 2.026$	

In 18 interest periods (or 9 years) the principal will have more than doubled, but in 17 periods it will not quite be double. ∎

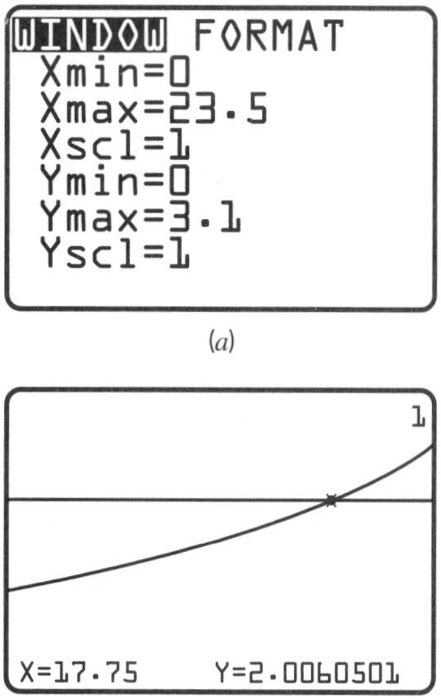

Calculator Note

We can obtain a graphical solution to this problem. We graph $y = (1.04)^x$ and $y = 2$ (Figure 11) and find where they intersect; that is, we find where $(1.04)^x = 2$. Using the Trace key (see page 3 of Appendix A), we find x to be 17.75 interest periods.

Practice Problem 4   You want to borrow $10,000 for 6 years. Which is a better interest rate:

a) $8\frac{1}{4}\%$ simple interest   or   b) 8% compounded quarterly?

ANSWER

(a) $A = 10{,}000(1 + 0.0825 \cdot 6) = \$14{,}950$
(b) $A = 10{,}000(1 + 0.02)^{6 \cdot 4} = \$16{,}084.37$
    The loan at $8\frac{1}{4}\%$ simple interest is the better rate.

```
WINDOW FORMAT
 Xmin=0
 Xmax=23.5
 Xscl=1
 Ymin=0
 Ymax=3.1
 Yscl=1
```

(a)

X=17.75   Y=2.0060501

(b)

Figure 11

Calculator Note

Practice Problem 4 suggests the following illustration. Let's compare graphically the value of $1 invested at 6% simple interest and at 6% interest compounded monthly. For x years we have

Simple interest: $y = (1 + 0.06x)$
Compound interest: $y = (1 + 0.005)^{12x}$

The value of $1 for various years is shown in Figure 12.

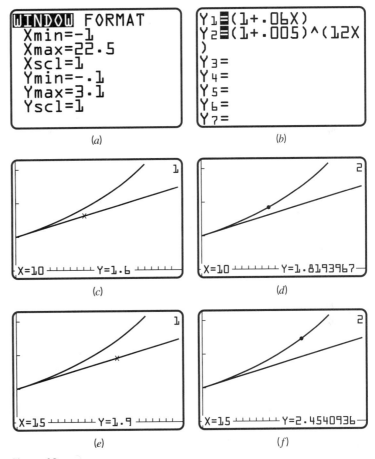

(a)

(b)

(c)

(d)

(e)

(f)

Figure 12

Many different types of problems can be considered using the compound-interest formula. For example, suppose that inflation is increasing at a rate of 5% per year. A $1.50 milkshake would cost $1.50(1.05)^6 = $2.01 at the end of 6 years, and a $10,000 automobile would cost $10,000(1.05)^8 = $14,774.55 at the end of 8 years.

SUMMARY

For simple interest

$$I = Prt$$

$$A = P(1 + rt)$$

where P = principal (or present value), r = annual simple-interest rate, t = time in number of years, and A = amount (or future value).

For compound interest

$$A = P\left(1 + \frac{r}{k}\right)^{kt} = P(1 + i)^n$$

where A = compound amount after n periods, P = principal invested (present value), r = annual rate, k = number of compound-interest periods per year, t = time in years, $i = r/k$, and $n = kt$.

Exercise Set 4.1

1. Compute the simple interest when the principal, rate, and time of the loan are given.
   (a) $P = \$500$, $r = 0.08$, $t = 2$ years
   (b) $P = \$300$, $r = 0.03$, $t = 4$ years
   (c) $P = \$500$, $r = 0.04$, $t = 5$ years

2. What is the amount to be repaid in (a), (b), and (c) of Exercise 1?

3. Find the interest and the amount of a loan for $3000 borrowed for 2 years at 6% simple interest.

Find the compound interest and compound amount for the investments in Exercises 4 through 6.

4. (a) $5000 at 8% compounded annually for 3 years
   (b) $5000 at 8% compounded semiannually for 3 years
   (c) $5000 at 8% compounded quarterly for 3 years

5. (a) $2000 at 8% compounded annually for 4 years
   (b) $2000 at 8% compounded semiannually for 4 years
   (c) $2000 at 8% compounded quarterly for 4 years

6. (a) $3000 at 6% compounded annually for 6 years
   (b) $3000 at 6% compounded semiannually for 6 years
   (c) $3000 at 6% compounded quarterly for 6 years

7. Compute the amount to be repaid when the principal, simple-interest rate, and time of the loan are given.
   (a) $P = \$4000$, $r = 0.06$, $t = 2$ years
   (b) $P = \$3000$, $r = 0.05$, $t = 6$ months
   (c) $P = \$100$, $r = 0.08$, $t = 3$ years

8. Find the simple-interest rate on the loan when the principal, the amount repaid, and the term of the loan are given.
   (a) Principal, $3500; amount repaid in 2 years, $4130
   (b) Principal, $1000; amount repaid in 120 days, $1046.67
   (c) Principal, $500; amount repaid in 45 days, $510

Find the present value of the money in Exercises 9 through 12.

9. $5000 due in 5 years if money is worth 6% compounded annually

10. $6000 due in 5 years 6 months if money is worth 8% compounded semiannually

11. $7000 due in 4 years if money is worth 6% compounded semiannually

12. $8000 due in 6.5 years if money is worth 7% compounded annually

Solve each formula for the indicated variable.

13. $I = Prt$ (r)

14. $A = P + Prt$ (P)

15. How long will it take for $125 to amount to $375 at 6% interest compounded quarterly?

16. How many years will it take to double $1000 at 4% interest compounded semiannually?

17. Graphically compare the accumulations of two loans, one at 9% simple interest and one at 9% interest compounded monthly, over a period of 20 years.

18. Work Exercise 17 for 8% simple interest and 8% compounded quarterly.

Applications (Business and Economics)

19. **Interest.** Find the interest on $2000 borrowed for 8 months at 10% simple interest.

20. **Interest.** Find the amount of a 90-day, $1500 loan at 6% simple interest. (Use a banker's year of 360 days in the computation.)

21. **Interest.** How much interest will you owe on a $1000 loan from March 3 to August 7 at 5% simple interest? (Use a 365-day year.)

22. **Interest.** What is the interest on a $1500 loan from June 15 to September 11 at a simple-interest rate of 8%? (Use a 365-day year.)

23. **Loans.** Some lending institutions have a minimum amount of interest they must collect on any loan. An institution with a minimum charge of $5 will expect to receive $5 interest on the loan, even though actual interest charges are only $3. How long must $1000 be borrowed at 8% interest to reach the minimum service charge of $5? (Use a banker's year of 360 days in the computation.)

24. **Loans.** A college student borrowed $700 and agreed to repay the principal with interest at 8% compounded semiannually. What will she owe at the end of 5 years?

25. **Loans.** On April 1, 1986, the owner of a small business borrowed $3000 at 8% compounded quarterly. What will she owe on October 1, 1998?

26. **Savings.** The sum of $1000 was deposited in a bank at an interest rate of 6% compounded semiannually. Five years later the rate increased to 8% compounded semiannually. If the money was not withdrawn, how much was in the account at the end of 6 years?

27. **Savings.** How much should parents invest for their daughter at 6% interest compounded seminannually in order to have $5000 at the end of 20 years?

28. **Cash Value.** A lot is sold for $750 cash and $600 a year for the next 3 years. Find the cash value of the lot if money is worth 6% compounded annually.

Applications (Social and Life Sciences)

The effects of inflation can be obtained using the formula for compound interest where i becomes the inflation rate per period.

29. **Inflation.** Find the cost of the following items in 10 years at an annual inflation rate of 6%.
   (a) A $100,000 house
   (b) A $1.90 hamburger
   (c) A $5.00 movie ticket
   (d) A $14.65 hourly labor rate.

30. **Inflation.** Rework Exercise 29 for a 3% annual inflation rate.

31. **Inflation.** Rework Exercise 29 for an 8% annual inflation rate.

32. **Inflation.** Find how long it will take a price to double with an average annual inflation rate of
   (a) 3%   (b) 4%
   (c) 5%   (d) 6%

33. **Population Growth.** The population of a city of 60,000 is expected to increase at a rate of 4% per year for the next 10 years. What is the population at the end of 10 years?

34. **Fish Population.** The number of fish in a lake is expected to increase at a rate of 6% per year for 5 years. How many fish will be in the lake in 5 years if 10,000 are placed in the lake today?

4.2 EFFECTIVE RATES, CONTINUOUS COMPOUNDING, AND GEOMETRIC PROGRESSIONS (Optional)

OVERVIEW   Effective rates and continuously compounded rates are useful extensions of our discussion on compound interest. Geometric progressions are introduced to facilitate an understanding of the next section on annuities. In this section we consider the following:

- A formula for effective rates
- A formula for continuous compounding
- The definition of a geometric progression
- A formula for the sum of a geometric progression

We have not yet established how to compare the interest rates promised by two institutions. For example, if the bank down the street offers an interest rate of 10% compounded 5 times a year and the bank 12 miles across town offers a rate of 12% compounded 3 times a year, should we undertake the long drive? To compare the two rates, we need to introduce the notion of an **effective rate of interest**. The effective rate of $i\%$ per period compounded k times a year is the simple-interest rate that gives the same amount due at the end of 1 year as the rate i per period compounded k times a year. That is,

$$P(1 + r \cdot 1) = P(1 + i)^k$$
$$1 + r = (1 + i)^k \qquad \text{Divide by } P$$
$$r = (1 + i)^k - 1$$

where r is the simple-interest rate, and i is compounded k times a year. From this relationship the effective rate can be obtained or the following formula can be used.

DEFINITION: EFFECTIVE ANNUAL RATE

A rate of i per period compounded k times a year produces an

$$\text{Effective rate} = (1 + i)^k - 1 \qquad i = r/k$$

EXAMPLE 10   Find the effective rate equivalent to a nominal rate of 6% compounded quarterly.

SOLUTION

$$\text{Effective rate} = (1 + 0.015)^4 - 1 \qquad i = \frac{0.06}{4} = 0.015$$
$$= 1.0614 - 1 \qquad k = 4$$
$$= 0.0614$$

A nominal rate of 6% compounded quarterly is equivalent to an effective rate of 6.14%. ■

EXAMPLE 11   For a savings account, which is the better rate, 6.5% compounded semiannually, or 6% compounded monthly?

SOLUTION   To compare the two rates, we first find and compare the effective rates:

$$\left(1 + \frac{0.065}{2}\right)^2 - 1 = (1 + 0.0325)^2 - 1 = 0.06606 \quad \text{or} \quad 6.606\%$$

$$\left(1 + \frac{0.06}{12}\right)^{12} - 1 = (1.005)^{12} - 1 = 0.06168 \quad \text{or} \quad 6.168\%$$

The effective rate for 6.5% compounded 2 times a year is greater than that for 6% compounded 12 times a year.   ■

Practice Problem 1 Find the effective rate equivalent to 8% compounded monthly.

ANSWER $r = \left(1 + \frac{0.08}{12}\right)^{12} - 1 = 0.083$ or 8.3%.

A comparison of the following interest formulas leads intuitively to a formula for compounding continuously.

Summary of Compound Interest

Amount at compound interest:

$$A = P(1 + i)^n$$

Annually:

$$A = P(1 + r)^t$$

Semiannually:

$$A = P\left(1 + \frac{r}{2}\right)^{2t}$$

Quarterly:

$$A = P\left(1 + \frac{r}{4}\right)^{4t}$$

Monthly:

$$A = P\left(1 + \frac{r}{12}\right)^{12t}$$

Where $i = r/k$
r = annual quoted rate (nominal rate)
k = number of compounding periods per year
$n = kt$
t = number of years
P = present value

Summary of Compound Interest *(Continued)*

Daily:

$$A = P\left(1 + \frac{r}{365}\right)^{365t}$$

Continuously:

$$A = Pe^{rt}$$

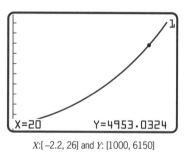

(a)

X:[–2.2, 26] and Y: [1000, 6150]

(b)

Figure 13

Suppose that you compound your interest every hour or 8760 times a year,

$$A = P\left(1 + \frac{r}{8760}\right)^{8760t}$$

or you could compound your interest every minute or 525,600 times a year,

$$A = P\left(1 + \frac{r}{525,600}\right)^{525,600t}$$

These compound amounts are approaching what we would get by using the formula for **continuous compounding**:

$$A = Pe^{rt}$$

where P is the present value, r is the annual interest rate, and t is the number of years. This function was introduced in Chapter 1. The graph of $A = 1000e^{rt}$ is shown in Figure 13 for $r = 0.08$ and for t varying from 0 to 26 years. Notice that the amount more than quadrupled in 20 years.

Calculator Note

The number e is an irrational number. Your graphing calculator should give a good approximation for e. Use

| 2nd | | LN | | 1 | | ENTER |

to get an approximate value for e, namely 2.718281828.

In Figure 14 note the rather significant increase in interest earned as the number of compound-interest periods per year changes from 1 to 2 to 4 to 12. Then note the insignificant change in interest earned from compounding daily to compounding continuously.

EXAMPLE 12   Find the amount obtained from an investment of $2000 compounded continuously for 5 years at 6%.

SOLUTION
$$A = Pe^{rt} = 2000e^{5(0.06)}$$
$$= 2000e^{0.3}$$

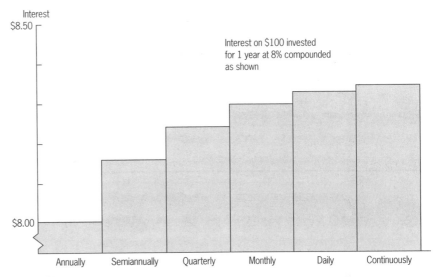

Figure 14

Using your calculator, press

$$2000 \quad \boxed{\times} \quad \boxed{\text{2nd}} \quad \boxed{\text{LN}} \quad .3 \quad \boxed{\text{ENTER}}$$

The answer is $2699.72. ■

Practice Problem 2 To what amount will $10,000 accumulate in 6 years if the 8% interest is compounded continuously?

ANSWER $A = \$10,000e^{6(0.08)} = \$16,160.74$

Geometric Progressions

The next two sections involve a sequence of payments or deposits that draw interest. In the preceding section, we learned that to accumulate a payment with compound interest we multiply by the factor $(1 + i)^n$. To find the sum of such terms, we introduce what is known as a geometric progression.

A **geometric progression** is characterized by the fact that each term is obtained from the preceding term by multiplying by the constant r, called the **common ratio**.

EXAMPLE 13 Determine whether the following is a geometric progression, find the first term, and find the common ratio:

$$12 + 36 + 108 + 324 + \cdots$$

SOLUTION   The first term is $a = 12$. Since $36 = 12 \cdot 3$, $108 = 36 \cdot 3 = 12 \cdot 3^2$, and $324 = 108 \cdot 3 = 12 \cdot 3^3$, this expression represents a geometric progression with $r = 3$.   ■

If the first term of a geometric progression is a and the common ratio is r, then the sum of the first n terms of a geometric progression can be expressed as

$$S_n = a + ar + ar^2 + \cdots + ar^{n-1}$$

$$\underset{\substack{1st \\ term}}{} \quad \underset{\substack{2nd \\ term}}{} \quad \underset{\substack{3rd \\ term}}{} \qquad \underset{\substack{nth \\ term}}{\phantom{ar^{n-1}}}$$

Let's multiply both sides of this equation by r and subtract S_n from rS_n.

$$\begin{aligned} rS_n &= ar + ar^2 + \cdots + ar^{n-1} + ar^n \\ S_n &= a + ar + ar^2 + \cdots + ar^{n-1} \\ \hline rS_n - S_n &= ar^n - a \\ (r-1)S_n &= a(r^n - 1) \end{aligned}$$

If $r \neq 1$, dividing by $r - 1$ gives

$$S_n = \frac{a(r^n - 1)}{r - 1}$$

Thus, we have the following:

Sum of a Geometric Progression

The sum S_n of the first n terms of a geometric progression with first term a and common ratio $r \neq 1$ is

$$S_n = \frac{a(r^n - 1)}{r - 1} \quad \text{or} \quad S_n = \frac{a(1 - r^n)}{1 - r}$$

EXAMPLE 14   Find the sum of six terms of the following geometric progression:

$$6 + 12 + 24 + \cdots$$

SOLUTION   Since this is a geometric progression, each term is obtained from the preceding term by multiplying by a constant r. The constant r may be found by dividing any term by the preceding term; that is,

$$r = \frac{12}{6} = 2 \quad \text{or} \quad r = \frac{24}{12} = 2$$

After r is found, other terms of the progression may be obtained by multiplying the preceding term by r. The sum of n terms is found by using the formula

$$S_n = \frac{a(r^n - 1)}{r - 1}$$

$$S_6 = \frac{6(2^6 - 1)}{2 - 1} \qquad 2^6 - 1 = 63$$

$$= 378 \qquad \blacksquare$$

EXAMPLE 15 Find the sum of

$$100 + 100(1.06) + 100(1.06)^2 + \cdots + 100(1.06)^4$$

SOLUTION Notice that S_n is now the sum of five terms of a geometric progression whose first term is 100 and whose common ratio is 1.06; hence,

$$S_5 = \frac{a(r^n - 1)}{r - 1} = \frac{100\,[(1.06)^5 - 1]}{1.06 - 1} \approx 563.709$$

On your calculator, use

100 $\boxed{\times}$ $\boxed{(}$ 1.06 $\boxed{\wedge}$ 5 $\boxed{-}$ 1 $\boxed{)}$ $\boxed{\div}$ $\boxed{(}$ 1.06 $\boxed{-}$ 1 $\boxed{)}$ $\boxed{\text{ENTER}}$ $\blacksquare$

In the next section we relate this example to a problem such as finding the amount of an investment of $100 at the end of each year for 5 years at an interest rate of 6% per year.

EXAMPLE 16 Find the sum of

$$100(1.06)^{-1} + 100(1.06)^{-2} + 100(1.06)^{-3} + 100(1.06)^{-4} + 100(1.06)^{-5}$$

SOLUTION First write this with positive exponents:

$$\frac{100}{(1.06)^1} + \frac{100}{(1.06)^2} + \frac{100}{(1.06)^3} + \frac{100}{(1.06)^4} + \frac{100}{(1.06)^5}$$

Then note that this is a geometric progression with $a = 100/1.06$ and common ratio 1/1.06. To verify, multiply the first term by 1/1.06 and see if you get the second term, and so on through the progression. The sum is

$$S_5 = \frac{\dfrac{100}{1.06}\left[1 - \dfrac{1}{(1.06)^5}\right]}{1 - \dfrac{1}{1.06}} \qquad \begin{array}{l} \text{Use} \\[4pt] S_n = \dfrac{a(1 - r^n)}{1 - r} \end{array}$$

$$= \frac{100}{1.06}\left[\frac{1 - (1.06)^{-5}}{\dfrac{1.06 - 1}{1.06}}\right] \qquad \frac{1}{(1.06)^5} = (1.06)^{-5}$$

$$= 100\left[\frac{1 - (1.06)^{-5}}{0.06}\right] \approx 421.24$$

On a calculator, use

$$100 \;\boxed{\text{x}}\; \boxed{(}\; 1 \;\boxed{-}\; 1.06 \;\boxed{\wedge}\;\boxed{(-)}\; 5 \;\boxed{)}\;\boxed{\div}\; .06 \;\boxed{\text{ENTER}}$$

to attain 421.24.   ■

In Section 4.4 we relate this example to the problem of finding the present value of a sequence of $100 payments made at the end of each year for 5 years.

Now let's consider abstract geometric progressions, which we will use in the next section.

EXAMPLE 17   Find the sum of

$$R + R(1 + i) + R(1 + i)^2 + \cdots + R(1 + i)^{n-2} + R(1 + i)^{n-1}$$

SOLUTION   Note that this expression is the sum of a geometric progression with the common ratio $1 + i$ and first term R. Thus,

$$S_n = \frac{a(r^n - 1)}{r - 1}$$

$$= R\left[\frac{(1 + i)^n - 1}{(1 + i) - 1}\right]$$

$$= R\left[\frac{(1 + i)^n - 1}{i}\right] \qquad\qquad ■$$

EXAMPLE 18   Find the sum of

$$R(1 + i)^{-1} + R(1 + i)^{-2} + R(1 + i)^{-3} + \cdots + R(1 + i)^{-n}$$

SOLUTION   First write this as

$$\frac{R}{(1+i)} + \frac{R}{(1+i)^2} + \frac{R}{(1+i)^3} + \cdots + \frac{R}{(1+i)^n}$$

Now

$$a = \frac{R}{1+i} \quad \text{and} \quad r = \frac{1}{(1+i)}$$

Using the formula for the sum of a geometric progression gives

$$S_n = \frac{R}{1+i}\left[\frac{1 - \left(\frac{1}{1+i}\right)^n}{1 - \frac{1}{1+i}}\right] \qquad \text{Use} \quad S_n = \frac{a(1-r^n)}{1-r}$$

$$= R\left[\frac{1 - (1+i)^{-n}}{(1+i) - 1}\right] \qquad (1+i)\left(1 - \frac{1}{1+i}\right) = (1+i) - 1$$

$$= R\left[\frac{1 - (1+i)^{-n}}{i}\right]$$

SUMMARY

1. The effective rate of a nominal (or annual) rate r compounded k times a year is

$$\text{Effective rate} = (1 + r/k)^k - 1$$

2. For continuous compounding the amount is given by

$$A = Pe^{rt}$$

where P is the principal, r is the annual interest rate, t is the number of years, and A is the amount.

3. The sum of n terms of a geometric progression is

$$S_n = a\frac{(r^n - 1)}{r - 1}$$

where r is the common ratio and a is the first term.

Exercise Set 4.2

Identify whether the following are geometric progressions. If so, find the common ratio.

1. 4, 12, 36, 108, . . .

2. 6, 14, 22, 38, 46, . . .

3. 12, 9, 6, 3, 0, -3, . . .

4. 12, 3, $\dfrac{3}{4}$, $\dfrac{3}{16}$, . . .

5. a, $a + r$, $a + 2r$, $a + 3r$, . . .

6. $\dfrac{1}{k}$, $\dfrac{r}{k}$, $\dfrac{r^2}{k}$, $\dfrac{r^3}{k}$, . . .

7. a, $a(1+i)^2$, $a(1+i)^3$, . . .

8. a, $a(1+i)^{-1}$, $a(1+i)^{-2}$, $a(1+i)^{-3}$, . . .

Applications (Business and Economics)

Find the effective rates equivalent to the nominal rates for Exercises 9–12.

9. 8% compounded semiannually

10. 6% compounded semiannually

11. 8% compounded monthly

12. 10% compounded monthly

13. $1000 is invested at 6% compounded
    (a) Annually   (b) Semiannually
    (c) Quarterly   (d) Monthly
    (e) Daily   (f) Continuously
    What is the amount after 10 years?

14. Find the compound amount for
    (a) $3000 at 6% compounded continuously for 5 years
    (b) $6000 at 5% compounded continuously for 6 years

15. Find the effective rate corresponding to 10% interest compounded continuously.

16. Approximately how many years will it take money to double at 10% interest compounded continuously?

17. Find the accumulated value of $10,000 invested for 10 years at 4% interest compounded daily.

18. How many years will it take money to double if 8% interest is compounded daily?

19. What is the effective rate of 8% compounded daily?

20. Three junk bonds pay interest at 14.6% compounded annually, 14.2% compounded monthly, and 14% compounded continuously. Show the accumulations of these three bonds graphically over a 20-year period.

21. Assume in Exercise 20 that the 14.6% is compounded semiannually, the 14.2% daily, and the 14% continuously. Show these 20-year accumulations graphically.

Applications (Social and Life Sciences)

22. **Fish Population.**   Suppose that the instantaneous rate of growth of the fish population in a lake is 10% per year. If you stock the lake with 100 fish today, how many fish should the lake contain 6 years from today? (**Hint**: $N = 100e^{6(0.10)}$. Find N and round to the nearest whole number.)

Review Exercises

23. Find the simple interest due on a loan of $5000 at 6% interest for 3 months.

24. Find the compound interest and compound amount for an investment of $4000 at 6% interest compounded semiannually for 10 years.

4.3 AMOUNT OF AN ORDINARY ANNUITY; SINKING FUNDS

OVERVIEW   In this section we are interested in a sequence of equal payments made at the ends of equal time intervals. Deposits (of equal amounts) in a savings account at the end of each quarter fit our pattern. For example, suppose that you deposit $500 at the end of each quarter in a savings and loan that pays interest at 8% compounded quarterly. You might wish to know if you will have enough money in this savings account at the end of 5 years to purchase the automobile of your dreams.
    In this section we find for an ordinary annuity

- Amounts
- Periodic deposits
- Sinking funds

    An **ordinary annuity** is a sequence of equal payments made at equal time intervals. For the formula we will develop, these payments, denoted as R, are made at the *ends* of equal successive payment periods, and the interest periods are the same as the payment periods. The sum of all payments R plus their interest is called the **amount of an annuity**, as illustrated in the following example.

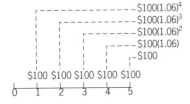

Figure 15

EXAMPLE 19   Suppose that at the end of each year you receive $100 and invest it at 6% compounded annually. How much would you have at the end of 5 years?

SOLUTION   Your first deposit, made at the end of the first year, accumulates compound interest for only 4 years. The value of this deposit at the 5-year mark is $100(1.06)^4$ (see Figure 15). The second deposit will accumulate interest for $5 - 2 = 3$ years and has a value of $100(1.06)^3$. At the 5-year mark the third deposit has a value of $100(1.06)^2$, the fourth deposit $100(1.06)^1$, and finally the last deposit is made on the 5-year mark and has a value of $100. You have in the bank at the end of 5 years the sum of the five accumulations.

$$\$S = \$100(1.06)^4 + \$100(1.06)^3 + \$100(1.06)^2 + \$100(1.06) + 100$$
$$= 100 + 100(1.06) + 100(1.06)^2 + 100(1.06)^3 + 100(1.06)^4 \quad \blacksquare$$

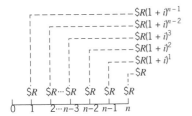

Figure 16

Using the preceding example as a pattern, we now derive an expression for the amount of an annuity of R deposits at the end of intervals for n intervals, as shown in Figure 16.

$$\$S = \$R + \$R(1 + i) + \$R(1 + i)^2 + \$R(1 + i)^3$$
$$+ \cdots + \$R(1 + i)^{n-2} + \$R(1 + i)^{n-1}$$

Using the formula for the sum of a geometric progression in Example 17 of the preceding section, this sum can be expressed as

$$S = R\left[\frac{(1 + i)^n - 1}{i}\right]$$

This is summarized as follows:

Amount of an Annuity

$$S = R\left[\frac{(1 + i)^n - 1}{i}\right]$$

where S = amount of the annuity, R = periodic payment of the annuity, i = rate *per period*, and n = number of payments (periods). (Payments are made at the end of each period.)

The use of the formula is illustrated by the following example.

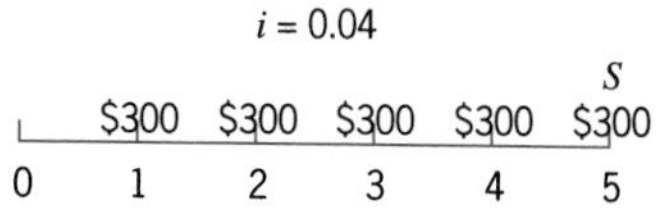

$i = 0.04$

$300 $300 $300 $300 $300 S

0   1   2   3   4   5

Figure 17

EXAMPLE 20 A nurse deposits $300 at the end of each year in a savings account that pays 4% interest compounded annually. How much money does she have just after the fifth deposit?

SOLUTION The time diagram is shown in Figure 17. Remember that S gives the amount just after a payment is made. Hence, it can be used to solve the problem.

$$S = R\left[\frac{(1 + i)^n - 1}{i}\right]$$

$i = 0.04$
$n = 5$
$R = \$300$

$$= \$300\left[\frac{(1 + 0.04)^5 - 1}{0.04}\right]$$

$$= \$1624.90$$

The nurse has $1624.90 after the fifth deposit. ■

Values of $[(1 + i)^n - 1]/i$ for various values of n and i are found in Table 2 of Appendix B. However, it is recommended that you work most of the problems using a calculator (without referring to the tables) because the tables are limited to given interest rates and all different rates can be used with the calculator approach.

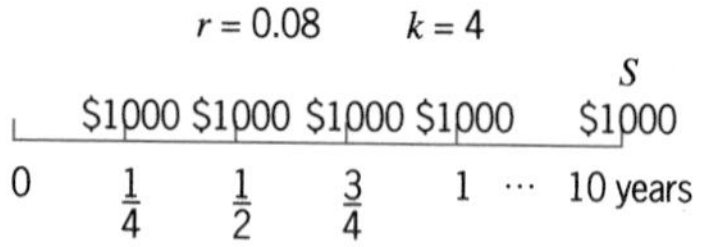

$r = 0.08$ $k = 4$

$1000 $1000 $1000 $1000 $1000 S

0    $\frac{1}{4}$ $\frac{1}{2}$ $\frac{3}{4}$ 1   ···   10 years

Figure 18

EXAMPLE 21 A mathematics teacher deposits $1000 in his savings and loan IRA at the end of each quarter for 10 years. How much money does he have at the end of 10 years if the savings and loan pays 8% interest compounded quarterly?

SOLUTION The time diagram is shown in Figure 18.

$$S = R\left[\frac{(1 + i)^n - 1}{i}\right]$$

$i = \frac{r}{k} = \frac{0.08}{4} = 0.02$

$$= \$1000\left[\frac{(1 + 0.02)^{40} - 1}{0.02}\right]$$

$n = kt = 4(10) = 40$

$$= \$60,401.98$$

$R = \$1000$

Thus, the mathematics teacher has $60,401.98 in his IRA at the end of 10 years. ■

Practice Problem 1 If you deposit $100 a month in a savings and loan paying 6% compounded monthly, how much money do you have at the end of 5 years?

ANSWER $\$100\left[\dfrac{(1.005)^{60} - 1}{0.005}\right] = \$6977$

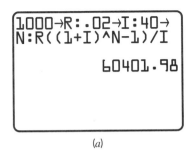

(a)

```
100→R:.005→I:60→
N:R((1+I)^N-1)/I
                6977.00
```

(b)

Figure 19

$r = 0.12$ $k = 12$

$$\begin{array}{ccccc} & & & & \$20,000 \\ \$R & \$R & \$R & \$R & \$R \\ \hline 0 & 1 & 2 & \cdots & 1 & \cdots & 7 & \cdots & 8 \\ & \tfrac{1}{12} & \tfrac{2}{12} & & & \end{array}$$

Figure 20

Calculator Note

To demonstrate repeated use of the formula for S we store the solution for Example 21 and then obtain the solution for Practice Problem 1. Press 1000 STO▶ ALPHA R 2nd : .02 STO▶ ALPHA I 2nd : 40 STO▶ ALPHA N 2nd : ALPHA R ((1 + ALPHA I) ^ ALPHA N − 1) ÷ ALPHA I ENTER to obtain 60,401.98 [see Figure 19(a)].

Then press 2nd ENTRY and replace 1000 with 100, .02 with .005, and 40 with 60 to obtain 6977.00 [see Figure 19(b)] to get the answer to Practice Problem 1.

EXAMPLE 22   A family decides to make monthly deposits into a college-education fund for a daughter so that she will have $20,000 at the end of 8 years. They locate a bond fund that pays 12% compounded monthly. How much must the family deposit each month?

SOLUTION   The time diagram is shown in Figure 20.

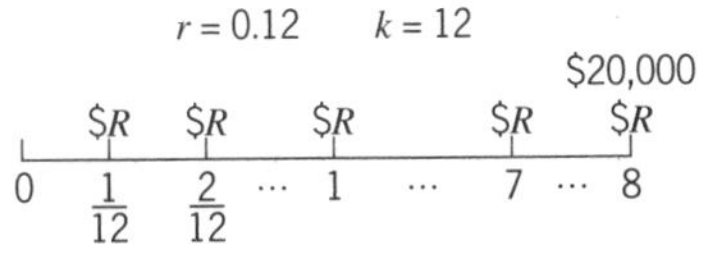

$$\$S = \$R \left[\frac{(1 + i)^n - 1}{i} \right]$$

$i = r/k = 0.12/12 = 0.01$
$n = kt = 12(8) = 96$
$S = \$20,000$

$$\$20,000 = \$R \left[\frac{(1 + 0.01)^{96} - 1}{0.01} \right]$$

On your calculator, use

(1.01 ^ 96 − 1) ÷ .01 ENTER x^{-1} × 20000 ENTER

which gives $R = \$125.06$. The family must deposit $125.06 each month in order to have $20,000 at the end of 8 years.   ■

Sometimes a fund is set up to accumulate money to pay a loan. Such a fund is called a **sinking fund**. For example, a hospital might accumulate money in a sinking fund to pay off a bond issue due in 10 years. Often there will be one interest rate for the sinking fund and another interest rate for the loan (or, in the example, bonds).

EXAMPLE 23   A businessman wishes to create a sinking fund to pay off his loan, which will amount to $2500 in 3 years. If his sinking fund pays 4% interest compounded semiannually, what is his semiannual deposit into the sinking fund?

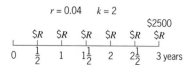

Figure 21

SOLUTION   Deposits of $R in the sinking fund are shown in Figure 21.

$$S = \$R \left[\frac{(1 + i)^n - 1}{i} \right]$$

$$i = r/k = 0.04/2 = 0.02$$
$$n = kt = 2(3) = 6$$
$$S = \$2500$$

$$\$2500 = \$R \left[\frac{(1 + 0.02)^6 - 1}{0.02} \right]$$

On a calculator, press

$$\boxed{(}\ 1.02\ \boxed{\wedge}\ 6\ \boxed{-}\ 1\ \boxed{)}\ \boxed{\div}\ .02\ \boxed{\text{ENTER}}\ \boxed{x^{-1}}\ \boxed{\times}\ 2500\ \boxed{\text{ENTER}}$$

to obtain $R = \$396.31$.

Now let's investigate how the businessman's sinking fund in the preceding example grows year by year. To accomplish this goal look at the first line of Table 3. Since the first deposit is made at the end of the first period (6 months), there is no accumulation of interest for the first 6 months. The increase in the fund for the first 6 months (1 period) is $396.31, and the amount in the fund at the end of the first period is $396.31. This amount will accumulate interest during the second period as 0.02 ($396.31) = $7.93. The deposit is constant, but the fund increases by $396.31 + $7.93 = $404.24. At the end of the second period the amount in the fund is $396.31 + $404.24 = $800.55. Now complete the next line of Table 3 using the same reasoning.

TABLE 3

Period	Interest	Deposit	Increase in Fund	Amount in Fund
1	$ 0	$396.31	$396.31	$ 396.31
2	7.93	396.31	404.24	800.55
3	16.01	396.31	412.32	1212.87
4	24.26	396.31	420.57	1633.43
5	32.67	396.31	428.98	2062.41
6	41.25	396.31	437.56	2499.97

To demonstrate how to complete the remainder of Table 3 we use the table capabilities of our calculator. The amount of money in the fund at the end of $n - 1$ periods can be written as

$$A_{n-1} = \$396.31 \left[\frac{(1.02)^{n-1} - 1}{0.02} \right]$$

The interest over the nth period will be $0.02(A_{n-1})$. The amount of money in the fund at the end of n periods is

$$A_n = \$396.31 \left[\frac{(1.02)^n - 1}{0.02} \right]$$

For the calculator solution we let Y_1 represent the amount in the fund at the end of X periods, and Y_2 the interest on the amount in the fund at the end of the $(X - 1)$st period as seen in Figure 22(c). Refer to page 8 of Appendix A for details.

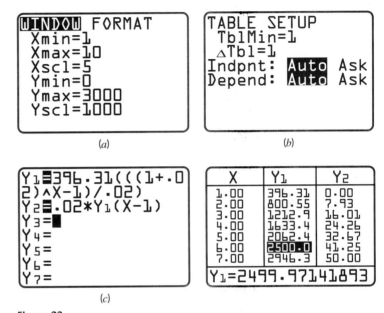

(a)

(b)

(c)

Figure 22

Practice Problem 2 A self-employed person is setting up a defined-benefit retirement plan with a goal of having $100,000 in the plan at the end of 5 years. What amount must be deposited in the fund each month if the fund accumulates interest at 9% compounded monthly?

ANSWER

$$\$R \left[\frac{(1.0075)^{60} - 1}{0.0075} \right] = \$100,000$$

$$\$R = \$1325.84$$

SUMMARY

The amount of an ordinary annuity of $R at the end of each period for n periods is

$$A = R \left[\frac{(1 + i)^n - 1}{i} \right]$$

where i is the interest rate per period.

Exercise Set 4.3

With a calculator, compute each of the following. Interpret what you have found.

1. $100 \left[\dfrac{(1.06)^{10} - 1}{0.06} \right]$ 2. $50 \left[\dfrac{(1.01)^{100} - 1}{0.01} \right]$

3. $200 \left[\dfrac{(1.08)^{14} - 1}{0.08} \right]$ 4. $1 \left[\dfrac{(1.12)^{17} - 1}{0.12} \right]$

Find the amount of the following annual annuities. Interest is compounded annually.

5. $R = \$100$, $i = 6\%$, $n = 10$
6. $R = \$1000$, $i = 8\%$, $n = 12$

Find the amount of each of the following annuities. Payments or deposits are made at the end of each interest period.

7. $R = \$100$, $r = 6\%$ compounded semiannually, $n = 10$ years

8. $R = \$1000$, $r = 9\%$ compounded quarterly, $n = 5$ years

9. $R = \$2000$, $r = 6\%$ compounded monthly, $n = 4$ years

10. $R = \$5000$, $r = 5.5\%$ compounded monthly, $n = 10$ years

11. Find the amount of the following annuities.
   (a) $1000 per year for 20 years at 8% interest compounded annually
   (b) $500 per quarter for 6 years at 8.5% compounded quarterly
   (c) $600 per half year for 5 years at 8% compounded semiannually

12. **Exam.**   A businessperson wants to invest a certain sum of money at the end of each year for 10 years. The investment will earn 6% compounded annually. At the end of 10 years, the businessperson will need $100,000. How should the person compute the required annual investment?

(a) $\dfrac{\$100,000}{\left[\dfrac{1 - (1.06)^{-10}}{0.06} \right]}$ (b) $\dfrac{\$100,000}{\left[\dfrac{(1.06)^{10} - 1}{0.06} \right]}$

(c) $\$100,000 \left[\dfrac{(1.06)^{10} - 1}{0.06} \right]$

(d) $\$100,000 \left[\dfrac{1 - (1.06)^{-10}}{0.06} \right]$

Applications (Business and Economics)

13. **Investments.**   A man is to receive $1000 at the end of each year for 5 years. If he invests each year's payment at 8% compounded annually, how much will he have at the end of his 5 years?

14. **Investments.**   Suppose that you deposit $500 each 6 months in a credit union that pays 8% interest compounded semiannually. How much would you have after 5 years?

15. **Retirement Account.**   A man deposited $2000 per year in a retirement account. Make a table showing how his money accumulates for the first 5 years and then find how much he will have at the end of 20 years if his bank pays
   (a) 8% compounded annually.
   (b) 10% compounded annually.

16. **Sinking Fund.**   Compute the quarterly deposit a woman must make to a sinking fund that pays 6% interest compounded quarterly to pay off a loan of $5000 due in 4 years. Construct three lines of a schedule for this fund.

17. **Sinking Fund.**   What deposit must be made to a sinking fund that pays 6% interest compounded quarterly to pay off a loan of $1500, due in 4 years, at 8% interest compounded annually? Construct three lines of a schedule for this fund.

18. **Sinking Fund.**   What deposit must be made to a sinking fund that pays 4% compounded quarterly to pay off in 5 years a loan of $3200 at 6% interest compounded semiannually?

Applications (Social and Life Sciences)

19. **Inflation.**   If inflation holds steady at 4.2% per year for 5 years, what will be the cost in 5 years of a $10,000 car today? How much will you need to deposit each year in a sinking fund earning 8% per year to purchase the new car in 5 years?

20. **Fish Population.** You place 1000 fish in your lake each year for 5 years. If the fish increase at a rate of 5% per year, how many will you have at the end of 5 years?

Review Exercises

21. Find the interest on $3000 at 6% simple interest for 6 months.

22. Compute the compound interest and compound amount for an investment of $4000 at 8% compounded quarterly for 10 years.

23. Find the effective rate equivalent to 4% interest compounded semiannually.

24. Find the present value of $4000 due in 5 years at 6% interest compounded semiannually.

4.4 PRESENT VALUE OF AN ORDINARY ANNUITY; AMORTIZATION

OVERVIEW   At some point in our lives most of us buy an automobile or a home. If we are typical Americans, we pay for these expensive items with monthly payments. If you know the cash value of your automobile or your home and the prevailing interest rate, you can compute the required monthly payments. In this section we

- Define the present value of an ordinary annuity
- Develop a formula for the present value of an annuity
- Find regular payments to amortize a debt
- Show what happens with each payment by constructing an amortization schedule

EXAMPLE 24   Compute the present value of an annuity of $100 per year for 5 years at 6% compounded annually. In other words, find the amount of money that must be invested now at 6% compounded annually so that payments of $100 per year can be made from this investment for 5 years.

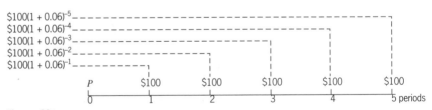

Figure 23

SOLUTION   The payments of $100 per year have been placed on the time diagram in Figure 23. Notice that the first payment is at the end of the first year. We learned in Section 4.1 that the present value of this payment (at time 0) is $100(1.06)^{-1}$. The second $100 payment is at the end of 2 years. The present value of this payment (at time 0) is $100(1.06)^{-2}$. The present value of the third payment is $\$100(1.06)^{-3}$; the fourth payment is $\$100(1.06)^{-4}$; and finally the present value of the fifth payment is $\$100(1.06)^{-5}$. The present value of the annuity is the sum of these terms or

$$
\begin{aligned}
\$P &= \$100(1.06)^{-1} + \$100(1.06)^{-2} + \$100(1.06)^{-3} \\
&\quad + \$100(1.06)^{-4} + \$100(1.06)^{-5} \\
&= \frac{\$100}{(1.06)^1} + \frac{\$100}{(1.06)^2} + \frac{\$100}{(1.06)^3} + \frac{\$100}{(1.06)^4} + \frac{\$100}{(1.06)^5}
\end{aligned}
$$

■

Before finding the value of the preceding expression, let's follow the same pattern for a general case. Consider $\$R$ payments at the ends of payment intervals for n intervals on the time diagram in Figure 24. The rate per interest period is i. The value at $t = 0$ of the first $\$R$ is $R(1 + i)^{-1}$. The second $\$R$ has a value at $t = 0$ of $R(1 + i)^{-2}$. The last $\$R$ payment has a value of $R(1 + i)^{-n}$ at $t = 0$.

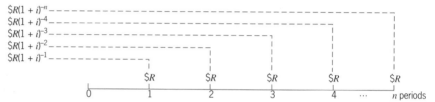

Figure 24

$$
\begin{aligned}
\$P &= \$R(1 + i)^{-1} + \$R(1 + i)^{-2} + \$R(1 + i)^{-3} + \$R(1 + i)^{-4} \\
&\quad + \cdots + \$R(1 + i)^{-n}
\end{aligned}
$$

In Example 18, we found the sum of these terms as the sum of a geometric progression. This sum was

$$
P = R\left[\frac{1 - (1 + i)^{-n}}{i}\right]
$$

This suggests the following formula for the present value of an annuity.

DEFINITION: PRESENT VALUE OF AN ANNUITY

The formula for the **present value of an ordinary annuity** P is

$$
P = R\left[\frac{1 - (1 + i)^{-n}}{i}\right]
$$

where P = present value of an annuity, R = periodic payment, i = rate per period, and n = number of payments (periods). Note that P is the present value at the beginning of the first period; payments are made at the end of each period.

Recall that the compound interest formula $A = P(1 + i)^n$ expresses a relationship between the present value P and the value at the end of n interest intervals, A. We can use this relationship to develop the formula for the present value of an ordinary annuity from the amount of an ordinary annuity.

$$P(1 + i)^n = A = R\left[\frac{(1 + i)^n - 1}{i}\right]$$

Annuity of $R per period

$$P = R(1 + i)^{-n}\left[\frac{(1 + i)^n - 1}{i}\right]$$

$$= R\left[\frac{1 - (1 + i)^{-n}}{i}\right]$$

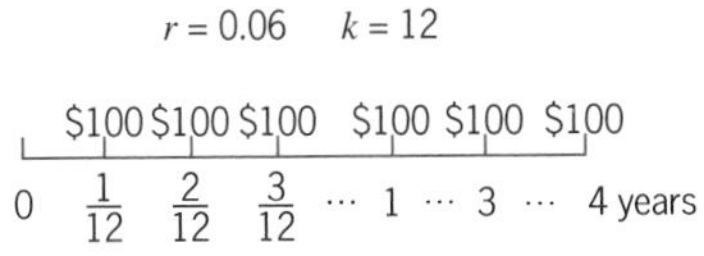

$r = 0.06$ $k = 12$

$100 \ $100 \ $100 \ \ $100 \ $100 \ 100

$0 \quad \frac{1}{12} \quad \frac{2}{12} \quad \frac{3}{12} \quad \cdots \quad 1 \quad \cdots \quad 3 \quad \cdots \quad 4\ \text{years}$

Figure 25

EXAMPLE 25 Compute the present value of an ordinary annuity that pays $100 each month for 4 years at 6% interest compounded monthly.

SOLUTION The time diagram is shown in Figure 25.

$$P = R\left[\frac{1 - (1 + i)^{-n}}{i}\right]$$

$$= \$100\left[\frac{1 - (1.005)^{-48}}{0.005}\right]$$

$i = r/k = 0.06/12 = 0.005$

$n = kt = 12 \cdot 4 = 48$

$R = \$100$

On a calculator, use

$100\ \boxed{\times}\ \boxed{(}\ 1\ \boxed{-}\ 1.005\ \boxed{\wedge}\ \boxed{(-)}\ 48\ \boxed{)}\ \boxed{\div}\ .005\ \boxed{\text{ENTER}}$

which gives $P = \$4258.03$. ∎

If you need practice using tables, values of $[1 - (1 + i)^{-n}]/i$ are found in Table 2 of Appendix B. However, it is recommended that you work most of the problems using a calculator, since tables may not always be available and all interest rates are not found in most tables.

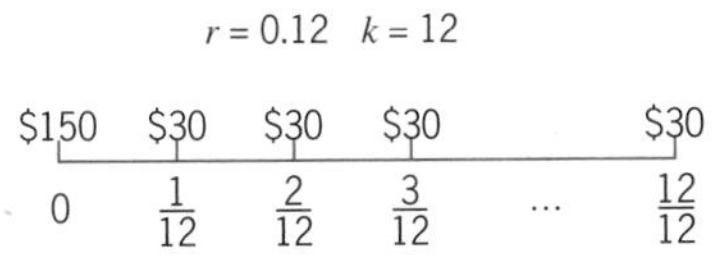

$r = 0.12$ $k = 12$

$150 \quad \$30 \quad \$30 \quad \$30 \qquad \$30$

$0 \quad \frac{1}{12} \quad \frac{2}{12} \quad \frac{3}{12} \quad \cdots \quad \frac{12}{12}$

Figure 26

EXAMPLE 26 The college students in Apartment D purchased a refrigerator for $150 down and $30 a month for 12 months. If the finance charge is 12% compounded monthly, find the cash price.

SOLUTION The time diagram is shown in Figure 26.

$$C = \$150 + P$$

$$= \$150 + \$30\left[\frac{1 - (1.01)^{-12}}{0.01}\right]$$

$$= \$487.65$$

$i = r/k = 0.12/12 = 0.01$

$n = kt = 12(1) = 12$

$R = \$30$

The cash price of the refrigerator is $487.65. ∎

Practice Problem 1 Jodi agrees to pay $500 down, $200 per month for 5 years for her new automobile. If the finance rate is 12% compounded monthly, what is the cash value of her car?

ANSWER   Cash value = $500 + Present value of a $200 annuity

$$= \$500 + \$200 \left[\frac{1 - (1 + 0.01)^{-60}}{0.01} \right]$$

$$= \$500 + \$8991.01$$

$$= \$9491.01$$

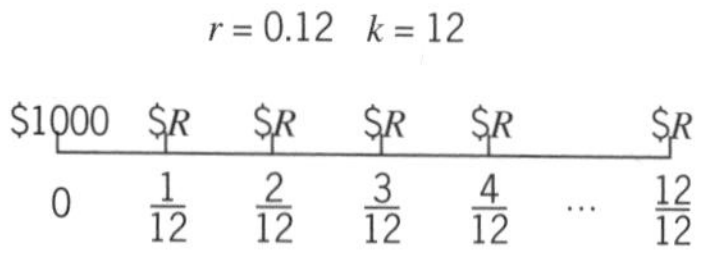

$r = 0.12$ $k = 12$

$1000   \$R   \$R   \$R   \$R   \quad\quad \$R$

0   $\frac{1}{12}$ $\frac{2}{12}$ $\frac{3}{12}$ $\frac{4}{12}$ $\cdots$ $\frac{12}{12}$

Figure 27

EXAMPLE 27   Find the payment needed each month for 1 year to pay off a debt of $1000 at 12% compounded monthly.

SOLUTION   The time diagram is shown in Figure 27.

$$P = R \left[\frac{1 - (1 + i)^{-n}}{i} \right] \qquad \begin{array}{l} i = 0.12/12 = 0.01 \\ n = kt = 12(1) = 12 \\ P = \$1000 \end{array}$$

$$\$1000 = \$R \left[\frac{1 - (1.01)^{-12}}{0.01} \right]$$

or

$$R = \frac{1000}{\left[\dfrac{1 - (1.01)^{-12}}{0.01} \right]} = \$88.85$$

Using a calculator, press

$\boxed{(}\ 1\ \boxed{-}\ 1.01\ \boxed{\wedge}\ \boxed{(-)}\ 12\ \boxed{)}\ \boxed{\div}\ .01\ \boxed{\text{ENTER}}\ \boxed{x^{-1}}\ \boxed{\times}\ 1000\ \boxed{\text{ENTER}}$

which gives $R = \$88.85$.   ■

An interest-bearing debt is **amortized** if both the principal and the interest are paid by a sequence of equal payments at equal periods of time. The amortization schedule for the $1000 debt of the preceding example is given in Table 4.

At the beginning of the first month the debt is $1000. For the first month, interest is $1000(0.01) = $10. A payment of $88.85 is made. The payment less interest ($88.85 − $10 = $78.85) gives that part of the payment that is applied to the debt. So, at the beginning of the second month, the debt is $1000 − $78.85 = $921.15. Can you find this on the first line of Table 4? See if you understand the second line.

TABLE 4

Months	Outstanding Principal	Interest Due	Payment	Principal Repaid Each Period
1	$1000.00	$10.00	$88.85	$78.85
2	921.15	9.21	88.85	79.64
3	841.51	8.42	88.85	80.43
4	761.08	7.61	88.85	81.24
5	679.84	6.80	88.85	82.05
6	597.79	5.98	88.85	82.87
7	514.92	5.15	88.85	83.70
8	431.22	4.31	88.85	84.54
9	346.68	3.47	88.85	85.38
10	261.30	2.61	88.85	86.24
11	175.06	1.75	88.85	87.10
12	87.96	0.88	88.84	87.96

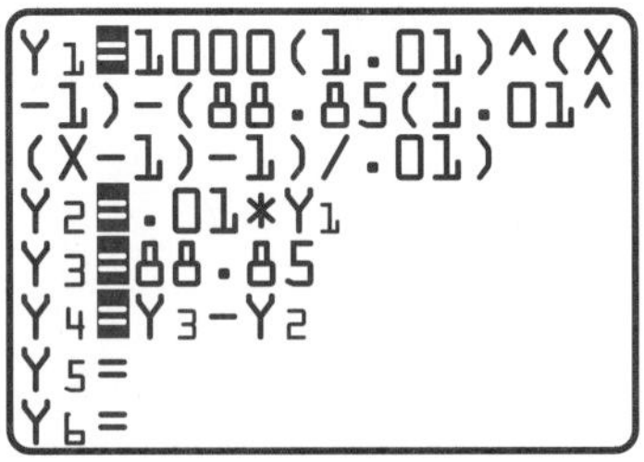

```
Y₁■1000(1.01)^(X
-1)-(88.85(1.01^
(X-1)-1)/.01)
Y₂■.01*Y₁
Y₃■88.85
Y₄■Y₃-Y₂
Y₅=
Y₆=
```

(a)

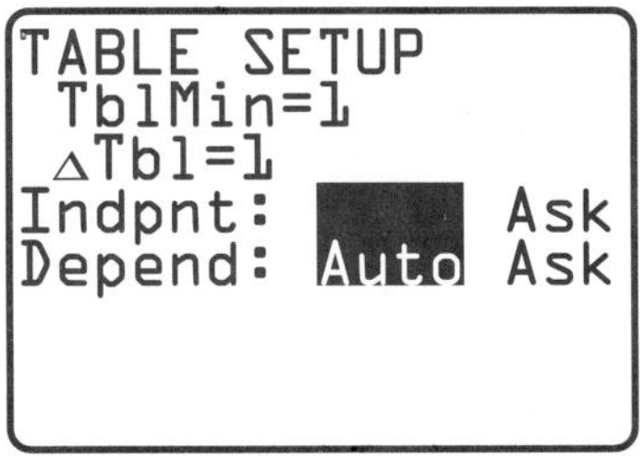

```
TABLE SETUP
 TblMin=1
 △Tbl=1
Indpnt: ███  Ask
Depend: Auto Ask
```

(b)

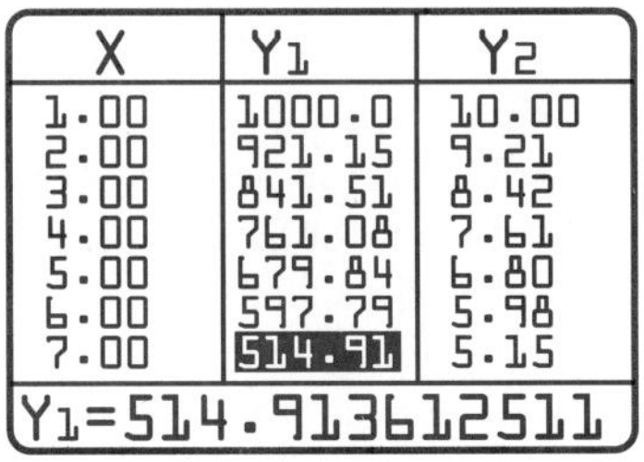

X	Y₁	Y₂
1.00	1000.0	10.00
2.00	921.15	9.21
3.00	841.51	8.42
4.00	761.08	7.61
5.00	679.84	6.80
6.00	597.79	5.98
7.00	514.91	5.15

Y₁=514.913612511

(c)

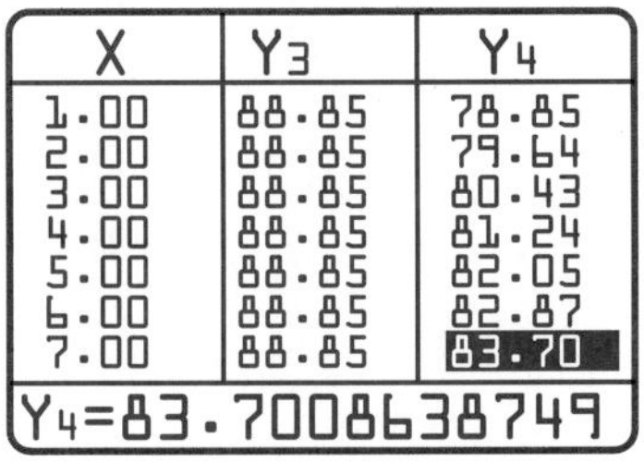

X	Y₃	Y₄
1.00	88.85	78.85
2.00	88.85	79.64
3.00	88.85	80.43
4.00	88.85	81.24
5.00	88.85	82.05
6.00	88.85	82.87
7.00	88.85	83.70

Y₄=83.7008638749

(d)

Figure 28

To demonstrate how to complete such a table on a calculator we reason as follows. The amount of debt outstanding at the end of the $(n-1)$st period is the debt accumulated for $n-1$ periods less the accumulated value of the payments for $n-1$ periods; that is,

$$A_{n-1} = \$1000(1.01)^{n-1} - \frac{\$88.85(1.01^{n-1} - 1)}{0.01}$$

This is the amount that accumulates interest at 1% over the nth period. Using a calculator we let Y_1 represent the amount of the debt at the beginning of the xth period [see Figure 28(a)]. The interest over the period is $Y_2 = .01 \cdot Y_1$. Y_3 is the periodic deposit or $88.85. The principal repaid, Y_4, is the payment minus the interest or $Y_4 = Y_3 - Y_2$. For details see page 8 of Appendix A.

Practice Problem 2   You purchase a house for $100,000, pay 20% down, and amortize your debt with monthly payments for 30 years. What is your monthly payment if your loan charges 9% compounded monthly?

ANSWER $100,000 - 0.20(\$100,000) = \$80,000$ (owe on the house)

$$\$80,000 = R \left[\frac{1 - (1 + 0.0075)^{-360}}{0.0075} \right]$$

$$R = \$643.70$$

Practice Problem 3   Show three lines of the amortization schedule in Practice Problem 2. That is, show how the payments contain both interest and payment on the principal.

ANSWER

Month	Outstanding Principal	Interest Due	Payment	Principal Repaid
1	$80,000	$600.00	$643.70	$43.70
2	79,956.30	599.67	643.70	44.03
3	79,912.27	599.34	643.70	44.36

SUMMARY

The present value of an ordinary annuity of $R at the end of each period for n periods is

$$P = R \left[\frac{1 - (1 + i)^{-n}}{i} \right]$$

where i is the interest rate per period.

Exercise Set 4.4

With a calculator, compute each of the following. Interpret what you have found.

1. $50 \left[\dfrac{1 - (1.06)^{-10}}{0.06} \right]$ 2. $50 \left[\dfrac{1 - (1.01)^{-100}}{0.01} \right]$

3. $200 \left[\dfrac{1 - (1.08)^{-14}}{0.08} \right]$ 4. $1 \left[\dfrac{1 - (1.12)^{-7}}{0.12} \right]$

Find the present value of each of the following annuities. Interest is compounded annually.

5. $R = \$100$, $i = 6\%$, $n = 10$

6. $R = \$1000$, $i = 8\%$, $n = 12$

7. Find the present value of the following annuities:
   (a) $1000 per year for 20 years at 8% interest compounded annually
   (b) $500 per quarter for $6\frac{1}{2}$ years at 8.5% interest compounded quarterly
   (c) $600 per half year for 5 years at 8% interest compounded semiannually

8. Find the present value of the following annuities:

(a) $100 per month for 8 years at 12% interest compounded monthly
(b) $500 per month for 3 years at 4.4% interest compounded monthly

9. **Exam.** A businessperson wants to receive $6000 (including principal) from a fund at the end of each year for 10 years. How should she compute her required initial investment at the beginning of the first year if the fund earns 6% compounded annually?

(a) $\dfrac{6000}{\left[\dfrac{(1.06)^{10} - 1}{0.06} \right]}$ (b) $6000 \left[\dfrac{(1.06)^{10} - 1}{0.06} \right]$

(c) $\dfrac{6000}{\left[\dfrac{(1 - (1.06)^{-10})}{0.06} \right]}$ (d) $6000 \left[\dfrac{1 - (1.06)^{-10}}{0.06} \right]$

Applications (Business and Economics)

10. **Payments.** Compute the monthly payment necessary to finance a used car for $3500 at 6% interest compounded monthly for 3 years.

11. **Payments**. Find the payment necessary each quarter for 2 years to amortize a debt of $2000 at 8% interest compounded quarterly.

12. **Amortization**.   Make an amortization schedule for Exercise 10.

13. **Amortization**.   Make an amortization schedule for Exercise 11.

14. **Equity**.   A woman bought a house for $60,000. She paid $10,000 down and amortized the balance with monthly payments for 8 years at 6% interest compounded monthly.
   (a) What is her equity after 30 payments? (**Hint**: $60,000 less the present value of a 66-payment annuity.)
   (b) What is her equity after 60 payments?

15. **Equity**. A fisherman bought a cabin for $32,000. He paid $5000 down and amortized the balance with monthly payments for 7 years at 8% interest compounded monthly.
   (a) What is his equity after his 50th payment? (**Hint**: $32,000 less the present value of a 34-payment annuity.)
   (b) What is his equity after his 80th payment?

Review Exercises

16. A man obtained $190 from the bank and signed a 3-month non-interest-bearing note for $200. Compute the simple-interest rate he was charged.

17. A doctor is contributing $75,000 at the end of each six months to endow a chair at the university. Compute the amount in the endowment fund at the end of 10 years if the fund is to be invested at 6% interest compounded semiannually.

18. Use a geometric progression to find the present value of the annuity

$$\frac{\$500}{1.08} + \frac{\$500}{(1.08)^2} + \frac{\$500}{(1.08)^3} + \cdots + \frac{\$500}{(1.08)^{12}}$$

19. Use geometric progressions to find the sum of the annuity

$$\$100 + \$100(1.06) + \$100(1.06)^2 + \cdots + \$100(1.06)^8$$

4.5 EQUATIONS OF VALUE; PERPETUITIES (Optional)

OVERVIEW   Not all annuities fit exactly the format of an ordinary annuity. A furniture company advertises a sale where your monthly payments start 1 year from the date of purchase. This is not an ordinary annuity. Also, in this section we study annuities that continue forever, called perpetuities.

In this section, we pay particular attention to where payments are located on a time scale. Then by either accumulating or discounting values to a selected time, we can form an equation relating the values. At the same time, we look at a sequence of payments that are not ordinary annuities as we defined them. However, we use our ordinary annuity formulas to find the amounts and present values of such payments. Therefore, in this section we study

- Equations of value
- Payments that are not ordinary annuities in standard form
- Perpetuities

We begin by summarizing material from the preceding sections. That is, we will learn to find values at a given position on a time diagram by using the formulas from the preceding sections.

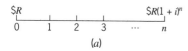

Figure 29

1. To accumulate a single payment $R for n periods at interest rate of i per period, multiply $R by

$$(1 + i)^n \qquad \text{Figure 29}(a)$$

2. To find the value of a payment $R, n periods before it is due at interest rate i per period, multiply $R by

$$(1 + i)^{-n} \qquad \text{Figure 29}(b)$$

3. To accumulate the payments of an annuity of n $R payments to the date of the last payment, multiply $R by

$$\left[\frac{(1 + i)^n - 1}{i} \right] \qquad \text{Figure 29}(c)$$

4. To find the value of an annuity of n $R payments 1 period before the first payment is made, multiply $R by

$$\left[\frac{1 - (1 + i)^{-n}}{i} \right] \qquad \text{Figure 29}(c)$$

That is, to accumulate a present value of $R from time 0 to time n at interest rate i, use $R(1 + i)^n$, [see Figure 29(a)]. Similiarly, if P is the present value of $R at time n, then $P(1 + i)^n = R$ or $P = R/(1 + i)^n$ or $P = R(1 + i)^{-n}$. A value of $R(1 + i)^{-n}$ at time 0 is equivalent to a value of R at time n under the assumption that interest is $i\%$ per period, [see Figure 29(b)].

Similarly, for a sequence of payments made at the end of periods for n periods, the value of this sequence of payments at time 0 is

$$\$R \left[\frac{1 - (1 + i)^{-n}}{i} \right]$$

and at time n is

$$\$R \left[\frac{(1 + i)^n - 1}{i} \right]$$

See Figure 29(c).

The preceding ideas are used in **equations of value**. In an equation of value, you select a given date. Then you accumulate [multiply by $(1 + i)^n$] or discount [multiply by $(1 + i)^{-n}$] all obligations to this date. Likewise, you accumulate or discount all payments to this date. When you set equal the value of the obligations and the value of the payments on this date, you have an equation of value.

EXAMPLE 28   Suppose that you owe $10,000. You want to pay this debt by making payments of $1000 each at the end of each year for 6 years. Then, at the end of 6 years, you will make a payment of $x to retire the remaining debt. If money is worth 8% interest compounded annually, what is the size of the $x payment you will make at the end of 6 years?

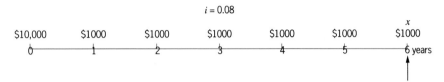

Figure 30

SOLUTION   The time diagram is shown in Figure 30. We choose the 6-year mark as the time to compare the debt and the payments. At the end of 6 years the debt has a value of $10,000(1 + 0.08)^6$. At the end of 6 years the $1000-a-year annuity has a value of

$$\$1000 \left[\frac{(1.08)^6 - 1}{0.08} \right] \qquad \text{Sum of the annuity}$$

x has a value of x at the 6-year mark. The equation of value at the 6-year mark is

$$\$10{,}000(1 + 0.08)^6 = \$1000 \left[\frac{(1.08)^6 - 1}{0.08} \right] + x \qquad \begin{array}{l}\text{Debts equal payments at}\\ \text{the 6-year mark.}\end{array}$$

$$x = \$8532.81$$

The one additional payment at the end of 6 years that will retire the debt is $8532.81.   ■

EXAMPLE 29   Use an equation of value at time 0 to find x for Example 28.

SOLUTION   Debt = Present value of an annuity + Present value of payment at 6-year mark; that is,

$$\$10{,}000 = \$1000 \left[\frac{1 - (1.08)^{-6}}{0.08} \right] + x(1.08)^{-6}$$

Using a calculator, press

$$\boxed{(}\ 10000\ \boxed{-}\ 1000\ \boxed{(}\ 1\ \boxed{-}\ 1.08\ \boxed{\wedge}\boxed{(-)}\ 6\ \boxed{)}\ \boxed{\div}\ .08\ \boxed{)}\ \boxed{\div}$$

$$1.08\ \boxed{\wedge}\boxed{(-)}\ 6\ \boxed{\text{ENTER}}$$

which gives $x = \$8532.81$. As expected, the value of x is the same regardless of the time selected for the equation of value.   ■

Previously we have classified our annuities as ordinary annuities. The reason for this classification is that there are other annuities called annuities certain and deferred annuities. With **annuities certain** the payments are made at the beginning of periods (instead of at the end), and with **deferred annuities** the payments do not start at time 0 but are deferred a given number of periods. These terms are not important in that we use equations of value and ordinary annuities to solve problems involving these concepts.

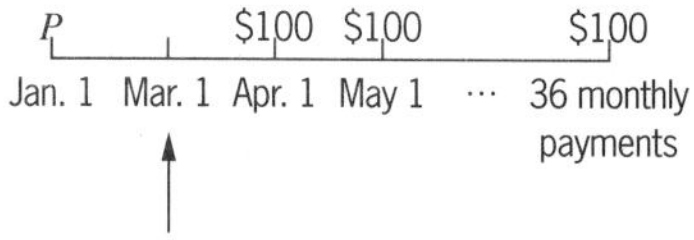

Figure 31

EXAMPLE 30 A car was bought on January 1 with the agreement that there would be 36 monthly payments of $100, the first of which would be due on April 1. Find the equivalent cash price if interest is 12% compounded monthly.

SOLUTION The time diagram is shown in Figure 31. Let's select March 1 as the date for an equation of value in order to use the formula for the present value of an ordinary annuity.

$$P(1.01)^2 = \$100 \left[\frac{1 - (1.01)^{-36}}{0.01} \right] \qquad \begin{array}{l} P \text{ must be} \\ \text{accumulated for} \\ \text{two periods.} \end{array}$$

$$P = \$2951.43$$

The equivalent cash price is $2951.43. ∎

Equations of value are useful when the payment and interest periods do not exactly fit our definition of an ordinary annuity.

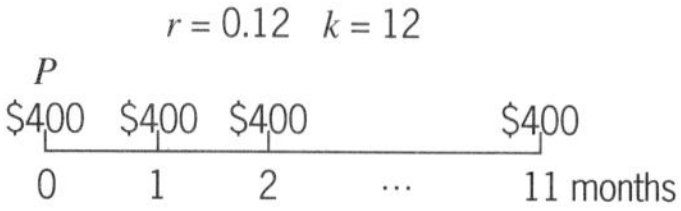

Figure 32

EXAMPLE 31 At an apartment near the college a student pays $400 rent each month, payable in advance. What would be the equivalent yearly rent at 12% interest compounded monthly if the student paid it in advance?

SOLUTION The $400 payment at time 0 does not belong in the annuity because for an ordinary annuity you must have interval then payment, interval then payment, and so forth. As seen in Figure 32, we can use an equation of value including an annuity within it. At time 0, the equation of value can be written as

$$P = \$400 + \$400 \left[\frac{1 - (1.01)^{-11}}{0.01} \right] \qquad \begin{array}{l} \text{Without the first} \\ \$400, \text{ there are only} \\ 11 \text{ payments of } \$400 \\ \text{in the annuity.} \end{array}$$

$$= \$4547.05 \qquad \blacksquare$$

Practice Problem 1 To pay for his car, Aaron is to give his brother $1000 at the end of each year for 6 years. Aaron fails to make the first two payments. At the end of 3 years, Aaron is going to borrow money to pay off his debt to his brother. If it is agreed that he should pay interest at 8% compounded annually, how much should Aaron pay his brother at the end of 3 years?

ANSWER

$$\$1000(1.08)^2 + \$1000(1.08)^1 + \$1000 + \$1000\left[\frac{1 - (1.08)^{-3}}{0.08}\right] = \$5823.50$$

Perpetuities

We conclude this section with a discussion of perpetuities. An annuity whose payments begin on a certain date and continue indefinitely is called a **perpetuity**. Since the payments continue indefinitely, it would be impossible to compute the amount of a perpetuity; however, the present value can be found. The present value P of a perpetuity that is payable at the end of each interest period at $i\%$ per period is the principal that would, in one interest period, earn the payment R. Thus,

$$Pi = R \quad \text{or} \quad P = \frac{R}{i}$$

Interest or income from a stock is called a **dividend**. If the dividends of a stock remained constant and the value of a stock remained constant, a stock would be a perpetuity.

EXAMPLE 32   A company is expected to pay a $3.00 dividend every 6 months on a share of its stock. What is the present value of this stock if money is worth 6% interest compounded semiannually?

SOLUTION
$$P = \frac{R}{i} = \frac{\$3.00}{0.03} = \$100 \qquad \blacksquare$$

Often, the payment periods and the interest periods for a perpetuity may not be the same. When this happens, the present value may be found by the following formula.

DEFINITION: PRESENT VALUE OF PERPETUITY

The present value of a perpetuity that yields the payment R at the end of n interest periods with money worth $i\%$ per interest period is

$$P = \frac{R}{(1 + i)^n - 1}$$

Notice that the preceding formula is $P = R/r$, where $r = (1 + i)^n - 1$ is the effective rate for the payment period.

EXAMPLE 33 Find the present value of a company's stock, which is expected to pay $3.00 every 6 months, if money is worth 8% interest compounded quarterly.

SOLUTION Since money is worth 8% interest compounded quarterly, the interest is 2% each quarter. The payment of $3.00 is made after two interest periods; therefore, $n = 2$. Substituting in the equation gives

$$P = \frac{R}{(1 + i)^n - 1}$$
$$= \frac{\$3.00}{(1.02)^2 - 1}$$
$$= \$74.26$$

∎

Practice Problem 2 An accountant plans to retire in 1 year. He wants to place in a trust fund, bearing interest at 9% compounded semiannually, enough money to receive $9000 every year forever (starting 1 year from today). How much money must he place in the bank?

ANSWER $P = \dfrac{\$9000}{(1.045)^2 - 1} = \$97,799.51$

SUMMARY

In this section we considered the following topics:

1. Equations of value.
2. The present value of an annuity certain and a deferred annuity.
3. $P = R/i$ to find the present value of a perpetuity, where R is the payment per period and i is the interest rate per period.
4. $P = R/[(1 + i)^n - 1]$ to find the present value of a perpetuity that yields the payment R at the end of n interest periods with money worth $i\%$ per interest period.

Exercise Set 4.5

Find the present value of the following perpetuities.

1. $10,000 a year, interest at 8% annually
2. $20,000 a year, interest at 5% annually
3. $10,000 a year, interest at 8% quarterly
4. $20,000 a year, interest at 4% semiannually

Find the payments of the following perpetuities.

5. Present value $200,000, annual payments, interest at 8% annually

6. Present value $1,000,000, monthly payments, interest at 6% monthly
7. Present value $200,000, annual payments, interest at 8% compounded quarterly
8. Present value $1,000,000, semiannual payments, interest at 4% monthly

Applications (Business and Economics)

9. *Stock Dividend.* A company is expected to pay $6.00 every 6 months on a share of its stock. What is the present value of this stock if money is worth 8% interest compounded semiannually?

10. **Cash Premium**. The annual premium for a 20-year-pay insurance policy is $100 payable at the beginning of each year for 20 years. What is the equivalent cash premium if money is worth 6% interest compounded annually?

11. **Cash Premium**. Find the equivalent cash premium if money is worth 8% interest compounded annually for a 20-year-pay insurance policy that has an annual premium of $100 payable at the beginning of each year for 20 years.

12. **Stock Dividend**. A stock pays $0.225 each quarter. What is the present value of the stock if money is worth 6% interest compounded quarterly?

13. **Payment of Debt**. A teacher has a debt of $8000 due in 5 years. He wants to cancel this debt by paying $3000 1 year from now, $1000 3 years from now, and a last payment 7 years from now. If money is worth 6% interest compounded semiannually, what will be the amount of the last payment?

14. **Cash Payment of Rent**. A student pays $300 rent each month, payable in advance. What would be her equivalent yearly rent at 9% interest compounded monthly if she paid it in advance?

15. **Deferred Payment**. Find the present value of an annuity of $1000 per year with the first payment due 3 years from now and the last occurring 12 years from now if the interest is 8% compounded annually.

16. **Equations of Value**. To cancel three loans of $2000 due now, $5000 due in 4 years, and $10,000 due in 6 years, I agree to pay $R at the end of each year for 10 years. If the interest rate is 8% compounded annually, what is my annual payment?

17. **Installment Purchase**. A stereo set sells for $20 down and $30 a month for 12 months. What is the cash price of the set if interest is 4% compounded monthly?

18. **Installment Purchase**. A car was bought on March 1 with the agreement that there would be 24 monthly payments of $150, the first of which is due on July 1. Find the equivalent cash price if interest is 6% compounded monthly.

19. **Stock Dividend**. A stock pays $4.00 every 6 months. What is the present value of the stock if money is worth 8% interest compounded quarterly?

20. **Installment Purchase**. A TV set that has a cash price of $337.26 is sold for $20 down and $30 a month for 12 months. What is the compound interest rate monthly?

Review Exercises

21. Compute the present value and the amount of an annuity of $400 a year for 10 years at 8% interest compounded annually.

22. A contract pays $200 at the end of each quarter for 4 years and $2000 additional at the end of the last quarter. What is the present value of the contract at 8% interest compounded quarterly?

23. What should be the semiannual deposit to a sinking fund established to pay off a loan of $300 at 6% interest compounded annually in 3 years if the fund pays 6% interest compounded semiannually?

24. You wish to borrow $10,000 today and $5000 5 years from now. You plan to repay these loans with equal payments at the end of each year for 10 years. If the interest is 8% compounded annually, what is your annual payment?

25. A $10,000 debt is to be repaid by equal payments at the end of each 6 months for 3 years. If money is worth 8% interest compounded semiannually, make an amortization table for this debt.

26. A deposit of $500 is made to a bank at the end of each 6 months for $2\frac{1}{2}$ years. If money is worth 8% interest compounded semiannually, make a table showing how much money accumulates in the sinking fund.

27. Find the compound amount the Native Americans would have if they had invested $24 for 300 years at 8% interest compounded annually.

28. A department store charges 2% interest per month service charge on unpaid balances. Assume that no payments are made for 1 year and compute the approximate effective rate of interest.

29. A house trailer was bought for $27,500, with $2500 down and the balance amortized with monthly payments at 12% interest compounded monthly for 8 years. What is the equity after the 50th payment?

30. An orchard will produce its first crop at the end of 6 years. If after 6 years it is expected to produce an annual income of $600 for 15 years, what is the cash value of the orchard? Assume that money is worth 6% interest compounded annually.

31. A medical student borrows $6000 today with interest at 6% computed annually. He agrees to pay $1000 in 1 year, $2000 in 2 years, and the balance 4 years from today. Compute the final payment.

32. You purchase an automobile for $500 down and $300 a month for 48 months. If interest is at 14.2% compounded monthly, what is the selling price of the car?

33. A company sets up a sinking fund with yearly deposits of $10,000 to replace a piece of equipment with a life of 8 years. At an interest rate of 8% compounded yearly, how much money will be available to replace the old equipment?

34. Kate's parents are saving money for her college education. How much must they save yearly in order to have $10,000 when Kate enters college at 17 years of age if the money can be invested at 8% compounded yearly?

Chapter Review

This chapter contains mathematics commonly referred to as the "mathematics of finance." You should now be familiar with the following terms.

Important Terms

Simple interest	Amount of an annuity
Principal	Sinking fund
Compound interest	Amortization
Compound period	Equation of value
Compound amount	Perpetuity
Effective rate	Deferred annuity
Annuity	Geometric progression
Present value of an annuity	

Important Formulas

Formulas that we have used in this chapter are as follows.
   Simple interest and amount:

$$I = Prt$$
$$A = P(1 + rt)$$

Compound amount:

$$A = P(1 + i)^n$$

Compound amount if a principal is compounded continuously:

$$A = Pe^{rt}$$

Present value:

$$P = \frac{A}{(1 + i)^n}$$

Amount of an ordinary annuity:

$$A = R\left[\frac{(1 + i)^n - 1}{i}\right]$$

Present value of an ordinary annuity:

$$P = R\left[\frac{1 - (1 + i)^{-n}}{i}\right]$$

Present value of a perpetuity:

$$P = \frac{R}{(1 + i)^n - 1}$$

The sum S_n of a geometric progression of n terms is

$$S_n = \frac{a(r^n - 1)}{r - 1}$$

where the first term is a and each term is obtained from the preceding terms by multiplying by the constant $r \neq 1$.

Chapter Test

1. Find the present value of $1000 due in 3 years at 8% compounded semi-annually.

2. How much money will you have in the bank at the end of 3 years if you deposit $500 at the end of each month for the 3 years? You draw 6% interest compounded monthly.

3. If $1000 accumulates to $1200 in 6 months, what simple-interest rate is being charged?

4. You borrow $2000 from a loan company. How much do you pay back in 2 years if they charge 12% compounded monthly?

5. On a new automobile you agree to pay $100 every 6 months for 5 years. If you are charged interest at a rate of 8% compounded semiannually, what is the cash value of the automobile?

6. You place your savings in a bank that pays a simple-interest rate of 5%. If you have $2400 at the end of 4 years, how much did you deposit?

7. What is the effective rate of 12% compounded monthly?

8. You purchase a house for $120,000, pay $20,000 down, and amortize the debt by monthly payments for 10 years. What are your monthly payments if money is worth 6% compounded monthly?

9. What is the present value of a perpetuity of $1000 every 6 months at 8% compounded semiannually?

10. You buy a house for $100,000. Your monthly payments are $1200. Make a schedule to show how you are reducing the loan each month for the first 3 months if you pay 12% interest compounded monthly.

11. A young man secures a loan of $10,000 to start a small business. At the end of 1 year he repays $2000. At the end of 2 years he pays $3000. At the end of 4 years he agrees to pay off the debt with 6 equal payments made semiannually starting at the end of 4 years. If the bank charges 8% compounded semiannually, find the payments necessary to pay off the loan.

12. How much do you owe at the end of 90 days, if you borrow $2000 at a simple-interest rate of 8%?

13. You agree to pay $2320 at the end of 2 years. If a bank charges 8% simple interest, how much do you obtain?

CHAPTER 5

Counting Techniques and Probability

rcheological artifacts indicate that many early peoples played some version of dice, either for recreation or to determine the will of the tribal deity. As more elaborate games were developed, the players began to observe certain patterns in the results, but they did not have the language of probability with which to describe and analyze them. Over time, outstanding mathematicians established probability as a legitimate field of inquiry, but the main application seemed to be games of chance for gamblers. Blaise Pascal (1623–1662) helped to develop probability theory through a series of letters with another mathematician, Pierre DeFermat (1601–1665).

As more time passed, it became clear that probability was much more than just a technique for gamblers. Managers in government and industry use probabilistic

311

techniques in decision-making processes. Physicists now use probability theory when studying various gas and heat laws as well as in the theory of atomic physics. Biologists apply the techniques of probability in genetics, the theory of natural selection, and learning theory. Furthermore, probability is the theoretical basis of statistics, a discipline that permeates modern thinking.

In this chapter, you will come to understand the basic concepts of probability. Counting techniques, such as the fundamental principle of counting, tree diagrams, permutations and combinations, and additional set terminology are used to assist you in computing probabilities.

5.1   THE LANGUAGE OF PROBABILITY

OVERVIEW   One of the significant characteristics of our increasingly complex society is that we must deal with questions for which there is no known answer; rather, there is one or more *probable* (or *improbable*) answers. We hear statements such as, "This surgery has a 90% chance of success." Chances of success and failure are often given in political campaigns, and, of course, we hear about the probability of rain with each weather forecast. The ability to measure the degree of uncertainty in an undetermined situation is a necessary skill in our ever-changing world. In this section, we discuss **probability**—the language of uncertainty. As the language of the undetermined and the uncertain, probability is an important tool in many phases of our uncertain modern life. In this section we learn about

- Experiments
- Outcomes
- Sample space
- Uniform sample space
- Events
- Properties of probability
- Classical definition of probability

We first introduce some of the language and terminology of probability and then we formally define probability. Since probability is a language of uncertainty, any discussion of probability presupposes a **process of observation**. A process that has observable results is called an **experiment**. Any possible result of an experiment is called an **outcome**. The order of the outcomes is not predictable. We define the terms sample space, sample point, and event as follows.

DEFINITION: SAMPLE SPACE, SAMPLE POINT, EVENT

1. **Sample space**: The set of elements that are all the possible outcomes of an experiment.
2. **Sample point**: Each element in the sample space.
3. **Event**: Any subset of the sample space of an experiment.

EXAMPLE 1   For each experiment stated, state the sample space.

(a)   A coin is tossed and the result is noted.

(b)   A pair of dice is tossed and the sum of the numbers on top is noted.

SOLUTION

(a)   The sample space is $S = \{$Head (H), Tail (T)$\}$.

(b)   The sample space is $S = \{2, 3, 4, 5, 6, 7, 8, 9, 10, 11, 12\}$. ■

In experiments like those in Example 1, where each sample point has the same chance of occurring as another, we say that each sample point is **equally likely**. This leads us to the following definition.

DEFINITION: UNIFORM SAMPLE SPACE

If each sample point of a sample space is equally likely to occur, the sample space is called a **uniform sample space**.

EXAMPLE 2   Ten blank cards are marked with the numbers 1 to 10. An experiment consists of shuffling the cards and then drawing one card.

(a)   Determine the sample space for the experiment.

(b)   How many sample points are in the sample space?

(c)   Is the sample space a uniform sample space?

SOLUTION

(a)   The sample space is $S = \{1, 2, 3, 4, 5, 6, 7, 8, 9, 10\}$.

(b)   There are 10 sample points.

(c)   Since each card has a equal chance of being drawn, each is equally likely and the sample space is a uniform sample space. ■

EXAMPLE 3   Using the same cards as in Example 2, consider the experiment of drawing one card and determining whether it is a 1 or is not a 1. What is the sample space and is it a uniform sample space?

SOLUTION   The sample space can be written as $S = \{1, \text{not } 1\}$. There are only two outcomes for this experiment, but they are not equally likely (i.e., they do not have the same chance of occurring); therefore, the sample space is not a uniform sample space. ■

Notice that in Examples 2 and 3, the sample space completely exhausted all possibilities of what could happen if the experiment were to be performed. In Example 2, the sample space is uniform, but in Example 3 the sample space is not uniform. You should also note from these two examples that the sample points are not repeated in two different outcomes. That is, the list of sample points for an experiment is distinct; there is no overlapping. From this we can see that when a sample space is being constructed we must observe the following:

1. The categories do not overlap.
2. No result is classified more than once.
3. The set of sample points is complete—all possibilities are exhausted.

Practice Problem 1 Five balls numbered 1 to 5 are placed in a bag and one ball is drawn. For each experiment, state the sample space and whether it is a uniform sample space.

(a) The number of the ball is noted.
(b) Whether the number drawn is odd or even is noted.

ANSWER

(a) $S = \{1, 2, 3, 4, 5\}$. This is a uniform sample space.
(b) $S = \{\text{odd, even}\}$. This is not a uniform sample space because there are three odd numbers and only two even numbers.

EXAMPLE 4 A fair coin is flipped two times in succession.

(a) If the result of each flip is recorded and the order is noted, what is the sample space?
(b) If only the number of heads and tails is recorded, without regard to order of occurrence, what is the sample space?

SOLUTION

(a) Since the order of the results is the required outcome, the sample space is

$$S = \{(HH), (HT), (TH), (TT)\}$$

Note that the way the elements in the sample space are listed is not unique. You could list the sample space as $S = \{(TH), (TT), (HH), (HT)\}$. The only requirement is that each possible outcome is a sample point.

(b) There are several ways this sample space could be listed. One way is $S = \{(2H), (2T), (1H \text{ and } 1T)\}$, which is clear in its intent. Another way for S to be listed is $S = \{2H, 1H, 0H\}$. ■

Practice Problem 2 A sack contains 5 chocolate candies, 3 butterscotch candies, and 1 peppermint candy. One candy is drawn from the sack and eaten and then a second candy is drawn and eaten. If the flavor of each candy is listed in order, what is the sample space and is it a uniform sample space?

ANSWER $S = \{(CC), (CB), (CP), (BC), (BB), (BP), (PC), (PB)\}$. Since there are different amounts of each flavor, each one does not have an equally likely chance of being drawn and the sample space is not uniform.

In Practice Problem 2, we see that there are a number of possible outcomes. When an experiment has more than one level, such as drawing a card from a deck of cards and then drawing another, it may be difficult to determine if the entire sample space has been found, since there may be a large number of possible outcomes. One tool that can be used for a relatively small number of elements and repetitions is a tree diagram. A tree diagram can help you organize the experiment in an orderly way and serves to give you an overall picture of the experiment and the possible outcomes. A tree diagram also proves useful when different events are being discussed for one experiment. Let's look at how to construct a tree diagram, and then we will see how to use it.

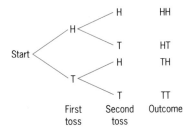

Figure 1

EXAMPLE 5   A fair coin is tossed two times. Construct a tree diagram to demonstrate the sample space if the order is noted.

SOLUTION   In Figure 1 we see that the first branches of the tree are the results of the first toss. The second level of branches shows each possibility that occurs after the first toss. To determine the sample space, you follow each path of the tree. You can see that the results are the same as those found in Example 4(a). ■

Now let's look at how to construct a tree diagram for the experiment in Practice Problem 2. At this point, we are not concerned with how many of each type of candy there is in the bag, but rather with the type of candy drawn. However, since there is only one peppermint candy, the sample point (PP) is not possible. The tree shown in Figure 2 demonstrates this. Later, when we are determining probabilities, the number of each type will be important. By following each path, we see that the results shown by the tree diagram in Figure 2 are the same as the sample points in the sample space given in Practice Problem 2.

Earlier in this section, we defined an event as a subset of a sample space. If an event has only one element, it is called a **simple event**. For example, when tossing a fair coin, the event that a tail is showing is {T} and is a simple event. If it has more than one element, it is called a **compound event**.

To see what a compound event looks like, refer to Example 2 where we looked at a set of cards numbered 1 to 10. Suppose that we would like to know if the card drawn has an even number on it. Let's define event E to be that an even-numbered card is drawn. From this definition we can now write

$$E = \{2, 4, 6, 8, 10\}$$

Event E is a subset of the sample space $S = \{1, 2, 3, 4, 5, 6, 7, 8, 9, 10\}$. Event E is a compound event because it contains more than one element. If we draw a card and an even number comes up, we say that event E has *occurred*. An event **occurs** when one of the elements in the event is the result of the experiment.

Suppose we let event N be that a number between 1 and 10 inclusive is drawn. From this definition, we have $N = \{1, 2, 3, 4, 5, 6, 7, 8, 9, 10\}$. Looking at N and the sample space S, we can see that they are equal sets. When a card is drawn, event N will occur because there are no other possible outcomes. Thus, event N is a **certain event**. *A certain event will occur when the experiment is performed.*

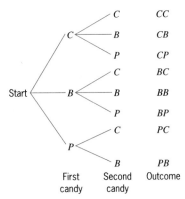

Figure 2

Now let's suppose we define event *I* to be that a 20 is drawn. Since the number 20 is not a sample point of *S*, event *I* cannot occur with this experiment. Event *I* is called an **impossible event** because it will never occur when the experiment is performed.

In our study of sets in Chapter 0, we determined that the empty set is a subset of every set. However, since the empty set contains no elements, it cannot occur when an experiment is performed. An experiment will always have a result. Therefore, the empty set is an impossible event. Likewise, since the sample space *S* contains all of the sample points, when an experiment is performed, *S* will occur. It is, therefore, a certain event.

EXAMPLE 6   Suppose that a bowl contains 5 blue balls, 5 red balls, and 5 green balls. One ball is drawn from the bowl and then another ball is drawn from the bowl. The color of each ball in the order drawn is noted. Make a tree diagram for the possible outcomes of the experiment and then find the elements in each event described below.

(a)   *A* = Exactly one red ball is drawn.
(b)   *B* = At least one red ball is drawn.
(c)   *C* = The balls are the same color.
(d)   *D* = The balls are not the same color.
(e)   *E* = Neither ball is blue.

SOLUTION   The tree diagram for this experiment is shown in Figure 3.

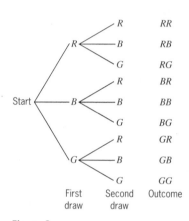

Figure 3

(a)   *A* = {(*RB*), (*BR*), (*RG*), (*GR*)}
(b)   *B* = {(*RR*), (*RB*), (*BR*), (*RG*), *GR*)}
(c)   *C* = {(*RR*), (*BB*), (*GG*)}
(d)   *D* = {(*RB*), (*RG*), (*BR*), (*BG*), (*GR*), (*GB*)}
(e)   *E* = {(*RR*), (*RG*), (*GR*), (*GG*)}

Practice Problem 3   Six cards are numbered 1 to 6 and one card is drawn.

(a) List three simple events associated with this experiment.
(b) List three compound events associated with this experiment.

ANSWER

(a) Some possible answers are as follows: observing a 1; observing a 2; observing a 3; and so on.
(b) Some compound events are as follows: the number is even; the number is odd; the number is less than 3, the number is prime, and so on.

Earlier in this section, we discussed the fact that the empty set is an impossible event and the sample space is a certain event. We begin our discussion of probability at this point. If a weatherman is absolutely certain that it is not going to rain, he will make the statement, "The probability of rain today is 0%." We could say that rain is an impossible event on that day. In the same manner, if the weather man is absolutely certain that it is going to rain, then he would make the statement, "The probability of rain today is 100%." In other words, the event of rain is a certain event. Now, 0% = 0 and 100% = 1. It is impossible for a probability to be less than 0 or more than 1, since every event is either an impossible event, a certain event, or somewhere in between.

Now let's associate with the outcomes of a sample space a probability (or, to be more precise, a probability function). A probability function is a function that assigns to each outcome in the sample space a real number between 0 and 1, inclusive.

PROPERTIES OF PROBABILITY

A probability function on a sample space must satisfy these properties:

(a)   For any event A, the probability of A, denoted $P(A)$, is a real number between 0 and 1 inclusive. That is,

$$0 \leq P(A) \leq 1$$

(b)   The sum of the probabilities of all outcomes, $P(s)$, in a sample space, S, must equal 1.

$$P(S) = 1$$

The following example demonstrates these properties.

EXAMPLE 7   If a fair coin is tossed, $P(T) = \frac{1}{2}$ and $P(H) = \frac{1}{2}$. Show that this satisfies both properties listed above.

SOLUTION   The sample space for tossing a coin is $S = \{H, T\}$.

(a) $0 \leq P(H) \leq 1$ and $0 \leq P(T) \leq 1$.
(b) $P(S) = P(H) + P(T) = \frac{1}{2} + \frac{1}{2} = 1$.   ■

Practice Problem 4   There are 5 red balls, 5 blue balls, and 5 green balls in a sack. One ball is drawn from the sack. Find the probabilities for the events in the sample space.

ANSWER $P(\text{red}) = \frac{1}{3}$, $P(\text{blue}) = \frac{1}{3}$, $P(\text{green}) = \frac{1}{3}$. The probability of each color is equal, so each has probability $\frac{1}{3}$ and the sum is 1.

If a sample space is a uniform sample space, the probability assignments are quite easy to determine. Since each sample point is equally likely, we can assign the same probability to each sample point.

PROBABILITY OF A SAMPLE POINT IN A UNIFORM SAMPLE SPACE

If a sample space S has m sample points, and each sample point is equally likely to occur, then

$$S = \{s_1, s_2, \ldots, s_m\}$$

is a uniform sample space and we can assign the probabilities of each sample point as

$$P(s_1) = P(s_2) = \cdots = P(s_m) = \frac{1}{m}$$

EXAMPLE 8   Eight identical balls numbered 1 to 8 are placed in a sack. Determine the sample space and the corresponding probabilities associated with drawing one ball from the sack.

SOLUTION   The sample space for this experiment is $S = \{1, 2, 3, 4, 5, 6, 7, 8\}$, with each number representing one ball. Since this is a uniform sample space with each sample point being equally likely, we can assign a probability of $\frac{1}{8}$ to each sample point.

$$P(1) = \frac{1}{8}, \quad P(2) = \frac{1}{8}, \quad \ldots, \quad P(8) = \frac{1}{8}$$

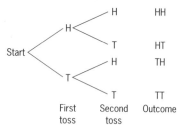

First toss   Second toss   Outcome

Figure 4

EXAMPLE 9   If a fair coin is tossed two times and the order of outcomes noted, state the sample space and determine a probability distribution.

SOLUTION   We looked at this experiment in Example 5. The outcomes are repeated in Figure 4. Each of the outcomes HH, HT, TH, and TT is equally likely, since on each toss of a fair coin the outcome of a head or tail is equally likely. The sample space is a uniform sample space and we can assign each outcome a probability of $\frac{1}{4}$.

Events often contain more than one sample point. For instance, in Example 9, we could let an event be that one or two heads are tossed in any order. Another example would be when a single die is tossed, we could let event E be that an even number is tossed, so E would contain $\{2, 4, 6\}$. We now extend our definition of probability for simple events to compound events in a uniform sample space.

DEFINITION: PROBABILITY OF EVENTS IN A UNIFORM SAMPLE SPACE

Let S represent a uniform sample space with m equally likely sample points, and let A be an event on sample space S.

1. If A is the empty set (impossible event), then $P(A) = 0$ [i.e., $P(\emptyset) = 0$].

DEFINITION: PROBABILITY OF EVENTS IN A UNIFORM SAMPLE SPACE (Continued)

2.   If A is the sample space (certain event), then $P(A) = 1$. [We can also write $P(S) = 1$.]
3.   If A is an event consisting of r sample points in S, then $P(A)$ is given by

$$P(A) = \frac{\text{Number of elements in } A}{\text{Number of elements in } S} = \frac{n(A)}{n(S)} = \frac{r}{m}$$

EXAMPLE 10   Ten identical balls numbered 1 to 10 are placed in a sack and a ball is drawn. Determine the probability for each event listed.

(a)   Let A denote the event that a number less than 5 is drawn.
(b)   Let B denote the event that an odd number is drawn.
(c)   Let C denote the event that a multiple of 3 is drawn.
(d)   Let D denote the event that a multiple of 15 is drawn.
(e)   Let E denote the event that an odd or an even number is drawn.

SOLUTION

(a) $A = \{1, 2, 3, 4\}$, so $n(A) = 4$ and since $n(S) = 10$, we have $P(A) = \frac{4}{10} = \frac{2}{5}$.
(b) $B = \{1, 3, 5, 7, 9\}$, so $n(B) = 5$ and $P(B) = \frac{5}{10} = \frac{1}{2}$.
(c) $C = \{3, 6, 9\}$, so $n(C) = 3$ and $P(C) = \frac{3}{10}$.
(d)   Since there are no multiples of 15 in the sample space, we have $D = \emptyset$ and $P(D) = 0$.
(e) $E = \{1, 2, 3, 4, 5, 6, 7, 8, 9, 10\}$, which is the entire sample space, so $P(E) = 1$. ■

EXAMPLE 11   A standard deck of 52 cards is shuffled and one card is drawn. Find the probability of each event listed.

(a)   Let A denote the event that an ace is drawn.
(b)   Let B denote the event that a club is drawn.
(c)   Let C denote the event that a red card is drawn.
(d)   Let D denote the event that a face card (jack, queen, or king) is drawn.

SOLUTION

(a)   There are four aces in a standard deck; therefore, from $n(A) = 4$ and $n(S) = 52$ we have $P(A) = \frac{4}{52} = \frac{1}{13}$.
(b)   There are 13 clubs in a deck of cards, so $n(B) = 13$ and $P(B) = \frac{13}{52} = \frac{1}{4}$.
(c)   The set of red cards consists of hearts and diamonds. Since there are 13 cards in each suit, there are 26 red cards and $n(C) = 26$, which gives $P(C) = \frac{26}{52} = \frac{1}{2}$.
(d)   Each suit has 3 face cards and there are 4 suits; therefore, $n(D) = 12$ and $P(D) = \frac{12}{52} = \frac{3}{13}$. ■

EXAMPLE 12 In Example 9 a fair coin was tossed two times. Let E denote the event that at least one head is tossed. Find $P(E)$.

SOLUTION Looking at the tree diagram in Figure 4, we see that at least 1 head is tossed 3 times, so $n(E) = 3$. Also, we can see that $n(S) = 4$ and we can calculate $P(E) = \frac{3}{4}$. ∎

Practice Problem 5 Again refer to Example 9. Find the probability that the results from the two tosses are not the same, that is, HT or TH.

ANSWER $P(E) = \frac{1}{2}$

SUMMARY

It is important to know the terminology used in probability. You should know the following terms that were introduced in this section: experiment, sample space, sample point, event, uniform sample space, impossible event, certain event, simple event, compound event, and probability function. We also introduced how to assign probabilities to events in a uniform sample space.

Exercise Set 5.1

1. State the sample space for each of the spinners shown below. Assume equal fractional sections.

(a)　　　(b)　　　(c)　　　(d)

2. Find the probabilities for the events in each sample space in Exercise 1.

3. A fair die is rolled one time and the number showing on top is noted. Find a sample space for this experiment and the corresponding probabilities.

4. Doug, Tom, Ben, Fred, and Louis place their names in a hat. One name is to be drawn to determine who will confess to breaking a window. Tabulate a sample space and the probabilities for the events in the sample space.

5. What is the probability of getting heads when a two-headed coin is tossed?

6. Refer to Exercise 5 and determine the probability of getting tails.

Suppose that there is an equally likely probability that the spinner shown in the figure will stop at any one of the six sections. Answer questions 7–11 using this figure.

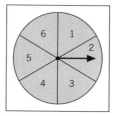

7. What is the probability of stopping on an even number?

8. What is the probability of stopping on a multiple of 3?

9. What is the probability of stopping on an even number or a multiple of 3?

10. What is the probability of stopping on an even number or an odd number?

11. What is the probability of stopping on a number that is both a multiple of 3 and an even number?

12. A card is drawn from an ordinary deck of 52 cards. What is the probability that the card is
    (a) a heart?   (b) an ace?
    (c) the jack of spades?   (d) a black card?
    (e) a black ace?   (f) a red face card?

13. A multiple-choice question has five possible answers. You have not studied and hence have no idea which answer is correct. So, you randomly choose one of the five answers. What is the probability that you select the correct answer? An incorrect answer?

14. List the elements in a sample space for the simultaneous tossing of a fair coin and the drawing of a card from a set of six cards numbered 1 to 6. (**Hint**: Make a tree diagram.)

15. Which of the following could not be a probability and why?
    (a) $-\dfrac{1}{2}$ (b) $\dfrac{17}{16}$ (c) .001   (d) 0

    (e) 1.03   (f) .01   (g) $\dfrac{5}{4}$ (h) 1

16. A box contains three red balls and four blue balls. Let R represent a red ball and B represent a blue ball. Find a sample space if
    (a) one ball is drawn.
    (b) two balls are drawn at a time. (Make a tree.)
    (c) three balls are drawn at a time. (Make a tree.)
    (d) Are these sample spaces uniform?

17. Three coins are tossed, and the number of heads is recorded. Which of the following sets is a sample space for this experiment? Why do the other sets fail to qualify as sample spaces?
    (a) {1, 2, 3}   (b) {0, 1, 2}
    (c) {0, 1, 2, 3, 4}   (d) {0, 1, 2, 3}

18. A box contains four balls numbered 1 to 4. Record a sample space for the following experiments.
    (a) A ball is drawn and the number is recorded. The ball is returned and a second ball is drawn and recorded.
    (b) A ball is drawn and recorded. Without replacing the first ball, a second is drawn and recorded.

19. A box contains 4 black balls, 7 white balls, and 3 red balls. If a ball is drawn, what is the probability of getting the following colors?
    (a) Black   (b) Red
    (c) White   (d) Red or White
    (e) Black or White   (f) Red or Black or White

20. Two identical sets of four cards are numbered 1 to 4. One card is drawn from each set. List a uniform sample space to show that the two cards can be drawn in 16 different ways. Find the probability that the sum of the numbers is
    (a) greater than 4.   (b) equal to 9.
    (c) equal to 7.   (d) greater than 7.

21. Two fair coins are tossed. Find the probability that the results are
    (a) 2 heads.   (b) exactly 1 head.
    (c) at most 1 head.

22. In drawing a card from a standard deck of 52 cards, you reason that you might get a heart (H) or you might not get a heart. Therefore, there are two outcomes, and you determine that $P(H) = \frac{1}{2}$. Is this reasoning correct? Why or why not?

23. Fifteen balls, numbered 1 to 15, are placed in a bag, and one ball is drawn. Find each probability.
    (a) The number is even.
    (b) The number is odd.
    (c) The number is less than 7.
    (d) The number is odd or even.
    (e) The number is a multiple of 20.

24. A box contains a yellow, a green, a black, and a blue marble.
    (a) How many outcomes are possible if you pick one marble from the box?
    (b) List the possible outcomes.
    (c) What is the probability of picking a blue marble?
    (d) Is each outcome equally likely?
    (e) What is the probability of not getting a yellow marble?

25. Each letter of the word *mathematics* is placed on a card, and the cards are placed in a basket. A card is randomly drawn from the basket.
    (a) List the possible outcomes.
    (b) Are the outcomes equally likely?
    (c) What is the probability of drawing a c?
    (d) What is the probability of drawing an e?

(e) What is the probability of drawing an *a*?

(f) What is the probability of drawing an *m* or an *s*?

(g) What is the probability of drawing a vowel?

(h) What is the probability of drawing a consonant?

26. The probability that a student will pick a blue button out of a jar is $\frac{1}{2}$. There are 16 buttons in the jar. How many of them are blue?

27. If you flipped a fair coin 15 times and got 15 heads, what would be the probability of getting a head on the 16th toss?

Applications (Business and Economics)

28. **Quality Control**.   A shipment is believed to contain 100 good articles, 5 articles with minor defects, and 3 articles with major defects. If one article is drawn from the shipment, what is the probability that it will

(a) not have a defect?

(b) have a defect?

(c) have a major defect?

29. **Executive Boards**.   An executive board of a corporation is made up of 5 members whom we shall call A, B, C, D, and E. A committee of 3 is chosen to select a president for the corporation. Find outcomes of the sample space that represent all the possible committees.

Applications (Social and Life Sciences)

30. **Drug Analysis**.   A medical research institute is experimenting with possible cures for cancer. The doctor in charge of the experiment initially selects 3 of 5 possible drugs—V, W, X, Y, Z—for concentrated research.

(a) List the sample outcomes where Z is one of the chosen drugs.

(b) List the sample outcomes where X and Y are both among the chosen drugs.

(c) Another doctor suddenly announces that drug V is definitely not a cure for cancer. If drug V is dropped, list the sample space for which the doctor in charge can choose three of four possible drugs.

(d) The doctor decides that a new drug, drug A, will definitely be 1 of the 3 drugs used for experimentation. She now needs to choose 2 of the 4 other possible drugs. List the sample outcomes where W is one of the chosen drugs.

5.2 EMPIRICAL PROBABILITY AND THE FUNDAMENTAL PRINCIPLE OF COUNTING

OVERVIEW   Subjective judgment may be less precise to use in assigning probabilities, but there are certain situations where subjective judgment is the only accessible tool. A sales manager might declare, "We have a probability of 60% of getting that contract." The man on the street might venture to say, "The Cowboys have an 80% chance of winning the Superbowl." Generally, such assignments are merely measures of the strength of the person's belief. However, if the person making the projection has a lot of experience in the area and a keen sense of either observation or the undefinable intuition, the probability assignments might prove to be of some practical use. This is evidenced daily in the decision-making centers of government, education, stock markets, industry, and weather forecasting.

In this section, we study a rule for assigning probabilities based on **empirical data**. Then we consider counting procedures that assist in assigning probabilities. We need to understand

- Relative frequency
- Empirical probability
- Tree diagrams
- The fundamental principle of counting

In the last section, we assigned probabilities to outcomes using a definition involving the number of ways an event could occur. In this section, we assign probabilities based on **empirical data**. To introduce this term, consider the following example.

EXAMPLE 13   A fair die is rolled 10,000 times. Table 1 itemizes the number of times a 1 has occurred at various stages of the process. Notice that as the number of rolls N increases, the *relative frequency* begins to stabilize around the number $.166 \approx \frac{1}{6}$. We can reasonably assign the probability

$$P(1) = \frac{1}{6}$$

TABLE 1

Number of Rolls (N)	Number of 1's Occurring (m)	Relative Frequency (m/N)
10	4	.4
100	20	.2
1,000	175	.175
3,000	499	.16633. . .
5,000	840	.168
7,000	1,150	.164285714. . .
10,000	1,657	.1657

The probability introduced in the preceding example describes the fraction of times one would *expect* the outcome A (getting a 1) to occur if the experiment were performed a *large number of times*. Suppose that a thumbtack is tossed in the air 10,000 times and lands with the point up 1000 times. The relative frequency for this outcome is $1000/10,000 = \frac{1}{10}$. If we continue the experiment 10,000 more times and find that the ratio is still approximately $\frac{1}{10}$, we are willing to assign this number as a measure of our degree of belief that it will land point up on the next toss. These examples suggest the following definition.

DEFINITION: EMPIRICAL PROBABILITY

Suppose that an experiment is performed N times, where N is a very large number. The probability of an outcome A should be approximately equal to the following ratio:

$$P(A) \approx \frac{\text{Number of times } A \text{ occurs}}{N}$$

EXAMPLE 14   A loaded die (one for which outcomes are not equally likely) is thrown 7000 times with the results shown in Table 2. Determine a rule for assigning a probability to each outcome. Is this a legitimate sample space?

TABLE 2

Outcome	Frequency	Relative Frequency
1	967	$\dfrac{967}{7000} \approx .14$
2	843	$\dfrac{843}{7000} \approx .12$
3	931	$\dfrac{931}{7000} \approx .13$
4	1504	$\dfrac{1504}{7000} \approx .21$
5	1576	$\dfrac{1576}{7000} \approx .23$
6	1179	$\dfrac{1179}{7000} \approx .17$

SOLUTION   Use $S = \{1, 2, 3, 4, 5, 6\}$ as a sample space to assign the following probabilities:

$$P(1) = .14, \qquad P(2) = .12, \qquad P(3) = .13$$
$$P(4) = .21, \qquad P(5) = .23, \qquad P(6) = .17$$

You can determine that the sum of the probabilities is 1, so this is a legitimate sample space. ∎

EXAMPLE 15   A poll was taken of a sample of 500 employees at a large car manufacturing plant to determine whether they wanted to go on strike. Table 3 shows the outcome of this poll. The three groups are divided according to salary.

TABLE 3

	In Favor of a Strike	Not in Favor of a Strike	No Opinion
Group A	150	50	10
Group B	100	80	8
Group C	30	70	2

(a) What is the probability that an employee selected at random from Group A is in favor of a strike? (To say that an employee is selected **at random** means that each employee has the same chance of being selected.)

(b) What is the probability that an employee selected at random from Group B has no opinion?

SOLUTION

(a) Using our definition of probability in Section 5.1, we find that the probability of an employee from Group A being in favor of a strike is

$$P = \frac{\text{Number in Group A in favor of strike}}{\text{Total number in Group A}} = \frac{150}{210}$$

(b) The probability of an employee from Group B having no opinion on the strike is

$$P = \frac{\text{Number in Group B with no opinion}}{\text{Total number in Group B}} = \frac{8}{188}$$

Practice Problem 1 Using the information from Example 15, find each probability.

(a) What is the probability that an employee selected at random is in Group C?

(b) What is the probability that an employee selected at random has no opinion?

(c) What is the probability that an employee selected at random favors a strike?

ANSWER

(a) P(employee is in Group C) $= \frac{102}{500}$.

(b) P(employee has no opinion) $= \frac{20}{500}$.

(c) P(employee is in favor of a strike) $= \frac{280}{500}$.

In Section 5.1, we introduced the use of a tree diagram to help us determine a sample space. We now use tree diagrams as we begin to look at counting techniques.

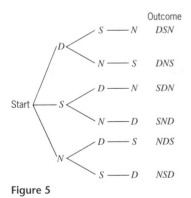

Figure 5

EXAMPLE 16 A college choral group is planning a concert tour with performances in Dallas, St. Louis, and New Orleans. In how many ways can their itinerary be arranged?

SOLUTION If there is no restriction on the order of the performances, any one of the three cities can be chosen as the first stop. We see this illustrated in Figure 5. After the first city is selected, either of the other cities can be chosen as the second stop, with the remaining city being the third stop on the tour. As we saw in Section 5.1, to determine all possible outcomes, we follow each path to its completion and then record the path. In Figure 5, we have designated Dallas as D, New Orleans as N, and St. Louis as S.

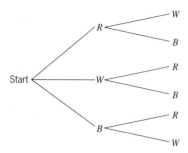

Figure 6

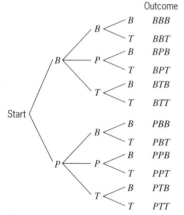

Figure 7

Whenever a task can be done in two or more stages and each stage can be done in a number of ways, such as planning a trip to several cities and then selecting a mode of transportation to each city, a tree diagram provides a good illustration of the choices involved and serves as an aid in determining the total number of ways that the entire task can be accomplished.

Practice Problem 2 Use a tree diagram to determine the total number of ways that a two-toned car can be painted with red (R), white (W), or black (B) paint.

ANSWER   As shown in Figure 6, there are six ways that this can be done.

Now let's return to the problem situation given in Example 16.

EXAMPLE 17   The members of the choral group decide on the itinerary of New Orleans, Dallas, and St. Louis. Now they must decide on the mode of transportation from one city to the next. They can travel from their campus to New Orleans by bus or plane and from New Orleans to Dallas by bus, plane, or train. From Dallas to St. Louis they can travel by bus or train. Use a tree diagram to determine the total number of different modes of transportation they can use.

SOLUTION   The tree diagram in Figure 7 organizes the information above. By counting the paths, we see that there are 12 different ways to arrange the modes of transportation. ■

Looking at Example 17 and Figure 7, we see that there are 2 ways that the first part of the trip can be made, 3 ways for the second part, and 2 ways for the last part. We counted 12 total ways, but we can also see that $2 \cdot 3 \cdot 2 = 12$. It is not simply a coincidence that this product is the same as the total number of paths in the tree. We can also see that if the number of choices in a multistage problem like this were to be large, a tree diagram would be of little use because of the time required to construct it. For example, suppose that the choral group were planning a 20-city tour. To construct a tree diagram with all the possible itineraries would, for all practical purposes, be impossible.

As we learn more about counting techniques, we will encounter problems where the total number of possible outcomes is very large. We need a way other than tree diagrams to count the total possible ways that a multistage task can be performed. As we progress, we will look at several counting techniques. The first we will study is given below.

The Fundamental Principle of Counting

If two experiments are performed in order with n_1 possible outcomes of the first experiment and n_2 possible outcomes of the second experiment, then there are

The Fundamental Principle of Counting (Continued)

$$n_1 \cdot n_2$$

combined outcomes of the first experiment followed by the second. In general, if k experiments are performed in order with possible number of outcomes n_1, n_2, n_3, . . . , n_k, respectively, then there are

$$n_1 \cdot n_2 \cdot n_3 \cdot \;\cdot\;\cdot\;\cdot\; \cdot n_k$$

possible outcomes of the experiments performed in order.

The fundamental principle of counting is helpful in solving problems such as those given in the following examples.

EXAMPLE 18   Texas has 12.6 million registered vehicles. If the license plates are made by placing three letters followed by three digits, will there be enough different license plates possible, if all letters and all digits may be used with repetition, when the number of vehicles reaches 14 million?

SOLUTION   There are 26 letters to choose for each of the 3-letter positions, and there are 10 digits (0–9) to choose for each of the digit positions. By the fundamental principle of counting, the number of combinations is given by

$$26 \cdot 26 \cdot 26 \cdot 10 \cdot 10 \cdot 10 = 17{,}576{,}000$$

Consequently, there will be enough license plates available.   ■

Practice Problem 3   A box contains 12 identical balls numbered 1 to 12. A ball is drawn, its number is noted, and the ball is not replaced. A second ball is then drawn and its number is noted. How many ways can 2 balls be drawn?

ANSWER   There are $12 \cdot 11 = 132$ ways of drawing the balls.

EXAMPLE 19   A fair coin is tossed 5 times. If each outcome is either a head or a tail, how many different outcomes (arrangements of heads and tails) are in the sample space?

SOLUTION   Since there are 5 steps, each with 2 possible outcomes, there are $2 \cdot 2 \cdot 2 \cdot 2 \cdot 2 = 32$ different outcomes.   ■

Practice Problem 4   If a fair coin is tossed 4 times, determine the possible outcomes and then determine each of the following probabilities.

(a) P(exactly 3 heads)   (b) P(at most 3 heads)
(c) P(at least 3 heads)

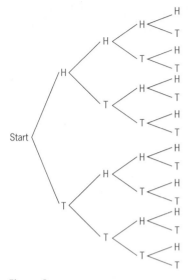

Figure 8

ANSWER   There are $2 \cdot 2 \cdot 2 \cdot 2 = 16$ total outcomes. These are shown in the tree diagram in Figure 8. The probabilities are

(a) $P(\text{exactly 3 heads}) = \frac{4}{16}$ (b) $P(\text{at most 3 heads}) = \frac{15}{16}$

(c) $P(\text{at least 3 heads}) = \frac{5}{16}$

As you can see, even for a relatively small number of outcomes, tree diagrams become quite difficult after only a few stages. In the next section, we look at two counting techniques that will enable us to determine probabilities such as those in Practice Problem 4 without having to construct a tree diagram.

NOTE: In Practice Problem 4, you were asked to find $P(\text{at least 3 heads})$ and $P(\text{at most 3 heads})$. This terminology is sometimes confusing to students. When a coin is tossed 4 times (or 4 coins are all tossed at the same time), there are 5 possibilities for the number of heads: 0 heads, 1 head, 2 heads, 3 heads, or 4 heads. Therefore, the phrase *at least 3 heads* is asking for 3 heads or more and those possibilities are 3 heads or 4 heads. Likewise, the phrase *at most 3 heads* is asking for no more than 3 heads and those possibilities are 0 heads, 1 head, 2 heads, or 3 heads.

EXAMPLE 20   How many 4-digit identification tags can be made using the digits 1, 2, 3, 4, and 5 (a) if repetitions are permitted and (b) if repetitions are not permitted?

SOLUTION

(a)   If repetitions are permitted, then each number can be used each time and there are $5 \cdot 5 \cdot 5 \cdot 5 = 5^4 = 625$ total tags possible.

(b)   If each number can be used only once, then there are $5 \cdot 4 \cdot 3 \cdot 2 \cdot 1 = 120$ total tags possible.   ∎

Practice Problem 5 How many 4-digit identification tags can be made using the digits 1, 2, 3, 4, and 5 if repetitions are permitted and the number must be even?

ANSWER   There are only two even numbers (2 and 4), so there are two choices for the last digit and five choices for each other position, giving $5 \cdot 5 \cdot 5 \cdot 2 = 250$ possible tags.

SUMMARY

We have seen that tree diagrams are useful in helping to determine the total number of outcomes in a multistage experiment or event. However, we also determined that for even a small number of choices, the extent of a tree diagram can very quickly get to be too much. Consequently, we must look for other methods of counting. In this section, we introduced the fundamental principle of counting, which is one of the counting techniques used in determining the total number of possible outcomes. In the next section, we introduce two more counting techniques.

Exercise Set 5.2

1. There are 6 roads from city A to city B and 4 roads from city B to city C.
   (a) In how many ways can Joy drive from city A to city C if she must go through city B?
   (b) In how many ways can Joy drive round trip from city A to city B to city C and return to city A through city B?

2. A woman wants to buy an automobile. She has a choice of 2 body styles (standard and sports model) and 4 colors (green, red, black, and blue). What is the probability she selects a red sports model? (Assume that each choice is equally likely.)

3. A restaurant offers the following menu.

Main Course	Vegetable	Beverage
Beef	Potatoes	Milk
Ham	Green beans	Coffee
Fried chicken	Green peas	Tea
Shrimp	Asparagus	

   If you choose 1 main course, 2 different vegetables, and 1 beverage, in how many ways could you order a meal?

4. A boy and his father play in a Ping-Pong tournament. The first player to win 4 games wins the tournament. If the father wins game 1, make a tree diagram showing the possible ways in which the tournament can turn out.

5. A fair die is rolled and a chip is drawn from a box containing three chips numbered 1, 2, 3. How many possible outcomes can be obtained from this experiment? Verify your answer with a tree diagram.

6. In how many ways can two speakers be arranged on a program? Three speakers? Four speakers?

7. A box contains 6 different-colored balls: red, white, blue, black, green, and yellow. If 2 balls are drawn at random, one at a time, and replaced, find each probability below. Assume that each outcome is equally likely.
   (a) P(a yellow ball followed by a red ball)
   (b) P(a red ball followed by a blue ball)
   (c) P(red ball and a blue ball in any order)
   (d) P(both balls being the same color)

8. Four coins are tossed at the same time and the number of heads and tails is noted. Find each probability.
   (a) P(4 tails)
   (b) P(exactly 2 tails)
   (c) P(at most 2 tails)
   (d) P(at least 2 tails)

9. A letter cube (with sides A, B, C, D, E, and F) is tossed and the number of times (frequency) that each letter occurred is recorded, as shown in the table.

Letter	Frequency
A	8
B	10
C	9
D	11
E	12
F	10

   (a) How many possible outcomes are there?
   (b) What are the possible outcomes?
   (c) From the Frequency column, assign a probability of tossing a D.
   (d) Assign a probability to the outcome D based on the assumption that each outcome is equally likely.
   (e) Why is there a difference?
   (f) Would you expect the two probabilities to have been closer if the cube had been tossed 1000 times? Why or why not?

10. Suppose that you toss a fair coin 40 times and get 19 heads.
   (a) On the basis of this experiment, assign a probability of getting a head on one toss of a coin.
   (b) From the definition of probability, compute the probability of getting a head on one toss of a fair coin.
   (c) Why do the two answers differ?

11. The manager of an appliance store has recorded the

color of all dishwashers sold in the past 6 months, as shown in the table.

Color	Number Sold
White	24
Tan	18
Blue	8
Yellow	22
Brown	28

(a) What is the probability that a person who bought a dishwasher bought a white one?

(b) What is the probability that a person who bought a dishwasher bought a brown one?

12. (a) Refer to Exercise 11 and find the probability that a customer selected a tan or blue or a yellow dishwasher.

(b) Find the sum of the five probabilities (probability for each color).

13. Social Security numbers are nine digits long. The first number cannot be 0, but that is the only restriction. How many different Social Security numbers can be issued?

14. An ordinary deck of 52 cards is shuffled and two cards are pulled from the deck, one after the other, without replacement. Determine how many different ways you can draw the following cards.

(a) Two diamonds

(b) Two red cards

(c) A red card and then a black card

(d) One red card and one black card

15. You have forgotten the combination to your locker. There are 30 numbers on the lock. To open the lock, you turn to the right, then to the left, and back to the right. The numbers cannot repeat. How many combinations are possible? What is the probability that you guess correctly the first time?

16. How many four-letter words (any combination of letters being considered a word) are possible using the complete alphabet if

(a) no repetition is allowed?

(b) adjacent letters cannot be alike?

(c) letters can be repeated?

17. A license plate consists of 3 digits, a letter, and then 3 more digits. How many plates are possible if

(a) there are no restrictions?

(b) no repetition of digits is allowed?

(c) the first digit must be odd and there are no other restrictions?

(d) the first and last digits must be even and there are no other restrictions?

18. A survey of 500 employees in a large plant asked whether the employee would be willing to give up three days of paid sick leave in exchange for one week of half paid vacation. The results are shown in the table. If an employee is selected at random, find each probability.

	In Favor of the Offer	Not in Favor of the Offer	No Opinion
Ages 18–24	150	30	5
Ages 25–40	75	120	3
Ages 41–65	20	90	7

(a) P(employee is between 18 and 24 and is in favor of the offer)

(b) P(employee is between 41 and 65 and is not in favor of the offer)

(c) P(employee is in favor of the offer)

(d) P(employee is not in favor of the offer)

19. **Exam.** A cooling and heating company established the following distribution of monthly service calls over the last 4 years.

Number of Calls	Number of Occurrences
801–850	4
851–900	10
901–950	80
951–1000	40
1001–1050	20
1051–1100	12
1101–1150	12
1151–1200	10
1201–1250	8
1251–1300	4
	200

The probability of experiencing more than 1150 calls for service during a given month would be

(a) .06   (b) .89   (c) .17   (d) .11

Applications (Business and Economics)

20. **Commissions**. A poll is taken among 100 salespeople of a corporation concerning how each wishes to be paid. The results of the poll are tabulated by categories—those in the top 25% in sales last year, the next 25%, and the lowest 50%.

	Number of Salespeople		
Method of Payment	Top 25% in Sales	Second 25% in Sales	Lowest 50% in Sales
Flat salary	1	8	23
All commission	20	15	5
Half salary/half commission	4	2	22

If one of the 100 salespeople is chosen at random, what is the probability that this person
(a) favors receiving all the salary by commission?
(b) wants a flat salary?
(c) favors half flat salary and half commission?
(d) is in the top 25% of sales?
(e) is in the second 25% of sales?

21. **Unemployment**. A sample of the employment status of the residents in a certain town is given in the following table.

	Employed	Unemployed
Male	1000	40
Female	800	180

Find each probability.
(a) P(an unemployed person selected at random is female)
(b) P(an unemployed person selected at random is male)

(c) P(a male selected at random is unemployed)
(d) P(a female selected at random is employed)

22. **Transportation**. A sales representative plans a trip from Atlanta to Boston to London. From Atlanta to Boston, he can travel by bus, train, or airplane. However, from Boston to London, he can travel only by ship or airplane.
(a) In how many ways can the trip be made?
(b) Verify your answer by drawing an appropriate tree diagram and counting the routes.

23. **Management Solution**. Applicants at a corporation are classified from tests as high IQ, average IQ, or low IQ, and from interviews as aggressive or passive. How many combinations will exist from the tests and interviews?

24. **Quality Control**. Suppose that in a shipment of 100 items there are 4 that are defective. Items from the shipment are drawn one at a time and tested. The testing terminates when 2 defective items are found or after five tests. Show the testing process with a tree diagram.

Applications (Social and Life Sciences)

25. **Family Planning**. Assume that a family wishes to have 4 children and that the chance for a boy is the same as the chance for a girl. Find each probability.
(a) P(all 4 will be girls)
(b) P(all the children will be the same sex)
(c) P(at least 3 will be boys)
(d) P(at most 3 will be boys)

26. **Politics**. Suppose that in a local election, only two parties are represented, Democrats (D) and Republicans (R). Draw a tree diagram illustrating four consecutive elections and determine how many possibilities result in each of the following.
(a) At least 1 party change
(b) No party changes
(c) Exactly 1 party change
(d) More than 2 party changes

27. **Medicine**. A clinic indicates a cure has been found for a certain blood disease. Out of 80,000 patients, 74,000 recovered after using the medication. Assign

a probability that a person suffering from the blood disease will recover using this medicine.

28. **On-the-Job Accidents.**   A sociology class made a study of the relationship between a worker's age and the number of on-the-job accidents. The following table summarizes the findings.

Age Group	Number of Accidents			
	0	1	2	3 or More
Under 20	18	22	8	12
20–39	26	18	8	10
40–59	34	14	8	6
60 and over	42	10	12	2

(a) What is the probability that an employee who had 1 accident is in the 20–39 age group?

(b) What is the probability that an employee under 20 years of age will have 2 accidents?

(c) What is the probability that an employee 40 years of age or older will have more than 2 accidents?

(d) What is the probability of an employee having an accident?

Review Exercises

29. A number x is selected at random from the set of numbers $\{1, 2, 3, \cdots, 10\}$. What is the probability that

(a) $x < 5$?

(b) x is even?

(c) x is divisible by 3?

(d) x is less than 5 and even?

(e) $x \geq 2$?

(f) $x < 7$?

(g) x is less than 5 or even?

(h) $x^2 > 9$?

30. A bag contains 6 red balls, 4 black balls, and 3 green balls. Tabulate a sample space for the following experiments.

(a) A single ball is drawn and the color is noted.

(b) A single ball is drawn and pocketed. A second ball is drawn. (You may want to use a tree diagram.)

(c) A ball is drawn, its color recorded, and it is replaced. A second ball is drawn.

(d) A ball is drawn and pocketed. A second ball is drawn and pocketed. A third ball is drawn and pocketed.

5.3   COUNTING TECHNIQUES USING PERMUTATIONS AND COMBINATIONS

OVERVIEW   The fundamental principle of counting can be used to study two extremely important concepts for counting: **permutations** and **combinations**. Both of these concepts are useful, not only in solving complicated probability problems but also in other types of applications. The distinguishing difference between the two concepts is **order**. Permutations involve the order of an arrangement of objects, and combinations do not involve order. In this section, we look at

- Permutations
- Factorial
- Number of permutations
- Permutations for indistinguishable objects
- Combinations

Suppose that four letters, A, B, C, D, are listed two at a time in every possible way, without repetition of letters such as AA. The list is as follows:

$$\{AB, AC, AD, BA, BC, BD, CA, CB, CD, DA, DB, DC\}$$

Note that *AB* is listed and so is *BA*. Since we wanted every possible way to arrange the letters, **order** is important in making this list. When order is important, an arrangement of distinct objects is called a **permutation**. Each element in the list above is a different permutation of two of the letters.

DEFINITION: PERMUTATION

An ordered arrangement of *r* objects selected from a set of *n* distinct objects ($r \leq n$) is called a **permutation** of the objects.

NOTE: Since a permutation involves an arrangement of objects in a particular order, we can see that repetition of an object in that arrangement is not allowed. For example, when we listed the permutations above, we did not include the arrangements *AA*, *BB*, *CC*, *DD* because there was only one *A*, one *B*, one *C*, and one *D*. Consequently, each letter could only be used once.

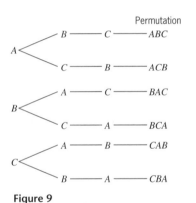

Figure 9

EXAMPLE 21 For the set $S = \{A, B, C\}$:

(a) Determine the number of permutations using the fundamental principle of counting.

(b) Draw a tree diagram to find all the permutations.

SOLUTION

(a) According to the fundamental principle of counting, we have three decisions to make to fill in three spaces with letters. Since in a permutation repetition of the same object is not allowed, there are 3 choices for the first letter, 2 choices for the second letter, and 1 choice for the third letter. Therefore, there are $3 \cdot 2 \cdot 1 = 6$ permutations of the letters in set *S*.

(b) The tree diagram in Figure 9 shows the permutations and confirms that there are 6 permutations of the letters in set *S*. The permutations are

$$ABC, ACB, BAC, BCA, CAB, CBA$$ ■

Practice Problem 1 Use the fundamental principle of counting to determine the number of permutations of the set $S = \{A, B, C, D\}$.

ANSWER Each letter can be used only once, so there are $4 \cdot 3 \cdot 2 \cdot 1 = 24$ permutations of *S*.

As the number of objects being arranged increases, the number of permutations becomes quite large. Because of this, we now introduce a useful notation and then use it with permutations.

We might ask ourselves the question, How many permutations are there for a set with *n* distinct objects? Following our strategy from the examples above, we could correctly assume that we must multiply $n(n - 1)(n - 2) \cdot \cdots \cdot 3 \cdot 2 \cdot 1$.

However, since this notation is cumbersome, we have a notation that solves this problem—it is $n!$.

DEFINITION: FACTORIAL

The notation $n!$ is called n **factorial** and is defined as

$$n! = n(n - 1)(n - 2) \cdot \cdots \cdot 3 \cdot 2 \cdot 1$$

for n a positive integer. We also define $0! = 1$.

The statement that $0! = 1$ may seem unusual, especially since $1! = 1$, but you will learn later in your work with factorials that this definition is reasonable and consistent with the factorial idea for positive integers.

Now that we have a convenient notation, we can define the number of permutations of a set with n distinct objects.

NUMBER OF PERMUTATIONS OF n OBJECTS

The number of permutations of n distinct objects is $n!$.

Calculator Note

On your graphing calculator, the factorial is part of the MATH PRB menu (press $\boxed{\text{MATH}}$ $\boxed{\blacktriangleleft}$). Use the factorial key on your calculator to find 8!. Press 8 $\boxed{\text{MATH}}$ $\boxed{\blacktriangleleft}$ (select 4:) $\boxed{\text{ENTER}}$. The answer is 40,320.

EXAMPLE 22   Six workers are assigned to six different jobs. In how many ways can the assignments be made?

SOLUTION   Each worker is distinct. The first worker has 6 choices of jobs. After the first worker has selected a job, there are 5 jobs left for the next worker. This process will continue for each worker, so there are $6 \cdot 5 \cdot 4 \cdot 3 \cdot 2 \cdot 1$ $= 6! = 720$ different ways the jobs can be assigned for the six workers. ■

EXAMPLE 23   In Section 5.2, we looked at an example of a college choral group planning a trip. If the choral group is planning a tour of 20 cities, how many possible ways can the itinerary be done?

SOLUTION   As we saw in the last section, a tree diagram would, for all practical purposes, be impossible. What is really being asked for here is the number of permutations of 20 objects. Therefore, there are 20! ways of planning the itinerary. This number is so large, that we will simply use the factorial notation. However, you should calculate the number on your calculator to see just how many choices there are! (2.43×10^{18}) ■

Calculator Note

Some graphing calculators have a limit to the size of the numbers that can be used with the factorial key. This is not a problem. It is not practical to write out these extremely large numbers. In many cases, the factorial notation is more descriptive and it is desirable to use that notation as the answer to a problem.

In order to be more efficient in our work and to make our work with large numbers easier, let's turn our attention to developing a formula for the number of permutations of n distinct objects taken r at a time, where $r \leq n$. For example, how many permutations of 6 objects taking only 3 at a time are there? The notation for a permutation of n distinct objects taken r at a time is $P(n, r)$.

EXAMPLE 24   A group of stockholders is to elect a president, a vice president, a secretary, and a treasurer from six board members who qualify. How many ways can the officers be selected?

SOLUTION   To designate different candidates, we use C1 for candidate 1, C2 for candidate 2, and so on. If we consider the order as president, vice president, secretary, and treasurer, then {C1, C2, C3, C4} is certainly a different set of officers than {C4, C2, C1, C3}. Order is important here and the answer to the problem is $P(6, 4)$, the number of permutations of 6 things taken 4 at time. Use the fundamental principle of counting and note that the position of president can be filled in 6 ways. After this occurs, the position of vice president can be filled in 5 ways, or the 2 positions can be filled in $6 \cdot 5$ ways. Then the secretary can be selected in 4 ways, or the 3 positions in $6 \cdot 5 \cdot 4$ ways. Finally, there are only 3 people left to be selected for treasurer. Hence, the number of ways that all 4 positions can be filled is $6 \cdot 5 \cdot 4 \cdot 3$ or

$$P(6, 4) = 6 \cdot 5 \cdot 4 \cdot 3 \quad \text{or} \quad \text{Keystrokes: 6 } \boxed{\text{MATH}} \, \boxed{\blacktriangleleft} \, \text{(Select 2) 4 } \boxed{\text{ENTER}}$$

To express $P(6, 4)$ in terms of factorials, multiply and divide by 2! (which is equivalent to multiplying by 1):

$$P(6, 4) = \frac{6 \cdot 5 \cdot 4 \cdot 3 \cdot 2!}{2!}$$

Now, since $6! = 6 \cdot 5 \cdot 4 \cdot 3 \cdot 2 \cdot 1$ and $2! = 2 \cdot 1$, we can write

$$P(6, 4) = \frac{6!}{2!} = \frac{6!}{(6 - 4)!}$$

to express everything in terms of the only numbers (6 and 4) that are given in the example.   ∎

Using the reasoning from Example 24, the number of n distinct objects taken r at a time is given by the following definition.

DEFINITION: NUMBER OF PERMUTATIONS

The **number of permutations** of n distinct objects taken r at a time is given by

$$P(n, r) = \frac{n!}{(n - r)!} \qquad (1 \le r \le n)$$

Calculator Note

To evaluate an expression such as $P(72, 28)$ you may notice that your calculator uses the notation $_nP_r$ instead of P(n, r). To evaluate $_{72}P_{28}$ we use the MATH PRB menu as follows: 72 | MATH | ◄ | (select 2) 28 | ENTER |. The answer is approximately $2.30\,(10)^{49}$.

EXAMPLE 25   A marketing survey lists 10 types of candy and asks those being surveyed to list their favorite, second favorite, and least favorite type on the list. How many ways can a survey form be filled out?

SOLUTION   There are 10 objects ($n = 10$) and 3 ($r = 3$) of them are to be selected. Therefore, there are $P(10, 3) = 720$ ways to fill out the survey.   ■

Practice Problem 2   In how many ways can a president, a vice president, and a secretary be selected from a group of 15 people?

ANSWER   There are $P(15, 3) = 2730$ ways.

Suppose that we wish to find the number of arrangements of n objects where not all of the objects are distinguishable. For example, suppose we permute the letters in the word *seem*.

seem	*seme*	*smee*	*mees*	*mese*	*msee*
eesm	*eems*	*emes*	*esem*	*esme*	*emse*

Thus, there are 12 arrangements (permutations) of the letters in the word *seem*; if two of the letters had not been alike, there would have been $4! = 24$ ways. When objects, such as two *e*'s, are not distinguishable, the number of permutations can be found using the following rule.

Number of Permutations with Indistinguishable Objects

If there is a set of n objects to be arranged when there are n_1 of one indistinguishable type, n_2 of a second indistinguishable type, continuing until there are n_k of the kth indistinguishable type, then the number of possible permutations of the n objects is

$$\frac{n!}{n_1!\, n_2! \cdots n_k!}$$

where $n_1 + n_2 + \cdots + n_k = n$.

Using this rule with the word *seem*, the number of arrangements is

$$\frac{4!}{2!\,1!\,1!} = 12$$

which is the total we found previously.

EXAMPLE 26   How many permutations are there for the letters in the word *Louisiana*?

SOLUTION   Using the rule above we have

$$\frac{9!}{2!\,2!\,1!\,1!\,1!\,1!\,1!} = 90{,}720$$

■

Practice Problem 3 Find the number of arrangements of the letters in the word *Mississippi*.

ANSWER $\dfrac{11!}{4!\,4!\,2!\,1!} = 34{,}650$

Let's turn our attention now to the other counting technique that we use in this section, **combinations**. This is a method of selecting subsets of a set in which order is not a consideration. For example, let's look at the set of letters {A, B, C, D}. Suppose that we want to select two letters at a time, but the order is not important. Therefore, the arrangements AB and BA are considered equivalent. So, from the given set we have

{AB, AC, AD, BC, BD, CD}

The notation for combinations that we use is $C(n, r)$, which is the number of combinations of n objects taken r at a time. This notation varies in books and on calculators. Some books use the notation $_nC_r$ or $\binom{n}{r}$. In Example 27, we develop the rule for calculating these values.

EXAMPLE 27   An economics club is to elect four class officers from six club members who qualify. How many sets of officers are possible?

SOLUTION   The key word here is *sets*. Since the arrangement of elements in a set does not matter in defining the set, we know that we are to use combinations in this problem. The answer is $C(6, 4)$, the number of combinations of 6 things taken 4 at a time. Since order is not important, {Gerry, Phoebe, Karen, Michelle} is the same as {Karen, Michelle, Gerry, Phoebe}. By writing down the different sets of 4 officers from 6 prospects, let's say {G, P, K, M, J, L}, we get the following possibilities.

{G, P, K, M}   {G, P, K, J}   {G, P, K, L}   {G, P, M, J}   {G, P, M, L}
{G, P, J, L}   {G, K, M, J}   {G, K, M, L}   {G, K, J, L}   {G, M, J, L}
{P, K, M, J}   {P, K, M, L}   {P, K, J, L}   {P, M, J, L}   {K, M, J, L}

We see that $C(6, 4) = 15$.

Now let's obtain the answer another way. Take each combination of officers and arrange the 4 officers as president, vice president, secretary, and treasurer. There would be $P(4, 4)$ arrangements for each combination. So, if we multiply this result by the number of selections of 4 officers, $C(6, 4)$, that number should be the same as the number of permutations of 6 things taken 4 at a time, or

$$P(6, 4) = C(6, 4) \cdot P(4, 4)$$

Thus,

$$
\begin{aligned}
C(6, 4) &= \frac{P(6, 4)}{P(4, 4)} = \frac{\dfrac{6!}{2!}}{4!} \\
&= \frac{6!}{2! \, 4!} \\
&= \frac{6 \cdot 5 \cdot 4 \cdot 3 \cdot 2 \cdot 1}{(2 \cdot 1)(4 \cdot 3 \cdot 2 \cdot 1)} \\
&= 15
\end{aligned}
$$

In general, by reasoning as we did in Example 27, the number of combinations of n objects taken r at a time, $C(n, r)$, relates to permutations as

$$P(n, r) = C(n, r)P(r, r)$$

So we can now compute

$$
\begin{aligned}
C(n, r) &= \frac{P(n, r)}{P(r, r)} \\
&= \frac{n!}{(n - r)! \, r!}
\end{aligned}
$$

COMBINATION OF n OBJECTS TAKEN r AT A TIME

The number of ways of selecting r objects from n objects without regard to order is the number of **combinations of n objects taken r at a time** and is

$$C(n, r) = \frac{n!}{r!(n - r)!} \qquad (0 \leq r \leq n)$$

Calculator Note

Find $_6C_4$ with your calculator. The keystrokes are 6 $\boxed{\text{MATH}}$ $\boxed{\blacktriangleleft}$ (select 3) 4 $\boxed{\text{ENTER}}$. Your calculator should give the result 15.

We now look at a number of counting problems that involve the use of combinations and how combinations can be used in probability problems. In some of these examples, we will see that a counting problem may involve more than one combination in conjunction with the fundamental principle of counting.

EXAMPLE 28   From a group of 20 employees a delegation of 3 members is to be chosen. How many ways can this be done?

SOLUTION   Since order is not important here, the total number of 3 member delegations is $C(20, 3) = 1140$. ■

EXAMPLE 29

(a)   Find the number of 5-card hands that can be drawn from an ordinary deck of cards.
(b)   Find the number of 5-card hands that contain all hearts.
(c)   What is the probability that a hand will contain all hearts?

SOLUTION

(a)   There are 52 cards in a standard deck. We will be selecting 5 cards from the deck to form a hand, so order is not important. Therefore, the number of 5 card hands is a combination of 52 objects taken 5 at a time.

$$C(52, 5) = \frac{52!}{(52 - 5)!\, 5!} = 2{,}598{,}960$$

(b)   There are 13 hearts in a standard deck. The number of ways that 5 hearts can be selected is $C(13, 5) = 1287$.
(c)   Going back to our definition of probability, we have

$$P(\text{all hearts in a 5-card hand}) = \frac{\text{Number of 5-card hands with all hearts}}{\text{Total number of 5-card hands}}$$

$$= \frac{1287}{2{,}598{,}960}$$

$$\approx .0005$$ ■

EXAMPLE 30   On a test, a student must answer any 7 of the first 10 questions, and any 5 of the next 8 questions. In how many ways can this done?

SOLUTION   The first 7 questions answered must come from the first 10 questions but the order in which the questions are answered is not important. Thus, there are $C(10, 7)$ ways of selecting the first set. From the next group of 8 questions, the student must choose 5 to answer, and again the order is not important, so we determine that there are $C(8, 5)$ ways of selecting this set. We can think of this problem as a two-stage problem—selecting the first set of questions and then selecting the next set of questions. Since there are $C(10, 7)$ ways of making the first decision and $C(8, 5)$ ways of making the second decision, by the fundamental principle of counting, there are $C(10, 7) \cdot C(8, 5) = 6720$ ways of making both decisions. ■

EXAMPLE 31   How many ways can 11 different jobs be distributed among four workmen if the first workman gets 3, the second 4, and the last two get 2 each?

SOLUTION   The order of the job selection by each workman does not matter. But after each selection, the number of jobs is reduced. The first workman has $C(11, 3)$ ways of selecting his jobs. Since there are only 8 jobs left, the second workman has $C(8, 4)$ ways of selecting his jobs. Now there are only 4 jobs left, so the third workman has $C(4, 2)$ ways to select his. The last workman has only $C(2, 2)$ ways to select his jobs. Now, following the reasoning in Example 30, we can now use the fundamental principle of counting to determine the total number of ways the selections can be made:

$$C(11, 3) \cdot C(8, 4) \cdot C(4, 2) \cdot C(2, 2) = 69{,}300$$ ■

Practice Problem 4   There are 15 patients in a ward to which 5 nurses are assigned. Each nurse is to be responsible for 3 patients. In how many different ways can the patients be distributed among the nurses?

ANSWER $C(15, 3) \cdot C(12, 3) \cdot C(9, 3) \cdot C(6, 3) \cdot C(3, 3) = 168{,}168{,}000$

NOTE: When calculating combinations, you may have noticed the following:

$$C(n, 0) = 1, \quad C(n, n) = 1, \quad \text{and} \quad C(n, 1) = n$$

EXAMPLE 32   A box contains 4 red balls and 6 blue balls. Two balls are drawn at random. Find each probability.
(a) $P(2 \text{ red balls})$
(b) $P(2 \text{ blue balls})$
(c) $P(1 \text{ red ball and 1 blue ball})$

SOLUTION
(a)   To determine the probability, we need two numbers: the number of ways that 2 red balls can be drawn and the total number of ways that 2 balls can be drawn. Since there are 4 red balls, there are $C(4, 2)$ ways of select-

ing 2 red balls. There are 10 balls in all, so there are $C(10, 2)$ total ways of selecting 2 balls from the 10. Consequently,

$$P(2 \text{ red balls}) = \frac{C(4, 2)}{C(10, 2)} = \frac{6}{45} = \frac{2}{15}$$

(b) To find the probability of two blue balls, we use the same strategy as in part (a):

$$P(2 \text{ blue balls}) = \frac{C(6, 2)}{C(10, 2)} = \frac{15}{45} = \frac{1}{3}$$

(c) We must find the number of ways to select 1 ball of each color and then determine the probability. This problem requires that we use the fundamental principle of counting.

$$P(1 \text{ red ball and 1 blue ball}) = \frac{C(4, 1) \cdot C(6, 1)}{C(10, 2)} = \frac{24}{45} = \frac{8}{15} \quad \blacksquare$$

Practice Problem 5 A quality control inspector at a factory randomly selects 5 bulbs from each lot of 100 bulbs that is produced and inspects them for defects. If a lot has 96 good bulbs and 4 defective bulbs, what is the probability that the inspector will find 3 good bulbs and 2 defective bulbs in her sample?

ANSWER

$$P(3 \text{ good and 2 defective bulbs}) = \frac{C(96, 3) \cdot C(4, 2)}{C(100, 5)} = \frac{857,280}{75,287,520} \approx .01$$

EXAMPLE 33 A 5-card hand is drawn from a standard deck of cards. Find each probability.

(a) $P(\text{exactly 3 aces})$
(b) $P(\text{exactly 2 clubs})$
(c) $P(\text{all cards are the same suit})$

SOLUTION

(a) In finding the probability of exactly 3 aces, we first determine that the other two cards are not aces. There are 4 aces in each deck and $C(4, 3)$ ways of selecting 3 of them. There are 48 cards that are not aces and the remaining 2 cards must come from those. Therefore, there are $C(48, 2)$ ways of completing the hand. We can now find the probability as follows:

$$P(\text{exactly 3 aces}) = P(\text{exactly 3 aces and 2 other cards})$$
$$= \frac{C(4, 3) \cdot C(48, 2)}{C(52, 5)}$$
$$= \frac{4 \cdot 1128}{2,598,960} \approx .00174$$

(b)   There are 13 clubs in a deck and $C(13, 2)$ ways of selecting 2 clubs. Also, there are 39 cards that are not clubs and $C(39, 3)$ ways of selecting 3 of them. So, we having the following calculation:

$$P(\text{exactly 2 clubs}) = P(\text{exactly 2 clubs and 3 nonclubs})$$
$$= \frac{C(13, 2) \cdot C(39, 3)}{C(52, 5)}$$
$$= \frac{78 \cdot 9139}{2{,}598{,}960} \approx 0.2743$$

(c)   There are 4 suits in a deck of cards, so there are $C(4, 1) = 4$ ways of selecting one suit. There are 13 cards in each suit and there are $C(13, 5) = 1287$ ways of selecting 5 cards from each suit. Therefore,

$$P(\text{all 5 cards in same suit}) = \frac{C(4, 1) \cdot C(13, 5)}{C(52, 5)} = \frac{5148}{2{,}598{,}960} \approx .002.$$

∎

EXAMPLE 34   A total of 14 people—8 men and 6 women—have applied for admission to business school. The school has five openings.

(a)   How many ways are there of selecting the students to be admitted if there are no restrictions based on gender?

(b)   How many ways are there of selecting the students to be admitted if exactly 3 of them must be men?

(c)   How many ways are there of selecting the students to be admitted if at least 3 of them must be women?

(d)   How many ways are there of selecting the students, if at least one of them must be a man?

SOLUTION

(a)   If there are no restrictions on the admissions, then the total selection comes from the entire group and there are $C(14, 5) = 2002$ ways of selecting the students.

(b)   If exactly 3 of the students must be men, then exactly 2 of the students must be women and there are $C(8, 3) \cdot C(6, 2) = 840$ ways of selecting the students.

(c)   If at least 3 of the students must be women, there is more than one possibility for the number of women selected and hence the number of men. If at least 3 are women, we have the following possibilities: 3 women and 2 men, 4 women and 1 man, 5 women and no men. We count the ways for each possibility and then add them together.

$$C(6, 3) \cdot C(8, 2) + C(6, 4) \cdot C(8, 1) + C(6, 5) \cdot C(8, 0) = 686$$

(d) This will be the total number of possible ways minus the number that have no men: $C(14, 5) - C(8, 0) \cdot C(6, 5) = 1996$. ■

NOTE: In this section, we learned two very useful counting techniques—permutations and combinations. These are used for counting as well as in probability problems. The problem that most students have is in determining whether to use a permutation or a combination. When you are trying to decide which technique to use, ask yourself whether a different order will result in a different outcome. In card problems, unless the order of selection is specifically mentioned, the order of cards in your hand will not be important. Five cards of one suit will yield the same result in any order. However, when we are dealing with different positions of objects, then the order is important and a permutation is used. You may wish to think of these two situations. (1) A teacher comes into the class and selects 5 students, all of whom will get an A. (2) A teacher comes into the class and selects 5 students; the first will get an A, the second will get a B, and so on. It is easy to see that the order of the selection in situation (1) does not matter, but the order in situation (2) would definitely matter to the students!

SUMMARY

Permutation: An ordered arrangement of r objects selected from a set of n distinct objects is

$$P(n, r) = \frac{n!}{(n - r)!} \qquad (1 \le r \le n)$$

Combination: The number of ways of selecting r objects from n objects without regard to order is the number of combinations of n objects taken r at a time and is

$$C(n, r) = \frac{n!}{r!(n - r)!} \qquad (0 \le r \le n)$$

Exercise Set 5.3

1. Consider set $S = \{W, X, Y, Z\}$.
   (a) How many permutations of two objects can be chosen from this set? List them.
   (b) How many combinations of two objects can be chosen from this set? List them.

2. Write in factorial notation.
   (a) $P(5, 3)$ (b) $C(4, 2)$
   (c) $C(44, 22)$ (d) $C(8, 8)$

   (e) $P(6, 6)$ (f) $P(7, 5)$
   (g) $C(7, 5)$ (h) $C(100, 89)$
   (i) $C(100, 11)$

3. Evaluate each of the following.
   (a) $C(10, 6)$ (b) $C(10, 0)$
   (c) $C(15, 1)$ (d) $C(4, 2)$
   (e) $C(r, 2)$ (f) $C(r, r - 1)$

4. Evaluate each of the following.
   (a) $P(19, 3)$ (b) $P(6, 5)$ (c) $P(8, 1)$
   (d) $P(9, 2)$ (e) $P(7, 2)$ (f) $P(8, 7)$

5. In how many ways can 5 speakers be arranged on a program?

6. A coach of a football team must choose a starting quarterback and a backup quarterback from a group of 8 aspiring superstars. In how many ways can the choice be made?

7. In how many ways can 7 students line up outside Professor Smith's door to complain about grades?

8. A company president is asked to appoint a committee consisting of 5 men and 3 women. A list of 12 men and 10 women is provided from which to make the appointment. How many different committees can be selected?

9. A 7-card hand is dealt from a standard deck.
   (a) How many hands are possible?
   (b) How many hands have exactly 5 hearts?
   (c) How many hands have all cards in the same suit?
   (d) How many hands have at most 2 aces?
   (e) How many have 4 aces and 3 kings?

10. A special committee of 3 persons must be selected from a 12-person board of directors. In how many ways can the committee be selected?

11. Simplify.
   (a) $P(r, 1)$ (b) $P(k, 2)$ (c) $C(n, 1)$
   (d) $C(n, 0)$ (e) $P(r, r)$ (f) $C(n, n)$

12. The license plates for a certain state display 3 letters followed by 3 numbers.
   (a) How many different license plates can be manufactured if no repetitions are allowed?
   (b) If repetitions are allowed?
   (c) If no repetitions and zero is not allowed?
   (d) If no repetitions and X, Y, Z are not allowed?

13. Employee ID numbers at a large factory consist of four-digit numbers such as 0133, 4499, and 0000.
   (a) How many badges are possible if repetition of digits is allowed?
   (b) How many badges are possible in which all four digits are different?
   (c) How many badges are possible if the first digit cannot be a zero? (Repetitions are allowed.)
   (d) How many badges are possible if the last number must be even? (Repetitions are allowed.)
   (e) What is the probability that a badge selected at random contains all four different digits? (Repetitions are allowed.)

14. A bowl contains 8 red marbles and 14 black marbles. Three marbles are selected at random from the bowl without replacement. Find each probability.
   (a) P(all 3 are black)
   (b) P(1 is red and 2 are black)
   (c) P(at least 2 are black)
   (d) P(at least 1 is red)

15. A hat contains 20 slips of paper numbered 1 to 20. If 3 are drawn without replacement, what is the probability that all are numbered less than 10?

16. A Social Security number has 9 digits. If a Social Security number is chosen at random, what is the probability that all the digits would be the same?

17. A seven-card hand is drawn from a standard deck of 52 cards. Find the probability of each hand.
   (a) Seven spades
   (b) Five clubs and two hearts
   (c) Four clubs, one spade, two hearts
   (d) Three clubs, two hearts, two diamonds

18. How many ways can 3 people be selected from a group of 5 (a) if order matters and (b) if order does not matter?

19. A disc jockey has 10 rock records and 20 jazz records. He has time to play 7 records on his program.
   (a) How many different programs can he broadcast?
   (b) How many different programs can he broadcast if he is to alternate jazz and rock records, beginning with jazz?

20. A thunderstorm has caused a total of 25 power outages, 11 to residential areas and 14 to commercial areas. An electrician has time to repair only 8 of them on her shift.
   (a) How many ways can she select the ones to be repaired?
   (b) How many ways can be chosen if she has to repair 5 outages in residential areas and 3 in commercial areas?

For Exercises 21–24, any combination of letters is considered a word.

21. How many distinct 6-letter words can be formed from the letters of the word *fiasco*?

22. How many distinct 6-letter words can be formed from the letters of the word *bubble*?

23. How many distinct 7-letter words can be formed from the letters of the word *winning*?

24. How many distinct 7-letter words can be formed from the letters of the word *Bahamas*?

25. How many ways can 3 people be assigned to 5 offices if
    (a) there are no restrictions?
    (b) two particular people must be adjacent?
    (c) two particular people must not be adjacent?

26. How many ways can 4 men and 3 women be seated in a row of 7 chairs if
    (a) there are no restrictions?
    (b) men and women must alternate?
    (c) there must be a man at each end?
    (d) two particular people must sit next to each other?
    (e) two particular people must not sit next to each other?
    (f) two particular people must be seated at each end?

27. How many ways can 4 contracts be awarded to 7 different firms (a) if no firm can have more than one contract and (b) if there are no restrictions?

28. How many ways can 4 contracts be awarded to 7 different firms if 2 particular contracts must be awarded to the same firm and
    (a) that firm can have no other contracts?
    (b) there are no further restrictions?

29. A coin is tossed 6 times. Find the number of ways it will land tails
    (a) exactly 1 time.   (b) exactly 3 times.
    (c) at most 1 time.   (d) at least 1 time.

30. Find the probability for each situation in Exercise 29.

31. **Exam.** An urn contains 10 red, 20 white, and 30 blue balls. If 6 balls are drawn at random without replacement, what is the probability that 1 ball is red, 2 are white, and 3 are blue?
    (a) $\dfrac{5}{36}$ (b) $\dfrac{100}{649}$ (c) $\dfrac{10}{59}$
    (d) $\dfrac{5}{18}$ (e) $\dfrac{549}{649}$

32. **Exam.** A box contains 12 varieties of candy and exactly two pieces of each variety. If 12 pieces of candy are selected at random, what is the probability that a given variety is represented?
    (a) $\dfrac{12}{12!}$ (b) $\dfrac{2}{24!}$ (c) $\dfrac{2}{C(24,\,12)}$
    (d) $\dfrac{11}{46}$ (e) $\dfrac{35}{46}$

33. **Exam.** A random sample of 6 balls is selected with replacement from an urn that contains 10 red, 5 white, and 5 blue balls. What is the probability that the sample contains 2 balls of each color?
    (a) $\dfrac{5}{1024}$ (b) $\dfrac{1}{646}$ (c) $\dfrac{45}{512}$
    (d) $\dfrac{75}{646}$ (e) $\dfrac{45}{64}$

34. **Exam.** In a particular softball league each team consists of 5 women and 5 men. In determining a batting order for the 10 players, a woman must bat first and successive batters must be of opposite sex. How many different batting orders are possible for a team?
    (a) 5!   (b) $2 \cdot 5!$ (c) $(5!)^2$
    (d) $\dfrac{10!}{2}$ (e) 10!

Applications (Business and Economics)

35. **Sales.** A sales force is made up of 4 people. There are 12 prospects on file, but each prospect needs a great deal of individual attention. In how many ways can the 12 prospects be considered by the 4 salespeople without regard to order if each salesperson takes 3 clients.

36. **Purchasing.** A firm buys material from three local companies and five out-of-state companies. Four orders are submitted at one time. In how many ways can two orders be submitted to two local firms and two orders to out-of-state firms?

37. **Quality Control.** In a quality control check at a tire company, 3 tires are randomly selected and inspected from each lot of 20 tires produced. Suppose that a lot contains 4 defective tires and 16 good tires.
    (a) How many ways can 3 tires be selected from 20?
    (b) How many different selections of 3 tires from 20 will contain exactly 1 defective tire?
    (c) What is the probability that exactly 1 of the 3 tires selected will be defective?

(d) What is the probability that no defective tires will be among the 3 selected?

38. **Management**.   A management company has 15 clients that need to be seen. In how many ways can the clients be seen by 5 salesmen if each salesman is to get 3 clients?

39. **Organizational Management**.   Among the 30 employees in a corporation, there are 20 women and 10 men. Of these, 18 are pollsters, 6 are sales employees, and 6 are management employees. Three employees are chosen to form a committee. In how many ways can a committee be selected to have
    (a) two women and 1 man?
    (b) 2 pollsters and 1 salesperson?
    (c) 1 salesperson, 1 pollster, and 1 management employee?
    (d) at least 1 man?

40. **Quality Control**.   From a shipment of 60 control boards, 5 of which are defective, a sample of 4 boards is selected at random.
    (a) How many samples are possible?
    (b) How many of the samples will not have any defective boards?
    (c) How many samples will contain at least 1 defective board?

Applications (Social and Life Sciences)

41. **Dispensing Drugs**.   The order of administering five different drugs is important in an experiment for a cure of a disease.
    (a) In how many ways can all 5 drugs be administered?
    (b) In how many ways can 3 of the 5 drugs be administered?

42. **Experimental Design**.   Ten rats are selected for an experiment. Each trial run is to involve 3 rats at a time. How many trial runs can be performed using different groups of three?

43. **Medicine**.   Assume there are 8 classifications of blood types. In samples of 5, how many possible samples exist so that all classifications are different and no two samples have exactly the same 5 classifications?

Review Exercises

44. At a charity bazaar, one booth sells 60 identically wrapped surprise packages, where 8 of the packages contain tape players, 12 of the packages contain electric razors, 16 of the packages contain rubber galoshes, and 24 contain boxes of tissue.
    (a) What is the probability of getting a tape player or an electric razor?
    (b) What is the probability of getting something other than a box of tissues?

45. A box contains 6 balls, numbered 1 to 6. Record a sample space for the following experiments.
    (a) A ball is drawn and the number is recorded. The ball is returned and a second ball is drawn.
    (b) A ball is drawn and the number is recorded. The ball is not replaced, but a second ball is drawn.

46. A deck of cards is shuffled and three cards are drawn one after the other without replacement. Determine how many ways you can draw
    (a) three red cards.
    (b) a red card, then a black card, and then a red card.

5.4   PROBABILITY FOR THE UNION AND INTERSECTION OF EVENTS

OVERVIEW   Consider the following problems. A card is drawn from a standard deck of cards. What is the probability that it is either an ace or a spade? As a car part comes off a production assembly line, it is tested for two defects. It is known that 4% of the parts fail test 1, 3.5% fail test 2, and 3% fail both tests. A quality control manager would like to know what percentage of parts fail only test 1 or only test 2. In both of the preceding examples, the first event has parts in common with the second event. In the first example, the ace of spades is common to both events. In the second example the parts that fail both tests are common to both events. The new notations introduced in this section help us find solutions to problems like these. In this section we study

- Intersection and union of sets
- Probability of A or B
- Probability of A and B
- Probability of A'
- Mutually exclusive events
- Odds

As an introduction to the study of the union and intersection of events, let's review the special notations that are used for discussing the relationship among members of two or more sets.

DEFINITION: INTERSECTION AND UNION OF SETS

1. The **intersection** of two sets A and B (denoted by $A \cap B$) is the set of all elements common to both sets A and B. Therefore,

$$A \cap B = \{x \mid x \in A \text{ and } x \in B\}$$

2. If A and B are any two sets, the **union** of A and B (denoted by $A \cup B$) is the set consisting of all the elements in set A or set B or in both A and B. Therefore,

$$A \cup B = \{x \mid x \in A \text{ or } x \in B\}.$$

The shaded regions in Figure 10 compare intersection and union under different situations that can occur for sets A and B. (Recall that U is the universal set.) Note in Figure 10(a) that A and B overlap—they have elements in common. In Figure 10(b), A is part of B (A is a subset of B). In Figure 10(c), A and B have no elements in common ($A \cap B = \emptyset$).

EXAMPLE 35   A cable company offers two packages of premium channels. Package A consists of channels 20, 21, 22, 30, 31, and 32 and package B consists of 21, 22, 25, 26, 30, and 31. Which channels are offered in both packages and which are offered in either package?

SOLUTION   The channels offered in both packages are those in the intersection: A ∩ B = {21, 22, 30, 31}. The channels offered in either package are those in the union: A ∪ B = {20, 21, 22, 25, 26, 30, 31, 32}.   ■

Practice Problem 1 If $A = \{1, 2, 3, 4\}$ and $B = \{4, 6, 7\}$, find $A \cup B$ and $A \cap B$.

ANSWER $A \cup B = \{1, 2, 3, 4, 6, 7\}$ and $A \cap B = \{4\}$

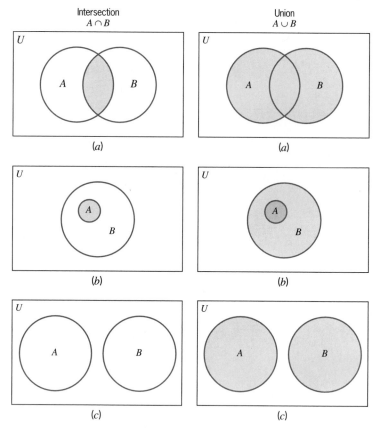

Intersection
$A \cap B$

Union
$A \cup B$

Figure 10

Recall that $A \cup B$ is the set of all elements that belong to A or B or to both A and B. If there are elements in common to both sets, they are listed only once in the union.

We are now ready to consider some additional configurations of events. Three of these relationships are of such importance that we list them as special events.

DEFINITION: AND, OR, COMPLEMENT

S is a sample space of an experiment.

1. The event $A \cap B$ is the set of all outcomes *in both event A and in event B.*
2. The event $A \cup B$ is the set of all outcomes that are *in event A or in event B* or in both events A and B.
3. The **complement** of an event A, denoted A', is the set of all outcomes that are *in S and are not in A.*

We now illustrate these concepts with examples involving the roll of a fair die.

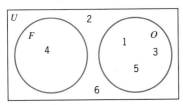

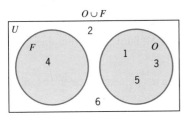

Figure 11

EXAMPLE 36   In rolling a fair die, what is the probability of rolling an odd number or a 4?

SOLUTION   We let O represent the event that an odd number is rolled and F represent the event that a 4 is rolled. These sets are shown in Figure 11. From the figure, we see that $S = \{1, 2, 3, 4, 5, 6\}$, $O = \{1, 3, 5\}$, $F = \{4\}$, and $O \cup F = \{1, 3, 4, 5\}$. Therefore,

$$P(O \cup F) = \frac{n(O \cup F)}{n(S)} = \frac{4}{6} = \frac{2}{3}$$

Note in Figure 11 that

$$P(O) = \frac{3}{6} \quad \text{and} \quad P(F) = \frac{1}{6}$$

Consequently,

$$P(O \cup F) = P(O) + P(F) = \frac{3}{6} + \frac{1}{6} = \frac{4}{6} = \frac{2}{3}$$

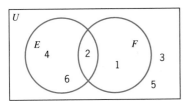

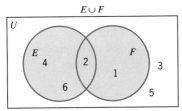

Figure 12

EXAMPLE 37   In the rolling of the same fair die, what is the probability of rolling an even number or a number that is less than 3?

SOLUTION   Let E represent the event that an even number is rolled and F represent a number less than 3 being rolled. We seek $P(E \cup F)$. Figure 12 shows the sets, and we see that $E = \{2, 4, 6\}$, $F = \{1, 2\}$, and $E \cup F = \{1, 2, 4, 6\}$. Therefore,

$$P(E \cup F) = \frac{n(E \cup F)}{n(S)} = \frac{4}{6} = \frac{2}{3}$$

We can also see in Figure 12 that

$$P(E) = \frac{3}{6} \quad \text{and} \quad P(F) = \frac{2}{6}$$

However,

$$P(E \cup F) \neq P(E) + P(F) \quad \text{because} \quad \frac{4}{6} \neq \frac{3}{6} + \frac{2}{6}$$

In Example 36, we were able to add the probability of each event and get the correct answer, but in Example 37, this did not result in the correct answer. Do you see the difference between the two problems? Events F and O have no sample points in common, but events E and F have a common element. This discussion suggests the following definition and property of probability.

MUTUALLY EXCLUSIVE EVENTS

1. Events A and B are **mutually exclusive** if they have no outcomes in common; that is, $A \cap B = \varnothing$.
2. If events A and B are mutually exclusive, then

$$P(A \cup B) = P(A) + P(B)$$

Practice Problem 2 From a standard deck of 52 cards, we draw one card. What is the probability of getting a spade or a red card?

ANSWER We will let SP be the event of drawing a spade and R be the event of drawing a red card. Since the events are mutually exclusive, we have

$$P(SP \cup R) = P(SP) + P(R) = \frac{13}{52} + \frac{26}{52} = \frac{3}{4}$$

Now let's return to Example 37 where we discovered that $P(E \cup F) \neq P(E) + P(F)$. In Figure 12, we can see that outcome 2 is in both event E and event F, so the events are not mutually exclusive. If we just add $P(E) + P(F)$, then outcome 2 is counted twice. Since $E \cap F$ is included twice in $P(E) + P(F)$, we subtract one of these and note that

$$P(E \cup F) = P(E) + P(F) - P(E \cap F)$$

$$\frac{4}{6} = \frac{3}{6} + \frac{2}{6} - \frac{1}{6}$$

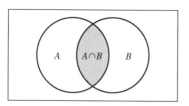

Figure 13

We can generalize this concept by realizing that in set theory the number of outcomes in event A or in event B is the number in A plus the number in B less the number in $A \cap B$, which has been counted in both A and B (see Figure 13). Thus,

$$n(A \cup B) = n(A) + n(B) - n(A \cap B)$$

Divide both sides of the equation by N, the number of elements in a sample space, to obtain

$$\frac{n(A \cup B)}{N} = \frac{n(A)}{N} + \frac{n(B)}{N} - \frac{n(A \cap B)}{N}$$

from which we obtain the inclusion–exclusion principle,

$$P(A \cup B) = P(A) + P(B) - P(A \cap B)$$

PROBABILITY OF *A* OR *B*

For any two events *A* and *B*, the **probability of *A* or *B*** is given by

$$P(A \cup B) = P(A) + P(B) - P(A \cap B)$$

EXAMPLE 38   In a manufacturing plant, parts are tested as they come off the assembly line. It is found that 12% fail the stress test (*S*), 8% fail a heat test (*H*), and 4% fail both tests. What percent failed the stress test or the heat test?

SOLUTION   The events are not mutually exclusive. We are given the probabilities

$$P(S) = .12, \quad P(H) = .08, \quad \text{and} \quad P(S \cap H) = .04$$

Therefore,

$$P(S \cup H) = P(S) + P(H) - P(S \cap H) = .12 + .08 - 04 = .16$$

This relationship is shown in Figure 14.   ■

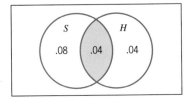

Figure 14

Practice Problem 3   In drawing a card from 8 plain cards numbered 1 to 8, what is the probability of getting an even number or a number less than 5?

ANSWER $P(\text{even number}) = \frac{1}{2}$, $P(\text{number less than 5}) = \frac{1}{2}$, and $P(\text{even number and a number less than 5}) = \frac{1}{4}$. Therefore, $P(\text{even or less than 5}) = \frac{1}{2} + \frac{1}{2} - \frac{1}{4} = \frac{3}{4}$.

The preceding discussion can be extended to three events, *A*, *B*, and *C*.

PROBABILITY OF *A* OR *B* OR *C*

For any three events *A*, *B*, and *C*, the **probability of *A* or *B* or *C*** is given by

$$P(A \cup B \cup C) = P(A) + P(B) + P(C) - P(A \cap B) - P(A \cap C)$$
$$- P(B \cap C) + P(A \cap B \cap C)$$

Venn diagrams are particularly useful when looking at the relationship among three groups. This is illustrated in Example 39.

EXAMPLE 39   A survey of 100 rivers shows that 55 are polluted with chemicals (*C*), 45 are polluted with pesticides (*P*), 25 are polluted with unrefined ore (*O*), 12 are polluted with chemicals and pesticides (*C* ∩ *P*), 10 with chemicals and ore (*C* ∩ *O*), 8 with pesticides and ore (*P* ∩ *O*), and 5 are polluted with all three (*C* ∩ *P* ∩ *O*). Each river has at least one pollutant. Use a Venn diagram to display the relationship among the three pollutants.

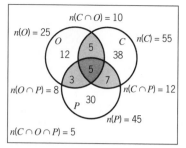

Figure 15

SOLUTION   The Venn diagram is shown in Figure 15. When constructing a Venn diagram of three groups of rivers, it is helpful to start with the intersection of all three groups if possible. Therefore, we put a 5 in the area that is the intersection of all three. Next, we work through all the intersections of 2 groups. It must be remembered, however, that the 5 rivers that are polluted with all three are also in the intersections of 2 groups. Those 5 must be counted. Consequently, when we are told that there are 8 rivers polluted with pesticides and ore, we deduct the 5 and find that there are 3 rivers polluted with only pesticides and ore. Likewise, since 10 rivers are polluted with chemicals and ore, we find that 5 are polluted with only chemicals and ore. In like manner, we find the remaining numbers. Notice that all the numbers total 100.

The number of pollutants and the associated probability is summarized as follows:

$$n(C) = 55 \qquad\qquad P(C) = \tfrac{55}{100}$$

$$n(P) = 45 \qquad\qquad P(P) = \tfrac{45}{100}$$

$$n(O) = 25 \qquad\qquad P(O) = \tfrac{25}{100}$$

$$n(C \cap P) = 12 \qquad\qquad P(C \cap P) = \tfrac{12}{100}$$

$$n(C \cap O) = 10 \qquad\qquad P(C \cap O) = \tfrac{10}{100}$$

$$n(O \cap P) = 8 \qquad\qquad P(O \cap P) = \tfrac{8}{100}$$

$$n(C \cap P \cap O) = 5 \qquad\qquad P(C \cap P \cap O) = \tfrac{5}{100}$$

$$P(C \cup P \cup O) = P(C) + P(O) + P(P) - P(C \cap P) - P(C \cap O) \\ - P(O \cap P) + P(C \cap P \cap O)$$

$$= \frac{55}{100} + \frac{25}{100} + \frac{45}{100} - \frac{12}{100} - \frac{10}{100} - \frac{8}{100} + \frac{5}{100}$$

$$= 1 \qquad\blacksquare$$

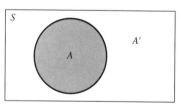

Figure 16

In performing an experiment, an event will either occur or it will not occur. We now look at the probability that an event does not occur. Earlier in this section we defined the complement of event A to be all the outcomes in S that are not in A. For example, if $S = \{1, 2, 3, 4, 5, 6\}$ and $A = \{6\}$, then $A' = \{1, 2, 3, 4, 5\}$. We can divide all of the events in a sample space into two mutually exclusive sets, A and A', as shown in Figure 16.

Note that

$$A \cap A' = \varnothing \quad \text{and} \quad A \cup A' = S$$

Thus,

$$P(A \cup A') = P(A) + P(A')$$

$$1 = P(A) + P(A')$$

$$P(A') = 1 - P(A) \quad \text{or}$$

$$P(A) = 1 - P(A')$$

PROBABILITY OF A COMPLEMENT

If A is any event in the sample space S, and if set A' denotes the complement of A, then

$$P(A') = 1 - P(A) \quad \text{or} \quad P(A) = 1 - P(A')$$

EXAMPLE 40   What is the probability of not getting an ace when drawing a card from a standard deck of cards?

SOLUTION $P(\text{no ace}) = 1 - P(\text{ace}) = 1 - \frac{4}{52} = \frac{48}{52} = \frac{12}{13}$ ∎

Practice Problem 4   Refer to Example 39 and Figure 15 to find each probability.

(a) P(river is polluted by ore or pesticides but not chemicals)
(b) P(river is polluted by exactly 1 of the pollutants)
(c) P(river is polluted by chemicals and ore but not pesticides)
(d) P(river is polluted by chemicals but not ore)

ANSWER

(a) $\dfrac{45}{100}$ (b) $\dfrac{80}{100}$ (c) $\dfrac{5}{100}$ (d) $\dfrac{45}{100}$

In set theory, there are two laws called DeMorgan's laws. These can be extended to probability.

DeMorgan's Laws

1. The law $A' \cap B' = (A \cup B)'$ extends to $P(A' \cap B') = P[(A \cup B)']$. (The probability of the intersection of the complements of two sets is equal to the probability of the complement of the set that is the union of the given two sets.)
2. The law $A' \cup B' = (A \cap B)'$ extends to $P(A' \cup B') = P[(A \cap B)']$. (The probability of the union of the complements of two sets is equal to the probability of the complement of the set that is the intersection of the given sets.)

EXAMPLE 41   Let A and B be two events that are mutually exclusive. Given that $P(A) = .4$ and $P(B) = .1$, find each probability.

(a) $P(A \cap B)$ (b) $P(A \cup B)$ (c) $P(A')$
(d) $P(A' \cup B')$ (e) $P(A' \cap B')$

SOLUTION

(a) Since the events are mutually exclusive, $P(A \cap B) = 0$.
(b) $P(A \cup B) = P(A) + P(B) - P(A \cap B) = .4 + .1 - 0 = .5$

(c) $P(A') = 1 - P(A) = 1 - .4 = .6$

(d) $P(A' \cup B') = P[(A \cap B)'] = 1 - P(A \cap B)$ By DeMorgan's law

$= 1 - 0 = 1$

(e) $P(A' \cap B') = P[(A \cup B)'] = 1 - P(A \cup B)$ By DeMorgan's law

$= 1 - .5 = .5$

Practice Problem 5   Given that A and B are two events from a sample space S and $P(A) = .2$, $P(B) = .1$, and $P(A \cap B) = .05$, find each probability.

(a) $P(A \cup B)$ (b) $P(A' \cap B')$

(c) $P(A' \cap B)$ (**Hint**: Use a Venn diagram.)

ANSWER

(a)   .25   (b)   .75   (c)   .05

We often hear the word *odds* given in association with an event. Sometimes people confuse odds with probability. Odds and probability are not the same. The probability of drawing a heart from a deck of cards is $\frac{1}{4}$, but the odds of drawing a heart are 1 to 3, because there is a ratio of 1 to 3 of hearts to nonhearts in a deck of cards. Probability can be used to compute odds. The odds in favor of an event or the odds against an event are actually comparisons of the probability of an event and the probability of its complement. We define odds as follows.

DEFINITION: ODDS

1.   The **odds in favor of event E** are found by calculating the ratio of the probability of event E to the probability of event E':

$$\frac{P(E)}{P(E')} \quad \text{or} \quad \frac{P(E)}{1 - P(E)}$$

2.   The **odds against event E** are found by calculating the ratio of the probability of event E' to the probability of event E:

$$\frac{P(E')}{P(E)} \quad \text{or} \quad \frac{1 - P(E)}{P(E)}$$

EXAMPLE 42   Find the probability of rolling a 6 and the odds in favor of rolling a 6 with a single die.

SOLUTION   We have $P(6) = \frac{1}{6}$ and $P(\text{not a } 6) = 1 - P(6) = \frac{5}{6}$. Therefore, we calculate

$$\text{Odds in favor of rolling a } 6 = \frac{P(6)}{P(\text{not a } 6)} = \frac{P(6)}{1 - P(6)} = \frac{\frac{1}{6}}{\frac{5}{6}} = \frac{1}{5}$$

The probability of a 6 is $\frac{1}{6}$, but the odds in favor of rolling a 6 are 1 to 5.

The probability of an event is a number between 0 and 1, inclusive. The odds in favor of an event or the odds against an event are given as a ratio using the probabilities, $P(E)$ and $P(E')$. Probabilities can be expressed as a fraction or as a decimal; however, odds are not given as a decimal.

Practice Problem 6 Ten identical balls numbered 1 to 10 are put in a sack and one ball is drawn.

(a) What is the probability of a 5 being drawn?
(b) What are the odds in favor of drawing a 5?
(c) What are the odds against drawing a 5?

ANSWER

(a) $P(5) = \frac{1}{10}$
(b) The odds in favor of drawing a 5 are 1 to 9 or $\frac{1}{9}$.
(c) The odds against drawing a 5 are 9 to 1 or $\frac{9}{1}$.

At times we are given the odds for an event, and from the odds we can obtain the probability that the event will occur.

PROBABILITY FROM ODDS

If the odds favoring an event E are m to n, then

$$P(E) = \frac{m}{m + n} \quad \text{and} \quad P(E') = \frac{n}{m + n}$$

EXAMPLE 43   The odds that it will rain today are 1 to 3. What is the probability that it will rain?

SOLUTION   For the given odds, m can be taken as 1 and n as 3. Thus,

$$P(R) = \frac{1}{1 + 3} = \frac{1}{4}$$

Practice Problem 7 What are the odds against two heads on two tosses of a coin?

ANSWER   The odds against two heads are 3 to 1.

SUMMARY

1. For events A and B:
    (a) $P(A \cup B) = P(A) + P(B) - P(A \cap B)$
    (b) $P(A \cup B) = P(A) + P(B)$ if A and B are mutually exclusive, i.e. $P(A \cap B) = 0$.
2. For events A, B, and C,

    $$P(A \cup B \cup C) = P(A) + P(B) + P(C) - P(A \cap B) - P(A \cap C)$$
    $$- P(B \cap C) + P(A \cap B \cap C)$$

3. If A' is the complement of A, then $P(A') = 1 - P(A)$.
4. The odds in favor of event E are

$$\frac{P(E)}{P(E')} \quad \text{or} \quad \frac{P(E)}{1 - P(E)}$$

5. The odds against event E are

$$\frac{P(E')}{P(E)} \quad \text{or} \quad \frac{1 - P(E)}{P(E)}$$

6. If the odds favoring event E are m to n, then

$$P(E) = \frac{m}{m + n} \quad \text{and} \quad P(E') = \frac{n}{m + n}$$

Exercise Set 5.4

1. If A and B are events with $P(A) = .6$, $P(B) = .3$, and $P(A \cap B) = .2$, find each of the following.
    (a) $P(A \cup B)$ (b) $P(A' \cap B)$
    (c) $P(A' \cap B')$

2. If A and B are events in a sample space such that $P(A) = .6$, $P(B) = .2$, and $P(A \cap B) = .1$, compute each of the following.
    (a) $P(A')$ (b) $P(B')$
    (c) $P(A \cup B)$ (d) $P(A' \cup B')$

3. If A and B are mutually exclusive events with $P(A) = .2$ and $P(B) = .5$, compute each of the following.
    (a) $P(A \cap B)$ (b) $P(A')$
    (c) $P(A \cup B)$ (d) $P(A' \cap B')$
    (e) $P(A' \cup B')$

4. If A and B are events with $P(A \cup B) = \frac{5}{8}$, $P(A \cap B) = \frac{1}{3}$, and $P(A') = \frac{1}{2}$, compute the following.
    (a) $P(A)$ (b) $P(B)$
    (c) $P(B')$ (d) $P(A' \cup B)$

5. At a college, 30% of the freshmen failed mathematics, 20% failed English, and 15% failed both mathematics and English. What is the probability that a freshman failed mathematics or English?

6. An experiment consists of tossing a coin 7 times. Describe in words the complement of each of the following.
    (a) Getting at least 2 heads
    (b) Getting exactly 3, or exactly 4, or exactly 5 heads
    (c) Getting exactly 1 tail
    (d) Getting no heads

7. A number x is selected at random from the set of numbers $\{1, 2, 3, \ldots, 8\}$. What is the probability that
    (a) x is less than 5?
    (b) x is even?
    (c) x is less than 5 and even?
    (d) x is less than 5 or is a 7?

8. A single card is drawn from a standard 52-card deck. What is the probability that the card is
    (a) either a heart or a club?
    (b) either a heart or a king?
    (c) not a jack?
    (d) either red or black?

9. From a bag containing 6 red balls, 4 black balls, and 3 green balls, 1 ball is drawn. What is the probability that the ball is
    (a) red or black?

(b) red or black or green?

(c) not black?

(d) not red or not black?

10. If the probability that it will rain tomorrow is .3, what are the odds that
    (a) it will rain tomorrow?
    (b) it will not rain tomorrow?

11. The probability of event A occurring is .8.
    (a) What are the odds in favor of A occurring?
    (b) What are the odds against A occurring?

12. The odds in favor of an event A are 10 to 7. What is the probability that A will occur?

13. The odds against an event A occurring are 3 to 2. Find $P(A')$.

14. ***Exam***. Let $P(A \cap B) = .2$, $P(A) = .6$, and $P(B) = .5$. Then $P(A' \cup B')$ equals which of the following?
    (a) 1   (b) .3   (c) .7
    (d) .8   (e) .9

15. ***Exam***. A card hand selected from a standard deck consists of two kings, a queen, a jack, and a ten. Three additional cards are selected at random, without replacement, from the remaining cards in the deck. What is the probability that the enlarged hand contains at least three kings?

    (a) $\dfrac{3}{1081}$ (b) $\dfrac{132}{1081}$ (c) $\dfrac{135}{1081}$

    (d) $\dfrac{264}{1081}$ (e) $\dfrac{267}{1081}$

Applications (Business and Economics)

16. ***Marketing Survey***. A recent survey found that 60% of the people in a given community drink a certain cola and 40% drink other soft drinks; 15% of the people interviewed indicated that they drink both the cola and other soft drinks. What percent of the people drink either the cola or other soft drinks?

17. ***Forecasting***.   In a survey of the presidents of leading banks by an economics consulting group, the following information was obtained relative to their forecast for the next year:

65% expect higher inflation

15% expect a recession

 5% expect both higher inflation and a recession

75% expect higher interest rates

50% expect higher inflation and higher interest rates

10% expect higher interest rates and a recession

 3% expect higher inflation, higher interest rates, and a recession

What is the probability that a bank president selected at random would forecast

(a) no recession or lower interest rates?

(b) no increase in inflation and no increase in interest rates?

(c) no recession, or no increase in interest rates, or no increase in inflation?

18. ***Reading Habits***.   A survey of 100 people in a library revealed the following:

40 read the *Wall Street Journal*

30 read *National Geographic*

25 read *Sports Illustrated*

15 read the *Wall Street Journal* and *National Geographic*

12 read the *Wall Street Journal* and *Sports Illustrated*

10 read *National Geographic* and *Sports Illustrated*

 4 read all three

If a person surveyed is selected at random, what is the probability that the person

(a) reads exactly two of the magazines?

(b) reads only one magazine?

(c) reads *Sports Illustrated* and *National Geographic* but not the *Wall Street Journal?*

(d) does not read any of the magazines?

19. ***Consumer Survey***. A survey was taken of 120 people concerning their preferences for toothpaste samples they were asked to try. The samples were labeled A, B, C. The results were as follows:

22 liked A

25 liked B

26 liked C

15 liked A and B

10 liked A and C

12 liked B and C

8 liked all three

If one of those surveyed is chosen at random, find the probability that the person

(a) liked at least one of the brands.

(b) liked A and B but not C.

(c) did not like any brand.

(d) liked only one brand.

20. *Advertising*. In Atlanta, 600,000 people read newspaper A, 450,000 read newspaper B, and 160,000 read both newspapers. How many read either newspaper A or newspaper B?

Applications (Social and Life Sciences)

21. *Prediction Relative to Children and Divorce*. In a survey, families were classified as C, children, and C', no children. At the same time, families were classified according to D, husband and wife divorced, and D', not divorced. Out of 200 families surveyed, the following results were obtained.

	C	C'	Total
D	60	20	80
D'	90	30	120
Total	150	50	200

(a) What is the probability that a family selected at random has children?

(b) What is the probability that in a family selected at random, the parents are not divorced?

(c) What is the probability that for a family selected at random, there are children or the parents are divorced?

(d) What is the probability that in a family selected at random the parents are not divorced or there are no children?

22. *Survey of Family Characteristics*. In a survey of 100 families of a school district in 1990, each family was asked the following questions: (1) Do you have children in public school? (2) Do you object to the modern approach of teaching mathematics? (3) Do you object to placing students in classes according to IQ tests? The yes answers to these questions were tabulated as follows, where A, B, and C represent questions (1), (2), (3), respectively:

$$n(A) = 70 \quad n(A \cap B) = 15$$

$$n(B) = 30 \quad n(A \cap C) = 6$$

$$n(C) = 10 \quad n(B \cap C) = 8 \quad n(A \cap B \cap C) = 5$$

(a) If a family is selected at random, what is the probability that the answer from this family was yes on either (1) or (2)?

(b) If a family is selected at random, what is the probability that it answered yes to exactly one of the three questions?

(c) If a family is selected at random, what is the probability that it did not object to modern mathematics or did not object to placing students in classes by IQ tests?

23. *Politics*. In a sample of 50 people, it was found that 28 planned to vote for the Democratic candidates for mayor and assistant to the mayor, 10 planned to vote for the Republican candidates for mayor and assistant to the mayor, and 5 planned to vote for a Republican mayor and a Democratic assistant to the mayor, and the remainder voted for Democratic mayor and Republican assistant to the mayor.

(a) What is the probability that a person plans to vote for a Republican mayor?

(b) What is the probability that a person plans to vote for at least one Republican?

(c) What is the probability of a person voting for a Republican mayor and a Democratic assistant mayor?

(d) What is the probability of a person voting for a Republican mayor or a Democratic assistant mayor?

Review Exercises

24. Two fair dice are cast.

(a) List a sample space for the numbers showing on each die.

(b) Let E be the event that an even number shows on exactly one of the dice. List the elements in E.

25. Two fair dice are cast. What is the probability that
    (a) the sum of the numbers is 8?
    (b) the sum of the numbers is odd?
    (c) only one of the dice is showing an odd?
    (d) the sum of the numbers is more than 7 or less than 4?
    (e) the sum of the numbers is more than 6 or odd?

Sum of Two Dice

First die

	1	2	3	4	5	6
1	2	3	4	5	6	7
2	3	4	5	6	7	8
3	4	5	6	7	8	9
4	5	6	7	8	9	10
5	6	7	8	9	10	11
6	7	8	9	10	11	12

(Second die)

5.5 CONDITIONAL PROBABILITY AND INDEPENDENT EVENTS

OVERVIEW   When dealing with an uncertain situation, as more information is obtained, the probabilities might change. Alternatively, we could say that as more information is available, the sample space can be modified. Suppose, for instance, that the top executives for a corporation are evaluating their chances of obtaining a large fabrication contract. They feel that their corporation and two others are equally likely to win the bidding. Hence, in their minds, the probability is $\frac{1}{3}$ that they will win the contract. Then comes information that one corporation has withdrawn from the bidding. Excitement reigns at the first corporation because in this modified sample space, their probability of success is reevaluated at $\frac{1}{2}$. Is this reasonable? You will be able to answer this question when you study

- Conditional probability: $P(A|B) = \dfrac{P(A \cap B)}{P(B)}$

- Independent events: $P(A \cap B) = P(A) \cdot P(B)$

- Multistage experiments

- Simulations

Conditional probability is the mathematical term used to describe probability with additional information. For example, the probability of drawing a heart from a standard deck of cards is $\frac{1}{4}$. However, if you know that the card drawn is red, then the probability that it is a heart is $\frac{1}{2}$. The sample space is changed when additional information is given. The symbol $P(A|B)$ denotes the problem of finding the probability that event A will occur, given the information or condition that event B has occurred. $P(A|B)$ is read "the probability of A, given B." In this section we consider two procedures for computing conditional probability.

EXAMPLE 44   In a sample of 120 customers, a restaurant manager finds that 80 customers like the lunch buffet, 60 like the lunch special menu, and 20 like both (Figure 17). What is the probability that a customer selected at random likes the buffet $P(B)$? What is the probability that a customer selected at random likes the buffet, given that the same customer also likes the lunch menu, $P(B|L)$?

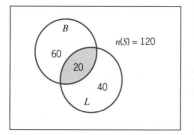

Figure 17

SOLUTION   The probability that a customer likes the buffet is

$$P(B) = \frac{80}{120}$$

The given condition of also liking the lunch menu reduces the number of possibilities to 60, of which 20 like the buffet; thus,

$$P(B \mid L) = \frac{20}{60} = \frac{1}{3}$$ ■

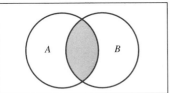

Figure 18

Looking at Figure 18, we can see that $P(A \mid B)$ means that we know event B has occurred, and the sample space is reduced to just event B; we write

$$P(A \mid B) = \frac{n(A \cap B)}{n(B)}$$

CONDITIONAL PROBABILITY

If A and B are any events, where $P(A) \neq 0$ and $P(B) \neq 0$, then the probability that event A occurs given that event B has occurred is the quotient of the probability that both A and B occur and the probability of B. This is given as

$$P(A \mid B) = \frac{P(A \cap B)}{P(B)}$$

Likewise, the probability that event B occurs given that event A has occurred is the quotient of the probability that both A and B occur and the probability of A. This is given as

$$P(B \mid A) = \frac{P(B \cap A)}{P(A)}$$

In the definition for $P(A \mid B)$ and $P(B \mid A)$, if we multiply by the denominators, we get the following rule.

Multiplication Rule

The probability that both of two events will occur is equal to the probability that the first event will occur multiplied by the conditional probability that the second event

---

Multiplication Rule *(Continued)*

will occur when it is known that the first event has occurred. That is,

$$P(A \cap B) = P(A) \cdot P(B|A)$$
$$P(B \cap A) = P(B) \cdot P(A|B)$$

We can observe from this rule that we now have a procedure for computing the probability of A and B [i.e., $P(A \cap B)$]. This is something that has been missing from our repertoire of skills. However, this relationship is helpful only if one of the relevant conditional probabilities is known or can be computed.

EXAMPLE 45   A poll is taken to determine whether 700 hourly employees of a company favor a strike. The 700 employees are divided into three groups, X, Y, and Z, according to compensation levels (see Table 4). Suppose that an hourly employee is selected at random. The probability that he or she is in favor of a strike is $\frac{380}{700}$. Now suppose that an hourly employee is selected at random from group X. What is the probability that he or she is in favor of a strike?

TABLE 4

	In Favor of a Strike	Not in Favor of a Strike	No Opinion	Total
Group X	150	50	10	210
Group Y	100	80	8	188
Group Z	130	170	2	302
Total	380	300	20	700

SOLUTION   This conditional probability is denoted by

$$P(\text{in favor of a strike} | \text{group X}) = \frac{150}{210}$$

210 in group X; of these, 150 favor a strike

Other conditional probabilities are

$$P(\text{not in favor of a strike} | \text{group Z}) = \frac{170}{302}$$

302 in group Z; of these, 170 not in favor of a strike

and

$$P(\text{group Y} | \text{in favor of a strike}) = \frac{100}{380}$$

380 in favor of a strike; of these, 100 in group Y

TABLE 5

	A	B	Total
C	40	30	70
D	10	20	30
Total	50	50	100

EXAMPLE 46　From the information given in Table 5, find
(a) $P(A\,|\,C)$, (b) $P(D\,|\,B)$, (c) $P(A \cap C)$, and (d) $P(D \cap B)$. (e) Now show that

$$P(A\,|\,C) = \frac{P(A \cap C)}{P(C)}$$

(f)　Show that

$$P(D\,|\,B) = \frac{P(D \cap B)}{P(B)}$$

SOLUTION

(a)　$P(A\,|\,C) = \dfrac{40}{70} = \dfrac{4}{7}$

(b)　$P(D\,|\,B) = \dfrac{20}{50} = \dfrac{2}{5}$

(c)　$P(A \cap C) = \dfrac{40}{100} = \dfrac{2}{5}$

(d)　$P(D \cap B) = \dfrac{20}{100} = \dfrac{1}{5}$

(e)　$P(A\,|\,C) = \dfrac{P(A \cap C)}{P(C)}$　because　$\dfrac{40}{70} = \dfrac{\frac{40}{100}}{\frac{70}{100}} = \dfrac{4}{7}$

(f)　$P(D\,|\,B) = \dfrac{P(D \cap B)}{P(B)}$　because　$\dfrac{20}{50} = \dfrac{\frac{20}{100}}{\frac{50}{100}} = \dfrac{2}{5}$ ■

Practice Problem 1　Given $P(B) = .8$, $P(A\,|\,B) = .4$, find $P(A \cap B)$.

ANSWER　$P(A \cap B) = P(B) \cdot P(A\,|\,B) = (.8)(.4) = .32$

If we know that event B has occurred, but that information does not yield any additional information about the occurrence or nonoccurrence of A, we say event A is **independent** of event B. For example, knowing that the first toss of a coin is heads adds nothing to our knowledge of what will happen on the second toss. In this instance, $P(A \cap B) = P(B) \cdot P(A\,|\,B) = P(B) \cdot P(A)$. We can use the relationship $P(A \cap B) = P(A) \cdot P(B)$ in situations in which two events are performed, and it is clear that what happens on the first trial has no influence on what occurs on the second trial.

PROBABILITY OF INDEPENDENT EVENTS

If two events, A and B, are independent, then

$$P(A \cap B) = P(A) \cdot P(B)$$

We can also state that events A and B are independent if

$$P(A|B) = P(A) \quad \text{or} \quad P(B|A) = P(B)$$

NOTE: Do not confuse independent events with mutually exclusive events. If events are independent, then the occurrence of one during one stage of an experiment has no bearing on the outcome of the next stage. If two events are mutually exclusive, they cannot both occur at the same time. This difference is demonstrated in the next example.

EXAMPLE 47   Let $S = \{a, b, c, d, e\}$ be a sample space with

$$P(a) = P(b) = P(c) = \frac{1}{4} \quad \text{and} \quad P(d) = P(e) = \frac{1}{8}$$

(a) Show that $A = \{a, b\}$ and $B = \{b, c\}$ are independent events but are not mutually exclusive.
(b) Show that $A = \{c, d\}$ and $B = \{d, e\}$ are not independent events and are not mutually exclusive.

SOLUTION

(a) To show that A and B are independent, we must show that

$$P(A \cap B) = P(A) \cdot P(B)$$

$$P(A \cap B) = P(\{b\}) = \frac{1}{4} \qquad A \cap B = \{b\}$$

$$P(A) \cdot P(B) = \left(\frac{1}{4} + \frac{1}{4}\right)\left(\frac{1}{4} + \frac{1}{4}\right)$$

$$= \left(\frac{1}{2}\right)\left(\frac{1}{2}\right)$$

$$= \frac{1}{4}$$

Therefore,

$$P(A \cap B) = P(A) \cdot P(B)$$

If A and B are mutually exclusive, then $A \cap B = \emptyset$. But we have already seen that $A \cap B = \{b\}$, so the events are independent but are not mutually exclusive.

(b)   To show that A and B are not independent, we show that $P(A \cap B) \neq P(A) \cdot P(B)$.

$$P(A \cap B) = P(\{d\}) = \frac{1}{8} \qquad A \cap B = \{d\}$$

$$P(A) \cdot P(B) = \left(\frac{1}{4} + \frac{1}{8}\right)\left(\frac{1}{8} + \frac{1}{8}\right)$$

$$= \left(\frac{3}{8}\right)\left(\frac{1}{4}\right)$$

$$= \frac{3}{32}$$

Therefore,

$$P(A \cap B) \neq P(A) \cdot P(B)$$

If A and B are mutually exclusive, then $A \cap B = \emptyset$. But we have already seen that $A \cap B = \{d\}$, so the events are not independent and are not mutually exclusive. ∎

EXAMPLE 48   A fair coin is tossed twice, and the outcomes are noted. Show that the event $A = \{\text{heads on first toss}\}$ is independent of event $B = \{\text{tails on second toss}\}$.

SOLUTION   The sample space is $S = \{HH, HT, TH, TT\}$. Event $A = \{HH, HT\}$, $B = \{HT, TT\}$, and $A \cap B = \{HT\}$. Therefore, we have

$$P(A \cap B) = \frac{1}{4}, \qquad P(A) = \frac{1}{2}, \qquad P(B) = \frac{1}{2}$$

So

$$P(A \cap B) = P(A) \cdot P(B)$$

$$\frac{1}{4} = \frac{1}{2} \cdot \frac{1}{2}$$

∎

TABLE 6

	D	D'	Total
E	1540	360	1900
E'	25	75	100
Total	1565	435	2000

EXAMPLE 49   A study of 2000 people classified each person according to employment status, E (employed) and E' (not employed) and college degree status, D (degree) and D' (no degree). The results are shown in Table 6.

(a)   Find $P(E|D)$ and $P(E)$.
(b)   Are E and D independent events?

SOLUTION

(a) $P(E|D) = \dfrac{P(E \cap D)}{P(D)} = \dfrac{n(E \cap D)}{n(D)} = \dfrac{1540}{1565} \approx .98$

$P(E) = \dfrac{1540 + 360}{2000} = \dfrac{1900}{2000} = .95$

(b) $P(E|D) \neq P(E)$, so the events are not independent.

Multistep Experiments

Some experiments are performed as a sequence of consecutive steps and are called **multistep experiments**. The probabilities of outcomes of multistep experiments are easily computed using tree diagrams and conditional probability. Consider the following example.

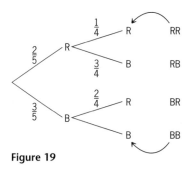

Figure 19

EXAMPLE 50   A basket contains 2 red balls and 3 black balls. A ball is drawn, set aside, and its color is noted. Then a second ball is drawn. Find a sample space for this experiment and assign probabilities.

SOLUTION   The tree diagram in Figure 19 makes it easy to determine the sample space, but it is clear that this is not a uniform sample space. The outcome *RR* (red and then red) is less likely than *BB* (black and then black). (Note that the paths in Figure 19 are conditional probabilities.)

The probability of obtaining a red ball on the first draw is $\frac{2}{5}$, but since the first ball is laid aside, the probability of the second draw producing a red ball is $\frac{1}{4}$. Thus, only $\frac{1}{4}$ of the times that the first ball is red will the second ball be red. Hence, it is reasonable to assign a probability of

$$\frac{2}{5} \cdot \frac{1}{4}$$

to the outcome *RR*.

Label each branch of the tree with the probability that the outcome on that branch will occur on the next step (see Figure 19). The probability of *RR* that we computed is the product of the probabilities along the path to *RR* in the tree diagram. Consequently, we can assign the following probabilities:

$$P(RR) = \left(\frac{2}{5}\right)\left(\frac{1}{4}\right) = \frac{1}{10}$$

$$P(RB) = \left(\frac{2}{5}\right)\left(\frac{3}{4}\right) = \frac{3}{10}$$

$$P(BR) = \left(\frac{3}{5}\right)\left(\frac{2}{4}\right) = \frac{3}{10}$$

$$P(BB) = \left(\frac{3}{5}\right)\left(\frac{2}{4}\right) = \frac{3}{10}$$

The scheme used above to compute probabilities in a multistep experiment can be used in general. Suppose that a tree diagram is used to find a sample space for a multistep experiment.

Probabilities on a Tree Diagram

The probability of an outcome described by a path through a tree diagram is equal to the product of the probabilities along that path.

Example 50 illustrated the preceding property when the probabilities along a path of the tree diagram are conditional probabilities. We now illustrate this property where the probabilities come from independent events.

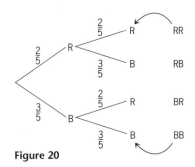

Figure 20

EXAMPLE 51   Let us use the basket from Example 50 that contains 2 red balls and 3 black balls. This time, after we draw the first ball and record the results, we place it back in the basket (Figure 20). In this circumstance, what are the probabilities of the outcomes RR and RB in the sample space?

SOLUTION   From Figure 20 we can see that $P(RR) = \left(\frac{2}{5}\right)\left(\frac{2}{5}\right) = \frac{4}{25}$ while $P(RB) = \left(\frac{2}{5}\right)\left(\frac{3}{5}\right) = \frac{6}{25}$. ■

Practice Problem 2 Find the probability of BR and BB from Example 51.

ANSWER $P(BR) = \dfrac{6}{25}$ and $P(BB) = \dfrac{9}{25}$

EXAMPLE 52   A card is drawn from a deck of cards. Then the card is replaced, the deck is reshuffled, and a second card is drawn. What is the probability of an ace on the first draw and a king on the second draw?

SOLUTION   We do not need to draw the whole tree to compute this probability. Consider only the path of interest: ace–king. $P(\text{ace then king}) = \frac{4}{52} \cdot \frac{4}{52} \approx .0059$. ■

Practice Problem 3 From Example 52, compute the probability if the first card is not replaced.

ANSWER $\dfrac{4}{52} \cdot \dfrac{4}{51} \approx .006$

Practice Problem 4 If you toss a coin and draw a card from a deck of cards, what is the probability of getting a head and drawing an ace?

ANSWER $P(H \cap A) = P(H) \cdot P(A) = \dfrac{1}{2} \cdot \dfrac{1}{13} = \dfrac{1}{26}$

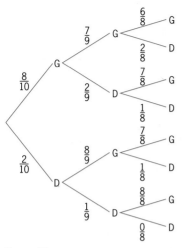

Figure 21

EXAMPLE 53   A box contains 10 radios; 2 of the radios are defective (D) and 8 are good (G). Suppose that 3 radios are selected in succession without replacing those already selected.

(a)   What is the probability that all 3 radios selected are good?
(b)   What is the probability that the third radio selected is good?
(c)   What is the probability that 2 are good and one is defective?
(d)   What is the probability that at least 2 are good?

SOLUTION   The tree diagram with associated path probabilities is shown in Figure 21.

(a) $P(GGG) = \dfrac{8}{10} \cdot \dfrac{7}{9} \cdot \dfrac{6}{8} = \dfrac{336}{720} \approx .47$. Note that this could also have been found using

$$\frac{C(8, 3)}{C(10, 3)} = \frac{56}{120} \approx .47$$

(b) $P(\text{third is good}) = P(GGG) + P(GDG) + P(DGG) + P(DDG)$

$$= \frac{8}{10} \cdot \frac{7}{9} \cdot \frac{6}{8} + \frac{8}{10} \cdot \frac{2}{9} \cdot \frac{7}{8} + \frac{2}{10} \cdot \frac{8}{9} \cdot \frac{7}{8} + \frac{2}{10} \cdot \frac{1}{9} \cdot \frac{8}{8}$$

$$= \frac{576}{720} = .8$$

(c) $P(2G, 1D) = P(GGD) + P(GDG) + P(DGG)$

$$= \frac{8}{10} \cdot \frac{7}{9} \cdot \frac{2}{8} + \frac{8}{10} \cdot \frac{2}{9} \cdot \frac{7}{8} + \frac{2}{10} \cdot \frac{8}{9} \cdot \frac{7}{8}$$

$$= \frac{336}{720} \approx .47$$

Again, since no particular order was asked for, combinations could be used.

$$P(2G, 1D) = \frac{C(8, 2) \cdot C(2, 1)}{C(10, 3)} = \frac{56}{120} \approx .47$$

(d) $P(\text{at least 2 good}) = P(2 \text{ good or 3 good})$

$$= \frac{336}{720} + \frac{336}{720} \qquad \text{From parts (a) and (c)}$$

$$\approx .93$$

We see from parts (a) and (c) in Example 53 that if a particular order is not asked for, combinations can be used. Combinations can also be used when there are many steps and a tree is not practical. Remember,

$$P(A\,|\,B) = \frac{P(A \cap B)}{P(B)} = \frac{n(A \cap B)}{n(B)}$$

EXAMPLE 54 A box contains 10 yellow, 6 red, and 8 blue balls. Five balls are selected at random and the colors are noted. Find each probability.

(a) P(all yellow balls | all the same color)
(b) P(exactly 2 red balls | no blue balls)
(c) P(at least 3 blue balls | no red balls)

SOLUTION

(a) There are $C(10, 5)$ ways of choosing all yellow balls. We are told that all the balls are the same color. There are $C(6, 5)$ ways of selecting all red and $C(8, 5)$ ways of selecting all blue.

$$P(\text{all yellow balls} \mid \text{all the same color}) = \frac{\text{Number of ways}}{\text{for all yellow}}$$
$$= \frac{\text{Number of ways for}}{\text{all the same color}}$$

$$= \frac{C(10, 5)}{C(10, 5) + C(6, 5) + C(8, 5)}$$

$$= \frac{252}{252 + 6 + 56} = \frac{252}{314} \approx .80$$

(b) We can interpret the problem as asking for 2 red balls and 3 yellow balls, since there are no blue balls. Therefore, the probability is found as

$$P(\text{exactly 2 red balls} \mid \text{no blue balls}) = \frac{C(6, 2) \cdot C(10, 3)}{C(16, 5)} = \frac{1800}{4368} \approx .412$$

(c) At least three blue balls means that there can be 3 blue balls and 2 others or there can be 4 blue balls and 1 other or there can be 5 blue balls and no other. Since we are told there are no red balls, the sample space is reduced to only yellow and blue, a total of 18 balls. Thus, we find the probability as

$$\frac{C(8, 3) \cdot C(10, 2) + C(8, 4) \cdot C(10, 1) + C(8, 5)}{C(18, 5)} = \frac{2520 + 700 + 56}{8568}$$

$$= \frac{3276}{8568}$$

$$\approx .38 \qquad \blacksquare$$

Practice Problem 5 If 5 cards are drawn from a deck without replacement, find each probability.

(a) P(exactly 3 kings | there are no aces)
(b) P(2 aces and 3 fours | all the numbers are less than 10) (Count the ace as a 1.)

ANSWER

(a) $\dfrac{C(4, 3) \cdot C(44, 2)}{C(48, 5)} \approx .002$ (b) $\dfrac{C(4, 2) \cdot C(4,3)}{C(36, 5)} \approx .00006$

Simulation

One of the most powerful tools available to both scientists and business analysts is **simulation**. When faced with a complicated system that is not well understood but is built from processes whose behavior is understood, simulations are often useful. In this way, rocket engines, marketing schemes, and weapons systems are tested before they are even built. We will certainly not test any rocket engines, but we can experience the flavor of simulations. For example, since the probability of having a boy is roughly the same as the probability of having a girl, we could use the flip of a coin to simulate the birth of a child. We could simulate a possible

TABLE 7

Random Numbers							
12135	65186	86886	72976	79885	07369	49031	45451
10724	95051	70387	53186	97116	32093	95612	93451
53493	56442	67121	70257	74077	66687	45394	33414
15685	73627	54287	42596	05544	76826	51353	56404
74106	66185	23145	46426	12855	48497	05532	36299
57126	99010	29015	65778	93911	37997	89034	79788
94676	32307	41283	42498	73173	21938	22024	76374
68251	71593	93397	26245	51668	47244	13732	48369
60907	17698	32865	24490	56983	81152	12448	00902
07263	16764	71561	52515	93269	61210	55526	70912
43501	10248	34219	83416	91239	45279	19382	82151
57365	84915	11497	98102	58168	61534	69495	85183
38161	22848	06673	35293	27893	58461	10404	17385
26760	51437	87751	41523	10816	54858	35715	47947
65592	93388	36555	21136	43900	89837	78093	28870
48651	16719	99032	86292	40668	72821	59266	44970
71495	84760	35193	06961	41211	33548	40026	63873
81242	06154	69109	60926	62177	72065	70225	86018
26574	84854	38915	83783	46780	08735	38781	94657
07736	70130	46808	18940	14795	34231	23671	05856
26533	06561	09049	67618	12560	59539	41937	18490
36335	84039	05960	38850	62976	65958	99682	64250
92074	87770	31924	99481	15505	55099	42072	57637
00243	48272	45390	24171	96173	98887	03335	45965
68900	91374	18868	45389	57567	89557	56764	59362
57663	88219	88929	03419	28838	89659	64710	60768
27715	05262	06208	96357	65700	82054	28590	95933
91798	54270	85403	30110	00426	19915	38883	43423
64221	42325	55273	68399	91856	76729	25130	64615
10852	21818	08641	82759	75389	96295	05934	53697

outcome of having three children by flipping a coin three times. By repeating this simulation 100 times, we could compute empirical probabilities for the experiment without having undergone the considerable difficulty of having 300 children.

A very useful tool in performing a simulation is a table of random digits, such as the one in Table 7. Tables of random numbers or random number generators can decrease the amount of time necessary to run a simulation. There is no bias in the numbers generated and outcomes can be assigned arbitrarily. The numbers in Table 7 are collections of digits that are randomly generated by a computer. We use this table in the simulation in Example 55.

EXAMPLE 55   Simulate the experiment of tossing a pair of coins 50 times.

SOLUTION   We use Table 7 to determine the result of tossing each coin. Because each digit is equally likely to be even or odd, we use the first two digits in each entry to represent the outcome for each coin. If the first digit is even, the first coin shows a head; if the first digit is odd, the first coin shows a tail. Similarly, the second digit will determine whether the second coin shows a head or a tail. We can start at any point in the table. In Table 7, the circle around a set of digits is where we start for this problem. Since the first two digits of 35293 are odd, the outcome of our first toss is two tails. Proceeding to the next number down the column (you could go across if you wanted), the 4 and 1 in 41523 indicate that the next pair of tosses results in a head and a tail. Continuing in this manner until we have simulated 50 tosses, we get the results shown in Table 8. How does this compare to the theoretical probabilities of .25?

TABLE 8

	Frequency	Empirical Probability
HH	12	12/50 = .24
HT	11	11/50 = .22
TH	14	14/50 = .28
HH	13	13/50 = .26

EXAMPLE 56   A basketball player hits 40% of her shots from three-point range. Use simulation to determine the probability that in a game in which she takes 5 long shots (from beyond the three-point line) that she will hit at least 3 of them.

SOLUTION   We let a single digit represent each attempted long shot. Since she hits 40% of her shots from this distance, we let the four digits 0, 1, 2, 3 represent successful shots and the digits 4, 5, 6, 7, 8, 9 represent missed attempts. (Note that 0, 1, 2, 3 are 40% of all shots.) A 5-digit entry from Table 7 represents a game in which she attempts five shots. We randomly pick a place to begin. We choose the number with the rectangle around it. The first entry is 93397. Since the 3's represent successes and the 9 and 7 represent misses, in the first game she made only 2 of her 5 shots. If we simulate this 50 times, we learn that in 14 of 50 games she hit 3 or more of her three-point goals. Thus, from the simulation, we assign a probability of .28 to this event. ∎

SUMMARY

1. Conditional probability of A given B is

$$P(A \mid B) = \frac{P(A \cap B)}{P(B)}$$

2. $P(A \cap B) = P(A) \cdot P(B \mid A)$.
3. $P(A \cap B) = P(A) \cdot P(B)$ if A and B are independent.
4. A tree and path probabilities can be used to organize a problem.
5. A table of random digits can be useful in a simulation.

Exercise Set 5.5

1. A single card is drawn at random from a standard deck. Let $B = \{$the card is black$\}$, $H = \{$the card is a heart$\}$, and $C = \{$the card is a club$\}$.
   (a) Describe in words a sample space for the experiment.
   (b) How is the sample space changed if we have the additional information that a black card is drawn?
   (c) Compute $P(H \mid B)$.
   (d) Compute $P(C \mid B)$.
   (e) Compute $P(B \mid C)$.

2. Roll a single fair die. Let $A = \{$the die shows less than 4$\}$ and $B = \{$the die shows an odd number$\}$. Compute (a) $P(A \mid B)$ and (b) $P(B \mid A)$.

3. If $P(A) = .6$, $P(B \mid A) = .7$, and $P(B) = .6$, compute each probability.
   (a) $P(A \cap B)$
   (b) $P(A \mid B)$
   (c) $P(B')$
   (d) $P(A \cup B)$

4. Given the following table, compute each probability.

	C	D	E	Total
A	.20	.10	.05	.35
B	.30	.20	.15	.65
Total	.50	.30	.20	1

   (a) $P(A)$ (b) $P(E)$
   (c) $P(B \cap C)$ (d) $P(A \cap E)$
   (e) $P(E')$ (f) $P(B')$
   (g) $P(C \mid A)$ (h) $P(A \mid C)$
   (i) $P(B \mid D)$

5. If $P(A) = .70$, $P(B) = .30$, and $P(A \cap B) = .20$, compute each probability.
   (a) $P(A \mid B)$ (b) $P(B \mid A)$
   (c) $P(A \cup B)$ (d) $P(A \mid B')$
   (e) $P((A \cup B)')$ (f) $P(A' \cap B')$
   (g) $P(A' \mid B)$ (h) $P(A \cup B')$

6. A baseball player is batting .300. In a typical game he will have three official times at bat.
   (a) Draw a three-stage tree diagram describing the possible outcomes.
   (b) What is the probability of 3 hits?
   (c) What is the probability of at least 2 hits?
   (d) What is the probability of no hits?

7. A card is drawn from a standard deck of cards. Find each probability.
   (a) $P(\text{jack} \mid \text{it is a face card})$
   (b) $P(\text{ace} \mid \text{it is not a face card})$

8. From a box containing 5 red balls and 3 white balls, 2 balls are drawn successively at random, without replacement.
   (a) What is the probability that the first is white and the second is red?
   (b) What is the probability that both are red if it is known that the first is red?

9. A new low-flying missile has a probability of .9 of penetrating the enemy defenses and a probability of .7 of hitting the target if it penetrates the defenses. What is the probability that the missile will penetrate the defenses and hit the target?

10. A box contains 3 red balls and 4 white balls. What is the probability of drawing 2 white balls
    (a) if the first is replaced before the second one is drawn?
    (b) if the first ball is not replaced?

11. In a certain college, 30% of the students failed mathematics, 20% failed English, and 15% failed both mathematics and English.
    (a) If a student failed English, what is the probability that she failed mathematics?
    (b) If a student failed mathematics, what is the probability that he failed English?
    (c) What is the probability that a student failed mathematics or English?
    (d) If a student did not fail mathematics, what is the probability that he failed English?
    (e) If a student did not fail English, what is the probability that she did not fail mathematics?

12. A box contains the following balls: 5 colored red and white, 3 colored black and white, 4 colored green and white, 6 colored red and black, 4 colored red and green, and 5 colored black and green.
    (a) Given that you have drawn a ball that is partly green, what is the probability that it is partly white?
    (b) Given that the ball you have drawn is partly white, what is the probability that it is partly red?

13. A candy jar contains 6 pieces of peppermint, 4 pieces of chocolate, and 12 pieces of butterscotch candy. A small boy reaches into the jar, snatches a piece, and eats it rapidly. He repeats this act quickly.
    (a) What is the probability that he eats a peppermint and then a chocolate?
    (b) What is the probability that he eats 2 chocolates?
    (c) What is the probability that he eats a chocolate and then a butterscotch?
    (d) What is the probability that he eats a chocolate and a butterscotch?

14. Suppose that the small boy in Exercise 13 is caught by his mother immediately after he snatches his first piece. She makes him return the candy to the jar. He waits an appropriate length of time and then again snatches a piece.
    (a) What is the probability that the frustrated thief snatches a peppermint and then a chocolate?
    (b) What is the probability that he gets chocolate on both tries?

15. Three cards are drawn without replacement from a well-shuffled deck of 52 playing cards. What is the probability that the third card drawn is a club?

16. John must take a four-question, true–false quiz, but he has failed to study. What is the probability that he will score 100% if he guesses on each question?

17. Suppose the quiz that John must take in Exercise 16 consists of four multiple choice questions with answers (a) to (d). What is the probability that John will make 100%?

18. Use Table 7 to simulate the experiment of flipping a coin three times. Use the first three digits of each entry in the table to determine respectively whether the first, second, and third flips show a head or a tail. Simulate the experiment 50 times and compute the appropriate empirical probabilities.

19. A new cereal company is placing 1 of 3 tiny dolls in each box of cereal. Each of the dolls is equally likely to occur in a given box of cereal. We are interested in the probability that if we open 5 boxes of cereal, we will have a collection of all 3 dolls.
    (a) How could we use simulation with rolls of a die to answer this question?
    (b) Use simulation to answer this question by rolling five dice 20 times.

20. Rework Exercise 6 using simulation. Let the first three digits of an entry in Table 7 represent 3 times at bat. If the entry is 0, 1, or 2, record a hit; otherwise, record a failure. Simulate 50 games, each with 3 times at bat and compute the probabilities requested in Exercise 6.

21. For events A and B, suppose that $P(A) = .5$ and $P(A \cup B) = .8$. Find x such that $P(B) = x$ if
    (a) A and B are independent.
    (b) A and B are mutually exclusive.

22. In a survey of 100 movie-goers, three films were listed. The results were as follows:

    32 people liked movie A
    18 people liked movie B
    50 people liked movie C
    4 people liked movies A and C
    5 people liked movies B and C
    7 people liked movies A and B
    2 people liked all three movies

    If a person is selected at random from the survey group and they like movie C, what is the probability that they also liked movie A?

23. If $P(A) = \frac{1}{6}$ and $P(B) = \frac{1}{5}$, find $P(A \cap B)$ if it is known that A and B are independent events.

24. Let A and B be events in a sample space S with $P(A) = .4$, $P(B) = .3$, and $P(A \cap B) = .12$.
    (a) Find $P(A|B)$ and $P(B|A)$.
    (b) Are A and B independent events?

25. Determine whether events A and B are independent.
    (a) $P(A) = .7$, $P(B) = .5$, $P(A \cup B) = .85$
    (b) $P(A \cap B) = .5$, $P(A) = .4$, $P(B) = .3$

26. Two cards are drawn without replacement from a standard deck of 52 cards.
    (a) What is the probability that the first card drawn is a diamond?
    (b) What is the probability that the second card drawn is a club, if the first card is a club?
    (c) What is the probability that the second card drawn is a king, given that the first card is not a king?
    (d) What is the probability that both cards are the same size?
    (e) What is the probability that both cards are the same suit?
    (f) What is the probability that both cards are not the same suit? [**Hint**: use complements and part (e).]

27. A bowl has 10 red balls, 10 blue balls, and 10 yellow balls. Three balls are drawn in succession without replacement. Use both tree diagrams and combinations to find the following probabilities.
    (a) P (exactly 2 are red)
    (b) P(there is 1 of each color)

    (c) P(all are the same color)
    (d) P(all are red | all are the same color)
    (e) P(at least 1 is red) [**Hint**: P (at least 1) $= 1 - P$(none).]

28. A coin is tossed 3 times. Find each probability.
    (a) It will land as tails at least 2 times.
    (b) It will be heads on the second toss if it landed as heads on the first toss.
    (c) It will be tails on the third toss, if the first two tosses were heads.

29. **Exam.** An urn contains 10 red, 20 white, and 30 blue balls. What is the probability that, of 6 balls drawn at random with replacement, 1 ball will be red, 2 white, and 3 blue?
    (a) $\dfrac{5}{36}$ (b) $\dfrac{100}{649}$ (c) $\dfrac{10}{59}$ (d) $\dfrac{5}{18}$

30. **Exam.** What is the probability that a three-card hand drawn at random and without replacement from an ordinary deck consists entirely of black cards?
    (a) $\dfrac{1}{17}$ (b) $\dfrac{2}{17}$ (c) $\dfrac{1}{8}$
    (d) $\dfrac{3}{17}$ (e) $\dfrac{4}{17}$

Applications (Business and Economics)

31. **Investments.** Of 100 businesspeople polled, 50 have investments in common stocks, 35 have investments in bonds, and 25 have investments in both stocks and bonds. What is the probability that a person chosen at random from the businesspeople polled
    (a) invests in common stocks and not in bonds?
    (b) invests in bonds and not in common stocks?
    (c) does not invest in stocks or does not invest in bonds?
    (d) invests in stocks or bonds?
    (e) invests in stocks, if you know she invests in bonds?
    (f) invests in bonds, if you know he invests in stocks?
    (g) invests in stocks, if you know he does not invest in bonds?
    (h) invests in bonds, if you know she does not invest in stocks?

32. **Quality Control.** A machine is assembled using components A and B. The two components are each built in separate fabricating plants. Experience indicates that the probability that A is defective is .01, and the probability that B is defective is .05. (**Hint:** Since A and B are fabricated in different plants, whether A is good or defective is independent of the quality of B.)

   (a) What is the probability that both components are defective?

   (b) What is the probability that both components are good?

33. **Quality Control.** You know that 4% of all light bulbs produced by a given company weigh less than specifications, and 2% of all bulbs are both defective and weigh less than specifications. What is the probability that a light bulb selected at random is defective, if you know that it weighs less than specifications?

34. **Quality Control.** A manufacturer receives a shipment of 20 articles. Unknown to him, 6 are defective. He selects 2 articles at random and inspects them. What is the probability that the first is defective and the second is satisfactory?

35. Plants A and B manufacture tires for a warehouse. The warehouse receives 40% of its tires from plant A and 60% from plant B. Of the tires made by plant A, 90% are good and 10% have some defect. Of the tires made by plant B, 85% are good and 15% have some defect. Suppose that one of the plants is chosen at random and two tires from that plant are selected.

   (a) What is the probability that neither tire is defective?

   (b) What is the probability that both tires are good if it is known that they came from Plant B?

36. **Product Testing.** The probability that a new, long-lasting AA battery will last at least 15 hours is .80 and the probability that it will last at least 20 hours is .15. A battery is selected at random and after 15 hours it is still good. What is the probability that it will last at least 20 hours?

37. **Quality Control.** A shipment of spotlights contains 2 defective bulbs and 13 bulbs with no defects. A stock clerk is told to test each bulb one at a time until the two defective bulbs are found. What is the probability that he will get lucky and the two defective bulbs will be found after only 3 tests?

38. A large firm on Wall Street did a survey of its personnel. The survey grouped each person by degree received and by incomed earned. The results are shown in the table.

	Income $\leq 30{,}000$	Income $> 30{,}000$	Total
College degree	75	25	100
Graduate degree	15	35	50
Total	90	60	150

Let G be the event that an employee has a graduate degree, D be the event that an employee has only one degree, L be the event that an employee makes at most $30,000, and M be the event that an employee makes more than $30,000. Find each probability.

   (a) $P(G \cap M)$ (b) $P(D|L)$

   (c) $P(M|D)$ (d) $P(L|G)$

Applications (Social and Life Sciences)

39. **Genetics.** In a study of genetics, a class used a sample of 100 people to obtain the following information.

	Male (M)	Female (F)	Total
Color blind (C)	4	1	5
Not color blind (C')	40	55	95
Total	44	56	100

What is the probability that a person is

   (a) color blind, given that the person is a female?

   (b) color blind, given that the person is a male?

   (c) not color blind, given that the person is a male?

   (d) not color blind, given that the person is a female?

40. **Genetics.** According to the genetic theories of Mendel, a parent with genes of type AA can transmit only an A gene to offspring. A parent with type aa can transmit either an A or an a gene, each with probability $\frac{1}{2}$. For each of the following mates, find

the probability of the offspring being type *AA*, type *aa*, and type *Aa*.

(a) Type *AA* mates with *AA*.

(b) Type *aa* mates with *aa*.

41. **Genetics**.   An animal with *BB* genes is crossed with one with *Bb* genes. Suppose that there is a litter of four. Find each probability.

(a) All will be *Bb*   (b) All will be *BB*

Review Exercises

42. A bag contains 6 red balls, 4 black balls, and 3 green balls. A ball is drawn from the bag. What is the probability that the ball is

(a) red or black?

(b) blue?

(c) red or black or green?

(d) not red and not green?

(e) not black?

(f) green?

(g) not red or not black?

(h) red and black?

(i) not red or not green?

(j) not green?

43. A ball is drawn from the bag in Exercise 42, and its color is recorded. The ball is replaced, and a second ball is drawn. What is the probability of drawing each of the following?

(a) Two red balls

(b) A red ball followed by a green ball

(c) Two black balls

44. Suppose   that $P(A) = .35,$ $P(B) = .51,$ and $P(A \cap B) = .17.$ Compute   (a) $P(A')$ and   (b) $P(A \cup B)$.

Chapter Review

Review the following terms to ensure that you understand their application to probability.

Important Terms

Experiment	$A \cup B$
Outcome	$A \cap B$
Sample space	Complement
Relative frequency	Probability rule on a sample space
Event	Compound event
Probability of an event	Mutually exclusive events
Conditional probability	Uniform sample space
Independent events	Simple event

Review the following concepts used in counting.

Important Concepts

Tree diagrams

Fundamental principal of counting

Permutations

Combinations

Permutations of *n* things taken *r* at a time, *P(n, r)*

Combinations of *n* things taken *r* at a time, *C(n, r)*

Success in solving probability problems depends on the ability to use the correct formula. Be sure you know the conditions that allow you to use each of the following formulas.

Important Formulas

$$P(A) = \frac{\text{Number of times } A \text{ occurs}}{N}$$

$$0 \leq P(A) \leq 1$$

$$P(A) = \frac{1}{n(S)}$$

$$P(A \cup B) = P(A) + P(B)$$

$$P(A \cup B) = P(A) + P(B) - P(A \cap B)$$

$$P(A') = 1 - P(A)$$

$$P(\text{at least one}) = 1 - P(\text{none})$$

$$P(A|B) = \frac{P(A \cap B)}{P(B)}$$

$$P(A) = \frac{n(A)}{n(S)}$$

$$P(A \cap B) = P(A) \cdot P(B|A)$$

$$n! = n(n - 1)(n - 2) \cdot \cdots \cdot 3 \cdot 2 \cdot 1$$

$$P(A \cap B) = P(B) \cdot P(A|B)$$

$$P(n, r) = \frac{n!}{(n - r)!}$$

$$P(A \cap B) = P(A) \cdot P(B)$$

$$P(n, n) = n!$$

$$C(n, r) = \frac{n!}{r!(n - r)!}$$

$$\text{Permutations} = \frac{n!}{n_1! \, n_2! \cdot \cdots \cdot n_k!}$$

Chapter Test

1. What is the probability of getting heads when a 2-headed coin is tossed? What is the probability of getting tails?

2. In how many ways can 6 books be arranged on a shelf?

3. A box contains 3 red balls and 4 white balls. What is the probability of drawing 2 white balls
   (a) if the first ball is replaced before the second one is drawn?
   (b) if the first ball is not replaced?

4. From a group of 7 people, how many committees of 4 can be selected?

5. In a certain college, 30% of the students failed mathematics, 20% failed English, and 15% failed both mathematics and English. If a student failed English, what is the probability that he failed mathematics?

6. A file contains 20 good sales contracts and 5 canceled contracts. In how many ways can 4 good contracts and 2 canceled contracts be selected?

7. If you toss 2 dice, what is the probability of getting a sum of 7 or 8?

8. A box contains 6 red and 4 black balls. Three balls are drawn at random without replacement.
   (a) What is the probability of getting 2 red balls and 1 black ball?

     (b) What are the odds in favor of getting all red balls?

     (c) What are the odds against getting 2 black and 1 red ball?

9.   In how many ways can a chairman, a treasurer, and a secretary be selected from a board of 12 persons?

10. Consider a family of 3 children. Find the probability that all 3 children are the same sex?

11. The probability that John will live at least 20 more years is $\frac{1}{5}$. The probability that Marie will live at least 20 more years is $\frac{1}{4}$. Find the probability that neither will live at least 20 more years.

12. A proposal to accept or reject a union is submitted to all employees of the Arnold Corporation. Of the employees, 30% are laborers, 35% are white-collar workers, and 35% are blue-collar workers. In the response, 20% of the laborers, 80% of the white-collar workers, and 83% of the blue-collar workers vote to reject the union. What is the probability that a laborer selected at random voted against the union?

13. Use Table 7 to simulate the experiment of flipping a coin 3 times. The coin is loaded so that the probability of a head is 0.4 and the probability of a tail is 0.6. Simulate the experiment 50 times and compute the appropriate empirical probabilities.

14. A team of 3 people is to be selected from a group consisting of 4 men and 4 women.

     (a) How many different teams can be selected?

     (b) How many different teams can be selected if the team is required to have at least one woman on it?

     (c) How many different teams can be selected if the team is required to have at least one man and at least one woman on it?

15. In one university 40% of the new freshmen take college algebra. The grade distribution for one fall semester was as follows: 5% made A's, 10% made B's, 50% made C's, 20% made D's, and 15% made F's. You meet a new freshman at the end of the fall semester. What is the probability that this person

     (a) made an A in college algebra?

     (b) passed college algebra (A, B, C, or D)?

     (c) failed college algebra?

16. In a classroom there are 75 students. Fifty-five of them went to see a movie last night, 25 ate a pizza last night, and 5 did neither. If you pick a student in the classroom at random, what is the probability that this student ate a pizza and saw a movie last night?

17. A deck of 52 cards is shuffled and 5 cards are selected from the deck.

     (a) Find the number of 5-card hands that are possible.

     (b) Find the probability that 3 kings and 2 aces are dealt.

     (c) Find the probability that all 5 cards are of the same suit.

18. If the odds in favor of event A occurring are 5 to 2, find the probability of A occurring.

19. How many 9-letter words (any combination of letters being considered a word) can be formed from the letters in the word *classroom*?

20. If $P(A) = .6$, $P(B) = .5$, and $P(A \cup B) = .9$, are A and B independent events?

21. A survey of 80 sophomores at a western college showed the following:

   36 take English
   32 take history
   32 take political science
   16 take political science and history
   16 take history and English
   14 take political science and English
   6 take all three

   (a) How many students take only English?
   (b) What is the probability that a student chosen at random from the 80 does not take any of the courses?
   (c) What is the probability that a student takes history or English, but not political science?

Important Applications of Probability

In this chapter we use the topics of probability introduced in the preceding chapter to solve other types of probability problems and to introduce special applications of probability. For example, when you complete this chapter you will be able to determine that it is not wise financially to play your local lottery. We introduce **expected value** and show how it is used in choosing a course of action such as whether to buy a lottery ticket.

In this chapter we reverse the typical probability problem and find the probability of an earlier event conditioned on the occurrence of a later one. In many tests in the medical profession, the probability that the test is accurate is often known. Suppose that you take a test and it indicates that you have a certain disease. You are certainly interested in the probability that you do not have the disease given that the test is positive. The mathematician Thomas Bayes (1702–1763) is supposed to have been trying to prove the existence of God when he developed a formula that we use to solve such probability problems today.

At the end of the chapter we introduce Markov chains, named for the Russian mathematician Andrei Markov (1856–1922), in which matrix theory and probability are combined to discuss physical systems and how they change over time.

6.1 BAYES' FORMULA

OVERVIEW   In this section we are given conditional probabilities in one direction and we need to find conditional probabilities in the opposite direction. For example, of the people in a given area, 10% are known to have some kind of cancer. A new method of detecting cancer is developed that gives a correct positive result 95% of the time when a person has cancer and gives a false positive result 10% of the time when a person does not have cancer. Suppose that the test is positive. What is the probability that the person has cancer? Such problems can be solved by using a procedure called Bayes' formula. In this section we find Bayes' probabilities

- From a formula
- Using a tree diagram

In the preceding chapter we discussed briefly the probability that two events will occur in a given order. This is equal to the probability that the first event will occur multiplied by the conditional probability that the second event will occur when it is known that the first event has occurred:

$$P(A \cap B) = P(A) \cdot P(B|A)$$

or

$$P(A \cap B) = P(B) \cdot P(A|B)$$

The sequence of experiments to obtain events A and B is called a **stochastic process**, which means that the outcome of one event depends on the outcome of a previous event.

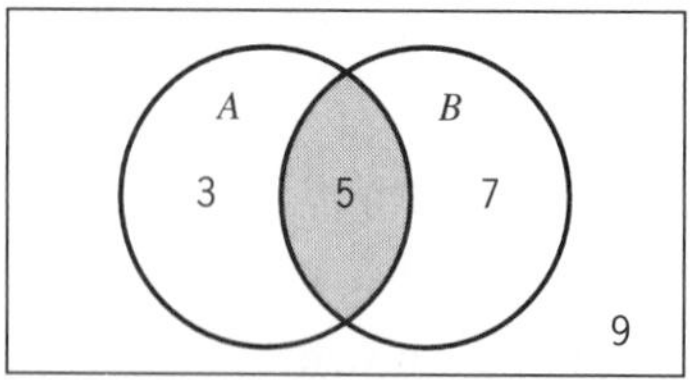

Figure 1 shows the number of ways events A and B can occur. That is, event A can occur in $3 + 5$ ways; event B in $5 + 7$ ways; event A and B in 5 ways; and 9 represents the ways that neither A nor B can occur. Thus we have

Figure 1

$$P(A) = \frac{8}{24} \qquad P(B) = \frac{12}{24} \qquad P(A \cap B) = \frac{5}{24}$$

$$P(A|B) = \frac{5}{12} \qquad P(B|A) = \frac{5}{8}$$

From these probabilities we note that

$$P(A \cap B) = P(A) \cdot P(B|A)$$
$$= \frac{8}{24} \cdot \frac{5}{8} = \frac{5}{24}$$

and

$$P(A \cap B) = P(B) \cdot P(A|B)$$
$$= \frac{12}{24} \cdot \frac{5}{12} = \frac{5}{24}$$

In the first probability, A occurs and then B; in the second probability, B occurs and then A. It is somewhat easier to understand these formulas by denoting with subscripts the order in which events occur:

$$P(A_1 \cap B_2) = P(A_1) \cdot P(B_2|A_1)$$

or

$$P(A_2 \cap B_1) = P(B_1) \cdot P(A_2|B_1)$$

In general, since A and B can occur in either order,

$$P(A \cap B) = P(A_1) \cdot P(B_2|A_1) + P(B_1) \cdot P(A_2|B_1)$$

EXAMPLE 1   From a box containing 6 red balls and 4 black balls, 2 balls are drawn. (This is equivalent to drawing 2 balls one at a time without replacement.) What is the probability of getting a red ball and a black ball?

SOLUTION   This event can happen in two mutually exclusive ways: a red ball followed by a black ball or a black ball followed by a red ball. Thus,

$$P(R \cap B) = P(R_1) \cdot P(B_2|R_1) + P(B_1) \cdot P(R_2|B_1)$$
$$= \frac{6}{10} \cdot \frac{4}{9} + \frac{4}{10} \cdot \frac{6}{9}$$
$$= \frac{48}{90} = \frac{8}{15}$$

The tree diagram in Figure 2 shows all possibilities in the drawing of 2 balls from the box. The second and third branches give the desired probability for this problem.

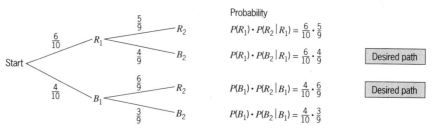

Figure 2

EXAMPLE 2

Box I contains 3 red and 4 black balls; box II contains 4 red and 5 black balls. A ball is drawn from box I and placed in box II; then a ball is drawn from box II. What is the probability that the second ball is red?

SOLUTION   The first experiment involves drawing a ball from box I with two possible outcomes: R_1, a red ball, or B_1, a black ball. The second experiment involves drawing a ball from box II after a ball has been drawn from box I and placed in box II. Four possibilities exist, as indicated in Figure 3. There are two mutually exclusive paths for getting a red ball on the second draw: a red ball on the first draw and a red ball on the second draw, or a black ball on the first draw and a red ball on the second draw. The probability of a particular path is the product of the probabilities along the path. Therefore,

$$P(\text{red ball on the second draw}) = P(R_1) \cdot P(R_2 | R_1) + P(B_1) \cdot P(R_2 | B_1)$$
$$= \frac{3}{7} \cdot \frac{5}{10} + \frac{4}{7} \cdot \frac{4}{10} = \frac{31}{70}$$

■

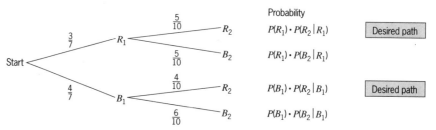

Figure 3

EXAMPLE 3

The probability that a chest X ray will be positive for a person with lung cancer or tuberculosis (TB) is .90, while the probability that the X ray will be positive for a person who does not have either disease is .05. Four percent of the people in a city have lung cancer or tuberculosis. If a person is selected at random, what is the probability that the X ray will be positive?

SOLUTION   As indicated in Figure 4, the X ray can be positive in two ways: (1) when the person has cancer or TB, denoted by C, and (2) when the person

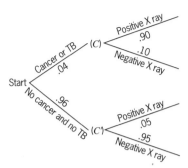

does not have either disease, denoted by C'. Thus,

$$P(\text{positive X ray}) = [P(C) \cdot P(\text{positive X ray}|C)]$$
$$+ [P(C') \cdot P(\text{positive X ray}|C')]$$
$$= .04(.90) + .96(.05)$$
$$= .084$$ ∎

Practice Problem 1 Three machines, A, B, and C, produce 50%, 25%, and 25%, respectively, of all the items produced in a given area of a factory. It has been found that defective items make up about 5% of the items produced from machine A, 3% of those from machine B, and 1% from machine C. If an item is selected out of a day's production from the three machines, what is the probability that it is defective?

ANSWER

$$P(\text{article is defective}) = P(A) \cdot P(D|A) + P(B) \cdot P(D|B) + P(C) \cdot P(D|C)$$
$$= .50(.05) + .25(.03) + .25(.01) = .035$$

Figure 4

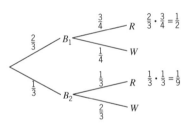

Figure 5

Bayes' Probabilities

To introduce Bayes' formula let's consider the following example. Colored balls are distributed in two boxes as follows. The first box contains 3 red balls and 1 white ball, while the second box contains 2 white balls and 1 red ball. A box is selected in such a manner that the probability of selecting box B_1 is $\frac{2}{3}$ and the probability of selecting box B_2 is $\frac{1}{3}$. Figure 5 summarizes these facts. From the first branch in Figure 5 we see that

$$P(B_1 \cap R) = \frac{2}{3} \cdot \frac{3}{4} = \frac{1}{2}$$

Also,

$$P(B_2 \cap R) = \frac{1}{3} \cdot \frac{1}{3} = \frac{1}{9} \quad \text{Third branch}$$

Now,

$$P(R) = (B_1 \cap R) \cup (B_2 \cap R)$$

So

$$P(R) = P(B_1 \cap R) + P(B_2 \cap R) \quad B_1 \cap R \text{ and } B_2 \cap R \text{ are mutually exclusive.}$$
$$= \frac{2}{3} \cdot \frac{3}{4} + \frac{1}{3} \cdot \frac{1}{3} = \frac{11}{18}$$

From Figure 5 we can see that

$$P(B_1 \cap R) = \frac{2}{3} \cdot \frac{3}{4} = \frac{1}{2}$$

We have learned that

$$P(B_1 | R) = \frac{P(B_1 \cap R)}{P(R)}$$

$$= \frac{1/2}{11/18} = \frac{9}{11}$$

So when a red ball is drawn, the probability that it came from B_1 is $\frac{9}{11}$.

A procedure to find $P(B_1 | R)$ in the preceding example can be stated as

$$P(B_1 | R) = \frac{\text{Probability of the path through } B_1 \text{ and } R}{\text{Sum of the probabilities of all paths leading to } R}$$

$$= \frac{1/2}{1/2 + 1/9} = \frac{9}{11}$$

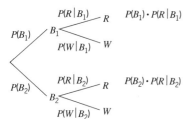

Figure 6

Now let's consider the same problem without a specific number of red and white balls and without specific probabilities for drawing each box. The corresponding tree diagram is shown in Figure 6.

The formula for finding the probability that if a red ball is drawn, it came from box B_1 can be formulated in the following manner. First note that in Figure 6,

$$R = (B_1 \cap R) \cup (B_2 \cap R)$$

$$P(R) = P(B_1 \cap R) + P(B_2 \cap R) \qquad B_1 \cap R \text{ and } B_2 \cap R \text{ are mutually exclusive.}$$

$$P(B_1 \cap R) = P(B_1) \cdot P(R | B_1) \qquad \text{First path}$$

$$P(B_2 \cap R) = P(B_2) \cdot P(R | B_2) \qquad \text{Third path}$$

So

$$P(R) = P(B_1) \cdot P(R | B_1) + P(B_2) \cdot P(R | B_2) \qquad \text{Substitution}$$

Also, from Figure 6,

$$P(R \cap B_1) = P(B_1) \cdot P(R | B_1) \qquad \text{First path}$$

Now, from the preceding chapter,

$$P(B_1|R) = \frac{P(R \cap B_1)}{P(R)}$$

$$= \frac{P(R \cap B_1)}{P(B_1) \cdot P(R|B_1) + P(B_2) \cdot P(R|B_2)} \qquad \text{Substitution for denominator}$$

$$= \frac{P(B_1)P(R|B_1)}{P(B_1) \cdot P(R|B_1) + P(B_2) \cdot P(R|B_2)} \qquad \text{Substitution for numerator}$$

In a similar manner, we have

$$P(R \cap B_2) = P(B_2) \cdot P(R|B_2) \qquad \text{Third path}$$

and

$$P(B_2|R) = \frac{P(B_2 \cap R)}{P(R)}$$

$$P(B_2|R) = \frac{P(B_2) \cdot P(R|B_2)}{P(B_1) \cdot P(R|B_1) + P(B_2) \cdot P(R|B_2)} \qquad \text{Substitution}$$

Again, let's obtain the same result with a tree-diagram approach.

$$P(B_1|R) = \frac{\text{Probability of path through } B_1 \text{ and } R}{\text{Sum of the probabilities of all paths to } R}$$

From the tree diagram in Figure 6,

$$P(B_1|R) = \frac{P(B_1) \cdot P(R|B_1)}{P(B_1) \cdot P(R|B_1) + P(B_2) \cdot P(R|B_2)}$$

In general we have the following formula.

BAYES' FORMULA

Let B_i ($i = 1, 2, \ldots, n$), with probabilities $P(B_i)$, be a finite set of disjoint events whose union is the sample space. Let A be an event that has occurred when the experiment was performed, where A is a subset of the union of the B_i. Then

$$P(B_i|A) = \frac{P(B_i) \cdot P(A|B_i)}{P(B_1) \cdot P(A|B_1) + P(B_2) \cdot P(A|B_2) + \ldots + P(B_n) \cdot P(A|B_n)}$$

EXAMPLE 4 If $P(A|B) = \frac{1}{2}$, $P(A|C) = \frac{2}{5}$, $P(A|D) = \frac{2}{3}$, $P(B) = \frac{1}{4}$, $P(C) = \frac{5}{8}$, and $P(D) = \frac{1}{8}$, find $P(C|A)$.

SOLUTION

$$P(C|A) = \frac{P(C) \cdot P(A|C)}{P(B) \cdot P(A|B) + P(C) \cdot P(A|C) + P(D) \cdot P(A|D)}$$

$$= \frac{\frac{5}{8} \cdot \frac{2}{5}}{\frac{1}{4} \cdot \frac{1}{2} + \frac{5}{8} \cdot \frac{2}{5} + \frac{1}{8} \cdot \frac{2}{3}} = \frac{6}{11}$$ ■

Now let's look at the example given in the Overview of this section.

EXAMPLE 5 Of the people in a given area, 10% are known to have some kind of cancer. A new method of detecting cancer is developed that gives a positive result 95% of the time when a person has cancer and gives a positive result 10% of the time when a person does not have cancer. Suppose that the test is positive. What is the probability that the person has cancer?

SOLUTION We want to find the probability that the person has cancer knowing that the test is positive, or $P(C|T)$.

$$P(C|T) = \frac{P(C) \cdot P(T|C)}{P(C) \cdot P(T|C) + P(C') \cdot P(T|C')}$$

We have $P(T|C) = .95$ and $P(T|C') = .10$. Likewise, $P(C) = .10$ and $P(C') = .90$.

$$P(C|T) = \frac{(.10)(.95)}{(.10)(.95) + (.90)(.10)} \approx .51$$

Thus, even though the test showed positive, the probability that the person has cancer is only .51. ■

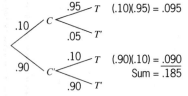

Figure 7

From the preceding example, we can construct a tree diagram as shown in Figure 7. Since we desire $P(C|T)$, we need to consider only those branches of the tree diagram that terminate in T (testing positive).

$$P(C|T) = \frac{\text{Probability of path through } C \text{ and } T}{\text{Sum of probabilities of all paths to } T}$$

$$= \frac{.095}{.185} = .51$$

Practice Problem 2 A parts company discovers a defective ignition system. Thirty percent of all ignition systems are purchased from company A, 30% from company B, and 40% from company C. Company A claims that less than 1% of

.30 A —.01— D (.30)(.01) = .0003

.30 B —.03— D (.30)(.03) = .0009

.40 C —.02— D (.40)(.02) = .0008
                    Sum = .0020

Figure 8

their parts are defective, B claims less than 3%, and C claims less than 2%. What is the probability that the defective part came from company C?

ANSWER   The tree diagram is shown in Figure 8.

$$P(D|C) = \frac{.4(.02)}{.3(.01) + .3(.03) + .4(.02)} = \frac{2}{5}$$

or

$$P(D|C) = \frac{.008}{.020} = \frac{8}{20} = \frac{2}{5}$$

SUMMARY

In this section we learned that a tree diagram is very helpful in finding Bayes' probabilities for events B_i:

$$P(B_i|A) = \frac{\text{Probability of path through } B_i \text{ and } A}{\text{Sum of the probabilities of all paths to } A}$$

Exercise Set 6.1

Find the following probabilities by referring to the tree diagram.

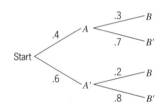

1. $P(A)$
3. $P(B|A)$
5. $P(A' \cap B)$
7. $P(B)$

2. $P(A')$
4. $P(B'|A')$
6. $P(A \cap B')$
8. $P(B')$

Find the following probabilities by referring to the tree diagram.

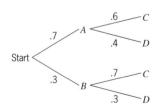

9. $P(C|A)$ 10. $P(C|B)$
11. $P(D|A)$ 12. $P(A \cap C)$
13. $P(A \cap D)$ 14. $P(B \cap C)$
15. $P(C) = P(A \cap C) + P(B \cap C)$
    $= P(A) \cdot P(C|A) + P(B) \cdot P(C|B)$
16. $P(D)$
17. $P(A|C) = \dfrac{P(A \cap C)}{P(C)} =$
    $$\frac{P(A) \cdot P(C|A)}{P(A) \cdot P(C|A) + P(B) \cdot P(C|B)}$$
18. $P(A|D)$
19. $P(B|C)$ 20. $P(B|D)$
21. Use the following tree diagram and Bayes' formula to find each probability.

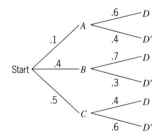

(a) $P(A|D)$ (b) $P(B|D')$
(c) $P(C|D)$

22. Box A contains 3 red chips and 4 black chips. Box B contains 5 red chips and 2 black chips. A chip is drawn from box A and placed in box B, and then a chip is drawn from box B. What is the probability that the chip is black?

23. Box A contains 6 cards numbered 1 through 6, and box B contains 4 cards numbered 1 through 4. A card is drawn from box A. If it is even, a card is then drawn from box B; if it is odd, a second card is drawn from box A. What is the probability that
(a) both cards are odd?
(b) both cards are even?
(c) one card is odd and the other is even?

24. A box contains 2 coins, one of which has two tails. A coin is selected at random and tossed. What is the probability of getting a tail?

25. Box I contains 6 red and 4 white balls, and box II contains 5 red, 3 white, and 2 green balls. An experiment consists of selecting a box (equally likely) and then randomly drawing a ball from the box selected.
(a) Find the probability of getting a red ball.
(b) Find the probability of getting a white ball.
(c) Find the probability of getting a green ball.
(d) If a red ball is drawn, find the probability that it came from box I.
(e) If a white ball is drawn, find the probability that it came from box I.
(f) If a green ball is drawn, find the probability that it came from box I.

26. An urn contains three coins; two of them are fair, and the other has heads on both sides. A coin is selected at random and tossed twice, coming up heads both times. What is the probability that the coin selected has two heads?

27. Box A contains 3 red and 5 black chips, box B has 2 red and 3 black chips, and box C has 1 red and 2 black chips. A box is selected at random and a chip is drawn. If the chip is red, what is the probability that it came from box B?

Applications (Business and Economics)

28. **Salaries.**   Of the employees in a corporation, 5% of the men and 7% of the women have salaries in excess of $40,000. Furthermore, 60% of the employees are men. If an employee who is selected at random earns more than $40,000, what is the probability that the employee is a man?

29. **Unions.**   A proposal to accept or reject a union is submitted to all employees of a corporation. Of the employees, 30% are laborers, 35% are white-collar workers, and 35% are blue-collar workers. In the response, 80% of the laborers, 15% of the white-collar workers, and 5% of the blue-collar workers vote to reject the union. What is the probability that a person selected at random who has voted against the union is
(a) a laborer?
(b) a blue-collar worker?

30. **Marketing.**   A given area is divided into submarkets according to the percentage of prospective customers: area I is 30%; area II, 45%; and area III, 25%. From a sampling of the customers, it is found in area I that 23% favor a given detergent. In area II, 20% favor the detergent, and in area III, 16%. If a customer is selected at random and favors the detergent, what is the probability that the customer came from area I? Area III?

31. **Quality Control.**   A dealer reports a major defect in a TV set. The set could have been manufactured at any one of three plants, A, B, or C. For the week in which the given set was manufactured, plant A made 30% of the sets, B made 45%, and C, 25%. During this week quality control located 2% defects at A, 3% defects at B, and 5% defects at C. What is the probability that the TV set came from C? From A?

32. **Stock Trends.**   Three proposed economic projections about the future have probabilities of .2, .5, and .3 of occurring. If the first projection occurs, the probability that the common stock of the Investors Corporation will increase by 5 points is .6; if the second projection occurs, the probability of the increase is .7; and if the third projection occurs, the probability is .8. During the next year the common

stock of the Investors Corporation does increase by 5 points. What is the probability that the first economic projection has occurred?

33. **Credit.** A credit-card company classifies its credit cards in three categories: golden cards (10% of customers), preferred cards (30% of customers), and regular cards (60% of customers). In the past, 1% of those with golden cards, 6% of the preferred customers, and 11% of the regular cardholders became delinquent each year. The company receives a nasty letter from a cardholder whose account was canceled because it had become delinquent. What is the probability that this person holds a preferred card?

Applications (Social and Life Sciences)

34. **Medicine.**   The probability that a person has disease D is $P(D) = .1$. The probability that a medical examination will indicate the disease if a person has the disease is $P(I|D) = .8$, and the probability that the examination will indicate the disease if a person does not have it is $P(I|D') = .02$. What is the probability that Patty has the disease if the medical examination indicates that she does?

35. **Lie Detection.**   Suppose that a lie-detector test is 90% accurate for guilty people and 95% accurate for innocent people. If T stands for the test indicating guilt, G stands for a person being guilty, and I stands for a person being innocent, $P(T|G) = .90$ and $P(T|I) = .05$ or $P(T'|I) = .95$. One of five people in an office is guilty of stealing money. A person is selected and given the lie-detector test.
   (a) If the test indicates he is guilty, what is the probability that he is guilty?
   (b) If the test indicates he is guilty, what is the probability that he is innocent?

36. **Diabetes.**   If a person has diabetes, a particular blood test shows positive 95% of the time. On the other hand, the test also shows positive 2% of the time for those who do not have diabetes. About 5% of the general public have diabetes. If you were to test positive, what would be the probability that you actually have diabetes?

37. **Family.**   A family has 2 girls out of 3 children. What is the probability that the first was a girl given that the last was a girl?

6.2   BERNOULLI TRIALS AND THE BINOMIAL DISTRIBUTION

OVERVIEW A new drug that causes side effects in 6% of the patients is being tested. What is the probability of no side effects if the drug is tested on 20 patients? This problem illustrates a general category of probability problems that are concerned with experiments in which an experiment (or trial) is repeated many times. For example, we might desire to find the probability of 5 heads in 10 coin tosses, or to find 1 defective item in a sample of 20 items. Probability problems of this nature are called problems with repeated experiments or **Bernoulli trials**, after John Bernoulli (1654–1705) who contributed significantly to the field of probability. In each problem, one outcome is designated as a success, and any other result on a single trial is considered a failure. The probability of success must remain constant from trial to trial. That is, in tossing a coin, the probability of getting a head on a single toss (or trial) is $\frac{1}{2}$, and this probability remains $\frac{1}{2}$ from toss to toss (or trial to trial). In this section, we

- Use a formula to find the probability of x successes in n trials
- Define a random variable
- Discuss the characteristics of a probability density function

In this section we are interested only in an outcome that can be classified as a success or a failure. For example, in tossing a coin, either a tail occurs or does not occur. In tossing a die, we either get a 6 or do not get a 6. What do these outcomes have in common? In each experiment there are only two outcomes: one we designate as a success, and its complement, which we call a failure.

EXAMPLE 6   Suppose that we toss a coin four times and we are interested in the probability of exactly two tails. Discuss the circumstances of this experiment in the context of successes and failures.

SOLUTION   In the first toss we get either a tail (called a success) or we get a head (the complement of a tail and called a failure). In the second toss the probability of a tail (namely, $\frac{1}{2}$) is the same as in the first toss. We need to keep in mind that each toss is completely independent of the other tosses.   ■

This discussion leads to the following definition.

DEFINITION: BERNOULLI TRIALS

Repeated trials of an experiment are called **Bernoulli trials** if

1. There are only two possible outcomes (success or failure) on each trial.
2. The probability of success p remains constant from trial to trial. (The probability of failure is $q = 1 - p$.)
3. All trials are independent.

Two examples of Bernoulli trials are as follows:

1. Tossing a die several times in succession. Consider getting a 6 as a success. (Note that $p = \frac{1}{6}$ is the probability of getting a 6 on one toss and this probability remains constant from toss to toss. In addition, the results of one toss are completely independent of another toss.)
2. Drawing a card from a deck several times, replacing it each time, and designating getting a heart as a success.

EXAMPLE 7   Consider the experiment of tossing a die four times. Getting a 6 will be a success, which we denote as S. (Do not confuse this S with the S we used in the preceding chapter to represent sample space.) What is the probability that the die will come up failure, success, success, success (FSSS)?

SOLUTION   We can write these outcomes as

$$F \cap S \cap S \cap S \quad \text{or} \quad FSSS$$

Now

$$P(S) = P(6) = \frac{1}{6}$$

and

$$P(F) = 1 - P(S) = 1 - \frac{1}{6} = \frac{5}{6}$$

The trials are independent, so

$$P(FSSS) = P(F \cap S \cap S \cap S) = P(F) \cdot P(S) \cdot P(S) \cdot P(S)$$
$$= \frac{5}{6} \cdot \frac{1}{6} \cdot \frac{1}{6} \cdot \frac{1}{6} = \frac{5}{1296} \qquad \blacksquare$$

EXAMPLE 8   Consider the experiment in Example 7, but this time find $P(SFSS)$.

SOLUTION $P(SFSS) = P(S \cap F \cap S \cap S)$
$$= P(S) \cdot P(F) \cdot P(S) \cdot P(S)$$
$$= \frac{1}{6} \cdot \frac{5}{6} \cdot \frac{1}{6} \cdot \frac{1}{6} = \frac{5}{1296} \qquad \blacksquare$$

We can see from the two preceding examples that, for two of the possibilities for obtaining three 6's out of four tosses of a die, the probability remains constant as $\frac{5}{1296}$. Since these possibilities are mutually exclusive, to find the probability of three 6's in four tosses, we add the probabilities of the individual possibilities. Let's list the number of ways of getting one failure and three successes. We could get a success on the first toss (S_1), or on the second (S_2), or on the third (S_3), or on the fourth (S_4). We can have only three successes. In how many ways can we select a combination of three successes from a possibility of four? The answer is the number of combinations of four things taken three at a time or $C(4, 3)$. Therefore,

$$P(\text{three 6's in 4 tosses}) = C(4, 3) \cdot \frac{5}{1296}$$
$$= \frac{4 \cdot 5}{1296} = \frac{5}{324}$$

In general, the preceding reasoning can be used to validate the following theorem.

PROBABILITY OF x SUCCESSES IN n BERNOULLI TRIALS

The probability of exactly x successes (and $n - x$ failures) in a sequence of n Bernoulli trials is given by

$$P(x \text{ successes}) = C(n, x)p^x q^{n-x}$$

where p and q are the probability of success and failure, respectively.

Thus, the probability of 10 successes in 30 Bernoulli trials is given by

$$P(10 \text{ successes}) = C(30, 10)p^{10}q^{20}$$

where $p + q = 1$.

EXAMPLE 9   An advertising agency believes that 2 out of every 3 people who respond to their advertisement will purchase the product involved. Five people who have read the advertisement are selected at random. What is the probability that exactly 3 people will purchase the product?

SOLUTION   Using $C(n, x)p^x q^{n-x}$, we note that

$$n = 5, \qquad p = \frac{2}{3}$$

$$x = 3, \qquad q = 1 - p = 1 - \frac{2}{3} = \frac{1}{3}$$

$$P(\text{exactly 3 out of 5}) = C(5, 3) \left(\frac{2}{3}\right)^3 \left(\frac{1}{3}\right)^2 = \frac{10}{1} \cdot \frac{8}{243} = \frac{80}{243} \quad \blacksquare$$

Calculator Note

In the preceding chapter we evaluated $C(n, r)$ using a calculator. Using the same procedures, P(exactly 3 successes) in the preceding example could be calculated as 5 $\boxed{\text{MATH}}$ $\boxed{\blacktriangleleft}$ (select 3: $\boxed{{}_n C_r}$) 3 $\boxed{\times}$ (2/3) $\boxed{\wedge}$ 3 $\boxed{\times}$ (1/3) $\boxed{\wedge}$ 2 $\approx$ .33.

EXAMPLE 10   In the production of radio parts, it is found that 1 out of 100 parts is defective. A sample of 8 parts is selected. What is the probability that 2 or fewer parts are defective?

SOLUTION   Let getting a defective part be a success. The probability of a defective part is $p = \frac{1}{100} = .01$. Then

$$q = 1 - \frac{1}{100} = \frac{99}{100} = .99$$

$$P(2 \text{ or fewer defects}) = P(\text{no defects}) + P(1 \text{ defect}) + P(2 \text{ defects})$$
$$= C(8, 0)p^0q^8 + C(8, 1)p^1q^7 + C(8, 2)p^2q^6$$

On a calculator use 8 ⌑MATH⌑ ⌑◀⌑ (Select 3) 0 ⌑×⌑.99 ∧ 8 ⌑+⌑ 8 ⌑MATH⌑ ⌑◀⌑ (select 3) 1 ⌑×⌑.01 ⌑∧⌑ 1 ⌑×⌑.99 ⌑∧⌑ 7 ⌑+⌑ 8 ⌑MATH⌑ ⌑◀⌑ (select 3) 2 ⌑×⌑.01 ⌑∧⌑ 2 ⌑×⌑.99 ⌑∧⌑ 6 ≈ .9999.

Thus, the probability of 2 or fewer defects is approximately .9999. ∎

Practice Problem 1 A coin is tossed 6 times. What is the probability of obtaining more than 1 head?

ANSWER $P(\text{more than 1 head}) = 1 - P(\text{no heads}) - P(1 \text{ head})$
$$= 1 - \left(\frac{1}{2}\right)^6 - 6\left(\frac{1}{2}\right)^6 = \frac{57}{64}$$

The sequence of Bernoulli trials is often called a **binomial experiment** because $C(n, x)$ also represents the coefficients in the binomial expansion of $(p + q)^n$. The number of successes can be called a binomial random variable.

In Chapters 0 and 1 we introduced the concept of a variable. When a variable represents the numerical values obtained from an experiment in probability, it is called a **random variable**, indicating that the variable is associated with a random experiment. When one performs experiments such as tossing a coin, rolling a die, or counting defective items, a random variable is used to represent the real numbers that exist for each outcome of the experiment. We consider the following abbreviated definition of a random variable.

DEFINITION: RANDOM VARIABLE

A **random variable** is a function or rule that assigns numerical values to the elements of a sample space.

A random variable, usually denoted by a capital letter, is said to be finite if it can take on only a fixed number (such as 20 or 100) of different values. The following is an example of a finite random variable.

EXAMPLE 11   Toss a pair of dice and define the random variable X to take on values that are the sum of the two numbers on top. What are the possible values of this random variable?

SOLUTION   The values of X that we associate with this experiment are 2, 3, 4, 5, 6, 7, 8, 9, 10, 11, and 12, that is, the sum of the two numbers on top of the two dice (see the first column of Table 1). Again, since X can take on only 11 values, it is said to be finite. ∎

TABLE 1

X	Sample Points	Probability
2	(1, 1)	$\frac{1}{36}$
3	(1, 2),(2, 1)	$\frac{2}{36}$
4	(1, 3),(2, 2),(3, 1)	$\frac{3}{36}$
5	(1, 4),(2, 3),(3, 2),(4, 1)	$\frac{4}{36}$
6	(1, 5),(2, 4),(3, 3),(4, 2),(5, 1)	$\frac{5}{36}$
7	(1, 6),(2, 5),(3, 4),(4, 3),(5, 2),(6, 1)	$\frac{6}{36}$
8	(2, 6),(3, 5),(4, 4),(5, 3),(6, 2)	$\frac{5}{36}$
9	(3, 6),(4, 5),(5, 4),(6, 3)	$\frac{4}{36}$
10	(4, 6),(5, 5),(6, 4)	$\frac{3}{36}$
11	(5, 6),(6, 5)	$\frac{2}{36}$
12	(6, 6)	$\frac{1}{36}$

Now it seems logical to associate a probability with each value of a random variable. For example, the probability that a 2 occurs in Example 11 is $\frac{1}{36}$, that a 3 occurs is $\frac{2}{36}$, that a 4 occurs is $\frac{3}{36}$, and so on. The function that assigns these probabilities is called a distribution function of the variable X.

DEFINITION: PROBABILITY DISTRIBUTION FUNCTION OF A DISCRETE RANDOM VARIABLE

Let X be a finite random variable. The **probability distribution function** of X is a function or rule $p(x_i)$ defined for $1 \le i \le n$ by

$$p(x_i) = p(X = x_i)$$

where n is the natural number of elements x_i.

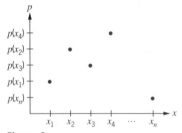

Figure 9

The preceding distribution is sometimes called a **discrete probability distribution function**. It is discrete because it is defined only for integral values, 0, 1, 2, 3, . . . , n. Graphically, a probability distribution function is defined by a finite number of points, as shown in Figure 9.

EXAMPLE 12   Graph the probability distribution function of the random variable in Example 11.

SOLUTION   The graph of the finite probability distribution function in Example 11 is shown in Figure 10. ∎

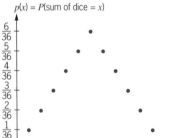

Figure 10

Calculator Note

The characteristics and graph of a probability distribution function can be displayed using a graphing calculator. See page 000 of Appendix A for a discussion of Example 11.

From the definition of a probability distribution function and the preceding examples, you can discover the following properties of a probability distribution function.

PROPERTIES OF A PROBABILITY DISTRIBUTION FUNCTION

A discrete probability distribution function (sometimes called a discrete probability density function) of a random variable satisfies
(a) $0 \leq p(x_i) \leq 1$, where $i = 1, 2, \ldots, n$.
(b) $p(x_1) + p(x_2) + \cdots + p(x_n) = 1$, where $x_1, x_2, \ldots, x_n$ are all the possible values assigned by the random variable.

Note in Example 11 that $0 \leq p(x_i) \leq 1$ because each $p(x_i)$ is a probability. The sum of all the $p(x_i)$ is 1 because the probability of getting either a 2, 3, 4, 5, 6, 7, 8, 9, 10, 11, or 12 in tossing two dice is 1.

Let's now consider a discrete probability distribution that represents the number of successes in n Bernoulli trials with corresponding probabilities $C(n, x)p^x q^{n-x}$. This probability distribution is called the **binomial distribution**.

EXAMPLE 13   The probability that a medical procedure causes a reaction is $\frac{1}{6}$. A doctor plans to use this procedure on 4 patients but desires to study possible reactions. Find the probability of getting x reactions and tabulate the values of the variable and corresponding probabilities of the discrete probability distribution function.

SOLUTION   We have

$$P(x \text{ successes in 4 tosses}) = C(4, x) \left(\frac{1}{6}\right)^x \left(\frac{5}{6}\right)^{4-x}$$

The probability distribution is defined in Table 2. ∎

Practice Problem 2   A family decides to have exactly 4 children. What is the probability of having exactly 2 girls and 2 boys?

ANSWER $P(\text{exactly 2 boys and 2 girls}) = C(4, 2) \left(\frac{1}{2}\right)^2 \left(\frac{1}{2}\right)^2 = \frac{3}{8}$

TABLE 2

x	$P(x)$	
0	$C(4, 0) \left(\dfrac{1}{6}\right)^0 \left(\dfrac{5}{6}\right)^4$	$\approx .482$
1	$C(4, 1) \left(\dfrac{1}{6}\right)^1 \left(\dfrac{5}{6}\right)^3$	$\approx .386$
2	$C(4, 2) \left(\dfrac{1}{6}\right)^2 \left(\dfrac{5}{6}\right)^2$	$\approx .116$
3	$C(4, 3) \left(\dfrac{1}{6}\right)^3 \left(\dfrac{5}{6}\right)^1$	$\approx .015$
4	$C(4, 4) \left(\dfrac{1}{6}\right)^4 \left(\dfrac{5}{6}\right)^0$	$\approx .001$
	Total	1.000

SUMMARY

In this section we introduced probability distributions and discussed their properties. In the repeated application of an experiment, the trials are called Bernoulli trials if

1.   There are only two possible outcomes (called a success or a failure).
2.   The probability of a success remains constant from trial to trial.
3.   All trials are independent.

The probability of x successes in n Bernoulli trials is given by

$$P(x \text{ successes}) = C(n, x)p^x q^{n-x}$$

Exercise Set 6.2

Find C(n, x)p^xq^{n-x} for the given values of n, x, and p

1. $n = 4$, $x = 3$, $p = \dfrac{1}{3}$

2. $n = 6$, $x = 4$, $p = .1$

3. $n = 8$, $x = 1$, $p = \dfrac{2}{3}$

4. $n = 10$, $x = 8$, $p = .9$

5. Assume that there are 4 independent repetitions of a Bernoulli experiment with the probability of success on a single trial equal to $\frac{1}{3}$. Find the following probabilities:
   (a) No successes
   (b) Exactly 1 success
   (c) Exactly 2 successes
   (d) Exactly 3 successes
   (e) Exactly 4 successes
   (f) Exactly 5 successes

6. What is the probability of obtaining exactly 3 heads if a coin is tossed 6 times?

7. An ordinary die is tossed 6 times. What is the probability of obtaining exactly three 5's?

8. Records show that 0.2% of the taxicabs in a firm have accidents each day. If 5 taxicabs are operating on a given day, what is the probability of exactly 1 accident?

9. Suppose that a family has 5 children. Find the probability of each of the following.
   (a) Exactly 3 girls
   (b) No boys
   (c) Exactly 5 boys
   (d) At least 4 girls
   (e) No more than 3 boys
   (f) At least 3 boys

10. Define a probability distribution function from Exercise 5 by making a table.

11. A family consists of 8 girls and no boys. What is the probability of such an occurrence?

12. In its training program, a company has a drop-out rate of 0.20. If 8 trainees start the program, what is the probability that 6 or more will finish?

13. A baseball player has a batting average of .200. What is the probability that he will get exactly 2 hits in the next 5 official times at bat?

14. On a true–false test consisting of 5 questions, what is the probability of passing (60% or better) by guessing?

15. Sam Selpeep is noted for his ability to sleep through an entire class. Following a psychology lecture during which he performed admirably by sleeping throughout the class period, he was given a 10-question, true–false test. What is his probability of getting better than 75% of the answers correct by guessing?

16. Let x be the number of successes in 10 independent repetitions of a Bernoulli experiment where the probability of success on a single trial is .4. Find each of the following:
   (a) $P(x = 6)$ (b) $P(x = 0)$
   (c) $P(x > 8)$ (d) $P(x < 6)$
   (e) $P(x > 9)$ (f) $P(x \leq 2)$

Applications (Business and Economics)

17. *Sales.*   A company has found that 25% of all customers contacted will buy its product. If 10 customers are contacted, what is the probability of more than 2 sales?

18. *Quality Control.*   Suppose that 10% of the items produced by a factory are defective. If 6 items are chosen at random, what is the probability that
   (a) exactly 1 is defective?
   (b) exactly 3 are defective?
   (c) at least 1 is defective?
   (d) fewer than 3 are defective?

19. *Quality Control.*   In the manufacture of a certain item, under normal conditions 4% of the items have a defect. If 10 items are selected from an assembly line, what is the probability that exactly 2 will be defective?

Applications (Social and Life Sciences)

20. *Fertility Drug.*   The probability of multiple births for women using a certain fertility drug is 20%. Ten women take the fertility drug. What is the probability

of more than 2 but fewer than 6 of the women having multiple births?

21. ***Drug Effectiveness***.   A drug manufacturer claims that a particular drug is effective 90% of the time. A physician prescribes the drug to 10 patients and 6 respond to this treatment. What is the probability of 6 or fewer successes if $p = .9$? What conclusion is the physician apt to draw?

22. ***Side Effects***.   A new drug is being tested that causes side effects in 6% of the patients. What is the probability of no side effects if the drug is tested on 20 patients?

23. ***Divorce***.   The probability that a couple in Colorado will get a divorce within the next 10 years is $\frac{1}{5}$. Six couples living in Colorado form a dinner club. What is the probability that at least one of the couples in the club will get a divorce in the next 10 years?

24. ***Politics***.   Two-thirds of the participants at a state Republican convention are conservatives. If 7 members of the convention are chosen at random to serve on a committee, what is the probability that the conservatives will be the majority?

Review Exercises

25. Supervisors indicate that 90% of the workers at an aluminum company are good workers. On a work-aptitude test, 80% of the good workers and 20% of those classified as not good workers made passing grades. If a worker makes a passing grade on the aptitude test, what is the probability that she is a good worker?

26. The probability that a person has diabetes is .06. The probability that the sugar-tolerance test will show diabetes if the disease is present is .90 and .02 if the person does not have diabetes. What is the probability that a person has diabetes if the sugar-tolerance test so indicates?

6.3   EXPECTED VALUE AND DECISION MAKING

OVERVIEW   Probabilities alone do not always supply all the information that is useful when making a decision in an uncertain situation. For example, what a businessperson stands to gain or lose in a transaction is important in making a decision. The tool that is used to determine such expected gain or loss is **expected value**, which was originally developed for games of chance. We discuss several games of chance in this section. In addition, we use a formula for finding the expected value for the binomial distribution discussed in the preceding section. In this section we

- Define expected value
- Use a formula for expected value
- Find expected value for a binomial distribution

An important property associated with probability is that of **expectation** or **expected value**. If we toss a coin 100 times, we would expect to get a head

$$100 \cdot \frac{1}{2} = 50 \text{ times}$$

If we spin a spinner with 10 equal sections 1000 times, we would expect the spinner to stop on any given section

$$1000 \cdot \frac{1}{10} = 100 \text{ times}$$

The concept is perhaps most easily explored in the analysis of a simple game of chance such as the following. A coin is tossed. If a head appears, we receive $5 from our opponent; if a tail appears, we pay our opponent a sum of $2.

We investigate what happens if we play the game 100 times. In 100 flips of the coin we can expect approximately 50 heads and 50 tails. Hence, we can expect a payoff of approximately (50) ($5) from the heads and a payoff of (50) (− $2) from the tails (we lose $2 each time a tail occurs). Our net profit would thus be (50) ($5) + (50) (− $2) = $150. Since the game was played 100 times, our average profit per game would be $\frac{150}{100}$ = $1.50.

It is very important to observe that there is an alternative way to compute this average gain per game.

$$P(\text{winning \$5}) = P(\text{H}) = \frac{1}{2}$$

$$P(\text{losing \$2}) = P(\text{T}) = \frac{1}{2}$$

$$(\$5)P(\text{winning \$5}) + (-\$2)P(\text{losing \$2}) = \$5 \cdot \frac{1}{2} + (-\$2) \cdot \frac{1}{2}$$
$$= \$2.50 - \$1 = \$1.50$$

DEFINITION: EXPECTED VALUE

If an experiment has a probability distribution

x	x_1	x_2	$\cdots$	x_n
$p(x)$	$p(x_1)$	$p(x_2)$	$\cdots$	$p(x_n)$

then the **expected value** E of the experiment is

$$E(x) = x_1 p(x_1) + x_2 p(x_2) + \cdots + x_n p(x_n)$$

To find expected value of an experiment:

1. Find all possible values of some variable x: $x_1, x_2, \ldots, x_n$.
2. Find the probabilities for each x_i. Make certain that the sum of the probabilities is 1.
3. Compute $E = x_1 p(x_1) + x_2 p(x_2) + \cdots + x_n p(x_n)$.

TABLE 3

x	$p(x)$	$xp(x)$
0	$\frac{1}{8}$	0
1	$\frac{3}{8}$	$\frac{3}{8}$
2	$\frac{3}{8}$	$\frac{6}{8}$
3	$\frac{1}{8}$	$\frac{3}{8}$
Total	1	$\frac{12}{8}$

EXAMPLE 14   Find the expected number of heads in the toss of 3 coins.

SOLUTION   Let x be a variable representing the number of heads that can appear in the toss of 3 coins. In this example x can assume values 0, 1, 2, or 3. Tabulating the results as in Table 3 assists in computing the expected value of x. The expected number of heads is $\frac{3}{2}$.

$$E(x) = \frac{12}{8} = \frac{3}{2}$$

It must be understood that expected value deals with what happens to the average in the long run when an experiment is performed again and again. Hence, in any given toss of the 3 coins, we could get 2 heads or even 3 heads. However, in repeated experimentation the *average* number of heads would be 1.5.

EXAMPLE 15   A nationwide lottery promises a first prize of $25,000, two second prizes of $5000, and four third prizes of $1000. A total of 950,000 persons enter the lottery.

(a)   What is the expected value if the lottery costs nothing to enter?
(b)   Is it worth the stamp required to mail the lottery form?

SOLUTION

(a)   Since

$$P(\$25,000) = \frac{1}{950,000}, \qquad P(\$5000) = \frac{2}{950,000}$$

$$P(\$1000) = \frac{4}{950,000}, \qquad P(\$0) = \frac{949,993}{950,000}$$

the expected value is

$$(\$25,000)\frac{1}{950,000} + (\$5000)\frac{2}{950,000} + (\$1000)\frac{4}{950,000} = .041$$

In repeated participation in this lottery, the average value of prizes won would be $0.04.

(b)   Not when the expected value is only approximately $0.04.

Expected value is useful in studying games of chance. In a **fair game**, the expected value of the game is 0. Casinos and lotteries understandably do not operate fair games; instead, they operate games in which the player has an expected value that is negative. That is, the player can expect to lose money on repeated playing of the game.

Consider a game in which a player pays $2 for the privilege of playing. Suppose that her probability of winning $10 is $\frac{1}{10}$ and her probability of losing is $\frac{9}{10}$. Actually, if she wins, her winnings are $10 − $2 = $8 because she pays $2 to play. If she loses, she will lose $2. The expected value of the game is

$$E(x) = \$8 \left(\frac{1}{10}\right) - \$2 \left(\frac{9}{10}\right) = -\$1$$

Thus, if the player should continue to play the game, her average earnings would be −$1. That is, she would lose, on the average, $1 per game.

It should be clear that expected value would be helpful in making decisions involving uncertainty. One method in choosing a course of action is to choose the action that gives the largest expected value. We will see throughout the remainder of this chapter how expected value plays a role in monetary situations and in various social science problems.

EXAMPLE 16   A company has the privilege of bidding on two contracts, A and B. It is estimated that, if the company should win contract A, a profit of $14,000 (above all costs) would be realized; however, it costs $500 to prepare a proposal in order to submit a bid. Contract B would give a profit of only $10,000 (above all costs), but the cost of preparing a proposal would be only $200. It is estimated that the probability that the company will win contract A is $\frac{1}{4}$, and the probability of winning contract B is $\frac{1}{3}$. If the company can submit a proposal for only one contract, which proposal should be submitted?

SOLUTION   The probability of winning contract A is $\frac{1}{4}$, so the probability of not winning this contract is $1 - \frac{1}{4} = \frac{3}{4}$. If the contract is won, the profit is $14,000; if the contract is not won, the loss is $500. Thus,

$$E(A) = \$14,000 \left(\frac{1}{4}\right) + (-\$500)\left(\frac{3}{4}\right)$$

$$= \$3125$$

Similarly, the expected value from B is

$$E(B) = \$10,000 \left(\frac{1}{3}\right) + (-\$200)\left(\frac{2}{3}\right)$$

$$= \$3200$$

If the expected value is considered as an appropriate criterion for a decision, the proposal should be for contract B.

TABLE 4

Minutes Late	Probability
1	.1
2	.4
3	.3
4	.1
5	.05
6	.05

EXAMPLE 17   Students over the years have observed that a certain professor is never on time and yet never misses class. Records kept by students in the back row indicate the probabilities shown in Table 4.

(a)   Find the expected number of minutes that the professor will be late?

(b)   If a class is to meet 45 times a semester, what is the expected number of minutes of enlightenment to be denied the students?

SOLUTION

(a) $E = 1P(1) + 2P(2) + 3P(3) + 4P(4) + 5P(5) + 6P(6)$
$= 1(.1) + 2(.4) + 3(.3) + 4(.1) + 5(.05) + 6(.05)$
$= 2.75$ minutes

(b) $(45)(2.75) = 123.75$ minutes

TABLE 5

Liability	Corresponding Probabilities
500,000	.0001
100,000	.001
50,000	.004
30,000	.01
5,000	.04
1,000	.06
0	.8849

Practice Problem 1   An insurance company insures 200,000 cars each year. Records indicate that during the year the company will make the liability payments shown in Table 5 for accidents. What amount could the company expect to pay per car insured?

ANSWER   The expected liability per car is $910.

We now find the expected value of a binomial distribution.

EXAMPLE 18   Consider rolling a die 4 times, where getting a 6 is a success. Find the expected number of successes.

TABLE 6

x	$p(x)$	$xp(x)$
0	.482	.000
1	.386	.386
2	.116	.232
3	.015	.045
4	.001	.004
Total		.667

SOLUTION   As in Example 14, Table 6 defines the probability distribution function. The last column will be used to find the expected value.
The expected value is .667. Note also that $4(\frac{1}{6}) = .667$, where 4 is the number of times the experiment was performed, and $\frac{1}{6}$ is the probability of success on a single trial.   ∎

The preceding example suggests that there is a faster method for finding the expected value of a binomial distribution. Expected value of a distribution (also called the mean of a distribution is sometimes denoted by the Greek letter μ or as μ_x.

EXPECTED VALUE OF A BINOMIAL DISTRIBUTION

The **expected value** (or the **mean**) **of a binomial distribution** of n independent trials, each with a probability of success p, is given by

$$E(x) = np$$

The binomial distribution is a special example of theoretical distributions that are studied in a statistics course. The probabilities and the mean are often computed to characterize the distribution. For example, the probability that a can of cola has a defect at a bottling company is .001. A sample of 6 cans is selected at random. We write the binomial distribution for the number of defective cans in the sample (see Table 7), and we compute the mean of the distribution. The mean of the distribution is

$$np = 6(.001) = .006$$

TABLE 7

x (defects)	$P(x)$
0	$C(6, 0)(.001)^0(.999)^6 \approx .99402$
1	$C(6, 1)(.001)^1(.999)^5 \approx .00597$
2	$C(6, 2)(.001)^2(.999)^4 \approx .00001$
3	$C(6, 3)(.001)^3(.999)^3 \approx 0$
4	$C(6, 4)(.001)^4(.999)^2 \approx 0$
5	$C(6, 5)(.001)^5(.999)^1 \approx 0$
6	$C(6, 6)(.001)^6(.999)^0 \approx 0$

EXAMPLE 19 The probability of a computer failure in the first year of a new Oldsmobile is .0001. The Southeast region sold 100,000 Oldsmobiles. Find the expected number of computer failures.

SOLUTION $E(x) = np$ $\qquad p = .0001,\ n = 100,000$
$= 100,000(.0001)$
$= 10$

Practice Problem 2 In a quality-control inspection, 1 out of 500 of all items inspected have a defect. In a shipment of 10,000 items, how many defective items are expected?

ANSWER $10,000 \left(\dfrac{1}{500} \right) = 20$

SUMMARY

In this section we introduced the formula for expected value:

$$E(x) = x_1 p(x_1) + x_2 p(x_2) + x_3 p(x_3) + \cdots + x_n p(x_n)$$

where $p(x_i)$ is the probability that x_i occurs. This formula can be used to find the mean of a finite probability distribution. The expected value of a binomial distribution is

$$E(x) = np.$$

Exercise Set 6.3

1. An alphabet block (six sides containing letters A, B, C, D, E and F) is rolled on the floor 360 times. How many times would you expect the A to be on the top?

Find the expected value for each of the following:

2.

Outcome	1	2	3	4	5
Probability	.1	.2	.4	.2	.1

3.

Outcome	2	4	6	8
Probability	.2	.3	.3	.2

4.

x_i	0	1	2	3	4
$p(x_i)$	$\frac{1}{8}$	$\frac{1}{4}$	$\frac{1}{4}$	$\frac{1}{4}$	$\frac{1}{8}$

5.

x_i	0	2	4
$p(x_i)$	$\frac{1}{6}$	$\frac{4}{6}$	$\frac{1}{6}$

6. Find the expected value of the following binomial distributions with given p and n:

   (a) $p = \dfrac{1}{2}, n = 60$

   (b) $p = \dfrac{1}{3}, n = 240$

   (c) $p = \dfrac{1}{6}, n = 180$

   (d) $p = .01, n = 1000$

7. A highway engineer knows that her crew can repair 4 miles of road a day in dry weather and 2 miles of road a day in wet weather. If the weather in her region is rainy 25% of the time, what is the average number of miles of repairs per day that she can expect?

8. In a lottery, 200 tickets are sold for $1 each. There are four prizes, worth $50, $25, $10, and $5. What is the expected value for someone who purchases 1 ticket?

9. In one version of the game of roulette, the wheel has 37 slots numbered 0 through 36. The player bets $1 on a number from 0 to 36. If the ball comes to rest on his number, he receives $36 including his stake. Otherwise, he loses his dollar.
   (a) What is the expected value for the player?
   (b) How much can he expect to lose in 100 games?

10. Several students decide to play the following game for points. A single die is rolled. If it shows an even number, the student receives points equal to twice the number of dots showing. If it shows an odd number, the student loses points equal to 3 times the number of dots showing. What is the expected value of the game?

11. A man who rides a bus to work each day determines that the probability that the bus will be on time is $\frac{7}{16}$. The probability that the bus will be 5 minutes late is $\frac{3}{16}$; 10 minutes late, $\frac{1}{4}$; and 15 minutes late, $\frac{1}{8}$. What is his expected waiting time if the man arrives at the bus stop at the scheduled time?

12. Two coins, each biased (or weighted) so that $P(H) = \frac{1}{3}$ and $P(T) = \frac{2}{3}$, are tossed. The payoff is $5 for matching heads, $3 for matching tails, and $-$2 if they don't match. What is the expected value of the game?

Find the expected value for the random variable whose probability function is graphed in Exercises 13 and 14.

13.

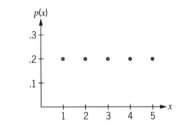

14.

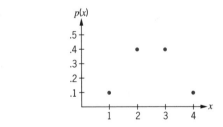

15. Records show that 0.2% of the taxicabs in New York City have accidents each day. If a company has 100 cabs operating on a given day, how many accidents can be expected?

16. Sam Selpeep is noted for his ability to sleep through an entire class. Following a psychology lecture during which he performed admirably by sleeping throughout the class period, he was given a 10-question, true–false test. What is his expected number of correct answers?

17. Find the expected value of the total number of dots that appear in the toss of 3 dice.

Applications (Business and Economics)

18. *Contracts*.   A corporation prepares a bid on a job at a cost of $7000. They estimate that if they get the job, they will make $250,000 in profits. If the probability of getting the job is .4, what is their expected profit or loss?

19. *Contracts*.   Find the expected values for the following contracts.
    (a) Estimated profit $5000, cost of proposal $500, probability of winning $\frac{1}{8}$
    (b) Estimated profit $1000, cost of proposal $200, probability of winning $\frac{1}{3}$
    (c) Estimated profit $10,000, cost of proposal $600, probability of winning $\frac{1}{3}$

20. *Sales*.   During a sale, an appliance dealer offers a chance on a $1200 motorcycle for each refrigerator sold. If he sells his refrigerators at $25 more than other dealers, and if he sells 120 units during the sale, what is the expected value of a purchase to the consumer?

21. *Sales*.   A company has found that 25% of all customers contacted will buy its product. If 20 customers are contacted, how many sales can be expected?

22. *Quality Control*.   During inspection of 1000 welded joints produced by a certain machine, 100 defective joints are found. Consider the random variable to be the number of defective joints that result when 5 joints are welded. How many defective joints are expected?

23. *Quality Control*.   The output of an automatic machine at a corporation is analyzed and found to be a binomial process with a probability of a defect of .04. Consider a sample of 6 units to be tested. What is the expected number of defects?

24. *Training Program*.   In a training program, a company has a dropout rate of 12%. If 50 trainees start the program, how many can the company expect to finish it?

Applications (Social and Life Sciences)

25. *Mortality Table*.   According to a mortality table, the probability that a 20-year-old woman will live 1 year is .994; the probability that she will die is, of course, .006. She buys a $1000, 1-year term life-insurance policy for $20. What is the expected loss or gain of the insurance company? (Assume no interest for the year.)

26. *Mortality Table*.   The probability that a man, age 40, will live for 1 year is .906. How large a premium should he be willing to pay for a 1-year, $2000 term policy? (Assume no interest for the year.)

27. *Polls*.   A public-opinion pollster finds that, in a mailing process costing $0.50 per questionnaire, she gets a 40% return. Her follow-up procedure to ensure a reply to most of the questionnaires costs $3 per questionnaire. She devises a scheme that costs $1 per questionnaire, but she thinks she will get an 80% return. Which scheme should she use for minimum expected cost?

28. *Unemployment*.   The probability that a steelworker will remain employed during the next year is .866. Each steelworker who loses his job is eligible for $2000 in unemployment benefits from his state. How much money should the state have in its budget for each steelworker?

29. *Testing Techniques*.   A psychology professor notes that a systematized review increases her students' scores on final examinations. Of the number of times she has given final examinations after systematized reviews, scores have increased $\frac{1}{4}$ of the time by 20%, $\frac{1}{2}$ of the time by 10%, $\frac{1}{8}$ of the time by 8%, and $\frac{1}{8}$ of the time by 4%. What is the expected increase in final-examination scores if the professor gives her students a systematized review?

30. *Politics*.   Two-thirds of the participants at a state Republican convention are conservatives. If 12 people are chosen at random to be the rules committee,

how many conservatives can you expect to be on this committee?

Review Exercises

31. An experiment involves two boxes. Box A contains 4 red chips and 3 black chips, box B contains 5 red and 2 black chips. One of the boxes is selected at random, and a chip is drawn. Suppose we know only that a red chip was drawn. What is the probability that it was drawn from box A?

32. A shipment contains 96 good items and 4 defective ones. Three items are drawn, one at a time, from the shipment.
    (a) What is the probability of 3 defective items?
    (b) What is the probability of 2 good items and 1 defective item?

33. Eight poker chips are numbered 10, 11, 12, 13, 14, 15, 16, and 17, respectively. Suppose that a chip is drawn and its number recorded as x. Find $E(x)$.

34. Eight poker chips are numbered 1, 2, 3, 4, 5, 6, 7, and 8, respectively. If x is the random variable denoting the sum of numbers on two chips drawn at random, tabulate the probability function; find $E(x)$.

6.4 MARKOV CHAINS

OVERVIEW   In this section, matrix theory and probability theory are combined to focus on a new application. In the first section of this chapter we introduced stochastic processes, processes or experiments in which outcomes depend on previous outcomes. If the outcome of an experiment is dependent on *only* the outcome immediately preceding it, the process is called a Markov process or **Markov chain**. Andrei Andreyevich Markov (1856–1922) was a Russian mathematician who first introduced the study of this type of stochastic process. Weather patterns, the behavior of animals in psychological tests, population studies, the study of price and market trends, as well as many other processes can be modeled using this technique. In this section we

- Write transition matrices for problems involving Markov chains
- Find steady-state matrices for Markov processes
- Determine when a Markov chain is regular

The United Way of Atlanta has determined that 90% of those who contribute one year will contribute the next year. (This of course implies that 10% do not contribute the next year.) Also, 20% of those who do not contribute one year will contribute the next year. This information can be represented by a **transition matrix**, where C represents those who contribute and DNC represents those who do not contribute:

$$
\begin{array}{c}
\text{Next State} \\
\begin{array}{cc} \text{C} & \text{DNC} \end{array} \\
\text{Present State} \begin{array}{c} \text{C} \\ \text{DNC} \end{array}
\begin{bmatrix} .9 & .1 \\ .2 & .8 \end{bmatrix}
\end{array}
$$

Note in the preceding discussion that today (called present state) a person is classified as a contributor (C) or a noncontributor (DNC) and the elements of the matrix give the probabilities that these will be contributors or noncontributors in the next state.

In this section we are interested in a process moving from one state to the next or a transition from the present state to the next state with the following property.

DEFINITION: MARKOV CHAIN

If the probabilities of the outcomes of any trial except the first depend on the outcome of the preceding trial only, the series of trials is called a Markov chain.

EXAMPLE 20 The following transition matrix represents the various probabilities of smoking or nonsmoking fathers having smoking or nonsmoking sons.

		Sons (Next State)	
		Smoking	Nonsmoking
Fathers	Smoking	$\begin{bmatrix} .8 & .2 \end{bmatrix}$	
(Present State)	Nonsmoking	$\begin{bmatrix} .4 & .6 \end{bmatrix}$	

We obtain the desired probabilities by locating the appropriate row and column. For example, the probability of a smoking father having a nonsmoking son is only .2, whereas the probability that a nonsmoking father will have a nonsmoking son is .6. ∎

We illustrate a transition matrix of a Markov chain by a simple experiment having only two possible outcomes, S_1 and S_2. If the present state is S_1 and the experiment is performed, then the next state can be either S_1 or S_2. Likewise, if the present state is S_2, then the next state can be S_1 or S_2. Note that there are four transition probabilities involved:

p_{11}, probability of going from S_1 to S_1 Staying in S_1

p_{12}, probability of going from S_1 to S_2

p_{21}, probability of going from S_2 to S_1

p_{22}, probability of going from S_2 to S_2 Staying in S_2

The tree diagram in Figure 11(a) and the transition diagram in Figure 11(b) show the possibilities of this experiment. As shown, at S_1 it is possible to go to only S_1 or S_2. Therefore,

$$p_{11} + p_{12} = 1$$

Likewise,

$$p_{21} + p_{22} = 1$$

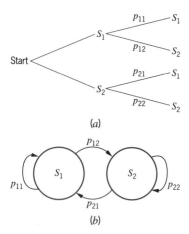

(a)

(b)

Figure 11

Thus, the matrix

$$\begin{bmatrix} p_{11} & p_{12} \\ p_{21} & p_{22} \end{bmatrix}$$

is called a **transition matrix for a Markov chain**. Such matrices have the following characteristics.

DEFINITION: TRANSITION MATRIX OF A MARKOV CHAIN

A **transition matrix of a Markov chain** having n states is the $n \times n$ matrix

New State

$$\text{Present State} \quad \begin{array}{c} S_1 \\ S_2 \\ \\ \\ \\ S_n \end{array} \begin{bmatrix} \begin{array}{cccc} S_1 & S_2 & \cdots & S_n \end{array} \\ p_{11} & p_{12} & \cdots & p_{1n} \\ p_{21} & p_{22} & \cdots & p_{2n} \\ \cdot & \cdot & \cdot & \cdot \\ \cdot & \cdot & \cdot & \cdot \\ \cdot & \cdot & \cdot & \cdot \\ p_{n1} & p_{n2} & \cdots & p_{nn} \end{bmatrix}$$

where p_{ij} is the probability of moving from state S_i to state S_j and where $p_{i1} + p_{i2} + \cdots + p_{in} = 1$ for $1 \le i \le n$ (the sum of probabilities in any row is 1).

EXAMPLE 21   A basketball player has discovered a very interesting fact about his free-throw accuracy. When he makes a free throw, he is 80% accurate on his next attempt. When he misses a free throw, he is only 60% accurate on the next attempt. Identify appropriate states for a Markov chain, and find the transition matrix.

SOLUTION   Let S_1 represent making a free throw and S_2 represent missing one. Assume that only the present state influences the next attempt. The transition diagram for this problem is given in Figure 12. The transition matrix for this Markov chain is

$$\mathbf{A} = \begin{array}{c} \\ S_1 \\ S_2 \end{array} \begin{array}{c} \begin{array}{cc} S_1 & S_2 \end{array} \\ \begin{bmatrix} .80 & .20 \\ .60 & .40 \end{bmatrix} \end{array}$$

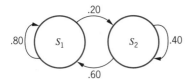

Figure 12

In Markov chains, an experiment may be repeated many times. Suppose that an experiment is repeated four times. Then $p_{13}(4)$ would represent the probability of going from state 1 to state 3 when the experiment has been performed four times. The following example illustrates an interesting way to obtain probabilities of repeated experiments.

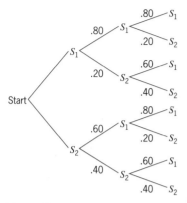

Figure 13

EXAMPLE 22   Use the data in Example 21 to find the transition probabilities after 2 free throws have been shot.

SOLUTION   The transition probabilities are shown in Figure 13, where S_1 represents the state of having made a free throw and S_2 represents the state of having missed one. Note that the transition can be made from state 1 to state 1 in two experiments in two ways: through either S_1 or S_2 at the end of one experiment. The same is true for the other transitions, the probability of each being the sum of two products.

$$p_{11}(2) = .80(.80) + .20(.60) = .76 \qquad S_1 \text{ to } S_1$$

$$p_{12}(2) = .80(.20) + .20(.40) = .24 \qquad S_1 \text{ to } S_2$$

$$p_{21}(2) = .60(.80) + .40(.60) = .72 \qquad S_2 \text{ to } S_1$$

$$p_{22}(2) = .60(.20) + .40(.40) = .28 \qquad S_2 \text{ to } S_2$$

Let's note what happens when we square the transition matrix **A**:

$$\mathbf{A}^2 = \begin{bmatrix} .80 & .20 \\ .60 & .40 \end{bmatrix}^2 = \begin{bmatrix} .76 & .24 \\ .72 & .28 \end{bmatrix}$$

A comparison of the probabilities within the square of the transition matrix and the transition probabilities after 2 free throws indicates that

$$\mathbf{A}^2 = \begin{bmatrix} p_{11}(2) & p_{12}(2) \\ p_{21}(2) & p_{22}(2) \end{bmatrix}$$

The preceding example illustrates the following theory.

Power of a Transition Matrix

If a transition matrix of a Markov chain is raised to the nth power, the element $p_{ij}(n)$ of the nth power of the matrix gives the probability that an experiment in state S_i will be in state S_j after n repetitions of the experiment.

EXAMPLE 23   Given the transition matrix

$$\begin{array}{cc} & S_1 \quad S_2 \end{array}$$

$$\mathbf{A} = \begin{array}{c} S_1 \\ S_2 \end{array} \begin{bmatrix} \frac{1}{2} & \frac{1}{2} \\ \frac{1}{3} & \frac{2}{3} \end{bmatrix}$$

suppose that an experiment is in state 1. Find the probability it is in state 2

(a) one step later.
(b) two steps later.
(c) three steps later.

SOLUTION

$$\mathbf{A}^2 = \begin{bmatrix} \frac{5}{12} & \frac{7}{12} \\ \frac{7}{18} & \frac{11}{18} \end{bmatrix} \quad \text{and} \quad \mathbf{A}^3 = \begin{bmatrix} \frac{29}{72} & \frac{43}{72} \\ \frac{43}{108} & \frac{65}{108} \end{bmatrix}$$

(a) From $\mathbf{A}$, $p_{12}(1) = \frac{1}{2}$.
(b) From $\mathbf{A}^2$, $p_{12}(2) = \frac{7}{12}$.
(c) From $\mathbf{A}^3$, $p_{12}(3) = \frac{43}{72}$. ∎

Calculator Note

In Chapters 2 and 3 we learned to find powers of square matrices using a calculator. For

$$\mathbf{A} = \begin{bmatrix} \frac{1}{2} & \frac{1}{2} \\ \frac{1}{3} & \frac{2}{3} \end{bmatrix}$$

on a calculator, we have

$$\mathbf{A}^2 = \begin{bmatrix} .41667 & .58333 \\ .38889 & .61111 \end{bmatrix}$$

$$\mathbf{A}^3 = \begin{bmatrix} .40278 & .59722 \\ .39815 & .60185 \end{bmatrix}$$

and

$$p_{1,2}(1) = .5$$
$$p_{1,2}(2) = .58333$$
$$p_{1,2}(3) = .59722$$

Practice Problem 1 A Markov chain has a transition matrix

$$\mathbf{A} = \begin{matrix} & \begin{matrix} S_1 & S_2 \end{matrix} \\ \begin{matrix} S_1 \\ S_2 \end{matrix} & \begin{bmatrix} .6 & .4 \\ 0 & 1 \end{bmatrix} \end{matrix}$$

Find $p_{12}(2)$, the probability of going from state 1 to state 2 after two applications of the transition matrix.

ANSWER $p_{12}(2) = .64$

We return now to the example at the beginning of this section, where

$$
\begin{array}{c}
 & \begin{array}{cc} \text{C} & \text{DNC} \end{array} \\
\begin{array}{c} \text{C} \\ \text{DNC} \end{array} &
\left[\begin{array}{cc} .9 & .1 \\ .2 & .8 \end{array} \right]
\end{array}
$$

represents the relationship between contributions to United Way from one year to the next. That is, 90% of those who contribute one year will contribute the next year. Now suppose that a new company moves into the city and in the first year 60% of the employees contribute. This information can be written in a form we call the **initial-state probability matrix**:

$$[.60 \quad .40]$$

Note that if 60% contribute, then 40% do not contribute.

After 1 year, let's see what the state of contributions will be at this company:

$$[.60 \quad .40] \left[\begin{array}{cc} .9 & .1 \\ .2 & .8 \end{array} \right] = [.62 \quad .38]$$

During the second campaign, 62% of this company will contribute to United Way. Under these circumstances, let's see what happens over a period of several years.

$$[.62 \quad .38] \left[\begin{array}{cc} .9 & .1 \\ .2 & .8 \end{array} \right] = [.634 \quad .366] \qquad \text{Third state}$$

$$[.634 \quad .366] \left[\begin{array}{cc} .9 & .1 \\ .2 & .8 \end{array} \right] = [.644 \quad .356] \qquad \text{Fourth state}$$

$$[.644 \quad .356] \left[\begin{array}{cc} .9 & .1 \\ .2 & .8 \end{array} \right] = [.651 \quad .349] \qquad \text{Fifth state}$$

$$[.651 \quad .349] \left[\begin{array}{cc} .9 & .1 \\ .2 & .8 \end{array} \right] = [.656 \quad .344] \qquad \text{Sixth state}$$

$$[.656 \quad .344] \left[\begin{array}{cc} .9 & .1 \\ .2 & .8 \end{array} \right] = [.659 \quad .341] \qquad \text{Seventh state}$$

If we were to continue multiplying in the same manner, we would see that the state probability matrices approach or converge to a fixed probability matrix:

$$[.666 \quad .334]$$

When this happens, the system is said to be in **equilibrium**, because additional repetition of the transition matrix will not change the state matrix very much. Thus, we will have found a probability matrix $\mathbf{X}$ such that $\mathbf{XA} = \mathbf{X}$.

FIXED OR STEADY-STATE PROBABILITY MATRIX

A state matrix **X** such that

$$\mathbf{XA} = \mathbf{X}$$

is a **steady-state probability matrix** for the transition matrix A.

EXAMPLE 24 Find the steady-state probability matrix for

$$\begin{bmatrix} 0 & 1 \\ \frac{3}{5} & \frac{2}{5} \end{bmatrix}$$

SOLUTION Using the preceding theorem, we are looking for a steady-state probability matrix **X** such that $\mathbf{XA} = \mathbf{X}$. Suppose that we denote **X** by $[p_1 \quad p_2]$. Then

$$[p_1 \quad p_2] \begin{bmatrix} 0 & 1 \\ \frac{3}{5} & \frac{2}{5} \end{bmatrix} = [p_1 \quad p_2]$$

$$\left[\left(\frac{3}{5}\right) p_2 \quad p_1 + \left(\frac{2}{5}\right) p_2 \right] = [p_1 \quad p_2] \qquad \text{Multiply matrices.}$$

so

$$\left(\frac{3}{5}\right) p_2 = p_1 \quad \text{and} \quad p_1 + \left(\frac{2}{5}\right) p_2 = p_2 \qquad \text{Set elements equal.}$$

Then

$$p_1 = \left(\frac{3}{5}\right) p_2$$

Recall that

$$p_1 + p_2 = 1 \qquad \text{p_1 and p_2 are probabilities with $p_1 + p_2 = 1$.}$$

Thus,

$$\left(\frac{3}{5}\right) p_2 + p_2 = 1 \qquad \text{Substitute for p_1.}$$

$$\left(\frac{8}{5}\right) p_2 = 1 \qquad \text{Add coefficients.}$$

$$p_2 = \frac{5}{8} \qquad \text{Divide by 8/5.}$$

$$p_1 = \frac{3}{8} \qquad \text{$p_1 + p_2 = 1$}$$

We can verify that $[\frac{3}{8} \quad \frac{5}{8}]$ is a steady-state matrix for our transition matrix as follows:

$$[\tfrac{3}{8} \quad \tfrac{5}{8}] \begin{bmatrix} 0 & 1 \\ \frac{3}{5} & \frac{2}{5} \end{bmatrix} = [\tfrac{3}{8} \quad \tfrac{5}{8}] \qquad \text{Multiplication of matrices}$$

Practice Problem 2 Find the steady-state probability matrix for the regular transition matrix

$$\begin{bmatrix} \frac{3}{4} & \frac{1}{4} \\ \frac{2}{5} & \frac{3}{5} \end{bmatrix}$$

ANSWER $[\frac{8}{13} \quad \frac{5}{13}]$

The question arises, "Can we always find a steady-state probability matrix by the procedure used in the preceding example?" The answer is "Yes, provided the steady-state matrix exists." However, the steady-state matrix may not exist. The following discussion should answer all of our questions.

REGULAR MARKOV CHAIN

1. A transition matrix is said to be *regular* if some power of it contains only positive elements. The chain is then called a **regular Markov chain**.
2. If the transition matrix **A** is regular, then a steady-state probability matrix **X** exists such that

$$\mathbf{XA} = \mathbf{X}$$

EXAMPLE 25   The following matrix is regular because the first power contains all positive elements.

$$\begin{bmatrix} .7 & .3 \\ .4 & .6 \end{bmatrix}$$

Likewise,

$$\mathbf{A} = \begin{bmatrix} .1 & .9 & 0 \\ .3 & .3 & .4 \\ .8 & .1 & .1 \end{bmatrix}$$

is regular, since

$$\mathbf{A}^2 = \begin{bmatrix} .28 & .36 & .36 \\ .44 & .40 & .16 \\ .19 & .76 & .05 \end{bmatrix}$$

contains all positive elements. However,

$$\mathbf{A} = \begin{bmatrix} 1 & 0 \\ .7 & .3 \end{bmatrix}$$

is not regular because the element in the upper right corner will be 0 for all $\mathbf{A}^n$, and zero is not a positive element. ∎

The following theorem, presented without proof, summarizes the important properties of regular transition matrices. Refer to Chapter 2 if you have forgotten the meaning of a vector.

PROPERTIES OF REGULAR TRANSITION MATRICES

If **A** is a **regular transition matrix**, then

(a) **A** has a unique steady-state probability vector **X**.
(b) The sequence $\mathbf{A}, \mathbf{A}^2, \mathbf{A}^3, \ldots$ approaches a steady-state matrix **M**. The rows of **M** are identical. The elements of a row are equal to the elements of **X**, the steady-state probability vector.
(c) If $\mathbf{V}_i$ is any vector, the sequence

$$\mathbf{V}_i\mathbf{A}, \; \mathbf{V}_i\mathbf{A}^2, \; \mathbf{V}_i\mathbf{A}^3, \ldots$$

approaches a fixed or steady-state vector **V**, or

$$\mathbf{V}_i\mathbf{M} = \mathbf{V} \quad \text{and} \quad \mathbf{V}\mathbf{A} = \mathbf{V}$$

This theorem states that no matter what the initial-state vector, a regular Markov chain tends to stabilize, and thus the long-term behavior becomes predictable. Powers of the transition matrix approach a matrix in which the elements in a column are approximately identical. Thus, no matter what the initial state, the long-term probability of being in state S_i from any state is constant.

EXAMPLE 26   Use Example 24 to find the matrix **M** that $\mathbf{A}, \mathbf{A}^2, \mathbf{A}^3$, $\mathbf{A}^4, \ldots$ approaches; find for the initial-state matrix $\mathbf{V}_0 = [100 \quad 200]$, the steady-state matrix **V**; and show that $\mathbf{V}\mathbf{A} = \mathbf{V}$.

SOLUTION   In Example 24 we found that the fixed probability matrix is $[\frac{3}{8} \quad \frac{5}{8}]$. Thus, we can form **M** as

$$\begin{bmatrix} \frac{3}{8} & \frac{5}{8} \\ \frac{3}{8} & \frac{5}{8} \end{bmatrix}$$

since each row of **M** is identical to the steady-state probability matrix **X**. The long-term probability of being in state 1 is $\frac{3}{8}$, regardless of whether the initial state is 1 or 2. Likewise, the long-term probability of being in state 2 is $\frac{5}{8}$.

Now the steady-state matrix for $\mathbf{V}_0 = [100 \quad 200]$ is

$$\mathbf{V}_0\mathbf{M} = [100 \quad 200] \begin{bmatrix} \frac{3}{8} & \frac{5}{8} \\ \frac{3}{8} & \frac{5}{8} \end{bmatrix} = [\frac{900}{8} \quad \frac{1500}{8}]$$

Hence, $\mathbf{V}_0\mathbf{A}, \mathbf{V}_0\mathbf{A}^2, \mathbf{V}_0\mathbf{A}^3, \ldots$ approaches $[\frac{900}{8} \quad \frac{1500}{8}]$. Also note that for steady-state matrix $\mathbf{V}$,

$$\mathbf{VA} = \mathbf{V}$$

$$[\frac{900}{8} \quad \frac{1500}{8}] \begin{bmatrix} 0 & 1 \\ \frac{3}{5} & \frac{2}{5} \end{bmatrix} = [\frac{900}{8} \quad \frac{1500}{8}] \qquad ■$$

Practice Problem 3 To serve additional customers, a company secures a nationally known advertising agency. After 1 year it is noted that 10% of those who had been using another product switched to the product being advertised and 90% did not switch. It was also noted that 95% of those using the advertised product continued to use the product while 5% changed.

(a) Write a transition matrix for this Markov chain.
(b) After two successive advertising campaigns, what percentage of people had changed to the advertised product from using other products?
(c) Find a steady-state matrix.

ANSWER (a) $\begin{bmatrix} .95 & .05 \\ .10 & .90 \end{bmatrix}$ (b) 18.5% (c) $\begin{bmatrix} \frac{2}{3} & \frac{1}{3} \\ \frac{2}{3} & \frac{1}{3} \end{bmatrix}$

SUMMARY

In this section we considered the following concepts:

1. The meaning of the transition matrix of a Markov chain.
2. Requirements that a Markov chain be regular.
3. Requirements for the existence of a steady-state probability vector $\mathbf{X}$ such that $\mathbf{XA} = \mathbf{X}$, where $\mathbf{A}$ is a transition matrix.

Exercise Set 6.4

Which of the following are probability matrices?

1. $[\frac{1}{3} \quad \frac{1}{3} \quad \frac{1}{3}]$
2. $[0 \quad 1]$
3. $[1 \quad 0 \quad 0]$
4. $[\frac{2}{3} \quad \frac{2}{3} \quad \frac{2}{3}]$

5. $[1 \quad 1]$
6. $[1 \quad 0 \quad 1]$

Which of the following are transition matrices?

7. $\begin{bmatrix} 0 & 1 \\ 1 & 0 \end{bmatrix}$
8. $\begin{bmatrix} \frac{3}{4} & \frac{1}{4} \\ \frac{1}{2} & \frac{1}{2} \end{bmatrix}$

9. $\begin{bmatrix} \frac{1}{6} & \frac{5}{6} \\ \frac{2}{3} & \frac{3}{4} \end{bmatrix}$

10. $\begin{bmatrix} 1 & 1 & 1 \\ 0 & 0 & 1 \\ 1 & 0 & 0 \end{bmatrix}$

11. $\begin{bmatrix} \frac{1}{3} & \frac{1}{3} & \frac{1}{3} \\ \frac{1}{2} & \frac{1}{2} & 0 \\ 1 & 0 & 1 \end{bmatrix}$

12. $\begin{bmatrix} -\frac{1}{2} & \frac{1}{2} & 0 \\ 1 & 0 & 0 \\ 0 & 0 & 1 \end{bmatrix}$

13. $\begin{bmatrix} .2 & .7 & .1 \\ .3 & .5 & .2 \\ .8 & -.2 & .4 \end{bmatrix}$

14. $\begin{bmatrix} \frac{1}{3} & \frac{2}{3} & 0 \\ 0 & \frac{1}{2} & \frac{1}{2} \end{bmatrix}$

15. $\begin{bmatrix} .6 & .4 & 0 \\ 0 & 1 & 0 \end{bmatrix}$

16. A Markov chain has the transition matrix

$$A = \begin{bmatrix} .7 & .3 \\ 0 & 1 \end{bmatrix}$$

(a) What is the probability of starting in state 1 and going to state 2?

(b) What is the probability of being in state 2 and staying in state 2?

17. A Markov chain has the transition matrix

$$A = \begin{bmatrix} .3 & .3 & .4 \\ .2 & .2 & .6 \\ .1 & .5 & .4 \end{bmatrix}$$

(a) What is the probability of starting in state 1 and going to state 3?

(b) What is the probability of being in state 3 and going to state 2?

(c) What is the probability of being in state 3 and staying in state 3?

18. A Markov chain has the following transition matrix. Draw a tree diagram to represent it.

$$A = \begin{bmatrix} .2 & .8 \\ .7 & .3 \end{bmatrix}$$

19. Repeat Exercise 18 for the transition matrix

$$A = \begin{bmatrix} .2 & .3 & .5 \\ .5 & .5 & 0 \\ 1 & 0 & 0 \end{bmatrix}$$

20. For the following transition matrices and initial-probability matrices, find the probability matrix for the second stage.

(a) $\begin{bmatrix} .4 & .6 \\ .5 & .5 \end{bmatrix}$, $[.2 \quad .8]$

(b) $\begin{bmatrix} .8 & .2 \\ .1 & .9 \end{bmatrix}$, $[.5 \quad .5]$

21. In Exercise 20, find the probability matrix after the third stage (two after the initial stage).

22. In Exercise 20, find for each a steady-state probability matrix.

23. In Exercise 20, find the matrix M that the powers of the transition matrix approach.

24. For the matrices (a) $V_0 = [100, 200]$ and (b) $V_0 = [60, 40]$, find, in Exercise 23, V_0M getting an answer V. Then show $VA = V$, where A is the original transition matrix.

Which of the following transition matrices are regular?

25. $\begin{bmatrix} \frac{1}{2} & \frac{1}{2} \\ 0 & 1 \end{bmatrix}$

26. $\begin{bmatrix} 0 & 1 \\ \frac{1}{2} & \frac{1}{2} \end{bmatrix}$

27. $\begin{bmatrix} 1 & 0 \\ \frac{1}{2} & \frac{1}{2} \end{bmatrix}$

28. $\begin{bmatrix} 0 & 1 \\ 1 & 0 \end{bmatrix}$

29. $\begin{bmatrix} \frac{1}{3} & \frac{2}{3} \\ 1 & 0 \end{bmatrix}$

30. $\begin{bmatrix} \frac{1}{3} & \frac{2}{3} \\ 0 & 1 \end{bmatrix}$

31. $\begin{bmatrix} 1 & 0 \\ \frac{1}{3} & \frac{2}{3} \end{bmatrix}$

32. $\begin{bmatrix} 0 & 1 \\ \frac{1}{3} & \frac{2}{3} \end{bmatrix}$

33. $\begin{bmatrix} 1 & 0 \\ \frac{3}{4} & \frac{1}{4} \end{bmatrix}$

34. $\begin{bmatrix} 0 & 1 \\ \frac{3}{4} & \frac{1}{4} \end{bmatrix}$

35. $\begin{bmatrix} \frac{2}{3} & \frac{1}{3} \\ \frac{3}{4} & \frac{1}{4} \end{bmatrix}$

36. $\begin{bmatrix} 1 & 0 & 0 \\ \frac{1}{3} & \frac{1}{3} & \frac{1}{3} \\ 0 & \frac{3}{4} & \frac{1}{4} \end{bmatrix}$

37. $\begin{bmatrix} \frac{1}{3} & \frac{2}{3} & 0 \\ \frac{1}{2} & 0 & \frac{1}{2} \\ 0 & 1 & 0 \end{bmatrix}$

38. $\begin{bmatrix} 0 & 1 & 0 \\ \frac{1}{2} & 0 & \frac{1}{2} \\ \frac{1}{3} & \frac{2}{3} & 0 \end{bmatrix}$

39. $\begin{bmatrix} 0 & .5 & .5 \\ .8 & .1 & .1 \\ .2 & .2 & .6 \end{bmatrix}$

For the following, find the steady-state probability matrix. Also find the matrix that each transition matrix approaches as it is raised to higher powers.

40. Exercise 29

41. Exercise 31

42. Exercise 35

43. Exercise 37

Determine the steady-state matrix **V** *for the given matrix and transition matrix.*

44. $[100 \quad 60]$, $\begin{bmatrix} \frac{1}{3} & \frac{2}{3} \\ 1 & 0 \end{bmatrix}$

(See Exercise 40.)

45. $[100 \quad 100]$, $\begin{bmatrix} \frac{2}{3} & \frac{1}{3} \\ \frac{3}{4} & \frac{1}{4} \end{bmatrix}$

(See Exercise 42.)

46. $[50 \quad 70]$, $\mathbf{A} = \begin{bmatrix} 1 & 0 \\ \frac{1}{2} & \frac{1}{2} \end{bmatrix}$

47. $[60 \quad 30]$, $\mathbf{A} = \begin{bmatrix} .7 & .3 \\ .5 & .5 \end{bmatrix}$

48. $[10 \quad 20 \quad 30]$, $\begin{bmatrix} \frac{1}{3} & \frac{2}{3} & 0 \\ \frac{1}{2} & 0 & \frac{1}{2} \\ 0 & 1 & 0 \end{bmatrix}$

(See Exercise 43.)

49. $[2, \quad 3, \quad 6]$, $\begin{bmatrix} 0 & .5 & .5 \\ .8 & .1 & .1 \\ .2 & .2 & .6 \end{bmatrix}$

(See Exercise 39.)

50. A Markov chain has the transition matrix

$$\mathbf{A} = \begin{bmatrix} \frac{1}{4} & \frac{3}{4} \\ \frac{2}{3} & \frac{1}{3} \end{bmatrix}$$

Find

(a) $\mathbf{A}^2$

(b) $\mathbf{A}^3$

51. In Exercise 50, find the following probabilities.
   (a) What is the probability of starting in state 1 and being in state 2 two steps later?
   (b) What is the probability of being in state 2 and remaining in state 2 two steps later? (Two repetitions of the experiment.)
   (c) What is the probability of being in state 2 and being in state 1 three steps later?

(d) What is the probability of being in state 1 and remaining in state 1 three steps later?

Using a calculator, multiply the transition matrix by the row matrix until you determine what you believe to be a steady-state matrix.

52. $[.2 \quad .3 \quad .4]$, $\begin{bmatrix} .3 & .3 & .4 \\ .2 & .4 & .4 \\ .6 & .2 & .2 \end{bmatrix}$

53. $[.1 \quad .3 \quad .4]$, $\begin{bmatrix} .5 & .2 & .3 \\ .3 & .5 & .2 \\ .2 & .2 & .6 \end{bmatrix}$

Applications (Business and Economics)

54. *Marketing.* Two newspapers in town, the *Star* and the *Times,* are competing for customers. A study shows the following: At the beginning of the study, the *Star* had $\frac{2}{3}$ of the customers; at the end of the year, it is found that the *Star* kept 60% of its customers and lost 40% to the *Times.* The *Times* kept 70% of its customers and lost 30% to the *Star.* What will be the customer distribution (a) at the end of next year and (b) two years after the study was made? (c) What is the long-range distribution prediction?

55. *Occupational Probabilities.* The following transition matrix gives occupational-change probabilities:

		Sons	
		White collar	Blue collar
Fathers	White	.7	.3
	Blue collar	.2	.8

(a) If the father is a blue-collar worker, what is the probability that a son is a white-collar worker?
(b) If a man is a blue-collar worker, what is the probability that his grandson is also a blue-collar worker?
(c) If a man is a blue-collar worker, what is the probability that his great-grandson is a white-collar worker?

56. *Advertising.* After an intensive advertising campaign, it was found that 90% of the people continued to use product X and 10% changed brands. Of those

not using product X, it was found that 40% switched to product X and 60% continued to use what they had been using.

(a) Write a transition matrix for this Markov chain.

(b) After 2 intensive advertising campaigns, what percentage of people changed to product X from other brands?

(c) After 2 campaigns, what percentage of those using product X decided not to use product X?

57. **Demographics.** Over a given year, the following shifts in employment are recorded:

Employment Shift	Percentage
From business (or industry) to business (or industry)	75%
From business (or industry) to unemployment	15
From business (or industry) to self-employment	10
From unemployment to business (or industry)	25
From unemployment to unemployment	60
From unemployment to self-employment	15
From self-employment to business (or industry)	10
From self-employment to unemployment	10
From self-employment to self-employment	80

How will people be distributed in the various employment categories in the long run?

Applications (Social and Life Sciences)

58. **Genetics.** A basic assumption in a simple genetics problem is that the offspring inherits one gene from each parent and that these genes are selected at random. Suppose that an inheritance trait is governed by a pair of genes, each of which is of type G or g. The possible combinations are gg, gG (same genetically as Gg), and GG. Those possessing the gg combination are called *recessive* (denoted by R); *hybrid* (denoted by H) is used to indicate those possessing gG; and *dominant* (denoted by D) indicates those possessing GG. Note that in the mating of two dominant parents, the offspring must be dominant. In the mating of two recessive parents, the offspring must be recessive. In the mating of a dominant parent with a recessive parent, the offspring must be hybrid. Suppose that a person of unknown genetic character is

crossed with a hybrid, the offspring is again crossed with a hybrid, and the process is continued, hence forming a Markov chain; the transition matrix is

$$\text{Parent} \begin{array}{c} \\ R \\ H \\ D \end{array} \overset{\begin{array}{ccc} & \text{Offspring} & \\ R & H & D \end{array}}{\begin{bmatrix} \frac{1}{2} & \frac{1}{2} & 0 \\ \frac{1}{4} & \frac{1}{2} & \frac{1}{4} \\ 0 & \frac{1}{2} & \frac{1}{2} \end{bmatrix}}$$

(a) What is the probability that a recessive parent will have a dominant child?

(b) What is the probability that a hybrid parent will have a recessive child?

(c) What is the probability that a recessive parent will have a hybrid grandchild?

59. **Communications.** A system transmits the digits 0 and 1. A digit must pass through several stages before reaching its destination. The probability that the digit 1 entering a stage will be unchanged is .80, and that the digit 0 will be unchanged is .90.

(a) Write a transition matrix for this system.

(b) What is the probability that 0 will be unchanged through two stages?

60. **Population Movement.** The transition matrix of a population-movement model is as follows:

	City	Suburb	Nonmetro Area
City	.86	.10	.04
Suburb	.02	.97	.01
Nonmetro Area	.005	.005	.99

(a) What is the probability of moving from the city to the suburbs?

(b) What is the probability of moving from the suburbs to a nonmetro area?

(c) What is the probability of moving from the suburbs to the city in 2 years?

(d) What is the probability of staying in the city for 2 years?

61. **Political-Party Change.** The transition matrix for changing political parties in a given election is

$$\begin{array}{c} \quad\quad\quad \text{Demo-} \quad \text{Repub-} \quad \text{Inde-} \\ \quad\quad\quad \text{crats} \quad\;\; \text{licans} \quad \text{pendents} \end{array}$$

$$\begin{array}{c} \text{Democrats} \\ \text{Republicans} \\ \text{Independents} \end{array} \begin{bmatrix} .7 & .2 & .1 \\ .1 & .8 & .1 \\ .4 & .2 & .4 \end{bmatrix}$$

(a) What is the probability of a Democrat voting Republican?

(b) What is the probability of a Republican voting Republican?

(c) What is the probability of a Republican voting Democratic in a second election?

(d) What is the probability of a Democrat voting for a Democrat in a second election?

62. **Genetics.** Assume that the transition matrix for an inheritance trait is given by the following matrix:

$$\begin{array}{c} \quad\quad\quad\quad\quad \text{Offspring} \\ \quad\quad\quad\quad R \quad\;\; H \quad\;\; D \end{array}$$

$$\text{Parent} \begin{array}{c} R \\ H \\ D \end{array} \begin{bmatrix} .25 & .75 & 0 \\ .25 & .50 & .25 \\ 0 & .75 & .25 \end{bmatrix}$$

Find the fixed probability matrix, and interpret your results.

63. **Political Parties.** In a town 3 political parties are competing for members. A survey shows that party I has 50,000 members, party II has 120,000 members, and party III has 30,000 members. The survey predicts the following transition matrix:

$$\begin{array}{c} \quad\quad\quad\quad \text{I} \quad\;\; \text{II} \quad\;\; \text{III} \end{array}$$

$$\begin{array}{c} \text{I} \\ \text{II} \\ \text{III} \end{array} \begin{bmatrix} .7 & .1 & .2 \\ .2 & .5 & .3 \\ .2 & .2 & .6 \end{bmatrix}$$

How will the members be distributed in the long run?

Review Exercises

64. A Markov chain has the transition matrix

$$\begin{bmatrix} \frac{1}{3} & \frac{1}{3} & \frac{1}{3} \\ \frac{1}{4} & \frac{1}{4} & \frac{1}{2} \\ 0 & \frac{1}{2} & \frac{1}{2} \end{bmatrix}$$

(a) If the initial state is $[\frac{2}{3} \;\; \frac{1}{6} \;\; \frac{1}{6}]$, what is the state matrix after one transition?

(b) After two transitions?

65. According to a mortality table, the probability John Salls will live for 1 year is .992. He buys a $10,000 1-year term life insurance policy for $40. What is the expected gain or loss of the insurance company? (Assume no interest rate.)

66. Explain why

$$\begin{bmatrix} \frac{1}{3} & \frac{1}{3} & \frac{1}{3} \\ 0 & 1 & 0 \\ \frac{3}{4} & \frac{1}{2} & \frac{1}{4} \end{bmatrix}$$

cannot be a transition matrix.

Chapter Review

Review the following concepts to ensure that you understand and can use them.

Important Concepts

Finite stochastic processes	Bayes' formula
Bernoulli trials	Binomial probability
Expected value	Random variable
Binomial distribution	Probability distribution

Markov Chains

Transition matrix	Initial state
Probability matrix	Steady-state matrix
Regular Markov chain	

Success in solving application problems with probability depends on the ability to use the correct formula. Review the conditions that allow you to use each of the following formulas:

Important Formulas

$$P(A \cap B) = P(A_1) \cdot P(B_2|A_1) + P(B_1) \cdot P(A_2|B_1)$$

$$P(B_i|A) = \frac{P(B_i) \cdot P(A|B_i)}{P(B_1) \cdot P(A|B_1) + P(B_2) \cdot P(A|B_2) + \cdots + P(B_n) \cdot P(A|B_n)}$$

$$E(x) = x_1 p(x_1) + x_2 p(x_2) + \cdots + x_n p(x_n)$$

$$P(x) = C(n, x) p^x q^{n-x}$$

Chapter Test

1. Consider the following tree diagram. Find $P(A|D)$.

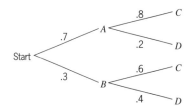

2. Three machines, A, B, and C, produce 50%, 30%, and 20%, respectively, of all the parts in a small motor. It has been found that 3% of the items produced by A, 2% of the items produced by B, and 1% of the items produced by C are defective. What is the probability that some part of a small motor is defective?

3. Find the steady-state probability matrix for

$$\begin{bmatrix} 0 & 1 \\ \frac{1}{2} & \frac{1}{2} \end{bmatrix}$$

4. Eight percent of all residents of Orange County have the flu virus. The probability that a medical examination will indicate the disease if a person has the virus is .8, and the probability that the examination will incorrectly indicate the disease is .3. What is the probability that Sue actually has the flu if her examination indicates that she is sick with the flu?

5. Discuss whether or not

$$\begin{bmatrix} 0 & .6 & .4 \\ .2 & .3 & .5 \\ .7 & 0 & .3 \end{bmatrix}$$

represents a regular Markov chain.

6. A Markov chain has a transition matrix

$$\begin{bmatrix} \frac{1}{3} & 0 & \frac{2}{3} \\ \frac{1}{2} & \frac{1}{4} & \frac{1}{4} \\ \frac{1}{3} & \frac{1}{3} & \frac{1}{3} \end{bmatrix}$$

If the initial state matrix is $[\frac{1}{3}, \frac{1}{3}, \frac{1}{3}]$, what will be the state matrix two transitions later?

7. Find the steady-state matrix for

$$\begin{bmatrix} 0 & 1 & 0 \\ \frac{1}{3} & 0 & \frac{2}{3} \\ \frac{1}{2} & \frac{1}{2} & 0 \end{bmatrix}$$

8. A lot of 12 tires contains 3 defective and 9 good tires. What is the probability that 3 good tires will be selected at random?

9. Rework Exercise 8 for the probability that 1 good and 2 defective tires will be selected at random.

10. The probability of success in a binomial distribution is $\frac{1}{50}$. If the experiment is performed 1000 times, find the expected number of successes.

Statistics

W hen we speak of *data* we usually refer to a collection of numbers. The branch of mathematics called **statistics** is the study of data for the purpose of obtaining information from the data. In this chapter we present a brief introduction to the study of statistics. In modern business a vast amount of data is collected about product acceptability, advertising effectiveness, production costs, sales costs, and profits. A statistical study of such data is important in today's business world. Life scientists collect data from experiments, and from these data they must formulate conclusions. Social scientists reach conclusions from a study of samples from a population of people. We have seen election results predicted when only a small percentage of the votes have been counted.

Statistics can be divided into two subdivisions: descriptive statistics and inferential statistics, as described below.

1. *Descriptive statistics:* Techniques are used to summarize and describe the characteristics of a set of data.
2. *Inferential statistics:* Generalizations or conclusions are made about the data in a large group (called a **population**) from a portion (called a **sample**).

In the first three sections of this chapter we study descriptive statistics: understanding a set of data by forming frequency distributions and drawing associated graphs, finding measures of central tendency (mean, median, and mode), and finding measures of the scattering of data. Inferential statistics is introduced by studying the normal distribution, which is the foundation of much of the research studies in business, economics, and the social and life sciences. In this chapter, we consider such diverse applications of statistics as marketing, pollution, psychology, politics, quality control, and drug effectiveness.

7.1 FREQUENCY DISTRIBUTIONS AND GRAPHICAL REPRESENTATIONS

OVERVIEW   We are immersed daily in a torrent of numbers flowing from our televisions, radios, newspapers, hair stylist, and our favorite uncle. Although data are a part of our daily lifestyle, it is evident that we often do not know how to organize, interpret, or understand the message being conveyed. In this section, we learn to organize and summarize data for better understanding of statistics. First, we introduce some basic techniques for classifying and summarizing a set of observed measurements. Then we represent the data with graphs. In this section, we construct

- Tables representing frequency distributions
- Relative frequency distributions
- Bar graphs
- Line graphs
- Histograms
- Frequency polygons
- Circle graphs

Organizing Data

The first objective of a statistician is to develop a plan to collect data for study. After the data have been collected, the second objective is to make sense out of

TABLE 1

7	1	1	0	3	4
5	5	3	2	3	3
6	6	2	4	2	1
0	0	3	4	5	6
3	1	4	1	3	4

TABLE 2

Number of Colds	Tally	Frequency
0					3
1							5
2					3
3									7
4							5
5					3
6					3
7			1
	Total	30

TABLE 3

Lengths of Engagements (months)

10	2	9	6	11
17	4	10	7	3
1	4	11	6	3
8	15	12	9	12
8	18	12	6	10
8	18	12	6	9

TABLE 4

Class Intervals	Tallies	Frequency
1–3	||||	4
4–6	|||||		6
7–9	||||| ||	7
10–12	||||| ||||	9
13–15			1
16–18	|||	3

this large mass of information. Suppose that you have collected the numbers shown in Table 1. The table contains your tabulation of the number of colds experienced during one winter by each of a group of 30 elementary-school children.

A quick glance at this array of numbers tells us very little about what the data imply about the group of people represented. Closer observation indicates that the largest number of colds experienced was 7, while the smallest number experienced was 0. The difference between the largest and smallest entry in the data is called the **range** of the data. In this case, the range is $7 - 0 = 7$. To organize this list of numbers more thoroughly, the data can be summarized as a **frequency distribution** in a frequency table. In such a table the number of times that a data entry occurs is tabulated as the frequency of the entry.

EXAMPLE 1   Make a frequency distribution for the data in Table 1.

SOLUTION   From Table 2 we can see that the number of colds most often reported was 3. We can also see how the number of colds was distributed among the 30 students. ∎

When numerous data are involved, a frequency distribution may become cumbersome. In this situation we consider **intervals** (or **classes**) of data and construct a grouped frequency distribution table for them.

In a grouped frequency distribution, we divide the range of data into intervals of equal length and record the number of pieces of data that fall into each interval.

EXAMPLE 2   Table 3 gives the lengths of engagements (by the number of months) of 30 newly married couples. Construct a grouped frequency distribution for the data in Table 3.

SOLUTION   The range of the data is

$$18 - 1 = 17$$

We can arbitrarily select six classes for our grouping. Since $17 \div 6$ is 2.833, the length of the classes (if the classes are to be of equal length) must be more than 2.833 in order to include all the data in six classes. Whenever feasible, classes should be of equal length. We arbitrarily select the following class limits: 1–3, 4–6, 7–9, 10–12, and so on. The grouped frequency distribution is found in Table 4. ∎

If in the preceding example we divide each frequency by the total number of items in the original data (in our case, 30), we obtain the **relative frequency** of each interval (see Table 5). We can interpret relative frequency to be a probability. That is, the probability that an engagement is of length 10 to 12 months is .30.

TABLE 5

Class Interval	Frequency	Relative Frequency
1–3	4	$\frac{4}{30} \approx .133$
4–6	6	$\frac{6}{30} \approx .200$
7–9	7	$\frac{7}{30} \approx .233$
10–12	9	$\frac{9}{30} \approx .300$
13–15	1	$\frac{1}{30} \approx .033$
16–18	3	$\frac{3}{30} \approx .100$
Total	30	$.999 \approx 1$

TABLE 6

Class Interval	Frequency
13–19	4
20–26	7
27–33	5
34–40	2
41–47	2

TABLE 7

Interval	Relative Frequency
13–19	$\frac{4}{20} = .2$
20–26	$\frac{7}{20} = .35$
27–33	$\frac{5}{20} = .25$
34–40	$\frac{2}{20} = .1$
41–47	$\frac{2}{20} = .1$
Total	1.0

EXAMPLE 3   From the grouped frequency distribution given in Table 6, find a relative frequency distribution.

SOLUTION   The relative frequency distribution is shown in Table 7. ■

Calculator Note

Grouped frequency distributions can also be obtained on a calculator. We study this procedure as we draw histograms later in this section.

Stem and Leaf Plots

Another interesting way of summarizing data is to use what is called a **stem and leaf plot**. To illustrate this procedure, let's consider the grades obtained by two classes who were tested on material from Chapter 6. These data are given in Table 8.

TABLE 8

Class I	Class II
56, 64, 72, 73, 84,	99, 81, 50, 64, 76,
98, 80, 86, 75, 68,	63, 71, 78, 81, 92,
46, 78, 75, 91, 63,	87, 79, 74, 60, 68,
84, 79, 69, 76, 58	92, 84, 86, 65, 78

The first digit serves as the stem, and the second digit as the leaf. For example, the stem of the 46 in Class I is 4, and the leaf is 6. Likewise, 56 and 58 have stems of 5 and leaves of 6 and 8, respectively. The data for Class I are listed with six stems—4, 5, 6, 7, 8, and 9—and with appropriate leaves in Table 9. In Table 10, the same leaves are arranged in increasing order.

TABLE 9

Class I	
Stems	Leaves
4	6
5	6, 8
6	4, 8, 3, 9
7	2, 3, 5, 8, 5, 9, 6
8	4, 0, 6, 4
9	8, 1

TABLE 10

Class I	
Stems	Leaves
4	6
5	6, 8
6	3, 4, 8, 9
7	2, 3, 5, 5, 6, 8, 9
8	0, 4, 4, 6
9	1, 8

Now let's compare Class I and Class II, using the same stems. In Table 11 the leaves of Class I increase from left to right, and the leaves of Class II increase from right to left.

TABLE 11

Class II		Class I
Leaves	Stems	Leaves
	4	6
0	5	6, 8
8, 5, 4, 3, 0	6	3, 4, 8, 9
9, 8, 8, 6, 4, 1	7	2, 3, 5, 5, 6, 8, 9
7, 6, 4, 1, 1	8	0, 4, 4, 6
9, 2, 2	9	1, 8

A quick inspection of how the leaves increase in Class I and in Class II suggests that the students in Class II did better on this test.

TABLE 12

Weights of Members of the Basketball Team	
Stems	Leaves
15	8
16	4, 7
17	5, 6, 8
18	3, 4, 6, 8
19	4, 5
20	3

Steps in Making a Stem and Leaf Plot

1. Decide on the number of digits in the data to be listed under stems (one-digit, two-digit, or three-digit numbers). Usually only one digit is given under leaves and the other digits are listed under stems.
2. List the stems in a column, from least to greatest.
3. List the remaining digits in each data entry as leaves. (You may wish to order these data from smallest to largest.)

It is always a good idea to describe or explain a stem and leaf plot with a heading as in Table 12. Likewise, it is a good idea to explain the notation. For example, 17 under stems and 6 under leaves in Table 12 represent 176 pounds.

Bar Graphs

We have seen throughout this book that we can greatly improve our understanding and our problem-solving ability if we can draw a graph, picture, or diagram. There are several ways to represent graphically a conglomeration of data. One such representation is a **bar graph**. To construct a bar graph, first construct a frequency distribution or a grouped frequency distribution, whichever is appropriate. Then plot the frequencies on the vertical axis and the data values or intervals on the horizontal axis. Finally, draw a bar to show the relationship between the values and the frequencies.

EXAMPLE 4 Draw a bar graph that represents the number of graduates per year, as shown in Table 13.

TABLE 13

Year	1980	1981	1982	1983	1984	1985	1986	1987	1988	1989	1990	1991
Number of Graduates	152	163	197	185	201	196	210	189	195	205	200	180

SOLUTION Figure 1 is a bar graph of the data in Table 13. Notice that the number of graduates is measured on the vertical axis and that the years are given on the horizontal axis. The break in the vertical axis, denoted by ⌇, indicates that the scale is not accurate from 0 to 150. The height of each bar represents the number of students who graduated in a given year. To determine from the bar graph the number of students who graduated in 1983, locate the bar labeled 1983 and draw a horizontal line from the top of the bar to the vertical axis. The point where this horizontal line meets the vertical axis identifies the number of graduating students. Thus, 185 students graduated in 1983. ■

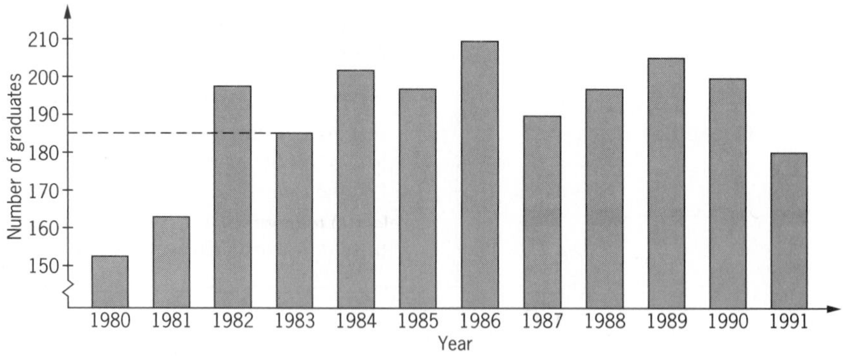

Figure 1

Line Graphs

In general, a **line graph** does a better job of showing fluctuations and emphasizing changes in the data than does a bar graph. The line graph in Figure 2 represents the distance (in meters) run in 6 minutes by a group of freshmen in a physical education class. Looking at this graph, you can readily see the variations in the numbers who ran given distances in 6 minutes.

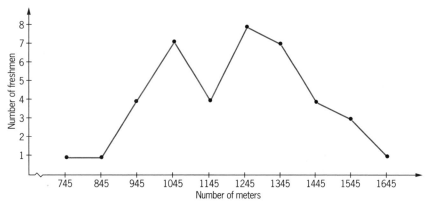

Figure 2

Calculator Note

To obtain a line graph on a calculator, consider the frequency table (Table 14) for points on the graph in Figure 2.

TABLE 14

x	745	845	945	1045	1145	1245	1345	1445	1545	1645
f	1	1	4	7	4	8	7	4	3	1

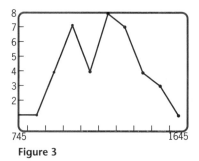

Figure 3

First list the x values under L_1 and the frequencies under L_2. For the WINDOW use X: [745, 1695] with a scale of 100 and Y: [0, 8] with a scale of 1. Press ⌊STAT⌋ (select 1) to input the values for L_1 and L_2. Press ⌊2nd⌋ ⌊STAT PLOT⌋ (select 1). Under Plot 1 select ON, under Type: select the second graph and press under the Xlist select L_1 and press ⌊ENTER⌋; and under the Ylist select L_2 and press ⌊ENTER⌋. Press ⌊GRAPH⌋ to get Figure 3. Note that the hash lines on the horizontal axis are not at 745, 845, and 945 but halfway between these values. The reason that this occurs will be explained when we draw histograms.

Histograms

A bar graph representing a grouped frequency distribution is called a **histogram**. To construct a histogram, we first construct a grouped frequency distribution. Then we represent each interval with a bar or a rectangle [Figure 4(*a*)]. The height of the rectangle indicates the frequency of the interval. To label each rectangle, we use the interval that the rectangle represents or the midpoint of the interval, called the **class mark** [Figure 4(*b*)]. Usually, the rectangles touch at a point halfway between each pair of class limits so that there are no gaps in the histogram. In Figure 4(*c*), the rectangles touch at 19.5, 24.5, 29.5, and 34.5. These are termed the **class boundaries**.

TABLE 15

Class	Tallies	Frequency
15–19	IIII	4
20–24	Ж II	7
25–29	Ж	5
30–34	II	2
35–39	II	2

EXAMPLE 5   Draw a histogram of the data in Table 15.

SOLUTION   The histograms in parts (*a*), (*b*), and (*c*) of Figure 4 all represent the same data. In part (*a*) each rectangle is labeled according to its class interval, in part (*b*) each rectangle is labeled according to the midpoint of its class interval (the class mark), and in part (*c*) each rectangle is labeled by class boundaries (the numbers that are halfway between class limits). ■

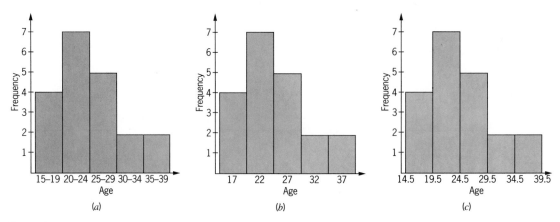

Figure 4

Calculator Note

The advantage of drawing a histogram on a calculator is that the step of making a frequency distribution is eliminated. The calculator performs that task for you.

EXAMPLE 6   Draw a histogram and from the histogram make a frequency distribution for the data in Table 3 on the lengths of engagements.

SOLUTION   In Table 3 the smallest value is 1 and the largest value is 18. Starting the first bar at .5 and ending the last bar at 18.5, there would be six

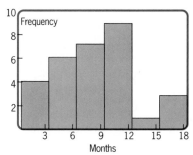

Figure 5

TABLE 16

Class Intervals	Frequency
1–3	4
4–6	6
7–9	7
10–12	9
13–15	1
16–18	3

bars of length 3. Thus we use X:[0.5, 18.5] with a scale of 3 and Y:[0, 10] with a scale of 2. Press ⌐STAT⌐ (select 1) to input the data under L_1. Press ⌐2nd⌐ ⌐STAT PLOT⌐ and select 1. Under Plot 1 select "ON" and press ⌐ENTER⌐. Under Type: select the histogram and press ⌐ENTER⌐. Under the Xlist select L_1 and press ⌐ENTER⌐, and under the Ylist select 1 and press ⌐ENTER⌐. Press ⌐GRAPH⌐ to get Figure 5. Using ⌐TRACE⌐ note that the first bar extends from .5 to 3.5, the second from 3.5 to 6.5 etc. although the hash lines are not at these points. The frequency distribution is tabulated in Table 16. The information in Table 16 is the same as that in Table 4. ■

Frequency Polygons

When a line graph is used to represent grouped data, it is called a **frequency polygon**. To draw a frequency polygon, we plot the frequency of the intervals versus the midpoints (class marks) of the intervals, and connect the resulting points with straight-line segments. Finally, we connect the first and last points to points on the horizontal axis that are located in the middle of the invervals beyond these first and last class marks.

EXAMPLE 7   Draw a frequency polygon for the data in Table 15.

SOLUTION   Figure 6 presents a frequency polygon for the grouped frequency distribution of Table 15. ■

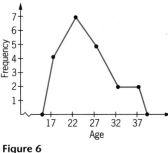

Figure 6

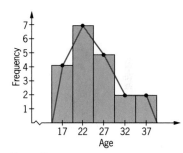

Figure 7

A frequency polygon can be obtained from a histogram by simply connecting the midpoints of the tops of the rectangles with straight-line segments and drawing the lines to the horizontal axis. For example, the frequency polygon shown in Figure 7 is easily obtained from Figure 4(*b*) by drawing straight-line segments.

Circle Graph

A graph that is often used in everyday situations is the **circle graph**, sometimes called a **pie chart**. It consists of a circle partitioned into sectors, each of which represents a percentage of the whole.

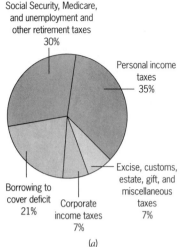

Social Security, Medicare,
and unemployment and
other retirement taxes
30%

Personal income
taxes
35%

Excise, customs,
estate, gift, and
miscellaneous
taxes
7%

Borrowing to
cover deficit
21%

Corporate
income taxes
7%

(a)

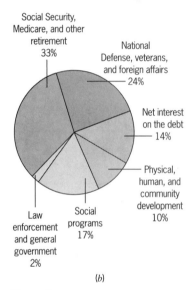

Social Security,
Medicare, and other
retirement
33%

National
Defense, veterans,
and foreign affairs
24%

Net interest
on the debt
14%

Physical,
human, and
community
development
10%

Law
enforcement
and general
government
2%

Social
programs
17%

(b)

Figure 8

EXAMPLE 8   On the IRS instructions for individual tax returns, each year we see circle graphs presenting information on income and expenses of the federal government. Look at the financial data in Table 17 and study the construction of Figure 8(*a*), which gives the relative sizes of income sources. Then refer to Figure 8(*b*) to answer each of the following. If expenditures in 1992 were 1380.9 billion dollars, calculate dollar expenditures for

(a)   social security, medicare, and other retirement expenses.
(b)   law enforcement and general government.
(c)   social programs.
(d)   physical, human, and community development.
(e)   defense, veterans, and foreign affairs.
(f)   interest on the nation's debt.

TABLE 17

1992 Income	
Source	Amount (billions of dollars)
Personal income taxes	483.32
Social Security, Medicare, unemployment, and other such taxes	414.26
Corporate income taxes	96.66
Excise, customs, estate, gift, and miscellaneous taxes	96.66
Borrowing to cover deficit	290.00
Total	1380.90

SOLUTION   Note that the personal income taxes were 483.32 billion dollars and this amount represented $483.32/1380.9 = 0.35$ or 35% of the whole. To meet budget expenditures it was necessary to borrow 290 billion dollars, representing $290/1380.9 = 0.21$ or 21% of the whole. The percentage of the whole for each income source is shown in Figure 8(*a*).

(a) $455.70 billion   (b) $27.62 billion   (c) $234.75 billion
(d) $138.09 billion   (e) $331.42 billion   (f) $193.33 billion   ■

SUMMARY

In this section we learned how to form frequency distributions, grouped frequency distributions, and stem and leaf plots. We also learned how to represent data graphically by using a circle graph, line graph, bar graph, histogram, and frequency polygon.

Exercise Set 7.1

1. In a transportation survey, bus riders on the Friday evening run were asked how many times they had ridden the bus that week. Summarize the following data in a frequency distribution.

4	8	6	4
7	2	2	8
2	5	8	1
7	9	8	3
8	2	4	8
10	3	3	9

2. Using the frequency distribution in Exercise 1, draw the line graph using a calculator.

3. The following is a tabulation of the ages of mothers of the first babies born in Morningside Hospital in 1990.

Class	Tally	Frequency
15–19	IIII	4
20–24	HHt II	7
25–29	HHt	5
30–34	II	2
35–39	II	2

   (a) Determine the number of mothers in the tabulation?
   (b) Find the number of mothers younger than 30.
   (c) Find the number of mothers who were at least 20 years of age.
   (d) Find the number of mothers whose ages were between 20 and 34, including the endpoints.
   (e) Find class marks.

4. In the given pie chart or circle graph,

   (a) what percentage are professionals?
   (b) what percentage are craftsmen?
   (c) what percentage are managers or clerical?
   (d) what percentage are neither managers nor professionals?

5. The following table gives the number of students who had a specific number of absences in a given semester.

Number of Absences	Frequency
0	25
1	18
2	20
3	31
4	34
5	14
6	13
7	12
8	8
9	3
10	1

   (a) Display the data with a bar graph.
   (b) Using a calculator, represent the data as a line graph.

6. Consider a grouped frequency distribution defined by the table.

Class	Frequency
20–29	1
30–39	2
40–49	4
50–59	5
60–69	9
70–79	6
80–89	3

   (a) Find the class marks.
   (b) Construct a histogram using a calculator.
   (c) In part (b) draw straight-line segments to obtain a frequency polygon.

7. A college has 1426 students. The table presents a tabulation of their ages.

Age	Number of Students
15–19	562
20–24	450
25–29	350
30–34	58
35–39	6

(a) With your calculator, draw a frequency polygon for these data.

(b) Use your calculator to draw a histogram.

8. (a) Indicate on a bar graph a comparison of the number of students in mathematics courses who are majoring in the following academic fields.

Academic Field	Number of Students
Business administration	110
Social sciences	100
Life sciences	60
Humanities	30
Physical sciences	60
Elementary education	140

(b) Display the data with a circle graph.

(c) Use your calculator to draw a frequency polygon.

9. A college has 1500 students. The table is a tabulation of their ages. Use relative frequencies to construct a probability distribution.

Age	Number of Students
15–19	500
20–24	400
25–29	300
30–34	200
35–39	100

(a) What is the probability that the age of a student is between 20 and 24?

(b) What is the probability that the age of a student is less than 35?

(c) What is the probability that the age of a student is more than 19?

(d) What is the probability that the age of a student is less than 40?

(e) What is the probability that the age of a student is greater than 14?

10. Park officials want to understand the use of a municipal park. One evening the officials interviewed 36 people and recorded their ages.

(a) Using a calculator, summarize the data in a grouped frequency distribution with 7 intervals of equal length. Let the first interval be 4–14.

(b) What trend can you see in the data?

7	18	35	73	18	28
15	19	41	61	16	24
51	65	12	65	61	26
16	62	14	73	72	48
17	59	16	62	43	68
21	16	17	19	32	72

For Exercises 11–13, *group the data on your calculator in the manner outlined in each exercise. Then use the tabulated data to complete parts* (a) *to* (c).

(a) *Make a grouped frequency distribution.*

(b) *What is the range?*

(c) *Find class marks.*

11. The 25 scores below were achieved by a group of high school seniors on a mathematics placement test. Tabulate this information into five groups of minimum integral length, with the first class beginning at 450.

477	485	527	483	582
567	513	609	596	525
566	540	451	519	530
576	656	525	621	603
648	555	535	528	546

12. The amounts (rounded to the nearest $1) that a sample of 50 freshmen spent on textbooks per class during a fall semester are listed in the table. Use 6 intervals of minimum integral length with the first class beginning at 30.

33	41	35	53	42	47	41	31	38	37
30	38	37	33	41	35	42	50	41	38
39	42	41	40	40	38	37	41	45	48
35	36	35	38	33	39	40	40	47	38
37	38	37	34	35	44	44	46	40	39

13. The grades of 60 students in a mathematics course were recorded as shown. Start the first class at 43 and use 7 intervals of minimum integral length.

96	71	43	77	74	73	87	81	91	79
78	72	82	81	87	93	95	53	64	66
83	71	58	97	53	74	61	68	67	63
55	81	62	87	76	74	71	65	93	71
74	71	77	83	85	84	94	56	63	65
75	48	89	84	75	76	75	61	91	90

14. The following stem and leaf plot records the distances in feet that 22 children in a recreation program could throw a softball.
    (a) Write the distances represented in the stem and leaf plot.
    (b) What is the shortest throw?
    (c) How many of the throws traveled more than 60 feet?

Lengths of Softball Throws

Stems	Leaves
3	1, 7
4	1, 2, 6
5	1, 2, 3, 3, 7, 8, 9
6	2, 8, 8, 9
7	1, 3, 6
8	2, 2
9	6

15. Using the first digit as a stem, make a stem and leaf plot for
    (a) the data in Exercise 12.
    (b) the data in Exercise 13.

16. Make a stem and leaf plot for the data in Exercise 10, and arrange the data in increasing order.

17. Given the following histogram, tabulate a frequency distribution and find the class marks.

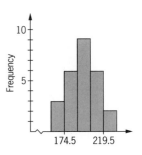

Applications (Business and Economics)

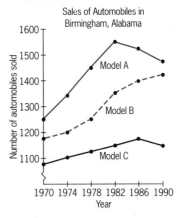

Use this graph in Exercises 18–21.

18. In 1990, how many more model A cars were sold than model B?

19. In which 4-year period did model C have the greatest decrease in sales? What was the decrease?

20. In which 4-year period did model B have the greatest increase in sales? What was the increase?

21. Compare the increase in sales of the three models from 1974 to 1990.

22. *Sales.* Tabulate a relative frequency distribution from the table, where the variable is the number of units sold per day.

Units Sold	Days
0	20
1	80
2	120
3	250
4	260
5	190
6	80

23. **Production.**   The new vice president of an oil company claims that production has doubled during the first 12 months of her tenure. To present this fact to the board of directors, she has the following graph prepared. What is misleading about this graph? (**Hint**: The viewer mentally compares volumes. What happens to the volume of a cylinder if you double its height and its radius?)

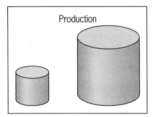

24. **Exports.**   A small foreign country exports 20 main products each year, ranging from iron ore to toy medical kits to surgical instruments. The value of each export in millions of dollars is given in the table. Group the values into 6 intervals of minimum integral length, starting the first class at 60.

86	62	239	290	207
285	232	214	131	195
424	343	476	140	398
363	348	156	222	370

25. **Common Stocks.**   The table gives the price–earnings ratio of 100 common stocks listed on the New York Stock Exchange. Let the variable be the price–earnings ratio and tabulate a probability distribution.

Interval	Frequency
0–4	6
5–9	46
10–14	30
15–19	10
20–24	4
25–29	2
30–34	2

(a) What is the probability that a price–earnings ratio is in the interval 10–14?
(b) What is the probability of a price–earnings ratio of less than 20?
(c) What is the probability of a price–earnings ratio of less than 35?
(d) What is the probability of a price–earnings ratio between 15 and 29, inclusive?

Applications (Social and Life Sciences)

26. **Classification of Occupations.**   Use a protractor to construct a pie chart showing the percentage of women who work in the following occupations.

Professional	16%
Managers	4%
Clerical	35%
Craftspeople	2%
Operative	14%
Service	17%
All other	12%

27. **Politics.**   The following circle graphs record the contributions to candidates for federal office in the late 1980s.
(a) Compare the percentage of contributions to Democratic and Republican candidates that came from labor organizations. From corporations.

(b) What dollar amount of support for Democratic candidates came from the party?

(c) What dollar amount of support for Republican candidates came from corporations?

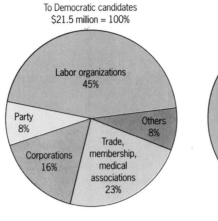

To Democratic candidates
$21.5 million = 100%

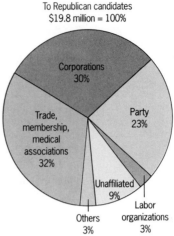

To Republican candidates
$19.8 million = 100%

7.2   MEASURES OF CENTRAL TENDENCY

OVERVIEW

In the preceding section we used histograms, frequency polygons, and circle graphs to summarize and explain sets of data. Sometimes we need a more concise procedure for characterizing a set of data. Suppose that a student receives 69, 71, 78, 82, and 73 on her five tests in Math 102. She gives her average grade as 74.6, but her friend claims her average is 73. Which average is correct? Actually, both averages are correct. The student found the mean and her friend found the median. In this section we introduce the concept of averages, also called **measures of central tendency.** Three measures are in general use—the **arithmetic mean,** the **median,** and the **mode.** The fact that there are these three (as well as others) often leads to misuses of statistics. One measure may be quoted, and the reader can automatically think that it means another. When a measure is quoted, immediately ask the question, ''Which one?'' In this section, we study

- The arithmetic mean
- The arithmetic mean for a frequency distribution
- The median
- The mode

The most widely used measure of central tendency is the **arithmetic mean,** sometimes called **arithmetic average.** The arithmetic mean of a set of n measurements is the sum of the measurements divided by n.

DEFINITION: ARITHMETIC MEAN

Consider n measurements $x_1, x_2, x_3, \ldots, x_n$. The formula for the **arithmetic mean**, denoted by $\bar{x}$, is

$$\bar{x} = \frac{x_1 + x_2 + x_3 + \cdots + x_n}{n}$$

Mathematicians employ a very useful notation to express complicated sums. When the Greek letter sigma, Σ, occurs in a mathematical expression, it means "add the indicated terms." For instance, the sum, $x_1 + x_2 + x_3 + \cdots + x_n$, can be represented as

$$\sum_{i=1}^{n} x_i$$

The index $i = 1$ at the bottom and the n at the top of the Σ indicate that the x terms should be added starting with the first one and stopping at the nth one. Using this notation, the formula for the mean can be expressed as follows:

$$\bar{x} = \frac{\sum_{i=1}^{n} x_i}{n}$$

We often omit the indices and simply write $\bar{x} = \dfrac{\sum x}{n}$

EXAMPLE 9   Find the arithmetic mean of 8, 16, 4, 12, and 10.

SOLUTION $\bar{x} = \dfrac{8 + 16 + 4 + 12 + 10}{5} = 10$ ■

EXAMPLE 10   Using a graphing calculator, find the arithmetic mean of 25, 25, 25, 25, 30, 30, 30, 40, 40, 40, 40, and 50.

SOLUTION   To find the mean of a set of data on a calculator, the data values are first stored (say, under L_1). Then we can use either the LIST MATH menu or the STAT CALC menu. This time we use the LIST MATH menu. Press $\boxed{\text{2nd}}$ $\boxed{\text{LIST}}$ $\boxed{\blacktriangleright}$ (select 3) $\boxed{\text{2nd}}$ $\boxed{\text{L}_1}$ $\boxed{)}$ $\boxed{\text{ENTER}}$. We immediately obtain $\bar{x}$ to be 33.33333. We could have listed the data in Mean ({ . . . }), instead of storing the data in L_1.   ■

Let's find the mean in the preceding example without the calculator in order to introduce the next concept.

$$\bar{x} = \frac{25 + 25 + 25 + 25 + 30 + 30 + 30 + 40 + 40 + 40 + 40 + 50}{12}$$

$$= \frac{25(4) + 30(3) + 40(4) + 50(1)}{4 + 3 + 4 + 1}$$

$$= \frac{400}{12} = 33\tfrac{1}{3}$$

Observe that the 4, 3, 4, and 1 are the frequencies of 25, 30, 40, and 50, respectively. The mean is obtained by multiplying each value by its frequency of occurrence and then dividing the sum of these products by the sum of the frequencies. Let's now generalize the formula for finding the arithmetic mean to include the frequencies of the observations.

DEFINITION: ARITHMETIC MEAN, FREQUENCY DISTRIBUTION

Let $x_1, x_2, \ldots, x_m$ be different measurements. Then the formula for the arithmetic mean $\bar{x}$ is

$$\bar{x} = \frac{x_1 f_1 + x_2 f_2 + x_3 f_3 + \cdots + x_m f_m}{f_1 + f_2 + f_3 + \cdots + f_m}$$

where f_i is the frequency of x_i for $i = 1, 2, 3, \ldots, m$.

Using the Greek letter Σ, the formula for the arithmetic mean of a grouped frequency distribution can be written as

$$\bar{x} = \frac{\sum_{i=1}^{m} x_i f_i}{n} \quad \text{or} \quad \bar{x} = \frac{\sum x f}{n}, \quad \text{where } n = \sum f$$

where the x_i represents class marks.

TABLE 18

x	4	14	24	34
f	2	8	20	10

EXAMPLE 11 Find the arithmetic mean of the data given in Table 18:

SOLUTION $\bar{x} = \dfrac{4(2) + 14(8) + 24(20) + 34(10)}{2 + 8 + 20 + 10} = \dfrac{940}{40} = 23.5$

Calculator Note

To find the mean of a frequency distribution using a calculator, the x values are stored in one list, say L_1, and the f values are stored in a second list, say L_2. To use the LIST MATH menu follow the procedures of Example 10 except use mean (L_1, L_2). To use the STAT CALC menu press $\boxed{\text{STAT}}$ $\boxed{\blacktriangleright}$ (select 1) $\boxed{\text{ENTER}}$ and read $\bar{x} = 23.5$. Use "3 SetUP" to make certain the frequency is in L_2.

Probability Distribution Mean

The formula for the mean of a frequency distribution assists in understanding the mean of a probability distribution. Suppose that we write

$$\bar{x} = \frac{f_1 x_1 + f_2 x_2 + \cdots + f_m x_m}{f_1 + f_2 + \cdots + f_m}$$

$$= \frac{f_1 x_1 + f_2 x_2 + \cdots + f_m x_m}{n} \qquad n = f_1 + f_2 + \cdots + f_m$$

$$= \frac{f_1}{n} x_1 + \frac{f_2}{n} x_2 + \cdots + \frac{f_m}{n} x_m \qquad \text{Divide each term by } n.$$

Since each f_i/n is a relative frequency, it can be interpreted as a probability. So

$$\bar{x} = p_1 x_1 + p_2 x_2 + \cdots + p_m x_m \qquad p_i = f_i/n$$

We have the same formula for the **mean** $\bar{x}$ or **expected value** $E(x)$ **of a probability distribution** as discussed in the preceding chapter. Recall that $\bar{x} = E(x)$.

TABLE 19

x	1	5	10	25
p	.3	.3	.3	.1

Practice Problem 1 A box contains 3 pennies, 3 nickels, 3 dimes, and 1 quarter. Let x be a variable representing the value (in cents) of a single coin drawn from the box. Verify the probability distribution shown in Table 19. Now find the expected value or mean value.

ANSWER 7.3 cents

TABLE 20

x	104	114	124	134
f	2	8	20	10

Practice Problem 2 Using a graphing calculator, find the arithmetic mean of the data given in Table 20.

ANSWER 123.5

Finding Medians

The **median** of a set of observations is the middle number when the observations are ranked according to size.

DEFINITION: MEDIAN

If $x_1, x_2, x_3, \ldots, x_n$ is a set of data placed in increasing or decreasing order, the **median** is the middle entry, if n is odd. If n is even, the median is the mean of the two middle entries.

EXAMPLE 12   Consider the set of five measurements 7, 1, 2, 1, and 3. Arranged in increasing order, they are written as

$$1, 1, 2, 3, 7$$
$$|$$
median

Hence, the median is 2.   ■

EXAMPLE 13   The array

$$25, 2, 5, 6, 5, 23, 7, 10, 22, 15, 21, 23$$

can be arranged in decreasing order as

$$25, 23, 23, 22, 21, 15, 10, 7, 6, 5, 5, 2$$
$$\diagdown \diagup$$
median

So the median is

$$\frac{15 + 10}{2} = 12.5$$

■

A small company has four employees with annual salaries of $15,500, $16,053, $17,144, and $17,553. The president of the company has an annual salary of $45,000. The mean of the 5 salaries is $22,250. This is a true average, as given by the arithmetic mean, but most people would not accept it as a meaningful measure of central tendency. The median salary, $17,144, is more representative, because the one large salary tends to weight the mean upward to a misleading extent.

Calculator Note

The median can be obtained on a calculator by using either the LIST MATH menu or the STAT CALC menu. To use the LIST MATH menu follow the same procedure as found in Example 10 for the mean except select 4: median; then use either median (list the data) or median (L_1) where the data are stored in L_1. Median ({6, 8, 19, 12}) gives an answer of 10. To use the STAT CALC menu follow the same procedure as found in the Calculator Note on finding the mean. Read from the screen Med = 10.

Practice Problem 3 Find the median of 12, 8, 20, 16, 5, 40, 45, 16, 13, and 5 using a calculator.

ANSWER   14.5

Practice Problem 4 The average salary of 6 office workers in a corporation is $24,000. John remembers 5 of the 6 salaries: $20,000, $23,000, $29,000, $26,000, and $22,000. If the average is the median, calculate for John the missing salary. Do the same if the average is the mean.

ANSWER   For the median a $25,000 salary is missing. For the mean a $24,000 salary is missing.

Finding Modes

The third measure of central tendency is called the **mode**. It refers to the measurement that appears most often in a given set of data.

DEFINITION: MODE

The **mode** of a set of measurements is the observation that occurs most often. If every measurement occurs only once, then there is no mode. If the two most common measurements occur with the same frequency, the set of data is *bimodal*. It may be the case that there are three or more modes.

EXAMPLE 14 Baseball caps with the following head sizes were sold in a week by a sporting goods store: 7, $7\frac{1}{2}$, 8, 6, $7\frac{1}{2}$, 7, $6\frac{1}{2}$, $8\frac{1}{2}$, $7\frac{1}{2}$, 8, and $7\frac{1}{2}$. Find the mode head size.

SOLUTION   The mode is $7\frac{1}{2}$; it occurs four times, more times than any other size.   ■

Practice Problem 5 Find the mode of 21, 23, 24, 22, 24, 20, 22, 24, 25, 20, 22, and 21.

ANSWER   There are two modes: 22 and 24.

The decision about which measure of central tendency to use in a given situation is not always easy. The mean is a good average of magnitudes, such as weights, test scores, and prices, provided that no extreme values are present to distort the data. When extraordinarily large or small values are included in the data set, the median is usually better than the mean. However, the mean is the average most often used, since it gives equal weight to the value of each measurement. The median is a positional average. The mode is used when the ''most common'' measurement is desired. The most appropriate measure for the price of pizzas in town would be the arithmetic mean. To select the best-tasting pizza in town, you could make a survey and use the mode (if a large number of persons were involved). Unfortunately, people with a stake in the outcome tend to use the measure of central tendency that best suits the objectives they hope to accomplish, and they quote the result as an accomplished (and exclusive) fact. This, of course, leads to a widespread mistrust of statistics.

EXAMPLE 15   In one group of games against the Dodgers, the Reds won six of seven games by the scores given in Table 21. Find the mean, median, and mode of the scores for each team.

TABLE 21

Dodgers	2	6	1	15	4	2	2
Reds	4	7	2	1	5	3	3

SOLUTION   When the mean scores are computed, the following results are obtained:

$$\text{Dodgers' mean score} = 4.57$$

$$\text{Reds' mean score} = 3.57$$

Although the Reds dominated the series, the Dodgers' mean score was substantially higher. In this case, the mean is not a good average to use, because the

Dodgers' extraordinarily high score in one game biased the mean. In such cases, it is often better to use the median.

Reds' scores (placed in order):   1 2 3 ③ 4 5 7

median

Dodgers' scores (placed in order):   1 2 2 ② 4 6 15

In this case, the median offers a better measure for comparing the scores. Coincidentally, the mode score for the Reds is 3, and the mode score for the Dodgers is 2.

Percentiles

One way of reporting a person's relative performance on a test is to identify the percentage of people taking the test who scored lower than the person under consideration. For example, someone who scores higher than 70% of those who take a test is said to be in the 70th percentile. Conversely, a percentile score of 85 means that the person scored higher than 85% of those in the sample. In the next several examples, we practice using percentiles.

DEFINITION: PERCENTILE

Let $x_1, x_2, x_3, \ldots, x_n$ be a set of n measurements arranged in order of magnitude. The pth **percentile** is the value of x such that $p\%$ of the measurements are less than the value of x and $(100 - p)\%$ are greater than x.

EXAMPLE 16   A student scored in the 68th percentile on the mathematics portion of the Iowa Test of Basic Skills. Can she conclude that she answered 68% of the questions correctly on this test?

SOLUTION   No, the student has no information about how many problems she answered correctly. She does know that she scored higher than 68% of those who took the mathematics portion of the test. We say that the *percentile rank* of the student's score is 68.

Because percentiles are difficult to compute accurately (except for large data sets), we are concerned with how to interpret them, not how to compute them. However, there are three percentiles to which we wish to give closer attention. The 25th percentile is also called the **first quartile**, the 50th percentile is the **second** or **middle quartile**, and the 75th percentile is the **third** or **upper quartile**. Notice that the second quartile or 50th percentile is the median. For small data sets we can get an easy approximation of the first quartile by computing the median of the scores below the median. Similarly, we can approximate the upper quartile by computing the median of the scores above the median.

EXAMPLE 17 Find the median and approximate the first and third quartiles of the following data:

8, 14, 12, 64, 7, 9, 42, 84, 76, 92, 41, 15, 17, 26, 47, 16, 21, 22, 23, 24

SOLUTION Arrange the data in order of increasing magnitude:

7, 8, 9, 12, 14,   15, 16, 17, 21, 22,   23, 24, 26, 41, 42,   47, 64, 76, 84, 92

There are 20 observations, so the median is the average of the tenth and eleventh scores. The median is 22.5, the average of 22 and 23. There are 10 scores below the median and their median is 14.5. This is the first quartile, Q_1. Likewise, there are 10 scores above the median and their median is 44.5. This is the third quartile, Q_3. ∎

Calculator Note

To illustrate the potential of a graphing calculator, let's store the data from Example 17 without arranging the numbers in order of increasing magnitude. Suppose that we store the data in L_1. On some graphing calculators you need to store corresponding 1's in L_2 to represent the frequency of each data point in L_1, or use "3:SetUP."

We arrange the data in order of increasing magnitude by using $\boxed{\text{STAT}}$ 2 followed by SortA(L_1). This arranges the data in L_1 in the same way that it is arranged in the solution of Example 17. To access the STAT CALC menu press $\boxed{\text{STAT}}$ $\boxed{\blacktriangleright}$, then select 1 and press $\boxed{\text{ENTER}}$. We obtain

$$\bar{x} = 33$$

$$n = 20$$

$$Q_1 = 14.5$$

$$\text{median} = 22.5$$

$$Q_3 = 44.5$$

SUMMARY

1. Formulas for the arithmetic mean are

$$\bar{x} = \frac{x_1 + x_2 + x_3 + \cdots + x_n}{n} \quad \text{and} \quad \bar{x} = \frac{x_1 f_1 + x_2 f_2 + \cdots + x_m f_m}{f_1 + f_2 + \cdots + f_m}$$

2. The expected value of a probability distribution is

$$E(x) = \sum_{i=1}^{n} x_i p_i$$

3. The median is the middle measurement.
4. The mode is the most common measurement.
5. The pth percentile is the score below which $p\%$ of the data lies.

Exercise Set 7.2

Use a graphing calculator whenever possible in this exercise set.

1. Compute the arithmetic mean, the median, and the mode (or modes if any) for the given sets of data.
   (a) 3, 4, 5, 8, 10
   (b) 4, 6, 6, 8, 9, 12
   (c) 3, 6, 2, 6, 5, 6, 4, 1, 1
   (d) 7, 1, 3, 1, 4, 6, 5, 2
   (e) 21, 13, 12, 6, 23, 23, 20, 19
   (f) 18, 13, 12, 14, 12, 11, 16, 15, 21

2. An elevator has a capacity of 15 people and a load limit of 2250 lb. What is the mean weight of the passengers if the elevator is loaded to capacity with people and weight?

3. Find the mean of the given distribution.

x	Frequency
10	2
20	6
30	8
40	4

4. Make up a set of data with 4 or more measurements, not all of which are equal, with each of the following characteristics:
   (a) The mean and median are equal.
   (b) The mean and mode are equal.
   (c) The mean, median, and mode are equal.
   (d) The mean and median have values of 8.
   (e) The mean and mode have values of 6.
   (f) The mean, median, and mode have values of 10.

5. The mean score of a set of 8 scores is 65. What is the sum of the 8 scores?

6. The mean score of 9 of 10 scores is 81. The tenth score is 100. What is the mean of the 10 scores?

7. Which of the three averages should be used for the following data?

(a) The average salary of 4 salesmen and the owner of a small store
(b) The average height of all male students in a high school
(c) The average dress size sold at an apparel outlet

8. At the initial meeting of an athletic club, the weights of the members were found to be 220, 275, 199, 246, 302, 333, 401, 190, 286, 254, 302, 323, 221.
   (a) Compute the mean, median, and mode of the data.
   (b) Which measure is most representative of the data?

9. The given grades were recorded for a test on this chapter. Find the arithmetic mean.

Score	Frequency
100	3
90	5
80	7
70	15
60	14
50	3
40	3

10. For each of the given sets of observations, find the median, Q_1, and Q_3.
    (a) 16, 14, 12, 13, 15, 18, 24, 8, 10, 4
    (b) 18, 47, 64, 32, 41, 92, 84, 27, 14, 12

11. In a class of 80 students, Jodi scored in the 80th percentile. How many students scored less than Jodi?

12. In a class of 80 students, Aaron's rank on the last test was 14th from the top. What is his percentile score?

13. For the following data, find the median, Q_1, and Q_3:

    17, 26, 34, 41, 52, 14, 13, 18, 27, 31,
    39, 43, 44, 47, 49

Find the (a) mean and (b) median of the distributions given by the following histograms. Use the class marks for the variable.

14.

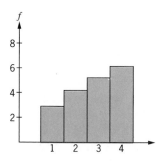

15.

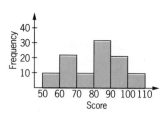

Find the mean or the expected value of the probability distributions defined by the following graphs.

16.

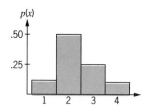

17.

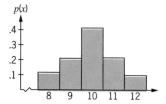

18. A student has an average of 89 on 9 tests. What will she need to make on the tenth test to have an average of 90?

19. An interesting property of the mean is that the sum of the differences in the mean and each observation (deviations of each score from the mean, considered as signed numbers) is 0. Show that this statement is true for the following data: 5, 8, 10, 12, 15.

20. The table shows the distribution of scores on a test administered to freshmen at a college. Find the arithmetic mean. Let x be the class mark.

Score	Frequency
140–149	3
130–139	4
120–129	8
110–119	13
100–109	4
90–99	2
80–89	0
70–79	1

Applications (Business and Economics)

21. *Salaries.* The table shows the salaries of the employees in a company. Find the mean salary of the employees.

Salary	Frequency
$ 8,000	4
10,000	3
18,000	2
30,000	1

22. *Salaries.* The president of a furniture factory draws a salary of $200,000 per year. Four supervisors have salaries of $40,000 each. Twenty blue-collar workers have salaries of $20,000 each. Discuss each of the following.
   (a) The president says the average salary is $30,400.
   (b) The union says the average salary is $20,000.
   (c) Which average is more representative of the factory salaries?

23. *Salaries.* If 99 people have a mean income of $18,000, how much is the mean income increased by the addition of a man with an income of $150,000?

24. *Salaries.* The following frequency distribution gives the weekly salaries by title of the employees of a company. Frank examined this list and concluded

that the mean salary is

$$\frac{550 + 450 + 350 + 250 + 150}{5} = \$350$$

Title	Number	Weekly Salary
Manager	1	$550
Supervisors	3	450
Inspectors	3	350
Line workers	21	250
Clerks	5	150

(a) Is he correct?

(b) If he is not correct, find what the mean should be.

25. **Price–Earnings Ratio**. Find the mean price–earnings ratio of 100 common stocks listed on the New York Stock Exchange. (See Exercise 20.)

Interval	Frequency
0–4	6
5–9	46
10–14	30
15–19	10
20–24	4
25–29	2
30–34	2

Applications (Social and Life Sciences)

26. **Psychology**. Find the mean of the following test scores (scores are class marks of a frequency distribution).

Test Score	Frequency
50	1
60	3
70	10
80	4
90	2

27. **Expense Accounts**. The following data have been

collected on the expenses (excluding travel and lodging) of 6 trips made by teachers in the mathematics department at a college. Let

$$\bar{x} = \frac{\$64.30}{20}$$

$$= \$3.22 \text{ average expense per day}$$

or

$$\bar{x} = \frac{\$148.00}{20}$$

$$= \$7.40 \text{ average expense per day}$$

or

$$\bar{x} = \frac{\$64.30}{5}$$

$$= \$12.86 \text{ average expense per day}$$

Which average is realistic?

Number of Days on Trip	Total Expenses	Expenses per Day
0.5	$ 13.50	$27.00
2.5	12.00	4.80
3	21.00	7.00
1	9.50	9.50
8	32.00	4.00
5	60.00	12.00
20	$148.00	$64.30

28. **Pollution**. A study of the number of oil spills into the nation's waterways in recent years gives the number of spills of various sizes:

Millions of Gallons of Oil	Number of Spills
1–3	6
4–6	9
7–9	13
10–12	10
13–15	7
16–18	3

Find the arithmetic mean.

29. **Vitamins.**   A pollster tabulated the ages of 30 users of a vitamin A designed to counteract colds. The results are shown in the table. Find the mean age.

Age	Frequency
20–29	1
30–39	2
40–49	4
50–59	5
60–69	9
70–79	6
80–89	3

Review Exercises

30. Group the following test scores, which were received by 24 students, into 10 classes (95–99, 90–94, 85–89, 80–84, 75–79, 70–74, 65–69, 60–64), 55–59, and 50–54.

    63, 71, 85, 96, 94, 90, 75, 72, 77, 71, 62, 84, 81, 76, 61, 54, 87, 94, 62, 81, 94, 77, 63, 60

31. For the data in Exercise 30, construct (a) a histogram and (b) a frequency polygon.

7.3   MEASURES OF VARIATION

OVERVIEW   The fact that the mean salary of management at one company exceeds the mean salary of management at another company does not mean that the salaries of the first company are superior to those of the second company. Four monthly salaries at one company are $3000, $3100, $3100, and $20,000, with a mean of $7300. Four monthly salaries at another company are $6800, $6900, $7000, and $7200, with a mean of $6975. The mean salary of the first company exceeds the mean salary of the second company, but the lowest salary of the second company is better than all the salaries except the largest salary of the first company. The average depth of the Cahaba River is 1 foot. This river should be a nice river in which to go wading. Wait a minute! There are many shallow areas, but there are also a number of holes 15 to 16 feet deep.

The preceding examples indicate the need for a measure of dispersion or scattering of data. That is, we need a measure to indicate whether the entries in a set of data are close to or not close to the average. In this section we consider

- The range
- Variance
- Standard deviation
- Standard deviation for a frequency distribution
- Standard deviation of a probability distribution
- z scores

There are several ways to measure dispersion of values within a set of data. The measure of scattering that is easiest to calculate is the range, which we introduced in Section 7.1.

EXAMPLE 18   For the set of data

$$7, 3, 1, 15, 41, 74, 35$$

the range is 73, since $74 - 1 = 73$.

Calculator Note

The range can be obtained in two ways using a calculator: Using the LIST MATH menu and using the STAT CALC menu followed by Var Stats. In both cases the calculator generates the maximum and minimum values of a set of data. The difference in these two values is the range.

Although the range is easy to obtain, it is not always a good measure of dispersion because it is so radically affected by a single extreme value. For example, suppose that the 74 in the set of data listed in Example 18 was miscopied and listed as 24 instead. Note that the range changes from 73 to 40.

Since the range is affected significantly by extreme values (either large or small), other measures of scattering are preferable. In this section we consider **variance**, denoted by σ_x^2, and **standard deviation**, denoted by σ_x (often abbreviated as σ), which measure scattering of the data from the mean.

Variance

Variance for a set of data can be obtained in four steps.

1. Compute the mean, $\bar{x}$.
2. Compute the difference between each observation and the arithmetic mean: $x - \bar{x}$.
3. Square each difference: $(x - \bar{x})^2$.
4. Divide the sum of the differences squared by n, where n is the number of observations.

Standard deviation is the square root of variance.

EXAMPLE 19 Find the variance for the data 5, 7, 1, 2, 3, and 6.

SOLUTION

1. Compute the mean of the data (see Table 22):

$$\bar{x} = \frac{24}{6} = 4$$

2. Determine the difference between each x and $\bar{x} = 4$. (See the second column of Table 22.)
3. Compute the square of each of these differences; that is, compute $(x - \bar{x})^2$. (See the third column of Table 22.)

TABLE 22

x	$x - \bar{x}$	$(x - \bar{x})^2$
5	1	1
7	3	9
1	-3	9
2	-2	4
3	-1	1
6	2	4
Total 24		28

4. Sum the squares of the differences (i.e., sum the third column of Table 22) and divide by $n = 6$:

$$\sigma^2 = \frac{28}{6} = 4.67$$

The variance is 4.67. ■

The preceding four-step calculation is equivalent to the following formula for variance.

Formula for Variance

Variance, denoted by σ^2, is calculated as

$$\sigma^2 = \frac{(x_1 - \bar{x})^2 + (x_2 - \bar{x})^2 + (x_3 - \bar{x})^2 + \cdots + (x_n - \bar{x})^2}{n}$$

where $\bar{x}$ is the mean of the observations and n is the number of observations.

This formula is sometimes written in summation notation as

$$\sigma^2 = \frac{\sum_{i=1}^{n} (x_i - \bar{x})^2}{n} \quad \text{or} \quad \frac{\sum (x - \bar{x})^2}{n}$$

To compute standard deviation, remember that standard deviation is simply

$$\sigma = \sqrt{\sigma^2} = \sqrt{\frac{\sum_{i=1}^{n} (x_i - \bar{x})^2}{n}}$$

EXAMPLE 20 Find the standard deviation of the data given in Example 19.

$$\sigma = \sqrt{\sigma^2} = \sqrt{4.67} \qquad \sigma^2 = 4.67 \text{ in Example 19.}$$
$$= 2.16$$
■

Calculator Note

To obtain the standard deviation of a set of data, we must first list the data in the calculator, say under L_1. We then press [STAT] [▶] to access the STAT CALC menu, where statistical calculations may be obtained. The first item under this menu is 1:1-Var

Stats, which calculates one-variable statistics. For example, list the data from Example 19 under L_1: 5, 7, 1, 2, 3, 6. Under L_2 we list that each of these data points has a frequency of 1 or use "3:SetUp." Selecting from the STAT CALC menu, the $1:1$-Var Stats gives

$$\bar{x} = 4$$

$$s_x = 2.366$$

$$\sigma_x = 2.160$$

$$n = 6$$

The standard deviation we have defined is obtained from $\sigma_x = 2.160$. Note that this is the same answer that we obtained in Example 20.

NOTE: Most calculators provide two standard deviations. On a graphing calculator the standard deviation of a population (the standard deviation we obtain in this section) is denoted by σ_x. The standard deviation of a sample is denoted by s_x. What is the difference in these two standard deviations? In inferential statistics, when computing the standard deviation of a sample in order to estimate the standard deviation of the population, we divide by $n - 1$ instead of n to get the variance. Because we are concerned primarily with descriptive statistics in this book, we will not need to be concerned with this subtlety.

The formulas given for variance and standard deviation are useful because they make clear that they measure variation of the data from the arithmetic mean. However, a different formula is more accessible for computations.

Computational Formula for Variance

$$\sigma^2 = \frac{x_1^2 + x_2^2 + \cdots + x_n^2 - n\bar{x}^2}{n} = \frac{\sum\limits_{i=1}^{n} x_i^2 - n\bar{x}^2}{n}$$

EXAMPLE 21 Use the computational formula for variance to find the variance of the data given in Example 19.

SOLUTION The first column of Table 23 displays the values of the variable x, and the second column displays the values of x^2. Therefore,

$$\bar{x} = \frac{24}{6} = 4$$

$$\sigma^2 = \frac{124 - 6(4)^2}{6} \qquad \sum_{i=1}^{n} x_i^2 = 124$$

$$\qquad\qquad\qquad\qquad n = 6$$

$$= \frac{28}{6} = 4.67$$

This is the same answer that we obtained in Example 19.

TABLE 23

x	x^2
5	25
7	49
1	1
2	4
3	9
6	36
24	124

When the data are presented in a frequency distribution, the formula for variance can be written as follows.

Variance of Frequency Distributions

Suppose that $x_1, x_2, \ldots, x_m$ have respective frequencies $f_1, f_2, \ldots, f_m$. Therefore,

$$\sigma^2 = \frac{x_1^2 f_1 + x_2^2 f_2 + \cdots + x_m^2 f_m - n\bar{x}^2}{n} = \frac{\sum_{i=1}^{m} x_i^2 f_i - n\bar{x}^2}{n}$$

where $n = f_1 + f_2 + \cdots + f_m$.

As before, to find the standard deviation compute the square root of the variance.

EXAMPLE 22   Find the variance and the standard deviation of the data tabulated in the first two columns of Table 24.

SOLUTION

$$\bar{x} = \frac{120}{20} = 6$$

$$\sigma^2 = \frac{848 - (20)6^2}{20} \qquad \sum_{i=1}^{n} x_i^2 f_i = 848$$

$$= 6.4 \qquad n = 20$$

$$\sigma = \sqrt{6.4} = 2.53$$

TABLE 24

x	f	xf	x^2	$x^2 f$
2	3	6	4	12
4	4	16	16	64
6	6	36	36	216
8	4	32	64	256
10	3	30	100	300
Total	20	120		848

Calculator Note

To obtain the standard deviation of a frequency distribution on a graphing calculator, follow the same steps that we discussed for a regular distribution except replace the 1's in L_2 by the actual frequency of each data point. For example, place {2, 4, 6, 8, 10} in L_1 and {3, 4, 6, 4, 3} in L_2. Use ''3:SetUp'' to make certain the frequency is in L_2. Then perform the steps as discussed previously for the STAT CALC menu to obtain

$$\bar{x} = 6$$

$$\sigma_x = 2.53$$

$$n = 20$$

These are the same answers that we obtained in Example 22.

The formula for the variance of a population can be put in a form for the variance of a probability distribution as follows:

$$\sigma_x^2 = \frac{x_1^2 f_1 + x_2^2 f_2 + \cdots + x_m^2 f_m - n\bar{x}^2}{n}$$

$$= x_1^2 \frac{f_1}{n} + x_2^2 \frac{f_2}{n} + \cdots + x_m^2 \frac{f_m}{n} - \bar{x}^2$$

$$= x_1^2 p(x_1) + x_2^2 p(x_2) + \cdots + x_m^2 p(x_m) - [E(x)]^2$$

where f_i/n is interpreted as $p(x_i)$ and $\bar{x}$ is replaced by its probability distribution counterpart,

$$E(x) = x_1 p(x_1) + x_2 p(x_2) + \cdots + x_n p(x_n)$$

Variance of a Probability Distribution

If the probabilities of outcomes $x_1, x_2, \ldots, x_n$ are $p(x_1), p(x_2), \ldots, p(x_n)$, then the variance of the probability distribution is

$$\sigma_x^2 = x_1^2 p(x_1) + x_2^2 p(x_2) + \cdots + x_n^2 p(x_n) - [E(x)]^2$$

where $E(x) = x_1 p(x_1) + x_2 p(x_2) + \cdots + x_n p(x_n)$.

TABLE 25

x_i	1	2	3	4	5
$p(x_i)$	$\frac{1}{8}$	$\frac{1}{4}$	$\frac{1}{4}$	$\frac{1}{4}$	$\frac{1}{8}$

EXAMPLE 23   Find the variance of the probability distribution given in Table 25.

SOLUTION

$$\mu_x = E(x) = 1\left(\frac{1}{8}\right) + 2\left(\frac{1}{4}\right) + 3\left(\frac{1}{4}\right) + 4\left(\frac{1}{4}\right) + 5\left(\frac{1}{8}\right)$$

$$= 3$$

$$\sigma_x^2 = 1^2\left(\frac{1}{8}\right) + 2^2\left(\frac{1}{4}\right) + 3^2\left(\frac{1}{4}\right) + 4^2\left(\frac{1}{4}\right) + 5^2\left(\frac{1}{8}\right) - 3^2$$

$$= \frac{84}{8} - 9$$

$$= 1.5$$

∎

An important property of standard deviation is that for many collections of data most of the data are located within 2 standard deviations of the mean of the data. At least 75% of given data lie within 2 standard deviations of the mean for most data sets, and we will learn in the next section that for certain collections of data, 95% of the data lie within 2 standard deviations of the mean.

EXAMPLE 24   Babe Ruth won 12 American League home run championships in his career. The number of home runs he hit to win those championships were

$$11, 29, 54, 59, 41, 46, 47, 60, 54, 46, 49, 46$$

The mean of these data is 45.17 and the standard deviation is 13.02. What percentage of the data lies within 2 standard deviations of the mean?

SOLUTION

Mean + 2 standard deviations = 45.17 + 2 (13.02) = 71.21

Mean − 2 standard deviations = 45.17 − 2 (13.02) = 19.13

Eleven of the twelve values lie within 2 standard deviations of the mean or 92% of the data lie in this interval.   ■

EXAMPLE 25   A student made scores of 90, 82, 70, 61, and 94 on 5 tests. Of course, the student did his best work relative to the rest of the class on the last test. Or did he? What do we know about how the other students scored? Maybe everyone in the class made a higher score than 90 on the last test.   ■

Example 25 illustrates a need for standard scores by which to make comparisons. Such scores are called **z scores**. These scores are measured from the mean and are adjusted for scattering by dividing by the standard deviation. Thus, if the student in Example 25 had a z score of 2.3 on the last test and a z score of 2.6 on the first test, the student did better on the first test relative to the remainder of the class.

DEFINITION: z SCORES

A score or measurement, denoted by x, from a distribution with mean $\bar{x}$ and standard deviation σ has a corresponding **z score** given by

$$z = \frac{x - \bar{x}}{\sigma}$$

representing the number of standard deviations from the mean.

EXAMPLE 26   The average weight of bags of potato chips is 10 oz, with a standard deviation of 1 oz. One bag of potato chips is weighed and has a weight of 9.8 oz. Convert this measurement into standard units.

SOLUTION   We know that $x = 9.8$ oz, $\bar{x} = 10$ oz, and $\sigma = 1$ oz. Therefore,

$$z = \frac{x - \bar{x}}{\sigma} = \frac{9.8 - 10}{1} = -0.2$$

■

The z score is a measurement expressed in standard units or without units. For example, if x and $\bar{x}$ are in feet, then σ is in feet, and the division eliminates the units. Consequently, z scores are useful for comparing two sets of data with different units.

EXAMPLE 27   Teresa scores a 76 on the entrance test at school X and an 82 at school Y. At which school did she have the better score?

SOLUTION   To answer this question, we need to know that the mean score at school X was 70 with a standard deviation of 12 while the mean score at school Y was 76 with a standard deviation of 16. The z scores are then as follows:

$$\text{School X:} \qquad z = \frac{76 - 70}{12} = 0.5$$

$$\text{School Y:} \qquad z = \frac{82 - 76}{16} = 0.375$$

Since 0.5 is greater than 0.375, Teresa's score at school X was better than her score at school Y in comparison to the scores of others who took the test.   ■

Interquartile Range

Another measure of variation is called the **interquartile range (IQR)**. To find the IQR first we need to compute the lower quartile Q_1, called the first quartile, and the upper quartile Q_3, called the third quartile. Remember that Q_1 is the 25th percentile and Q_3 is the 75th percentile of a given set of data. The interquartile range is the difference between the lower quartile and the upper quartile. It is therefore the interval that contains the middle half (50%) of the data.

DEFINITION: INTERQUARTILE RANGE

The **interquartile range** (IQR) equals

$$Q_3 - Q_1$$

EXAMPLE 28   The following are 16 grades received on a test, arranged in increasing order. Find the median, Q_1, Q_3, and the interquartile range.

SOLUTION   Remember that we can approximate Q_1 by computing the median of the scores below the median. Similarly, we can approximate Q_3 by computing the median of the scores above the median.

64 66 68 71 73 75 76 78 82 83 84 90 92 95 96 99

$Q_1 = 72$ median $= 80$ $Q_3 = 91$

$$\text{IQR} = 91 - 72 = 19$$

Therefore, a span of 19 points includes the middle 50% of the grades.   ■

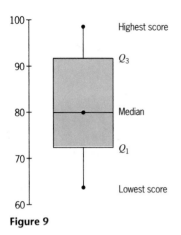

Figure 9

Box and Whisker Plots

Sometimes the median, Q_1, Q_3, and the range are displayed in a diagram, called a **box and whisker plot**. Figure 9 shows such a plot for the data in Example 28. The interquartile range is represented as the box. The lines extending from Q_1 and Q_3 to the lowest and highest scores are the whiskers.

Calculator Note

Let's draw a box and whisker plot on a calculator. We first insert the data in Example 28 in L_1. Then under WINDOW use X : [60, 100] with a scale of 5. Press 2nd | STAT PLOT | 1 and under Plot 1 select in order ON | ENTER |, under Type: the picture of a box plot | ENTER |; under Xlist: L_1 | ENTER | and under the Ylist select 1 and press | ENTER |. Press | GRAPH | to obtain Figure 10.

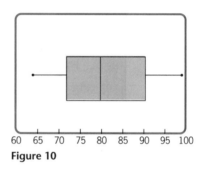

Figure 10

Data that are much larger (or much smaller) than the mean can adversely affect the usefulness of the mean as a measure of central tendency. One way to locate such data is to find the **outliers** of a set of data.

DEFINITION: OUTLIER

An **outlier** is any data point farther than 1.5 IQRs above Q_3 or farther than 1.5 IQRs below Q_1.

For the data in Example 28 we compute

$$Q_3 + 1.5(\text{IQR}) = 91 + 1.5(19) = 119.5$$

$$Q_1 - 1.5(\text{IQR}) = 72 - 1.5(19) = 43.5$$

For these data, there are no outliers. If the data contained a test grade of 40, this value would be classified as an outlier because it would be less than 43.5.

For the data in Example 17 the IQR $= 44.5 - 14.5 = 30.$

$$Q_3 + 1.5(\text{IQR}) = 89.5$$
$$Q_1 - 1.5(\text{IQR}) < 0$$
$$= -30.5$$

There is one outlier, namely 92.

SUMMARY

1. In this section we introduced the following measures of dispersion: range, variance, standard deviation, and interquartile range.
2. The variance is the average of squared deviations from the mean:

$$\sigma^2 = \frac{(x_1 - \bar{x})^2 + (x_2 - \bar{x})^2 + \cdots + (x_n - \bar{x})^2}{n}$$

$$\sigma^2 = \frac{(x_1 - \bar{x})^2 f_1 + (x_2 - \bar{x})^2 f_2 + \cdots + (x_m - \bar{x})^2 f_m}{n}$$

3. The standard deviation is the square root of the variance.
4. The standardized score or z score of a value x is given by

$$z = \frac{x - \bar{x}}{\sigma}$$

5. The box and whisker plot gives a geometric view of the relationship between the median, the first and third quartiles, and the largest and smallest values in the data set.

Exercise Set 7.3

Work the problems in this exercise set using a graphing calculator.

For the given sets of observations, find the range, the variance, and the standard deviation.

1. 6, 8, 8, 14
2. 10, 12, 13, 14, 16
3. 1, 4, 5, 7, 13
4. 80, 75, 80, 70, 80
5. 15, 17, 19, 23, 26
6. 1, 1, 4, 7, 7
7. 16, 14, 12, 13, 15, 18, 24, 8, 10, 4
8. 18, 47, 64, 32, 41, 92, 84, 27, 14, 12

9. 9, 7, 16, 14, 12, 13, 14, 18, 24, 8, 10, 4
10. 10, 17, 18, 47, 64, 32, 41, 92, 84, 27, 14, 12
11. For Exercise 7, construct a box and whisker plot. List all outliers.
12. For Exercise 8, construct a box and whisker plot. List all outliers.
13. For Exercise 9, construct a stem and leaf chart. From this chart, find $Q_3 + 1.5(\text{IQR})$ and $Q_1 - 1.5(\text{IQR})$. Are there any outliers?
14. For Exercise 10, construct a stem and leaf chart. From this chart, find $Q_3 + 1.5(\text{IQR})$ and $Q_1 - 1.5(\text{IQR})$. Are there any outliers?

15. Compute σ^2 for the following frequency distribution.

x	10	14	18	22
f	4	6	8	2

16. Find the standard deviation of the given sample data.

x	Frequency
1	10
4	20
7	30
10	40

17. The mean of a population is 100 with a standard deviation of 10. Convert the following to z scores.
(a) 110 (b) 80 (c) 71
(d) 120 (e) 140 (f) 40

18. Two instructors gave the same test to their classes. Both classes had a mean score of 72, but the scores of class A showed a standard deviation of 4.5 and class B showed a standard deviation of 9. Discuss the difference in the two classes.

19. Find the variance for the following probability distribution.

x_1	16	18	20	22	24
$p(x_i)$	$\frac{1}{8}$	$\frac{1}{4}$	$\frac{1}{4}$	$\frac{1}{4}$	$\frac{1}{8}$

20. It can be shown that, for any set of data, most of the values lie within two standard deviations on either side of the mean. Examine the data from the following table in view of this fact. First find the standard deviation of each class.

Height in Centimeters					
Class I			Class II		
156	158	182	168	180	183
178	159	176	180	187	190
160	176	174	176	176	178
166	160	172	188	186	174
189	187	154	179	192	188
153	180	198	176	179	181
159	162	176	173	174	180
180	166	192	178	176	175

(a) How many of the values from class I lie within 2 standard deviations on either side of the mean? What percentage of class I is this?
(b) How many of the values from class II lie within 2 standard deviations on either side of the mean? What percentage of class II is this?

21. **Exam.** A student received a grade of 80 on a math final where the mean grade was 72 and the standard deviation was σ. In the statistics final, he received a 90, where the mean grade was 80 and the standard deviation was 15. If the standardized scores (i.e., scores adjusted to a mean of 0 and a standard deviation of 1) were the same in each case, then σ equals which of the following?
(a) 10 (b) 12 (c) 16
(d) 18 (e) 20

Applications (Business and Economics)

22. **Quality Assessment.** The following data show the miles per gallon reported by owners of 5 eight-cylinder automobiles from different manufacturers.

Manufacturer				
A	B	C	D	E
18	18	24	21	18
19	18	16	18	18
20	20	18	19	19
21	21	20	18	27
22	24	22	20	18
22	19	24	21	18

(a) Which sample suggests the best gasoline mileage?
(b) Which sample has the lowest standard deviation?

23. **Salaries.** When Tran entered his profession in 1975, the average salary in the profession was $11,500, with a standard deviation of $1000. In 1985, the average salary in the profession was $21,000 with a standard deviation of $3000. Tran made $11,000 in 1975 and $20,000 in 1985. In which year did he do better in comparison with the rest of the profession?

24. *Salary Range*.   The salaries of the 10 supervisors at an automobile company are as follows: $30,000, $33,000, $32,000, $34,000, $36,500, $31,500, $35,000, $37,000, $36,000, and $31,000. How many of the 10 supervisors have salaries within 1 standard deviation of the mean? How many have salaries within 2 standard deviations of the mean?

25. *Price–Earnings Ratio*.   Find the standard deviation of the price–earnings ratio given in Exercise 25 of Exercise Set 7.2.

Applications (Social and Life Sciences)

26. *Test Scores*.   The following table shows the distribution of scores on a test administered to first-year students at a college. Find the standard deviation.

Score	Frequency
140–149	3
130–139	4
120–129	8
110–119	13
100–109	4
90–99	2
80–89	0
70–79	1

27. *Vitamins*.   A pollster tabulated the ages of 30 users of a vitamin pill designed to make one feel young. The results are shown in the following table. Find the standard deviation.

Age	Frequency
20–29	1
30–39	2
40–49	4
50–59	5
60–69	9
70–79	6
80–89	3

28. *Test Scores*.   Joan decides to join the New Army to seek her fortune. She takes a battery of tests to determine placement into the appropriate corps. She makes 75 on the office work test and 80 in the outdoor activity test. The office work test has a mean of 60 and a standard deviation of 20, whereas the outdoor activity test has a mean of 75 and a standard deviation of 10. Into which group should Joan be placed?

Review Exercises

29. Consider the set of scores 1, 8, 16, 18, 20, 20, 21, 23, 24, 29. Find the following.
    (a) Mean   (b) Median
    (c) Mode   (d) Range
    (e) Standard deviation

30. The mean salary of all employees at the Brown Corporation is $30,000. Make up an example to show how this statistic may be misleading.

7.4   THE NORMAL DISTRIBUTION

OVERVIEW   The grades on a certain standardized test are normally distributed with a mean of 70 and a standard deviation of 10. What is the probability that a randomly selected student, who took the test, scored between 70 and 85? The key to answering this question is an understanding of the phrase *normally distributed*.

One of the fortunate surprises in statistics is that many line graphs and bar graphs are approximately bell-shaped. In fact, if we modify some line graphs to indicate probability rather than frequency, the resulting graphs closely approximate a smooth, bell-shaped curve called the **normal probability curve**. If this is true, the data involved are said to be normally distributed. In this section we consider

- The normal distribution
- Properties of the normal curve

- How to use the normal curve
- Applications of the normal curve

Normal Probability Curve

To appreciate the characteristics of the normal probability curve, consider the line graph and histogram in Figure 11, which are based on the data in Table 26. Notice that the frequencies have been converted into relative frequencies or probabilities.

The height of each rectangle in Figure 11 represents the probability that the variable falls in the associated interval. For example, the probability that the variable falls between 19.5 and 24.5 is $\frac{6}{20}$ = .30. For a probability curve we define the graph as representing a continuous probability density function if and only if the area under the curve is 1, and the area under the curve between two values of the variable is the probability that the variable is between the two values. Since we draw probability curves as approximations to histograms, we want each area of a rectangle, rather than the height of the rectangle, to represent probability. Therefore, we can revise the graph in Figure 11 so that the area of each rectangle is equal to the probability by dividing the height of each rectangle by 5. The resulting graph is shown in Figure 12.

TABLE 26

Class	Frequency	Relative Frequency
5–9	1	$\frac{1}{20}$ = 0.05
10–14	2	$\frac{2}{20}$ = 0.10
15–19	4	$\frac{4}{20}$ = 0.20
20–24	6	$\frac{6}{20}$ = 0.30
25–29	4	$\frac{4}{20}$ = 0.20
30–34	2	$\frac{2}{20}$ = 0.10
35–39	1	$\frac{1}{20}$ = 0.05
	20	

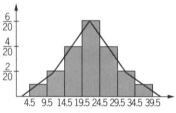

Figure 11

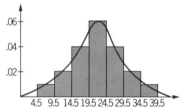

Figure 12

In Figure 12, the area of the rectangle over the interval from 9.5 to 14.5 is .10 and represents the probability that a randomly selected data value from the data set in Table 26 falls between 9.5 and 14.5. Said another way, the area in this rectangle is 10% of the total area of the rectangles in Figure 12 and represents 10% of the data in Table 26.

A smooth curve is sketched over the bar graph in Figure 12. This smooth curve is an example of a normal curve. We can observe some interesting properties of the normal probability curve (often called a normal distribution) by studying the approximating histogram.

Recall that for a discrete probability density function (see the preceding chapter), the $\Sigma\, p(x_i)$ is equal to 1. In a similar manner, the area under the graph of a continuous probability density function is equal to 1. Note that the sum of all the areas of the rectangles in the histogram in Figure 12 is equal to 1. The mean of the data is 22, and the curve is symmetric about a vertical line through the mean. The standard deviation of the data is slightly less than 7.5. The interval about the mean that extends for 1 standard deviation on either side of the mean is the interval from $(22 - 7.5)$ to $(22 + 7.5)$. This interval from 14.5 to 29.5 contains $\frac{4}{20} + \frac{6}{20} + \frac{4}{20} = \frac{14}{20}$, or 70%, of the data (see Figure 13). The interval about the mean that extends for 2 standard deviations on either side of the mean is the interval from $(22 - 15)$ to $(22 + 15)$, or from 7 to 37. Assume that half of the frequency in the first and last class intervals belongs in this range from 7 to 37. Then the interval (from mean $-$ 2 standard deviations to mean $+$ 2 standard deviations) contains $\frac{19}{20}$, or 95%, of the data (see Figure 14).

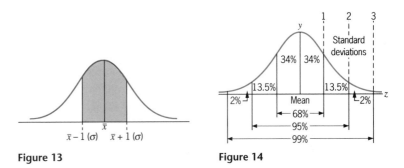

Figure 13   **Figure 14**

These properties for the histogram and the approximating smooth curve suggest the following well-known properties of a normal curve.

PROPERTIES OF A NORMAL CURVE

(a)   The area under a normal curve is equal to 1.

(b)   The normal curve is symmetric about a vertical line through the mean of the set of data.

(c)   The interval extending from 2 standard deviations to the left of the mean to 2 standard deviations to the right of the mean contains approximately 95% of the area; the corresponding interval extending 1 standard deviation on each side of the mean contains approximately 68% of the area; the corresponding interval extending 3 standard deviations on each side of the mean contains 99% of the area.

(d)   If x is a data value from a normally distributed set of data, then the probability that x is greater than a and less than b is the area under the normal curve between a and b.

In the preceding section, we discussed the process of converting data into standard units. Recall that to express x in standard units, we subtract the mean and then divide by the standard deviation:

$$z = \frac{x - \bar{x}}{\sigma}$$

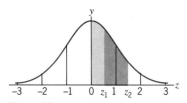

Figure 15

When a data value from a normal distribution is standardized, the resulting data value lies in a special normal distribution called the **standard normal distribution**. The standard normal distribution has a mean of 0 and a standard deviation of 1.

The curve in Figure 15 is the standard normal distribution. We use z to represent the standard normal variable and y to represent the probability. The maximum value of the curve is attained at the mean, $z = 0$. The standard normal curve has perfect symmetry. Because of this characteristic, the mean and median of the distribution have the same value, namely 0. The range is not bounded because values occur as far out as you wish to go; that is, the curve never intersects the horizontal axis.

The area under the standard normal curve is equal to 1. To find the probability that z is between z_1 and z_2, $P(z_1 \leq z \leq z_2)$, we obtain the area under the curve between z_1 and z_2 (the shaded region in Figure 15). Table 27 gives the area under the normal curve less than or equal to $z = z_1$, and greater than or equal to $z = 0$. That is, the area indicated by the light shading in Figure 15 is given in Table 27 at $z = z_1$. The area from $z = 0$ to $z = z_1$ is the same as the probability that z is less than or equal to z_1 and greater than or equal to 0, $P(0 \leq z \leq z_1)$. The table stops at $z = 3.09$, since the additional area under the curve beyond $z = 3.09$ is negligible.

The fact that the standard normal curve is symmetric about $z = 0$ means that the area under the curve on either side of 0 is 0.5. This symmetry allows us to compute probabilities that do not specifically occur in the table.

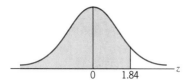

Figure 16

EXAMPLE 29   Find $P(z \leq 1.84)$.

SOLUTION   To solve such problems as this one, it is important to sketch the area in question. Give careful attention to the shaded area in Figure 16. We first find $P(0 \leq x \leq 1.84)$ by locating $z = 1.84$ in Table 27 and observing that the probability is .4671. Since the area under either half of the curve is .5000,

$$P(z \leq 1.84) = .5000 + P(0 \leq z \leq 1.84)$$
$$= .5000 + .4671$$
$$= .9671$$

For the normal curve, $P(z \leq a) = P(z < a)$, and $P(z \geq a) = P(z > a)$, for all a. In general, for distributions (called **continuous distributions**) like the standard normal curve, the probability that z is less than or equal to a number is the same as the probability that z is less than the number.

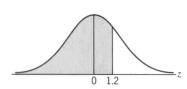

Figure 17

EXAMPLE 30   Find $P(z < 1.2)$.

SOLUTION   In Figure 17 we see that the area in question can be broken into two parts. The area to the left of 0 is .5, while the shaded area to the right of 0 can be found in Table 27 to be .3849. Thus,

$$P(z < 1.2) = P(z \leq 1.2) = .3849 + .5000 = .8849$$

TABLE 27

z	.00	.01	.02	.03	.04	.05	.06	.07	.08	.09
0.0	.0000	.0040	.0080	.0120	.0160	.0199	.0239	.0279	.0319	.0359
0.1	.0398	.0438	.0478	.0517	.0557	.0596	.0636	.0675	.0714	.0753
0.2	.0793	.0832	.0871	.0910	.0948	.0987	.1026	.1064	.1103	.1141
0.3	.1179	.1217	.1255	.1293	.1331	.1368	.1406	.1443	.1480	.1517
0.4	.1554	.1591	.1628	.1664	.1700	.1736	.1772	.1808	.1844	.1879
0.5	.1915	.1950	.1985	.2019	.2054	.2088	.2123	.2157	.2190	.2224
0.6	.2257	.2291	.2324	.2357	.2389	.2422	.2454	.2486	.2517	.2549
0.7	.2580	.2611	.2642	.2673	.2704	.2734	.2764	.2794	.2823	.2852
0.8	.2881	.2910	.2939	.2967	.2995	.3023	.3051	.3078	.3106	.3133
0.9	.3159	.3186	.3212	.3238	.3264	.3289	.3315	.3340	.3365	.3389
1.0	.3413	.3438	.3461	.3485	.3508	.3531	.3554	.3577	.3599	.3621
1.1	.3643	.3665	.3686	.3708	.3729	.3749	.3770	.3790	.3810	.3830
1.2	.3849	.3869	.3888	.3907	.3925	.3944	.3962	.3980	.3997	.4015
1.3	.4032	.4049	.4066	.4082	.4099	.4115	.4131	.4147	.4162	.4177
1.4	.4192	.4207	.4222	.4236	.4251	.4265	.4279	.4292	.4306	.4319
1.5	.4332	.4345	.4357	.4370	.4382	.4394	.4406	.4418	.4429	.4441
1.6	.4452	.4463	.4474	.4484	.4495	.4505	.4515	.4525	.4535	.4545
1.7	.4554	.4564	.4573	.4582	.4591	.4599	.4608	.4616	.4625	.4633
1.8	.4641	.4649	.4656	.4664	.4671	.4678	.4686	.4693	.4699	.4706
1.9	.4713	.4719	.4726	.4732	.4738	.4744	.4750	.4756	.4761	.4767
2.0	.4772	.4778	.4783	.4788	.4793	.4798	.4803	.4808	.4812	.4817
2.1	.4821	.4826	.4830	.4834	.4838	.4842	.4846	.4850	.4854	.4857
2.2	.4861	.4864	.4868	.4871	.4875	.4878	.4881	.4884	.4887	.4890
2.3	.4893	.4896	.4898	.4901	.4904	.4906	.4909	.4911	.4913	.4916
2.4	.4918	.4920	.4922	.4925	.4927	.4929	.4931	.4932	.4934	.4936
2.5	.4938	.4940	.4941	.4943	.4945	.4946	.4948	.4949	.4951	.4952
2.6	.4953	.4955	.4956	.4957	.4959	.4960	.4961	.4962	.4963	.4964
2.7	.4965	.4966	.4967	.4968	.4969	.4970	.4971	.4972	.4973	.4974
2.8	.4974	.4975	.4976	.4977	.4977	.4978	.4979	.4979	.4980	.4981
2.9	.4981	.4982	.4982	.4983	.4984	.4984	.4985	.4985	.4986	.4986
3.0	.4987	.4987	.4987	.4988	.4988	.4989	.4989	.4989	.4990	.4990

If it is known that a set of data values is normally distributed but not described by the standard normal distribution, then we can compute their probabilities by using the z scores discussed in Section 7.3.

EXAMPLE 31   Find the probability that the normal variable x, with mean 175 and standard deviation 20, is less than or equal to 215.

SOLUTION   The z value corresponding to $x = 215$ is

$$z = \frac{215 - 175}{20} = 2$$

Therefore,

$$P(x \leq 215) = P(z \leq 2) = .9772$$

Can you sketch the picture that shows the following?

$$P(z \leq 2) = .5 + .4772 = .9772 \qquad \blacksquare$$

Practice Problem 1 Find $P(z \leq 0.8)$.

ANSWER   .7881

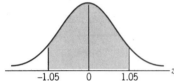

Figure 18

Recall that the standard normal curve is symmetric about $z = 0$. This fact is important to our discussion of areas under the curve. The fact that the standard normal curve is symmetric about the origin means that the area under the curve extending an equal distance on either side of 0 is the same. In Figure 18 we see that

$$P(-1.05 \leq z \leq 1.05) = 2P(0 \leq z \leq 1.05)$$
$$= 2(.3531)$$
$$= .7062$$

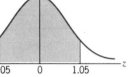

Figure 19

Since the total area under the curve is 1, the area to the right of $z = 1.66$ is 1 minus the area to the left of 1.66 (see Figure 19). In symbolic notation,

$$P(z \geq 1.66) = 1 - P(z < 1.66)$$
$$= 1 - [.5000 + P(0 < z < 1.66)]$$
$$= 1 - [.5000 + .4515]$$
$$= 1 - .9515$$
$$= .0485$$

Practice Problem 2 Find $P(z > 1.28)$.

ANSWER   .1003

EXAMPLE 32   The mileage data on a new-model automobile is normally distributed. The mean in-city mileage is 32 miles per gallon and the standard deviation is 1.5. What is the probability that the automobile you purchase will average less than 30 miles per gallon?

SOLUTION   To solve this problem we find the associated z score for 30 miles per gallon:

$$z = \frac{30 - 32}{1.5} = -1.33$$

$$P(x < 30) = P(z < -1.33)$$

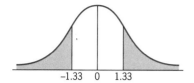

Figure 20

Careful attention to Figure 20 reveals that

$$P(z < -1.33) = P(z > 1.33)$$
$$= 1 - P(z \leq 1.33)$$
$$= 1 - .9082$$
$$= .0918$$

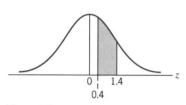

Figure 21

Sometimes we need to compute the probability that z is in a certain range—say, between 0.4 and 1.4 (see Figure 21). This probability is indicated by $P(0.4 < z < 1.4)$. It can be obtained by finding the probability that z is less than 1.4 and subtracting from this the probability that z is less than 0.4.

$$P(0.4 < z < 1.4) = P(z < 1.4) - P(z < 0.4)$$
$$= .9192 - .6554$$
$$= .2638$$

Practice Problem 3 Find $P(-0.51 \leq z \leq 1.98)$.

ANSWER   .6711

EXAMPLE 33   A security agency has uniforms to fit men ranging in height from 68 to 74 inches. The heights of adult males are normally distributed with a mean of 70 inches and a standard deviation of 2.5 inches. What percentage of male applicants to this agency can be fitted in their existing uniforms?

SOLUTION   The z values that correspond to 68 and 74 are

$$z = \frac{68 - 70}{2.5} = -0.8 \quad \text{and} \quad z = \frac{74 - 70}{2.5} = 1.6$$

Therefore,

$$P(68 \leq x \leq 74) = P(-0.8 \leq z \leq 1.6)$$
$$= P(-0.8 \leq z \leq 0) + P(0 \leq z \leq 1.6)$$
$$= P(0 \leq z \leq 0.8) + .4452$$
$$= .2881 + .4452$$
$$= .7333$$

Hence, the probability that a given applicant can be fitted in a uniform is .73, so 73% of the applicants can be fitted.

To this point we have been computing the probability that a data value from a normal distribution is less than some specific value. We can turn the question around and ask, "What is a z score, z_0, with the property that $P(z \leq z_0)$ is a specific number?"

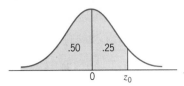

Figure 22

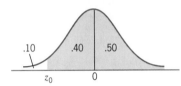

Figure 23

EXAMPLE 34   Find the third quartile, Q_3, for the standard normal distribution.

SOLUTION   We need to find a z score, z_0, so that $P(z < z_0) = .75$. Looking at Figure 22, we see that $P(0 \leq z \leq z_0) = .25$. Looking in Table 27 for probabilities near .25, we see that z_0 is about .67. This is the third quartile for the standard normal curve.   ∎

EXAMPLE 35   A company produces lightbulbs. The lifetimes of these bulbs are normally distributed with a mean of 800 hours and a standard deviation of 25 hours. The company wishes to guarantee a lifetime for their bulbs so that 90% of the bulbs will burn longer than the guaranteed lifetime. What lifetime should they advertise?

SOLUTION   We must first find a z score, z_0, above which 90% of the area lies and below which 10% of the area lies. Then we must find the corresponding score in a normal distribution with mean 800 and standard deviation of 25. Looking at Figure 23 we see that the desired z score is a negative number but that its opposite is the score that appears in Table 27 with an area of .4. Hence, z is approximately -1.28.

If we let x be the lightbulb life associated with the z score of -1.28, then

$$-1.28 = \frac{x - 800}{25}$$

Solving for x we learn that the advertised lifetime should be 768 hours.   ∎

A Comparison of the Binomial and Normal Distributions

A **binomial distribution** is a discrete distribution with two possible outcomes, which we call success and failure. The **binomial distribution function** is a discrete distribution function because it is defined for only integral values, 0, 1, 2, 3, . . . , n. For each variable x is given a probability

$$C(n, x)p^x q^{n-x}, \quad \text{where } q = p - 1 \qquad \text{p is the probability of success}$$

The mean and standard deviation for this distribution are given by

$$\mu_x = np$$
$$\sigma_x = \sqrt{npq}$$

The normal distribution function is a continuous distribution function (the variable can take on all real values) defined by

$$f(x) = \frac{1}{\sigma_x \sqrt{2\pi}} e^{-(x - \mu_x)^2/2\sigma_x^2}$$

where μ_x is the mean of the distribution and σ_x is the standard deviation. By making the transformation $z = (x - \mu_x)/\sigma_x$, we obtain the standard normal distribution function with mean 0 and standard deviation 1.

To demonstrate the use of these two distribution functions, consider the following example. Seventy percent of the participants at a state Republican convention are conservatives. If 12 people are chosen at random to be the rules committee, what is the probability of getting 8, 9, or 10 conservatives?

$$P(8) = C(12, 8) \cdot (.7)^8(.3)^4 \quad = .2311$$
$$P(9) = C(12, 9) \cdot (.7)^9(.3)^3 \quad = .2397$$
$$P(10) = C(12, 10) \cdot (.7)^{10}(.3)^2 = \underline{.1678}$$
$$.6386$$

$p = .7$
$q = .3$

So the probability of getting 8, 9, or 10 conservatives is .6386.

For many practical problems the methods of using the binomial distribution become very cumbersome. For this reason practical problems are usually worked with a normal curve approximation to the binomial distribution. The mean and standard deviation of the preceding distribution are

$$\mu_x = np = 12(.7) = 8.4$$
$$\sigma_x = \sqrt{npq} = \sqrt{12(.7)(.3)} = 1.587$$

Figure 24 shows the normal curve approximation to the histogram representing this binomial distribution. The probability of 8, 9, or 10 conservatives would be

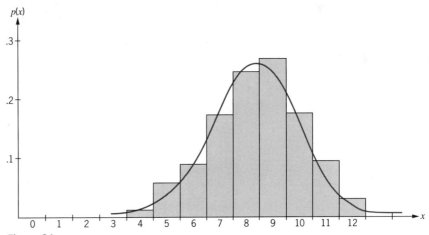

Figure 24

the area under the normal curve from 7.5 to 10.5. The z scores for these values are

$$z = \frac{7.5 - 8.4}{1.587} = -.57$$
8.4 is the mean and 1.587 the standard deviation of the normal distribution.

and

$$z = \frac{10.5 - 8.4}{1.587} = 1.32$$

The area under the normal curve from $z = -.57$ to $z = 1.32$ is .6223. This normal distribution area is a good approximation to the answer, .6386.

Consider another example. During inspection of 1000 welded joints produced by a certain machine, 100 defective joints were found. Consider the random variable to be the number of defective joints that result when 50 joints are welded. Compute the mean value of x, the standard deviation, and the probability of getting more than $\mu_x + 2\sigma_x$ defective parts, where $\mu_x + 2\sigma_x$ stands for 2 standard deviations above the expected value.

$$p = \frac{100}{1000} = \frac{1}{10}$$

Thus,

$$\mu_x = np = 50 \left(\frac{1}{10} \right) = 5$$

$$\sigma_x = \sqrt{npq} = \sqrt{50 \left(\frac{1}{10} \right) \left(\frac{9}{10} \right)} = 2.12$$

$$\mu_x + 2\sigma_x = 5 + 2(2.12) = 9.24$$

Therefore,

$$P(x > 9.24) = P(10) + P(11) + P(12) + \cdots + P(50)$$

This requires a great deal of computation using binomial probabilities. However, using the normal distribution

$$z = \frac{9.5 - 5}{2.12} = 2.12$$

and $P(z > 2.12) = .017$. Thus the probability of 10 or more defective parts is approximately .017.

The normal curve approximation to the binomial distribution is usually quite accurate, especially when both np and nq are 5 or more.

SUMMARY

1. A normal curve has the following properties:
   (a) The area under the curve is 1.
   (b) The curve is symmetric about the vertical line through its mean.
   (c) If x is a data value from a normally distributed set of data, $P(a < x < b)$ is the area under the curve from a to b.
2. The standard normal curve has a mean of 0 and a standard deviation of 1.
3. To find each of the following kinds of areas under the standard normal curve, we use Table 27 in different ways.

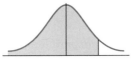 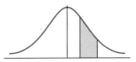 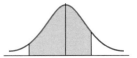

4. We can compute probabilities from normal distributions with means other than 0 and standard deviations other than 1 by using

$$z = \frac{x - \bar{x}}{\sigma}$$

Exercise Set 7.4

Whenever possible use a graphing calculator to obtain answers in this exercise set.

1. Find the area under the standard normal curve that lies between the following pairs of z values.
   (a) $z = 0$ to $z = 2.40$
   (b) $z = 0$ to $z = 0.41$
   (c) $z = 0$ to $z = 1.67$
   (d) $z = -0.36$ to $z = 0.36$

2. Assuming that the following sketches represent the standard normal curve, compute the shaded areas.
   (a)

   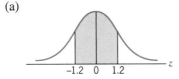

   (b)

   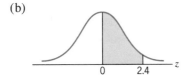

   (c)

   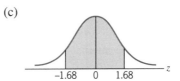

   (d)

   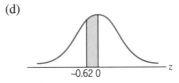

3. Sketch the area that is equal to each of the following probabilities and then find the probabilities using Table 27.
   (a) $P(z \le 1.7)$ (b) $P(z \le .6)$

4. If x is a variable having a normal distribution, with $\bar{x} = 12$ and $\sigma = 4$, find the probability that x assumes the following values.
   (a) $x \le 16$ (b) $x \le 14$

5. The scores on a national achievement test are normally distributed with a mean of 50 and a standard

deviation of 10. What fraction of the students taking the test make below 65 on the test?

6. Sketch the areas equal to each of the following probabilities and then find the probabilities using Table 27.
   (a) $P(z \le -2.1)$ (b) $P(z \ge -1.4)$
   (c) $P(z \le 0.1)$ (d) $P(z < -1.6)$
   (e) $P(z > 1.5)$ (f) $P(z > 2.4)$
   (g) $P(z > -2.1)$
   (h) $P(z > -1.8)$
   (i) $P(1.3 \le z \le 2.4)$
   (j) $P(2.1 < z < 2.8)$
   (k) $P(-1.2 \le z \le 0.3)$
   (l) $P(-2.6 \le z \le 1.4)$

7. Given that x is normally distributed, with mean 50 and standard deviation 10, find the following probabilities.
   (a) $P(x \ge 50)$ (b) $P(x \le 50)$
   (c) $P(x \ge 60)$ (d) $P(x \le 70)$
   (e) $P(x \le 40)$ (f) $P(x \ge 46)$
   (g) $P(38 \le x \le 54)$ (h) $P(32 \le x \le 61)$

8. The mean of a normal distribution is 35 with a standard deviation of 4. What percentage of the data should be in the following intervals?
   (a) 31.5 to 38.5   (b) 24.5 to 45.5
   (c) 17.5 to 52.5   (d) 35 to 38.5
   (e) 35 to 45.5   (f) 35 to 52.5

9. A large set of measurements is closely approximated by a normal curve with mean 30 and standard deviation 4.
   (a) What percentage of the measurements can be expected to lie in the interval from 20 to 32?
   (b) Find the probability that a measurement will differ from the mean by more than 5.

10. A student was noted for his ability to sleep through an entire class. Following a psychology lecture during which he performed admirably by sleeping throughout the class period, he was given a 10-question, true–false test. What is his expected number of correct answers? What is his probability of getting better than 75% of the answers correct by guessing?

Applications (Business and Economics)

In each of the following problems assume that the data are normally distributed.

11. **Sales.** A company has found that 25% of all customers contacted will buy its product. If 20 customers are contacted, how many sales can be expected? What is the probability of more than 2 sales? How many standard deviations below the mean are 2 sales?

12. **Sales.** It is known from experience that the number of telephone orders made daily to a company approximates a normal curve with mean 350 and standard deviation 20. What percentage of the time will there be more than 400 telephone orders per day?

13. **Quality Control.** A tire company manufactures a superior quality tire (sold at a superior price) that they guarantee for 50,000 miles. They know that the average life of one of their tires is 55,000 miles (with a standard deviation of 4000 miles), and they also know that occasionally a tire fails in less than 50,000 miles. What percentage of the tires will fail before 50,000 miles?

14. **Quality Control.** The life of a certain brand of batteries has a mean of 1200 days and a standard deviation of 100 days. If the manufacturer does not want to replace more than 12% of the batteries, for how long should the batteries be guaranteed?

Applications (Social and Life Sciences)

15. **Radar.** Radar is used to check the speed of traffic on Interstate 75 north of Atlanta. If the mean speed of the traffic is 70 miles per hour with a standard deviation of 4 miles per hour, what percentage of the cars are exceeding the legal speed of 65 miles per hour?

16. **Diets.** Young rabbits placed on a certain high-protein diet for a month show a weight gain with a mean of 120 grams and a standard deviation of 12 grams.
   (a) What is the probability that a given rabbit will gain at least 100 grams in weight?
   (b) If 15,000 rabbits are placed on this diet, how many can be expected to gain at least 140 grams?

17. *Heights.* The heights of men in a certain army regiment are normally distributed with mean 177 centimeters and standard deviation 4 centimeters.
    (a) What percentage of the men are between 173 and 181 centimeters in height?
    (b) What percentage of the men are between 169 and 185 centimeters in height?

18. *Grades.* The grades in a certain class are normally distributed with mean 76 and standard deviation 6. The lowest D is 61, the lowest C is 70, the lowest B is 82, and the lowest A is 91. What percentage of the class will make A's? What percentage will make B's? C's? D's? F's?

19. *Achievement Tests.* A nationally administered achievement test is known to have a mean score of 500 and a standard deviation of 100. What is the probability that a score is less than 300?

20. *Aptitude Tests.* The average time required for completing an aptitude test is 80 minutes with a standard deviation of 10 minutes. Assuming that the lengths of times necessary for completing the test are normally distributed, when should you stop the test to make certain that 90% of the people taking it have had time to complete the test?

21. *Medicine.* The pulse rate per minute of American males aged 18 to 25 has a mean of 72 and a standard deviation of 9.7. Requirements for employment by a law-enforcement agency are that the applicant have a pulse rate of less than 95 beats per minute. What percentage of males aged 18 to 25 would meet the requirements?

22. *Testing.* A test given to kindergarten students involves putting together a jigsaw puzzle. The mean time of completion of the puzzle is 150 seconds with a standard deviation of 20 seconds.

(a) If the test were given to 1000 children, about how many would finish the puzzle in less than 200 seconds?
(b) What percentage of the children would finish the puzzle in 120 seconds or less?

23. *Multiple Births.* The probability of multiple births for women using a certain fertility drug is 20%. Considering 10 pregnancies of women who have previously taken the fertility drug, what is the probability of more than 2 but fewer than 6 of the pregnancies resulting in multiple births?

24. *Drugs.* A drug manufacturer claims that a particular drug is effective 90% of the time. A physician prescribes the drug to 10 patients and 6 respond to this treatment. What is the probability of 6 or fewer successes if $p = .9$? What conclusion is the physician apt to draw? How many standard deviations below the mean are 6 or fewer?

Review Exercises

25. The following test scores were received by 24 students:

    63, 71, 85, 96, 94, 90, 75, 72, 77, 71, 62, 84,
    61, 54, 87, 94, 32, 81, 94, 77, 63, 60, 81, 76

    Find the following.
    (a) Mean   (b) Median
    (c) Mode   (d) First quartile
    (e) Range   (f) Standard deviation
    (g) IQR   (h) Outliers

26. For the data in Exercise 25 construct a stem and leaf plot.

27. For the data in Exercise 25 construct a box plot without whiskers.

Chapter Review

Organizing data

As you organize data make certain you understand the meaning of the following terms:

(a) Frequency   (b) Frequency distribution
(c) Grouped frequency distribution   (d) Class boundaries
(e) Class marks   (f) Stem and leaf plot

Graphing data

Make certain you understand the techniques for visually presenting statistical data:

(a) Bar graph   (b) Histogram
(c) Line graph   (d) Frequency polygon
(e) Circle graph

Measures of central tendency

(a) Formulas for the arithmetic mean:

$$\bar{x} = \frac{x_1 + x_2 + x_3 + \cdots + x_n}{n}$$

$$\bar{x} = \frac{x_1 f_1 + x_2 f_2 + \cdots + x_m f_m}{f_1 + f_2 + \cdots + f_m}$$

(b) The median is the middle measurement.
(c) The mode is the most common measurement.

Measures of scattering

(a) Variance is given by the following formulas.

$$\sigma^2 = \frac{(x_1 - \bar{x})^2 + (x_2 - \bar{x})^2 + \cdots + (x_n - \bar{x})^2}{n}$$

$$\sigma^2 = \frac{(x_1 - \bar{x})^2 f_1 + (x_2 - \bar{x})^2 f_2 + \cdots + (x_m - \bar{x})^2 f_m}{n}$$

where $n = \Sigma f$
(b) Standard deviation is defined by

$$\sigma = \sqrt{\text{variance}}$$

Interquartile range (IQR)
(a) Percentiles are values below which a given percent of the data is found.
(b) Q_1 represents the value below which 25% of the data are located.
(c) Q_3 represents the value below which 75% of the data are located.
(d) IQR = $Q_3 - Q_1$
(e) Box and whisker plot: Make sure you can draw a box and whisker plot and locate outliers [data above $Q_3 + 1.5$ (IQR) or below $Q_1 - 1.5$(IQR)].

Standard variable

To change from the variable x to the standard variable z, use

$$z = \frac{x - \bar{x}}{\sigma}$$

Normal distribution

The normal distribution has a bell-shaped graph with the following characteristics:

(a) 68% of data lie within 1 standard deviation of the mean.
(b) 95% of data lie within 2 standard deviations of the mean.
(c) 99% of data lie within 3 standard deviations of the mean.

Chapter Test

For Questions 1–5, consider the following set of values.

$$\{3, 7, 11, 15, 19, 23\}$$

1. Find the mean.
2. Find the median.
3. Find the variance of x.
4. Find the interquartile range.
5. Draw a box and whisker plot.

Questions 6–9 refer to the frequency distribution given in the following table.

Interval	Frequency
1–4	5
5–8	7
9–12	10
13–16	5
17–20	3

6. Draw a histogram.
7. Draw a frequency polygon.
8. Find the mean.
9. Find the standard deviation.

Questions 10–18 refer to the following scores on a mathematics test:

33, 37, 42, 48, 52, 53, 54, 55, 57, 59, 62, 62,
64, 64, 64, 68, 69, 71, 72, 72, 73, 73, 74, 74,
78, 79, 82, 83, 85, 87, 88, 89, 93, 96, 98

10. Approximate the interquartile range.
11. Find the mean.
12. Find the median.
13. Find the mode.
14. Make a stem and leaf chart.
15. Make a box and whisker plot.
16. Are there any outliers? Name them.
17. Make a group frequency distribution with class intervals 31–40, 41–50, ...
18. Find the standard deviation of your grouped frequency distribution in Exercise 17.
19. A variable is normally distributed with mean 25 and standard deviation 5. Find $P(25 < x < 30)$.
20. A set of measurements is approximately normally distributed, with a mean of 100 and a standard deviation of 10. Find the percentage of measurements larger than 110.

Differential Calculus

W hat is calculus? Simply put, calculus is the branch of mathematics that is used to study change. Since we live in a changing world, the study of calculus is very important. There are changes in inventory, wages, population, temperature, and so on. Quite often, we are not just interested in the fact that something is changing but also with how fast it is changing. For example, if your wages increase by $100, it is important to know whether that increase is per year, per month, or per week.

It may be surprising to learn that only a knowledge of algebra is needed for this study. One of our major objectives is to present calculus concepts with a minimum of prerequisite study. Isaac Newton (1642–1727) and Gottfried Leibniz (1646–1716) are given credit as being the creators of calculus. Two general problems characterize the study of elementary calculus: the tangent line problem and the area under a curve problem. How do we find the slope of the tangent line to the graph of a function $f(x)$ at the point P in Figure 1(a)? We consider this question in the study of differential calculus, which involves a multitude of ideas, all having to do with the rate of change of a function. How do we find the area of the region R bounded by the graph of $f(x)$ and the x-axis for $a \leq x \leq b$ in Figure 1(b)? We answer this question in the study of integral calculus with related topics in later chapters.

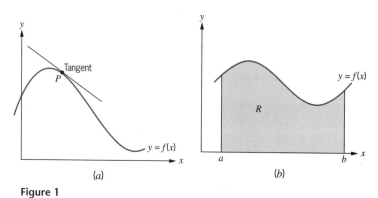

Figure 1

8.1 LIMITS

OVERVIEW   In this section we introduce a concept that opens the way for us to consider each of the two major goals of calculus. Whether we begin our study with differential calculus or with integral calculus, the notion of a limit must be considered. That is, we must investigate what happens to the value of a function as the variable gets closer and closer to some number. This concept helps distinguish between average rates of change and instantaneous rates of change. For example, in traveling from New York to Chicago you might compute your average speed. However, a glance at your speedometer gives an instantaneous speed. This instantaneous speed can be obtained as a **derivative**, which is defined in terms of a limit. We will find that derivatives (giving us instantaneous changes) are useful in economics and business management as well as in the social and life sciences. In this section we

- Introduce an intuitive concept of a limit
- Consider limits from the right and from the left
- Discuss the properties of limits

We introduce the concept of a limit by considering the graph of $f(x) = x + 1$ as x gets closer and closer to 3, but not equal to 3 (Figure 2). Recall that to determine $f(x)$ at a value x, proceed vertically from the x value on the x-axis to the graph of f,

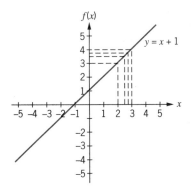

Figure 2

and then proceed horizontally to the corresponding value, $f(x)$, on the y-axis, as indicated by the dashed line for $x = 2$. In Figure 2, $f(2) = 3$.

What happens to $f(x)$ when x is chosen closer and closer to 3 for values of x just smaller than 3? If you let $x = 2.5$, then $f(x) = 3.5$; if $x = 2.7$, then $f(x) = 3.7$; and if $x = 2.9$, then $f(x) = 3.9$ (see Figure 2). In fact, if $x = 2.9999$, then $f(x) = 3.9999$. We see that $f(x)$ approaches 4 as x approaches 3 through values of x less than 3. In a similar manner we can demonstrate that $f(x)$ approaches 4 as x approaches 3 through values of x greater than 3.

EXAMPLE 1   Consider the function $f(x) = 3x + 4$ and a sequence of x values 1, 1.9, 1.99, 1.999, . . . , which are approaching the x-coordinate 2 from the left. That is, the x values are all less than 2 but are getting closer and closer to 2, as shown in the table. (The symbol $\rightarrow$ is read "approaches.")

x	1.000	1.900	1.990	1.999	$\cdots$	$\rightarrow$	2
$f(x)$	7.000	9.700	9.970	9.997	$\cdots$	$\rightarrow$	10

The sequence of corresponding $f(x)$ values 7, 9.7, 9.97, 9.997, . . . seems to be approaching (getting closer and closer to) 10. We say that the limit of $f(x)$ as x approaches 2 from the left (from the "less than 2" side) is 10, and we write this symbolically as

$$\lim_{x \to 2^-} f(x) = \lim_{x \to 2^-} (3x + 4) = 10$$

With your calculator, draw the graph of $y = f(x) = 3x + 4$, for $0 \le x \le 4$ (see Figure 3). For any value of x just less than 2 (no matter how close to 2), y is just less than 10. That is, as $x \to 2^-$, $y \to 10$. ■

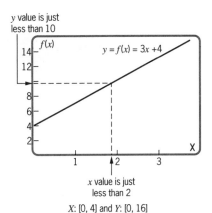

Figure 3

Limit
$$x \to a^-$$

We write

$$\lim_{x \to a^-} f(x) = L$$

if the functional value of $f(x)$ is close to the single real number L whenever $x < a$ is close to a.

EXAMPLE 2   Use the ⌐TRACE⌐ feature of your calculator to discover what happens as you let x approach 2 from the left for the function $f(x) = 3x + 4$. Use the Zoom feature for x to approach the 2 rapidly. Now let x approach 2 from the right and see what happens to the y coordinates of $f(x)$.

SOLUTION   Note that numbers will vary with different calculators and with different original ranges.

x	1.9574468	1.9972838	1.9995377	1.9999947	$\to$	2
$f(x)$	9.8723404	9.9812695	9.9997986	9.9999283	$\to$	10
x	2 $\leftarrow$	2.0003945	2.004045	2.0059054	2.0425532
$f(x)$	10 $\leftarrow$	10.001222	10.002215	10.022893	10.127666

As x approaches 2 from the right, the sequence of corresponding $f(x)$ values is approaching 10. We say that the limit of $f(x)$ as x approaches 2 from the right (from the ''greater side of 2'') is 10, and we write this as

$$\lim_{x \to 2^+} f(x) = \lim_{x \to 2^+} (3x + 4) = 10$$

Limit
$$x \to a^+$$

We write

$$\lim_{x \to a^+} f(x) = L$$

if the functional value of $f(x)$ is close to the single real number L whenever $x > a$ is close to a.

In the simple example above, the corresponding y-coordinates of f, as $x \to 2$, approach 10, which is the same value you would get if you merely evaluated $f(x)$

at $x = 2$, namely $f(2) = 3(2) + 4 = 10$. But it is not always this simple. Let's consider a case where the function does not exist at the x value that is being approached.

EXAMPLE 3   Discover what value the function $g(x) = (x^2 - 4)/(x - 2)$ approaches as x approaches 2 (a) from the left, and (b) from the right.

SOLUTION

(a)   From the table we can see that

$$\lim_{x \to 2^-} g(x) = \lim_{x \to 2^-} \frac{x^2 - 4}{x - 2} = 4$$

x	1.9	1.9900	1.9990	1.9999	$1.99999 \cdots \to 2$
$g(x)$	3.9	3.9900	3.9990	3.9999	$3.99999 \cdots \to 4$

(b)   From the following table we can see that

$$\lim_{x \to 2^+} g(x) = \lim_{x \to 2^+} \frac{x^2 - 4}{x - 2} = 4$$

x	$2 \leftarrow \cdots$	2.0001	2.0010	2.010	2.1
$g(x)$	$4 \leftarrow \cdots$	4.0001	4.0010	4.010	4.1

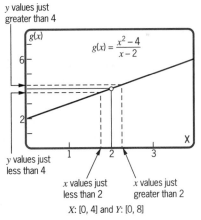

y values just greater than 4

$g(x)$

$g(x) = \dfrac{x^2 - 4}{x - 2}$

y values just less than 4

x values just less than 2

x values just greater than 2

$X: [0, 4]$ and $Y: [0, 8]$

Figure 4

Calculator Note

With your calculator, draw the graph of $g(x) = (x^2 - 4)/(x - 2)$ for $0 \leq x \leq 4$ and $0 \leq y \leq 8$ in Figure 4. Use the Trace key to locate the value of y at $x = 2$. On your calculator, the value of y is left blank. This means that there is no value of y for $x = 2$. In fact, there is a hole in the graph at $x = 2$ because $(2^2 - 4)/(2 - 2)$ is not a number [the denominator is zero and $g(2)$ is undefined]. So the point $(2, 4)$ is not a point on the graph, indicated by an empty circle. However, this does not change the fact that 4 is the value that the y-coordinate is approaching. As x approaches 2 from the left and right, the limit from the left and the limit from the right are both 4, as we see in Figure 4.

NOTE: In Figure 4, as the values of x approach 2, the values of $y = (x^2 - 4)/(x - 2)$ approach 4. The word *approach* means that we can make all y values be within a certain distance of 4 (and this distance can be made as small as we wish) by keeping the x values within a certain distance of 2.

Practice Problem 1 On a calculator, draw the graph of $y = (x^2 - 9)/(x + 3)$ and use the Zoom feature to discover the values of

$$\lim_{x \to -3^-} \frac{x^2 - 9}{x + 3} \quad \text{and} \quad \lim_{x \to -3^+} \frac{x^2 - 9}{x + 3}$$

ANSWER -6 and -6

Calculator Note

We will form tables many times to explore the possibility of a limit. Practice using the table capabilities of your calculator by forming the table in Example 3. See page 7 of Appendix A.

Limits do not always equal the same value from the left and from the right. When the limits from the left and from the right are equal, we define a limit as follows.

DEFINITION: LIMIT

When both the left limit, $\lim_{x \to a^-} f(x)$, and the right limit, $\lim_{x \to a^+} f(x)$, exist and are equal to some real number L, we say the limit of $f(x)$ as x approaches a is equal to L, or

$$\lim_{x \to a} f(x) = L$$

If the limit from the left and from the right are not equal or do not exist, we say the limit of $f(x)$ as x approaches a does not exist.

Using this definition, in Examples 1 and 2, $\lim_{x \to 2} (3x + 4) = 10$, and in Example 3,

$$\lim_{x \to 2} \frac{x^2 - 4}{x - 2} = 4$$

As we will discover, limits do not always exist or equal the same value from the left and from the right.

EXAMPLE 4   Find $\lim\limits_{x \to 1} \dfrac{1}{x - 1}$.

SOLUTION

x	0.9	0.99	0.999	0.9999	$\to$	1^-
$f(x)$	-10	-100	-1000	$-10{,}000$	$\to$	$-$ (large)

x	1.1	1.01	1.001	1.0001	$\to$	1^+
$f(x)$	10	100	1000	10,000	$\to$	(large)

Intuitively, as x is close to 1 but less than 1, $f(x)$ becomes numerically large but negative; likewise, as x is close to 1 but greater than 1, $f(x)$ gets numerically large but positive. Thus,

$$\lim\limits_{x \to 1^-} \frac{1}{x - 1} \quad \text{and} \quad \lim\limits_{x \to 1^+} \frac{1}{x - 1}$$

do not exist. We say that $f(x)$ is **unbounded** at $x = 1$. The calculator graph of this function is shown in Figure 5 for $-2 \le x < 1$, $1 < x \le 4$, and $-4 \le y \le 4$. ∎

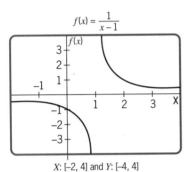

$f(x) = \dfrac{1}{x-1}$

X: [−2, 4] and Y: [−4, 4]

Figure 5

Calculator Note

Graph $f(x) = 1/(x - 1)$ again. This time use a very small horizontal viewing window, such as [0.9, 1.1], and a very large vertical window, [−1000, 1000]. Do you observe what we discussed in Example 4?

Practice Problem 2 With the Zoom feature of your graphing calculator, investigate what happens when $x \to 1$ on the graph of the function $y = 1/(x - 1)$ from the left and from the right. See page 2 of Appendix A.

ANSWER   The graph is unbounded as $x \to 1$ from both the left and the right. ∎

From the table in Example 4, the graph in Figure 5, and Practice Problem 2, we see that

$$\lim\limits_{x \to 1^-} \frac{1}{x - 1} \quad \text{and} \quad \lim\limits_{x \to 1^+} \frac{1}{x - 1}$$

do not exist.

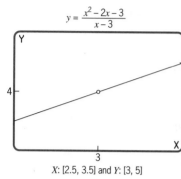

$$y = \frac{x^2 - 2x - 3}{x - 3}$$

X: [2.5, 3.5] and Y: [3, 5]

Figure 6

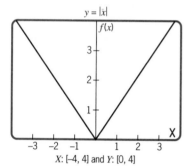

$y = |x|$

X: [−4, 4] and Y: [0, 4]

Figure 7

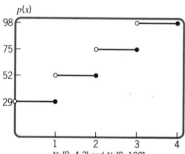

X: [0, 4.3] and Y: [0, 100]

Figure 8

Practice Problem 3 Use a calculator to investigate

$$\lim_{x \to 3} \frac{x^2 - 2x - 3}{x - 3}$$

by decreasing the size of the interval about $x = 3$ (or on some calculators let the box containing the limit point decrease in size).

ANSWER   The graph is shown in Figure 6.

$$\lim_{x \to 3} \frac{x^2 - 2x - 3}{x - 3} = 4$$

Sometimes graphs are in sections, with a different formula for each section. Such a graph represents what is called a **piecewise function**. The graph of $f(x) = |x|$ is the graph of a piecewise function because its two sections are $f(x) = -x$ when $x < 0$ and $f(x) = x$ when $x \geq 0$. On a calculator, with $-4 \leq x \leq 4$ and $0 \leq y \leq 4$, verify the graph in Figure 7. For this function $\lim_{x \to 0^-} |x| = 0$ and $\lim_{x \to 0^+} |x| = 0$, so $\lim_{x \to 0} |x| = 0$.

EXAMPLE 5   At one time postage for a first-class letter was $0.29 for the first ounce, or fraction thereof, and $0.23 more for each additional ounce, or fraction thereof. The piecewise function (using whole numbers for cents) is

$$p(x) = \begin{cases} 29, & \text{if } 0 < x \leq 1 \\ 52, & \text{if } 1 < x \leq 2 \\ 75, & \text{if } 2 < x \leq 3 \\ 98, & \text{if } 3 < x \leq 4 \\ & \text{etc.} \end{cases}$$

As shown in Figure 8, the graph of this function can be obtained by using the greatest integer function on your calculator. First select the dot mode. Press MODE and select "Dot"; then press 2nd QUIT. To graph press Y = 29 + 23 × MATH ▶ (select 4) X GRAPH to obtain Figure 7. Note that the dots and circles will not appear on your calculator screen; use the trace key to see the values change. In Figure 8, notice that $\lim_{x \to 2^-} p(x) = 52$ but $\lim_{x \to 2^+} p(x) = 75$.

That is, they are not equal. However, $\lim_{x \to 1.5^-} p(x) = 52$ and $\lim_{x \to 1.5^+} p(x) = 52$, so $\lim_{x \to 1.5} p(x) = 52$. ∎

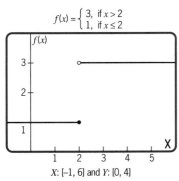

$$f(x) = \begin{cases} 3, & \text{if } x > 2 \\ 1, & \text{if } x \le 2 \end{cases}$$

X: [-1, 6] and Y: [0, 4]

Figure 9

EXAMPLE 6   Suppose that $f(x) = 1$ when $x \le 2$ and $f(x) = 3$ when $x > 2$. The graph of this function is given in Figure 9 for $-1 \le x \le 6$ and $0 \le y \le 4$. Find each of the following.

(a) $\displaystyle\lim_{x \to 0^-} f(x)$ (b) $\displaystyle\lim_{x \to 0^+} f(x)$ (c) $\displaystyle\lim_{x \to 0} f(x)$

(d) $\displaystyle\lim_{x \to 2^-} f(x)$ (e) $\displaystyle\lim_{x \to 2^+} f(x)$ (f) $\displaystyle\lim_{x \to 2} f(x)$

(g) $\displaystyle\lim_{x \to 4^-} f(x)$ (h) $\displaystyle\lim_{x \to 4^+} f(x)$ (i) $\displaystyle\lim_{x \to 4} f(x)$

SOLUTION

Since $f(x)$ is a constant 1 for all x close to 0, from Figure 9 it is clear that

(a, b, c) $\displaystyle\lim_{x \to 0^-} f(x) = 1$, $\displaystyle\lim_{x \to 0^+} f(x) = 1$,   and $\displaystyle\lim_{x \to 0} f(x) = 1$

(d)   From Figure 9, we see that as x approaches 2 from the left, $f(x)$ approaches 1; that is,

$$\lim_{x \to 2^-} f(x) = 1 \qquad f(x) = 1 \text{ to the left of } x = 2, \text{ no matter how close to } x = 2.$$

(e)   As x approaches 2 from the right, $f(x)$ approaches 3; that is,

$$\lim_{x \to 2^+} f(x) = 3 \qquad f(x) = 3 \text{ to the right of } x = 2.$$

(f)   Since $f(x)$ does not approach the same number as x approaches 2 from the left as it does from the right, $f(x)$ does not approach a unique limit as $x \to 2$.

(g, h, i)   From Figure 9, we see that as x approaches 4 from either direction, $f(x)$ approaches 3; that is,

$$\lim_{x \to 4^-} f(x) = 3 \qquad \text{As } x \text{ gets close to 4, } f(x) \text{ is a constant 3.}$$

$$\lim_{x \to 4^+} f(x) = 3$$

$$\lim_{x \to 4} f(x) = 3 \qquad \blacksquare$$

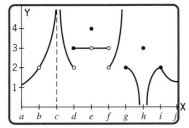

Figure 10

EXAMPLE 7   Find the limits at x for $x = b, c, d, e, f, g, h,$ and i for the function $g(x)$ in Figure 10, or state why the limit does not exist. Again, the black dots indicate the values of the function for the given values of x.

SOLUTION   At x values b, e, and i, both one-sided limits exist and agree, so the following limits exist:

$$\lim_{x \to b} g(x) = 2, \quad \lim_{x \to e} g(x) = 3, \quad \text{and} \quad \lim_{x \to i} g(x) = 2$$

At x values c, g, and h, at least one of the one-sided limits does not exist, so the limit does not exist at c, g, or h. At x values d and f, both one-sided limits exist, but they are unequal, so the limit does not exist. ■

In general, we can say that a function $f(x)$ has a finite limit L as x approaches the value a from either direction, which is written as

$$\lim_{x \to a} f(x) = L$$

if we can cause $f(x)$ to be as close to L as desired by restricting x to a sufficiently small interval surrounding a, but not necessarily including a. L is the y value we are *expecting* while on our approach to a, even though L may not turn out to be $f(a)$ when we arrive at a.

The following properties of limits are listed without proofs. It may be easier to remember the properties as stated in words; consequently, we give both the statement of the property and the symbolic form.

PROPERTIES OF LIMITS

Assume that a and c are constants, n is a positive integer, and

$$\lim_{x \to a} f(x) = L \quad \text{and} \quad \lim_{x \to a} g(x) = M$$

1. The limit of a constant function is that constant.

$$\lim_{x \to a} c = c$$

2. The limit of x^n is a^n

$$\lim_{x \to a} x^n = a^n$$

3. The limit of a constant times a function is the constant times the limit of the function.

$$\lim_{x \to a} cf(x) = c \lim_{x \to a} f(x) = cL$$

4. The limit of a sum (or difference) is equal to the sum (or difference) of the limits.

$$\lim_{x \to a} [f(x) \pm g(x)] = \lim_{x \to a} f(x) \pm \lim_{x \to a} g(x) = L \pm M$$

5. The limit of a product of two functions is the product of the limits of the functions provided that they both exist.

$$\lim_{x \to a} [f(x) \cdot g(x)] = \lim_{x \to a} f(x) \cdot \lim_{x \to a} g(x) = LM$$

PROPERTIES OF LIMITS *(Continued)*

6. The limit of a quotient of two functions is the quotient of the limits of the functions whenever both limits exist and the limit of the denominator is not zero.

$$\lim_{x \to a} \frac{f(x)}{g(x)} = \frac{\lim_{x \to a} f(x)}{\lim_{x \to a} g(x)} = \frac{L}{M}, \quad \text{if } M \neq 0$$

7. The limit of the *n*th power is the *n*th power of the limit.

$$\lim_{x \to a} [f(x)]^n = L^n, \quad \text{where } n \text{ is a positive integer}$$

8. [The domain of $f(x)$ is restricted so that $\sqrt[n]{f(x)}$ is always real.] The limit of a root is the root of the limit.

$$\lim_{x \to a} \sqrt[n]{f(x)} = \sqrt[n]{\lim_{x \to a} f(x)} = \sqrt[n]{L}$$

9. Let $P(x)$ be a polynomial function. Then $\lim_{x \to a} P(x) = P(a)$.

For practice in using these properties, Examples 8 through 15 illustrate separate properties rather than property 9, which could also be applied. Each example can be verified by using a calculator and an appropriate window.

EXAMPLE 8

$$\lim_{x \to 2} 7 = 7 \qquad \text{Property 1}$$

EXAMPLE 9

$$\lim_{x \to 3} 9x^2 = 9 \lim_{x \to 3} x^2 = 9 \cdot 3^2 = 81 \qquad \text{Properties 2, 3}$$

EXAMPLE 10

$$\lim_{x \to 2} (8x^3 - 3x) = \lim_{x \to 2} 8x^3 - \lim_{x \to 2} 3x \qquad \text{Property 4}$$
$$= 8 \lim_{x \to 2} x^3 - 3 \lim_{x \to 2} x \qquad \text{Property 3}$$
$$= 8 \cdot 2^3 - 3 \cdot 2 \qquad \text{Property 2}$$
$$= 64 - 6 = 58$$

EXAMPLE 11

$$\lim_{x \to 1} (3x + 5)^2 = \left[\lim_{x \to 1} (3x + 5) \right]^2 = 8^2 = 64 \qquad \text{Property 7}$$

EXAMPLE 12

$$\lim_{x \to 4} [(3x^2)(2x^3)] = \left(\lim_{x \to 4} 3x^2\right) \cdot \left(\lim_{x \to 4} 2x^3\right) \qquad \text{Property 5}$$

$$= \left(3 \lim_{x \to 4} x^2\right) \cdot \left(2 \lim_{x \to 4} x^3\right) \qquad \text{Property 3}$$

$$= (3 \cdot 4^2) \cdot (2 \cdot 4^3) \qquad \text{Property 2}$$
$$= 6144$$

EXAMPLE 13

$$\lim_{x \to 3} \frac{5x^3 + 5}{3x} = \frac{\lim_{x \to 3}(5x^3 + 5)}{\lim_{x \to 3} 3x} \qquad \text{Property 6}$$

$$= \frac{5 \lim_{x \to 3} (x^3 + 1)}{3 \lim_{x \to 3} x} \qquad \text{Property 3}$$

$$= \frac{5 \cdot 28}{3 \cdot 3} \qquad \text{Property 2}$$

$$= \frac{140}{9}$$

EXAMPLE 14

$$\lim_{x \to 2} \sqrt{(3x + 5)} = \sqrt{\lim_{x \to 2} (3x + 5)} \qquad \text{Property 8}$$

$$= \sqrt{\lim_{x \to 2} 3x + \lim_{x \to 2} 5} \qquad \text{Property 4}$$

$$= \sqrt{3 \lim_{x \to 2} x + \lim_{x \to 2} 5} \qquad \text{Property 3}$$

$$= \sqrt{3 \cdot 2 + 5} \qquad \text{Property 1, 2}$$
$$= \sqrt{11}$$

EXAMPLE 15

$$\lim_{x \to 2} (3x^3 - 4x^2 + x - 2) = 3(2)^3 - 4(2)^2 + 2 - 2 = 8 \qquad \text{Property 9}$$

Practice Problem 4 Find

$$\lim_{x \to 2} \frac{x^2 - 2x + 5}{2x - 3}$$

ANSWER

$$\frac{\lim_{x \to 2} (x^2 - 2x + 5)}{\lim_{x \to 2} (2x - 3)} = \frac{2^2 - 2 \cdot 2 + 5}{2 \cdot 2 - 3} = 5$$

SUMMARY

We introduced the following important concepts in this section:

1. The limit of function $f(x)$ as x approaches a is denoted by $\lim_{x \to a} f(x)$.
2. The limit of $f(x)$ as x approaches a from the left is denoted by $\lim_{x \to a^-} f(x)$ and as x approaches a from the right by $\lim_{x \to a^+} f(x)$.
3. If $\lim_{x \to a^-} f(x) = L$ and $\lim_{x \to a^+} f(x) = L$, then $\lim_{x \to a} f(x)$ exists and equals L.

Exercise Set 8.1

Use your calculator to complete the tables and obtain the limits intuitively.

1. $f(x) = 2x - 3$; find $\lim_{x \to 2} f(x)$.

x	1.900	1.990	1.999	$\to 2 \leftarrow$	2.001	2.010	2.100
$f(x)$	____	____	____	$\to ? \leftarrow$	____	____	____

2. $g(x) = \dfrac{3}{x^2}$; find $\lim_{x \to 0} g(x)$.

x	-0.100	-0.010	-0.001	$\to 0 \leftarrow$	0.0010	0.0100	0.1000
$g(x)$	____	____	____	$\to ? \leftarrow$	____	____	____

3. $h(x) = \dfrac{x^2 - 16}{x - 4}$; find $\lim_{x \to 4} h(x)$.

x	3.900	3.990	3.999	$\to 4 \leftarrow$	4.001	4.010	4.100
$h(x)$	____	____	____	$\to ? \leftarrow$	____	____	____

4. $r(x) = \dfrac{x}{x - 1}$; find $\lim_{x \to 1} r(x)$.

x	0.9000	0.9900	0.9990	$\to 1 \leftarrow$	1.001	1.010	1.100
$r(x)$	____	____	____	$\to ? \leftarrow$	____	____	____

Use the Zoom feature of your graphing calculator to find the limits of the following functions intuitively.

5. $\lim\limits_{x \to 1} \dfrac{3x^2 + 4x}{x + 2}$

6. $\lim\limits_{x \to 2} \dfrac{x^2 - 4}{x - 2}$

Find the following limits.

7. $\lim\limits_{x \to 4} -8$

8. $\lim\limits_{x \to 3} 0$

9. $\lim\limits_{x \to 3} 2x$

10. $\lim\limits_{x \to 2} 5x$

11. $\lim\limits_{x \to 2} \dfrac{3x}{2}$

12. $\lim\limits_{x \to -1} -\dfrac{2x}{3}$

In Exercises 13–22, use the following graph.

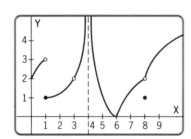

13. $\lim\limits_{x \to 1^-} f(x)$

14. $\lim\limits_{x \to 4} f(x)$

15. $\lim\limits_{x \to 1^+} f(x)$

16. $f(1)$

17. $f(8)$

18. $\lim\limits_{x \to 8^-} f(x)$

19. $f(3)$

20. $\lim\limits_{x \to 8^+} f(x)$

21. $\lim\limits_{x \to 3} f(x)$

22. $\lim\limits_{x \to 8} f(x)$

Find the following limits.

23. $\lim\limits_{x \to 3} (2x^2 - 4)$

24. $\lim\limits_{x \to 5} (3x + 5)$

25. $\lim\limits_{x \to 2} (3x^2 - 5)(x + 4)$

26. $\lim\limits_{x \to 2} (2x + 7)(3x^2 - 1)$

27. $\lim\limits_{x \to -3} \dfrac{7x}{2x + 3}$

28. $\lim\limits_{x \to 0} \dfrac{9x^2 - x + 1}{x^2 + 2x + 5}$

29. $\lim\limits_{x \to 1} \dfrac{5x}{2 + x^2}$

30. $\lim\limits_{x \to 2} (x + 1)^2(2x - 1)$

31. $\lim\limits_{x \to 2} x^2(x^2 + 1)^3$

32. Determine whether $\lim\limits_{x \to 2} f(x)$ exists for the following functions. If the limit exists, find it.

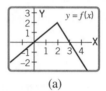
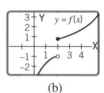
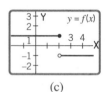

(a)   (b)   (c)

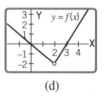
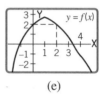
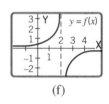

(d)   (e)   (f)

33. Determine whether $\lim\limits_{x \to 3} f(x)$ exists for the following functions. If the limit exists, find it.

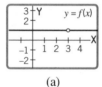
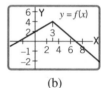
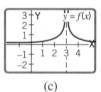

(a)   (b)   (c)

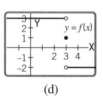
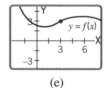
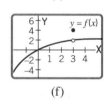

(d)   (e)   (f)

For Exercises 34–39, use the following graphs.

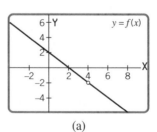
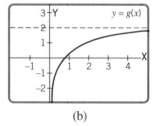

(a)   (b)

34. $\lim\limits_{x \to 0} g(x)$

35. $\lim\limits_{x \to 0} f(x)$

36. $\lim\limits_{x \to 2} f(x)$

37. $\lim\limits_{x \to 4} f(x)$

38. $\lim\limits_{x \to -1} f(x)$

39. $\lim\limits_{x \to 2} g(x)$

In Exercises 40–43, find the limit, if it exists, for each function as $x \to 2$. (Note that parts of the graphs of the functions are found in Exercise 32.)

40. $y = \begin{cases} x, & \text{for } x \le 2 \\ 6 - 2x, & \text{for } x > 2 \end{cases}$

41. $y = \begin{cases} -2/x, & \text{for } x < 2 \\ (x - 1)^2, & \text{for } x > 2 \end{cases}$

42. $y = \begin{cases} 1, & \text{for } x \le 2 \\ -1, & \text{for } x > 2 \end{cases}$

43. $y = \begin{cases} -x, & \text{for } x < 2 \\ 2x - 6, & \text{for } x > 2 \end{cases}$

Use the Zoom feature on your graphing calculator to find the following limits if they exist.

44. $\lim\limits_{x \to -2} \dfrac{x^3 + 8}{x + 2}$

45. $\lim\limits_{x \to 3} \dfrac{x^3 - 27}{x - 3}$

46. $\lim\limits_{x \to 2} \dfrac{x^3 - x^2 - 4x + 4}{x - 2}$

47. $\lim\limits_{x \to -3} \dfrac{x^2 - 9}{x + 3}$

48. Graph

$$f(x) = \frac{(x + 3)^2 - 9}{x}$$

Magnify the graph around $x = 0$. What do you discover? Validate your observation algebraically.

49. Graph

$$y = \frac{x^3 - 8}{x - 2}$$

Magnify the graph near $x = 2$. What do you ob-

serve? How can you reach the same conclusion algebraically?

50. Graph

$$y = \frac{x^3 + 8}{x - 1}$$

Use a very small horizontal viewing window and a very large vertical window near $x = 1$. What do you observe? Can you verify your conjecture algebraically?

51. Graph

$$y = \frac{x^3 + 1}{x - 2}$$

Follow the instructions for Exercise 50 near $x = 2$.

Applications (Business and Economics)

52. ***Telephone Rates.*** The cost of a direct call between Atlanta and Nashville is $0.64 for the first 3 minutes and $0.23 for each additional minute or fraction thereof. Sketch this function, where x is the number of minutes, and determine the following.
    (a) $\lim\limits_{x \to 3} f(x)$ (b) $\lim\limits_{x \to 5.5} f(x)$

53. ***Cost function.*** A cost function C is given as

$$C(x) = \begin{cases} 5, & \text{for } 0 \le x < 4 \\ 6, & \text{for } 4 \le x < 6 \\ x, & \text{for } 6 \le x < 10 \end{cases}$$

where x is the number of units. Find the following limits if they exist.
    (a) $\lim\limits_{x \to 1} C(x)$ (b) $\lim\limits_{x \to 4} C(x)$
    (c) $\lim\limits_{x \to 6} C(x)$ (d) $\lim\limits_{x \to 8} C(x)$

54. ***Demand Function.*** A price for a product is given as a function of x, the number of units available:

$$p(x) = \frac{900}{x^2} + \frac{800}{x}$$

Find each of the following.

(a) $\lim\limits_{x \to 1} p(x)$ (b) $\lim\limits_{x \to 0^+} p(x)$

(c) $\lim\limits_{x \to 0} p(x)$

Applications (Social and Life Sciences)

55. ***Growth function***.   A growth function is given as

$$n(t) = \frac{80,000t}{100 + t}$$

in terms of time t in minutes. Find the value of each of the following.

(a) $\lim\limits_{t \to 100} n(t)$

(b) $\lim\limits_{t \to 0} n(t)$

(c) $\lim\limits_{t \to -100} n(t)$

8.2 LIMITS AND CONTINUITY

OVERVIEW   A continuous function over an open interval is a function whose graph does not have any breaks in it. Another way of stating this concept is to say that a function is continuous on an interval if we can draw the graph of the function on the complete interval without taking our pen off the paper. These intuitive concepts are explained and defined in terms of limits in this section. In this section we

- Study additional techniques for finding limits
- Define what is meant by a function being continuous at a point
- Consider examples of discontinuous functions
- List properties of continuous functions

In our study of $\lim\limits_{x \to a} f(x) = L$, we were not concerned with the value of the function at $x = a$. We were interested in what happens to the function values as x approaches a. We now shift our interest to the value of the function at $x = a$. If the function is defined at $x = a$, if $\lim\limits_{x \to a} f(x)$ exists, and if $\lim\limits_{x \to a} f(x) = f(a)$, then the function is **continuous** at $x = a$.

DEFINITION: CONTINUOUS FUNCTION

A function $f(x)$ is **continuous** at $x = a$ if and only if

1. $f(a)$ is defined.
2. $\lim\limits_{x \to a} f(x)$ exists.
3. $\lim\limits_{x \to a} f(x) = f(a)$.

If any one of these conditions is not satisfied, the function $f(x)$ is discontinuous at $x = a$. The point $x = a$ is then called a **point of discontinuity**.

DEFINITION: CONTINUOUS ON AN OPEN INTERVAL

A function is **continuous on an open interval** if it is continuous at each point of the interval.

That is, a function is continuous on an open interval if there are no points of discontinuity in the interval.

EXAMPLE 16 State why the following functions are discontinuous at point P.

(a)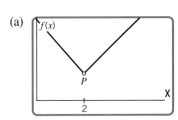

(b)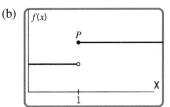

(c)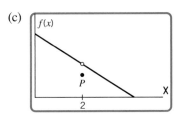

(d)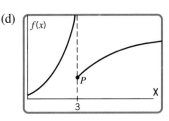

SOLUTION

(a) $f(2)$ is not defined.

(b) $\lim\limits_{x \to 1} f(x)$ does not exist because $\lim\limits_{x \to 1^-} f(x) \neq \lim\limits_{x \to 1^+} f(x)$.

(c) $\lim\limits_{x \to 2} f(x) \neq f(2)$.

(d) $\lim\limits_{x \to 3} f(x)$ does not exist because $\lim\limits_{x \to 3^-} f(x) \neq \lim\limits_{x \to 3^+} f(x)$. ■

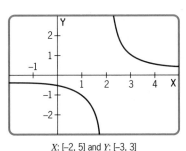

X: [–2, 5] and Y: [–3, 3]

Figure 11

EXAMPLE 17 Graph $f(x) = 1/(x - 2)$. Is $f(x)$ continuous at $x = 2$?

SOLUTION A calculator graph of $f(x) = 1/(x - 2)$ is given in Figure 11. Since $f(x)$ is not defined at $x = 2$, condition 1 of the definition is not satisfied. Thus, $f(x)$ is discontinuous at $x = 2$. It is continuous at all other values of x; hence, $f(x)$ is continuous on the interval $(-\infty, 2)$ and on the interval $(2, \infty)$. ■

Practice Problem 1 With a graphing calculator, draw the graph of

$$y = \frac{1}{(x - 1)^2}$$

on the intervals $-6 \leq x \leq 6$ and $0 \leq y \leq 6$. Is the function continuous at $x = 1$?

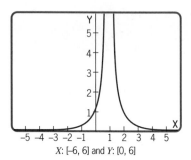

X: [-6, 6] and Y: [0, 6]

Figure 12

ANSWER   The graph is shown in Figure 12. The function is not continuous because a point of discontinuity exists at $x = 1$.

EXAMPLE 18   Graph

$$f(x) = \begin{cases} 1, & \text{if } x \leq 2 \\ 2, & \text{if } x > 2 \end{cases}$$

SOLUTION   The graph is shown in Figure 13. Note that $f(2) = 1$ but $\lim_{x \to 2} f(x)$ does not exist. Since condition 2 of the definition is not satisfied, $f(x)$ is not continuous at $x = 2$. It is continuous at all other values of x; consequently, $f(x)$ is continuous on the open interval $(-\infty, 2)$ and on the open interval $(2, \infty)$. ■

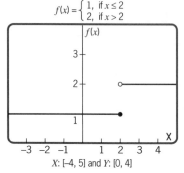

X: [-4, 5] and Y: [0, 4]

Figure 13

Practice Problem 2 Discuss the continuity of f at $x = c$ for each of the following.

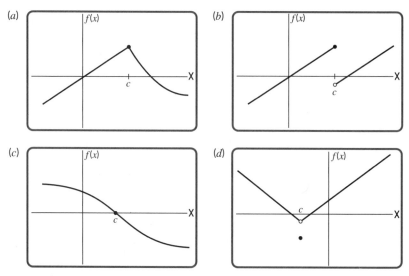

ANSWER

(a) Continuous at $x = c$.

(b) Discontinuous at $x = c$ because $\lim_{x \to c} f(x)$ does not exist.

(c) Continuous at $x = c$.

(d) Discontinuous at $x = c$ because $\lim_{x \to c} f(x) \neq f(c)$.

Practice Problem 3 Locate all points of discontinuity of

$$f(x) = \begin{cases} 2 + x, & \text{if } x \le 1 \\ 6 - x, & \text{if } x > 1 \end{cases}$$

ANSWER   A point of discontinuity is at $x = 1$, since $\lim_{x \to 1} f(x)$ does not exist.

$$\lim_{x \to 1^-} f(x) = \lim_{x \to 1^-} (2 + x) = 3 \quad \text{but} \quad \lim_{x \to 1^+} f(x) = \lim_{x \to 1^+} (6 - x) = 5$$

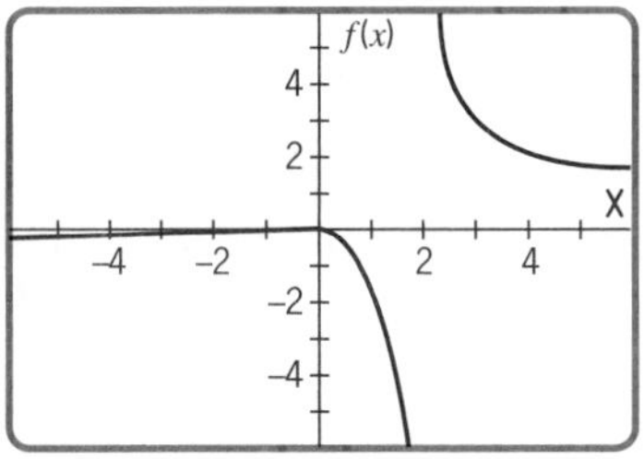

Practice Problem 4 Using a graphing calculator find all points of discontinuity of

$$f(x) = \frac{|x|}{x - 2} \qquad \text{for } -6 \le x \le 6$$

X: [–6, 6] and Y: [–6, 6]

Figure 14

ANSWER   As shown in Figure 14, the function is discontinuous at $x = 2$.

For the next two examples on continuity, we need to discuss an additional technique for finding limits. Property 6 on limits in Section 8.1 does not always give an answer, as we see in the following example.

EXAMPLE 19   Find $\lim\limits_{x \to 3} \dfrac{x^2 - 9}{x - 3}$.

SOLUTION   If we apply property 6, we have

$$\lim_{x \to 3} \frac{x^2 - 9}{x - 3} = \frac{\lim\limits_{x \to 3} (x^2 - 9)}{\lim\limits_{x \to 3} (x - 3)} = \frac{0}{0}$$

The form 0/0 is called an **indeterminate form**. When this occurs you might perform an algebraic simplification to determine whether the limit exists.

$$\lim_{x \to 3} \frac{x^2 - 9}{x - 3} = \lim_{x \to 3} \frac{(x + 3)(x - 3)}{x - 3}$$

If $x \ne 3$, then $(x - 3)/(x - 3) = 1$. Therefore

$$\lim_{x \to 3} \frac{x - 3}{x - 3} = 1$$

and the limit of a product is the product of the limits. Thus,

$$\lim_{x \to 3} \frac{x^2 - 9}{x - 3} = \lim_{x \to 3} \left[\frac{x - 3}{x - 3} \cdot \frac{x + 3}{1} \right]$$

$$= \lim_{x \to 3} \left[\frac{x - 3}{x - 3} \right] \cdot \lim_{x \to 3} (x + 3)$$

$$= 1 \cdot \lim_{x \to 3} (x + 3) = 6$$

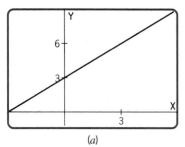

(a)

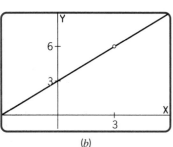

(b)

X: [−3, 6] and Y: [−1, 9]

Figure 15

The graph of $y = x + 3$ is shown in Figure 15(a), and the graph of $y = \dfrac{x^2 - 9}{x - 3}$ is shown in Figure 15(b). Actually, they are the same except that the point (3, 6) is not on the graph of $y = (x^2 - 9)/(x - 3)$.   ■

Calculator Note

Your calculator may show a hole in the graph at (3, 6), but if it does not, you can use the TRACE key and note that no value for y is given at x = 3.

Practice Problem 5 Find $\lim_{x \to -2} \dfrac{x^2 - 4}{x + 2}$.

ANSWER $\lim_{x \to -2} \dfrac{(x + 2)(x - 2)}{x + 2} = \lim_{x \to -2} (x - 2) = -4$

EXAMPLE 20 Find

$$\lim_{x \to 3} \frac{\dfrac{1}{x} - \dfrac{1}{3}}{x - 3}$$

SOLUTION In this example, we again use algebra to write an equivalent expression for one that results in an indeterminate form. First we clear the numerator of fractions by multiplying both numerator and denominator by $3x$ ($3x$ is the lcd).

$$\lim_{x \to 3} \frac{\dfrac{3x}{x} - \dfrac{3x}{3}}{3x(x - 3)} = \lim_{x \to 3} \frac{3 - x}{3x(x - 3)}$$

$$= \lim_{x \to 3} \frac{-(x - 3)}{3x(x - 3)}$$

$$= \lim_{x \to 3} \frac{-1}{3x}$$

$$= \frac{-1}{3(3)} = \frac{-1}{9}$$

■

Practice Problem 6 Find

$$\lim_{x \to -4} \frac{\dfrac{1}{x} + \dfrac{1}{4}}{x + 4}$$

ANSWER

$$\lim_{x \to -4} \frac{\dfrac{4x}{x} + \dfrac{4x}{4}}{4x(x + 4)} = \lim_{x \to -4} \frac{4 + x}{4x(x + 4)} = \lim_{x \to -4} \frac{1}{4x} = \frac{-1}{16}$$

EXAMPLE 21 Find $\displaystyle\lim_{x \to 4} \frac{\sqrt{x} - 2}{x - 4}$.

SOLUTION We rationalize the numerator by multiplying the numerator and denominator by $\sqrt{x} + 2$.

$$\lim_{x \to 4} \frac{\sqrt{x} - 2}{x - 4} = \lim_{x \to 4} \frac{(\sqrt{x} - 2)(\sqrt{x} + 2)}{(x - 4)(\sqrt{x} + 2)}$$

$$= \lim_{x \to 4} \frac{x - 4}{(x - 4)(\sqrt{x} + 2)}$$

$$= \lim_{x \to 4} \frac{x - 4}{x - 4} \cdot \lim_{x \to 4} \frac{1}{\sqrt{x} + 2} \qquad x \ne 4$$

$$= (1) \cdot \frac{1}{\sqrt{4} + 2} = \frac{1}{4}$$

This problem could also be worked by factoring $x - 4 = (\sqrt{x} + 2)(\sqrt{x} - 2)$. ■

Practice Problem 7 Find $\displaystyle\lim_{x \to 9} \frac{x - 9}{\sqrt{x} - 3}$. (**Hint**: Rationalize the denominator.)

ANSWER $\displaystyle\lim_{x \to 9} \frac{x - 9}{x - 9} \cdot \lim_{x \to 9} (\sqrt{x} + 3) = 1 \cdot (6) = 6$

The technique of using algebraic manipulation to produce equivalent expressions for finding limits will be useful as we continue our discussion of continuity.

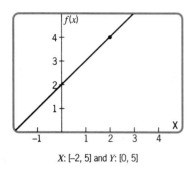

X: [-2, 5] and Y: [0, 5]

Figure 16

EXAMPLE 22 Graph

$$f(x) = \begin{cases} \dfrac{x^2 - 4}{x - 2}, & \text{if } x \neq 2 \\ 4, & \text{if } x = 2 \end{cases}$$

Is this function continuous at $x = 2$?

SOLUTION The calculator graph of the function is given in Figure 16. Note that if $x \neq 2$, then

$$\lim_{x \to 2} \frac{x^2 - 4}{x - 2} = \lim_{x \to 2} \frac{(x + 2)(x - 2)}{x - 2} = \lim_{x \to 2} (x + 2) = 4$$

Since $f(2) = 4$ and $\lim_{x \to 2} f(x)$ exists and $\lim_{x \to 2} f(x) = f(2)$, this function is continuous at $x = 2$. It is also continuous at all other values of x. ∎

EXAMPLE 23 The cost of a telephone call between two cities on a pay phone is $2.70 for the first 3-minute period (or fraction thereof) and $0.60 for each additional minute (or fraction thereof). Let $C(x)$ be the cost of a telephone call that lasts x minutes.

$$C(x) = \begin{cases} \$2.70, & \text{if } 0 < x \leq 3 \\ \$3.30, & \text{if } 3 < x \leq 4 \\ \$3.90, & \text{if } 4 < x \leq 5 \\ \$4.50, & \text{if } 5 < x \leq 6 \\ \$5.10, & \text{if } 6 < x \leq 7 \end{cases}$$

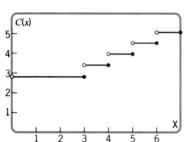

Figure 17

Determine whether this function is continuous at each of the following.

(a) $x = 4\frac{1}{2}$ minutes   (b) $x = 3$ minutes
(c) $x = 5$ minutes   (d) $x = 6.3$ minutes

SOLUTION The graph is shown in Figure 17. Note the breaks at $x = 3$, $x = 4$, $x = 5$, and $x = 6$.

(a) Since $\lim_{x \to 4\frac{1}{2}} C(x) = C(4\frac{1}{2})$, $C(x)$ is continuous at $x = 4\frac{1}{2}$.

(b) Since $\lim_{x \to 3} C(x)$ does not exist, $C(x)$ is not continuous at $x = 3$.

(c) Since $\lim_{x \to 5} C(x)$ does not exist, $C(x)$ is not continuous at $x = 5$.

(d) Since $\lim_{x \to 6.3} C(x) = C(6.3)$, $C(x)$ is continuous at $x = 6.3$.

We observe that $C(x)$ is continuous on the open interval $0 < x < 3$, on the open interval $3 < x < 4$, on the open interval $4 < x < 5$, on the open interval $5 < x < 6$, and on the open interval $6 < x < 7$. ∎

Some of the properties or theorems of continuous functions are listed below.

PROPERTIES OF CONTINUOUS FUNCTIONS

(a)   *Constant function:* $f(x) = c$, where c is any constant, is continuous for all real values of x.

(b) $f(x) = x^n$ is continuous for all real values of x.

If we assume that $f(x)$ and $g(x)$ are continuous functions, then

(c) $f(x) \pm g(x)$ is continuous.
(d) $f(x) \cdot g(x)$ is continuous.
(e) $f(x)/g(x)$ is continuous, except where $g(x) = 0$.
(f) $[f(x)]^n$, where n is a positive integer, is continuous.
(g) $\sqrt{f(x)}$ is continuous when x is restricted to those values that make $f(x) \geq 0$.

Practice Problem 8   Assume that $f(x) = 2x^2 + 4$ and $g(x) = x + 3$ are both continuous functions. Graph the following functions and determine whether they are continuous or discontinuous.

(a) $y = f + g = (2x^2 + 4) + (x + 3)$ (b) $y = f \cdot g = (2x^2 + 4)(x + 3)$

(c) $y = \dfrac{g}{f} = \dfrac{x + 3}{2x^2 + 4}$ (d) $y = f^3 = (2x^2 + 4)^3$

(e) $y = \sqrt{f} = \sqrt{2x^2 + 4}$

Note the graphs that have no breaks (i.e., graphs that represent continuous functions).

ANSWER   (a), (b), (c), (d), and (e) are continuous.

Two commonly used functions, $f(x) = c$, where c is any constant, and $f(x) = x^n$, where n is any positive integer, are continuous over $(-\infty, \infty)$. From the preceding properties, we can see that all polynomials are continuous functions for all real numbers; all rational functions are continuous except where the denominators are zero; for n, an odd positive integer greater than 1, $\sqrt[n]{f(x)}$ is continuous wherever $f(x)$ is continuous, and for n, an even positive integer, $\sqrt[n]{f(x)}$ is continuous whenever $f(x)$ is continuous and nonnegative.

SUMMARY

For a function f on an open interval, f is defined to be continuous at $x = a$ if and only if $\lim\limits_{x \to a} f(x) = f(a)$. This implies that

1. $\lim\limits_{x \to a} f(x)$ exists.
2. $f(a)$ is defined.
3. $\lim\limits_{x \to a} f(x)$ must equal $f(a)$.

In order to show that a function is continuous at $x = a$, we need to verify that the three conditions stated above are satisfied.

Exercise Set 8.2

Determine whether the following functions are continuous at the given points.

1. $f(x) = 5$;   at $x = -1$, $x = 0$, $x = 2$
2. $f(x) = -3$;   at $x = 1$, $x = 2$
3. $f(x) = 3x$;   at $x = -2$, $x = 0$, $x = 1$
4. $f(x) = -2x$;   at $x = -1$, $x = 3$
5. $f(x) = 2x^2$;   at $x = -1$, $x = 0$, $x = 2$
6. $f(x) = -2x^2$;   at $x = -2$, $x = 3$
7. $f(x) = \dfrac{1}{x}$;   at $x = -1$, $x = 0$, $x = 2$
8. $f(x) = \dfrac{2}{x}$;   at $x = 0$, $x = 1$
9. $f(x) = \dfrac{1}{x + 2}$;   at $x = -2$, $x = 0$, $x = 1$
10. $f(x) = \dfrac{x}{x + 2}$;   at $x = -2$, $x = 1$
11. $f(x) = \dfrac{x - 1}{x - 2}$;   at $x = -1$, $x = 0$, $x = 2$
12. $f(x) = \dfrac{1 - 2x}{x - 2}$;   at $x = 1$, $x = 2$
13. $f(x) = \dfrac{x^2 - 9}{x - 3}$;   at $x = -3$, $x = 0$, $x = 3$
14. $f(x) = \dfrac{x^2 - 9}{x + 3}$;   at $x = -3$, $x = 3$
15. $f(x) = \dfrac{1}{x(x - 1)}$;   at $x = -1$, $x = 0$, $x = 1$
16. $f(x) = \dfrac{1}{2x(x - 1)}$;   at $x = 0$, $x = 1$

Are the following functions continuous or discontinuous at $x = 2$? Explain your answer.

17. (a)

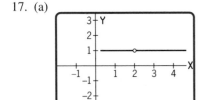

(b)

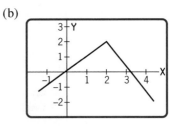

18. (a)

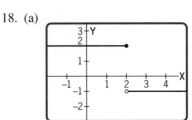

(b)

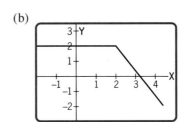

Find the following limits if they exist.

19. $\lim\limits_{x \to 3} \dfrac{x^2 - 9}{x - 3}$

20. $\lim\limits_{x \to 3} \dfrac{x^2 - 2x - 3}{x - 3}$

21. $\lim\limits_{x \to 0} \dfrac{4x^2 - 3x}{x}$

22. $\lim\limits_{x \to 0} \dfrac{(1/x) - 1}{1/x}$

23. $\lim\limits_{x \to 4} \dfrac{x - 4}{\sqrt{x} - 2}$

24. $\lim\limits_{x \to 16} \dfrac{x - 16}{\sqrt{x} - 4}$

25. $\lim\limits_{x \to 0} \dfrac{(2/x) + 1}{3/x}$

Find all points where the graphed functions are discontinuous.

26.

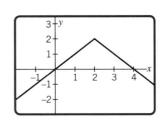

27.

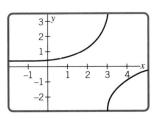

28.

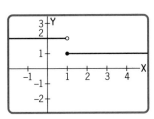

29.

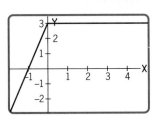

30. Discuss whether the function is continuous at each of the eight points.

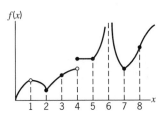

Use your graphing calculator to find all points of discontinuity for each function.

31. $g(x) = \dfrac{1}{3x}$

32. $f(x) = \dfrac{1}{x - 3}$

33. $g(x) = \dfrac{x}{x + 2}$

34. $f(x) = \dfrac{1}{x(x - 2)}$

35. $g(x) = \dfrac{x + 1}{x^2 - 4}$

36. $f(x) = \begin{cases} x^2, & \text{if } -1 \le x < 1 \\ 3x, & \text{if } x \ge 1 \end{cases}$

37. $f(x) = \dfrac{x - 1}{x^2 - 4}$

38. Where are the following functions continuous and discontinuous?

(a) $y = \begin{cases} 2, & \text{if } x < 2 \\ 2, & \text{if } x > 2 \end{cases}$

(b) $y = \begin{cases} x, & \text{if } x \le 2 \\ 6 - 2x, & \text{if } x > 2 \end{cases}$

39. Where are the following functions continuous and discontinuous?

(a) $y = \begin{cases} 2, & \text{if } x \le 2 \\ -1, & \text{if } x > 2 \end{cases}$

(b) $y = \begin{cases} 2, & \text{if } x < 2 \\ 6 - 2x, & \text{if } x \ge 2 \end{cases}$

40. ***Exam.***   Find $\displaystyle \lim_{x \to 3} \dfrac{x^2 - x - 6}{x^2 - 9}$.

(a) 0   (b) $\dfrac{5}{6}$ (c) 1   (d) $\dfrac{5}{3}$

(e) Increases without bound

41. Using a calculator, locate points of discontinuity of

$$f(x) = \dfrac{|x^2 + 3x + 2|}{x - 2}$$

42. Locate points of discontinuity of

$$g(x) = \dfrac{x^2 - 4x + 3}{x^2 - 1}$$

Applications (Business and Economics)

43. ***Cost Functions.***   A cost function $C(x)$ has the following graph for $0 \le x \le 8$.

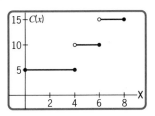

(a) Where is the cost function discontinuous?

(b) Find $\displaystyle \lim_{x \to 2} C(x)$.   (c) Find $C(2)$.

(d) Find $\lim\limits_{x \to 4} C(x)$.   (e) Find $C(4)$.

(f) Find $\lim\limits_{x \to 7} C(x)$.   (g) Find $C(7)$.

44. **Revenue Functions.**   A revenue function $R(x)$, giving revenue in terms of the number of items sold, x, is found to be

$$R(x) = 10x - \frac{x^2}{100}, \qquad 0 \le x \le 1000$$

(a) Find $R(10)$.   (b) Find $R(100)$.
(c) Find $R(500)$.   (d) Draw the graph of $R(x)$.
(e) Is $R(x)$ continuous on $0 < x < 1000$?

45. **Demand Functions.**   The demand function for an item is given by

$$p(x) = \frac{80}{x - 16}, \qquad 16 < x \le 96$$

where p is the price per unit and x is the number of items. Draw the graph of the demand function and discuss the continuity of $p(x)$.

Applications (Social and Life Sciences)

46. **Agronomy.**   The height of a plant is given by

$$h(t) = 3\sqrt{t}$$

where h is measured in inches and t in days of growth. Graph this function and find any points of discontinuity.

47. **Laboratory Experiment.**   A laboratory uses mice in its experiments. The number N of mice available is a function of time t. The following chart was drawn for 6 days.

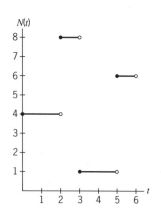

(a) Where is this function discontinuous?
(b) Find $N(1)$.   (c) Find $\lim\limits_{t \to 2^-} N(t)$.

(d) Find $\lim\limits_{t \to 2^+} N(t)$.

48. **Learning Functions.**   A person learns $L(x)$ items in x hours according to the function

$$L(x) = 40\sqrt{x}, \qquad 0 \le x \le 16$$

(a) Find $L(4)$.   (b) Find $L(9)$.
(c) Find $L(16)$.   (d) Draw the graph of $L(x)$.
(e) Discuss the continuity of $L(x)$.

49. **Learning Functions.**   Suppose that a learning function is given as

$$L(x) = 30x^{2/3}, \qquad 0 \le x \le 8$$

(a) Draw the graph of $L(x)$.
(b) Discuss the continuity of $L(x)$.

50. **Voting.**   The registered voting population of a city, $P(t)$, in thousands, is given for the next 8 years as

$$P(t) = 40 + 9t^2 - t^3, \qquad 0 \le t \le 8$$

(a) Find $P(2)$.   (b) Find $P(4)$.
(c) Find $P(6)$.   (d) Draw the graph of $P(t)$.
(e) Discuss the continuity of $P(t)$.

Review Exercises

Find the following limits if they exist.

51. $\lim\limits_{x \to 2} \dfrac{x}{x + 1}$

52. $\lim\limits_{x \to -1} \dfrac{x + 1}{x}$

53. $\lim\limits_{x \to 0} \dfrac{5}{x^2}$

54. $\lim\limits_{x \to 1/3} \left(6 - \dfrac{2}{x} \right)$

55. $\lim\limits_{x \to 2} \dfrac{4}{x - 2}$

56. $\lim\limits_{x \to -1} \dfrac{2}{x + 1}$

8.3   AVERAGE AND INSTANTANEOUS CHANGES

OVERVIEW   A person planning to buy a new house either now or in the future is interested in the rate at which prices are changing. Likewise, a manufacturer is interested in changes in manufacturing that could lead to increased profits. As previously stated, calculus is the mathematics used to study change. The concept used is usually the derivative of a function, which we consider in the next section. To introduce this concept we consider **average change** and **instantaneous change**. The average change of a graph over an interval may be interpreted by the slope of the secant line over that interval. The instantaneous change of a graph at a point is the slope of the line that is tangent to the graph at that point. A study of these concepts in this section will motivate a thorough study in the next section on derivatives. In this section we study

- Average rates
- Increments
- Instantaneous rates of change
- Secant lines
- Tangent lines

If you are driving on an interstate at 55 mph, and 4 miles down the road your speed increases to 65 mph, the average change in your speed per mile traveled is

$$\frac{65 - 55}{4} = 2.5$$

That is, in this 4-mile stretch your speed increased by 10 mph, or it has increased 2.5 mph per mile.

In general, the **average rate of change** of any function on an interval is the change in the function values at the endpoints of the interval divided by the change in the independent variable.

TABLE 1

x-Toys	C(x)
1	$ 75
2	135
3	185
4	231
5	275

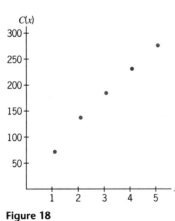

Figure 18

EXAMPLE 24   The cost of manufacturing a given toy includes the setup costs (use of equipment) plus the cost of labor and materials for each toy produced. In general, after a number of toys have been produced, the production cost per toy decreases. We consider a very simplified example of such a change. Table 1 gives the cost, $C(x)$, for the manufacture of x toys. The graph of this function is shown in Figure 18. Find the average rate of change of the cost function $C(x)$ as production increases from

(a)   1 to 2 toys.   (b) 1 to 3 toys.
(c)   1 to 4 toys.   (d) 2 to 5 toys.

SOLUTION

(a)   The average rate of change of the cost function between 1 and 2 toys is

$$\frac{C(2) - C(1)}{2 - 1} = \frac{135 - 75}{1} = 60$$

or the average rate of change of the cost is $60. That is, when one toy has been produced, the cost of producing one more toy is $60.

(b)   The average rate of change of the cost function between 1 and 3 toys is

$$\frac{C(3) - C(1)}{3 - 1} = \frac{185 - 75}{2} = 55$$

or the average rate of change of the cost is $55 as x changes from 1 to 3. That is, the average cost of producing each additional toy is $55 as production increases from 1 to 3 toys.

(c)   The average rate of change of the cost function between 1 and 4 toys is

$$\frac{C(4) - C(1)}{4 - 1} = \frac{231 - 75}{3} = 52$$

or the average rate of change of $C(x)$ is $52 as x changes from 1 to 4. That is, the average cost of producing each additional toy is $52 as production increases from 1 to 4 toys.

(d)   The average rate of change of $C(x)$ is

$$\frac{C(5) - C(2)}{5 - 2} = \frac{275 - 135}{3} = 46.67$$

or the average rate of change of $C(x)$ is $46.67 as x changes from 2 to 5. That is, the average cost of producing each extra toy is $46.67 as production increases from 2 to 5 toys.   ■

This example suggests the following formula.

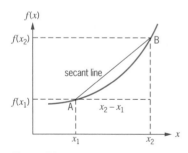

Figure 19

DEFINITION: AVERAGE RATE OF CHANGE

Assuming that the function is defined at x_1 and x_2, the **average rate of change** of $f(x)$ with respect to x as x changes from x_1 to x_2 is the ratio

$$\frac{f(x_2) - f(x_1)}{x_2 - x_1}$$

If we look at the graph of a function such as that in Figure 19, we note that

$$\frac{f(x_2) - f(x_1)}{x_2 - x_1}$$

is the slope of the line through the points $(x_1, f(x_1))$ and $(x_2, f(x_2))$, which are the points A and B in Figure 19. This straight line through A and B is called the **secant line** through A and B of the graph f.

EXAMPLE 25 Use your calculator to graph the function $y = f(x) = x^2 - 1$ and find the average rates of change as x changes from each of the following.

(a) 0 to 2   (b) 0 to 1   (c) 0 to $\frac{1}{2}$

SOLUTION The graph of the function $y = x^2 - 1$ is shown in Figure 20. Locate the three secants (actually segments).

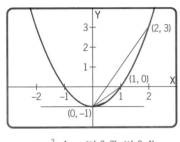

$y = x^2 - 1$ X:[-3, 3] Y:[-2, 4]

Figure 20

(a) The average rate of change from $x = 0$ to $x = 2$ (or the slope of the secant) is

$$\frac{f(x_2) - f(x_1)}{x_2 - x_1} = \frac{f(2) - f(0)}{2 - 0} = \frac{[2^2 - 1] - [0^2 - 1]}{2 - 0} = \frac{3 - (-1)}{2} = 2$$

(b) The average rate of change from $x = 0$ to $x = 1$ (or the slope of the secant) is

$$\frac{f(x_2) - f(x_1)}{x_2 - x_1} = \frac{f(1) - f(0)}{1 - 0} = \frac{[1^2 - 1] - [0^2 - 1]}{1 - 0} = \frac{0 - (-1)}{1 - 0} = 1$$

(c) The average rate of change from $x = 0$ to $x = \frac{1}{2}$ (or the slope of the secant) is

$$\frac{f(x_2) - f(x_1)}{x_2 - x_1} = \frac{f(\frac{1}{2}) - f(0)}{\frac{1}{2} - 0} = \frac{[(\frac{1}{2})^2 - 1] - [0^2 - 1]}{\frac{1}{2} - 0}$$

$$= \frac{-\frac{3}{4} - (-1)}{\frac{1}{2}} = \frac{\frac{1}{4}}{\frac{1}{2}} = \frac{1}{2}$$ ∎

Practice Problem 1 Find the average rate of change in the function $f(x) = 2x^2$ as x changes from 1 to 4.

ANSWER $\dfrac{f(4) - f(1)}{4 - 1} = \dfrac{2(4)^2 - 2(1)^2}{4 - 1} = \dfrac{32 - 2}{3} = 10$

Increments

Before we work toward instantaneous rates of change, let's simplify our notation. If $y = f(x)$, the change in x and the corresponding change in y, called **increments** in x and y, can be denoted by h and $f(x + h) - f(x)$, respectively. For example, in Figure 21, for $y = f(x)$, we have

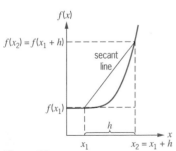

Figure 21

$$h = x_2 - x_1 \quad \text{or} \quad x_2 = x_1 + h$$

Likewise,

$$f(x_2) - f(x_1) = f(x_1 + h) - f(x_1)$$

This is the change in y corresponding to a change of h in x.

EXAMPLE 26   For the function $y = f(x) = x^2$, find

(a) h and $f(x_1 + h) - f(x_1)$ for $x_1 = 0.6$ and $x_2 = 1$.

(b) $\dfrac{f(x_1 + h) - f(x_1)}{h}$ given $h = 0.5$ and $x_1 = 1$.

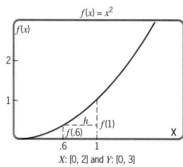

Figure 22

SOLUTION

(a)   As shown in Figure 22, h is the change in x from $x_1 = 0.6$ to $x_2 = 1$, or

$$h = x_2 - x_1 = 1 - 0.6 = 0.4$$

But $f(x_1 + h) - f(x_1)$ is the change in y as x changes from 0.6 to 1, or

$$f(x_1 + h) - f(x_1) = f(x_2) - f(x_1) = f(1) - f(0.6) = 1^2 - (0.6)^2 = 0.64$$

(b)   If we let $h = 0.5$ when $x_1 = 1$, then $x_1 + h = 1 + 0.5 = 1.5$.

$$\begin{aligned}
\frac{f(x_1 + h) - f(x_1)}{h} &= \frac{f(1.5) - f(1)}{0.5} \\
&= \frac{(1.5)^2 - 1^2}{0.5} \\
&= \frac{2.25 - 1}{0.5} \\
&= 2.5
\end{aligned}$$

Practice Problem 2 Let $y = f(x) = 2x + 3$. Find

(a) h and $f(x_1 + h) - f(x_1)$ for $x_1 = 2$ and $x_2 = 3$.

(b) $\dfrac{f(x_1 + h) - f(x_1)}{h}$ for $h = 1$ and $x_1 = 1$.

ANSWER

(a) $h = 3 - 2 = 1$ and $f(3) - f(2) = 9 - 7 = 2$

(b) $\dfrac{f(2) - f(1)}{1} = \dfrac{7 - 5}{1} = 2$

Looking at Figure 23, we see that the *slope of the secant line* from x_1 to $x_2 = x_1 + h$ or the *average rate of change* of f from x_1 to x_2 is

$$\frac{f(x_1 + h) - f(x_1)}{h}$$

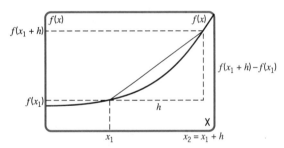

Figure 23

EXAMPLE 27   For $f(x) = x^2 + 2$, find the average rate of change when

(a) $x_1 = 1, h = 0.5$.   (b) $x_1 = 3, h \to 0$.

SOLUTION

(a)   The average rate of change is

$$\frac{f(x_1 + h) - f(x_1)}{h} = \frac{f(1.5) - f(1)}{0.5} \qquad x_1 + h = 1 + 0.5 = 1.5$$

$$= \frac{(1.5)^2 + 2 - (1^2 + 2)}{0.5}$$

$$= \frac{4.25 - 3}{0.5} = 2.5$$

(b)   The average rate of change is

$$\frac{f(3 + h) - f(3)}{h} = \frac{(3 + h)^2 + 2 - [(3)^2 + 2]}{h}$$

$$= \frac{9 + 6h + h^2 + 2 - 11}{h}$$

$$= \frac{6h + h^2}{h} \qquad h \neq 0$$

$$= 6 + h$$

As $h \to 0$, the average rate of change is 6.

These examples suggest the following definitions.

DEFINITION: AVERAGE RATE AND INSTANTANEOUS RATE

If $y = f(x)$, at $x = x_1$ and h is any number (usually small) but not equal to zero, then

$$\textbf{Average rate} = \frac{f(x_1 + h) - f(x_1)}{h}$$

$$\textbf{Instantaneous rate} = \lim_{h \to 0} \frac{f(x_1 + h) - f(x_1)}{h}$$

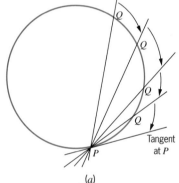

(a)
Intuitively, as Q approaches P along the circle, the secant line rotates into the tangent line.

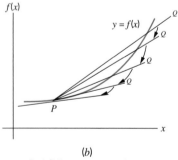

(b)
By definition, as Q approaches P along the graph of $y = f(x)$ from either side of P, the tangent line is the limiting position of the secant line.

Figure 24

Tangent Line to a Graph

What do we mean by the tangent line to the graph of a function at a point? In high school geometry the tangent line to the graph of a circle is defined as the line that intersects the circle in only one point. However, for curves other than a circle, the tangent line to a curve at a point might intersect the curve at other points. Suppose that we define the tangent line to a circle in terms of secant lines. In Figure 24(a), let points P and Q determine a secant line. Now consider point Q as moving toward point P on the circle as shown in the figure. As the point Q moves toward P, the secant rotates into the position of the tangent line of the circle at point P. This is the idea we use to define the tangent line to the graph of any function. [See Figure 24(b).]

Let's write the slope of the secant line in Figure 25 in terms of h. We have

$$\text{Slope of secant line} = \frac{f(x_1 + h) - f(x_1)}{h}$$

As we let h approach zero, B approaches A and the secant lines approach a limiting position. The line that the secant lines approach is the line that is **tangent**

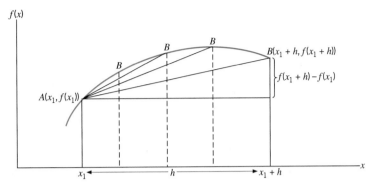

Figure 25

to the curve at point A, and the limit of the slopes of the secant lines is the slope of the tangent line.

TANGENT LINE

For the graph of the function $f(x)$, the line that is **tangent** to $y = f(x)$ at the point $(x_1, f(x_1))$ is the line that passes through this point and has a slope of

$$\lim_{h \to 0} \frac{f(x_1 + h) - f(x_1)}{h}$$

if this limit exists.

Note that the slope of a line that is tangent to a graph at $x = x_1$ is sometimes referred to as the **slope of the curve** (or graph) at $x = x_1$.

EXAMPLE 28 Find the slope of the graph of $y = x^2 - 1$ at $x = 0$.

SOLUTION

$$\begin{aligned}
\text{Slope of tangent line} &= \lim_{h \to 0} \frac{f(x_1 + h) - f(x_1)}{h} \\
&= \lim_{h \to 0} \frac{f(0 + h) - f(0)}{h} \qquad \text{At } x = 0 \\
&= \lim_{h \to 0} \frac{[(0 + h)^2 - 1] - [0^2 - 1]}{h} \\
&= \lim_{h \to 0} \frac{(h)^2}{h} \\
&= \lim_{h \to 0} h = 0 \qquad h \neq 0
\end{aligned}$$

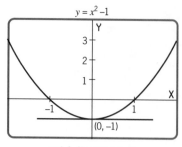

$y = x^2 - 1$

X: [−2, 2] and Y: [−2, 4]

Figure 26

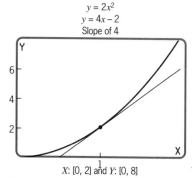

$y = 2x^2$
$y = 4x - 2$
Slope of 4

X: [0, 2] and Y: [0, 8]

Figure 27

Note that this answer agrees with the slope of what was thought to be the tangent line as drawn in Figure 20. For convenience, we repeat the pertinent parts of Figure 20 in Figure 26.

EXAMPLE 29 Find an equation of the tangent line to the graph of $y = 2x^2$ at $(1, 2)$.

SOLUTION First we need to find the slope of the tangent line to $y = 2x^2$ at $(1, 2)$ (see Figure 27).

$$\text{Slope} = \lim_{h \to 0} \frac{f(1 + h) - f(1)}{h}$$

$$= \lim_{h \to 0} \frac{2(1 + h)^2 - 2(1)^2}{h}$$

$$= \lim_{h \to 0} 2\frac{1 + 2h + h^2 - 1}{h}$$

$$= \lim_{h \to 0} 2\frac{2h + h^2}{h} \qquad \text{Divide by } h \neq 0$$

$$= \lim_{h \to 0} 2(2 + h) = 4$$

Using the point-slope equation of a line, $y - y_1 = m(x - x_1)$, we have $y - 2 = 4(x - 1)$ or $y = 4x - 2$.

SUMMARY

We learned two important concepts in this section:

$$\lim_{h \to 0} (\text{Average change}) = (\text{Instantaneous change})$$

$$\lim_{h \to 0} (\text{Secant slopes}) = (\text{Tangent slope})$$

We use these concepts in the next three sections.

Exercise Set 8.3

For Exercises 1–7, find each of the following for the function $y = f(x) = 3x^2$.

1. $\dfrac{f(x_2) - f(x_1)}{x_2 - x_1}$ when $x_1 = 2, x_2 = 4$

2. $f(x_1 + h)$ when $x_1 = 2, h = 1$

3. $\dfrac{f(x_1 + h) - f(x_1)}{h}$ when $x_1 = 2, h = 1$

4. The average rate of change when x changes from 3 to 5

5. $\dfrac{f(3 + h) - f(3)}{h}$

6. Find $\lim\limits_{h \to 0} \dfrac{f(3 + h) - f(3)}{h}$ and interpret your answer.

7. Find $\lim\limits_{h \to 0} \dfrac{f(2 + h) - f(2)}{h}$ and interpret your answer.

Find the average rate of change for the following functions.

8. $y = 3x + 4$ between $x = 1$ and $x = 4$

9. $y = x^2 + 2x$ between $x = 0$ and $x = 3$

10. $y = 2x^3 - 6$ between $x = 1$ and $x = 3$

11. $y = \sqrt{x}$ between $x = 1$ and $x = 4$

Find the instantaneous rate of change for the following functions.

12. Exercise 8 at $x = 2$ 13. Exercise 9 at $x = 0$

14. Find the average rate of change of $f(x) = 3x^2 + 2$ from
    (a) $x = 1$ to $x = 2$.   (b) $x = 1$ to $x = 1.5$.
    (c) $x = 1$ to $x = 1.1$.   (d) $x = 1$ to $x = 1.01$.

15. Find the instantaneous rate of change of $f(x) = 3x^2 + 2$ at $x = 1$.

16. Find the average rate of change of $f(x) = -2x^2 + 3$ from
    (a) $x = 2$ to $x = 3$.   (b) $x = 2$ to $x = 2.5$.
    (c) $x = 2$ to $x = 2.1$.   (d) $x = 2$ to $x = 2.01$.

17. Find the instantaneous rate of change of $f(x) = -2x^2 + 3$ at $x = 2$.

18. **Exam**. What is the y-coordinate of the point on the curve $y = 2x^2 - 3x$ at which the slope of the tangent line is the same as that of the slope of the secant line between $x = 1$ and $x = 2$?
    (a) -1 (b) 0 (c) 1
    (d) 3 (e) 9

Applications (Business and Economics)

19. **Consumer Price Index**. The consumer price index (CPI), compiled by the U.S. Department of Labor, gives the price today of $100 worth of food, clothing, housing, fuel, and so on in 1967.

Year	CPI
1982	$289
1983	297
1984	308
1985	320
1986	333
1987	348
1988	363

Find the average change in the consumer price index from
(a) 1983 to 1985.   (b) 1986 to 1988.
(c) 1984 to 1988.   (d) 1982 to 1988.

20. **Demand Function**. The price of an article is given as a function of the number of the articles available, x, as

$$p(x) = \frac{10,000}{x^2} + \frac{8000}{x}$$

Find the average change in $p(x)$ as
(a) x changes from 2 to 4.
(b) x changes from 1 to 2.
(c) x changes from $\frac{1}{2}$ to 1.

21. **Profit Functions**. A company lists its profit function as

$$P(x) = 8x - 0.02x^2 - 500$$

(a) Find the average rate of change in profit from $x = 100$ to $x = 200$.
(b) Find the instantaneous rate of change in profit at $x = 100$.
(c) Find the instantaneous rate of change in profit at $x = 200$.
(d) Find the profit at $x = 200$.

22. **Cost Functions**. A cost function is given as

$$C(x) = 40,000 - 300x + x^2$$

Graph the function and find the instantaneous rate of change of cost when $x = 100$, when $x = 150$, and when $x = 200$.

Applications (Social and Life Sciences)

23. **Bacteria**. The number of bacteria N, in thousands, present x hours after being treated by an antibiotic is

$$N(x) = x^2 - 6x + 10, \qquad 0 \le x \le 6$$

Find the average rate of change of the number present from $x = 1$ to $x = 3$. Find the instantaneous rate of change at $x = 1$.

24. **Births**. At Sunnyside Hospital the number of births is recorded by year as follows:

Year	1987	1988	1989	1990	1991	1992	1993
Births	1040	1102	1230	1308	1402	1460	1480

Find the average rate of change in the number of births from
(a) 1987 to 1990.   (b) 1990 to 1993.
(c) 1989 to 1993.

Review Exercises

Find the following limits.

25. $\lim\limits_{x \to 2} \dfrac{x^2 - 3x + 2}{x - 2}$

26. $\lim\limits_{x \to -1} \dfrac{x + 1}{x^2 + 3x + 2}$

27. $\lim\limits_{x \to 1} \dfrac{3x^2 + 4x - 1}{5x^2 + 2x + 3}$

28. $\lim\limits_{x \to 2} (x - 2)(x^2 - 2x - 2)$

29. $\lim\limits_{x \to 3} \dfrac{2x^2 - 2x - 12}{x - 3}$

30. $\lim\limits_{x \to 0} \dfrac{5x^3 + 3x}{x}$

8.4 DEFINITION OF THE DERIVATIVE

OVERVIEW   In the last section, we used an expression that we will call the **difference quotient** for function f,

$$\frac{f(x_1 + h) - f(x_1)}{h}$$

which gave the average change in $f(x)$ from x_1 to $x_1 + h$. If the limit of this difference,

$$\lim_{h \to 0} \frac{f(x_1 + h) - f(x_1)}{h}, \qquad h \neq 0$$

exists, it can be interpreted as the *slope of the tangent line* to the graph of the function at $x = x_1$. It can also be interpreted as the *instantaneous rate of change* of $f(x)$ at $x = x_1$. In this section we use this notation to define the **derivative** of $f(x)$, and we investigate some of its many applications. In economics, if $C(x)$ represents a cost function, the derivative of $C(x)$ is the rate of change of cost and is called the **marginal cost** function. The derivative of the profit function $P(x)$ is called the **marginal profit** function (the rate of change of profit); the derivative of the demand function $D(x)$ is the **marginal demand** function; and the derivative of the supply function $S(x)$ is the **marginal supply** function. In medicine, if $P(x)$ represents systolic blood pressure, the derivative of $P(x)$ may be thought of as the sensitivity to a drug. In this section, we

- Define the derivative of a function
- Outline a procedure for finding the derivative
- Illustrate the use of derivatives

In the difference quotient above, let's replace the fixed number x_1 by the more general symbol x to get

$$\frac{f(x + h) - f(x)}{h}, \qquad h \neq 0$$

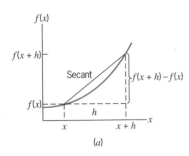

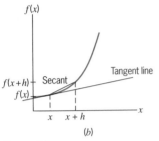

Figure 28

This difference quotient is pictured in Figure 28(*a*). Geometrically, this quotient gives the slope of the secant line as shown. As $h \to 0$ in Figure 28(*b*), the slopes of the secant lines approach the slope of the tangent line at *x*.

This difference quotient as $h \to 0$ provides for us one of the basic ideas of calculus, the derivative.

DEFINITION: DERIVATIVE OF A FUNCTION

For $y = f(x)$ and *h*, any number such that $h \neq 0$, the **derivative** of $f(x)$, which is denoted by $f'(x)$, is defined by

$$f'(x) = \lim_{h \to 0} \frac{f(x + h) - f(x)}{h}$$

provided that this limit exists.

Other symbols that can be used to denote the derivative are

$$y', \quad D_x f, \quad \text{and} \quad \frac{dy}{dx}$$

which can be read as "the derivative of *y*," or "the derivative of a function of *x* with respect to *x*."

NOTE: The *dy/dx* should not be regarded, at this time, as a quotient of two entities, *dy* and *dx*.

EXAMPLE 30   Use the definition of the derivative to find $f'(x)$ for the function $f(x) = x^2 - 1$.

SOLUTION   First, we set up the difference quotient:

$$\frac{f(x + h) - f(x)}{h} = \frac{[(x + h)^2 - 1] - [x^2 - 1]}{h}$$

Second, we write the limit as $h \to 0$:

$$\lim_{h \to 0} \frac{[(x + h)^2 - 1] - [x^2 - 1]}{h}$$

Third, we simplify this expression as much as possible:

$$\lim_{h \to 0} \frac{x^2 + 2xh + h^2 - 1 - x^2 + 1}{h} = \lim_{h \to 0} \frac{2xh + h^2}{h}$$

$$= \lim_{h \to 0} \frac{h(2x + h)}{h} \qquad h \neq 0$$

$$= \lim_{h \to 0} (2x + h)$$

Fourth, to evaluate the limit as $h \to 0$, we use the properties of limits that allow us to replace h with 0, since $2x + h$ is a continuous function:

$$\lim_{h \to 0} (2x + h) = 2x + 0 = 2x$$

Therefore, $f'(x) = 2x$ gives the derivative of the function $f(x) = x^2 - 1$ at any real number x. ■

Procedure for Finding a Derivative

1. Set up the difference quotient.
2. Write as a limit.
3. Simplify this limit so that h is no longer a factor of the denominator.
4. Evaluate the limit by replacing h with zero.

EXAMPLE 31   Given $f(x) = 2x^2 + 3$, compute $f'(x)$.

SOLUTION

1. Set up the difference quotient.

$$\frac{f(x + h) - f(x)}{h} = \frac{[2(x + h)^2 + 3] - [2x^2 + 3]}{h}$$

2. Write as a limit.

$$\lim_{h \to 0} \frac{[2(x + h)^2 + 3] - [2x^2 + 3]}{h}$$

3. Simplify the limit.

$$= \lim_{h \to 0} \frac{2x^2 + 4xh + 2h^2 + 3 - 2x^2 - 3}{h}$$

$$= \lim_{h \to 0} \frac{4xh + 2h^2}{h} = \frac{h(4x + 2h)}{h}$$

$$= \lim_{h \to 0} (4x + 2h) \qquad h \neq 0$$

4. Evaluate the limit.

$$= 4x + 2(0)$$

Thus, we have $f'(x) = 4x$. ■

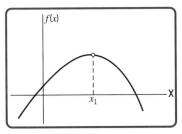

(a) Function is not defined at $x = x_1$

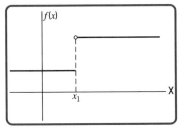

(b) Limit of the function does not exist at $x = x_1$

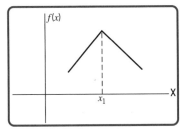

(c) No unique tangent line at $x = x_1$

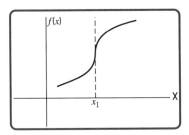

(d) Vertical tangent line at $x = x_1$

Figure 29

Calculator Note

Most calculators have a menu for finding the value of the derivative at a point or the value of a function at a point. We illustrate as follows: Press $\boxed{\text{MATH}}$, (select 8), and then insert the expression or function, the variable, and the value of the variable:

$$\text{nDeriv(expression, variable, value)}$$

To find $f'(1)$ in Example 31, we use

$$\text{nDeriv}(2x^2 + 3, x, 1)$$

to obtain a value of 4. From Example 31, $f'(1) = 4 \cdot 1 = 4$.

Practice Problem 1 Find the derivative of $f(x) = 3x^2$.

ANSWER

$$\lim_{h \to 0} \frac{3(x + h)^2 - 3x^2}{h} = \lim_{h \to 0} \frac{6xh + 3h^2}{h} = \lim_{h \to 0} \frac{h(6x + 3h)}{h}$$
$$= \lim_{h \to 0} (6x + 3h) = 6x + 3(0) = 6x$$

Nonexistence of Derivatives

The existence of a derivative at $x = x_1$ depends on the existence of the limit

$$\lim_{h \to 0} \frac{f(x_1 + h) - f(x_1)}{h}$$

If the limit does not exist, we say that the function is **not differentiable** at $x = x_1$ or that $f'(x_1)$ does not exist. A derivative will not exist at a point, x_1, if:

1. The function is not defined at the point x_1.
2. The limit of the function as x approaches x_1 does not exist.
3. There is no unique tangent line at $(x_1, f(x_1))$.
4. There is a vertical tangent line at $x = x_1$.

Examples of these cases are shown in Figure 29.

If the derivative of $f(x)$ exists at $x = a$, what can we say about the continuity of the function at $x = a$? If the derivative of $f(x)$ exists at $x = a$, then $f(x)$ is continuous at $x = a$. The converse of this statement is not true. If a function is continuous at $x = a$, then its derivative at $x = a$ may or may not exist; Figures 29(c) and 29(d) show examples in which f is continuous at x_1 but $f'(x_1)$ does not exist. As we try to find derivatives of functions, it is important to remember that if a function is not continuous at a point, the function does not have a derivative at that point.

Here is an example of a function that has a derivative everywhere except at $x = 0$, because the function does not even exist at $x = 0$.

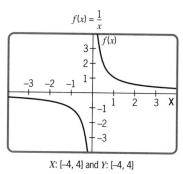

$f(x) = \dfrac{1}{x}$

X: [-4, 4] and Y: [-4, 4]

Figure 30

EXAMPLE 32 Find the derivative of $f(x) = 1/x$, which is shown in Figure 30.

SOLUTION

1. Set up the difference quotient.

$$\frac{f(x + h) - f(x)}{h} = \frac{\dfrac{1}{x + h} - \dfrac{1}{x}}{h}$$

2. Write as a limit.

$$\lim_{h \to 0} \frac{\dfrac{1}{x + h} - \dfrac{1}{x}}{h}$$

3. Simplify the limit by finding a common denominator.

$$= \lim_{h \to 0} \frac{\dfrac{x}{x(x + h)} - \dfrac{(x + h)}{x(x + h)}}{h}$$

$$= \lim_{h \to 0} \frac{x - (x + h)}{x(x + h)h} = \lim_{h \to 0} \frac{-h}{x(x + h)h}$$

$$= \lim_{h \to 0} \frac{-1}{x(x + h)}$$

4. Evaluate the limit.

$$= \frac{-1}{x(x + 0)}$$

Therefore,

$$f'(x) = \frac{-1}{x^2}$$

Notice that $f'(0)$ does not exist because there is no number $-1/0^2$. ■

> ## Calculator Note
>
> In the preceding example, $f'(x) = -1/x^2$, so $f'(1) = -1$. Using nDeriv($1/x$, x, 1) on your graphing calculator gives an answer of -1.000001. Why is there a discrepancy? The value of the derivative from the calculator is the result of a numerical procedure where the variable changes by increments. The answer is not accurate to the sixth decimal place. Consider the answer as -1 to five decimal places. If we evaluate the derivative at $x = 0$, we get $f'(0) \approx f'(.0010001) = -4{,}999{,}750{,}012$. This is as accurately as the calculator can express such numbers that increase without bound.

EXAMPLE 33 Find the derivative of $f(x) = \sqrt{x}$ and check your answer at $x = 9$ with a calculator.

SOLUTION   To simplify this expression, we multiply both the numerator and denominator by $\sqrt{x + h} + \sqrt{x}$.

$$f'(x) = \lim_{h \to 0} \frac{\sqrt{x + h} - \sqrt{x}}{h}$$

$$= \lim_{h \to 0} \frac{(\sqrt{x + h} - \sqrt{x})(\sqrt{x + h} + \sqrt{x})}{h(\sqrt{x + h} + \sqrt{x})}$$

$$= \lim_{h \to 0} \frac{(\sqrt{x + h})^2 - (\sqrt{x})^2}{h(\sqrt{x + h} + \sqrt{x})}$$

$$= \lim_{h \to 0} \frac{h}{h(\sqrt{x + h} + \sqrt{x})}$$

$$= \lim_{h \to 0} \frac{1}{\sqrt{x + h} + \sqrt{x}}$$

$$= \frac{1}{\sqrt{x + 0} + \sqrt{x}}$$

$$= \frac{1}{2\sqrt{x}}$$

> ✓ Calculator Check: $f'(9) = \dfrac{1}{2\sqrt{9}} = \dfrac{1}{6}$; nDeriv($\sqrt{x}$, x, 9) = .1666666669 and
>
> $\dfrac{1}{6} \approx .1666666667$ ∎

Economists make use of the derivative of a function to determine **marginal cost**.

EXAMPLE 34   A small manufacturer of boats (without motors) finds that to produce and sell x units of his product per month, it costs him $10x - 0.1x^2$ hundreds of dollars. That is, his monthly total-cost function is given by $C(x) = 10x - 0.1x^2$ for $0 \le x \le 10$. The graph of this function is shown in Figure 31. Find the marginal cost function.

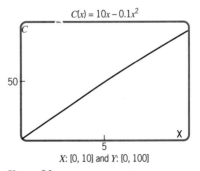

$C(x) = 10x - 0.1x^2$

X: [0, 10] and Y: [0, 100]

Figure 31

SOLUTION The derivative of $C(x)$ is the marginal cost:

$$\text{Marginal cost} = C'(x)$$

$$= \lim_{h \to 0} \frac{C(x + h) - C(x)}{h}$$

$$= \lim_{h \to 0} \frac{[10(x + h) - 0.1(x + h)^2] - [10x - 0.1x^2]}{h}$$

$$= \lim_{h \to 0} \frac{10h - 0.2xh - 0.1h^2}{h}$$

$$= \lim_{h \to 0} \frac{h(10 - 0.2x - 0.1h)}{h}$$

$$= \lim_{h \to 0} (10 - 0.2x - 0.1h)$$

$$= 10 - 0.2x - 0$$

$$= 10 - 0.2x \qquad \blacksquare$$

Practice Problem 2 A revenue function is given as $R(x) = 200x - 0.01x^2$. Find an expression for marginal revenue: $\text{MR} = R'(x)$. What is the approximate additional revenue for one more unit when $x = 10$?

ANSWER We have $\text{MR} = R'(x) = 200 - 0.02x$ and $R'(10) = 200 - 0.02(10) = 199.80$, so the additional revenue for one more unit when $x = 10$ will be \$199.80.

SUMMARY

The derivative of a function f, at a number x in its domain, is defined to be

$$f'(x) = \lim_{h \to 0} \frac{f(x + h) - f(x)}{h}$$

Since the difference quotient

$$\frac{[f(x + h) - f(x)]}{h}$$

is the slope of the line joining $(x, f(x))$ and $((x + h), f(x + h))$, the derivative may be interpreted as the slope of a tangent line to the curve. Economists often refer to the derivative of a function as the marginal value of that function.

Exercise Set 8.4

Find the derivative of the following functions.

1. $f(x) = 3x$
2. $f(x) = 2x$
3. $f(x) = 3x + 5$
4. $f(x) = -2x + 5$
5. $f(x) = \dfrac{5x - 2}{3}$
6. $f(x) = \dfrac{-3x - 5}{2}$

Find the derivative of each function, and then find $f'(0)$, $f'(1)$, and $f'(-1)$. Check $f'(1)$ using a graphing calculator.

7. $f(x) = 3x^2 + 2$
8. $f(x) = x^2$
9. $f(x) = 7x + x^2$
10. $f(x) = x^2 - 4x$
11. $f(x) = \dfrac{4}{x}$
12. $f(x) = \dfrac{5}{x}$
13. $f(x) = \sqrt{3x}$
14. $f(x) = \sqrt{5x}$
15. $f(x) = \dfrac{1}{\sqrt{x + 1}}$
16. $f(x) = \dfrac{1}{\sqrt{x + 4}}$
17. $f(x) = \dfrac{2}{x^2}$
18. $f(x) = \dfrac{3}{x^2}$
19. $f(x) = x^3$
20. $f(x) = 2x^3$
    [**Hint:** $(x + h)^3 = x^3 + 3x^2h + 3xh^2 + h^3$.]

21. Using your calculator, draw the graph of $y = x^2 - 3$ and evaluate the derivative at $x = 1$ on the graph. (Use ZOOM 4 for your window.) Now, find the derivative of $y = x^2 - 3$ and evaluate at $x = 1$. Are the answers approximately the same?

22. Work Exercise 21 using $y = 8/x$.

23. Find the slope of the tangent line to the graph of $f(x) = 2x^2 + 2x$ at $x = 1$, and then find the equation of the tangent line. [Use $y - y_1 = m(x - x_1)$.]

24. Find the slope of the tangent line to the graph of $f(x) = 3x^2$ at $x = 2$ and then find the equation of the tangent line.

25. For the function $f(x) = x^2 + 1$,
    (a) find the equation for the slope of the tangent lines to this curve.
    (b) find the slope at $x = 2$.
    (c) find an equation of the tangent line drawn to this curve at $x = 2$.
    (d) graph the function and draw the tangent line at $x = 2$.

Applications (Business and Economics)

26. **Manufacturing.** On an assembly line the rate at which articles are produced decreases with the number of hours on the line. For example, the number of articles produced at any given time t (t is time at work on the assembly line) is $f(t) = 400t - 3t^2$.
    (a) How many articles have been produced at the end of 2 hours?
    (b) What is the instantaneous rate of producing articles at the end of 2 hours?
    (c) Repeat part (a) at the end of 6 hours.
    (d) Repeat part (b) at the end of 6 hours.
    (e) Compare parts (b) and (d) and see if they agree with the first sentence of the problem.

27. **Cost Functions.** A cost function is given as

    $$C(x) = 900 + 300x + x^2$$

    (a) Find the marginal cost function, $C'(x)$.
    (b) Find $C'(2)$. Check with a graphing calculator.
    (c) Find $C'(3)$. Check with a graphing calculator.

28. **Demand Functions.** A demand function is given as

    $$p(x) = 144 - x^2, \quad 1 \le x \le 12$$

    (a) Find the instantaneous rate of change, $p'(x)$.
    (b) Find $p'(2)$. Check with a graphing calculator.
    (c) Find $p'(6)$. Check with a graphing calculator.

29. **Demand Functions.** A demand function is given as

    $$p(x) = 169 - x^2, \quad 1 \le x \le 13$$

    (a) Find the instantaneous rate of change, $p'(x)$.
    (b) Find $p'(3)$. Check with a graphing calculator.
    (c) Find $p'(5)$. Check with a graphing calculator.

Applications (Social and Life Sciences)

30. **Human Sensitivity.** The systolic blood pressure of a patient an hour after receiving a drug is given by

    $$P(x) = 136 + 18x - 8x^2, \quad 0 \le x \le 3$$

    (a) What is the sensitivity dP/dx of the patient to this drug?
    (b) What is the sensitivity when $x = 2$? Check with a graphing calculator.

31. **Human Sensitivity.**   A doctor administers x milligrams of a drug and records a patient's blood pressure. The systolic pressure is approximated by

$$P(x) = 140 + 10x - 5x^2, \qquad 0 \le x \le 4$$

(a) What is the sensitivity dP/dx of the patient to the
drug?
(b) What is the sensitivity when 2 milligrams are
administered? Check with a graphing calculator.
(c) What is the sensitivity when 3 milligrams are
administered? Check with a graphing calculator.

32. **Learning Functions.** A learning function for the
number of words learned in a foreign language class
after t hours of study is given as

$$N(t) = 15t - t^2, \qquad 0 \le t \le 8$$

(a) Find the rate of change of this function.
(b) Find $N'(2)$ and $N'(4)$. Check with a graphing
calculator.

33. **Bacteria.**   When an antibiotic is introduced into a
culture of bacteria, the number of bacteria present
after t hours is given by $N(t) = 2000 + 10t - 5t^2$,

where $N(t)$ is the number (in thousands) of bacteria
present at the end of t hours.
(a) Find the number of bacteria present after 2
hours.
(b) Find the rate of change in the number of bacteria
present at the end of 2 hours. Check with a
graphing calculator.
(c) Repeat part (a) at the end of 4 hours. Check with
a graphing calculator.
(d) Repeat part (b) at the end of 4 hours. Check with
a graphing calculator.

Review Exercises

Find the following limits if they exist.

34. $\displaystyle \lim_{x \to 1/2} \frac{2x - 3}{4x - 2}$

35. $\displaystyle \lim_{x \to -2/3} \frac{3x^2 + 4x - 1}{3x + 2}$

36. Find the average rate of change of $f(x) = x^2 - 4$
from $x = 1$ to $x = 2$.

8.5 TECHNIQUES FOR FINDING DERIVATIVES

OVERVIEW   In the preceding section, we found a derivative by using the definition formula

$$f'(x) = \lim_{h \to 0} \frac{f(x + h) - f(x)}{h}$$

In this section and the sections that follow, we discover some rules or formulas
from this definition so that we can take the derivative of a very large number of
functions without using the definition formula. In fact, these rules or formulas will
make the derivatives of many functions easy to obtain. In this section we focus
our attention on

- Derivative of x to a power
- Derivative of a constant
- Derivative of a constant times a function
- Derivative of a sum
- Marginal analysis

Geometrically, the graph of $y = f(x) = C$, where C is any constant, is a horizontal line; that is, it has a slope of zero. Intuitively, we know that the derivative

of $f(x) = C$ must be zero, since there is no change in the slope. However, we verify this fact with our four-step process given in the preceding section.

1. $\dfrac{f(x + h) - f(x)}{h} = \dfrac{C - C}{h}$

2. $\lim\limits_{h \to 0} \dfrac{C - C}{h}$

3. $= \lim\limits_{h \to 0} \dfrac{0}{h} = \lim\limits_{h \to 0} 0$

4. $= 0$

Thus we have a proof that the derivative of a constant function is zero.

DERIVATIVE OF A CONSTANT FUNCTION

If $f(x) = C$, then $f'(x) = 0$.

EXAMPLE 35

(a)   If $f(x) = -5$, then $f'(x) = 0$.
(b)   If $f(x) = 3$, then $f'(x) = 0$.

Consider now the derivative of a constant times a function. Suppose that $f(x) = C \cdot g(x)$, where C is a constant and $g'(x)$ exists. Then

1. $\dfrac{f(x + h) - f(x)}{h} = \dfrac{C \cdot g(x + h) - C \cdot g(x)}{h}$

2. $\lim\limits_{h \to 0} \dfrac{C \cdot g(x + h) - C \cdot g(x)}{h}$

3. $= \lim\limits_{h \to 0} C \left[\dfrac{g(x + h) - g(x)}{h} \right] = C \lim\limits_{h \to 0} \left[\dfrac{g(x + h) - g(x)}{h} \right]$

4. $= Cg'(x)$

Therefore we have proved that $f'(x) = Cg'(x)$.

DEFINITION: DERIVATIVE OF A CONSTANT TIMES A FUNCTION

Let C be any real number and $g(x)$ be any function that is differentiable. The derivative of

$$f(x) = Cg(x) \quad \text{is} \quad f'(x) = Cg'(x)$$

That is, the **derivative of a constant times a function** is the constant times the derivative of the function.

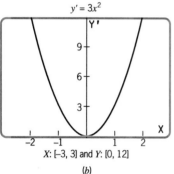

(a)

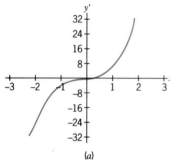

$y' = 3x^2$

X: [-3, 3] and Y: [0, 12]

(b)

Figure 32

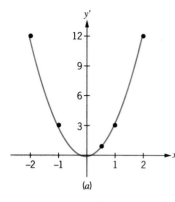

(a)

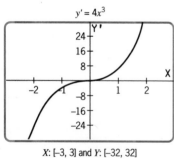

$y' = 4x^3$

X: [-3, 3] and Y: [-32, 32]

(b)

Figure 33

EXAMPLE 36 Find the derivative of $y = 4x^2$.

SOLUTION

Consider y as the product of 4 and x^2. As we found in the exercise set in the last section, the derivative of x^2 is $2x$. Therefore,

$$y' = 4(2x) = 8x$$

Functions of the form $y = x^n$ occur often in application problems. Next we examine some special cases for this type of function and then state the power rule without proof. For example, we have already shown that if $f(x) = x$, then $f'(x) = 1 \cdot x^0 = 1$, and if $f(x) = x^2$, then $f'(x) = 2x^{2-1} = 2x$. It would appear that if $f(x) = x^n$, then $f'(x) = nx^{n-1}$.

Let's continue to explore this conjecture with our calculator.

First, with our calculator, we obtain values of the derivative of $f(x) = x^3$ at a number of points, and then we draw a smooth curve through these points to represent the graph of $f'(x)$; see Figure 32(a). Next, we draw the graph of our conjectured derivative, $f'(x) = 3x^2$; see Figure 32(b). Are the two graphs in agreement? The two graphs appear to coincide. Thus, if $f(x) = x^3$, then $f'(x)$ appears to be $3x^2$.

Similarly, compare the graphs of the derivative values of $f(x) = x^4$ and $f'(x) = 4x^3$ shown in Figure 33. Again, the curves seem to coincide and it appears that if $f(x) = x^4$, then $f'(x) = 4x^3$.

In general, for any positive integer n, it is true that if $f(x) = x^n$, then $f'(x) = nx^{n-1}$. A similar formula can be developed when n is a negative integer. Apply the formula in Example 37, and then show that the answer is correct by using the definition of a derivative.

EXAMPLE 37 Use the definition of a derivative given in the preceding section to verify that if

$$f(x) = x^{-1}, \quad \text{then} \quad f'(x) = -1x^{-1-1} = -x^{-2} = -\frac{1}{x^2}$$

SOLUTION

$$\lim_{h \to 0} \frac{(x + h)^{-1} - x^{-1}}{h} = \lim_{h \to 0} \frac{\dfrac{1}{x + h} - \dfrac{1}{x}}{h}$$

$$= \lim_{h \to 0} \frac{x - (x + h)}{x(x + h)h} \qquad \text{Multiply by } \frac{x(x + h)}{x(x + h)}.$$

$$= \lim_{h \to 0} \frac{-h}{x(x + h)h}$$

$$= \lim_{h \to 0} \frac{-1}{x(x + h)}$$

$$= \frac{-1}{x(x + 0)} = \frac{-1}{x^2}$$

Do you think the formula for the derivative of x^n holds when n is not an integer? Apply the formula in Example 38, and then show that the answer is correct by using the definition of the derivative.

EXAMPLE 38   Use the definition of a derivative in the preceding section to verify that if

$$f(x) = x^{1/2}, \quad \text{then} \quad f'(x) = \frac{1}{2}x^{(1/2)-1} = \frac{1}{2}x^{-1/2}$$

SOLUTION

$$\lim_{h \to 0} \frac{(x+h)^{1/2} - x^{1/2}}{h} = \lim_{h \to 0} \frac{\sqrt{x+h} - \sqrt{x}}{h} \cdot \frac{\sqrt{x+h} + \sqrt{x}}{\sqrt{x+h} + \sqrt{x}}$$

$$= \lim_{h \to 0} \frac{(x+h) - x}{h(\sqrt{x+h} + \sqrt{x})} \qquad \text{Rationalize the numerator.}$$

$$= \lim_{h \to 0} \frac{h}{h(\sqrt{x+h} + \sqrt{x})}$$

$$= \lim_{h \to 0} \frac{1}{\sqrt{x+h} + \sqrt{x}}$$

$$= \frac{1}{\sqrt{x+0} + \sqrt{x}}$$

$$= \frac{1}{2\sqrt{x}}$$

$$= \frac{x^{-1/2}}{2} \qquad \blacksquare$$

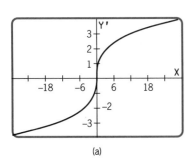

(a)

$$y' = \frac{4}{3}x^{\frac{1}{3}}$$

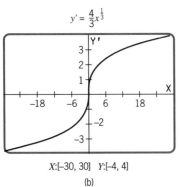

X:[-30, 30] Y:[-4, 4]

(b)

Figure 34

Calculator Note

Show that if $f(x) = x^{4/3}$, then $f'(x) = \dfrac{4x^{1/3}}{3}$. This time instead of finding the value of the derivative at several points and then drawing the graph, we will accomplish both on the calculator in one step. Press $\boxed{\text{Y} =}$ $\boxed{\text{MATH}}$ (Select 8) and insert nDeriv ((x $\boxed{\wedge}$ 4) $\boxed{\wedge}$ (1/3), x, x) $\boxed{\text{GRAPH}}$. The graph is shown in Figure 34(a). Then the calculator graph of our assumed answer $y' = \dfrac{4}{3}x^{1/3}$ is found in Figure 34(b).

Since the two graphs in Figure 34 seem to be identical, it appears that the derivative of $f(x) = x^n$ is $f'(x) = nx^{n-1}$ when n is a rational number.

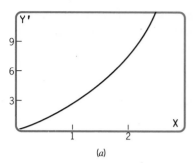

(a)

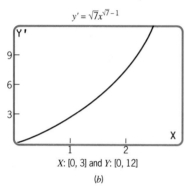

$$y' = \sqrt{7}x^{\sqrt{7}-1}$$

X: [0, 3] and Y: [0, 12]

(b)

Figure 35

Calculator Note

Using a calculator we demonstrate that the power rule holds for n, an irrational number such as $\sqrt{7}$, consider $f(x) = x^{\sqrt{7}}$. Figure 35(a) shows the graph of y' using the numerical procedure for finding a derivative on a calculator. Press $\boxed{Y =}$ $\boxed{\text{MATH}}$ (select 8) and insert nDeriv($x^{\sqrt{7}}$, x, x). Our conjecture will be that $f'(x) = \sqrt{7}x^{\sqrt{7}-1}$. Figure 35(b) is the graph of our assumed derivative. Note that they are identical.

A formal proof of the following theorem is beyond the scope of this book.

POWER RULE

Let $f(x) = x^n$, where n is any real number. Then $f'(x) = nx^{n-1}$.

EXAMPLE 39

$$\text{If} \quad f(x) = x^4, \quad \text{then} \quad f'(x) = 4x^3 \qquad {\scriptstyle n = 4;\, n - 1 = 3}$$

EXAMPLE 40

$$\text{If} \quad y = x^{-3}, \quad \text{then} \quad \frac{dy}{dx} = -3x^{-4} \qquad {\scriptstyle n = -3;\, n - 1 = -4}$$

EXAMPLE 41

$$\text{If} \quad f(x) = x^{1/3}, \quad \text{then} \quad f'(x) = \frac{1}{3}x^{(1/3)-1} = \frac{1}{3}x^{-2/3} \qquad {\scriptstyle n = \frac{1}{3};\, n-1 = -\frac{2}{3}}$$

The notation

$$\frac{d}{dx}[f(x)]$$

means *find the derivative of the function inside the brackets.*

EXAMPLE 42   If $y = \sqrt[3]{x^2}$, find y'. Recall that $\sqrt[3]{x^2}$ can be written equivalently as $(x^2)^{1/3} = x^{2/3}$.

SOLUTION $y' = \dfrac{d}{dx}[x^{2/3}] = \dfrac{2}{3}x^{(2/3)-1} = \dfrac{2}{3}x^{-1/3} = \dfrac{2}{3\sqrt[3]{x}}$

Practice Problem 1 If $y = \sqrt{x^3}$, find y'.

ANSWER $y' = \dfrac{3\sqrt{x}}{2}$

EXAMPLE 43   Find the derivative of $y = 7x^3$.

SOLUTION   Consider y as a product of 7 and x^3. Then

$$y' = 7\frac{d}{dx}[x^3] \qquad \text{Derivative of constant times a function}$$

$$= 7 \cdot 3x^2 \qquad \text{Power rule}$$

$$= 21x^2$$

EXAMPLE 44   Find $\dfrac{d}{dx}\left[\dfrac{x^3}{4}\right]$.

SOLUTION   We can write $x^3/4$ as $\frac{1}{4}x^3$. Thus,

$$\frac{d}{dx}\left[\frac{1}{4}x^3\right] = \frac{1}{4}\frac{d}{dx}[x^3] = \frac{1}{4}[3x^2] = \frac{3x^2}{4}$$

EXAMPLE 45   Find the derivative of $f(x) = x^\pi$.

SOLUTION   For this example $n = \pi$, so $f'(x) = \pi x^{\pi - 1}$.

Practice Problem 2   Find y' for $y = \dfrac{3}{\sqrt{x}}$. (**Hint:** Write $1/\sqrt{x}$ using a negative exponent.)

ANSWER $y' = \dfrac{-3}{2x\sqrt{x}}$

The next rule is a great time-saver, especially when taking the derivative of a polynomial. To differentiate the sum of two functions, differentiate each function individually and add the two derivatives. That is, the derivative of a sum (or difference) is the sum (or difference) of the derivatives.

DERIVATIVE OF A SUM OF TWO FUNCTIONS

Let $f(x)$ and $g(x)$ be two functions whose derivatives exist. The derivative of their sum (or difference) is the sum (or difference) of their derivatives:

$$\frac{d}{dx}[f(x) \pm g(x)] = \frac{d[f(x)]}{dx} \pm \frac{d[g(x)]}{dx} = f'(x) \pm g'(x)$$

To prove this theorem, we apply our four-step procedure. Let

$$s(x) = f(x) + g(x)$$

1. $\dfrac{s(x + h) - s(x)}{h} = \dfrac{[f(x + h) + g(x + h)] - [f(x) + g(x)]}{h}$

2. $s'(x) = \lim\limits_{h \to 0} \dfrac{[f(x + h) + g(x + h)] - [f(x) + g(x)]}{h}$

3. $= \lim\limits_{h \to 0} \dfrac{f(x + h) - f(x)}{h} + \lim\limits_{h \to 0} \dfrac{g(x + h) - g(x)}{h}$

4. $= f'(x) + g'(x)$

Therefore, $s'(x) = f'(x) + g'(x)$.

EXAMPLE 46 Find the derivative of $5x^3 + 3x^2$.

SOLUTION Let $f(x) = 5x^3$ and $g(x) = 3x^2$. Then

$$\frac{d}{dx}[f(x) + g(x)] = \frac{d[f(x)]}{dx} + \frac{d[g(x)]}{dx}$$

$$\frac{d}{dx}[5x^3 + 3x^2] = \frac{d[5x^3]}{dx} + \frac{d[3x^2]}{dx}$$

$$= 5\frac{d(x^3)}{dx} + 3\frac{d(x^2)}{dx} \qquad \text{Constant times a function}$$

$$= 5(3x^2) + 3(2x) \qquad \text{Power rule}$$

$$= 15x^2 + 6x$$

By the repeated application of the sum formula, we see that the derivative of a polynomial is simply the sum of the derivatives of its terms.

EXAMPLE 47 Find the derivative of $f(x) = 7x^4 - 5x^3 + 3x^2 - 2$.

SOLUTION

$$f'(x) = \frac{d}{dx}(7x^4) + \frac{d}{dx}(-5x^3) + \frac{d}{dx}(3x^2) + \frac{d}{dx}(-2) \qquad \text{Derivative of a sum}$$

$$= 7\frac{d}{dx}(x^4) - 5\frac{d}{dx}(x^3) + 3\frac{d}{dx}(x^2) + 0 \qquad \begin{array}{l}\text{Derivative of a constant}\\\text{times a function and}\\\text{derivative of a constant}\end{array}$$

$$= 7(4x^3) - 5(3x^2) + 3(2x) \qquad \text{Power rule}$$

$$= 28x^3 - 15x^2 + 6x$$

EXAMPLE 48 Find the derivative of $f(x) = 5x^3 - \dfrac{2}{3}x^{-2} + x^{1/2} - 3$.

SOLUTION

$$f'(x) = 5 \cdot 3x^{(3-1)} + \left(\frac{-2}{3}\right) \cdot (-2)x^{(-2-1)} + \frac{1}{2}x^{(1/2)-1} + 0$$

$$= 15x^2 + \frac{4}{3}x^{-3} + \frac{1}{2}x^{-1/2}$$

$$= 15x^2 + \frac{4}{3x^3} + \frac{1}{2\sqrt{x}}$$

EXAMPLE 49

(a)   Find an equation of the line tangent to $y = 4x^2 + 2x$ at the point $(1, 6)$.
(b)   Where is $y' = 0$?

SOLUTION

(a)   The slope m is the derivative of the function evaluated at $(1, 6)$:

$$f'(x) = 8x + 2$$
$$f'(1) = 8(1) + 2$$
$$= 10$$

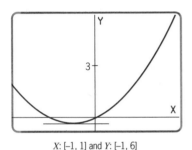

X: [–1, 1] and Y: [–1, 6]

Figure 36

The slope m is 10 at the point $(1, 6)$ and an equation of the line is

$$y - 6 = 10(x - 1) \quad \text{or} \quad y = 10x - 4$$

(b) $y' = 0$ wherever $8x + 2 = 0$, that is, where $x = -\frac{1}{4}$ (see Figure 36).

Practice Problem 3 Find an equation of the line that is tangent to the graph of $f(x) = 25 - 3x^2$ at the point $(2, 13)$.

ANSWER   Since $f'(x) = -6x$, $f'(2) = -12$. The tangent line has a slope of -12 and contains the point $(2, 13)$. The tangent line is $y - 13 = -12(x - 2)$ or $y = -12x + 37$.

EXAMPLE 50   If $f(x) = (x + 1)(x + 2)$, find $f'(x)$. (**Hint:** Multiply the two factors.)

SOLUTION $f(x) = x^2 + 3x + 2; f'(x) = 2x + 3.$

EXAMPLE 51   If $f(x) = \dfrac{3x^2 + 3x - 8}{x}$, find $f'(x)$. (**Hint:** Divide by x and write as three terms.)

SOLUTION $f(x) = 3x + 3 - 8x^{-1}$; therefore, $f'(x) = 3 + 0 - 8(-1)x^{-2} = 3 + 8x^{-2}.$

The symbols x and y can represent any variables that have a functional relationship, and $f'(x)$ or dy/dx means the rate of change of y with respect to x. Other letters, such as $h = s(t)$, could be used to represent the same idea. In this case the derivative would be $s'(t)$ or dh/dt and would mean the rate of change of h with respect to t.

Practice Problem 4 The total population of a city at time t is given by $p(t) = 1000 + \frac{1}{30}t^2$. What is the rate of change of the population with respect to time at $t = 45$?

ANSWER $p'(t) = 0 + \dfrac{1}{30} \cdot 2t = \dfrac{t}{15}$ and $p'(45) = 3$ persons per year.

Marginal Analysis

Recall from the preceding section that *marginal* refers to rate of change, that is, to a derivative. (The term *marginal* comes from the Marginalist School of Economic Thought, which originated in Austria with mathematics being applied to economics.) We can think of **marginal cost** as the approximate cost of producing one more item after x items have already been produced. Management must be careful to keep track of marginal costs. If the marginal cost of producing an extra unit exceeds the revenue received from the sale, the company will lose money on that additional unit.

EXAMPLE 52 Suppose that the total cost to produce x hundred cases (24 large bottles per case) of a cola under given conditions is given by

$$C(x) = 600 + 1000x - 2x^2$$

(a) Find the cost of producing 500 cases.
(b) Find an expression for the marginal cost.
(c) Find the marginal cost when $x = 5$.
(d) When 500 cases have been produced, what is the approximate cost of producing 100 more cases?

SOLUTION

(a) The cost of producing 500 cases of cola is $C(5)$ or

$$C(5) = 600 + 1000(5) - 2(5)^2 = \$5550$$

(b) The marginal cost is the derivative with respect to x of the cost function. That is, marginal cost is

$$C'(x) = 1000 - 4x$$

(c)   The marginal cost when $x = 5$ is

$$C'(5) = 1000 - 4(5) \quad \text{or} \quad 1000 - 20 = \$980$$

(d)   If 500 cases is equivalent to $x = 5$, then 100 more cases is equivalent to a change in x of 1. Since the marginal cost is $980, the approximate cost of one more x when $x = 5$ is $980, or the cost of 100 additional cases when the number of cases is 500 is $980.   ■

Calculator Note

Again, we need to take time to emphasize the usefulness of the calculator for problems such as Example 52. The manager of the cola plant has her cost function, which she places in her graphing calculator as $y = 600 + 1000x - 2x^2$. Under the CALC menu she can obtain the cost for any production and the marginal cost for any level of production as she moves the cursor along the graph. Also, she can use TABLE with $Y_1 = 600 + 1000X - 2X^2$ and $Y_2 = 1000 - 4X$ to obtain a table similar to the following.

Production (number of cases of cola)	5	25	50	100
Approximate Cost	$5550	$24,350	$45,600	$80,600
Approximate Marginal Cost	$980	$900	$800	$600

Then, if the manager wishes, she can change parts of the cost function and note changes in production costs and marginal costs.

In the preceding example, if we classify the terms of $C(x)$ as

$$C(x) = \underbrace{600}_{\text{Fixed cost}} + \underbrace{100x - 2x^2}_{\text{Variable cost}}$$

then only the part associated with variable cost is found in the derivative (the derivative of 600 is 0). This is in agreement with the economic principle which states that fixed costs of a company have no effect on marginal costs. Generally, marginal cost decreases as more units are produced until it reaches some minimum value, and then it starts to increase. Often this increase is due to the need to add more machinery, more labor, or overtime pay.

It is often appropriate in business to consider average costs and average profits. To take the average of several numerical values, add the values and then divide the total by the number of values. For example, the average of 50, 47, 53, 62, and 58 is computed as

$$\frac{50 + 47 + 53 + 62 + 58}{5} = \frac{270}{5} = 54$$

It is often enlightening to compute and compare the average cost per unit, the average revenue per unit, and the average profit per unit with the marginal functions. These averages per unit are defined as follows.

DEFINITION: AVERAGE COST, REVENUE, PROFIT

If x is the number of units produced, then

$$\text{Average cost} = \overline{C}(x) = \frac{C(x)}{x} \qquad \text{Cost per unit}$$

$$\text{Average revenue} = \overline{R}(x) = \frac{R(x)}{x} \qquad \text{Revenue per unit}$$

$$\text{Average profit} = \overline{P}(x) = \frac{P(x)}{x} \qquad \text{Profit per unit}$$

EXAMPLE 53   For $C(x) = 500 + 20x^2$, find the average cost and the marginal cost and sketch these on the same coordinate system.

SOLUTION   We have

$$\overline{C}(x) = \frac{C(x)}{x} = \frac{500}{x} + 20x \quad \text{and} \quad C'(x) = 40x$$

In Figure 37, $C'(x)$ and $\overline{C}(x)$ appear to intersect at the point where $\overline{C}(x)$ is a minimum. ∎

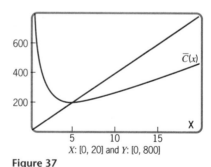

Figure 37

X: [0, 20] and Y: [0, 800]

NOTE: In economic theory, it is well known that at the level of production where the average cost is at a minimum, the average cost equals the marginal cost. We will be able to verify this in the next chapter.

Practice Problem 5 Use a graphing calculator for $C(x) = x^3 - 4x^2 + 8x$ to show that the marginal cost function and the average cost function intersect at the minimum value of the average cost function. (Use $0 \le x \le 8$ and $0 \le y \le 10$.)

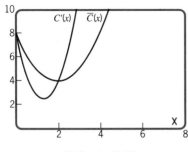

X: [0, 8] and Y: [0, 10]

Figure 38

ANSWER   We have $C'(x) = 3x^2 - 8x + 8$ and $\overline{C}(x) = x^2 - 4x + 8$. As shown in Figure 38, the graph for the marginal cost function appears to intersect the graph for the average cost function at the point $(2, 4)$, which is the point at which there is a minimum value of the average cost function.

SUMMARY

In this section we introduced the following concepts involving derivatives.

1. The derivative of a constant is zero.

$$\frac{d}{dx}C = 0$$

2. The derivative of x^n, where n is any real number, is nx^{n-1}.

$$\frac{d}{dx}x^n = nx^{n-1}$$

3. The derivative of a sum (or difference) is the sum (or difference) of the derivatives.

$$\frac{d}{dx}[f(x) \pm g(x)] = f'(x) \pm g'(x) \quad \text{if } f'(x) \text{ and } g'(x) \text{ exist}$$

4. The derivative of a constant times a function of x is the constant times the derivative of the function.

$$\text{If } f = cg(x), \quad \text{then} \quad f'(x) = cg'(x)$$

Exercise Set 8.5

Find each of the following.

1. $f'(x)$ for $f(x) = 6$

2. $f'(x)$ for $f(x) = 6x + 2$

3. $\dfrac{dy}{dx}$ for $y = -x + 3$

4. $\dfrac{dy}{dx}$ for $y = \sqrt{5}$

5. $D_x[3x^2 + 5]$ (D_x is d/dx.)

6. $D_x[5x^2 - 3x]$

7. y' for $y = 5x^3 - 3x^2 + 7$

8. y' for $y = -3x^3 + 2x^2 - x$

9. $\dfrac{d}{dx}(2x^{-4})$

10. $\dfrac{d}{dx}\left(\dfrac{4}{x^2}\right)$

11. $f'(x)$ for $f(x) = \dfrac{2}{\sqrt{x}}$

12. $f'(x)$ for $f(x) = \dfrac{2}{\sqrt[3]{x}}$

Find the derivative $f'(x)$, and then find, if they exist, $f'(0)$, $f'(2)$, and $f'(-3)$ for the following functions. Check the values for $f'(-3)$ using a graphing calculator.

13. $f(x) = 6x^4 - 3x$

14. $f(x) = 5x^4 - 3x^3 + 2x^2$

15. $f(x) = \dfrac{3}{x} + 4x$

16. $f(x) = -3x^2 - \dfrac{4}{x}$

17. $f(x) = 4x^2 - \dfrac{4}{x}$

18. $f(x) = \dfrac{-3}{x} + 4x$

19. $f(x) = 4x^{1/2}$

20. $f(x) = 6x^{2/3}$

21. $f(x) = 6x^{2/3} + 4x^{1/2}$

22. $f(x) = 9x^{-2/3} - \dfrac{4}{\sqrt{x}}$

23. $f(x) = 3x^2 - \dfrac{4}{\sqrt{x}}$ 24. $f(x) = 3x^4 - \dfrac{6}{\sqrt[3]{x}}$

25. $f(x) = 7\sqrt{x} + 9\sqrt[3]{x}$ 26. $f(x) = \dfrac{6}{\sqrt{x}} + \dfrac{3}{\sqrt[3]{x}}$

For the following exercises, find the points on the graph of each where the tangent line is horizontal (i.e., $y' = 0$).

27. $y = 3x^2$ 28. $y = 5x^2$

29. $y = 4x^2 - x$

30. $y = 0.01x^2 + 0.4x + 30$

For Exercises 31–34, find the equation of the tangent line to the graph at $x = 1$.

31. $y = 3x^2 + 2$ 32. $y = \dfrac{8}{x} + 3$

33. $y = -0.02x^2 + 0.2x$ 34. $y = \dfrac{1}{3}x^3 + 4$

35. ***Exam.*** Reading Company manufactures and sells an industrial-strength cleaning fluid. The equations presented below represent the revenue (R) and cost (C) functions for the company, where x is equal to a thousand gallons of fluid.

$R(x) = -80 + 26x - 0.05x^2$ and

$C(x) = 40 + 8x + 0.01x^2$

Which of the following functions represents the marginal cost of 1 gallon of cleaning fluid?
(a) $8x + 0.01x^2$ (b) $8 + 0.01x$
(c) $8 + 0.02x$ (d) $\dfrac{8}{0.02x}$

(e) Some function other than those given above.

36. ***Exam.*** For $C(x)$ and $R(x)$ given in Exercise 35, what is the marginal profit?
(a) $R(x) - C(x)$ (b) $18 - 1.2x$
(c) $\dfrac{R'(x)}{C'(x)}$ (d) $18x - 1.2x^2$

(e) Some function other than those given above.

Applications (Business and Economics)

37. ***Cost Functions.*** A cost function is given as

$C(x) = 800 + 400x - x^2$

(a) Find the marginal cost function, $C'(x)$.
(b) Find $C'(1)$.   (c) Find $C'(2)$.
(d) Find $\bar{C}(2)$.
(e) Check parts (b), (c), and (d) with a graphing calculator.

38. ***Demand Functions.*** A demand function is given as

$$p(x) = 138 - x^2, \qquad 1 \le x \le 11$$

(a) Find the instantaneous rate of change, $p'(x)$.
(b) Find $p'(2)$.
(c) Find $p'(5)$.
(d) Check parts (b) and (c) with a graphing calculator.

39. ***Revenue Functions.*** A revenue function is given as

$$R(x) = 20x - \dfrac{x^2}{500}, \qquad 0 \le x \le 10{,}000$$

(a) Find the marginal revenue function.
(b) Find the value of x that makes the marginal revenue equal to zero.
(c) Find the value of the revenue at the value of x found in part (b).
(d) Find $\bar{R}(x)$.

40. ***Profit Functions.*** A profit curve is given by

$$P(x) = \dfrac{x^2}{2} + 4x$$

(a) Find the derivative.
(b) What values of x make the derivative positive?
(c) What value of x makes the derivative zero?
(d) What values of x make the derivative negative?
(e) Find $\bar{P}(x)$.

41. ***Cost, Revenue, Profit.*** Cost and revenue functions are given as

$C(x) = 6x + 100$ and $R(x) = 40x - 0.01x^2$

(a) Find the marginal cost at $x = 4$. At $x = 10$. Check with a graphing calculator.
(b) What can you say about marginal cost?
(c) Find the marginal revenue at $x = 5$. Check with a graphing calculator.

(d) Find the marginal profit at $x = 4$. Check with a graphing calculator.

(e) What is the profit when $x = 3$?

(f) Find $\overline{P}(x)$.

Applications (Social and Life Sciences)

42. **Pulse Rates.** A doctor has found the relation between the pulse rate y and a person's height x, in inches, to be approximately

$$y = \frac{600}{\sqrt{x}}, \qquad 34 \le x \le 74$$

Find the rate of change of the pulse rate at the following heights.

(a) 36 inches   (b) 49 inches

(c) 64 inches

43. **Rate of Pollution Change.**   Suppose that the concentration of a pollutant in the air is given in parts per million by

$$P(x) = 0.2x^{-2}$$

where x is the distance in miles from a factory.

(a) Find the instantaneous rate of change of pollutant.

(b) Find $P'(1)$ and check.

(c) Find $P'(2)$ and check.

44. **Learning Functions.**   Suppose that a person learns $f(x)$ instructions in x hours according to the function

$$f(x) = 48\sqrt{x}, \qquad 0 \le x \le 10$$

(a) Find the instantaneous rate of change of learning.

(b) Find $f'(1)$.

(c) Find $f'(4)$.

(d) Find $f'(9)$.

(e) Check parts (b), (c), and (d) using a graphing calculator.

45. **Learning Functions.** Suppose that the learning function in Exercise 44 is changed to

$$f(x) = 72\sqrt[3]{x}, \qquad 0 \le x \le 10$$

(a) Find the instantaneous rate of change of learning.

(b) Find $f'(1)$.

(c) Find the instantaneous rate of change of learning at the end of 8 hours; that is, find $f'(8)$.

(d) Find the number of items learned at the end of 8 hours, $f(8)$.

(e) Check parts (b), (c), and (d) with a graphing calculator.

Review Exercises

Find the following limits if they exist.

46. $\displaystyle \lim_{x \to 1/\sqrt{3}} \frac{6x^3 + 4}{3x^2 - 1}$ 47. $\displaystyle \lim_{x \to 2} \frac{x^2 + x + 4}{x - 2}$

Use the definition of average change for $h = 0.1$ to find the average change in y when $x = 2$ in the following functions.

48. $y = 6x^2 + 5$ 49. $y = 4x^2 + 3x + 2$

Use the definition of a derivative,

$$\lim_{h \to 0} \frac{f(x + h) - f(x)}{h}$$

to find the derivatives of the following functions.

50. $y = 6x^2 + 5$ 51. $y = 4x^2 + 3x + 2$

8.6   PRODUCTS AND QUOTIENTS

OVERVIEW   Since the derivative of the sum of two functions is the sum of their derivatives, if these derivatives exist, we might expect the derivative of the product of two functions to be the product of their derivatives. In this section we show that this statement is not true. We develop formulas for finding the derivative of the product of two functions and the quotients of two functions. When these formulas are combined with formulas already obtained, we can find the derivatives of many new functions. These two formulas significantly increase our ability to compute and apply derivatives. In this section we develop and use

- The product formula
- The quotient formula

There are situations where a function can be considered the product of two other functions.

EXAMPLE 54   For $y = (x^2 + x - 1)(3x + 2)$, find dy/dx.

SOLUTION   One way we can solve this problem is to multiply the two functions and then take the derivative term by term.

$$y = (x^2 + x - 1)(3x + 2)$$
$$= 3x^3 + 5x^2 - x - 2$$

Then we have

$$y' = 9x^2 + 10x - 1$$

Now let's consider this example without multiplying the two factors to get a polynomial. Think of the problem as a product of two functions,

$$y = f(x) \cdot g(x)$$

where $f(x) = (x^2 + x - 1)$ and $g(x) = 3x + 2$. We use the following formula for problems of this nature.

PRODUCT FORMULA

Let $f(x)$ and $g(x)$ be two functions whose derivatives with respect to x exist. Then

$$\frac{d}{dx} [f(x)g(x)] = f(x)g'(x) + g(x)f'(x)$$

Sometimes students use a memory scheme for the product formula. Let's write the formula as *First factor times derivative of the Second plus the Second factor times the derivative of the First,* abbreviated as *FS′ + SF′.*

EXAMPLE 55   For

$$y = \overbrace{(x^2 + x - 1)}^{F}\overbrace{(3x + 2)}^{S}$$

find dy/dx.

SOLUTION

$$\frac{dy}{dx} = \overbrace{(x^2 + x - 1)}^{F \;\cdot} \overbrace{(3)}^{S' \;+} + \overbrace{(3x + 2)}^{S \;\cdot} \overbrace{(2x + 1)}^{F'}$$

$$= 3x^2 + 3x - 3 + 6x^2 + 7x + 2$$
$$= 9x^2 + 10x - 1$$

Note that this is the same answer that we obtained in Example 54.

EXAMPLE 56   For $f(x) = x^2(x + 1)$, find $f'(x)$.

SOLUTION

$$f(x) = \overbrace{x^2}^{F \;\cdot} \overbrace{(x + 1)}^{S}$$

$$f'(x) = \overbrace{x^2(1)}^{F \cdot S' \;+} + \overbrace{(x + 1)(2x)}^{S \cdot F'}$$
$$= 3x^2 + 2x$$

Calculator Note

A formal proof of the product formula is given later in this section. For now, we support the product rule graphically. In Figure 39(a) is shown the graph of the numerical values of the derivative of $f(x) = x^2 (x + 1)$. (See page 519.) Compare this graph with the graph in Figure 39(b) of $y' = 3x^2 + 2x$, obtained using the product formula in the preceding example. Note that the graphs in Figure 39 appear to be identical.

Practice Problem 1 For $y = (3x^2 + 5x)(x^2)$, find y'.

ANSWER $y' = 12x^3 + 15x^2$

EXAMPLE 57   For $y = \dfrac{2x + 1}{x^2}$, find y'.

SOLUTION

$$y = (2x + 1) \cdot \frac{1}{x^2} \qquad \text{Rewrite as a product.}$$

$$= \overbrace{(2x + 1) \cdot x^{-2}}^{F \cdot S}$$

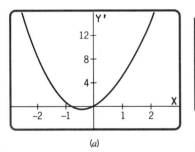

(a)

$y' = 3x^2 + 2x$

$X: [-3, 3]$ and $Y: [-4, 16]$

(b)

Figure 39

$$\overbrace{y' = (2x + 1)(-2x^{-3}) + x^{-2}(2)}^{F \cdot S' + S \cdot F'}$$

$$= \frac{-2(2x + 1)}{x^3} + \frac{2}{x^2}$$

$$= \frac{-4x - 2 + 2x}{x^3}$$

$$= \frac{-2x - 2}{x^3}$$

For practice, let's find this derivative another way. Write the equation as

$$y = \frac{2x}{x^2} + \frac{1}{x^2} = 2x^{-1} + x^{-2}$$

This results in

$$y' = -2x^{-2} - 2x^{-3} = \frac{-2}{x^2} - \frac{2}{x^3} = \frac{-2x - 2}{x^3}$$

which is the same answer.

Practice Problem 2 If $h(x) = (\sqrt[3]{x} - 3x^2)(9x^{-1})$, find $h'(x)$.

ANSWER $h'(x) = \dfrac{-6\sqrt[3]{x} - 27x^2}{x^2}$

Calculator Note

You can catch careless mistakes by doing a quick check with a calculator. Let's evaluate $h'(x)$ from Practice Problem 2 at $x = 1$.

$$h'(1) = \frac{-6 - 27}{1} = -33$$

Using nDeriv, $h'(1) \approx -33$. Although evaluation at a single point is not proof that an accurate derivative has been obtained, such a check can help you locate mistakes and should increase your confidence level.

We now give a formal proof of the product formula using our four-step procedure for finding a derivative (from the definition of a derivative). Let $y(x) = f(x) \cdot g(x)$. Then

1. $\dfrac{y(x + h) - y(x)}{h} = \dfrac{f(x + h) \cdot g(x + h) - f(x)g(x)}{h}$

2. $y' = \lim\limits_{h \to 0} \dfrac{f(x + h) \cdot g(x + h) - f(x)g(x)}{h}$

Now add and subtract $f(x + h)g(x)$ as needed to get

3. $\displaystyle\lim_{h\to 0} \frac{[f(x+h)g(x+h) - f(x+h)g(x)] + [f(x+h)g(x) - f(x)g(x)]}{h}$

$\displaystyle = \lim_{h\to 0} \frac{f(x+h)g(x+h) - f(x+h)g(x)}{h} + \lim_{h\to 0} \frac{f(x+h)g(x) - f(x)g(x)}{h}$

$\displaystyle = \lim_{h\to 0} f(x+h) \cdot \lim_{h\to 0} \frac{g(x+h) - g(x)}{h} + \lim_{h\to 0} g(x) \cdot \lim_{h\to 0} \frac{f(x+h) - f(x)}{h}$

4. $\quad = f(x)g'(x) + g(x)f'(x)$

Quotient Formula

Let's now consider situations where the function is the quotient of two functions, $f(x)$ and $g(x)$.

EXAMPLE 58 Find the derivative of $y = \dfrac{x^3 + 4}{x^2}$.

SOLUTION Without a formula for the derivative of a quotient, we write the function as two fractions.

$$y = \frac{x^3 + 4}{x^2} = \frac{x^3}{x^2} + \frac{4}{x^2} = x + 4x^{-2}$$

Therefore,

$$y' = 1 - 8x^{-3} = 1 - \frac{8}{x^3} = \frac{x^3 - 8}{x^3}$$ ∎

Many quotients cannot be written as a sum or difference of two fractions, so we need a formula for the derivative of a quotient of two functions. You might be tempted to think that the derivative of a quotient, $f(x)/g(x)$, is the quotient of the derivatives $f'(x)$ and $g'(x)$. We can see that this is incorrect from the fact that $x^2 = x^5/x^3$, but

$$\frac{d}{dx} x^2 \neq \frac{\dfrac{d}{dx}[x^5]}{\dfrac{d}{dx}[x^3]}$$

We know that they are unequal because

$$\frac{d}{dx} x^2 = 2x \quad \text{and} \quad \frac{\dfrac{d}{dx} x^5}{\dfrac{d}{dx} x^3} = \frac{5x^4}{3x^2} = \frac{5}{3} x^2$$

The correct formula for the derivative of a quotient is as follows.

QUOTIENT FORMULA

Let $f(x)$ and $g(x)$ be two differentiable functions with respect to x. If $g(x) \neq 0$, then

$$\frac{d}{dx}\left[\frac{f(x)}{g(x)}\right] = \frac{g(x)f'(x) - f(x)g'(x)}{[g(x)]^2}$$

In other words, the quotient formula states that the derivative of a quotient is **the denominator times the derivative of the numerator minus the numerator times the derivative of the denominator, all divided by the denominator squared:**

$$\frac{DN' - ND'}{D^2}$$

EXAMPLE 59   Using the quotient formula, find the derivative of

$$y = \frac{x^3 + 4}{x^2}$$

SOLUTION

$$y = \frac{\overbrace{x^3 + 4}^{N}}{\underbrace{x^2}_{D}}$$

$$y' = \frac{x^2 \cdot (3x^2) - (x^3 + 4) \cdot 2x}{(x^2)^2} \qquad \frac{DN' - ND'}{D^2}$$

$$= \frac{3x^4 - 2x^4 - 8x}{x^4}$$

$$= \frac{x^4 - 8x}{x^4}$$

$$= \frac{x^3 - 8}{x^3} \qquad \text{Divide by } x.$$

Notice that this is the same answer that we found in Example 58.   ■

EXAMPLE 60   If $y = \dfrac{3x - 2}{5x + 3}$, find $\dfrac{dy}{dx}$.

SOLUTION

$$y = \dfrac{\overbrace{3x - 2}^{N}}{\underbrace{5x + 3}_{D}}$$

$$\begin{aligned}
\dfrac{dy}{dx} &= \dfrac{(5x + 3)(3) - (3x - 2)(5)}{(5x + 3)^2} \qquad \dfrac{DN' - ND'}{D^2} \\[2mm]
&= \dfrac{15x + 9 - 15x + 10}{(5x + 3)^2} \\[2mm]
&= \dfrac{19}{(5x + 3)^2}
\end{aligned}$$

■

The formal proof of the quotient formula is very similar to that of the product formula and will not be given at this time. However, we give support graphically to the quotient formula as follows.

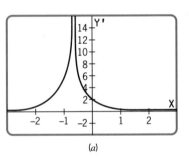

(a)

$$f'(x) = \dfrac{19}{(5x + 3)^2}$$

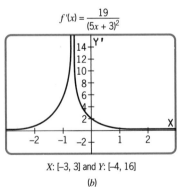

X: [-3, 3] and Y: [-4, 16]

(b)

Figure 40

Calculator Note

For the function $f(x) = \dfrac{3x - 2}{5x + 3}$, we obtain in Figure 40(a) the graph of the numerical values of $f'(x)$. (See page 519.) Then using $f'(x) = \dfrac{19}{(5x + 3)^2}$ from the preceding example we obtain the graph shown in Figure 40(b). These two graphs appear to be identical.

Practice Problem 3   Find $y' = \dfrac{6x - 1}{3x + 2}$.

ANSWER $y' = \dfrac{15}{(3x + 2)^2}$

EXAMPLE 61   Find the derivative of $y = \dfrac{2x + 1}{\sqrt{x}}$.

SOLUTION

$$y = \dfrac{\overbrace{2x + 1}^{N}}{\underbrace{\sqrt{x}}_{D}}$$

$$y' = \frac{\sqrt{x}(2) - (2x+1)\left(\dfrac{1}{2\sqrt{x}}\right)}{(\sqrt{x})^2} \qquad \frac{DN' - ND'}{D^2}$$

$$= \frac{2\sqrt{x} - \dfrac{x}{\sqrt{x}} - \dfrac{1}{2\sqrt{x}}}{x} \qquad \text{Multiply and simplify.}$$

$$= \frac{4x - 2x - 1}{2x\sqrt{x}} \qquad \text{Multiply numerator and denominator by } 2\sqrt{x}.$$

$$= \frac{2x - 1}{2x\sqrt{x}} \qquad \blacksquare$$

Calculator Check: $y'(1) = \dfrac{2 \cdot 1 - 1}{2 \cdot 1\sqrt{1}} = \dfrac{1}{2}$ and nDeriv$((2x + 1)/(\sqrt{x}), x, 1)$

$$= .4999998126.$$

Practice Problem 4 Find the derivative of $y = \dfrac{x^2}{2x + 1}$.

ANSWER $y' = \dfrac{2x^2 + 2x}{(2x + 1)^2}$

Whenever the numerator of a quotient is a constant, the quotient rule may not be the rule that you should use. Although you get the correct answer, often the problem can be worked more easily by rewriting the function and using the power rule. This is demonstrated in the next example.

EXAMPLE 62 Find the derivative of $y = \dfrac{2}{x^2}$.

SOLUTION Using the quotient rule we have

$$y' = \frac{x^2 \cdot \dfrac{d}{dx}(2) - 2 \cdot \dfrac{d}{dx}(x^2)}{(x^2)^2} = \frac{x^2 \cdot 0 - 4x}{x^4} = \frac{-4}{x^3}$$

If we rewrite the function as $y = 2/x^2 = 2x^{-2}$, we can use the power rule and arrive at the derivative much more quickly.

$$y' = -2(2)x^{-3} = \frac{-4}{x^3}$$

$\blacksquare$

EXAMPLE 63   Find the equation of the tangent line to the graph of $y = x/(x + 1)$ when $x = 1$.

SOLUTION   When $x = 1$,

$$y = \frac{1}{1 + 1} = \frac{1}{2}$$

so the equation of the tangent line will be $y - \frac{1}{2} = m(x - 1)$, where $m = y'(1)$.

$$y'(x) = \frac{(x + 1)(1) - x \cdot 1}{(x + 1)^2} = \frac{1}{(x + 1)^2} \qquad \frac{DN' - ND'}{D^2}$$

Evaluating $y'(1) = \frac{1}{4}$, we now have the slope and can find an equation of the tangent line.

$$y - \frac{1}{2} = \frac{1}{4}(x - 1) \quad \text{or} \quad y = \frac{1}{4}x + \frac{1}{4} \qquad \blacksquare$$

EXAMPLE 64   If $y = \dfrac{3x^2 - 1}{2x^2 + 2}$, find $\dfrac{dy}{dx}$ and check your answer at $x = 1$.

SOLUTION

$$\frac{dy}{dx} = \frac{(2x^2 + 2)(6x) - (3x^2 - 1)(4x)}{(2x^2 + 2)^2} \qquad \frac{DN' - ND'}{D^2}$$

$$= \frac{12x^3 + 12x - 12x^3 + 4x}{(2x^2 + 2)^2}$$

$$= \frac{16x}{(2x^2 + 2)^2}$$

✔ Calculator Check: $\dfrac{dy}{dx}$ at $x = 1$ is $\dfrac{16 \cdot 1}{(2 \cdot 1 + 2)^2} = \dfrac{16}{16} = 1.$

nDeriv$((3x^2 - 1)/(2x^2 + 2), x, 1) = 1.$ $\blacksquare$

Now that we have a quotient formula, we can introduce a new concept in our discussion of marginal analysis. The average cost per item can be found by dividing the total cost by the number of items, $C(x)/x$, and the derivative of this expression is the **marginal average cost**.

DEFINITION: MARGINAL AVERAGE COST

The **marginal average cost** is the derivative of the average cost.

$$\text{Marginal average cost} = \frac{d}{dx}\left[\frac{C(x)}{x}\right] = \frac{d}{dx}[\overline{C}(x)] = \overline{C}'(x)$$

Industry is interested in making the average cost as small as possible. In the next chapter we will learn how to accomplish this by finding the minimum average cost.

EXAMPLE 65

The total cost to manufacture x automobile engines in a week is

$$C(x) = 1000 + 500x - x^3$$

(a) Find the average cost per engine function.
(b) Find the marginal average cost function.

SOLUTION

(a) $\overline{C}(x) = \dfrac{C(x)}{x} = \dfrac{1000 + 500x - x^3}{x}$. By dividing through by x we have

$$\overline{C}(x) = 1000x^{-1} + 500 - x^2$$

(b) The marginal average cost function is

$$\frac{d}{dx}[\overline{C}(x)] = -1000x^{-2} - 2x = \frac{-1000}{x^2} - 2x$$

SUMMARY

1. *Product formula:* The derivative of a product is the first factor times the derivative of the second factor plus the second factor times the derivative of the first factor. If $f'(x)$ and $g'(x)$ exist, then

$$\frac{d}{dx}[f(x)g(x)] = f(x)g'(x) + g(x)f'(x)$$

2. *Quotient formula:* The derivative of a quotient is the denominator times the derivative of the numerator minus the numerator times the derivative of the denominator, all divided by the denominator squared. That is,

$$\frac{d}{dx}\left[\frac{f(x)}{g(x)}\right] = \frac{g(x)f'(x) - f(x)g'(x)}{[g(x)]^2}$$

Exercise Set 8.6

Find dy/dx by the product formula; then check your answer by multiplying the two factors and taking the derivative term by term.

1. $y = (3x + 2)(x - 3)$ 2. $y = (2x - 3)(x + 2)$
3. $y = (x^2 - 1)(x^2 + 1)$
4. $y = (3x^2 - 2)(3x^2 + 2)$

5. $y = (x^2 + x + 1)(x - 1)$
6. $y = (x^2 - x + 1)(x + 1)$

Find the derivatives of the following functions by using the quotient formula.

7. $y = \dfrac{5x}{2x - 3}$ 8. $y = \dfrac{3x - 2}{5x}$

9. $y = \dfrac{3x + 2}{2x - 3}$

10. $y = \dfrac{3x + 4}{5x - 1}$

Find the following and check each answer by finding the value of the derivative at $x = 1$ on a graphing calculator.

11. $\dfrac{d}{dx}[(3x^2 - 2x + 1)(4x + 5)]$

12. $\dfrac{d}{dx}[(4x^2 - 3x + 2)(3x - 2)]$

13. $D_x\left[\dfrac{5}{x^2 + 1}\right]$

14. $D_x\left[\dfrac{7}{2x^2 - 1}\right]$

15. $\dfrac{d}{dx}\left(\dfrac{3x - 2}{x^2 - 3}\right)$

16. $\dfrac{d}{dx}\left(\dfrac{2x + 3}{x^2 - 2}\right)$

17. Find y' for $y = \dfrac{3\sqrt{x}}{x^2 - 4}$.

18. Find y' for $y = \dfrac{4x^{1/2}}{\sqrt{x} - 3}$.

Find dy/dx for the following functions and check with a graphing calculator.

19. $y = 5(x - 2)\left(\dfrac{4x}{x + 1}\right)$

20. $y = (3x + 2)(2x + 7)x^{-1}$

21. $y = (3x - 2)^{-1}(x^3 + 2)$

22. $y = (2x - 3)^{-1}(3x^2 + 2)$

23. $y = 3(2x + 5)^{-1} + \sqrt{x}\,(3x - 5)$

24. $y = \dfrac{2x^2 - 3}{3x + 2} + \dfrac{3}{x}$

25. $y = \dfrac{2x^2 - 1}{x^3 + 1} + \dfrac{1}{\sqrt{x}}$

26. $y = \sqrt{x}\,(x^2 + x)$

27. Find an equation of the tangent line to the graph of $y = 5/(x^2 - 2x)$ at $(-2, \frac{5}{8})$.

Applications (Business and Economics)

28. **Average Cost.** A cost function is given by $C(x) = 7x^2 - 3x + 10$.
   (a) Find the average cost, $\overline{C}(x) = C(x)/x$.
   (b) Find the marginal cost, $C'(x)$.

   (c) Find the marginal average cost, $\overline{C}'(x)$.
   (d) Find $\overline{C}'(5)$.

29. **Demand Function.** A demand function is given by

$$p(x) = \dfrac{4000(25 - x)}{x^2}, \qquad 1 \le x \le 25$$

   (a) Find the marginal demand function.
   (b) Find the marginal demand when $x = 2$. Check with your graphing calculator.
   (c) Find the marginal average demand.

30. **Average Cost.** A cost function is given by $C(x) = 500x^2 + 2$.
   (a) Find the marginal cost function.
   (b) Find the marginal average cost function.

31. **Average Revenue.** A revenue function is given by $R(x) = 1000(x^2 - 3)$.
   (a) Find the marginal revenue function.
   (b) Find the marginal average revenue function.

32. **Average Profit**
   (a) Using Exercises 30 and 31, find the average profit function.
   (b) Find the marginal profit function.
   (c) Find the marginal average profit function.

Applications (Social and Life Sciences)

33. **Learning Functions.** A learning function giving the proportion of a set of activities performed without a mistake after t hours is given as

$$L = \dfrac{80t}{90t + 85}$$

   (a) Find the instantaneous rate of learning.
   (b) Find the rate of learning at $t = 1$ and check with a graphing calculator.
   (c) Find the rate of learning at $t = 8$ and check.

34. **Rate of Change of Drug Concentration.** The concentration of a drug in a person's bloodstream t hours after injection is

$$c(t) = \dfrac{t}{16t^2 + 10t + 63}$$

   (a) Find the rate of change of the concentration.
   (b) What is the concentration after 1 hour?
   (c) What is the rate of change of the concentration after 1 hour? Check with a graphing calculator.

Review Exercises

Find dy/dx for the following functions.

35. $y = 7x^2 - 3x + 2$ 36. $y = 10x^2 - 5x + 3$
37. $y = 3x^2 - x^3 + 2x^4$ 38. $y = 5 - 3x + 6x^2$
39. $y = 5x^3 - 2x^{-1}$ 40. $y = 4x^3 - 6x^{-2}$
41. $y = 6x^4 - 5x^{-2} + 3$ 42. $y = 7x^{-3} + 2x - 3$

Use the definition of the average change to find the average change in y for the following functions when x = 2 and h = 0.1.

43. $y = 3x^2$ 44. $y = x^2 + 4$

Use the definition of the derivative,

$$\lim_{h \to 0} \frac{f(x + h) - f(x)}{h}$$

to find the derivatives of the following functions.

45. $y = 3x^2$ 46. $y = x^2 + 4$

Find f'(x) for the following functions.

47. $f(x) = 2x^{1/2} + 3x^2$ 48. $f(x) = 3x^{1/3} - 5x^3$
49. $f(x) = 2x^{1/2} - 4x^{-1/2}$ 50. $f(x) = 3x^{1/3} - 3x^{-1/3}$
51. $f(x) = 4x^{-1/2} - 6x^{-2/3}$
52. $f(x) = 9x^{-2/3} + 8x^{-3/4}$

8.7 THE CHAIN RULE

OVERVIEW   In many practical situations, the quantity under consideration is described as a function of one variable, which in turn is a function of a second variable. In this section we develop a formula for the derivative of the original quantity with respect to the second variable. The formula we develop for such derivatives is called the **chain rule**. The chain rule is important because it describes how the derivative of a composition of a function is obtained. For example, a company may know

$$\frac{dC}{dP} = \text{Rate of change of cost with respect to production}$$

$$\frac{dP}{dt} = \text{Rate of change of production with respect to time}$$

Since cost also varies with respect to time, we would like to know dC/dt. In this section we learn how to find dC/dt in terms of dC/dP and dP/dt.

A useful way of combining functions $f(x)$ and $g(x)$ is to let the y-coordinates of $g(x)$ be used as the x-coordinates of $f(x)$. This results in a new function $f(g(x))$, a function of a function, which is called the **composite** of $f(x)$ and $g(x)$. In this section we

- Define the composite of two functions
- Introduce the chain rule
- Develop the general power rule

We begin this section by considering functions of functions.

EXAMPLE 66   If $f(x) = x^2 - 7$ and $g(x) = 1/3x$, then what would we mean by $f(g(x))$? Since $f(x)$ is the function that squares its x values and then subtracts 7, then f would square $g(x)$ and subtract 7:

$$f(g(x)) = [g(x)]^2 - 7$$

or

$$f(g(x)) = \left(\frac{1}{3x}\right)^2 - 7 \qquad \text{After substituting for } g(x)$$

$$= \frac{1}{9x^2} - 7$$

Let's call the new function

$$h(x) = \frac{1}{9x^2} - 7$$

The function h is said to be a **composite** of the two functions f and g.

DEFINITION: COMPOSITE FUNCTION

A function h is a **composite** of functions f and g if

$$h(x) = f[g(x)]$$

The domain of h is the set of all numbers x such that x is in the domain of g and $g(x)$ is in the domain of f.

Similarly, to decide what is meant by $g(f(x))$, we note that $g(x)$ is the function that triples its x values and then takes the reciprocal. Therefore, g would triple $f(x)$ and then take the reciprocal:

$$g(f(x)) = \frac{1}{3 \cdot f(x)}$$

or

$$g(f(x)) = \frac{1}{3 \cdot (x^2 - 7)} \qquad \text{After substituting for } f(x)$$

$$= \frac{1}{3x^2 - 21}$$

We observe that $g(f(x)) \neq f(g(x))$, so the order of the composition makes a difference.

EXAMPLE 67   Find $f(g(x))$ if $f(x) = 5x^2 + 3x$ and $g(x) = x - \sqrt{x}$. Each x in the equation for $f(x)$ may be replaced by $g(x)$ to obtain

$$f(g(x)) = 5 \cdot (g(x))^2 + 3 \cdot (g(x))$$
$$= 5 \cdot (x - \sqrt{x})^2 + 3 \cdot (x - \sqrt{x})$$

Similarly,

$$g(f(x)) = f(x) - \sqrt{f(x)}$$
$$= (5x^2 + 3x) - \sqrt{5x^2 + 3x}$$

■

EXAMPLE 68   If $f(x) = x^2 + 4$ and $g(x) = 3x + 6$, find $f[g(x)]$.

SOLUTION   Replace each x in $f(x)$ by $g(x)$ to get

$$f[g(x)] = [g(x)]^2 + 4 = (3x + 6)^2 + 4 = 9x^2 + 36x + 40$$

■

Practice Problem 1 If $f(x) = \sqrt{x + 5}$ and $g(x) = x^2$, find $f[g(x)]$. Also find $g[f(x)]$.

ANSWER $f[g(x)] = \sqrt{x^2 + 5}$ and $g[f(x)] = (\sqrt{x + 5})^2 = x + 5$ for $x \geq -5$

Sometimes it is easier to think of a composite function in terms of two variables, u and x. Let's write the problem in Example 68 as $f(u) = u^2 + 4$ and $u = 3x + 6$. Then $f(3x + 6)$ becomes $(3x + 6)^2 + 4$, which is the same answer that we found for $f[g(x)]$.

EXAMPLE 69   Given $y(u) = u^2$ and $u(x) = x^2 + 1$, find $y[u(x)]$.

SOLUTION $y[u(x)] = (x^2 + 1)^2 = x^4 + 2x^2 + 1$

■

EXAMPLE 70   Let $u(x) = x^2 + 1$ and $y(u) = u^2$. Then $y[u(x)]$ is a composite function, and it is simple enough so that we can find its derivative with methods we already know.

$$y[u(x)] = (x^2 + 1)^2$$
$$= x^4 + 2x^2 + 1$$
$$\frac{d}{dx}[y(u(x))] = 4x^3 + 4x$$

This can be separated into components as follows:

$$\frac{d}{dx}[y(u(x))] = 2(x^2 + 1) \cdot (2x) \qquad \text{Factor } 4x^3 + 4x.$$

$$= 2u \cdot 2x$$

$$= \frac{d}{du}[u^2] \cdot \frac{d}{dx}[x^2 + 1]$$

$$= \frac{dy}{du} \cdot \frac{du}{dx}$$

■

In the preceding example we can see that the derivative of a composite is equal to the product of the derivatives of its components. This is an example of the general rule for the derivative of a composite function, called the **chain rule**. We state it without proof.

THE CHAIN RULE

If y is a function of u and u is a function of x, whose range is in the domain of y, and if dy/du and du/dx exist, then y is a function of x and

$$\frac{dy}{dx} = \frac{dy}{du} \cdot \frac{du}{dx}$$

NOTE: These derivatives are *not* considered as fractions where du in the numerator and denominator divide out.

Alternative notations often used for the chain rule are

$$\frac{d}{dx} y(u(x)) = \frac{dy}{du} \cdot \frac{du}{dx} \quad \text{and} \quad [y(u)]'(x) = y'(u(x)) \cdot u'(x)$$

EXAMPLE 71   If $y = 5u^2 + 2u - 1$ and $u = 3x + 2$, then

$$\frac{dy}{du} = 10u + 2 \quad \text{and} \quad \frac{du}{dx} = 3$$

So, by the chain rule, $\dfrac{dy}{dx} = \dfrac{dy}{du} \cdot \dfrac{du}{dx} = (10u + 2)(3)$

Therefore, $dy/dx = 30u + 6$. Notice that when $3x + 2$ is substituted for u we obtain

$$\frac{dy}{dx} = 30(3x + 2) + 6 = 90x + 66$$

If the function y is found in terms of x in the preceding example, we obtain

$$\begin{aligned}
y(x) &= 5(3x + 2)^2 + 2(3x + 2) - 1 \\
&= 5(9x^2 + 12x + 4) + 6x + 4 - 1 \\
&= 45x^2 + 60x + 20 + 6x + 3 \\
&= 45x^2 + 66x + 23
\end{aligned}$$

Now dy/dx is $\dfrac{dy}{dx} = 90x + 66$

which agrees with the derivative found by using the chain rule.

EXAMPLE 72   If $P = s^{3/4}$ and $s = 2t^3 - t + 1$, find dP/dt.

SOLUTION

$$\frac{dP}{dt} = \frac{dP}{ds} \cdot \frac{ds}{dt}$$

$$= \frac{3}{4} s^{-1/4} (6t^2 - 1)$$

$$= \frac{3(6t^2 - 1)}{4s^{1/4}}$$

$$= \frac{3(6t^2 - 1)}{4(2t^3 - t + 1)^{1/4}} \qquad \text{Substitute } s = 2t^3 - t + 1.$$

$$= \frac{3(6t^2 - 1)}{4\sqrt[4]{2t^3 - t + 1}}$$

Practice Problem 2 If $y = u^2 + 4u - 3$ and $u = x + 6$, find dy/dx by the chain rule.

ANSWER $\dfrac{dy}{du} = 2u + 4$ and $\dfrac{du}{dx} = 1$; so

$$\frac{dy}{dx} = \frac{dy}{du} \cdot \frac{du}{dx} = (2u + 4) \cdot 1 = 2(x + 6) + 4 = 2x + 16$$

EXAMPLE 73   Find dy/dx if $y = (3x^2 + 2x - 5)^5$.

SOLUTION   This derivative could be found by expanding the trinomial and then differentiating each term. However, this method can be very time consuming and not practical for exponents greater than 3. Let's use the chain rule instead. We notice that $y = (3x^2 + 2x - 5)^5$ is in the form $y = u^5$, where $u = 3x^2 + 2x - 5$. Now

$$\frac{dy}{du} = 5u^4 \quad \text{and} \quad \frac{du}{dx} = 6x + 2$$

Substituting these into the chain rule form

$$\frac{dy}{dx} = \frac{dy}{du} \cdot \frac{du}{dx}$$

gives us
$$\frac{dy}{dx} = 5u^4 \cdot (6x + 2)$$

$$= 5(3x^2 + 2x - 5)^4 \cdot (6x + 2)$$

Suppose that $y(u) = u^n$ and u is a function of x. We have

$$\frac{dy}{du} = n \cdot u^{n-1} \quad \text{and} \quad \frac{du}{dx} = u'(x)$$

Then

$$\frac{dy}{du} \cdot \frac{du}{dx} = n \cdot u^{n-1} \cdot u'(x)$$

Therefore,

$$\frac{d}{dx}[y(u(x))] = n \cdot [u(x)]^{n-1} \cdot u'(x)$$

which is called the **generalized power rule**.

THE GENERALIZED POWER RULE

If u is a function whose derivative exists at x, n is any real number, and

$$y(u(x)) = [u(x)]^n$$

then

$$\frac{d}{dx}[y(u(x))] = n \cdot [u(x)]^{n-1} \cdot u'(x)$$

That is,

$$\frac{dy}{dx} = n[u(x)]^{n-1} \cdot \frac{du}{dx}$$

EXAMPLE 74   If $y = (3x + 2)^4$, find dy/dx.

SOLUTION   First, we recognize that y is of the form u^n, where $u = 3x + 2$. Then

$$\frac{dy}{dx} = nu^{n-1} \frac{du}{dx}$$

$$\frac{dy}{dx} = 4(3x + 2)^3 \cdot 3 \qquad u = 3x + 2; \frac{du}{dx} = 3$$

$$= 12(3x + 2)^3$$

EXAMPLE 75   Find dy/dx for $y = \sqrt{4x^2 + 3}$ and check the answer with a graphing calculator.

SOLUTION   Write $y = \sqrt{4x^2 + 3}$ as $y = (4x^2 + 3)^{1/2}$. Recognize the form $u^{1/2}$, where $u = 4x^2 + 3$. Then

$$\frac{dy}{dx} = nu^{n-1}\frac{du}{dx}$$

$$\frac{dy}{dx} = \frac{1}{2}(4x^2 + 3)^{-1/2}(8x) \qquad u = 4x^2 + 3;\; \frac{du}{dx} = 8x$$

$$= \frac{4x}{\sqrt{4x^2 + 3}}$$

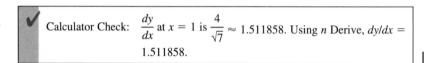

Calculator Check: $\dfrac{dy}{dx}$ at $x = 1$ is $\dfrac{4}{\sqrt{7}} \approx 1.511858$. Using n Derive, $dy/dx =$ 1.511858.

Now let's work some similar problems without the "in between" variable u.

EXAMPLE 76   If $y = (4x^3 - 3x^2 + x - 2)^3$, find dy/dx and check with a graphing calculator at $x = 1$.

SOLUTION   We use the following formula:

$$\text{If} \quad y = [g(x)]^n, \quad \text{then} \quad \frac{dy}{dx} = n[g(x)]^{n-1}\, g'(x)$$

Now, $g(x) = 4x^3 - 3x^2 + x - 2$, so $g'(x) = 12x^2 - 6x + 1$. Substituting in the formula we now have

$$\frac{d}{dx}[g(x)]^n = n[g(x)]^{n-1}\, g'(x)$$

$$\frac{dy}{dx} = 3(4x^3 - 3x^2 + x - 2)^2\, g'(x) \qquad n = 3; n - 1 = 2$$

$$= 3(4x^3 - 3x^2 + x - 2)^2(12x^2 - 6x + 1)$$

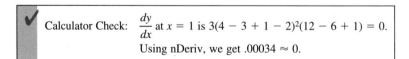

Calculator Check: $\dfrac{dy}{dx}$ at $x = 1$ is $3(4 - 3 + 1 - 2)^2(12 - 6 + 1) = 0$. Using nDeriv, we get $.00034 \approx 0$.

EXAMPLE 77   If $y = (3x^2 + 2)^{1/2}$, find dy/dx and check with a calculator at $x = 1$.

SOLUTION   We have $g(x) = 3x^2 + 2$ so $g'(x) = 6x$, and

$$\frac{dy}{dx} = \frac{1}{2}(3x^2 + 2)^{-1/2}(6x) = \frac{3x}{(3x^2 + 2)^{1/2}} \qquad n = \tfrac{1}{2}$$

> **Calculator Check**   The derivative dy/dx at $x = 1$ is $3/\sqrt{5} \approx 1.34164$, which is the same answer that you get using nDeriv on your calculator.

EXAMPLE 78   If $y = x^2\sqrt{3x^2 + 2}$, find dy/dx and check at $x = 1$ with a graphing calculator.

SOLUTION   First, you must recognize this as a product so that you can use the product formula. If $y = f(x)g(x)$, then $y' = f(x)g'(x) + f'(x)g(x)$. This gives

$$y' = x^2 \frac{d}{dx}[\sqrt{3x^2 + 2}] + 2x\sqrt{3x^2 + 2} \qquad f(x) = x^2; f'(x) = 2x$$

From Example 77, we know that

$$\frac{d}{dx}\sqrt{3x^2 + 2} = \frac{3x}{(3x^2 + 2)^{1/2}}$$

Substituting this value in the expression for y' gives

$$y' = x^2 \cdot \frac{3x}{\sqrt{3x^2 + 2}} + 2x\sqrt{3x^2 + 2}$$

$$= \frac{3x^3 + 2x(3x^2 + 2)}{\sqrt{3x^2 + 2}} \qquad \text{Using a common denominator}$$

$$= \frac{9x^3 + 4x}{\sqrt{3x^2 + 2}}$$

> **Calculator Check:**   At $x = 1$, $\dfrac{dy}{dx} = \dfrac{13}{\sqrt{5}} \approx 5.81378$
>
> Using nDeriv, it is ≈ 5.81378.

Practice Problem 3   If $y = (2x^3 + 4)^5$, find dy/dx.

ANSWER $\dfrac{dy}{dx} = 5(2x^3 + 4)^{5-1} \cdot \dfrac{d}{dx}(2x^3 + 4) = 5(2x^3 + 4)^4(6x^2)$

EXAMPLE 79   If $y = \dfrac{x}{(x^2 + 3)^2}$, find $\dfrac{dy}{dx}$.

SOLUTION   This is in the form of a quotient, so we start with the quotient formula.

$$\frac{dy}{dx} = \frac{(x^2 + 3)^2 \cdot \dfrac{d}{dx}(x) - x \cdot \dfrac{d}{dx}(x^2 + 3)^2}{[(x^2 + 3)^2]^2}$$

$$= \frac{(x^2 + 3)^2 \cdot (1) - x \cdot [2(x^2 + 3)^1 \cdot 2x]}{(x^2 + 3)^4}$$

$$= \frac{(x^2 + 3)^2 - 4x^2(x^2 + 3)}{(x^2 + 3)^4}$$

$$= \frac{(x^2 + 3)(x^2 + 3 - 4x^2)}{(x^2 + 3)^4}$$

$$= \frac{(3 - 3x^2)}{(x^2 + 3)^3}$$

Practice Problem 4   Differentiate $y = \dfrac{\sqrt{x^2 - 5}}{x + 1}$.

ANSWER $y' = \dfrac{x + 5}{(x + 1)^2 \cdot \sqrt{x^2 - 5}}$

COMMON ERROR   Many times students will differentiate like this:

$$y(x) = (3x^2 + 4x)^5, \quad \text{so} \quad y'(x) = 5(3x^2 + 4x)^4$$

What is wrong with this? The answer is $5(3x^2 + 4x)^4(6x + 4)$.

SUMMARY

1. The derivative of a function of x to the nth power is equal to n times the function to the $(n - 1)$st power times the derivative of the function of x with respect to x.

$$\text{If} \quad f(x) = [u(x)]^n, \quad \text{then} \quad f'(x) = n[u(x)]^{n-1} \cdot \frac{du}{dx}$$

2. If $f(x) = y[u(x)]$, then $f'(x) = \dfrac{dy}{du} \cdot \dfrac{du}{dx}$.

Exercise Set 8.7

In each case find f[g(x)] and g[f(x)].

1. $f(x) = x^2 + 3$, $g(x) = x - 1$
2. $f(x) = x^2 - 7$, $g(x) = x + 5$
3. $f(x) = \sqrt{x - 7}$, $g(x) = x^3$
4. $f(x) = x^4$, $g(x) = \dfrac{1}{x}$

Express y in terms of x only.

5. $y = u^2$, $u = x^2 + 1$
6. $y = \sqrt{u + 1}$, $u = x^2 - 1$
7. $y = \dfrac{1}{\sqrt{u + 3}}$, $u = x^4$
8. $y = \dfrac{1}{u^2}$, $u = \sqrt{x + 3}$

Write each of the following functions in the form y = f(u) and u = g(x).

9. $y = (3x + 4)^5$
10. $y = (x + 7)^4$
11. $y = \dfrac{1}{x + 8}$
12. $y = \sqrt[3]{x - 13}$

Use the chain rule to find dy/dx. To check your results, find y as a function of x and then find dy/dx.

13. $y = 3u$, $u = 3x + 5$
14. $y = 4u$, $u = 3x - 2$
15. $y = -4u$, $u = 3x^2$
16. $y = 2u + 4$, $u = 3x^2$
17. $y = 4u^2$, $u = 3x + 5$
18. $y = -5u^2$, $u = 3x - 2$

Find dy/dx using the chain rule for powers:

$$\frac{dy}{dx} = nu^{n-1}\frac{du}{dx}$$

Check your answers using a graphing calculator.

19. $y = (3x + 4)^2$
20. $y = (4x - 7)^5$
21. $y = (6 - 2x^2)^3$
22. $y = \sqrt{4 - x^2}$

Find each of the following and check with a graphing calculator.

23. $D_x\left[\dfrac{1}{2x + 4}\right]$
24. $D_x\sqrt{5x - 1}$
25. $\dfrac{d}{dx}(2x^3 - 1)^{1/3}$
26. $\dfrac{d}{dx}(1 - 8x^2)^{-1/4}$
27. y' for $y = (3x^2 - 2x + 1)^2$
28. y' for $y = (2x^2 - x)^4$
29. $\dfrac{dy}{dx}$ for $y = -5(2x + 1)^3 - 7$
30. $\dfrac{dy}{dx}$ for $y = 3 - 4(3x - 2)^3$
31. $D_x[(x - 1)^3\sqrt{x}]$
32. $D_x[(x + 5)^2\sqrt[3]{x}]$
33. $\dfrac{d}{dx}\left(\dfrac{1}{\sqrt{2x^2 - 1}}\right)$
34. $\dfrac{d}{dx}\left[\dfrac{2x + 1}{(2x^2 + 1)^3}\right]$

Find the derivative of each function.

35. $f(x) = (3x + 5)(2x - 3)^2$
36. $f(x) = (3x - 2)^2(2x + 3)$
37. $f(x) = 4(2x^2 + x - 1)^{1/2}$
38. $f(x) = 9(3x^2 - x + 1)^{1/3}$
39. $f(x) = 4x^{1/2}(3x - 1)^2$
40. $f(x) = 6x^{-1/2}(4x - 3)^2$
41. $f(x) = \dfrac{4x^2}{(2x - 3)^2}$
42. $f(x) = \dfrac{3x^2 - 2}{(2x + 3)^2}$
43. $f(x) = [(x + 3)\sqrt{x}]^4$
44. $f(x) = \left(\dfrac{x + 1}{x - 1}\right)^5$
45. $f(x) = \left(\dfrac{x - 3}{x + 2}\right)^2$
46. $f(x) = [(x - 7)\sqrt{x}]^3$

Applications (Business and Economics)

47. **Cost Functions.** A cost function is found to be $C(x) = (3x - 10)^2 + 24$.
    (a) Find the marginal cost function.
    (b) Find $C'(3)$.
    (c) Find $C'(4)$.
    (d) Find $\overline{C}'(x)$.

48. **Revenue Functions.**   A revenue function for the sale of stereo sets is

$$R(x) = 4608 - 372 \left(12 - \frac{x}{3} \right) - \left(12 - \frac{x}{3} \right)^2$$

where $0 \leq x \leq 36$.
   (a) Find the marginal revenue function.
   (b) Find $R'(9)$.
   (c) Find $R'(18)$.
   (d) Find $\overline{R}(x)$.

Applications (Social and Life Sciences)

49. **Learning Functions.**   For the learning function giving the number of actions learned after x hours of practice,

$$L(x) = 36(2x - 1), \qquad \tfrac{1}{2} \leq x \leq 15$$

find the rate of learning at the end of   (a) 1 hour and   (b) 14 hours.

50. **Pollution.**   The pollution from a factory is given by

$$P(x) = 0.4(2x + 3)^{-2}$$

where x is measured in miles. Find the rate of change of the pollution. What is the rate of change of pollution at   (a) 1 mile and   (b) 3 miles?

Review Exercises

Find the derivatives of the following functions.

51. $y = \dfrac{2x + 5}{3x - 2}$

52. $y = \dfrac{4x + 3}{2x - 3}$

53. $y = (4x^2 - 2)(3x + 2)$

54. $y = (5x^2 + 2)(3x + 5)$

55. $y = 3(2x + 5)^3$

56. $y = 4(3x - 2)^2$

Chapter Review

Review the following definitions and concepts to ensure that you understand and can use each of them.

Slope of a function	Derivative of a function
Limit of a function	Average rate of change
Tangent to a graph	Four-step procedure for derivatives
Continuity	Instantaneous rate of change
Marginal cost function	Marginal supply function
Marginal profit function	Average cost function
Marginal demand function	Marginal average cost function
Power rule	Quotient rule
Product rule	General power rule

Make sure you understand and know when and how to use the following formulas.

$$\frac{dy}{dx} = \lim_{h \to 0} \frac{f(x + h) - f(x)}{h}$$

$$\frac{d}{dx} x^n = nx^{n-1}$$

$$\text{Average change in } f(x) = \frac{f(x + h) - f(x)}{h}$$

$$\frac{d}{dx} C = 0$$

$$\frac{d}{dx} [C f(x)] = C \frac{d}{dx} [f(x)]$$

$$\frac{dy}{dx} = \frac{dy}{du} \cdot \frac{du}{dx}$$

$$\frac{d}{dx}[f(x) \pm g(x)] = \frac{d}{dx}[f(x)] \pm \frac{d}{dx}[g(x)]$$

$$\frac{d}{dx}[g(x)]^n = n[g(x)]^{n-1}\, g'(x)$$

$$\frac{d}{dx}[f(x)g(x)] = f(x)g'(x) + g(x)f'(x)$$

$$\frac{d}{dx}\left[\frac{f(x)}{g(x)}\right] = \frac{g(x)f'(x) - f(x)g'(x)}{[g(x)]^2}$$

Chapter Test

1. Define $f'(x)$ in terms of limits.
2. Using the preceding definition find the derivative of $f(x) = 3x^2$.
3. If $y = 2x^2 + 4x + 1$, find the average change in y as x changes from 2 to 3.
4. For $f(x) = 4x - 3x^2$, find $f'(0)$ and $f'(-1)$.
5. For $y = \sqrt{3}$, find y'.
6. Given $y = -4u^2 + 3$ and $u = x^2$, find dy/dx using the chain rule.
7. A cost function is given as $1000 + 20x + x^2$. Find the marginal average cost at $x = 2$.

8. Find

$$\lim_{x \to 0} \frac{3x - 2}{4x + 1}$$

9. Find

$$\lim_{x \to 1} \frac{x^2 - 3x + 3}{x - 1}$$

10. Find $f'(x)$ for $f(x) = (3/x) + 4\sqrt{x}$.
11. A profit function is given as

$$P(x) = \frac{x^2}{2} + 4x$$

Find where $P'(x) > 0$ and where $P'(x) < 0$.

12. Find y' for $y = 2(x^2 - 3)^{3/2}$. Check with a graphing calculator at $x = 4$.
13. Find y' for $y = (3x + 2)^{1/2}(2x - 3)$. Check with a graphing calculator at $x = 1$.

14. Find y' for

$$y = \frac{2x + 5}{\sqrt{x^2 - 1}}$$

Check with a graphing calculator at $x = 2$.

15. A demand function is given as

$$p(x) = \frac{6000(20 - x)}{x^2}, \qquad 1 \le x \le 20$$

(a) Find the marginal demand function.
(b) Find the marginal average demand function.

Applications of Differentiation and Additional Derivatives

In this chapter we consider applications that are important in the business world. A CEO desires to optimize profits, but many factors must be considered. What will be the result of increased production? Will the increase in revenue exceed the additional cost? To accomplish this goal of optimizing profits, one of the vice presidents is given the responsibility of minimizing inventory costs. In a like manner, a physician might be interested in maximizing the amount of medicine in the bloodstream after a given time, or an agricultural researcher might be interested in how long it will take a plant to reach its maximum height. In this chapter, we learn to find maximum and minimum values of functions. This study leads to the consideration of graphing functions. In the process of applying calculus concepts to the graphing of functions, we find a need for additional procedures in taking derivatives. Thus, we consider higher-order derivatives.

Throughout this chapter, and those that follow, we will be looking at revenue, cost, and profit functions. These functions are actually models that fit patterns determined from empirical data. In some cases we will state the domain. However, even if a domain is not stated, you should be aware that these functions will usually involve limited domains.

9.1   HOW DERIVATIVES ARE RELATED TO GRAPHS

OVERVIEW   As indicated previously, finding the greatest and least values attained by a function has extensive applications. A business owner wants to find the greatest value of a profit function or perhaps the least value attained by a cost function. The management of a restaurant needs to be able to minimize waste. To locate values that maximize or minimize, we apply principles of calculus to graphs. Recall that the slope of a curve at a point is the slope of the tangent line to the curve at the point. Also remember that this slope can be found by evaluating the derivative of the function at the point. By studying changes in the slope (changes in the sign of the derivative), we obtain interesting information that is relevant to finding maximum and minimum values of the function. In this section we find

- Intervals where a function is increasing
- Intervals where a function is decreasing
- Critical points

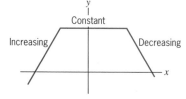

Figure 1

If the graph of a function rises from left to right on an interval [$f(x)$ increases], we say that the function is **increasing** on the interval. If the graph falls [$f(x)$ decreases] from left to right on an interval, we say that the function is **decreasing** on the interval. If the graph of a function is horizontal on an interval, we say that the function is **constant** on the interval. (See Figure 1.) We describe these terms mathematically as follows.

DEFINITION: INCREASING, DECREASING, CONSTANT

1.   A function is **increasing** on an open interval I if, for every a and b in I,

$$f(a) < f(b), \quad \text{when } a < b$$

2.   A function is **decreasing** on an open interval I if, for every a and b in I,

$$f(a) > f(b), \quad \text{when } a < b$$

3.   A function is **constant** on an open interval I if, for every a and b in I,

$$f(a) = f(b), \quad \text{when } a < b$$

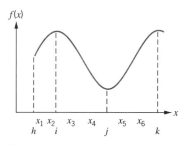

f(x)

Figure 2

EXAMPLE 1 Using the preceding definition, locate the intervals on an open interval I where $f(x)$ is increasing and the intervals where $f(x)$ is decreasing on the interval (h, k), as shown in Figure 2.

SOLUTION

The function is increasing on (h, i) because $f(x_1) < f(x_2)$ for all x values on this interval where $x_1 < x_2$.

The function is decreasing on (i, j) because $f(x_3) > f(x_4)$ for all x values on this interval where $x_3 < x_4$.

The function is increasing on (j, k) because $f(x_5) < f(x_6)$ for all x values on this interval where $x_5 < x_6$.

Practice Problem 1 On a graphing calculator, graph the function $f(x) = x^3 - 3x^2$ over the interval $(-1, 4)$. Determine the intervals where the function is increasing and the intervals where the function is decreasing.

ANSWER By using the Trace function (see page 3 of Appendix A), you can see in Figure 3 that the function increases on the intervals $(-1, 0)$ and $(2, 4)$ and decreases on the interval $(0, 2)$.

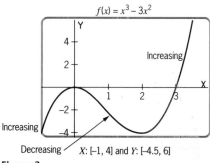

$f(x) = x^3 - 3x^2$

Increasing

Increasing

Decreasing   X: [–1, 4] and Y: [–4.5, 6]

Figure 3

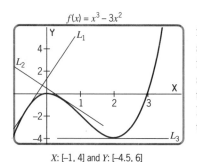

$f(x) = x^3 - 3x^2$

X: [–1, 4] and Y: [–4.5, 6]

Figure 4

Now let's examine the graphs of functions and the slopes of tangent lines to the functions we examine. Using the function $f(x) = x^3 - 3x^2$, let's see how the slopes of the tangent lines relate to the curve. Looking at Figure 4, we see that the first tangent line drawn (L_1) is rising from left to right and therefore has a positive slope. The second tangent line drawn (L_2) is going down from left to right and therefore has a negative slope. The third tangent line drawn (L_3) is a horizontal line and therefore has slope 0. What relationship do you see between the slopes of the tangent lines and whether the function is increasing or decreasing?

Calculator Note

1. You can draw a tangent line with your calculator. For example in Figure 4, to draw the tangent line at $x = -0.5$ use the DRAW DRAW menu as follows. From the home screen, press [2nd] [DRAW] (Select 5) and insert Tangent ($x^3 - 3x^2, -.5$) [ENTER]. Draw several tangent lines on the graph of $f(x) = x^3 - 3x^2$.

Do these lines confirm what we have just discussed? Are the tangent lines rising from left to right when f is increasing? If you were to draw a tangent line anywhere on the interval $(-\infty, 0)$ would it have a positive or negative slope? What about a tangent line on the interval $(0, 2)$?

2. With your calculator draw the graph of another polynomial, such as $y = 4x^4 - 5x^2 + x$ with X:$[-2, 2]$, Y:$[-3, 3]$ and then draw some tangent lines along the curve. What can you determine about the slope of the tangent lines as they relate to the function?

3. With your graphing calculator draw the graph of $y = 3x^3 - 2x^2 - 1$ and then $y' = 9x^2 - 4x$. Use the intervals X:$[-3, 3]$ and Y:$[-15, 15]$. Can you see the relationship between y and y' when y is increasing? What about when y is decreasing?

It would seem that when the derivative is positive over an interval (the slopes of the tangent lines are positive), the function is increasing on the interval. Likewise, when the derivative is negative over an interval (the slopes of the tangent lines are negative) the function is decreasing on that interval. When the derivative is 0, the function is not increasing or decreasing. We state this property as follows.

POSITIVE AND NEGATIVE SLOPES

(a) If $f'(x) > 0$ for all x in an open interval I, and if $y = f(x)$ is a differentiable function on I, then $y = f(x)$ is increasing for all x in the interval.

(b) If $f'(x) < 0$ for all x in an open interval I, and if $y = f(x)$ is a differentiable function on I, then $y = f(x)$ is decreasing for all x in the interval.

(c) If $f'(x) = 0$ for all x in an open interval I, and if $y = f(x)$ is a differentiable function on I, then $y = f(x)$ is constant for all x in the interval.

A function is increasing (the graph is rising) when its derivative is positive and decreasing (the graph is falling) when its derivative is negative. If the derivative of a function is defined for all points in an interval, to change from positive to negative or from negative to positive, $f'(x)$ must assume the value of 0 on the interval. (Think of a number line. What number divides the positive numbers from the negative numbers? Zero.) That is, if the graph of f changes from rising to falling or falling to rising, at this change the slope of the tangent line is 0 and the tangent line is horizontal, as we saw in Figure 4.

EXAMPLE 2   With your calculator, draw the graph of $f(x) = 5x^4 - 4x^5$ and $f'(x) = 20x^3 - 20x^4$. Determine the intervals where $f(x)$ is increasing and where it is decreasing. Compare these intervals to the intervals where $f'(x) > 0$ and where $f'(x) < 0$.

SOLUTION   The graphs shown in Figure 5 indicate that $f(x)$ is increasing and $f'(x) > 0$ on $(0, 1)$ and $f(x)$ is decreasing and $f'(x) < 0$ on $(-\infty, 0)$, $(1, \infty)$. ■

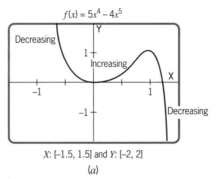

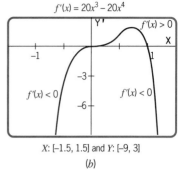

X: [–1.5, 1.5] and Y: [–2, 2]   X: [–1.5, 1.5] and Y: [–9, 3]

(a)   (b)

Figure 5

Calculator Note

Have your calculator draw tangent lines at $(0, 0)$ and $(1, 0)$ on the curve $y = 5x^4 - 4x^5$. What do you see? Can you see the first tangent line? Note that both are horizontal.

Tangent lines to the graph of $f(x)$ at the points $(0, 0)$ and $(1, 1)$ in Figure 5(a) are horizontal lines and have slope $= 0$. Note in Figure 5(b) that $f'(x) = 0$ at these two points. When a derivative equals 0 or the derivative is undefined, we classify the points as **critical points**.

DEFINITION: CRITICAL POINT

If c is in the domain of f and if either $f'(c) = 0$ or $f'(c)$ is undefined, then c is called a **critical value** of f. If c is a critical value, then the point $(c, f(c))$ is called a **critical point**.

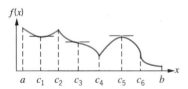

Figure 6

EXAMPLE 3   Find the critical values on (a, b) for the function shown in Figure 6.

SOLUTION   Critical values occur at $x = c_1, c_2, c_3, c_4, c_5, c_6$. At $x = c_1, x = c_3, x = c_5$, the graph shows horizontal tangents. $f'(x) = 0$ at each of these points. What about the other critical values? Note that even though at c_3 the tangent appears to be horizontal, the function remains decreasing. At c_2 and c_4 it is possible to draw more than one tangent line, and at c_6 there is a vertical tangent; therefore, $f'(x)$ is undefined at c_2, c_4, and c_6.   ■

Steps for Finding Critical Points

1. Determine the domain of the function.
2. Compute the derivative.
3. Factor the derivative, if possible, and determine the points at which the function is defined and the derivative is 0 or is undefined.

NOTE: Even though a critical point is found, it does not necessarily mean that the function changes from increasing to decreasing or from decreasing to increasing. For example, draw a calculator graph of the function $f(x) = x^{1/3}$ and note that the graph of the function reveals that f is always increasing. However, there is a critical value at $x = 0$, where the derivative $f'(x) = 1/(3x^{2/3})$ is undefined. Have your calculator draw the graph of $f'(x) = 1/(3x^{2/3})$ and confirm this.

EXAMPLE 4   Find all critical points for $y = 7x^2 - 14x + 3$.

SOLUTION   Critical points for a function occur where the derivative is 0 or does not exist but the value is within the domain of the function. To find the critical points, we follow the steps listed above. Note that the function is defined for all real values.

1.   We have

$$\frac{dy}{dx} = y' = 14x - 14 = 14(x - 1)$$

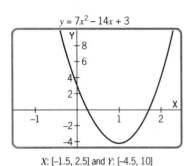

$y = 7x^2 - 14x + 3$

X: [-1.5, 2.5] and Y: [-4.5, 10]

(a)

(The derivative is defined for all real values of x.)

2.   Setting the derivative equal to 0 and solving for x we have

$$14(x - 1) = 0 \quad \text{and} \quad x = 1$$

3.   The only critical value occurs at $x = 1$. At $x = 1$ we have

$$y = 7(1)^2 - 14(1) + 3 = -4$$

The only critical point for this function is $(1, -4)$.   ■

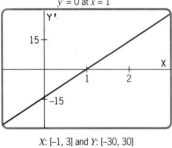

$y' = 14x - 14$
$y' = 0$ at $x = 1$

X: [-1, 3] and Y: [-30, 30]

(b)

Figure 7

Practice Problem 2 Use your calculator to draw the graph of y and y' from Example 4 and verify that $(1, -4)$ is the only critical point.

ANSWER   The graphs are shown in Figure 7. Note that y' changes signs only once, at $(1, 0)$, and y has only the one critical point.

Practice Problem 3 Find the critical points algebraically for the function $f(x) = 2x^3 - 9x^2 + 12x - 10$.

ANSWER $(1, -5)$ and $(2, -6)$

Practice Problem 4 With a calculator, draw the graph of the function $f(x) = 2x^3 - 9x^2 + 12x - 10$ from Practice Problem 3 and $f'(x) = 6x^2 - 18x + 12$ to validate graphically your answer in Practice Problem 3.

ANSWER   From the graphs in Figure 8, we can see that the critical points are at $(1, -5)$ and $(2, -6)$ and that $f'(x)$ changes signs at $x = 1$ and $x = 2$. This confirms the algebraic work done in Practice Problem 3.

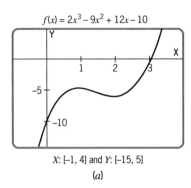

$f(x) = 2x^3 - 9x^2 + 12x - 10$

X: [-1, 4] and Y: [-15, 5]

(a)

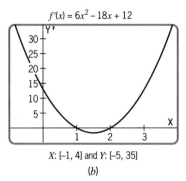

$f'(x) = 6x^2 - 18x + 12$

X: [-1, 4] and Y: [-5, 35]

(b)

Figure 8

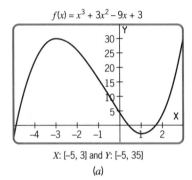

$f(x) = x^3 + 3x^2 - 9x + 3$

X: [-5, 3] and Y: [-5, 35]

(a)

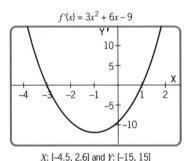

$f'(x) = 3x^2 + 6x - 9$

X: [-4.5, 2.6] and Y: [-15, 15]

(b)

Figure 9

In application problems it is important to know the exact value of the critical points. Therefore, we use algebraic techniques to find the critical points and we then validate our work with a graphical analysis of the function.

EXAMPLE 5   Determine the intervals where the function $f(x) = x^3 + 3x^2 - 9x + 3$ is increasing and where it is decreasing by finding the critical points algebraically and then verifying by using the graphs of f and f'.

SOLUTION   For $f(x) = x^3 + 3x^2 - 9x + 3$, we have $f'(x) = 3x^2 + 6x - 9$. We note here that both f and f' are defined for all real numbers, so all critical points will occur where $f'(x) = 0$. Thus, we examine the equation $f'(x) = 0$.

$$f'(x) = 3x^2 + 6x - 9 = 3(x + 3)(x - 1) = 0$$

From this we determine that $f'(x) = 0$ only when $x = -3$ or $x = 1$; so we have two critical values. Looking at the graphs in Figure 9, we see that f is increasing [and $f'(x) > 0$] over $(-\infty, -3)$, $(1, \infty)$ and f is decreasing [and $f' < 0$] over $(-3, 1)$. ∎

Even though the turning points of a function can be found by looking at the graph of a function or its derivative, these values are often decimal approximations. That is why we usually use both the graphing calculator and algebraic techniques of factoring and reduction to find our answers. It is helpful to support our algebraic work with graphing. Likewise, graphical work can be supported with algebraic work.

EXAMPLE 6   Determine the intervals where $f(x) = \frac{2}{3}x^{3/2} - 2x^{1/2}$ is increasing and where it is decreasing for $x \geq 0$. Use both algebraic techniques and the graphs of f and f'.

SOLUTION   The domain of the function is $[0, \infty)$. For $f(x) = \frac{2}{3}x^{3/2} - 2x^{1/2}$, we have

$$f'(x) = \frac{2}{3} \cdot \frac{3}{2} x^{1/2} - 2 \cdot \frac{1}{2} x^{-1/2}$$

$$= x^{1/2} - x^{-1/2}$$

$$= \sqrt{x} - \frac{1}{\sqrt{x}} \qquad\qquad \text{Convert to radical form.}$$

$$= \frac{x - 1}{\sqrt{x}} \qquad\qquad \text{Combine the terms.}$$

Notice that $f(0)$ is defined even though $f'(0)$ is undefined and if $x = 1$, $f'(x) = 0$. Therefore, our critical values are $x = 0$ (derivative is undefined) and $x = 1$ (derivative is zero). Looking at the graphs in Figure 10 we see that $f(x)$ is decreasing on $(0, 1)$ and is increasing on $(1, \infty)$. We also see that f' changes from negative to positive at $x = 1$.   ■

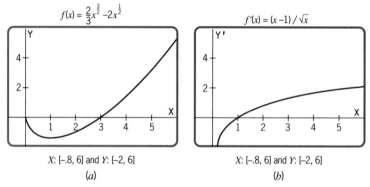

$$f(x) = \frac{2}{3}x^{\frac{3}{2}} - 2x^{\frac{1}{2}}$$

$$f'(x) = (x-1)/\sqrt{x}$$

X: [–.8, 6] and Y: [–2, 6]   X: [–.8, 6] and Y: [–2, 6]

(a)   (b)

Figure 10

For some problems in the exercise set, we ask you to locate intervals where a function is increasing and where it is decreasing. If you have critical values at c_1, c_2, and c_3 for a function with $c_1 < c_2 < c_3$ on a closed interval $[a, b]$, consider the closed interval as being divided into the following subintervals.

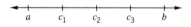

Then, for a continuous curve between any two of these consecutive points, the function is either decreasing or increasing. That is, find the sign of $f'(x)$ for some x on $a < x < c_1$. If the sign is positive, the function is increasing on (a, c_1). If the sign is negative, the function is decreasing on (a, c_1). This procedure can be generalized for any number of critical points on a closed interval. In the next section, we relate critical points to points associated with relative maxima and relative minima.

Practice Problem 5 For $y = 1/(x - 2)^2$ find the intervals where the function is increasing and where it is decreasing. Using your graphing calculator, graph both y and y' to confirm your answer.

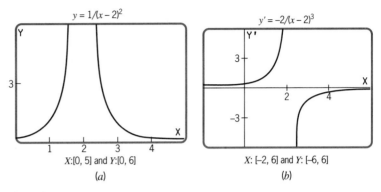

$y = 1/(x - 2)^2$

$X:[0, 5]$ and $Y:[0, 6]$

(a)

$y' = -2/(x - 2)^3$

$X: [-2, 6]$ and $Y: [-6, 6]$

(b)

Figure 11

ANSWER $y' = -2/(x - 2)^3$. From Figure 11 we have $y' < 0$ when $x > 2$ and $y' > 0$ when $x < 2$. Therefore, the function is decreasing on $(2, \infty)$ and increasing on $(-\infty, 2)$. Notice that y is not defined for $x = 2$, and therefore, $x = 2$ is not a critical value.

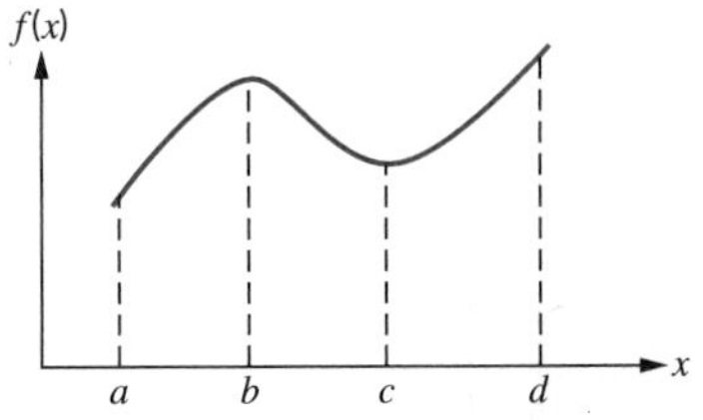

Practice Problem 6 For the function defined by the graph in Figure 12,

(a) find the intervals where the function is increasing over (a, d).
(b) find all the critical values over (a, d).

ANSWER

(a) The function is increasing on (a, b) and (c, d).
(b) Critical values occur at $x = b$ and $x = c$.

$f(x)$

a b c d x

Figure 12

EXAMPLE 7   For $f(x) = x^3(\sqrt{x} + 1)$, find any critical points and the intervals where f is increasing and where it is decreasing. Use f' and the calculator graphs of both f and f'.

SOLUTION   Note that the domain of the function is $[0, \infty)$, and the function will always be positive. We now find the derivative.

$$f(x) = x^3(\sqrt{x} + 1) = x^3(x^{1/2} + 1) = x^{7/2} + x^3 \qquad \text{Rewrite the function to find } f'.$$

$$f'(x) = \frac{7}{2} x^{5/2} + 3x^2 \qquad \text{Find the derivative.}$$

$$= x^2 \left(\frac{7}{2} x^{1/2} + 3 \right) \qquad \text{Factor the derivative.}$$

The only critical value is $x = 0$, since there is no real number that satisfies the equation $(\frac{7}{2}x^{1/2} + 3) = 0$. Therefore, the only critical point is $(0, f(0)) = (0, 0)$ and the function is increasing over $(0, \infty)$. The graphs are shown in Figure 13.

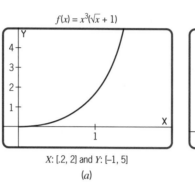

$f(x) = x^3(\sqrt{x} + 1)$

X: [.2, 2] and Y: [-1, 5]

(a)

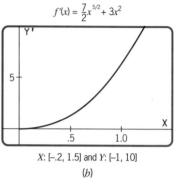

$f'(x) = \frac{7}{2}x^{5/2} + 3x^2$

X: [-.2, 1.5] and Y: [-1, 10]

(b)

Figure 13

COMMON ERROR　Students sometimes substitute a critical value in the first derivative, $(c, f'(c))$, to get a critical point. Why is this incorrect?

SUMMARY

When the derivative of a function is positive, the function is increasing. Likewise, when the derivative is negative, the function is decreasing. We can determine these intervals by locating the critical points. Remember, a critical value occurs at c if $f(c)$ is defined and if $f'(c) = 0$ or if $f'(c)$ is undefined. An analysis of the graph in conjunction with the knowledge of the values of the critical points enables you to find the intervals where a function is increasing or decreasing.

Exercise Set 9.1

Find all critical points algebraically for the following functions and determine the intervals where the function is increasing.

1. $y = 5$
2. $y = -3x$
3. $f(x) = 3x + 5$
4. $f(x) = -5x + 3$
5. $f(x) = x^2$
6. $f(x) = -3x^2$
7. $y = x^2 + 2x$
8. $y = -3x^2 + 2$
9. $y = 4x^2 + 3x - 2$
10. $y = -5x^2 - 3x + 8$
11. $y = 3x^2 + 12x - 5$
12. $y = 5x^2 - 4x + 3$
13. $y = 12x^{2/3} + x$
14. $y = 4x^{1/2} - 2x$

Determine the intervals where the following functions are increasing or decreasing. The functions continue in the directions indicated with no more critical points.

15.

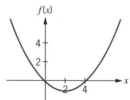

16.

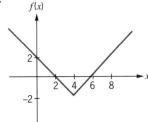

17.

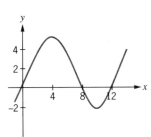

18.

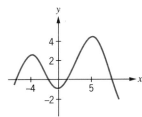

In Exercises 19–22, assume that the function is continuous. From the given information, sketch the graph of a function that meets the given requirements.

19. The derivative is negative over $(-\infty, 1)$ and positive over $(1, \infty)$. The function contains the points $(0, 3)$, $(1, 1)$, $(2, 2)$, and $(4, 6)$.

20. The derivative is negative over $(-\infty, -1)$ and positive over $(-1, \infty)$. The function contains the points $(-2, 2)$, $(-1, 1)$, $(0, 2)$, and $(1, 5)$.

21. The derivative is positive over $(-\infty, 3)$ and negative over $(3, \infty)$. The function contains the points $(1, 7)$, $(2, 10)$, $(3, 11)$, and $(4, 10)$.

22. The derivative is positive over $(-\infty, -2)$ and negative over $(-2, \infty)$. The function contains the points $(-4, 1)$, $(-2, 5)$, $(0, 1)$, and $(1, -4)$.

23. Refer to the function $f(x)$ represented by the following graph to answer each question.

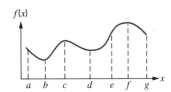

(a) Find all intervals where $f(x)$ is increasing.
(b) Find all intervals where $f(x)$ is decreasing.
(c) Find all the critical values on the interval (a, k).

24. Refer to the function $f(x)$ represented by the following graph to answer each question.

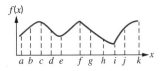

(a) Find all intervals where $f(x)$ is increasing.
(b) Find all intervals where $f(x)$ is decreasing.
(c) Find all the critical values on the interval (a, g).

Find all critical points, intervals where the functions are increasing, and intervals where the functions are decreasing. On a graphing calculator verify your answer from the graphs of the functions and their derivatives.

25. $y = 2x^2$

26. $y = 3x^2$

27. $y = -2x^2 + 3$

28. $y = -2x^2 + 4x$

29. $y = 3x^2 + 3x + 2$

30. $y = -3x^2 - 3x + 2$

31. $y = x^3$

32. $y = -x^3$

33. $y = x^3 + 3x$

34. $y = x^3 - 3x$

35. $y = x^3 + 6x^2$

36. $y = x^3 - 6x^2$

37. $y = 2x^3 - 6x^2$

38. $y = 2x^3 + 6x^2$

39. $y = x^3 - 6x^2 + 9x$

40. $y = x^3 + 6x^2 + 3x - 5$

41. $y = 3x^{1/3} + 8x$

42. $y = 3x^{5/3}$

43. $y = 2\sqrt{x}$

44. $y = \sqrt{x} - \frac{1}{2}x$

45. $f(x) = \dfrac{x^3 - 5x^2 + 3x + 9}{x - 3}$

46. $f(x) = (x + 2)^{3/2} + 4$

Applications (Business and Economics)

47. **Profit Function**.   Profit P is related to selling price S by the formula

$$P = 10{,}000S - 250S^2$$

For what range of prices is the profit increasing and for what range of prices is it decreasing?

48. **Demand Function**.   A price per unit $p(x)$ for a product is given as a function of the number of units x by

$$p(x) = 800x^{-2} + 400x^{-1}$$

for $x > 0$. Is this function increasing or decreasing?

49. **Cost of Production**.   For $x \geq 1$, find the production values of x for which the cost of production is decreasing and the values of x for which it is increasing given

$$C(x) = x^3 - 6x^2 + 9x + 150$$

50. **Average Cost**.   The cost of producing x units of a given product is given by

$$C(x) = 2000 + 10x + 0.01x^2$$

(a) Find the intervals where the cost function is increasing or decreasing.
(b) Find the intervals where the average cost function $C(x)/x$ is increasing or decreasing.

51. **Marginal Analysis**.   A manufacturer's profit function is given as $P(x) = R(x) - C(x)$, where $R(x)$ is the revenue function and $C(x)$ is the cost function. Suppose that the profit function has a critical value at a production level of $x = c$. Argue that the marginal cost and the marginal revenue at this production level are equal.

Applications (Social and Life Sciences)

52. **Drug Concentration**.   The concentration $c(t)$ of a drug in a patient's bloodstream t hours after a drug is taken is given by

$$c(t) = \frac{0.2t}{t^2 + 4t + 9}$$

For what interval of time will the concentration be increasing and for what interval will it be decreasing?

53. **Population**.   A population study has been made for 8 years in a community and the following population function P has been obtained:

$$P = 1000t^2 - 6000t + 29{,}000$$

where t is time measured in years from the beginning date of the study. When is the population increasing and when is it decreasing?

9.2 FINDING RELATIVE MAXIMA AND RELATIVE MINIMA

OVERVIEW   When the graph of a continuous function changes from rising to falling, a local high point occurs. In a like manner, when the graph changes from falling to rising, a local low point occurs. In this section we define or name these points and demonstrate how they facilitate curve tracing. The calculus techniques of this section can be applied to a wide variety of application problems (finding maximum velocity or time of peak efficiency of a worker). When we obtain the function that represents a mathematical model of some problem, we can then analyze the model relative to maximum and minimum values. In this section we study

- Relative maxima
- Relative minima
- The first derivative test

When a "high point" or "low point" occurs on a graph, we refer to that point as an *extremum*. In this chapter, we consider two kinds of *extrema:* **absolute extrema** and **relative extrema**. The greatest value of a function (if one exists) over a specified interval or the entire domain is called the **absolute maximum**. The least value of a function (if one exists) over a specified interval or the entire domain is called the **absolute minimum**. We will discuss these in a later section. In this section we turn our attention to the other kind of extrema, called relative extrema.

DEFINITION: RELATIVE MAXIMUM AND RELATIVE MINIMUM

1. $f(c)$ is a **relative maximum** if there exists an interval (a, b) with $(a < c < b)$ such that $f(x) \leq f(c)$ for all x in (a, b).
2. $f(c)$ is a **relative minimum** if there exists an interval (a, b) with $(a < c < b)$ such that $f(x) \geq f(c)$ for all x in (a, b).

This definition is not as complicated as it may seem. **A relative maximum is simply a high point on the graph where the function changes from *increasing* to *decreasing*. Similarly, a relative minimum is a low point on the graph where the function changes from *decreasing* to *increasing*.** There can be more than one relative maximum or relative minimum on a given interval or over the entire domain of the function.

EXAMPLE 8 Locate points of relative maxima and relative minima in Figure 14.

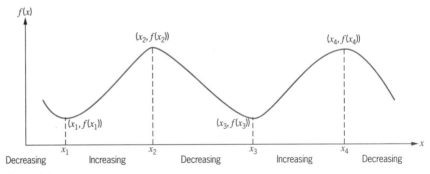

Figure 14

SOLUTION   By inspection we see that

$f(x_1)$ is a relative minimum. The function changes from decreasing to increasing at x_1.

$f(x_2)$ is a relative maximum. The function changes from increasing to decreasing at x_2.

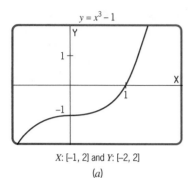

X: [–1, 2] and Y: [–2, 2]

(a)

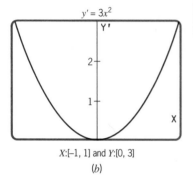

X:[–1, 1] and Y:[0, 3]

(b)

Figure 15

$f(x_3)$ is a relative minimum. The function changes from decreasing to increasing at x_3.

$f(x_4)$ is a relative maximum. The function changes from increasing to decreasing at x_4. ■

By examining the points where relative maxima and relative minima occur in Figure 14, we note that all points where relative maxima and relative minima occur are critical points. It is possible to prove the following theorem.

THEOREM

If $f(x)$ is continuous on the interval (a, b) and $f(c)$ is a relative maximum or relative minimum where $a < c < b$, then either $f'(c) = 0$ or $f'(c)$ does not exist. That is, c is a critical point.

The preceding theorem states that relative extrema can occur only at critical points. However, be careful, because the existence of a critical point does not always mean that there is an extremum. There are times when a critical point is present but there is no extremum at that point.

With your graphing calculator draw the graphs of $y = x^3 - 1$ and $y' = 3x^2$ shown in Figure 15.

We can determine that $(0, -1)$ is a critical point; however, there is neither a relative maximum nor a relative minimum at $(0, -1)$, since y' is always positive (except when it is 0 at $x = 0$), indicating that the curve is increasing on each side of $x = 0$. What is different about this critical point? (The function does not change from increasing to decreasing.) This difference leads to the following test for relative maxima and relative minima.

FIRST DERIVATIVE TEST FOR RELATIVE EXTREMA

At critical value c:

(a)   The function has a relative maximum at $(c, f(c))$ if the derivative is positive everywhere in an interval just to the left of c and negative in an interval just to the right of c.

(b)   The function has a relative minimum at $(c, f(c))$ if the derivative is negative everywhere in an interval just to the left of c and positive just to the right of c.

(c)   If the derivative does not change signs just to the right and just to the left of c, then $f(c)$ is not a relative extremum.

This is summarized in Figure 16. Notice that in the last two cases listed in the figure we see no extrema, but it does seem that some change occurs. This is called an **inflection point** [f(critical value) = 0, but there is no change of sign of the first derivative] and will be discussed in the next section. Let's now look at the steps we will follow in determining the exact value of a relative extremum. We

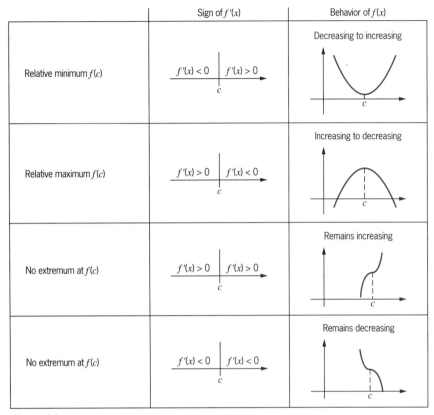

	Sign of $f'(x)$	Behavior of $f(x)$
Relative minimum $f(c)$	$f'(x) < 0 \quad \vert \quad f'(x) > 0$ c	Decreasing to increasing
Relative maximum $f(c)$	$f'(x) > 0 \quad \vert \quad f'(x) < 0$ c	Increasing to decreasing
No extremum at $f(c)$	$f'(x) > 0 \quad \vert \quad f'(x) > 0$ c	Remains increasing
No extremum at $f(c)$	$f'(x) < 0 \quad \vert \quad f'(x) < 0$ c	Remains decreasing

Figure 16

will use both calculus and the graph of the function. The use of graphing calculators along with the knowledge gained from the first derivative help to locate the extremum.

Steps for Determining Relative Extrema of a Function

1. Find the derivative.
2. Factor the derivative, if possible, and determine critical values.
3. Graph the function and its derivative. If the function changes from increasing to decreasing or from decreasing to increasing at a critical point, you have found an extremum. You should graph the derivative as well, to confirm your answer with the first derivative test.
4. Evaluate the function at the critical value to determine the point where the relative extremum is located.

$f(x) = 3x^2 - 6x$

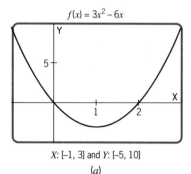

X: [–1, 3] and Y: [–5, 10]

(a)

$f'(x) = 6x - 6$

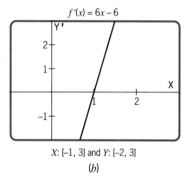

X: [–1, 3] and Y: [–2, 3]

(b)

Figure 17

EXAMPLE 9   Find any relative extrema for $f(x) = 3x^2 - 6x$ by following the steps listed above.

SOLUTION

1.   Find the derivative. The derivative of $f(x) = 3x^2 - 6x$ is $f'(x) = 6x - 6$.
2.   Factor the derivative. Factoring f', we have $f'(x) = 6(x - 1)$, and we can determine that the only critical value is $x = 1$, since $f'(1) = 0$.
3.   Graph the function and the derivative. The graphs of $f(x) = 3x^2 - 6x$ and $f'(x) = 6x - 6$ are shown in Figure 17.
4.   Evaluate the function at the critical values. We can determine from the graph that the function changes from decreasing to increasing and the derivative changes from negative to positive at $x = 1$. Since $f(1) = -3$, there is a relative minimum at the point $(1, -3)$.   ∎

EXAMPLE 10   Given the function $f(x) = 2x^3 + 3x^2 - 12x - 4$, find any relative extrema.

SOLUTION   We first determine that $f'(x) = 6x^2 + 6x - 12$ and then solve the equation $f'(x) = 0$. Since $f'(x) = 6x^2 + 6x - 12 = 6(x + 2)(x - 1)$, the critical values are $x = -2$ and $x = 1$. This is confirmed by the graph of f' shown with the graph of f in Figure 18. Looking at the graph, we determine that $f(-2) = 16$ is a relative maximum and $f(1) = -11$ is a relative minimum.   ∎

$f(x) = 2x^3 + 3x^2 - 12x - 4$

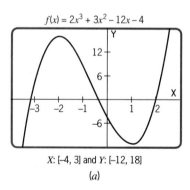

X: [–4, 3] and Y: [–12, 18]

(a)

$f'(x) = 6x^2 + 6x - 12$

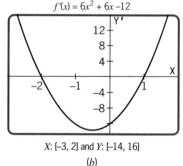

X: [–3, 2] and Y: [–14, 16]

(b)

Figure 18

Practice Problem 1   For the function $f(x) = 2x^3 - 3x^2 - 12x + 8$, locate all relative extrema.

ANSWER   We have $f'(x) = 6x^2 - 6x - 12 = 6(x + 1)(x - 2)$. A relative maximum occurs at $(-1, 15)$ and a relative minimum occurs at $(2, -12)$. This is verified by the graphs in Figure 19.

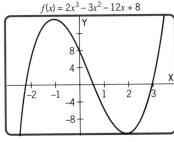

$f(x) = 2x^3 - 3x^2 - 12x + 8$

X: [−3, 4] and Y: [−12, 16]

(a)

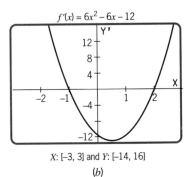

$f'(x) = 6x^2 - 6x - 12$

X: [−3, 3] and Y: [−14, 16]

(b)

Figure 19

Let's now consider the four cases where $f'(c)$ is not defined but $f(c)$ is defined. If $f'(x)$ changes signs, then a relative extremum occurs. If $f'(x)$ does not change signs, then there is a vertical tangent. This is summarized in Figure 20.

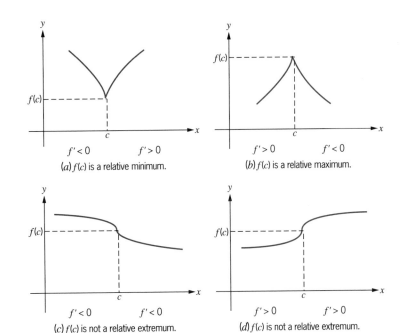

$f' < 0 \qquad f' > 0$

(a) $f(c)$ is a relative minimum.

$f' > 0 \qquad f' < 0$

(b) $f(c)$ is a relative maximum.

$f' < 0 \qquad f' < 0$

(c) $f(c)$ is not a relative extremum.

$f' > 0 \qquad f' > 0$

(d) $f(c)$ is not a relative extremum.

Figure 20

EXAMPLE 11   Draw the graph of $f(x) = 6(x - 1)^{2/3} + 4$ and determine any relative extrema.

SOLUTION   The function is defined for all real numbers, but the derivative,

$$f'(x) = 6 \cdot \tfrac{2}{3}(x - 1)^{-1/3} = \frac{4}{(x - 1)^{1/3}}$$

is not defined at $x = 1$. However, $x = 1$ is a critical value since $f(1)$ is defined. The graph of the function is shown in Figure 21. Be careful about how you enter the function in your calculator. Just using the exponent 2/3 may not work. You will most likely have to enter ⎡Y=⎤ 6 ⎡(⎤ ⎡(⎤ X ⎡−⎤ 1 ⎡)⎤ ⎡∧⎤ 2 ⎡)⎤ ⎡∧⎤ ⎡(⎤ 1 ⎡÷⎤ 3 ⎡)⎤ ⎡+⎤ 4. We can determine from Figure 21 that a relative minimum occurs at (1, 4).

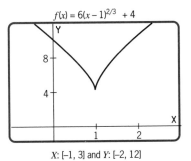

$f(x) = 6(x − 1)^{2/3} + 4$

X: [−1, 3] and Y: [−2, 12]

Figure 21

Practice Problem 2   Find the critical points of each function and draw the graph, then determine where any relative extrema occur.

(a) $f(x) = (x + 2)^{2/3} - 1$
(b) $f(x) = -x^{1/3}$
(c) $f(x) = -(2 - x)^{2/3}$
(d) $f(x) = (x - 2)^{1/3}$

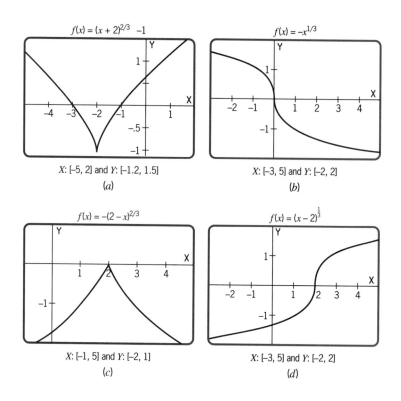

Figure 22

ANSWER

(a) Critical value $x = -2$. There is a relative minimum at $(-2, -1)$; see Figure 22(a).
(b) Critical value $x = 0$. There are no relative extrema; see Figure 22(b).
(c) Critical value $x = 2$. There is a relative maximum at $(2, 0)$; see Figure 22(c).
(d) Critical value $x = 2$. There are no relative extrema; see Figure 22(d).

Now let's turn our attention to a function that has a restricted domain.

EXAMPLE 12   For the function $f(x) = \sqrt{x}\,(x - 4)$, $x > 0$, locate any critical values and relative extrema. Use your graphing calculator to draw the graph and verify the answer.

SOLUTION   We first note that the function is only defined for $x \geq 0$. To find $f'(x)$ for this function, it will be easier to rewrite the function as $f(x) =$

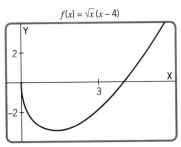

$f(x) = \sqrt{x}\,(x - 4)$

X: [–1.8, 6] and Y: [–4, 4]

Figure 23

$x^{1/2}(x - 4) = x^{3/2} - 4x^{1/2}$. From this we determine that

$$f'(x) = \frac{3}{2}x^{1/2} - 2x^{-1/2}$$

$$= \frac{1}{2}x^{-1/2}(3x - 4)$$

$$= \frac{3x - 4}{2x^{1/2}}$$

We can now determine the only critical value is $x = 4/3$ and from the graph in Figure 23 we see that there is a relative minimum at $(\frac{4}{3}, -3.08)$. ∎

Practice Problem 3 On your graphing calculator draw the graphs of each function and determine where any relative extrema occur.

(a) $f(x) = -\sqrt{x}(x - 2)$ (b) $f(x) = |x^2 - 2x - 3|$

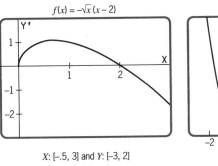

$f(x) = -\sqrt{x}\,(x - 2)$

X: [–.5, 3] and Y: [–3, 2]

(a)

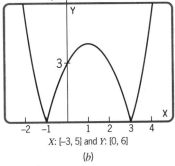

$f(x) = |x^2 - 2x - 3|$

X: [–3, 5] and Y: [0, 6]

(b)

Figure 24

ANSWER   The graphs are shown in Figure 24.

(a) For $x > 0$ there is a relative maximum at $(\frac{2}{3}, 1.09)$.

(b) There are relative minima at $(-1, 0)$ and $(3, 0)$ and there is a relative maximum at $(1, 4)$.

COMMON ERROR   Students sometimes classify $x = 2$ as a critical value of $f(x) = 1/(x - 2)$. Why is this incorrect? (It is incorrect because $x = 2$ is not in the domain of f.)

SUMMARY

In this section we added to our ability to analyze the graphs of functions by using the graphs generated by a graphing calculator and the first derivative test. Sometimes we obtained the graph of the derivative of a function as a means of applying the first derivative test. Remember, if a function is defined at c and either $f'(c) = 0$ or $f'(c)$ is undefined, then a possible extremum has been located.

Exercise Set 9.2

Locate all relative maxima and minima.

1.

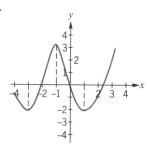

2.

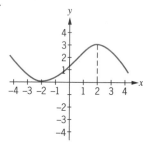

3. Classify $f(c)$ as either a relative maximum, a relative minimum, or neither. Assume the function is continuous over (a, b) and $a < c < b$.

	$f'(c)$	$f'(x)$ over (a, c)	$f'(x)$ over (c, b)
(a)	0	+	−
(b)	0	−	−
(c)	0	−	+
(d)	0	+	+
(e)	Not defined	−	+
(f)	Not defined	−	−
(g)	Not defined	+	−
(h)	Not defined	+	+

For Exercises 4–6, refer to the function $f(x)$ represented by the following graph. Answer the questions from your observations of the graph.

4. Where are there relative maxima on the interval (a, k)?

5. Where are there relative minima on the interval (a, k)?

6. Name any point(s) where the derivative is undefined on the interval (a, k).

Find the critical points, relative extrema, and intervals over which the function is increasing and decreasing. First find approximate answers with your graphing calculator and then validate or find exact answers with calculus procedures.

7. $y = 2x^2$

8. $y = 3x^2$

9. $f(x) = -2x^2 + 3$

10. $f(x) = -2x^2 + 4x$

11. $f(x) = 3x^2 + 3x + 2$

12. $f(x) = -3x^2 - 3x + 2$

13. $f(x) = x^3$

14. $f(x) = -x^3$

15. $f(x) = x^3 + 4x$

16. $f(x) = x^3 + 4x^2$

17. $f(x) = x^3 - 6x^2$

18. $f(x) = 2x^2 + 4x^4$

19. $f(x) = x(1 - x)^{1/2}$

20. $f(x) = x^2(1 - x)^{1/2}$

21. $f(x) = 4(x + 2)^{2/3}$

22. $f(x) = 2x^{1/3}$

23. $f(x) = 4\sqrt{x}(x - 2)$

24. $f(x) = -2\sqrt{x}(x - 4)$

25. $f(x) = \dfrac{x}{x - 1}$

26. $f(x) = \dfrac{x}{x + 1}$

27. $f(x) = x + \dfrac{4}{x}$

28. $f(x) = x^2 - \dfrac{4}{x}$

29. **Exam.** What values of x produce a relative minimum and a relative maximum, respectively, for $f(x) = 2x^3 + 3x^2 - 12x - 5$?
   (a) $-5, 0$ (b) $-2, 1$ (c) $1, -3$
   (d) $1, -2$ (e) $2, -1$

30. **Exam.** Vulcan Screw and Bolt Company manufactures and sells screws and bolts. The equations below represent the revenue (R) and cost (C) for a given bolt where x is 1000 cases of bolts.

$$R(x) = 100 + 26x - 0.05x^2 \quad \text{and}$$
$$C(x) = 50 + 8x + 0.01x^2$$

What quantity should be produced to maximize profit?
   (a) 520,000 cases   (b) 260,000 cases

(c) 150,000 cases

(d) Some amount other than those given.

31. Find any points at which there are relative extrema for $y = |x - 1|^3$.

Applications (Business and Economics)

All application problems have been selected so that the maxima and minima requested are at relative maxima and minima.

32. **Cost Function.** A cost function is given by $C(x) = 400 - 12x + x^2$, where x is the number of units produced. Find the minimum cost.

33. **Cost Function.** A cost function is given by $C(x) = 2x^2 - 12x + 200$, where x is the number of units produced. Find the minimum cost.

34. **Demand Function.** A unit price $p(x)$ for a product is given as a function of the number of units for $x > 0$ and is defined as

$$p(x) = \frac{60,000 - x}{20,000}$$

Is there a maximum or minimum price?

35. **Profit Function.** Profit P is related to selling price S by the formula

$$P(S) = 10,000S - 250S^2$$

What selling price provides a maximum profit?

36. **Maximum Profit.** The cost and revenue from sales curves are given by $C(x) = 2x^2 - x$ and $R(x) = x^2 + 5x$. Find the number of units for which the profit (in hundreds) will be maximum. What is the maximum profit?

Applications (Social and Life Sciences)

37. **Population.** A population study has been made for 8 years in a community, and the following population function P has been obtained.

$$P(t) = 1000t^2 - 6000t + 29,000$$

where t is time measured in years from the beginning date of the study. What is the minimum population?

38. **Drug Concentration.** The concentration $c(t)$ of a drug in a patient's bloodstream t hours after a drug is taken is given by

$$c(t) = \frac{0.2t}{t^2 + 4t + 9}$$

(a) When will the concentration be a maximum?

(b) What is the maximum concentration?

Review Exercises

Find all critical points and the intervals where the functions are increasing and decreasing. Draw each graph with your graphing calculator to check your answers.

39. $y = x^2 - 2x + 3$

40. $y = x^2 - 4x + 3$

41. $y = 2x^2 - 4x + 3$

42. $y = 3x^2 - 3x + 2$

9.3 HIGHER-ORDER DERIVATIVES AND CONCAVITY

OVERVIEW   Consider a function such as $f(x) = x^3 + 2x^2$. The derivative of this function is $f'(x) = 3x^2 + 4x$. Since $f'(x)$ is also a function of x, there is no reason we cannot take the derivative of it, which we denote as $f''(x) = 6x + 4$. In this section we take derivatives of derivatives. The derivative of a derivative is useful in discussing concavity, which is another tool we can use to analyze and sketch the graph of a function. In this section we study

- Higher-order derivatives
- Concavity
- Points of inflection

Higher-Order Derivatives

The derivative of the derivative of a function is called the second derivative of the function. The derivative of the second derivative gives the third derivative. We could continue this process getting fourth derivatives, fifth derivatives, and so on. These are called **higher-order derivatives**. Many different notations are used to indicate derivatives and higher-order derivatives. Some of the notations are given in Table 1.

TABLE 1

Function	First Derivative	Second Derivative	Third Derivative	nth Derivative
$y = f(x)$	y'	y''	y'''	$y^{(n)}$
$y = f(x)$	$f'(x)$	$f''(x)$	$f'''(x)$	$f^{(n)}(x)$
$y = f(x)$	$D_x f(x)$	$D_x^2 f(x)$	$D_x^3 f(x)$	$D_x^n f(x)$
$y = f(x)$	$\dfrac{dy}{dx}$	$\dfrac{d^2y}{dx^2}$	$\dfrac{d^3y}{dx^3}$	$\dfrac{d^ny}{dx^n}$
$y = f(x)$	$\dfrac{d}{dx}[f(x)]$	$\dfrac{d^2}{dx^2}[f(x)]$	$\dfrac{d^3}{dx^3}[f(x)]$	$\dfrac{d^n}{dx^n}[f(x)]$

EXAMPLE 13   Given $y = 3x^4 + 2x^3 + 5x^2 - x + 2$, find each of the following.

(a) $\dfrac{dy}{dx}$ (b) $\dfrac{d^2y}{dx^2}$ (c) $\dfrac{d^3y}{dx^3}$ (d) $\dfrac{d^4y}{dx^4}$ (e) $\dfrac{d^5y}{dx^5}$

SOLUTION

(a) $\dfrac{dy}{dx} = 12x^3 + 6x^2 + 10x - 1$ First derivative

(b) $\dfrac{d^2y}{dx^2} = 36x^2 + 12x + 10$ Second derivative (derivative of the first derivative)

(c) $\dfrac{d^3y}{dx^3} = 72x + 12$ Third derivative (derivative of the second derivative)

(d) $\dfrac{d^4y}{dx^4} = 72$ Fourth derivative (derivative of the third derivative)

(e) $\dfrac{d^5y}{dx^5} = 0$ All other derivatives will be zero.

EXAMPLE 14   Given $f(x) = 3x^{2/3}$, find   (a) $f'(x)$,   (b) $f''(x)$, and   (c) $f'''(x)$.

SOLUTION

(a) $f'(x) = 3 \cdot \frac{2}{3} x^{-1/3} = 2 x^{-1/3}$ First derivative

(b) $f''(x) = 2 \cdot (-\frac{1}{3}) x^{-4/3} = (-\frac{2}{3}) x^{-4/3}$ Second derivative (derivative of the first derivative)

(c) $f'''(x) = (-\frac{2}{3}) \cdot (-\frac{4}{3}) x^{-7/3} = \frac{8}{9} x^{-7/3}$ Third derivative (derivative of the second derivative) ■

Practice Problem 1 For $y = x^3 + \dfrac{1}{x}$, find $y'''(x)$.

ANSWER $y'''(x) = 6 - 6x^{-4}$

For functions that involve products, such as $f(x) = x(x - 2)^2$, there are two approaches to finding higher-order derivatives. One approach would be to expand the function and then differentiate. However, for a function that is not easily expanded, such as $f(x) = \sqrt{x}(x - 4)^5$, this is not a reasonable approach. For functions such as these, we use the product rule.

EXAMPLE 15 Find $f''(x)$ if $f(x) = x(x - 2)^2$.

SOLUTION To find $f'(x)$ we use the product formula:

$$f'(x) = x \cdot 2(x - 2) + (x - 2)^2 \cdot 1$$
$$= (x - 2)(2x + x - 2) = (x - 2)(3x - 2) \qquad FS' + SF'$$

Now find the derivative of $f'(x)$ by using the product rule again.

$$f''(x) = (x - 2)(3) + (3x - 2)(1) = 3x - 6 + 3x - 2 = 6x - 8 \qquad ■$$

For functions that involve quotients, we again have a choice of techniques. For example, $f(x) = x/(x - 1)$ can be written as $f(x) = x(x - 1)^{-1}$. Therefore, we can use the quotient rule or the product rule. The choice of techniques will usually depend on the complexity of the function. There are times when both techniques can be used. Often, after a first derivative is taken, this derivative can be rewritten and the product rule or the general power rule can be used. This is advisable when the numerator is a constant, as illustrated in the next example.

EXAMPLE 16 Find $f''(x)$ for $f(x) = \dfrac{x}{x - 1}$.

SOLUTION We use the quotient formula to find the first derivative:

$$f'(x) = \frac{1(x - 1) - x(1)}{(x - 1)^2} \qquad \frac{N'D - ND'}{D^2}$$
$$= \frac{x - 1 - x}{(x - 1)^2} = \frac{-1}{(x - 1)^2}$$

To find the second derivative, we rewrite the first derivative and use the general power rule.

$$f'(x) = -(x-1)^{-2}$$

$$f''(x) = 2(x-1)^{-3} \cdot 1 = \frac{2}{(x-1)^3}$$

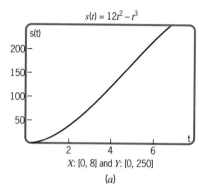

$s(t) = 12t^2 - t^3$

X: [0, 8] and Y: [0, 250]

(a)

Practice Problem 2 Find y'' for each function.

(a) $y = \sqrt{x}(x-2)$

(b) $y = x^2(x-3)^2$

(c) $y = \dfrac{2x}{3x-2}$

ANSWER

(a) $y'' = \frac{3}{4}x^{-1/2} + \frac{1}{2}x^{-3/2}$

(b) $y'' = 12x^2 - 36x + 18$

(c) $y'' = 24(3x-2)^{-3}$

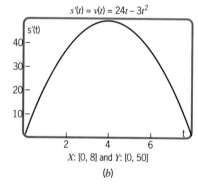

$s'(t) = v(t) = 24t - 3t^2$

X: [0, 8] and Y: [0, 50]

(b)

Applications

A derivative measures the rate of change of a function. Therefore, f'' measures the rate of change of f' in the same way that f' measures the rate of change of f at some particular point. For example, if $s(t) = 12t^2 - t^3$ is a position function for an object at time t for $0 \le t \le 8$, then $s'(t) = v(t)$ is the *velocity* of the object at time t in feet per second as defined by $v(t) = 24t - 3t^2$. Likewise, $v'(t) = a(t) = 24 - 6t$ is the *acceleration* of the object at time t. We see then that at $t = 2$, the velocity of the object is $v(2) = 36$ and $a(2) = 12$, which tells us that at 2 seconds the object's velocity is still increasing. The object will have increasing velocity when the derivative of the velocity function (acceleration) is positive.

EXAMPLE 17 Using the position function $s(t) = 12t^2 - t^3, 0 \le t \le 8$, find the interval(s) of time when the velocity is increasing. The object will have increasing velocity when the derivative of the velocity function (acceleration) is positive.

SOLUTION We found above that $v(t) = 24t - 3t^2$ and $v'(t) = a(t) = 24 - 6t$. The critical value for v occurs at $t = 4$. Over the interval $(0, 4)$, $a(t) > 0$ and over the interval $(4, 8)$, $a(t) < 0$. Therefore, velocity increases over the time interval 0 seconds to 4 seconds. This is verified by looking at the graphs in Figure 25.

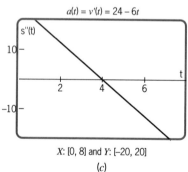

$a(t) = v'(t) = 24 - 6t$

X: [0, 8] and Y: [−20, 20]

(c)

Figure 25

Another useful application of higher-order derivatives is determining the **concavity** of the graph of a function. Look at the two graphs in Figure 26. The graph of $f_1(x)$ can be described as turning upward (concave up) and the graph of $f_2(x)$ turns downward (concave down).

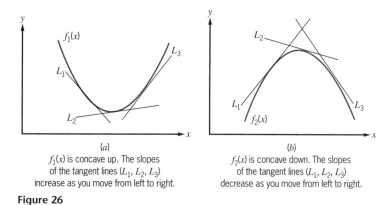

(a).
$f_1(x)$ is concave up. The slopes
of the tangent lines (L_1, L_2, L_3)
increase as you move from left to right.

(b).
$f_2(x)$ is concave down. The slopes
of the tangent lines (L_1, L_2, L_3)
decrease as you move from left to right.

Figure 26

In Figure 26 notice that the tangent lines lie below the graph of $f_1(x)$ and the slopes of the tangent lines of $f_1(x)$ increase from left to right (starting as negative and ending as positive). This means that $f_1'(x)$ is increasing and, therefore, its derivative is positive, $f_1''(x) > 0$. Likewise, the tangent lines lie above the graph of $f_2(x)$ and the slopes of the tangent lines of $f_2(x)$ are decreasing from left to right (starting as positive and ending as negative). This means that the derivative of $f_2'(x)$ must be negative, $f_2''(x) < 0$. This discussion is summarized as follows.

DEFINITION: CONCAVITY OF A FUNCTION

1. If $f''(x) > 0$ on an open interval I, then $f(x)$ is turning upward on I and the graph is said to be **concave upward** over I.
2. If $f''(x) < 0$ on an open interval I, then $f(x)$ is turning downward on I and the graph is said to be **concave downward** over I.

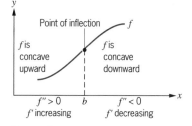

Figure 27

Notice that where the graph is concave upward, $f''(x) > 0$ so that $f'(x)$ is increasing, and where the graph is concave downward, $f''(x) < 0$ so that $f'(x)$ is decreasing. Now, what happens at the point where a graph changes from concave upward to concave downward (Figure 27)? Remember that $f''(x)$ must either assume a value of 0 or it does not exist at this point. The point $(b, f(b))$ where concavity changes is called a **point of inflection**.

DEFINITION: INFLECTION POINT

An **inflection point** is a point on the graph of a function where concavity changes from upward to downward or from downward to upward. At this point, $f''(x) = 0$ or it does not exist.

NOTE: Where $f''(x) = 0$ or where it does not exist a *possible* inflection point occurs. However, if $f''(x)$ does not change sign, there is no inflection point. The graph of the second derivative will help you see what is occurring at the point.

EXAMPLE 18 Determine where the graph of the function $f(x) = x^3 - 6x^2 + 9x + 4$ is concave up, is concave down, and has points of inflection. Also, determine where f is increasing, where it is decreasing, and the values of any relative extrema. Use critical values, possible inflection points, and the graphs of f, f', and f'' to find the information.

SOLUTION

1. Find the first derivative:

$$f'(x) = 3x^2 - 12x + 9 = 3(x^2 - 4x + 3) = 3(x - 3)(x - 1)$$

From this we see that the critical values are $x = 1$ and $x = 3$.

2. Find the second derivative:

$$f''(x) = 6x - 12 = 6(x - 2)$$

From this we determine that the only possible inflection point is at $x = 2$.

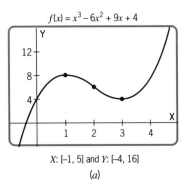

$f(x) = x^3 - 6x^2 + 9x + 4$

X: [-1, 5] and Y: [-4, 16]

(a)

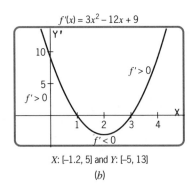

$f'(x) = 3x^2 - 12x + 9$

X: [-1.2, 5] and Y: [-5, 13]

(b)

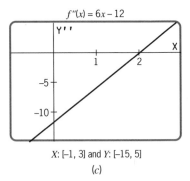

$f''(x) = 6x - 12$

X: [-1, 3] and Y: [-15, 5]

(c)

Figure 28

3. Using the information found in (1) and (2), we look at the graphs of the function and the derivatives in Figure 28 to obtain the following:

Increasing on $(-\infty, 1)$, $(3, \infty)$ where $f' > 0$.
Decreasing on $(1, 3)$ where $f' < 0$.
Relative maximum at $(1, 8)$ where $f' = 0$ and changes from positive to negative.
Relative minimum at $(3, 4)$ where $f' = 0$ and changes from negative to positive.
Concave down on $(-\infty, 2)$ where $f'' < 0$.
Concave up on $(2, \infty)$ where $f'' > 0$.
Inflection point at $(2, 6)$ where $f'' = 0$ and concavity changes. ∎

Practice Problem 3 Draw the graph of $f(x) = x^3 - 12x^2$. Find all extrema and points of inflection.

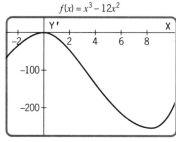

$f(x) = x^3 - 12x^2$

X: [-3, 10] and Y: [-280, 40]

Figure 29

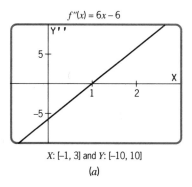

$f''(x) = 6x - 6$

X: [-1, 3] and Y: [-10, 10]

(a)

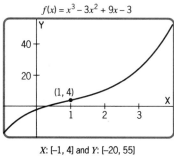

$f(x) = x^3 - 3x^2 + 9x - 3$

X: [-1, 4] and Y: [-20, 55]

(b)

Figure 30

ANSWER   The graph is shown in Figure 29.

$f'(x) = 3x^2 - 24x = 3x(x - 8)$. Relative maximum at $(0, 0)$ and relative minimum at $(8, -256)$.

$f''(x) = 6x - 24$. Inflection point at $(4, -128)$.

EXAMPLE 19   For $f(x) = x^3 - 3x^2 + 9x - 3$:

(a)   Find the intervals where the graph is concave up and the intervals where it is concave down.

(b)   Find the slope of the tangent line at the point of inflection.

SOLUTION

(a)   We must first find the second derivative.

$$f(x) = x^3 - 3x^2 + 9x - 3$$
$$f'(x) = 3x^2 - 6x + 9$$
$$f''(x) = 6x - 6 = 6(x - 1)$$

We see from this work that the only possible inflection point occurs at $x = 1$. By drawing the graph of $f''(x)$, we can see that it does indeed change from negative to positive and, therefore, the function is concave down on the interval $(-\infty, 1)$ and concave up on the interval $(1, \infty)$. The graphs of $f''(x)$ and $f(x)$ are shown in Figure 30.

(b)   To find the slope of the tangent line at the inflection point, we evaluate $f'(1)$ and determine the slope as $f'(1) = 6$ at $(1, 4)$ ■

EXAMPLE 20   For $f(x) = x^4 - 2x^3$, find any relative extrema and inflection points using the first and second derivatives and the graphs of the functions.

SOLUTION   First, we find the derivatives.

$$f(x) = x^4 - 2x^3$$
$$f'(x) = 4x^3 - 6x^2 = 2x^2(2x - 3) \qquad \text{Critical values are } x = 3/2 \text{ and } x = 0.$$
$$f''(x) = 12x^2 - 12x = 12x(x - 1) \qquad \text{Possible inflection points are } x = 0 \text{ and } x = 1.$$

Using the information just found together with the graphs shown in Figure 31, we can determine that there is a relative minimum at $(\frac{3}{2}, -1.7)$ and inflection points at $(0, 0)$ and $(1, -1)$. Notice that even though $f'(0) = 0$ and is a critical value, the derivative does not change signs and there are no relative extrema at that point.   ■

Now, let's use our knowledge of the relationship between a function and its first and second derivative to sketch a graph of a function without knowing the definition of the function or its derivatives.

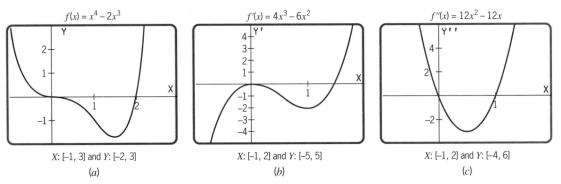

$f(x) = x^4 - 2x^3$

$f'(x) = 4x^3 - 6x^2$

$f''(x) = 12x^2 - 12x$

X: [-1, 3] and Y: [-2, 3]

(a)

X: [-1, 2] and Y: [-5, 5]

(b)

X: [-1, 2] and Y: [-4, 6]

(c)

Figure 31

EXAMPLE 21 The graphs of f' and f'' are shown in Figure 32 along with the value of the function f at several points. Use the graphs, what you know about a function and its derivatives, and the values given to sketch the graph of a function f that fits the given information.

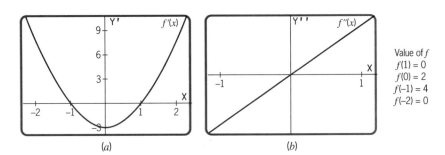

$f'(x)$

$f''(x)$

Value of f
$f(1) = 0$
$f(0) = 2$
$f(-1) = 4$
$f(-2) = 0$

(a)

(b)

Figure 32

SOLUTION Since $f'(x) > 0$ on $(-\infty, -1)$, $(1, \infty)$, we know that f is increasing on those intervals.

Since $f'(x) < 0$ on $(-1, 1)$, we know that f is decreasing on that interval. Also, f' changes from positive to negative at $x = -1$ and f changes from increasing to decreasing; therefore, there is a relative maximum at $x = -1$. Likewise, we can see that a relative minimum occurs at $x = 1$ because f' changes from negative to positive and f changes from decreasing to increasing.

The graph of f'' tells us that f is concave down on the interval $(-\infty, 0)$, since $f'' < 0$ there. Also, f is concave up where $f'' > 0$, which is on the interval $(0, \infty)$. Therefore, the only inflection point occurs at $x = 0$.

We now use the preceding information to sketch a graph of a function meeting the above requirements (Figure 33).

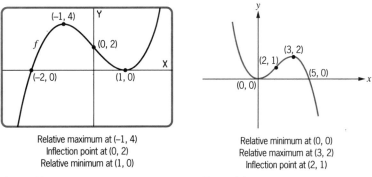

Relative maximum at (−1, 4)
Inflection point at (0, 2)
Relative minimum at (1, 0)

Figure 33

Relative minimum at (0, 0)
Relative maximum at (3, 2)
Inflection point at (2, 1)

Figure 34

Practice Problem 4 Given the graph of a function in Figure 34, sketch the graphs of f' and f''.

ANSWER The graphs are shown in Figure 35.

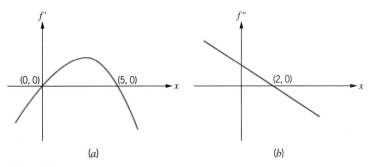

(a)

(b)

Figure 35

Practice Problem 5 Use the given information on $f'(x)$ and $f''(x)$ and the ordered pairs to the right to sketch a rough graph of the function. Assume that the function is continuous.

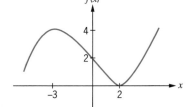

Figure 36

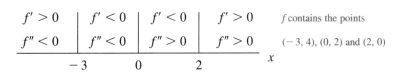

$f' > 0$	$f' < 0$	$f' < 0$	$f' > 0$	f contains the points
$f'' < 0$	$f'' < 0$	$f'' > 0$	$f'' > 0$	$(-3, 4), (0, 2)$ and $(2, 0)$

ANSWER The graph is shown in Figure 36.

SUMMARY

In this section we learned that derivatives can have derivatives. These higher-order derivatives are useful in many ways. The use that we concentrated on most in this section was finding the concavity of a graph. Discontinuities may occur in derivatives even though a function is continuous. This fact does not affect the usefulness of the derivative in determining possible inflection points and concavity. The conclusions are summarized in Table 2.

TABLE 2

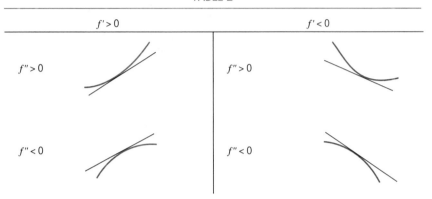

	$f' > 0$		$f' < 0$	
$f'' > 0$			$f'' > 0$	
$f'' < 0$			$f'' < 0$	

Exercise Set 9.3

Find

$$\frac{dy}{dx}, \quad \frac{d^2y}{dx^2}, \quad and \quad \frac{d^3y}{dx^3}$$

for each function.

1. $y = 25$
2. $y = -5$
3. $y = 3x + 5$
4. $y = 3x^2 - 5$
5. $y = 35x^2 - 27x + 5$
6. $y = -3x^3 + 4x + 2$
7. $y = 2x^3 + 3x^2 - x + 2$
8. $y = 4x^3 - 3x^2 + 5x - 3$
9. $y = 2x^6 + 3x^3 - 2x + 7$
10. $y = -3x^8 + 2x^5 - 3x^2$
11. $y = \frac{1}{2}x^4 + \frac{1}{3}x^3$
12. $y = -\frac{2}{3}x^3 + \frac{1}{2}x^2$
13. $y = 3x^{1/3}$
14. $y = 3x^{2/3}$
15. $y = x\sqrt[3]{x}$
16. $y = \sqrt{16 - x}$
17. $y = \frac{3}{x - 1}$
18. $y = \frac{4}{(x - 1)^4}$

Use the figure to answer Exercises 19–21.

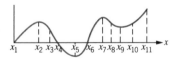

19. For the points named on the figure, list all x_i's over the interval (x_1, x_{11}) for each concept.
    (a) Critical point
    (b) Inflection point
    (c) x-intercept
    (d) Relative maximum
    (e) Relative minimum

20. Does it appear that there are points in the interval (x_1, x_{11}) where $f'(x)$ is undefined?

21. For each blank, write one of the following: positive, negative, or zero.
    (a) At x_2, f is _____, f' is _____, and f'' is _____.
    (b) At x_5, f is _____, f' is _____, and f'' is _____.
    (c) At x_6, f is _____, f' is _____, and f'' is _____.

(d) At x_{10}, f is _____ , f' is _____ , and f'' is _____ .

(e) At x_8, f is _____ , f' is _____ , and f'' is _____ .

(f) On the interval (x_3, x_6), f'' is _____ .

(g) On the interval (x_8, x_{10}), f'' is _____ .

(h) Wherever f is increasing, f' is _____ .

(i) Wherever f is concave up, f'' is _____ .

(j) List the intervals where f is decreasing.

(k) List the intervals where f is concave down over the interval (x_1, x_{11}).

Discuss whether the function described below is increasing or decreasing, is concave upward or concave downward, and draw a small section of the curve in some interval containing c.

22. $f'(c) > 0$, $f''(c) > 0$

23. $f'(c) > 0$, $f''(c) < 0$

24. $f'(c) < 0$, $f''(c) > 0$

25. $f'(c) < 0$, $f''(c) < 0$

26. $f'(c) = 0$, $f''(c) > 0$

27. $f'(c) = 0$, $f''(c) < 0$

28. Find an equation of the tangent line at the point of inflection for $y = x^3 + 3x^2 + 2$.

In Exercises 29, 30, and 31, use the given information on $f'(x)$ and $f''(x)$ and the table of values to sketch a rough graph of the function. Assume that the function is continuous.

29.

$f' > 0$ $f'' < 0$	$f' < 0$ $f'' < 0$	$f' < 0$ $f'' > 0$	$f' > 0$ $f'' > 0$
-1	0	1	

x	-2	-1	0	1	2
$f(x)$	-2	2	0	-2	2

30.

$f' < 0$ $f'' > 0$	$f' > 0$ $f'' > 0$
	1

x	0	1	2
$f(x)$	0	-3	0

31.

$f' < 0$ $f'' > 0$	$f' > 0$ $f'' > 0$	$f' > 0$ $f'' < 0$	$f' < 0$ $f'' < 0$
-2	-1	0	

x	-2	-1	0
$f(x)$	-16	-5	0

Find $f''(x)$ for each of the following functions

32. $f(x) = (x^2 - 3)(x - 1)$

33. $f(x) = (x - 3)(x + 4)^2$

34. $f(x) = \dfrac{x + 1}{x + 3}$

35. $f(x) = \dfrac{x}{(x + 3)^2}$

36. $f(x) = (x^2 + 4)^3$

37. $f(x) = \dfrac{1}{\sqrt{x^2 + 4}}$

Find the intervals over which the graph of the given function is concave downward and concave upward, the inflection points, and all relative extrema. Use calculator graphs of the function and its first and second derivatives.

38. $y = x^2$

39. $y = -x^2$

40. $y = 3x^2$

41. $y = -2x^2$

42. $y = 4x^2 + 3x$

43. $y = -3x^2 + 2x$

44. $y = x^3$

45. $y = -x^3$

46. $y = x^3 - 3x^2$

47. $y = x^3 + 4x^2$

48. $y = x^4 - 2x^2$

49. $y = x^4 + x^2$

50. $y = 3x^{5/3} - 15x^{2/3}$

51. $y = x^{5/3} + 5x^{2/3}$

Given the graph of f, make a reasonable sketch of f' and f''.

52.

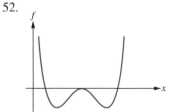

x-intercepts: 1, 4, 7

Relative minima at $x = 2$, $x = 6$

Relative maximum at $x = 4$

Inflection points at $x = 3$, $x = 5$

53.

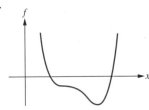

x-intercepts: 4, 14
Relative minima at $x = 6$, $x = 12$
Relative maximum at $x = 8$
Inflection points at $x = 7$, $x = 10$

Given the graphs of f' and f'' along with values of the function in the table, make a reasonable sketch of a function fitting the given requirements.

54.

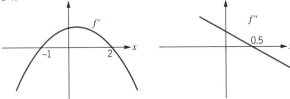

x	-1	0	0.5	2
$f(x)$	-5	2	8.5	22

55.

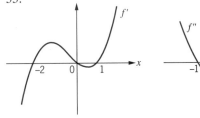

x	-2	-1	0	0.5	1
$f(x)$	-4	-2	1	0	-1

56. A helicopter rises vertically, and after t seconds its height $s(t)$ above the ground is given by the formula $s(t) = 6t^2 - t^3$ feet for $0 \le t \le 6$.
 (a) Find the velocity function and the acceleration function.

 (b) Find the velocity and acceleration after 2 seconds.

57. A missile rises to a height of $s(t) = t^3 + 0.5t^2$ feet in t seconds.
 (a) Find the velocity function and the acceleration function.
 (b) Find the velocity and acceleration after 8 seconds.

Applications (Business and Economics)

All application problems have been selected so that the maxima and minima requested are at relative maxima and minima. Use your graphing calculator to find approximate values of relative extrema and then use calculus techniques to find exact values.

58. Cost Function. A cost function is given by $C(x) = (3x - 10)^2 + 24$ for $x \ge 3$. Find the minimum value of the cost function.

59. Maximum Profit. Profit P is related to selling price S by the formula $P = 15{,}000S - 500S^2$ for $0 \le S \le 30$.
 (a) Describe the rate of change of P at $S = 7$.
 (b) For what range of values of S is the profit increasing?
 (c) At what selling price is the profit a maximum?
 (d) What is the maximum profit?

60. Maximum Profit. The cost and revenue from sales curves are given by $C = 2x^2 - 8x$ and $R = x^2 - 6x$, where x is given in millions.
 (a) For what production is the revenue function a minimum?
 (b) For what production is the cost function a minimum? Discuss.
 (c) Find the number of units for which the profit will be a maximum.
 (d) What is the maximum profit?

61. A revenue function for the sale of x stereo sets per year is

$$R(x) = 460x - 372\left(12 - \frac{x}{3}\right) - \left(12 - \frac{x}{3}\right)^2$$

for $0 \le x \le 5000$.

(a) For what production do you have maximum revenue?

(b) Find the maximum revenue.

Applications (Social and Life Sciences)

62. ***Drug Concentration.*** The concentration of a drug in a person's blood x hours after administration is

$$c = \frac{x}{20(x^2 + 6)}$$

(a) When will the concentration be a maximum?

(b) What is the maximum concentration?

63. ***Bacteria Count.*** Assume that the number of bacteria, in millions, present in a culture at time t is given by $B(t) = t^2(t - 8) + 400$. Find the minimum population.

64. ***Contagious Disease.*** Suppose that the number of people who are affected by the introduction of a flu virus in a state is given by $N(t) = 40t + 12 - t^2$, where t is the number of days after the virus is detected. On what day will the maximum number of people be affected, and how many people will be affected?

Review Exercises

Find the critical points, the intervals where the graphs are increasing, and the intervals where the graphs are decreasing. Draw each graph by hand and then check your work with a graphing calculator.

65. $y = x^2 + 3x$

66. $y = x^3 - 3x^2$

67. $y = x^3 - 3x + 2$

68. $y = x^3 + x - 1$

69. $y = x^4 - 4x + 1$

70. $y = x^4 - 2x^2 + 1$

9.4 THE SECOND DERIVATIVE TEST AND ABSOLUTE EXTREMA

OVERVIEW   Our study of concavity leads naturally into another test that helps determine whether a critical point is a relative maximum or relative minimum. A relative maximum may not be an absolute maximum; likewise, a relative minimum may not be an absolute minimum. We consider the differences in absolute and relative extrema. In this section we study

- The second derivative test
- Absolute maximum
- Absolute minimum

In Section 9.2, the first derivative test was given for relative maxima or relative minima (relative extrema). Another test for relative extrema uses the second derivative; this test is used when the second derivative is easy to obtain. Just as the first derivative gives the slope of the tangent line to a curve, the second derivative indicates the concavity of the curve. Suppose that a function is concave down in an open interval and $f'(c) = 0$ for some point c in this interval. Intuitively, a relative maximum exists at $(c, f(c))$; see Figure 37(a). Likewise, if a function is concave up in an open interval and $f'(c) = 0$ for some c in this interval, then a relative minimum exists at $(c, f(c))$; see Figure 37(b). This leads to the following test.

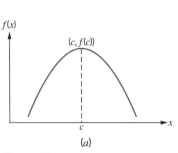
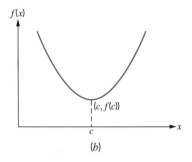

Figure 37

TABLE 3

$f'(c)$	$f''(c)$	$f(c)$
0	$-$	Relative maximum
0	$+$	Relative minimum
0	0	Test fails; use first derivative test or graph

SECOND DERIVATIVE TEST

Let $f'(c) = 0$ and suppose that $f''(x)$ exists in an open interval containing c.

(a)   If $f''(c) < 0$, then $f(c)$ is a relative maximum.
(b)   If $f''(c) > 0$, then $f(c)$ is a relative minimum.
(c)   If $f''(c) = 0$, then the test gives no information.

The second derivative test is summarized in Table 3.

EXAMPLE 22   Use the second derivative test to find the points on $f(x) = x^3 - 3x^2 - 9x + 5$ that are relative extrema.

SOLUTION   We have

$$f'(x) = 3x^2 - 6x - 9 = 3(x^2 - 2x - 3) = 3(x - 3)(x + 1)$$

The critical values of the function are $x = -1$ and $x = 3$, since those are the values that make $f'(x) = 0$. Let's test these critical values in the second derivative, $f''(x) = 6x - 6$.

$$f''(-1) = -6 - 6 = -12 < 0 \qquad \text{So } f(-1) = 10 \text{ is a relative maximum.}$$

$$f''(3) = 18 - 6 = 12 > 0 \qquad \text{So } f(3) = -22 \text{ is a relative minimum.}$$

This is confirmed by the graph of f shown in Figure 38.   ■

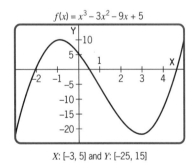

$f(x) = x^3 - 3x^2 - 9x + 5$

X: [-3, 5] and Y: [-25, 15]

Figure 38

EXAMPLE 23   Use the second derivative test to test $f(x) = 2x^3 - 12x^2 + 24x + 12$ for relative extrema.

SOLUTION   We have

$$f'(x) = 6x^2 - 24x + 24 = 6(x^2 - 4x + 4) = 6(x - 2)^2$$

We can see that the only critical value is $x = 2$. Next we apply the second derivative test.

$$f''(x) = 12x - 24 \quad \text{and} \quad f''(2) = 12(2) - 24 = 0$$

The second derivative test fails here, so we must use the first derivative test or a graph. We can see by looking at the graph of $f(x)$ in Figure 39 that the function appears to be increasing for all x. We can confirm this by looking at the graph of f', which is nonnegative for all x; hence, $f(x)$ is increasing for all x. If we look at the graph of f'', we can see that there is an inflection point at $(2, 28)$. ∎

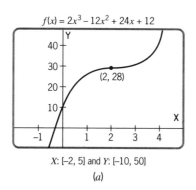

$f(x) = 2x^3 - 12x^2 + 24x + 12$

X: [-2, 5] and Y: [-10, 50]

(a)

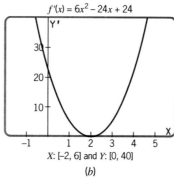

$f'(x) = 6x^2 - 24x + 24$

X: [-2, 6] and Y: [0, 40]

(b)

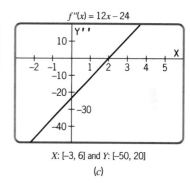

$f''(x) = 12x - 24$

X: [-3, 6] and Y: [-50, 20]

(c)

Figure 39

Practice Problem 1 The critical values for the function $f(x) = 2x^3 + 3x^2 - 12x + 8$ are $x = -2$ and $x = 1$. Use the second derivative test to classify as relative maxima, relative minima, or no information.

ANSWER $f''(x) = 12x + 6$. A relative maximum occurs at $x = -2$ ($f''(-2) < 0$) and a relative minimum at $x = 1$ ($f''(1) > 0$).

EXAMPLE 24 Determine where the graph of the function $f(x) = x^3 - 3x^2$:
(a) Is concave upward.
(b) Is concave downward.
(c) Has a point of inflection.
(d) Is Increasing.
(e) Is Decreasing.
(f) Has a relative maximum.
(g) Has a relative minimum.
(h) Find the slope of the tangent line at the point of inflection.

SOLUTION

$$f'(x) = 3x^2 - 6x = 3x(x - 2) \qquad \text{Critical values are } x = 0, x = 2.$$

$$f''(x) = 6x - 6 = 6(x - 1) \qquad \text{Possible inflection point at } x = 1$$

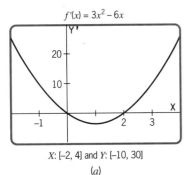

$f'(x) = 3x^2 - 6x$

X: [-2, 4] and Y: [-10, 30]

(a)

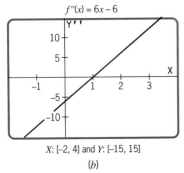

$f''(x) = 6x - 6$

X: [-2, 4] and Y: [-15, 15]

(b)

Figure 40

From the information above and the graphs shown in Figure 40 we can conclude the following.

(a) f is concave upward on $(1, \infty)$.

(b) f is concave downward on $(-\infty, 1)$.

(c)   There is a point of inflection at $(1, -2)$.

(d) f is increasing on $(-\infty, 0)$ and $(2, \infty)$.

(e) f is decreasing on $(0, 2)$.

(f) $f'(0) = 0$ and $f''(0) = -6 < 0$; therefore, there is a relative maximum at $(0, 0)$.

(g) $f'(2) = 0$ and $f''(2) = 6 > 0$; therefore, there is a relative minimum at $(2, -4)$.

(h)   The point of inflection occurs at $x = 1$. The slope of the tangent at this point is $f'(1) = -3$.

All of the preceding information is summarized in Figure 41.   ■

Up until this point, we have discussed relative extrema. Now let's turn our attention to **absolute extrema**. The **absolute maximum** of a function is the greatest value that the function attains in a closed interval, and the **absolute minimum** is the least value attained in a closed interval. A function that is continuous on a closed interval must have an absolute maximum and absolute minimum on that interval.

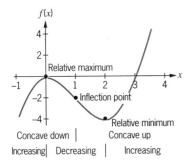

$f(x)$

Relative maximum

Inflection point

Relative minimum

Concave down | Concave up

Increasing| Decreasing | Increasing

Figure 41

DEFINITION: ABSOLUTE MAXIMUM AND ABSOLUTE MINIMUM

(a)   A function f has an absolute maximum at $x = c$ if $f(c) \geq f(x)$ for all x in the selected domain of f.

(b)   A function f has an absolute minimum at $x = c$ if $f(c) \leq f(x)$ for all x in the selected domain of f.

To determine where the absolute maximum and the absolute minimum occur, study Figure 42. Note that for values in the intervals between the endpoints and the critical points, a continuous function is either increasing or decreasing. So, to find the absolute maximum and the absolute minimum of a continuous function that is not constant on a closed interval, we evaluate the function at each endpoint and at each critical point *within the interval* and simply choose the greatest value for the absolute maximum and the least value for the absolute minimum.

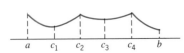

$a \quad c_1 \quad c_2 \quad c_3 \quad c_4 \quad b$

Figure 42

Procedure for Finding Absolute Maximum and Absolute Minimum

For $f(x)$ defined and continuous on $[a, b]$:

Procedure for Finding Absolute Maximum and Absolute Minimum (*Continued*)

1. Find all critical values $c_1, c_2, \ldots, c_n$ on (a, b).
2. Evaluate $f(a), f(b), f(c_1), f(c_2), \ldots, f(c_n)$.
3. The greatest value found in step 2 is the absolute maximum on $[a, b]$. The least value found in step 2 is the absolute minimum on $[a, b]$.

EXAMPLE 25   Draw the graph of $f(x) = x^3 - 6x^2 + 8$ and find the absolute maximum and the absolute minimum on each given interval.

(a) $[-1, 3]$ (b) $[-1, 5]$ (c) $[-3, 7]$

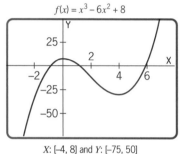

$f(x) = x^3 - 6x^2 + 8$

X: [-4, 8] and Y: [-75, 50]

Figure 43

SOLUTION   The graph of $f(x) = x^3 - 6x^2 + 8$ is given in Figure 43. Since $f'(x) = 3x^2 - 12x = 3x(x - 4)$, the critical values are at $x = 0$ and $x = 4$.

(a)   For the interval $[-1, 3]$ we have $a = -1$ and $b = 3$. The critical value in the interval is $x = 0$.

$$f(-1) = (-1)^3 - 6(-1)^2 + 8 = 1 \qquad \text{Endpoint}$$

$$f(0) = (0)^3 - 6(0)^2 + 8 = 8 \qquad \text{Critical point}$$

$$f(3) = (3)^3 - 6(3)^2 + 8 = -19 \qquad \text{Endpoint}$$

Therefore, for the interval $[-1, 3]$ the absolute maximum occurs at $x = 0$ and is 8. The absolute minimum occurs at $x = 3$ and is -19.

(b)   For the interval $[-1, 5]$ we have $a = -1$ and $b = 5$. The critical values are $x = 0$ and $x = 4$.

$$f(-1) = (-1)^3 - 6(-1)^2 + 8 = 1 \qquad \text{Endpoint}$$

$$f(0) = (0)^3 - 6(0)^2 + 8 = 8 \qquad \text{Critical point}$$

$$f(4) = (4)^3 - 6(4)^2 + 8 = -24 \qquad \text{Critical point}$$

$$f(5) = (5)^3 - 6(5)^2 + 8 = -17 \qquad \text{Endpoint}$$

For the interval $[-1, 5]$ the absolute maximum occurs at $x = 0$ and is 8. The absolute minimum occurs at $x = 4$ and is -24.

(c)   For the interval $[-3, 7]$ we have $a = -3$ and $b = 7$. The critical values are $x = 0$ and $x = 4$.

$$f(-3) = (-3)^3 - 6(-3)^2 + 8 = -73 \qquad \text{Endpoint}$$

$$f(0) = (0)^3 - 6(0)^2 + 8 = 8 \qquad \text{Critical point}$$

$$f(4) = (4)^3 - 6(4)^2 + 8 = -24 \qquad \text{Critical point}$$

$$f(7) = (7)^3 - 6(7)^2 + 8 = 57 \qquad \text{Endpoint}$$

For the interval $[-3, 7]$, the absolute maximum occurs at $x = 7$ and is 57. The absolute minimum occurs at $x = -3$ and is -73. ∎

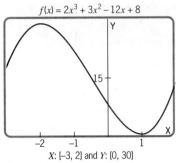

$f(x) = 2x^3 + 3x^2 - 12x + 8$

X: [-3, 2] and Y: [0, 30]

Figure 44

Practice Problem 2 For $f(x) = 2x^3 + 3x^2 - 12x + 8$, find the absolute maximum and absolute minimum on the interval $[-3, 2]$. Verify your answer with a graph.

ANSWER   We have

$$f'(x) = 6x^2 + 6x - 12 \quad \text{and} \quad f''(x) = 12x + 6$$

The absolute maximum on $[-3, 2]$ is 28 at $x = -2$ and the absolute minimum is 1 at $x = 1$. The graph is shown in Figure 44.

   If we look at the function in Practice Problem 2 as it is defined for all real numbers, there is no absolute maximum and no absolute minimum because the function can become infinitely large in magnitude in both directions. This illustrates the importance of the interval for which an absolute maximum or an absolute minimum is desired.

   When there is only one critical value of f on (a, b), and the second derivative is negative at the critical value, the absolute maximum of f on $[a, b]$ occurs at the relative maximum. Likewise, when there is only one critical value of f on (a, b), and the second derivative is positive at the critical value, the absolute minimum of f on $[a, b]$ occurs at the relative minimum. This is summarized in Table 4.

TABLE 4

Test for Absolute Maximum or Minimum			
Let $f(x)$ be continuous on $a < x < b$, and let there be only one critical value c on this interval.			
$f'(c)$	$f''(c)$	$f(c)$	Graph
0	$-$	Absolute maximum	
0	$+$	Absolute minimum	

EXAMPLE 26   Find the absolute maximum and the absolute minimum of $f(x) = 3x^2 - 6x$ on $[0, 4]$.

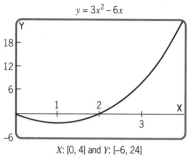

$y = 3x^2 - 6x$

X: [0, 4] and Y: [-6, 24]

Figure 45

SOLUTION   We have $f'(x) = 6x - 6 = 6(x - 1)$; therefore, the only critical value is $x = 1$, with $f'(1) = 0$. Since $x = 1$ is the only critical value on $[0, 4]$, the second derivative test can be used.

$$f''(x) = 6 \quad \text{and} \quad f''(1) = 6$$

Since $f''(x)$ is positive, the absolute minimum of $f(x)$ occurs at $x = 1$ and is $f(1) = 3(1)^2 - 6(1) = -3$. The absolute maximum is found by evaluating $f(x)$ at the two endpoints of the interval $x = 0$ and $x = 4$.

$$f(0) = 3(0)^2 - 6(0) = 0 \quad \text{and} \quad f(4) = 3(4)^2 - 6(4) = 48 - 24 = 24$$

Thus, the absolute maximum of f occurs at $x = 4$ and is 24.   ∎

Practice Problem 3 With your graphing calculator, draw the graph of $y = 3x^2 - 6x$ on the intervals $0 \le x \le 4$ and $-6 \le y \le 24$. Then from the graph find x for the absolute maximum and the absolute minimum. Do your answers agree with those in Example 26?

ANSWER   The graph is shown in Figure 45. The absolute maximum on $[0, 4]$ is 24 and the absolute minimum is -3, which agree with Example 26.

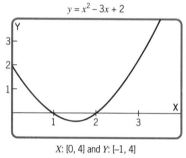

$y = x^2 - 3x + 2$

X: [0, 4] and Y: [-1, 4]

Figure 46

Practice Problem 4 Using your graphing calculator, draw the graph and find the absolute minimum for $y = x^2 - 3x + 2$ on $[0, 4]$.

ANSWER   The graph is shown in Figure 46. The absolute minimum is $-\frac{1}{4}$ at $x = \frac{3}{2}$.

The absolute extrema can occur at more than one point on a closed interval. For example, $f(x) = x^2$ on $[-1, 1]$ has an absolute maximum at $x = -1$ and $x = 1$. This is demonstrated in Figure 47.

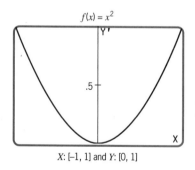

$f(x) = x^2$

X: [-1, 1] and Y: [0, 1]

Figure 47

SUMMARY

The second derivative test gives us a method of determining relative extrema without having to see a graph. This is useful in applications, as we will see in Section 9.6. Over a closed interval, a continuous function that is not constant has an absolute maximum and absolute minimum. These occur at the endpoints of the interval or at critical points that are within the closed interval.

Exercise Set 9.4

Find the location of all absolute maxima and absolute minima (if any) for the following functions.

1.

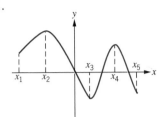

2.

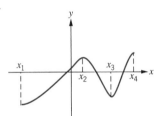

3.

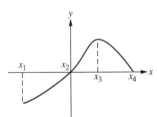

4.

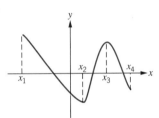

Find the absolute maximum and the absolute minimum on the given intervals.

5. $y = x^2 - x$
   (a) $[0, 1]$ (b) $[0, 3]$ (c) $[-2, 3]$
6. $y = x^2 + 2x$
   (a) $[-2, 0]$ (b) $[-3, 1]$ (c) $[-4, 4]$
7. $y = 2x^3$
   (a) $[-1, 1]$ (b) $[-2, 2]$ (c) $[-3, 3]$

8. $y = x^3 - 3x^2$
   (a) $[-1, 3]$ (b) $[-2, 4]$ (c) $[-3, 6]$
9. $y = x^4 + 2x^2$
   (a) $[-1, 1]$ (b) $[-2, 2]$ (c) $[-3, 2]$
10. $y = 2x + \dfrac{1}{x}$
    (a) $[1, 2]$ (b) $[-2, -1]$ (c) $[1, 3]$

Using the second derivative test (if possible), find relative maxima and relative minima for each function. Verify your answers with your graphing calculator.

11. $y = 6x^2 + 4x^3$ 12. $y = 9x^4 - 4x^2$
13. $y = 3x^4 - 4x^3$ 14. $y = 3x^4 - 6x^3$

15. $y = x^3 + \dfrac{19}{2} x^2 + 20x$ 16. $y = 6(x - 2)^{2/3}$

Find the critical points, relative extrema, absolute extrema, and inflection points of the following functions. Use calculator graphs to verify your work.

17. $y = 4x^2 - x$ 18. $y = x^2 + 2x$
19. $y = 2x^3$ 20. $y = -2x^3$
21. $y = x^3 - 3x^2$ 22. $y = x^3 + 3x^2$
23. $y = x^4 + 2x^2$ 24. $y = 3x - x^{-1}$
25. $y = 2x + x^{-1}$

Find the intervals where the function is increasing or decreasing, and where the graph is concave upward or downward. Then find all points of relative maxima, relative minima, and points of inflection. First find all of these using a graphing calculator. Then, using calculus techniques, indicate all of these on a coordinate system as you graph the function.

26. $f(x) = x^2(x - 1)$ 27. $f(x) = \sqrt{x}$
28. $f(x) = \sqrt[3]{x}$ 29. $f(x) = x(x - 1)^2$
30. $f(x) = x - 3x^{1/3}$ 31. $f(x) = x^{4/3} + 4x^{1/3}$
32. **Exam.**   The five graphs below have equal scales on the x- and y-axis. Which one could be the graph of $y = 3x^4 - 4x^3$?

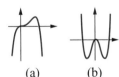

(a)   (b)   (c)   (d)   (e)

33. **Exam**.   On which of the following open intervals is the function $f(x) = (x^3/6) - 2x$ decreasing and concave upward?

(a) $(-2, 2)$ (b) $(-2, 0)$ (c) $(0, 2)$
(d) $(2, 3)$ (e) $(-2, 4)$

34. **Exam**.   What is the maximum value of a function defined by $f(x) = x^3 + x^2$ over $[-2, 1]$?

(a) -2 (b) 0 (c) $\dfrac{4}{27}$ (d) $\dfrac{20}{27}$

(e) 2

35. **Exam**.   To find the point at which a relative minimum cost occurs given a total cost equation, the steps are as follows: (1) Find the first derivative, (2) set the first derivative equal to zero, and (3) solve the equation. Using the solution(s) so derived, which additional steps must be taken to determine if a minimum exists?

(a) Substitute the solution(s) in the first derivative of the cost function, and a negative result indicates a minimum.
(b) Substitute the solution(s) in the first derivative of the cost function, and a positive result indicates a minimum.
(c) Substitute the solution(s) in the second derivative of the cost function, and a negative result indicates a minimum.
(d) Substitute the solution(s) in the second derivative of the cost function, and a positive result indicates a minimum.

36. Use a graphing calculator to find the absolute minimum of

$$y = \frac{4}{|x^2 + 2x - 3|}$$

over the interval $[-2, 0]$.

Applications (Business and Economics)

37. **Average Cost Function**.   If a cost function is given by $C(x) = 2x^2 + 50$, for $0 \le x \le 100$, find the minimum value of the average cost function,

$$\overline{C}(x) = \frac{C(x)}{x}$$

38. **Average Revenue**.   A revenue function is given by $R(x) = 3x^2 - 10x + 75$, for $0 \le x \le 100$. Find the minimum value of the average revenue function,

$$\overline{R}(x) = \frac{R(x)}{x}$$

39. **Average Profit**.   Using Exercises 37 and 38, find the minimum value of the average profit function,

$$\overline{P}(x) = \frac{P(x)}{x}$$

40. **Average Cost Function**.   Given that the cost function of producing x items of a given product is

$$C(x) = 40 + 10x + \frac{x^3}{200}$$

find the maximum value of the average cost function,

$$\overline{C}(x) = \frac{C(x)}{x}$$

and compare the value of $\overline{C}(x)$ with $C'(x)$, the marginal cost at the value of x where $\overline{C}(x)$ is a minimum.

41. **Revenue Function**.   The cost and revenue from sales are

$$C(x) = x^3 - 15x^2 + 72x + 25$$
$$\text{and} \quad R(x) = 51x - 3x^2$$

Draw the cost function and the revenue function at the same time. Where does profit occur and where does loss occur? Find the point at which there is maximum profit.

Applications (Social and Life Sciences)

42. **Maximum Work**.   A psychiatrist has found that for $t \ge 0$, the equation $y(t) = c_0 + c_1 t + c_2 t^2 + c_3 t^3$, where $c_0, c_1, c_2,$ and c_3 are constants, describes the relationship between amount of work output $y(t)$ and elapsed time t for $0 \le t \le 5$. Find any relative max-

ima or relative minima, and sketch the curve when $c_0 = 0$, $c_1 = 92$, $c_2 = 10$, and $c_3 = -6$.

Review Exercises

Find the critical points, intervals where the graph is increasing and where it is decreasing, relative maxima and minima, points of inflection, intervals where the

curve is concave upward and concave downward, and sketch the graphs of the following functions.

43. $y = x^{-1} + x^{-2}$

44. $y = x + x^{2/3}$

45. $y = \dfrac{x^3}{9(x + 2)}$

46. $y = \left(\dfrac{1}{7}\right)(2x^3 - 9x^2 + 12x + 3)$

47. $y = x^{5/3} + 5x^{2/3}$

9.5 INFINITE LIMITS; LIMITS AT INFINITY; CURVE TRACING

OVERVIEW

The real number system is unbounded. There is no *largest* real number and no *smallest* real number. Geometrically, this means that the real number line extends indefinitely to the right and to the left (Figure 48). We have already used the symbols ∞ and $-\infty$ (read "infinity" and "negative infinity") to denote this unboundedness. The symbol ∞ was introduced by the English mathematician John Wallis (1616–1703). The symbol does not represent a number and cannot be manipulated algebraically. It has meaning primarily in connection with limits. In this section we give additional attention to limits that do not exist. That is, we consider functions that increase without bound. In addition, we sometimes need to determine limits when the independent variable is getting larger and larger. This new concept of a limit will be very useful in graphical analysis such as that used in studying growth of bacteria or the average cost function in business. In this section we

Figure 48

- Define vertical and horizontal asymptotes
- Evaluate limits at infinity
- Analyze graphs of rational functions

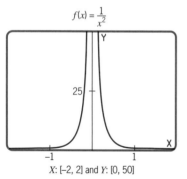

X: [-2, 2] and Y: [0, 50]

Figure 49

Consider the graph of $f(x) = 1/x^2$ shown in Figure 49. Consider the function and $\lim_{x \to 0} f(x)$. We see that $f(x)$ gets larger and larger as x gets closer and closer to 0. We say that the limit of the function increases without bound as x approaches 0, or

$$\lim_{x \to 0} \frac{1}{x^2} \to \infty$$

The notation $\to \infty$ is read "approaches infinity," which means the function is increasing without bound as x approaches 0.

Generally, if $f(x)$ increases without bound as x gets closer and closer to some number a from both sides of a, then $\lim_{x \to a} f(x) \to \infty$.

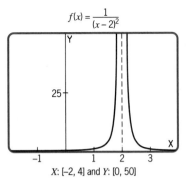

$f(x) = \dfrac{1}{(x-2)^2}$

X: [-2, 4] and Y: [0, 50]

Figure 50

EXAMPLE 27 Find $\lim\limits_{x \to 2} \dfrac{1}{(x-2)^2}$.

SOLUTION Study the following table along with the graph in Figure 50.

x	1.9	1.99	1.999	$\to$ 2 $\leftarrow$	2.001	2.01	2.1
$\dfrac{1}{(x-2)^2}$	100	10,000	1,000,000	$\to \infty \leftarrow$	1,000,000	10,000	100

It seems that as $x \to 2$, $1/(x-2)^2$ gets larger and larger, or, increases without bound. That is,

$$\lim_{x \to 2} \frac{1}{(x-2)^2} \to \infty$$

This is verified in the graph shown in Figure 50. ∎

Calculator Note

In order to get a look at how the values of the function are changing as x gets closer and closer to 2, you can use the $\boxed{\text{TRACE}}$ key. However, you can also have your calculator evaluate the function at values of x very close to $x = 2$. Unlike the $\boxed{\text{TRACE}}$ key, the evaluation function allows you to select the values of x.

Practice Problem 1 Draw the graph of

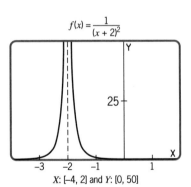

$f(x) = \dfrac{1}{(x+2)^2}$

X: [-4, 2] and Y: [0, 50]

Figure 51

$$f(x) = \frac{1}{(x+2)^2}$$

and determine what happens as x gets closer and closer to -2.

ANSWER The graph is shown in Figure 51. As x gets closer and closer to -2, the function becomes increasingly large. We can say that

$$\lim_{x \to -2} \frac{1}{(x+2)^2} \to \infty$$

Sometimes a curve is unbounded in a negative direction. Sometimes a curve is unbounded in a positive direction as x approaches a number from one side and is unbounded in a negative direction as x approaches the number from the other side. We use the following terminology for infinite limits.

INFINITE LIMITS

(a) If as $x \to a^-$, $f(x)$ increases without bound, we indicate this behavior by

$$\lim_{x \to a^-} f(x) \to \infty$$

(b) If as $x \to a^-$, $f(x)$ decreases without bound, we indicate this behavior by

$$\lim_{x \to a^-} f(x) \to -\infty$$

(c) If as $x \to a^+$, $f(x)$ increases without bound, we indicate this behavior by

$$\lim_{x \to a^+} f(x) \to \infty$$

(d) If as $x \to a^+$, $f(x)$ decreases without bound, we indicate this behavior by

$$\lim_{x \to a^+} f(x) \to -\infty$$

It should be emphasized that ∞ is not a number. When we use the notation $\lim_{x \to a} f(x) \to \infty$ or $\lim_{x \to a} f(x) \to -\infty$, the limit *does not exist*. We use ∞ or $-\infty$ to give us an indication of the unbounded behavior of the function.

EXAMPLE 28   Find $\lim_{x \to 3} \dfrac{x}{x - 3}$

as x approaches 3 from the left and from the right.

SOLUTION

x	2.9	2.99	2.999	$\to$	3	$\leftarrow$	3.001	3.01	3.1
$\dfrac{x}{x-3}$	-29	-299	-2999	$\to -\infty$		$\infty \leftarrow$	3001	301	31

As x approaches 3 from the left, the function approaches $-\infty$ (gets smaller and smaller or is unbounded in a negative direction). We say that

$$\lim_{x \to 3^-} \frac{x}{x - 3} \to -\infty$$

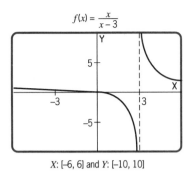

$f(x) = \dfrac{x}{x-3}$

X: [–6, 6] and Y: [–10, 10]

Figure 52

Also, as x approaches 3 from the right, the function approaches ∞ (gets larger and larger or is unbounded in a positive direction). We say that

$$\lim_{x \to 3^+} \frac{x}{x-3} \to \infty$$

The graph is shown in Figure 52 and confirms the behavior we saw indicated in the table of values. The line $x = 3$ is a vertical asymptote. ∎

We have been looking at functions that exhibit **asymptotic behavior**. A function exhibits asymptotic behavior around a vertical asymptote $x = a$ when it becomes arbitrarily close to the asymptote as x approaches a. This leads us to the following definition.

DEFINITION: VERTICAL ASYMPTOTES

If $\lim\limits_{x \to a^+} f(x) \to \pm \infty$ and/or $\lim\limits_{x \to a^-} f(x) \to \pm \infty$, then the line $x = a$ is called a **vertical asymptote**. For a rational function, a vertical asymptote can be found by locating any value of x that will make the denominator (but not the numerator) equal to 0. If the numerator is also 0 at that value, then $x = a$ may not be a vertical asymptote.

In Example 27, $x = 2$ is a vertical asymptote of $f(x) = 1/(x - 2)^2$. In Example 28, $x = 3$ is a vertical asymptote of $f(x) = x/(x - 3)$. A function such as

$$f(x) = \frac{x(x - 1)}{x(x + 1)}$$

will have a "hole" in the graph at $x = 0$, but it will not have a vertical asymptote at $x = 0$ because x is a factor of both the numerator and the denominator.

EXAMPLE 29   Find the vertical asymptotes and draw the graph of $y = 3/(x - 2)$.

SOLUTION

$$\lim_{x \to 2^-} \frac{3}{x-2} \to -\infty \quad \text{and} \quad \lim_{x \to 2^+} \frac{3}{x-2} \to \infty$$

Thus $x = 2$ is a vertical asymptote. We note that y increases without bound as x approaches this asymptote from the right. As $x \to 2$ from the left $y \to -\infty$. Note this on the graph in Figure 53. With your graphing calculator, duplicate this graph and use the Trace feature to note how the values change rather dramatically as $x \to 2^+$ and $x \to 2^-$. ∎

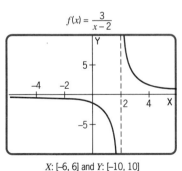

$f(x) = \dfrac{3}{x-2}$

X: [–6, 6] and Y: [–10, 10]

Figure 53

You can determine the behavior of a function as it approaches a vertical asymptote by analyzing the sign of the function very close to that value. It is not necessary to actually calculate the function, but only whether it will be positive or negative. When a function demonstrates asymptotic behavior, that behavior will remain consistent very close to the asymptote. The function will not suddenly change directions. Notice that if $x = 2.1$, then $3/(2.1 - 2) > 0$; therefore,

$$\lim_{x \to 2^+} \frac{3}{x - 2} \to \infty$$

Likewise, if $x = 1.9$, then $3/(1.9 - 2) < 0$ and we have

$$\lim_{x \to 2^-} \frac{3}{x - 2} \to -\infty$$

Calculator Note

As we learned earlier, the evaluation function on your calculator allows you to determine the value of the function at any value for x within the domain of x entered into the calculator. Using this function, you can quickly determine the sign of the function very close to the vertical asymptote. You can also use the Table feature to examine the behavior of the function close to the vertical asymptote.

Limit as $x \to \infty$

In addition to being interested in what happens to function values as the independent variable approaches a particular number, we are often concerned about what happens as the independent variable gets larger and larger or increases without bound. For example, consider the function

$$f(x) = \frac{x - 1}{x} = 1 - \frac{1}{x}$$

and determine what happens as x gets larger and larger. Values for $f(x)$ are tabulated in Table 5. It appears that $f(x) = 1 - 1/x$ approaches 1 as x gets larger and larger. To indicate this, we write

$$\lim_{x \to \infty} \left(1 - \frac{1}{x} \right) = 1$$

where $x \to \infty$ (read "x approaches infinity") means that x is increasing without bound.

TABLE 5

x	1	2	3	5	10	100	1000	100,000	$\to$	∞
$f(x)$	0	$\frac{1}{2}$	$\frac{2}{3}$	$\frac{4}{5}$	0.9	0.99	0.999	0.99999	$\to$	1

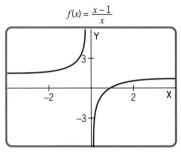

$f(x) = \dfrac{x-1}{x}$

X: [-4, 4] and Y: [-6, 6]

Figure 54

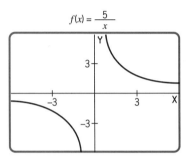

$f(x) = \dfrac{5}{x}$

X: [-6, 6] and Y: [-6, 6]

Figure 55

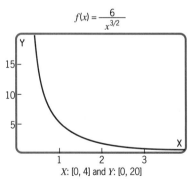

$f(x) = \dfrac{6}{x^{3/2}}$

X: [0, 4] and Y: [0, 20]

Figure 56

Similarly, as x becomes more and more negative, the function also approaches 1. We indicate this by writing $\lim\limits_{x \to -\infty} f(x) = 1$. Note the graph in Figure 54.

For real numbers L and M, the notation $\lim\limits_{x \to \infty} f(x) = L$ means that $f(x)$ approaches L as x gets larger and larger; the notation $\lim\limits_{x \to -\infty} f(x) = M$ means that $f(x)$ approaches M as x gets smaller and smaller. (The number is negative but the absolute value is large.)

All the limit properties listed previously hold for $\lim\limits_{x \to \infty} f(x)$, with the exception of the limit of a quotient. The limit of a quotient is replaced by the following property.

SPECIAL LIMIT

For any constant c,

$$\lim\limits_{x \to \infty} \frac{c}{x^n} = 0 \quad \text{and} \quad \lim\limits_{x \to -\infty} \frac{c}{x^n} = 0 \quad \text{for } n > 0$$

EXAMPLE 30

$$\lim\limits_{x \to \infty} \frac{5}{x} = 0$$

The graph is shown in Figure 55. We see also that

$$\lim\limits_{x \to -\infty} \frac{5}{x} = 0$$

EXAMPLE 31   Find $\lim\limits_{x \to \infty} \dfrac{6}{x^{3/2}}$.

SOLUTION   Since $n = 3/2 > 0$, we can say that

$$\lim\limits_{x \to \infty} \frac{6}{x^{3/2}} = 0$$

See the graph in Figure 56. Note that the function is not defined for $x \le 0$.

EXAMPLE 32   Find $\lim\limits_{x \to \infty} \dfrac{3x^2 - 5}{2x^2 + x - 3}$.

SOLUTION   This function is a quotient of two other functions (both the numerator and denominator are second-degree functions). However, we cannot use the formula for the limit of a quotient because

$$\lim\limits_{x \to \infty} (3x^2 - 5) \to \infty \quad \text{and} \quad \lim\limits_{x \to \infty} (2x^2 + x - 3) \to \infty$$

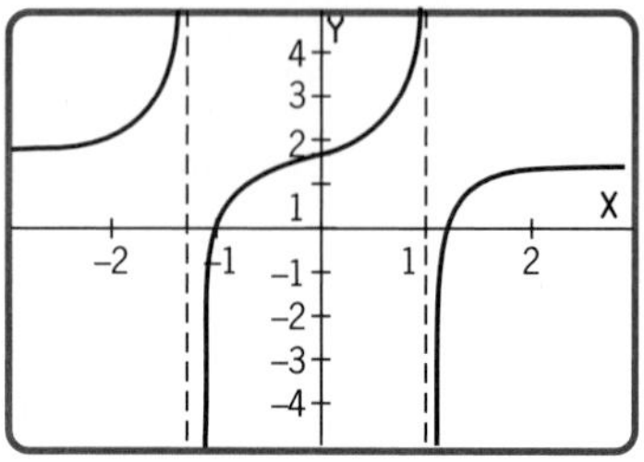

$f(x) = \dfrac{3x^2 - 5}{2x^2 + x - 3}$

X:[-3, 3] and Y:[-5, 5]

Figure 57

In this situation we use algebra to transform the quotient into a form where the numerator and denominator do have limits. The algebraic manipulation is to divide the numerator and the denominator by x raised to the largest exponent that occurs in the expression. In this case, we divide each term in the expression by x^2 and then use the special limit property for a quotient (see Figure 57).

$$\lim_{x \to \infty} \frac{3x^2 - 5}{2x^2 + x - 3} = \lim_{x \to \infty} \frac{3 - \dfrac{5}{x^2}}{2 + \dfrac{1}{x} - \dfrac{3}{x^2}}$$

$$= \frac{\lim_{x \to \infty} \left(3 - \dfrac{5}{x^2}\right)}{\lim_{x \to \infty} \left(2 + \dfrac{1}{x} - \dfrac{3}{x^2}\right)} = \frac{3 - 0}{2 + 0 - 0} = \frac{3}{2}$$

We arrived at the above answer because

$$\lim_{x \to \infty} \frac{-5}{x^2} = 0, \quad \lim_{x \to \infty} \frac{1}{x} = 0, \quad \text{and} \quad \lim_{x \to \infty} \frac{-3}{x^2} = 0 \qquad \blacksquare$$

Calculator Note

You must use some care when graphing the function

$$f(x) = \frac{3x^2 - 5}{2x^2 + x - 3}$$

If you do not use **dot mode**, you may get a rather strange looking graph if the window is too large. Press the MODE key and moved the cursor down to the line that has "Connected Dot." Move the cursor to the right so that it is flashing over the word "Dot." Press ENTER. The calculator will not now try to connect parts of the graph that should not be connected. Use the ZOOM menu and you will get a good picture of the function.

Practice Problem 2 Find $\lim_{x \to \infty} \dfrac{5x^2 + 2}{3x^2 + 2x}$.

ANSWER $\dfrac{5}{3}$

EXAMPLE 33 Find $\lim_{x \to \infty} \dfrac{3x^2 + x - 5}{2x + 5}$.

SOLUTION   Divide the numerator and denominator by x^2, since the largest exponent in the expression is 2.

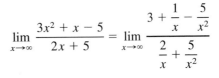

$$\lim_{x \to \infty} \frac{3x^2 + x - 5}{2x + 5} = \lim_{x \to \infty} \frac{3 + \dfrac{1}{x} - \dfrac{5}{x^2}}{\dfrac{2}{x} + \dfrac{5}{x^2}}$$

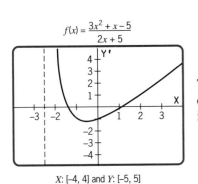

$f(x) = \dfrac{3x^2 + x - 5}{2x + 5}$

X: [–4, 4] and Y: [–5, 5]

Figure 58

The limit of the numerator is 3, and the limit of the denominator is 0; consequently, the quotient becomes very large and the limit as x approaches infinity increases without bound. This is indicated by

$$\lim_{x \to \infty} \frac{3x^2 + x - 5}{2x + 5} \to \infty$$

Figure 58 shows the graph of the function

$$f(x) = \frac{3x^2 + x - 5}{2x + 5}$$

NOTE: In Examples 30, 31 and 32, the limits exist. In Example 33, the limit fails to exist. With this in mind, we state the following:

1. If the degree of the polynomial in the numerator of a rational function is greater than the degree of the denominator, the limit of the fraction as $x \to \infty$ is $\pm \infty$. (That is, it fails to exist and exhibits unbounded behavior as in Example 33.)
2. If the degree of the numerator and denominator are the same, the limit as $x \to \pm \infty$ is a number that is the quotient of the coefficients of the highest-degree terms in the numerator and in the denominator. (See Example 32.)
3. If the degree of the denominator is greater than the degree of the numerator, the limit of the function as $x \to \pm \infty$ is 0. (See Examples 30 and 31.)

As $|x|$ gets very large, we see in Examples 30 and 31 that the function gets very close to the horizontal line $y = 0$ but does not reach it. In Example 32, the function gets very close to the line $y = \frac{3}{2}$ as $|x|$ gets very large, but it does not reach the line. The line $y = 3/2$ is called a **horizontal asymptote**.

DEFINITION: HORIZONTAL ASYMPTOTE

If $\lim\limits_{x \to \infty} f(x) = c$ and/or $\lim\limits_{x \to -\infty} f(x) = c$, then the line $y = c$ is called a **horizontal asymptote**.

Be aware of the fact that we are dealing with very large values of x, $x \rightarrow \pm \infty$. A function may cross a horizontal asymptote when x is relatively small, but it will not cross a horizontal asymptote as $x \rightarrow \pm \infty$ (see Example 32) for these functions.

EXAMPLE 34   Find the horizontal and vertical asymptotes for

$$y = \frac{2x + 1}{x - 1}$$

SOLUTION

$$\lim_{x \to \infty} \frac{2x + 1}{x - 1} = \lim_{x \to \infty} \frac{2 + \dfrac{1}{x}}{1 - \dfrac{1}{x}} = 2 \qquad \text{Remember that } \lim_{x \to \infty} \frac{1}{x} = 0.$$

The line $y = 2$ is a horizontal asymptote.

By inspection, we see that $x = 1$ makes the denominator zero and does not make the numerator zero. Thus, $x = 1$ is a vertical asymptote. Looking at the first derivative we have

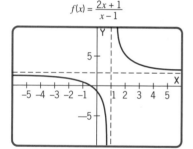

$f(x) = \frac{2x + 1}{x - 1}$

X: [–6, 6] and Y: [–10, 10]

Figure 59

$$y' = \frac{2(x - 1) - (2x + 1) \cdot 1}{(x - 1)^2} = \frac{-3}{(x - 1)^2} \qquad \frac{N'D - ND'}{D^2}$$

We see that $y' < 0$ for all values of x, the function is always decreasing. (Draw a graph of y' to confirm this.)

Likewise,

$$y'' = \frac{6}{(x - 1)^3}$$

When $x < 1$, we see that $y'' < 0$ and the graph will be concave downward. When $x > 1$, then $y'' > 0$ and the graph will be concave upward. (Draw the graph of y'' to confirm this.) These conditions are satisfied in the graph of the function in Figure 59. ∎

Practice Problem 3 Find the horizontal and vertical asymptotes of

$$y = \frac{3x - 1}{1 - x}$$

ANSWER   There is a vertical asymptote at $x = 1$ and a horizontal asymptote at $y = -3$. The graph is shown in Figure 60.

$f(x) = \frac{3x - 1}{1 - x}$

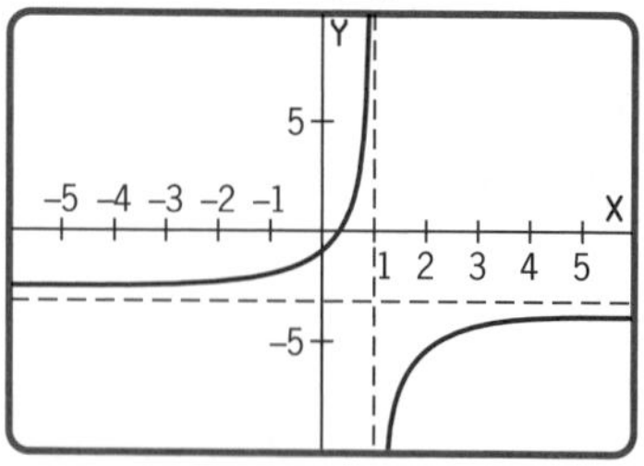

X: [–6, 6] and Y: [–10, 10]

Figure 60

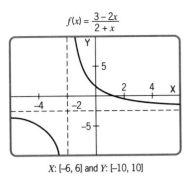

$f(x) = \dfrac{3 - 2x}{2 + x}$

X: [-6, 6] and Y: [-10, 10]

Figure 61

Practice Problem 4 Find the horizontal and vertical asymptotes of

$$y = \frac{3 - 2x}{2 + x}$$

ANSWER   There is a vertical asymptote at $x = -2$ and a horizontal asymptote at $y = -2$ (see Figure 61).

EXAMPLE 35   Discuss the graph of the function

$$y = f(x) = \frac{1}{x^2 - 1}$$

relative to each of the following.

(a) Vertical asymptotes
(b) Horizontal asymptotes
(c) Intervals where the function is increasing and where it is decreasing
(d) Relative extrema
(e) Concavity

SOLUTION

(a) Vertical asymptotes: Since

$$f(x) = \frac{1}{x^2 - 1} = \frac{1}{(x - 1)(x + 1)}$$

we see that $x = -1$ and $x = 1$ are vertical asymptotes.

(b) Horizontal asymptotes: Since

$$\lim_{x \to \infty} \frac{1}{x^2 - 1} = \lim_{x \to \infty} \frac{\dfrac{1}{x^2}}{1 - \dfrac{1}{x^2}} = 0$$

we know that $y = 0$ (the x-axis) is a horizontal asymptote.

(c) Increasing and decreasing intervals: We have

$$f'(x) = \frac{-2x}{(x^2 - 1)^2} = \frac{-2x}{(x - 1)^2(x + 1)^2}$$

The only critical value is $x = 0$, since the function is not defined for $x = 1$ and $x = -1$. Look at the graph in Figure 62. From the graph of f', we see that $f' > 0$ over $(-\infty, -1)$, $(-1, 0)$. Thus, $f(x)$ is increasing there.

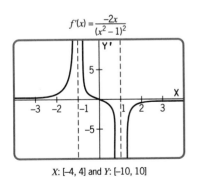

$f'(x) = \dfrac{-2x}{(x^2 - 1)^2}$

X: [-4, 4] and Y: [-10, 10]

Figure 62

$$f''(x) = \frac{6x^2 + 2}{(x^2 - 1)^3}$$

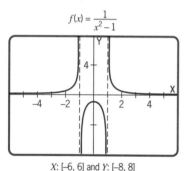

X: [-6, 6] and Y: [-10, 10]

Figure 63

$$f(x) = \frac{1}{x^2 - 1}$$

X: [-6, 6] and Y: [-8, 8]

Figure 64

Likewise, $f' < 0$ over $(0, 1)$, $(1, \infty)$; therefore, f is decreasing on those intervals.

(d)   Relative extrema: Since $x = 0$ is the only critical point, we can see from the graph that $f'(0) = 0$ and f' changes from positive to negative; therefore, there is a relative maximum at $(0, -1)$ [i.e., $f(0) = -1$].

(e)   Concavity: To determine concavity, we must find $f''(x)$.

$$f''(x) = \frac{-2(x^2 - 1)^2 - (-2x) \cdot 2(x^2 - 1)2x}{(x^2 - 1)^4}$$

$$= \frac{(x^2 - 1)(-2x^2 + 2 + 8x^2)}{(x^2 - 1)^4} = \frac{6x^2 + 2}{(x^2 - 1)^3}$$

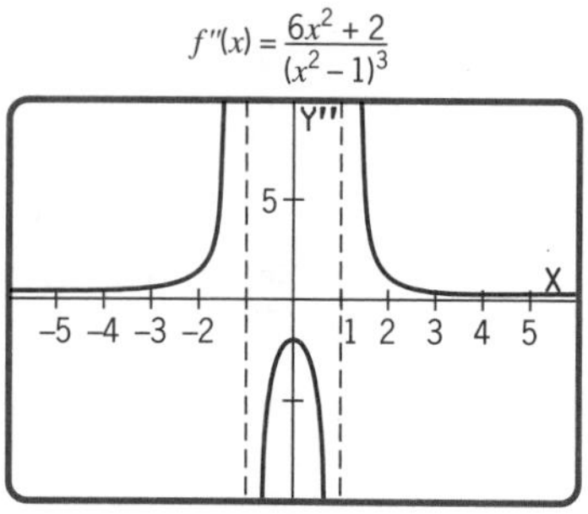

Looking at the graph of f'' in Figure 63, we determine that $f'' > 0$ on $(-\infty, -1)$ and $(1, \infty)$; therefore, f is concave up on those intervals. Likewise, $f'' < 0$ on $(-1, 1)$, and f is concave down on that interval. The graph of f is shown in Figure 64. ■

When finding horizontal asymptotes, instead of evaluating the limits you can use the following guidelines derived from taking $\lim_{x \to \infty} [f(x)/g(x)]$ (see the note on page 601).

For a rational function $y = f(x)/g(x)$, if the degree of $f(x) = m$ and the degree of $g(x) = n$, then we have the following:

$$\lim_{x \to \pm\infty} \frac{f(x)}{g(x)} = \begin{cases} 0, & \text{if } m < n \\ \dfrac{\text{Coefficient of } x^m}{\text{Coefficient of } x^n}, & \text{if } m = n \\ \pm\infty & \text{if } m > n \end{cases}$$

EXAMPLE 36   Find each of the following limits using the properties above.

(a) $\displaystyle\lim_{x \to \infty} \frac{3x}{x^2 + 2}$

(b) $\displaystyle\lim_{x \to \infty} \frac{x^2 - 3}{4x^2 + 2}$

(c) $\displaystyle\lim_{x \to \infty} \frac{x^4 + 1}{x^2 + 2x + 1}$

(d) $\displaystyle\lim_{x \to -\infty} \frac{x^2}{x + 1}$

SOLUTION

(a) $\displaystyle\lim_{x \to \infty} \frac{3x}{x^2 + 2} = 0$ Degree of numerator < Degree of denominator

(b) $\displaystyle\lim_{x \to \infty} \frac{x^2 - 3}{4x^2 + 2} = \frac{1}{4}$ Degree of numerator = Degree of denominator

(c) $\displaystyle\lim_{x \to \infty} \frac{x^4 + 1}{x^2 + 2x + 1} \to \infty$ Degree of numerator > Degree of denominator

(d) $\displaystyle\lim_{x \to -\infty} \frac{x^2}{x + 1} \to -\infty$ Degree of numerator > Degree of denominator   ■

$f'(x) = \dfrac{-6-x}{x^3}$

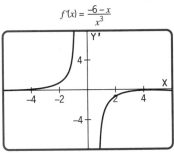

X: [–6, 6] and Y: [–8, 8]

(a)

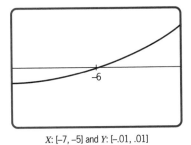

X: [–7, –5] and Y: [–.01, .01]

(b)

Figure 65

Remember, a function may cross a horizontal asymptote, but a function will not cross a vertical asymptote. Be careful when analyzing a graph drawn on your graphing calculator, because inflection points may not be apparent when you are looking at the screen as we will discover in Example 37.

EXAMPLE 37   Analyze the graph of $f(x) = \dfrac{x+3}{x^2}$.

SOLUTION

1. By examining the function, we can see that it has a vertical asymptote at $x = 0$ and a horizontal asymptote at $y = 0$.

2. We have

$$f'(x) = \frac{-6-x}{x^3}$$

so the only critical value is $x = -6$. Note that $x = 0$ is not a critical value because the function is not defined there (Figure 65).

3. We have

$$f''(x) = \frac{18+2x}{x^4}$$

so there is a possible inflection point at $x = -9$ (Figure 66).

$f''(x) = \dfrac{18+2x}{x^4}$

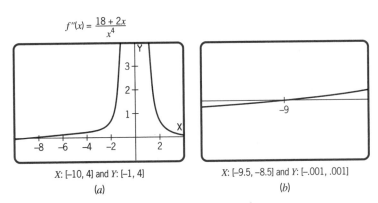

X: [–10, 4] and Y: [–1, 4]

(a)

X: [–9.5, –8.5] and Y: [–.001, .001]

(b)

Figure 66

4. The graphs in Figure 65 will help, but you must use care when using a graphing calculator. There is a sign change in f' at $x = -6$, but it may not be readily apparent on your calculator screen. Look at the graph in Figure 65(b) with a much more narrow range. When a sign change, or lack of change, is not apparent at a critical value or a possible inflection point, you must investigate by using either the Zoom feature or a more restricted range.

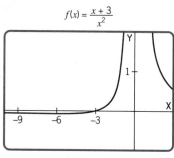

$f(x) = \dfrac{x+3}{x^2}$

X: [-10, 3] and Y: [-1, 2]

Figure 67

$f(x) = \dfrac{x-1}{x^2}$

X: [-4, 4] and Y: [-6, 2]

Figure 68

$f(x) = \dfrac{2x^2}{1+x^2}$

X: [-4, 4] and Y: [-.5, 2.5]

Figure 69

5. We see that there is a relative minimum at $(-6, -\frac{1}{12})$ and an inflection point at $(-9, -\frac{2}{27})$ and that the function is concave down on the interval $(-\infty, -9)$ and concave up on the intervals $(-9, 0)$ and $(0, \infty)$.
6. The graph of f is shown in Figure 67. Note that the point of inflection is very subtle and is hard to see, so you must examine that part of the graph very carefully.   ■

Practice Problem 5 For $f(x) = (x-1)/x^2$ find all extrema, asymptotes, and inflection points.

ANSWER We have

$$f'(x) = \frac{2-x}{x^3} \quad \text{and} \quad f''(x) = \frac{2x-6}{x^4}$$

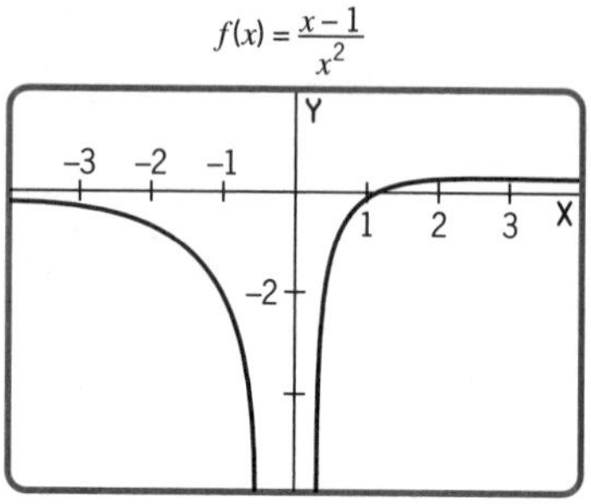

From Figure 68 we see that there is a vertical asymptote at $x = 0$ and a horizontal asymptote at $y = 0$. There is a relative maximum at $(2, \frac{1}{4})$ and an inflection point at $(3, \frac{2}{9})$.

Practice Problem 6 Analyze the graph of $f(x) = \dfrac{2x^2}{1+x^2}$.

ANSWER Looking at the graph in Figure 69, we see that there is a horizontal asymptote at $y = 2$ and there are no vertical asymptotes. By using the first and second derivatives, we find that there is a relative minimum at $(0, 0)$ and there are inflection points at $(0.577, 0.5)$ and $(-0.577, 0.5)$.

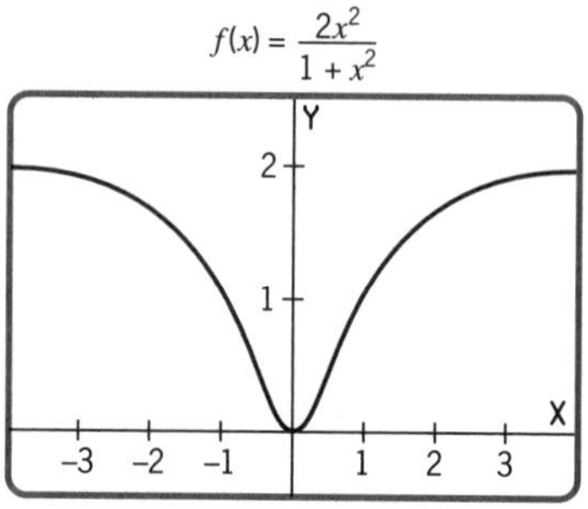

EXAMPLE 38 Analyze the graph of

$$f(x) = \frac{2x}{\sqrt{x^2+2}}$$

by using the Trace key on your calculator.

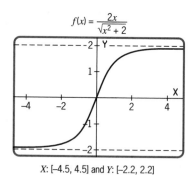

$f(x) = \dfrac{2x}{\sqrt{x^2+2}}$

X: [-4.5, 4.5] and Y: [-2.2, 2.2]

Figure 70

SOLUTION   The graph is shown in Figure 70. This is an example of a function that has two horizontal asymptotes.

$$\lim_{x \to +\infty} f(x) = 2 \quad \text{and} \quad \lim_{x \to -\infty} f(x) = -2$$

Draw the curve with the lines $y = 2$ and $y = -2$ to demonstrate this.   ■

SUMMARY

When the limit of a function fails to exist as x approaches some number a either from the left or the right, we use the notation $\lim_{x \to a} f(x) \to \pm\infty$ to indicate this unbounded behavior. Remember that ∞ is not a number and cannot be manipulated like one; it is an indication of a behavior. If, for very large values of x, a function approaches a limiting value L, we use the notation $\lim_{x \to \infty} f(x) = L$. The line $y = L$ is a horizontal asymptote. If there is a value of the variable, c, that makes the denominator of a rational function zero, but does not make the numerator zero, then the line $x = c$ may be a vertical asymptote.

Exercise Set 9.5

Use a calculator to evaluate each function at $x = 10$, $x = 100$, $x = 1000$, and $x = 10,000$. Estimate $\lim_{x \to \infty} f(x)$ for each of the following functions.

1. $f(x) = \dfrac{4}{x + 2}$

2. $f(x) = \dfrac{x + 2}{2x - 2}$

3. $f(x) = \dfrac{x^2}{x + 2}$

4. $f(x) = \dfrac{x + 2}{x}$

Use a calculator to evaluate each function at $x = -10$, $x = -100$, $x = -1000$, and $x = -10,000$. Estimate $\lim_{x \to -\infty} f(x)$ for each of the following functions.

5. $f(x) = \dfrac{4}{x + 2}$

6. $f(x) = \dfrac{x + 2}{2x - 2}$

7. $f(x) = \dfrac{x^2}{x + 2}$

8. $f(x) = \dfrac{x + 2}{x}$

Use a calculator to evaluate each function at $x = 0.9$, $x = 0.99$, $x = 0.999$, and $x = 0.9999$. Estimate

$\lim_{x \to 1^-} f(x)$ *for each of the following functions from your calculations.*

9. $f(x) = \dfrac{2}{x - 1}$

10. $f(x) = \dfrac{2x}{x - 1}$

11. $f(x) = \dfrac{1}{|x - 1|}$

12. $f(x) = \dfrac{2}{(x - 1)^2}$

Use a calculator to evaluate each function at $x = 1.1$, $x = 1.01$, $x = 1.001$, and $x = 1.0001$. Estimate $\lim_{x \to 1^+} f(x)$ for each of the following functions from your calculations.

13. $f(x) = \dfrac{2}{x - 1}$

14. $f(x) = \dfrac{2x}{x - 1}$

15. $f(x) = \dfrac{1}{|x - 1|}$

16. $f(x) = \dfrac{2}{(x - 1)^2}$

Study the following graphs and identify each of the following, if they exist.

(a) *Vertical asymptotes*
(b) *Horizontal asymptotes*

(c) $\lim_{x \to \infty} f(x)$

(d) $\lim_{x \to -\infty} f(x)$

17.

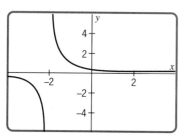

18.

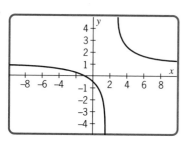

Find the following limits, if they exist. Use both the properties of limits and graphs of the functions.

19. $\lim_{x \to \infty} \dfrac{3}{x}$

20. $\lim_{x \to \infty} \dfrac{5}{x}$

21. $\lim_{x \to \infty} \left(7 - \dfrac{5}{x}\right)$

22. $\lim_{x \to \infty} \left(\dfrac{2}{x^2} - \dfrac{3}{x} + 5\right)$

23. $\lim_{x \to \infty} \dfrac{3x - 2}{3x + 5}$

24. $\lim_{x \to \infty} \dfrac{4x^2 - 2}{4x + 2}$

25. $\lim_{x \to \infty} \dfrac{3x^2 - 5x + 7}{4x^2 + x + 1}$

26. $\lim_{x \to \infty} \dfrac{5x^2 + 3x - 7}{6x^2 + 3x + 2}$

27. $\lim_{x \to \infty} \dfrac{3x^3 - 2x + 5}{4x + 5}$

28. $\lim_{x \to \infty} \dfrac{4x + 5}{3x^2 - 2x + 5}$

29. $\lim_{x \to \infty} \dfrac{x^3 + 2x}{x^2 + 4}$

30. $\lim_{x \to \infty} \dfrac{4x^4}{x^2 + 1}$

31. $\lim_{x \to \infty} \dfrac{1}{\sqrt{1 + x}}$

32. $\lim_{x \to \infty} \dfrac{x}{\sqrt{x - 2}}$

For each function below, find the horizontal and vertical asymptotes, relative extrema, inflection points,

and sketch the graph. Use both your graphing calculator and calculus techniques.

33. $f(x) = \dfrac{x - 3}{x + 2}$

34. $f(x) = \dfrac{1}{(x - 2)^3}$

35. $f(x) = \dfrac{2x}{1 + x}$

36. $f(x) = \dfrac{x}{x^2 - 1}$

37. $f(x) = \dfrac{2x + 3}{x - 1}$

38. $f(x) = \dfrac{5 - 2x}{3 + x}$

39. $f(x) = \dfrac{2x^2 - 6}{(x - 1)^2}$

40. $f(x) = 1 + \dfrac{1}{x} + \dfrac{1}{x^2}$

41. $f(x) = \dfrac{6 - x}{x^2}$

42. $f(x) = \dfrac{x^2}{(x + 2)^2}$

43. $f(x) = \dfrac{1 - 2x}{x^2}$

44. $f(x) = \dfrac{x + 1}{x^2}$

Find any horizontal and vertical asymptotes.

45. $y = \dfrac{x^3}{x^2 - 1}$

46. $y = \dfrac{-4x}{x^2 + 4}$

47. $y = 2 + \dfrac{x^2}{x^4 + 1}$

48. $y = \dfrac{2x^2 - 3x + 5}{x^2 + 1}$

49. Find $\lim_{x \to 0} \sqrt{\dfrac{1 + x}{x^2} - \dfrac{1}{x}}$

Applications (Business and Economics)

50. **Average Cost.** The cost function for manufacturing x hair brushes is $C(x) = 4000 + 2.50x$. Given that the average cost function is

$$\overline{C}(x) = \dfrac{C(x)}{x}$$

find $\lim_{x \to \infty} \overline{C}(x)$.

51. **Average Profit.** If the hair brushes in Exercise 50 sell for \$6 each and the average profit function is

$$\overline{P}(x) = \dfrac{P(x)}{x}$$

find $\lim_{x \to \infty} \overline{P}(x)$.

52. **Supply Function.** A supply function is given as

$$p(x) = \frac{900}{x} - \frac{800}{x^2}, \qquad x \geq 1$$

where x is the number of available items. Find $\lim\limits_{x \to \infty} p(x)$.

53. **Cost Function.** A cost function is given as

$$C(x) = 5000 + 10x + \frac{40}{x}$$

where x is the number of items produced. Find $\lim\limits_{x \to \infty} C(x)$.

54. **Profit.** A company finds that the cost of manufacturing x electric saws in a week is given by

$$C(x) = 0.16x^2 - 57.8x + 6220$$

The company is able to sell the saws at $20 each minus $x/1000$ (discount for large purchases).
(a) Find the revenue function.
(b) Find the maximum profit.
(c) Draw graphs showing $C(x)$ and $R(x)$ and locate the regions for profit and loss.
(d) Mark where the maximum profit occurs on the graph in part (c).

Applications (Social and Life Sciences)

55. **Voting Trends.** Suppose that the probability P that a person will vote yes as the nth voter on an issue is

$$P = \frac{1}{3} + \frac{1}{10}\left(-\frac{1}{2}\right)^n$$

Find the value of $\lim\limits_{n \to \infty} P$.

56. **Growth.** A growth function is given by

$$N(t) = \frac{80{,}000t}{100 + t}$$

where t is the time in minutes. Find the value of $\lim\limits_{t \to \infty} N(t)$.

57. **Bacteria.** When bacteria are introduced into a medium, they often increase according to the graph shown below. Explain the significance of each of the following.
(a) $x = a$ (b) $(a, f(a))$ (c) $y = b$

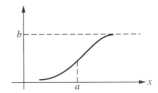

Review Exercises

Find points of absolute maxima and absolute minima on the given closed intervals.

58. $f(x) = x^2(x - 3)$ on $[-3, 3]$
59. $f(x) = x(x - 2)^2$ on $[-3, 3]$

9.6 OPTIMIZATION

OVERVIEW   Many practical problems require determining maximum or minimum values. For example, businesspeople wish to maximize profit and minimize cost. Builders wish to maximize the strength of their structures. The government has to be concerned about maximizing tax revenue, and the retailer wants to minimize inventory cost. Such problems are called optimization problems. They require the determination of absolute maxima or absolute minima. In this section we apply the theory developed in the preceding sections to help determine the absolute extrema.

Although it is impossible to describe mathematical procedures that can solve all optimization problems, it is possible to state some general rules.

> ## Steps in Optimization
>
> 1. Use any or all of the techniques and strategies for solving problems you have learned in previous chapters.
> 2. Work toward a functional relationship between variables to be optimized and one other variable. (If two or more independent variables arise, it will be necessary in this chapter to eliminate all but one.)
> 3. Find critical values and locate absolute maxima and minima.

For the problems in this section, you should check your work by drawing the graph of the function being optimized. The first two examples should give some understanding of optimization before we consider the more important applications of this section.

EXAMPLE 39   The sum of two positive numbers is 220. Find the two numbers such that the product of the two numbers is a maximum.

SOLUTION   Since the sum of the two numbers is 220, if one number is 100, the other number is $220 - 100 = 120$. The product of the two numbers is $100(120) = 12,000$. This is probably not the maximum value, but we can use this as we proceed. Instead of 100, let the first of the two numbers be x. Then the second number is $220 - x$. The product of these two numbers is

$$P = x(220 - x) = 220x - x^2$$

Furthermore, since the numbers must be positive, we look only at the interval $[0, 220]$. We want P (the product) to be a maximum, so we take the first derivative to locate the critical value of x.

$$\frac{dP}{dx} = 220 - 2x$$

Setting $220 - 2x$ equal to zero yields

$$220 - 2x = 0$$

$$2x = 220$$

$$x = 110 \qquad \text{Critical value}$$

$$\frac{d^2P}{dx^2} = -2 \qquad \text{Second derivative}$$

The second derivative is negative and we now know that there is a relative maximum at $x = 110$. At $x = 0$ and at $x = 220$, the endpoints of the interval, the value of the product is 0. Therefore, at $x = 110$ we see that

$$P(110) = 110(220 - 110) = 12,100$$

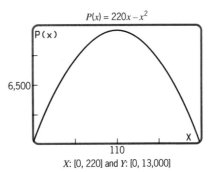

Figure 71

is the maximum product. So the two numbers are 110 and 110. The graph of the product function over [0, 220] is shown in Figure 71. ∎

Practice Problem 1 Find two numbers whose sum is 72 and whose product is a maximum.

ANSWER 36 and 36

EXAMPLE 40 Using 120 feet of fencing, a farmer wishes to contain a cow in a rectangular plot of land that has one side along the bank of a river. If no fencing is needed along the river, what should be the dimensions of the rectangular field to provide the cow with maximum grazing area?

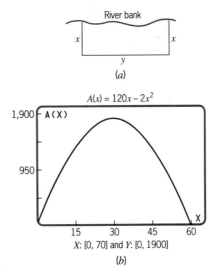

Figure 72

SOLUTION Whenever possible, it is a good idea to draw a figure to help set up a problem. The situation is shown in Figure 72(*a*). By looking at the figure, let's first make a guess of the dimensions we will select. If we were to choose $x = 10$, then we would have $y = 120 - 2(10) = 100$, since a total of 120 feet is available. With these dimensions, the area would be, $A = 100 \cdot 10 = 1000$. Although this may seem like a good guess, we will need a function to maximize to find the dimensions for maximum area. Let's use our guess to help us set up an area function.

First, we must establish a relationship between the two variables. Since we know the amount of fence available, we let y be the length of the side that is parallel to the river and x be the length of the other two sides, resulting in the equation $2x + y = 120$. However, we want the area function to be in terms of one variable so that we can maximize it. By solving for y, we have $y = 120 - 2x$ and now we have a relationship established between the two variables. This will enable us to substitute one for the other as needed.

We now write the area of our rectangular field as a function of x, find the derivative and critical value(s), and then find the second derivative to use the second derivative test for extrema.

Area $= x \cdot y$	Now substitute $y = 120 - 2x$.
$A(x) = x(120 - 2x) = 120x - 2x^2$	This is area as a function of x and only x.
$A'(x) = 120 - 4x$	First derivative
$120 - 4x = 0$	Set the derivative to zero and solve.
$x = 30$	This is the critical value.
$A''(x) = -4$	Second derivative is negative.

The second derivative is negative; therefore, we have found a relative maximum. For $x = 30$, we have $y = 120 - 2(30) = 60$. The dimensions should be 30 by 60 feet to produce the maximum area of 1800 square feet. The graph of the area function is shown in Figure 72(b). ■

Practice Problem 2 A farmer wants to fence a rectangular area next to a mountain. Find the dimensions with the largest area possible using 3000 feet of fence if no fence is needed on the mountain side. (Use a figure similar to Figure 72.)

ANSWER 1500 feet by 750 feet

EXAMPLE 41 Design a box with a square base and maximum volume using 216 square inches of cardboard. The box will have a top and bottom.

SOLUTION We will let the dimensions of the base of the box be x by x and the height be h (see Figure 73). The volume of this box is $V = x^2h$. Before we try to work with variables, let's make a guess of some measurements to help us see how to proceed with the problem. If we let $x = 10$, then the base and top will each measure $10 \cdot 10 = 100$ square inches. That would be a total of 200 square inches, leaving only 16 square inches for the sides! This is quite obviously not a good guess, but we now have an idea of how to continue.

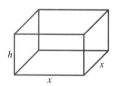

Figure 73

Note that the volume V is a function of two variables, x and h (length or width and height). Just as in Example 40, we will solve for one of the variables in terms of the other. The 216 square inches of cardboard are utilized in the four sides and two ends as follows:

$$\text{Area of four sides} = 4xh \qquad \text{Area of each side is } xh.$$

$$\text{Area of top} + \text{Area of bottom} = 2x^2$$

$$4xh + 2x^2 = 216 \qquad \text{Total area}$$

$$h = \frac{216 - 2x^2}{4x} \qquad \text{Solve for } h.$$

Substituting for h in the expression for V we obtain

$$V = x^2 h$$

$$V = x^2 \left[\frac{216 - 2x^2}{4x} \right]$$

$$V = 54x - \frac{1}{2} x^3$$

$$\frac{dV}{dx} = 54 - \frac{3}{2} x^2 \qquad \text{Differentiate volume function.}$$

Now find the critical value(s).

$$54 - \frac{3}{2} x^2 = 0$$

$$\frac{3}{2} x^2 = 54$$

$$x^2 = 36 \qquad \text{Dimensions of square base}$$

$$x = 6 \qquad \begin{array}{l} -6 \text{ cannot be a critical value, since } x \\ \text{cannot be negative.} \end{array}$$

$$h = \frac{216 - 2(6)^2}{4 \cdot 6}$$

$$= \frac{216 - 72}{24}$$

$$= 6 \text{ inches} \qquad \text{This is the height.}$$

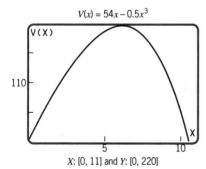

$V(x) = 54x - 0.5x^3$

V(X)

110

5   10   X

X: [0, 11] and Y: [0, 220]

Figure 74

Since $d^2V/dx^2 = -3x$ and at $x = 6$ we have $d^2V/dx^2 = -18$, there is a relative maximum at the critical value. Looking at the graph of V in Figure 74, note that the relative maximum is also an absolute maximum. Maximum $V = x^2 h = 6^2 \cdot 6 = 216$ cubic inches.

Practice Problem 3 A box with no top is to be made from a piece of metal that measures 20 feet by 20 feet. It is to be made by cutting out squares from each corner of the metal and then folding up the sides. What should the dimensions be to create a box with the maximum volume possible? What is the maximum volume?

ANSWER A square of width $\frac{10}{3} = 3\frac{1}{3}$ feet should be cut out (Figure 75). The dimension of the box would be $13\frac{1}{3}$ by $13\frac{1}{3}$ by $3\frac{1}{3}$ feet, yielding a volume of 16,000/27 cubic feet.

Figure 75

Maximum Profit

The next five examples in this section involve methods of optimization of business problems. Before we continue, make note of the following:

Marginal cost, $C'(x)$: Rate of change of cost
Marginal revenue, $R'(x)$: Rate of change of revenue
Marginal profit, $P'(x)$: Rate of change of profit

Furthermore, we have

$$\text{Profit} = \text{Revenue} - \text{Cost} \qquad P = R - C$$

and

$$\text{Marginal Profit} = \text{Marginal Revenue} - \text{Marginal Cost} \qquad P' = R' - C'$$

Since $P' = 0$ when $R' = C'$, **maximum profit** occurs when

$$\text{Marginal cost} = \text{Marginal revenue}$$

EXAMPLE 42 The revenue function and the cost function for an item are given by

$$R(x) = 12x - \frac{x^2}{12} - \frac{x^3}{6} \quad \text{and} \quad C(x) = 2 + 4x - \frac{x^2}{12}$$

where $0 \le x \le 6$ and x is in thousands. For what value of x is the profit a maximum?

SOLUTION To maximize profit, we must find the derivative and critical values, and then test with the second derivative.

$$P(x) = R(x) - C(x)$$

$$= 12x - \frac{x^2}{12} - \frac{x^3}{6} - \left[2 + 4x - \frac{x^2}{12} \right]$$

$$P(x) = \frac{-x^3}{6} + 8x - 2 \qquad \text{Profit function}$$

$$P'(x) = \frac{-x^2}{2} + 8 \qquad \text{Marginal profit function}$$

$$\frac{-x^2}{2} + 8 = 0 \qquad \text{Find critical values.}$$

$$16 - x^2 = 0$$

$$x = 4 \qquad \text{Reject } -4, \text{ since } x \ge 0.$$

$$P(4) = \frac{58}{3} \qquad \text{Evaluate the profit at } x = 4.$$

$$P''(x) = -x \qquad \text{Second derivative}$$

$$P''(4) < 0 \qquad \text{By the second derivative test, it is a maximum.}$$

By looking at the graph in Figure 76, we see that $x = 4$ is the absolute maximum. Thus, the profit is a maximum at $x = 4$ (or 4000 items). ■

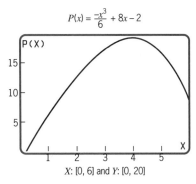

Figure 76

Profit can be maximized by solving the equation $MR = MC$. A producer can estimate marginal revenue and marginal cost at a given level of production and can raise production if $MR > MC$ or reduce production if $MR < MC$.

Practice Problem 4 A company makes and sells purses. The daily cost function is $C(x) = 80 + 8x - 1.65x^2 + 0.1x^3$ and the revenue function is $R(x) = 32x$, both in dollars. Assuming that the company can produce at most 30 bags per day, what production level x will yield the maximum daily profit?

ANSWER We have

$$P(x) = -80 + 24x + 1.65x^2 - 0.1x^3 \quad \text{and} \quad P'(x) = 24 + 3.3x - 0.3x^2$$

The critical value is $x = 16$, and a production level of 16 bags will yield the maximum daily profit.

EXAMPLE 43 If $C = 0.01x^2 + 5x + 100$ is a cost function, find the average cost function and the level of production x where average cost is a minimum.

SOLUTION Average cost is

$$\overline{C}(x) = \frac{C(x)}{x}$$

This is the function that we will minimize.

$$\overline{C}(x) = 0.01x + 5 + \frac{100}{x} \qquad \text{Find the average cost function.}$$

$$\overline{C}'(x) = 0.01 - \frac{100}{x^2} \qquad \text{Differentiate.}$$

$$0 = 0.01 - \frac{100}{x^2} \qquad \text{Find the critical values.}$$

$$0.01x^2 = 100$$

$$x^2 = 10{,}000$$

$$x = 100 \qquad \text{Use } x = 100 \text{ and reject } x = -100.$$

$$\overline{C}''(x) = \frac{200}{x^3} \qquad \text{Find the second derivative.}$$

$$\overline{C}''(100) > 0 \qquad \text{Minimum by the second derivative test}$$

Therefore, $x = 100$ is the level of production that yields the minimum average cost (see Figure 77). ■

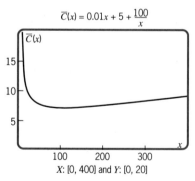

$$\overline{C}(x) = 0.01x + 5 + \frac{100}{x}$$

$\overline{C}(x)$

X: [0, 400] and Y: [0, 20]

Figure 77

Practice Problem 5 If $C(x) = x^2 + 160x + 6400$ is a cost function, find the value of x that gives minimum average cost. Graph the average cost function on your graphing calculator to verify the result.

ANSWER We have

$$\overline{C}(x) = x + 160 + \frac{6400}{x}$$

so $x = 80$ is the minimum average cost. The graph is shown in Figure 78.

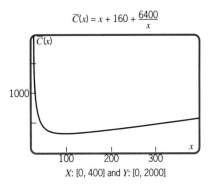

$\overline{C}(x) = x + 160 + \frac{6400}{x}$

X: [0, 400] and Y: [0, 2000]

Figure 78

EXAMPLE 44 A company sells a product for $1000 per set of 100 units. The cost, in dollars, of making x sets of 100 units in 1 year is $C(x) = 6 + 2x + 0.01x^2$. Write an expression for profit in terms of x. Find the number of sets of 100 units that would give a maximum profit.

SOLUTION The revenue for x sets of 100 units would be $R(x) = xp = 1000x$. Since Profit = Revenue − Cost, the profit on x sets of 100 units would be

$$P = R - C$$
$$P = 1000x - (6 + 2x + 0.01x^2)$$
$$= 1000x - 6 - 2x - 0.01x^2$$
$$= 998x - 6 - 0.01x^2$$

The first-derivative condition for a maximum profit is $dP/dx = 998 - 0.02x = 0$. We solve this and find that $x = 49{,}900$. Likewise, $d^2P/dx^2 = -0.02$, which shows that the function is concave down and is therefore a maximum by the second derivative test. Thus 49,900 sets of 100 units would yield a maximum profit. The maximum profit would be $24,900,094. The graph is shown in Figure 79. ■

NOTE: The first derivative of C with respect to x, called marginal cost, is the rate of change of cost when one more set of 100 units is produced. For the given example, $dC/dx = 2 + 0.02x$. In like manner, the rate of change of revenue with

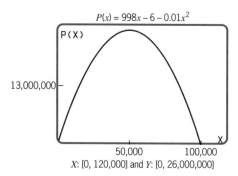

$P(x) = 998x - 6 - 0.01x^2$

X: [0, 120,000] and Y: [0, 26,000,000]

Figure 79

respect to the number of sets of units produced is called marginal revenue, and for this example is $dR/dx = 1000$. We see that the marginal cost increases with increasing x, whereas the marginal revenue is a constant. If production is continued until marginal cost equals marginal revenue, the number of sets of 100 units would be obtained from

$$1000 = 2 + 0.02x$$

$$x = 49{,}900$$

which is the same answer that we obtained in Example 44.

Practice Problem 6 A firm producing baseball gloves determines that in order to sell x gloves per week, the price per glove must be set at $p = 150 - 0.5x$. It is also determined that the cost of producing x number of gloves will be $C(x) = 2500 + 0.25x^2$. How many gloves must be sold to maximize profit each week?

ANSWER We have $R(x) = xp$ and $P'(x) = R'(x) - C'(x) = 150 - 1.5x$. The critical value is $x = 100$, which is value of x that produces maximum profit.

EXAMPLE 45 A 100-room budget motel is filled to capacity every night at $20 per room. For each $1 increase in rent, two fewer rooms are rented. If each rented room costs $4 to service each day, how much should the management charge for each room to maximize profit?

SOLUTION The change in the number of rooms rented depends on each $1 increase in rate. Therefore, we will let $x =$ the number of $1 increases in the rent. The remaining functions are then found as follows:

New number of rooms rented $= 100 - 2x$, since the number decreases by 2 for each $1 increase.
New rent $= 20 + x$.

Revenue = (Number of rooms rented)(Rent per room) = $(100 - 2x)(20 + x)$.

Cost = $4 · Number of rooms rented = $4(100 - 2x) = 400 - 8x$.

Profit = Revenue - Cost = $(100 - 2x)(20 + x) - 4(100 - 2x) = 1600 + 68x - 2x^2$.

Marginal profit = $P'(x) = 68 - 4x$.

From $P'(x) = 68 - 4x = 0$, we find that $x = 17$ is the only critical value. Since $P''(x) = -4 < 0$, the function is concave down and we have found that at $x = 17$ a rent of $20 + $17 = $37 should be charged to obtain maximum profit. The graph is shown in Figure 80. ■

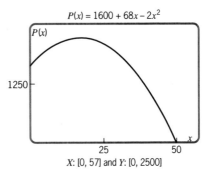

$P(x) = 1600 + 68x - 2x^2$

X: [0, 57] and Y: [0, 2500]

Figure 80

Practice Problem 7 An electronics supplier sells 50 CD players at $260 each. For every player over 50, the price of the player is reduced $2. If the supplier buys the CD players from the manufacturer for $120 each, how many CD players must be sold for the supplier to reach maximum profit each day?

ANSWER 60 players

EXAMPLE 46 A fruit grower has been planting 30 peach trees per acre in his orchard. The average yield is 400 peaches per tree. It is predicted that for each additional tree planted per acre, the yield will be reduced by 10 peaches per tree. How many trees should be planted per acre in order to have maximum yield?

SOLUTION To understand the problem, we summarize the data as follows. In the past, 30 trees have been planted per acre. The yield has been 400 peaches per tree. Each additional tree planted will reduce the yield per tree by 10 peaches. We can summarize the problem as follows by letting x = the number of trees planted.

New number of trees per acre = $30 + x$
Decrease in yield per tree = $10x$

New yield per tree $= 400 - 10x$
New total yield $= Y(x) = (30 + x)(400 - 10x) = 12,000 + 100x - 10x^2$
Marginal yield $= Y'(x) = 100 - 20x$

By setting $Y'(x) = 0$, we have $100 - 20x = 0$ or $x = 5$. Since $Y''(x) = -20 < 0$, $x = 5$ produces a relative maximum. By planting 5 additional trees production will be

$$Y(5) = 12,000 + 100(5) - 10(5^2) = 12,250$$

peaches per acre [the absolute maximum yield since $Y(x)$ is always concave down]. Note that we have worked this problem as if the function is continuous. We need to realize that if the answer had been 5.3 trees, we could not have 0.3 of a tree. The graph is shown in Figure 81. ■

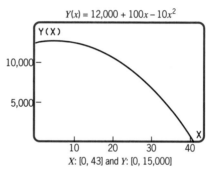

$Y(x) = 12,000 + 100x - 10x^2$

X: [0, 43] and Y: [0, 15,000]

Figure 81

Practice Problem 8 A fruit grower has been planting 25 trees per acre in his orchard. For this design, the average yield is 495 apples per tree. It is predicted that for each additional tree planted per acre the yield will be reduced by 15 apples per tree. How many additional trees should be planted to maximize the yield?

ANSWER 4 trees

Inventories

Every large retail outlet is concerned about the cost of storing the inventory. For example, suppose that a department store sells 3650 refrigerators a year. Of course, it could order all of these at one time, but then the store would have to pay very high **storage costs** (for space, insurance, etc.). If the store makes too many small orders, then costs can increase due to delivery charges, office records, manpower, and so on. We classify these as **reorder costs**. Inventory cost is calculated by adding the storage cost and the reorder cost. We will be trying to minimize

Inventory costs = Storage costs + Reorder costs

To obtain the function representing inventory costs, we simplify the problem by assuming that the number demanded is uniform per day. That is, 10 refrigerators will be sold each day for 365 days. Now let x be the number ordered originally and at each reordering period. A reorder is made so that the next order arrives just as the inventory is depleted. Thus, we have x refrigerators on hand at the beginning of a period and 0 at the end of the period. Hence the average number in storage over a period is

$$\frac{x + 0}{2} = \frac{x}{2}$$

Since this average is the same for each reorder period, we say that $x/2$ is the average number in storage over the year (see Figure 82).

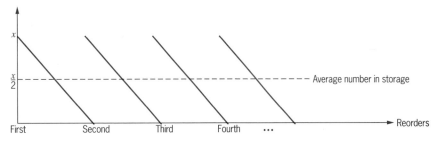

Figure 82

The yearly storage cost is usually found by multiplying the cost of storing 1 unit times $x/2$ (average number in storage). The number of reorders would be the total number needed in a year divided by x; that is, the number of reorders for refrigerators is $3650/x$. Usually, some expression can be found in terms of x for the cost of each order. Therefore, we can now express inventory costs as follows.

DEFINITION: INVENTORY COSTS

$$\text{Inventory costs} = (\text{Cost of storing 1 unit})\left(\frac{x}{2}\right) + (\text{Cost of each reorder})\left(\frac{N}{x}\right)$$

where N is the number of items needed per year and x is the number obtained in each order.

EXAMPLE 47   A department store sells 3600 refrigerators per year. It costs $20 per unit to store a refrigerator for 1 year. The cost of reorders has been estimated to be $10 plus $5 for each refrigerator. How many orders should be placed per year (or how many should be ordered per order) to minimize inventory costs?

SOLUTION   Let x be the number of refrigerators ordered in each reorder. Then, by the preceding discussion, $x/2$ is the average number in storage and $3600/x$ is the number of orders.

$$\text{Inventory costs} = C(x)$$

$$C(x) = 20\left(\frac{x}{2}\right) + (10 + 5x)\left(\frac{3600}{x}\right)$$

$$C(x) = 10x + \frac{36,000}{x} + 18,000$$

$$C'(x) = 10 - \frac{36,000}{x^2}$$

$$C''(x) = \frac{72,000}{x^3}$$

Setting $C'(x)$ equal to 0 yields $x = 60$ as a positive critical point. Also, $C''(x)$ is positive for positive values of x; therefore, there is a relative minimum at $x = 60$. Since $C''(x)$ is positive for positive values of x, $x = 60$ provides an absolute minimum. Hence, $3600/60 = 60$ orders should be placed each year and the minimum inventory cost is

$$C(60) = 10 \cdot 60 + \frac{36,000}{60} + 18,000 = \$19,200 \qquad \blacksquare$$

Practice Problem 9   An automobile dealership sells 2000 automobiles each year. It costs \$200 to store a car for one year. The cost of reorders is estimated at \$125. How many orders should be placed each year to minimize inventory costs?

ANSWER   40 orders of 50 each

EXAMPLE 48   Suppose that it costs \$1000 to prepare a factory to produce a batch or run a certain item. After the preparation is complete, it costs \$40 to produce each item. After the items are produced, it costs an average of \$10 per year for each item held in inventory. If a company needs 5000 items per year, how many units should be run off in each batch to minimize cost?

SOLUTION   Let $x \geq 0$ be the number of items produced in each run or batch. Then

$$\text{Number of runs per year} = \frac{5000}{x}$$

$$\text{Cost to prepare factory for production} = \frac{1000(5000)}{x}$$

$$\text{Cost for 5000 items} = 40(5000)$$

We assume that all the production, or run, is put in inventory, the items are withdrawn at a uniform rate, and the inventory is depleted before a second run is made. Thus, the average number in inventory would be $x/2$. Cost for inventory storage is then $10 \cdot (x/2)$. The total cost for 5000 items is

$$C = 1000 \left(\frac{5000}{x} \right) + 40(5000) + \frac{10x}{2}$$ Cost function

$$\frac{dC}{dx} = \frac{-5,000,000}{x^2} + 5$$ Differentiate the cost function.

$$0 = \frac{-5,000,000}{x^2} + 5$$ Find the critical values.

$$= \frac{-5,000,000 + 5x^2}{x^2}$$

$$= \frac{5(-1,000,000 + x^2)}{x^2}$$

$$x = 1000$$ Since x must be positive, this is the only solution.

$$\frac{dC^2}{dx^2} = \frac{10,000,000}{x^3}$$ Second derivative will be positive for positive x.

Thus, the minimum cost occurs when 1000 items are produced in each batch. This minimum cost is $210,000. ■

SUMMARY

Solving optimization problems requires practice and development of reasoning skills. In each problem you must identify each function that is to be maximized or minimized. This involves finding the derivative and critical value(s) along with function values at any endpoints, and then using the second derivative to test that you have found an absolute maximum or minimum. It is wise to let your variable be the amount around which the problem changes. For example, let $x = $ the number of $5 increases in a rent problem, or let $x = $ the additional number of trees in an orchard. If you have to develop the function to be optimized, then it is important to break the problem down into small units and see how the variable is used in each unit. Then put the units together to form the proper function. You can use a graphing calculator to sketch the graphs of the functions you have found to see if they seem to be reasonable functions. You should always make sure that your answer is reasonable within the context of the problem.

Exercise Set 9.6

1. Find two numbers whose sum is 52 and whose product is a maximum.

2. The product of two positive numbers is 36. Find the numbers that will make their sum a minimum.

3. The product of two positive numbers is 49. Find the numbers that will make their sum a minimum.

4. Find the dimensions of the rectangle of area 16 for which the perimeter is a minimum.

5. A box with square ends and rectangular sides is to be made from 600 square inches of cardboard. What is the maximum volume of the box? The box will have a top.

6. Find the dimensions of a box with a square top and bottom and rectangular sides with a volume of 27 cubic inches that can be constructed with the minimum amount of cardboard.

7. A rectangular plot of ground containing 576 square feet is to be fenced. Find the dimensions that require the least amount of fence.

8. Work Exercise 7 for a rectangle containing 100 square feet.

9. A rectangular plot of ground containing 1350 square feet is to be fenced, and an additional fence is to be constructed in the middle of the longest side to divide the plot into two equal parts. Find the dimensions that require the least amount of fence.

10. Suppose that the fence used to enclose the plot in Exercise 9 costs $12 per foot, and the fence used to divide the plot costs $6 per foot. Find the dimensions that make the cost a minimum.

11. From a piece of cardboard 60 centimeters by 60 centimeters, square corners are cut out so that the sides can be folded up to form a box. What size squares should be cut out to give a box of maximum volume? What is the maximum volume?

12. Redo Exercise 11 for a piece of cardboard 50 centimeters by 50 centimeters.

13. A company has set aside $3000 to fence in a rectangular portion of land adjacent to their building, using the building as one side of the enclosed area. The cost of the fencing running parallel to the building is $5 per foot installed, and the fencing for the remaining 2 sides will cost $3 per foot installed. Find the dimensions of the maximum enclosed area.

14. **Exam.** A furniture company uses the following model for the replenishment of square feet of white pine lumber used in the manufacture of furniture. If the monthly cost of purchasing, carrying costs, and storage is given by

$$y = \frac{200{,}000}{x} + 0.05x$$

where x is the number of square feet in each order, the quantity ordered that minimized this cost is
(a) 633 square feet   (b) 2000 square feet
(c) 6325 square feet   (d) 4000 square feet
(e) Some number other than those given.

Applications (Business and Economics)

15. **Cost Function.** A cost function is given as $C(x) = 10x + 30 + 0.01x^2$.
   (a) Find the marginal cost, $C'(x)$.
   (b) Find the average cost,

$$\overline{C}(x) = \frac{C(x)}{x}$$

   (c) Find the marginal average cost, $\overline{C}'(x)$.
   (d) Show that the average cost $\overline{C}(x)$ is a minimum when the marginal cost is the same as the average cost.

16. **Maximum Revenue.** For a charter boat to operate, a minimum of 75 people paying $125 each is necessary. For each person in excess of 75, the fare is reduced $1 per person. Find the number of people that will make the revenue a maximum. What is the maximum revenue?

17. **Cost and Revenue Functions.** The cost and revenue from sales are

$C = x^3 - 15x^2 + 76x + 25$ and

$$R = 55x - 3x^2$$

Find the number of units for which the profit will be a maximum.

18. Verify for Exercise 17 that the marginal cost and marginal revenue are equal when the profit is a maximum.

19. **Maximum Profit.**   A furniture maker can produce 25 tables per week. If $p = 110 - 2x$ is the demand equation for x tables at price p and the cost of producing the tables is $C(x) = 600 + 10x + x^2$ dollars, how many tables should be made each week to give the largest profit? What is the largest profit?
Hint: $R(x) = xp$.

20. **Maximum Revenue.**   If $R(x) = x\sqrt{800 - x^2}$ is the revenue function for the sale of x tennis rackets per day, how many tennis rackets should be produced and sold to maximize revenue?

21. **Maximum Profit.**   A company manufactures and sells x CD players per month. The monthly cost and demand equations are given as

$$C(x) = 30{,}000 + 30x \quad \text{and} \quad p = 300 - \frac{x}{30}$$

where $0 \le x \le 9000$. Find the production level that will realize the maximum profit. What is the maximum profit? **Hint:** $R(x) = xp$.

22. **Minimum Cost.**   For safety reasons, a company plans to fence in a 10,800 square-foot rectangular employee parking lot adjacent to the building by using the building as one side of the enclosed area. The fencing parallel to the building faces a highway and will cost $3 per foot installed and the fencing on the other two sides will cost $2 per foot installed. Find the amount of each type of fence so that the total cost of the fence will be a minimum. What is the minimum cost?

23. **Maximum Profit.**   A restaurant is being planned on the basis of the following information. For a seating capacity of 50 to 100 persons, the weekly profit is approximately $6 per seat. As the seating capacity increases beyond 100 chairs, the weekly profit on each chair in the restaurant decreases by $0.05 times the excess above 100 chairs. What seating capacity would yield the maximum profit?

24. **Maximum Income.**   A rental agency has a problem in determining the rent to charge for each of 100 apartments to obtain a maximum income. It is estimated that if the rent is set at $100 per month, all units will be occupied. On the average, one unit will remain vacant for each $5-per-month increase in rent. What should the rent be in order to maximize the income?

25. **Inventory.**   A company has a contract to supply 500 units per month at a uniform daily rate. Since it costs $100 to start production and the production cost is $5 per unit, it is decided to produce a large quantity at one time, storing the excess units until time for delivery. Storage costs run $10 per item per month. How many items should be made per run to minimize cost?

26. **Advertising.**   ABC Auto has determined that profits are related to x thousands of dollars spent on advertising by

$$P(x) = 8 + 30x - \frac{x^2}{2}$$

What amount should be spent on advertising to attain maximum profit?

27. **Inventory.**   ABC Auto sells 400 new automobiles a year. It costs $100 to store one automobile for a year. To reorder automobiles there is a fixed cost of $60 and a cost per automobile of $50. How many automobiles must be ordered each time to minimize inventory costs?

28. **Inventory.**   An appliance dealer sells 3000 VCRs a year. It costs $12 to store one VCR for a year. To reorder there is a fixed fee of $10 plus $6 for each VCR. How many times a year should the store reorder VCRs, and what should be the size of the order so as to minimize costs?

29. **Maximum Income.**   If a crop of oranges is harvested now, the average yield of 80 pounds per tree can be sold at $0.40 per pound. From past experience, the owners expect the crop yield to increase at a rate of 10 pounds per week per tree and the price to decrease at a rate of $0.02 per pound per week. When should the oranges be picked to attain maximum sale?

30. **Maximum Profit.**   An apartment complex has 24 units. Upkeep and utilities come to $900 per month. The manager estimates that he can keep all the apartments rented at $150 per month rent, but he has one vacancy for each $20 per month added to the rent. However, for each vacant apartment, he would save

$30 per month out of the $900 expenses. What should the rent be to maximize the profit? How many apartments will he have rented at this rent? What will be the maximum profit?

31. **Maximum Yield.** If 25 pear trees are planted per acre in an orchard, the yield will be 525 pears per tree. For each additional tree planted per acre, the yield per tree will be reduced by 15 pears. What is the optimal number of trees to plant per acre?

32. **Maximum Revenue.** A company handles an apartment building with 50 units. Experience has shown that if the rent for each of the units is $160 per month, all of the units will be filled, but one unit will become vacant for each $5 increase in the monthly rent. What rent should be charged to maximize the total revenue from the building?

33. **Maximum Revenue.** A bus company charges $10 for a trip to an historical landmark if 30 people travel in a group. But for each person above the 30, the charge will be reduced by $0.20. How many people in a group will maximize the total revenue for the bus company?

34. **Maximum Revenue.** A consulting company will hold a workshop if at least 30 people sign up at a cost of $50 per person. The company will agree to a reduction of $1.25 per person for each person over the 30 minimum. (All attending receive the discount.) In order for the company to maximize its revenue, how many people should attend the workshop? Assume the maximum number allowed is 40.

35. **Maximum Yield.** A pecan grower has found that if 20 trees are planted per acre, each tree will produce an average of 60 pounds of pecans per year. If, for each additional tree planted per acre (up to 15 additional trees) the average yield per tree will drop by 2 pounds, how many additional trees (if any) should be planted to maximize the yield?

36. **Maximum Revenue.** When a travel agency charges $600 for a fantasy baseball weekend with a professional baseball team, it attracts 1000 people. For each $20 decrease in the charge, an additional 100 people will sign up. What price should be charged to maximize the revenue?

Applications (Social and Life Sciences)

37. **Mosquito Population.** Assume that the number of mosquitoes $N(x)$, in thousands, depends on the rainfall, in inches, according to the function

$$N(x) = 60 - 45x + 12x^2 - x^3, \qquad 0 \le x \le 6.4$$

(a) Find the amount of rainfall that will produce the minimum number of mosquitoes.
(b) Find the amount of rainfall that will produce the maximum number of mosquitoes.

38. **Bacteria Population.** Suppose that the bacteria count t days after a treatment is given by

$$C(t) = 30t^2 - 180t + 700$$

(a) When will the count be minimum?
(b) What is the minimum count?

39. **Voter Registration.** The number of registered voters, in thousands, is estimated to grow according to the function for the time in years as

$$N(t) = 12 + 3t - t^3, \qquad 0 \le t \le 3$$

(a) Find the rate of increase.
(b) When is the number a maximum?

Review Exercises

Draw the graph of each function and find the critical points, the intervals where the graph is increasing and where decreasing, the relative maxima and minima, points of inflection, the intervals where the graph is concave upward and where concave downward, and the horizontal and vertical asymptotes if they exist.

40. $y = \dfrac{4x}{x - 2}$

41. $y = \dfrac{4}{x - 2}$

42. $y = (4 - x^{2/3})^{3/2}$

43. $y = \dfrac{x}{2}(16 - x^2)^{1/2}$

9.7 ELASTICITY AS IT AFFECTS BUSINESS DECISIONS (Optional)

OVERVIEW In this section we are interested in relative change. By relative change, we mean a comparison of change and the value of a variable before the change takes place. For example, a salary increase of $2000 per year for a person whose salary is $10,000 is a 20% increase. A $2000-a-year increase for a person whose salary is $200,000 is a 1% increase. In this section we

- Introduce average elasticity (the rate of the relative changes in two variables)
- Introduce point elasticity, which is the limit of average elasticity as the change approaches 0
- Use these concepts to study revenue, cost, demand, other business applications, and applications in the social and life sciences

Relative Change

Consider the function $y = f(x)$ as x changes by Δx (the change in the value of x). We will define the change in y to be

$$\Delta y = f(x + \Delta x) - f(x)$$

Now let's define the relative change in x and y.

Relative change in x at $x = x_0$ is $\dfrac{\Delta x}{x_0}$ (x_0 is the initial value of x).

Relative change in $y = y_0$ is $\dfrac{\Delta y}{y_0}$ (y_0 is the initial value of y).

EXAMPLE 49

(a) For the function $y = x^2$, find the relative changes in x and y when x changes from 4 to 5.
(b) For the function $y = x$, find the relative changes in x and y as x changes from 4 to 5.

SOLUTION

(a) For x, we have

$$\frac{\Delta x}{x_0} = \frac{5 - 4}{4} = \frac{1}{4} = 0.25 \quad \text{or} \quad 25\%$$

For y, we have

$$\Delta y = f(5) - f(4) = 5^2 - 4^2 = 25 - 16 = 9$$

Therefore,

$$\frac{\Delta y}{y_0} = \frac{5^2 - 4^2}{4^2} = \frac{9}{16} = 0.5625 \quad \text{or} \quad 56\frac{1}{4}\%$$

(b)   We have

$$\text{Relative change in } x = \frac{5 - 4}{4} = \frac{1}{4} = 25\%$$

$$\text{Relative change in } y = \frac{5 - 4}{4} = \frac{1}{4} = 25\%$$

This time the relative change in y is the same as the relative change in x.

∎

In the preceding example, we say that the function y in part (a) is more sensitive or more responsive to a change in x than the function in part (b). We define average elasticity as a measure of this responsiveness. In part (a) the average elasticity is calculated as

$$\frac{\text{Relative change in } y}{\text{Relative change in } x} = \frac{\dfrac{\Delta y}{y}}{\dfrac{\Delta x}{x}} = \frac{0.5625}{0.25} = 2.25$$

and in part (b) the average elasticity is

$$\frac{\dfrac{\Delta y}{y}}{\dfrac{\Delta x}{x}} = \frac{0.25}{0.25} = 1$$

The average elasticity of 2.25 in part (a) is greater than the average elasticity of 1 in part (b) and is, therefore, more sensitive or more responsive to a change in x.

DEFINITION: AVERAGE ELASTICITY

For a function $y = f(x)$ the **average elasticity** of $f(x)$ relative to x is defined to be

$$\frac{\dfrac{\Delta y}{y}}{\dfrac{\Delta x}{x}} = \frac{\dfrac{f(x + \Delta x) - f(x)}{y}}{\dfrac{\Delta x}{x}} \qquad \text{(Relative change in } y)/(\text{Relative change in } x)$$

$$= \frac{x}{y}\left[\frac{f(x + \Delta x) - f(x)}{\Delta x}\right]$$

EXAMPLE 50   Find the average elasticity of $y = f(x) = x^3$ as x changes from 2 to 3; that is, $\Delta x = 1$.

SOLUTION

$x = 2$ therefore; $\Delta x = 3 - 2 = 1$

$y = f(2) = 2^3 = 8$ therefore; $\Delta y = f(x + \Delta x) - f(x) = f(2 + 1) - f(2) =$
$$3^3 - 2^3 = 27 - 8 = 19$$

$$\frac{\dfrac{\Delta y}{y}}{\dfrac{\Delta x}{x}} = \frac{\dfrac{19}{8}}{\dfrac{1}{2}} = \frac{19}{4}$$

The average elasticity of this function when $x = 2$ and $\Delta x = 1$ is 19/4. ■

Recall that at the beginning of our study of calculus, we went from average slope to instantaneous (or point) slope by taking the limit as the change in x (Δx) approaches 0. We apply the same reasoning to average elasticity.

$$\lim_{\Delta x \to 0} \frac{\dfrac{\Delta y}{y}}{\dfrac{\Delta x}{x}} = \lim_{\Delta x \to 0} \frac{x}{y} \left[\frac{f(x + \Delta x) - f(x)}{\Delta x} \right]$$

$$= \frac{x}{y} \lim_{\Delta x \to 0} \frac{f(x + \Delta x) - f(x)}{\Delta x} \qquad x/y \text{ is constant relative to } \Delta x.$$

$$= \frac{x}{y} f'(x) = \frac{xf'(x)}{f(x)} \qquad\qquad \text{Definition of derivative}$$

Thus, point elasticity (simply called elasticity) is the limit of average elasticity if the limit exists.

DEFINITION: ELASTICITY

For the function $y = f(x)$ the **elasticity** at x is

$$E(x) = \frac{xf'(x)}{f(x)}$$

provided that $f'(x)$ exists and $f(x) \neq 0$.

If $|E(x)| > 1$, then y is elastic relative to x.
If $|E(x)| = 1$, then y has unit elasticity with respect to x.
If $|E(x)| < 1$, then y is inelastic relative to x.

When a function is elastic, the relative change in output is greater than the relative change in input. When a function is inelastic, a relative change in output is less than the relative change in input.

In the business world, Alfred Marshall, a British economist, used the preceding ideas when he introduced what he called price elasticity of demand. This concept provided a measure of the relative amount by which demand would change in response to a change in price.

Suppose that a demand D is given as a function of price p.

$$D = f(p)$$

For most products, demand is a decreasing function of price. As the price (input) increases, the demand (output) decreases. Substituting in the general elasticity formula,

$$E(x) = \frac{xf'(x)}{f(x)} \quad \text{becomes} \quad E(p) = \frac{pf'(p)}{f(p)}$$

Note that $E(p)$ will always be negative. Most management books define $E(p)$ to be

$$\frac{-pf'(p)}{f(p)}$$

so that E will be easier to work with because it will always be positive.

DEFINITION: ELASTICITY OF DEMAND

Suppose that the demand and price are related by

$$D = f(p)$$

The **point elasticity of demand** is

$$E(p) = \frac{-pf'(p)}{f(p)}$$

1. If $0 \le E(p) < 1$, demand is inelastic.
   (a) Increase in unit price $\Rightarrow$ Increase in revenue.
   (b) Decrease in unit price $\Rightarrow$ Decrease in revenue.
2. If $E(p) > 1$, demand is elastic.
   (a) Increase in unit price $\Rightarrow$ Decrease in revenue.
   (b) Decrease in unit price $\Rightarrow$ Increase in revenue.
3. If $E(p) = 1$, demand has unit elasticity.
   (a) Increase in price $\Rightarrow$ Revenue remains the same.

Note that the absolute value sign on $|E(x)|$ was removed for $E(p)$, since it is assumed that $f(p)$ is a decreasing function; that is, $f'(p)$ is negative at the same time p and $f(p)$ are positive. Thus $E(p)$ is positive.

EXAMPLE 51   If $D = f(p) = 1600 - 80p$, with $0 < p < 20$, find $E(p)$ and give one interpretation at

(a) $p = \$2$ (b) $p = \$10$ (c) $p = \$12$

SOLUTION

$$E(p) = \frac{-pf'(p)}{f(p)} = \frac{p \cdot (80)}{1600 - 80p}, \qquad f'(p) = -80$$

(a)   At $p = 2$, we substitute $p = 2$ into $E(p)$ and get

$$E(2) = \frac{2(80)}{1600 - 80 \cdot 2} = \frac{1}{9} \approx 0.111, \qquad 0 \le E(p) < 1$$

The demand is inelastic (i.e., the demand is not overly sensitive to a change in price). Actually, a price increase of 10% would result in a

$$0.111(10\%) = 1.1\%$$

increase in demand.

(b)   At $p = 10$, we substitute $p = 10$ into $E(p)$ and get

$$E(10) = \frac{10(80)}{1600 - 80(10)} = 1$$

In this case, a percentage change in price will result in approximately the same percentage change in demand. When $E(p) = 1$, the demand has unit elasticity.

(c)   At $p = 12$, we substitute $p = 12$ into $E(p)$ and get

$$E(12) = \frac{12(80)}{1600 - 80(12)} = 1.5$$

Since $E(p) > 1$, demand is elastic (i.e., there is a change in demand when price changes.) For example, a 10% change in price would result in a

$$1.5(10\%) = 15\%$$

change in demand.   ■

EXAMPLE 52 Given $D = f(p) = 3840 - 20p^2$ for $0 < p < \sqrt{192}$.

(a) Determine the interval on p where the demand is inelastic and the interval where it is elastic.
(b) Interpret the result for a 10% increase in price at $p = \$6$.
(c) Interpret the result for a 10% increase in price at $p = \$10$.

SOLUTION

(a) We have

$$E(p) = -\frac{p \cdot (-40p)}{3840 - 20p^2} \qquad \frac{dD}{dp} = -40p$$

$$= \frac{2p^2}{192 - p^2} \qquad \text{Factor 20 from numerator and denominator and reduce.}$$

Now, if $E(p) = 1$, then

$$\frac{2p^2}{192 - p^2} = 1$$

$$2p^2 = 192 - p^2$$

$$3p^2 = 192$$

$$p^2 = 64$$

$$p = 8 \qquad p \text{ cannot be negative.}$$

When $p = 8$ the demand has unit elasticity. In a similar manner, if $E(p) < 1$, then

$$\frac{2p^2}{192 - p^2} < 1$$

$$2p^2 < 192 - p^2 \qquad 192 - p^2 > 0$$

$$3p^2 < 192$$

$$p^2 < 64$$

$$p < 8 \qquad p \text{ cannot be negative.}$$

For $0 < p < 8$, the demand is inelastic. If $E(p) > 1$, then

$$\frac{2p^2}{192 - p^2} > 1$$

$$2p^2 > 192 - p^2 \qquad 192 - p^2 > 0$$

$$3p^2 > 192$$

$$p^2 > 64$$

$$p > 8 \qquad p \text{ cannot be negative.}$$

For $8 < p < \sqrt{192}$ demand is elastic.

(b) At $p = 6$, we know the demand is inelastic. A 10% increase in p would result in

$$\left(\frac{2p^2}{192 - p^2}\right) \cdot 10\% = \left(\frac{2 \cdot 36}{192 - 36}\right) \cdot 10\% \approx 4.6\% \text{ increase in demand}$$

(c) At $p = 10$, we know the demand is elastic. A 10% increase in p would result in

$$\left(\frac{2p^2}{192 - p^2}\right) \cdot 10\% = \left(\frac{2 \cdot 100}{192 - 100}\right) \cdot 10\%$$
$$\approx 21.7\% \text{ decrease in demand} \qquad ■$$

Practice Problem 1 With your graphing calculator draw the graph of

$$y = \frac{2p^2}{192 - p^2}$$

to check the preceding discussion. Use the range $0 \le p \le 20$ and $-10 \le y \le 10$.

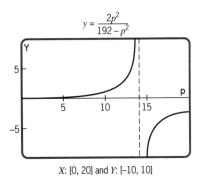

$$y = \frac{2p^2}{192 - p^2}$$

X: [0, 20] and Y: [–10, 10]

Figure 83

ANSWER The graph in Figure 83 shows that $E(p)$ is inelastic when $0 < p < 8$, and $E(p)$ is elastic for $8 < p < \sqrt{192}$. The graph below the axis where $x > \sqrt{192}$ is not in the domain of p. The graph appears to agree with Example 52.

EXAMPLE 53 Returning to Example 51, where $D = f(p) = 1600 - 80p$ for $0 < p < 20$, we noted that

$$E(p) = \frac{80p}{1600 - 80p} = \frac{p}{20 - p}$$

Let's discover where $E(p) < 1$ and $E(p) > 1$.

SOLUTION The inequality $E(p) < 1$ means that

$$\frac{p}{20 - p} < 1$$

$$p < 20 - p \qquad \text{20 − p is given as positive.}$$

$$2p < 20$$

$$p < 10$$

For $0 \leq p < 10$, $E(p) < 1$ and the demand is inelastic.
For $E(p) > 1$,

$$\frac{p}{20 - p} > 1$$

$$p > 20 - p \qquad \text{20 − p is given as positive.}$$

$$2p > 20$$

$$p > 10$$

For $10 \leq p < 20$, $E(p) > 1$ and the demand is elastic. ■

Revenue Function

Now let's examine where a revenue function is increasing and where it is decreasing. Using the preceding example,

$$R = p \cdot D \qquad \text{Price · Demand}$$
$$= pf(p)$$
$$= p(1600 - 80p)$$
$$= 1600p - 80p^2$$

$$\frac{dR}{dp} = 1600 - 160p$$

If $1600 - 160p = 0$ $\qquad$ Setting $\dfrac{dR}{dp} = 0$

then $p = 10$

For $0 < p < 10$, dR/dp is positive and the function is increasing; therefore, we have increasing revenue. For $10 < p < 20$, dR/dp is negative and the function is decreasing; therefore we have decreasing revenue.

For the preceding example, we note that revenue is increasing precisely when demand is inelastic and revenue is decreasing when demand is elastic. This is further emphasized by Figure 84.

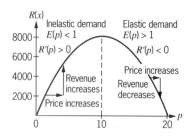

Figure 84

Practice Problem 2 $E(p) = \dfrac{p^2}{80 - p^2}$. Is the demand elastic or inelastic at

(a) $p = 4$?
(b) $p = 8$?

ANSWER

(a) At $p = 4$, $E(p) = 0.25$, so demand is inelastic.
(b) At $p = 8$, $E(p) = 4$, so demand is elastic.

SUMMARY

In business it is very important to understand the relationship between revenue, cost, profit, and the changes in these as prices change.

1. When demand is elastic at price p [i.e., $E(p) > 1$], an increase in the price per unit will cause a decrease in revenue. However, a decrease in the price per unit will result in an increase in revenue.
2. When demand is inelastic at price p [i.e., $E(p) < 1$], then increasing the price per unit results in an increase in revenue. Likewise, a decrease in the price per unit causes the total revenue to decrease.
3. If the demand has unit elasticity at price p [i.e., $E(p) = 1$], an increase in the price per unit will result in very little change in the revenue.

Exercise Set 9.7

1. Consider the function $y = x^2$.
   (a) Find the relative change in x as x changes from 2 to 4.
   (b) Find the relative change in y.
   (c) Find the average elasticity.

2. Consider the function $y = x^3$.
   (a) Find the relative change in x as x changes from 1 to 3.
   (b) Find the relative change in y.
   (c) Find the average elasticity.

Applications (Business and Economics)

Find the elasticity (point elasticity) of the given function $y = f(x)$ at the designated point. Tell whether the function is elastic, inelastic, or has unit elasticity.

3. $y = x^2 + 2$ at $x = 1$
4. $y = x(x + 1)$ at $x = 2$

5. $y = \dfrac{1}{x}$ at $x = 3$
6. $y = 3x^2 + 2$ at $x = 3$
7. $y = 3x(x^2 + 1)$ at $x = 1$
8. $y = \dfrac{1}{x^2}$ at $x = 2$

9. *Elasticity of Demand.* The following demand equation is given:

$$D(p) = 200(40 - p), \qquad 0 \le p \le 40$$

   (a) Find the elasticity of demand, $E(p)$.
   (b) What is the elasticity when $p = \$10$?
   (c) Classify part (b) as elastic, inelastic, or as having unit elasticity.
   (d) What will be the approximate change in demand if p is increased by 5% when $p = \$10$?
   (e) What is the elasticity when $p = \$30$?
   (f) Classify part (e) as elastic, inelastic, or as having unit elasticity.

(g) What will be the approximate change in demand if p is increased by 5% when $p = \$30$?

10. **Elasticity of Demand.** The following demand equation is given:

$$D(p) = 20(10 - p), \qquad 0 \le p \le 10$$

(a) Find the elasticity of demand, $E(p)$.
(b) What is the elasticity when $p = \$3$?
(c) Classify part (b) as elastic, inelastic, or as having unit elasticity.
(d) What will be the approximate change in demand if p is decreased by 10% when $p = \$3$?
(e) What is the elasticity when $p = \$8$?
(f) Classify part (e) as elastic, inelastic, or as having unit elasticity.
(g) What will be the approximate change in demand if p is decreased by 5% when $p = \$8$?

For the following demand equations, determine if demand is elastic, inelastic, or has unit elasticity for the given values of p.

11. $D(p) = 100 - p^2$ for $p = 8$
12. $D(p) = 100 - p^2 - p$ for $p = 6$
13. $D(p) = 1000 - p^2$ for $p = 10$
14. $D(p) = 1000 - p^2 - 4p$ for $p = 10$

For each of the following demand equations, determine intervals of p that produce a demand that is elastic and the intervals that produce a demand that is inelastic.

15. $D(p) = 200(40 - p), 0 \le p \le 40$
16. $D(p) = 20(10 - p), 0 \le p \le 10$
17. $D(p) = 20(100 - p)^2, 0 \le p \le 100$
18. $D(p) = 10(64 - 2p)^2, 0 \le p \le 32$

19. **Elasticity of Cost.** Suppose that the cost of producing x units weekly is

$$C(x) = \frac{1}{2}x^2 + 6x + 200$$

Find the elasticity of cost when $x = 4$, $x = 10$, and $x = 20$. Would you classify the elasticity of cost as being elastic, inelastic, or as having unit elasticity at these three points? An increase of 10% in production at these points produces what change in cost?

For each of the following demand equations, find the revenue equation, sketch the graph of the revenue equation, label where the revenue graph is increasing and where it is decreasing, and find the regions of elastic and inelastic demand and mark these on the graph of the revenue equation.

20. $D(p) = 40(20 - p), 0 \le p \le 20$
21. $D(p) = 100(100 - p), 0 \le p \le 100$
22. $D(p) = 40(20 - p)^2, 0 \le p \le 20$
23. $D(p) = 100(100 - p)^2, 0 \le p \le 100$

Review Exercises

Find the critical points, intervals where the graph is increasing and where decreasing, relative maxima and minima, points of inflection, intervals where the graph is concave upward and concave downward, horizontal and vertical asymptotes if they exist, and sketch the graph of the following functions.

24. $y = x + x^{-1}$
25. $y = \frac{1}{3}(x^4 - 4x^3)$
26. $y = 2x\sqrt{x - 1}$
27. $y = 2x(x - 2)^2$

Chapter Review

Review the following concepts to ensure that you understand and can use them.

Important Terms

Inflection points	Concavity
Curve sketching	First derivative test
Relative maxima	Second derivative test

Relative minima	Average elasticity
Absolute maxima	Point elasticity
Absolute minima	Elastic
Asymptote	Inelastic
Critical point	Unit elasticity
Increasing and decreasing function	Elasticity of demand

Make sure you understand the following concepts and formulas.

$f'(c) = 0$ or $f'(c)$ is undefined when c is a critical value

$f'(x) > 0$; graph is increasing

$f'(x) < 0$; graph is decreasing

$f''(x) > 0$; graph concave upward

$f''(x) < 0$; graph concave downward

$\lim\limits_{x \to \infty} f(x) = c$; then $y = c$ is a possible horizontal asymptote

$\lim\limits_{x \to c} f(x) = \infty$; then $x = c$ is a possible vertical asymptote

$$\text{Average elasticity} = \frac{x}{y}\left[\frac{f(x + \Delta x) - f(x)}{\Delta x}\right]$$

$$E(x) = \frac{xf'(x)}{f(x)}$$

$$E(p) = \frac{-pf'(p)}{f(p)}$$

Chapter Test

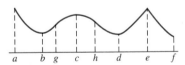

Use the figure to answer questions 1–4. Give answers as the largest possible intervals if possible (otherwise as points).

1. From the figure determine over (a, f).
   (a) Where is the curve increasing?
   (b) Where is the curve decreasing?
   (c) For what value or values of x are there relative maxima?

2. From the figure determine over (a, f).
   (a) For what value(s) are there relative minima?
   (b) On $a \le x \le h$, where is there an absolute minimum?
   (c) On $a \le x \le h$, where is there an absolute maximum?

3. From the figure determine over (a, f).
   (a) Where are the points of inflection?
   (b) Intervals where the second derivative is positive?
   (c) Intervals where the second derivative is negative?

4. From the figure determine over (a, f).
   (a) Where is the curve concave upward?
   (b) Where is the curve concave downward?
   (c) Where is the first derivative undefined?

5. Given $y = -2x^2 + 4x$.
   (a) Where is the curve increasing and where is it decreasing?
   (b) Find all points of relative extrema.
   (c) Sketch the curve.

6. For $y = x^4 - x^3$ use both your graphing calculator and calculus techniques to draw the graphs of y, y', and y'' to answer each question.
   (a) Where is the curve concave upward? Concave downward?
   (b) Where are the points of inflection?
   (c) Find all points of relative extrema.

7. For $y = x + x^{-1}$ use both your graphing calculator and calculus techniques to draw the graphs of y, y', and y'' to answer each question.
   (a) Where is the curve increasing? Decreasing?
   (b) Where is the curve concave upward? Concave downward?
   (c) Find points of relative extrema.
   (d) Find a vertical asymptote.
   (e) Sketch the curve.

8. An electronics store sells 5000 radios each year. It costs $10 to store a radio for a year. To reorder there is a fixed fee of $10 plus $6 for each radio on inventory. How many times a year should the store reorder radios in order to minimize costs, and what should the size of the orders be?

9. When 200 cars are sold per month at an automobile dealership, the profit per car is $400. For each sale above 200 cars, the profit decreases by $1 per car. How many cars should be sold for maximum profit?

10. For the demand equation $D(p) = -p + 6$, is demand elastic or inelastic at $p = 2$? Interpret this result for a price increase of 5%.

CHAPTER 10

Additional
Derivative Topics

In this chapter, we introduce procedures for differentiation of exponential functions and logarithmic functions that are useful in the discussion of applications of calculus. These functions serve as models for population growth, growth of investments, depreciation of capital goods, decay of radioactive material, the rate of learning, the spread of epidemics, and compound interest. You may need to review exponents and exponential functions in Chapter 1.

Sometimes functions cannot be given with one variable expressed explicitly as a function of the other. When this occurs, a procedure for finding a derivative, called implicit differentiation, is most useful. Similarly, sometimes two variables are functions of a third variable. In this chapter, we investigate the relationship between the derivatives of the two variables in terms of the derivative of the third variable.

639

10.1 FINDING DERIVATIVES OF EXPONENTIAL FUNCTIONS

OVERVIEW This is an important section because exponential functions are used extensively in applications. Growth and decay models are based on exponentials. Exponential functions are used by economists to study the growth rate of the money supply. They are needed by businesspeople to study the rate of change in sales, by biologists to study the rate of growth of organisms, and by social scientists to study the rate of population growth. In this section we

- Find the derivative of exponential functions
- Use the chain rule with exponential functions
- Study graphing techniques
- Introduce applications of exponential functions

We introduced exponential functions in Chapter 1. Most of our work with exponential functions will involve the base e.

$$e = \lim_{n \to \infty} \left[1 + \frac{1}{n} \right]^n$$

if this limit exists. This concept of the constant e, defined as a limit, is in agreement with our use of e in Chapter 1.

To get a feel for the size of e, we use the Table feature of our graphing calculator to approximate e. We see that e does appear to exist and is approximately 2.7182818 (see Table 1).

In calculus, e is an excellent choice for the base of an exponential function because of the simplicity of the derivative. If $y = e^x$, then $dy/dx = e^x$. This fact will be proved after we apply this formula in several examples. First consider the more general case in which the exponent of e is a function of x. If $y = e^{u(x)}$, we apply the chain rule to obtain

$$\frac{dy}{dx} = \frac{dy}{du} \cdot \frac{du}{dx}$$

$$= e^u u'(x) \qquad \frac{d}{du} e^u = e^u$$

These ideas are summarized as follows.

TABLE 1

n	$\left(1 + \dfrac{1}{n}\right)^n$
100	2.7048138
500	2.7155685
1,000	2.7169239
10,000	2.7181459
100,000	2.7182682
1,000,000	2.7182805
100,000,000	2.7182818

DIFFERENTIATION FORMULAS FOR EXPONENTIAL FUNCTIONS

$$\frac{d}{dx}(e^x) = e^x$$

If u is a differentiable function of x, then

$$\frac{d}{dx}[e^{u(x)}] = e^{u(x)} \frac{du}{dx} = u'(x)e^{u(x)}$$

The following examples illustrate the use of these formulas.

EXAMPLE 1   If $y = e^{3x}$, find $\dfrac{dy}{dx}$.

SOLUTION   Let $u(x) = 3x$. Then $u'(x) = 3$, and

$$\frac{dy}{dx} = e^{u(x)}u'(x)$$
$$= e^{3x} \cdot 3$$
$$= 3e^{3x}$$

■

EXAMPLE 2   If $y = e^{3x^2}$, find $\dfrac{dy}{dx}$.

SOLUTION   Let $u(x) = 3x^2$.

$$\frac{dy}{dx} = e^u u'(x)$$
$$= e^{3x^2} \cdot 6x \qquad u = 3x^2, \text{ so } u' = 6x.$$
$$= 6xe^{3x^2}$$

■

EXAMPLE 3   If $y = e^{\sqrt{x}}$, find $\dfrac{dy}{dx}$.

SOLUTION   Let $u(x) = \sqrt{x}$.

$$\frac{dy}{dx} = e^u u'$$
$$= e^{\sqrt{x}} \frac{1}{2\sqrt{x}} \qquad u = \sqrt{x}, \text{ so } u' = \frac{1}{2\sqrt{x}}.$$
$$= \frac{1}{2\sqrt{x}} e^{\sqrt{x}}$$

■

EXAMPLE 4   If $y = e^{3x^2 + x - 2}$, find $\dfrac{dy}{dx}$.

SOLUTION   Let $u(x) = 3x^2 + x - 2$.

$$\frac{dy}{dx} = e^u u'$$
$$= e^{3x^2 + x - 2}(6x + 1) \qquad u = 3x^2 + x - 2, \text{ so } u' = 6x + 1.$$
$$= (6x + 1)e^{3x^2 + x - 2}$$

■

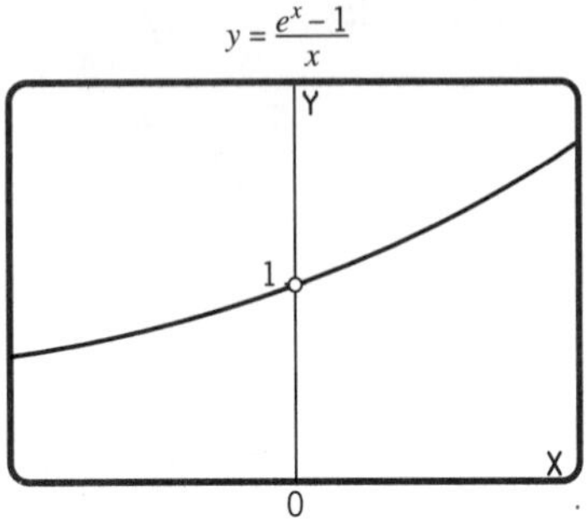

$y = \dfrac{e^x - 1}{x}$

X: [-1, 1] and Y: [0, 2]

Figure 1

In the process of developing the formula for the derivative of e^x, we will need $\lim\limits_{h \to 0} \dfrac{e^h - 1}{h}$. To get this limit we investigate with our calculator what happens to the function $y = (e^h - 1)/h$ as h approaches 0 from both the left and the right. To find this limit, consider the graph of $y = (e^x - 1)/x$ in Figure 1. This function is not defined at $x = 0$, but it seems that

$$\lim_{x \to 0} \frac{e^x - 1}{x} = 1$$

as we use the $\boxed{\text{TRACE}}$ key to look at x values that are very close to 0.

Using the Table feature of our calculator, Table 2 presents a picture of $(e^x - 1)/x$ approaching 1 as x approaches 0 from both the left and right.

TABLE 2

x	0.01	0.001	0.0001	$\to 0 \leftarrow$	-0.0001	-0.001	-0.01
$(e^x - 1)/x$	1.005	1.0005	$1.00005 \to$	$1 \leftarrow$	0.99995	0.9995	0.995

Since

$$\lim_{h \to 0} \frac{e^h - 1}{h} \quad \text{is the same as} \quad \lim_{x \to 0} \frac{e^x - 1}{x}$$

we use the fact that

$$\lim_{h \to 0} \frac{e^h - 1}{h} = 1$$

in the following derivation of the derivative of e^x. We use the four-step procedure from Chapter 8.

$$f(x) = e^x$$

$$f(x + h) = e^{x+h} = e^x \cdot e^h$$

$$f(x + h) - f(x) = e^x \cdot e^h - e^x$$

$$\frac{f(x + h) - f(x)}{h} = \frac{e^x \cdot e^h - e^x}{h} \qquad h \neq 0$$

$$\frac{dy}{dx} = \lim_{h \to 0} \frac{f(x + h) - f(x)}{h} = \lim_{h \to 0} \frac{e^x(e^h - 1)}{h}$$

$$= e^x \lim_{h \to 0} \frac{e^h - 1}{h}$$

$$= e^x \cdot 1 = e^x$$

Consequently, if $y = e^x$, then $dy/dx = e^x$.

EXAMPLE 5   If $y = e^x$, find $\dfrac{d^2y}{dx^2}$.

SOLUTION   Since $\dfrac{dy}{dx} = e^x$, then $\dfrac{d^2y}{dx^2} = e^x$ and, in general,

$$\frac{d^n y}{dx^n} = e^x$$

EXAMPLE 6   If $y = xe^x$, find $\dfrac{dy}{dx}$.

SOLUTION   Notice that y is a product of two functions, $f(x) = x$ and $g(x) = e^x$. Therefore, by the derivative formula for a product we have

$$\frac{dy}{dx} = x \cdot \frac{d}{dx}(e^x) + \frac{d}{dx}(x) \cdot e^x \qquad FS' + F'S$$
$$= xe^x + 1 \cdot e^x$$
$$= xe^x + e^x$$
$$= e^x(x + 1)$$

Practice Problem 1   Given $f(x) = 3xe^{x^2}$, find $f'(x)$.

ANSWER $f'(x) = 6x^2 e^{x^2} + 3e^{x^2} = e^{x^2}(6x^2 + 3)$

EXAMPLE 7   Find y' where $y = \sqrt[3]{e^{2x} - 3}$.

SOLUTION

$$y = (e^{2x} - 3)^{1/3}$$

$$y' = \frac{1}{3}(e^{2x} - 3)^{-2/3} \cdot \frac{d}{dx}(e^{2x} - 3) \qquad \frac{d}{dx}(u^n) = nu^{n-1}\frac{du}{dx}$$

$$= \frac{1}{3}(e^{2x} - 3)^{-2/3} \cdot e^{2x} \cdot (2) \qquad \frac{d}{dx}(e^{2x} - 3) = e^{2x}(2)$$

$$= \frac{2e^{2x}}{3(e^{2x} - 3)^{2/3}}$$

How can we tell whether the graphs obtained from data represent mathematical models for growth or decay? Since many growth and decay models can be represented by graphs of exponential functions, a knowledge of the shape and characteristics of graphs of exponential functions helps us to answer this question. We now investigate characteristics of the graph of $y = e^x$.

The derivative of $y = e^x$ is $y' = e^x$, which is positive for all finite values of x. Thus, the graph of $y = e^x$ is always increasing. Since y'' is also e^x, the graph is concave upward for all x. When $x = 0$, $y = e^0 = 1$ (the y-intercept). A calculator

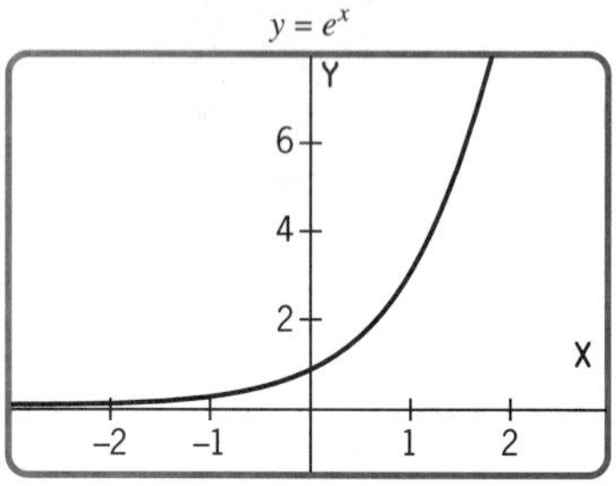

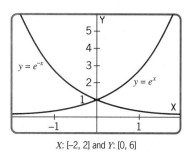

X: [–3, 3] and Y: [–1.6, 8]

Figure 2

X: [–2, 2] and Y: [0, 6]

Figure 3

graph of $y = e^x$ is shown in Figure 2. By shifting the viewing window to the left, y seems to approach 0 as x gets smaller and smaller. In fact, $\lim\limits_{x \to -\infty} e^x = 0$ and the x-axis is an asymptote of the graph.

Practice Problem 2 With your graphing calculator, draw the graph of both $y = e^x$ and $y = e^{-x}$. Do these graphs seem to be mirror images of each other about some line?

ANSWER The graph is shown in Figure 3. The graphs appear to be mirror images about the y-axis.

Compounding Continuously

Recall that if interest is compounded continuously, then the amount of a loan, P, at the end of t years is

$$A = Pe^{rt}$$

where r is the rate expressed in terms of a year.

EXAMPLE 8 Suppose that a loan is for $1000 and the interest rate is 8% compounded continuously. Find the rate of change of the loan amount with respect to time.

SOLUTION

$$A = 1000e^{0.08t} \quad \text{and} \quad \frac{dA}{dt} = 1000 \frac{d}{dt}[e^{0.08t}]$$

Now

$$\frac{d}{dt}(e^u) = e^u \frac{du}{dt}$$

and we have

$$\frac{dA}{dt} = 1000e^{0.08t}(0.08) \quad u = 0.08t, \text{ so } \frac{du}{dt} = 0.08$$

$$= 80e^{0.08t}$$

Therefore, $80e^{0.08t}$ is the rate of change of the loan amount at time t. ∎

Growth equations and decay equations (dependent variable decreasing) are important in business, economics, biology, and the social sciences. At this time we discuss three of the many growth equations, often named for the people who first used them.

The **logistic curve** (discussed in Exercise 46) is defined by

$$y = \frac{A}{1 + Be^{-kt}}$$

and is a suitable model for defining natural phenomena where there is at first rapid growth and then a slow down of growth because of overcrowding, scarcity of food, and other factors.

One family of curves that has been used to describe both growth and deterioration is called the **Gompertz curve**, whose equation is of the form

$$N = Ca^{kt}$$

where N is the possible number of individuals at a given time t. For example, a business executive may predict the number of employees of his company by the Gompertz curve $N = 1000(0.5)^{0.6t}$, where t represents the number of years after starting the new business.

In Exercise 47 we discuss the characteristics of the Von Bertalanffy curve, which is useful in biology. A discussion of this and other exponential models is found in *Mathematical Models and Applications* by D. Maki and M. Thompson (Englewood Cliffs, N.J.: Prentice Hall, 1973), pp. 312–317.

EXAMPLE 9   The following equation is an example of a Gompertz curve.

$$N(t) = 1000e^{-0.05t}$$

where t is measured in days. Find the rate of decay (or decrease) after 4 days.

SOLUTION   The rate of decay is the derivative of the decay function with respect to time or $N'(t)$.

$$N'(t) = (-0.05)(1000)e^{-0.05t} \qquad \frac{d}{dt}(-0.05t) = -0.05$$

$$N'(4) = -50e^{-0.05(4)} = -50e^{-0.2}$$
$$= -40.94$$

The population is decreasing at a rate of approximately 41 per day at the end of 4 days. In fact, the population is always decreasing, but the rate of decrease approaches zero for large t.   ■

Calculator Note

If all that is needed is the rate of decay after 4 days, finding the numerical value of a derivative on a calculator saves a great deal of time. On page 511 of Chapter 8 we learned to obtain the numerical value of a derivative using nDeriv ($1000\, e^{-0.05x}$, x, 4), which equals -40.9365.

SUMMARY

In this section we introduced two important derivative formulas:

1. If $y = e^x$, then $\dfrac{dy}{dx} = e^x$.

2. If $y = e^{u(x)}$, then $\dfrac{dy}{dx} = e^{u(x)}u'(x)$.

These derivatives are important when working with growth and decay models.

Exercise Set 10.1

Find dy/dx for the following functions.

1. $y = e^{4x}$
2. $y = 4e^{3x}$
3. $y = 5 + e^{2x}$
4. $y = 3x^2 - 2e^{4x}$
5. $y = 4x - e^{-3x}$
6. $y = 8x^2 + e^{-5x}$
7. $y = 4x - e^{3x}$
8. $y = e^{3x} + e^{4x}$
9. $y = e^{x^2}$
10. $y = 3 \cdot e^{x+4}$
11. $y = \sqrt{e^x + 4}$
12. $y = \sqrt{e^{4x+1}}$
13. $y = (3 + e^x)(2 - e^{-x})$
14. $y = \dfrac{3 + e^x}{3 - e^{-x}}$
15. $y = 3x^2e^{4x}$
16. $y = (3 + 5x)e^{2x}$
17. $y = e^{3x} - e^x$
18. $y = e^{3x} - 2e^{4x}$
19. $y = \sqrt{e^{2x} + e^x}$
20. $y = x^2e^x - 5xe^{2x} + 2e^x$

Find f''(x) for each of the following.

21. $y = xe^{x-1}$
22. $y = xe^{x^2}$
23. $y = e^{x^2+x}$
24. $y = (2x + 5)e^{-3x}$
25. Find the equation of the tangent line to the graph of $y = e^{3x-2}$ at the point $(\frac{2}{3}, 1)$.
26. Find the equation of the tangent line to the graph of $y = e^{x^2}$ at the point $(0, 1)$.

Find dy/dx for the following functions.

27. $y = (e^{2x} + 5)e^x$
28. $y = \dfrac{e^{4x} - 1}{e^x - 5}$
29. $y = xe^{2x} + e^{2x^2}$
30. $y = e^{3x} + e^{x^2} - 5$
31. $y = e^{3x} + e^{x^2}$
32. $y = \dfrac{e^{5x}}{e^{3x} + 2}$

Applications (Business and Economics)

33. **Salvage Value.** If the salvage value of an airplane after t years is

$$V(t) = 400,000e^{-0.1t}$$

what is the rate of depreciation, dV/dt, after (a) 1 year, (b) 3 years, and (c) 10 years?

34. **Marginal Revenue Function.** If the price demand and revenue functions for x units of an item are, respectively,

$$p(x) = 100e^{-0.06x} \quad \text{and}$$
$$R(x) = xp(x) = 100xe^{-0.06x}$$

find the marginal revenue function. What number of units gives the maximum revenue?

35. **Sales Decay.** Sales at an automobile dealership were excellent when the 1995 cars were introduced, but then leveled off with time according to the model

$$S(t) = 160 - 80e^{-0.2t}$$

where t represents time in weeks. Find the rate of change of sales at the end of (a) 4 weeks, (b) 10 weeks, and (c) 20 weeks.

36. **Compounding Continuously.** If $P = \$1000$ and interest is compounded continuously at a rate of 10%, find the rate of change of A with respect to time (a) at the end of 2 years and (b) at the end of 10 years.

Applications (Social and Life Sciences)

Sometimes exponential graphs are classified as unlimited growth, limited growth, logistic growth, and exponential decay. In Exercises 37–44, use your cal-

culator to compare the graphs of the functions. First use $0 < t < 20$ and $0 < y < 500$. Change your window if you do not feel you are getting a useful section of each graph.

37. Unlimited growth: $y = 500e^{0.15t}$
38. Unlimited growth: $y = 1000e^{0.06t}$
39. Limited growth: $y = 500(1 - e^{-0.15t})$
40. Limited growth: $y = 1000(1 - e^{-0.06t})$
41. Logistic growth: $y = \dfrac{500}{1 + 200e^{-0.15t}}$
42. Logistic growth: $y = \dfrac{1000}{1 + 400e^{-0.06t}}$
43. Exponential decay: $y = 500e^{-0.15t}$
44. Exponential decay: $y = 1000e^{-0.06t}$
45. **Learning Function.**   A learning function is given by

$$N(t) = 100(-e^{-0.03t})$$

Find the rate of learning, $N'(t)$. What is the rate (a) after 10 hours and   (b) after 20 hours?

46. **Logistic Growth Curve.**   The equation of a logistic growth curve of bacteria is of the form

$$w(t) = \frac{1000}{1 + 50e^{-0.2t}}$$

where t is measured in days and $w(t)$ is the weight after t days. Find the rate of change of weight w with respect to time t.

47. **Von Bertalanffy Curve.**   The number of bacteria in a culture after t hours can be given by

$$N(t) = 1000(1 - .4e^{-0.2t})^3$$

Find the rate of change of $N(t)$ with respect to time t in hours.

10.2   LOGARITHMIC FUNCTIONS AND THEIR PROPERTIES

OVERVIEW
Approximately 400 years ago John Napier discovered an ingenious way to multiply by adding and to divide by subtracting. He did this by developing **logarithms**. In 1614 Napier published his discovery in a paper entitled *A Description of the Marvelous Rule of Logarithms*. For many years logarithms were used to facilitate the multiplication and division of large numbers. Today, with modern calculators we no longer need logarithms for computation. However, logarithmic functions are important because they are used to express relationships in such areas as advertising, archeology, and biochemistry. In business, some cost functions and demand functions are expressed in terms of logarithms. The well-known Richter scale for earthquakes is expressed in terms of logarithms. As you will see, logarithmic functions are closely related to exponential functions. In this section we

- Define a logarithm in terms of exponents
- State the properties of logarithms
- Examine the characteristics of logarithmic functions
- Draw graphs of logarithmic functions
- Solve exponential equations

In Chapter 1 we determined that one bacteria can divide into two new ones. If the split occurs each minute, how long will it take for 1 bacteria to become 256 bacteria? To solve this problem, we observe the pattern that occurs. After 1 minute we have 2 or 2^1 bacteria; after 2 minutes we have 4 or 2^2 bacteria; and after

3 minutes we have 8 or 2^3 bacteria. In general, after y minutes we have 2^y bacteria. For our problem we have $2^y = 256$. Thus it takes $y = 8$ minutes for 1 bacteria to become 256 bacteria since $2^8 = 256$.

The exponential equation $2^8 = 256$ is often rewritten in logarithmic form as

$$\log_2 256 = 8$$

The symbol $\log_2 256$ is read the "logarithm, base 2, of 256." Another example of an exponential equation and its logarithmic equivalent is $10^3 = 1000$ and $\log_{10} 1000 = 3$. In general, if $x = b^y$, we say that y is the log of x to the base b and denote this by

$$y = \log_b x$$

That is, in $x = b^y$ the exponent y can be written in terms of another function called a logarithm. Remember that logarithms are exponents.

LOGARITHM

If x and b are positive numbers and b is not equal to 1, then the **logarithm** of x to the base b is equal to y, which is written as

$$\log_b x = y, \qquad x \geq 0$$

if and only if $x = b^y$.

We read $\log_b x$ as "the logarithm of x to the base b." The function defined is called the **logarithmic function**. Table 3 illustrates a number of logarithmic expressions and their exponential equivalents.

TABLE 3

Logarithmic Form	Exponential Form
$\log_{10} 100 = 2$	$10^2 = 100$
$\log_{10} 0.01 = -2$	$10^{-2} = 0.01$
$\log_2 0.25 = -2$	$2^{-2} = 0.25$
$\log_{10} 1 = 0$	$10^0 = 1$
$\log_b 1 = 0$	$b^0 = 1$
$\log_b b = 1$	$b^1 = b$

EXAMPLE 10

(a) Write $81 = 3^4$ in logarithmic form.
(b) Write $\log_2 \frac{1}{32} = -5$ in exponential form.
(c) If $3 = \log_2 x$, find x.

SOLUTION

(a) $\log_3 81 = 4$ (b) $2^{-5} = \frac{1}{32}$ (c) $x = 2^3 = 8$

Practice Problem 1 Find $\log_3 \frac{1}{81}$.

ANSWER -4

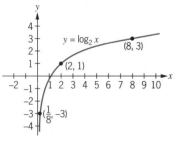

Figure 4

EXAMPLE 11   Sketch the graph of $y = \log_2 x$.

SOLUTION   We know that $\log_2 8 = 3$ because $2^3 = 8$, so the point $x = 8$, $y = 3$ is on the graph. In a similar manner, $(1, 0)$, $(2, 1)$, $(\frac{1}{2}, -1)$, $(\frac{1}{8}, -3)$ are points on the graph. The graph of $y = \log_2 x$ is shown in Figure 4.

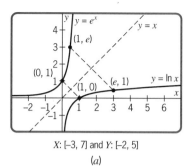

X: [-3, 7] and Y: [-2, 5]

(a)

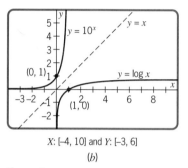

X: [-4, 10] and Y: [-3, 6]

(b)

Figure 5

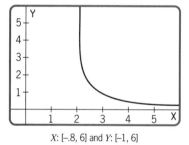

X: [-.8, 6] and Y: [-1, 6]

Figure 6

In the preceding example, notice that the domain of $y = \log_2 x$ is the set of *positive* real numbers and the range is the set of all real numbers such that

$$y = \log_2 x \quad \begin{cases} < 0, & \text{for } 0 < x < 1 \\ = 0, & \text{for } x = 1 \\ > 0, & \text{for } x > 1 \end{cases}$$

The x-intercept is at $(1, 0)$. The graph approaches the y-axis as an asymptote.

As we have noted, the base of a logarithm may be any positive number except 1. However, most logarithms use the base 10 (called **common logarithms**) or the base e (called **natural logarithms**). On your calculator you will see a ⌐LOG⌐ key representing common logarithms and an ⌐LN⌐ key representing natural logarithms.

With your graphing calculator draw the graphs of $y = e^x$ and $y = \ln x$ on the same coordinate system, as shown in Figure 5(a). Notice that the graphs of $y = e^x$ and $y = \ln x$ are reflections of each other about the line $y = x$. Likewise, draw the graphs of $y = 10^x$ and $y = \log x$ on the same coordinate system, as shown in Figure 5(b). Notice that $y = 10^x$ and $y = \log x$ are also reflections of each other about $y = x$.

The pairs of functions in Figure 5 have special names; $y = e^x$ and $y = \ln x$ are called inverse functions. In general, we say that $y = f(x)$ and $y = g(x)$ are **inverse functions** if, whenever the pair (a, b) satisfies $y = f(x)$, the pair (b, a) satisfies $y = g(x)$. Furthermore, because the values of the x- and y-coordinates are interchanged for inverse functions, their graphs are reflections of each other about the line $y = x$.

Practice Problem 2 Use a calculator to identify the domain of

$$y = \ln \frac{|x|}{x - 2}$$

ANSWER   As shown in Figure 6, the domain is $x > 2$.

Logarithmic functions have several useful properties that follow directly from their definitions and from the properties of exponents.

PROPERTIES OF LOGARITHMS

For $b > 0$, $b \neq 1$, and n a real number, if M and N are positive numbers, then the following properties are true.

1. $\log_b(M \cdot N) = \log_b M + \log_b N$
2. $\log_b \dfrac{M}{N} = \log_b M - \log_b N$
3. $\log_b M^N = N \log_b M$
4. $\log_b 1 = 0$
5. $\log_b b = 1$
6. $\log_b M = \log_b N$ if and only if $M = N$

These properties are stated in words to assist you in remembering them.

1. The logarithm of a product is the sum of the logarithms of the factors.
2. The logarithm of a quotient is the difference of the logarithm of the numerator and the logarithm of the denominator.
3. The logarithm of a number to a power is the power times the logarithm of the number.
4. The logarithm of 1 to any base is 0.
5. The logarithm of a number which equals the base is one.
6. If the logarithms of two numbers to the same base are equal, then the numbers are equal.

We can demonstrate the first three properties by comparing the graphs in Figure 7. In part (a) we have the graph of $y = \ln 6x$. Note that this is exactly the same graph as $y = \ln 6 + \ln x$ in part (b); that is, $\ln 6x = \ln 6 + \ln x$. The graph of $y = \ln (x/6)$ in part (c) is the same as the graph of $y = \ln x - \ln 6$ in part (d); that is, $\ln (x/6) = \ln x - \ln 6$. In part ($e$) the graph of $y = \ln x^3$ is the same as the graph of $y = 3 \ln x$ in part (f); that is, $\ln x^3 = 3 \ln x$.

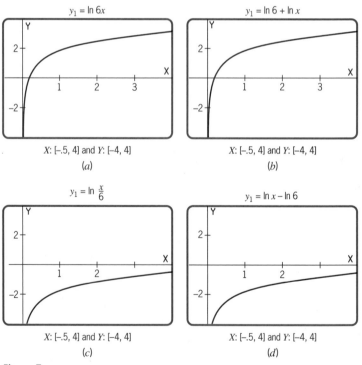

$y_1 = \ln 6x$

X: [−.5, 4] and Y: [−4, 4]

(a)

$y_1 = \ln 6 + \ln x$

X: [−.5, 4] and Y: [−4, 4]

(b)

$y_1 = \ln \frac{x}{6}$

X: [−.5, 4] and Y: [−4, 4]

(c)

$y_1 = \ln x - \ln 6$

X: [−.5, 4] and Y: [−4, 4]

(d)

Figure 7

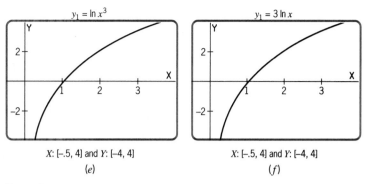

Figure 7

EXAMPLE 12   Using $\ln 2 = 0.693147$ and $\ln 3 = 1.098612$, find $\ln 6$ using the properties of logarithms.

SOLUTION

$$\ln 6 = \ln (2 \cdot 3) = \ln 2 + \ln 3 \qquad \text{Property 1}$$

$$\ln 6 = 0.693147 + 1.098612 = 1.791759$$

EXAMPLE 13   Find $\ln e$.

SOLUTION $\ln e = 1$ by property 5.

Calculator Note

To find ln e use ⎡LN⎤ ⎡e⎤ ⎡∧⎤ 1.

EXAMPLE 14   Find $\ln 8$ using the properties of logarithms. (See Example 12.)

SOLUTION

$$\ln 8 = \ln 2^3 = 3 \ln 2 \qquad \text{Property 3}$$

$$\ln 8 = 3(0.693147) = 2.079441$$

EXAMPLE 15   Solve for x in $3^x = 4$ given that $\log 2 = 0.301030$ and $\log 3 = 0.477121$. (Verify these values with your calculator.)

SOLUTION Take the logarithm of both sides to obtain

$$\log 3^x = \log 4$$
$$= \log 2^2$$

$$x \log 3 = 2 \log 2 \qquad \text{Property 3}$$

$$x = \frac{2 \log 2}{\log 3}$$

$$= \frac{0.602060}{0.477121}$$

$$= 1.261860 \qquad \blacksquare$$

EXAMPLE 16 Express

$$\log \left(\frac{x^2 y^{1/2}}{z^3} \right)$$

as a sum or difference of logarithms without exponents.

SOLUTION

$$\log \left(\frac{x^2 y^{1/2}}{z^3} \right) = \log x^2 + \log y^{1/2} - \log z^3 \qquad \text{Properties 1 and 2}$$

$$= 2 \log x + \frac{1}{2} \log y - 3 \log z \qquad \text{Property 3} \qquad \blacksquare$$

EXAMPLE 17 Solve for x in $\log_5 x + \log_5(x - 4) = 1$.

SOLUTION

$$\log_5 x + \log_5(x - 4) = 1$$

$$\log_5 x(x - 4) = 1 \qquad \log M + \log N = \log MN$$

$$x(x - 4) = 5^1 \qquad \text{Definition of a logarithm}$$

$$x^2 - 4x - 5 = 0$$

$$(x - 5)(x + 1) = 0$$

$$x = 5 \quad \text{or} \quad x = -1$$

It would seem that $x = 5$ and $x = -1$ are solutions of the given equation, but a check of these values in the original equation shows that only $x = 5$ is a solution. Why is $x = -1$ not a solution? $\qquad \blacksquare$

Practice Problem 3 Use your calculator to find the value of $\log \frac{120}{62}$.

ANSWER   0.286790

Logarithms are very useful in solving for exponents in exponential equations. For example, we can use logarithms to compute the length of time necessary for an investment to double using the formula $A = P(1 + i)^n$.

EXAMPLE 18   How long will it take for an investment of $10,000 to double if a bank pays 8% interest compounded annually?

SOLUTION   If P dollars are invested at $i\%$ compounded yearly, the investment will accumulate to A dollars in n years by the formula

$$A = P(1 + i)^n$$

If $P = \$10,000$ is to double, then $A = 2P = \$20,000$, and i is 0.08. The problem is to find n.

$$A = P(1 + i)^n$$
$$20,000 = 10,000(1 + 0.08)^n \qquad \text{Substitution}$$
$$2 = (1.08)^n$$
$$\log 2 = \log(1.08)^n \qquad \text{Solve by taking log or ln of both sides.}$$
$$\qquad = n \log 1.08 \qquad \text{Property 3}$$
$$n = \frac{\log 2}{\log 1.08}$$
$$\qquad = 9.0065$$

Thus, in approximately 9 years the investment will have doubled.

NOTE: Compare the answer in the preceding example with the Rule of 72 often used in the business world. This rule gives the doubling time of an investment at $i\%$ as being approximately $72/i$, where i is not changed to a decimal. For Example 18 the doubling time would be estimated to be $72/8 = 9$ years.

SUMMARY

In this section we defined a logarithm in terms of exponents and learned how to use logarithmic functions and properties of logarithms. We also solved logarithmic equations. All of these will be important as we learn to take derivatives of logarithmic functions in the next section.

Exercise Set 10.2

1. Write the following equations in logarithmic notation.
   (a) $3^4 = 81$ (b) $2^7 = 128$
   (c) $3^{-2} = \dfrac{1}{9}$

2. Write the following equations in logarithmic notation.
   (a) $5^4 = 625$ (b) $8^{1/3} = 2$
   (c) $16^{1/4} = 2$

3. Write the following in exponential notation.
   (a) $\log_7 49 = 2$ (b) $\log_6 36 = 2$
   (c) $\log_3 \dfrac{1}{9} = -2$

4. Write the following in exponential notation.
   (a) $\log_3 \dfrac{1}{27} = -3$ (b) $\log_{10} 1 = 0$
   (c) $\log_{10} 1000 = 3$

5. Find the following logarithms.
   (a) $\log_2 16$ (b) $\log_2 \dfrac{1}{8}$
   (c) $\log_{1/2} \dfrac{1}{8}$ (d) $\log_3 27$

6. Find the following logarithms.
   (a) $\log_{10} 10$ (b) $\log_{10} 1$
   (c) $\log_{10} 0.1$ (d) $\log_{10} 0.01$

7. Find the base of the following logarithms.
   (a) $\log_b 2 = \dfrac{1}{2}$ (b) $\log_b 2 = -1$

8. Find the base of the following logarithms.
   (a) $\log_b \dfrac{1}{8} = -3$ (b) $\log_b 100 = 2$

9. Determine the solution for x by inspection or by first writing the equation in exponential form.
   (a) $\log_3 9 = x$ (b) $\log_{10} x = 3$

10. Determine the solution for x by inspection or by first writing the equation in exponential form.
    (a) $\log_{10} x = -3$ (b) $\log_3 27 = x$

Express each of the following as a sum or difference of logarithms without exponents.

11. $\ln \dfrac{x^2 z^4}{y}$

12. $\ln \dfrac{x^2 y^{-2}}{z^3}$

13. $\log_b \sqrt[3]{\dfrac{x^2}{y}}$

14. $\log_b \sqrt{\dfrac{x}{y^3}}$

15. $\ln \sqrt{\dfrac{x^2 y}{z}}$

16. $\ln \sqrt{\dfrac{x}{z^3 y}}$

Use your calculator to find the value of each of the following.

17. $\log \sqrt[3]{165}$

18. $\log \sqrt{16^3 \cdot 71}$

19. $\log \dfrac{\sqrt{131}}{\sqrt[3]{9}}$

20. $\ln \sqrt[3]{7}$

21. $\ln \sqrt{18^3 \cdot 27}$

22. $\ln \sqrt[5]{\dfrac{161}{5}}$

Solve the following equations for x.

23. $\log_6 1 = x$ 24. $\log_4 x = 1$
25. $\log x = 2 \log 3 - 3 \log 2$
26. $\log \sqrt{x} = 2 \log 4 - \dfrac{1}{2} \log 9$
27. $\ln x^2 = 2$
28. $\ln x^2 + \ln 4 = \ln x + \ln 8$
29. $\ln e^{-0.01x} = 4$ 30. $\ln e^{x^2} = 25$
31. $\log x + \log(x + 1) = \log 2$
32. $\ln(x + 2) - \ln 3 = \ln 4x$

Use your calculator to solve the following equations. Round your answers to four decimal places.

33. $8^x = 61$ 34. $14^y = 0.1$
35. $74^x = 16$ 36. $e^{2x} = 6$
37. $5e^{-3x} = 8$ 38. $e^{-x/2} = 0.4$

Applications (Business and Economics)

39. **Doubling of Sales.**   A corporation has formulated the model

    $$y = 1{,}000{,}000 e^{0.06t}$$

    to predict sales growth, where y is annual sales in dollars in year t when t starts at 0 in 1990. In what year will the annual sales reach $2,000,000?

40. **Compound Interest.**   Brooke has placed $10,000 in a savings and loan that pays 9% interest compounded annually. When will she have $25,000 in her account? [**Hint:** $A = P(1 + i)^n$.]

41. **Compounded Continuously.** How long will it take to double an investment if a bank pays 8% interest compounded continuously?

42. **Cost Function.** The cost function for manufacturing a certain commodity is given as

$$C(x) = 10,000 + 600 \ln (2x^2 + 1)$$

where x is the number of items to be manufactured.
   (a) What is the fixed cost (the cost when $x = 0$)?
   (b) What is the cost of manufacturing 10,000 items?
   (c) What is the cost of manufacturing 1000 items?

Applications (Social and Life Sciences)

43. **Earthquake.** A Richter scale measurement of the intensity of an earthquake is given as

$$RS = \log \frac{I}{I_0}$$

where I_0 is the intensity used for comparison.
   (a) Find RS if I is 1,000,000 times I_0.
   (b) The 1983 earthquake measured 7.7 on the Richter scale. Find the intensity in terms of I_0.

Review Exercises

Find dy/dx for the following functions.

44. $y = e^{x^2}$

45. $y = e^{5x^3}$

46. $y = x^2 e^{2x^2}$

10.3   DERIVATIVES OF LOGARITHMIC FUNCTIONS

OVERVIEW   Many functions in business and economics are expressed in terms of logarithms. In order to obtain marginal functions, such as marginal revenue and marginal cost, we need to know the derivative of a logarithmic function. In this section we use formulas for dy/dx when $y = \ln x$ and $y = \ln u(x)$. The derivation of dy/dx when $y = \ln x$ using our four-step procedure is rather cumbersome. So we postpone this derivation until Section 10.5 where the derivation is simple using implicit differentiation. In this section we study

- Formulas for the derivative of logarithmic functions
- The chain rule
- Graphing techniques

The derivative formula for $y = \ln x$ is very simple.

$$\text{If } y = \ln x, \text{ then } \frac{dy}{dx} = \frac{1}{x}.$$

With your calculator, on some range $x > 0$, sketch the graph of $y = $ nDeriv $(\ln x, x, x)$ and then sketch $y = 1/x$ and note that the two graphs are identical for $x > 0$. We postpone the derivation of this derivative formula until Section 10.5, as implicit derivatives facilitate an easy derivation. The derivative formula for $y = \ln x$, along with the properties of logarithms, enable us to differentiate a wide variety of functions.

EXAMPLE 19   Find $\dfrac{dy}{dx}$ if $y = \ln x^3$.

SOLUTION

$$y = \ln x^3$$
$$= 3 \ln x \qquad \ln M^N = N \ln M$$

$$\frac{dy}{dx} = 3 \cdot \frac{1}{x} = \frac{3}{x}$$

For $y = \ln u$, where u is a function of x, we use the chain rule to find dy/dx, where $u(x) > 0$.

$$y = \ln u(x)$$

$$\frac{dy}{dx} = \frac{d}{du} \ln u \cdot \frac{du}{dx}$$

$$= \frac{1}{u} \cdot \frac{du}{dx} \qquad \frac{d}{du} \ln u = \frac{1}{u}$$

Now let's rework Example 19 using this formula.

$$y = \ln x^3 \qquad \text{Let } u = x^3.$$

$$\frac{dy}{dx} = \frac{1}{u} \cdot \frac{du}{dx}$$

$$= \frac{1}{x^3} \cdot (3x^2) \qquad \frac{du}{dx} = 3x^2$$

$$= \frac{3}{x}$$

We state these formulas as follows.

DERIVATIVES OF LOGARITHMIC FUNCTIONS

1. If $f(x) = \ln x$, then

$$f'(x) = \frac{1}{x}$$

for all $x > 0$.

2. If $f(x) = \ln u(x)$, where $u(x) > 0$ and has a derivative with respect to x, then

$$f'(x) = \frac{u'(x)}{u(x)}$$

EXAMPLE 20   If $f(x) = \ln(2x + 5)$, find $f'(x)$.

SOLUTION   Let $u(x) = 2x + 5$. Then $u'(x) = 2$, and

$$f'(x) = \frac{u'(x)}{u(x)} = \frac{2}{2x + 5}$$

■

EXAMPLE 21   If $f(x) = \ln(3x^2 + x - 2)$, find $f'(x)$.

SOLUTION   Let $u(x) = 3x^2 + x - 2$. Then $u'(x) = 6x + 1$, and

$$f'(x) = \frac{u'(x)}{u(x)} = \frac{6x + 1}{3x^2 + x - 2}$$

■

EXAMPLE 22   If $f(x) = \ln(x^{3/2} - 2x^{5/2})$, find $f'(x)$.

SOLUTION

$$f'(x) = \frac{u'(x)}{u(x)} = \frac{u'(x)}{x^{3/2} - 2x^{5/2}}$$

If $u(x) = x^{3/2} - 2x^{5/2}$, then $u'(x) = \frac{3}{2}x^{1/2} - 5x^{3/2}$. Therefore,

$$f'(x) = \frac{\frac{3}{2}x^{1/2} - 5x^{3/2}}{x^{3/2} - 2x^{5/2}}$$

$$= \frac{x^{1/2}[\frac{3}{2} - 5x]}{x^{1/2}[x - 2x^2]} \qquad \text{Factor } x^{1/2} \text{ from the numerator and denominator.}$$

$$= \frac{\frac{3}{2} - 5x}{x - 2x^2} \qquad \text{Divide out } x^{1/2}.$$

$$= \frac{3 - 10x}{2x - 4x^2}$$

■

EXAMPLE 23   If $f(x) = x \ln(x + 1)$, find $f'(x)$.

SOLUTION

$$f'(x) = x\frac{d}{dx}\ln(x + 1) + \frac{d}{dx}(x) \ln(x + 1) \qquad \text{Product formula}$$

$$= x \cdot \frac{1}{x + 1} + 1 \cdot \ln(x + 1)$$

$$= \frac{x + (x + 1) \ln(x + 1)}{x + 1}$$

■

Practice Problem 1   Given $f(x) = x^2 \ln(x^2 + 2)$, find $f'(x)$.

ANSWER $f'(x) = 2x \ln(x^2 + 2) + \dfrac{2x^3}{x^2 + 2}$

Whenever the equation is given, such as the revenue function in Example 24, calculus procedures can be used to maximize profit.

EXAMPLE 24 A company can produce men's ties at a cost of $4 per tie. The revenue equation is

$$R(x) = 10x - 0.8x \ln x$$

where x is the number of ties manufactured each week. Find the weekly production that will maximize profit.

SOLUTION

$$\begin{aligned} P(x) &= R(x) - C(x) \\ &= 10x - 0.8x \ln x - 4x = 6x - 0.8x \ln x \qquad C(x) = 4x \end{aligned}$$

Then we have

$$P'(x) = 6 - \frac{0.8x}{x} - 0.8 \ln x \qquad \text{Derivative of a product}$$

$$0 = 5.2 - 0.8 \ln x \qquad \text{Set } P'(x) = 0.$$

$$\ln x = \frac{5.2}{0.8} = 6.5$$

$$x = e^{6.5} \approx 665 \qquad \text{Definition of } \ln x$$

Since $P''(x) = -0.8/x$, which is negative for $x > 0$, there is a relative maximum at $x = 665$. Therefore, a production of 665 ties per week would produce a maximum profit. ■

Sometimes it is helpful to apply the properties of logarithms before taking derivatives. This procedure saves time when working problems such as Example 25.

EXAMPLE 25 If $f(x) = \ln[(x + 2)^2(x^3 + 4)]$, find $f'(x)$.

SOLUTION

$$\begin{aligned} f(x) &= \ln[(x + 2)^2(x^3 + 4)] \\ &= \ln(x + 2)^2 + \ln(x^3 + 4) \qquad \ln MN = \ln M + \ln N \\ &= 2 \ln(x + 2) + \ln(x^3 + 4) \qquad \ln M^N = N \ln M \end{aligned}$$

$$f'(x) = \frac{2}{x + 2} + \frac{3x^2}{x^3 + 4} \qquad \frac{d}{dx}(x^3 + 4) = 3x^2$$ ■

SUMMARY

In this section we learned to use two formulas involving logarithms.

1. If $y = \ln x$, then

$$y' = \frac{1}{x} \quad \text{for } x > 0$$

2. If $y = \ln u$, where u is a function of x, then

$$y' = \frac{u'}{u} \quad \text{for } u > 0$$

These formulas can be combined with the rules of differentiation in Chapter 8 to differentiate a wide variety of functions.

Exercise Set 10.3

Find dy/dx for the following functions and simplify.

1. $y = 3 \ln x$
2. $y = \ln x^2$
3. $y = \ln(2x + 3)$
4. $y = \ln(3x + 10)$
5. $y = 4 \ln(3x + 2)$
6. $y = 7 \ln(2x + 5)$
7. $y = \ln(3x^2 + 2x + 5)^2$
8. $y = 3 \ln(2x^2 + x + 3)^2$
9. $y = 3x^2 + 2x + \ln(2x + 5)$
10. $y = 2x \ln(3x + 7)$
11. $y = 3x^2 \ln(3x + 2)$
12. $y = (2x^2 + 1) \ln 5x^3$
13. $f(x) = (\ln x)^3$
14. $f(x) = (\ln 2x^2)^2$
15. $y = 10x^2 \ln(x^2 + 1)^3$
16. $y = 2x^3 \ln(3x^2 + x)$
17. $y = \ln[3 \sqrt{x}(2x + 1)^3]$
18. $y = \ln[2x^3 \sqrt{3x + 2}]$
19. $y = [\ln(3x^2 - 2x + 5)]^2$
20. $y = \ln\left[\dfrac{3x^2 + 4x + 5}{(2x + 3)^4}\right]$

Applications (Business and Economics)

21. **Maximum Profit.**   If the revenue function of an industry is given as

$$R(x) = 400 \ln(2x - 10)$$

and the cost function is $C(x) = 40x$ for the production of x items, find the number of items that must be produced to have a maximum profit.

22. **Demand Equation.**   The market research of a company indicates that the demand equation is $p = D(x) = 20 - 4 \ln x$, where p is the price and x is the number of units in demand. Find dp/dx.

23. **Cost Function.**   The cost function for selling a product is given as

$$C(x) = 100 + 60 \ln(2x^2 - 2x + 1)$$

where x is the number of units of 10,000 pounds sold. Find the number of pounds that should be sold to keep cost at a minimum. First, approximate by using your graphing calculator.

24. **Business Advertising.**   The number of responses (automobiles sold) to x thousands of dollars of advertising by an automobile dealership seems to follow the model

$$N(x) = 200x - 2000 \ln x.$$

Find the amount of advertising that would produce the maximum number of responses. First, approximate using your graphing calculator.

Applications (Social and Life Sciences)

25. ***Earthquakes***. Earthquakes are reported in units R on the Richter scale with R being defined as

$$R = \frac{0.67}{\ln 10} \ln(E) - 7.9$$

where E is the energy in ergs released by the earthquake. Find an expression for the rate of change of R with respect to E.

26. ***Archaeology***. The formula

$$\ln(P) = \frac{1}{2} \ln(A)$$

relates the population of a site to the area of the site. Find the rate of change of the population with respect to the area.

27. ***pH Factor***. Recall that the pH of a solution is defined to be

$$pH = \frac{\ln \left(\dfrac{1}{H^+} \right)}{\ln 10}$$

where H^+ represents the concentration of hydrogen ions per liter. Compute the derivative of pH with respect to H^+, and show that pH is decreasing for $H^+ > 0$.

Review Exercises

28. Write as a sum:

$$y = \ln \left(\frac{\sqrt{x} \cdot y^{-1/3}}{z^{-1/3}} \right)$$

29. Write $\log_5 x = 3$ as an exponential expression and then find x.

For the following functions, find dy/dx.

30. $y = e^{x-2}$

31. $y = xe^{-3/x}$

32. $y = (4 + 2x^2)(3 - e^{-2x})$

33. $y = \dfrac{4 - 2e^{2x}}{3 + e^{-3x}}$

10.4 GRAPHING EXPONENTIAL AND LOGARITHMIC FUNCTIONS (Optional)

OVERVIEW   An easy way to graph exponential and logarithmic functions is to use a calculator with graphing capabilities. This will be our first step in this section. However, the main objective of this section is to apply the calculus concepts of Chapters 8 and 9 to graphing exponential and logarithmic functions. Our objective is to show how calculus can give more specific information about graphs than that obtained from calculator graphs. In this section we

- Discuss the characteristics of a graph using a graphing calculator
- Discuss the characteristics of a graph using calculus

Our first step in working problems in this section is to draw a given curve with a calculator. Then we discuss intuitively what seem to be the characteristics of the curve.

EXAMPLE 26   With a calculator, draw the graphs of y, y', and y'' for the function $y = f(x) = xe^x$ and discuss each of the following.

(a) Intercepts of $f(x)$

(b) Points where maxima or minima occur for $f(x)$

(c) Where the graph of $f(x)$ is increasing or decreasing

(d)   What happens to $f(x)$ when $x \to \infty$ and when $x \to -\infty$
(e)   Changes in concavity of $f(x)$

SOLUTION   The graphs of y, y', and y'' in Figure 8 suggest the following characteristics of $y = xe^x$.

(a)   Both the x- and y-intercepts of $f(x)$ are at $(0, 0)$, as shown in Figure 8(a).
(b)   A relative minimum occurs when $x = -1$, since we can see in Figure 8(b) that y' changes from negative to positive at $x = -1$.
(c)   The graph of $f(x)$ is increasing on $(-1, \infty)$ and decreasing on $(-\infty, -1)$.
(d)   As $x \to \infty$, $y \to \infty$; as $x \to -\infty$, $y \to 0$ as a horizontal asymptote.
(e)   The second derivative is 0 at $x = -2$ in Figure 8(c). At this point there is a point of inflection or a change in concavity. The curve is concave upward for $x > -2$ and concave downward for $x < -2$.   ∎

Of course, the solutions in the preceding example could have been obtained to greater accuracy; however, the procedure we are following at this time is to obtain approximate answers quickly using a graphing calculator and then to obtain exact answers using calculus procedures.

EXAMPLE 27   For the curve $y = xe^x$, use calculus techniques to

(a)   find the intercepts.
(b)   locate points of relative extrema.
(c)   discuss where the graph is increasing or decreasing.
(d)   discuss concavity.
(e)   determine what happens when $x \to \infty$ and when $x \to -\infty$.
(f)   sketch the graph of $y = xe^x$.

SOLUTION

(a)   When $x = 0$, $y = 0$, so the curve goes through the origin.
(b)   First we find y' to be $y' = xe^x + e^x$. To find the critical points, find where $y' = 0$.

$$y' = e^x(x + 1) = 0$$

Since e^x cannot be 0, $x = -1$ is the only solution and the only critical value. Next we have $y(-1) = -1 \cdot e^{-1} = -1/e$. So, $(-1, -1/e)$ is a critical point and we apply the second derivative test to determine whether this critical point is an extremum.

$$\begin{aligned} y'' &= e^x(1 + x) + e^x \qquad \text{Product formula} \\ &= e^x(2 + x) \end{aligned}$$

$$\begin{aligned} y''(-1) &= e^{-1}[2 + (-1)] \\ &= e^{-1} \cdot 1 \\ &= \frac{1}{e} > 0 \end{aligned}$$

Therefore, there is a relative minimum at $(-1, -1/e)$.

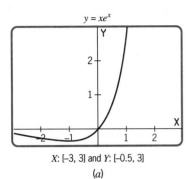

$y = xe^x$

X: [-3, 3] and Y: [-0.5, 3]

(a)

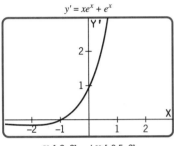

$y' = xe^x + e^x$

X: [-3, 3] and Y: [-0.5, 3]

(b)

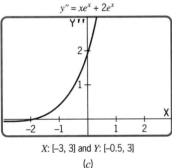

$y'' = xe^x + 2e^x$

X: [-3, 3] and Y: [-0.5, 3]

(c)

Figure 8

(c)   To determine where the function is increasing and where it is decreasing, we examine the first derivative $y' = e^x(1 + x)$. The slope changes at $x = -1$. Looking at the calculator graph of y', the derivative seems to be negative for $x < -1$ and positive for $x > -1$, or since e^x is always positive, $e^x(1 + x)$ is negative when $x < -1$ and positive when $x > -1$. Therefore, the function is decreasing for x on $(-\infty, -1)$ and increasing for x on $(-1, \infty)$.

(d)   To determine concavity, we find the second derivative and determine where is it zero.

$$y'' = e^x(2 + x)$$

Set y'' equal to zero and solve for x.

$$e^x(2 + x) = 0$$
$$2 + x = 0 \qquad \text{e^x cannot be zero.}$$
$$x = -2$$

Since $y'' = 0$ at $x = -2$, concavity can change only at $x = -2$. For $(-\infty, -2)$, e^x is always positive and $e^x(2 + x) < 0$; the curve is concave downward. For $(-2, \infty)$, $e^x(2 + x) > 0$ and the curve is concave upward.

(e)   We can use a calculator to verify that as x gets larger, y gets larger, or as $x \to \infty$, $y \to \infty$ and as $x \to -\infty$, $y \to 0$.

(f)   To sketch the graph shown in Figure 9, we use the preceding information for $y = xe^x$:

The y-intercept is $(0, 0)$.
A minimum point exists at $(-1, -1/e)$.
It is decreasing over $(-\infty, -1)$ and increasing over $(-1, \infty)$.
It is concave downward over $(-\infty, -2)$ and upward over $(-2, \infty)$.
$\lim\limits_{x \to -\infty} y = 0$ and $\lim\limits_{x \to \infty} y \to \infty$.   ■

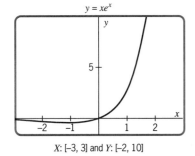

$y = xe^x$

X: [-3, 3] and Y: [-2, 10]

Figure 9

In a similar manner, we discuss the graph of a function involving logarithms by drawing the graph on a calculator and then obtaining characteristics of the graph intuitively.

EXAMPLE 28   With your graphing calculator draw the graph of $y = x \ln x$, the graph of the slope function $y' = \ln x + 1$, and the graph of $y'' = 1/x$. Then discuss each of the following intuitively.

(a)   Intercepts of y
(b)   Domain of y
(c)   What happens to y as $x \to 0$ and as $x \to \infty$

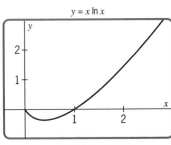

$y = x \ln x$

X: [-.5, 3] and Y: [-1, 3]

(a)

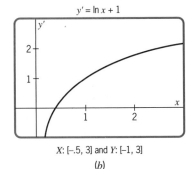

$y' = \ln x + 1$

X: [-.5, 3] and Y: [-1, 3]

(b)

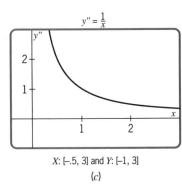

$y'' = \frac{1}{x}$

X: [-.5, 3] and Y: [-1, 3]

(c)

Figure 10

(d) Points of relative extrema of y
(e) Where the graph of y is increasing and decreasing
(f) Where the graph of y is concave up and concave down

SOLUTION   A study of the graphs of y, y' and y'' in Figure 10 suggest the following.

(a) From Figure 10(a) we see that an x-intercept is at $x = 1$. The graph is approaching (0, 0); however, (0, 0) may not be an intercept.
(b) · The domain is $x > 0$.
(c) As $x \to 0$, the graph seems to approach $y = 0$. As $x \to \infty$, y seems to approach ∞.
(d) There is a relative minimum at x slightly less than $x = \frac{1}{2}$ because the slope is 0 at this point; see Figures 10(a) and (b).
(e) The graph in Figure 10(a) is decreasing for x between 0 and some x slightly less than $x = \frac{1}{2}$ (also $y' < 0$ in (b)) and increasing from there on ($y' > 0$).
(f) The graph is concave up for all x in the domain [$y'' > 0$ for all x in Figure 10(c)].

EXAMPLE 29   For $y = x \ln x$, use calculus to discuss each of the following.

(a) Intercepts
(b) Domain
(c) What happens as $x \to 0$, and as $x \to \infty$
(d) Points of relative extrema
(e) Where the function is decreasing and where it is increasing
(f) Where the graph is concave up and where it is concave down
(g) Sketch the graph.

SOLUTION

(a) When $x = 1$, $\ln 1 = 0$, so an x-intercept is at $x = 1$. There is no y-intercept because $\ln 0$ is undefined.
(b) Since $\ln x$ is defined only for $x > 0$, the domain is $x > 0$.
(c) By substituting values closer and closer to $x = 0$, it seems that as $x \to 0$, $y \to 0$. (We cannot be certain at this time.) As $x \to \infty$, we see from the following table that y seems to get very large or $y \to \infty$.

x	100	1000	10,000
y	461	6908	92,103

(d) We determine the critical points by finding the first derivative and setting it equal to 0.

$$y = x \ln x$$

$$y' = x \left(\frac{1}{x}\right) + \ln x \qquad \text{Product formula}$$

$$= 1 + \ln x$$

Set the first derivative to zero and solve.

$$\ln x + 1 = 0$$

$$\ln x = -1$$

$$x = e^{-1} \qquad \text{Definition of } \ln x$$
$$\approx 0.368$$

When $x \approx 0.368$, $y = 0.368 \ln 0.368 \approx -0.368$. Thus, a critical point occurs at approximately $(0.368, -0.368)$. We use the second derivative to determine whether we have a relative maximum or minimum.

$$y'' = \frac{1}{x}$$

Since $1/x$ is always positive for $x > 0$, a relative minimum occurs at approximately $(0.368, -0.368)$.

(e) Since $y' = 1 + \ln x$ and $1 + \ln x$ is 0 only for x at approximately 0.368, we can easily verify that y' is negative for $x < 0.368$ and positive for $x > 0.368$. Therefore, the function is decreasing on $(0, 0.368)$ and increasing on $(0.368, \infty)$.

(f) The second derivative, $y'' = 1/x$, is positive for all $x > 0$. Therefore, the graph is concave up for all x in the domain.

(g) Using these characteristics, we can sketch the graph shown in Figure 11.

■

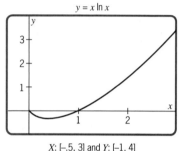

$y = x \ln x$

X: [–.5, 3] and Y: [–1, 4]

Figure 11

SUMMARY

In this section, we used the calculus concepts of Chapters 8 and 9 along with the graphing calculator to investigate graphs of exponential and logarithmic functions.

Exercise Set 10.4

With a graphing calculator, graph the following functions.

1. $y = e^{x^2}$
2. $y = e^{-(4x+2)}$
3. $y = \ln(2x^2 + 1)$
4. $y = \ln(3x - 4)$

With a graphing calculator, graph the given functions and their derivatives and then find each of the following intuitively.

(a) *Intercepts*
(b) *Domain*
(c) *Where the function is increasing and where it is decreasing*
(d) *Points where there are relative extrema*
(e) *Where the graph is concave up and where it is concave down*
(f) *What happens when $x \to \infty$ and when $x \to -\infty$*

5. $y = e^{x-4}$
6. $y = e^{-(3x+2)}$
7. $y = \ln(x - 4)$
8. $y = \ln(x^2 + 2)$

For each of the following use algebra and calculus techniques to

(a) *find the intercepts.*
(b) *find where the function is increasing or decreasing and locate points where there are relative extrema.*

(c) *find where the curve is concave up and where it is concave down.*

(d) *discuss what happens to y as $x \to \infty$ and as $x \to -\infty$.*

(e) *sketch the graph.*

9. $y = e^{x-4}$. (Check Exercise 5.)

10. $y = e^{-(3x+2)}$. (Check Exercise 6.)

11. $y = \ln(x - 4)$. (Check Exercise 7.)

12. $y = \ln(x^2 + 2)$. (Check Exercise 8.)

13. $y = xe^{-x}$

14. $y = \dfrac{e^x}{x}$

15. $y = x^2 \ln x$

16. $y = e^x \ln x$

17. Use a graphing calculator to find the points at which there are relative maxima or minima for $y = e^{-|x^2 - 1|}$.

Applications (Business and Economics)

18. **Sales Decay.** Automobile sales at an automobile dealership were excellent when the 1995 cars were introduced but then leveled off with time according to the model

$$S(t) = 160 - 80e^{-0.2t}$$

where t represents time in weeks. Sketch the graph of this function and verify with a graphing calculator.

19. **Compounding Continuously.** Sketch the graph of the rate of change of A with respect to time where $P = \$1000$ and $i = 10\%$. (Recall that $A = Pe^{rt}$.)

20. **Marginal Revenue Function.** The price–demand and revenue functions for x units of an item are, respectively,

$$p(x) = 100e^{-0.06x} \quad \text{and}$$
$$R(x) = xp(x) = 100xe^{-0.06x}$$

Find the marginal revenue function. On the same coordinate system, sketch the graphs of $R(x)$ and $R'(x)$. Discuss the relationship between these two graphs.

Applications (Social and Life Sciences)

21. **Standard Normal Curve.** The equation of a standard normal curve is given by

$$y = \frac{1}{\sqrt{2\pi}} e^{-x^2/2}$$

(a) Find the points of relative extrema and the points of inflection.

(b) Sketch the curve.

22. **Logistic Growth Curve.** The equation of a logistic growth curve of bacteria is of the form

$$w(t) = \frac{1000}{1 + 50e^{-0.2t}}$$

where t is measured in days and $w(t)$ is the weight after t days.

(a) Discuss where the curve is increasing and decreasing.

(b) Discuss where the curve is concave up and concave down.

(c) What happens when $x \to \infty$? (Use your calculator.)

(d) Sketch the curve.

23. **Von Bertalanffy Curve.** The number of bacteria in a culture after t hours can be given by

$$N = 1000(1 - .4e^{-0.2t})^3$$

(a) Discuss where the curve is increasing and decreasing.

(b) Discuss where the curve is concave up and concave down.

(c) What happens when $x \to \infty$? (Use your calculator.)

(d) Sketch the curve.

24. **Learning Function.** A learning function is given by

$$N(t) = 100(1 - e^{-0.03t})$$

Find the rate of learning, $N'(t)$. On the same coordinate system, sketch the graphs of $N(t)$ and $N'(t)$. Discuss the relationship between these two graphs.

Review Exercises

Find dy/dx for each function.

25. $y = xe^{3x^2}$

26. $y = \ln(1 + x^2)$

27. $y = x^2 \ln x$

28. $y = e^{1/x^3}$

29. $y = \ln\left[\dfrac{1 + x^2}{x}\right]$

30. $y = 4x^3 e^{2x}$

10.5 IMPLICIT DIFFERENTIATION

OVERVIEW   In Chapter 8 the functions for which we found derivatives were given in the form

$$y = f(x)$$

which are called **explicit functions**. For example, $y = -4x^3 + 3$ defines explicitly a function $f(x)$ with x as the independent variable and y as the dependent variable. This equation can be written as

$$f(x, y) = y + 4x^3 - 3 = 0$$

The y in this equation is the same y as in $y = -4x^3 + 3$. The equation $y + 4x^3 - 3 = 0$ gives **implicitly** y as a function of x. To find dy/dx when y is an implicit function of x is called **implicit differentiation**. In this section we

- Do implicit differentiation
- Develop a formula for y' where $y = \ln x$
- Find derivatives of $y = a^x$, where $a > 0$

To introduce the procedures for this differentiation, we find dy/dx by differentiating both sides of the equation with respect to the independent variable x.

$$y + 4x^3 - 3 = 0$$

$$\frac{d}{dx}(y + 4x^3 - 3) = \frac{d}{dx}(0)$$

By using the fact that the derivative of a sum is the sum of the derivatives, we have

$$\frac{d}{dx}(y) + \frac{d}{dx}(4x^3) + \frac{d}{dx}(-3) = \frac{d}{dx}(0)$$

$$\frac{d}{dx}(y) + 12x^2 + 0 = 0$$

Since y is a function of x, we write

$$\frac{d}{dx}(y) \quad \text{as} \quad \frac{dy}{dx} \text{ or } y'$$

Solving

$$\frac{dy}{dx} + 12x^2 = 0$$

for dy/dx gives

$$\frac{dy}{dx} = -12x^2$$

Note that this is the same result that is obtained by rewriting the equation as an explicit function and then differentiating.

$$y = -4x^3 + 3$$

$$\frac{dy}{dx} = -12x^2$$

Of greater importance is finding dy/dx for a function that cannot be expressed explicitly as a function of the independent variable. Sometimes it may be difficult or even impossible to express a relation as an explicit function, and yet the relation may still be such that it defines y as a function of x. Such a function is called an **implicit function**. The derivative of an implicit function can be found, if it exists, as in the preceding example, by differentiating both sides of the equation with respect to the independent variable. To illustrate implicit differentiation we consider

$$y^3 + y^2 + x^2 = 0$$

where y is an implicit function of x. Note that when we differentiate each term with respect to x we encounter

$$\frac{d}{dx}(y^3) \quad \text{and} \quad \frac{d}{dx}(y^2)$$

Recall the formula in Chapter 8 for the derivative with respect to x of a function of x raised to the nth power. We can write this derivative as

$$\frac{d}{dx}u^n = nu^{n-1}\frac{du}{dx}$$

where u is a function of x. If we consider y as a function of x, then

$$\frac{d}{dx}(y^n) = ny^{n-1}\frac{dy}{dx}$$

Therefore,

$$\frac{d}{dx}(y^2) = 2y\frac{dy}{dx} = 2yy'$$

$$\frac{d}{dx}(y^3) = 3y^2\frac{dy}{dx} = 3y^2y'$$

$$\frac{d}{dx}(y^4) = 4y^3\frac{dy}{dx} = 4y^3y'$$

Returning to our example, we differentiate both sides of the equation $y^3 + y^2 + x^2 = 0$ with respect to x.

$$\frac{d}{dx}(y^3 + y^2 + x^2) = \frac{d}{dx}(0)$$

$$\frac{d}{dx}(y^3) + \frac{d}{dx}(y^2) + \frac{d}{dx}(x^2) = 0 \qquad \text{Derivative of a sum}$$

$$3y^2y' + 2yy' + 2x = 0 \qquad \frac{d}{dx}(y^2) = 2y \cdot y'; \frac{d}{dx}(y^3) = 3y^2y'$$

$$(3y^2 + 2y)y' = -2x \qquad \text{Factor.}$$

$$y' = \frac{-2x}{3y^2 + 2y} \qquad \text{Solve for } y'.$$

EXAMPLE 30   Find y' for the function defined implicitly by

$$y^2 - x^2 - 3y = 0$$

SOLUTION   Differentiate both sides with respect to x to obtain

$$\frac{d}{dx}(y^2 - x^2 - 3y) = \frac{d}{dx}(0)$$

$$\frac{d}{dx}(y^2) + \frac{d}{dx}(-x^2) + \frac{d}{dx}(-3y) = 0$$

$$2yy' - 2x - 3y' = 0 \qquad \frac{d}{dx}(y^2) = 2yy'$$

$$(2y - 3)y' = 2x$$

$$y' = \frac{2x}{2y - 3} \qquad \text{Solve for } y'. \qquad ■$$

Calculator Note

As support for implicit differentiation, let's explore what happens when we separate $x^2 + y^2 = 25$ into two explicit functions by solving for y.

$$y = \pm\sqrt{25 - x^2}$$

We draw $y_1 = \sqrt{25 - x^2}$ and then $y_2 = -\sqrt{25 - x^2}$ on the same window and note what happens. Do you get the complete curve for $x^2 + y^2 = 25$? Using your calculator, find the slope of each graph when $x = 3$ and when $x = -3$. Do you get a slope of $-\frac{3}{4}$ at $(3, 4)$ and a slope of $\frac{3}{4}$ at $(-3, 4)$ on $y_1 = \sqrt{25 - x^2}$? Do you get a slope of $\frac{3}{4}$ at $(3, -4)$ and a slope of $-\frac{3}{4}$ at $(-3, -4)$ on $y_2 = -\sqrt{25 - x^2}$? The corresponding tangent lines are shown in Figure 12.

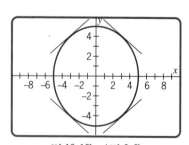

X:[-10, 10] and Y:[-6, 6]
Figure 12

Now we find the derivative of $x^2 + y^2 = 25$ implicitly.

$$\frac{d}{dx}(x^2) + \frac{d}{dx}(y^2) = \frac{d}{dx}(25)$$

$$2x + 2yy' = 0$$

$$y' = -\frac{x}{y}$$

Thus the slope of the graph at $(3, 4)$ is $-\frac{3}{4}$; at $(-3, 4)$ is $\frac{3}{4}$; at $(3, -4)$ is $\frac{3}{4}$; and at $(-3, -4)$ is $-\frac{3}{4}$. Note these slopes are in agreement with our calculator answers for explicit functions.

Practice Problem 1 Given $y^3 + x^2 = y - 1$, find dy/dx.

ANSWER $\dfrac{dy}{dx} = \dfrac{-2x}{3y^2 - 1}$

COMMON ERROR For equations involving a product such as $x^2 y$, students often forget that the product rule must be used when taking the derivative.

$$\frac{d}{dx}(x^2 y) \neq 2xy' \qquad \frac{d}{dx}(x^2 y) = 2xy + x^2 y'$$

EXAMPLE 31 If $x^2 y = x^3 + 4y$, find $\dfrac{dy}{dx}$.

SOLUTION We first take the derivative with respect to x of each term to obtain

$$\frac{d}{dx}(x^2 y) = \frac{d}{dx}(x^3) + \frac{d}{dx}(4y)$$

Now, as mentioned above, $x^2 y$ is a product and we must use the product formula.

$$x^2 \frac{d}{dx}(y) + 2xy = 3x^2 + 4\frac{d}{dx}(y) \qquad \text{Derivative of a product}$$

$$x^2 \frac{dy}{dx} - 4\frac{dy}{dx} = 3x^2 - 2xy$$

$$\frac{dy}{dx}(x^2 - 4) = 3x^2 - 2xy \qquad \text{Factor.}$$

$$\frac{dy}{dx} = \frac{3x^2 - 2xy}{x^2 - 4} \qquad \text{Divide by } x^2 - 4 \neq 0.$$

Practice Problem 2 Given $xy + y^2 = 4x$, find y'.

ANSWER $y' = \dfrac{4 - y}{x + 2y}$

EXAMPLE 32   Find dy/dx and the slope of the function
$$y^3 + y + 3x^2 + 2x + 1 = 0 \quad \text{at} \quad (-1, -1)$$

SOLUTION   Differentiating both sides with respect to x gives

$$\frac{d}{dx}(y^3 + y + 3x^2 + 2x + 1) = \frac{d}{dx}(0)$$

$$3y^2\frac{dy}{dx} + \frac{dy}{dx} + 6x + 2 = 0 \qquad\qquad \frac{d}{dx}y^3 = 3y^2 \cdot \frac{dy}{dx}$$

$$3y^2\frac{dy}{dx} + \frac{dy}{dx} = -6x - 2 \qquad\qquad \text{Subtraction}$$

$$(3y^2 + 1)\frac{dy}{dx} = -6x - 2 \qquad\qquad \text{Factoring}$$

$$\frac{dy}{dx} = \frac{-6x - 2}{3y^2 + 1} \qquad\qquad \text{Division}$$

By substituting $(-1, -1)$, we have

$$\left.\frac{dy}{dx}\right|_{(-1,\,-1)} = \frac{-6(-1) - 2}{3(-1)^2 + 1} = \frac{6 - 2}{3 + 1} = \frac{4}{4} = 1$$

Note that

$$\left.\frac{dy}{dx}\right|_{(-1,\,-1)}$$

means to evaluate dy/dx at $x = -1$, $y = -1$. The slope of the function is 1 at $(-1, -1)$. ∎

EXAMPLE 33   Find y'' for $xy + y^2 = 4$.

SOLUTION   First we find y' by implicit differentiation.

$$\frac{d}{dx}(xy) + \frac{d}{dx}(y^2) = \frac{d}{dx}(4)$$

$$xy' + y + 2yy' = 0 \qquad\qquad \text{Note the product.}$$

$$(x + 2y)y' = -y$$

$$y' = \frac{-y}{x + 2y} \qquad\qquad \text{Solve for } y'.$$

Now

$$\frac{d}{dx}(y') = y''$$

but the derivative of the right side of the equation involves a quotient.

$$y'' = \frac{\frac{d}{dx}(-y) \cdot (x + 2y) - \frac{d}{dx}(x + 2y) \cdot (-y)}{(x + 2y)^2}$$

$$= \frac{-xy' - 2yy' + y + 2yy'}{(x + 2y)^2}$$

$$= \frac{-xy' + y}{(x + 2y)^2}$$

$$= \frac{-x\left(\dfrac{-y}{x + 2y}\right) + y}{(x + 2y)^2} \qquad\qquad y' = \frac{-y}{x + 2y}$$

$$= \frac{xy + (xy + 2y^2)}{(x + 2y)^3}$$

$$= \frac{2(xy + y^2)}{(x + 2y)^3}$$

∎

In Section 10.3 we introduced a formula for the derivative of $y = \ln x$ without proof that the formula was valid. We now use implicit differentiation to prove that if $y = \ln x$, then

$$\frac{dy}{dx} = \frac{1}{x}$$

First, we write $y = \ln x$ as $e^y = x$. Then we take the derivative of both sides of the equation with respect to x.

$$\frac{d}{dx}(e^y) = \frac{d}{dx}(x)$$

$$e^y \frac{dy}{dx} = 1 \qquad\qquad d[e^{u(x)}] = e^{u(x)}\frac{dy}{dx}$$

$$\frac{dy}{dx} = \frac{1}{e^y} = \frac{1}{x} \qquad\qquad \text{Divide by } e^y.$$

In a like manner, we use implicit differentiation to develop the formula for the derivative of

$$y = a^x$$

$$\ln y = \ln a^x \qquad \text{Take ln of both sides.}$$

$$= x \ln a \qquad \ln M^N = N \ln M$$

Then we differentiate both sides with respect to x.

$$\frac{1}{y} \cdot \frac{dy}{dx} = \ln a \qquad \frac{d}{dx}(\ln y) = \frac{1}{y}\frac{dy}{dx}$$

$$\frac{dy}{dx} = y \ln a \qquad \text{Multiply both sides by } y.$$

$$\text{or} \quad \frac{dy}{dx} = a^x \ln a \qquad \text{Substitute } a^x \text{ for } y.$$

Note that if $a = e$, then $\ln e = 1$, and the differentiation formula becomes

$$\frac{dy}{dx} = e^x \qquad \text{When } y = e^x$$

Out of all the possible choices for bases of logarithmic and exponential functions, the simplest derivative formulas occur when the base is e.

Implicit differentiation of logarithmic functions gives a procedure for simplifying the derivative of products and quotients.

EXAMPLE 34 Find dy/dx if

$$y = x^3 \left(\frac{x^2 + 1}{\sqrt{x - 1}} \right)$$

SOLUTION Before taking the derivative we take the natural logarithm of both sides of the equation. Then we find dy/dx implicitly.

$$y = x^3 \left(\frac{x^2 + 1}{\sqrt{x - 1}} \right)$$

$$\ln y = \ln \left[x^3 \left(\frac{x^2 + 1}{\sqrt{x - 1}} \right) \right]$$

$$= \ln x^3 + \ln(x^2 + 1) - \ln \sqrt{x - 1}$$

$$= 3 \ln x + \ln(x^2 + 1) - \frac{1}{2} \ln(x - 1)$$

$$\frac{d}{dx} \ln y = \frac{d}{dx}(3 \ln x) + \frac{d}{dx}[\ln(x^2 + 1)] - \frac{d}{dx}\left[\frac{1}{2} \ln(x - 1) \right]$$

$$\frac{y'}{y} = \frac{3}{x} + \frac{2x}{x^2 + 1} - \frac{\frac{1}{2}}{x - 1}$$

$$y' = y \left[\frac{3}{x} + \frac{2x}{x^2 + 1} - \frac{\frac{1}{2}}{x - 1} \right] \qquad \text{Solve for } y'$$

$$= x^3 \left(\frac{x^2 + 1}{\sqrt{x - 1}} \right) \left[\frac{3}{x} + \frac{2x}{x^2 + 1} - \frac{\frac{1}{2}}{x - 1} \right]$$

■

SUMMARY

In this section we introduced the following procedure for differentiating implicitly.

1. Differentiate both sides of the equation with respect to the independent variable.
2. If y is the dependent variable, use the chain rule for terms that involve y to a power.
3. Solve for the derivative, y', in terms of the other variables.

Exercise Set 10.5

Find dy/dx by implicit differentiation. Solve for the explicit function $y = f(x)$, and then differentiate to check your results.

1. $y - 4x^2 + 3x = 0$
2. $x - y + 3x^3 = 0$
3. $x + 2y + 3x^2 = 0$
4. $3x - 2y + 5x^3 = 0$
5. $4x^2 - 3x + 7y = 2$
6. $9y - 3x + 4x^3 - 2 = 0$
7. $x^2 - 3x + 2 + 5y = 0$
8. $x^2 + 4x - 3y = 7$
9. $x^2 - 3x^3 + x - 2y = 0$
10. $2x^2 - 4x - 7y = 5$

Find dy/dx by implicit differentiation.

11. $3xy - 4x^2 = 0$
12. $4xy + 5x^2 = 0$
13. $2x^2y - 3x + 5 = 0$
14. $3x^2y + 4x^2 - 2 = 0$
15. $y + 3xy - 4 = 0$
16. $y + 2xy - 3x^3 = 0$
17. $y + 2x^2y + 3x = 0$
18. $y + 3x^2y - 4x = 0$
19. $y - 3x^2y - 2x = 0$
20. $y - 4x^2y + 5x = 0$

Find dy/dx by implicit differentiation.

21. $y^2 + y - 3x = 0$
22. $y^3 + y + 4x = 0$
23. $3y^2 - y + 4x^2 = 0$
24. $4y^2 - 2y + 3x = 0$
25. $y^2 + xy - 4x = 5$
26. $y^2 + 3xy - x^3 = 4$
27. $3xy^2 - 2y - 3 = 0$
28. $4x^2y^2 - 3y + 6 = 0$
29. $2xy^2 - 3xy + x^3 = 3$
30. $3xy^2 - 2xy + 4x^2 = 0$

Find the equation(s) of the tangent(s) to the graphs at the indicated value of x.

31. $3xy - x - 2 = 0$ at $x = 1$

32. $y^2 + 2y - x = 0$ at $x = 0$
33. $y + 3xy - 7 = 0$ at $x = 2$
34. $y^2 - 2xy - 8 = 0$ at $x = -1$
35. $x^3y + 3x^2 + 4 = 0$ at $x = -2$

Find dy/dx and the value of the slope of the graph at the indicated point.

36. $(1 + y)^3 + y = 2x + 7$ at $(1, 1)$
37. $(y - 2)^2 + x = y$ at $(2, 3)$
38. $(x + y)^2 + 3y = -3$ at $(1, -1)$
39. $(2x + y)^2 + 2x = -1$ at $(-1, 3)$
40. $(x + 2y)^2 - x^2 + 8 = 0$ at $(3, -1)$
41. $xy = 6$ at $(3, 2)$

Find dy/dx for the following exponential functions.

42. $y = 4^{x^2}$
43. $y = 10^{3x+4}$

Find the derivatives of the following functions by first taking the ln of both sides.

44. $y = \dfrac{x^3(x^2 + 1)}{x + 3}$
45. $y = \dfrac{\sqrt{x + 1}}{x^2}$
46. $y = \dfrac{(3x + 2)^{1/3}}{(2x - 4)^2}$
47. $y = \dfrac{3x^2(2x + 3)^{1/3}}{(4x^3 - 3x + 1)^2}$

Find y" for the following and leave the answer in terms of y'.

48. $xy + x^2 + y^2 = 4$
49. $x^3 + y^3 = xy$

50. Separate $3x^2 + 4y^2 = 16$ into two explicit functions and draw both with a common window. Two of the following points are on one graph and the other two are on the second graph: $(2, 1)$, $(-2, 1)$, $(2, -1)$, and $(-2, -1)$. With your calculator, find slopes of

the graphs at the appropriate points. Draw the tangent lines. Then find the derivative implicitly and find the slopes of the tangent lines at the points. Is there agreement?

51. Follow the instructions in Exercise 50 for

$$5x^2 - 4y^2 = 4$$

at $(2, 2)$, $(-2, 2)$, $(2, -2)$, and $(-2, -2)$.

52. **Exam.** What is the slope of the line tangent to the curve $y^3 - x^2y + 6 = 0$ at the point $(1, -2)$?

(a) $\dfrac{-2}{5}$ (b) $\dfrac{-4}{11}$ (c) $\dfrac{4}{11}$

(d) $\dfrac{11}{4}$ (e) 8

53. **Exam.** $\dfrac{d}{dx}(2^{-x^3}) = ?$

(a) $-3x^2 \ln 2$ (b) $-6x^{-2x^3}$
(c) $-3x^2 \cdot 2^{-x^3} \ln 2$ (d) $6x^{2-x^3}$
(e) $6x^{2-x^3} \ln 2$

Applications (Business and Economics)

54. **Instantaneous Change in Sales.** Suppose that a company's sales S, in hundreds of thousands of dollars, is related to the amount x, in thousands of dollars spent on training, by

$$Sx + S = 900 + 40x$$

(a) Find dS/dx, the instantaneous rate of change of S.
(b) Find

$$\left. \frac{dS}{dx} \right|_{(1,470)}$$

55. **Instantaneous Demand.** If x is the number of

items that can be sold at a price of p dollars in the demand equation

$$x^3 + p^3 = 1200$$

find dp/dx.

56. Rework Exercise 55 for the demand equation

$$p^3 - 3p^2 - x + 300 = 0$$

Applications (Social and Life Sciences)

57. **Pollution.** Suppose that pollution P, in parts per million, x yards away from the source, is given by

$$P + 2xP + x^2P = 600$$

Find the instantaneous rate of pollution 10 yards away.

58. **Learning Function.** A learning function L is given in terms of t hours as

$$L^2 - 256t = 0$$

(a) Find the instantaneous rate of learning.
(b) Find the instantaneous rate of learning at the end of 1 hour.
(c) Find the instantaneous rate of learning at the end of 9 hours.

Review Exercises

59. If $y = \ln(x^2 + 3)^2 \sqrt{5x + 1}$, find dy/dx.
60. Using logarithms, find dy/dx for

$$y = \frac{\sqrt[3]{3x + 2}}{\sqrt{5x^2 + 1}}$$

10.6 DERIVATIVES AND RELATED RATES (Optional)

OVERVIEW   The rate of increase of the radius of a balloon is related to the rate of pumping helium into the balloon. If the bottom of a ladder is pulled from the wall, the rate of descent of the top is related to the rate at which the bottom is pulled away. Production changes with time; that is, production is a function of time. Cost, revenue, and profit are all functions of time. The rate of change of cost with respect to time is related to the rate of change of production with respect to time. In this section we

- Define related rates
- Outline a procedure for finding related rates
- Apply related rates to application problems

Suppose that y is a function of x, say $y = f(x)$, and x and y vary with time t. That is, both x and y are functions of time t. The chain rule gives

$$\frac{dy}{dt} = \frac{dy}{dx} \cdot \frac{dx}{dt}$$

Thus the rate of change of y with respect to time is related to the rate of change of x with respect to t. That is, if $y = f(x)$ and both y and x are functions of t, by the chain rule we have

$$\frac{dy}{dt} = \frac{d}{dx}[f(x)] \cdot \frac{dx}{dt}$$

Now let's use the chain rule to solve problems.

EXAMPLE 35   If $y = 3x^2$, find dy/dt when $dx/dt = \frac{1}{2}$ and $x = 2$.

SOLUTION

$$\frac{dy}{dt} = 6x\frac{dx}{dt} \qquad \frac{d}{dt}(3x^2) = \frac{d}{dx}(3x^2)\frac{dx}{dt}$$

$$= 6 \cdot 2 \cdot \frac{1}{2} = 6 \qquad x = 2; \frac{dx}{dt} = \frac{1}{2}$$

∎

Often, two or more variables may be differentiable functions of another variable, such as time, and yet the explicit function may not be given. Suppose that x and y are related by the equation $x^2 + y^2 = 36$ and that both x and y are functions of t. If we differentiate both sides implicitly with respect to t, we obtain

$$\frac{d}{dt}(x^2 + y^2) = \frac{d}{dt}(36)$$

$$2x\frac{dx}{dt} + 2y\frac{dy}{dt} = 0 \qquad x \text{ and } y \text{ are functions of } t.$$

Practice Problem 1   Given $y = x^2 + 2x$, find dy/dt in terms of dx/dt where x and y are functions of t.

ANSWER $\dfrac{dy}{dt} = (2x + 2)\dfrac{dx}{dt}$

EXAMPLE 36   Given $x^2 + y^2 = 169$ and $dy/dt = 2$, find dx/dt at $(5, 12)$, where x and y are functions of t.

SOLUTION   The equation relating the variables is $x^2 + y^2 = 169$. The rate dy/dt is known, and the rate dx/dt is to be found. Differentiate both sides of the given equation with respect to t.

$$\frac{d}{dt}(x^2 + y^2) = \frac{d}{dt}(169) \qquad \text{Implicit differentiation}$$

$$2x\frac{dx}{dt} + 2y\frac{dy}{dt} = 0$$

$$2(5)\frac{dx}{dt} + 2(12)(2) = 0 \qquad x = 5,\ y = 12,\ \frac{dy}{dt} = 2$$

$$10\frac{dx}{dt} + 48 = 0$$

$$\frac{dx}{dt} = -\frac{48}{10} = -\frac{24}{5}$$

Practice Problem 2   Given $xy + y^2 = 3$, find dy/dt at $(2, 1)$ where $dx/dt = 5$.

ANSWER $\dfrac{dy}{dt} = -\dfrac{5}{4}$

The procedure for solving application problems involving related rates is outlined as follows.

Procedure for Solving Related Rates Problems

1. First read the problem and draw a picture if possible.
2. Carefully identify all the variables involved. This step is one of the most important steps in the procedure and is the one that is most often neglected or only partly done.
3. Relate the variables by some equation that holds generally and not just at some particular time. A sketch will often help you find this equation.
4. List all the rates involved in the problem, those that are known and those that are to be found.
5. If the variables are functions of time, differentiate, with respect to time, both sides of the equation relating the variables.
6. Substitute the values of the known rates and the known variables at the given time, and solve for the unknown rate.

EXAMPLE 37   Suppose that the radius of a circular oil slick is increasing at the rate of 12 yards per hour. Find the rate at which the area is increasing when $r = 40$ yards.

SOLUTION   Let r be the radius and A be the area. We know that the area of a circle is $A = \pi r^2$. We also know that $dr/dt = 12$ yd/hr and dA/dt is unknown. Then

$$\frac{dA}{dt} = \frac{d}{dt}(\pi r^2) = \frac{d}{dr}(\pi r^2)\frac{dr}{dt} \qquad \text{Chain rule}$$

$$= 2\pi r \cdot \frac{dr}{dt}$$

$$= 2\pi(40 \text{ yd}) \cdot (12 \text{ yd/hr}) \qquad r = 40; \frac{dr}{dt} = 12$$

$$= 960\pi \text{ yd}^2/\text{hr}$$

EXAMPLE 38   A toy manufacturer has found that cost, revenue, and profit functions can be expressed as functions of production. If x is the number of toys produced in a week and C, R, and P represent cost, revenue, and profit, respectively, then

$$C(x) = 6000 + 2x, \qquad R(x) = 20x - \frac{x^2}{2000}, \qquad P(x) = R(x) - C(x)$$

Suppose that production is increasing at the rate of 200 toys per week from a production level of 1000 toys. Find the rate of increase in   (a) cost,   (b) revenue, and   (c) profit.

SOLUTION   Since production is changing with respect to time, production must be a function of time. Hence, cost, revenue, and profit are all functions of time. To find the rate of increase, we differentiate each with respect to time.

(a)   Cost:

$$C(x) = 6000 + 2x$$

$$\frac{dC}{dt} = \frac{d}{dt}(6000) + \frac{d}{dt}(2x) \qquad \text{Implicit differentiation}$$

$$= 0 + 2\frac{dx}{dt}$$

$$\frac{dC}{dt} = 2(200) = 400 \qquad \frac{dx}{dt} = 200$$

Cost is increasing at a rate of $400 per week.

(b) Revenue:

$$R = 20x - \frac{x^2}{2000}$$

$$\frac{dR}{dt} = \frac{d}{dt}(20x) - \frac{d}{dt}\left(\frac{x^2}{2000}\right) \qquad \text{Implicit differentiation}$$

$$= 20\frac{dx}{dt} - \left(\frac{x}{1000}\right)\frac{dx}{dt} \qquad \text{Chain rule}$$

$$= 20(200) - \frac{1000}{1000}(200) \qquad x = 1000;\ \frac{dx}{dt} = 200$$

$$= 3800$$

Revenue is increasing at a rate of $3800 per week.

(c) Profit:

$$P = R - C$$

$$\frac{dP}{dt} = \frac{dR}{dt} - \frac{dC}{dt}$$

$$= 3800 - 400$$

$$= 3400$$

Profit is increasing at a rate of $3400 per week.   ■

EXAMPLE 39   The amount of time A, in minutes, required to perform an operation on an assembly line is given as a function of the number of trials x by the equation

$$A = 9 + 9x^{-1/2}$$

Find dA/dt if $dx/dt = 4$ when $x = 25$.

SOLUTION

$$A = 9 + 9x^{-1/2}$$

$$\frac{dA}{dt} = \frac{d}{dt}(9) + \frac{d}{dt}(9x^{-1/2}) \qquad \text{Implicit differentiation}$$

$$= 0 + 9\frac{d}{dt}(x^{-1/2})$$

$$= 9\left(\frac{-1}{2}\right)x^{-3/2}\frac{dx}{dt} \qquad \text{Chain rule}$$

$$= \left(\frac{-9}{2}\right) \cdot (25)^{-3/2} \cdot (4) \qquad x = 25,\ \frac{dx}{dt} = 4$$

$$= \frac{-18}{125}$$

SUMMARY

The chain rule is very important in this section. If $y = f(x)$ and both x and y are functions of t, then

$$\frac{dy}{dt} = \frac{d}{dx}f(x) \cdot \frac{dx}{dt}$$

If x and y are functions of t, then

$$\frac{d}{dt}(x^2) = 2x \cdot \frac{dx}{dt} \quad \text{and} \quad \frac{d}{dt}(y^3) = 3y^2 \cdot \frac{dy}{dt}$$

Exercise Set 10.6

For the following equations assume that x and y are functions of t. Find dy/dt given that dx/dt = 3 when x = 2.

1. $x + y = 3$
2. $2x + y = 5$
3. $3x - 2y = 4$
4. $4x - 3y = 2$
5. $y - 2\sqrt{x} = 0$
6. $x^2 - 2y = 0$
7. $y - 3\sqrt{x} = 0$
8. $y - 2\sqrt[3]{x} = 0$
9. $y - x^2 = 4$
10. $y - 2x^2 + x = 0$
11. $xy = 4$
12. $3xy = 1$
13. $x + xy = 6$
14. $2x - xy = 4$
15. $y + xy = 5$
16. $y - 2xy = 3$
17. $y - xy + x^2 = 5$
18. $y + 2xy - x^2 = 3$
19. $y + x^2y - 3x = 2$
20. $y - x^2y - 4x = 4$

21. A particle travels along the curve $y = \sqrt{x}$ so that $dx/dt = 4$ centimeters per second. How fast is y changing when (a) $x = 2$ and (b) $x = 9$?

22. The radius of a circle is increasing at the rate of 2 centimeters per second. At what rate is the area of the circle increasing when the radius is (a) 10 centimeters and (b) 15 centimeters?

23. The length of a rectangle is 4 times the width. If the length is increasing at the rate of 8 centimeters per second, how fast is the area changing when the width is (a) 1 centimeter and (b) 2 centimeters?

24. The edges of a cube are increasing at a rate of 2 centimeters per second. At what rate is the volume changing when the edge is 3 centimeters?

25. At what rate is the surface area of the cube in Exercise 24 increasing when the edge is 3 centimeters?

Applications (Business and Economics)

26. **Cost, Revenue, and Profit Rates.** Suppose that the revenue from the sale of x stereos is given by $R(x) = 500x - x^2$ and the cost is given by $C(x) = 2000 + 30x$. If the company is selling six stereos per day, then when 50 stereos are sold find each of the following.
    (a) Rate of change in revenue.
    (b) Rate of change in cost.
    (c) Rate of change in profit.

27. **Demand Rates.** Suppose that a demand equation is $p^2 + p + 3x = 39$. Find dp/dt if $dx/dt = 3$ when $p = 2$.

Applications (Social and Life Sciences)

28. **Pollution.** Oil is leaking from a tanker and has formed a circular oil slick 3 centimeters thick. To estimate the rate of leakage, the radius was measured and found to be 100 meters and increasing at the rate of 5 centimeters per minute. Assume that the depth is constant.
    (a) Use these results to find the rate of leakage.
    (b) After finding the rate of leakage, assume that the rate of leakage is constant and find how fast the radius is changing 5 hours after the leakage began.

29. **Learning Function.** A learning function is given as required time R after x tries,

$$R(x) = 7 + 7x^{-1/2}$$

Find dR/dt if $dx/dt = 5$ when $x = 49$.

30. **Medicine.** The cross section area of a tumor is given by

$$A = \pi r^2$$

The radius is increasing at a rate of 0.4 millimeters per day. At a radius of 20 millimeters, how fast is the area of the tumor increasing? If

$$V = \frac{4}{3}\pi r^3$$

how fast is the volume increasing per day?

Review Exercises

Find dy/dx at $x = 1$, $y = 3$.

31. $xy = 3$

32. $x^2 + 3xy = 10$

33. $2x^2 + y^2 = 11$

Find dy/dx.

34. $x^2y + y^2 = 4$

35. $2xy^2 + y = 7$

36. $4x^3 + 5xy + 7y^3 = 2$

37. $ye^x + x = y$

38. $ye^{x^2} + x^4 = 4$

39. $x \ln(x^2 + 2) = xy$

40. $xy - \ln x = 4$

Chapter Review

Review the following concepts to ensure that you understand and can use them.

Logarithmic Functions

Graphs $\log_b b = 1$ and $\log_b 1 = 0$

$\log x$ $\log_b b^x = x$

$\ln x$

Implicit derivatives   Related rates

Make sure you understand the following relations.

$\log_b x = y$ if and only if $b^y = x$ $\log_b M^N = N \log_b M$

$\log_b M \cdot N = \log_b M + \log_b N$ $\log_b x = \log_b y$ if and only if $x = y$

$\log_b \dfrac{M}{N} = \log_b M - \log_b N$

Can you use each of the following formulas?

$$\frac{d}{dx}(e^u) = e^u \frac{du}{dx}, \quad u > 0 \qquad \frac{d}{dx}(\ln u) = \frac{1}{u} \cdot \frac{du}{dx}, \quad u > 0$$

Chapter Test

1. If $3xy - 4x^2 = 0$, find dy/dx.
2. If $3xy - 2x = 4$, find dy/dt if $dx/dt = 4$ when $x = 1$.
3. If $w = 4e^{z^2+5}$, find dw/dz.

4. If $r = 6 \ln(s^2 + 4s)$, find dr/ds.

5. If $y = x \ln x^2$, find dy/dx.

6. (a) With your graphing calculator, find an approximation to the maximum value of the function

$$y = \frac{\ln x}{x}$$

   (b) Using calculus, find the exact value of the maximum value of

$$y = \frac{\ln x}{x}$$

7. If $y = \ln(1 + e^x)$, find y'.

8. (a) Using your graphing calculator, for $y = xe^{-x}$ find each of the following and round to two decimal places if necessary.
   (i) Find the intervals over which y is increasing or where it is decreasing.
   (ii) Find the intervals where y is concave up and where it is concave down.
   (iii) Sketch the graph.
   (b) Repeat part (a) using calculus techniques.

9. If $y = x^2 \ln x + xe^y$, find y'.

10. The amount of money in a savings account is $A = 10{,}000e^{0.08t}$, where t is the time measured in years. What is the rate of increase of the account at the end of 5 years?

11. If $x^2 e^y + \ln y = \ln(x + 1)$, find y'.

12. Make a calculator sketch of $y = x^2 e^x$.

13. For Problem 12, find the relative maxima and minima.

14. For Problem 12, discuss where the curve is increasing or decreasing.

15. For Problem 12, discuss concavity.

CHAPTER 11

Integral Calculus

As indicated earlier, calculus is divided into two parts, differential calculus, discussed in the three preceding chapters, and integral calculus, which we introduce in this chapter. We considered the derivative of a function as a rate of change and found many ways to apply that concept. In integral calculus the rate of change is known and our problem is to find the function. Again, many interesting applications arise from this type of problem. For example:

A businessperson knows the rate of change of profit in terms of items produced and would like to know the profit for a given production.

A biologist introduces a bactericide into a culture and knows the rate of change of the number of bacteria present. The biologist would like to know the number of bacteria present at a given time.

This chapter begins with a discussion of antiderivative problems in which information about the derivative of a function is given and the function is unknown. The set of antiderivatives is called the indefinite integral of a function. Another concept, the definite integral of a function, seems quite different from the indefinite integral; however, a remarkable theorem, the fundamental theorem of calculus, will show how the two are intimately related.

11.1 THE ANTIDERIVATIVE (INDEFINITE INTEGRAL)

OVERVIEW   Many of the operations we have studied so far have inverse operations. For example, subtraction is the inverse operation for addition, division is the inverse operation for multiplication, and taking the square root is the inverse operation for squaring ($x \geq 0$). Inverse functions exist for many functions. Thus, it seems natural to consider an inverse operation for differentiation. If the derivative of a function is known, can the function be found? We answer this question by introducing the concept of **antiderivative**. In this section we learn

- The meaning of antiderivative
- Ways to find antiderivatives
- Relationships between indefinite integrals and antiderivatives
- Notation for indefinite integrals

We learned in the preceding chapters that the derivative of x^2 is $2x$. Now let's perform an inverse operation of $f(x) = 2x$, called **antidifferentiation**. Instead of going forward to find the derivative

$$x^2 \rightarrow 2x \qquad \text{Differentiation}$$

we will work backward to find a function whose derivative is $2x$.

$$x^2 \leftarrow 2x \qquad \text{Antidifferentiation}$$

That is, an antiderivative of $2x$ is x^2.

ANTIDERIVATIVE

A function $F(x)$ is an **antiderivative** of $f(x)$ on the interval (a, b) if

$$\frac{d}{dx} F(x) = f(x)$$

for all x in (a, b).

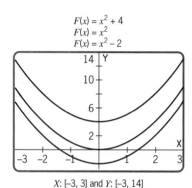

$F(x) = x^2 + 4$
$F(x) = x^2$
$F(x) = x^2 - 2$

X: [−3, 3] and Y: [−3, 14]
A few antiderivatives of $f(x) = 2x$

Figure 1

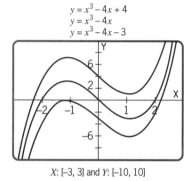

$y = x^3 - 4x + 4$
$y = x^3 - 4x$
$y = x^3 - 4x - 3$

X: [−3, 3] and Y: [−10, 10]

Figure 2

In the definition above, we could have also used the notation $D_x[F(x)] = F'(x) = f(x)$.

Did you note the word *an* in the preceding definition? This word implies that there is more than one antiderivative of a function. As we shall see, there is a set of functions that are all antiderivatives of a given function.

EXAMPLE 1   Let's continue to look at the antiderivatives of $2x$. Note that

$$\frac{d}{dx}x^2 = 2x, \qquad \frac{d}{dx}(x^2 + 4) = 2x, \qquad \frac{d}{dx}(x^2 - 2) = 2x$$

Thus, x^2, $x^2 + 4$, and $x^2 - 2$ are all antiderivatives of $2x$. We can see that the only difference in each function is the constant. This is demonstrated in Figure 1. Since the only difference in the antiderivatives is the constant, we can use the notation $F(x) = x^2 + C$, where C is any constant, as an antiderivative of x, to indicate the entire set of antiderivatives of $f(x) = 2x$. ■

Practice Problem 1 With a graphing calculator, draw the graphs of $y = x^3 - 4x$, $y = x^3 - 4x + 4$, and $y = x^3 - 4x - 3$. Then take the derivative of each function. The functions represented by the three graphs are antiderivatives of what function?

ANSWER   The graphs are shown in Figure 2. $y' = 3x^2 - 4$. The graphs are antiderivatives of $3x^2 - 4$.

In general, if $F(x)$ is an antiderivative of $f(x)$, then each of the functions $F(x) + C$ is also an antiderivative, where C is any real number. It can be shown that there are no other antiderivatives, so that this is the entire set. This property can be stated as follows.

PROPERTY OF ANTIDERIVATIVES

If the derivatives of two functions are equal, then the functions differ at most by a constant.

The preceding property suggests that we need a notation or a symbol to represent all antiderivatives of $f(x)$. Such a symbol is called the **indefinite integral** and is denoted by

$$\int f(x)\, dx$$

where $\int$ is called the **integral sign**, $f(x)$ is the **integrand**, and dx is the **differential**.

The importance of dx will be discussed in Section 11.3. At this time we consider dx simply as a part of the indefinite integral, which can be used to designate the independent variable.

INDEFINITE INTEGRAL

Let $F(x)$ be an antiderivative of the function $f(x)$. The **indefinite integral** of $f(x)$ is defined to be

$$\int f(x)\, dx = F(x) + C$$

The $\int$ and dx indicate that antidifferentiation is to be performed with respect to x on the function $f(x)$, and C is called the constant of integration. Hence, $F'(x) = f(x)$.

Before we introduce the rules of integration, let's look at several examples and use our knowledge of derivatives to find some antiderivatives.

EXAMPLE 2 Find the indefinite integral $\int 5x^4\, dx$.

SOLUTION Since the derivative of x^5 is $5x^4$, x^5 is an antiderivative of $5x^4$; hence, $x^5 + C$ is the indefinite integral of $5x^4$.

$$\int 5x^4\, dx = x^5 + C \quad \text{because} \quad D_x[x^5 + C] = 5x^4 \qquad \blacksquare$$

EXAMPLE 3 Find $\int 2x^3\, dx$.

SOLUTION The derivative of x^4 is $4x^3$ and not $2x^3$. Since $2x^3$ is one-half of $4x^3$, we should try $\frac{1}{2}x^4$ as an antiderivative of $2x^3$. The derivative of $\frac{1}{2}x^4$ is $2x^3$; therefore, $\frac{1}{2}x^4 + C$ is the indefinite integral of $2x^3$.

$$\int 2x^3\, dx = \frac{1}{2}x^4 + C \quad \text{because} \quad D_x\left[\frac{x^4}{2} + C\right] = 2x^3 \qquad \blacksquare$$

EXAMPLE 4 Find the indefinite integral $\int 3\, dx$.

SOLUTION The derivative of $3x$ is 3. Thus $3x$ is an antiderivative of 3. The indefinite integral of 3 is $3x + C$.

$$\int 3\,dx = 3x + C \quad \text{because} \quad D_x[3x + C] = 3$$

The preceding examples should suggest the following indefinite integral formulas.

INDEFINITE INTEGRAL FORMULAS

For k and C constants:

1. $\displaystyle\int k\,dx = kx + C$

2. $\displaystyle\int x^n\,dx = \frac{x^{n+1}}{n+1} + C, \qquad n \neq -1$ Power rule for antiderivatives

To verify the power rule for antiderivatives, all we need to do is take the derivative of the right side and show that this is the integrand. Since

$$\frac{d\left(\dfrac{x^{n+1}}{n+1}\right)}{dx} = \left(\frac{1}{n+1}\right)\frac{d(x^{n+1})}{dx} \qquad \frac{1}{n+1} \text{ is a constant}$$

$$= \left(\frac{1}{n+1}\right)[(n+1)x^{n+1-1}]$$

$$= x^n$$

then

$$\int x^n\,dx = \frac{x^{n+1}}{n+1} + C, \qquad n \neq -1$$

A special formula for $n = -1$ is discussed in the next section. The present formula states that to find the indefinite integral of x^n with respect to x, you must increase the exponent of x by 1 and divide by the new exponent. (You may have noticed that this is the reverse of the power rule for differentiation. When we take the derivative of a term x^n, we *subtract* 1 from n and *multiply*. When we integrate, we *add* 1 to n and *divide*.)

EXAMPLE 5

$$\int x^4\,dx = \frac{x^5}{5} + C \qquad n = 4; n + 1 = 5$$

EXAMPLE 6

$$\int x^{2/3}\,dx = \frac{x^{(2/3)+1}}{\frac{5}{3}} + C = \frac{3}{5}x^{5/3} + C \qquad n = \frac{2}{3}; n + 1 = \frac{5}{3}$$

EXAMPLE 7 Find $\int \dfrac{1}{x^2}\,dx$.

SOLUTION First, we rewrite the function so that we can apply the power rule.

$$\int \frac{1}{x^2}\,dx = \int x^{-2}\,dx \qquad\qquad \text{Now apply the power rule.}$$

$$= \frac{x^{-2+1}}{-1} + C \qquad\qquad n = -2;\, n+1 = -1$$

$$= \frac{x^{-1}}{-1} + C = -\frac{1}{x} + C \qquad\qquad \blacksquare$$

The properties of derivatives allow us to develop and use the following properties for antiderivatives.

PROPERTIES OF INDEFINITE INTEGRALS

If both f and g have antiderivatives and k is any constant, then

1. Constant rule: $\displaystyle\int kf(x)\,dx = k\int f(x)\,dx$

2. Sum rule: $\displaystyle\int [f(x) + g(x)]\,dx = \int f(x)\,dx + \int g(x)\,dx$

3. Difference rule: $\displaystyle\int [f(x) - g(x)]\,dx = \int f(x)\,dx - \int g(x)\,dx$

EXAMPLE 8 Find $\int 10x^3\,dx$.

SOLUTION

$$\int 10x^3\,dx = 10\int x^3\,dx = 10\left(\frac{x^4}{4}\right) + C = \frac{5x^4}{2} + C \qquad \text{Property 1} \quad \blacksquare$$

EXAMPLE 9 Find $\int 30x^5\,dx$.

SOLUTION

$$\int 30x^5\,dx = 30\int x^5\,dx = 30\left(\frac{x^6}{6}\right) + C = 5x^6 + C \qquad \text{Property 1} \quad \blacksquare$$

NOTE: It is important to remember that property 1 states that a *constant* can be factored from the integrand; a *variable* cannot be factored in this manner.

COMMON ERROR Students sometimes try to treat a variable like a constant.

Correct   Incorrect

$$\int 5\sqrt{x}\,dx = 5\int \sqrt{x}\,dx \qquad \int x\sqrt{x}\,dx = x\int \sqrt{x}\,dx$$

EXAMPLE 10

$$\int (x^3 + x^{1/2})\,dx = \int x^3\,dx + \int x^{1/2}\,dx = \frac{x^4}{4} + \frac{2}{3}x^{3/2} + C \qquad \text{Property 2} \quad \blacksquare$$

EXAMPLE 11

$$\begin{aligned}
\int (x^3 + x^2 + 2x + 3)\,dx &= \int x^3\,dx + \int (x^2 + 2x + 3)\,dx && \text{Property 2}\\
&= \int x^3\,dx + \int x^2\,dx + \int (2x + 3)\,dx && \text{Property 2}\\
&= \int x^3\,dx + \int x^2\,dx + \int 2x\,dx + \int 3\,dx && \text{Property 2}\\
&= \frac{x^4}{4} + \frac{x^3}{3} + x^2 + 3x + C
\end{aligned}$$

The C's for the four integrands are combined into one C, since the constant is arbitrary. $\blacksquare$

This example shows that by repeated application of the sum formula, the indefinite integral of a function that is the sum of a finite number of functions can be obtained by taking the sum of the indefinite integrals of the functions.

EXAMPLE 12

$$\begin{aligned}
\int (3x^4 - 2x^3 + x - 3)\,dx &= \int 3x^4\,dx + \int (-2)x^3\,dx + \int x\,dx + \int (-3)\,dx\\
&= 3\int x^4\,dx - 2\int x^3\,dx + \int x\,dx + \int (-3)\,dx\\
&= \frac{3x^5}{5} - \frac{x^4}{2} + \frac{x^2}{2} - 3x + C
\end{aligned}$$
$\blacksquare$

Practice Problem 2 Find $\int (3x^2 + 4x)\,dx$.

ANSWER $x^3 + 2x^2 + C$

Some algebraic manipulation or simplification will often be necessary before the integral can be found. This is demonstrated in the following examples.

EXAMPLE 13   Find $\int \dfrac{x^3 + 2x^2 + 3}{x^2}\,dx.$

SOLUTION   Before the power rule or the other rules can be used, the expression must first be simplified. We rewrite the expression so that each term can be reduced.

$$
\int \frac{x^3 + 2x^2 + 3}{x^2}\,dx = \int \frac{x^3}{x^2}\,dx + \int \frac{2x^2}{x^2}\,dx + \int \frac{3}{x^2}\,dx
$$

$$
= \int x\,dx + \int 2\,dx + \int 3x^{-2}\,dx
$$

$$
= \frac{x^2}{2} + 2x + \frac{3x^{-1}}{-1} + C \qquad \text{Only one constant is needed.}
$$

$$
= \frac{x^2}{2} + 2x - \frac{3}{x} + C \qquad\qquad ■
$$

Practice Problem 3   Find $\int \dfrac{x^5 - 2}{x^2}\,dx.$

ANSWER $\dfrac{x^4}{4} + \dfrac{2}{x} + C$

EXAMPLE 14   Find each of the following.

(a) $\int x\,(x^2 + 2x + 1)\,dx$

(b) $\int (x + 1)(2x - 1)\,dx$

(c) $\int \sqrt{x}\,(x^2 + 3)\,dx$

SOLUTION   Each integral is multiplied before integrating.

(a) $\displaystyle\int x(x^2 + 2x + 1)\,dx = \int (x^3 + 2x^2 + x)\,dx = \frac{x^4}{4} + \frac{2x^3}{3} + \frac{x^2}{2} + C$

(b) $\displaystyle\int (x + 1)(2x - 1)\,dx = \int (2x^2 + x - 1)\,dx = \frac{2x^3}{3} + \frac{x^2}{2} - x + C$

(c) $\displaystyle\int \sqrt{x}\,(x^2 + 3)\,dx = \int x^{1/2}(x^2 + 3)\,dx = \int (x^{5/2} + 3x^{1/2})\,dx$

$$
= \frac{x^{7/2}}{\frac{7}{2}} + \frac{3x^{3/2}}{\frac{3}{2}} + C = \frac{2}{7}x^{7/2} + 2x^{3/2} + C \qquad ■
$$

Practice Problem 4 Find each of the following.

(a) $\displaystyle\int x^2(x^3 - 2x)\, dx$ (b) $\displaystyle\int (x + 2)(x - 1)\, dx$

(c) $\displaystyle\int \sqrt{x}(3 - 2x)\, dx$

ANSWER

(a) $\dfrac{x^6}{6} - \dfrac{x^4}{2} + C$

(b) $\dfrac{x^3}{3} + \dfrac{x^2}{2} - 2x + C$

(c) $2x^{3/2} - \dfrac{4}{5}x^{5/2} + C$

COMMON ERROR Do not try to integrate a product as you would a sum. In the same way that the derivative is not the product of the derivatives, the integral of a product is not the product of the integrals.

| Correct | Incorrect |

$$\int x(x^2 + x)\, dx = \int (x^3 + x^2)\, dx \qquad \int x(x^2 + x)\, dx = \int x\, dx \cdot \int (x^2 + x)\, dx$$

Finding a Particular Antiderivative

We have observed that there is more than one, in fact there are an infinite number of antiderivatives associated with a function. By requiring that the graph of an antiderivative go through a given point or by requiring the antiderivative $F(x)$ to be equal to some value for a fixed value of x, we can evaluate the constant of integration and attain a particular antiderivative.

EXAMPLE 15 Find the particular function $F(x)$ whose derivative is

$$F'(x) = f(x) = x^2 + 3x + 2$$

and $F(0) = 4$.

SOLUTION First, we find the set of antiderivatives.

$$F(x) = \int (x^2 + 3x + 2)\, dx$$

$$= \int x^2\, dx + \int 3x\, dx + \int 2\, dx \qquad \text{Property 2}$$

$$= \frac{x^3}{3} + \frac{3x^2}{2} + 2x + C$$

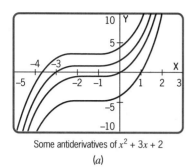

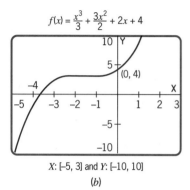

Some antiderivatives of $x^2 + 3x + 2$

(a)

$f(x) = \frac{x^3}{3} + \frac{3x^2}{2} + 2x + 4$

(0, 4)

X: [-5, 3] and Y: [-10, 10]

(b)

Figure 3

Looking at the graphs in Figure 3(a), we see some of the antiderivatives. However, we can see that only one of the graphs contains the point (0, 4). Let's now find that particular function.

$$F(0) = \frac{0^3}{3} + \frac{3(0)^2}{2} + 2(0) + C = 4 \qquad \text{Substitute } x = 0; F(0) = 4$$

$$C = 4 \qquad\qquad \text{Solve for } C.$$

The specific function sought is

$$f(x) = \frac{x^3}{3} + \frac{3x^2}{2} + 2x + 4$$

and is graphed in Figure 3(b).

An **initial condition** is often used to find the constant of integration. In Example 15, the initial condition given was $F(0) = 4$.

EXAMPLE 16   Given $f'(x) = 8x - 6$ and $f(1) = 6$, find $f(4)$.

SOLUTION   The function $f(1) = 6$ is the initial condition and is used to find the constant of integration. Since $f'(x) = 8x - 6$, we know that

$$f(x) = \int (8x - 6)\, dx = 4x^2 - 6x + C$$

We were given that $f(1) = 6$; therefore, $f(1) = 4(1)^2 - 6 \cdot 1 + C = 6$, so we have that $C = 8$. Thus,

$$f(x) = 4x^2 - 6x + 8 \quad \text{and} \quad f(4) = 4(4)^2 - 6 \cdot 4 + 8 = 48$$

The graph is shown in Figure 4.

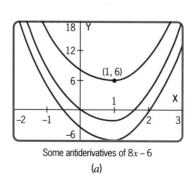

Some antiderivatives of $8x - 6$

(a)

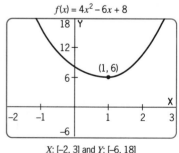

$f(x) = 4x^2 - 6x + 8$

X: [-2, 3] and Y: [-6, 18]

(b)

Figure 4

Practice Problem 5 Given $F'(x) = 2x + 3$ and $F(2) = 4$. Find the particular function $F(x)$.

ANSWER $F(x) = x^2 + 3x - 6$

COMMON ERROR   Students often make careless mistakes when working with negative exponents. They subtract instead of add.

Correct	Incorrect

$$\int x^{-4}\, dx = \frac{x^{-3}}{-3} + C = -\frac{1}{3x^3} + C \qquad \int x^{-4}\, dx = \frac{x^{-5}}{-5} + C = -\frac{1}{5x^5} + C$$

SUMMARY

You must understand the meaning of antiderivative to be successful in the remainder of this chapter. An antiderivative is a function that is found from its derivative. We use the process of antidifferentiation to accomplish this. Before using the constant rule or the general power rule, it is sometimes necessary to rewrite the integrand so that the rule can be applied. You can always check your work by differentiating the antiderivative to make certain your answer is identical to the integrand. Checking your work with differentiation also reinforces your differentiation skills. When you find a particular antiderivative, you are finding only one of a set of antiderivatives all sharing the integrand as their derivative.

Exercise Set 11.1

Find each indefinite integral and check by differentiation.

1. $\displaystyle\int 5\, dx$

2. $\displaystyle\int 3\, dx$

3. $\displaystyle\int \sqrt{3}\, dx$

4. $\displaystyle\int \pi\, dx$

5. $\displaystyle\int -2x\, dx$

6. $\displaystyle\int 7x\, dx$

7. $\displaystyle\int \sqrt{2x}\, dx$

8. $\displaystyle\int \pi x\, dx$

9. $\displaystyle\int 6x^2\, dx$

10. $\displaystyle\int 9x^2\, dx$

11. $\displaystyle\int 3x^3\, dx$

12. $\displaystyle\int 5x^3\, dx$

Find an expression for all the antiderivatives of each of the following derivatives.

13. $\dfrac{dy}{dx} = \dfrac{6}{x^2}$

14. $\dfrac{dy}{du} = 4u^3$

15. $\dfrac{dy}{du} = 16u$

16. $\dfrac{dy}{dt} = 4$

17. $\dfrac{dy}{dt} = 4\sqrt{t}$

18. $\dfrac{dy}{du} = \dfrac{3}{u^2} + 4u$

Find each indefinite integral and check by differentiation.

19. $\int x^{1/2}\, dx$

20. $\int x^{-1/2}\, dx$

21. $\int x^{-3}\, dx$

22. $\int 2x^{1/3}\, dx$

23. $\int 3x^{-3}\, dx$

24. $\int x^{2/3}\, dx$

25. $\int x^{-2}\, dx$

26. $\int 3x^{-2}\, dx$

27. $\int \left(\dfrac{1}{x^2} - 2x \right) dx$

28. $\int \left(6u^2 - \dfrac{7}{u^2} \right) du$

29. $\int \left(\dfrac{4}{\sqrt{u}} - \sqrt{u} \right) du$

30. $\int x^2 \left(x + 2x^2 - \dfrac{3}{x} \right) dx$

31. $\int \dfrac{1}{x^3} \left(x^4 + x^3 + \dfrac{4}{x} \right) dx$

32. $\int (x^2 - 1)(x - 3)\, dx$

Find the particular antiderivative that satisfies each condition. Use your graphing calculator to graph the function.

33. $f'(x) = 3x + 2, \quad f(0) = 10$

34. $f'(u) = 3u - 2, \quad f(1) = 7$

35. $\dfrac{dy}{du} = \dfrac{10}{u^2}, \quad y(5) = 2$

36. $\dfrac{dy}{dx} = \dfrac{10}{\sqrt{x}}, \quad y(4) = 2$

37. $C'(x) = 0.2x + 4x^2, \quad C(0) = 10$

38. $C'(x) = x + \dfrac{1}{x^2}, \quad C(4) = 8$

Find each indefinite integral.

39. $\int \sqrt{x}\,(x^2 + 2x - 1)\, dx$

40. $\int \sqrt[3]{x}\,(x^2 - 3x + 2)\, dx$

41. $\int \dfrac{4x^5 - 2x^4 + 3x^2 - 1}{x^2}\, dx$

42. $\int \dfrac{5x^4 - 2x^3 + 2x - 4}{x^3}\, dx$

43. $\int (4x^{1/2} - 3)\, dx$

44. $\int \dfrac{x^3 + 7}{\sqrt{x}}\, dx$

45. $\int \dfrac{x^4 - 7x^3 + x^2 - 3}{x^2}\, dx$

46. $\int \dfrac{x^2 - x}{x^{4/3}}\, dx$

47. $\int \dfrac{2x^3 - 3x}{\sqrt[3]{x}}\, dx$

48. $\int \dfrac{3x^2 - 5}{\sqrt[4]{x}}\, dx$

49. $\int \dfrac{x^2 - 2x + 3}{\sqrt[3]{x}}\, dx$

50. $\int \left(\dfrac{x^{4.3} - x^{2.1}}{x^2} \right) dx$

51. $\int \left(\dfrac{-3}{\sqrt[5]{x^2}} + \dfrac{2}{3\sqrt{x}} \right) dx$

52. $\int \left(\dfrac{2}{\sqrt[4]{x^3}} - \dfrac{1}{2\sqrt{x}} \right) dx$

Applications (Business and Economics)

53. **Sales Function.** The rate of sales of an item is

$$\dfrac{dS}{dt} = 8t + 6$$

Find the sales function and the number of sales at $t = 2$ if $S = 0$ when $t = 0$.

54. **Cost Function.** The marginal cost for producing x items is given by

$$C'(x) = -0.002x^2 - 0.6x + 100$$

Find the cost function if $C(0) = 60$.

55. **Profit Function.** The marginal profit for producing x items is given by

$$P'(x) = 500 - 4x$$

Find the profit function if $P = 0$ when $x = 0$.

56. **Revenue Function.** A marginal revenue function is given by

$$R'(x) = 400 - 0.8x$$

Find the revenue function if $R(0) = 0$, and find the revenue for a sale of 300 items.

Applications (Social and Life Sciences)

57. **Bacteria.** After introducing a bactericide into a culture, a biologist gives the rate of change of the number of bacteria present as

$$\frac{dN}{dt} = 60 - 12t$$

If $N(0) = 1200$, find $N(t)$, $N(5)$, and $N(8)$. When will the number of bacteria be 0?

58. **Flu Epidemic.** A city has a flu epidemic. The health department gives the rate of change of the number of people with the flu to be

$$\frac{dN}{dt} = 500t + 10$$

where t is the number of days after the start of the epidemic. If $N(0) = 600$, find a function $N(t)$ for the number of people with the flu in terms of t. Use this function to find the number of people with the flu 20 days after the epidemic began.

59. **Learning Rate.** A learning rate is given by

$$\frac{dL}{dt} = 0.06t - 0.0006t^2$$

where L is the number of words learned and t is time in minutes. Find $L(30)$ and $L(40)$ if $L = 0$ when $t = 0$.

60. **Population.** The change in the population of a certain area is estimated in terms of time t years as

$$\frac{dP}{dt} = 600 + 500\sqrt{t}, \qquad 0 \le t \le 5$$

If the current population is 8000, what will the population be in 4 years?

11.2 INTEGRATION FORMULAS AND MARGINAL ANALYSIS

OVERVIEW   The integration formulas introduced in this section are important. The functions that we study in this section are of the form $y = e^{kx}$ and $y = \ln |x|$. Such functions are useful in studying the growth rate of money supply, the rate of decline or the rate of increase in sales, the rate of growth of organisms in biology, and human population growth. In this section we

- Introduce new integration formulas
- Take a new look at marginal analysis

In the preceding chapter we learned that

$$\frac{d(e^{kx})}{dx} = ke^{kx}$$

The antiderivatives of the expressions on the left side should equal the functions on the right side; that is,

$$\int ke^{kx} \, dx = e^{kx} + C_1$$

$$k \int e^{kx} \, dx = e^{kx} + C_1 \qquad \text{Property 1}$$

$$\int e^{kx} \, dx = \frac{e^{kx}}{k} + C \qquad C = \frac{C_1}{k}$$

When $k = 1$, this formula becomes

$$\int e^x \, dx = e^x + C$$

Also recall that for $x > 0$,

$$\frac{d}{dx}(\ln |x|) = \frac{d(\ln x)}{dx} = \frac{1}{x}$$

If $x < 0$, then $|x| = -x$ or for this case

$$\frac{d(\ln |x|)}{dx} = \frac{d(\ln (-x))}{dx}$$

$$= \frac{1}{-x} \frac{d(-x)}{dx} \qquad \text{Chain rule: } D_x \ln |u| = \frac{1}{u} D_x u$$

$$= \frac{1}{-x} \cdot (-1) = \frac{1}{x} \qquad \frac{d(-x)}{dx} = -1$$

So $\quad \dfrac{d}{dx} \ln |x| = \dfrac{1}{x}$

Therefore $\quad \displaystyle\int \frac{1}{x} \, dx = \ln |x| + C$

These indefinite integrals are stated as follows.

INTEGRATION FORMULAS

1. $\displaystyle\int e^x \, dx = e^x + C$

2. $\displaystyle\int e^{kx} \, dx = \frac{e^{kx}}{k} + C = \frac{1}{k} e^{kx} + C$

3. $\displaystyle\int \frac{dx}{x} = \ln |x| + C$

COMMON ERROR

Correct	Incorrect

$$\int \frac{dx}{x} = \ln |x| + C \qquad \int \frac{dx}{x} = \int x^{-1} \, dx = \frac{x^{-1+1}}{-1+1} + C = \frac{x^0}{0} + C$$

Division by 0 is undefined.

The following examples utilize these integration formulas along with the properties and formulas from Section 11.1.

EXAMPLE 17   Find $\int 5e^x \, dx$.

SOLUTION

$$\int 5e^x \, dx = 5 \int e^x \, dx = 5e^x + C \qquad \text{Formula 1}$$

EXAMPLE 18   Find $\int 3x^{-1} \, dx$.

SOLUTION

$$\int 3x^{-1} \, dx = 3 \int \frac{dx}{x} = 3 \ln |x| + C \qquad \text{Formula 3}$$

EXAMPLE 19   Find $\int e^{3x} \, dx$.

SOLUTION

$$\int e^{3x} \, dx = \frac{e^{3x}}{3} + C \qquad k = 3 \text{ in Formula 2}$$

EXAMPLE 20   Find $\int \left(\frac{1}{x} - e^{2x} \right) dx$.

SOLUTION

$$\int \left(\frac{1}{x} - e^{2x} \right) dx = \int \frac{dx}{x} - \int e^{2x} \, dx \qquad \text{Property 3 and Formula 2}$$

$$= \ln |x| - \frac{e^{2x}}{2} + C$$

Now let's look at some examples that involve using multiple formulas and properties.

EXAMPLE 21   Find $\int \left(e^{3x} - \frac{1}{x^2} + \frac{3}{x} \right) dx$.

SOLUTION

$$\int \left(e^{3x} - \frac{1}{x^2} + \frac{3}{x} \right) dx = \int e^{3x} \, dx - \int \frac{1}{x^2} \, dx + 3 \int \frac{dx}{x} \qquad \text{Properties 2 and 3}$$

Now we have

$$\int e^{3x} dx = \frac{e^{3x}}{3} + C_1 \qquad \text{Formula 2}$$

$$\int \frac{dx}{x^2} = \int x^{-2} dx = \frac{x^{-1}}{-1} + C_2 = \frac{-1}{x} + C_2 \qquad \int x^n dx$$

$$3 \int \frac{dx}{x} = 3 \ln |x| + C_3 \qquad \text{Formula 3}$$

Substituting for these three integrals and combining the three constants into one constant called C yields

$$\int \left(e^{3x} - \frac{1}{x^2} + \frac{3}{x} \right) dx = \frac{e^{3x}}{3} + \frac{1}{x} + 3 \ln |x| + C \qquad \blacksquare$$

EXAMPLE 22   Find $\int \dfrac{dx}{e^{4x}}$.

SOLUTION

$$\int \frac{dx}{e^{4x}} = \int e^{-4x} dx = \frac{-1}{4} e^{-4x} + C = \frac{-1}{4e^{4x}} + C \qquad \text{Formula 2} \qquad \blacksquare$$

Practice Problem 1   Find each indefinite integral.

(a) $\int \dfrac{dx}{e^{3x}}$

(b) $\int \left(e^{2x} - \dfrac{1}{e^x} - \dfrac{3}{x} \right) dx$

ANSWER

(a) $\dfrac{-1}{3e^{3x}} + C$

(b) $\dfrac{1}{2} e^{2x} + \dfrac{1}{e^x} - 3 \ln |x| + C$

EXAMPLE 23   Find $\int \left(\dfrac{5xe^{3x} - 4 + x^2}{x} \right) dx$.

SOLUTION   This expression must be rewritten and then simplified.

$$\int \left(\frac{5xe^{3x} - 4 + x^2}{x} \right) dx = \int \left(5e^{3x} - \frac{4}{x} + x \right) dx \qquad \text{Simplify each term.}$$

$$= \int 5e^{3x}\, dx - \int \frac{4}{x}\, dx + \int x\, dx \qquad \text{Property 2}$$

$$= 5 \int e^{3x}\, dx - 4 \int \frac{dx}{x} + \int x\, dx \qquad \text{Property 1}$$

$$= \frac{5e^{3x}}{3} - 4 \ln |x| + \frac{x^2}{2} + C$$

Practice Problem 2   Find each indefinite integral.

(a) $\displaystyle \int \frac{x^2 - 2x + 1}{x}\, dx$

(b) $\displaystyle \int \left(e^{-2x} - \frac{1}{x} \right) dx$

ANSWER

(a) $\dfrac{x^2}{2} - 2x + \ln |x| + C$

(b) $\dfrac{-1}{2e^{2x}} - \ln |x| + C$

A New Look at Marginal Analysis

In the preceding chapters we have seen that cost, average cost, revenue, profit, and productivity functions (sometimes called *total functions*) have derivatives that we call marginal functions. Now that we understand indefinite integrals, we can work backward from marginal functions to the total functions. In practice, economists often obtain marginal functions from empirical data and then obtain total functions by integration.

To illustrate, assume that a corporation has the information given in Table 1. The company finds the marginal productivity function to be

$$MPD = \frac{d(PD)}{dx} = 5x - x^2$$

where x is the number of units ($100,000 each) of investment capital. (Verify the values in Table 1 by substituting the inputs in the expression for marginal productivity.) **Marginal productivity**, *MPD*, is the rate at which productivity changes (increases or decreases) with changes in capital (money available). From marginal productivity we can obtain productivity, *PD*, as a function of capital available.

TABLE 1

Input	Marginal Productivity
1	4
2	6
3	6
4	4
5	0
6	−6

EXAMPLE 24   Let the marginal productivity of a corporation be expressed as $5x - x^2$, where x is investment capital in hundreds of thousands of dollars. If productivity is zero when investment capital is zero, express productivity as a function of investment capital.

SOLUTION   Since

$$\frac{d}{dx}[PD(x)] = 5x - x^2$$

$$PD(x) = \int (5x - x^2)\, dx \qquad \text{Indefinite integral}$$

$$= \frac{5x^2}{2} - \frac{x^3}{3} + C$$

$$PD(0) = \frac{5(0)^2}{2} - \frac{0^3}{3} + C = 0 \qquad \text{Set } x = 0.$$

$$C = 0$$

Therefore,

$$PD(x) = \frac{5x^2}{2} - \frac{x^3}{3}$$

■

EXAMPLE 25

(a)   Find the total revenue function $R(x)$ if the marginal revenue is

$$R'(x) = 500 - 0.6x$$

(b)   Find the total revenue for a sale of 800 items.

SOLUTION

(a)   Since the marginal revenue function is the derivative of the total revenue function, the total revenue function is an antiderivative of the marginal revenue function.

$$R'(x) = 500 - 0.6x$$

$$\int R'(x)\, dx = \int (500 - 0.6x)\, dx \qquad \text{Integrate both sides of the equation.}$$

$$R(x) = \int (500 - 0.6x)\, dx \qquad \int R'(x) = R(x)$$

$$= 500x - 0.3x^2 + C$$

Now we find the constant C by setting $R = 0$ when $x = 0$. (The total revenue is 0 when the number of items sold is 0.)

$$0 = 500(0) - 0.3(0)^2 + C \qquad \text{Substitute } x = 0.$$

$$0 = C \qquad \text{Solve for } C.$$

Hence, the total revenue function is

$$R(x) = 500x - 0.3x^2$$

(b) Setting $x = 800$ gives

$$R = 500(800) - 0.3(800)^2$$
$$= \$208{,}000$$

■

EXAMPLE 26

(a) Find the total profit function if the marginal profit function is

$$P'(x) = 600 - 4x$$

where x is the number of items produced. Assume that there is a loss of $100 when no items are produced.

(b) Find the maximum profit.

SOLUTION

(a) Since the marginal profit function is the derivative of the total profit function, the total profit function is an antiderivative of the marginal profit function.

$$P'(x) = 600 - 4x \qquad \text{Marginal profit function}$$

$$P(x) = \int (600 - 4x)\, dx \qquad \text{Integrate both sides.}$$

$$= 600x - 2x^2 + C$$

Substituting $P(x) = -100$ when $x = 0$ gives

$$-100 = 600(0) - 2(0)^2 + C \qquad \text{Solve for } C.$$
$$= C$$

Hence,

$$P(x) = 600x - 2x^2 - 100 \qquad \text{Total profit function}$$

(b) To find the maximum profit, first set $P'(x)$ equal to 0, and find the critical value(s).

$$P'(x) = 600 - 4x$$
$$0 = 600 - 4x$$
$$x = 150 \qquad \text{Critical value}$$

Since $P''(x) = -4 < 0$, the graph is concave downward and the absolute maximum occurs at $x = 150$. The absolute maximum value of the total profit function is then

$$P(150) = 600(150) - 2(150)^2 - 100 \qquad \text{Substitute } x = 150 \text{ in } P(x).$$
$$= \$44{,}900$$

The graph is shown in Figure 5. From the graph we see that the maximum profit occurs when $x = 150$. Use a range of $0 \leq x \leq 200$ with the increments set at 50 and a range of $0 \leq y \leq 50{,}000$ with the increments set at 5000. ■

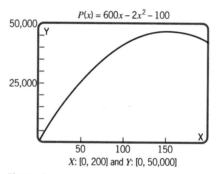

Figure 5

SUMMARY

Finding the antiderivatives of functions involving the natural exponent and those problems where the natural logarithm is the antiderivative are fairly straightforward. With some practice, these should not be difficult for you. Be sure to be careful when working with negative exponents.

Exercise Set 11.2

Find the following indefinite integrals.

1. $\int 6x^{-1}\, dx$

2. $\int 8x^{-1}\, dx$

3. $\int 4e^x\, dx$

4. $\int 10e^x\, dx$

5. $\int (6e^x + 2x)\, dx$

6. $\int (7e^x + 3x^2 - 5)\, dx$

7. $\int (2x^{-1} + e^x)\, dx$

8. $\int (5x^{-1} + x^2)\, dx$

9. $\int \left(\frac{3}{x}\right) dx$

10. $\int e^{-3t}\, dt$

11. $\int (e^{2t} - t^{-2})\, dt$

12. $\int (x^{-1} + e^{-x})\, dx$

13. $\int \left(\frac{2}{x} - e^{-10x}\right) dx$

14. $\int (-3x^{-2} + 4x^{-1})\, dx$

15. $\int 3e^{-0.02t}\, dt$

16. $\int \frac{e^{-x} - 3x^{-1}}{6}\, dx$

17. $\int \frac{e^{-2x} - 2x^{-1}}{4}\, dx$

18. $\int (e^{-x} - e^{-3x})\, dx$

19. $\int \frac{(xe^x - 2x)e^x}{x}\, dx$

20. $\int \frac{(xe^{2x} - 3x)e^x}{2x}\, dx$

21. $\int \frac{x^2 + 1}{x}\, dx$

22. $\int \frac{3x^2 + x + 2}{x}\, dx$

23. $\int (e^{-0.1x} + 2x^{-1} + e^{-0.3x})\, dx$

24. $\int (e^{-0.2x} - e^{-0.1x})\, dx$

Find the particular antiderivative of each derivative that satisfies the given condition.

25. $f'(x) = e^{2x} + 1, \quad f(0) = 6$

26. $f'(x) = e^{-x}, \quad f(0) = 6$

27. $f'(x) = \dfrac{3 + x}{x}, \quad f(1) = 4$

28. $f'(x) = \dfrac{3x^2 + 4}{x^2}, \quad f(1) = 2$

Applications (Business and Economics)

29. **Total Revenue.** A marginal revenue function is given by

$$R'(x) = 800 - 0.4x$$

Find the revenue function and the revenue for a sale of 1000 items. What is the maximum revenue? (**Hint:** $R = 0$ when $x = 0$.)

30. **Profit Function.** ABC Company has determined its marginal profit function to be

$$MP = 200 - 5x$$

If ABC Company loses $50 when no items are produced, find the company's total profit function. What is ABC Company's profit when 30 items are produced? What number of items should ABC Company produce in order to have maximum profit? What is the maximum profit?

31. **Price–Demand Equation.** The marginal price dp/dx at x units demand per month for a car is given by

$$\frac{dp}{dx} = -300e^{-0.05x}$$

Find the price–demand equation if at a price of $10,000 each, the demand is 10 cars per month.

32. **Cost Function.** The marginal average cost for producing x items is given by

$$\overline{C}'(x) = \frac{-600}{x^2}$$

where $\overline{C}(x)$ is the average cost in dollars. If $\overline{C}(50) = 20$, find the average cost function.

33. **Revenue.** Find the revenue function given the marginal revenue

$$R'(x) = 4000 - 5x$$

and knowing that $R(0) = 0$.

34. **Cost.** Find the cost for 100 units given the marginal cost

$$C'(x) = 3000 - 4x$$

when $C(0) = \$5000$.

35. **Profit.** Use the marginal functions in Exercises 33 and 34 to answer the following.
    (a) Write an expression for marginal profit.
    (b) If $P(0) = -\$5000$, find the profit for 100 units.
    (c) Find the profit when $x = 10$.

Applications (Social and Life Sciences)

36. **Population.** The rate of change of a population of bacteria in terms of time t in hours is given in thousands by

$$\frac{dP}{dt} = 60 - 0.06t$$

If $P(0) = 100$, find $P(t)$ and $P(60)$.

Review Exercises

Find the following indefinite integrals.

37. $\displaystyle\int 5x^{1/2} \, dx$

38. $\displaystyle\int \left(\frac{2}{x^2} - 3x\right) dx$

39. $\displaystyle\int \left(\frac{1}{\sqrt{x}} - x^{3/2}\right) dx$

11.3 DIFFERENTIALS AND INTEGRATION

OVERVIEW   When we studied differentiation, we did not treat dy/dx as the quotient of two separate quantities. In this section we give meaning to both dy and dx and study how they are related to integral calculus. We begin this section with a review of increments. Then we define differentials and show how the two are related. In this section we study

- Increments
- Differentials
- Integration

Increments

Recall from previous work that for two points $P(x_1, y_1)$ and $Q(x_2, y_2)$ in Figure 6 that

$$\Delta x = x_2 - x_1 \qquad \text{Increment: change in } x$$
$$\Delta y = y_2 - y_1 \qquad \text{Increment: change in } y$$

We are sometimes interested in how y is affected by a small change in x. For example, let

$$y = f(x) = x^2$$

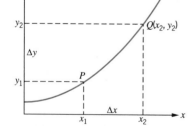

Figure 6

What is the change in y as x changes from 2 to 2.1? We have

$$\Delta x = 2.1 - 2 = 0.1 \qquad \text{Change in } x$$

$$\begin{aligned} \Delta y &= f(2.1) - f(2) \\ &= (2.1)^2 - (2)^2 \\ &= 4.41 - 4 = 0.41 \qquad \text{Corresponding change in } y \end{aligned}$$

Differentials

An understanding of increments leads to the definition of differentials. We have defined dy/dx to be one symbol. However, by themselves the symbols dy and dx may be given distinct meanings.

DIFFERENTIALS

Let $y = f(x)$ be a differentiable function of x. Then,

(a) dx, the differential of the independent variable x, is an arbitrary increment of x.

$$dx = \Delta x$$

(b) dy, the differential of the dependent variable y, is a function of x and dx given by

$$dy = f'(x)\, dx$$

Now let's use these definitions to find differentials.

EXAMPLE 27

(a) If $u = g(x) = 2x^3$, find du.
(b) If $y = h(x) = x^{-1}$, find dy.
(c) If $w = f(t) = \sqrt{t^2 + 1}$, find dw.

SOLUTION

(a)
$$u = g(x) = 2x^3$$

$$du = g'(x)\, dx \qquad \text{Definition of differential}$$

$$= 6x^2\, dx \qquad \dfrac{d(2x^3)}{dx} = 6x^2$$

(b)
$$y = h(x) = x^{-1}$$

$$dy = h'(x)\, dx \qquad \text{Definition of differential}$$

$$= -x^{-2}\, dx \qquad \frac{d(x^{-1})}{dx} = -x^{-2}$$

$$= \frac{-1}{x^2}\, dx$$

(c)
$$w = f(t) = \sqrt{t^2 + 1}$$

$$dw = f'(t)\, dt \qquad \text{Definition of differential}$$

$$= \frac{1}{2}(t^2 + 1)^{-1/2}(2t)\, dt \qquad \frac{d(t^2 + 1)^{1/2}}{dt} = \frac{1}{2}(t^2 + 1)^{-1/2}(2t)$$

$$= \frac{t}{\sqrt{t^2 + 1}}\, dt$$

Practice Problem 1 If $u = f(x) = x^2 + 3x$, find du.

ANSWER $du = (2x + 3)\, dx$

EXAMPLE 28

(a) If $u = g(x) = e^{-3x}$, find du.
(b) If $y = h(x) = \ln (1 + 2x)$, find dy.
(c) If $w = f(t) = e^{3t^2}$, find dw.

SOLUTION

(a)
$$u = g(x) = e^{-3x}$$

$$du = g'(x)\, dx \qquad \frac{d}{dx}(-3x) = -3$$

$$= -3e^{-3x}\, dx$$

(b)
$$y = h(x) = \ln (1 + 2x)$$

$$dy = h'(x)\, dx \qquad D_u(\ln u) = \frac{1}{u} u'$$

$$= \frac{1}{1 + 2x} \cdot 2\, dx \qquad \frac{d}{dx}(1 + 2x) = 2$$

$$= \frac{2}{1 + 2x}\, dx$$

(c)
$$w = f(t) = e^{3t^2}$$

$$dw = f'(t)\, dt \qquad D_u e^u = e^u u'$$

$$= e^{3t^2}(6t)\, dt \qquad \frac{d}{dy}(3t^2) = 6t$$

$$= (6t)e^{3t^2}\, dt$$

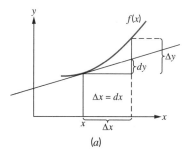

(a)

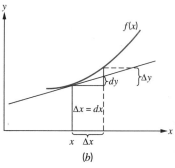

(b)

$m = f'(x) = dy/dx$, which now can be considered as either a derivative or a quotient of differentials.

Figure 7

Practice Problem 2   If $w = f(z) = e^{-4z^2}$, find dw.

ANSWER $dw = -8ze^{-4z^2}\, dz$

We can represent dy and dx pictorially as in Figure 7. Let x change by Δx. The corresponding change in the function is Δy. Now let $dx = \Delta x$. The differential dy is the change in the tangent line to the curve at x. Note that $dy \neq \Delta y$. However, for small values of dx, $dy \approx \Delta y$.

Integration

The integration formulas presented in the preceding two sections were expressed in terms of the variable x; however, they could have been expressed in terms of any variable. For instructional purposes we repeat the formulas in terms of a variable u, where we consider u as a function of x.

INTEGRATION FORMULAS

1. $\displaystyle \int u^n\, du = \frac{u^{n+1}}{n+1} + C, \quad n \neq -1$

2. $\displaystyle \int e^u\, du = e^u + C$

3. $\displaystyle \int \frac{du}{u} = \ln|u| + C$

Integration is not as straightforward as differentiation. When we differentiate $y = (2x^2 + 3)^4$ by the chain rule, we obtain

$$\frac{dy}{dx} = 4(2x^2 + 3)^3(4x) = (16x)(2x^2 + 3)^3$$

So, if we are to integrate $y = 16x(2x^2 + 3)^3$, we must think about using the chain rule in reverse. Suppose that we wish to integrate

$$\int 2(2x + 3)^2\, dx$$

Note that a function of x, namely $(2x + 3)$, is raised to the second power. If we call this function u, so that we have $u = 2x + 3$, then u is raised to the second power. To use integration formula 1, we must have an expression corresponding to du. Now, if $u = 2x + 3$, then

$$du = u'\, dx \qquad \text{Definition of the differential}$$

$$du = 2\, dx$$

Thus,

$$\int 2(2x + 3)^2 \, dx = \int (2x + 3)^2 \, 2 \, dx$$

would be in the form $\int u^2 \, du$ if $(2x + 3)$ were replaced by u and $2 \, dx$ were replaced by du, so that the expression can now be written as

$$\int 2(2x + 3)^2 \, dx = \int \overset{u^2}{\overbrace{(2x + 3)^2}} \overset{du}{\overbrace{2 \, dx}} = \int u^2 \, du = \frac{u^3}{3} + C = \frac{(2x + 3)^3}{3} + C$$

Now, suppose that the integral above had been $\int (2x + 3)^2 \, dx$, with $du = 2 \, dx$ not within the integrand. An adjustment can be made to solve a problem like this, as we see in Example 29.

EXAMPLE 29   Find $\displaystyle\int (3x + 5)^4 \, dx$.

SOLUTION   To use $\int u^n \, du$ you must consider $3x + 5$ as u, that is,

$$u = 3x + 5$$

$$\text{so} \quad du = 3 \, dx \qquad \frac{d(3x + 5)}{dx} = 3$$

$$\text{Therefore,} \quad dx = \frac{du}{3}$$

We now proceed to find the antiderivative.

$$\int (3x + 5)^4 \, dx = \int \overset{u^4}{\overbrace{(3x + 5)^4}} \cdot \overset{\frac{du}{3}}{\overbrace{dx}}$$

$$= \int u^4 \, \frac{du}{3} \qquad\qquad \text{Substituting } u = 3x + 5 \text{ and } dx = \frac{du}{3}$$

$$= \frac{1}{3} \int u^4 \, du \qquad\qquad \text{Constant rule}$$

$$= \frac{1}{3} \cdot \frac{u^5}{5} + C$$

$$= \frac{(3x + 5)^5}{15} + C$$

Practice Problem 3 Find $\int (5x - 1)^3 \, dx$.

ANSWER $\dfrac{(5x - 1)^4}{20} + C$

EXAMPLE 30 Find $\int e^{3x} \, dx$.

SOLUTION   We already have a formula for this integral. However, let's consider the integral as $\int e^u \, du$.

$$\text{Let} \quad u = 3x$$

$$\text{Then} \quad du = 3 \cdot dx$$

$$dx = \frac{du}{3}$$

$$\text{Therefore,} \quad \int e^{3x} \, dx = \int e^u \frac{du}{3}$$

$$= \frac{1}{3} \int e^u \, du$$

$$= \frac{1}{3} e^u + C$$

$$= \frac{1}{3} e^{3x} + C$$

This is exactly the same answer you would get using the formula on page 695. ■

EXAMPLE 31 Find $\int \dfrac{dx}{6x + 7}$.

SOLUTION   The integral seems to be of the form

$$\int \frac{du}{u}$$

If this is true, then

$$u = 6x + 7$$

$$du = 6 \cdot dx \qquad \frac{d(6x + 7)}{dx} = 6$$

$$dx = \frac{du}{6}$$

To put the integral in the form

$$\int \frac{du}{u}$$

we substitute and get

$$\int \frac{dx}{6x + 7} = \int \frac{1}{6x + 7}\, dx$$

$$= \int \frac{1}{u}\frac{du}{6}$$

$$= \frac{1}{6}\int \frac{du}{u}$$

$$= \frac{1}{6}\ln |u| + C$$

$$= \frac{1}{6}\ln |6x + 7| + C$$

Practice Problem 4 Find $\displaystyle\int \frac{dx}{7x - 3}$.

ANSWER $\dfrac{1}{7}\ln |7x - 3| + C$

SUMMARY

Much of what has been done in this section is used and amplified in the next section, when we begin to integrate using a technique called substitution. Therefore, make certain that you have a good grasp of the integration formulas and how to use them before proceeding to the next section.

Exercise Set 11.3

In Exercises 1–6, find dy and Δy if x = 2, Δx = 0.1, and dx = 0.1.

1. $y = 2x + 3$

2. $y = 3 - 2x$

3. $y = x^2 - 2$

4. $y = 3 - x^2$

5. $y = x + \dfrac{1}{x}$

6. $y = \dfrac{1}{x - 1}$

In Exercises 7–12, find dy and Δy if x = 2, Δx = 0.01 and dx = 0.01. How do the differences in Δy and dy compare to the differences found in Exercises 1–6?

7. $y = 2x + 3$

8. $y = 3 - 2x$

9. $y = x^2 - 2$

10. $y = 3 - x^2$

11. $y = x + \dfrac{1}{x}$

12. $y = \dfrac{1}{x - 1}$

In Exercises 13–26, find the differentials specified.

13. $u = t^2 + 4t + 1$; du

14. $w = \sqrt{x}$; dw

15. $u = e^{0.3x}$; du

16. $w = e^{3x^2}$; dw

17. $v = \dfrac{1}{x - 1}$; dv

18. $u = \sqrt{7x^2 + 2}$; du

19. $p = \ln(1 + t^2)$; dp

20. $u = \ln(1 + e^{2x})$; du

21. $u = e^{0.3x^2 + 6}$; du

22. $w = e^{3x^2 + 4x}$; dw

23. $v = \dfrac{1}{\sqrt{x^2 - 1}}$; dv

24. $u = \sqrt{7x^2 + 2}$; du

25. $p = \ln\sqrt{1 + t^2}$; dp

26. $u = \ln\sqrt{1 + e^{2x}}$; du

Find the following indefinite integrals.

27. $\displaystyle\int (x - 3)^2 \, dx$

28. $\displaystyle\int (x + 4)^2 \, dx$

29. $\displaystyle\int (2x - 3)^2 \, dx$

30. $\displaystyle\int (2x + 4)^2 \, dx$

31. $\displaystyle\int (3x + 2)^2 \, dx$

32. $\displaystyle\int e^{3x} \, dx$

33. $\displaystyle\int e^{5x} \, dx$

34. $\displaystyle\int \frac{dx}{2x + 1}$

35. $\displaystyle\int \frac{dx}{3x + 2}$

Review Exercises

Find the following indefinite integrals.

36. $\displaystyle\int 6x^{-2} \, dx$

37. $\displaystyle\int (3x^2 + 5x - 7) \, dx$

38. $\displaystyle\int (2x^3 - 3x + 5) \, dx$

11.4   INTEGRATION BY SUBSTITUTION

OVERVIEW   In this section we work with integration by substitution. We will not be as concerned with building up the differential as we did in Section 11.3; instead, we proceed with the method of substitution. The work in this section is similar to that in the last section but with much more varied functions. In this section we

- Learn the process of formal substitution
- Learn to use algebraic simplification in integration

The key step in using the method of substitution with the integration formulas listed in the preceding section is recognizing which function of x to set equal to u. This will involve some practice, but you will soon find that it is not too difficult to recognize the avenue of substitution. Remember, so far we have only three key formulas of integration. These are on page 707 in Section 11.3.

When you are trying to find an integral, you must "think backwards" because you are looking at the result of differentiation. Therefore, you must decide which formula is a "match" for the particular integration problem you are working with and then decide on the proper substitution.

EXAMPLE 32   Find $\int (2x + 3)^{1/2}\, dx$.

SOLUTION   This integration problem is of the form $\int u^n\, du$, so we use formula 1. If we let $u = 2x + 3$, then

$$du = \frac{d}{dx}(2x + 3) \cdot dx$$

$$= 2 \cdot dx$$

$$2\, dx = du$$

Solving for dx yields

$$dx = \frac{1}{2}\, du$$

Now we substitute into the original problem and integrate as follows.

$$\int (2x + 3)^{1/2}\, dx = \int u^{1/2} \cdot \frac{1}{2}\, du \qquad \text{Substitute } u = 2x + 3 \text{ and } dx = \tfrac{1}{2} du.$$

$$= \frac{1}{2} \int u^{1/2}\, du$$

$$= \frac{\frac{1}{2} u^{3/2}}{\frac{3}{2}} + C$$

$$= \frac{1}{3}\, u^{3/2} + C$$

$$= \frac{1}{3}(2x + 3)^{3/2} + C \qquad \text{Substitute } u = 2x + 3. \qquad \blacksquare$$

Practice Problem 1   Find $\int (3x + 2)^{1/3}\, dx$ by substituting $u = 3x + 2$.

ANSWER $\dfrac{(3x + 2)^{4/3}}{4} + C$

EXAMPLE 33   Find $\int \sqrt{5 - x}\, dx$.

SOLUTION   Since $\sqrt{5 - x} = (5 - x)^{1/2}$, we use formula 1.

$$\int (5 - x)^{1/2}\, dx = \int u^{1/2}(-du) = -\int u^{1/2}\, du \qquad \text{Substitution: } u = 5 - x; \text{ so } du = -dx \text{ and } dx = -du.$$

$$= \frac{-u^{3/2}}{\frac{3}{2}} + C = -\frac{2}{3} u^{3/2} + C$$

$$= -\frac{2}{3}(5 - x)^{3/2} + C \qquad \text{Substitute } u = 5 - x. \qquad \blacksquare$$

Practice Problem 2 Find $\int \sqrt[3]{x - 3} \, dx$.

ANSWER $\frac{3}{4}(x - 3)^{4/3} + C$

In the previous examples, there was really only one substitution possible. However, some problems may be a product or quotient of functions and the proper substitution must be carefully considered. Several problems of this nature are demonstrated in the next examples.

EXAMPLE 34 Find $\int 2x^2 \sqrt{3 + 5x^3} \, dx$.

SOLUTION Here we must decide which part to use for substitution. Since $\sqrt{3 + 5x^3} = (3 + 5x^3)^{1/2}$ is of the form u^n and is the more complicated term, we use formula 1 and let $u = 3 + 5x^3$.

$$\int 2x^2 \sqrt{3 + 5x^3} \, dx = 2 \int \sqrt{3 + 5x^3} \, (x^2) \, dx$$

$$= 2 \int \sqrt{u} \, \frac{du}{15} \qquad \begin{array}{l} \text{Substitution: } u = 3 + 5x^3; \\ \text{so } du = 15x^2 \, dx \text{ and} \\ du/15 = x^2 \, dx. \end{array}$$

$$= \frac{2}{15} \int u^{1/2} \, du$$

$$= \frac{2}{15} \frac{u^{3/2}}{\frac{3}{2}} + C$$

$$= \frac{4}{45} (3 + 5x^3)^{3/2} + C \qquad \text{Substitute } u = 3 + 5x^3.$$

Practice Problem 3 Find $\int 6x^2 \sqrt{x^3 + 5} \, dx$.

ANSWER $\frac{4}{3}(x^3 + 5)^{3/2} + C$

EXAMPLE 35 Find $\int x^2 e^{x^3} \, dx$.

SOLUTION

$$\int x^2 e^{x^3} \, dx = \int e^{x^3} x^2 \, dx \qquad \begin{array}{l} \text{Substitution: } u = x^3; \text{ so} \\ du = 3x^2 \, dx \text{ and } x^2 \, dx = \dfrac{du}{3}. \end{array}$$

$$= \int e^u \frac{du}{3}$$

$$= \frac{1}{3} \int e^u \, du$$

$$= \frac{e^u}{3} + C$$

$$= \frac{e^{x^3}}{3} + C \qquad \text{Substitute } u = x^3.$$

Practice Problem 4 Find $\displaystyle\int xe^{-4x^2} \, dx$.

ANSWER $\dfrac{-1}{8} e^{-4x^2} + C$

EXAMPLE 36 Find $\displaystyle\int \frac{x}{1 + x^2} \, dx$.

SOLUTION

$$\int \frac{x}{1 + x^2} \, dx = \int \frac{x \, dx}{1 + x^2} \qquad \begin{array}{l} \text{Substitution: } u = 1 + x^2; \text{ so} \\ du = 2x \, dx \text{ and } x \, dx = \dfrac{du}{2}. \end{array}$$

$$= \int \frac{du/2}{u}$$

$$= \frac{1}{2} \int \frac{du}{u}$$

$$= \frac{1}{2} \ln |u| + C$$

$$= \frac{1}{2} \ln |1 + x^2| + C \qquad \text{Substitution: } u = 1 + x^2.$$

$$= \frac{1}{2} \ln (1 + x^2) + C \qquad \text{Since } x^2 + 1 > 0$$

Practice Problem 5 Find $\displaystyle\int \frac{(\ln x)^2}{x} \, dx$.

ANSWER $\dfrac{(\ln x)^3}{3} + C$

EXAMPLE 37 Find $\displaystyle\int \frac{1}{\sqrt{x}} e^{\sqrt{x}} \, dx$.

SOLUTION   The integral seems to be of the form $\int e^u \, du$. If this is true, then $u = \sqrt{x}$ or $x = u^2$.

$$\int \frac{1}{\sqrt{x}} e^{\sqrt{x}} \, dx = \int e^{\sqrt{x}} \frac{dx}{\sqrt{x}}$$

Substitution: $x = u^2$; so $u = \sqrt{x}$

and $du = \dfrac{dx}{2\sqrt{x}} \Rightarrow 2 \, du = \dfrac{dx}{\sqrt{x}}$.

$$= \int e^u \, 2 \, du$$

$$= 2 \int e^u \, du$$

$$= 2e^u + C$$

$$= 2e^{\sqrt{x}} + C \qquad \text{Substitute } u = \sqrt{x}.$$ ∎

Practice Problem 6 Find $\displaystyle\int \frac{1}{x^2} e^{1/x} \, dx$.

ANSWER $-e^{1/x} + C$

Sometimes an integration problem will not seemingly match any of the three formulas. When this occurs, you must make an initial substitution to simplify the problem and then use other techniques as demonstrated in the following examples to complete the process of integration. Note that there may be times when more than one substitution is possible, and if the substitution you choose does not seem to be working, go back and try another substitution. There will, of course, be many functions for which substitution simply will not work and other methods must be used.

EXAMPLE 38 Find $\displaystyle\int \frac{x}{x + 1} \, dx$.

SOLUTION   There is no clear-cut substitution in this problem. Therefore, let's try an initial substitution of $u = x + 1$. If $u = x + 1$, then $x = u - 1$ and $du = dx$. Now let's substitute these into the problem.

$$\int \frac{x}{x + 1} \, dx = \int \frac{u - 1}{u} \, du$$

Substitution: $u = x + 1$, $x = u - 1$, and $du = dx$

$$= \int \left(\frac{u}{u} - \frac{1}{u} \right) du$$

Algebraic simplification

$$= \int \left(1 - \frac{1}{u} \right) du$$

$$= u - \ln |u| + C$$

$$= (x + 1) - \ln |x + 1| + C \qquad \text{Substitute } u = x + 1.$$

$$= x - \ln |x + 1| + C \qquad \text{The 1 is unnecessary, since } C \text{ is arbitrary.}$$ ∎

Practice Problem 7 Find $\int \dfrac{x}{2 - x}\, dx$.

ANSWER $-x - 2 \ln |2 - x| + C$

EXAMPLE 39   Find $\int x\sqrt{x - 3}\, dx$.

SOLUTION   At first it appears that we should substitute with $u = \sqrt{x - 3}$, but this would not help us with this problem because $u' = \frac{1}{2}(x - 3)^{-1/2}$ is not part of the expression to be integrated.

For our substitution we use $u = x - 3$ so that we have $x = u + 3$ and $du = dx$.

$$\int x\sqrt{x - 3}\, dx = \int x(x - 3)^{1/2}\, dx \qquad \text{Substitution: } u = x - 3;$$

$$\text{therefore, } du = dx \text{ and}$$
$$x = u + 3.$$

$$= \int (u + 3)u^{1/2}\, du$$

$$= \int (u^{3/2} + 3u^{1/2})\, du \qquad \text{Simplify algebraically.}$$

$$= \frac{u^{5/2}}{\frac{5}{2}} + 3\,\frac{u^{3/2}}{\frac{3}{2}} + C$$

$$= \frac{2}{5}(x - 3)^{5/2} + 2(x - 3)^{3/2} + C \qquad \text{Substitute } u = x - 3. \quad \blacksquare$$

Practice Problem 8 Find $\int \dfrac{x}{\sqrt{x + 4}}\, dx$.

ANSWER $\frac{2}{3}(x + 4)^{3/2} - 8(x + 4)^{1/2} + C$

NOTE: You may have noticed a pattern to the work we have done. When using substitution, try to let the denominator be u, or the term that is under a radical, or the term raised to the highest power. These are guidelines that usually lead to a successful substitution.

EXAMPLE 40   Find $\int \dfrac{x\, dx}{\sqrt{4 - x^2}}$.

SOLUTION   We let the term under the radical in the denominator be u.

$$\int \frac{x\, dx}{\sqrt{4 - x^2}} = \int \frac{\dfrac{-1}{2}\, du}{\sqrt{u}} \qquad \text{Substitution: } u = 4 - x^2 \text{ and}$$

$$du = -2x\, dx, \ \frac{-1}{2}\, du = x\, dx$$

$$= \frac{-1}{2} \int \frac{du}{u^{1/2}} = \frac{-1}{2} \int u^{-1/2}\, du$$

$$= -\frac{1}{2}(2u^{1/2}) + C = -u^{1/2} + C$$
$$= -\sqrt{4 - x^2} + C$$

Practice Problem 9 Find $\displaystyle\int \frac{(x + 1)\, dx}{\sqrt[3]{3x^2 + 6x + 5}}$.

ANSWER $\frac{1}{4}(3x^2 + 6x + 5)^{2/3} + C$

**COMMON
ERROR**   When using substitution in finding an integral, do not forget to substitute back. The variable in your answer should be the same as in the original problem.

SUMMARY

It is important to realize that it is not always necessary or advantageous to use substitution. In the two integrals listed below, it would be better to multiply the first one out and then use the simple integration techniques. In the second integral, substitution would be a good choice.

$$\int x(3x^2 + 4)^2\, dx \qquad \text{Multiply first before integrating.}$$

$$\int x(3x^2 + 4)^3\, dx \qquad \text{Use substitution.}$$

It is also important to be able to distinguish between problems requiring you to use the formula involving ln x and the formula involving a term such as u^{-n}, as shown below.

$$\int \frac{1}{3x + 4}\, dx \qquad \text{requires} \qquad \int \frac{1}{u}\, du = \ln |u| + C \qquad \text{Why?}$$

$$\int \frac{1}{(3x + 4)^2}\, dx \quad \text{requires} \quad \int u^n\, du = \frac{u^{n+1}}{n + 1} + C, \quad n \neq -1 \qquad \text{Why?}$$

When integrating using the technique of substitution, you may realize that certain substitutions are obvious while others are not. When a substitution is not obvious, some algebraic simplification will usually be necessary after the initial substitution. Since these problems involve more steps and are somewhat more involved than those in previous sections, it is wise to check your solutions by differentiating your answer.

Exercise Set 11.4

Find the following indefinite integrals.

1. $\displaystyle\int 6\sqrt{x} + 3 \; dx$

2. $\displaystyle\int 5\sqrt{x} - 4 \; dx$

3. $\displaystyle\int 6\sqrt{4 - 3x} \; dx$

4. $\displaystyle\int 5\sqrt{2x - 3} \; dx$

5. $\displaystyle\int 2xe^{3x^2} \; dx$

6. $\displaystyle\int \frac{x \; dx}{1 - x^2}$

7. $\displaystyle\int 3xe^{-5x^2} \; dx$

8. $\displaystyle\int \frac{-3x \; dx}{2x^2 + 3}$

9. $\displaystyle\int \sqrt[3]{3 + 7x} \; dx$

10. $\displaystyle\int \sqrt[3]{x + 3} \; dx$

11. $\displaystyle\int x(2x^2 - 7)^{3/4} \; dx$

12. $\displaystyle\int x\sqrt{x^2 + 1} \; dx$

13. $\displaystyle\int x\sqrt{x^2 + 4} \; dx$

14. $\displaystyle\int 2x(3x^2 - 4)^{2/3} \; dx$

15. $\displaystyle\int x^2\sqrt{1 + x^3} \; dx$

16. $\displaystyle\int x^2\sqrt[3]{1 + x^3} \; dx$

17. $\displaystyle\int \frac{x + 1}{x - 1} \; dx$

18. $\displaystyle\int \frac{2x}{\sqrt{x + 2}} \; dx$

19. $\displaystyle\int \frac{-2x}{3 - x} \; dx$

20. $\displaystyle\int \frac{4x}{x + 2} \; dx$

21. $\displaystyle\int x(x + 1)^4 \; dx$

22. $\displaystyle\int 2x(x - 3)^3 \; dx$

23. $\displaystyle\int 2x\sqrt{x - 5} \; dx$

24. $\displaystyle\int x\sqrt[3]{x + 3} \; dx$

25. $\displaystyle\int \frac{e^{2x}}{1 + e^{2x}} \; dx$

26. $\displaystyle\int 3x(e^{x^2})^4 \; dx$

27. $\displaystyle\int \frac{x}{1 - 3x} \; dx$

28. $\displaystyle\int (x + 2)\sqrt{2 - x} \; dx$

29. $\displaystyle\int (3 - x)\sqrt[3]{x + 1} \; dx$

30. $\displaystyle\int \frac{x}{(x - 3)^4} \; dx$

31. $\displaystyle\int \frac{e^{\sqrt{x - 1}}}{\sqrt{x - 1}} \; dx$

32. $\displaystyle\int \frac{1/\sqrt{x}}{\sqrt{x} - 1} \; dx$

33. $\displaystyle\int \frac{(\sqrt{x} + 3)^5}{\sqrt{x}} \; dx$

34. $\displaystyle\int \frac{(2x + 1)}{x^2 + x} \; dx$

35. Why does

$$\int \frac{1}{3x + 4} \; dx$$

require the use of the formula

$$\int \frac{1}{u} \; du$$

while

$$\int \frac{1}{(3x + 4)^2} \; dx$$

requires the use of the formula

$$\int u^n du?$$

Applications (Business and Economics)

36. **Revenue.** If the marginal revenue for the sale of x units per week is

$$R'(x) = x(x^2 + 1)^2 \quad \text{and} \quad R(0) = 0$$

find $R(10)$.

37. **Cost.** If the marginal cost of x items is

$$C'(x) = 40 + 1000e^{-2x}$$

with $C(0) = 1000$, find $C(10)$.

38. **Profit.** Use Exercises 36 and 37 to find an expression for marginal profit. Then with $P(0) = -1000$, find $P(10)$. Is $P(10) = R(10) - C(10)$?

39. **Revenue Function.** Suppose that a marginal revenue function is given by

$$R'(x) = \frac{x}{\sqrt{x^2 + 16}}$$

Find the revenue function. Find the revenue for a production of three items. [**Hint:** $R(0) = 0$.]

Applications (Social and Life Sciences)

40. **Pollution.** Suppose that the radius R, in feet, of an oil slick is increasing at the rate of

$$\frac{dR}{dt} = 80(t + 16)^{-1/2}$$

where t is time in minutes. If $R = 0$ when $t = 0$, find the radius of the slick after 20 minutes.

41. **Learning Function.** Suppose that the rate of change of a learning function is given as a function of time t in hours as

$$N'(t) = -1.6e^{0.02t}$$

If $N(0) = 80$, find the learning function as a function of time.

Review Exercises

Find the following indefinite integrals.

42. $\displaystyle\int (x + 3)^{-2}\, dx$

43. $\displaystyle\int e^{-4x}\, dx$

44. $\displaystyle\int e^{-2x}\, dx$

45. $\displaystyle\int \frac{dx}{x + 1}$

11.5   DEFINITE INTEGRALS

OVERVIEW In the previous sections, we have been concerned with finding the indefinite integral (antiderivative) $F(x)$ of some function $f(x)$ denoted by

$$F(x) = \int f(x)\, dx$$

In this section, we are interested in finding a function that gives the total change over an interval in an antiderivative of a function. For example, if $f(x) = 2x$, then $F(x) = x^2 + C$. The total change in F over the interval $[1, 4]$ will be

$$F(4) - F(1) = [4^2 + C] - [1^2 + C] = 15$$

This total change over an interval in the antiderivative is called the **definite integral** and is one of the important concepts of calculus. A definite integral can be used to evaluate the total depreciation of a machine over a period of time or the total amount of money that a machine will generate over several years. Our approach in this section is intuitive and informal. In Section 11.7 we consider the definite integral as a Riemann integral defined as the limit of a sum. This more formal approach is needed to develop many of the applications of the definite integral. In this section we study

- Definite integrals
- Properties of definite integrals
- Use of substitution in definite integrals
- Applications

We illustrate the usefulness of the definite integral with the following example.

EXAMPLE 41   The marginal cost in producing a certain model of a radio at an electronics corporation is

$$\frac{dC}{dx} = 6 - 0.04x$$

where x is the number of radios produced in a day. If the number produced in a day changes from 50 to 100 radios, what is the change in cost?

SOLUTION   Recall that $C(x)$ is the cost function. We seek

$$C(100) - C(50)$$

Using the indefinite integral, we find

$$C(x) = \int (6 - 0.04x) \, dx$$
$$= 6x - 0.02x^2 + C \qquad \text{Antiderivative of } 0.04x = 0.02x^2.$$

Therefore,

$$C(100) - C(50) = [6(100) - 0.02(100)^2 + C] - [6(50) - 0.02(50)^2 + C]$$
$$= 150$$

We see that the constant C in each term is eliminated by the subtraction process and does not play a part in the computation.   ◼

Notice in the preceding example that the change in $C(x)$ was computed using an antiderivative of $C'(x)$. Since the antiderivative is symbolized by

$$C(x) = \int C'(x) \, dx$$

we can indicate the change in $C(x)$ by the following notation:

$$\text{Change in } C(x) = C(600) - C(400) = \int_{400}^{600} C'(x) \, dx$$

This integral form is called the definite integral.

DEFINITE INTEGRAL

The **definite integral** of a nonnegative, continuous function $f(x)$ over the interval from $x = a$ to $x = b$ is the net change in an antiderivative of $f(x)$ over the interval. This fact is symbolized by

$$\int_a^b f(x) \, dx = F(x) \Big|_a^b = F(b) - F(a)$$

where $F'(x) = f(x)$, for all x in $[a, b]$.

In Example 41 we noted that the constant of integration subtracted out. Since the definite integral was defined for any antiderivative, it is customary to take the simplest antiderivative where $C = 0$. In the symbolic form

$$\int_a^b f(x)\, dx$$

b is called the **upper limit** of integration, a is the **lower limit** of integration, and $f(x)$ is called the **integrand**.

Even though many graphing calculators can evaluate definite integrals, it is important for you to understand the definition and what is being found when a definite integral is used. A definite integral calculates the change in a function over an interval.

The following examples are done using the traditional calculus methods. Before you use a graphing calculator exclusively to evaluate definite integrals, you should be able to do them as shown in the examples.

EXAMPLE 42   Evaluate $\displaystyle\int_2^6 x^2\, dx$.

SOLUTION

$$\int_2^6 x^2\, dx = \frac{x^3}{3}\Big|_2^6 = \frac{6^3}{3} - \frac{2^3}{3} = 69\tfrac{1}{3} \qquad \text{Simplest antiderivative of } x^2 \text{ is } \frac{x^3}{3}. \qquad \blacksquare$$

EXAMPLE 43   Evaluate $\displaystyle\int_0^3 e^{2x}\, dx$.

SOLUTION

$$\int_0^3 e^{2x}\, dx = \frac{e^{2x}}{2}\Big|_0^3 = \frac{e^6}{2} - \frac{e^0}{2}$$

$$= \frac{e^6 - 1}{2} \approx 201.21 \qquad \text{Antiderivative of } e^{2x} = \frac{e^{2x}}{2}. \qquad \blacksquare$$

Practice Problem 1 Evaluate $\displaystyle\int_1^2 (x - x^{-1})\, dx$.

ANSWER $\frac{3}{2} - \ln 2 \approx 0.807$

Practice Problem 2 Evaluate $\displaystyle\int_1^2 e^{3x}\, dx$.

ANSWER $\dfrac{e^6 - e^3}{3} \approx 127.78$

NOTE: We should take time now to note the distinction between the definite integral and the indefinite integral. The **definite integral is a real number** (a value), whereas the **indefinite integral is a set of functions** [all of the antiderivatives of $f(x)$].

The following is a list of properties of definite integrals. Properties 1 and 2 parallel properties of indefinite integrals. Properties 3, 4, and 5 follow from the definition of the integral given in this section. Even when using a graphing calculator, it is important to know these properties.

DEFINITE INTEGRAL PROPERTIES

Assume that $f(x)$ and $g(x)$ are continuous functions on the indicated intervals and k is a constant. Then

1. $\displaystyle\int_a^b kf(x)\,dx = k\int_a^b f(x)\,dx$

2. $\displaystyle\int_a^b [f(x) \pm g(x)]\,dx = \int_a^b f(x)\,dx \pm \int_a^b g(x)\,dx$

3. $\displaystyle\int_a^b f(x)\,dx = -\int_b^a f(x)\,dx$

4. $\displaystyle\int_a^a f(x)\,dx = 0$

5. $\displaystyle\int_a^b f(x)\,dx = \int_a^c f(x)\,dx + \int_c^b f(x)\,dx, \quad a \le c \le b$

Calculator Note

Now let's look at how to evaluate a definite integral with your graphing calculator. On the MATH menu you will find a key for evaluating the integral. It may be fnInt(or something similar to that notation. When evaluating the definite integral, you enter the function, the variable, and the limits of integration. For example, to find the integral from Practice Problem 1,

$$\int_1^2 (x - x^{-1})\,dx$$

your entry and answer would look like

$$\text{fnInt}(x - x \wedge -1, x, 1, 2) = .8068528194$$

If your calculator has a slightly different notation, it should work the same way.

We now give some examples to illustrate properties of definite integrals. We evaluate definite integrals using both the properties and then with the graphing calculator to check our work.

EXAMPLE 44

1. $\displaystyle\int_a^b kf(x)\,dx = k\int_a^b f(x)\,dx$

$$\int_2^3 6x\,dx = 6\int_2^3 x\,dx = 6\left(\frac{x^2}{2}\right)\Big|_2^3$$

$$= 6\left(\frac{3^2}{2}\right) - 6\left(\frac{2^2}{2}\right) \qquad F(3) - F(2)$$

$$= 6\left(\frac{9}{2} - \frac{4}{2}\right) = 15$$

✓ Calculator Check: fnInt(6x, x, 2, 3) = 15

2. $\displaystyle\int_a^b [\,f(x) \pm g(x)]\,dx = \int_a^b f(x)\,dx \pm \int_a^b g(x)\,dx$

$$\int_2^3 (6x + 4)\,dx = \int_2^3 6x\,dx + \int_2^3 4\,dx$$

$$= \left(\frac{6x^2}{2}\right)\Big|_2^3 + (4x)\Big|_2^3$$

$$= \left[\frac{6\cdot 3^2}{2} - \frac{6\cdot 2^2}{2}\right] \qquad F(3) - F(2) \text{ for each part}$$

$$+ [4\cdot 3 - 4\cdot 2]$$

$$= 15 + 4 = 19$$

✓ Calculator Check: fnInt(6x + 4, x, 2, 3) = 19

3. $\displaystyle\int_a^b f(x)\,dx = -\int_b^a f(x)\,dx$

$$\int_2^3 6x\,dx = -\int_3^2 6x\,dx$$

$$3x^2\Big|_2^3 = -3x^2\Big|_3^2$$

$$27 - 12 = -(12 - 27)$$

$$15 = -(-15)$$

✔ Calculator Check: $(-)$fnInt$(6x, x, 3, 2) = 15$

4. $\displaystyle\int_a^a f(x)\, dx = 0$

$$\int_2^2 6x\, dx = 3x^2 \Big|_2^2 = 12 - 12 = 0$$

✔ Calculator Check: fnInt$(6x, x, 2, 2) = 0$

5. $\displaystyle\int_a^b f(x)\, dx = \int_a^c f(x)\, dx + \int_c^b f(x)\, dx$

$$\int_1^3 6x\, dx = \int_1^2 6x\, dx + \int_2^3 6x\, dx \qquad C = 2 \text{ is selected arbitrarily.}$$

$$3x^2 \Big|_1^{3.} = 3x^2 \Big|_1^2 + 3x^2 \Big|_2^3$$

$$27 - 3 = 12 - 3 + 27 - 12$$

$$24 = 9 + 15$$

✔ Calculator Check: fnInt$(6x, x, 1, 2) +$ fnInt$(6x, x, 2, 3) = 24$. Note: Try this same problem with your calculator using another value for C.

Practice Problem 3 Evaluate each integral using the properties and then check your answer with your graphing calculator.

(a) $\displaystyle\int_1^4 5x^2\, dx$

(b) $\displaystyle\int_2^4 (3x^2 - 2x)\, dx$

(c) $\displaystyle\int_1^3 \left(\frac{4}{x^2} - 2\right) dx$

ANSWER

(a) 105   (b) 44   (c) $-4/3$

Sometimes algebraic simplification is used to find the definite integral.

EXAMPLE 45   Evaluate $\int_4^9 \dfrac{x+1}{\sqrt{x}}\, dx$.

SOLUTION

$$\int_4^9 \frac{x+1}{\sqrt{x}}\, dx = \int_4^9 \left(\frac{x}{\sqrt{x}} + \frac{1}{\sqrt{x}} \right) dx = \int_4^9 (x^{1/2} + x^{-1/2})\, dx$$

$$= \left(\frac{x^{3/2}}{\frac{3}{2}} + \frac{x^{1/2}}{\frac{1}{2}} \right) \Bigg|_4^9 = \left(\frac{2}{3} x^{3/2} + 2x^{1/2} \right) \Bigg|_4^9$$

$$= \left[\frac{2}{3} \cdot 9^{3/2} + 2 \cdot 9^{1/2} \right] - \left[\frac{2}{3} \cdot 4^{3/2} + 2 \cdot 4^{1/2} \right]$$

$$= (18 + 6) - \left(\frac{16}{3} + 4 \right) = 14\tfrac{2}{3}$$

✔ Calculator Check: fnInt$((x + 1)/\sqrt{x}, x, 4, 9) = 14.66666667$

Practice Problem 4 Evaluate $\int_0^4 \sqrt{x}\,(3 + x)\, dx$.

ANSWER $\dfrac{144}{5} = 28\tfrac{4}{5}$.

Integration by Substitution

A definite integral in which substitution is required, can be evaluated using the properties of integration with two different methods. Method 1 involves actually changing the limits of integration by finding corresponding values for the new variable (u). Method 2 involves obtaining the antiderivative in terms of the original variable before evaluating the definite integral. We demonstrate both methods in Example 46. Generally, Method 1 results in less complicated calculations. As before, after we find the integral, we will check our answer with a calculator.

EXAMPLE 46   Evaluate $\int_1^5 \sqrt{2x - 1}\, dx$.

SOLUTION

Method 1   This method involves substitution and changing the limits of integration to correspond to the new variable. For the integral

$$\int_1^5 \sqrt{2x - 1}\, dx$$

we use the following substitutions:

$$u = 2x - 1 \quad \text{gives} \quad du = 2\,dx, \quad \text{which yields} \quad dx = \frac{1}{2}\,du$$

We now change the limits of integration to match the new variable of integration, u. When $x = 1$, $u = 2(1) - 1 = 1$; and when $x = 5$, $u = 2(5) - 1 = 9$.

$$
\begin{aligned}
\int_1^5 \sqrt{2x - 1}\ dx &= \int_1^9 u^{1/2}\,\frac{1}{2}\,du \qquad \text{Substitution} \\[2mm]
&= \frac{1}{2}\int_1^9 u^{1/2}\,du \\[2mm]
&= \frac{u^{3/2}}{3}\,\bigg|_1^9 \qquad\qquad \text{Integration} \\[2mm]
&= \frac{9^{3/2}}{3} - \frac{1^{3/2}}{3} \\[2mm]
&= \frac{26}{3}
\end{aligned}
$$

✓ Calculator Check: fnInt($\sqrt{}(2x - 1)$, x, 1, 5) = 8.666666667 and (1/2)fnInt ($x \wedge (1/2)$, x, 1, 9) = 8.666666667.

Method 2 For this method, we find the antiderivative in terms of x (the original variable). We use the same substitutions as above, but notice that we do not place the limits of integration on the indefinite integral.

$$
\begin{aligned}
\int \sqrt{2x - 1}\ dx &= \int u^{1/2}\,\frac{1}{2}\,du \qquad\qquad\qquad \text{Substitutions} \\[2mm]
&= \frac{u^{3/2}}{3} = \frac{(2x - 1)^{3/2}}{3} \qquad\quad \text{Substitute back.} \\[4mm]
\int_1^5 \sqrt{2x - 1}\ dx &= \frac{(2x - 1)^{3/2}}{3}\,\bigg|_1^5 \\[2mm]
&= \frac{(2 \cdot 5 - 1)^{3/2}}{3} - \frac{(2 \cdot 1 - 1)^{3/2}}{3} \qquad F(5) - F(1) \\[2mm]
&= \frac{27}{3} - \frac{1}{3} = \frac{26}{3}
\end{aligned}
$$

From this point on, we will show only Method 1 when evaluating definite integrals requiring substitution. Method 1 reduces the amount of actual calculation that has to be done because you deal with simpler terms.

Calculator Note

We will no longer show the Calculator Check, but you should continue to check all work with your calculator.

EXAMPLE 47   Evaluate $\displaystyle\int_0^2 x(2x^2 + 1)^3 \, dx$.

SOLUTION

$$\int_0^2 x(2x^2 + 1)^3 \, dx = \int_1^9 u^3 \frac{1}{4} \, du$$

Substituting $u = 2x^2 + 1$ yields $du = 4x \, dx$ and $du/4 = x \, dx$.

$$= \frac{1}{4} \int_1^9 u^3 \, du$$

Therefore, when $x = 0$, $u = 1$; and when $x = 2$, $u = 9$.

$$= \frac{u^4}{16} \Big|_1^9$$

$$= \frac{9^4}{16} - \frac{1^4}{16} = \frac{6560}{16} = 410$$

Practice Problem 5   Evaluate $\displaystyle\int_0^2 x^2(3x^3 + 2)^2 \, dx$.

ANSWER $650\frac{2}{3}$

COMMON ERROR   When evaluating a definite integral by using substitution and the methods of changing the limits, students often go back to the original terms to evaluate the limit. Once the limits have been changed, you no longer use the variable x. You use only the variable u.

Practice Problem 6   Evaluate $\displaystyle\int_0^3 \frac{dx}{(3x + 1)^2}$.

ANSWER $\dfrac{3}{10}$

Integration is used in many important applications in many fields. In applied problems we may know the rate of change of the function, but we may be interested in computing the actual change in the function as the value of the variable changes. Let's look at some examples here, and then we will expand the topic in Section 11.8.

EXAMPLE 48   Suppose that a company's marginal cost, marginal revenue, and marginal profit are given in thousands of dollars in terms of the number x of units produced as

$$C'(x) = 1$$

$$R'(x) = 12 - 2x \quad \text{for } 0 \le x < 12$$

$$P'(x) = R'(x) - C'(x)$$

If production changes from 3 units to 6 units, find the change in (a) cost, (b) revenue, and (c) profit.

SOLUTION

(a)   The change in cost is

$$\int_3^6 C'(x)\, dx = \int_3^6 1\, dx = x \Big|_3^6 = 6 - 3 = 3$$

Therefore, the change in cost by the increase in production from 3 units to 6 units is $3000.
(b)   The change in revenue is

$$\int_3^6 R'(x)\, dx = \int_3^6 (12 - 2x)\, dx = (12x - x^2) \Big|_3^6$$
$$= (72 - 36) - (36 - 9) = 9$$

Therefore, revenue increases by $9000 with the change in production.
(c)   The change in profit is

$$\int_3^6 P'(x) = \int_3^6 [R'(x) - C'(x)]\, dx$$
$$= \int_3^6 R'(x)\, dx - \int_3^6 C'(x)\, dx = 9 - 3 = 6$$

Therefore, profit increases by $6000 by the change in production.

Note that we are not calculating the total amount of the cost, revenue, and profit functions. We are calculating the change in the amount of the function resulting from a change in production. ∎

EXAMPLE 49   The marginal revenue that a manufacturer receives is given by $R'(x) = 2 - 0.02x + 0.003x^2$ dollars per unit sold. How much additional revenue is received if sales increase from 100 to 200 units?

SOLUTION

$$\int_{100}^{200} (2 - 0.02x + 0.003x^2)\, dx = 2x - 0.01x^2 + 0.001x^3 \Big|_{100}^{200}$$
$$= [2 \cdot 200 - 0.01 \cdot 200^2 + 0.001 \cdot 200^3]$$
$$- [2 \cdot 100 - 0.01 \cdot 100^2 + 0.001 \cdot 100^3]$$
$$= 8000 - 1100$$
$$= 6900$$

Therefore, the revenue will increase by a total of $6900 if production is increased from 100 to 200 units. ■

SUMMARY

Make certain that you understand the difference between an indefinite integral (antiderivative) and a definite integral. When you are determining an indefinite integral (antiderivative), you are actually finding an entire set of functions. When you evaluate a definite integral, you are calculating a change in the antiderivatives as the variable changes from one value to another. The constant of integration does not play a role in the definite integral, since it subtracts out during the process of evaluation. Therefore, we always choose the simplest antiderivative, with $C = 0$, when evaluating a definite integral. Remember, if you use a graphing calculator to evaluate a definite integral, you should first make certain that you know how to do the problem using the properties of integration.

Exercise Set 11.5

Evaluate the following definite integrals using the properties of integration.

1. $\int_{1}^{3} dx$

2. $\int_{3}^{5} e^2\, dx$

3. $\int_{2}^{3} 7\, dx$

4. $\int_{1}^{6} \frac{4}{x}\, dx$

5. $\int_{1}^{3} 3x\, dx$

6. $\int_{2}^{5} 4x\, dx$

7. $\int_{2}^{5} (2x + 3)\, dx$

8. $\int_{1}^{3} (4x - 1)\, dx$

9. $\int_{0}^{4} e^{2t}\, dt$

10. $\int_{0}^{2} (x^2 + x)\, dx$

11. $\int_{2}^{5} (x^2 - x + 1)\, dx$

12. $\int_{0}^{1} e^{-t/3}\, dt$

13. $\int_{1}^{3} (x - 3)^2\, dx$

14. $\int_{0}^{2} (x - 3)^2\, dx$

15. $\int_{3}^{6} \sqrt[4]{x - 3}\, dx$

16. $\int_{3}^{5} \sqrt{2x - 3}\, dx$

17. $\int_{-2}^{2} x^{2/3}\, dx$

18. $\int_{-1}^{2} 3x^{2/3}\, dx$

Evaluate using the properties of integration and then use a graphing calculator to check your work.

19. $\int_{3}^{6} 2x^{-1}\, dx$

20. $\int_{1}^{2} \frac{dt}{4t + 6}$

21. $\int_{0}^{4} \frac{dt}{e^{4t}}$

22. $\int_{1}^{3} xe^{2x^2}\, dx$

23. $\int_{0}^{1} x^2 e^{x^3}\, dx$

24. $\int_{0}^{4} (1 + xe^{-x^2})\, dx$

25. $\int_0^1 x\sqrt{x^2 + 1}\ dx$

26. $\int_0^1 x\sqrt[3]{x^2 + 2}\ dx$

27. $\int_2^4 \dfrac{x + 1}{x - 1}\ dx$

28. $\int_4^9 \dfrac{x + 4}{\sqrt{x}}\ dx$

29. $\int_0^2 3x(x + 1)^4\ dx$

30. $\int_0^2 \sqrt{x}(x + 1)\ dx$

If $\int_2^5 f(x)\ dx = 6$ *and* $\int_2^5 g(x) = -4$ *use the properties of definite integrals to find the following.*

31. $\int_2^5 [f(x) - g(x)]\ dx$

32. $\int_5^2 [2f(x) + g(x)]\ dx$

33. $\int_2^5 [4f(x) - 3g(x)]\ dx$

If $\int_1^3 f(x)\ dx = 10$ *and* $\int_3^7 f(x) = 8$ *use the properties of definite integrals to find the following.*

34. $\int_1^7 f(x)\ dx$

35. $\int_7^1 f(x)\ dx$

36. $\int_1^7 f(x)\ dx - \int_3^7 f(x)\ dx$

Applications (Business and Economics)

37. **Assembly Function.** From assembling several units of a product, the production manager obtained the following rate of assembly function:

$$f(x) = 120x^{-1/2}$$

where $f(x)$ represents the rate of labor hours required to assemble x units. The company plans to bid on a new order of 12 additional units. Help the manager estimate the total labor requirements for assembling the additional 12 units, if the current production level is 50 units.

38. **Manufacturing Costs.** A corporation has determined that the marginal cost per week of manufacturing electric shavers is given by

$$C'(x) = 40 - 0.2x + \dfrac{x^2}{10}$$

where x is the number of units manufactured per week. What will be the total cost change if the company decides to increase weekly production from 400 to 450 razors per week?

39. **Revenue.** A corporation determines a marginal revenue function associated with selling x shavers to be

$$R'(x) = 80 - 0.1x$$

Find the additional revenue generated by increasing sales from 200 to 300 shavers.

40. **Cost.** The daily marginal cost function associated with producing x large-screen televisions is given by

$$C'(x) = 0.0009x^2 - 0.004x + 4$$

where $C'(x)$ is measured in dollars per unit and x denotes the number of units produced. The daily fixed cost of production is $250.00. Find the total cost involved in the production of the first 1000 units.

41. **Cost, Revenue, Profit.** A company's marginal cost, revenue, and profit equations (in thousands of dollars per day) where x is the number of units produced per day are

$$C'(x) = 1$$
$$R'(x) = 10 - 2x \quad \text{for } 0 \le x \le 10$$
$$P'(x) = R'(x) - C'(x)$$

Find the change in cost, revenue, and profit in going from a production level of 3 units per day to 5 units per day.

42. **Cost, Revenue, Profit.** The marginal cost and marginal revenue functions (in dollars per item) of a manufacturer of small computers are found to be

$$C'(x) = 300 - 0.2x \quad \text{and} \quad R'(x) = 400 + 0.01x$$

where x is the level of production. Assume that each computer manufactured is sold.
   (a) Find the change in revenue received when the level of production increases from 20 to 40 computers per week.
   (b) Find the revenue received from the manufacture of 50 computers.

(c) If the fixed cost is $200, find the cost function and the cost of producing 50 computers.
(d) Find the profit received from the manufacture and sale of 50 computers.

Applications (Social and Life Sciences)

43. **Temperature of a Patient.** The rate of change of a patient's temperature 1 hour after x milligrams of a drug is administered is

$$\frac{dT}{dx} = 0.3x - \frac{x^2}{4}, \qquad 0 \le x \le 4$$

What total change of temperature occurs when the dosage changes from (a) 0 to 3 milligrams and (b) 1 to 4 milligrams?

44. **Poiseville's Law.** Suppose that V represents the total flow in cubic units per second of blood through an artery whose radius is R, and suppose that r is the distance of a particle of blood from the center of the artery. Assume that

$$\frac{dV}{dr} = 2\pi C(Rr - r^2)$$

where C is a constant that depends on the units used.

(a) Compute

$$V = \int_{R}^{1.1R} 2\pi C(Rr - r^2)\, dr$$

to find the increase in blood flow when the radius of the artery is increased by 10%.
(b) Find the increase in blood flow when the radius of the artery is increased by 20%.

45. **Learning Function.** A person's learning rate of new words per minute is given by

$$\frac{dL}{dt} = 30t^{-1/2}, \qquad 1 \le t \le 16$$

Find the number of words learned (i.e., find L) from $t = 4$ to $t = 9$ minutes.

Review Exercises

Find the following indefinite integrals.

46. $\displaystyle\int \sqrt{3x - 7}\, dx$ 47. $\displaystyle\int \sqrt[3]{2x + 3}\, dx$

48. $\displaystyle\int \frac{x + 1}{x - 2}\, dx$ 49. $\displaystyle\int x\sqrt{x - 1}\, dx$

11.6 THE DEFINITE INTEGRAL AND AREA

OVERVIEW   The definite integral can be used to find the area under a curve [see Figure 8(a)]. In this section, we illustrate how the indefinite integral, the definite integral, and the area under a curve are related. As one application, we note that the area between the marginal revenue curve and marginal cost curve can be interpreted as profit over an interval [Figure 8(b)]. In this section we consider

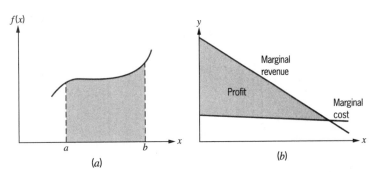

(a)

(b)

Figure 8

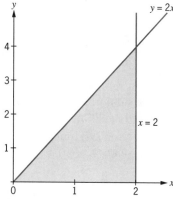

Figure 9

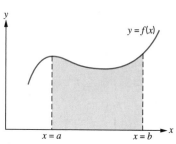

Figure 10

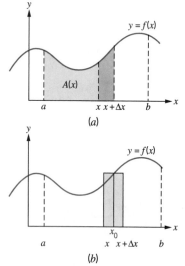

Figure 11

- The area under a curve
- Integrands with negative values
- Area between curves

To introduce the concept of the area under a curve, consider the following example which we will work in two ways.

EXAMPLE 50   Find the area of the shaded region in Figure 9 that is bounded by the line $y = 2x$, the vertical line $x = 2$, and the x-axis.

SOLUTION   The shaded area is a right triangle with base = 2 and altitude = 4, so the area is

$$\text{Area of triangle} = \frac{1}{2} \cdot 2 \cdot 4 = 4$$

Now suppose that we find the definite integral of the function $y = f(x) = 2x$ as x changes from 0 to 2. We have

$$\int_0^2 2x \, dx = x^2 \Big|_0^2 = 2^2 - 0^2 = 4$$

The definite integral of the bounding function over $[0, 2]$ gives the area under the curve from $x = 0$ to $x = 2$.   ∎

Practice Problem 1 Find the area of the region that is bounded by the line $y = 4x$, the vertical line $x = 3$, and the x-axis. Use the area of the triangle formula and then evaluate the definite integral

$$\int_0^3 4x \, dx$$

ANSWER   18

You should always draw the graphs involved in an area problem. Generally, the area under a curve cannot be found with the area formula of a geometric figure. For example, the shaded area of Figure 10 is given by

$$\int_a^b f(x) \, dx$$

Let's investigate now why the definite integral gives the exact area under a curve. In Figure 11(a), let $A(x)$ be the area under the curve from a to x. We show first that $A(x)$ is an antiderivative of $f(x)$, for $A'(x) = f(x)$. Now, $A(x + \Delta x)$ is the area under the curve from a to $x + \Delta x$ in Figure 11(a).

In Figure 11(a), $A(x + \Delta x) - A(x)$ is the darker shaded area. This area can be approximated by a rectangle in Figure 11(b). The width of the rectangle is $(x + \Delta x) - x = \Delta x$ and the height of the rectangle where $x \leq x_0 \leq (x + \Delta x)$ is $f(x_0)$. Thus, the area of the rectangle is approximately $f(x_0) \, \Delta x$. Hence,

$$A(x + \Delta x) - A(x) \approx f(x_0) \, \Delta x$$

This approximation of the area of a rectangle improves as Δx becomes smaller and smaller because the area of the rectangle being calculated will get closer and closer to the actual area. Dividing both sides by Δx gives

$$\frac{A(x + \Delta x) - A(x)}{\Delta x} \approx f(x_0) \frac{\Delta x}{\Delta x} \qquad \Delta x \neq 0$$

$$\lim_{\Delta x \to 0} \frac{A(x + \Delta x) - A(x)}{\Delta x} = \lim_{\Delta x \to 0} f(x_0) \qquad \frac{\Delta x}{\Delta x} = 1$$

$$A'(x) = \lim_{\Delta x \to 0} f(x_0) \qquad \text{Definition of a derivative}$$

$$A'(x) = \lim_{\Delta x \to 0} f(x_0) = f(x) \qquad f(x) \text{ is continuous and } x \leq x_0 \leq (x + \Delta x).$$

Thus, $A(x)$ is an antiderivative of $f(x)$. Hence,

$$A(b) - A(a) = \int_a^b f(x) \, dx \qquad \text{Definition of definite integral}$$

Since $A(b) - A(a)$ represents the area under a curve from $x = a$ to $x = b$,

$$\int_a^b f(x) \, dx$$

is the area under the curve defined by $y = f(x)$ from $x = a$ to $x = b$.

AREA UNDER A CURVE

The area bounded by the continuous function $f(x) \geq 0$ on $a \leq x \leq b$, the x-axis, and the lines $x = a$ and $x = b$ is given by

$$A = \int_a^b f(x) \, dx$$

Later, this area will be expressed as a limit of a sum leading to the fundamental theorem of integral calculus.

EXAMPLE 51 Find the area of the region bounded by $f(x) = 3x^2$, the x-axis, and the lines $x = 2$ and $x = 4$. This region is illustrated in Figure 12.

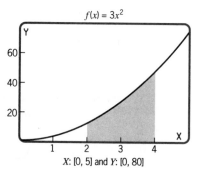

$f(x) = 3x^2$

X: [0, 5] and Y: [0, 80]

Figure 12

SOLUTION This area is

$$\int_a^b f(x)\, dx = \int_2^4 3x^2\, dx = 3 \int_2^4 x^2\, dx$$

$$= 3 \left(\frac{x^3}{3} \right) \Big|_2^4 = x^3 \Big|_2^4$$

$$= 4^3 - 2^3 = 64 - 8 = 56 \qquad \blacksquare$$

Practice Problem 2 Find the area under $y = 2x^2 - x + 1$ from $x = 1$ to $x = 4$. First look at the area on your graphing calculator and make an estimate for the area.

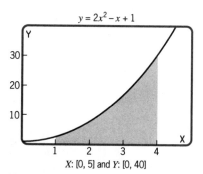

$y = 2x^2 - x + 1$

X: [0, 5] and Y: [0, 40]

Figure 13

ANSWER The answer is 37.5. The graph is shown in Figure 13.

The definite integral for area was given for $f(x) \geq 0$. Now let's look at what happens if the graph of $f(x)$ falls below the x-axis between $x = a$ and $x = b$

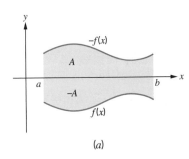

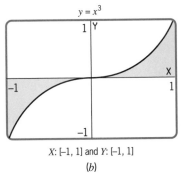

$y = x^3$

X: [-1, 1] and Y: [-1, 1]

(b)

Figure 14

[Figure 14(a)]. The definite integral is the total change of f. If f is below the x-axis, the change is negative, the integral is negative, and the "area" is negative. Therefore, to calculate the area we use $-f(x)$ in our formula and we have

$$\int_a^b f(x)\, dx = -\int_a^b -f(x)\, dx = -A$$

To get the area bounded by the x-axis, where $f(x) \le 0$ on $[a, b]$ and $x = a$ and $x = b$, we take the absolute value of the definite integral:

$$\left| \int_a^b f(x)\, dx \right|$$

Calculator Note

Your graphing calculator will shade an area under a curve and then state the value of $\int f(x)\, dx$ for the function being graphed with the established lower and upper bounds.

However, the calculator function that gives you a value for a definite integral is not calculating area. A definite integral can be negative or zero, but area must be positive. It is important to know when to use the absolute value of a function in calculating the definite integral, especially when you are using a graphing calculator. For example, have your graphing calculator draw x^3 over X:$[-1, 1]$ and Y:$[-1, 1]$ and then use $\int f(x)\, dx$ by using $x = -1$ and $x = 1$ as the lower and upper bounds. The calculator will show

$$\int_{-1}^1 x^3\, dx = 0$$

but this is certainly not the area shaded in Figure 14(b). This is one example of why you must look at the graph and use the properties of integrals to calculate the area correctly using a graphing calculator. To do this, on the home screen enter $\boxed{\text{2nd}}$ $\boxed{\text{ABS}}$ $\boxed{\text{MATH}}$ (select 9) (fnInt($x \wedge 3, x, -1, 0$) + fnInt($x \wedge 3, x, 0, 1$) = 0.5. Also note that when we are working the examples, we will often indicate the answer as a fraction, but your graphing calculator will give answers in decimal form unless you have it convert to fraction form using $\boxed{\text{MATH}}$ (select ▶Frac).

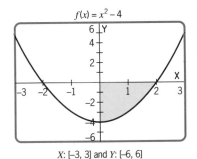

$f(x) = x^2 - 4$

X: [-3, 3] and Y: [-6, 6]

Figure 15

EXAMPLE 52 Find the area bounded by the x-axis and $f(x) = x^2 - 4$ from $x = 0$ to $x = 2$. The graph of the region bounded by this function, the x-axis, and the lines $x = 0$ to $x = 2$ is shaded in Figure 15.

SOLUTION   The bounded area is given by the absolute value of the definite integral:

$$\int_a^b f(x)\, dx = \int_0^2 (x^2 - 4)\, dx = \left(\frac{x^3}{3} - 4x\right)\Big|_0^2$$

$$= \left[\frac{2^3}{3} - 4(2)\right] - \left[\frac{0^3}{3} - 4(0)\right]$$

$$= \frac{8}{3} - 8 = -\frac{16}{3} \approx -5.3$$

Thus, the area of this region is $\left|-\frac{16}{3}\right| = \frac{16}{3}$. ∎

Calculator Note

When you have your calculator shade the area under a function to calculate $\int f(x)\, dx$, you may not be able to set the lower and upper boundaries exactly where you want by using $\boxed{\text{TRACE}}$. You can get around this problem by setting the Xmin and Xmax to the values you want as boundaries for the integration. Now the proper area can be shaded. Be careful, however, because the integral being calculated by the calculator is the definite integral and not the area. You must use absolute value at the proper places to calculate the area. For instance, in Example 52 you could go to the home screen and enter $\boxed{\text{2nd}}$ $\boxed{\text{ABS}}$ $\boxed{\text{MATH}}$ (select 9)fnInt $(x^2 - 4, x, 0, 2)$ $\boxed{\text{MATH}}$ (select ▶Frac) $\boxed{\text{ENTER}}$ and get the result 16/3.

Practice Problem 3 Find the area bounded by the x-axis and $f(x) = x^2 - 4$ from $x = -2$ to $x = 2$.

ANSWER $\dfrac{32}{3} \approx 10.67$

EXAMPLE 53   Find the area of the region bounded by $f(x) = x^2 - 4$ and the x-axis from $x = 2$ to $x = 4$. This area is shaded in Figure 16(a).

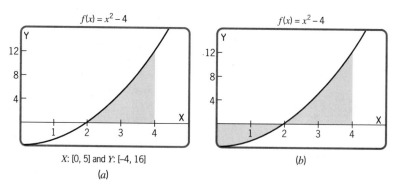

X: [0, 5] and Y: [−4, 16]

(a)

(b)

Figure 16

SOLUTION   Notice that since this area lies above the x-axis, we do not need to use the absolute value to evaluate the area.

$$
\begin{aligned}
\int_2^4 (x^2 - 4)\, dx &= \left(\frac{x^3}{3} - 4x \right) \Bigg|_2^4 \\
&= \left[\frac{4^3}{3} - 4(4) \right] - \left[\frac{2^3}{3} - 4(2) \right] \\
&= \left(\frac{64}{3} - 16 \right) - \left(\frac{8}{3} - 8 \right) \\
&= \frac{16}{3} + \frac{16}{3} = \frac{32}{3}
\end{aligned}
$$

EXAMPLE 54   Find the area of the region bounded by the y-axis, the x-axis, the graph of $f(x) = x^2 - 4$, and the line $x = 4$. This region is shaded in Figure 16(b).

SOLUTION   Since part of the shaded region is below the x-axis and part is above, the area of the total region must be found by finding the area of the region below the x-axis and the area of the region above the x-axis and adding their absolute values. The area of the region below the x-axis was found in Example 52 to be $\left| -\frac{16}{3} \right|$. The area of the region above the x-axis was found in Example 53 to be $\frac{32}{3}$. Therefore, the total area is

$$
\left| -\frac{16}{3} \right| + \left| \frac{32}{3} \right| = \frac{16}{3} + \frac{32}{3} = \frac{48}{3} = 16
$$

NOTE: Notice that the answer in Example 54 could not have been found by evaluating the definite integral

$$
\int_0^4 (x^2 - 4)\, dx = \frac{16}{3}
$$

since the definite integral combines the values $-\frac{16}{3} + \frac{32}{3} = \frac{16}{3}$. With your graphing calculator, you can use absolute value and determine the correct answer.

COMMON ERROR   When calculating areas that lie above and below the x-axis, students often neglect to separate the integrals and to use the absolute value. You should always draw the function being integrated with your graphing calculator so that you can avoid this error.

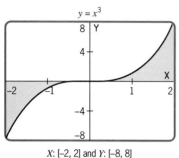

$y = x^3$

X: [-2, 2] and Y: [-8, 8]

Figure 17

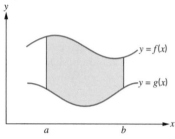

$y = f(x)$

$y = g(x)$

Figure 18

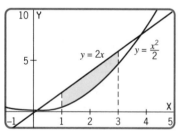

$y = 2x$ $y = \frac{x^2}{2}$

X: [-1, 5] and Y: [-1.4, 10]

Figure 19

Practice Problem 4 On your calculator, shade the area bounded by the graph of $f(x) = x^3$ and the x-axis from $x = -2$ to $x = 2$ and use this graph to help you evaluate the area.

ANSWER The graph is shown in Figure 17. The area = 8.

Up until this point, we have only considered areas that were bounded by a function and the x-axis from one value of x to another. Now let's consider the area bounded by $f(x)$ and $g(x)$, with $f(x) \geq g(x)$, between $x = a$ and $x = b$. Using definite integrals, we determine this area by finding the area under the curve $f(x)$ and subtracting the area under $g(x)$. This situation can be seen in Figure 18. The difference in total change of $f(x)$ and $g(x)$ is the area between the two curves.

$$\text{Area of shaded region} = \text{Area under } f(x) - \text{Area under } g(x)$$
$$= \int_a^b f(x)\,dx - \int_a^b g(x)\,dx$$
$$= \int_a^b [f(x) - g(x)]\,dx \qquad \text{Property of definite integrals}$$

AREA BETWEEN TWO CURVES

If $y = f(x)$ lies above $y = g(x)$, where $f(x) \geq g(x)$, for $a \leq x \leq b$, the area of the region bounded by $f(x)$ and $g(x)$ from $x = a$ to $x = b$ is

$$\int_a^b [f(x) - g(x)]\,dx$$

EXAMPLE 55 Find the area between the curves $y = 2x$ and $y = x^2/2$ from $x = 1$ to $x = 3$ (see Figure 19).

SOLUTION The graph shows the $2x \geq x^2/2$ over [1, 3]; therefore, by the preceding property, the area is given by

$$A = \int_1^3 \left(2x - \frac{x^2}{2}\right) dx = x^2 - \frac{x^3}{6}\Big|_1^3 \qquad \text{Antiderivative}$$
$$= \left(3^2 - \frac{3^3}{6}\right) - \left(1^2 - \frac{1^3}{6}\right) \qquad F(3) - F(1)$$
$$= \left(9 - \frac{9}{2}\right) - \left(1 - \frac{1}{6}\right)$$
$$= 3\tfrac{2}{3} \qquad \blacksquare$$

Calculator Note

How would you calculate the problem in Example 55 using your calculator? It is necessary to look at the graphs to determine which function is above the other over the designated interval. After that is done, you can use the fnInt function by subtracting the lower function from the one above.

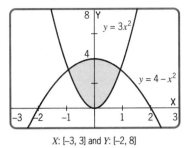

X: [-3, 3] and Y: [-2, 8]

Figure 20

Practice Problem 5 Using your calculator, find the area of the region bounded by the functions in Example 55 from $x = 0$ to $x = 4$.

ANSWER $\dfrac{16}{3}$. Enter fnInt$(2x, x, 0, 4)$ − fnInt$(x^2/2, x, 0, 4)$.

EXAMPLE 56 Find the area of the region bounded by $y = 4 - x^2$ and $y = 3x^2$. This region is illustrated in Figure 20.

SOLUTION Note that these graphs intersect at the points $(1, 3)$ and $(-1, 3)$. You can locate these points with your calculator using the intersect function. However, if the points of intersection given by the calculator are decimal approximations, you can find the exact points of intersection algebraically by setting the two functions equal and solving the resulting equation as we have done below. (Your calculator will convert some decimals to fraction form, but there may be a restriction on how many places it will allow in the denominator.)

$$3x^2 = 4 - x^2$$
$$4x^2 = 4$$
$$x^2 = 1$$
$$x = 1 \quad \text{or} \quad x = -1$$

Using the preceding property, the area is given by

$$A = \int_{-1}^{1} [(4 - x^2) - 3x^2]\, dx = \int_{-1}^{1} (4 - 4x^2)\, dx$$

$$= \left(4x - \frac{4}{3}x^3 \right) \Big|_{-1}^{1} \qquad \text{Antiderivative}$$

$$= \left[4 \cdot 1 - \frac{4}{3} \cdot 1 \right] - \left[4 \cdot (-1) - \frac{4}{3}(-1) \right] \qquad F(1) - F(-1)$$

$$= \frac{16}{3}$$

Calculator Note

It is not necessary to simplify expressions before entering them in the calculator. Thus, an expression such as $4 - x^2 - 3x^2$ is allowed. However, if the second function has more than one term, you will need to use parentheses or distribute the negative sign.

Practice Problem 6 Use a calculator to draw the graphs of $y = x^2 - 2$ and $y = x + 4$. Then find the points of intersection and the area bounded by the two curves.

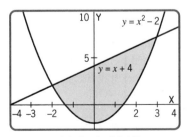

X: [-4, 4] and Y: [-3, 10]

Figure 21

ANSWER As shown in Figure 21, the points of intersection are at $x = -2$ and $x = 3$. Area $= 125/6$.

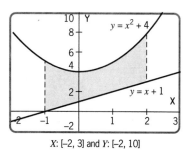

X: [-2, 3] and Y: [-2, 10]

Figure 22

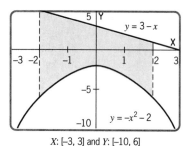

X: [-3, 3] and Y: [-10, 6]

Figure 23

EXAMPLE 57   Find the area of the region bounded by the functions $y = x^2 + 4$ and $y = x + 1$ from $x = -1$ to $x = 2$.

SOLUTION   Before we calculate this area, let's look at the graphs in Figure 22 so that we can accurately set up the definite integral. Since the function $y = x^2 + 4$ lies above $y = x + 1$, we set up and evaluate the integral as follows:

$$A = \int_{-1}^{2} [(x^2 + 4) - (x + 1)]\, dx = \int_{-1}^{2} (x^2 - x + 3)\, dx$$

$$= \frac{x^3}{3} - \frac{x^2}{2} + 3x \Big|_{-1}^{2} \qquad \text{Antiderivative}$$

$$= \left[\frac{2^3}{3} - \frac{2^2}{2} + 3(2) \right] - \left[\frac{(-1)^3}{3} - \frac{(-1)^2}{2} + 3(-1) \right] \qquad F(2) - F(-1)$$

$$= 10\tfrac{1}{2}$$

Practice Problem 7   Use your calculator to draw the graphs of $y = -x^2 - 2$ and $y = 3 - x$. Find the area bounded by these functions from $x = -2$ to $x = 2$.

ANSWER   The graphs are shown in Figure 23. Area = 76/3

EXAMPLE 58   Find the area between the curves $y = x^2$ and $y = 2x$ from $x = 1$ to $x = 3$.

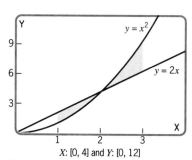

X: [0, 4] and Y: [0, 12]

Figure 24

SOLUTION   From the graph in Figure 24, we can see that the functions intersect and that neither function lies above the other over the entire interval $1 \le x \le 3$. Therefore, we must find the point of intersection within the given interval and use that point to find the area. You can use your calculator to find the point of intersection or you can find the point of intersection algebraically by setting the two functions equal and solving the resulting equation.

$$x^2 = 2x$$

$$x^2 - 2x = x(x - 2) = 0$$

$$x = 0 \quad \text{or} \quad x = 2$$

The point of intersection within the desired interval is at $x = 2$. Notice that in Figure 24, $y = 2x$ lies above $y = x^2$ over the interval $1 \le x \le 2$, and $y = x^2$ lies above $y = 2x$ over the interval $2 \le x \le 3$. Therefore, the area of the region between the two curves is

$$A = \int_1^2 [2x - x^2]\, dx + \int_2^3 [x^2 - 2x]\, dx$$

$$= \left[x^2 - \frac{x^3}{3} \bigg|_1^2 \right] + \left[\frac{x^3}{3} - x^2 \bigg|_2^3 \right] \qquad \text{Antiderivative}$$

$$= \left[2^2 - \frac{2^3}{3} \right] - \left[1^2 - \frac{1^3}{3} \right]$$

$$+ \left[\frac{3^3}{3} - 3^2 \right] - \left[\frac{2^3}{3} - 2^2 \right] \qquad F(b) - F(a) \text{ for each}$$

$$= 4 - \frac{8}{3} - \frac{2}{3} + \frac{4}{3}$$

$$= 2$$

$\blacksquare$

EXAMPLE 59 Find the area bounded by the curves $y = x^3$ and $y = x^2 + 2x$.

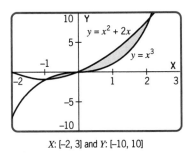

X: [-2, 3] and Y: [-10, 10]

Figure 25

SOLUTION From the graphs shown in Figure 25, we see that the curves intersect at three points. We find the points of intersection algebraically by setting the two functions equal and solving the resulting equation. (They can be found on your calculator using the intersect function.)

$$x^3 = x^2 + 2x$$

$$x^3 - x^2 - 2x = 0$$

$$x(x^2 - x - 2) = x(x - 2)(x + 1) = 0$$

$$x = -1 \quad \text{or} \quad x = 0 \quad \text{or} \quad x = 2$$

Therefore, we set up and solve the integral using the interval $-1 \le x \le 0$ where $y = x^3$ is the top function and the interval $0 \le x \le 2$ where $y = x^2 + 2x$ is the top function.

$$A = \int_{-1}^{0} [x^3 - (x^2 + 2x)] \, dx + \int_{0}^{2} [(x^2 + 2x) - x^3] \, dx$$

$$= \left[\frac{x^4}{4} - \frac{x^3}{3} - x^2 \right]\Big|_{-1}^{0} + \left[-\frac{x^4}{4} + \frac{x^3}{3} + x^2 \right]\Big|_{0}^{2} \qquad \text{Antiderivatives}$$

$$= \left[\left(\frac{0^4}{4} - \frac{0^3}{3} - 0^2 \right) - \left(\frac{(-1)^4}{4} - \frac{(-1)^3}{3} - (-1)^2 \right) \right]$$

$$+ \left[\left(-\frac{2^4}{4} + \frac{2^3}{3} + 2^2 \right) - \left(-\frac{0^4}{4} + \frac{0^3}{3} + 0^2 \right) \right]$$

$$= \frac{37}{12}$$

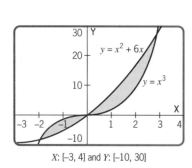

30   Y
20   $y = x^2 + 6x$
10   $y = x^3$
X
-3 -2 -1   1   2   3   4
-10

X: [-3, 4] and Y: [-10, 30]

Figure 26

Practice Problem 8 Use your calculator to draw the graphs of $y = x^3$ and $y = x^2 + 6x$ and find the area bounded by the two curves.

ANSWER The graphs are shown in Figure 26. Area = 253/12

SUMMARY

It is important to draw the graphs of functions when evaluating the area bounded by their curves. There are times when the points of intersection are obvious or when the limits of integration are given. However, there will also be problems for which you must find the point or points of intersection algebraically or use your calculator to find the intersection points. Consequently, you will need a graph to assist you in setting up the definite integrals with proper limits of integration.

Exercise Set 11.6

Find the area of the region bounded by the following curves using techniques of integration. Use your graphing calculator to draw the curves and check your work.

1. $f(x) = 5$, $f(x) = 0$, $x = 2$, $x = 5$
2. $f(x) = x$, $f(x) = 0$, $x = 1$, $x = 3$
3. $f(x) = 2x$, $f(x) = 0$, $x = 2$, $x = 4$
4. $f(x) = x^2$, the x-axis on $1 \le x \le 3$
5. $f(x) = x^3$, the x-axis on $0 \le x \le 2$
6. $f(x) = x^2 - 4$, $f(x) = 0$, $x = -3$, $x = 4$
7. $f(x) = x^2 - x$, $f(x) = 0$, $x = 1$, $x = 3$
8. $f(x) = \sqrt{x}$, the x-axis on $1 \le x \le 4$
9. $f(x) = 1 + \sqrt{x}$, the x-axis on $1 \le x \le 9$

10. $f(x) = \dfrac{1}{\sqrt{x}}$, the x-axis on $1 \le x \le 4$
11. $f(x) = e^{2x}$, the x-axis on $0 \le x \le 2$
12. $f(x) = \dfrac{1}{x}$, the x-axis on $1 \le x \le 2$
13. $f(x) = \dfrac{1}{2x + 1}$, the x-axis on $\dfrac{1}{2} \le x \le 1$
14. $f(x) = 3x$, $f(x) = 1$, $x = 1$, $x = 4$
15. $f(x) = 7x$, $f(x) = 2$, $x = 2$, $x = 3$
16. $f(x) = (x - 2)^2$, the x-axis, $x = 0$, $x = 5$
17. $y = x^2 + 2x + 2$, $y = 2x + 3$
18. $f(x) = x^2$, $f(x) = x$
19. $y = \sqrt{x}$, $x = 9$
20. $y = x^2$, $y = \sqrt{x}$

Find the area of the region bounded by the following curves using either techniques of integration, your graphing calculator, or both methods.

21. $y = 16 - x^2$, $y = 6x$
22. $y = x^2 - 4$, the x-axis, $x = -1$, $x = 2$
23. $y = x^3$, the x-axis, $x = -2$, $x = 1$
24. $y = 2xe^{-x^2}$, the x-axis, $x = 0$, $x = 2$
25. $y = x^2 - 4x$, $y = 2$, $x = 0$, $x = 2$
26. $y = x^2 - 1$, $y = 0$, $x = -1$, $x = 3$
27. $y = 2x + 2$, $y = x^2 + 2$
28. $y = x + 2$, $y = x^2 + x - 7$
29. $y = 3x + 3$, $y = x^2 + 2x + 1$
30. $y = x^2$, $y = x^3 - 6x$
31. $y = x^3 - 3x^2 + 3x$, $y = x^2$
32. $f(x) = 2x^2$, $f(x) = x^4 - 2x^2$
33. $y = -x^2 + 4x + 3$, $y = x^2 - 4x + 3$
34. $y = x^2 - x$, $y = 2x + 4$

35. $y = 2x^2$, $y = x^3 - 8x$
36. $y = x + 4$, $y = -x^2 - 2$, $x = -2$, $x = 4$
37. $y = x - 4$, $y = -x^2 + 3$, $x = -2$, $x = 2$
38. $y = x^2 + 3$, $y = -x^2 - 3$, $x = -1$, $x = 2$
39. $y = 2x^2 + 1$, $y = 2 - x^2$, $x = -5$, $x = 5$

Review Exercises

Find the following indefinite integrals.

40. $\displaystyle\int xe^{-x^2}\,dx$
41. $\displaystyle\int x^2 e^{-x^3}\,dx$

Find the value of the following definite integrals using your graphing calculator.

42. $\displaystyle\int_0^3 (9x - x^3)^2\,dx$
43. $\displaystyle\int_2^3 \frac{2x\,dx}{(1 + x^2)^2}$

11.7 RIEMANN SUMS AND THE DEFINITE INTEGRAL (Optional)

OVERVIEW   In the preceding section, we used the definite integral

$$\int_a^b f(x)\,dx$$

to find the area between the x-axis and $y = f(x) > 0$ from $x = a$ to $x = b$. In this section, we look at an alternative approach to the definition of a definite integral. This procedure was developed by George Bernhard Riemann, a nineteenth century mathematician. In this section we

- Approximate the area under a curve using the sum of the areas of approximating rectangles
- Define the definite integral as the limit of such sums as the number of rectangles increases without bound
- Use the definition of the definite integral as an infinite sum to give meaning to such applications as the average value of a function over a closed interval

First, we consider how to approximate the area under a curve by dividing it into subregions that are *almost* rectangles. In Figure 27(a), $[a, b]$ has been divided into four subintervals of equal width, Δx.

$$\Delta x = \frac{b - a}{4}$$

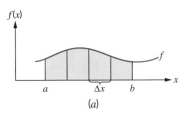

(a)

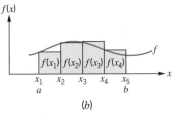

(b)

Figure 27

Note the four regions that are *almost* rectangles.

In Figure 27(b), we approximate the area under the curve with rectangles. We select the left edge of each subinterval (we could have selected the right edge or any point in the subinterval) and use $f(x_1), f(x_2), f(x_3)$, and $f(x_4)$ as the heights of four rectangles. The sum of the areas of these four rectangles approximates the area under the curve. Thus, the area under the curve is approximately

$$f(x_1) \, \Delta x + f(x_2) \, \Delta x + f(x_3) \, \Delta x + f(x_4) \, \Delta x$$

We can simplify this by using summation notation

$$\sum_{i=1}^{4} f(x_i) \, \Delta x = f(x_1) \, \Delta x + f(x_2) \, \Delta x + f(x_3) \, \Delta x + f(x_4) \, \Delta x$$

For notational purposes we will use the following definitions:

1.   Let s_{min} be defined as the sum of the areas of the rectangles formed using the minimum value of $f(x)$ for x on the base of each rectangle.
2.   Let s_{max} be defined as the sum of the areas of the rectangles formed using the maximum value of $f(x)$ for x on the base of each rectangle.

EXAMPLE 60   Approximate the area bounded by $y = x^2$, the x-axis, $x = 1$, and $x = 3$ using eight intervals.

SOLUTION   The graph of $y = x^2$ is shown in Figure 28. The width of each rectangle is Δx and

$$\Delta x = \frac{3 - 1}{8} = \frac{1}{4} = 0.25$$

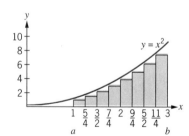

Figure 28

so the sides of rectangles occur at

$$x = 1, \frac{5}{4}, \frac{6}{4}, \frac{7}{4}, 2, \frac{9}{4}, \frac{10}{4}, \frac{11}{4}, 3$$

To obtain s_{min} we use the minimum height of the function, which occurs at the left endpoint in each interval, to find the area of each rectangle. The area of the first rectangle is $f(1)$ times the width Δx; the area of the second rectangle is $f(\frac{5}{4})$ times the width Δx. Thus,

$$s_{\min} = f(1)\,\Delta x + f\left(\frac{5}{4}\right)\Delta x + f\left(\frac{6}{4}\right)\Delta x + f\left(\frac{7}{4}\right)\Delta x$$

$$+ f(2)\,\Delta x + f\left(\frac{9}{4}\right)\Delta x + f\left(\frac{10}{4}\right)\Delta x + f\left(\frac{11}{4}\right)\Delta x$$

$$= 1\left(\frac{1}{4}\right) + \frac{25}{16}\left(\frac{1}{4}\right) + \frac{9}{4}\left(\frac{1}{4}\right) + \frac{49}{16}\left(\frac{1}{4}\right) \qquad \Delta x = \frac{1}{4}$$

$$+ 4\left(\frac{1}{4}\right) + \frac{81}{16}\left(\frac{1}{4}\right) + \frac{25}{4}\left(\frac{1}{4}\right) + \frac{121}{16}\left(\frac{1}{4}\right)$$

$$= \left(\frac{16 + 25 + 36 + 49 + 64 + 81 + 100 + 121}{16}\right)\left(\frac{1}{4}\right)$$

$$\approx 7.69$$

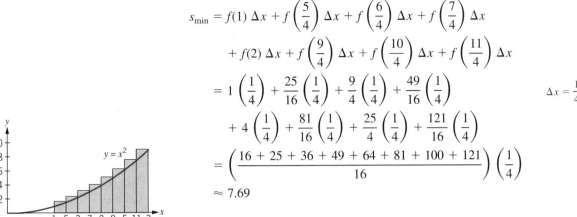

y = x²

Figure 29

The graph of $y = x^2$ and associated rectangles using the maximum heights of the rectangles (here the right endpoints of each interval) is shown in Figure 29. We have

$$s_{\max} = f\left(\frac{5}{4}\right)\Delta x + f\left(\frac{6}{4}\right)\Delta x + f\left(\frac{7}{4}\right)\Delta x + f(2)\,\Delta x$$

$$+ f\left(\frac{9}{4}\right)\Delta x + f\left(\frac{10}{4}\right)\Delta x + f\left(\frac{11}{4}\right)\Delta x + f(3)\,\Delta x$$

$$= \frac{25}{16}\left(\frac{1}{4}\right) + \frac{9}{4}\left(\frac{1}{4}\right) + \frac{49}{16}\left(\frac{1}{4}\right) + 4\left(\frac{1}{4}\right)$$

$$+ \frac{81}{16}\left(\frac{1}{4}\right) + \frac{25}{4}\left(\frac{1}{4}\right) + \frac{121}{16}\left(\frac{1}{4}\right) + 9\left(\frac{1}{4}\right)$$

$$= \left(\frac{25 + 36 + 49 + 64 + 81 + 100 + 121 + 144}{16}\right)\left(\frac{1}{4}\right)$$

$$\approx 9.69$$

The actual area is a number between $s_{\min}$ and $s_{\max}$, so we know that

$$7.69 \le A \le 9.69 \qquad \blacksquare$$

From the preceding discussion, two facts become evident about the approximation of the area under a curve using rectangles.

1. If a point is taken to get a height of each rectangle so that there is about as much area of the rectangle above the curve as there is missing below the curve, each rectangle gives a better approximation for the area under the curve (see Figure 30).

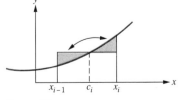

Figure 30

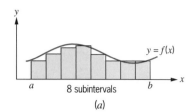

(a)

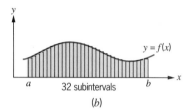

(b)

Figure 31

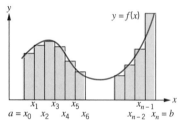

Figure 32

2. As the number of rectangles increases or as Δx gets smaller and smaller, the sum of the areas of the rectangles gets closer and closer to the area under the curve (see Figure 31).

Practice Problem 1 On your graphing calculator, draw the graph of $y = x^2$, using $2 \le x \le 3$ and $4 \le y \le 9$.

(a) What characteristic of the curve do you see on this portion of the graph?
(b) If the endpoints for s_{min} are at $x = 2, 2.2, 2.4, 2.6, 2.8$ and the endpoints for s_{max} are at $x = 2.2, 2.4, 2.6, 2.8, 3$, calculate bounds for the area using rectangles of width 0.2 and interpret this result.

ANSWER

(a) The curve appears to be almost linear.
(b) $s_{min} = 5.84$ and $s_{max} = 6.84$. The actual area is somewhere between these values.

Now let's consider an area A bounded by $y = f(x)$, the x-axis, $x = a$, and $x = b$ as shown in Figure 32. This time let's divide the interval $[a, b]$ into n subintervals that may or may not be of equal length. Let $\Delta x_1 = x_1 - x_0$ and let c_1 be a point where $x_0 \le c_1 \le x_1$ such that $f(c_1) \Delta x_1$ gives a fairly good approximation for the area under the curve from x_0 to x_1. Likewise, $f(c_2) \Delta x_2$, where $\Delta x_2 = x_2 - x_1$, gives a good approximation for the area from x_1 to x_2. After forming n such rectangles, the area under the curve can be expressed as

$$A \approx f(c_1) \Delta x_1 + f(c_2) \Delta x_2 + \cdots + f(c_n) \Delta x_n$$

where $x_{i-1} \le c_i \le x_i$. Of course, as n gets larger (with the understanding that as n increases the largest Δx_i decreases), this expression gets closer and closer to the exact area under the curve. This discussion leads to the following definition.

THE DEFINITE INTEGRAL USING RIEMANN SUMS

Assume that $y = f(x)$ is a function defined on $a \le x \le b$ and that

1. $a = x_0 \le x_1 \le \cdots \le x_{n-1} \le x_n = b$
2. $\Delta x_i = x_i - x_{i-1}$ for $i = 1, 2, \ldots n$
3. $\Delta x_i \to 0$ as $n \to \infty$
4. $x_{i-1} \le c_i \le x_i$ for $i = 1, 2, \ldots n$

If

$$\lim_{n \to \infty} [f(c_1) \Delta x_1 + f(c_2) \Delta x_2 + \cdots + f(c_n) \Delta x_n]$$

exists, then this limit is called the **definite integral** of $f(x)$ from a to b. Symbolically, this is written as

$$\lim_{n \to \infty} \sum_{i=1}^{n} f(c_i) \Delta x_i = \int_{a}^{b} f(x) \, dx$$

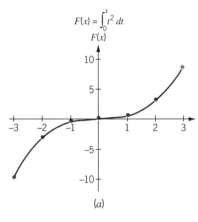

$$F(x) = \int_0^x t^2 \, dt$$

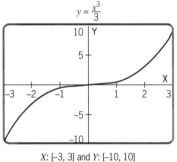

$$y = \frac{x^3}{3}$$

X: [−3, 3] and Y: [−10, 10]

(b)

Figure 33

Calculator Note

You should be aware that

$$F(x) = \int_0^x t^2 \, dt$$

is really the antiderivative of x^2, namely $x^3/3$. On your calculator, evaluate $F(x)$ at several points. Then, on paper, plot the points and connect them to make a graph as in Figure 33(a). Compare your graph with the graph of $y = x^3/3$ in Figure 33(b). Do these two graphs appear to be the same? We have plotted the points $F(-3) = -9$, $F(-2) = \frac{-8}{3}$, $F(-1) = \frac{-1}{3}$, $F(0) = 0$, $F(1) = \frac{1}{3}$, $F(2) = \frac{8}{3}$, $F(3) = 9$.

Practice Problem 2 Use the technique described in the preceding Calculator Note to plot the graph of

$$F(x) = \int_0^x t^3 \, dt$$

Then plot the graph of $y = x^4/4$. Are they the same graph?

ANSWER After plotting $F(x)$ for $x = -2, -1, 0, 1, 2$ to obtain the graph in Figure 34(a) and comparing this to the calculator graph of $y = x^4/4$ in Figure 34(b), the graphs appear to be the same.

Recall from the preceding section that if $F(x)$ is an antiderivative of $f(x)$, then

$$F(b) - F(a) = \int_a^b f(x) \, dx$$

Note that for $f(x) \geq 0$, $F(b) - F(a)$ represents the area under a curve from $x = a$ to $x = b$. Since

$$\lim_{n \to \infty} [f(c_1) \, \Delta x_1 + f(c_2) \, \Delta x_2 + \cdots + f(c_n) \, \Delta x_n]$$

also represents the area under a curve from $x = a$ to $x = b$, these results can be combined to give the **fundamental theorem of integral calculus.**

FUNDAMENTAL THEOREM OF INTEGRAL CALCULUS

If $F(x)$ is any antiderivative of $f(x)$ over $a \leq x \leq b$, then

$$\lim_{n \to \infty} [f(c_1) \, \Delta x_1 + f(c_2) \, \Delta x_2 + \cdots + f(c_n) \, \Delta x_n] = \int_a^b f(x) \, dx$$
$$= F(b) - F(a)$$

where $x_{i-1} \leq c_i \leq x_i$ for $i = 1$ to n.

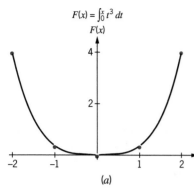

$F(x) = \int_0^x t^3\, dt$

(a)

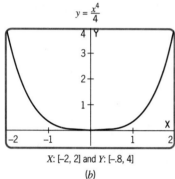

$y = \frac{x^4}{4}$

X: [-2, 2] and Y: [-.8, 4]

(b)

Figure 34

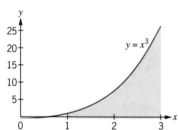

$y = x^3$

Figure 35

This theorem provides a way to evaluate a definite integral. The notation

$$\int_a^b f(x)\, dx = F(x)\ \Big|_a^b = F(b) - F(a)$$

is helpful. Also, notice that it is not necessary to include the constant of integration since

$$\int_a^b f(x)\, dx = [F(b) + C] - [F(a) + C] = F(b) - F(a)$$

EXAMPLE 61　Consider $f(x)$ on $[1, 3]$ and let $\Delta x = (3 - 1)/n$. Find the value of

$$\lim_{n \to \infty} \sum_{i=1}^n x_i^3\, \Delta x$$

as a definite integral.

SOLUTION　As $n \to \infty$, $\Delta x = (3 - 1)/n \to 0$. Furthermore, x_i^3 is the value of $f(x)$ for a value of x on the ith interval. Therefore, $f(x)$ could be x^3 (see Figure 35). Thus, we have

$$\lim_{n \to \infty} \sum_{i=1}^n x_i^3\, \Delta x = \int_1^3 x^3\, dx = \frac{x^4}{4}\ \Big|_1^3$$

$$= \frac{3^4}{4} - \frac{1^4}{4}$$

$$= \frac{80}{4} = 20$$

Practice Problem 3　Consider $f(x) = x^3$ on $[2, 4]$ and let $\Delta x = (4 - 2)/n$. Find the value of

$$\lim_{n \to \infty} \sum_{i=1}^n x_i^3\, \Delta x$$

as a definite integral. To reinforce your work visually, draw $f(x) = x^3$ on your calculator using $2 \le x \le 4$ and $0 \le y \le 64$.

ANSWER   60. The graph is shown in Figure 36.

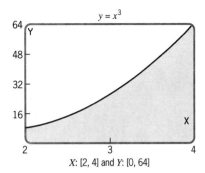

X: [2, 4] and Y: [0, 64]

Figure 36

Average Value

To illustrate one of the many ways that the Riemann sum definition of a definite integral can be used in application problems, let's consider the problem of finding the average value for a function over a desired interval. Let $y = f(x)$ and let $\bar{y}$ denote the average value of $y_1, y_2, \ldots, y_n$. That is,

$$\bar{y} = \frac{y_1 + y_2 + y_3 + \cdots + y_n}{n} = \frac{1}{n}[f(x_1) + f(x_2) + \cdots + f(x_n)]$$

This expression for $\bar{y}$ gives the average value of the function for n values. To find the average value of a continuous function over a closed interval $[a, b]$, it seems natural to generalize this expression for $\bar{y}$ by taking the limit of the expression as n increases without bound.

$$\bar{y} = \lim_{n \to \infty} \frac{1}{n}[f(x_1) + f(x_2) + \cdots + f(x_n)], \qquad y_i = f(x_i), \quad i = 1, 2, \ldots, n$$

To use a definite integral to evaluate this limit, we multiply the numerator and denominator by $b - a$ and then use summation and limit properties to put the expression in the following form:

$$\bar{y} = \lim_{n \to \infty} \left(\frac{1}{n}\right)\left(\frac{b - a}{b - a}\right)[f(x_1) + f(x_2) + \cdots + f(x_n)]$$

$$= \frac{1}{b - a} \lim_{n \to \infty} [f(x_1) + f(x_2) + \cdots + f(x_n)]\left(\frac{b - a}{n}\right)$$

Now let $\Delta x = (b - a)/n$ and substitute this value in the preceding equation.

$$\bar{y} = \frac{1}{b-a} \lim_{n\to\infty} [\, f(x_1)\,\Delta x + f(x_2)\,\Delta x + \cdots + f(x_n)\,\Delta x\,]$$

$$= \frac{1}{b-a} \lim_{n\to\infty} \sum_{i=1}^{n} f(x_i)\,\Delta x$$

$$= \frac{1}{b-a} \int_a^b f(x)\,dx$$

AVERAGE VALUE

If $y = f(x)$ is a continuous function over $a \le x \le b$, then the **average value** $\bar{y}$ of the function over $a \le x \le b$ is

$$\bar{y} = \frac{1}{b-a} \int_a^b f(x)\,dx$$

EXAMPLE 62 The average cost of a certain auto part is expected to increase over the next 2 years. Find the average cost of the part over the next 2 years if $C(x) = 8 + x^2$.

SOLUTION

$$\bar{y} = \frac{1}{b-a} \int_a^b f(x)\,dx = \frac{1}{2-0} \int_0^2 (8 + x^2)\,dx$$

$$= \frac{1}{2}\left[8x + \frac{x^3}{3} \right]\Bigg|_0^2 = \frac{1}{2}\left[8(2) + \frac{2^3}{3} \right] - \frac{1}{2}\left[8(0) + \frac{0^3}{3} \right]$$

$$= \frac{1}{2}\left(16 + \frac{8}{3} \right) - \frac{1}{2}(0 + 0) = \frac{1}{2}\left(\frac{48}{3} + \frac{8}{3} \right) - 0$$

$$= \frac{1}{2}\left(\frac{56}{3} \right) = \frac{28}{3}$$

As shown in Figure 37, the average value is 28/3 over the interval [0, 2]. ■

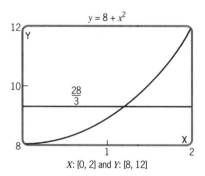

$y = 8 + x^2$

$\frac{28}{3}$

X: [0, 2] and Y: [8, 12]

Figure 37

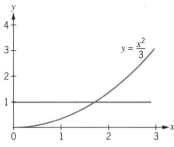

y = $\frac{x^2}{3}$

Average value is 1 over the interval [0, 3]

Figure 38

Practice Problem 4 Find the average of $f(x) = x^2/3$ from $x = 0$ to $x = 3$. Draw the graph of f on $[0, 3]$ and a horizontal line at the average value.

ANSWER $\dfrac{1}{3-0} \displaystyle\int_0^3 \dfrac{x^2}{3}\, dx = 1$. The graph is shown in Figure 38.

EXAMPLE 63   Suppose that the temperature, in Celsius degrees, t hours after midnight at a certain station, is given by the function

$$C(t) = t^3 - 3t + 15, \qquad 0 \le t \le 3$$

(a) What was the average temperature over this period?
(b) What was the average temperature over $1 \le t \le 3$?

SOLUTION

(a)
$$\overline{C} = \frac{1}{b-a}\int_a^b f(x)\,dx = \frac{1}{3-0}\int_0^3 (t^3 - 3t + 15)\,dt$$

$$= \frac{1}{3}\left[\frac{t^4}{4} - \frac{3t^2}{2} + 15t\right]\Bigg|_0^3 = \frac{1}{3}\left[\frac{3^4}{4} - \frac{3}{2}(3)^2 + 15(3)\right]$$

$$= \frac{1}{3}\left(\frac{81}{4} - \frac{27}{2} + 45\right) = \frac{1}{3}\left(\frac{27}{4} + \frac{180}{4}\right) = \frac{1}{3}\left(\frac{207}{4}\right)$$

$$= \frac{69}{4} = 17.25°C$$

(b)
$$\overline{C} = \frac{1}{b-a}\int_a^b f(x)\,dx = \frac{1}{3-1}\int_1^3 (t^3 - 3t + 15)\,dt$$

$$= \frac{1}{2}\left[\frac{t^4}{4} - \frac{3t^2}{2} + 15t\right]\Bigg|_1^3$$

$$= \frac{1}{2}\left[\frac{3^4}{4} - \frac{3}{2}(3)^2 + 15(3)\right] - \frac{1}{2}\left[\frac{1^4}{4} - \frac{3}{2}(1)^2 + 15(1)\right]$$

$$= \frac{1}{2}\left(\frac{81}{4} - \frac{27}{2} + 45\right) - \frac{1}{2}\left(\frac{1}{4} - \frac{3}{2} + 15\right)$$

$$= \frac{207}{8} - \frac{1}{2}\left(\frac{1}{4} - \frac{6}{4} + \frac{60}{4}\right) = \frac{207}{8} - \frac{1}{2}\left(\frac{55}{4}\right)$$

$$= \frac{207}{8} - \frac{55}{8} = \frac{152}{8} = 19°C$$

Calculator Note

There are functions, such as

$$f(x) = e^{-x^2} \quad \text{and} \quad f(x) = \frac{\log(1 + 2x)}{x}$$

that do not have antiderivatives. However, we can obtain values of definite integrals and graphical antiderivatives.

SUMMARY

Evaluating definite integrals using Riemann sums is more rigorous than what was done in Section 11.6. Even though it is not necessary to calculate a Riemann sum in order to evaluate a definite integral, you should work through several problems to reinforce your understanding of the definite integral. Also, being able to find the average value of a function is useful in many application areas such as the average cost of equipment.

Exercise Set 11.7

1. Use Riemann sums to approximate

$$\int_{7}^{37} f(x)\, dx$$

where $f(x)$ is defined by the following table. (**Hint:** $\Delta x = (37 - 7)/5 = 6$.)

x	10	16	22	28	34
$f(x)$	14	20	26	32	38

2. Use Riemann sums to approximate

$$\int_{1}^{11} f(x)\, dx$$

where $f(x)$ is defined by the following table. (**Hint:** $\Delta x = (11 - 1)/5 = 2$.)

x	2	4	6	8	10
$f(x)$	2	8	18	32	50

3. Find an approximation of the area of the region bounded above by $f(x)$ and below by the x-axis. Use the rectangles as constructed in the figure. (**Hint:** The curve goes through the middle of the top of each rectangle.)

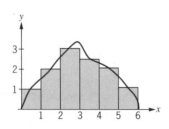

4. Given $f(x) = x^2$, compute the Riemann sum of f over the interval [0, 1] by partitioning the interval into five subintervals of the same length.

5. Repeat Exercise 4 using 10 subintervals of the same length.

Find the average value of each of the following functions over the indicated interval. Draw each function with your graphing calculator and try to estimate the average value before you actually evaluate it.

6. $f(x) = 5, \quad 1 \le x \le 4$
7. $f(x) = x, \quad 2 \le x \le 4$
8. $f(x) = 10 - 2x, \quad 3 \le x \le 4$
9. $f(t) = 3t^2, \quad 0 \le t \le 4$
10. $c(t) = t^2 - t + 3, \quad 0 \le t \le 2$
11. $f(x) = 3x^2 - 2x + 1, \quad 1 \le x \le 3$

Find an approximation for the areas of the following regions using the midpoint of each interval as the point at which to evaluate $f(x)$.

12. $y = 3x + 4$, [1, 5], $n = 4$
13. $y = \dfrac{x^2}{4}$, [1, 4], $n = 3$
14. $y = x^2 - 1$, [1, 7], $n = 6$
15. $y = e^x$, [0, 4], $n = 8$
16. $y = e^x$, [0, 2], $n = 4$

For F(x), calculate at least five points with your graphing calculator, plot the points on paper, and draw the graph of F. Then have your graphing calculator draw the graph of $f(x)$ to verify your work.

17. $F(x) = \displaystyle\int_{0}^{x} \sqrt{t}\, dt, \quad f(x) = \dfrac{2}{3}x^{3/2}$

18. $F(x) = \int_0^x e^{2t} \, dt, \quad f(x) = \dfrac{1}{2} e^{2x}$

19. $F(x) = \int_0^x (t^3 - 2t) \, dt, \quad f(x) = \dfrac{x^4}{4} - x^2$

20. $F(x) = \int_0^x \dfrac{1}{t} \, dt, \quad f(x) = \ln x \quad (x > 0)$

21. **Exam.** Which of the following is the average value of $1/x$ for $e \le x \le e^e$?

   (a) $\dfrac{2(e^{e-1} - 1)}{e^e}$ (b) $2(e^{e-1} - 1)$

   (c) $\dfrac{e - 1}{e^e - e}$ (d) $\dfrac{e - 1}{2}$

   (e) $e - 1$

Applications (Business and Economics)

22. **Cost of Production.** A corporation finds the marginal cost of a product at various production levels to be

Production (units)	20	24	28	32	36
Marginal Cost ($)	300	310	320	300	240

   Approximate the cost of increasing production from 18 to 38 units (use midpoints to obtain intervals). [**Hint**: $\Delta x = (38 - 18)/5 = 4$.]

23. **Increase in Production.** Suppose that production at a new company is increasing monthly at a rate of

   $$p'(t) = 100e^{0.5t}$$

   How much has production increased from the sixth

to the twelfth month? Approximate using the midpoint of six intervals.

Applications (Social and Life Sciences)

24. **Concentration of a Drug.** The amount of a certain drug in the body of a patient after t hours is

   $$A(t) = 6e^{-0.4t}$$

   Find the average amount of the drug present over a 3-day period after the drug has been administered.

Review Exercises

Find each indefinite integral.

25. $\displaystyle\int \dfrac{2x^2 + x - 3}{\sqrt{x}} \, dx$ 26. $\displaystyle\int \sqrt{x}(x - 1)^2 \, dx$

27. $\displaystyle\int \dfrac{e^x}{e^x + 1} \, dx$ 28. $\displaystyle\int 3x^2 - 2e^{4x} \, dx$

29. $\displaystyle\int \dfrac{5x}{\sqrt{x + 2}} \, dx$ 30. $\displaystyle\int \dfrac{3x - 1}{x + 1} \, dx$

Evaluate.

31. $\displaystyle\int_0^3 \dfrac{x + 1}{x - 4} \, dx$ 32. $\displaystyle\int_2^3 \dfrac{2x}{x^2 + 1} \, dx$

Find the area bounded by the following curves.

33. $f(x) = x^2, \quad f(x) = 0, \quad 4 \le x \le 7$

34. $f(x) = -x^2, \quad f(x) = x$

11.8 APPLICATIONS OF THE DEFINITE INTEGRAL

OVERVIEW   The number of application problems that can be solved using integral calculus is inexhaustible. Many interesting applications in economics involve what we will call consumers' surplus and producers' surplus. Other interesting problems involve continuous money flow. We also consider problems that determine who gets what proportion of the income in a company or in a country. To solve such problems we learn to compute a coefficient of inequality. In this section, we use definite integrals to find

- Consumers' surplus
- Producers' surplus

- The amount of a continuous money flow
- The coefficient of inequality

Figure 39(*a*) shows a demand curve for a product. The equation $p = D(x)$ indicates the price p per unit at which consumers will demand (or purchase) x units. Of course, the higher the price the smaller the demand, and the lower the price the greater the demand. Figure 39(*b*) shows a supply curve for a given product. The equation $p = S(x)$ indicates the price p per unit at which the manufacturer will supply x units. The greater the price the more units the manufacturer will supply. In Figure 39(*a*) at price p_1, customers demand x_1 units. Likewise, at price p_1, manufacturers supply x_1 units.

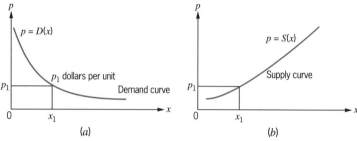

(*a*)   (*b*)

Figure 39

As we can see in Figure 39, some consumers are willing to pay more than the price of p_1 dollars per unit for the product. The consumers who would have paid higher prices have saved money. However, when an industry sets the market value at p_1, it loses the revenue of those who would have paid a higher price. The revenue received at price p_1 is $R = p_1 x_1$ (the area of the rectangle in Figure 40). The income that could have been received is the area under the curve from $x = 0$ to $x = x_1$. The loss of income is the shaded area. Since the loss by the industry is a gain by the consumer, the shaded area is called the **consumers' surplus**, *CS*, at market price p_1. Since this shaded area is the area between two curves, $p = D(x)$ and $p = p_1$, we have

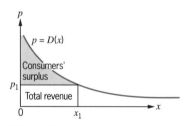

Figure 40

$$CS = \int_0^{x_1} [D(x) - p_1]\, dx$$

CONSUMERS' SURPLUS

If (x_1, p_1) is a point on the price–demand curve for $p = D(x)$, then the **consumers' surplus** at p_1 is

$$CS = \int_0^{x_1} [D(x) - p_1]\, dx$$

EXAMPLE 64   Find the consumer's surplus at a price level of $5 for the demand function

$$p = D(x) = 10 - \frac{5}{6}x$$

SOLUTION   Find the demand for given price, $p_1 = 5$.

$$5 = 10 - \frac{5}{6}x_1 \qquad \text{Replace } p_1 \text{ with 5.}$$

$$-30 = -5x_1 \qquad \text{Multiply by 6.}$$

$$x_1 = 6$$

Copy the calculator graph and shade in the area of consumers' surplus (see Figure 41).

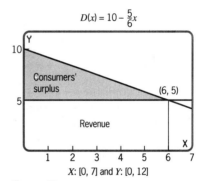

Figure 41

Find the consumers' surplus (the shaded region in Figure 41).

$$CS = \int_0^6 \left[10 - \frac{5}{6}x - 5 \right] dx \qquad \text{Substitute in formula.}$$

$$= 5 \int_0^6 \left(1 - \frac{x}{6} \right) dx$$

$$= 5 \left(x - \frac{x^2}{12} \right) \bigg|_0^6 = 5(6 - 3) = 15 \qquad \text{Definite integral}$$

EXAMPLE 65   Solve Example 64 at price level $4.

SOLUTION   When $p_1 = \$4$, x_1 is obtained from

$$4 = 10 - \frac{5}{6}x_1 \qquad \text{Replace } p_1 \text{ with } 4.$$

$$24 - 60 = -5x_1 \qquad \text{Multiply by 6.}$$

$$-36 = -5x_1$$

$$x_1 = \frac{36}{5} \qquad \text{Solve for } x_1.$$

So the consumer's surplus is

$$CS = \int_0^{36/5} \left[10 - \frac{5}{6}x - 4 \right] dx \qquad \text{Substitute in formula.}$$

$$= \int_0^{36/5} \left[6 - \frac{5}{6}x \right] dx$$

$$= 6x - \frac{5}{12}x^2 \Big|_0^{36/5} \qquad \text{Definite integral}$$

$$= 6 \cdot \frac{36}{5} - \frac{5}{12}\left(\frac{36}{5} \right)^2$$

$$= \frac{108}{5} = 21.6$$

We see that for a $1 decrease in price, the consumers' surplus increases from 15 to 21.6, an increase of 44%. ◼

For a given market price, say p_1, some producers are willing to sell a product for less than this price. These producers actually gain by selling at the given market price. In Figure 42, the rectangle gives the total revenue at fixed price p_1, and the area under the curve gives the revenue with the price varying with the price–supply curve. The additional money that the producers gain from the higher price is called the **producers' surplus**, PS (the dark shaded region in Figure 42), and can be expressed in terms of a definite integral, as we did with consumers' surplus. This integral gives the area between the two curves $p = S(x)$ and $p = p_1$ from $x = 0$ to $x = x_1$.

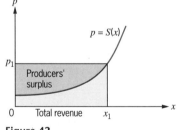

Figure 42

PRODUCERS' SURPLUS

If (x_1, p_1) is a point on the price–supply curve representing $p = S(x)$, then the **producers' surplus** at price p_1 is

$$PS = \int_0^{x_1} [p_1 - S(x)] \, dx$$

EXAMPLE 66 Find the producers' surplus at a price level of $5 for the price–supply equation

$$p = S(x) = \frac{x}{2} + 2$$

SOLUTION Find the supply x_1 when the price is $5.

$$5 = \frac{x_1}{2} + 2 \qquad P_1 = 5$$

$$x_1 = 6 \qquad \text{Solve for } x_1.$$

Copy the calculator graph and shade in the area of producers' surplus (see Figure 43).

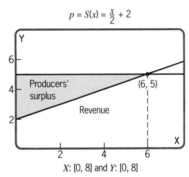

$$p = S(x) = \frac{x}{2} + 2$$

Producers' surplus

(6, 5)

Revenue

X: [0, 8] and Y: [0, 8]

Figure 43

Now find the producers' surplus.

$$PS = \int_0^6 \left[5 - \left(\frac{x}{2} + 2 \right) \right] dx \qquad \text{Substitute in formula.}$$

$$= 3x - \frac{x^2}{4} \Big|_0^6 \qquad \text{Definite integral}$$

$$= 18 - 9 = 9$$

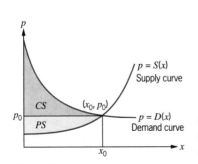

Figure 44

Typical demand and supply curves are shown in Figure 44. The supply curve is usually increasing and the demand curve decreasing. In general, there is a tendency for supply and demand to balance. Let (x_0, p_0) be the point where these two curves intersect. This point where the supply is equal to the demand is called the **equilibrium point**, and p_0 is called the **equilibrium price**. The regions representing consumers' surplus and producers' surplus are shown in the Figure 44.

EXAMPLE 67   Sketch the calculator graphs of the price–demand function

$$p = D(x) = 10 - \frac{5x}{6}$$

and the price–supply function $p = S(x) = \frac{x}{2} + 2$ on the same coordinate system. Locate graphically and find algebraically the equilibrium point. Shade with lighter shading the region representing the producers' surplus and with darker shading the region representing the consumers' surplus.

SOLUTION

$$D(x) = S(x) \qquad \text{Solve for } x_0.$$

$$10 - \frac{5x_0}{6} = \frac{x_0}{2} + 2 \qquad \text{Multiply by 6.}$$

$$60 - 5x_0 = 3x_0 + 12$$

$$48 = 8x_0$$

$$x_0 = 6$$

$$p_0 = \frac{x_0}{2} + 2 = \frac{6}{2} + 2 = 5 \qquad \text{Find } p_0.$$

So $(6, 5)$ is the equilibrium point. The shaded regions for the producers' surplus and consumers' surplus are shown in Figure 45. The actual computations of consumers' surplus and producers' surplus are found in Examples 64 and 66.   ■

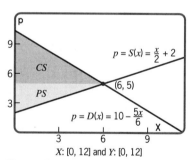

Figure 45

Practice Problem 1 With your graphing calculator, graph on the same coordinate system $y = 8 - x$ (demand function) and $y = x^2 + 2$ (supply function) and locate the point of intersection, y_0. Then graph $y = y_0$. Can you locate the region of consumers' surplus and producers' surplus? Calculate *CS* and *PS*.

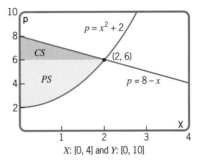

Figure 46

ANSWER   The graph is shown in Figure 46. The equilibrium point is $(2, 6)$: $CS = 2$; and $PS = \frac{16}{3}$.

Continuous Money Flow

A viable business will continually receive income from its activities. Most of the money a business receives is reinvested. For our purposes, we will assume that the income is received continuously over a year and is immediately reinvested at a certain rate of interest compounded continuously. This can be thought of as a continuous money flow.

If we select an initial time to be $t = 0$, then at a later time t, we define the rate of the money flowing in to be

$$\text{Rate of flow} = f(t)$$

This money is immediately invested at rate r and compounded continuously $[f(t)e^{rt}]$. Now, if $A(t)$ is the total amount of money at time t (t can be any unit of time: day, month, year), then

$$\frac{dA}{dt} = f(t)e^{rt} \quad \text{and} \quad A(T) - A(0) = \int_0^T f(t)e^{rt}\,dt$$

AMOUNT OF A CONTINUOUS MONEY FLOW

If $f(t)$ is the rate of flow of money into an investment, then the accumulated value of the amount of continuous money flow over time T at an interest rate r compounded continuously is given by

$$A = \int_0^T f(t)e^{rt}\,dt$$

EXAMPLE 68   An amount of \$5000 a year is being invested at 10% compounded continuously for 12 years. Find the amount of the continuous money flow.

SOLUTION   For this problem $f(t)$ is a constant function, $f(t) = \$5000$, $r = 0.10$, and $T = 12$ years.

$$\int_0^{12} 5000e^{0.10t}\,dt = \frac{5000e^{0.10t}}{0.10}\Bigg|_0^{12}$$

Antiderivative of e^{kx} is $\dfrac{e^{kx}}{k}$.

$$= \frac{5000e^{(0.10)(12)}}{0.10} - \frac{5000e^{(0.10)(0)}}{0.10}$$

$$= \frac{5000e^{1.2}}{0.10} - \frac{5000}{0.10} = 116{,}005.85$$

The investment will amount to \$116,005.85 in 12 years.   ■

Practice Problem 2   An amount of \$4000 a year is being invested at 4.5% compounded continuously for 10 years. Find the amount of the continuous money flow.

ANSWER   \$50,516.64

If \$5000 is invested at 10% compounded continuously for 5 years, the amount will be \$8243.61. We say that \$5000 is the present value of \$8243.61 in 5 years. Present value is the principal that will grow to the given sum at a specified future date at a constant rate of interest.

Now if $A = Pe^{rt}$ is the formula for the amount where a fixed P dollars has been invested at an interest rate of r compounded continuously, then the present value P is

$$P = Ae^{-rt}   \text{Divide by } e^{-rt}.$$

Suppose that the sum of money A is to be received continuously (in a series of frequent payments). That is, let $f(t)$ be the rate of flow of money from the fund. The accumulation of all the present values can be obtained from the following integral.

PRESENT VALUE OF A CONTINUOUS MONEY FLOW

If $f(t)$ is the rate of flow of a continuous money flow, then the present value from now until some time T in the future is

$$\int_0^T f(t)e^{-rt}\,dt$$

where r is the interest rate compounded continuously.

EXAMPLE 69   Find the present value of an investment over a 10-year period if there is a continuous money flow of \$1000 yearly and interest is at 8% compounded continuously.

SOLUTION   The function $f(t)$ is a constant function equal to $1000 yearly.

$$\int_0^{10} 1000e^{-0.08t}\, dt = \frac{1000e^{-0.08t}}{-0.08}\bigg|_0^{10} \qquad \text{Antiderivative of } e^{kx} \text{ is } \frac{e^{kx}}{k}.$$

$$= \frac{-1000e^{-0.8} + 1000}{0.08}$$

$$= 6883.39$$

The present value of this money flow is $6883.39.

Practice Problem 3 Find the present value of an investment over a 6-year period if there is a continuous money flow of $3000 yearly and interest is at 3.5% compounded continuously.

ANSWER $16,235.64

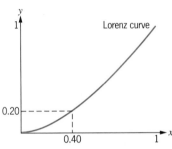

This curve shows that the lowest 40% of income recipients receive 20% of the total income.

Figure 47

Coefficient of Inequality

A **Lorenz curve**, named for the American statistician, M. D. Lorenz, is used by economists to study the distribution of income. It shows what percent of income is received by x percent of the population of a country or state, or perhaps the employees of a corporation. In Figure 47, x represents the cumulative proportion of income recipients, ranked from those with lowest income to those with highest income. The point $x = 0.40$ indicates that 40% of the people have incomes below this point. The function $f(x)$ represents the proportion of total income received. The Lorenz curve in Figure 47 indicates that the lowest 40% of income recipients receive about 20% of the total income. The function has domain $[0, 1]$, range $[0, 1]$, $f(0) = 0$, $f(1) = 1$, and $f(x) \le x$ for all x in $[0, 1]$.

EXAMPLE 70   The income distribution of a developing country is described by

$$f(x) = \frac{9}{10}x^2 + \frac{1}{10}x$$

(a) Use your graphing calculator to draw the Lorenz curve for the given function. Use $X:[0, 1]$ with Xscl$=.2$ and $Y:[0, 1]$ with Yscl $= .2$.

(b) The lowest 20% of income recipients receive what percent of the total income?

(c) The top 10% of income recipients receive what percent of the total income?

SOLUTION

(a) The graph is shown in Figure 48.

(b) $f(0.20) = \frac{9}{10}(0.20)^2 + \frac{1}{10}(0.20) = 0.056$. The lowest 20% of income recipients receive 5.6% of the total income.

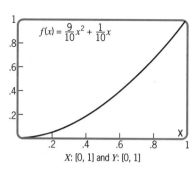

$$f(x) = \frac{9}{10}x^2 + \frac{1}{10}x$$

X: [0, 1] and Y: [0, 1]

Figure 48

(c) $f(0.90) = 0.819$, so the bottom 90% of income recipients receive 81.9% of the total income, leaving $1 - 0.819 = 0.181$ or 18.1% of the total income for the top 10%.   ■

This measure is defined as follows.

Equality line
y = x

Lorenz curve
y = f(x)

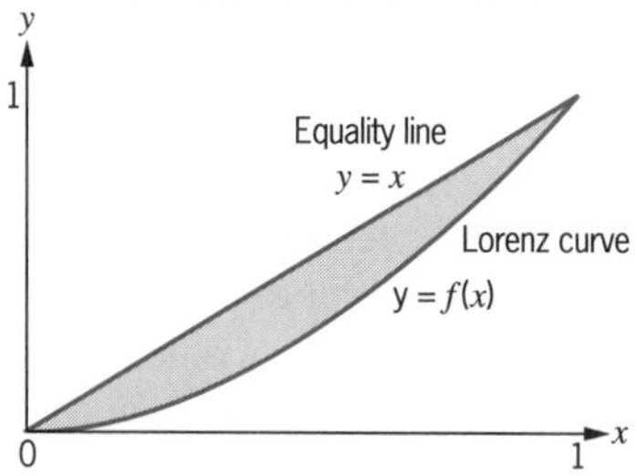

The smaller the area between the two curves, the more equitable the distribution of income.

Figure 49

One of the most interesting applications of the area between two curves when one of the curves is a Lorenz curve is the **Gini index** or **coefficient of inequality**. The function $y = f(x)$ represents **equality of income** if the lowest 25% of income recipients receive 25% of the total income, the lowest 50% of income recipients receive 50% of the total income, and so on. The area between the equality line and a Lorenz curve (the shaded region in Figure 49), gives a measure of inequality. This measure is defined as follows.

GINI INDEX OR COEFFICIENT OF INEQUALITY

If $y = f(x)$ is the equation of a Lorenz curve then

$$2 \int_0^1 [x - f(x)]\, dx$$

is called the **Gini index** or the **coefficient of inequality**.

EXAMPLE 71   Find the coefficient of inequality for the Lorenz curve defined in Example 70.

SOLUTION

$$\text{Coefficient of inequality} = 2 \int_0^1 [x - f(x)]\, dx$$

$$= 2 \int_0^1 \left[x - \frac{9}{10}x^2 - \frac{1}{10}x \right] dx \qquad \text{Substitute for } f(x).$$

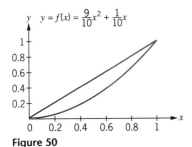
Figure 50

$$= 2 \int_0^1 \left(\frac{9}{10} x - \frac{9}{10} x^2 \right) dx$$

$$= 2 \left[\frac{9}{20} x^2 - \frac{9}{30} x^3 \right] \Big|_0^1 \qquad \text{Definite integral}$$

$$= 2 \left[\frac{9}{20} - \frac{9}{30} \right] = \frac{9}{30} = \frac{3}{10}$$

See Figure 50.   ■

The coefficient of inequality can be used to compare income distributions for various countries, or to check to see if there is a change in the income distribution in a given country. For example, a few years ago the coefficient of inequality for the United States was 0.26 while that for Sweden was 0.18 and that for Brazil was 0.34. From these values we can see that among these three countries income was more equitably distributed in Sweden than in the United States and was the least equitably distributed in Brazil.

According to the *1993 World Almanac*, the gap between rich and poor did not widen in 1991, but the long-term trend pointed to an increasing inequality of income. The richest 20% of all households got 46.5% of all household income in 1991, up from 43.5% in 1971 and 44.4% in 1981. The poorest 20% got 3.8% of all income in 1991, contrasted with 4.1% in 1971 and 1981. (Note that most government assistance to the poor is not included and does affect the income inequity.)

Income distribution can be applied to situations other than nations and states.

Practice Problem 4 The salary or pay distribution of a corporation is

$$f(x) = 0.95x^2 + 0.05x$$

Compute the coefficient of inequality.

ANSWER $2 \int_0^1 (x - 0.95x^2 - 0.05x) \, dx \approx 0.32$

SUMMARY

We have looked at only a few of the applications of the definite integral. Many are beyond the scope of this book but are used throughout economics and business. You should investigate this topic by doing some research in the areas of economics and business.

Exercise Set 11.8

Applications (Business and Economics)

1. **Coefficient of Inequality.** Find the coefficient of inequality for each of the Lorenz curves.
   (a) $f(x) = x^2$ (b) $f(x) = x^3$

2. **Coefficient of Inequality.** Find the coefficient of inequality for each of the Lorenz curves.
   (a) $f(x) = x^{1.5}$ (b) $f(x) = x^{5/2}$

3. **Continuous Money Flow.** Find the total income produced in 6 years by the continuous money flow if the rate of flow is $f(t) = 2000$ per year and $r = 8\%$ compounded continuously.

4. **Continuous Money Flow.** Find the total income in 10 years by a continuous money flow with a rate of $f(t) = e^{0.04t}$ and $r = 10\%$.

5. **Continuous Money Flow.** Find the total income in 8 years by a continuous money flow with a rate of $f(t) = e^{0.06t}$ and $r = 10\%$.

6. **Continuous Money Flow.** Find the present value at 8% compounded continuously for 6 years for a continuous money flow where $f(t) = 2000$.

7. **Continuous Money Flow.** Find the present value at 8% compounded continuously for 6 years for a continuous money flow where $f(t) = e^{-0.02t}$.

8. **Continuous Money Flow.** Find the present value at 10% compounded continuously for 8 years for a continuous money flow $f(t) = e^{-0.06t}$.

For each pair of equations: (a) Use your graphing calculator to graph the pair on the same screen, (b) find the equilibrium point, (c) add the graph of $y = p$ (equilibrium price) and (d) find the consumers' surplus and the producers' surplus.

9. $D(x) = -2x + 6$, $S(x) = x + 2$

10. $D(x) = -3x + 4$, $S(x) = 2x + 1$

11. $D(x) = -2x + 16$, $S(x) = x^2 + 7$

12. $D(x) = -3x + 4$, $S(x) = x^2$

13. $D(x) = 9 - x$, $S(x) = x + 3$

14. $D(x) = 8 - 2x$, $S(x) = x + 2$

15. $D(x) = 6 - x$, $S(x) = \frac{1}{4}x + 1$

16. $D(x) = 7 - x$, $S(x) = \frac{1}{4}x + 2$

17. $D(x) = 8 - \frac{1}{4}x^2$, $S(x) = \frac{1}{8}x^2 + 2$

18. $D(x) = 9 - \frac{1}{12}x^2$, $S(x) = \frac{1}{12}x^2 + 3$

19. **Lorenz Curve.** The wages of a given industry are described by a Lorenz curve

$$f(x) = 0.8x^2 + 0.2x$$

(a) What percent of the total wages of the industry are earned by workers classified as the lowest-paid 20% of the work force of the industry?
(b) What percent of the total wages of the industry are earned by the workers classified as the lowest-paid 80% of the work force of the industry?
(c) Find the coefficient of inequality for the Lorenz curve.

20. **Coefficient of Inequality.** The Lorenz curve for a laborer in a plant is $f(x) = \frac{11}{12}x^2 + \frac{1}{12}x$ and that of a supervisor is $f(x) = \frac{8}{11}x^4 + \frac{3}{11}x$.
(a) Compute the coefficient of inequality for each Lorenz curve.
(b) Which has a more equitable income distribution?

21. **Consumers' Surplus.** The demand for electric shavers in x hundred units per week in terms of the wholesale price p in dollars is given as

$$p = D(x) = \sqrt{400 - 6x}$$

Determine the consumers' surplus if the wholesale price is set at $16.

22. **Producers' Surplus.** The supply function for the electric shavers in Exercise 21 is

$$p = S(x) = \sqrt{40 + 2x}$$

Determine the producers' surplus if the wholesale price is set at $10.

23. **Price Equilibrium**
(a) Using the demand function from Exercise 21 and the supply function from Exercise 22, find the point of equilibrium.
(b) Find the consumers' surplus at this point.

(c) Find the producers' surplus at this point.

(d) Show parts (a), (b), and (c) on a graph.

24. **Continuous Money Flow.** Find the amount of a continuous money flow if $1000 is being invested each year at 10% compounded continuously for 12 years.

25. **Continuous Money Flow.** The Formans want to establish a fund for the education of their children. If they deposit $5000 yearly for 10 years in a savings account that is paying 8% compounded continuously, find how much is available for the college education of their children. What is the present value of this fund?

26. **Lorenz Curve.** The income distribution in the United States in 1986 was approximated to be the Lorenz curve $f(x) = x^{2.2}$. Find the coefficient of inequality.

27. **Lorenz Curve.** The income distribution of a certain country is given by

$$f(x) = \frac{19}{20} x^2 + \frac{1}{20} x$$

(a) Sketch the Lorenz curve for this function.

(b) Compute $f(0.60)$ and interpret your answer.

(c) Compute the coefficient of inequality.

28. **Coefficient of Inequality.** The Lorenz curve for the distribution of U.S. income in 1929 was $f(x) = x^{3.2}$; find the coefficient of inequality.

29. **Coefficient of Inequality.** An approximation for the income distribution in Argentina in 1986 is given by

$$f(x) = 0.4x - 2.2x^2 + 2.8x^3$$

Compute the coefficient of inequality.

30. **Consumer Demand.** For a certain product, the consumer demand is $p = D(x) = 600 - 10x$ and market price is $100.

(a) How many units will be sold at this price?

(b) What is the total cost of these units?

(c) Find the total value of these units (area under the curve) to the consumer.

(d) Find the consumers' surplus.

(e) Is (d) = (c) − (b)?

31. **Consumer Demand.** For a certain product, the consumer demand is given by $p = D(x) = 356.25 - 0.01x^2$ and the market price is $200.

(a) How many units will be sold at this price?

(b) What is the total cost of these units?

(c) Find the total value of these units (area under the curve) to the consumer.

(d) Find the consumers' surplus.

(e) Is (d) = (c) − (b)?

Review Exercises

Find the values of the following definite integrals.

32. $\displaystyle\int_0^7 \sqrt[3]{x + 1}\, dx$

33. $\displaystyle\int_0^3 \sqrt{4 - x}\, dx$

Chapter Review

Make certain that you understand the following important terms.

Antiderivative	Indefinite integral
Integration constant	Integral sign
Integrand	Definite integral
Lower limit of integration	Upper limit of integration
Area under a curve	Area between curves
Integration by substitution	Consumers' surplus
Continuous money flow	Producers' surplus
Coefficient of inequality	Lorenz curve

Be sure that you are proficient in working with the following formulas.

$$\int x^n \, dx = \frac{x^{n+1}}{n+1} + C, \quad n \neq -1$$

$$\int e^{kx} \, dx = \frac{e^{kx}}{k} + C$$

$$\int kf'(x) \, dx = k \int f'(x) \, dx = kf(x) + C$$

$$\int [f(x) + g(x)] \, dx = \int f(x) \, dx + \int g(x) \, dx$$

$$\int \frac{1}{x} \, dx = \ln |x| + C \qquad \int_a^b f(x) \, dx = F(b) - F(a)$$

Chapter Test

Find the following indefinite integrals.

1. $\int \left(x^3 + 4x^2 + \frac{1}{2} \right) dx$

2. $\int \frac{dx}{(x+1)^{1/3}}$

3. $\int \left(\frac{-4x^2 - 7x + 6}{x^4} \right) dx$

4. $\int x(1 - x^2)^{1/2} \, dx$

5. $\int 2(4x + 1)^{-1} \, dx$

6. $\int (20e^{4x} + 4x^{-3}) \, dx$

7. $\int 12e^{3x^2} x \, dx$

8. $\int \frac{2x + 1}{x - 1} \, dx$

9. $\int x\sqrt{x^2 - 3} \, dx$

10. $\int x(x + 3)^4 \, dx$

Find the value of the following definite integrals using both the properties of integrals and your graphing calculator.

11. $\int_1^4 (6x^2 - 4x + 3) \, dx$

12. $\int_{-2}^0 (4 - 6x)^{1/2} \, dx$

13. Find the differential du if $u = \sqrt{3x^2 - 1}$.

14. Find the area bounded by the curves $y = x^2 - 1$ and $y = 0$. Use your graphing calculator to draw the graph.

15. The Goad Coal Company's marginal cost function is $C'(x) = 12x + 30x^{1/2}$. Find the cost of increasing production from 25 tons to 40 tons.

16. Find the producers' surplus at a price level of $17 for the price–supply equation

$$p = S(x) = 1 + \frac{1}{100} x^2$$

17. If a Lorenz curve is given by $f(x) = 0.9x^2 + 0.1x$, find the coefficient of inequality.

Techniques of Integration
and Numerical Integration

I n the preceding chapter, we defined the definite integral $\int_a^b f(x)\, dx$ to be the total change in any derivative of $f(x)$ from $x = a$ to $x = b$. We developed or applied this idea to areas under curves. In fact, our discussion of topics relating to the fundamental theorem of integral calculus was associated with the area under a curve. This theorem, however, is important in a multitude of applications. The definition of a definite integral, as the limit of a sum, allows us to apply the fundamental theorem of integral calculus to various application problems.

Sometimes it is difficult or even impossible to find an antiderivative for a given integrand. We introduce two procedures to approximate the value of a definite integral when it cannot be evaluated by any other method. Also in this chapter, we introduce three new techniques for integration. A technique known as integration by parts assists in finding an antiderivative when the function can be considered as a product. We expand our notion of the definite integral to encompass cases in which the interval of integration is infinite. We include in this chapter a discussion of the use of a table of integrals. Infinite (improper) integrals and integration by parts enable us to compute the capital value of a continuous money flow where interest is compounded continuously and where time increases without bound. Improper integrals enable us to study statistical distributions and probability density functions.

<u>12.1</u> INTEGRATION BY PARTS

OVERVIEW Recall that the derivative of a product is not the product of the derivatives. Hence, you would expect that the integral of a product is not the product of the integrals. We introduce in this section a very powerful method of integration called **integration by parts**. This method of integration is based on the formula for the derivative of a product. This method is particularly useful in finding integrals involving the product of an exponential (or logarithmic) function and an algebraic function. The need for such a procedure is quite evident. Suppose that we desire to find

$$\int xe^x \, dx$$

Prior to this section we have had no procedure for integrating such a product. In this section we

- Find indefinite integrals using integration by parts
- Evaluate definite integrals using integration by parts

To derive a formula for the antiderivative of a product, recall that

$$\frac{d(uv)}{dx} = u \cdot \frac{dv}{dx} + v \frac{du}{dx}$$
  Where u and v are functions of x. Now rewrite the formula.

$$\frac{d(uv)}{dx} = uv' + vu'$$
  $v' = \dfrac{dv}{dx}$ and $u' = \dfrac{du}{dx}$

$$uv + C = \int (uv' + vu') \, dx$$
  Take antiderivatives of both sides of this equation.

Hence $$uv + C = \int uv' \, dx + \int vu' \, dx$$
  Integral of a sum

or $$\int uv' \, dx = uv - \int vu' \, dx + C$$

INTEGRATION-BY-PARTS FORMULA

If $u(x)$ and $v(x)$ are functions whose derivatives exist, then

$$\int uv' \, dx = uv - \int vu' \, dx + C$$

or

$$\int u \, dv = uv - \int v \, du + C$$

The second part of the preceding formula can be obtained by substituting

$$dv = v' \, dx \quad \text{and} \quad du = u' \, dx \qquad dv = \frac{dv}{dx} \cdot dx$$

When an antiderivative is found by using the formula just derived, the process is called **integration by parts**. The factors u and dv should be chosen so that $\int v \, du$ is simpler to evaluate than the original problem.

In Example 1, we return to the problem that we stated at the beginning of the section.

EXAMPLE 1   Find $\int xe^x \, dx$.

SOLUTION   We use the following substitutions:

$$x = u \quad \text{and} \quad e^x \, dx = dv$$

We can see that $dx = du$ and we now find v by integrating:

$$dv = e^x \, dx$$

$$v = \int e^x \, dx$$

$$v = e^x \qquad \text{Constant is added later.}$$

Therefore,

$$\int \overset{u}{x} \, \overset{dv}{\overbrace{e^x \, dx}} = \overset{u}{x} \, \overset{v}{e^x} - \int \overset{v}{e^x} \, \overset{du}{dx} + C = xe^x - e^x + C$$

For some functions, it does not matter which factor is chosen for u or which is selected for dv in $\int u\, dv$; however, for many functions the choice of u and dv is critical in the integration process. For example, suppose that in Example 1 we had chosen $u = e^x$ and $dv = x\, dx$. We would then have $du = e^x\, dx$ and $v = \int x\, dx = x^2/2$. Then by using the formula we would have

$$\int e^x x\, dx = \frac{x^2}{2} \cdot e^x - \int \frac{x^2}{2} e^x\, dx$$

which is more complicated than the original integral. Thus, our first choice for u and dv was the appropriate choice in order to perform the integration. Undoubtedly you are asking, How do I know which factor to call u and which to call dv? To a large extent, it is a matter of some trial and error and practice. You should, however, work toward making the second integral, $\int v\, du$, one which you can readily integrate. Also, you must be able to find v, given dv. After we have done some examples and you have had the opportunity to see some problems solved, we will give some helpful hints for selecting the proper substitutions.

NOTE: In Example 2, we could use substitution as we did in Section 11.4. However, we integrate by parts to illustrate that there are times when you have a choice of techniques of integration to use in a particular problem.

EXAMPLE 2 Find $\int x\sqrt{1 + x}\, dx$.

SOLUTION We use the following substitutions because dv can be integrated to find v.

$$u = x \quad \text{and} \quad dv = \sqrt{1 + x}\, dx$$

From these substitutions we obtain

$$du = dx \quad \text{and} \quad v = \int dv = \int (1 + x)^{1/2}\, dx = \frac{2}{3}(1 + x)^{3/2}$$

Substitution in the integration-by-parts formula

$$\int u\, dv = uv - \int v\, du$$

yields

$$\int \underbrace{x}_{u} \; \underbrace{\sqrt{1 + x} \; dx}_{dv} = \underbrace{x}_{u} \; \underbrace{\frac{2}{3} (1 + x)^{3/2}}_{v} - \int \underbrace{\frac{2}{3} (1 + x)^{3/2}}_{v} \; \underbrace{dx}_{du} + C$$

Since

$$\int (1 + x)^{3/2} \; dx = \frac{2}{5} (1 + x)^{5/2}$$

we now have

$$\int x\sqrt{1 + x} \; dx = \frac{2}{3} x(1 + x)^{3/2} - \frac{4}{15} (1 + x)^{5/2} + C$$ ■

Calculator Note

Use your graphing calculator to find several values for

$$F(x) = \int_0^x t\sqrt{t + 1} \; dt$$

and then plot these points on a graph and draw a curve through the points. Then have your calculator draw the graph of $y(x) = \frac{2}{3}x(1 + x)^{3/2} - \frac{4}{15}(1 + x)^{5/2}$. How do these graphs compare? Are you surprised that they differ only by the constant of integration, C?

Practice Problem 1 Find $\int xe^{2x} \; dx$.

ANSWER $\dfrac{xe^{2x}}{2} - \dfrac{e^{2x}}{4} + C$

EXAMPLE 3 Find $\int \ln x \; dx$.

SOLUTION By just looking at this integrand, it would appear that there is only one factor. However, let's use $(\ln x)$ as one factor and (dx) as the other factor because $\ln x$ can be differentiated and dx can be integrated. Since we cannot integrate $(\ln x)$, we choose the following substitutions and work as follows:

$$u = \ln x \qquad\qquad dv = dx$$

$$du = \frac{1}{x} \; dx \qquad\quad v = \int dv = \int dx$$

$$v = x$$

$$\int \overbrace{(\ln x)}^{u} \overbrace{(1 \ dx)}^{dv} = \overbrace{(\ln x)}^{u} \overbrace{(x)}^{v} - \int x \overbrace{\frac{1}{x} dx}^{v \, du} + C$$

$$= x \ln x - \int dx + C$$

$$= x \ln x - x + C \qquad \blacksquare$$

Repeated Integration by Parts

Sometimes it is necessary to repeat the integration by parts in order to integrate a function. When this occurs, you usually select for u a type of function that will become a constant after several differentiations. Let's take a look at this situation in the next example.

EXAMPLE 4   Find $\int x^2 e^x \, dx$.

SOLUTION   Since x^2 will become a constant after two differentiations, we select our substitutions and work as follows:

$$u = x^2 \qquad\qquad dv = e^x \, dx$$

$$du = 2x \, dx \qquad\qquad v = \int dv = \int e^x \, dx$$

$$v = e^x$$

$$\int \overbrace{(x^2)}^{u} \overbrace{(e^x \, dx)}^{dv} = \overbrace{(x^2)}^{u} \overbrace{(e^x)}^{v} - \int \overbrace{(e^x)}^{v} \overbrace{(2x \, dx)}^{du} + C$$

$$= x^2 e^x - 2 \int x e^x \, dx + C$$

Note that $\int x^2 e^x \, dx$ is now expressed in terms of $\int x e^x \, dx$, which we found in Example 1. Therefore, using the substitution $\int x e^x \, dx = x e^x - e^x$ from that example, we now have

$$\int x^2 e^x \, dx = x^2 e^x - 2[x e^x - e^x] + C = x^2 e^x - 2x e^x + 2e^x + C \qquad \blacksquare$$

We now list some hints for using the integration-by-parts formula.

Hints for Integration by Parts

1. You must be able to integrate dv. It will usually be the most complicated part that fits an integration formula.
2. An application of the formula should produce an integral that is easier to integrate.
3. For $\int x^p e^{kx}\, dx$, $p > 0$, let $u = x^p$ and $dv = e^{kx}\, dx$.
4. For $\int x^p (\ln x)^q\, dx$, $p \geq 0$, let $u = (\ln x)^q$ and $dv = x^p\, dx$.

EXAMPLE 5 Find $\int x \ln x\, dx$.

SOLUTION Using hint 4, we let

$$u = \ln x \qquad dv = x\, dx$$

$$du = \frac{1}{x}\, dx \qquad v = \int dv = \int x\, dx = \frac{x^2}{2}$$

$$\int \overbrace{(\ln x)}^{u}\, \overbrace{(x\, dx)}^{dv} = \overbrace{(\ln x)}^{u} \overbrace{\left(\frac{x^2}{2}\right)}^{v} - \int \overbrace{\left(\frac{x^2}{2}\right)}^{v} \overbrace{\left(\frac{1}{x}\, dx\right)}^{du} + C$$

$$= \frac{x^2 \ln x}{2} - \frac{1}{2} \int x\, dx + C$$

$$= \frac{x^2 \ln x}{2} - \frac{1}{2} \cdot \frac{x^2}{2} + C$$

$$= \frac{x^2 \ln x}{2} - \frac{x^2}{4} + C$$

EXAMPLE 6 Evaluate $\int \frac{xe^x}{(x + 1)^2}\, dx$.

SOLUTION We use the following substitutions:

$$u = xe^x \qquad\qquad dv = \frac{1}{(x + 1)^2}$$

$$du = (xe^x + e^x)\, dx \qquad v = \int dv = \int \frac{1}{(1 + x)^2}\, dx$$

$$= e^x(1 + x)\, dx \qquad\qquad v = \frac{-1}{x + 1}$$

Therefore, we have

$$
\int \frac{xe^x}{(x+1)^2}\,dx = \overbrace{(xe^x)}^{u}\overbrace{\left(\frac{-1}{x+1}\right)}^{v} - \int \overbrace{\left(\frac{-1}{x+1}\right)}^{v}\overbrace{[e^x(x+1)\,dx]}^{du}
$$

$$
= \frac{-xe^x}{x+1} + \int e^x\,dx
$$

$$
= \frac{-xe^x}{x+1} + e^x + C
$$

$$
= \frac{e^x}{x+1} + C
$$

We can evaluate definite integrals as well as indefinite integrals using the integration-by-parts formula.

DEFINITE INTEGRAL: INTEGRATION BY PARTS

$$
\int_a^b u\,dv = uv\,\Big|_a^b - \int_a^b v\,du
$$

where the limits of integration are for x since both u and v are functions of x.

In Chapter 11 we stated that definite integrals can be found on most graphing calculators. However, we will continue to show the techniques involved in the calculation of the definite integral and check with a graphing calculator.

EXAMPLE 7 Find

$$
\int_1^e \ln x\,dx
$$

using integration by parts and check with numerical integration on your graphing calculator.

SOLUTION From Example 3 we know that

$$
\int \ln x\,dx = x \ln x - x + C
$$

Therefore, we now have

$$
\int_1^e \ln x\,dx = (x \ln x - x)\,\Big|_1^e = (e \ln e - e) - (1 \ln 1 - 1)
$$

$$
= (e - e) - (0 - 1) = 1
$$

> ✓ Calculator Check: fnInt(ln x, x, 1, e) = 1   ■

We know that definite integrals are associated with the area under a curve. Draw the graph of $f(x) = \ln x$ with your calculator and shade the area from $x = 1$ to $x = e$. (Use e^1 on your calculator to get e.) See Figure 1.

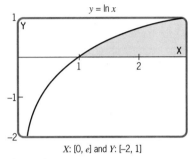

X: [0, e] and Y: [–2, 1]

Figure 1

Practice Problem 2 Find

$$\int_1^e 2x \ln x \, dx$$

using integration by parts and check with numerical integration on your graphing calculator.

ANSWER $\displaystyle\int_1^e 2x \ln x \, dx = \frac{e^2}{2} + \frac{1}{2} \approx 4.2$

> ✓ Calculator Check: fnInt($2x$ ln x, x, 1, $e \wedge 1$) ≈ 4.2

EXAMPLE 8 Find $\displaystyle\int_0^4 x^2 \sqrt{1 + 2x} \, dx$ and check.

SOLUTION Let

$$u = x^2 \qquad\qquad dv = (1 + 2x)^{1/2} \, dx$$

$$du = 2x \, dx \qquad v = \int dv = \int (1 + 2x)^{1/2} \, dx = \frac{(1 + 2x)^{3/2}}{3}$$

We now substitute into the formula to get

$$\int_0^4 x^2 \sqrt{1 + 2x} \, dx = x^2 \, \frac{(1 + 2x)^{3/2}}{3} \bigg|_0^4 - \int_0^4 \frac{(1 + 2x)^{3/2}}{3} \, 2x \, dx$$

Before we proceed, we must find

$$\int \frac{(1 + 2x)^{3/2}}{3} \, 2x \, dx$$

Let

$$u = 2x \qquad dv = \frac{(1 + 2x)^{3/2}}{3} \, dx$$

$$du = 2 \, dx \qquad v = \int dv = \frac{(1 + 2x)^{5/2}}{15}$$

We can now complete the problem.

$$\int_0^4 x^2 \sqrt{1 + 2x} \, dx = \frac{x^2(1 + 2x)^{3/2}}{3} \bigg|_0^4 - \frac{2x(1 + 2x)^{5/2}}{15} \bigg|_0^4$$

$$+ \int_0^4 \frac{(1 + 2x)^{5/2}}{15} \cdot 2 \, dx$$

$$= \left[\frac{x^2(1 + 2x)^{3/2}}{3} - \frac{2x(1 + 2x)^{5/2}}{15} + \frac{2(1 + 2x)^{7/2}}{105} \right] \bigg|_0^4$$

$$= \left[\frac{16(9)^{3/2}}{3} - \frac{8(9)^{5/2}}{15} + \frac{2(9)^{7/2}}{105} \right] - \left[\frac{2}{105} \right]$$

$$= 144 - \frac{648}{5} + \frac{1458}{35} - \frac{2}{105}$$

$$= \frac{5884}{105} \approx 56.04$$

✓ Calculator Check: fnInt($x \wedge 2 \sqrt{(1 + 2x)}, x, 0, 4$) ≈ 56.04

Draw the graph of $y = x^2 \sqrt{1 + 2x}$ with your graphing calculator and shade the area being found (see Figure 2). ∎

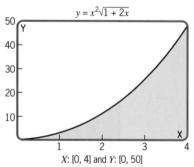

$y = x^2\sqrt{1 + 2x}$

X: [0, 4] and Y: [0, 50]

Figure 2

EXAMPLE 9   The rate at which natural gas will be produced from a well t years after production begins is estimated to be

$$P(t) = 100te^{-0.1t}$$

thousand cubic feet per year. How many cubic feet will be produced in the first 2 years that the well is in production? Draw the graph of the production function with your calculator and shade the area being calculated.

SOLUTION

$$\int_0^2 100te^{-0.1t}\, dt = 100 \int_0^2 te^{-0.1t}\, dt$$

Let

$$u = t \qquad dv = e^{-0.1t}\, dt$$

$$du = dt \qquad v = \int dv = \int e^{-0.1t}\, dt$$

$$v = -\frac{1}{0.1}e^{-0.1t} = -10e^{-0.1t}$$

Thus

$$100\int_0^2 te^{-0.1t}\, dt = 100\left[(-10te^{-0.1t}) \Big|_0^2 + 10\int_0^2 e^{-0.1t}\, dt \right]$$

$$= 100[-10te^{-0.1t} - 100e^{-0.1t}] \Big|_0^2$$

$$= -1000te^{-0.1t} - 10{,}000e^{-0.1t} \Big|_0^2$$

$$= [(-1000)(2)e^{-0.2} - 10{,}000e^{-0.2}] - (0 - 10{,}000e^0)$$

$$= 175.23$$

The graph of the production function is shown in Figure 3.

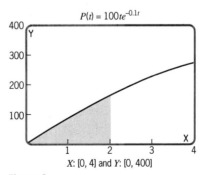

$$P(t) = 100te^{-0.1t}$$

X: [0, 4] and Y: [0, 400]

Figure 3

$$\boxed{\checkmark \quad \text{Calculator Check: fnInt}(100xe^{-.1x}, x, 0, 2) \approx 175.23}$$

Therefore, $175.23 \cdot 1000 = 175,230$ cubic feet will be produced in the first 2 years. ∎

SUMMARY

Knowing *how* to use an integration technique is important, but you must also know *when* to use a particular technique. When do you use substitution, or a formula, or integration by parts? You must practice *recognizing* which technique to use as much as how to use the technique. Just a slight change in the integrand calls for a different technique.

Integrand	Technique
$\int x \ln x \, dx$	Integration by parts
$\int \dfrac{\ln x}{x} \, dx$	Substitution and power rule
$\int \dfrac{1}{x \ln x} \, dx$	Substitution and ln rule

Exercise Set 12.1

Find the following indefinite integrals.

1. $\int 3xe^x \, dx$

2. $\int xe^{3x} \, dx$

3. $\int 2xe^{-x} \, dx$

4. $\int xe^{-2x} \, dx$

5. $\int (x - 1)e^x \, dx$

6. $\int (x + 1)e^x \, dx$

7. $\int (2x - 1)e^{3x} \, dx$

8. $\int (3x + 2)e^{4x} \, dx$

9. $\int (3x + 1)e^{-2x} \, dx$

10. $\int (4x - 3)e^{-3x} \, dx$

11. $\int 5x \ln x \, dx$

12. $\int 3x^2 \ln x \, dx$

13. $\int x^2 \ln x \, dx$

14. $\int 3x^3 \ln x \, dx$

15. $\int x^3 \ln x \, dx$

16. $\int 2x^3 \ln x \, dx$

17. $\int 3x\sqrt{1 + x} \, dx$

18. $\int 3x(1 + x)^{-1/2} \, dx$

19. $\int (3x - 1)(x + 2)^{-1/2} \, dx$

20. $\int (xe^{2x} - \ln x) \, dx$

21. $\int \dfrac{x}{(x + 4)^5} \, dx$

22. $\int \dfrac{x}{\sqrt{2 + 3x}} \, dx$

23. $\int \dfrac{1}{x(\ln x)^3} \, dx$

24. $\int \dfrac{(\ln x)^2}{x} \, dx$

25. $\int \dfrac{xe^{2x}}{(2x + 1)^2} \, dx$

26. $\int \dfrac{x^3 e^{x^2}}{(x^2 + 1)^2} \, dx$

Evaluate the definite integrals using both the integration-by-parts formula and your graphing calculator.

27. $\displaystyle\int_1^3 xe^{4x}\,dx$

28. $\displaystyle\int_0^4 xe^{-2x}\,dx$

29. $\displaystyle\int_{1/2}^{e/2} \ln{(2x)}\,dx$

30. $\displaystyle\int_0^2 \frac{x^2}{e^x}\,dx$

31. $\displaystyle\int_0^1 \ln{(1+2x)}\,dx$

32. $\displaystyle\int_1^e x^4 \ln{x}\,dx$

33. $\displaystyle\int_0^1 x^2 e^x\,dx$

Locate with your graphing calculator the areas bounded by the following and use calculus techniques to find the areas. Check your answers with your graphing calculator.

34. $y = xe^x$, $y = 0$, $x = 1$, and $x = 2$

35. $y = \ln{x}$, $y = 0$, $x = 1$, and $x = 2$

36. $y = x\sqrt{2x+1}$, $y = 0$, $x = 0$, and $x = 4$

37. $y = \dfrac{x}{(x+2)^2}$, $y = 0$, $x = 0$, and $x = 1$

The following exercises may require the use of the integration-by-parts formula more than once, or they may be integrable by previous methods. Check your answers with your graphing calculator.

38. $\displaystyle\int x^2 e^{3x}\,dx$

39. $\displaystyle\int xe^{3x^2}\,dx$

40. $\displaystyle\int \frac{3\ln{x}}{x}\,dx$

41. $\displaystyle\int \frac{2\ln{x}}{x^2}\,dx$

42. $\displaystyle\int x^3 e^{x^2}\,dx$

Use your graphing calculator to approximate the integral to four decimal places.

43. $\displaystyle\int_0^1 e^{x^3}\,dx$

44. $\displaystyle\int_0^4 \frac{1}{\sqrt{x^3+2}}\,dx$

45. $\displaystyle\int_{-1}^1 \sqrt{x^6+1}$

46. **Exam**

$$\int_{-1}^0 x\sqrt{x+1}\,dx =$$

(a) $-\dfrac{2}{3}$ (b) $-\dfrac{4}{15}$ (c) $-\dfrac{1}{30}$

(d) 0 (e) $\dfrac{1}{3}$

Applications (Business and Economics)

47. **Total Revenue.**   Find the total revenue for a product over the next 5 years if the demand is given by

$$D(t) = 1000(1 - e^{-t})$$

and the price is given by

$$P(t) = 1.04t$$

[**Hint:** $R(t) = P(t)D(t)$]

48. **Marginal Profit.**   A company gives its marginal profit function as

$$MP = 3x - xe^{-x}$$

If $P(0) = 0$, find $P(x)$.

49. **Production.**   The instantaneous rate of production of a company is given by

$$\frac{dP}{dt} = 20te^{-0.2t}$$

Find the total production for a year, if t is measured in months.

50. **Continuous Cash Flow.**   Find the present value at 10% compounded continuously for 6 years for a continuous cash flow of

$$f(t) = 10{,}000 - 400t$$

51. **Total Cost.**   The marginal cost for producing x electric shavers at a manufacturing company is

$$C'(x) = 0.040x \ln{x}$$

What is the total cost of producing 100 shavers?

Applications (Social and Life Sciences)

52. ***Population***.   Suppose that the population P, in thousands, of a town is approximated by

$$P = 30 + 5t - 4te^{-0.1t}$$

where t is time in years after 1980. Find the average population from $t = 1$ to $t = 5$.

53. ***Medicine***.   The rate, dA/dt, of assimilation of a drug after t minutes is

$$\frac{dA}{dt} = te^{-0.3t}$$

Find the total amount assimilated after 8 minutes.

12.2   MORE TECHNIQUES FOR INTEGRATION

OVERVIEW   In Section 11.4, we used the method of substitution to integrate when a direct application of a formula was not possible. In this section we

- Look at the process of substitution again
- Examine graphically what happens in a substitution problem when dealing with the definite integrals and the area under a curve

You should have a list of the basic integration formulas with you as a quick reference when you are using the technique of substitution.

Integration Formulas	
$\int k \, dx = kx + C$	Constant
$\int x^n \, dx = \dfrac{x^{n+1}}{n+1} + C, \quad n \neq -1$	Simple power rule
$\int u^n \, du = \dfrac{u^{n+1}}{n+1} + C, \quad n \neq -1$	General power rule
$\int e^{kx} \, dx = \dfrac{e^{kx}}{k} + C$	Exponential rule
$\int e^u \, du = e^u + C$	General exponential rule
$\int \dfrac{1}{x} \, dx = \ln |x| + C$	Log rule
$\int \dfrac{1}{u} \, du = \ln |u| + C$	General log rule
$\int u \, dv = uv - \int v \, du + C$	Integration by parts

When using substitution for indefinite integrals, you can generally follow these guidelines.

Guidelines for Substitution

1. Let u be some function of x. (It will usually be an expression that is raised to a power, an expression under a radical, or an expression in the denominator.)
2. Solve for x and dx in terms of u and du.
3. Substitute into the integral and determine which basic formula to use. You may find it necessary to use more than one substitution.
4. After integrating, you must substitute back and rewrite the antiderivative in terms of x.

EXAMPLE 10   Find $\int \dfrac{x}{(x + 1)^4}\, dx$.

SOLUTION   Let u equal the expression in the denominator that is raised to the fourth power: $u = x + 1$ (guideline 1). Therefore, $du = dx$ and $x = u - 1$ (guidelines 2 and 3). We are now ready to substitute. We integrate using u and then substitute back for u in terms of x.

$$
\begin{aligned}
\int \frac{x}{(x + 1)^4}\, dx &= \int \frac{u - 1}{u^4}\, du \\[2mm]
&= \int \left(\frac{u}{u^4} - \frac{1}{u^4} \right) du \\[2mm]
&= \int (u^{-3} - u^{-4})\, du \\[2mm]
&= \frac{u^{-2}}{-2} - \frac{u^{-3}}{-3} + C \qquad \text{Substitute back.} \\[2mm]
&= \frac{-1}{2(x + 1)^2} + \frac{1}{3(x + 1)^3} + C
\end{aligned}
$$

■

EXAMPLE 11   Find $\int \dfrac{e^{4x}}{1 + e^{4x}}\, dx$.

SOLUTION   Let $u = 1 + e^{4x}$. From this we obtain $du = 4e^{4x}\, dx$, which gives us $\frac{1}{4}\, du = e^{4x}\, dx$. The key here is to recognize that e^{4x} is part of the expression for du. Now we substitute, integrate, and substitute back in terms of x.

$$
\begin{aligned}
\int \frac{e^{4x}}{1 + e^{4x}}\, dx &= \int \left(\frac{1}{1 + e^{4x}} \right) (e^{4x}\, dx) \\[2mm]
&= \int \left(\frac{1}{u} \right) \left(\frac{1}{4}\, du \right) \\[2mm]
&= \frac{1}{4} \int \frac{1}{u}\, du
\end{aligned}
$$

$$= \frac{1}{4} \ln |u| + C$$

$$= \frac{1}{4} \ln (1 + e^{4x}) + C \qquad \text{Absolute value is not needed here,}$$
since $1 + e^{4x} > 0$ for all x. ■

Calculator Note

With your calculator, compare the values of

$$y_1 = \int_0^x \frac{e^{4t}}{1 + e^{4t}} \, dt \quad \text{and} \quad y_2 = \frac{1}{4} \ln (1 + e^{4x})$$

for $x = -1$, $x = 0$, $x = 1$, $x = 2$. Do you see that the values are the same *constant* apart? How would the two graphs compare?

Practice Problem 1

(a) Find $\displaystyle \int \frac{2x}{(x-1)^5} \, dx.$　　　(b) Find $\displaystyle \int \frac{e^{-2x}}{1 - e^{-2x}} \, dx.$

ANSWER

(a) $\displaystyle \frac{1 - 4x}{6(x-1)^4} + C$　　　(b) $\displaystyle \frac{\ln |1 - e^{-2x}|}{2} + C$

COMMON ERROR　When working substitution problems, students often forget to substitute back to the original variable. You must substitute back to the original variable in an indefinite integral.

In some integration problems, there can be more than one way to perform a substitution. There are times when the specific substitution will make little difference, but in others, one way can be much more efficient than the other. With experience and practice, you should be able to determine which substitution will be the most efficient.

EXAMPLE 12　Find $\displaystyle \int \frac{3}{\sqrt{x} - 2} \, dx.$

SOLUTION　None of the basic formulas can be used here. Let's try the substitution

$$u = \sqrt{x} - 2$$

In order to find dx in terms of u and du, we must first solve for x.

$$\sqrt{x} = u + 2$$
$$x = (u + 2)^2 = u^2 + 4u + 4 \qquad \text{Square both sides.}$$
$$dx = (2u + 4)\, du \qquad \qquad \text{Differentiate with respect to } u.$$

Now we can substitute into the original integral, perform the integration, and substitute back to x.

$$\int \frac{3}{\sqrt{x} - 2}\, dx = \int \left(\frac{3}{u}\right)(2u + 4)\, du$$
$$= \int \frac{6u + 12}{u}\, du$$
$$= \int \left(6 + \frac{12}{u}\right) du$$
$$= 6u + 12 \ln |u| + C$$
$$= 6(\sqrt{x} - 2) + 12 \ln |\sqrt{x} - 2| + C$$
$$= 6\sqrt{x} + 12 \ln |\sqrt{x} - 2| + C \qquad -12 \text{ is part of the constant, } C.$$

$\blacksquare$

Practice Problem 2 Find $\displaystyle\int \frac{4}{2 - \sqrt{x}}\, dx$.

ANSWER $-8\sqrt{x} - 16 \ln |2 - \sqrt{x}| + C$

Substitution and Definite Integrals

In Section 11.6 we saw that definite integrals are related to the area under a curve. In Section 11.5 we looked at definite integrals and substitution. We now combine these two and see graphically what happens when substitution is used and the limits of integration are changed in a definite integral.

Calculator Note

On page 725 we demonstrated two methods for evaluating a definite integral using substitution. In Method 2, we kept the original limits of integration and obtained the integration in terms of the original variable. In Method 1, we changed the limits of integration to the new variable. We now give a calculator verification that these procedures give the same answers. Then we use Method 1 for the examples in this section. For example, using substitution and changing the limits in the second integral, we have

$$\int_1^4 \frac{dx}{2\sqrt{x}\,(1 + \sqrt{x})^2} = \int_2^3 \frac{du}{u^2} \qquad \text{Using } u = 1 + \sqrt{x} \text{ and } du = \frac{dx}{2\sqrt{x}}$$

With a calculator, we get an answer of $\frac{1}{6}$ for both integrals.

EXAMPLE 13   Find the area bounded by

$$y = \frac{x}{\sqrt{x+1}}$$

and the x-axis from $x = 3$ to $x = 8$.

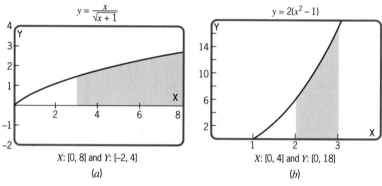

X: [0, 8] and Y: [−2, 4]

(a)

X: [0, 4] and Y: [0, 18]

(b)

Figure 4

SOLUTION   In Figure 4(a) we see the area to be calculated in the shaded portion of the graph. We know that

$$A = \int_3^8 \frac{x}{\sqrt{x+1}}\, dx$$

To evaluate the integral we let

$$u = \sqrt{x+1}$$

$$u^2 = x + 1 \qquad \text{Square both sides and solve for } x.$$

$$x = u^2 - 1$$

$$dx = 2u\, du \qquad \text{Differentiate with respect to } u.$$

We must now change the limits of integration to match the new variable of integration, u.

Change the lower limit: $x = 3$ gives $u = \sqrt{3+1} = 2$
Change the upper limit: $x = 8$ gives $u = \sqrt{8+1} = 3$

We are now ready to integrate.

$$\int_3^8 \frac{x}{\sqrt{x+1}}\, dx = \int_2^3 \frac{u^2 - 1}{u}\, 2u\, du$$

$$= 2 \int_2^3 (u^2 - 1)\, du$$

$$= 2 \left(\frac{u^3}{3} - u \right) \Big|_2^3$$

$$= 2 \left[\left(\frac{3^3}{3} - 3 \right) - \left(\frac{2^3}{3} - 2 \right) \right]$$

$$= \frac{32}{3}$$

Now, we know that

$$\int_3^8 \frac{x}{\sqrt{x+1}}\, dx = \int_2^3 2(u^2 - 1)\, du = \frac{32}{3}$$

Geometrically, this means that the two regions have the same area [see Figure 4(b)]. ■

Practice Problem 3 Find $\displaystyle\int_0^5 \frac{x}{(x+5)^2}\, dx$.

ANSWER $\ln 2 - \dfrac{1}{2} \approx 0.193$

EXAMPLE 14   A company has fixed costs of $150. It is found that the marginal cost is given by $C'(x) = x\sqrt{x^2 + 9}$. Find the company's total cost function.

SOLUTION   Fixed costs of $150 means that $C(0) = 150$. Therefore,

$$C(x) = \int C'(x)\, dx = \int x\sqrt{x^2 + 9}\, dx$$

$$= \frac{1}{2} \int u^{1/2}\, du \qquad \text{Let } u = x^2 + 9 \text{ so that } du = 2x\, dx$$
$$\text{and } \tfrac{1}{2}\, du = x\, dx.$$

$$= \frac{\frac{1}{2} u^{3/2}}{\frac{3}{2}} + C$$

$$= \frac{u^{3/2}}{3} + C$$

$$= \frac{(x^2 + 9)^{3/2}}{3} + C$$

$$C(0) = 150 \qquad \text{We must now find } C.$$

$$\frac{9^{3/2}}{3} + C = 150$$

$$C = 150 - 9 = 141 \qquad \text{The constant is } 141.$$

Thus, we now have

$$C(x) = \frac{(x^2 + 9)^{3/2}}{3} + 141$$

■

SUMMARY

You may come across a problem where, after substitution, the lower limit is greater than the upper limit of integration. Do not rearrange the limits. It is not unusual for this to occur and it is not incorrect. Also, if you encounter a problem where one substitution does not work or seems to be too involved, try a different substitution.

Exercise Set 12.2

Find each indefinite integral. (Some may require integration by parts.)

1. $\displaystyle\int \frac{12x + 2}{3x^2 + x}\, dx$

2. $\displaystyle\int \frac{6x^2 + 2}{x^3 + x}\, dx$

3. $\displaystyle\int \frac{x^2}{x + 2}\, dx$

4. $\displaystyle\int \frac{2x}{x + 2}\, dx$

5. $\displaystyle\int e^{8x}\, dx$

6. $\displaystyle\int \frac{e^x}{1 + e^x}\, dx$

7. $\displaystyle\int \frac{x}{(x + 3)^3}\, dx$

8. $\displaystyle\int \frac{x^2}{(x + 1)^4}\, dx$

9. $\displaystyle\int \frac{1}{6 + \sqrt{2x}}\, dx$

10. $\displaystyle\int \frac{2\sqrt{x} + 3}{x}\, dx$

11. $\displaystyle\int \frac{1}{1 + \sqrt{x}}\, dx$

12. $\displaystyle\int \frac{1 - \sqrt{x}}{1 + \sqrt{x}}\, dx$

13. $\displaystyle\int x^2(x^3 + 1)^8\, dx$

14. $\displaystyle\int (4x - 1)(4x^2 - 2x + 1)^{1/3}\, dx$

15. $\displaystyle\int \frac{x - 1}{x^2 - 2x + 5}\, dx$

16. $\displaystyle\int \frac{x}{x^2 + 1}\, dx$

17. $\displaystyle\int \left(x + \frac{1}{2}\right) e^{x^2 + x + 1}\, dx$

18. $\displaystyle\int \frac{e^{-x} - 1}{(e^{-x} + x)^2}\, dx$

19. $\displaystyle\int \frac{\sqrt{\ln x}}{x}\, dx$

20. $\displaystyle\int \frac{(\ln x)^5}{x}\, dx$

21. $\displaystyle\int \frac{x}{e^x}\, dx$

22. $\displaystyle\int x \ln(x + 1)\, dx$

Evaluate each definite integral using both the properties of integration and your graphing calculator. (Some may require integration by parts.)

23. $\displaystyle\int_1^4 x\sqrt{3x^2 + 1}\, dx$

24. $\displaystyle\int_0^6 \frac{1}{\sqrt{2x + 4}}\, dx$

25. $\displaystyle\int_2^5 \frac{x}{\sqrt{x - 1}}\, dx$

26. $\displaystyle\int_1^3 \frac{x}{x^2 + 1}\, dx$

27. $\displaystyle\int_0^5 \frac{4x}{x^2 + 9}\, dx$

28. $\displaystyle\int_0^4 \frac{1}{x + 1}\, dx$

29. $\int_0^2 x(4 - x^2)^{1/2} \, dx$ 30. $\int_0^4 \sqrt{3x + 4} \, dx$

31. $\int_1^{e^2} \frac{4(\ln x)^3}{x} \, dx$ 32. $\int_2^4 \frac{1 + 3x^2}{5(x + x^3)} \, dx$

33. $\int_0^6 \frac{x}{\sqrt{5x + 6}} \, dx$

34. $\int_1^3 (x + 1)(\sqrt{3} - x) \, dx$

35. $\int_0^1 \frac{xe^x}{(x + 1)^2} \, dx$ 36. $\int_2^4 \frac{\ln x}{x^2} \, dx$

Applications (Business and Economics)

37. **Demand.** A marginal demand function is defined to be

$$D'(x) = \frac{-2000x}{\sqrt{25 - x^2}}$$

Find the demand function, $D(x)$, if $D(3) = 13,000$.

38. **Profit.** A distributing company has found that the marginal profit function for their company is

$$P'(x) = \frac{9000 - 3000x}{(x^2 - 6x + 10)^2}$$

If $P(3) = 1500$, find $P(x)$.

39. **Production.** A toy manufacturing company finds that the marginal production t hours after a shift begins is

$$p'(t) = \frac{t}{\sqrt{t^2 + 1}}, \qquad 0 \le t \le 8$$

What is the total volume of production during an 8-hour shift?

40. **Revenue.** If a marginal revenue function is given as

$$R'(x) = \frac{20x}{x^2 + 1}$$

what would be the total change in revenue if x is increased from 3 to 7?

Review Exercises

Integrate using integration by parts.

41. $\int 4x^2 \ln x \, dx$ 42. $\int 3xe^x \, dx$

12.3   IMPROPER INTEGRALS AND STATISTICAL DISTRIBUTIONS

OVERVIEW   Some applications of integral calculus in business, in statistics, and the social and life sciences involve finding areas that extend infinitely to the right or left or extend infinitely in both directions, as shown by the shaded regions in Figure 5.

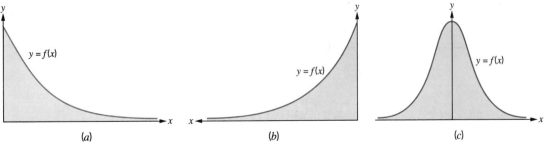

(a)   (b)   (c)

Figure 5

In this section, we show how to determine if the area of such a region exists and how to find the area if it does exist. To do this we must define improper integrals. In this section we

- Determine improper integrals
- Determine whether an improper integral does or does not exist
- Use improper integrals to compute capital values

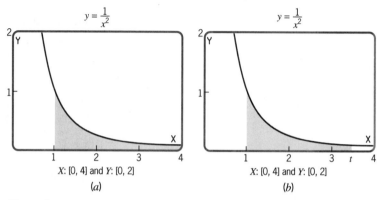

X: [0, 4] and Y: [0, 2]   X: [0, 4] and Y: [0, 2]

(a)   (b)

Figure 6

We introduce improper integrals by finding the area between $y = 1/x^2$ and the x-axis to the right of $x = 1$ [see Figure 6(a)]. Notice that this region is not bounded. Since we have not yet considered how to find the area of such an unbounded region, we consider the following problem. Find the area under this curve from $x = 1$ to t [see Figure 6(b)].

$$A = \int_1^t \frac{1}{x^2}\, dx = -\frac{1}{x}\Big|_1^t \qquad \text{Antiderivative of } \frac{1}{x^2}.$$

$$= -\frac{1}{t} - \frac{-1}{1}$$

$$= 1 - \frac{1}{t}$$

Now as t becomes large, $1/t$ becomes very small and actually approaches 0. [Remember from Chapter 9 that $\lim\limits_{x \to \infty} (1/x) = 0$.] We can see that the area gets closer and closer to 1. You can verify the above conclusion by substituting values for t and finding A.

$$t = 1000 \qquad A = 0.999$$

$$t = 100{,}000 \qquad A = 0.99999$$

$$t = 1{,}000{,}000 \qquad A = 0.999999$$

Thus, intuitively,

$$\lim_{t \to \infty} \int_1^t \frac{1}{x^2}\, dx = 1$$

This example suggests the following definition.

IMPROPER INTEGRALS

If $f(x)$ is continuous over the interval of integration and if the limit exists, then

1. $\displaystyle\int_a^\infty f(x)\, dx = \lim_{t \to \infty} \int_a^t f(x)\, dx$; see Figure 7(a).

2. $\displaystyle\int_{-\infty}^b f(x)\, dx = \lim_{t \to -\infty} \int_t^b f(x)\, dx$; see Figure 7(b).

3. $\displaystyle\int_{-\infty}^\infty f(x)\, dx = \int_{-\infty}^c f(x)\, dx + \int_c^\infty f(x)\, dx$ [see Figure 7(c)], where c is any real number provided that both improper integrals exist.

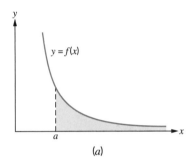

(a)

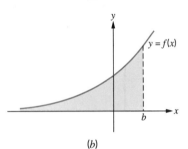

(b)

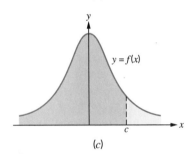

(c)

Figure 7

Calculator Note

With your calculator using fnInt, find values for

$$\int_0^t \frac{dx}{\sqrt{x^4 + 1}}$$

by using increasingly large values for t. What do your calculations tell you about

$$\lim_{t \to \infty} \int_0^t \frac{dx}{\sqrt{x^4 + 1}}\,?$$

Be careful here. Some calculators may give incorrect information if t gets too large. The answer is 1.85407, and if you use 10,000 as the upper limit you get very close to this answer. However, if you use 90,000,000, the answer given is incorrect. Choose your upper limits carefully and use enough limits to see a pattern.

If the indicated limit of an improper integral exists, then the improper integral is said to **converge**; if the limit does not exist, then the improper integral is said to **diverge**.

EXAMPLE 15   Find

$$\int_1^\infty e^{-x}\, dx$$

if the value of this integral exists. With your calculator draw the graph of $y = e^{-x}$ over $[0, 10]$ (see Figure 8).

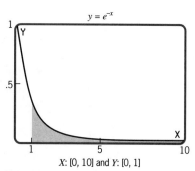

$$y = e^{-x}$$

X: $[0, 10]$ and Y: $[0, 1]$

Figure 8

SOLUTION

$$\int_1^\infty e^{-x}\, dx \quad \text{equals the value of} \quad \int_1^t e^{-x}\, dx$$

as t becomes very large, so

$$\int_1^t e^{-x}\, dx = -e^{-x}\bigg|_1^t = -e^{-t} + e^{-1}$$

As t becomes very large, e^{-t} approaches 0; hence, $-e^{-t} + e^{-1}$ approaches e^{-1}. Thus

$$\int_1^\infty e^{-x}\, dx = e^{-1} = \frac{1}{e}$$

EXAMPLE 16   Find

$$\int_{-\infty}^{-3} \frac{dx}{(1 - x)^{3/2}}$$

if it exists. Draw the graph of $y = 1/(1 - x)^{3/2}$ over $[-9, 1]$ (see Figure 9).

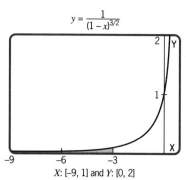

$$y = \frac{1}{(1 - x)^{3/2}}$$

X: $[-9, 1]$ and Y: $[0, 2]$

Figure 9

SOLUTION

$$\int_{-\infty}^{-3} \frac{dx}{(1-x)^{3/2}} = \lim_{t \to -\infty} \int_{t}^{-3} \frac{dx}{(1-x)^{3/2}}$$

$$= \lim_{t \to -\infty} \int_{t}^{-3} (1-x)^{-3/2} \, dx$$

$$= \lim_{t \to -\infty} \left[\frac{-(1-x)^{-1/2}}{-\frac{1}{2}} \right] \Big|_{t}^{-3} \qquad \int u^n \, du = \frac{u^{n+1}}{n+1}$$

$$= \lim_{t \to -\infty} [2(4)^{-1/2} - 2(1-t)^{-1/2}]$$

$$= \lim_{t \to -\infty} \left[\frac{2}{\sqrt{4}} - \frac{2}{\sqrt{1-t}} \right]$$

$$= \lim_{t \to -\infty} \left[1 - \frac{2}{\sqrt{1-t}} \right] = 1 \qquad \lim_{t \to -\infty} \frac{2}{\sqrt{1-t}} = 0 \quad \blacksquare$$

Calculator Note

Evaluating improper integrals is not as easy to do on your calculator as definite integrals. The numbers you may have to use as the limits of integration to get a fair estimate of the integral may be very large. With this in mind, it is really better to use the properties of integrals for calculating an improper integral and use the calculator for drawing the graph and perhaps getting an estimate of the answer. However, you should be aware that even numbers as large as 250,000 used as a limit of integration may not give a really accurate answer. Also, you will need to see the graph and use care in working with integrals that are divergent. It may not be readily apparent from the graph that the integral diverges. For example, fnInt($1/x$, x, 1, 1000) = 6.9077 but fnInt($1/x$, x, 1, 1,000,000) = 13.816.

EXAMPLE 17 Find

$$\int_{1}^{\infty} \frac{x \, dx}{(1+x^2)^2}$$

if it exists. Use your graphing calculator to draw the graph of $y = x/(1+x^2)^2$ over [1, 11] (see Figure 10).

SOLUTION

$$\lim_{t \to \infty} \int_{1}^{t} \frac{x \, dx}{(1+x^2)^2} = \lim_{t \to \infty} \int_{1}^{t} (1+x^2)^{-2} x \, dx$$

Now let's find the indefinite integral $\int(1+x^2)^{-2}x \, dx$, which can be obtained by a change of variable. Let $u = 1 + x^2$. Then

$$du = 2x \, dx \quad \text{and} \quad x \, dx = \tfrac{1}{2} \, du$$

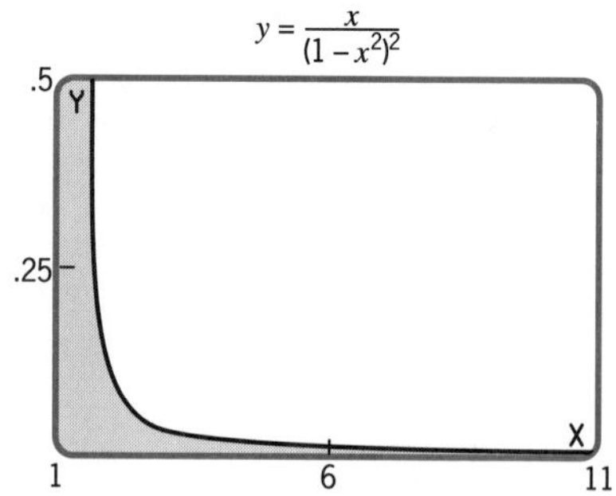

$$y = \frac{x}{(1 - x^2)^2}$$

X: [1, 11] and Y: [0, .5]

Figure 10

Substituting gives

$$\int (1 + x^2)^{-2} x \, dx = \int u^{-2} \left(\frac{1}{2} \right) du = \frac{1}{2} \int u^{-2} \, du$$

$$= -\frac{1}{2} u^{-1} + C = -\frac{1}{2(1 + x^2)} + C$$

Hence

$$\lim_{t \to \infty} \int_1^t (1 + x^2)^{-2} x \, dx = \lim_{t \to \infty} \frac{-1}{2(1 + x^2)} \Bigg|_1^t = \lim_{t \to \infty} \left[\frac{-1}{2(1 + t^2)} + \frac{1}{4} \right]$$

Since $-1/[2(1 + t^2)]$ approaches 0 as t becomes very large,

$$\int_1^\infty \frac{x \, dx}{(1 + x^2)^2} = 0 + \frac{1}{4} = \frac{1}{4}$$

Practice Problem 1 Evaluate each of the following.

(a) $\displaystyle\int_2^\infty \frac{dx}{(1 + x)^{3/2}}$ (b) $\displaystyle\int_{-\infty}^2 e^{2x} \, dx$

ANSWER

(a) $\dfrac{2}{\sqrt{3}}$ (b) $\dfrac{e^4}{2}$

EXAMPLE 18 Graph $f(x) = 1/x$ and find the area of the region under this curve and above the x-axis to the right of $x = 1$, if this area exists.

SOLUTION

$$A = \int_1^\infty \frac{dx}{x}$$

$$\int_1^t \frac{dx}{x} = \ln x \Bigg|_1^t = \ln t - \ln 1 \qquad \text{See Figure 11.}$$

Now, $\ln 1 = 0$; hence,

$$\int_1^t \frac{dx}{x} = \ln t$$

$$\int_1^\infty \frac{dx}{x} = \lim_{t \to \infty} \int_1^t \frac{dx}{x}$$

$$\int_1^\infty \frac{dx}{x} = \lim_{t \to \infty} (\ln t)$$

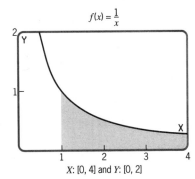

$$f(x) = \frac{1}{x}$$

X: [0, 4] and Y: [0, 2]

Figure 11

Now with your calculator (see Figure 11), you can substitute values and intuitively determine that $\ln t$ becomes large as t becomes large and the limit does not exist. That is,

$$\int_1^\infty \frac{dx}{x}$$

does not exist or the integral diverges. Be careful, however, when you use the calculator for a problem like this. Some large upper bounds require a lot of time for your calculator to compute. Numerical integration without some thought given to the problem can be deceiving.

Calculator Note

You can approximate infinite limits with your graphing calculator using the numerical integration function. Look at

$$\int_1^\infty \frac{dx}{\sqrt{x^2 + x + 4}}$$

Take the upper bound to be increasingly larger numbers and use this to determine whether

$$\lim_{t \to \infty} \int_1^\infty \frac{dx}{\sqrt{x^2 + x + 4}}$$

converges or diverges. What do you see as you take the upper bound from $x = 10$ to $x = 1000$?

We have already discussed that the present value of a continuous income stream over a fixed number of years can be found by integrating a definite integral. When the number of years is extended indefinitely, this present value is called a **capital value of a money flow** or the **present value of a perpetuity**.

CAPITAL VALUE OF MONEY FLOW

$$\text{Capital value} = \int_0^\infty f(t)e^{-rt} \, dt$$

where $f(t)$ is the annual rate of flow at time t, and r is the annual interest rate, compounded continuously.

EXAMPLE 19 How much capital must you invest to provide a continuous money flow with an annual flow given by $f(t) = 10,000$ if the interest rate is 8% compounded continuously?

SOLUTION The capital value of the fund is given by

$$\int_0^\infty 10,000e^{-0.08t} \, dt = \lim_{b \to \infty} \int_0^b 10,000e^{-0.08t} \, dt$$

$$= \lim_{b \to \infty} \frac{10,000e^{-0.08t}}{-0.08} \Big|_0^b$$

$$= \lim_{b \to \infty} \left[\frac{10,000e^{-0.08b}}{-0.08} - \frac{10,000}{-0.08} \right] \qquad e^0 = 1$$

$$= \lim_{b \to \infty} \left[\frac{10,000}{0.08} \left(1 - \frac{1}{e^{0.08b}} \right) \right]$$

$$= \frac{10,000}{0.08} = \$125,000 \qquad\qquad \lim_{b \to \infty} \left(\frac{1}{e^{0.08b}} \right) = 0$$

Do you see why

$$\lim_{b \to \infty} \left(\frac{1}{e^{0.08b}} \right) = 0?$$

If not, you can show that this is a reasonable answer with your calculator. ■

Probability Density Functions

A study of improper integrals in this chapter enables us to consider probability density functions of a continuous variable. These functions are very important in the study of statistics. A continuous probability density function is used to measure probabilities for a continuous random variable. For example, the normal distribution is the foundation for many statistical tests. First we look at five characteristics of probability density functions, commonly called a **pdf of a distribution**. The continuous variable is called a **random variable**. Then we consider problems relative to these characteristics.

Characteristics of Probability Density Functions

If X is a continuous random variable and $f(x)$ is a probability density function, then:

1. $f(x) \geq 0$ for $(-\infty, \infty)$ The function is always nonnegative.

2. $\displaystyle\int_{-\infty}^{\infty} f(x)\,dx = 1$ The area under the curve is 1.

3. The probability of x in $[a, b]$ is

$$P(a \leq X \leq b) = \int_{a}^{b} f(x)\,dx$$

     This is the area under the curve from $x = a$ to $x = b$.

4. The expected value or mean of x is defined to be

$$\mu = E(x) = \int_{-\infty}^{\infty} xf(x)\,dx$$

5. The variance of x (square of standard deviation σ) is defined to be

$$\sigma^2 = \int_{-\infty}^{\infty} (x - \mu)^2 f(x)\,dx \quad \text{or}$$

$$= \int_{-\infty}^{\infty} x^2 f(x)\,dx - \mu^2 \qquad \text{Measure of dispersion or scattering.}$$

Four useful probability density functions of a random variable X are the **uniform** probability density function, the **exponential** probability density funtion, the **normal** probability density function, and the **standard** normal density function (see Figure 12). μ is the mean of the distribution and δ is the standard deviation.

(a)

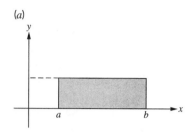

The **uniform probability density function,** (Figure 12(a)), is given by

$$f(x) = \begin{cases} \left(\dfrac{1}{b-a}\right), & a \le x \le b \\ 0, & \text{otherwise} \end{cases}$$

$$\mu = \frac{a+b}{2}$$

$$\sigma = \frac{b-a}{\sqrt{12}}$$

(b)

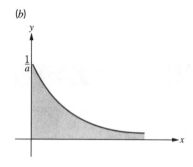

The **exponential probability density function,** (Figure 12(b)), is given by

$$f(x) = \begin{cases} \dfrac{1}{(a)}\, e^{-x/a}, & x \ge 0 \\ 0, & \text{otherwise} \end{cases}$$

$$\mu = a$$

$$\sigma = a$$

(c)

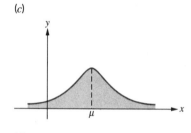

The **normal probability density function** (Figure 12(c)), is defined

$$f(x) = \frac{1}{\sigma\sqrt{2\pi}}\, e^{-(x-\mu)^2/2\sigma^2}$$

$$\mu = \mu \quad \text{(of the formula)}$$

$$\sigma = \sigma \quad \text{(of the formula)}$$

(d)

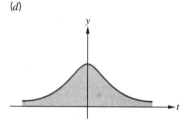

Figure 12

The standard normal density function, (Figure 12(d)), is defined by

$$f(x) = \frac{1}{\sqrt{2\pi}}\, e^{-t^2/2}$$

$$\mu = 0$$

$$\sigma = 1$$

In the normal probability density function and the standard normal density function, the value of μ gives the x-coordinate of the point at which there is a relative maximum and σ determines the spread of the curve.

EXAMPLE 20   Suppose that a random variable X has a pdf defined by

$$f(x) = \begin{cases} 1 - \tfrac{1}{2}x, & 0 \le x \le 2 \\ 0, & \text{elsewhere} \end{cases}$$

See Figure 13. Find   (a) $P(1 \le X \le 2)$ and   (b) $P(X \le 1)$.

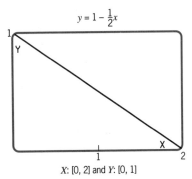

$$y = 1 - \tfrac{1}{2}x$$

X: [0, 2] and Y: [0, 1]

Figure 13

SOLUTION

(a) $P(1 \le X \le 2) = \int_1^2 \left(1 - \frac{1}{2}x\right) dx = \left(x - \frac{x^2}{4}\right)\Big|_1^2 = 1 - \frac{3}{4} = \frac{1}{4}$

(b)   The random variable X assumes nonzero values only in the interval $[0, 2]$ because the pdf is 0 outside of this interval. Therefore, we have

$$P(X \le 1) = P(0 \le X \le 1) = \int_0^1 \left(1 - \frac{1}{2}x\right) dx = \left(x - \frac{x^2}{4}\right)\Big|_0^1$$

$$= \left(\frac{3}{4} - 0\right) = \frac{3}{4}$$

Practice Problem 2   Using the pdf from Example 20, find $P(\frac{1}{2} \le x)$.

ANSWER

$$P\left(\frac{1}{2} \le x\right) = P\left(\frac{1}{2} \le X \le 2\right) = \int_{1/2}^2 \left(1 - \frac{1}{2}x\right) dx = \frac{9}{16}$$

EXAMPLE 21   Because of unpredictable variations in weather and air traffic, the 9:45 A.M. flight from New Orleans to Houston takes off at various times from 9:45 A.M. to 10:00 A.M. What is the probability that a passenger will experience a delay of at most 10 minutes?

SOLUTION   We let X be the delay in minutes; therefore, X is in the interval $[0, 15]$. Thus X is uniformly distributed over the interval and its pdf is

$$f(x) = \begin{cases} \frac{1}{15}, & 0 \le x \le 15 \\ 0, & \text{elsewhere} \end{cases}$$

Thus, the probability of a delay of at most 10 minutes is

$$P(0 \le X \le 10) = \int_0^{10} \frac{1}{15} \, dx = \frac{1}{15}x \Big|_0^{10} = \frac{10}{15} - 0 = \frac{2}{3}.$$

EXAMPLE 22   A company has found that the average length of each business call is 7 minutes. If the length of a call has an exponential distribution

$$f(x) = \frac{1}{a} e^{-x/a}$$

find the following probabilities.

(a)   The probability that a call will be no more than 5 minutes.
(b)   The probability that a call will be longer than 5 minutes.

SOLUTION

(a)   Since the average call is 7 minutes, $a = 7$, so the pdf for X is $f(x) = \frac{1}{7}e^{-x/7}$, where $x \geq 0$. We now find the probability.

$$P(0 \leq X \leq 5) = \int_0^5 \frac{1}{7} e^{-x/7} \, dx$$

$$= -e^{-x/7} \Big|_0^5$$

$$= -e^{-5/7} - (-e^0)$$

$$= 1 - e^{-5/7}$$

$$\approx 0.510458$$

Therefore, approximately 51% of the calls are 5 minutes or less.

(b)   The probability that a call will be longer than 5 minutes is

$$P(x > 5) = \int_5^\infty \frac{1}{7} e^{-x/7} \, dx$$

However, from the information in part (a) and the fact that the total area under the curve is 1 (why?), we can find the probability as follows:

$$P(x > 5) = 1 - P(0 \leq X \leq 5) = 1 - 0.510458 = 0.489542 \quad \blacksquare$$

Practice Problem 3 Let X be an exponentially distributed random variable with an average value of 2. Find $P(1 \leq X \leq 4)$.

ANSWER

$$P(1 \leq X \leq 4) = \int_1^4 \frac{1}{2} e^{-x/2} \, dx = -e^{-x/2} \Big|_1^4 = 0.4712$$

SUMMARY

Improper integrals allow us to solve a number of important applications in all areas, from cash flow to probability distributions. Remember, it is not always possible to integrate an improper integral. If a limit can be found, the integral converges, otherwise it is divergent.

Exercise Set 12.3

Find the value of the following improper integrals if they exist.

1. $\displaystyle\int_2^\infty \frac{dx}{x^2}$

2. $\displaystyle\int_{-\infty}^{-1} \frac{dx}{x^2}$

3. $\displaystyle\int_1^\infty \frac{dx}{x^2}$

4. $\displaystyle\int_1^\infty \frac{dx}{x^{3/2}}$

5. $\displaystyle\int_1^\infty \frac{dx}{(1+x)^{3/2}}$

6. $\displaystyle\int_3^\infty \frac{dx}{(1+x)^{3/2}}$

7. $\displaystyle\int_0^\infty e^x\, dx$

8. $\displaystyle\int_{-\infty}^0 e^x\, dx$

9. $\displaystyle\int_0^\infty e^{-x}\, dx$

10. $\displaystyle\int_{-\infty}^{-1} e^{-x}\, dx$

11. $\displaystyle\int_0^\infty e^{-2x}\, dx$

12. $\displaystyle\int_{-\infty}^0 e^{3x}\, dx$

13. $\displaystyle\int_1^\infty \frac{x\, dx}{1+x^2}$

14. $\displaystyle\int_{-\infty}^{-1} \frac{x\, dx}{1+x^2}$

15. $\displaystyle\int_0^\infty xe^{-x^2}\, dx$

16. $\displaystyle\int_{-\infty}^0 xe^{-2x}\, dx$

17. $\displaystyle\int_2^\infty \frac{x\, dx}{(1+x^2)^2}$

18. $\displaystyle\int_{-\infty}^{-1} \frac{x\, dx}{(1+x^2)^2}$

19. $\displaystyle\int_0^\infty x^2 e^{-x^3}\, dx$

20. $\displaystyle\int_{-\infty}^{-1} 2x^2 e^{x^3}\, dx$

21. $\displaystyle\int_{-\infty}^\infty \frac{x\, dx}{(1+x^2)^2}$

Determine whether the following integrals are convergent or divergent. If convergent, find the value. Check your answers with your graphing calculator.

22. $\displaystyle\int_{-\infty}^\infty x\, dx$

23. $\displaystyle\int_{-\infty}^\infty \frac{2x\, dx}{1+x^2}$

24. $\displaystyle\int_{-\infty}^\infty xe^{-x^2}\, dx$

25. $\displaystyle\int_0^\infty xe^{-x}\, dx$

26. $\displaystyle\int_{-\infty}^0 e^{3x}\, dx$

27. $\displaystyle\int_{-\infty}^0 xe^{-x^2}\, dx$

Use your graphing calculator to find each limit, if it exists.

28. $\displaystyle\lim_{t\to\infty} \int_1^t \frac{dx}{x^4}$

29. $\displaystyle\lim_{t\to\infty} \int_1^t \frac{dx}{\sqrt[3]{x}}$

30. $\displaystyle\lim_{t\to\infty} \int_1^t \frac{dx}{\sqrt{x}}$

Applications (Business and Economics)

31. **Perpetuity Trust.** Find the present value of a perpetuity trust that pays a yearly amount given by

$$f(t) = \$2000e^{-0.06t}$$

if interest is compounded continuously.

32. **Perpetuity Trust.** What is the value of a trust that pays $10,000 a year in perpetuity (forever) if interest is compounded continuously at a rate of 6% per year? See Exercise 31.

33. **Capital Value of Perpetuity.** A donor wishes to provide a gift for a university that will provide a continuous money flow of $15,000 per year for scholarships. If money is worth 8% compounded continuously, find the capital value of this gift.

34. **Production.** Suppose that the rate of production of an oil company per month is given in thousands of barrels as

$$\frac{dP}{dt} = 26te^{-0.04t}$$

How much oil is produced during the first 12 months of operation? If the production is continued indefinitely, what would be the total amount produced?

35. **Investment.** An investment gives

$$\frac{dA}{dt} = 2000e^{-0.12t}$$

dollars per year. Assume that the investment is continued indefinitely, and find the total amount A returned from this investment.

36. **Capital Value.** How much capital must you invest to provide a continuous money flow of $f(t) = 10,000 + 500t$, if the interest rate is 8% compounded continuously?

37. **Present Value of a Perpetuity.** If the continuous money flow for a perpetuity is given by $f(t) = R$ dollars per year, show that the present value of the perpetuity is R/r, where r is the rate compounded continuously.

38. **pdf.** Show that

$$\int_{-\infty}^{\infty} f(x)\, dx = 1$$

for the uniform density function.

39. **pdf.** Show that

$$\int_{-\infty}^{\infty} f(x)\, dx = 1$$

for the exponential density function.

40. **pdf.** Using the definition, find the mean of the uniform density function.

41. **pdf.** Using the definition, find the standard deviation of the uniform density function.

42. **pdf.** Using the definition, find the mean of the exponential density function.

43. **pdf.** Using the definition, find the standard deviation of the exponential density function.

44. **pdf.** If $a = 5$ and $b = 35$, find $P(10 \leq X \leq 20)$ for the uniform density function.

45. **pdf.** If $a = 10$ for the exponential density function, find $P(X \geq 20)$.

46. **pdf.** In Exercise 45, find $P(1 \leq X \leq 20)$.

47. **pdf.** In Exercise 45, find $P(X \leq 10)$.

48. **pdf.** For $f(x)$ to be a probability density function,

$$\int_{-\infty}^{\infty} f(x)\, dx$$

must equal 1. Find k so that this condition is satisfied for

$$f(x) = \begin{cases} kx^2, & \text{for } 0 < x < 2 \\ 0, & \text{elsewhere.} \end{cases}$$

49. **pdf.** Find k so that

$$\int_{-\infty}^{\infty} f(x)\, dx = 1 \quad \text{for } f(x) = \begin{cases} ke^{-x/2}, & \text{for } x > 0 \\ 0, & \text{elsewhere.} \end{cases}$$

50. **pdf.** Let X be uniformly distributed over the interval $[1, 6]$. Find each of the following.
   (a) $P(2 \leq X \leq 3)$
   (b) $P(X \geq 2)$
   (c) $P(X \leq 4)$
   (d) $P(X \geq 1)$
   (e) Find the expected value (mean).

51. **pdf.** The voltage in a 220-volt line varies randomly between 210 and 230 volts.
   (a) What is the probability that the voltage will be between 215 and 225 volts?
   (b) If a power surge can cause damage to an electrical system if the voltage is more than 225 volts, what percentage of the time will there be a danger of damage?
   (c) What is the expected voltage?

52. **pdf.** Let X be an exponentially distributed random variable with an average value of 5.
   (a) Find the pdf for X.
   (b) Find $P(X \leq 5)$.
   (c) Find $P(0 \leq X \leq 3)$.
   (d) Use the result in part (b) to find $P(X > 5)$.
   (e) Find $P(X > 5)$ by evaluating an appropriate improper integral.

53. **pdf.** Suppose that the average distance between barges on a river is 300 feet and the distance is exponentially distributed.
   (a) What is the probability that the distance between two successive barges will be no more than 150 feet?
   (b) What is the probability that the distance will be between 250 and 350 feet?

54. Given the pdf

$$f(x) = \begin{cases} 1.5x - 0.75x^2, & \text{for } 0 \leq x \leq 2 \\ 0, & \text{elsewhere} \end{cases}$$

use your graphing calculator to graph the function and approximate to three decimal places the necessary value of x to make the equation true.
   (a) $P(0 \leq X \leq x) = 0.3$
   (b) $P(x \leq X \leq 2) = 0.5$

55. Given the pdf

$$f(x) = \begin{cases} 0.75 - 0.375\sqrt{x}, & \text{for } 0 \leq x \leq 4 \\ 0, & \text{elsewhere} \end{cases}$$

use your graphing calculator to graph the function and approximate to three decimal places the necessary value of x to make the equation true.

(a) $P(0 \le X \le x) = 0.5$

(b) $P(x \le X \le 4) = 0.3$

Applications (Social and Life Sciences)

56. **Drug Elimination.** When a person takes a drug, the body does not absorb all of the drug. One way to determine the amount of drug absorbed is to measure the rate at which the drug is eliminated from the body. A doctor finds that the rate of elimination of a drug in milliliters per minute, t, is given by

$$\frac{dE}{dt} = te^{-0.3t}$$

Assume that the elimination continues indefinitely, and find the total amount eliminated, E.

Review Exercises

Find the following integrals.

57. $\displaystyle\int \frac{5e^x\, dx}{\sqrt[3]{e^x + 2}}$

58. $\displaystyle\int 3xe^{-2x}\, dx$

59. $\displaystyle\int (2x - 3)e^{2x}\, dx$

60. $\displaystyle\int 4x^2 \ln x\, dx$

61. $\displaystyle\int x^4 \ln x\, dx$

62. $\displaystyle\int 3x\sqrt{x + 1}\, dx$

12.4   INTEGRATION USING TABLES

OVERVIEW   We have studied several techniques for finding the antiderivatives of a function. You have probably noticed that finding antiderivatives is not as straightforward and is somewhat more difficult than finding derivatives. Because of this, formulas for integration have been compiled into tables of integrals in most mathematics handbooks. Our goal in this section is to practice properly matching a given integral with a formula found in a table. To accomplish this goal, we have included a very abbreviated table of integrals. It is customary in integral tables to omit the constant of integration. Therefore, you must remember to include it in your answer. In this section, we

- Practice recognizing formulas
- Use substitution to match an integral formula
- Use reduction formulas

In our abbreviated table of integrals we have not included formulas that have already been given. The table is divided into four sections:

Integrals involving $au + b$
Integrals involving $\sqrt{au + b}$
Integrals involving $u^2 \pm a^2$ and $a \pm u^2$
Integrals involving e^{au} and $\ln u$.

This is not, by any means, an exhaustive table, but it will serve to introduce you to the technique. Remember to add the constant of integration. There are handbooks that contain extensive tables of integration.

A Brief Table of Integrals

Integrals Involving $au + b$

1. $\displaystyle \int \frac{u}{au + b} \, du = \frac{u}{a} - \frac{b}{a^2} \ln |au + b|$

2. $\displaystyle \int \frac{u^2}{au + b} \, du = \frac{1}{a^3} \left[\frac{1}{2} (au + b)^2 - 2b(au + b) + b^2 \ln |au + b| \right]$

3. $\displaystyle \int \frac{u}{(au + b)^2} \, du = \frac{1}{a^2} \left[\frac{b}{au + b} + \ln |au + b| \right]$

4. $\displaystyle \int \frac{1}{(au + b)(cu + d)} \, du = \frac{1}{bc - ad} \ln \left| \frac{cu + d}{au + b} \right| \quad (bc - ad > 0)$

5. $\displaystyle \int \frac{1}{u^2(au + b)} \, du = -\frac{1}{bu} + \frac{a}{b^2} \ln \left| \frac{au + b}{u} \right|$

6. $\displaystyle \int \frac{1}{u(au + b)^2} \, du = \frac{1}{b(au + b)} + \frac{1}{b^2} \ln \left| \frac{u}{au + b} \right|$

Integrals Involving $\sqrt{au + b}$

7. $\displaystyle \int u\sqrt{au + b} \, du = \frac{2(3au - 2b)(au + b)^{3/2}}{15a^2}$

8. $\displaystyle \int \frac{u}{\sqrt{au + b}} \, du = \frac{2}{3a^2} (au - 2b) \sqrt{au + b}$

9. $\displaystyle \int \frac{u^n}{\sqrt{au + b}} \, du = \frac{2u^n \sqrt{au + b}}{a(2n + 1)} - \frac{2bn}{a(2n + 1)} \int \frac{u^{n-1}}{\sqrt{au + b}} \, du \quad (n \geq 2)$

Integrals Involving $u^2 \pm a^2$ and $a^2 \pm u^2$ $(a > 0)$

10. $\displaystyle \int \frac{du}{u^2 - a^2} = \frac{1}{2a} \ln \left| \frac{u - a}{u + a} \right| \quad (u^2 \neq a^2)$

11. $\displaystyle \int \frac{du}{a^2 - u^2} = \frac{1}{2a} \ln \left| \frac{a + u}{a - u} \right| \quad (u^2 \neq a^2)$

12. $\displaystyle \int \frac{du}{\sqrt{u^2 \pm a^2}} = \ln |u + \sqrt{u^2 \pm a^2}|$

13. $\displaystyle \int \frac{du}{u\sqrt{a^2 \pm u^2}} = -\frac{1}{a} \ln \left| \frac{a + \sqrt{a^2 \pm u^2}}{u} \right|$

14. $\displaystyle \int \frac{du}{u^2\sqrt{a^2 \pm u^2}} = -\frac{\sqrt{a^2 \pm u^2}}{a^2 u}$

15. $\displaystyle \int \frac{du}{u^2\sqrt{u^2 - a^2}} = \frac{\sqrt{u^2 - a^2}}{a^2 u}$

16. $\displaystyle \int \sqrt{u^2 \pm a^2} \, du = \frac{1}{2} (u\sqrt{u^2 \pm a^2} \pm a^2 \ln |u + \sqrt{u^2 \pm a^2}|)$

17. $\displaystyle \int u^2\sqrt{u^2 \pm a^2} \, du = \frac{1}{8} [u(2u^2 \pm a^2) \sqrt{u^2 \pm a^2} - a^4 \ln |u + \sqrt{u^2 \pm a^2}|]$

18. $\displaystyle\int \frac{\sqrt{a^2 \pm u^2}}{u}\, du = \sqrt{a^2 \pm u^2} - a \ln \left| \frac{a + \sqrt{a^2 \pm u^2}}{u} \right|$

19. $\displaystyle\int \frac{\sqrt{u^2 \pm a^2}}{u^2}\, du = -\frac{\sqrt{u^2 \pm a^2}}{u} + \ln \left| u + \sqrt{u^2 \pm a^2} \right|$

20. $\displaystyle\int \frac{u^2}{\sqrt{u^2 \pm a^2}}\, du = \frac{1}{2}\left(u\sqrt{u^2 \pm a^2} \mp a^2 \ln \left| u + \sqrt{u^2 \pm a^2} \right|\right)$

Integrals Involving e^{au} and $\ln u$

21. $\displaystyle\int u e^{au}\, du = \frac{e^{au}}{a^2}(au - 1)$

22. $\displaystyle\int u^n e^{au}\, du = \frac{u^n e^{au}}{a} - \frac{n}{a}\int u^{n-1} e^{au}\, du \qquad n > 0$

23. $\displaystyle\int u^n \ln u\, du = \frac{u^{n+1}}{(n+1)^2}[(n+1)\ln u - 1]$

24. $\displaystyle\int \frac{1}{u \ln u}\, du = \ln |\ln u|$

EXAMPLE 23   Find $\displaystyle\int \frac{5\, dx}{x^2 - 36}$.

SOLUTION

$$\int \frac{5\, dx}{x^2 - 36} = 5 \int \frac{dx}{x^2 - 36}$$

The denominator seems to involve $u^2 - a^2$. Letting $u = x$ and $du = dx$ and comparing with formula 10, we have

$$\int \frac{du}{u^2 - a^2} = \frac{1}{2a} \ln \left| \frac{u - a}{u + a} \right| + C \qquad u^2 \ne a^2,\ a^2 = 36$$

$$\int \frac{dx}{x^2 - 36} = \frac{1}{2 \cdot 6} \ln \left| \frac{x - 6}{x + 6} \right| + C \qquad \text{for } x^2 \ne 36$$

So $\displaystyle\int \frac{5\, dx}{x^2 - 36} = \frac{5}{12} \ln \left| \frac{x - 6}{x + 6} \right| + C$

To check this result with your calculator, calculate

$$\int \frac{5\, dx}{x^2 - 36}$$

as a definite integral, say

$$\int_1^3 \frac{5\, dx}{x^2 - 36}$$

Use the answer you got for the integral and substitute $x = 1$ and $x = 3$. Thus, we have

$$\frac{5}{12} \ln \left| \frac{x-6}{x+6} \right| \Big|_1^3 = \frac{5}{12} \ln \left| \frac{3-6}{3+6} \right| - \frac{5}{12} \ln \left| \frac{1-6}{1+6} \right| \approx -0.317558$$

Likewise, using the numeric integration on the calculator, you get

$$\text{fnInt}(5/(x^2 - 36), x, 1, 3) \approx -.317558$$

Calculator Note

For each example in this section you should continue to check your work as we did in the last example.

EXAMPLE 24 Find $\displaystyle\int \frac{dx}{6x^2 + x - 15}$.

SOLUTION There does not appear to be a formula in the table for this expression. Note, however, that the denominator may be factored.

$$\int \frac{dx}{6x^2 + x - 15} = \int \frac{dx}{(3x + 5)(2x - 3)}$$

It seems that formula 4 can be used with $u = x$ and $du = dx$.

$$\int \frac{1}{(au + b)(cu + d)} \, du = \frac{1}{bc - ad} \ln \left| \frac{cu + d}{au + b} \right| + C \qquad \text{Formula 4}$$

We will use $a = 3$, $b = 5$, $c = 2$, and $d = -3$.

$$\int \frac{1}{(3x + 5)(2x - 3)} \, dx = \frac{1}{\underbrace{5 \cdot 2 - (3)(-3)}_{bc - ad}} \ln \left| \frac{2x - 3}{3x + 5} \right| + C \qquad bc - ad = 19 > 0$$

$$= \frac{1}{19} \left[\ln |2x - 3| - \ln |3x + 5| \right] + C$$

EXAMPLE 25 Find $\displaystyle\int \frac{2x^2}{\sqrt{25x^2 - 36}} \, dx$.

SOLUTION To put this integral into a form involving $\sqrt{u^2 - a^2}$, let $u = 5x$. Then $x = \frac{1}{5}u$ and $dx = \frac{1}{5} \, du$.

$$\int \frac{2x^2}{\sqrt{25x^2 - 36}} \, dx = \int \frac{2(\frac{1}{25})u^2}{\sqrt{u^2 - 36}} \frac{1}{5} \, du = \frac{2}{125} \int \frac{u^2}{\sqrt{u^2 - 36}} \, du$$

Using formula 20 gives

$$\frac{2}{125} \int \frac{u^2}{\sqrt{u^2 - 36}} \, du$$

$$= \left(\frac{2}{125}\right)\left(\frac{1}{2}\right)\left(u\sqrt{u^2 - 36} + 36 \ln\left|u + \sqrt{u^2 - 36}\right|\right) + C$$

Thus

$$\int \frac{2x^2}{\sqrt{25x^2 - 36}} \, dx = \frac{1}{125}\left(5x\sqrt{25x^2 - 36} + 36 \ln\left|5x + \sqrt{25x^2 - 36}\right|\right) + C$$

■

EXAMPLE 26   Find $\int x^2 e^{4x} \, dx$.

SOLUTION   Formula 22 can be used as a first step in this problem. Let $u = x$ and $du = dx$.

$$\int u^n e^{au} \, du = \frac{u^n e^{au}}{a} - \frac{n}{a}\int u^{n-1} e^{au} \, du + C \qquad \text{Formula 22}$$

$$\int x^2 e^{4x} \, dx = \frac{x^2 e^{4x}}{4} - \frac{2}{4}\int x^{2-1} e^{4x} \, dx + C \qquad n = 2, \, a = 4$$

Now using formula 21, let $u = x$ and $du = dx$.

$$\int u e^{au} \, du = \frac{e^{au}}{a^2}(au - 1) \qquad\qquad \text{Formula 21}$$

$$\int x e^{4x} \, dx = \frac{e^{4x}}{4^2}(4x - 1) \qquad\qquad a = 4$$

So $\int x^2 e^{4x} \, dx = \frac{x^2 e^{4x}}{4} - \frac{2}{4}\left[\frac{e^{4x}}{16}(4x - 1)\right] + C$

$$= e^{4x}\left[\frac{x^2}{4} - \frac{1}{8}x + \frac{1}{32}\right] + C$$

$$= e^{4x}\left[\frac{8x^2 - 4x + 1}{32}\right] + C$$

■

Formulas such as formula 22 used in the preceding example are called **reduction formulas**. These may be applied repeatedly until the integral is in a form that you can integrate.

Practice Problem 1 Find $\int \dfrac{x^2}{3x + 2}\, dx$.

ANSWER

$$\frac{1}{27}\left[\frac{1}{2}(3x + 2)^2 - 4(3x + 2) + 4 \ln |3x + 2|\right] + C$$

EXAMPLE 27 Find

$$\int_9^{12} \frac{1}{x^2\sqrt{225 - x^2}}\, dx$$

Use your calculator to check your work.

SOLUTION Use formula 14 with $u = x$ and $a = 15$. Then

$$\int_9^{12} \frac{1}{x^2\sqrt{225 - x^2}}\, dx = \int_9^{12} \frac{1}{u^2\sqrt{225 - u^2}}\, du$$

$$= -\frac{\sqrt{225 - u^2}}{225u}\Bigg|_{u=9}^{u=12}$$

$$= -\frac{\sqrt{225 - 144}}{(225)(12)} + \frac{\sqrt{225 - 81}}{(225)(9)}$$

$$= -\frac{9}{(225)(12)} + \frac{12}{(225)(9)}$$

$$= -\frac{9}{2700} + \frac{16}{2700} = \frac{7}{2700}$$ ■

Practice Problem 2 Using a calculator, graph

$$y = \frac{1}{x^2\sqrt{225 - x^2}}$$

for $9 \le x \le 12$ and determine whether the answer in Example 27 can be considered as an area.

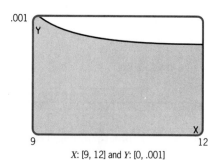

X: [9, 12] and Y: [0, .001]

Figure 14

ANSWER   Yes. (see Figure 14).

EXAMPLE 28   Find the consumers' surplus at a price level of $8 for the price–demand equation

$$p = D(x) = \frac{20x - 5000}{x - 400}$$

SOLUTION

1.   The first step is to find the production for a price level of $8.

$$8 = \frac{20x_1 - 5000}{x_1 - 400} \qquad p_1 = 8$$

$$8x_1 - 3200 = 20x_1 - 5000 \qquad \text{Multiply by } x_1 - 400.$$

$$12x_1 = 1800$$

$$x_1 = 150$$

2.   The second step is to show the consumers' surplus by a shaded region on a graph (see Figure 15).

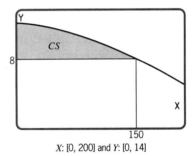

X: [0, 200] and Y: [0, 14]

Figure 15

3.   The consumers' surplus, the shaded region of the graph, is the area between

$$p_1 = 8 \quad \text{and} \quad p = D(x) = \frac{20x - 5000}{x - 400}$$

Thus

$$CS = \int_0^{x_1} [D(x) - p_1]\, dx$$

$$= \int_0^{150} \left[\frac{20x - 5000}{x - 400} - 8 \right] dx$$

$$= \int_0^{150} \left[\frac{12x - 1800}{x - 400} \right] dx$$

$$= 12 \int_0^{150} \frac{x}{x - 400} \, dx - 1800 \int_0^{150} \frac{dx}{x - 400}$$

$$= 12[x + 400 \ln |x - 400|] \Big|_0^{150} - 1800 \ln |x - 400| \Big|_0^{150} \qquad \text{Formula 1}$$

$$= 12[150 + 400 \ln 250 - 400 \ln 400]$$
$$\quad - 1800 \ln 250 + 1800 \ln 400$$

$$= 1800 + 3000 \ln 250 - 3000 \ln 400$$

$$\approx 389.99$$

The consumers' surplus is $389.99.

SUMMARY

Integration tables can be extensive. We have included only a few entries to introduce you to the use of an integration table. Using a table requires practice, since
you must become familiar with the various forms that the table has and how a
formula may fit the problem you are trying to solve.

Exercise Set 12.4

Using a table of integrals, find each of the following.

1. $\displaystyle \int \frac{dx}{x^2 - 25}$

2. $\displaystyle \int \frac{3 \, dx}{x^2 - 36}$

3. $\displaystyle \int \frac{dx}{36 - x^2}$

4. $\displaystyle \int \frac{4 \, dx}{16 - x^2}$

5. $\displaystyle \int \frac{dx}{\sqrt{x^2 + 36}}$

6. $\displaystyle \int \frac{dx}{\sqrt{x^2 - 36}}$

7. $\displaystyle \int \frac{dx}{x \sqrt{x^2 + 25}}$

8. $\displaystyle \int \frac{dx}{x \sqrt{49 - x^2}}$

9. $\displaystyle \int \frac{dx}{x^2 \sqrt{49 + x^2}}$

10. $\displaystyle \int \frac{dx}{x^2 \sqrt{49 - x^2}}$

11. $\displaystyle \int \frac{dx}{x^2 \sqrt{x^2 - 49}}$

12. $\displaystyle \int \sqrt{x^2 - 49} \, dx$

13. $\displaystyle \int x^2 \sqrt{x^2 + 25} \, dx$

14. $\displaystyle \int \frac{\sqrt{x^2 + 25}}{x^2} \, dx$

15. $\displaystyle \int \frac{\sqrt{x^2 + 25}}{x} \, dx$

16. $\displaystyle \int \frac{x^2}{\sqrt{x^2 - 16}} \, dx$

17. $\displaystyle \int \frac{x^2}{\sqrt{x^2 + 49}} \, dx$

18. $\displaystyle \int \frac{\sqrt{x^2 + 16}}{x^2} \, dx$

19. $\displaystyle \int \frac{dx}{(x + 3)(x - 2)}$

20. $\displaystyle \int \frac{dx}{(2x + 3)(3x - 2)}$

21. $\displaystyle \int x^2 e^{x/2} \, dx$

22. $\displaystyle \int x^3 \ln x \, dx$

23. $\displaystyle \int \frac{x}{\sqrt{2x + 3}} \, dx$

*Find the value of each of the following and check your
work with your calculator.*

24. $\displaystyle \int_3^5 \frac{1}{x^2 - 1} \, dx$

25. $\displaystyle \int_4^6 \frac{1}{\sqrt{x^2 - 9}} \, dx$

26. $\displaystyle \int_2^4 \frac{1}{x^2(2x + 1)} \, dx$

Find the following integrals.

27. $\displaystyle \int \frac{dx}{\sqrt{25x^2 + 16}}$

28. $\displaystyle \int \frac{3 \, dx}{\sqrt{16x^2 - 25}}$

29. $\displaystyle\int \frac{8 \, dx}{x\sqrt{25 - 4x^2}}$

30. $\displaystyle\int \frac{5 \, dx}{x^2\sqrt{16 + 9x^2}}$

31. $\displaystyle\int 3x^2\sqrt{9x^2 + 16} \, dx$

32. $\displaystyle\int \frac{\sqrt{16 - 25x^2}}{3x} \, dx$

33. $\displaystyle\int \frac{\sqrt{16 + 25x^2}}{3x^2} \, dx$

34. $\displaystyle\int \frac{4x^2}{\sqrt{36x^2 - 9}} \, dx$

35. $\displaystyle\int \frac{dx}{6x^2 - 5x - 6}$

36. $\displaystyle\int \frac{2 \, dx}{8x^2 - 10x - 3}$

Applications (Business and Economics)

37. **Producers' Surplus.** Find the producers' surplus for a price level of \$10 given the price–demand equation

$$p = S(x) = \frac{5x}{30 - x}$$

Show the producers' surplus on a graph.

38. **Consumers' Surplus.** Find the consumers' surplus for

$$D(x) = \frac{360}{x^2 + 4x + 3} \quad \text{and} \quad S(x) = \frac{5x}{x + 3}$$

39. **Cash Reserves.** The cash reserves of a company for a year are given in thousands of dollars by

$$c = 2 + \frac{x^2\sqrt{1 + x^2}}{144}, \qquad 0 \le x \le 12$$

where x represents the number of months. Find the average cash reserve for (a) the first quarter, (b) the first half of the year, and (c) the last quarter.

Applications (Social and Life Sciences)

40. **Learning Rates.** A rate of learning is given by

$$\frac{dL}{dt} = \frac{64}{\sqrt{t^2 + 36}}, \qquad t \ge 0$$

where L is the number of items learned in t hours. Find the number of items learned in 8 hours.

41. **Pollution.** The radius R of an oil slick is increasing at the rate

$$\frac{dR}{dt} = \frac{144}{\sqrt{t^2 + 16}}, \qquad t \ge 0$$

where t is time in minutes. Find the radius after 5 minutes if $R = 0$ when $t = 0$.

Review Exercises

42. Suppose that the rate of production of an oil company per month is given in thousands of barrels as

$$\frac{dP}{dt} = 24te^{-0.06t}$$

How much oil is produced during the first 6 months? If production is continued indefinitely, what would be the total amount produced?

Find the value of the improper integrals if they exist.

43. $\displaystyle\int_2^\infty 3x^{-3/2} \, dx$

44. $\displaystyle\int_2^\infty \frac{4 \, dx}{(1 + x)^{3/2}}$

45. $\displaystyle\int_0^\infty 3e^{-2x} \, dx$

46. $\displaystyle\int_{-\infty}^\infty e^{2x} \, dx$

12.5   APPROXIMATE INTEGRATION

OVERVIEW   We have defined the definite integral $\int_a^b f(x) \, dx$ over the closed interval $[a, b]$ to be $F(b) - F(a)$, where $F(x)$ is an antiderivative of $f(x)$. Sometimes it is difficult or maybe impossible to find $F(x)$. For example,

$$\int_a^b e^{-x^2} \, dx$$

cannot be evaluated using antiderivatives because no antiderivatives exist in

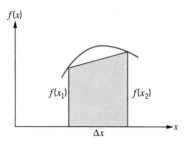

Figure 16

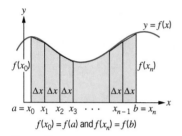

$f(x_0) = f(a)$ and $f(x_n) = f(b)$

Figure 17

terms of functions we have studied. Numerical procedures exist to help us evaluate such integrals. In this section we study two techniques:

- Trapezoidal rule
- Simpson's rule

The Riemann sum procedures that we have used to define a definite integral are sometimes the procedures we use to approximate the value of the definite integral. The two methods of numerical integration that we consider (the trapezoidal rule and Simpson's rule) are based on the interpretation of an integral as an area. For example, one can approximate the area under a curve as a trapezoid instead of as a rectangle, as seen in Figure 16. The area of the trapezoid is

$$\left[\frac{f(x_1) + f(x_2)}{2} \right] \Delta x$$

An approximation for the definite integral $\int_a^b f(x)\, dx$ can be found by calculating the area of n trapezoids. In Figure 17, the area under the curve from $x = a$ to $x = b$ is approximately

$$\overbrace{\frac{f(x_0) + f(x_1)}{2} \cdot \Delta x}^{\text{Area of 1st trapezoid}} + \overbrace{\frac{f(x_1) + f(x_2)}{2} \cdot \Delta x}^{\text{Area of 2nd trapezoid}} + \cdots + \overbrace{\frac{f(x_{n-1}) + f(x_n)}{2} \cdot \Delta x}^{\text{Area of }n\text{th trapezoid}}$$

$$= \left[\frac{f(x_0)}{2} + 2\frac{f(x_1)}{2} + 2\frac{f(x_2)}{2} + \cdots + 2\frac{f(x_{n-1})}{2} + \frac{f(x_n)}{2} \right] \Delta x$$

This allows us to employ the trapezoidal rule for approximating the value of a definite integral.

TRAPEZOIDAL RULE

Let $f(x)$ be a continuous function on $[a, b]$. Then

$$\int_a^b f(x)\, dx \approx \left[\frac{f(x_0)}{2} + f(x_1) + f(x_2) + \cdots + f(x_{n-1}) + \frac{f(x_n)}{2} \right] \Delta x$$

where

$$\Delta x = \frac{b - a}{n}, \quad x_0 = a, \quad x_1 = x_0 + \Delta x, \quad x_2 = x_1 + \Delta x,$$

$$\ldots, \quad x_n = x_{n-1} + \Delta x = b$$

The number of approximations will affect the accuracy of the approximation. As n increases, the approximation becomes more accurate.

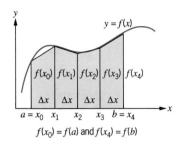

Figure 18

EXAMPLE 29 Write the trapezoidal rule to approximate the area under the curve defined by $y = f(x)$ in Figure 18 from $x = a$ to $x = b$ using $n = 4$ subdivisions.

SOLUTION In the formula

$$\Delta x = \frac{b - a}{n} = \frac{b - a}{4}, \quad a = x_0, \quad b = x_4$$

$$\int_a^b f(x)\, dx \approx \left[\frac{f(x_0)}{2} + f(x_1) + f(x_2) + f(x_3) + \frac{f(x_4)}{2} \right] \Delta x$$

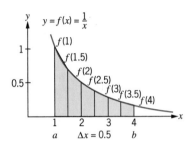

Figure 19

EXAMPLE 30 Use the trapezoidal rule with six intervals to approximate

$$\int_1^4 \frac{1}{x}\, dx$$

Compare this answer to the answer you get by numeric integration on your calculator (see Figure 19).

SOLUTION We first find Δx and identify all other parts to be used in the formula.

$$\Delta x = \frac{b - a}{n} = \frac{4 - 1}{6} = 0.5 \qquad x_0 = a = 1$$

$$x_1 = 1 + 0.5 = 1.5 \qquad\qquad x_2 = 1.5 + 0.5 = 2$$

$$x_3 = 2 + 0.5 = 2.5 \qquad\qquad x_4 = 2.5 + 0.5 = 3$$

$$x_5 = 3 + 0.5 = 3.5 \qquad\qquad x_6 = b = 4 \qquad \text{See Figure 19}$$

$$\int_1^4 \frac{1}{x}\, dx \approx \left[\frac{f(1)}{2} + f(1.5) + f(2) + f(2.5) + f(3) + f(3.5) + \frac{f(4)}{2} \right] (0.5)$$

$$= \left[\frac{1}{2} + \frac{1}{1.5} + \frac{1}{2} + \frac{1}{2.5} + \frac{1}{3} + \frac{1}{3.5} + \frac{1}{8} \right] (0.5)$$

$$= 1.41$$

With a calculator, we get the answer 1.386. How could we get the answer using the trapezoidal rule to be closer to the answer we get with the calculator?

When we use the trapezoidal rule, the approximation of the area under $f(x)$ is being done, in effect, by a linear function. Intuition suggests that the accuracy of a numerical integration procedure can be improved by replacing a linear approximation with something that would better fit the curve, a second-degree polynomial (Figure 20).

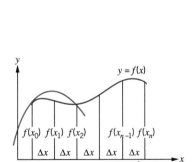

Figure 20

It can be shown (but will be omitted here) that the area under the parabola between $x = x_0$ and $x = x_2$ is given by

$$\frac{\Delta x}{3} [f(x_0) + 4f(x_1) + f(x_2)]$$

This argument can be repeated for $[x_2, x_4]$, $[x_4, x_6]$, and so on. Therefore, if n is even, the area under the curve from $x = a$ to $x = b$ is approximately the sum of the areas under $n/2$ approximating parabolas and we obtain Simpson's rule.

SIMPSON'S RULE

Let $f(x)$ be a continuous function on $[a, b]$. Then,

$$\int_a^b f(x)\, dx \approx \frac{\Delta x}{3} [f(x_0) + 4f(x_1) + 2f(x_2) + 4f(x_3)$$
$$+ \cdots + 2f(x_{n-2}) + 4f(x_{n-1}) + f(x_n)]$$

where $\Delta x = (b - a)/n$ is the width of each subinterval, n is even, and

$$x_0 = a, \quad x_1 = x_0 + \Delta x, \quad x_2 = x_1 + \Delta x, \quad \ldots, \quad x_n = b = x_{n-1} + \Delta x$$

EXAMPLE 31 Approximate the definite integral

$$\int_0^1 e^{-x^2}\, dx$$

using Simpson's rule with $n = 4$ subintervals.

SOLUTION For this problem $a = 0$, $b = 1$, $n = 4$, and $f(x) = e^{-x^2}$.

$$\Delta x = \frac{b - a}{n} = \frac{1 - 0}{4} = 0.25 \qquad \text{Width of the subintervals}$$

$$x_0 = a = 0 \qquad\qquad\qquad x_1 = 0 + 0.25 = 0.25$$

$$x_2 = 0.25 + 0.25 = 0.50 \qquad\qquad x_3 = 0.50 + 0.25 = 0.75$$

$$x_4 = b = 1$$

$$\int_0^1 e^{-x^2}\, dx \approx \frac{0.25}{3} [f(x_0) + 4f(x_1) + 2f(x_2) + 4f(x_3) + f(x_4)]$$

$$= \frac{0.25}{3} [f(0) + 4f(0.25) + 2f(0.50) + 4f(0.75) + f(1)]$$

$$= \frac{0.25}{3} [e^0 + 4e^{-(0.25)^2} + 2e^{-(0.50)^2} + 4e^{-(0.75)^2} + e^{-(1)^2}]$$

$$\approx \frac{0.25}{3} [1 + 3.7577 + 1.5576 + 2.2791 + 0.3679]$$

$$\approx 0.7469$$

You will find that the answer above is the same as the one you will get with numeric integration on your calculator to 4 decimal places. In fact most calculators use Simpson's rule for numeric integration, but use more terms. ■

Practice Problem 1 Using Simpson's rule with eight intervals to integrate

$$\int_1^5 \left(\frac{1}{x^2}\right) dx$$

find the following.

(a) Δx (b) $f(x_0)$ (c) $4f(x_1)$

ANSWER

(a) $\Delta x = \dfrac{5 - 1}{8} = \dfrac{1}{2}$ (b) $x_0 = 1, f(x_0) = \dfrac{1}{1^2} = 1$

(c) $x_1 = 1 + \dfrac{1}{2} = \dfrac{3}{2}, f(x_1) = \dfrac{4}{9}, 4f(x_1) = \dfrac{16}{9} = 1.7778$

NOTE: In more advanced courses, it is proven that the error in using the trapezoidal rule to approximate the integral $\int_a^b f(x)\, dx$ with n subdivisions has the following upper bound:

$$|\text{Error}| \le \frac{(b - a)^3 M}{12n^2}$$

where M is the maximum value of $|f''(x)|$ on $[a, b]$.
    For Simpson's rule it is

$$|\text{Error}| \le \frac{(b - a)^5 M}{180n^4}$$

where M is the maximum value of $|f^{(4)}(x)|$ on $[a, b]$.
    In Example 30, the maximum error is calculated as follows:

$$f(x) = \frac{1}{x}, \quad f'(x) = -\frac{1}{x^2}, \quad \text{and} \quad f''(x) = \frac{2}{x^3}$$

The maximum value of $\dfrac{2}{x^3}$ on $[1, 4]$ is 2; so $M = 2$.

$$|\text{Error}| \le \frac{(4 - 1)^3 \cdot 2}{12 \cdot 6^2} = \frac{1}{8} \qquad n = 6$$

SUMMARY

For the trapezoidal rule, errors are approximately proportional to $1/n^2$ and for a given n, the size of the error depends on the size of f''. For Simpson's rule, errors are approximately proportional to $1/n^4$, and, for a given n, the size of the error depends on the size of the fourth derivative, $f^{(4)}$. Thus, we can see that the definite integral $\int_a^b f(x)\, dx$ can be computed fairly quickly and quite accurately in most cases using Simpson's rule. Difficulties will arise when f' or a higher derivative of f does not exist or gets very large over the interval $a \leq x \leq b$.

Exercise Set 12.5

Evaluate the given integrals using the trapezoidal rule. Compare your answer with the answer you get by numeric integration on your graphing calculator.

1. $\displaystyle\int_1^2 (x^2 + 1)\, dx; \quad n = 4$

2. $\displaystyle\int_1^2 (x^3 - 4)\, dx; \quad n = 4$

3. $\displaystyle\int_0^1 \sqrt{x + 1}\, dx; \quad n = 4$

4. $\displaystyle\int_0^1 x(x^2 + 1)^2\, dx; \quad n = 6$

Evaluate the given integrals using Simpson's rule. Compare your answer with the answer you get by numeric integration on your graphing calculator.

5. $\displaystyle\int_1^2 (x^2 + 1)\, dx; \quad n = 4$

6. $\displaystyle\int_1^2 (x^3 - 4)\, dx; \quad n = 4$

7. $\displaystyle\int_0^1 \sqrt{x + 1}\, dx; \quad n = 4$

8. $\displaystyle\int_0^1 x(x^2 + 1)^2\, dx; \quad n = 6$

9. Using the trapezoidal rule, find the maximum error in Exercise 3.

10. Using Simpson's rule, find the maximum error in Exercise 7.

Applications (Business and Economics)

11. **Total Cost.** A corporation determines its marginal costs at various production levels to be as follows:

Production (units of 100)	Marginal Costs ($)
20	400
25	420
30	440
35	450
40	460

Use the trapezoidal rule to approximate the cost of increasing production from 20 units to 40 units.

12. **Production.** The rate of production of an oil company per month is given in thousands of barrels as

$$\frac{dp}{dt} = 30e^{-0.06t}$$

How much oil is produced during the first 6 months of operation? Approximate the answer using the trapezoidal rule with $n = 6$.

13. **Consumers' Surplus.** For a demand function

$$p = D(x) = 85 - e^{0.01x^2}$$

find the consumers' surplus using Simpson's rule with six subintervals when the price level is 30.

Applications (Social and Life Sciences)

14. **Learning Function.** The number of words learned in a minute is given at the following times.

Time After Start (minutes)	Number of Words Learned
4	15
5	14.5
6	13.2
7	12
8	11
9	10

Using the trapezoidal rule, find the number of words learned from $t = 4$ to $t = 9$.

Review Exercises

15. Find each integral if it exists.

(a) $\displaystyle\int_{-\infty}^{2} xe^{-3x}\, dx$ (b) $\displaystyle\int_{-\infty}^{\infty} \frac{2x}{\sqrt{x^2 + 1}}\, dx$

16. Find each indefinite integral.

(a) $\displaystyle\int \frac{-3\sqrt{\ln x}}{x}\, dx$ (b) $\displaystyle\int \frac{1 + x}{2 - x}\, dx$

Chapter Review

Review to make certain that you understand the following concepts.

Integration by parts   Improper integrals
Simpson's rule   Trapezoidal rule
Integration by use of tables   Convergent integral
Divergent integral   Probability density function

Make sure that you can use the following formulas.

$$\int u\, dv = uv - \int v\, du$$

$$\int_{a}^{\infty} f(x)\, dx = \lim_{t \to \infty} \int_{a}^{t} f(x)\, dx$$

$$\int_{-\infty}^{a} f(x)\, dx = \lim_{t \to -\infty} \int_{t}^{a} f(x)\, dx$$

$$\int_{-\infty}^{\infty} f(x)\, dx = \lim_{t \to -\infty} \int_{t}^{c} f(x)\, dx + \lim_{t \to \infty} \int_{c}^{t} f(x)\, dx$$

Chapter Test

1. Determine whether $\displaystyle\int_{0}^{\infty} \frac{dx}{\sqrt{x}}$ is convergent or divergent.

2. Find $\displaystyle\int x\sqrt{x + 2}\, dx$.

3. Use $\displaystyle\int \frac{du}{u^2 - a^2} = \frac{1}{2a} \ln \left| \frac{u - a}{u + a} \right| + C$ to find $\displaystyle\int \frac{dx}{25x^2 - 36}$.

4. Evaluate $\displaystyle\int_0^2 xe^{4x}\, dx$ and check with your graphing calculator.

5. Approximate

$$\int_0^2 x^3\, dx$$

by using Simpson's rule with four subintervals and compare this answer to the one you get by using your graphing calculator.

6. Evaluate $\displaystyle\int_2^4 x^2 \ln x\, dx$. Check your work with your graphing calculator.

7. Approximate

$$\int_0^4 x^2\, dx$$

using the trapezoidal rule with four subintervals and compare to the answer given by your graphing calculator.

8. Evaluate $\displaystyle\int x^2 e^x\, dx$ and check with your graphing calculator.

9. Evaluate $\displaystyle\int_0^1 15x^2 \sqrt{5x^3 + 4}\, dx$ using substitution.

10. Let X be an exponentially distributed random variable with an average value of 7. Find the pdf for X and find $P(X \le 8)$.

Multivariable Calculus

T he goal of this chapter is to extend the theory of calculus to real-valued functions that involve more than one independent variable. Examples of such functions abound in both business and economics and the social and life sciences. In economics and business, one encounters revenue as a function of both price, p, and the number of items sold, x. That is, $R = f(p, x)$. You need to know the number of items sold and the price per item to calculate the revenue. In this chapter, we study three-dimensional coordinates, functions of several variables, partial derivatives, relative maxima and minima for functions of two or more variables, and Lagrange multipliers.

13.1 FUNCTIONS OF SEVERAL VARIABLES

OVERVIEW   We have already encountered many examples of functions of several variables. In the formula for simple interest, $A = P(1 + rt)$, A is a function of P, r, and t, and in the formula for compound interest, $A = P(1 + i)^n$, A is a function of P, i, and n. In an earlier exercise, the cost function $C(x)$ for the number x of items produced was given as

$$C(x) = 0.25x + 70$$

Suppose now that the manufacturer decides to expand and produce an additional, different item. Let y represent the number of these additional items produced. The cost function $C(x, y)$ for producing both items would depend upon both x and y, and hence is a function of two variables. For example, suppose that

$$C(x, y) = 130 + 0.25x + 0.30y$$

If the manufacturer decides to produce even more items, the cost function would involve more variables. We could define functions of three variables, four variables, or, in general, n variables. In this section, we consider

- Functions of two or more independent variables
- Applications of functions of several variables
- The three-dimensional coordinate system
- Traces of graphs in three dimensions

A company produces pens and pencils. The cost of producing a pen is $2.40 and a pencil is $1.80 with fixed costs of $130. The cost function for producing x pens and y pencils is

$$C(x, y) = 130 + 2.40x + 1.80y$$

We say that C is a function of x and y. The domain of (x, y) is the set of ordered pairs where the elements are positive integers. To understand this concept, we need the following definition of a function with two independent variables.

FUNCTION OF TWO INDEPENDENT VARIABLES

The function $f(x, y)$ describes a function of two independent variables if for each ordered pair (x, y) from the domain of f there is one and only one value of $f(x, y)$ in the range of f. Unless stated otherwise, the domain of $f(x, y)$ is the set of all ordered pairs of real numbers (x, y) such that $f(x, y)$ is also a real number.

An equation such as

$$f(x, y) = 2x^2 + xy + 3y^2$$

defines f as a function of two independent variables x and y. The domain of this function is the set of ordered pairs (a, b), where a and b are any real numbers. The range of f is the set of real numbers.

EXAMPLE 1   Let $f(x, y) = 2x^2 + xy + 3y^2$. Then

$$f(0, 0) = 2(0)^2 + (0)(0) + 3(0)^2 = 0$$

$$f(1, 2) = 2(1)^2 + (1)(2) + 3(2)^2 = 16$$

$$f(-1, -2) = 2(-1)^2 + (-1)(-2) + 3(-2)^2 = 16$$

$$f(h, k + 2) = 2(h)^2 + h(k + 2) + 3(k + 2)^2$$
$$= 2h^2 + hk + 2h + 3k^2 + 12k + 12$$

Practice   Problem   1 Let $f(x, y) = 2x^3 - 3x^2y + 2xy^2 + y^3$.   Find
(a) $f(0, 0)$, (b) $f(-1, -2)$, and (c) $f(1, 2)$.

ANSWER

(a) 0
(b) -12
(c) 12

Similarly, a function defined by an equation such as

$$f(x, y, z) = 3x^2 - 2xy + y^2 - 3xyz + 2z^2$$

is a function of three variables. The domain of such a function is the set of ordered triples (a, b, c), where a, b, and c are real numbers.

EXAMPLE 2   If $f(x, y, z) = 3x^2 - 2xy + y^2 - 3xyz + 2z^2$, find
(a) $f(1, 2, 3)$ and (b) $f(-1, 2, -3)$.

SOLUTION

(a) $f(1, 2, 3) = 3(1)^2 - 2(1)(2) + (2)^2 - 3(1)(2)(3) + 2(3)^2$
$$= 3 - 4 + 4 - 18 + 18$$
$$= 3$$
(b) $f(-1, 2, -3) = 3(-1)^2 - 2(-1)(2) + (2)^2 - 3(-1)(2)(-3) +$
$$2(-3)^2$$
$$= 3 + 4 + 4 - 18 + 18$$
$$= 11$$

Practice Problem 2 For $f(x, y, z) = 3x^2 - 2xy + y^2 - 3xyz + 2z^2$, find
(a) $f(1, -1, 2)$ and (b) $f(h, h + 2, \Delta h)$.

ANSWER

(a) 20
(b) $2h^2 + 4 - 3h^2 \Delta h - 6h \Delta h + 2(\Delta h)^2$

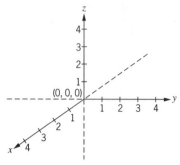

Figure 1

Functions with two independent variables can be represented on a coordinate system in much the same manner as functions with one independent variable. A function with one independent variable is drawn on a two-dimensional coordinate system. For a function with two independent variables, a three-dimensional coordinate system is needed. A three-dimensional Cartesian coordinate system is shown in Figure 1. In this figure the x-axis, the y-axis, and the z-axis intersect and are perpendicular to one another at the origin, $(0, 0, 0)$. The dashed lines indicate the negative portion of each of these coordinate lines. Any ordered triple (a, b, c) can be plotted as a point using this coordinate system (a is the x-coordinate, b is the y-coordinate, and c is the z-coordinate).

The plane formed by the x-axis and the y-axis is called the **xy-coordinate plane**; the plane formed by the x-axis and the z-axis is called the **xz-coordinate plane**; and the plane formed by the y-axis and the z-axis is called the **yz-coordinate plane**. In a three-dimensional system, the three axes divide the space into eight parts; each part is called an **octant**. In octant 1, the x-, y-, and z-coordinates are all positive.

The point $(3, 2, 4)$ is located by starting at the origin, $(0, 0, 0)$, and moving 3 units in the direction of the positive x-axis, 2 units in the direction of the positive y-axis, and 4 units in the direction of the positive z-axis [see Figure 2(a)]. Any point can be plotted in this manner. First move along the x-axis, then move parallel to the y-axis, and then parallel to the z-axis.

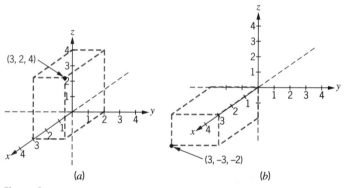

(a)   (b)

Figure 2

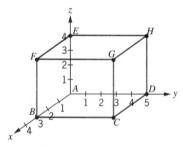

Figure 3

EXAMPLE 3   Plot the point $(3, -3, -2)$ on the coordinate system in Figure 2(b).

SOLUTION   First move 3 units in a positive direction on the x-axis. Then move 3 units in a negative direction parallel to the y-axis, and finally move 2 units in a negative direction parallel to the z-axis and mark the point $(3, -3, -2)$. ∎

Practice Problem 3 Find the coordinates of the vertices (A to H) of the rectangular box shown in Figure 3.

ANSWER $A(0, 0, 0)$, $B(3, 0, 0)$, $C(3, 5, 0)$, $D(0, 5, 0)$, $E(0, 0, 4)$, $F(3, 0, 4)$, $G(3, 5, 4)$, $H(0, 5, 4)$

The graph of a function $z = f(x, y)$ consists of all the points $(x, y, f(x, y))$ in a three-dimensional coordinate system, in other words, all points (x, y, z) where (x, y) is in the domain of f and $z = f(x, y)$. The graph of all such points is a **surface** in space. It is difficult or even impossible to sketch the graph of some three-variable equations by hand. Computers and math programs have the capabilities of graphing these surfaces. Figure 4 shows a graph generated by a computer. We will not try to sketch difficult figures, but we will consider some techniques for graphing equations with three variables.

To draw graphs of functions in three dimensions, we make use of traces. The intersections of a graph of a function with the coordinate planes of the three-dimensional coordinate system are called **traces** of the graph of the function. Since in each coordinate plane one variable is always zero, the equations of the traces can be obtained by setting $x = 0$, $y = 0$, and $z = 0$ one at a time.

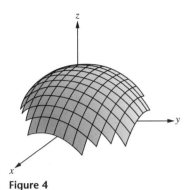

Figure 4

TRACES OF FUNCTIONS OF TWO VARIABLES

Let $z = f(x, y)$ be a function of two variables.

(a)   If we set $x = 0$, the curve $z = f(0, y)$ is the trace of the graph of f in the yz-plane.

(b)   If we set $y = 0$, the curve $z = f(x, 0)$ is the trace of the graph of f in the xz-plane.

(c)   If we set $z = 0$, the curve $f(x, y) = 0$ is the trace of the graph of f in the xy-plane.

One type of surface in space is a plane. The general equation of a plane is $Ax + By + Cz = D$. By forming the intersection of a plane with each coordinate plane, we can see that the resulting traces are lines.

For example, to graph the plane $3y + 4z + 2x = 16$, we draw three traces.

1.   Setting $z = 0$ results in the equation $3y + 2x = 16$. This is the trace on the xy-plane. The x-intercept is the point $(8, 0, 0)$ and the y-intercept is the point $(0, \frac{16}{3}, 0)$.

2.   Setting $y = 0$ results in the equation $4z + 2x = 16$, which is the trace on the xz-plane. The x-intercept is the point $(8, 0, 0)$ and the z-intercept is the point $(0, 0, 4)$.

3.   Setting $x = 0$ results in the equation $3y + 4z = 16$, which is the trace on the yz-plane. The y- and z-intercepts are the same as those found above, $(0, \frac{16}{3}, 0)$ and $(0, 0, 4)$.

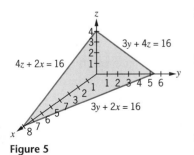

Figure 5

We draw the three traces of this plane as shown in Figure 5 and shade the region bounded by the traces. The shaded region shows a portion of the plane in the first octant—the octant where x, y, and z are all positive.

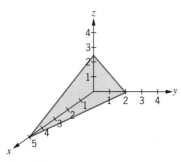

Figure 6

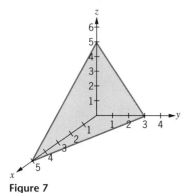

Figure 7

EXAMPLE 4   Sketch the traces of $2x + 5y + 4z = 10$ by finding the intercepts, and then shade the plane.

SOLUTION   By letting $y = 0$ and $z = 0$, $2x = 10$ or $x = 5$ (the x-intercept). From $z = 0$ and $x = 0$, $5y = 10$ or $y = 2$ (the y-intercept). The z-intercept is obtained by setting $y = 0$ and $x = 0$; that is, $4z = 10$, so $z = \frac{5}{2}$. The three intercepts are plotted in Figure 6; the lines connecting these points are the traces of the plane, the first octant of which is shaded.   ■

Practice Problem 4   Sketch the traces of $3x + 5y + 3z = 15$ by finding the intercepts, and then shade the plane.

ANSWER   See Figure 7.

Special planes are obtained when some of the variables are missing in the three-dimensional linear equation. Consider the surface that is the graph of $z = 4$. In this case x and y can have any value but z is always 4. Thus $z = 4$ is the equation of a plane parallel to the xy-plane and four units above the xy-plane (see Figure 8).

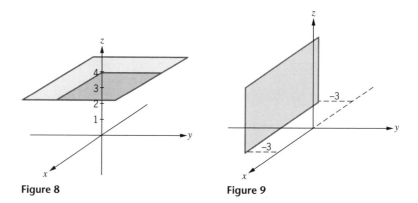

Figure 8   **Figure 9**

EXAMPLE 5   Sketch the plane $y = -3$ on a three-dimensional coordinate system.

SOLUTION   This plane is parallel to the xz-plane and 3 units in a negative direction from the xz-plane (see Figure 9). Remember: $y = -3$ *regardless of the values of x and z*.   ■

The planes shown in Figures 8 and 9 have only one intercept. It is possible for a plane to have two intercepts. Let's look at such a plane in the next example.

EXAMPLE 6   Sketch the graph of $2x + y = 2$ on a three-dimensional coordinate system.

SOLUTION Since the variable z is missing in the equation, z can assume any value. The trace in the xy-plane is $2x + y = 2$. The x-intercept is 1 and the y-intercept is 2. In Figure 10, vertical lines are drawn to the intercepts to show that z can assume any value.

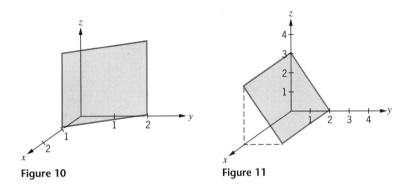

Figure 10   **Figure 11**

Practice Problem 5 Sketch the plane $3y + 2z = 6$ in three-space.

ANSWER

See Figure 11. Note that x can be anything.

Contour Maps

Your calculator is valuable in visualizing a three-dimensional surface using a **contour map**. To illustrate, let's consider the graph of the surface defined by $z = x^2 + y^2$. Suppose that we let $z = c$, where $c \neq 0$. The equation $z = c$ is a plane parallel to the xy-plane. If we draw the graph of $x^2 + y^2 = c$, we will have the trace of the surface on the $z = c$ plane. By considering a number of such traces, as z takes on various values, we can often visualize the surface that such traces would yield. These traces projected onto the xy-plane constitute a contour map. With your calculator, make a contour map for $z = x^2 + y^2$, considering traces when $z = 4, 9, 16$, and 25.

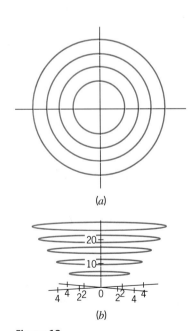

(a)

(b)

Figure 12

For $z = 4$, let $Y_1 = \sqrt{4 - x^2}$ and $Y_2 = -Y_1$.
For $z = 9$, let $Y_3 = \sqrt{9 - x^2}$ and $Y_4 = -Y_3$.
For $z = 16$, let $Y_5 = \sqrt{16 - x^2}$ and $Y_6 = -Y_5$.
For $z = 25$, let $Y_7 = \sqrt{25 - x^2}$ and $Y_8 = -Y_7$.

You will get a contour plot that looks something like that in Figure 12(a). In Figure 12(b) the graph shown is a contour plot done by a computer program.

We can visualize the traces as z takes on five values as we draw the sketch of the surface in Figure 13(a). In Figure 13(b), a computer generated plot of $z = x^2 + y^2$ is shown.

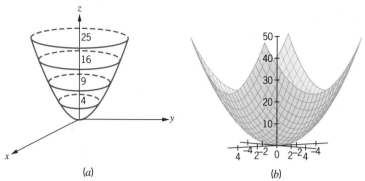

(a)   (b)

Figure 13

SUMMARY

When evaluating functions of more than one independent variable, the same techniques that we use with functions of one independent variable apply. Graphing, however, requires that we use a three-dimensional axis. Your calculator can help you with contour maps and then you can try to visualize the three-dimensional figure. In addition, traces on the coordinate planes help to visualize three-dimensional figures.

Exercise Set 13.1

Find the values of the following functions at the indicated points of their domain.

$$f(x, y) = 3x^2 - 2xy + y^3$$

$$g(x, y) = 7 - 3x + 2y$$

$$h(x, y) = 4x^3 - 3xy + 2y^2$$

1. $f(0, 1)$ 2. $f(1, 0)$
3. $f(-1, 0)$ 4. $f(-1, 1)$
5. $f(2, 1)$ 6. $f(1, 2)$
7. $g(0, 0)$ 8. $g(3, 2)$
9. $g(2, 3)$ 10. $g(-3, -1)$
11. $h(1, 2)$ 12. $h(2, -1)$

13. Plot the following points in a three-space coordinate system.
    (a) $(2, 4, 3)$ (b) $(3, -1, -2)$
    (c) $(-2, 0, 1)$ (d) $(-1, 0, -2)$
    (e) $(-1, 2, -3)$ (f) $(0, 0, -1)$

Obtain the equations of the traces, find the coordinate intercepts, and then shade parts of the following planes.

14. $x + y + z = 3$ 15. $3x + 7y + 2z = 14$
16. $x - y + z = 4$ 17. $5x + 2y + 3z = 15$
18. $-x + y - z = 3$ 19. $2x - 3y - 4z = 12$

20. Classify the following statements as true or false.
    (a) $3x + y = 7$ is a line in three-space.
    (b) In three-space, $y = 0$ is the equation of the x-axis.
    (c) One trace of $x - 3y + 4z = 6$ is $3y - 4z = -6$.
    (d) In three-space, $z = 0$ represents the xz-plane.
    (e) In three-space, every linear equation represents a plane.
    (f) The trace of $6z + 4x = 12$ in the yz-plane is $z = 2$.
    (g) The x-intercept of $y + 6z + 4x = 12$ is 3.
    (h) The point $(1, 3, -2)$ is a point on the plane $x - y + 2z = -6$.

(i) The graph of $y = 4$ in three-space is a plane perpenicular to the y-axis.

(j) The plane $2x + 3y = 6$ never intersects the z-axis.

Shade the following planes in three-space.

21. $z = 5$ 22. $x = -2$

23. $y = -4$ 24. $x + 3y = 6$

25. $2x - y = 4$ 26. $3y - 4x = 8$

For Exercises 27–31, make contour maps (possibly using a graphing calculator). (Optional: sketch the surface.)

27. $z = 4x^2 + y^2$ 28. $y = 4z^2 + x^2$

29. $z = x^2 - y^2$ 30. $y = x^2 - z^2$

Applications (Business and Economics)

31. **Profit Function.**   A profit function $P(x, y)$ is given by

$$P(x, y) = 2x^2 - xy + 3y^2 + 4x + 2y + 4$$

Find (a) $P(3, 2)$, (b) $P(4, 1)$, and (c) $P(3, 4)$.

32. **Revenue and Cost Functions.**   Suppose that the revenue and cost functions for a firm producing two items are given in units of thousands by

$$R(x, y) = 3x + 4y$$

$$C(x, y) = x^2 - 3xy + 3y^2 + 2x + 4y + 7$$

Find each of the following.
(a) $P(x, y) = R(x, y) - C(x, y)$ (b) $R(3, 2)$
(c) $C(2, 3)$ (d) $P(3, 2)$

33. **Prince–Earnings Ratio.**   The price–earnings ratio of a stock is given by

$$R(P, E) = \frac{P}{E}$$

where P is the price of a stock, and E is the earnings per share for a year. Find and interpret $R(110, 10)$ for IBM.

34. **Stock Yield.**   The yield of a stock is given by

$$Y(d, P) = \frac{d}{P}$$

where d is the yearly dividends and P is the price. Find and interpret $Y(33, 110)$ for IBM.

35. **Production Model.**   The production of a given company is given by

$$P = 0.8l^{1/2}C^{1/2}$$

where P is the number of units produced, l is the number of hours of labor available, and C is the number of dollars of capital available. Find the production when $l = 400$ and $C = \$10,000$.

36. **Production Model.**   Find the production for $P = 10l^{3/4}C^{1/4}$ if $l = 1296$ work-hours and $C = \$625$ in capital.

Applications (Social and Life Sciences)

37. **Scuba Diving.**   The time of a scuba dive is estimated by

$$T(v, d) = \frac{35v}{d + 35}$$

where v is the volume of air in the diver's tanks and d is the depth in feet of the dive. Find (a) $T(965, 30)$ and (b) $T(75, 45)$.

38. **Intelligence Quotient.**   The function for determining the intelligence quotient is given as

$$I(M, C) = \frac{100M}{C}$$

where I represents the intelligence quotient, M represents mental age, and C represents chronological age.
(a) If $M = 14$ and $C = 10$, find $I(14, 10)$.
(b) If $M = 11$ and $C = 10$, find $I(11, 10)$.
(c) If $M = 10$ and $C = 10$, find $I(10, 10)$.

13.2 PARTIAL DERIVATIVES

OVERVIEW   A function of more than one variable may have derivatives with respect to each independent variable. These derivatives in turn may have derivatives with respect to each independent variable. Such derivatives are called **partial derivatives.** We consider first a function of two variables. Since we already know how to differentiate a function of a single variable, it would seem reasonable to reduce our function of two variables to a function of one variable by holding one variable as constant. This leads to the definition of a partial derivative. In this section we

- Introduce partial derivatives
- Evaluate partial derivatives at points
- Interpret partial derivatives as slopes of tangents to surfaces.

The definition of a partial derivative is introduced for a function with two independent variables.

PARTIAL DERIVATIVE OF $z = f(x, y)$

The **partial derivative of** $z = f(x, y)$ **with respect to** x—denoted by $\partial z/\partial x, f_x$, or $f_x(x, y)$—is defined to be

$$\frac{\partial z}{\partial x} = \lim_{h \to 0} \frac{f(x + h, y) - f(x, y)}{h} \qquad y \text{ is held constant.}$$

The **partial derivative of** $z = f(x, y)$ **with respect to** y—denoted by $\partial z/\partial y, f_y$, or $f_y(x, y)$—is defined to be

$$\frac{\partial z}{\partial y} = \lim_{k \to 0} \frac{f(x, y + k) - f(x, y)}{k} \qquad x \text{ is held constant.}$$

if these limits exist.

NOTE: This definition states that partial derivatives of a function of two variables are determined by treating one variable as a constant. For example, for $z = f(x, y)$, to determine $\partial z/\partial x$, y is treated like a constant in the differentiation process and x is treated as a variable.

Since partial derivatives are simply ordinary derivatives with respect to one variable while keeping the other variable constant, partial derivatives can be found by using our previously obtained differentiation formulas instead of using the limit process.

EXAMPLE 7   If $z = f(x, y) = 2x^2 + y$, find $\partial z/\partial x$ and $\partial z/\partial y$.

SOLUTION   To find $\partial z/\partial x$, we differentiate with respect to x while holding y constant.

$$\frac{\partial z}{\partial x} = 4x \qquad \begin{array}{l} \text{Since } y \text{ is a constant, } (\partial/\partial x)(y) = 0 \\ \text{and } (\partial/\partial x)(2x^2) = 4x. \end{array}$$

Likewise, to find $\partial z/\partial y$, we differentiate with respect to y while holding x constant.

$$\frac{\partial z}{\partial y} = 1 \qquad \text{Since } x \text{ is a constant, } (\partial/\partial y)(2x^2) = 0 \\ \text{and } (\partial/\partial y)(y) = 1.$$

Practice Problem 1 For $z = f(x, y) = 4x^3 + 3x - y^2 + 2y$, find $\partial z/\partial x$ and $\partial z/\partial y$.

ANSWER $\partial z/\partial x = 12x^2 + 3$ and $\partial z/\partial y = -2y + 2$

EXAMPLE 8 If $z = e^{x^2 + y^2}$, find $\partial z/\partial y$.

SOLUTION

$$\frac{\partial z}{\partial y} = 2ye^{x^2 + y^2} \qquad \text{Since } x \text{ is a constant, } \frac{\partial}{\partial y}(x^2 + y^2) = 2y.$$

In the next example, the function has both independent variables in the same term. In this situation, you still treat one variable as a constant and differentiate as before.

EXAMPLE 9 If $z = f(x, y) = 3x^2 - 2xy + 5xy^2 - 3y$, find each of the following.

(a) $\dfrac{\partial z}{\partial y}$ (b) $f_y(3, 2)$

SOLUTION

(a) $\dfrac{\partial z}{\partial y} = -2x + 10xy - 3$ Hold x constant.

(b) $f_y(3, 2) = -2(3) + 10(3)(2) - 3$
$\qquad\qquad = -6 + 60 - 3$
$\qquad\qquad = 51$

Practice Problem 2 Given $z = f(x, y) = e^{xy}$, find $\partial f/\partial x$ and $\partial f/\partial y$.

ANSWER $\dfrac{\partial f}{\partial x} = ye^{xy}$ and $\dfrac{\partial f}{\partial y} = xe^{xy}$

EXAMPLE 10 If $z = \ln(3x^2 + 2y)$, find each of the following.

(a) $\dfrac{\partial z}{\partial x}$ (b) $f_x(1, 0)$

(c) $\dfrac{\partial z}{\partial y}$ (d) $f_y(1, 0)$

SOLUTION

(a) $\dfrac{\partial z}{\partial x} = \dfrac{6x}{3x^2 + 2y}$ Hold y constant.

(b) $f_x(1, 0) = \dfrac{6(1)}{3(1)^2 + 2(0)} = \dfrac{6}{3 + 0} = 2$

(c) $\dfrac{\partial z}{\partial y} = \dfrac{2}{3x^2 + 2y}$ Hold x constant.

(d) $f_y(1, 0) = \dfrac{2}{3(1)^2 + 2(0)} = \dfrac{2}{3 + 0} = \dfrac{2}{3}$ ∎

Practice Problem 3 For $z = \ln(5x^2y - 2x)$ find $\partial z/\partial x$ and $\partial z/\partial y$.

ANSWER

$$\frac{\partial z}{\partial x} = \frac{10xy - 2}{5x^2y - 2x} \quad \text{and} \quad \frac{\partial z}{\partial y} = \frac{5x^2}{5x^2y - 2x}$$

The definition of partial derivatives can be extended for functions of more than two independent variables. To be specific, if z is a function of more than two independent variables, such as $z = f(w, x, y) = x^2yw + ye^w$, consider all independent variables except one as fixed and take the derivative with respect to this one variable. For example, to find $\partial z/\partial w = f_w(w, x, y)$, treat both x and y as constants and w as a variable.

EXAMPLE 11 Given $z = x^2yw + ye^w$, find $\partial z/\partial x$, $\partial z/\partial y$, and $\partial z/\partial w$.

SOLUTION

$\dfrac{\partial z}{\partial x} = 2xyw$ w, y treated as constants.

$\dfrac{\partial z}{\partial y} = x^2w + e^w$ w, x treated as constants.

$\dfrac{\partial z}{\partial w} = x^2y + ye^w$ x, y treated as constants. ∎

Practice Problem 4 Given $f(x, y, z) = (x - y + 2z)^2$, find f_x, f_y, and f_z.

ANSWER $f_x = 2(x - y + 2z)$

$f_y = -2(x - y + 2z)$

$f_z = 4(x - y + 2z)$

In Section 13.1 we considered $z = f(x, y)$ as a surface in three dimensions. The partial derivative of z with respect to x at a point (x_1, y_1, z_1) may be thought of as the slope of the tangent to the surface in the direction of the x-axis at the point

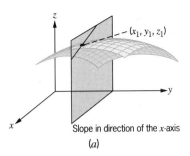

Slope in direction of the x-axis

(a)

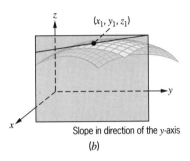

Slope in direction of the y-axis

(b)

Figure 14

(x_1, y_1, z_1) on the surface. That is, if $\partial z/\partial x$ is evaluated at (x_1, y_1, z_1), then it represents the slope of the tangent to the surface at (x_1, y_1, z_1) parallel to the x-axis [Figure 14(a)]. Similarly, if $\partial z/\partial y$ is evaluated at (x_1, y_1, z_1), then it represents the slope of the tangent to the surface at (x_1, y_1, z_1) parallel to the y-axis [Figure 14(b)].

EXAMPLE 12   Find the slope of the tangent line to $z = x^2 + 2y^2$ at $(1, -1, 3)$ (a) in the direction of the x-axis and (b) in the direction of the y-axis.

SOLUTION

(a)   We have

$$\frac{\partial z}{\partial x} = 2x \quad \text{and} \quad \frac{\partial z}{\partial x}(1, -1, 3) = 2 \cdot 1 = 2$$

Thus the slope of the tangent line to the surface in the direction of the x-axis at the point $(1, -1, 3)$ is 2 [see Figure 15(a)].

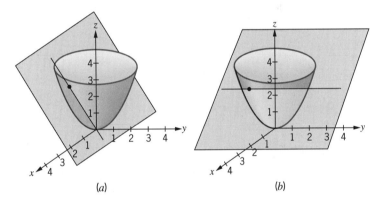

(a)   (b)

Figure 15

(b)   We have

$$\frac{\partial z}{\partial y} = 4y \quad \text{and} \quad \frac{\partial z}{\partial y}(1, -1, 3) = -4$$

Thus the slope of the tangent line to the surface in the direction of the y-axis at $(1, -1, 3)$ is -4 [see Figure 15(b)]. ■

SUMMARY

Evaluating partial derivatives for functions of more than one independent variable involves the same rules that were used in earlier chapters involving functions with only one independent variable. The difference is that when evaluating a partial derivative, only one variable is differentiated at a time and the others are held constant. If z is defined to be a function of two variables, x and y, $z = f(x, y)$, then $\partial z/\partial x$ is found by taking y to be constant while we differentiate with respect to x. Likewise, $\partial z/\partial y$ is found by differentiating z with respect to y while x is held constant. This work can be extended to a function, z, of any number of variables. To differentiate z with respect to any variable, you just hold all the others constant while differentiating with respect to that variable.

Exercise Set 13.2

Find $\partial z/\partial x$, $\partial z/\partial y$, $f_x(3, 2)$, and $f_y(3, 2)$ for the following functions.

1. $z = f(x, y) = 7 + 2x$
2. $z = f(x, y) = 7 - 3y$
3. $z = f(x, y) = 7 + 3x - 2y$
4. $z = f(x, y) = 5 - 2x + 3y$
5. $z = f(x, y) = 10 - 3x^2 - 2xy + y^3$
6. $z = f(x, y) = 4xy^2 - 3y^2$
7. $z = f(x, y) = 3e^x - 2e^y$
8. $z = f(x, y) = 3xy^2 - 2e^y$
9. $z = f(x, y) = 3 \ln (xy + x^2)$
10. $z = f(x, y) = 2 \ln (3x^2 + 2y^2)$

Find z_x, z_y, and z_w if they exist.

11. $z = e^{xyw}$
12. $z = e^{3x^2yw^2}$
13. $z = \ln (x^2 + xy + w^2)$
14. $z = \ln (x^2 + w^2 + yw)$

Find the slope of the tangent line in an x direction and in a y direction at the point given.

15. $z = x^3 + y^2$, (1, 1, 2)
16. $z = 4x^2 + 6y^2$, (1, 1, 10)
17. $z = xy$, (2, $\frac{1}{2}$, 1)
18. $z = x^2 + xy$, (2, 1, 6)

Applications (Business and Economics)

19. **Compound Interest.** If a principal P is invested at an interest rate i compounded annually for n years, then the amount after n years is given by $A =$

$P(1 + i)^n$. Find the rate of change of A with respect to i with P and n held constant.

20. **Revenue and Cost Functions.** The revenue and cost functions for a company are given by

$$R(x, y) = 3x + 5y$$

$$C(x, y) = x^2 - 2xy + 3y^2 + 3x + 2y + 5$$

where x and y represent the number of items produced. Find each of the following.
  (a) $P(x, y) = R(x, y) - C(x, y)$
  (b) $R_y(3, 2)$ (c) $C_x(2, 3)$
  (d) $C_y(2, 3)$ (e) $P_x(2, 3)$

21. **Profit Function.** A profit function is given in thousands of dollars as

$$P(x, y) = 3x^2 - xy + 2y^3 + 2x + 3y + 3$$

where x and y represent the number of items produced. Find each of the following.
  (a) $P(1, 2)$ (b) $P(2, 1)$
  (c) $P_x(1, 2)$ (d) $P_y(1, 2)$

22. **Production Model.** The production model of a given company is given by $P = 0.8l^{1/2}C^{1/2}$, where P is the number of items produced, l is the number of hours of available labor, and C is the number of dollars of capital available. Find $\partial P/\partial l$ and $\partial P/\partial C$.

23. **Production Model.** For the production model in Exercise 22, find $\partial P/\partial l$ and $\partial P/\partial C$ when $l = 400$ hours and $C = \$10,000$.

Applications (Social and Life Sciences)

24. **Safety.** The length L of skid marks is given by

$$L(w, s) = 0.000014ws^2$$

where w is the weight of the car and s is the speed of the car. Find and interpret the results for each of the following.

(a) $L(2000, 50)$ (b) $L(3000, 55)$
(c) $L_w(3000, 55)$ (d) $L_s(3000, 55)$

25. **Intelligence Quotient.** IQ, represented by I, is given by the following equation:

$$I(M, C) = \frac{100M}{C}$$

where M represents mental age and C represents chronological age.

(a) Find $I(13, 10)$.
(b) Find $I(10, 13)$.
(c) Find $I_M(13, 10)$.
(d) Interpret your results.

26. **Scuba Diving.** The time of a scuba dive is estimated by

$$T(v, d) = \frac{36v}{d + 36}$$

where v is the volume of air in the diver's tanks and d is the depth of the dive.

(a) Find $T(65, 30)$.
(b) Find $T_v(65, 30)$
(c) Find $T_d(65, 30)$.
(d) Interpret your results.

Review Exercises

Sketch the graphs of the following functions.

27. $z = 4z + 2y + 6$
28. $z = 6x - y + 12$

13.3   APPLICATION OF PARTIAL DERIVATIVES

OVERVIEW   Recall that partial derivatives represent the instantaneous rate of change of the dependent variable with respect to one independent variable (the others are held fixed). As an independent variable increases, a positive partial derivative indicates an increase in the dependent variable and a negative partial derivative indicates a decrease in the dependent variable. Keep these concepts in mind as we study marginal analysis (rates of change) with two or more independent variables. In this section we consider

- Marginal productivity
- Marginal cost, revenue, and profit for joint cost functions
- Marginal demand functions using related products

Companies that produce more than one product can be classified as multicommodity firms. Although most of these companies produce several products, for mathematical simplicity we discuss the cost, revenue, and profit in terms of two products, item 1 and item 2. We will discuss $C(x_1, x_2)$, $R(x_1, x_2)$, and $P(x_1, x_2)$, for x_1 units of item 1 and x_2 units of item 2. The following partial derivatives give marginal costs (C_{x_1} and C_{x_2}), marginal revenues (R_{x_1} and R_{x_2}), and marginal profits (P_{x_1} and P_{x_2}), which are useful approximations for the management of a multicommodity company.

When more than one product is produced by a company, the cost function involves all the products and is called a joint cost function.

C_{x_1}:   Cost of making one *additional* unit of item 1 for given values of x_1 and x_2.

C_{x_2}:   Cost of making one *additional* unit of item 2 for given values of x_1 and x_2.

R_{x_1}:   Increase in revenue from selling one *additional* unit of item 1 for given values of x_1 and x_2.

R_{x_2}:   Increase in revenue from selling one *additional* unit of item 2 for given values of x_1 and x_2.

P_{x_1}:   Change in profit from selling one *additional* unit of item 1 for given values of x_1 and x_2.

P_{x_2}:   Change in profit from selling one *additional* unit of item 2 for given values of x_1 and x_2.

EXAMPLE 13   If the joint cost function of a multicommodity firm is

$$C(x_1, x_2) = 100 + x_1^2 + 4x_1x_2 + 2x_2^2$$

find the following.

(a) C_{x_1} at $x_1 = 20$, $x_2 = 30$
(b) C_{x_2} at $x_1 = 20$, $x_2 = 30$

SOLUTION

(a) $C_{x_1} = 2x_1 + 4x_2$; $C_{x_1}(20, 30) = 2 \cdot 20 + 4 \cdot 30 = 160$
(b) $C_{x_2} = 4x_1 + 4x_2$; $C_{x_2}(20, 30) = 4 \cdot 20 + 4 \cdot 30 = 200$

Practice Problem 1   A revenue function is given by

$$R(x_1, x_2) = 200x_1 + 200x_2 - 4x_1^2 - 8x_1x_2 - 4x_2^2$$

If $x_1 = 4$ and $x_2 = 12$, find (a) R_{x_1} and (b) R_{x_2}.

ANSWER

(a) $R_{x_1} = 200 - 8x_1 - 8x_2$; $R_{x_1}(4, 12) = 72$
(b) $R_{x_2} = 200 - 8x_1 - 8x_2$; $R_{x_2}(4, 12) = 72$

Practice Problem 2   If a cost function is defined as

$$C(x_1, x_2) = 32\sqrt{x_1x_2} + 175x_1 + 205x_2 + 1050$$

find the marginal costs, C_{x_1} and C_{x_2}, when $x_1 = 80$ and $x_2 = 20$.

ANSWER $C_{x_1}(80, 20) = 183$; $C_{x_2}(80, 20) = 237$

Production Model

Production depends on many factors, such as labor, capital, type of machinery, age of machinery, and so on. However, for the purpose of simplicity, we consider production (or output) as a function of only two variables, labor and capital. A function

$$P = f(l, C)$$

that gives the output P for l units of labor (such as the number of hours of labor) and C units of capital is called a **production function.**

$\dfrac{\partial P}{\partial l}$ gives the marginal productivity with respect to labor units.

$\dfrac{\partial P}{\partial C}$ gives the marginal productivity with respect to capital.

EXAMPLE 14 A furniture manufacturing company can produce

$$P = 0.8l^{1/2}C^{1/2}$$

custom chairs per week. Here l is the number of hours of labor and C is the number of dollars of capital. For example, if 400 work-hours and \$10,000 in capital are available, the production would be

$$P(400, 10,000) = 0.8(400)^{1/2}(10,000)^{1/2}$$
$$= 1600 \text{ chairs}$$

Then the marginal productivities are

$$P_l = \frac{0.4C^{1/2}}{l^{1/2}}$$

$$P_l(400, 10,000) = \frac{0.4(10,000)^{1/2}}{(400)^{1/2}} = 2$$

$$P_C = \frac{0.4l^{1/2}}{C^{1/2}}$$

$$P_C(400, 10,000) = \frac{0.4(400)^{1/2}}{(10,000)^{1/2}} = 0.08$$

Since marginals give an approximation of the change in a function, $P_l = 2$ tells us that if capital remains fixed at \$10,000 per week, the availability of an *additional* work-hour each week would result in an *approximate* increase in production of 2 chairs. On the other hand, $P_C = 0.08$ tells us that if labor remains fixed at 400 work-hours per week, the availability of one additional dollar of capital would increase production by *approximately* 0.08 of a chair. ∎

This example introduces what is known as the Cobb–Douglas production function that is used extensively in business and economics to calculate the number of units produced by different amounts of capital and labor.

COBB–DOUGLAS PRODUCTION FUNCTION

The Cobb–Douglas production function is given by

$$P(l, C) = Al^a C^{1-a}, \qquad A > 0, 0 < a < 1$$

where P is the number of units produced with l units of labor and C units of capital.

EXAMPLE 15 A company has the following production model for a product:

$$P(l, C) = 10l^{3/4}C^{1/4}$$

(a) Find the production that will result from 1296 work-hours of labor and $625 in capital.

(b) Find the marginal productivity of labor and the marginal productivity of capital.

(c) Evaluate the marginal productivity of labor and capital when $l = 64$ work-hours and $C = \$256$.

(d) Interpret each result from part (c).

SOLUTION

(a) When $l = 1296$ and $C = 625$,

$$P(l, C) = 10(1296)^{3/4}(625)^{1/4} = 10{,}800$$

Production is 10,800 units.

(b) Given $P(l, C) = 10l^{3/4}C^{1/4}$, the marginal productivity of labor and the marginal productivity of capital are, respectively,

$$\frac{\partial P}{\partial l} = \frac{7.5C^{1/4}}{l^{1/4}} \quad \text{and} \quad \frac{\partial P}{\partial C} = \frac{2.5l^{3/4}}{C^{3/4}}$$

(c) We have

$$\frac{\partial P}{\partial l}(64,256) = \frac{7.5(256)^{1/4}}{(64)^{1/4}} = 10.6066$$

and

$$\frac{\partial P}{\partial C}(64,256) = \frac{2.5(64)^{3/4}}{(256)^{3/4}} = 0.8839$$

(d) From part (c), $\partial P/\partial l = 10.6066$ tells us that if capital is kept fixed at $256, the availability of an additional hour of labor would *increase* production by approximately 10.6066 units when $l = 64$ hours. Furthermore, $\partial P/\partial C = 0.8839$ tells us that if work-hours are kept constant, the availability of an extra dollar of capital would *increase* production by approximately 0.8839 units when $C = \$256$ and $l = 64$ hours. ■

Joint Demand Functions

Sometimes two products are so related that a change in the price of one affects the demand for the other. Typical examples are pork and beef, chicken and fish, IBM computers and MacIntosh computers. If such a relationship exists, then the demand for each product is a function of the price of *both* products. Let p_1 and q_1 and p_2 and q_2 be the prices and numbers demanded, respectively, for two products. Then

$$q_1 = f(p_1, p_2)$$

$$q_2 = g(p_1, p_2)$$

We can find four partial derivatives.

$\dfrac{\partial q_1}{\partial p_1}$: Marginal demand for the first product with respect to the price of the first product (rate of change in demand for product 1 as the price of product 1 changes).

$\dfrac{\partial q_1}{\partial p_2}$: Marginal demand for the first product with respect to the price of the second product (rate of change in demand for product 1 as the price of product 2 changes).

$\dfrac{\partial q_2}{\partial p_1}$: Marginal demand for the second product with respect to the price of the first product (rate of change in demand for product 2 as the price of product 1 changes).

$\dfrac{\partial q_2}{\partial p_2}$: Marginal demand for the second product with respect to the price of the second product (rate of change in demand for product 2 as the price of product 2 changes).

Two products are said to be **competitive** if a decrease in the quantity demanded of one product can lead to an increase in the quantity demanded of the other product. For example, IBM computers and MacIntosh computers are competitive. On the other hand, if a decrease in the quantity demanded of one product leads to a decrease in the quantity demanded of a second product, then the products are said to be **complementary.** For example, motorboats and motorboat engines are complementary. The marginal demands can be used to classify products as competitive or complementary.

COMPETITIVE OR COMPLEMENTARY OR NEITHER

If p_1 and p_2 and q_1 and q_2 represent, respectively, the price and demand for two products, then the two products are said to be **competitive** if

$$\frac{\partial q_1}{\partial p_2} > 0 \quad \text{and} \quad \frac{\partial q_2}{\partial p_1} > 0$$

and **complementary** if

$$\frac{\partial q_1}{\partial p_2} < 0 \quad \text{and} \quad \frac{\partial q_2}{\partial p_1} < 0$$

All other cases are neither competitive nor complementary.

EXAMPLE 16   Given the following two demand functions for product 1 and product 2, classify the two products as competitive or complementary.

$$q_1 = 1000 - 6p_1 + 8p_2$$
$$q_2 = 600 + 12p_1 - 10p_2$$

SOLUTION

$$\frac{\partial q_1}{\partial p_2} = 8 \quad \text{and} \quad \frac{\partial q_2}{\partial p_1} = 12$$

Since $\partial q_1/\partial p_2$ and $\partial q_2/\partial p_1$ are both positive, the two products are competitive.

SUMMARY

We can see that the concepts and uses of partial derivatives relative to marginal analysis are much the same as we encountered in Chapter 9 when we dealt with only one independent variable. Remember that these ideas can be used with any number of independent variables, and that often, in real cases, there are numerous independent variables. Partial derivatives are very useful in the discussion of production models such as the Cobb–Douglas model.

Exercise Set 13.3

Applications (Business and Economics)

1. **Marginal Profit.** A corporation makes and sells x units of product A and y units of product B. The cost and revenue functions are given by

$$C(x, y) = 400 + 30x + 20y$$

and $$R(x, y) = 50x + 30y$$

Find and interpret each of the following.
   (a) Marginal cost with respect to x
   (b) Marginal cost with respect to y
   (c) Marginal revenue with respect to x
   (d) Marginal revenue with respect to y
   (e) Marginal profit with respect to x
   (f) Marginal profit with respect to y

2. **Marginal Cost.** Suppose that a company produces x units of product A and y units of product B with a cost function of

$$C(x, y) = 1000 + 40x + 30y + 2x^2 + y^2$$

Find and interpret $C_x(8, 10)$ and $C_y(8, 10)$.

3. **Marginal Revenue.** Relative to the company in Exercise 2, the revenue function is

$$R(x, y) = 100x + 60y - 0.02xy$$

Find and interpret $R_x(8, 10)$ and $R_y(8, 10)$.

4. **Marginal Profit.** Relative to the company in Exercises 2 and 3, find and interpret the following for the profit function: $P_x(8, 10)$ and $P_y(8, 10)$.

5. **Marginal Profit.** A company uses the following cost and revenue functions:

$$C(x, y) = 10x^2 + 4y^2 + 2xy$$

and $$R(x, y) = 12x^2 + 6y^2$$

Find and interpret the marginal profit functions P_x and P_y.

Analyze the demand functions in Exercises 6–11 to determine whether product 1 and product 2 are competitive, complementary, or neither. p_1 and p_2 are the prices and q_1 and q_2 are the demands for products 1 and 2, respectively.

6. $q_1 = 160 - 0.10p_1 - 0.05p_2$
   $q_2 = 200 - 0.08p_1 - 0.01p_2$

7. $q_1 = 250 - 0.04p_1 + 0.01p_2$
   $q_2 = 200 - 0.03p_1 + 0.005p_2$

8. $q_1 = 6000 - \dfrac{400}{p_1 + 2} - 40p_2$

   $q_2 = 8000 + 50p_1 - \dfrac{400}{p_2 + 2}$

9. $q_1 = \dfrac{200}{p_1\sqrt{p_2}}, \quad q_2 = \dfrac{400}{p_2\sqrt{p_1 + 3}}$

10. $q_1 = \dfrac{60\sqrt{p_1}}{p_2}, \quad q_2 = \dfrac{40p_2}{p_1}$

11. $q_1 = 600 - 0.04p_1^2 - 0.02p_2^2$
    $q_2 = 400 - 0.01p_1^2 - 0.04p_2^2$

12. **Production Function.** The production function for a given product of an electronics company is given by

$$f(l, C) = 400l^{1/3}C^{2/3}$$

where l is labor in terms of work-hours per week, C is capital expenditures in dollars spent per week, and the output f is in units manufactured per week. Find each of the following and interpret your results.
   (a) $f(512, 27{,}000)$
   (b) $f_l(512, 27{,}000)$
   (c) $f_C(512, 27{,}000)$

13. **Production Function.** Suppose that the production function for a product is

$$f = 60l^{3/4}C^{1/4}$$

where l is in work-hours and C is capital expenditures in dollars.
   (a) Find the marginal productivity of l.
   (b) Find the marginal productivity of C.

Review Exercises

Find $\partial z/\partial x$, $\partial z/\partial y$, $f_x(1, 2)$, and $f_y(1, 2)$ for the following functions.

14. $z = f(x, y) = 3x^2 + 2xy - 5y^2$

15. $z = f(x, y) = 4x^2y - 3xy^3$

16. $z = f(x, y) = 4x^2y + 2e^x$

17. $z = f(x, y) = 4xy + 5e^{-y}$

13.4 HIGHER-ORDER PARTIAL DERIVATIVES

OVERVIEW

When working with functions of one variable, we found that it was easy to extend the concept of the first derivative to obtain a second derivative, a third derivative, and so on. Although such an extension with partial derivatives is somewhat more involved, in this section we

• Extend the partial derivative concept to higher-order partial derivatives
• Find mixed partial derivatives

Just as with ordinary derivatives, it is possible to take the second partial derivative of a function. We can also take the third, fourth, and so on partial derivatives. The definition is given for second-order partial derivatives but can be extended to higher derivatives as needed.

SECOND-ORDER PARTIAL DERIVATIVES

Suppose that $z = f(x, y)$. Then

$$\frac{\partial^2 z}{\partial x^2} = \frac{\partial}{\partial x}\left(\frac{\partial z}{\partial x}\right) \qquad \text{Take the partial derivative with respect to } x \text{ of } \frac{\partial z}{\partial x}.$$

$$\frac{\partial^2 z}{\partial y^2} = \frac{\partial}{\partial y}\left(\frac{\partial z}{\partial y}\right) \qquad \text{Take the partial derivative with respect to } y \text{ of } \frac{\partial z}{\partial y}.$$

$$\frac{\partial^2 z}{\partial x \partial y} = \frac{\partial}{\partial x}\left(\frac{\partial z}{\partial y}\right) \qquad \text{Take the partial derivative with respect to } x \text{ of } \frac{\partial z}{\partial y}.$$

The following notations are for second-order derivatives.

NOTATION OF SECOND-ORDER PARTIAL DERIVATIVES

Suppose that $z = f(x, y)$; then

$$\frac{\partial^2 z}{\partial x^2} = \frac{\partial}{\partial x}\left(\frac{\partial z}{\partial x}\right) = f_{xx}(x, y) = f_{xx}$$

$$\frac{\partial^2 z}{\partial y \partial x} = \frac{\partial}{\partial y}\left(\frac{\partial z}{\partial x}\right) = f_{xy}(x, y) = f_{xy}$$

$$\frac{\partial^2 z}{\partial x \partial y} = \frac{\partial}{\partial x}\left(\frac{\partial z}{\partial y}\right) = f_{yx}(x, y) = f_{yx}$$

$$\frac{\partial^2 z}{\partial y^2} = \frac{\partial}{\partial y}\left(\frac{\partial z}{\partial y}\right) = f_{yy}(x, y) = f_{yy}$$

EXAMPLE 17 If $z = f(x, y) = 2x^3 - 3xy^2 + y^3 - 2$, find each of the following.

(a) $\dfrac{\partial^2 z}{\partial x^2}$ (b) $\dfrac{\partial^2 z}{\partial y \partial x}$ (c) $\dfrac{\partial^2 z}{\partial x \partial y}$ (d) $\dfrac{\partial^2 z}{\partial y^2}$

(e) $f_{xx}(1, 2)$ (f) $f_{xy}(1, 2)$ (g) $f_{yx}(2, 1)$ (h) $f_{yy}(5, 2)$

SOLUTION

(a) $\dfrac{\partial z}{\partial x} = 6x^2 - 3y^2$

$\dfrac{\partial^2 z}{\partial x^2} = 12x$

(b) $\dfrac{\partial z}{\partial x} = 6x^2 - 3y^2$

$\dfrac{\partial^2 z}{\partial y \partial x} = -6y$

(c) $\dfrac{\partial z}{\partial y} = -6xy + 3y^2$

$\dfrac{\partial^2 z}{\partial x \partial y} = -6y$

(d) $\dfrac{\partial z}{\partial y} = -6xy + 3y^2$

$\dfrac{\partial^2 z}{\partial y^2} = -6x + 6y$

(e) $f_{xx}(x, y) = \dfrac{\partial^2 z}{\partial x^2} = 12x$

$f_{xx}(1, 2) = 12(1) = 12$

(f) $f_{xy}(x, y) = \dfrac{\partial^2 z}{\partial y \partial x} = -6y$

$f_{xy}(1, 2) = -6(2) = -12$

(g) $f_{yx}(x, y) = \dfrac{\partial^2 z}{\partial x \partial y} = -6y$

$f_{yx}(2, 1) = -6(1) = -6$

(h) $f_{yy}(x, y) = \dfrac{\partial^2 z}{\partial y^2} = -6x + 6y$

$f_{yy}(5, 2) = -6(5) + 6(2)$
$= -30 + 12 = -18$ ∎

Practice Problem 1 For $f(x, y) = 7x^2y^3 + 3x^2y - 2y + 4x^4$, find f_{xx}, f_{xy}, f_{yy}, and f_{yx}.

ANSWER $f_{xx} = 48x^2 + 6y + 14y^3$, $f_{xy} = 6x + 42xy^2$, $f_{yy} = 42x^2y$, and $f_{yx} = 6x + 42xy^2$

EXAMPLE 18 Find $f_{xx}(1, 2), f_{yy}(1, 2)$, and $f_{xy}(1, 2)$ for $f(x, y) = e^{x^2 + y^2}$.

SOLUTION

$$f(x, y) = e^{x^2 + y^2}$$

$$f_x = e^{x^2 + y^2}(2x)$$ $\qquad \dfrac{\partial(x^2 + y^2)}{\partial x} = 2x$

$$f_{xx} = 2e^{x^2 + y^2} + e^{x^2 + y^2}(4x^2)$$ $\qquad$ Product formula
$$= (4x^2 + 2)e^{x^2 + y^2}$$

$$f_y = e^{x^2 + y^2}(2y)$$ $\qquad \dfrac{\partial(x^2 + y^2)}{\partial y} = 2y$

$$f_{yy} = 2e^{x^2 + y^2} + e^{x^2 + y^2}(4y^2)$$ $\qquad$ Product formula
$$= (4y^2 + 2)e^{x^2 + y^2}$$

$$f_{xy} = e^{x^2 + y^2}(4xy)$$ $\qquad \dfrac{\partial(f_x)}{\partial y} = \dfrac{\partial[e^{x^2+y^2}(2x)]}{\partial y}$

$$f_{xx}(1, 2) = [4(1)^2 + 2]e^{1^2 + 2^2} = 6e^5$$

$$f_{yy}(1, 2) = [4(2)^2 + 2]e^{1^2 + 2^2} = 18e^5$$

$$f_{xy}(1, 2) = 4(1)(2)e^{1^2 + 2^2} = 8e^5$$ ∎

Practice Problem 2 Given $f = e^{xy}$, find f_{xx}, f_{yy}, and f_{xy}.

ANSWER $f_{xx} = y^2 e^{xy}$, $f_{yy} = x^2 e^{xy}$, $f_{xy} = e^{xy} + xye^{xy}$

You may have already discovered in many examples that

$$\frac{\partial^2 z}{\partial y \partial x} = \frac{\partial^2 z}{\partial x \partial y}$$

Actually, all of the functions with which we shall be dealing satisfy this relationship.

There are four ways to find a second-order partial derivative of $z = f(x, y)$.

1. f_{xx} Differentiate twice with respect to x.
2. f_{yy} Differentiate twice with respect to y.
3. f_{xy} Differentiate with respect to x and then with respect to y.
4. f_{yx} Differentiate with respect to y and then with respect to x.

The derivatives f_{xy} and f_{yx} are called **mixed partial derivatives.**

We now extend the concept of a second-order partial derivative to a higher-order partial derivative. We use the following notation:

$$\frac{\partial^3 f}{\partial x^3} = f_{xxx} = \frac{\partial(f_{xx})}{\partial x}$$

$$\frac{\partial^3 f}{\partial y^3} = f_{yyy} = \frac{\partial(f_{yy})}{\partial y}$$

$$\frac{\partial f}{\partial x^2 \partial y} = f_{xxy} \quad \text{or} \quad f_{xyx} \quad \text{or} \quad f_{yxx}$$

$$f_{xxy} = \frac{\partial(f_{xx})}{\partial y}$$

$$\text{and} \quad f_{xyx} = \frac{\partial(f_{xy})}{\partial x}$$

$$\text{and} \quad f_{yxx} = \frac{\partial(f_{yx})}{\partial x}$$

EXAMPLE 19   If $f(x, y) = x^4 y^3 + x^6 + y^4$, find f_{xxy}, f_{xyx}, and f_{xxx}

SOLUTION

$$f(x, y) = x^4 y^3 + x^6 + y^4$$

$$f_x = 4x^3 y^3 + 6x^5$$

$$f_{xx} = 12x^2 y^3 + 30x^4$$

$$f_{xxy} = 36x^2 y^2 \qquad f_{xxy} = \frac{\partial(f_{xx})}{\partial y}$$

$$f_{xy} = 12x^3y^2 \qquad f_{xy} = \frac{\partial(f_x)}{\partial y}$$

$$f_{xyx} = 36x^2y^2 \qquad f_{xyx} = \frac{\partial(f_{xy})}{\partial x}$$

$$f_{xxx} = 24xy^3 + 120x^3 \qquad f_{xxx} = \frac{\partial(f_{xx})}{\partial x}$$

Practice Problem 3 For $f(x, y) = x^4 - 3x^2y^2 + y^4$, find f_{xxy}, f_{xyx}, and f_{xxx}.

ANSWER $f_{xxy} = -12y, f_{xyx} = -12y, f_{xxx} = 24x$

SUMMARY

In this section, we have expanded what we previously learned about partial derivatives. Partial derivatives can be expanded to any level to give second, third, or higher derivatives. These derivatives may all be with respect to the same variable, or they can be mixed.

Exercise Set 13.4

Find $\partial^2z/\partial x^2$, $\partial^2z/\partial y^2$, and $f_{xy}(1, 3)$ for the following functions.

1. $z = f(x, y) = 7 + 5x + 3y$

2. $z = f(x, y) = 5x^2 - 3y^3$

3. $z = f(x, y) = 10x^2y^3$

4. $z = f(x, y) = 5x^3y^2$

5. $z = f(x, y) = 5x^2y - 3xy^3$

6. $z = f(x, y) = 4xy^2 - 3x^2y^3$

7. $z = f(x, y) = 3e^x - 2e^y$

8. $z = f(x, y) = 4xe^x - 2xe^y$

9. $z = f(x, y) = 3e^{xy}$

10. $z = f(x, y) = \ln(3x + y)$

11. $z = f(x, y) = x \ln(3x^2 + 2y^2)$

12. $z = f(x, y) = 3e^{x^2} + y^2$

13. $z = f(x, y) = e^x \ln(3x + 2y)$

14. $z = f(x, y) = (e^x + e^y) \ln x$

15. $z = f(x, y) = \sqrt{2x + 3y}$

16. $z = f(x, y) = \sqrt[3]{x^2 + 3y^2}$

17. $z = f(x, y) = \dfrac{x}{y}$

18. $z = f(x, y) = \dfrac{1}{2x + 3y^2}$

19. $z = f(x, y) = e^{x-y}(y^2)$ 20. $z = f(x, y) = xe^{x^2-y^2}$

Find $f_{xxx}, f_{yyy}, f_{xyx}$, and f_{yxy} for each of the following:

21. $f(x, y) = x^3y + y^3x$

22. $f(x, y) = y^2x + x^2y$

23. $f(x, y) = e^{x+y}$

24. $f(x, y) = e^{x^2+3y^2}$

25. $f(x, y) = \ln(x + y)$

26. $f(x, y) = \ln(x^2 + y)$

Applications (Business and Economics)

27. **Production Function.** The production function for a given product of an electronics company is given by

$$f(l, C) = 400l^{1/3}C^{2/3}$$

where l is labor in terms of work-hours per week, C is the dollars spent per week, and the output f is in units manufactured per week. Find f_{ll}, f_{CC}, and f_{lC}.

28. **Production Function.** Suppose that the production function for a product is

$$f(l, C) = 60l^{3/4}C^{1/4}$$

where l is in work-hours and C is capital expenditures in dollars. Find f_{ll}, f_{CC}, and f_{Cl}.

Review Exercises

Find $\partial z/\partial x$, $\partial z/\partial y$, $f_x(1, 2)$, and $f_y(1, 2)$.

29. $z = f(x, y) = 3xy + \ln(x^2 + 2y^2)$

30. $z = f(x, y) = e^{xy} + \ln(2x + y)$

13.5 RELATIVE MAXIMA AND MINIMA FOR FUNCTIONS OF TWO VARIABLES

OVERVIEW   In this section, we show how partial derivatives may be used to find relative maxima or relative minima for functions of two variables. The development of the theory is somewhat similar to the development of the theory for functions of one variable. Since we have demonstrated that there are two first partial derivatives and three or more second partial derivatives for functions of two variables, you may expect the procedure of finding and testing critical points to be more complicated. To simplify the discussion, we are going to assume that all higher-order partial derivatives of $z = f(x, y)$ exist in some circular region of the xy-plane. In this section we

- Define a relative maximum and relative minimum for two independent variables
- Define a saddle point
- Find necessary conditions for a relative maximum and relative minimum
- Find sufficient conditions for a relative maximum, relative minimum, and a saddle point

The following definition extends the definition of relative maximum and relative minimum to a function of two variables.

RELATIVE MAXIMUM AND RELATIVE MINIMUM FOR FUNCTIONS OF TWO VARIABLES

(a) If there exists a circular region with (a, b) as the center in the domain of $z = f(x, y)$ such that

$$f(a, b) \geq f(x, y)$$

for all (x, y) in the region, then $f(a, b)$ is a **relative maximum**.

(b) If there exists a circular region with (a, b) as the center in the domain of $z = f(x, y)$ such that

$$f(a, b) \leq f(x, y)$$

for all (x, y) in the region, then $f(a, b)$ is a **relative minimum**.

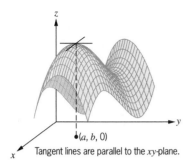

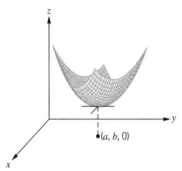

(a, b, 0)

Tangent lines are parallel to the xy-plane.

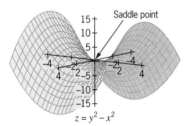

(a, b, 0)

Figure 16

Recall that at a relative maximum or relative minimum for a function of one variable, the first derivative is 0. For a function of two variables, Figure 16 suggests that both partial derivatives must be 0 at a relative maximum or relative minimum. This is true and is stated as a *necessary* condition for a relative maximum or relative minimum.

A NECESSARY CONDITION FOR A RELATIVE MAXIMUM OR RELATIVE MINIMUM

If f_x and f_y exist at (a, b), then a *necessary* condition for $f(a, b)$ to be a relative maximum or a relative minimum is that

$$f_x(a, b) = f_y(a, b) = 0$$

The point (a, b) is called a **critical point** of the domain.

The converse of the theorem is not true. If $f_x(a, b) = f_y(a, b) = 0$, then $f(a, b)$ may or may not be a relative maximum or relative minimum. Figure 17 shows a point (a, b) at which both partial derivatives equal 0, but the point is not a relative maximum or a relative minimum; this is called a **saddle point**.

The following test is similar to the second derivative test in one variable and gives *sufficient* conditions for a relative minimum, a relative maximum, or a saddle point.

Saddle point

$z = y^2 - x^2$

Figure 17

Sufficient Conditions for a Relative Minimum, a Relative Maximum, or a Saddle Point

Assume that $z = f(x, y)$ and all higher-order partial derivatives exist in some circular region with (a, b) as the center. Suppose that $f_x(a, b) = f_y(a, b) = 0$ and let

$$A = f_{xx}(a, b), \quad B = f_{xy}(a, b), \quad \text{and} \quad C = f_{yy}(a, b)$$

1. If $AC - B^2 > 0$, and $A > 0$, then $f(a, b)$ is a relative minimum.
2. If $AC - B^2 > 0$, and $A < 0$, then $f(a, b)$ is a relative maximum.
3. If $AC - B^2 < 0$, then $f(x, y)$ has a saddle point at (a, b).
4. If $AC - B^2 = 0$, then no conclusion can be reached about $f(a, b)$.

We now outline the procedure that should be followed in locating points where there are relative maxima, relative minima, or saddle points.

> ### Four-Step Procedure for Finding Relative Maxima, Relative Minima, or Saddle Points
>
> Step 1. Find f_x and f_y and set each equal to zero.
> Step 2. Solve the equations in step 1 to get the critical points, which we denote as (a, b).
> Step 3. Compute $A = f_{xx}(a, b)$, $B = f_{xy}(a, b)$, and $C = f_{yy}(a, b)$.
> Step 4. Evaluate $AC - B^2$ and determine which sufficient condition applies.

Step 2 may require solving the pair of simultaneous equations

$$\frac{\partial z}{\partial x} = 0$$

$$\frac{\partial z}{\partial y} = 0$$

EXAMPLE 20   Find all critical points and classify them for $z = f(x, y) = x^2 + y^2$.

SOLUTION   We will follow the four-step process outlined above.

Step 1.   Find f_x and f_y and set each equal to zero.

$$f_x(x, y) = 2x$$
$$f_y(x, y) = 2y$$

Setting both of these equal to 0 gives

$$f_x(x, y) = 2x = 0$$
$$f_y(x, y) = 2y = 0$$

Step 2.   Solve the equations in step 1 to get the critical points, which we denote as (a, b). In this step, we may obtain two equations in two unknowns in which we desire all common solutions. For this example, the only common solution is $x = 0$ and $y = 0$ and $(0, 0)$ is the only critical point of the domain.

Step 3.   Compute $A = f_{xx}(a, b)$, $B = f_{xy}(a, b)$, and $C = f_{yy}(a, b)$.

$$f_{xx}(x, y) = 2 \qquad \text{Generally, second partials are not constants.}$$
$$f_{xy}(x, y) = 0$$
$$f_{yy}(x, y) = 2$$

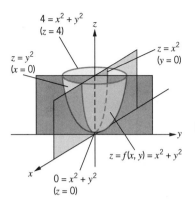

Figure 18

Therefore,

$$A = f_{xx}(0, 0) = 2$$

$$B = f_{xy}(0, 0) = 0$$

$$C = f_{yy}(0, 0) = 2$$

Step 4.   Evaluate $AC - B^2$ and determine which sufficient condition applies. $AC - B^2 = 2(2) - (0)^2 = 4 > 0$ and $A = 2 > 0$; so, condition 1 applies and $f(0, 0)$ is a relative minimum. The minimum is $f(0, 0) = 0$. This fact agrees with the graph shown in Figure 18.   ■

EXAMPLE 21   Find and classify all critical points of

$$z = f(x, y) = x^2 + xy + y^2 - 7x - 8y + 10$$

SOLUTION

Step 1.

$$f_x(x, y) = 2x + y - 7 = 0$$

$$f_y(x, y) = x + 2y - 8 = 0$$

Step 2.   Solving this system gives $(2, 3)$ as the only critical point of the domain.

Step 3.

$$f_{xx}(x, y) = 2 = f_{xx}(2, 3) = A$$

$$f_{xy}(x, y) = 1 = f_{xy}(2, 3) = B$$

$$f_{yy}(x, y) = 2 = f_{yy}(2, 3) = C$$

Step 4. $AC - B^2 = (2)(2) - (1)^2 = 3 > 0$. Since $A = 2 > 0$, condition 1 tells us that $f(2, 3)$ is a relative minimum. The minimum is $f(2, 3) = -9$.   ■

Practice Problem 1   Find and classify all critical points for

$$z = f(x, y) = 2x^2 + y^2 + 8x - 6y + 20$$

ANSWER   Relative minimum at $(-2, 3)$; $f(-2, 3) = 3$

Practice Problem 2   Find and classify all critical points for $z = x - \frac{1}{3}x^3 - \frac{1}{2}y^2$.

ANSWER   Relative maximum at $(-1, 0)$; $f(-1, 0) = -\frac{2}{3}$; saddle point at $(1, 0)$

EXAMPLE 22   Find all critical points and classify them for

$$z = f(x, y) = xy - \frac{x^4}{4} - \frac{y^2}{2} + 10$$

SOLUTION

Step 1.

$$f_x(x, y) = y - x^3 = 0$$

$$f_y(x, y) = x - y = 0$$

Step 2.   To solve these two equations, we obtain $y = x$ from the second equation and substitute into the first equation to give

$$x - x^3 = 0$$

$$x(1 - x^2) = 0$$

$$x(1 + x)(1 - x) = 0$$

$$x = 0 \quad \text{or} \quad x = -1 \quad \text{or} \quad x = 1$$

There are three critical points of the domain: $(0, 0)$, $(-1, -1)$, and $(1, 1)$. We will test each of these by repeating steps 3 and 4.

Step 3.

$$f_{xx}(x, y) = -3x^2$$

$$f_{xy}(x, y) = 1$$

$$f_{yy}(x, y) = -1$$

For $(0, 0)$ we have

$$A = f_{xx}(0, 0) = -3(0)^2 = 0$$

$$B = f_{xy}(0, 0) = 1$$

$$C = f_{yy}(0, 0) = -1$$

Step 4. $AC - B^2 = 0(-1) - (1)^2 = -1 < 0$. Condition 3 tells us that $f(x, y)$ has a saddle point at $(0, 0)$.

Step 3.   For $(-1, -1)$ we have

$$A = f_{xx}(-1, -1) = -3(-1)^2 = -3$$

$$B = f_{xy}(-1, -1) = 1$$

$$C = f_{yy}(-1, -1) = -1$$

Step 4. $AC - B^2 = (-3)(-1) - (1)^2 = 2 > 0$. Since $A = -3 < 0$, condition 2 tells us that $f(-1, -1)$ is a relative maximum.

Step 3.   For (1, 1) we have

$$A = f_{xx}(1, 1) = -3(1)^2 = -3$$
$$B = f_{xy}(1, 1) = 1$$
$$C = f_{yy}(1, 1) = -1$$

Step 4. $AC - B^2 = (-3)(-1) - (1)^2 = 2 > 0$. Since $A = -3 < 0$, condition 2 tells us that $f(1, 1)$ is a relative maximum.   ■

Practice Problem 3 Find the points at which there are relative maxima, relative minima, or saddle points for $z = 3xy - x^3 - y^3$.

ANSWER   Relative maximum of 1 at (1, 1); saddle point at (0, 0)

There are times when a restriction is placed on a function. These restrictions may be given as equations and are called **constraint equations**.

EXAMPLE 23   Find three numbers x, y, and z such that $x + y + z = 10$ and x^2yz is a relative maximum. (The equation $x + y + z = 10$ is a constraint equation.)

SOLUTION   Although x^2yz is not a function of two variables, we can solve $x + y + z = 10$ for z to obtain $z = 10 - x - y$. Substituting this value of z gives a function of two variables.

$$f(x, y) = x^2y(10 - x - y) = 10x^2y - x^3y - x^2y^2$$

Step 1.

$$f_x(x, y) = 20xy - 3x^2y - 2xy^2 = xy(20 - 3x - 2y) = 0$$
$$f_y(x, y) = 10x^2 - x^3 - 2x^2y = x^2(10 - x - 2y) = 0$$

Step 2.   Although $(0, y)$ is a critical point for any y, it is obvious that none of these yield a maximum because for each of these $f(0, y) = 10(0)^2y - (0)^3y - (0)^2y^2 = 0$. An additional critical point $(5, \frac{5}{2})$ is obtained from solving the system

$$20 - 3x - 2y = 0$$
$$10 - x - 2y = 0$$

Step 3.

$$f_{xx}(x, y) = 20y - 6xy - 2y^2$$
$$f_{xy}(x, y) = 20x - 3x^2 - 4xy$$
$$f_{yy} = (x, y) = -2x^2$$

For $(5, \frac{5}{2})$ we have

$$A = f_{xx}(5, \tfrac{5}{2}) = 20(\tfrac{5}{2}) - 6(5)(\tfrac{5}{2}) - 2(\tfrac{5}{2})^2 = 50 - 75 - \tfrac{25}{2} = -\tfrac{75}{2}$$

$$B = f_{xy}(5, \tfrac{5}{2}) = 20(5) - 3(5)^2 - 4(5)(\tfrac{5}{2}) = 100 - 75 - 50 = -25$$

$$C = f_{yy}(5, \tfrac{5}{2}) = -2(5)^2 = -50$$

Step 4. $AC - B^2 = -\tfrac{75}{2}(-50) - (-25)^2 = 1250 > 0.$ Since $A = -\tfrac{75}{2} < 0$, condition 2 tells us that

$$f(5, \tfrac{5}{2}) = 10(5)^2(\tfrac{5}{2}) - (5)^3(\tfrac{5}{2}) - (5)^2(\tfrac{5}{2})^2 = \tfrac{625}{4}$$

is a relative maximum. So x^2yz is a maximum at $(5, \frac{5}{2}, \frac{5}{2})$. ■

SUMMARY

To locate relative extrema on the surface $z = f(x, y)$, we begin by finding the critical points. The critical points are found by solving the pair of simultaneous equations

$$\frac{\partial z}{\partial x} = 0 \quad \text{and} \quad \frac{\partial z}{\partial y} = 0$$

In this section we stated the necessary and sufficient conditions for a relative maximum, relative minimum, and a saddle point. This can be used in marginal analysis to determine maximum profit, minimum cost, and maximum revenue.

Exercise Set 13.5

In Exercises 1–5, find the critical points of each function.

1. $f(x, y) = x^2 + y^2 - 2y + 4$
2. $f(x, y) = x^2 + 2y^2 - 2xy - 3x + 5y + 8$
3. $f(x, y) = x^2 - 2x + y^4 - 2y^2 + 6$
4. $f(x, y) = 4x^2 - 4xy + 3y^2 + 8x - 4y + 6$
5. $f(x, y) = x^2 - 2xy + 2y^2 + 3x + 8$

In Exercises 6–10, classify each critical point as a relative maximum, relative minimum, or saddle point.

6. Exercise 1

7. Exercise 2

8. Exercise 3

9. Exercise 4

10. Exercise 5

Find all critical points and classify each as a relative maximum, relative minimum, or saddle point.

11. $f(x, y) = x^2 + 2y^2$
12. $f(x, y) = 3x^2 + y^2$
13. $f(x, y) = 4 - x^2 - y^2$
14. $f(x, y) = 9 - x^2 - y^2$
15. $f(x, y) = 10 - x^2 - 2y^2$
16. $f(x, y) = y + x^2 + 3y^2$
17. $f(x, y) = x^2 - y^2$
18. $f(x, y) = xy$

19. $f(x, y) = 5 - x^2 + 4x - y^2$

20. $f(x, y) = x^2 + y^2 - 6x + 2y + 10$

21. $f(x, y) = xy - 3x + 2y - 3$

22. $f(x, y) = xy + 5x - 3y + 7$

23. $f(x, y) = x^2 + y^2 - 6xy$

24. $f(x, y) = 8xy - x^2 - y^2$

25. $f(x, y) = x^2 + xy + 2y^2 - 3x + 2y + 2$

26. $f(x, y) = x^2 + 4xy + y^2 + 6y + 1$

27. $f(x, y) = -2x^2 + xy - y^2 + 10x + y - 3$

28. $f(x, y) = 3x^2 - 2xy + 2y^2 - 8x - 4y + 10$

29. $f(x, y) = 3xy - x^2y - xy^2$

30. $f(x, y) = e^{xy}$

31. $f(x, y) = x^3 + y^3 - 6x^2 - 3y^2 - 9y$

32. $f(x, y) = 2x^3 - x^2 - 4x + y^2 - 4y + 2$

33. Divide a straight line of length L into three parts such that the sum of the squares of their lengths is a minimum.

Applications (Business and Economics)

34. **Profit Function.** A company has the following profit function (in thousands of dollars):

$$P(x, y) = 2xy - x^2 - 2y^2 - 4x + 12y - 5$$

where x is the number of thousands of item I and y is the number of thousands of item II produced. How many of each type should be produced to maximize (relative maximum) profit? What is the maximum profit?

35. **Cost Function.** A cost function (in thousands of dollars) is given by

$$C(x, y) = 3x^2 + 2xy + 2y^2 - 18x - 16y + 180$$

Find the critical point and the minimum (relative minimum) cost if x and y represent the number of items produced.

36. **Maximum Revenue.** The revenue function (in thousands of dollars) associated with Exercise 35 is

$$R(x, y) = -5x^2 + 42x - 8y^2 - 2xy + 102y$$

Find the point that will give maximum (relative maximum) revenue and find the maximum revenue.

37. **Maximum Profit.** Using the cost and revenue functions of Exercises 35 and 36, find the point at which there is a maximum (relative maximum) profit and find the maximum profit.

Review Exercises

38. Find f_{xx}, f_{ww}, f_{xy}, and f_{yw} for $f(x, y, w) = e^{xyw}$.

39. Find f_{xx} and f_{xy} for $f(x, y) = \ln(x^2 + 3y^2)$.

13.6 LAGRANGE MULTIPLIERS

OVERVIEW   In Example 23 of Section 13.5, we found three numbers x, y, and z, which made $f(x, y, z) = x^2yz$ a maximum subject to the constraint $x + y + z = 10$. In the solution to that example, we solved the constraint equation for z in terms of x and y, that is,

$$z = 10 - x - y$$

and substituted this value of z into the function x^2yz to obtain a function of two variables,

$$x^2yz = x^2y(10 - x - y)$$

We then found the critical points for this function and used the critical points to find the relative maximum.

Sometimes it is difficult or even impossible to solve a constraint equation for one variable in terms of the others. In that case, we may try a method called the **method of Lagrange multipliers**. In this method, the introduction of another variable allows us to solve the constrained optimization problem without first solving the constraint equation for one of the variables. The method can be used for functions of two variables, three variables, four variables, or more. For simplicity, we state the method for two independent variables. The method was invented by the French mathematician Joseph Louis Lagrange (1736–1813). In this section we

- Learn to use Lagrange multipliers
- Use Lagrange multipliers to solve optimization problems

The method of Lagrange multipliers is used to solve constrained optimization problems. For example, a soft drink distributor may wish to design the least expensive can that will hold a certain number of ounces, or a company may need to maximize production subject to a strict budget. The method of Lagrange multipliers helps to solve these problems.

We will consider problems such as, find the points that maximize or minimize the function $f(x, y)$ subject to the constraint $g(x, y) = 0$. The example we mentioned in the overview could be stated as follows: Find the point on the line $x + y = 10$ for which $f(x, y) = x^2 y$ is a maximum. That is, maximize $f(x, y) = x^2 y$ subject to $x + y = 10$.

The following procedure is very helpful in solving such problems.

THE METHOD OF LAGRANGE MULTIPLIERS

To find the relative maxima or relative minima of a function $f(x, y)$ subject to a constraint $g(x, y) = 0$, introduce a new variable, λ (called a Lagrange multiplier). Then:

Step 1. Form

$$F(x, y, \lambda) = f(x, y) + \lambda \cdot g(x, y)$$

Step 2. Form the system

$$F_x(x, y, \lambda) = 0$$

$$F_y(x, y, \lambda) = 0$$

$$F_\lambda(x, y, \lambda) = 0$$

We assume all indicated partial derivatives exist.

Step 3. Solve the system found in step 2 for values of x, y, and λ that satisfy the system. The desired extrema will be found among the points (x, y) that satisfy the system.

The steps in solving a problem by the method of Lagrange multipliers are illustrated in the next example.

EXAMPLE 24   Find the minimum value of $f(x, y) = x^2 + y^2$ subject to $2x + y = 5$.

SOLUTION   Note that $f(x, y) = x^2 + y^2$ and $g(x, y) = 2x + y - 5 = 0$.

Step 1.   Form

$$F(x, y, \lambda) = f(x, y) + \lambda \cdot g(x, y)$$

$$F(x, y, \lambda) = \underbrace{x^2 + y^2}_{\text{Objective function}} + \underbrace{\lambda \cdot (2x + y - 5)}_{\text{Constraint function}}$$

Step 2.   Form the system.

$$F_x(x, y, \lambda) = 2x + 2\lambda = 0$$

$$F_y(x, y, \lambda) = 2y + \lambda = 0$$

$$F_\lambda(x, y, \lambda) = 2x + y - 5 = 0$$

Step 3.   The system may be solved by any of several methods, but we choose to use the substitution method for this particular example. The first equation gives $x = -\lambda$. The second equation gives $y = -\frac{1}{2}\lambda$. Substituting these values into the third equation gives

$$2(-\lambda) + \left(-\frac{1}{2}\lambda\right) - 5 = 0$$

$$-\frac{5}{2}\lambda = 5$$

$$\lambda = -2$$

Hence $x = -\lambda = -(-2) = 2$

and $y = -\frac{1}{2}\lambda = -\frac{1}{2}(-2) = 1$

The point $(x, y) = (2, 1)$ gives a relative minimum of $f(2, 1) = 2^2 + 1^2 = 5$ for the function subject to the constraint $2x + y = 5$.   ■

Note that in the preceding example we did not prove that $f(2, 1) = 5$ is a relative minimum subject to the constraint. Sufficient conditions may be found in textbooks on mathematical analysis, but you can usually judge by substituting some point close to the critical point and comparing the values obtained.

Practice Problem 1   Minimize $f(x, y) = x^2 + y^2$ subject to $x + 2y = 10$.

ANSWER   The minimum is 20, which occurs at $x = 2$ and $y = 4$.

Practice Problem 2 Use the method of Lagrange multipliers to find two numbers whose sum is 76 and whose product is a maximum.

ANSWER $x = 38$ and $y = 38$

The method of Lagrange multipliers can be applied to functions of three or more variables, as the following example illustrates.

EXAMPLE 25   Use the method of Lagrange multipliers to find a maximum of $f(x, y, z) = x^2yz$ subject to the constraint $x + y + z = 10$. (This is a Lagrange multiplier solution to the example mentioned in the Overview.)

SOLUTION

Step 1.   Form

$$F(x, y, z, \lambda) = x^2yz + \lambda(x + y + z - 10)$$

Step 2.   Form the system.

$$F_x(x, y, z, \lambda) = 2xyz + \lambda = 0$$
$$F_y(x, y, z, \lambda) = x^2z + \lambda = 0$$
$$F_z(x, y, z, \lambda) = x^2y + \lambda = 0$$
$$F_\lambda(x, y, z, \lambda) = x + y + z - 10 = 0$$

Step 3.   From the first equation,

$$\lambda = -2xyz$$

Substituting in the second equation gives

$$x^2z - 2xyz = 0$$

which gives

$$xz(x - 2y) = 0$$
$$x = 0, \quad z = 0, \quad \text{or} \quad x = 2y$$

Substituting in the third equation gives

$$x^2y - 2xyz = 0$$

Therefore,

$$xy(x - 2z) = 0$$
$$x = 0, \quad y = 0, \quad \text{or} \quad x = 2z$$

From the fourth equation,

$$z = 10 - x - y$$

If $x = 0$, $y = 0$, or $z = 0$, the function $f(x, y, z) = 0$ and is not a maximum. So,

$$x = 2y$$
$$x = 2z$$
$$z = 10 - x - y$$

or

$$\frac{x}{2} = 10 - x - \frac{x}{2} \qquad \left(z = \frac{x}{2}, y = \frac{x}{2}\right)$$

$$2x = 10$$

$$x = 5, \quad y = \tfrac{5}{2}, \quad \text{or} \quad z = \tfrac{5}{2}$$

Thus, $f(5, \tfrac{5}{2}, \tfrac{5}{2}) = (5)^2(\tfrac{5}{2})(\tfrac{5}{2}) = \tfrac{625}{4}$ is a relative maximum subject to the constraint. This is the same value that we obtained in Example 23 of the last section. ■

Practice Problem 3 Using the method of Lagrange multipliers, maximize the function $f(x, y) = x^2 y$ subject to $x + y = 10$.

ANSWER $x = \tfrac{20}{3}, y = \tfrac{10}{3}, f(\tfrac{20}{3}, \tfrac{10}{3}) = \tfrac{4000}{27}$

EXAMPLE 26 A manufacturing company has a production function

$$P(l, C) = 200l^{1/2}C^{1/2}$$

where l denotes the number of units of labor and C denotes the number of units of capital. In addition, there is a budget constraint of \$30,000 where each unit of labor costs \$10 and each unit of capital \$100. Find the l and C that will maximize production.

SOLUTION We want to maximize

$$P(l, C) = 200l^{1/2}C^{1/2}$$

subject to the constraint

$$10l + 100C = 30,000$$

$$F(l, \lambda) = 200l^{1/2}C^{1/2} + \lambda(10l + 100C - 30,000)$$

$$F_l = \frac{100C^{1/2}}{l^{1/2}} + 10\lambda$$

$$F_C = \frac{100l^{1/2}}{C^{1/2}} + 100\lambda$$

$$F_\lambda = 10l + 100C - 30,000$$

Set each partial equal to zero.

$$\frac{100C^{1/2}}{l^{1/2}} + 10\lambda = 0$$

$$\frac{100l^{1/2}}{C^{1/2}} + 100\lambda = 0$$

$$10l + 100C - 30,000 = 0$$

Solving for λ in the first two equations gives

$$\lambda = \frac{-10C^{1/2}}{l^{1/2}} = \frac{-l^{1/2}}{C^{1/2}}$$

Multiplying by $l^{1/2}C^{1/2}$ gives

$$-10C = -l \quad \text{or} \quad l = 10C$$

Substituting this value in the third equation gives

$$10(10C) + 100C = 30,000$$

$$C = 150$$

So $l = 10(150) = 1500$ and we see that 1500 units of labor and 150 units of capital maximize production. The maximum production is

$$P(1500, 150) = 200\sqrt{150 \cdot 1500} = 94,868 \quad \text{units of production}$$

At $C = 120$, $l = 1800$ and production is 92,952.
At $C = 180$, $l = 1200$ and production is 92,952.

It seems that 94,868 is indeed a maximum.

EXAMPLE 27 A utility function is used to measure the satisfaction, called the utility, that one gets from using products in a given time. Suppose that $U(x, y)$ is such a function for products A and B. Let p_1 and p_2 be the prices of products A and B, respectively. Now suppose that there is a limit or budget that restricts the amount I that can be spent on products A and B. For example, suppose that

$$I = p_1 x + p_2 y$$

We are interested in maximizing the function $U(x, y)$ subject to the budget constraint $I = p_1 x + p_2 y$. For example, if $U = x^2 y^2$, find the maximum satisfaction if $40 = 2x + 3y$.

SOLUTION

$$F(x, y, \lambda) = x^2 y^2 + \lambda(2x + 3y - 40)$$
$$F_x = 2xy^2 + 2\lambda$$
$$F_y = 2x^2 y + 3\lambda$$
$$F_\lambda = 2x + 3y - 40$$

Setting each partial equal to zero yields

$$2xy^2 + 2\lambda = 0$$
$$2x^2 y + 3\lambda = 0$$
$$2x + 3y - 40 = 0$$

From the first two equations we get $\lambda = -xy^2$ and $\lambda = -\frac{2}{3}x^2 y$ or $-xy^2 = -\frac{2}{3}x^2 y$. We now have

$$2x^2 y - 3xy^2 = xy(2x - 3y) = 0$$

so

$$x = 0, \quad y = 0, \quad \text{or} \quad x = \frac{3y}{2}$$

Either $x = 0$ or $y = 0$ makes $U(x, y) = x^2 y^2$ equal to 0. A maximum value of $U(x, y)$ does not occur at either $x = 0$ or $y = 0$. [Note that $U(x, y)$ is always greater than or equal to 0.]

Substituting $x = 3y/2$ in $2x + 3y - 40 = 0$ yields

$$2\left(\frac{3y}{2}\right) + 3y - 40 = 0$$

$$6y = 40 \quad \text{or} \quad y = \frac{20}{3}$$

and $2x + 3(\frac{20}{3}) = 40$ or $x = 10$. Then

$$U\left(10, \frac{20}{3}\right) = 10^2 \left(\frac{20}{3}\right)^2 = \frac{40,000}{9}$$

Use your calculator to evaluate $U(x, y)$ at two or three points as evidence that $\frac{40,000}{9}$ is a maximum.

At $(5, 10)$, which satisfies $2x + 3y = 40$,

$$U(5, 10) = 5^2(10)^2 = 2500 < \frac{40,000}{9}$$

At $(11, 6)$, which again satisfies $2x + 3y = 40$,

$$U(11, 6) = 11^2(6)^2 = 4356 < \frac{40,000}{9}$$

SUMMARY

In this section we learned that it is often necessary to maximize or minimize a function subject to a certain constraint. This is especially true in business. To find a relative maximum or minimum value of $f(x, y, z)$ subject to the constraint $g(x, y, z) = 0$, we form

$$F(x, y, z, \lambda) = f(x, y, z) + \lambda g(x, y, z)$$

We then set all four partial derivatives of F equal to zero and solve for x, y, and z. The partial derivatives give

$$F_x(x, y, z, \lambda) = \frac{\partial f}{\partial x} + \lambda \frac{\partial g}{\partial x} = 0$$

$$F_y(x, y, z, \lambda) = \frac{\partial f}{\partial y} + \lambda \frac{\partial g}{\partial y} = 0$$

$$F_z(x, y, z, \lambda) = \frac{\partial f}{\partial z} + \lambda \frac{\partial g}{\partial z} = 0$$

$$g(x, y, z) = 0$$

Exercise Set 13.6

Use the method of Lagrange multipliers to solve the following exercises.

1. Find a relative maximum of $f(x, y) = 3xy$ subject to $x + y = 10$.

2. Find a relative maximum of $f(x, y) = 4xy$ subject to $x + y = 8$.

3. Find a relative maximum of $f(x, y) = 4xy$ subject to $x + 2y = 6$.

4. Find a relative maximum of $f(x, y) = 2xy + 3$ subject to $x + y = 12$.

5. Find a relative maximum of $f(x, y) = 3xy + 2$ subject to $3x + y = 4$.

6. Find x and y such that $x + y = 22$ and xy^2 is a maximum.

7. Find x and y such that $x + y = 34$ and $x^2 y$ is a maximum.

8. Find a relative maximum of $f(x, y) = 3x^2 y$ subject to $x + 2y = 5$.

9. Find a relative maximum of $f(x, y) = 2xy^2$ subject to $2x + 3y = 5$.

10. Find two numbers whose sum is 26 and whose product is a maximum.

11. Find two numbers whose sum is 94 and whose product is a maximum.

12. Find a relative minimum of $f(x, y) = 3x^2 + 2y^2 - xy$ subject to $x + y = 4$.

13. Find a relative minimum of $f(x, y) = x^2 + 3y^2 - 2xy$ subject to $2x + y = 5$.

14. Find three numbers whose sum is 80 and whose product is a maximum.

15. Find three numbers whose sum is 146 and whose product is a maximum.

Applications (Business and Economics)

16. **Production.** If x thousand dollars are spent on labor and y thousand dollars are spent on equipment, the production of a factory would be

$$P(x, y) = 40x^{1/3} y^{2/3}$$

How should \$100,000 be allocated to obtain the maximum production?

17. **Change in Production.** Use the method of Lagrange multipliers to estimate the change in the maximum production that would occur if the money available was increased by \$1000 in Exercise 16.

18. **Cost Function.** The cost function for x units of item I and y units of item II is

$$C(x, y) = 8x^2 + 14y^2$$

If it is necessary that $x + y = 99$, how many of each item should be produced for minimum cost? What is the minimum cost?

19. **Maximum Utility.** If $U(x, y) = 14x - x^2 + 20y - y^2$ is a utility function with constraint $5x + 4y = 100$, where $p_1 = \$5$, $p_2 = \$4$, and $I = \$100$, find the maximum utility.

20. **Maximum Utility.** If the utility function for two products is $U(x, y) = xy^2$, with $p_1 = \$4$, $p_2 = \$6$, and $I = \$60$, find the maximum utility or satisfaction.

21. **Warehouse.** An electric utility needs a warehouse that contains 1 million cubic feet. It is estimated that the floor and ceiling will cost \$3 per square foot and the walls will cost \$7 per square foot. Find the dimensions of the most economical building.

Applications (Social and Life Sciences)

22. **Feed Mixture.** A feed company mixed three feeds (called A, B, and C) to obtain a desired feed for calves. Let x be the number of tons of A in the mixture, y the number of tons of B, and z the number of tons of C in the mixture. To obtain the desired amount of mixture (due to cost)

$$x + 2y + 3z = 140$$

In addition, the company desires to maximize the units of iron in the mixture. If 2% of A are units containing iron, 4% of B, and 6% of C, determine the desired mixture to maximize the number of units of iron.

23. **Farming**.   A farmer has 300 feet of fence. Find the dimensions of the rectangular field of maximum area he can enclose.

24. Redo Exercise 23 if the farmer does not need to fence one side because of a building.

25. **Construction**.   A rectangular box with no top is to be built with 600 square feet of material. Find the dimensions of the box that would have maximum volume.

26. Redo Exercise 25 if the box has a top.

Review Exercises

Find the relative maxima and relative minima for the following functions.

27. $f(x, y) = 4x^2 + y^2 - 4y$

28. $f(x, y) = 4x^2 + y^2 + 8x - 2y - 1$

29. $f(x, y) = x^2 + xy + y^2 - 6x - 2y$

30. $f(x, y) = x^3 - 3x - y^2$

Chapter Review

Review to ensure that you understand the following concepts.

Function of two variables	Saddle point
Function of several variables	Partial derivative
Relative maxima	Second-order partial derivative
Relative minima	Lagrange multiplier
Critical point	Lagrange method

Can you use the following formulas?

$$\frac{\partial^2 z}{\partial x^2} = \frac{\partial}{\partial x}\left(\frac{\partial z}{\partial x}\right)$$

$$\frac{\partial^2 z}{\partial y^2} = \frac{\partial}{\partial y}\left(\frac{\partial z}{\partial y}\right)$$

$$\frac{\partial^2 z}{\partial x\,\partial y} = \frac{\partial}{\partial y}\left(\frac{\partial z}{\partial x}\right) \quad \text{or} \quad \frac{\partial}{\partial x}\left(\frac{\partial z}{\partial y}\right)$$

Chapter Test

1.   If $f(x, y, z) = x^3 + y^3 + z^3 - xyz$, find $f(1, 2, -1)$.

2.   Plot the following points on a three-dimensional coordinate system.
     (a) $(0, 2, 1)$ (b) $(-3, 2, -1)$

3.   Sketch the graph of $z = 4x + 3y + 12$.

4.   If $z = x^2 + 4xy^2$, find $\partial^2 z/\partial y^2$.

5.   Given $z = 3 \ln(x^2 + 3y^2)$, find $\partial^2 z/\partial x^2$.

6.   If $f = e^{xy}$, find f_y.

7.   If $z = x^2 y + 4x + xy^2$, find $\partial^2 z/\partial x\,\partial y$.

8. Sketch the three dimensional graph of $x + y = 4$.

9. Find all critical points for

$$f(x, y) = x^2 + xy - x + \frac{3y^2}{2} + 7y$$

10. In Problem 9, classify the critical points as relative maxima, relative minima, or neither.

11. Make a contour map for $z = 4x^2 + y^2$ and then draw a rough sketch of the surface defined by $z = 4x^2 + y^2$.

12. Use the method of Lagrange multipliers to find a maximum of $f(x, y, z) = 9xyz$ subject to $6x + 4y + 3z = 24$.

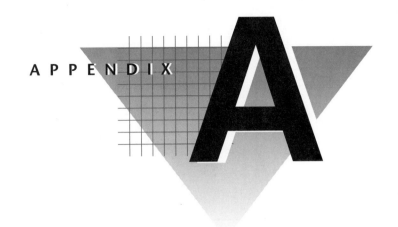

APPENDIX A

Using Graphing Calculators

INSTRUCTIONS FOR TI-82

Graphing Functions

The first step in graphing a function is to press the $\boxed{\text{Y}=}$ key. The screen shown in Figure 1(a) will appear in the calculator window. Press the down arrow to see the remaining Y = locations [Figure 1(b)].

Suppose that you want to graph $y = -x + 4$. With the arrow keys ($\boxed{\blacktriangledown}$, $\boxed{\blacktriangleleft}$, $\boxed{\blacktriangleright}$, $\boxed{\blacktriangle}$) you can move the cursor to any Y = location you choose. Move the cursor to the Y_1 position. Press the negative key $\boxed{(-)}$ (grey key, row 10, column 4), the $\boxed{\text{X, T, }\theta}$ key (row 3, column 2), the addition operation sign $\boxed{+}$, and the number 4. On the screen you have $Y_1 = -X + 4$ [Figure 2(a)]. You could have entered the X with the $\boxed{\text{ALPHA}}$ key and the $\boxed{\text{X}}$ key (row 9, column 1), but the $\boxed{\text{X, T, }\theta}$ key is easier. Note also that we used the negative key $\boxed{(-)}$ (grey key, row 10, column 4) instead of the subtraction key $\boxed{-}$ (blue key, row 8, column 5).

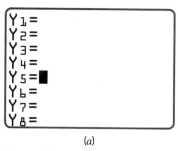

(a)   (b)

Figure A1

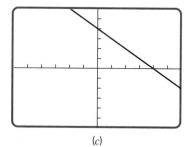

(a)

(b)

(c)

Figure A2

When you graph a function with paper and pencil, you must draw coordinate axes and decide what values you want to see on the axes. Similarly, with a graphing calculator you must decide what viewing window to use. Press $\boxed{\text{WINDOW}}$ (row 1, column 2). The Ymin, Ymax, Xmin, and Xmax define what is called a viewing window. You can change the size of your viewing window by changing the window settings. Move the arrow down to Xmin and input -6. Each time you press $\boxed{\text{ENTER}}$ or the down arrow the cursor moves down one line. Let Xmin $= -6$, Xmax $= 6$, Xscl $= 1$, Ymin $= -6$, Ymax $= 6$, and Yscl $= 1$ [Figure 2(b)]. When you have input all of the entries, press $\boxed{\text{Y=}}$ (to make certain of the function to be graphed) and then press $\boxed{\text{GRAPH}}$. When you press $\boxed{\text{GRAPH}}$, you obtain the graph shown in Figure 2(c).

The hash lines (called tick marks) on both the x-axis and y-axis represent 1 unit. If we count the tick marks in Figure 2(c), we see that the maximum y value on the screen is 6 and the minimum y value is -6. Press the $\boxed{\text{WINDOW}}$ key and change the minimum and maximum values of x and y to -10 and 10 and examine the graph in Figure 3(a). Note this time that there are 10 hash lines, each representing 1 unit. This particular range setting is the built-in window setting on the TI-82 that is attained by pressing $\boxed{\text{ZOOM}}$ and then choosing 6 : ZStandard [Figure 3(b)]. To clear the screen of a graph, simply press $\boxed{\text{CLEAR}}$ when the graph is on the screen. This will return you to the home screen.

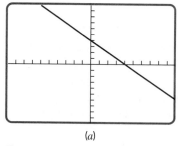

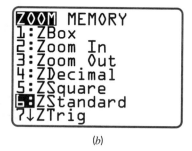

(a)   (b)

Figure A3

Press $\boxed{\text{Y=}}$ to input the function

$$y = \frac{320 - 4x}{3}$$

[see Figure 4(*b*)]. Note that we need to use parentheses:

$$\boxed{\text{Y=}}\,\boxed{(}\,320\,\boxed{-}\,4\,\boxed{\text{X, T, }\theta}\,3\,\boxed{)}\,\boxed{\div}\,3$$

Now when you press $\boxed{\text{GRAPH}}$ the screen is blank. What is the difficulty? The viewing window is too small. When $x = 0$, y is 320/3 or 106.6. The range of y values is too small. Suppose that we are interested in the graph for only positive values of x and large values of y. We change our window to Xmin = 0, Xmax = 8, Xscl = 2, Ymin = 80, Ymax = 120, and Yscl = 10 [Figure 4(*a*)]. The graph is shown in Figure 4(*c*).

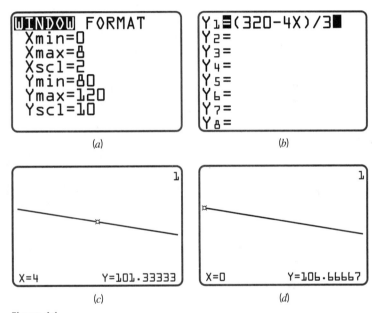

Figure A4

Trace Features

There are two cursors that can be moved to different positions on the screen. The free-moving cursor (+) is activated by pressing one of the arrow keys ($\boxed{\blacktriangledown}$, $\boxed{\blacktriangleleft}$, $\boxed{\blacktriangleright}$, $\boxed{\blacktriangle}$). The x- and y-coordinates of the position of the cursor are given at the bottom of the screen. The trace cursor (¤) is activated by pressing the $\boxed{\text{TRACE}}$ key. The trace cursor moves along graphs of functions that have been activated and placed on the screen. The trace cursor gives the current x- and the $f(x)$-coordinates of the selected function along the bottom of the screen. A small number at the upper right side of the screen indicates which function is being traced [see Figures 4(*c*) and 4(*d*)]. The up and down arrows ($\boxed{\blacktriangle}$, $\boxed{\blacktriangledown}$) move between Y = functions, and the left and right arrows ($\boxed{\blacktriangleleft}$, $\boxed{\blacktriangleright}$) move the cursor along the selected function. To deactivate the trace cursor, press $\boxed{\text{CLEAR}}$ or press $\boxed{\text{GRAPH}}$.

To graph $Y_1 = 2X + 1$ and $Y_2 = -X + 7$, enter the functions on the $\boxed{Y=}$ screen [Figure 5(a)]. Press the $\boxed{\text{ZOOM}}$ key, select 6: ZStandard, and then press $\boxed{\text{TRACE}}$ [Figure 5(b)]. Now press $\boxed{\blacktriangleright}$ and move the cursor to what seems to be the intersection of the two lines [Figure 5(c)]. The actual intersection is $x = 2$ and $y = 5$. However, the most accurate answer we can obtain by moving the cursor is $x = 2.1276596$ and $y = 4.8723404$ or $x = 1.9148936$ and $y = 4.8297872$.

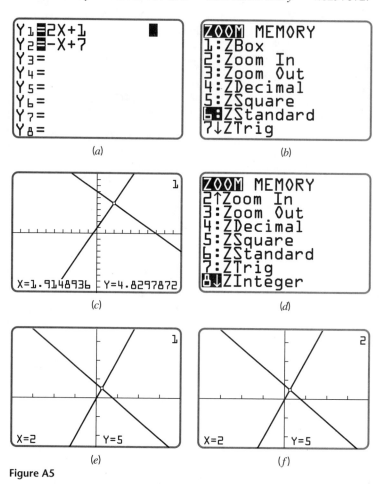

Figure A5

The answer is not exact because the cursor has width and length and moves a pixel distance; thus, it cannot be placed exactly on the intersection of the two lines. The viewing window 8: ZInteger [Figure 5(d)] under the $\boxed{\text{ZOOM}}$ menu makes the pixel distance such that the cursor will be on or close to the x- and y-coordinates. This procedure enables us to obtain a more accurate answer for the coordinates of intersection, especially if they are integral coordinates. Press the $\boxed{\text{ZOOM}}$ key, choose 8: ZInteger, and then press $\boxed{\text{ENTER}}$. The calculator gives you a chance to change the center of the coordinates to something other than

(0, 0). For now, let's use (0, 0), so press ENTER again. Press TRACE ; notice now that the trace cursor can be moved to exactly $x = 2$ and $y = 5$ (left and right arrows) whether you specify function 1 [Figure 5(e)] or change to view function 2 (up and down arrows) [Figure 5(f)].

The point of intersection can also be found by using the CALC menu. This will be discussed later in this appendix. Using the TABLE setup, we can also show where the two functions have the same x and y values. This will also be discussed later in this appendix.

Friendly Windows

A "friendly window" is one where the TRACE key moves in increments of .1 for the x-values. You can get a friendly window by using the ZDecimal option on the ZOOM menu (ZOOM 4:). When you use this key the WINDOW is set at Xmin = −4.7, Xmax = 4.7, Xscl = 1, Ymin = −3.1, Ymax = 3.1. Graph $y = x^2 − 2$ and use this window. Now use the TRACE key and you will see that the values of x change in increments of .1.

The incremental movement of the cursor is set by a variable, Δx. Δx is evaluated as $\Delta x = (\text{Xmax} − \text{Xmin})/94$. By setting the values of Xmin and Xmax appropriately, you can have $\Delta x = .1$. For example if you set Xmax = 5, then Xmin = −4.4. With some practice you can set these values so that you have a friendly window covering the area of the graph you wish to examine.

There is another variable, $\Delta y = (\text{Ymax} − \text{Ymin})/62$, but this variable does not affect the TRACE . Both Δx and Δy can be found using VARS : WINDOW. Δx is VARS 1 : 7 and Δy is VARS 1 : 8. You can store a value in these variables, but the resulting range on the graph screen may not be suitable.

Box Command

Suppose that we wish to graph $y = x^2$ and $y = 2x + 0.21$ on the standard screen. To graph $y = x^2$, press Y = , select Y_1 press X, T, θ , and x^2 . Similarly graph $Y_2 = 2X + .21$ [Figure 6(a)]. Using the standard viewing rectangle [Figure 6(b)], the point of intersection is close to $x = 2.13$, $y = 4.47$ [Figure 6(c)].

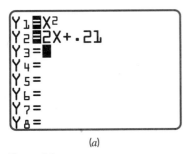

(a)

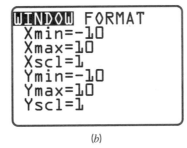

(b)

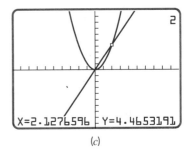

(c)

Figure A6

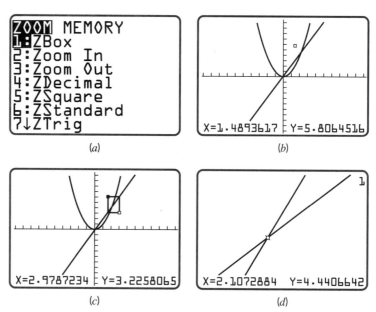

(a)

(b)

(c)

(d)

Figure A7

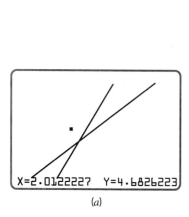

(a)

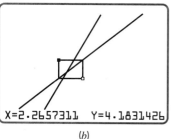

(b)

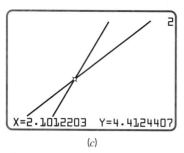

(c)

Figure A8

To improve the answer, place a box around the intersection and then change the screen to that which is in the box; that is, enlarge the picture of the intersection. Press ZOOM, select 1 : ZBox [Figure 7(a)], and move the cursor to a point where the upper left corner of the box could be, for example, $x = 1.4893617$ and $y = 5.8064516$ [Figure 7(b)]. Press ENTER to fix the corner. Then move the cursor to the lower right corner, for example, $x = 2.9787234$ and $y = 3.2258065$. A box is formed that visually contains the intersection [Figure 7(c)]. Press ENTER to obtain the interior of the box as the image on the screen [Figure 7(d)]. Repeat the process to obtain the answer $x = 2.10$, $y = 4.41$ to the accuracy desired [Figures 8(a), 8(b), and 8(c)]. (Note that we could also use the intersect option under the CALC menu or the ΔTbl increment using the TblSet key and TABLE setup to find the same solution.)

Solving Equations

In this section we illustrate two techniques for solving equations. The first technique gives an approximation of the answer to the desired accuracy. The second technique is helpful when the solution is an integer. Find the solution to $2x - 2 = 4$. Perform the necessary operations to get everything on one side of the equation, that is, $0 = 2x - 6$. The value of x that will make this expression equal to 0 is the value that will make y equal to 0 in $y = 2x - 6$. Of course, $y = 0$ is where the graph crosses the x-axis. This point is called the x-intercept. The first method consists of putting a box around the x-intercept on the graph and then

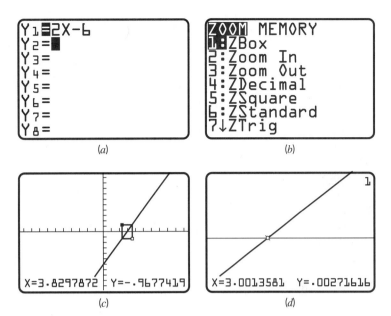

Figure A9

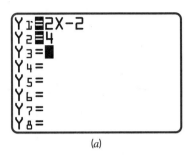

(a)

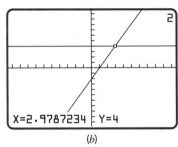

(b)

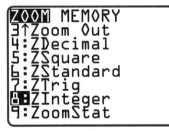

(c)

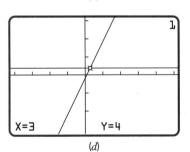

(d)

Figure A10

viewing the *x*-intercept when the box becomes the screen. That is, enlarge the picture so that the coordinates of the *x*-intercept can be read easily. Graph $y = 2x - 6$ in the Y_1 position [Figure 9(a)]. Press ZOOM, select 6:ZStandard, and then press ZOOM; next select 1:ZBox [Figure 9(b)], move the cursor to the upper left corner, and press ENTER to fix one corner of the box. Next move the cursor to the lower right corner to fix the box. Make sure that the box contains the *x*-intercept. Then press ENTER and the new graphing screen becomes the region that we enclosed in the box [Figure 9(c)]. Now move the cursor to the position that seems to be the location of the *x*-intercept [Figure 9(d)]. Continue this process until you are satisfied with the accuracy of the answer. The *x*-intercept is at $x = 3$; that is, the solution to the equation $2x - 2 = 4$ is $x = 3$.

Now let's solve the same equation, $2x - 2 = 4$, by considering two equations made up of the two sides of our given equation. Let $Y_1 = 2X - 2$ (left side) and $Y_2 = 4$ (right side) [Figure 10(a)]. Now press ZOOM, choose 6:ZStandard, and graph the two equations. This time let's use a different method to locate the point of intersection. Press TRACE and move the cursor to what seems to be the point of intersection. Notice at the bottom of the screen that $x = 2.9787234$, $y = 4$ [Figure 10(b)]. To determine if these are integral values, press ZOOM, select 8:ZInteger, and then press ENTER and TRACE [Figure 10(c)]. Move the cursor until you see that the coordinates of the intersection are $x = 3$, $y = 4$ [Figure 10(d)]. The solution to the equation is $x = 3$.

This problem could also be solved using the TABLE key or the CALC intersect key.

TABLE Key

The TABLE key displays the numerical values similar to the Graph command but in table format. Let's return to the earlier example of finding the intersection of $Y_1 = 2X + 1$ and $Y_2 = -X + 7$ [Figure 11(a)], but this time we will use the TABLE key. Press TblSet using 2nd WINDOW [Figure 11(b)]. Set TblMin $= -3$ and ΔTbl $= 1$, and then press TABLE using 2nd GRAPH. Notice that in the table at $x = 2$, both Y_1 and Y_2 have values of 5 [Figure 11(c)].

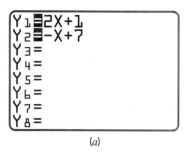

(a)

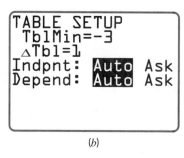

(b)

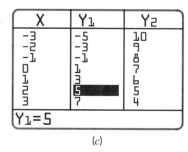

(c)

Figure A11

Table Setup

The selections we make on the TABLE SETUP screen determine which cells contain values when we press the TABLE key.

Indpnt: Auto Depend: Auto	Values appear in all cells in the table automatically.
Indpnt: Ask Depend: Auto	Table is empty. When a value is entered for the independent variable, the dependent values are calculated automatically.
Indpnt: Auto Depend: Ask	Values appear for the independent variable. To generate a value for a dependent variable, move to the specific cell and press ENTER.
Indpnt: Ask Depend: Ask	Table is empty. Enter values for independent variable. To generate a value for the dependent variable, move to the specific cell and press ENTER.

We use the functions in Y_1 and Y_2 and change the TblSet. Notice the effect of changing from Auto to Ask in each of the screens in Figure 12.

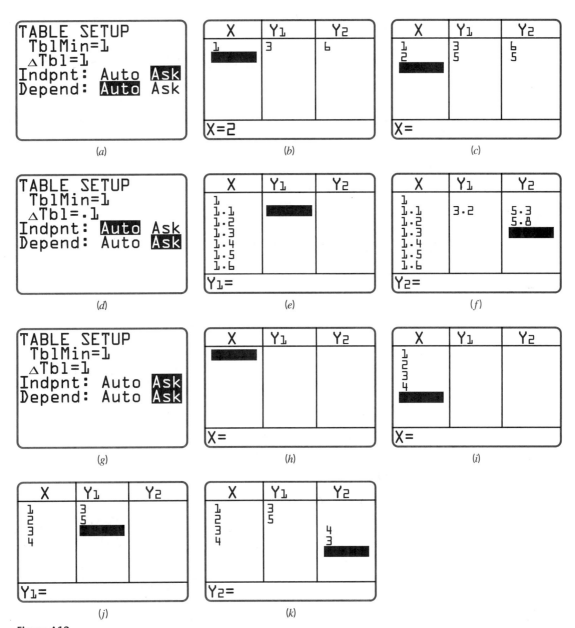

Figure A12

Now let's return to another example: $Y_1 = X^2$ and $Y_2 = 2X + .21$. Input the functions using the $\boxed{Y=}$ key [Figure 13(a)]. Then press the $\boxed{\text{TblSet}}$ (or $\boxed{\text{2nd}}$ $\boxed{\text{window}}$) key and set the TblMin to -1 and ΔTbl to 1 [Figure 13(b)]. Press $\boxed{\text{2nd}}$ $\boxed{\text{TABLE}}$; notice that at $x = 2$, both Y_1 and Y_2 have values that seem close [Figure 13(c)]. Now set TblMin to 1.9 and ΔTbl to .1 [Figure 13(d)]. At $x = 2.1$, the y

```
Y1=X²
Y2=2X+.21
Y3=
Y4=
Y5=
Y6=
Y7=
Y8=
```

(a)

```
TABLE SETUP
 TblMin=-1
 △Tbl=1
Indpnt: Auto Ask
Depend: Auto Ask
```

(b)

X	Y1	Y2
-1	1	-1.79
0	0	.21
1	1	2.21
2	4	4.21
3	9	6.21
4	16	8.21
5	25	10.21

X=2

(c)

```
TABLE SETUP
 TblMin=1.9
 △Tbl=.1
Indpnt: Auto Ask
Depend: Auto Ask
```

(d)

X	Y1	Y2
1.9	3.61	4.01
2	4	4.21
2.1	4.41	4.41
2.2	4.84	4.61
2.3	5.29	4.81
2.4	5.76	5.01
2.5	6.25	5.21

Y2=4.41

(e)

Figure A13

values agree exactly [Figure 13(e)]. If need be, we could continue to change the minimum value of the table and the increment of change in the table until the two y values reached an approximation close to the desired accuracy.

CALC Key

We can use the value command to calculate a function value for a given x [see Figure 14(c)]. Let $Y_1 = X^2$ and $Y_2 = 2X + .21$. Press the GRAPH key to see the graphs of the functions [Figure 14(d)]. Press 2nd TRACE. Use 1 : Value. Input an x value (in this case, 3) and press ENTER. Notice that the x and y values are displayed at the bottom of the screen [Figure 14(e)]. Press the up arrow or the down arrow and the x and y values for the other function will be displayed [Figure 14(f)].

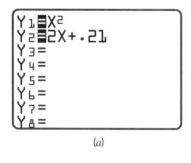

(a)

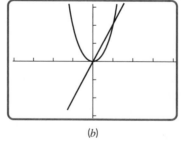

(b)

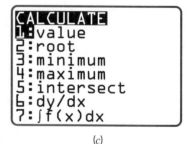

(c)

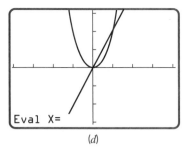

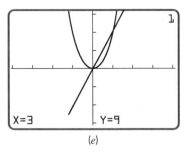

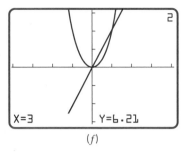

(d)

(e)

(f)

Figure A14

To solve for the intersection of $Y_1 = X^2$ and $Y_2 = 2X + .21$ using the ⎡CALC⎤ key, graph the functions using ⎡Y=⎤ [Figure 15(a)] and ⎡GRAPH⎤. To obtain the CALC menu press ⎡2nd⎤ ⎡TRACE⎤ and select 5 : intersect [Figure 15(b)]. In response to the question "First curve?" press ⎡ENTER⎤ if the cursor is on curve 1 close to the point of intersection [Figure 15(c)]. In response to the question "Second curve?" press ⎡ENTER⎤ if the cursor is on curve 2 close to the point of intersection [Figure 15(d)]. In response to "Guess?" move the cursor close to the point of intersection and press ⎡ENTER⎤ [Figure 15(e)]. The calculator will respond on the bottom of the screen with Intersection X = 2.1, Y = 4.41 [Figure 15(f)].

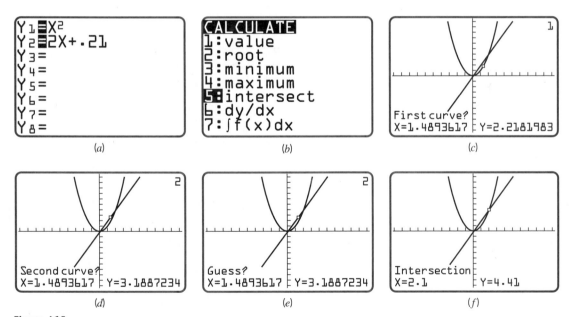

(a)

(b)

(c)

(d)

(e)

(f)

Figure A15

Matrices

The use of the TI-82 in the algebra of matrices is discussed thoroughly in Chapter 2 and the advantages of using a graphing calculator are demonstrated in Chapter 3 on linear programming.

The Simplex Program

This program will reduce a simplex tableau that has been stored in matrix **A**. Press ENTER to see each stage of reduction. The answer will be stored in matrix **E**. The program is shown in Figure 16.

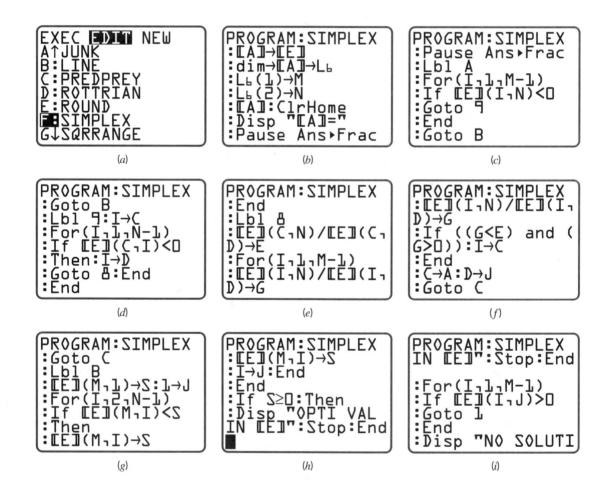

```
EXEC EDIT NEW
A↑JUNK
B:LINE
C:PREDPREY
D:ROTTRIAN
E:ROUND
F■SIMPLEX
G↓SQRRANGE
```

(a)

```
PROGRAM:SIMPLEX
:[A]→[E]
:dim→[A]→Lь
:Lь(1)→M
:Lь(2)→N
:[A]:ClrHome
:Disp "[A]="
:Pause Ans▸Frac
```

(b)

```
PROGRAM:SIMPLEX
:Pause Ans▸Frac
:Lbl A
:For(I,1,M-1)
:If [E](I,N)<0
:Goto 9
:End
:Goto B
```

(c)

```
PROGRAM:SIMPLEX
:Goto B
:Lbl 9:I→C
:For(I,1,N-1)
:If [E](C,I)<0
:Then:I→D
:Goto 8:End
:End
```

(d)

```
PROGRAM:SIMPLEX
:End
:Lbl 8
:[E](C,N)/[E](C,
D)→E
:For(I,1,M-1)
:[E](I,N)/[E](I,
D)→G
```

(e)

```
PROGRAM:SIMPLEX
:[E](I,N)/[E](I,
D)→G
:If ((G<E) and (
G>0)):I→C
:End
:C→A:D→J
:Goto C
```

(f)

```
PROGRAM:SIMPLEX
:Goto C
:Lbl B
:[E](M,1)→S:1→J
:For(I,2,N-1)
:If [E](M,I)<S
:Then
:[E](M,I)→S
```

(g)

```
PROGRAM:SIMPLEX
:[E](M,I)→S
:I→J:End
:End
:If S≥0:Then
:Disp "OPTI VAL
IN [E]":Stop:End
■
```

(h)

```
PROGRAM:SIMPLEX
IN [E]":Stop:End

:For(I,1,M-1)
:If [E](I,J)>0
:Goto 1
:End
:Disp "NO SOLUTI
```

(i)

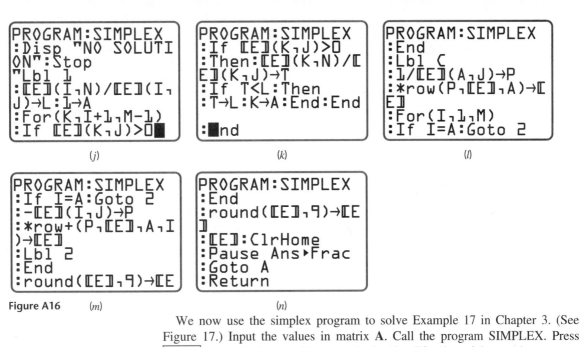

Figure A16

We now use the simplex program to solve Example 17 in Chapter 3. (See Figure 17.) Input the values in matrix **A**. Call the program SIMPLEX. Press ENTER to see the intermediate steps in the simplification of the matrix.

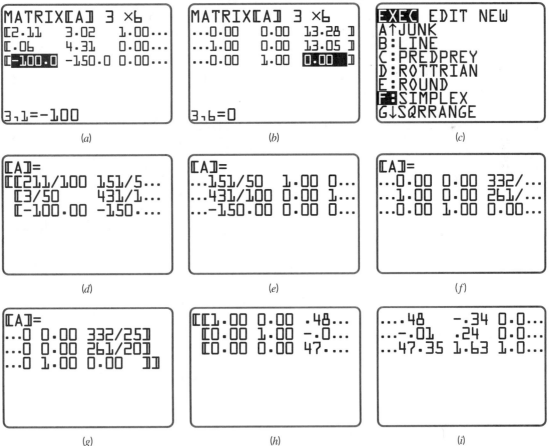

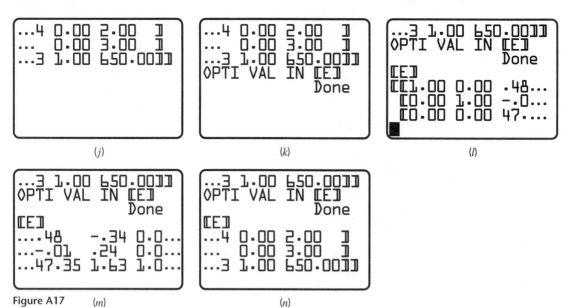

Figure A17 (m)   (n)

Probability and Statistics

The finite probability distribution function in Example 11, Chapter 6, can be displayed using a calculator. Press ⟨STAT⟩ and choose 1 : Edit [Figure 18(a)]. In the L_1 column enter the numbers 2 through 12, in the L_2 column enter the corresponding probabilities 1/36 to 6/36 to 1/36 [Figure 18(b)]. Use STAT PLOT by pressing ⟨2nd⟩ ⟨Y =⟩ [Figure 18(c)]. Choose 1 : Plot 1 and then select On; choose the first section under Type (scatterplot) and select L_1 for Xlist and L_2 for Ylist [Figure 18(d)]. Press the ⟨ZOOM⟩ key and select 9 : ZoomStat [Figure 18(e)].

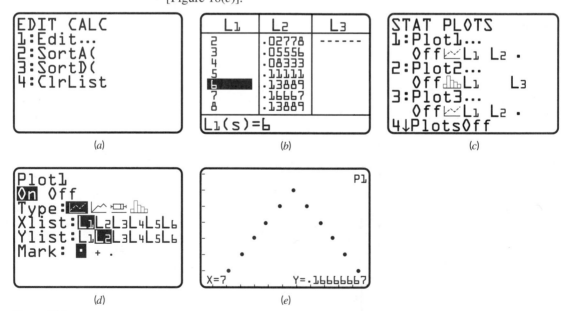

Figure A18

One way to visualize a set of data is to draw a histogram. This also facilitates making a frequency distribution from a set of data. Suppose that we wish to draw a histogram and also make a frequency distribution for the following data:

$$10, 17, 1, 8, 8, 8, 2, 4, 4, 15, 18, 18, 9, 10, 11,$$
$$12, 12, 12, 6, 7, 6, 9, 6, 6, 11, 3, 3, 12, 10, 9$$

The smallest value of the data is 1 and the largest value is 18. Suppose that we decide to have 6 classes into which we group the data.

$$\text{Class width} = \frac{18 - 1}{6} = \frac{17}{6}$$

If we round off to a whole number 3, we have 6 classes of width 3. Now the interval from the beginning of the first class to the end of the last class is $1 + 6(3) = 19$. Use Xmin = 1 and Xmax = 19. Xscl will be 3 (the width of the classes). We do not know the frequency in any class. However, with only 30 values, there possibly will not be more than 10 in any class. Select Ymax = 10. (We can easily change the range later if the histogram will not fit in the window of the screen.) Use Ymin = -1 to see the x-axis on the screen; then select Yscl = 1 to count the frequency easily.

Be sure that the functions in the $\boxed{Y =}$ menu are turned off. If the equal sign of a function is highlighted, put the cursor on the equal sign and press $\boxed{\text{ENTER}}$. This will deselect that function [Figures 19(a) and 19(b)]. Enter the data. Set the viewing window by pressing $\boxed{\text{WINDOW}}$ and set Xmin = 1, Xmax = 19, .Xscl = 3, Ymin = -1, Ymax = 10, and Yscl = 1 [Figure 19(d)]. Press $\boxed{\text{2nd}}$ $\boxed{Y =}$ and select Plot 1 [Figure 19(e)]. Highlight On and press $\boxed{\text{ENTER}}$. For Type choose the histogram icon, for the Xlist choose L_1, for Freq choose 1, and then press $\boxed{\text{GRAPH}}$ [Figure 19(f)]. The first rectangle of the histogram extends from 1 to 4 with the calculator programmed so that a data value 4 would be placed in the second class [Figure 19(g)]. This is the same as manually tabulating all values in a class from 1 to 3. From the histogram, we can read the frequencies and tabulate a frequency distribution. However, a data value 7 would be placed in the third rectangle. Again, this is equivalent to tabulating the frequency of the data from 4 to 6 [see Figures 19(h), 19(i), and 19(j)]. Press the $\boxed{\text{TRACE}}$ key to move the cursor along the histogram.

Class	Frequency
1–3	4
4–6	6
7–9	7
10–12	9
13–15	1
16–18	3
	30

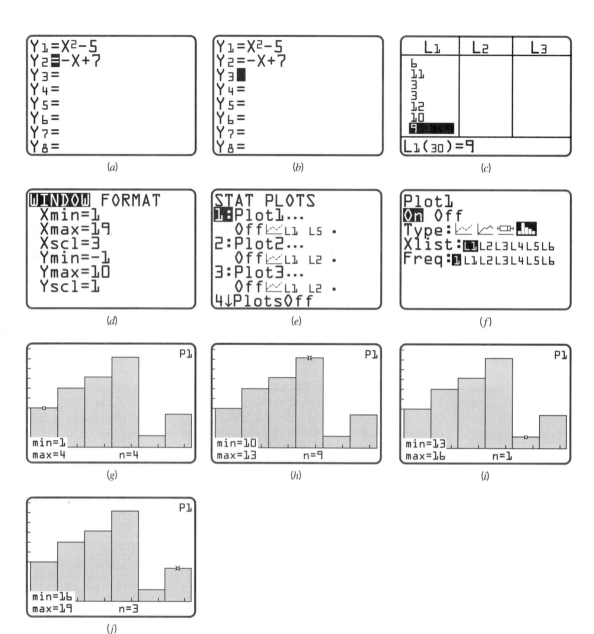

Figure A19

Calculus

The use of the TI-82 in the study of calculus is explained in the first five chapters on calculus.

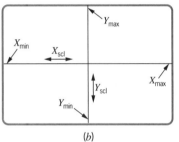

Cursor

```
Range
Xmin:█5.
 max:5.
 scl:2.
Ymin:-10.
 max:10.
 scl:5.
```

(a)

Figure A20

INSTRUCTIONS FOR THE CASIO 7000G AND 7700GB

Two types of graphs can be generated using this calculator: built-in function graphs and user-generated graphs. The calculator contains the following built-in graphs:

$$\sin \quad \cos \quad \tan \quad x^2 \quad \log \quad \ln$$
$$10 \quad e \quad x^{-1} \quad \sqrt{} \quad \sqrt[3]{}$$

We can draw such graphs as follows:

| Graph | x^{-1} | EXE | sketches $y = \dfrac{1}{x}$

| Graph | SHIFT | e^x | EXE | sketches $y = e^x$

Any time a built-in graph is executed, the ranges are set to their optimum values. User-generated graphs are not set automatically; instead, the user must decide on the range. The parts of a graph outside of the selected range do not appear on the display. The range contents are given in Figure 20(a); the maximum and minimum on each axis and the scale on each axis are given in Figure 20(b). To change the range use the keys | ← | | ⇒ | | ↓ | | ↑ | to move the cursor to the number you wish to change. Then insert the appropriate numbers. After you enter | Range |, the cursor is located at Xmin. After making a change, the cursor moves automatically down the scale by keying | EXE | as follows:

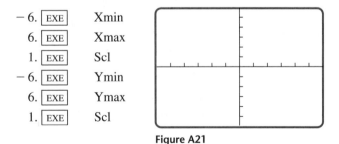

$-6.$	EXE	Xmin
$6.$	EXE	Xmax
$1.$	EXE	Scl
$-6.$	EXE	Ymin
$6.$	EXE	Ymax
$1.$	EXE	Scl

Figure A21

Now punch | G ↔ T | and you will see the coordinate system given in Figure 21. Again, punch | G ↔ T | to transfer the window from graphics back to text. Unless stated otherwise or unless built-in functions are used, the graphs in this appendix will use the above range of values. Enter | AC | and you are ready to draw a graph.

User-generated graphs can be drawn by entering the formula after entering [Graph]. An unknown is entered by using [ALPHA][X]. (The X appears in red and is the [+] key.) For example,

$$y = 4x^3 - 2x^2$$

is entered as

[Graph] 4 [ALPHA][X][x^y] 3 [−] 2 [ALPHA][X][x^2][EXE]

You will use the following keys in working problems.

[AC]   Clears the screen
[SHIFT][Cls][EXE]   Clears the memory of previously used graphs
[RANGE] followed by [RANGE]   Will show your coordinate system, then return to text
[SHIFT][■] (key [:])   Enables one to key in one graph followed by another; press [EXE] twice

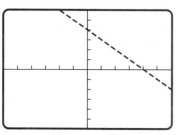

Figure A22

To assist you in using the graphing calculator as a tool throughout the book, the following program is given. To graph $y = -x + 4$, use [Graph][(−)][ALPHA][X][+] 4 [EXE] (see Figure 22).

Two graphs can be graphed on the screen at the same time and the calculator can be used to obtain an approximate solution of the point of intersection (see Figure 23). To find the intersection of $y = \frac{320 - 4x}{3}$ and $y = 20x$ first change the range on x to $2 \le x \le 6$ with a scale of 1, and on y to $80 \le y \le 120$ with a scale of 10:

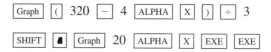

[Graph][(] 320 [−] 4 [ALPHA][X][)][÷] 3

[SHIFT][■][Graph] 20 [ALPHA][X][EXE][EXE]

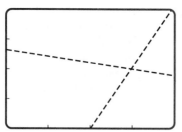

Figure A23

To read the point of intersection, use [SHIFT][Trace]. Locate the blinking pixel at the left of the screen. Use the arrow keys [⇒] and [⇐] to move it to the point of intersection. Then enter [SHIFT][X↔Y] and read the y-coordinate of the point of intersection. For greater accuracy we will use an automatic zoom-in feature. Press [SHIFT][$\hat{x}$] (the [×] key). You can also zoom in by changing the range settings, but the preceding procedure is much faster. The first graph appears automatically. The second graph appears when you press [EXE]. Again use [SHIFT][Trace]. Now, move the pixel to what seems to be the intersection and read x and y. This whole process can be repeated until you obtain the accuracy you desire. In two repetitions, it was found that $x = 5$ and $y = 100$.

For practice, let's obtain the intersections of $y = x^2 - 4x + 3$ and $y = -x + 1$. For $-6 \le x \le 6$ and $-6 \le y \le 6$,

[Graph][ALPHA][X][x^2][−] 4 [ALPHA][X][+] 3 [SHIFT][■]

[Graph][(−)][ALPHA][X][+] 1 [EXE][EXE]

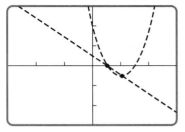

Figure A24

See Figure 24. To find a point of intersection for minimum y, use [SHIFT] [Trace]. Use the arrow keys [⇒] and [⇐] to move the blinking pixel to the desired intersection. Read the x value of this intersection from the screen. [SHIFT] [X ↔ Y] gives the y value associated with the x value of the intersection. For greater accuracy, [SHIFT] [$\hat{x}$] will zoom in on the intersection by telescoping the range. This procedure regraphs the first function. Use [EXE] to regraph the second function. Now start with [SHIFT] [Trace] and repeat the process. Continue to repeat the process until you have attained the desired accuracy ($x = 1.9997$, $y = -0.9997$).

Matrices

To enter the matrix mode use [MODE] 0. The following are the operations that are available from this menu. Press the function key as shown in Figure 25 for the operation that you want to perform.

[F1] (A)	Displays matrix **A** contents
[F2] (B)	Displays matrix **B** contents
[F3] (+)	Adds matrix **A** and matrix **B**
[F4] (−)	Subtracts matrix **B** from matrix **A**
[F5] (×)	Multiplies matrix **A** by matrix **B**
[F6] (C)	Displays matrix **C** contents

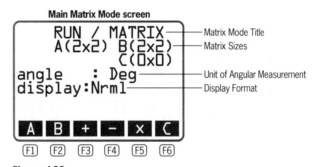

Figure A25

To clear matrix memory use [SHIFT] [CLR] [F3] (ARR). Press [F1] (YES) to clear matrix memory or [F6] (NO) (or [PRE]) to abort the operation without clearing anything (see Figure 26). You should clear matrix memory if you want to perform any nonmatrix calculations that use memories. Note that the above operation is not required if you have specified a new matrix size, because the size specification automatically clears matrix memory.

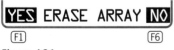

Figure A26

To input the following data into matrix **A**(3×4),

$$\begin{bmatrix} 1 & 0 & 3 & 4 \\ 2 & 1 & 0 & 1 \\ 3 & 1 & -2 & -3 \end{bmatrix}$$

input each value and press $\boxed{\text{EXE}}$.

$$1\ \boxed{\text{EXE}}\ 0\ \boxed{\text{EXE}}\ 3\ \boxed{\text{EXE}}\ 4\ \boxed{\text{EXE}}$$

$$2\ \boxed{\text{EXE}}\ 1\ \boxed{\text{EXE}}\ 0\ \boxed{\text{EXE}}\ 1\ \boxed{\text{EXE}}$$

$$3\ \boxed{\text{EXE}}\ 1\ \boxed{\text{EXE}}\ \boxed{-}\ 2\ \boxed{\text{EXE}}\ \boxed{-}\ 3\ \boxed{\text{EXE}}$$

You can use matrix **A** and matrix **B** contents in addition, subtraction, and multiplication operations. The examples of these operations presented here are based on the following two matrices:

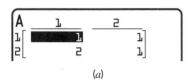

(a)

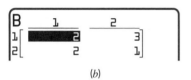

(b)

Figure A27

Matrix A   **Matrix B**

$$\begin{bmatrix} 1 & 1 \\ 2 & 1 \end{bmatrix} \quad \text{and} \quad \begin{bmatrix} 2 & 3 \\ 2 & 1 \end{bmatrix}$$

Create these matrices in memory using the following procedure. To input matrix **A** data [see Figure 27(a)], use

$$\boxed{\text{MODE}}\ 0\ \boxed{\text{F1}}\ (\text{A})\ \boxed{\text{F6}}\ (\heartsuit)\ \boxed{\text{F1}}\ (\text{DIM})$$

$$2\ \boxed{\text{EXE}}\ 2\ \boxed{\text{EXE}}$$

$$1\ \boxed{\text{EXE}}\ 1\ \boxed{\text{EXE}}$$

$$2\ \boxed{\text{EXE}}\ 1\ \boxed{\text{EXE}}$$

(a)

To input matrix **B** data [see Figure 27(b)], use

$$\boxed{\text{PRE}}\ \boxed{\text{F2}}\ (\text{B})\ \boxed{\text{F6}}\ (\heartsuit)\ \boxed{\text{F1}}\ (\text{DIM})$$

$$2\ \boxed{\text{EXE}}\ 2\ \boxed{\text{EXE}}$$

$$2\ \boxed{\text{EXE}}\ 3\ \boxed{\text{EXE}}$$

$$2\ \boxed{\text{EXE}}\ 1\ \boxed{\text{EXE}}$$

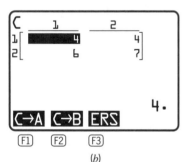

(b)

Figure A28

To multiply matrix **A** by matrix **B** [Figure 28(a)], use (A) $\boxed{\text{F5}}$ (B) to get matrix **C** shown in Figure 28(b). Matrix C shows the product of the values in the cells of matrix **A** and matrix **B**. The following are the operations that are available from the function display at the bottom of the screen. Press the function key below the operation that you want to perform.

$\boxed{\text{F1}}$ (C → A)   Transfers matrix **C** contents to matrix **A** (deleting matrix **A** contents)

$\boxed{\text{F2}}$ (C → B)   Transfers matrix **C** contents to matrix **B** (deleting matrix **B** contents)

$\boxed{\text{F3}}$ (ERS)   Deletes the matrix

Appendix A   A21

Figure A29

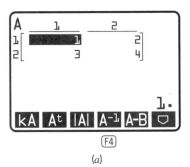

Figure A30

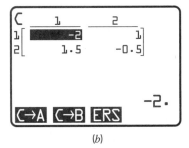

(a)

(b)

Figure A31

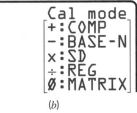

(a)

(b)

Figure A32

A matrix function menu provides calculation of the scalar product, transposition, calculation of the determinant, and calculation of the inverse matrix. A matrix editing menu lets you make changes to the configuration of a matrix after you already have it set up.

To display the matrix function menu for matrix **A**, use [F1] (A) (see Figure 29). The following are the operations that are available from the function display at the bottom of the screen. Press the function key below the operation that you want to perform.

[F1] (kA)	Returns the scalar product of matrix **A**
[F2] (Aᵗ)	Transposes matrix **A**
[F3] (⎮A⎮)	Returns the determinant of matrix **A**
[F4] (A⁻¹)	Returns the inverse matrix for matrix **A**
[F5] (A ⇌ B)	Exchanges the contents of matrix **A** and matrix **B**
[F6] (▽)	Matrix editing menu

Performing the above operations on the contents of matrix **A** stores the results in matrix **C**.

To display the matrix editing menu for matrix **A**, use [F6] (▽) (see Figure 30). The following are the operations that are available from the function display at the bottom of the screen. Press the function key below the operation that you want to perform.

[F1] (DIM)	For specification of the size of the matrix
[F2] (ERS)	Deletes the matrix
[F3] (CLR)	Clears the matrix
[F4] (ROW)	Adds, inserts, and deletes rows
[F5] (COL)	Adds, inserts, and deletes columns

To calculate the inverse matrix of the following data,

Matrix A

$$\begin{bmatrix} 1 & 2 \\ 3 & 4 \end{bmatrix}$$

use [F1] (A) [see Figure 31(a)], select [F4], and then press [F4] [A⁻¹] to get the answer given in Figure 31(b). This operation calculates the inverse of square matrix **A** or **B** and stores the results in matrix **C**.

Statistical Calculations

You should use the standard deviation mode to perform single-variable statistical calculations. In this mode, you can calculate the population standard deviation, the sample standard deviation, the mean, the sum of squares of the data, the sum

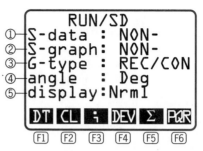

Figure A33

of the data, and the number of data items. To enter the standard deviation mode without data storage, press [MODE] [SHIFT]. In Figure 32(*a*), 2 specifies nonstorage of data. Press [MODE] [Figure 32(*b*)] and select [×] (SD) to get the screen shown in Figure 33.

① Indicates storage (STO) or nonstorage (NON-) of statistical data
② Indicates drawing (DRAW) or nondrawing (NON-) of a statistical graph
③ Graph type
④ Unit of angular measurement
⑤ Display format

The following are the operations that are available from the function display at the bottom of the screen. Press the function key below the operation that you want to perform.

[F1] (DT)	Input data
[F2] (CL)	Clears data
[F3] (;)	Used to input the number of data items
[F4] (DEV)	Displays a standard deviation function menu
[F5] (Σ)	Displays a data sum function menu
[F6] (PQR)	Displays a probability distribution function menu

Data Input To input the data 10, 20, 30, use

$$10 \; [F1] \; (DT) \; 20 \; [F1] \; (DT) \; 30 \; [F1] \; (DT)$$

To input the data 10, 20, 20, 20, 20, 20, 20, 30, use

$$10 \; [F1] \; (DT) \; 20 \; [F3] \; (;) \; 6 \; [F1] \; (DT) \; 30 \; [F1] \; (DT)$$

Note that you can input multiple data items by entering the data, pressing [F3] (;), and then entering the number of data items.

Integration   The following is the input format for integrations using Simpson's rule:

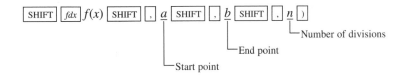

Number of divisions

End point

Start point

INSTRUCTIONS FOR SHARP EL-9200/9300

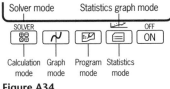

Solver mode   Statistics graph mode

SOLVER   OFF   ON

Calculation Graph Program Statistics
mode   mode   mode   mode

Figure A34

To turn on the calculator, press $\boxed{\text{ON}}$. To turn the calculator off, press $\boxed{\text{2ndF}}$ $\boxed{\text{OFF}}$. The top row (Figure 34) contains keys that determine the operation mode of the calculator.

The simplest mode to begin with is calculation mode (real). To operate in calculation mode, press $\boxed{\text{88}}$. Make sure the calculator is set to real mode and press $\boxed{\text{MENU}}$ $\boxed{1}$.

NOTE: Unless otherwise noted, all examples in this manual assume that you are in floating point display format.

Enter numbers using the number keys $\boxed{0}$ through $\boxed{9}$, the decimal point key $\boxed{.}$, and the change-sign key $\boxed{(-)}$.

Example: Enter the number 123.4.

Press: 1 2 3 $\boxed{.}$ 4 $\boxed{\text{ENTER}}$

Result: 123.4

Graphing

The general procedure for graphing equations is as follows:

1. Enter graph mode.
2. Key-in one or more functions.
3. Specify the range of the *x*- and *y*-axis.
4. Press $\boxed{\text{↵}}$ or $\boxed{\text{2ndF}}$ $\boxed{\text{AUTO}}$.
5. Press $\boxed{\text{SET-UP}}$ $\boxed{\text{E}}$.
6. When the menu appears, select 1 for rectangular coordinates. Then press $\boxed{\text{ENTER}}$.

After entering graph mode, use the keyboard and $\boxed{\text{MATH}}$ menus to enter a function (an equation or a number) and then press $\boxed{\text{ENTER}}$. The cursor will move to start of the next function. If there are any previous functions, use $\boxed{\text{CL}}$ to clear them. If you want to move between functions, press $\boxed{\blacktriangledown}$ or $\boxed{\blacktriangle}$ (if you are in equation edit mode, you must press $\boxed{\text{2ndF}}$ $\boxed{\blacktriangledown}$ or $\boxed{\text{2ndF}}$ $\boxed{\blacktriangle}$. Pressing $\boxed{\text{MENU}}$ displays a menu that lets you jump directly to any equation.

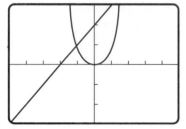

Figure A35

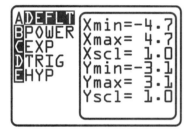

Figure A36

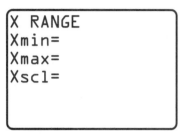

Figure A37

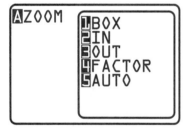

Figure A38

Before you graph a function, you must set a range. The easiest way to set the range is to use the auto scaling feature. Press [RANGE], enter a minimum, maximum, and scale value for *x*, and press [2ndF] [AUTO].

The following example shows some of the powerful graphing features: To graph $y = x^4$ and $y = x + 2$, press

[⟳] [SET UP] [E] [1] [ENTER]	Set up the calculator.
[X/θ/T] [aᵇ] 4 [ENTER]	
[X/θ/T] [+] 2 [ENTER]	
[RANGE] [MENU] [ENTER] [⟳]	Use the default range.

to get the graphs shown in Figure 35.

Press [RANGE] to get the screen shown in Figure 36. If rectangular coordinates are selected, the X RANGE screen lets you change the minimum (Xmin) and maximum (Xmax) points of the displayed *x*-axis. Xscl (*x*-scale) sets the scale of the tick marks on the *x*-axis. You can change any of these settings. Move to the Y RANGE screen by pressing [▶]. The Y RANGE screen displays the minimum (Ymin) and maximum (Ymax) points of the *y*-axis. Yscl sets the scale of the tick marks on the *y*-axis.

While in the range screen, press [MENU] and the screen shown in Figure 37 appears. This menu lets you select predefined ranges that are appropriate for common functions graphed in rectangular coordinates. To select the default press [ENTER].

Zooming

The zoom feature lets you change the range of the graph without entering specific numbers for each coordinate. Press [ZOOM] and the menu shown in Figure 38 is displayed. [1] BOX lets you draw a box around an area of interest. The boxed area then fills the entire screen, distorting the graph as necessary. After BOX is selected move the cursor to a starting point for the box and press [ENTER]. Move the cursor diagonally to draw a box and press [ENTER].

[2] IN zooms in on the graph by an amount determined by FACTOR. If you use the trace feature to select a point on the curve (before you zoom in), that point becomes the screen center.

[3] OUT zooms out on the graph by an amount determined by FACTOR. If you use the trace feature to select a point on the curve (before you zoom out), that point becomes the screen center.

[4] FACTOR lets you set the zoom factor (the number that the range is divided by when zooming in, or multiplied by when zooming out). The *x*-factor can differ from the *y*-factor.

[5] AUTO performs the same function as pressing [2ndF] [AUTO].

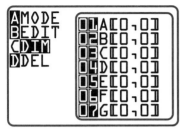

Figure A39

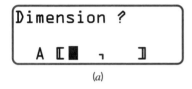

(a)

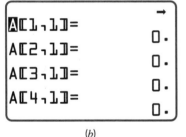

(b)

Figure A40

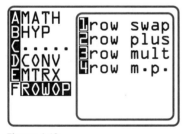

Figure A41

Figure A42

Matrix Algebra

Matrix mode lets you perform matrix operations and define matrix elements. While in calculation mode, press MENU A. Enter matrix mode by pressing 3, or use the cursor keys and press ENTER.

After entering matrix mode, press MENU C and the menu shown in Figure 39 appears. This option lets you define (and store) up to 26 different matrices. The matrices are stored even after the mode has been changed or the calculator has been turned off. To define the dimensions of a matrix, press 01 or use the cursor keys to select a matrix and press ENTER). The display shown in Figure 40(a) appears. Enter the number of rows and the number of columns. For example, to define a matrix with four rows and three columns, press 4 ENTER 3 ENTER. The first column of the matrix appears [see Figure 40(b)]. Use the keypad to enter the matrix elements and press ENTER. Each element of the matrix is labeled with its associated coordinate [matrix index (row number, column number)]. Use the cursor keys to move around the matrix (through the row and column elements). Arrow indicators show you which directions you can move. ▼ and ▲ move through the rows. ◄ and ► move through the columns. You can enter numbers, variables, and equations for each matrix element. (The final calculated value is stored.) Press QUIT to exit the matrix.

Deleting a Matrix   After entering matrix mode, press MENU D and the menu shown in Figure 41 appears. Select D to delete the matrix.

Matrix Function Keys   MAT is the matrix identifier function. This function key tells the calculator that a matrix is being referenced. After pressing MAT, you do not need to press ALPHA to select the matrix index. Use the matrix identifier to specify matrices used in calculations.

Example: Calculate matrix **A**: $\begin{bmatrix} 11 & 3 \\ 2 & 4 \end{bmatrix}$ + matrix **B**: $\begin{bmatrix} 1 & 16 \\ 31 & 41 \end{bmatrix}$.

(First, create the above matrices using the steps described earlier.)
Press: MAT A + MAT B ENTER

Two of the function keys, (x^{-1} and x^2) perform operations that are very different from their operation in other modes. x^{-1} calculates the inverse of a square matrix. x^2 calculates the square of a square matrix. Matrix **A** divided by matrix **B** is the same as matrix $\mathbf{A} \times$ matrix $\mathbf{B}^{-1}$.

Pressing MATH F displays the screen shown in Figure 42. 3 row mult multiplies a row by a scalar. The syntax is row mult (*scalar, matrix letter, row*).

Example: Multiply 4 times the first row of matrix **C**: $\begin{bmatrix} 1 & 2 & 3 \\ 0 & 5 & 6 \\ 0 & 0 & 9 \end{bmatrix}$.

Press: MATH F 3 4 ALPHA , ALPHA C ALPHA , 1) ENTER

4 row m.p. multiplies a row by a scalar, adds the row to a second row, and stores the result in the second row specified. The syntax is row m.p. (*scalar, matrix letter, row*1, *row*2).

Example: Multiply 2 times the first row of matrix **C**: $\begin{bmatrix} 1 & 2 & 3 \\ 0 & 5 & 6 \\ 0 & 0 & 9 \end{bmatrix}$ and

add the result to the second row.

Press: MATH F 4 2 ALPHA , ALPHA C ALPHA , 1 ALPHA ,
2) ENTER

NOTE: You can store the result of a matrix calculation into another matrix by pressing STO MAT *letter* instead of pressing ENTER .

Descriptive Statistics

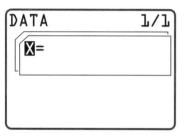

Figure A43

Press ▤ 1 to select one-variable statistics, and the first "card" of data appears as shown in Figure 43. Each card contains one data value (often called an *observation* by statisticians). Use the keypad to enter the data values. After pressing ENTER the calculator displays the next card (the next data value). To display the preceding data value, press ◄ . To display the next data value press ► . Use 2ndF ◄ or 2ndF ► to jump to the first or last card. The card/count number is displayed at the top right of the card. For example, 2/5 means that the second of five cards is currently displayed.

Example: What is the mean of the following test scores: 75, 85, 90, 82, and 77?

Press: ▤ 1 75 ENTER 85 ENTER 90 ENTER 82 ENTER 77 ENTER

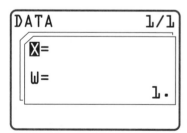

Figure A44

Frequency Distributions If your sample data contains a value that is repeated a number of times, enter it as a weighted variable. For example, entering a variable with a weight of 5 is the same as entering the value five times. Weighted one-variable data cards look like the one shown in Figure 44. Key-in the x value, press ENTER , and enter the weight.

When entering statistical data you can enter numbers, variables, or equations, but only the result is stored. If you make a mistake while entering data, press CL (before you press ENTER) and the preceding entry will be returned. 2ndF CA deletes an entire card.

Statistics Menu The MENU key lets you view statistical results; move through the data cards, mask, delete, or sort data, and store or retrieve data in matrices.

Figure A45

After entering the data, view statistical results by pressing [MENU] [A] and the menu is displayed (see Figure 45).

Example: What is the mean of the following test scores: 75, 75, 75, 75, 90, 90, 77?

Press: [▤] [MENU] [D] [2] [ENTER] Deletes any previous data.

Press: [2]   Selects one-variable with weight.

75 [ENTER] 4 [ENTER] 90 [ENTER] 2 [ENTER] 77 [ENTER] [MENU] [A] [1]

X VARS displays the statistical results shown in Figure 46.

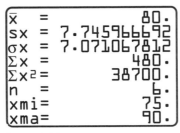

Figure A46

Figure A47

Ordering Data   You can sort the data cards in four different ways. Press [MENU] [E] and the menu shown in Figure 47 is displayed.

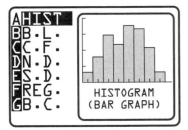

Figure A48

[1] X ASCEND sorts the cards from the lowest *x* value to the highest.

[2] X DESCEN sorts the cards from the highest *x* value to the lowest.

[3] Y ASCEND sorts the cards from the lowest *y* value to the highest. This function is not available in one-variable format.

[4] Y DESCEN sorts the cards from the highest *y* value to the lowest. This function is not available in one-variable format.

Statistical Graphs   To select statistics graph mode, press [2ndF] [⌐]. The menu shown in Figure 48 is displayed. Graphs can be drawn over each other for comparisons. Press [2ndF] [CA] to clear all previously drawn graphs.

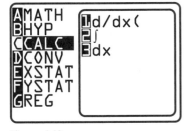

Figure A49

Calculus Functions

Your calculator can integrate and differentiate many types of functions using numerical estimations. Pressing [MATH] [C] displays the screen shown in Figure 49. [1] d/dx(selects the derivative function. This function estimates the first derivative of a function at a given value. The syntax of the derivative function is d/dx(*function*, *x*[*value*], Δ*x* [*change in x*]). (Δ*x* is optional.)

Example: If $F(x) = x^2 + x$, find $F'(4)$.

Press: MATH   C   1   X/θ/T x^2 +   X/θ/T   ALPHA   ,   4   ALPHA   ,
     0.00001   )   ENTER

Result: 9.

2 ∫ selects the integrate function. This calculates the area under a curve between two points. The calculator uses Simpson's rule to partition the area into a number of even subintervals and estimate the answer. The calculator doubles the number of subintervals that you specify. The answer is an estimate, and therefore will not be exactly correct.

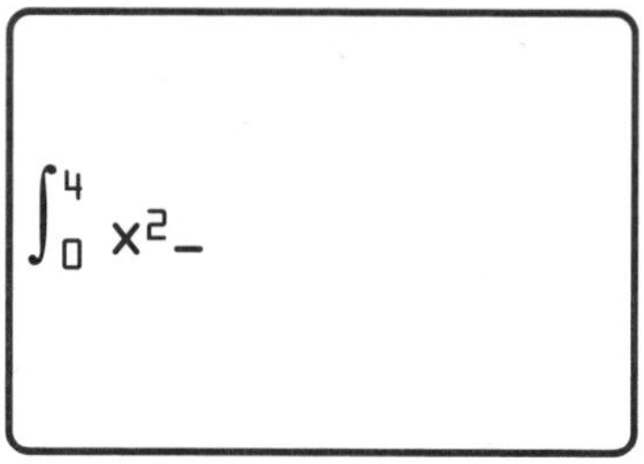

Example: What is $\int_0^4 x^2\, dx$ calculated over 16 subintervals?

Press: SET UP   F   1   ENTER   MATH   C   2   0   ▲   4   ►   X/θ/T x^2

Displays: See Figure 50.

Press: ALPHA   ,   8   (The comma marks the end of the equation, the 8 is one-half the number of desired subintervals.)

$$\int_0^4 x^2\_$$

Figure A50

Appendix B

Table 1 Compound Interest

$i = 1\%$ (interest rate per period) n = number of periods

n	$(1 + i)^n$	$(1 + i)^{-n}$	n	$(1 + i)^n$	$(1 + i)^{-n}$
1	1.010000	0.990099	51	1.661078	0.602019
2	1.020100	0.980296	52	1.677689	0.596058
3	1.030301	0.970590	53	1.694466	0.590157
4	1.040604	0.960980	54	1.711410	0.584313
5	1.051010	0.951466	55	1.728525	0.578528
6	1.061520	0.942045	56	1.745810	0.572800
7	1.072135	0.932718	57	1.763268	0.567129
8	1.082857	0.923483	58	1.780901	0.561514
9	1.093685	0.914340	59	1.798710	0.555954
10	1.104622	0.905287	60	1.816697	0.550450
11	1.115668	0.896324	61	1.834864	0.545000
12	1.126825	0.887449	62	1.853212	0.539604
13	1.138093	0.878663	63	1.871744	0.534261
14	1.149474	0.869963	64	1.890462	0.528971
15	1.160969	0.861349	65	1.909366	0.523734
16	1.172579	0.852821	66	1.928460	0.518548
17	1.184304	0.844377	67	1.947745	0.513414
18	1.196147	0.836017	68	1.967222	0.508331
19	1.208109	0.827740	69	1.986894	0.503298
20	1.220190	0.819544	70	2.006763	0.498315
21	1.232392	0.811430	71	2.026831	0.493381
22	1.244716	0.803396	72	2.047099	0.488496
23	1.257163	0.795442	73	2.067570	0.483660
24	1.269735	0.787566	74	2.088246	0.478871
25	1.282432	0.779768	75	2.109128	0.474130
26	1.295256	0.772048	76	2.130220	0.469435
27	1.308209	0.764404	77	2.151522	0.464787
28	1.321291	0.756836	78	2.173037	0.460185
29	1.334504	0.749342	79	2.194767	0.455629
30	1.347849	0.741923	80	2.216715	0.451118
31	1.361327	0.734577	81	2.238882	0.446651
32	1.374941	0.727304	82	2.261271	0.442229
33	1.388690	0.720103	83	2.283884	0.437851
34	1.402577	0.712973	84	2.306723	0.433515
35	1.416603	0.705914	85	2.329790	0.429223
36	1.430769	0.698925	86	2.353088	0.424974
37	1.445076	0.692005	87	2.376619	0.420766
38	1.459527	0.685153	88	2.400385	0.416600
39	1.474122	0.678370	89	2.424389	0.412475
40	1.488864	0.671653	90	2.448633	0.408391
41	1.503752	0.665003	91	2.473119	0.404348
42	1.518790	0.658419	92	2.497850	0.400344
43	1.533978	0.651900	93	2.522829	0.396380
44	1.549318	0.645445	94	2.548057	0.392456
45	1.564811	0.639055	95	2.573537	0.388570
46	1.580459	0.632728	96	2.599273	0.384723
47	1.596263	0.626463	97	2.625266	0.380914
48	1.612226	0.620260	98	2.651518	0.377142
49	1.628348	0.614119	99	2.678033	0.373408
50	1.644632	0.608039	100	2.704814	0.369711

(continued)

Table 1   Compound Interest *(continued)*

i = 2% (interest rate per period)   *n* = number of periods

n	$(1 + i)^n$	$(1 + i)^{-n}$	n	$(1 + i)^n$	$(1 + i)^{-n}$
1	1.020000	0.980392	51	2.745420	0.364243
2	1.040400	0.961169	52	2.800328	0.357101
3	1.061208	0.942322	53	2.856335	0.350099
4	1.082432	0.923845	54	2.913461	0.343234
5	1.104081	0.905731	55	2.971731	0.336504
6	1.126162	0.887971	56	3.031165	0.329906
7	1.148686	0.870560	57	3.091788	0.323437
8	1.171659	0.853490	58	3.153624	0.317095
9	1.195093	0.836755	59	3.216697	0.310878
10	1.218994	0.820348	60	3.281031	0.304782
11	1.243374	0.804263	61	3.346651	0.298806
12	1.268242	0.788493	62	3.413584	0.292947
13	1.293607	0.773033	63	3.481856	0.287203
14	1.319479	0.757875	64	3.551493	0.281572
15	1.345868	0.743015	65	3.622523	0.276051
16	1.372786	0.728446	66	3.694973	0.270638
17	1.400241	0.714163	67	3.768873	0.265331
18	1.428246	0.700159	68	3.844250	0.260129
19	1.456811	0.686431	69	3.921135	0.255028
20	1.485947	0.672971	70	3.999558	0.250028
21	1.515666	0.659776	71	4.079549	0.245125
22	1.545980	0.646839	72	4.161140	0.240319
23	1.576899	0.634156	73	4.244363	0.235607
24	1.608437	0.621722	74	4.329250	0.230987
25	1.640606	0.609531	75	4.415835	0.226458
26	1.673418	0.597579	76	4.504152	0.222017
27	1.706886	0.585862	77	4.594235	0.217664
28	1.741024	0.574375	78	4.686120	0.213396
29	1.775845	0.563112	79	4.779842	0.209212
30	1.811362	0.552071	80	4.875439	0.205110
31	1.847589	0.541246	81	4.972948	0.201088
32	1.884541	0.530633	82	5.072407	0.197145
33	1.922231	0.520229	83	5.173855	0.193279
34	1.960676	0.510028	84	5.277332	0.189490
35	1.999889	0.500028	85	5.382878	0.185774
36	2.039887	0.490223	86	5.490536	0.182132
37	2.080685	0.480611	87	5.600347	0.178560
38	2.122299	0.471187	88	5.712354	0.175059
39	2.164745	0.461948	89	5.826601	0.171627
40	2.208040	0.452890	90	5.943133	0.168261
41	2.252200	0.444010	91	6.061995	0.164962
42	2.297244	0.435304	92	6.183235	0.161728
43	2.343189	0.426769	93	6.306900	0.158557
44	2.390053	0.418401	94	6.433038	0.155448
45	2.437854	0.410197	95	6.561699	0.152400
46	2.486611	0.402154	96	6.692933	0.149411
47	2.536343	0.394268	97	6.826791	0.146482
48	2.587070	0.386538	98	6.963327	0.143610
49	2.638812	0.378958	99	7.102594	0.140794
50	2.691588	0.371528	100	7.244645	0.138033

(continued)

Table 1 Compound Interest *(continued)*

$i = 3\%$ (interest rate per period) $n =$ number of periods

n	$(1 + i)^n$	$(1 + i)^{-n}$	n	$(1 + i)^n$	$(1 + i)^{-n}$
1	1.030000	0.970874	51	4.515423	0.221463
2	1.060900	0.942596	52	4.650886	0.215013
3	1.092727	0.915142	53	4.790412	0.208750
4	1.125509	0.888487	54	4.934125	0.202670
5	1.159274	0.862609	55	5.082148	0.196767
6	1.194052	0.837484	56	5.234613	0.191036
7	1.229874	0.813092	57	5.391651	0.185472
8	1.266770	0.789409	58	5.553401	0.180070
9	1.304773	0.766417	59	5.720003	0.174825
10	1.343916	0.744094	60	5.891603	0.169733
11	1.384234	0.722421	61	6.068351	0.164789
12	1.425761	0.701380	62	6.250402	0.159990
13	1.468534	0.680951	63	6.437914	0.155330
14	1.512590	0.661118	64	6.631051	0.150806
15	1.557967	0.641862	65	6.829982	0.146413
16	1.604706	0.623167	66	7.034882	0.142149
17	1.652848	0.605016	67	7.245928	0.138009
18	1.702433	0.587395	68	7.463306	0.133989
19	1.753506	0.570286	69	7.687205	0.130086
20	1.806111	0.553676	70	7.917822	0.126297
21	1.860295	0.537549	71	8.155356	0.122619
22	1.916103	0.521893	72	8.400017	0.119047
23	1.973586	0.506692	73	8.652017	0.115580
24	2.032794	0.491934	74	8.911578	0.112214
25	2.093778	0.477606	75	9.178925	0.108945
26	2.156691	0.463695	76	9.454293	0.105772
27	2.221289	0.450189	77	9.737922	0.102691
28	2.287928	0.437077	78	10.030060	0.099700
29	2.356565	0.424346	79	10.330961	0.096796
30	2.427262	0.411987	80	10.640890	0.093977
31	2.500080	0.399987	81	10.960117	0.091240
32	2.575083	0.388337	82	11.288920	0.088582
33	2.652335	0.377026	83	11.627588	0.086002
34	2.731905	0.366045	84	11.976416	0.083497
35	2.813862	0.355383	85	12.335708	0.081065
36	2.898278	0.345032	86	12.705779	0.078704
37	2.985227	0.334983	87	13.086953	0.076412
38	3.074783	0.325226	88	13.479561	0.074186
39	3.167027	0.315754	89	13.883948	0.072026
40	3.262038	0.306557	90	14.300466	0.069928
41	3.359899	0.297628	91	14.729480	0.067891
42	3.460696	0.288959	92	15.171365	0.065914
43	3.564517	0.280543	93	15.626506	0.063994
44	3.671452	0.272372	94	16.095301	0.062130
45	3.781596	0.264439	95	16.578160	0.060320
46	3.895044	0.256737	96	17.075505	0.058563
47	4.011895	0.249259	97	17.587770	0.056858
48	4.132252	0.241999	98	18.115403	0.055202
49	4.256219	0.234950	99	18.658865	0.053594
50	4.383906	0.228107	100	19.218631	0.052033

(continued)

Table 1   Compound Interest *(continued)*

i = 4% (interest rate per period)   *n* = number of periods

n	$(1 + i)^n$	$(1 + i)^{-n}$	n	$(1 + i)^n$	$(1 + i)^{-n}$
1	1.040000	0.961538	51	7.390950	0.135301
2	1.081600	0.924556	52	7.686588	0.130097
3	1.124864	0.888996	53	7.994052	0.125093
4	1.169859	0.854804	54	8.313814	0.120282
5	1.216653	0.821927	55	8.646366	0.115656
6	1.265319	0.790315	56	8.992221	0.111207
7	1.315932	0.759918	57	9.351910	0.106930
8	1.368569	0.730690	58	9.725986	0.102817
9	1.423312	0.702587	59	10.115026	0.098863
10	1.480244	0.675564	60	10.519627	0.095060
11	1.539454	0.649581	61	10.940412	0.091404
12	1.601032	0.624597	62	11.378028	0.087889
13	1.665073	0.600574	63	11.833149	0.084508
14	1.731676	0.577475	64	12.306475	0.081258
15	1.800943	0.555265	65	12.798734	0.078133
16	1.872981	0.533908	66	13.310684	0.075128
17	1.947900	0.513373	67	13.843111	0.072238
18	2.025816	0.493628	68	14.396835	0.069460
19	2.106849	0.474642	69	14.972709	0.066788
20	2.191123	0.456387	70	15.571617	0.064219
21	2.278768	0.438834	71	16.194482	0.061749
22	2.369919	0.421955	72	16.842261	0.059374
23	2.464715	0.405726	73	17.515951	0.057091
24	2.563304	0.390121	74	18.216589	0.054895
25	2.665836	0.375117	75	18.945253	0.052784
26	2.772470	0.360689	76	19.703063	0.050754
27	2.883368	0.346817	77	20.491186	0.048801
28	2.998703	0.333477	78	21.310833	0.046924
29	3.118651	0.320651	79	22.163266	0.045120
30	3.243397	0.308319	80	23.049797	0.043384
31	3.373133	0.296460	81	23.971789	0.041716
32	3.508059	0.285058	82	24.930660	0.040111
33	3.648381	0.274094	83	25.927887	0.038569
34	3.794316	0.263552	84	26.965002	0.037085
35	3.946089	0.253415	85	28.043602	0.035659
36	4.103932	0.243669	86	29.165346	0.034287
37	4.268090	0.234297	87	30.331960	0.032969
38	4.438813	0.225285	88	31.545238	0.031701
39	4.616366	0.216621	89	32.807048	0.030481
40	4.801020	0.208289	90	34.119330	0.029309
41	4.993061	0.200278	91	35.484103	0.028182
42	5.192784	0.192575	92	36.903467	0.027098
43	5.400495	0.185168	93	38.379606	0.026056
44	5.616515	0.178046	94	39.914790	0.025053
45	5.841175	0.171198	95	41.511381	0.024090
46	6.074822	0.164614	96	43.171836	0.023163
47	6.317815	0.158283	97	44.898710	0.022272
48	6.570528	0.152195	98	46.694658	0.021416
49	6.833349	0.146341	99	48.562444	0.020592
50	7.106683	0.140713	100	50.504942	0.019800

(continued)

Table 1   Compound Interest *(continued)*

$i = 6\%$ (interest rate per period) n = number of periods

n	$(1 + i)^n$	$(1 + i)^{-n}$	n	$(1 + i)^n$	$(1 + i)^{-n}$
1	1.060000	0.943396	51	19.525363	0.051215
2	1.123600	0.889996	52	20.696885	0.048316
3	1.191016	0.839619	53	21.938698	0.045582
4	1.262477	0.792094	54	23.255020	0.043001
5	1.338226	0.747258	55	24.650321	0.040567
6	1.418519	0.704961	56	26.129340	0.038271
7	1.503630	0.665057	57	27.697100	0.036105
8	1.593848	0.627412	58	29.358926	0.034061
9	1.689479	0.591898	59	31.120462	0.032133
10	1.790848	0.558395	60	32.987690	0.030314
11	1.898299	0.526788	61	34.966951	0.028598
12	2.012196	0.496969	62	37.064968	0.026980
13	2.132928	0.468839	63	39.288866	0.025453
14	2.260904	0.442301	64	41.646198	0.024012
15	2.396558	0.417265	65	44.144970	0.022653
16	2.540352	0.393646	66	46.793668	0.021370
17	2.692773	0.371364	67	49.601288	0.020161
18	2.854339	0.350344	68	52.577365	0.019020
19	3.025599	0.330513	69	55.732007	0.017943
20	3.207135	0.311805	70	59.075928	0.016927
21	3.399564	0.294155	71	62.620483	0.015969
22	3.603537	0.277505	72	66.377712	0.015065
23	3.819750	0.261797	73	70.360375	0.014213
24	4.048935	0.246979	74	74.581997	0.013408
25	4.291871	0.232999	75	79.056917	0.012649
26	4.549383	0.219810	76	83.800332	0.011933
27	4.822346	0.207368	77	88.828352	0.011258
28	5.111687	0.195630	78	94.158053	0.010620
29	5.418388	0.184557	79	99.807536	0.010019
30	5.743491	0.174110	80	105.795988	0.009452
31	6.088101	0.164255	81	112.143748	0.008917
32	6.453387	0.154957	82	118.872372	0.008412
33	6.840590	0.146186	83	126.004715	0.007936
34	7.251025	0.137912	84	133.564997	0.007487
35	7.686087	0.130105	85	141.578897	0.007063
36	8.147252	0.122741	86	150.073631	0.006663
37	8.636087	0.115793	87	159.078049	0.006286
38	9.154252	0.109239	88	168.622731	0.005930
39	9.703507	0.103056	89	178.740095	0.005595
40	10.285718	0.097222	90	189.464501	0.005278
41	10.902861	0.091719	91	200.832371	0.004979
42	11.557032	0.086527	92	212.882312	0.004697
43	12.250454	0.081630	93	225.655252	0.004432
44	12.985482	0.077009	94	239.194566	0.004181
45	13.764610	0.072650	95	253.546240	0.003944
46	14.590487	0.068538	96	268.759014	0.003721
47	15.465916	0.064658	97	284.884555	0.003510
48	16.393871	0.060998	98	301.977628	0.003312
49	17.377504	0.057546	99	320.096286	0.003124
50	18.420154	0.054288	100	339.302062	0.002947

(continued)

Table 1 Compound Interest *(continued)*

$i = 8\%$ (interest rate per period) n = number of periods

n	$(1 + i)^n$	$(1 + i)^{-n}$	n	$(1 + i)^n$	$(1 + i)^{-n}$
1	1.080000	0.925926	51	50.653740	0.019742
2	1.166400	0.857339	52	54.706039	0.018280
3	1.259712	0.793832	53	59.082522	0.016925
4	1.360489	0.735030	54	63.809124	0.015672
5	1.469328	0.680583	55	68.913854	0.014511
6	1.586874	0.630170	56	74.426962	0.013436
7	1.713824	0.583490	57	80.381119	0.012441
8	1.850930	0.540269	58	86.811608	0.011519
9	1.999005	0.500249	59	93.756537	0.010666
10	2.158925	0.463193	60	101.257060	0.009876
11	2.331639	0.428883	61	109.357625	0.009144
12	2.518170	0.397114	62	118.106234	0.008467
13	2.719624	0.367698	63	127.554733	0.007840
14	2.937194	0.340461	64	137.759112	0.007259
15	3.172169	0.315242	65	148.779841	0.006721
16	3.425943	0.291890	66	160.682228	0.006223
17	3.700018	0.270269	67	173.536806	0.005762
18	3.996019	0.250249	68	187.419750	0.005336
19	4.315701	0.231712	69	202.413330	0.004940
20	4.660957	0.214548	70	218.606396	0.004574
21	5.033834	0.198656	71	236.094908	0.004236
22	5.436540	0.183941	72	254.982500	0.003922
23	5.871464	0.170315	73	275.381101	0.003631
24	6.341181	0.157699	74	297.411588	0.003362
25	6.848475	0.146018	75	321.204515	0.003113
26	7.396353	0.135202	76	346.900876	0.002883
27	7.988061	0.125187	77	374.652946	0.002669
28	8.627106	0.115914	78	404.625181	0.002471
29	9.317275	0.107328	79	436.995196	0.002288
30	10.062657	0.099377	80	471.954811	0.002119
31	10.867669	0.092016	81	509.711196	0.001962
32	11.737083	0.085200	82	550.488090	0.001817
33	12.676049	0.078889	83	594.527138	0.001682
34	13.690133	0.073045	84	642.089308	0.001557
35	14.785344	0.067635	85	693.456453	0.001442
36	15.968171	0.062625	86	748.932968	0.001335
37	17.245625	0.057986	87	808.847606	0.001236
38	18.625275	0.053690	88	873.555413	0.001145
39	20.115297	0.049713	89	943.439846	0.001060
40	21.724521	0.046031	90	1018.915031	0.000981
41	23.462483	0.042621	91	1100.428236	0.000909
42	25.339481	0.039464	92	1188.462491	0.000841
43	27.366640	0.036541	93	1283.539492	0.000779
44	29.555971	0.033834	94	1386.222649	0.000721
45	31.920449	0.031328	95	1497.120463	0.000668
46	34.474084	0.029007	96	1616.890095	0.000618
47	37.232011	0.026859	97	1746.241303	0.000573
48	40.210572	0.024869	98	1885.940604	0.000530
49	43.427418	0.023027	99	2036.815854	0.000491
50	46.901611	0.021321	100	2199.761119	0.000455

Table 2 Annuities

$i = 1\%$ (interest rate per period) n = number of periods

$$s_{\overline{n}|i} = \frac{(1 + i)^n - 1}{i} \qquad a_{\overline{n}|i} = \frac{1 - (1 + i)^{-n}}{i}$$

n	$s_{\overline{n}|i}$	$a_{\overline{n}|i}$	n	$s_{\overline{n}|i}$	$a_{\overline{n}|i}$
1	1.000000	0.990099	51	66.107810	39.798135
2	2.010000	1.970395	52	67.768888	40.394193
3	3.030100	2.940985	53	69.446577	40.984349
4	4.060401	3.901965	54	71.141043	41.568663
5	5.101005	4.853431	55	72.852453	42.147191
6	6.152015	5.795476	56	74.580977	42.719991
7	7.213535	6.728194	57	76.326787	43.287120
8	8.285670	7.651677	58	78.090055	43.848633
9	9.368527	8.566017	59	79.870956	44.404587
10	10.462212	9.471304	60	81.669665	44.955037
11	11.566834	10.367628	61	83.486361	45.500037
12	12.682502	11.255077	62	85.321225	46.039640
13	13.809327	12.133740	63	87.174437	46.573901
14	14.947421	13.003702	64	89.046181	47.102872
15	16.096895	13.865052	65	90.936643	47.626606
16	17.257864	14.717873	66	92.846010	48.145155
17	18.430442	15.562251	67	94.774470	48.658569
18	19.614747	16.398268	68	96.722214	49.166900
19	20.810894	17.226008	69	98.689436	49.670198
20	22.019003	18.045552	70	100.676330	50.168513
21	23.239193	18.856982	71	102.683094	50.661894
22	24.471585	19.660379	72	104.709924	51.150390
23	25.716301	20.455820	73	106.757024	51.634049
24	26.973463	21.243386	74	108.824594	52.112920
25	28.243198	22.023155	75	110.912840	52.587050
26	29.525630	22.795203	76	113.021968	53.056485
27	30.820886	23.559607	77	115.152188	53.521272
28	32.129095	24.316442	78	117.303709	59.981457
29	33.450386	25.065784	79	119.476747	54.437087
30	34.784890	25.807707	80	121.671514	54.888205
31	36.132739	26.542284	81	123.888229	55.334856
32	37.494066	27.269588	82	126.127111	55.777085
33	38.869006	27.989691	83	128.388382	56.214936
34	40.257696	28.702665	84	130.672266	56.648451
35	41.660273	29.408579	85	132.978988	57.077674
36	43.076876	30.107504	86	135.308778	57.502648
37	44.507645	30.799509	87	137.661866	57.923414
38	45.952721	31.484662	88	140.038484	58.340014
39	47.412248	32.163032	89	142.438869	58.752489
40	48.886371	32.834685	90	144.863258	59.160880
41	50.375234	33.499688	91	147.311890	59.565228
42	51.878987	34.158107	92	149.785009	59.965572
43	53.397776	34.810007	93	152.282859	60.361952
44	54.931754	35.455452	94	154.805687	60.754408
45	56.481072	36.094507	95	157.353744	61.142978
46	58.045882	36.727235	96	159.927281	61.527701
47	59.626341	37.353698	97	162.526554	61.908615
48	61.222604	37.973958	98	165.151819	62.285758
49	62.834830	38.588077	99	167.803338	62.659166
50	64.463178	39.196116	100	170.461370	63.028877

(continued)

Table 2   Annuities (continued)

$i = 2\%$ (interest rate per period) n = number of periods

$$s_{\overline{n}|i} = \frac{(1 + i)^n - 1}{i} \qquad a_{\overline{n}|i} = \frac{1 - (1 + i)^{-n}}{i}$$

n	$s_{\overline{n}|i}$	$a_{\overline{n}|i}$	n	$s_{\overline{n}|i}$	$a_{\overline{n}|i}$
1	1.000000	0.980392	51	87.270984	31.787848
2	2.020000	1.941561	52	90.016403	32.144949
3	3.060400	2.883883	53	92.816731	32.495046
4	4.121608	3.807729	54	95.673065	32.838282
5	5.204040	4.713459	55	98.586527	33.174787
6	6.308121	5.601431	56	101.558257	33.504693
7	7.434283	6.471991	57	104.589422	33.828130
8	8.582969	7.325481	58	107.681210	34.145226
9	9.754628	8.162236	59	110.834834	34.456104
10	10.949720	8.982585	60	114.051531	34.760886
11	12.168715	9.786848	61	117.332562	35.059692
12	13.412089	10.575341	62	120.679212	35.352639
13	14.680331	11.348373	63	124.092797	35.639842
14	15.973937	12.106248	64	127.574652	35.921414
15	17.293416	12.849263	65	131.126145	36.197465
16	18.639284	13.577709	66	134.748668	36.468103
17	20.012070	14.291871	67	138.443642	36.733434
18	21.412311	14.992031	68	142.212514	36.993563
19	22.840558	15.678462	69	146.056764	37.248591
20	24.297369	16.351433	70	149.977899	37.498619
21	25.783316	17.011209	71	153.977457	37.743744
22	27.298982	17.658048	72	158.057006	37.984062
23	28.844962	18.292204	73	162.218146	38.219669
24	30.421861	18.913925	74	166.462508	38.450656
25	32.030298	19.523456	75	170.791759	38.677114
26	33.670904	20.121035	76	175.207593	38.899131
27	35.344322	20.706897	77	179.711746	39.116795
28	37.051208	21.281272	78	184.305980	39.330191
29	38.792232	21.844384	79	188.992100	39.539403
30	40.568077	22.396455	80	193.771941	39.744513
31	42.379438	22.937701	81	198.647380	39.945601
32	44.227027	23.468334	82	203.620327	40.142746
33	46.111568	23.988563	83	208.692734	40.336025
34	48.033799	24.498591	84	213.866588	40.525515
35	49.994475	24.998619	85	219.143920	40.711289
36	41.994364	25.488842	86	224.526798	40.893421
37	54.034251	25.969453	87	230.017333	41.071981
38	56.114936	26.440640	88	235.617680	41.247040
39	58.237235	26.902588	89	241.330033	41.418667
40	60.401979	27.355478	90	247.156633	41.586929
41	62.610019	27.799489	91	253.099766	41.751891
42	64.862219	28.234793	92	259.161761	41.913618
43	67.159464	28.661562	93	265.344996	42.072175
44	69.602653	29.079962	94	271.651895	42.227622
45	71.892706	29.490159	95	278.084933	42.380022
46	74.330560	29.892313	96	284.646631	42.529433
47	76.817171	30.286581	97	291.339564	42.675915
48	79.353514	30.673119	98	298.166354	42.819524
49	81.940584	31.052077	99	305.129682	42.960318
50	84.579396	31.423605	100	312.232275	43.098351

(continued)

Table 2   Annuities *(continued)*

$i = 3\%$ (interest rate per period) n = number of periods

$$s_{\overline{n}|i} = \frac{(1 + i)^n - 1}{i} \qquad a_{\overline{n}|i} = \frac{1 - (1 + i)^{-n}}{i}$$

n	$s_{\overline{n}|i}$	$a_{\overline{n}|i}$	n	$s_{\overline{n}|i}$	$a_{\overline{n}|i}$
1	1.000000	0.970874	51	117.180769	25.951227
2	2.030000	1.913470	52	121.696192	26.166240
3	3.090900	2.828611	53	126.347078	26.374990
4	4.183627	3.717098	54	131.137490	26.577660
5	5.309136	4.579707	55	136.071615	26.774427
6	6.468410	5.417191	56	141.153763	26.965464
7	7.662462	6.230283	57	146.388376	27.150935
8	8.892336	7.019692	58	151.780027	27.331005
9	10.159106	7.786109	59	157.333428	27.505830
10	11.463879	8.530203	60	163.053431	27.675564
11	12.807795	9.252624	61	168.945034	27.840353
12	14.192029	9.954004	62	175.013384	28.000343
13	15.617790	10.634955	63	181.263786	28.155672
14	17.086324	11.296073	64	187.701699	28.306478
15	18.598913	11.937935	65	194.332750	28.452891
16	20.156881	12.561102	66	201.162733	28.595040
17	21.761587	13.166118	67	208.197614	28.733049
18	23.414435	13.753513	68	215.443543	28.867038
19	25.116868	14.323799	69	222.906849	28.997124
20	26.870374	14.877575	70	230.594054	29.123421
21	28.676485	15.415024	71	238.511875	29.246040
22	30.536780	15.936916	72	246.667232	29.365087
23	32.452883	16.443608	73	255.067249	29.480667
24	34.426469	16.935542	74	263.719266	29.592881
25	36.459263	17.413147	75	272.630844	29.701826
26	38.553041	17.876842	76	281.809769	29.807598
27	40.709632	18.327031	77	291.264062	29.910290
28	42.930921	18.764108	78	301.001983	30.009990
29	45.218849	19.188454	79	311.032043	30.106786
30	47.575414	19.600441	80	321.363004	30.200763
31	50.002677	20.000428	81	332.003894	30.292003
32	52.502757	20.388765	82	342.964010	30.380586
33	55.077840	20.765792	83	354.252930	30.466588
34	57.730175	21.131836	84	365.880518	30.550085
35	60.462080	21.487220	85	377.856933	30.631151
36	63.275942	21.832252	86	390.192641	30.709855
37	66.174221	22.167235	87	402.898420	30.786267
38	69.159447	22.492461	88	415.985373	30.860454
39	72.234231	22.808215	89	429.464934	30.932479
40	75.401258	23.114772	90	443.348881	31.002407
41	78.663295	23.412400	91	457.649348	31.070298
42	82.023194	23.701359	92	472.378828	31.136212
43	85.483890	23.981902	93	487.550192	31.200206
44	89.048406	24.254274	94	503.176698	31.262336
45	92.719858	24.518712	95	519.271998	31.322656
46	96.501454	24.775449	96	535.850158	31.381219
47	100.396498	25.024708	97	552.925662	31.438077
48	104.408392	25.266706	98	570.513433	31.493279
49	108.540644	25.501657	99	588.628835	31.546872
50	112.796863	25.729764	100	607.287700	31.598905

(continued)

Table 2   Annuities *(continued)*

$i = 4\%$ (interest rate per period) n = number of periods

$$s_{\overline{n}|i} = \frac{(1 + i)^n - 1}{i} \qquad a_{\overline{n}|i} = \frac{1 - (1 + i)^{-n}}{i}$$

n	$s_{\overline{n}|i}$	$a_{\overline{n}|i}$	n	$s_{\overline{n}|i}$	$a_{\overline{n}|i}$
1	1.000000	0.961538	51	159.773756	21.617485
2	2.040000	1.886095	52	167.164706	21.747582
3	3.121600	2.775091	53	174.851294	21.872675
4	4.246464	3.629895	54	182.845345	21.992956
5	5.416322	4.451822	55	191.159159	22.108612
6	6.632975	5.242137	56	199.805525	22.219819
7	7.898294	6.002055	57	208.797746	22.326749
8	9.214226	6.732745	58	218.149655	22.429567
9	10.582795	7.435331	59	227.875641	22.528429
10	12.006107	8.110896	60	237.990667	22.623490
11	13.486351	8.760477	61	248.510293	22.714894
12	15.025805	8.385074	62	259.450704	22.802783
13	16.626837	9.985648	63	270.828732	22.887291
14	18.291911	10.563123	64	282.661881	22.968549
15	20.023587	11.118387	65	294.968356	23.046682
16	21.824530	11.652295	66	307.767089	23.121809
17	23.697511	12.165669	67	321.077773	23.194048
18	25.645412	12.659297	68	334.920883	23.263507
19	27.671228	13.133939	69	349.317718	23.330295
20	29.778077	13.590326	70	364.290426	23.394515
21	31.969200	14.029160	71	379.862043	23.456264
22	34.247968	14.451115	72	396.056524	23.515639
23	36.617887	14.856841	73	412.898785	23.572730
24	39.082602	15.246963	74	430.414735	23.627625
25	41.645906	15.622080	75	448.631325	23.680408
26	44.311742	15.982769	76	467.576577	23.731162
27	47.084212	16.329585	77	487.279640	23.779963
28	49.967580	16.663063	78	507.770825	23.826888
29	52.966284	16.983714	79	529.081656	23.872007
30	56.084935	17.292033	80	551.244922	23.915392
31	59.328332	17.588493	81	574.294718	23.957107
32	62.701465	17.873551	82	598.266505	23.997219
33	66.209524	18.147645	83	623.197166	24.035787
34	69.857905	18.411197	84	649.125051	24.072872
35	73.652221	18.664613	85	676.090053	24.108531
36	77.598309	18.908282	86	704.133654	24.142818
37	81.702242	19.142579	87	733.299000	24.175787
38	85.970331	19.367864	88	763.630957	24.207487
39	90.409144	19.584485	89	795.176195	24.237969
40	95.025510	19.792774	90	827.983241	24.267278
41	99.826530	19.993052	91	862.102572	24.295459
42	104.819591	20.185627	92	897.586673	24.322557
43	110.012375	20.370795	93	934.490139	24.348612
44	115.412870	20.548841	94	972.869744	24.373666
45	121.029384	20.720040	95	1012.784531	24.397756
46	126.870560	20.884653	96	1054.295909	24.420919
47	132.945382	21.042936	97	1097.467747	24.443191
48	139.263197	21.195131	98	1142.366453	24.464607
49	145.833724	21.341472	99	1189.061112	24.485199
50	152.667073	21.482184	100	1237.623554	24.504999

(continued)

Table 2   Annuities *(continued)*

$i = 6\%$ (interest rate per period) n = number of periods

$$s_{\overline{n}|i} = \frac{(1 + i)^n - 1}{i} \qquad a_{\overline{n}|i} = \frac{1 - (1 + i)^{-n}}{i}$$

n	$s_{\overline{n}|i}$	$a_{\overline{n}|i}$	n	$s_{\overline{n}|i}$	$a_{\overline{n}|i}$
1	1.000000	0.943396	51	308.756049	15.813076
2	2.060000	1.833393	52	328.281411	15.861393
3	3.183600	2.673012	53	348.978296	15.906974
4	4.374616	3.465106	54	370.916993	15.949976
5	5.637093	4.212364	55	394.172013	15.990543
6	6.975318	4.917324	56	418.822333	16.028814
7	8.393838	5.582381	57	444.951673	16.064919
8	9.897468	6.209794	58	472.648773	16.098980
9	11.491316	6.801692	59	502.007700	16.131113
10	13.180795	7.360087	60	533.128160	16.161428
11	14.971642	7.886875	61	566.115851	16.190026
12	16.869941	8.383844	62	601.082800	16.217006
13	18.882137	8.852683	63	638.147769	16.242458
14	21.015066	9.294984	64	677.436635	16.266470
15	23.275970	9.712249	65	719.082832	16.289123
16	25.672528	10.105895	66	763.227802	16.310493
17	28.212879	10.477260	67	810.021470	16.330654
18	30.905652	10.827603	68	859.622755	16.349673
19	33.759991	11.158116	69	912.200122	16.367617
20	36.785591	11.469921	70	967.932127	16.384544
21	39.992726	11.764077	71	1027.008055	16.400513
22	43.392289	12.041582	72	1089.628537	16.415578
23	46.995827	12.303379	73	1156.006250	16.429791
24	50.815576	12.550357	74	1226.366622	16.443199
25	54.864511	12.783356	75	1300.948621	16.455848
26	59.156381	13.003166	76	1380.005534	16.467781
27	63.705764	13.210534	77	1463.805867	16.479039
28	68.528110	13.406164	78	1552.634216	16.489659
29	73.639797	13.590721	79	1646.792271	16.499679
30	79.058184	13.764831	80	1746.599804	16.509131
31	84.801676	13.929086	81	1852.395794	16.518048
32	90.889776	14.084043	82	1964.539537	16.526460
33	97.343163	14.230230	83	2083.411913	16.534396
34	104.183752	14.368141	84	2209.416623	16.541883
35	111.434777	14.498246	85	2342.981621	16.548947
36	119.120864	14.620987	86	2484.560511	16.555610
37	127.268116	14.736780	87	2634.634147	16.561896
38	135.904202	14.846019	88	2793.712188	16.567827
39	145.058455	14.949075	89	2962.334920	16.573421
40	154.761961	15.046297	90	3141.075010	16.578699
41	165.047679	15.138016	91	3330.539515	16.583679
42	175.950540	15.224543	92	3531.371874	16.588376
43	187.507572	15.306173	93	3744.254194	16.592808
44	199.758026	15.383182	94	3969.909436	16.596988
45	212.743508	15.455832	95	4209.104005	16.600932
46	226.508118	15.524370	96	4462.650238	16.604653
47	241.098605	15.589028	97	4731.409255	16.608163
48	256.564521	15.650027	98	5016.293804	16.611475
49	272.958392	15.707572	99	5318.271438	16.614599
50	290.335895	15.761861	100	5638.367708	16.617546

(continued)

Table 2   Annuities (continued)

$i = 8\%$ (interest rate per period) n = number of periods

$$s_{\overline{n}|i} = \frac{(1 + i)^n - 1}{i} \qquad a_{\overline{n}|i} = \frac{1 - (1 + i)^{-n}}{i}$$

n	$s_{\overline{n}|i}$	$a_{\overline{n}|i}$	n	$s_{\overline{n}|i}$	$a_{\overline{n}|i}$
1	1.000000	0.925926	51	620.671751	12.253227
2	2.080000	1.783265	52	671.325489	12.271506
3	3.246400	2.577097	53	726.031528	12.288432
4	4.506112	3.312127	54	785.114049	12.304103
5	5.866601	3.992710	55	858.923175	12.318614
6	7.335929	4.622880	56	917.837026	12.332050
7	8.922803	5.206370	57	992.263989	12.344491
8	10.636628	5.746639	58	1072.645104	12.356010
9	12.487558	6.246888	59	1159.456715	12.366676
10	14.486562	6.710081	60	1253.213249	12.376552
11	16.645487	7.138964	61	1354.470310	12.385696
12	18.977126	7.536078	62	1463.827931	12.394163
13	21.495296	7.903776	63	1581.934169	12.402003
14	24.214920	8.244237	64	1709.488897	12.409262
15	27.152114	8.559479	65	1847.248009	12.415983
16	30.324283	8.851369	66	1996.027847	12.422207
17	33.750225	9.121638	67	2156.710075	12.427969
18	37.450243	9.371887	68	2330.246878	12.433305
19	41.446263	9.603599	69	2517.666628	12.438245
20	45.761964	9.818147	70	2720.079956	12.442820
21	50.422921	10.016803	71	2938.686356	12.447055
22	55.456754	10.200744	72	3174.781254	12.450977
23	60.893295	10.371059	73	3429.763757	12.454608
24	66.764758	10.528758	74	3705.144851	12.457971
25	73.105939	10.674776	75	4002.556443	12.461084
26	79.954414	10.809978	76	4323.760948	12.463967
27	87.350767	10.935165	77	4670.661830	12.466636
28	95.338828	11.051078	78	5045.314764	12.469107
29	103.965934	11.158406	79	5449.939954	12.471396
30	113.283209	11.257783	80	5886.935134	12.473514
31	123.345866	11.349799	81	6358.889949	12.475476
32	134.213535	11.434999	82	6868.601137	12.477293
33	145.950617	11.513888	83	7419.089235	12.478975
34	158.626667	11.586934	84	8013.616357	12.480532
35	172.316800	11.654568	85	8655.705661	12.481974
36	187.102144	11.717193	86	9349.162105	12.483310
37	203.070315	11.775179	87	10098.095077	12.484546
38	220.315940	11.828869	88	10906.942661	12.485691
39	238.941216	11.878582	89	11780.498081	12.486751
40	259.056512	11.924613	90	12723.937900	12.487732
41	280.781033	11.967235	91	13742.852951	12.488641
42	304.243515	12.006699	92	14843.281143	12.489482
43	329.582997	12.043240	93	16031.743671	12.490261
44	356.949636	12.077074	94	17315.283134	12.490983
45	386.505607	12.108401	95	18701.505798	12.491651
46	418.426055	12.137409	96	20198.626205	12.492269
47	452.900140	12.164267	97	21815.516296	12.492842
48	490.132150	12.189137	98	23561.757575	12.493372
49	530.342722	12.212163	99	25447.698181	12.493863
50	573.770138	12.233485	100	27484.514007	12.494318

Answers

Exercise Set 0.1, page 11

1. (a) T   (b) F
    (c) F   (d) T
    (e) T   (f) F

3. (a) {4, 8}   (b) Yes   (c) No

5. (a) H, R   (b) Q, R
    (c) Q, R   (d) I, Q, R
    (e) Q, R   (f) H, R
    (g) H, R   (h) H, R
    (i) Q, R   (j) H, R

7. (a) 14.0712528 (14.07)
    (b) 7.139497 (7.14)

9. 11

11. 11

13. 7

15. $\dfrac{1}{5}$

17. $\dfrac{80}{3}$

19. Undefined

21. $\dfrac{1}{2}$

23. 0

25. $\dfrac{-7}{40}$

27. $2xy$

29. -1

31. (a) 17.2129
    (b) 130.0841
    (c) 355.4321
    (d) The number with the most decimal places determines the number of decimal places in the answer.

33. (a) 7000
    (b) 2
    (c) 0.2
    (d) Consider the division as a fraction. Move the decimal point in the denominator so it becomes a whole number, and move the decimal point in the numerator the same number of places. Then divide and keep the same number of decimal places in the answer as in the numerator.
    (e) Yes, as long as you change the number of places in the mode setting.

35. $H' = \{A, F, G\}$

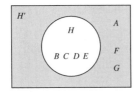

37.

Year	Depreciation
1	$2727.27
2	2454.55
3	2181.82
4	1909.09
5	1636.36
6	1363.64
7	1090.91
8	818.18
9	545.45
10	272.73

39. 0.5, 0.36, 0.01538, 0

Exercise Set 0.2, page 22

1. $x = 5$
    Check: $2(5) - 7 = 3$
          $10 - 7 = 3$
             $3 = 3$

3. $x = 3$
    Check: $4x - 7 = 5$
          $4(3) - 7 = 5$
           $12 - 7 = 5$
              $5 = 5$

5. $x = -1$
    Check: $4 - 2(-1) = [8 + 3(-1)] + 1$
          $4 + 2 = (8 - 3) + 1$
             $6 = 5 + 1$
             $6 = 6$

7. $x = 1$
    Check: $2(1) + 5 = 6 + 1$
          $2 + 5 = 7$
             $7 = 7$

9. $x = 5$
    Check: $\dfrac{5}{5} - 3 = -2$
          $1 - 3 = -2$
           $-2 = -2$

11. $x = -4$
    Check: $\dfrac{-4}{5} - \dfrac{1}{3} = \dfrac{-4}{3} + \dfrac{1}{5}$
        $\dfrac{-12}{15} - \dfrac{5}{15} = \dfrac{-20}{15} + \dfrac{3}{15}$
            $\dfrac{-17}{15} = \dfrac{-17}{15}$

13. $(3, \infty)$

15. $(-\infty, -21)$

17. $\left(\dfrac{28}{5}, \infty\right)$

19. $\dfrac{A - P}{Pt} = r$ $(Pt \neq 0)$

21. $y - mx = b$

23. $r = \dfrac{S - a}{S}$ $(S \neq 0)$

25. $x = 2$

Check: $-[2(2) - (3 - 2)] = 4(2) - 11$
$-[4 - 1] = 8 - 11$
$-3 = -3$

27. $x = \dfrac{2}{3}$

Check: $2 - \dfrac{2}{3} - \left(\dfrac{2}{3}\right)^2 = 1 - \left(\dfrac{2}{3} - 1\right)^2$

$2 - \dfrac{2}{3} - \dfrac{4}{9} = \dfrac{8}{9}$

$\dfrac{8}{9} = \dfrac{8}{9}$

29. $x = 9$

Check: $\dfrac{9 - 5}{4} = 1 + \dfrac{9 - 9}{12}$

$\dfrac{4}{4} = 1 + \dfrac{0}{12}$

$1 = 1 + 0$

$1 = 1$

31. $x = 5$ or $x = 3$

Check: $|4 - 5| = 1, \quad |4 - 3| = 1$
$|-1| = 1 \qquad |1| = 1$
$1 = 1 \qquad 1 = 1$

33. $x = -2, x = 1$

Check: $|2(-2) + 1| = 3, \quad |2(1) + 1| = 3$
$|-4 + 1| = 3 \qquad |2 + 1| = 3$
$|-3| = 3 \qquad |3| = 3$
$3 = 3 \qquad 3 = 3$

35. $[-6, 1]$

37. $\left(\dfrac{-25}{2}, -8\right]$

39. $(-5, 9)$

41. $(-\infty, -3]$ or $[1, \infty)$

43. $P = \dfrac{A}{1 + rt}$

45. $y = \dfrac{1 + 3x}{x^2 - 2z^3}$

47. $x < 1.69618$ 49. $x \le 6.64022$

51. $x = 0.20034$

53. (a) $3x - 6 = 21$ (b) $4x - 8 = 61$
(c) $x - 42 < 20$

Exercise Set 0.3, page 30

1. 10   3. Width = 4.4 inches
Length = 14.6 inches

5. 82 kg   7. $\dfrac{2}{3}$ gallon

9. (a) 9375 ties

11. $I = 0.07x + 0.10(27{,}000 - x)$

13. $R = \$5x$ 15. $p = 2x + 20$

17. $P > \$620{,}000$

19. $P = 200x + 400x = 600x$

21. \$1000 at 8%, \$1000 at 10%

23. \$1.36 per pound

25. $X = 2g_1 + 4g_2$ milligrams

27. $IQ = \left(\dfrac{x}{10}\right)100$

29. $P = 20{,}000 + 0.10(20{,}000)t$

31. $C = \dfrac{5}{9}(F - 32)$ 33. 200 grams

35. (a) 2   (b) $-2, 0, 2$ (c) $-2, -\dfrac{1}{2}, 0, 2$

(d) $-\sqrt{2}, \dfrac{1}{\sqrt{2}}, \sqrt{2}$ (e) $-2, -\dfrac{1}{2}, -\sqrt{2}$

(f) $-2, -\dfrac{1}{2}, -\sqrt{2}, 0, \dfrac{1}{\sqrt{2}}, \sqrt{2}, 2$

Exercise Set 0.4, page 39

1. (a) and (b)   (c) and (d)

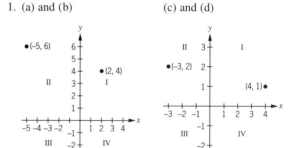

3. (a) $(0, 2), \left(1, \dfrac{4}{3}\right), \left(-1, \dfrac{8}{3}\right), \left(2, \dfrac{2}{3}\right), \left(-2, \dfrac{10}{3}\right)$

(b) $(-2, -1), (-2, -4), (0, 1), (1, 1), (2, 3)$

(c) $(0, 5), \left(1, \dfrac{10}{3}\right), \left(2, \dfrac{5}{3}\right), \left(-1, \dfrac{20}{3}\right), \left(-2, \dfrac{25}{3}\right)$

(d) $\left(\dfrac{13}{5}, -2\right)$, $(2, -1)$, $\left(\dfrac{7}{5}, 0\right)$, $\left(\dfrac{4}{5}, 1\right)$, $\left(\dfrac{1}{5}, 2\right)$

(e) $(0, 1)$, $\left(1, \dfrac{7}{4}\right)$, $\left(2, \dfrac{5}{2}\right)$, $(4, 4)$, $(-4, -2)$

(f) $(0, 0)$, $(-1, 0)$, $(-4, 2)$, $(5, 2)$, $(-2, 3)$

5. (a)

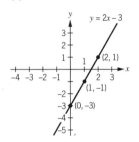

(b)

(c)

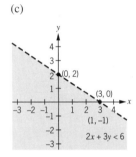

(d)

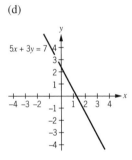

(e)

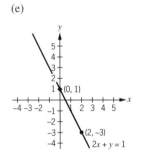

(f)

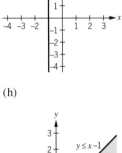

(g)

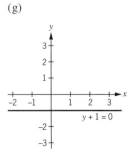

(h)

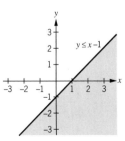

(i)

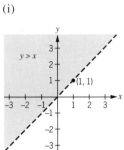

(j)

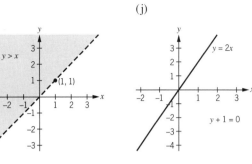

(k)

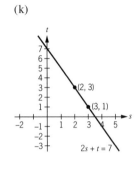

(l)

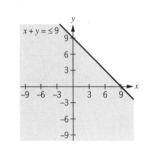

7. $y = 2x - 3$, $-1 \le x \le 4$

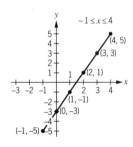

9. $y = -2x + 4$, $-2 \le x \le 4$

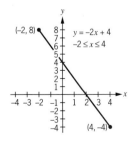

11. $y \leq 3x - 2$, $-2 \leq x \leq 2$

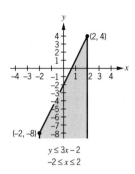

$y \leq 3x - 2$
$-2 \leq x \leq 2$

13.

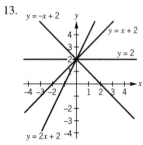

15.

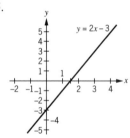

17.

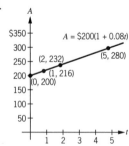

19. (a) $25   (b) $15.00   (c) Goes to 0

(d)

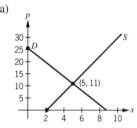

21. (a)

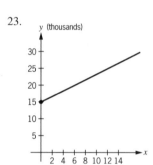

(b) (5, 11)   (c) $x < 5$

23.

Let 0 represent 1980 and 6 represent 1986. In 1990, the population was approximately 23,333. In 1994, approximately 26,666.

25. (a) $y \leq \dfrac{24}{7}$ (b) $x \geq \dfrac{135}{2}$ (c) $x \geq -8$

(d) $x < \dfrac{-8}{9}$

Exercise Set 0.5, page 54

1. $\dfrac{3}{x^7}$ 3. $\dfrac{5x^4}{2}$

5. $\dfrac{2}{x^3}$ 7. $\dfrac{1}{x^3}$

9. $\dfrac{4}{x^6}$ 11. 1

13. $\dfrac{4}{x^4}$ 15. $\dfrac{25}{x^2}$

17. $\dfrac{1}{49}$

19. $\dfrac{1}{729}$

21. $\dfrac{1}{27}$

23. $\sqrt{3}y$

25. $x^2 y \sqrt[3]{xy}$

27. $4x \sqrt[3]{y^2}$

29. 3

31. -2

33. $x^3 y^2$

35. $\dfrac{5x}{2y^2}$

37. 3

39. 4

41. 16

43. 5

45. Does not exist

47. $24x^4 y^3 z^5$

49. $\dfrac{5b}{a}$

51. $\dfrac{az^4}{3xy^2}$

53. $2x^2\sqrt{2x}$

55. $xy\sqrt[3]{2y}$

57. $\dfrac{\sqrt{2}}{2}$

59. $\dfrac{2\sqrt[3]{4x^2}}{x}$

61. $3x^{-1} = \dfrac{3}{x}$

63. $\dfrac{3x}{2}$

65. $\dfrac{3x^{-1/3}}{2} = \dfrac{3}{2x^{1/3}}$

67. -5

69. $5\sqrt{2}$

71. 6

73. $2y$

75. $\dfrac{5}{4x^{1/4}y^{1/4}}$

77. $\dfrac{b^5}{4c^3}$

79. $(1.3)^5 = 3.71293$

81. $(1.001)^{-1} = 0.999000999$

83. $2a^2 b \sqrt[3]{5ab^2}$

85. $2\sqrt{y}$

87. $\dfrac{2y^2\sqrt{3x}}{3x}$

89. (a) \$8000   (b) \$6400   (c) \$5120   (d) \$1677.72

91. \$9110.03

93. $P(0) = 100$; $P(1) = 60.656$; $P(2) = 36.792$;
$P(4) = 13.536$; $P(6) = 4.980$; $P(8) = 1.832$;
$P(10) = 0.674$

Exercise Set 0.6, page 65

1. $3x^2 - 10x + 9$

3. $-2x - z$

5. $-9x^2 + 12x - 4$

7. $6x^3 + 12x^2$

9. $x^2 - x - 6$

11. $3x^2 + 5x - 2$

13. $24x^2 + 44x - 28$

15. $4x^2 - 9$

17. $16x^2 + 2.4x + 0.09$

19. $12x^2 - 5x - 2$

21. $-16x^2 + 0.01$

23. $x(x - 5)$

25. $3(x + 5)(x - 5)$

27. $3x^2(y - 4)$

29. $(x + 5)^2$

31. $(x + 1)(x - 3)$

33. $2(x + 4)(x - 7)$

35. $\dfrac{2x + 3}{3x - 4}$

37. $\dfrac{3y - 2}{y + 2}$

39. $(x + 2)(x + 5)$

41. $y(y + 2)(x - 2)$

43. $2(x + 2)(x + 1)$

45. $(4x + 5y)(16x^2 - 20xy + 25y^2)$

47. $(x - 3)(x^2 + 3)$

49. $\dfrac{9(y + 2)}{20(y - 2)}$

51. $\dfrac{17}{12(x - 2)}$

53. $\dfrac{2(4y + 13)}{(y + 2)(y - 5)(y + 4)}$

55. $\dfrac{2x(x - 11)}{(3x - 2)(x + 3)(x - 4)}$

57. $\dfrac{(2x - 3)(4x + 3)}{(2x + 3)(4x - 3)}$

59. $\dfrac{\sqrt{2} + 1}{2}$

61. $\dfrac{4(\sqrt{a} + \sqrt{b})}{a - b}$

63. $\dfrac{3(\sqrt{x} + 2)}{2(x - 4)}$

65. $A = P(1 + rt)$

67. $-0.004(x^2 - 51x - 200)$

69. x^4

71. $\dfrac{15}{49xz}$

Chapter 0 Test, page 68

1. $\{3, 6, 9, 10\}$

2. $\dfrac{-55}{96}$

3. $x = 4$

4. Alice's age $= 5(n + 1) - 3$

5. $\dfrac{1}{x^4 y^3}$

6. $\dfrac{3x\sqrt{10y}}{5y}$

7. $\dfrac{3x\sqrt[3]{2x^2 y^2}}{y}$

8. $(3r + 5s)(3r - 5s)$

9. $(3k + 7)(k - 5)$

10. $(x^2 + 4)(y + x)$

11.

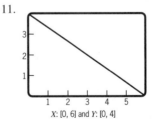

X: [0, 6] and Y: [0, 4]

12. $-\dfrac{19}{2}$ or $\dfrac{11}{2}$

13. $r = \dfrac{S - a}{S}$ $S \neq 0$

14. $x + 3y > 9$

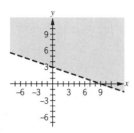

15. $x > -1$ 16. $-3 < x < 2$

17. $x \geq 3$ or $x \leq -5/3$

18.

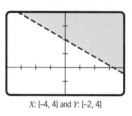

X: [-4, 4] and Y: [-2, 4]

19. \$1500   20. \$45,454.55 at 12%
                                  \$75,454.55 at 10%

21. $R = 50x$; $C = 30x + 10,000$; $P = 20x - 10,000$

22. $(134/7, 124/7)$

Exercise Set 1.1, page 77

1. -5 3. 1

5. -1 7. Not defined

9. 0

11.

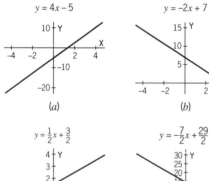

$y = 4x - 5$ $y = -2x + 7$

(a)   (b)

$y = \frac{1}{2}x + \frac{3}{2}$ $y = -\frac{7}{2}x + \frac{29}{2}$

(c)   (d)

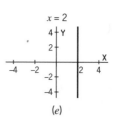

$x = 2$

(e)

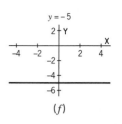

$y = -5$

(f)

13. $m = -2, b = 1$

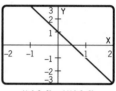

X: [-2, 2] and Y:[-3, 3]

15. $m = 5, b = 0$

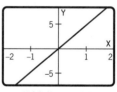

X: [-2, 2] and Y:[-8, 8]

17. $m = 3, b = 1$

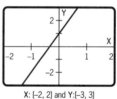

X: [-2, 2] and Y:[-3, 3]

19. $m = \frac{3}{2}, b = -\frac{5}{2}$

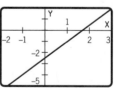

X: [-2, 3] and Y:[-5, 2]

21. $y = \frac{1}{2}x + \frac{5}{2}$ 23. $y = -\frac{1}{3}x - \frac{7}{3}$

25. $y = 4x - 3$ 27. $x = 1$

29. $y = -6$ 31. $y = 0$

33. $a; b$ 35. $b = -8$

37. (a) $P = 0.8x - 50$

                       (b) 63   (c) 110

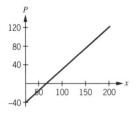

35. (a)

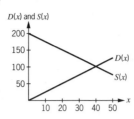

(b) (40, 100)

(c) $p = 100$

(d) $x > 40$

37. $y = 19x$ for 19 mpg. More miles per gallon will cause the slope of the line to increase, and the line will be steeper.

Exercise Set 1.2, page 86

1. (a), (b), (c), (d), (e), (g), (h), (j)

3. (a) Function   (b) Function   (c) Not a function
   (d) Function

5. $10, 4, -1$

7. $-1; 224; w^3 - 2w; 8z^3 - 4z; t^3 + 6t^2 + 10t + 4;$
   $27x^3 - 27x^2 + 3x + 1$

9. $[0, \infty)$

11. $(-\infty, -1), (-1, 5), (5, \infty)$

13. $[-\sqrt{3}, \sqrt{3}]$

15. $(-\infty, 0), (0, 3), (3, \infty)$

17. $(-\infty, -4], [0, \infty)$ 19. $(-\infty, -1), [1, \infty)$

21. $[-2, 1], [3, \infty)$ 23. $(-\infty, \infty)$

25. (a) 3   (b) 2
    (c) $4(4 + h)$ (d) $4 + h$
    (e) $\dfrac{\sqrt{2 + h} - \sqrt{2}}{h}$ (f) $\dfrac{-1}{2(2 + h)}$

27. (d)   29. (b)

31. $2, 3.50, 15.50, 36.50$; \$2 for first hour and \$1.50 for each additional hour.

33. (a) $\dfrac{25}{2}$ (b) $\dfrac{15}{2}$

(c) 0; it would take a negative price to generate a demand > 10.

(d)

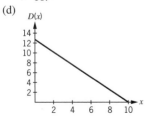

Exercise Set 1.3, page 96

1. $y = 4x^2$ magnifies $y = x^2$ by 4. $y = 4(x + 1)^2$ shifts $y = x^2$ to the left 1 and magnifies it by 4. $y = 4(x - 1)^2$ shifts $y = x^2$ to the right 1 and magnifies it by 4.

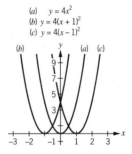

(a) $y = 4x^2$
(b) $y = 4(x + 1)^2$
(c) $y = 4(x - 1)^2$

3. $y = -x^2$ reflects $y = x^2$ across the x-axis. $y = -(x - 3)^2$ shifts $y = x^2$ to the right 3 and reflects it across the x-axis. $y = -(x + 3)^2$ shifts $y = x^2$ to the left 3 and reflects it across the x-axis.

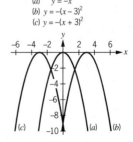

(a) $y = -x^2$
(b) $y = -(x - 3)^2$
(c) $y = -(x + 3)^2$

5. $y = x^2 - 2$ lowers $y = x^2$ down 2. $y = (x - 3)^2$ shifts $y = x^2$ to the right 3. $y = (x + 3)^2 + 2$ shifts $y = x^2$ to the left 3 and up 2.

(a) $y = x^2 - 2$
(b) $y = (x - 3)^2$
(c) $y = (x + 3)^2 + 2$

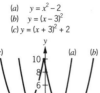

7. Magnify by 2
9. Shift to the left 1
11. Vertical shift up 3
13. Magnify by 2, horizontal shift right 1, vertical shift up 3
15. (a) $x = -0.5$
 (b) $(-0.5, -6.25)$
 (c) -6
 (d)

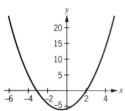

 (e) $x = -3, 2$
 (f) Minimum
17. (a) $x = -1$
 (b) $(-1, 25)$
 (c) 24
 (d)

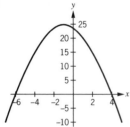

 (e) $x = -6, 4$
 (f) Maximum
19. (a) $x = -1.5$
 (b) $(-1.5, -18.75)$
 (c) -12

(d)

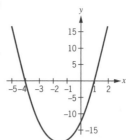

 (e) $x = -4, 1$
 (f) Minimum
21. (a) $x = -3$
 (b) $(-3, -9)$
 (c) 0
 (d)

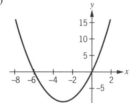

 (e) $x = -6, 0$
 (f) Minimum
23. (a) $x = 0$
 (b) $(0, -4)$
 (c) -4
 (d)

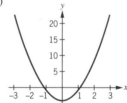

 (e) $x = -1.1547, 1.1547$
 (f) Minimum
25. $D = (-\infty, -4/3], [4/3, \infty), R = [0, \infty)$
27. $D = (-\infty, 0), (0, \infty), R = (-\infty, 1), (1, \infty)$

Exercise Set 1.4, page 103

1. $x = \pm 4$
3. $x = \pm \sqrt{7}$
5. $x = 0, 5$
7. $x = \pm 5$
9. $x = -1, 3$
11. $x = -4, 7$
13. $x = -2, 1$
15. $x = -3, 2$
17. $x = 1, -9$
19. $x = -5, -6$

21. (2, 3), (3, 4) 23. (−1, 2), (2/3, 2)

25. (−∞, −2], [2, ∞) 27. (−∞, −3), (−2, ∞)

29. [−1/2, 2/3]

31. $x = 0, -2, 2$; y is negative on $(-\infty, -2)$, positive on $(-2, 0)$, negative on $(0, 2)$, positive on $(2, \infty)$.

33. $x = 0, -1$; y is negative on $(-\infty, -1)$, negative on $(-1, 0)$, positive on $(0, \infty)$.

35. $x = -3, 0, 3$; y is positive on $(-\infty, -3)$, negative on $(-3, 0)$, negative on $(0, 3)$, positive on $(3, \infty)$.

37. $x = -2, 0, 1$; y is positive on $(-\infty, -2)$, negative on $(-2, 0)$, negative on $(0, 1)$, positive on $(1, \infty)$.

39. At $x = 193$, the maximum revenue is $37,249.

41. Equilibrium point is (8, 16) and the price is $16.

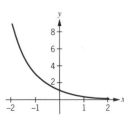

X: [4, 19] and Y: [0, 100]

43. Maximum profit occurs at $x = 14$ and is $242.

45.

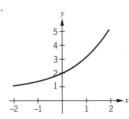

47. $r = 0$ and $r = 100$; $r = 76.83$

49. $y = -\dfrac{5}{9}x - \dfrac{1}{9}$

Exercise Set 1.5, page 112

1. 0.00137174 3. 123.144

5. 0.716531 7. −3.00417

9. 41.4666 11. 1; 0.707107; 4

13. 0; 2.16228; −0.99 15. 3; 2.12132; 12

17. 2; 2.25525; 7.52439

19.

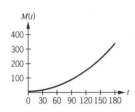

21.

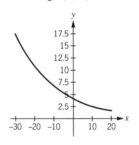

23.

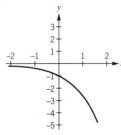

25.

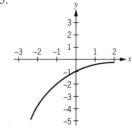

27. Range (0, ∞) 29. Range (−∞, 5)

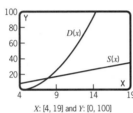 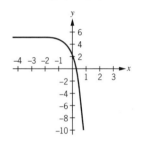

31. Range (0, ∞)

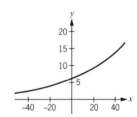

33. $x > 0$

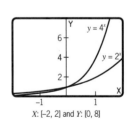

X: [−2, 2] and Y: [0, 8]

35. Minimum at (0, 0); Maximum at (2, 0.54)

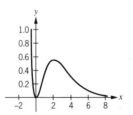

37. Minimum at (0, 0.368)

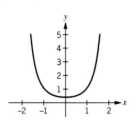

39. $x = -0.2$ 41. $x = -9$

43. Sales increase, but level off and are never above 600.

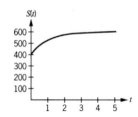

45. (a) $891.16, $410.57
    (b) Within the context of the problem the domain is $(0, \infty)$
        and the range is $(0, 1125)$.

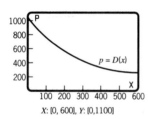

X: [0, 600], Y: [0,1100]

47. (a) $V(2) = 15,750$ (b) $V(5) = 6644.53$
    (c) $V(t) \approx 14,000$ when $x = 2.4$ years

49. $x = 0.75$ hours

51. (a) 0   (b) 6.71399
    (c) Around $x = 60$

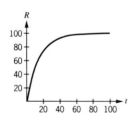

53. (a) 2412   (b) 3572

55. (a) $x = 4, -1$ (b) $x = 3, -0.5$

57. Vertex: $\left(\dfrac{2}{3}, \dfrac{4}{3}\right)$; x-intercepts: $0, \dfrac{4}{3}$; y-intercept: 0

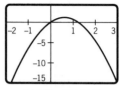

X: [-2, 3] and Y: [-15, 5]

59. Vertex: (1, 1); y-intercept: 3

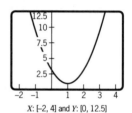

X: [-2, 4] and Y: [0, 12.5]

Chapter 1 Test, page 114

1. $y = x - 6$ 2. $y = -\dfrac{1}{2}x + \dfrac{5}{2}$

3. $x = \pm\sqrt{7}$ 4. $x = -3 \pm \sqrt{5}$

5. $x = -\dfrac{1}{2}, \dfrac{1}{3}$ 6. $x = .333424$

7. $x = 4.420696, x = -0.7540291$

8. $x = -1.61803399, x = 0.61803399, x = 1$

9. (a) -7 (b) $2x + h + 2$

10. $D = $ all real values of x except -1 and 5

11. $D = [-5, 5]$

12. $D = (-\infty, \infty)$

13. $D = (-\infty, \infty)$

14. (a) $x = 1$ (b) Up   (c) minimum at $(1, -4)$
    (d) y-intercept $= -3$, x-intercepts $= -1, 3$
    (e) $y = (x - 1)^2 - 4$
    (f)

15.

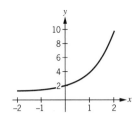

16. Function

17. Not a function

18. $x = 44$

19. $D(100) = 599.19$

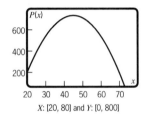

X: [20, 80] and Y: [0, 800]

Exercise Set 2.1, page 128

1. 2×3, $a_{12} = 1$

3. 3×3, $c_{32} = 2$

5. Impossible

7. $\begin{bmatrix} 18 & 24 & 30 \\ 6 & 12 & 18 \end{bmatrix}$

9. Impossible, the two matrices are not the same size.

11. $\begin{bmatrix} 20 & 10 & 15 & 15 \\ -5 & 0 & 10 & -5 \end{bmatrix}$

13. $\begin{bmatrix} 18 & -7 & 17 \\ 1 & 18 & 19 \\ -9 & 14 & -11 \end{bmatrix}$

15. $\begin{bmatrix} -2 & -1 \\ 0 & 3 \end{bmatrix}$

17. $\begin{bmatrix} 35 & -5 & 9 \\ -2 & 28 & -21 \end{bmatrix}$

19. $\begin{bmatrix} 2 & 1 \\ 3 & 0 \\ 4 & 5 \\ 5 & 2 \end{bmatrix}$

21. $x = -4$, $y = 8$, $z = 20$

23. $x = 0$, $y = 0$, $z = 0$

25. $\mathbf{X} = \begin{bmatrix} -12 \\ -1 \end{bmatrix}$

27. $\begin{bmatrix} 70.5 & 28.5 & -43 \\ -11.5 & 76 & -50.5 \\ 86 & 80.5 & -21.5 \end{bmatrix}$

29. $\begin{bmatrix} -14.08 & -7.4 & 10.9 \\ 4 & -17.9 & 9 \\ -20.6 & -11 & 0 \end{bmatrix}$

31. $\begin{bmatrix} 42.4 & 25.6 & -42.8 \\ -15.4 & 32.1 & -4.8 \\ 63.1 & 42.8 & -34.4 \end{bmatrix}$

33. (a)

	A	H	C
W34	100	50	200
G47	60	40	70
B71	13	40	24

(b)

	A	H	C
W34	10	11	15
G47	5	8	30
B71	8	16	10

(c) $\begin{bmatrix} 90 & 39 & 185 \\ 55 & 32 & 40 \\ 5 & 24 & 14 \end{bmatrix}$

(d) $\begin{bmatrix} 270 & 117 & 555 \\ 165 & 96 & 120 \\ 15 & 72 & 42 \end{bmatrix} = 3 \begin{bmatrix} 90 & 39 & 185 \\ 55 & 32 & 40 \\ 5 & 24 & 14 \end{bmatrix}$

35. $\begin{bmatrix} 0.60 & 0.40 \\ 0.30 & 0.70 \end{bmatrix}$

37. (a)

	V_1	V_2	V_3
V_1	1	1	0
V_2	1	0	1
V_3	0	1	0

(b) $\begin{bmatrix} 0 & 1 & 1 \\ 1 & 1 & 1 \\ 1 & 1 & 0 \end{bmatrix}$

(c) $\begin{bmatrix} 1 & 1 & 1 \\ 1 & 0 & 1 \\ 1 & 1 & 1 \end{bmatrix}$

(d) $\begin{bmatrix} 1 & 1 & 0 & 1 \\ 1 & 0 & 1 & 0 \\ 0 & 1 & 1 & 1 \\ 1 & 0 & 1 & 0 \end{bmatrix}$

39. Receivers 2 and 4 receive communication of sender 1. 1 and 4 receive from 2. 1 and 2 from sender 3. Receivers 1, 2, and 3 receive communication from sender 4.

41. (a)

	A	B	C	D
A	0	0	1	0
B	1	0	1	0
C	0	0	0	0
D	0	0	1	0

(b)

	A	B	C	D
A	0	1	1	0
B	1	0	1	0
C	0	0	0	1
D	0	1	0	0

43. If the parents are hybrid, 25% of the children will be recessive, 50% will be hybrid, and 25% will be dominant. If the parents are dominant, 75% of the children will be hybrid and 25% dominant.

Exercise Set 2.2, page 141

1. Undefined

3. -4

5. Undefined

7. 2×4

9. Undefined

11. Undefined

13. 3×3

15. 2×3

17. $\begin{bmatrix} 5 \\ -2 \\ 1 \end{bmatrix}$

19. $\begin{bmatrix} 24 & 2 & 38 \\ 6 & -4 & 25 \end{bmatrix}$

21. $\begin{bmatrix} 3 & 19 & 13 \\ 1 & 5 & -3 \\ 2 & 12 & 10 \end{bmatrix}$

23. $\begin{bmatrix} 0 & 0 & 0 \\ 0 & 0 & 0 \\ 0 & 0 & 0 \end{bmatrix}$

25. $\begin{bmatrix} 11 & 1 & 17 \\ 16 & 2 & 25 \\ 21 & 3 & 33 \end{bmatrix}$

27. $\begin{bmatrix} -3 & -2 \\ 2 & -4 \end{bmatrix}$

29. $\begin{bmatrix} 24 & 2 \\ 6 & -4 \end{bmatrix}$ 31. $\begin{bmatrix} 0 & 0 & 0 \\ 4 & 4 & 4 \\ 0 & 0 & 0 \end{bmatrix}$

33. $\begin{bmatrix} 4 & 3 \\ 6 & -1 \end{bmatrix}\begin{bmatrix} x \\ y \end{bmatrix} = \begin{bmatrix} 7 \\ 10 \end{bmatrix}$

35. $\begin{bmatrix} 4 & -1 \\ 1 & -2 \end{bmatrix}\begin{bmatrix} x \\ y \end{bmatrix} = \begin{bmatrix} -3 \\ -7 \end{bmatrix}$

37. $\begin{bmatrix} 1 & 0 \\ 1 & 2 \end{bmatrix}\begin{bmatrix} x \\ y \end{bmatrix} = \begin{bmatrix} 4 \\ 7 \end{bmatrix}$

39. $\begin{bmatrix} 6 & 3 & 7 \\ 5 & 4 & 1 \\ 2 & -1 & 1 \end{bmatrix}\begin{bmatrix} x \\ y \\ z \end{bmatrix} = \begin{bmatrix} 2 \\ 5 \\ 6 \end{bmatrix}$

41. $\begin{bmatrix} 1 & 0 & 0 \\ 0 & 1 & 0 \\ 0 & 0 & 1 \end{bmatrix}\begin{bmatrix} x \\ y \\ z \end{bmatrix} = \begin{bmatrix} 4 \\ 7 \\ 9 \end{bmatrix}$

43. $\begin{bmatrix} 25.13 & 56.77 & 65.54 \\ -12.83 & -44.36 & -57.52 \\ 12.66 & 49.71 & 65.76 \end{bmatrix}$

45. $\begin{bmatrix} 13{,}627 & -5{,}378 & 4{,}982 \\ -4{,}990 & 10{,}751 & -11{,}142 \\ 3{,}299 & -14{,}517 & 15{,}364 \end{bmatrix}$

47. $\begin{bmatrix} 4.58 \\ 3.04 \\ 1.96 \end{bmatrix}$ in millions of dollars

49. $\mathbf{A}^2 = \begin{bmatrix} 0.605 & 0.2075 & 0.1875 \\ 0.345 & 0.405 & 0.25 \\ 0.095 & 0.0825 & 0.8225 \end{bmatrix}$

From business to business, 60.5%
From business to unemployment, 20.75%
From business to self-employment, 18.75%
From unemployment to business, 34.5%
From unemployment to unemployment, 40.5%
From unemployment to self-employment, 25%
From self-employment to business, 9.5%
From self-employment to unemployment, 8.25%
From self-employment to self-employment, 82.25%

51. Birmingham and Nashville (by New Orleans and by Cincinnati)
Birmingham and Louisville (by New Orleans and by Cincinnati)
Nashville and Louisville (by New Orleans and by Cincinnati)

53. Process A   7.6
Process B   4.4
Process C   7.8
Process D   12.2

Exercise Set 2.3, page 152

1.

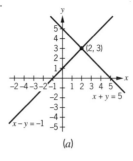

(a)

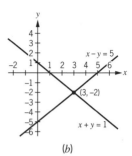

(b)

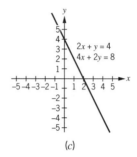

(c)

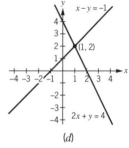

(d)

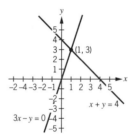

(e)

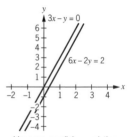

Lines are parallel; no solution
(f)

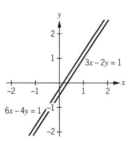

Lines are parallel; no solution
(g)

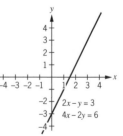

Lines are coincident: an infinite number of solutions
(h)

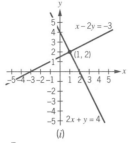

(i)

3. (a) $x = 5, y = -7$
   (b) No solutioh: lines are parallel
   (c) $x = 12, y = -5$
   (d) Infinite number of solutions: lines coincide
   (e) $x = 0, y = 2$
   (f) $x = 2, y = 1$

5. (93, 82)

7. (a) $x = 1.3417, y = 0.0761$
   (b) $x = -2.23, y = 1.79$
   (c) $x = 87, y = 103$
   (d) $x = -43, y = 117$

9. 32 pounds of $3.00 per pound candy
   48 pounds of $4.00 per pound candy

11. 1400 items at a cost of $350

13. Yes. The break-even point will be reached at a lower level of production.

15. Number of women: 360
    Number of men: 640

17. $a = 5, b = 9, c = 3, d = 2, e = 14, f = 20$

Exercise Set 2.4, page 165

1. $3x + y = 13$
   $2x - y = \ 2$

3. $x = 4, y = 6$

5. $\begin{bmatrix} 2 & 3 & | & 5 \\ 4 & -1 & | & 3 \end{bmatrix}$

7. $\begin{bmatrix} 5 & 2 & | & 3 \\ 1 & 3 & | & -2 \end{bmatrix}$

9. $\begin{bmatrix} -2 & 2 & | & -6 \\ 1 & -2 & | & -5 \end{bmatrix}$

11. (a) $\begin{bmatrix} 3 & 2 & | & 5 \\ 14 & 8 & | & 8 \end{bmatrix}$

   (b) $\begin{bmatrix} 7 & 4 & | & 4 \\ 3 & 2 & | & 5 \end{bmatrix}$

   (c) $\begin{bmatrix} 3 & 2 & | & 5 \\ 16 & 10 & | & 19 \end{bmatrix}$

   (d) $\begin{bmatrix} 12 & 8 & | & 20 \\ 7 & 4 & | & 4 \end{bmatrix}$

   (e) $\begin{bmatrix} 17 & 10 & | & 13 \\ 7 & 4 & | & 4 \end{bmatrix}$

   (f) $\begin{bmatrix} 3 & 2 & | & 5 \\ 0 & -\frac{2}{3} & | & -\frac{23}{3} \end{bmatrix}$

13. $\begin{bmatrix} 1 & 0 & | & 3 \\ 0 & 1 & | & 4 \end{bmatrix}$

15. $\begin{bmatrix} 1 & 0 & | & 1 \\ 0 & 1 & | & 1 \end{bmatrix}$

17. $\begin{bmatrix} 1 & 0 & | & 1 \\ 0 & 1 & | & -1 \end{bmatrix}$

19. $\begin{bmatrix} 1 & 0 & | & 11 \\ 0 & 1 & | & 8 \end{bmatrix}$

21. Infinite number of solutions

23. No solution

25. $x = -6.7857, y = -5.7857$

27. 47.5 (i.e., 47 tables), 4 chairs

29. 7 runs of best paper
    13 runs of good paper

31. $x = 100$ at a cost of $200

33. 2 units of food I
    3 units of food II

Exercise Set 2.5, page 175

1. Reduced form

3. Not reduced form

5. Reduced form

7. Not reduced form

9. Unique solution
   $x = 3, y = -1, z = 4$

11. Infinite number of solutions
    $x = 3 - 2c$
    $y = 1 - c$
    $z = c$

13. No solution

15. Unique solution
    $x = -2, y = 4, z = 6, w = 1$

17. No solution

19. $x = 1 - c, y = 3 + 2c, z = c$

21. $x = 10, y = 3, z = 15$ 23. $x = 1, y = -1, z = 1$

25. $x = 1, y = 3, z = 0$

27. $x = -2 + 3c, y = 4 - c, z = c$

29. $x = \dfrac{c + 10}{3}, y = \dfrac{4c - 2}{3}, z = c$

31. (a) No solution
    (b) Infinite number of solutions. For any real z, $x = -1 + 2z, y = 2 + z$.
    (c) Infinite number of solutions. For any values y and z, $x = 2y + 3z - 3$
    (d) $(4, -5, 6)$

33. $x_1 = 3/7, x_2 = 6/7$

35. $x_1 = -1, x_2 = 2$

37. $x_1 = -2r + s + 1, x_2 = r, x_3 = s.$ r and s are any real numbers.

39. $(1 + (5/11)c, 1 - (18/11)c, c)$

41. $(40 - 2k, 9k - 20, 32 - 6k, k)$ or 30 A, 25 B, 2 C, 5 D or 32 A, 16 B, 8 C, 4 D or 34 A, 7 B, 14 C, 3 D

43. $A = 200, B = 100, C = 300$

45. Food I: 3 servings

    Food II: $\frac{3}{2}$ servings

    Food III: $\frac{1}{2}$ serving

Exercise Set 2.6, page 186

1. (a) $\begin{bmatrix} 1 & 2 \\ 1 & 3 \end{bmatrix}\begin{bmatrix} 3 & -2 \\ -1 & 1 \end{bmatrix} = \begin{bmatrix} 3-2 & -2+2 \\ 3-3 & -2+3 \end{bmatrix}$

    $= \begin{bmatrix} 1 & 0 \\ 0 & 1 \end{bmatrix}$

    $\begin{bmatrix} 3 & -2 \\ -1 & 1 \end{bmatrix}\begin{bmatrix} 1 & 2 \\ 1 & 3 \end{bmatrix} = \begin{bmatrix} 3-2 & 6-6 \\ -1+1 & -2+3 \end{bmatrix}$

    $= \begin{bmatrix} 1 & 0 \\ 0 & 1 \end{bmatrix}$

  (b) $\begin{bmatrix} 4 & -6 \\ 2 & 2 \end{bmatrix}\begin{bmatrix} \frac{1}{10} & \frac{3}{10} \\ -\frac{1}{10} & \frac{2}{10} \end{bmatrix} = \begin{bmatrix} 1 & 0 \\ 0 & 1 \end{bmatrix}$

    $\begin{bmatrix} \frac{1}{10} & \frac{3}{10} \\ -\frac{1}{10} & \frac{2}{10} \end{bmatrix}\begin{bmatrix} 4 & -6 \\ 2 & 2 \end{bmatrix} = \begin{bmatrix} 1 & 0 \\ 0 & 1 \end{bmatrix}$

  (c) $\begin{bmatrix} 5 & 7 \\ 3 & 4 \end{bmatrix}\begin{bmatrix} -4 & 7 \\ 3 & -5 \end{bmatrix} = \begin{bmatrix} -20+21 & 35-35 \\ -12+12 & 21-20 \end{bmatrix}$

    $= \begin{bmatrix} 1 & 0 \\ 0 & 1 \end{bmatrix}$

    $\begin{bmatrix} -4 & 7 \\ 3 & -5 \end{bmatrix}\begin{bmatrix} 5 & 7 \\ 3 & 4 \end{bmatrix} = \begin{bmatrix} -20+21 & -28+28 \\ 15-15 & 21-20 \end{bmatrix}$

    $= \begin{bmatrix} 1 & 0 \\ 0 & 1 \end{bmatrix}$

  (d) $\begin{bmatrix} 2 & 0 & 2 \\ 4 & 2 & 0 \\ 2 & -2 & 0 \end{bmatrix}\begin{bmatrix} -\frac{1}{4} & \frac{1}{4} & \frac{1}{4} \\ \frac{1}{2} & 0 & -\frac{1}{2} \\ \frac{3}{4} & -\frac{1}{4} & -\frac{1}{4} \end{bmatrix} = \begin{bmatrix} 1 & 0 & 0 \\ 0 & 1 & 0 \\ 0 & 0 & 1 \end{bmatrix}$

    $\begin{bmatrix} -\frac{1}{4} & \frac{1}{4} & \frac{1}{4} \\ \frac{1}{2} & 0 & -\frac{1}{2} \\ \frac{3}{4} & -\frac{1}{4} & -\frac{1}{4} \end{bmatrix}\begin{bmatrix} 2 & 0 & 2 \\ 4 & 2 & 0 \\ 2 & -2 & 0 \end{bmatrix} = \begin{bmatrix} 1 & 0 & 0 \\ 0 & 1 & 0 \\ 0 & 0 & 1 \end{bmatrix}$

3. $2x - y = 4$
   $3x + 4y = 1$

5. $x + 3y + 2z = 5$
    $-x + 2y + z = 1$
        $y + 2z = 4$

7. $\begin{bmatrix} 5 & 3 \\ 4 & -1 \end{bmatrix}\begin{bmatrix} x \\ y \end{bmatrix} = \begin{bmatrix} 7 \\ 3 \end{bmatrix}$

9. $\begin{bmatrix} 3 & 4 & 5 \\ 2 & -1 & 0 \\ 7 & 0 & 2 \end{bmatrix}\begin{bmatrix} x \\ y \\ z \end{bmatrix} = \begin{bmatrix} 9 \\ 4 \\ 5 \end{bmatrix}$

11. (a) $A^{-1} = \begin{bmatrix} 3 & -1 \\ -2 & 1 \end{bmatrix}$
    $x = 4, y = -2$

  (b) $A^{-1} = \begin{bmatrix} -\frac{1}{2} & \frac{1}{2} \\ \frac{3}{2} & -\frac{1}{2} \end{bmatrix}$
    $x = 3, y = -2$

  (c) $A^{-1} = \begin{bmatrix} \frac{1}{2} & \frac{1}{2} \\ -\frac{1}{2} & \frac{1}{2} \end{bmatrix}$
    $x = 3, y = 2$

  (d) $A^{-1} = \begin{bmatrix} \frac{3}{5} & -\frac{1}{5} \\ -\frac{1}{5} & \frac{2}{5} \end{bmatrix}$
    $x = 1, y = 1$

13. (a) $A^{-1} = \begin{bmatrix} -\frac{1}{2} & \frac{7}{6} & -\frac{5}{6} \\ 0 & -\frac{1}{3} & \frac{2}{3} \\ \frac{1}{2} & -\frac{1}{2} & \frac{1}{2} \end{bmatrix}$
    $x = 3, y = -2, z = 1$

  (b) $A^{-1} = \begin{bmatrix} -1 & 0 & 1 \\ \frac{1}{2} & -\frac{1}{2} & 0 \\ \frac{3}{2} & \frac{1}{2} & -1 \end{bmatrix}$
    $x = 4, y = -1, z = 2$

  (c) $A^{-1} = \begin{bmatrix} -7 & -8 & 5 \\ 4 & 5 & -3 \\ -1 & -1 & 1 \end{bmatrix}$
    $x = 1, y = -1, z = 2$

  (d) $A^{-1} = \begin{bmatrix} \frac{1}{2} & 0 & -\frac{1}{2} \\ \frac{1}{2} & -\frac{1}{2} & 0 \\ 0 & \frac{1}{2} & \frac{1}{2} \end{bmatrix}$
    $x = 3, y = -2, z = 1$

15. $x_1 = 0.0435, x_2 = -2.652, x_3 = -3.217, x_4 = -1.739$

17. (a)

19. Type A: \$32,000   21. 5 of type I
    Type B: \$68,000   10 of type II

23. 5 of I and 4 of II

25. (a) No solution   (b) (0, 1)
    (c) $(3, c)$; an infinite number of solutions
    (d) $(2, 3, c)$; an infinite number of solutions
    (e) No solution   (f) (1, 0, 0)

Exercise Set 2.7, page 195

1. (a) $x_1 = 1.75x_3,\ x_2 = 1.625x_3$ or $x_1 = 14,000,\ x_2 = 13,000, x_3 = 8,000$

  (b) $x_1 = \frac{43}{41}x_3,\ x_2 = \frac{24}{41}x_3$ or $x_1 = 43,\ x_2 = 24, x_3 = 41$

3. (a) 0.2 unit   (b) 0.2 unit
  (c) 30 units or \$30,000,000   (d) B
  (e) A is least dependent.

5. $X = \begin{bmatrix} 20 \\ 28 \end{bmatrix}$ (rounded)   7. $X = \begin{bmatrix} 80 \\ 55 \end{bmatrix}$

9. $X = \begin{bmatrix} 32 \\ 29 \\ 18 \end{bmatrix}$ (rounded)   11. $\begin{bmatrix} 3 \\ 0.2 \\ 0.2 \end{bmatrix}$

13. $A = \begin{bmatrix} 0.3 & 0.5 & 0.1 \\ 0.2 & 0.2 & 0.3 \\ 0.3 & 0.1 & 0.2 \end{bmatrix}$

    $D = \begin{bmatrix} 3.0 \\ 2.1 \\ 0.3 \end{bmatrix}$

15. 71.2 units of service
    43.2 units of manufacturing

17. 93.82 units of electricity
     217.96 units of oil
      86.69 units of coal

19. (a) $x = 1, y = 1, z = 1$ (b) $x = 2, y = 3, z = 1$

21. $\begin{bmatrix} 5 \\ 10 \end{bmatrix}$

Chapter 2 Test, page 198

1.

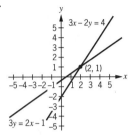

2. $x = 2, y = 1$

3. $a = -2, b = -2, c = 2, d = 0, e = 8, f = 14$

4. (b)

5. $\begin{bmatrix} 0 & 1 \\ 6 & -3 \end{bmatrix}$

6. $\begin{bmatrix} 1 & 2 \\ 1 & 3 \end{bmatrix}$

7. $x = 9, y = 7$

8. $\begin{bmatrix} 3 & -2 & | & 4 \\ -2 & 3 & | & -1 \end{bmatrix} \rightarrow \begin{bmatrix} 1 & -0.67 & | & 1.33 \\ -2 & 3 & | & -1 \end{bmatrix}$

$\rightarrow \begin{bmatrix} 1 & -0.67 & | & 1.33 \\ 0 & 1.67 & | & 1.67 \end{bmatrix} \rightarrow \begin{bmatrix} 1 & 0 & | & 2 \\ 0 & 1 & | & 1 \end{bmatrix}$

$x = 2, y = 1$

9. $\mathbf{A}^{-1} = \begin{bmatrix} \frac{9}{15} & \frac{2}{5} \\ \frac{2}{5} & \frac{3}{5} \end{bmatrix}$

$x = 2, y = 1$

10. $x = 1, y = 1, z = 1$

11. $\mathbf{A}^{-1} = \begin{bmatrix} -1 & -3 \\ -1 & -2 \end{bmatrix}$

$x = 9, y = 7$

12. $(-0.6, 3.1, -1.7)$

13. $1000 at 6%, $3000 at 8%, $6000 at 10%

14. $\begin{bmatrix} 3 \\ 2.1 \\ 0.3 \end{bmatrix}$ 15. $(2, 3)$

Exercise Set. 3.1, page 207

1.

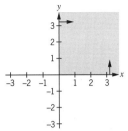

Unbounded; corner point: $(0, 0)$

3.

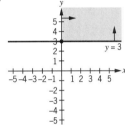

Unbounded; corner point: $(0, 3)$

5.

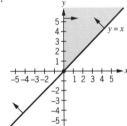

Unbounded; corner point: $(0, 0)$

7.

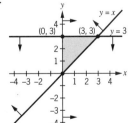

Bounded; corner points: $(0, 0), (0, 3), (3, 3)$

9.

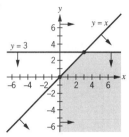

Unbounded; corner points: (0, 0), (3, 3)

11.

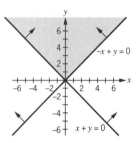

Unbounded; corner point: (0, 0)

13.

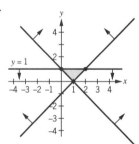

Bounded; corner points: (1, 0), (2, 1), (0, 1)

15.

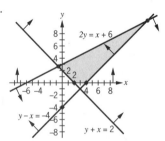

Bounded; corner points: $\left(\dfrac{-2}{3}, \dfrac{8}{3}\right)$, (14, 10), (2, 0), (4, 0)

17.

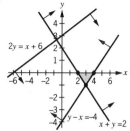

Bounded; corner points: (2, 0), (4, 0), (3, −1)

19.

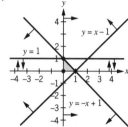

Bounded; corner points: (0, 0), (0, 1), (1, 0)

21. $x \geq 0$; $y \geq 0$; $x + 2y \leq 8$; $2x + 2y \leq 8$

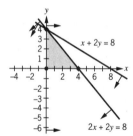

Corner points: (0, 0), (4, 0), (0, 4)

23. $x + y \leq \$30,000$; $x \geq 2y$; $x \geq 0$; $y \geq 0$

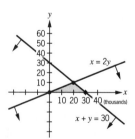

Corner points: (0, 0), ($30,000, 0), ($20,000, $10,000)

25.

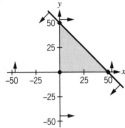

Corner points: (0, 0), (50, 0), (0, 50)

27. $x \geq 0$; $y \geq 0$, $2x + 3y \leq 150$; $40x + 30y \leq 1800$
    $x + y \leq 50$

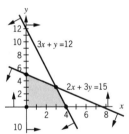

Corner points: (0, 0), (45, 0), (30, 20), (0, 50)

Exercise Set 3.2, page 214

1.

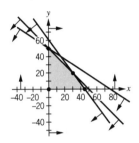

Corner points: (0, 0), (0, 5), (4, 0), (3, 3). Maximum value of P is 180 at (3, 3).

3.

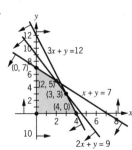

Corner points: (0, 0), (0, 7),   (2, 5),   (3, 3), (4, 0). Maximum value of F is 245 at (0, 7).

5.

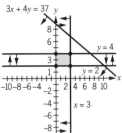

Maximum value of F is 52 at (3, 2).
Minimum value of F is 12 at (0, 4).

7.

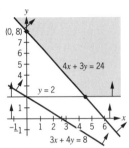

Minimum value of 20 at (0, 8)
No maximum value: unbounded

9.

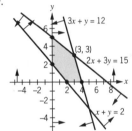

Maximum value of P is 180 at (3, 3).
Minimum value of P is 50 at (2, 0).

11.

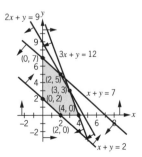

Maximum value of P is 245 at (0, 7).
Minimum value of P is 50 at (2, 0).

13.

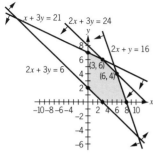

$x + 3y = 21$ $2x + 3y = 24$
$2x + y = 16$
$2x + 3y = 6$ (3, 6)
(6, 4)

Maximum value of P is 400 at (8, 0).
Minimum value of P is 40 at (0, 2).

15.

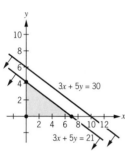

$3x + 5y = 30$
$3x + 5y = 21$

Corner points: $(0, 0)$, $(7, 0)$, $\left(0, \dfrac{21}{5}\right)$; bounded

Exercise Set 3.3, page 220

1. (a)

Type I	Type II	
2	3	
$2 \cdot 14$	$+ \ 3 \cdot 13 = 67$	
$14x$	$+ \ 13y = P$	
4	2	
$4x$	$+ \ 2y \le 12$	
3	3	
$3x$	$+ \ 3y \le 12$	

(b) $4x + 2y \le 12$
$3x + 3y \le 12$
$P = 14x + 13y$
$x \ge 0, y \ge 0$

(c)

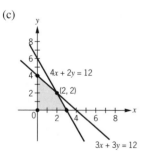

$4x + 2y = 12$
(2, 2)
$3x + 3y = 12$

Corner points: $(0, 0)$, $(0, 4)$, $(3, 0)$, $(2, 2)$

(d) Maximum profit of $54 occurs at $(2, 2)$.

3. (a)

Refinery I	Refinery II
4	5
$4 \cdot 3500$	$+ \ 5 \cdot 2000 = 24{,}000$
x	y
$3500x$	$+ \ 2000y = C$
$200x$	$+ \ 100y \ge 800$
$300x$	$+ \ 200y \ge 1400$
$100x$	$+ \ 100y \ge 500$

(b) $200x + 100y \ge 800$
$300x + 200y \ge 1400$
$100x + 100y \ge 500$
$x \ge 0, y \ge 0$
$C = 3500x + 2000y$

(c)

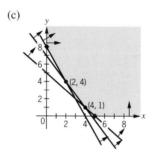

(2, 4)
(4, 1)

Corner points: $(5, 0)$, $(4, 1)$, $(2, 4)$, $(0, 8)$

(d) I operates 2 days and II operates 4 days for a minimum cost of $15,000.

5. Ship 25 tons from warehouse A to store I, 75 tons from warehouse A to store II, and 25 tons from warehouse B to store I for a minimum cost of $875.

7. $37,500 in AA bonds
12,500 in B bonds
Maximum income: $5625

9. 100 small saws
200 medium saws
Maximum revenue: $11,000

11. 400 boxes of lettuce and 600 boxes of celery for a maximum profit of $2300.

13. 35 instructors
6 graduate assistants
Minimum cost: $6608

15. 3 units of food I
1 unit of food II
Minimum: 130 calories

17. 0 bags of peanut hulls
4 bags of soybean meal
Minimum cost: $48

19. (c)   21. (b)

23. Minimum is 20. No maximum

Exercise Set 3.4, page 233

1.
$$3x + 2y + r = 8$$
$$2x + 4y + s = 8$$
$$-14x - 12y + P = 0$$
$$x = 0 \qquad r = 8$$
$$y = 0 \qquad s = 8$$
$$ P = 0$$

3.
$$2x + 4r + s = 7$$
$$3x + y + 2r = 5$$
$$4x - 3r + P = 12$$
$$x = 0 \qquad y = 5$$
$$r = 0 \qquad s = 7$$
$$ P = 12$$

5.

x	y	r	s	P	
2	3	1	0	0	15
3	1	0	1	0	12
-25	-35	0	0	1	0

7.

x	y	r	s	P	
2	3	1	0	0	15
3	1	0	1	0	12
-35	-25	0	0	1	0

9.

x	y	r	s	t	P	
1	1	1	0	0	0	7
2	1	0	1	0	0	9
3	1	0	0	1	0	12
-35	-25	0	0	0	1	0

11.

x	y	r	s	t	P	
1	1	1	0	0	0	7
2	1	0	1	0	0	9
3	1	0	0	1	0	12
-35	-25	0	0	1		0

13.

x	y	r	s	t	u	P	
2	3	1	0	0	0	0	6
1	3	0	1	0	0	0	21
2	3	0	0	1	0	0	24
2	1	0	0	0	1	0	16
-25	-35	0	0	0	0	1	0

15.
$$\tfrac{1}{2}R_1 \to R_1$$
$$-4R_1 + R_2 \to R_2$$
$$12R_1 + R_3 \to R_3$$

x	y	r	s	p	
$\frac{3}{2}$	1	$\frac{1}{2}$	0	0	4
-4	0	-2	1	0	-8
4	0	6	0	1	48

17.
$$\tfrac{1}{2}R_2 \to R_2$$
$$-4R_2 + R_1 \to R_1$$
$$3R_2 + R_3 \to R_3$$

x	y	r	s	p	
-4	-2	0	1	0	-3
$\frac{3}{2}$	$\frac{1}{2}$	1	0	0	$\frac{5}{2}$
$\frac{17}{2}$	$\frac{3}{2}$	0	0	1	$\frac{39}{2}$

19.

x	y	r	s	P	
0	$-\frac{7}{4}$	1	$-\frac{5}{4}$	0	$\frac{5}{2}$
1	$\frac{3}{4}$	0	$\frac{1}{4}$	0	$\frac{3}{2}$
0	$\frac{10}{4}$	0	$\frac{10}{4}$	1	15

$$\frac{-7}{4}y + r - \frac{5}{4}s = \frac{5}{2}$$
$$x + \frac{3}{4}y + \frac{1}{4}s = \frac{3}{2}$$
$$P = 15 - \frac{10}{4}y - \frac{10}{4}s$$

21.

x	y	r	s	P	
1	0	$\frac{1}{2}$	$-\frac{1}{4}$	0	2
0	1	$-\frac{1}{4}$	$\frac{3}{8}$	0	1
0	0	4	1	1	40

$$x + \frac{1}{2}r - \frac{1}{4}s = 2$$
$$y - \frac{1}{4}r + \frac{3}{8}s = 1$$
$$P = 40 - 4r - s$$

23.

x	y	r	s	t	P	
$\frac{5}{2}$	0	1	0	$-\frac{1}{2}$	0	$\frac{1}{2}$
$\frac{5}{4}$	0	0	1	$-\frac{3}{4}$	0	$\frac{1}{4}$
$\frac{1}{4}$	1	0	0	$\frac{1}{4}$	0	$\frac{5}{4}$
$-\frac{5}{2}$	0	0	0	$\frac{3}{2}$	1	$\frac{15}{2}$

$$\frac{5}{2}x + r - \frac{1}{2}t = \frac{1}{2}$$
$$\frac{5}{4}x + s - \frac{3}{4}t = \frac{1}{4}$$
$$\frac{1}{4}x + y + \frac{1}{4}t = \frac{5}{4}$$
$$P = \frac{5}{2}x - \frac{3}{2}t + \frac{15}{2}$$

25.

4	10	1	0	0	0	40
1	0	0	1	0	0	8
0	1	0	0	1	0	2
-20	-60	0	0	0	1	0

27.

1	2	1	0	0	48
2	3	0	1	0	84
-10	-25	0	0	1	0

29.

C	H	S			P	
30	45	15	1	0	0	5000
$\frac{1}{3}$	1	$\frac{1}{5}$	0	1	0	6000
-20	-15	-10	0	0	1	0

Exercise Set 3.5, page 243

1. $P = 9$ at $x = 3, y = 0$ or at $x = 1, y = 3$ or at any point on the line segment joining these two points

3. $P = 17$ at $x = 1, y = 3$ 5. $P = 175$ at $(0, 7)$

7. $P = 1920$ at $x = 120, y = 0, z = 200$

9. $P = 49$ at $(12, 0, 5)$

11. $P = 132.82$ at $x = 0, y = 56.52$

13. $x = 0, y = 12, z = 17, w = 3, P = 2460$

15. 16 of type I, 8 of type II for maximum profit of $1616.00

17. 18 of A and 10 of B to maximize profit at $1048

19. $18,000 invested in good quality; $9000 invested in high risk; maximum income: $2160

21. 5 newspaper and 2 television advertisements for a maximum audience of 220,000

23. 90 of type I, 0 of type II, 40 of type III for a profit of $9200

25. 500 liters by old process; 800 liters by new process; maximum income: $344

27. No horses, no cows, 333 sheep for a maximum profit of $3330

Exercise Set 3.6, page 252

1. Minimize $C = 15y_1 + 12y_2$ subject to
$$2y_1 + 3y_2 \geq 25$$
$$3y_1 + \quad y_2 \geq 35$$
$$y_1 \geq 0$$
$$y_2 \geq 0$$

3. Maximize $P = 5x_1 + 21x_2$ subject to
$$x_1 + 3x_2 \leq 4$$
$$x_1 + \quad x_2 \leq 7$$
$$x_1 \geq 0$$
$$x_2 \geq 0$$

5. Minimize $C = 4y_1 + 9y_2$
$$y_1 + 3y_2 \geq 3$$
$$y_1 + 2y_2 \geq 2$$
$$y_1, y_2 \geq 0$$

7. Minimize $C = 4y_1 + 9y_2$
$$y_1 + 3y_2 \geq 5$$
$$y_1 + 2y_2 \geq 4$$
$$y_1, y_2 \geq 0$$

9. Maximize $P = 7x_1 + 4x_2$ subject to
$$3x_1 + 2x_2 \leq 5$$
$$x_1 + 3x_2 \leq 8$$
$$x_1 \geq 0$$
$$x_2 \geq 0$$
Minimize $C = 5y_1 + 8y_2$ subject to
$$3y_1 + \quad y_2 \geq 7$$
$$2y_1 + 3y_2 \geq 4$$
$$y_1 \geq 0$$
$$y_2 \geq 0$$

11. Maximize $P = 6x_1 + 8x_2$ subject to
$$2x_1 + 3x_2 \leq 4$$
$$x_1 + 2x_2 \leq 7$$
$$3x_1 + \quad x_2 \leq 6$$
$$x_1 \geq 0$$
$$x_2 \geq 0$$
Minimize $C = 4y_1 + 7y_2 + 6y_3$ subject to
$$2y_1 + \quad y_2 + 3y_3 \geq 6$$
$$3y_1 + 2y_2 + \quad y_3 \geq 8$$
$$y_1 \geq 0$$
$$y_2 \geq 0$$
$$y_3 \geq 0$$

13. $C = 23$ at $y_1 = 2, y_2 = 3$

15. $C = 28$ at $y_1 = 7, y_2 = 0$

17. $C = 180$ at $y_1 = 3, y_2 = 3$

19. $C = 260.56$ at $y_1 = 7.78, y_2 = 1.89$

21. $10,000 at $y_1 = 0, y_2 = 9.879, y_3 = .121$

23. None from A to I; 40 tons from A to II; 50 tons from B to I; 35 tons from B to II. Minimum cost $970.

25. Produce 30 gallons of product 1 and 25 gallons of product 2 at a minimum cost of $55,000.

27. He should eat no units of food I and 5 units of food II for a total of 20 grams of carbohydrates.

29. The minimum is 21 grams at 3 units of I and 1 unit of II.

31. Three salmon patties and 2 slices of bread satisfy requirements with a minimum of 450 calories.

33. Maximum value of 21 at $(7, 0)$

Exercise Set 3.7, page 262

1.

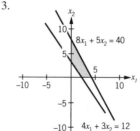

Corner points: (5, 0), (3, 0), (0, 8), (0, 4); P is a maximum of 80 at (0, 8).

3.

Corner points: (3, 0), (5, 0), (0, 4), (0, 8); P is a minimum of 12 at (3, 0).

5.

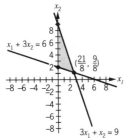

Corner points: $(0, 2)$, $(0, 9)$, $(\frac{21}{8}, \frac{9}{8})$; $P = 16$ at $(0, 2)$ is the minimum.

7. $\begin{bmatrix} 1 & 2 & 1 & 0 & 0 & 0 & | & 8 \\ 1 & 1 & 0 & -1 & 1 & 0 & | & 2 \\ \hline -8-k & -4-k & 0 & k & 0 & 1 & | & -2k \end{bmatrix}$

9. $\begin{bmatrix} -1 & 2 & -1 & 0 & 1 & 0 & | & 4 \\ 1 & 1 & 0 & 1 & 0 & 0 & | & 6 \\ \hline -5+k & -7-2k & k & 0 & 0 & 1 & | & -4k \end{bmatrix}$

11. $\begin{bmatrix} 1 & -1 & 1 & 0 & 0 & 0 & | & 2 \\ 1 & 1 & 0 & -1 & 1 & 0 & | & 4 \\ \hline -4-k & -6-k & 0 & k & 0 & 1 & | & -4k \end{bmatrix}$

13. $\begin{bmatrix} -1 & 3 & 1 & 0 & 0 & | & 2 \\ -1 & 2 & 0 & 1 & 0 & | & 3 \\ \hline 8 & -6 & 0 & 0 & 1 & | & 0 \end{bmatrix}$

15. $P = 80$ at $(0, 8)$

17. $C = 12$ at $(3, 0)$

19. $C = 16$ at $(0, 2)$

21. \$28,000 at $(7, 0)$

23. Maximize with 400 lb of A and 1200 lb of B, for a maximum of 260.

25. 8 kg algae and 12 kg of eggshells

27. $P = 20$ at $(0, 5)$

Chapter 3 Test, page 265

1.

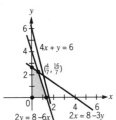

Corner points: $(0, 0)$, $\left(\frac{4}{7}, \frac{16}{7}\right)$, $\left(0, \frac{8}{3}\right)$, $\left(\frac{4}{3}, 0\right)$; bounded

2. Maximum value $= \frac{160}{3}$ at $\left(0, \frac{8}{3}\right)$; minimum value 0 at $(0, 0)$

3. $\begin{bmatrix} x & y & r & s & t & P & \\ 2 & 3 & 1 & 0 & 0 & 0 & | & 8 \\ 4 & 1 & 0 & 1 & 0 & 0 & | & 6 \\ 6 & 2 & 0 & 0 & 1 & 0 & | & 8 \\ \hline -10 & -20 & 0 & 0 & 0 & 1 & | & 0 \end{bmatrix}$

4. First row, second column; 3

5. Basic variables: t, x, z, P
$$x = 21$$
$$z = 28$$
$$t = 12$$
$$P = 50$$

6. $\begin{bmatrix} x & y & z & r & s & t & P & \\ 0 & 0 & 0 & 1 & \frac{1}{2} & \frac{1}{2} & 0 & | & 6 \\ 1 & 0 & 0 & 0 & -\frac{3}{2} & -\frac{3}{2} & 0 & | & 3 \\ 0 & 6 & 1 & 0 & 0 & -2 & 0 & | & 4 \\ \hline 0 & -1 & 0 & 0 & \frac{11}{2} & \frac{5}{2} & 1 & | & 80 \end{bmatrix}$

No, there is still a negative number in the last row.

7. Yes. All values in the last row are positive or zero.
$$54 \text{ at } x = 18, y = 6$$

8. $\begin{bmatrix} x_1 & x_2 & x_3 & y_1 & y_2 & P & \\ 3 & 1 & 1 & 1 & 0 & 0 & | & 5 \\ 2 & 4 & 1 & 0 & 1 & 0 & | & 2 \\ \hline -8 & -6 & -4 & 0 & 0 & 1 & | & 0 \end{bmatrix}$

9. Minimum value of 16 at $y_1 = 4$, $y_2 = 2$

10. 5 newspaper and 20 radio advertisements to reach 850,000 people

Exercise Set 4.1, page 278

1. (a) \$80   (b) \$36   (c) \$100

3. $I = \$360$, $A = \$3360$

5. (a) $I = \$720.98$, $A = \$2720.98$
   (b) $I = \$737.14$, $A = \$2737.14$
   (c) $I = \$745.57$, $A = \$2745.57$

7. (a) $A = \$4480$ (b) $A = \$3075$ (c) $A = \$124$

9. \$3736.29   11. \$5525.86

13. $r = \dfrac{I}{Pt}$

15. Between 18.25 and 18.5 years; ≈ 18.45 years

17. Accumulation of \$1

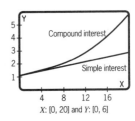

19. $133.33

21. $I = 21.64

23. 23 days

25. $8074.76

27. $P = 1532.78

29. (a) $179,084.77   (b) $3.40
    (c) $8.95   (d) $26.24

31. (a) $215,892.50   (b) $4.10   (c) $10.79
    (d) $31.63

33. 88,814

Exercise Set 4.2, page 287

1. Geometric progression with $r = 3$

3. Not a geometric progression

5. Not a geometric progression

7. Not a geometric progression

9. 8.16%   11. 8.3%

13. (a) $1790.85   (b) $1806.11   (c) $1814.02
    (d) $1819.40   (e) $1822.03   (f) $1822.12

15. 10.52%   17. $A = $14,917.92$

19. 8.33%

21.

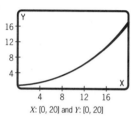

X: [0, 20] and Y: [0, 20]

The three graphs are so close together that they look like one graph.

23. $I = 75

Exercise Set 4.3, page 294

1. $1318.08. Amount of an annuity of $100 a year at 6% compounded annually for 10 years. (one possibility)

3. $4842.98. Amount of an annuity of $200 a year at 8% compounded annually for 14 years. (one possibility)

5. $1318.08   7. $2687.04

9. $A = $108,195.66$

11. (a) $45,761.96   (b) $15,445.10   (c) $7203.66

13. $5866.60

15. (a)

Year	Deposit	Interest	Amount
1	$2000	$ 0	$ 2,000
2	2000	160	4,160
3	2000	332.80	6,492.80
4	2000	519.42	9,012.22
5	2000	720.98	11,733.20

At the end of 20 years he will have $91,523.93.

(b)

Year	Deposit	Interest	Amount
1	$2000	$ 0	$ 2,000
2	2000	200	4,200
3	2000	420	6,620
4	2000	662	9,282
5	2000	928.20	12,210.20

At the end of 20 years he will have $114,550.

17.

Period	Interest	Deposit	Amount
1	$ 0	$113.80	$113.80
2	$1.71	113.80	229.31
3	3.44	113.80	346.55
⋮	⋮	⋮	⋮

19. $12,283.97; $2093.88

21. $I = 90

23. 4.04%

Exercise Set 4.4, page 300

1. $368.00. Present value of an annuity of $50 a year at an interest rate of 6% compounded annually for 10 years. (one possibility)

3. $1648.85. Present value of an annual annuity of $200 at an interest rate of 8% compounded annually for 14 years. (one possibility)

5. $736.01

7. (a) $9818.15   (b) $9909.41   (c) $4866.54

9. (d)

11. $273.02

13.

Quarter	Outstanding Principal	Interest Due	Payment	Principal Repaid
1	$2000.00	$40.00	$273.02	$233.02
2	1766.98	35.34	273.02	237.68
3	1529.30	30.59	273.02	242.43
4	1286.87	25.74	273.02	247.28
5	1039.59	20.79	273.02	252.23
6	787.36	15.75	273.02	257.27
7	530.09	10.60	273.02	262.42
8	267.67	5.35	273.02	267.67

15. (a) $19,235.36   (b) $30,344.37

17. $2,015,278.09

19. $989.75

Exercise Set 4.5, page 306

1. $125,000
3. $121,311.88
5. $16,000
7. $16,486.43
9. $150
11. $1060.36
13. $3460.02
15. $5752.81
17. $372.32
19. $99.01
21. $P = $2684.03
   $A = $5794.62
23. $55.24

25.

Period	Outstanding Principal	Interest Due	Payment	Principal Repaid Each Period
1	$10,000.00	$400	$1907.62	$1507.62
2	8492.38	339.70	1907.62	1567.92
3	6924.46	276.98	1907.62	1630.64
4	5293.82	211.75	1907.62	1695.87
5	3597.95	143.92	1907.62	1763.70
6	1834.25	73.37	1907.62	1834.25

27. $255,468,811,600 (rounded)
29. $12,576.99
31. $4136.65
33. $106,366.28

Chapter 4 Test, page 309

1. $790.31
2. $19,668.05
3. 40%
4. $2539.47
5. $811.09
6. $2000
7. 12.68%
8. $1110.21
9. $25,000

10.

Month	Outstanding Principal	Interest Due	Payment	Principal Repaid Each Period
1	100,000	1000	1200	200
2	99,800	998	1200	202
3	99,598	995.98	1200	204.02
4	99,393.98			

11. $1402.37
12. $2039.45
13. $2000

Exercise Set 5.1, page 320

1. (a) $S = \{A, B, C\}$ (b) $S = \{R, G\}$
   (c) $S = \{1, 2, 3, 4, 5, 6\}$ (d) $S = \{10, 11, 12, 13\}$
3. $S = \{1, 2, 3, 4, 5, 6\}$, $P(1) = \frac{1}{6}$, $P(2) = \frac{1}{6}$, $P(3) = \frac{1}{6}$,
   $P(4) = \frac{1}{6}, P(5) = \frac{1}{6}, P(6) = \frac{1}{6}$

5. 1
7. $\frac{1}{2}$
9. $\frac{4}{6} = \frac{2}{3}$
11. $\frac{1}{6}$
13. $\frac{1}{5}, \frac{4}{5}$
15. (a) No, negative   (b) No, greater than 1
   (c) Yes   (d) Yes   (e) No, greater than 1
   (f) Yes   (g) No, greater than 1
17. (a) Does not have a 0   (b) Does not have a 3
   (c) 4 is not an outcome   (d) Is a sample space
19. (a) $\frac{4}{14}$ (b) $\frac{3}{14}$ (c) $\frac{7}{14}$ (d) $\frac{10}{14}$
   (e) $\frac{11}{14}$ (f) 1
21. (a) $\frac{1}{4}$ (b) $\frac{1}{2}$ (c) $\frac{3}{4}$
23. (a) $\frac{7}{15}$ (b) $\frac{8}{15}$ (c) $\frac{6}{15}$ (d) 1   (e) 0
25. (a) $S = \{m, a, t, h, e, i, c, s\}$ (b) No   (c) $\frac{1}{11}$
   (d) $\frac{1}{11}$ (e) $\frac{2}{11}$ (f) $\frac{3}{11}$ (g) $\frac{4}{11}$ (h) $\frac{7}{11}$
27. $\frac{1}{2}$
29. $S = \{ABC, ABD, ABE, ACD, ACE, ADE, BCD, BCE, BDE, CDE\}$

Exercise Set 5.2, page 329

1. (a) $6 \cdot 4 = 24$ (b) $6 \cdot 4 \cdot 4 \cdot 6 = 576$
3. 72
5. $6 \cdot 3 = 18$

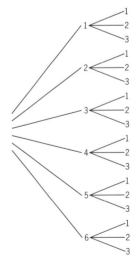

7. (a) $\dfrac{1}{36}$ (b) $\dfrac{1}{36}$ (c) $\dfrac{2}{36}$ (d) $6 \cdot \dfrac{1}{36} = \dfrac{1}{6}$

9. (a) 6 (b) {A, B, C, D, E, F} (c) $\dfrac{11}{60}$ (d) $\dfrac{1}{6}$

(e) 11/16 is assigned from empirical data and 1/6 is assigned as theoretical probability.
(f) Yes, with more repetitions, the result will come closer to theoretical or long-term relative frequency.

11. (a) $\dfrac{24}{100}$ (b) $\dfrac{28}{100}$ 13. $9 \cdot 10^8$

15. $24,360, \dfrac{1}{24,360}$

17. (a) 26,000,000 (b) 3,931,200 (c) 13,000,000
(d) 6,500,000

19. (d) 0.11

21. (a) $\dfrac{180}{220}$ (b) $\dfrac{40}{220}$ (c) $\dfrac{40}{1040}$ (d) $\dfrac{800}{980}$

23. $3 \cdot 2 = 6$

25. (a) $\dfrac{1}{16}$ (b) $\dfrac{2}{16} = \dfrac{1}{8}$ (c) $\dfrac{5}{16}$ (d) $\dfrac{15}{16}$

27. $\dfrac{74,000}{80,000} = 0.925$

29. (a) $\dfrac{4}{10}$ (b) $\dfrac{5}{10}$ (c) $\dfrac{3}{10}$ (d) $\dfrac{2}{10}$
(e) $\dfrac{9}{10}$ (f) $\dfrac{6}{10}$ (g) $\dfrac{7}{10}$ (h) $\dfrac{7}{10}$

Exercise Set 5.3, page 343

1. (a) $P(4, 2) = 12$; {WX, XW, WY, YW, WZ, ZW, XY, YX, XZ, ZX, YZ, ZY}
(b) $C(4, 2) = 6$; {WX, WY, WZ, XY, XZ, YZ}

3. (a) 210 (b) 1 (c) 15 (d) 6
(e) $\dfrac{r!}{2!(r-2)!} = \dfrac{r(r-1)}{2}$ (f) r

5. $5! = 120$ 7. $7! = 5040$

9. (a) $C(52, 7) = 133,784,560$
(b) $C(13, 5) \cdot C(39, 2) = 953,667$
(c) $C(4, 1) \cdot C(13, 7) = 6864$
(d) $C(4, 0) \cdot C(48, 7) + C(4, 1) \cdot C(48, 6) +$
$C(4, 2) \cdot C(48, 5) = 132,988,944$
(e) $C(4, 4) \cdot C(4, 3) = 4$

11. (a) r (b) $k(k-1)$ (c) n (d) 1
(e) $r!$ (f) 1

13. (a) $10^4 = 10,000$ (b) $P(10, 4) = 5040$
(c) $9 \cdot 10^3 = 9000$ (d) $10^3 \cdot 5 = 5000$
(e) $\dfrac{P(10, 4)}{10^4} = \dfrac{5040}{10,000} = \dfrac{63}{125}$

15. $\dfrac{C(9, 3)}{C(20, 3)} = \dfrac{7}{95}$

17. (a) $\dfrac{C(13, 7)}{C(52, 7)} \approx .00001283$
(b) $\dfrac{C(13, 5) \cdot C(13, 2)}{C(52, 7)} \approx .0007504$
(c) $\dfrac{C(13, 4) \cdot C(13, 1) \cdot C(13, 2)}{C(52, 7)} \approx .005419$
(d) $\dfrac{C(13, 3) \cdot C(13, 2) \cdot C(13, 2)}{C(52, 7)} \approx .013006$

19. (a) $P(30, 7) = 1.02(10)^{10}$
(b) $P(20, 4) \cdot P(10, 3) = 83,721,600$

21. $P(6, 6) = 6!$ 23. $\dfrac{7!}{3!\,2!\,1!\,1!} = 420$

25. (a) $P(5, 3) = 60$ (b) $4 \cdot 2 \cdot 3 = 24$
(c) $P(5, 3) - 24 = 36$

27. (a) $P(7, 4) = 840$ (b) $7^4 = 2401$

29. (a) $C(6, 1) = 6$ (b) $C(6, 3) = 20$
(c) $C(6, 0) + C(6, 1) = 7$ (d) $2^6 - 1 = 63$

31. (b)   33. (e)

35. $C(12, 3) \cdot C(9, 3) \cdot C(6, 3) \cdot C(3, 3)$

37. (a) $C(20, 3) = 1140$ (b) $C(4, 1) \cdot C(16, 2) = 480$
(c) $\dfrac{C(4, 1) \cdot C(16, 2)}{C(20, 3)} = \dfrac{8}{19}$ (d) $\dfrac{C(16, 3)}{C(20, 3)} = \dfrac{28}{57}$

39. (a) $C(20, 2) \cdot C(10, 1) = 1900$
(b) $C(18, 2) \cdot C(6, 1) = 918$
(c) $C(6, 1) \cdot C(18, 1) \cdot C(6, 1) = 648$
(d) Total: No men $= C(30, 3) - C(20, 3) = 2920$

41. (a) $5! = 120$ (b) $P(5, 3) = 60$

43. $C(8, 5) = 56$

45. (a) {(1, 1), (1, 2), (1, 3), (1, 4), (1, 5), (1, 6), (2, 1), (2, 2), (2, 3), (2, 4), (2, 5), (2, 6), (3, 1), (3, 2), (3, 3), (3, 4), (3, 5), (3, 6), (4, 1), (4, 2), (4, 3), (4, 4), (4, 5), (4, 6), (5, 1), (5, 2), (5, 3), (5, 4), (5, 5), (5, 6), (6, 1), (6, 2), (6, 3), (6, 4), (6, 5), (6, 6)}
(b) {(1, 2), (1, 3), (1, 4), (1, 5), (1, 6), (2, 1), (2, 3), (2, 4), (2, 5), (2, 6), (3, 1), (3, 2), (3, 4), (3, 5), (3, 6), (4, 1), (4, 2), (4, 3), (4, 5), (4, 6), (5, 1), (5, 2), (5, 3), (5, 4), (5, 6), (6, 1), (6, 2), (6, 3), (6, 4), (6, 5)}

Exercise Set 5.4, page 356

1. (a) .7 (b) .1 (c) .3

3. (a) 0 (b) .8 (c) .7 (d) .3 (e) 1

5. .35

7. (a) $\dfrac{1}{2}$ (b) $\dfrac{1}{2}$ (c) $\dfrac{2}{8} = \dfrac{1}{4}$ (d) $\dfrac{5}{8}$

9. (a) $\dfrac{10}{13}$ (b) 1 (c) $\dfrac{9}{13}$ (d) 1

11. (a) 4 to 1 (b) 1 to 4

13. $\dfrac{3}{5}$ 15. (c)

17. (a) .90 (b) .1 (c) .97

19. (a) $\dfrac{44}{120} = \dfrac{11}{30}$ (b) $\dfrac{7}{120}$

(c) $\dfrac{76}{120} = \dfrac{19}{30}$ (d) $\dfrac{23}{120}$

21. (a) $\dfrac{150}{200} = \dfrac{3}{4}$ (b) $\dfrac{120}{200} = \dfrac{3}{5}$

(c) $\dfrac{170}{200} = \dfrac{17}{20}$ (d) $\dfrac{140}{200} = \dfrac{7}{10}$

23. (a) $\dfrac{15}{50} = \dfrac{3}{10}$ (b) $\dfrac{11}{25}$

(c) $\dfrac{5}{50} = \dfrac{1}{10}$ (d) $\dfrac{43}{50}$

25. (a) $\dfrac{5}{36}$ (b) $\dfrac{1}{2}$ (c) $\dfrac{18}{36} = \dfrac{1}{2}$

(d) $\dfrac{18}{36} = \dfrac{1}{2}$ (e) $\dfrac{27}{36} = \dfrac{3}{4}$

Exercise Set 5.5, page 371

1. (a) Sample space consists of {H, D, C, S} or {R, B}.
   (b) Only cards in clubs and spades are listed.

   (c) 0 (d) $\dfrac{1}{2}$ (e) 1

3. (a) .42 (b) .7 (c) .4 (d) .78

5. (a) $\dfrac{.2}{.3} = \dfrac{2}{3}$ (b) $\dfrac{.2}{.7} = \dfrac{2}{7}$ (c) 0.8 (d) $\dfrac{.5}{.7} = \dfrac{5}{7}$

   (e) .2 (f) .2 (g) $\dfrac{.1}{.3} = \dfrac{1}{3}$ (h) .9

7. (a) $\dfrac{4}{12} = \dfrac{1}{3}$ (b) $\dfrac{4}{40} = \dfrac{1}{10}$

9. .63

11. (a) $\dfrac{3}{4}$ (b) $\dfrac{1}{2}$ (c) .35 (d) $\dfrac{1}{14}$

   (e) $\dfrac{.65}{.80} = \dfrac{13}{16}$

13. (a) $\dfrac{4}{77}$ (b) $\dfrac{2}{77}$ (c) $\dfrac{8}{77}$ (d) $\dfrac{16}{77}$

15. $\dfrac{1}{4}$ 17. $\dfrac{1}{256}$

19. (a) Doll 1 = 1 or 2, Doll 2 = 3 or 4, Doll 3 = 5 or 6
    (b) Answers will vary.

21. (a) $x = .6$ (b) $x = .3$

23. $\dfrac{1}{30}$ 25. (a) Yes (b) No

27. (a) $\dfrac{45}{203}$ (b) $\dfrac{50}{203}$ (c) $\dfrac{18}{203}$ (d) $\dfrac{1}{3}$ (e) $\dfrac{146}{203}$

29. (a)

31. (a) $\dfrac{1}{4}$ (b) $\dfrac{1}{10}$ (c) $\dfrac{3}{4}$ (d) $\dfrac{3}{5}$

   (e) $\dfrac{5}{7}$ (f) $\dfrac{1}{2}$ (g) $\dfrac{5}{13}$ (h) $\dfrac{1}{5}$

33. $\dfrac{1}{2}$ 35. (a) .75750 (b) .7225

37. $\dfrac{2}{105}$

39. (a) $\dfrac{1}{56}$ (b) $\dfrac{1}{11}$ (c) $\dfrac{10}{11}$ (d) $\dfrac{55}{56}$

41. (a) $\dfrac{1}{16}$ (b) $\dfrac{1}{16}$

43. (a) $\dfrac{36}{169}$ (b) $\dfrac{18}{169}$ (c) $\dfrac{16}{169}$

Chapter 5 Test, page 376

1. 1, 0 2. 6! = 720

3. (a) $\dfrac{16}{49}$ (b) $\dfrac{2}{7}$ 4. $C(7, 4) = 35$

5. $\dfrac{3}{4}$

6. $C(20, 4) \cdot C(5, 2) = 48{,}450$

7. $\dfrac{11}{36}$

8. (a) $\dfrac{1}{2}$ (b) 1 to 5 (c) 7 to 3

9. $P(12, 3) = 1320$ 10. $\dfrac{1}{4}$

11. $\dfrac{3}{5}$ 12. .06

13. Answers will vary.

14. (a) 56 (b) 52 (c) 48

15. (a) .02 (b) .34 (c) .06

16. $\dfrac{10}{75} = \dfrac{2}{15}$

17. (a) $C(52, 5) = 2{,}598{,}960$

    (b) $\dfrac{C(4, 3) \cdot C(4, 2)}{C(52, 5)} \approx .000009$

    (c) $4 \cdot \dfrac{C(13, 5)}{C(52, 5)} \approx .002$

18. $\dfrac{5}{7}$

19. $\dfrac{9!}{2!2!} = 90{,}720$

20. Not independent

21. (a) 12   (b) $\dfrac{1}{4}$ (c) $\dfrac{7}{20}$

Exercise Set 6.1, page 387

1. $P(A) = .4$

3. $P(B\,|\,A) = .3$

5. $P(A' \cap B) = .12$

7. $P(B) = .24$

9. .6

11. .4

13. .28

15. .63

17. $\dfrac{2}{3}$

19. $\dfrac{1}{3}$

21. (a) $\dfrac{1}{9}$ (b) .26   (c) $\dfrac{10}{27}$

23. (a) $\dfrac{1}{5} = .2$ (b) .25   (c) $\dfrac{11}{20} = .55$

25. (a) $\dfrac{11}{20} = .55$ (b) .35   (c) $\dfrac{1}{10} = .1$

    (d) .545   (e) $\dfrac{4}{7} = .57$ (f) 0

27. .36

29. (a) $\dfrac{24}{31} = .77$ (b) .06

31. $\dfrac{25}{64}; \dfrac{3}{16}$

33. $\dfrac{18}{85}$

35. (a) .82   (b) .18

37. $\dfrac{1}{2}$

Exercise Set 6.2, page 396

1. $\dfrac{8}{81}$

3. $\dfrac{16}{6561}$

5. (a) $\dfrac{16}{81} = .198$ (b) .395   (c) $\dfrac{8}{27} = .297$

   (d) .099   (e) $\dfrac{1}{81} = .012$ (f) 0

7. .054

9. (a) $\dfrac{5}{16} = .3125$ (b) .03125   (c) $\dfrac{1}{32} = .03125$

   (d) .1875   (e) $\dfrac{13}{16} = .8125$ (f) $\dfrac{1}{2}$

11. $\dfrac{1}{256}$

13. .2048

15. $\dfrac{7}{128}$

17. .474

19. .0519

21. .013. The drug does not seem to be 90% effective.

23. .738

25. $\dfrac{36}{37}$

Exercise Set 6.3, page 403

1. 60

3. 5

5. 2

7. 3.5

9. (a) -2.70 cents   (b) $2.70

11. 5.31 minutes

13. $E(x) = 3$

15. .2 (less than 1)

17. 10.5

19. (a) \$187.50   (b) \$40   (c) \$2933.33

21. 5

23. .24 (less than 1)

25. \$14.00 gain

27. The scheme that costs \$1 per questionnaire

29. 11.5%

31. .44

33. 13.5

Exercise Set 6.4, page 414

1. Is a probability matrix

3. Is a probability matrix

5. Is not a probability matrix

7. Transition matrix

9. Not a transition matrix

11. Not a transition matrix

13. Not a transition matrix

15. Not a transition matrix

17. (a) .4   (b) .5   (c) .4

19.

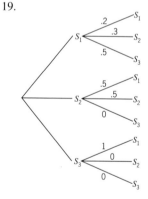

21. (a) $[.452 \quad .548]$ (b) $[.415 \quad .585]$

23. (a) $\begin{bmatrix} \frac{5}{11} & \frac{6}{11} \\ \frac{5}{11} & \frac{6}{11} \end{bmatrix}$ (b) $\begin{bmatrix} \frac{1}{3} & \frac{2}{3} \\ \frac{1}{3} & \frac{2}{3} \end{bmatrix}$

25. Not regular

27. Not regular

29. Regular

31. Not regular

33. Not regular

35. Regular

37. Regular

39. Regular

41. $[1 \quad 0]$

$$\mathbf{M} = \begin{bmatrix} 1 & 0 \\ 1 & 0 \end{bmatrix}$$

43. $[\frac{1}{3} \quad \frac{4}{9} \quad \frac{2}{9}]$

$$\begin{bmatrix} \frac{1}{3} & \frac{4}{9} & \frac{2}{9} \\ \frac{1}{3} & \frac{4}{9} & \frac{2}{9} \\ \frac{1}{3} & \frac{4}{9} & \frac{2}{9} \end{bmatrix}$$

45. $[\frac{1800}{13} \quad \frac{800}{13}]$

47. $[\frac{225}{4} \quad \frac{135}{4}]$

49. $[\frac{187}{57} \quad \frac{165}{57} \quad \frac{275}{57}]$

51. (a) $\dfrac{7}{16}$ (b) $\dfrac{11}{18}$ (c) $\dfrac{109}{216}$ (d) $\dfrac{83}{192}$

53. $[.261 \quad .228 \quad .310]$

55. (a) .2   (b) .7   (c) .35

57. $[.388 \quad .239 \quad .373]$

59. (a)
$$\begin{array}{c} \\ 0 \\ 1 \end{array} \begin{bmatrix} 0 & 1 \\ .9 & .1 \\ .2 & .8 \end{bmatrix}$$
(b) .83

61. (a) .2   (b) .8   (c) .19   (d) .55

63. The long-term prediction will be 80,000 members for party I, 74,286 members for III, and 45,714 members for II.

65. A loss of $40.00

Chapter 6 Test, page 419

1. $\dfrac{7}{13}$

2. .023

3. $[\frac{1}{3} \quad \frac{2}{3}]$

4. .188

5. The matrix represents a regular Markov chain, because $\mathbf{A}^2$ contains only positive elements.

6. $[\frac{79}{216} \quad \frac{3}{16} \quad \frac{193}{432}]$

7. $\begin{bmatrix} \frac{2}{7} & \frac{3}{7} & \frac{2}{7} \\ \frac{2}{7} & \frac{3}{7} & \frac{2}{7} \\ \frac{2}{7} & \frac{3}{7} & \frac{2}{7} \end{bmatrix}$

8. $\dfrac{63}{165} = \dfrac{21}{55}$

9. $\dfrac{27}{220}$

10. 20

Exercise Set 7.1, page 431

1.

Number of Times Ridden per Week	f
1	1
2	4
3	3
4	3
5	1
6	1
7	2
8	6
9	2
10	1
	24

3. (a) 20   (b) 16   (c) 16   (d) 14
   (e) 17, 22, 27, 32, 37

5. (a)

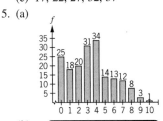

(b)

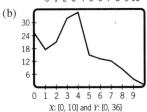

X: [0, 10] and Y: [0, 36]

7. (a), (b)

X: [14.5, 39.5], Y: [0, 600]

9.

Age	Relative Frequency
15–19	.333
20–24	.267
25–29	.200
30–34	.133
35–39	.067
	1.000

(a) .267   (b) .933   (c) .667
(d) 1.00   (e) 1.00

11. (a), (c)

Class	Class Marks	f
450–491	470.5	4
492–533	512.5	7
534–575	554.5	6
576–617	596.5	5
618–659	638.5	3

(b) Range = 205

13. (a), (c)

Class	ClassMarks	f
43–50	46.5	2
51–58	54.5	5
59–66	62.5	9
67–74	70.5	13
75–82	78.5	13
83–90	86.5	10
91–98	94.5	8

(b) The range is 54.

15. (a)

Stems	Leaves
3	0, 1, 3, 3, 3, 4, 5, 5, 5, 5, 5, 6, 7, 7, 7, 7, 7, 8, 8, 8, 8, 8, 8, 8, 9, 9, 9
4	0, 0, 0, 0, 0, 1, 1, 1, 1, 1, 1, 2, 2, 2, 4, 4, 5, 6, 7, 7, 8
5	0, 3

(b)

Stems	Leaves
4	3, 8
5	3, 3, 5, 6, 8
6	1, 1, 2, 3, 3, 4, 5, 5, 6, 7, 8
7	1, 1, 1, 1, 1, 2, 3, 4, 4, 4, 4, 5, 5, 5, 6, 6, 7, 7, 8, 9
8	1, 1, 1, 2, 3, 3, 4, 4, 5, 7, 7, 7, 9
9	0, 1, 1, 3, 3, 4, 5, 6, 7

17.

Class Marks	Class	Frequency
167	160–174	3
182	175–189	6
197	190–204	9
212	205–219	6
227	220–234	2

19. Decreased about 30 cars from 1986 to 1990

21. Model B increased the most, followed by model A, and then model C.

23. It is doubled in both radius and height, or has 8 times the original volume.

25.

Interval	Probability
0–4	.06
5–9	.46
10–14	.30
15–19	.10
20–24	.04
25–29	.02
30–34	.02

(a) $\dfrac{3}{10}$ (b) .92   (c) 1   (d) .16

27. (a) From labor, 45% was contributed to Democratic while only 3% to Republican candidates. From corporations, 16% was contributed to Democratic while 30% to Republican candidates.
   (b) 1.72 million   (c) 5.94 million

Exercise Set 7.2, page 443

1. (a) Mean = 6   (b) Mean = 7.5
      Median = 5   Median = 7
      No mode   Mode = 6
   (c) Mean = 3.78   (d) Mean = 3.625
      Median = 4   Median = 3.5
      Mode = 6   Mode = 1
   (e) Mean = 17.125   (f) Mean = 14.67
      Median = 19.5   Median = 14
      Mode = 23   Mode = 12

3. 27   5. 520

7. (a) Median   (b) Mean   (c) Mode

9. 69.4   11. 64

13. $Q_1 = 18$, median = 34, $Q_3 = 44$

15. (a) 81   (b) Between 80 and 90

17. 10

19. Mean = 10; $(10 - 5) + (10 - 8) + (10 - 10) + (10 - 12) + (10 - 15) = 0$

21. 12,800

23. Increases by 1320   25. 10.7

27. The $7.40 average expense per day

29. 60.83

31.

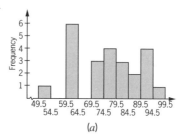

(a)

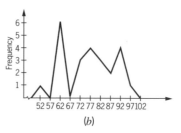

(b)

Exercise Set 7.3, page 455

1. Range = 8, variance = 9, standard deviation = 3
3. Range = 12, variance = 16, standard deviation = 4
5. Range = 11, variance = 16, standard deviation = 4
7. Range = 20, variance = 27.44, standard deviation = 5.24
9. Range = 20, variance = 26.74, standard deviation = 5.17
11.   No outliers

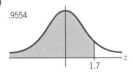

13.

Stems	Leaves
0	4, 7, 8, 9
1	0, 2, 3, 4, 4, 6, 8
2	4

IQR = 6.5
No outliers

15. 13.44
17. (a) 1   (b) -2 (c) -2.9 (d) 2   (e) 4
    (f) -6
19. 6   21. (b)
23. 1985   25. 5.90
27. 15.16
29. (a) 18   (b) 20   (c) 20   (d) 28   (e) 7.69

Exercise Set 7.4, page 467

1. (a) .4918   (b) .1591   (c) .4525   (d) .2812
3. (a)

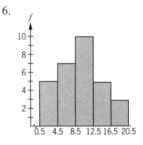

(b)

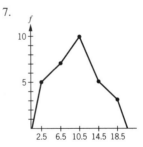

5. .9332 or 93.32%
7. (a) .5000   (b) .5000   (c) .1587
   (d) .9772   (e) .1587   (f) .6554
   (g) .5403   (h) .8284

9. (a) 68.53%   (b) 21.12%
11. 5; $p = .9087$ (normal, .9015); 1.55 standard deviations
    below the mean
13. 10.56%
15. 89.44%
17. (a) 68.26%   (b) 95.44%
19. 2.28%   21. 99.11%
23. .3158 (normal, .3419)
25. (a) 75   (b) 76.5   (c) 94   (d) 63
    (e) 64   (f) 14.95   (g) 23   (h) None
27.

Chapter 7 Test, page 471

1. 13   2. 13
3. 46.67   4. 12
5.

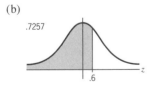

6.

7.

8. 9.7
9. 4.78
10. 25
11. 68.857
12. 71
13. 64

14.
Stems	Leaves
3	3, 7
4	2, 8
5	2, 3, 4, 5, 7, 9
6	2, 2, 4, 4, 4, 8, 9
7	1, 2, 2, 3, 3, 4, 4, 8, 9
8	2, 3, 5, 7, 8, 9
9	3, 6, 8

15.

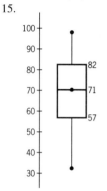

16. No outliers

17.
Class	Frequency
31–40	2
41–50	2
51–60	6
61–70	7
71–80	9
81–90	6
91–100	3
	35

18. 15.71

19. .3413

20. 15.87%

Exercise Set 8.1, page 485

1.
x	1.900	1.990	1.999 → 2 ← 2.001	2.010	2.100
$f(x)$	0.800	0.980	0.998 → 1 ← 1.002	1.020	1.200

3.
x	3.900	3.990	3.999 → 4 ← 4.001	4.010	4.100
$h(x)$	7.900	7.990	7.999 → 8 ← 8.001	8.010	8.100

5. $\lim\limits_{x \to 1} \dfrac{3x^2 + 4x}{x + 2} = \dfrac{7}{3}$

7. $\lim\limits_{x \to 4} -8 = -8$

9. $\lim\limits_{x \to 3} 2x = 6$

11. $\lim\limits_{x \to 2} \dfrac{3x}{2} = 3$

13. $\lim\limits_{x \to 1^-} f(x) = 3$

15. $\lim\limits_{x \to 1^+} f(x) = 1$

17. $f(8) = 1$

19. $f(3)$ is not defined.

21. $\lim\limits_{x \to 3} f(x) = 2$

23. $\lim\limits_{x \to 3} (2x^2 - 4) = 14$

25. $\lim\limits_{x \to 2} (3x^2 - 5)(x + 4) = 42$

27. $\lim\limits_{x \to -3} \dfrac{7x}{2x + 3} = 7$

29. $\lim\limits_{x \to 1} \dfrac{5x}{2 + x^2} = \dfrac{5}{3}$

31. $\lim\limits_{x \to 2} x^2(x^2 + 1)^3 = 500$

33. (a) $\lim\limits_{x \to 3} f(x) = 1$ (b) $\lim\limits_{x \to 3} f(x) = 4$

(c) $\lim\limits_{x \to 3} f(x)$ does not exist.

(d) $\lim\limits_{x \to 3} f(x)$ does not exist.

(e) $\lim\limits_{x \to 3} f(x) = 3$ (f) $\lim\limits_{x \to 3} f(x) = 2$

35. $\lim\limits_{x \to 0} f(x) = 2$

37. $\lim\limits_{x \to 4} f(x) = -2$

39. $\lim\limits_{x \to 2} g(x) = 1$

41. $\lim\limits_{x \to 2} y$ does not exist.

43. $\lim\limits_{x \to 2} y = -2$

45. $\lim\limits_{x \to 3} \dfrac{x^3 - 27}{x - 3} = 27$

47. $\lim\limits_{x \to -3} \dfrac{x^2 - 9}{x + 3} = -6$

49.

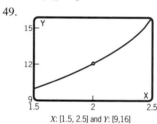

X: [1.5, 2.5] and Y: [9,16]

$\lim\limits_{x \to 2^-} f(x) = 12$; $\lim\limits_{x \to 2^+} f(x) = 12$; but $f(2)$ is not defined.

51.

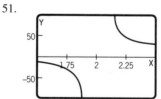

X: [1.5, 2.5] and Y: [–100, 100]

$\lim\limits_{x \to 2} f(x)$ is undefined.

53. (a) $\lim\limits_{x \to 1} C(x) = 5$ (b) $\lim\limits_{x \to 4} C(x)$ does not exist.

(c) $\lim\limits_{x \to 6} C(x) = 6$ (d) $\lim\limits_{x \to 8} C(x) = 8$

55. (a) $\lim\limits_{t \to 100} n(t) = 40{,}000$ (b) $\lim\limits_{t \to 0} n(t) = 0$

(c) $\lim\limits_{t \to -100} n(t)$ does not exist.

Exercise Set 8.2, page 496

1. Continuous at $-1, 0, 2$
3. Continuous at $-2, 0, 1$
5. Continuous at $-1, 0, 2$
7. Continuous at $-1, 2$; not continuous at 0
9. Continuous at $0, 1$; not continuous at -2
11. Continuous at $-1, 0$; not continuous at 2
13. Continuous at $-3, 0$; not continuous at 3
15. Continuous at -1; not continuous at $1, 0$
17. (a) Discontinuous because $\lim\limits_{x \to 2} f(x) = 1$ but $f(2) \neq 1$.
   (b) Continuous because $\lim\limits_{x \to 2} f(x) = 2$ and $f(2) = 2$.

19. $\lim\limits_{x \to 3} \dfrac{x^2 - 9}{x - 3} = 6$

21. $\lim\limits_{x \to 0} \dfrac{4x^2 - 3x}{x} = -3$

23. $\lim\limits_{x \to 4} \dfrac{x - 4}{\sqrt{x} - 2} = 4$

25. $\lim\limits_{x \to 0} \dfrac{(2/x) + 1}{3/x} = \dfrac{2}{3}$

27. Discontinuous at $x = 3$
29. Continuous everywhere
31. Discontinuous at $x = 0$
33. Discontinuous at $x = -2$
35. Discontinuous at $x = 2$ and $x = -2$
37. Discontinuous at $x = 2$ and $x = -2$
39. (a) Discontinuous at $x = 2$
   (b) Continuous everywhere
41. Discontinuous at $x = 2$
43. (a) Discontinuous at $x = 4$ and $x = 6$
   (b) $\lim\limits_{x \to 2} C(x) = 5$
   (c) $C(2) = 5$
   (d) $\lim\limits_{x \to 4} C(x)$ does not exist.
   (e) $C(4) = 5$
   (f) $\lim\limits_{x \to 7} C(x) = 15$
   (g) $C(7) = 15$
45. Continuous on $(16, 96)$

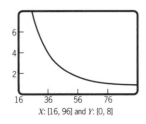
X: [16, 96] and Y: [0, 8]

47. (a) Discontinuous at $x = 2, 3, 5, 6$ (b) $N(1) = 4$
   (c) $\lim\limits_{t \to 2^-} N(t) = 4$ (d) $\lim\limits_{t \to 2^+} N(t) = 8$
49. (a)

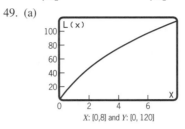
X: [0,8] and Y: [0, 120]

   (b) Continuous on $(0, 8)$

51. $\lim\limits_{x \to 2} \dfrac{x}{x + 1} = \dfrac{2}{3}$

53. $\lim\limits_{x \to 0} \dfrac{5}{x^2}$ does not exist.

55. $\lim\limits_{x \to 2} \dfrac{4}{x - 2}$ does not exist.

Exercise Set 8.3, page 506

1. 18
3. 15
5. $18 + 3h$
7. 12; this is the slope of the line tangent to the graph of $f(x) = 3x^2$ at $x = 2$.
9. 5
11. 1/3
13. 2
15. 6
17. -8
19. (a) 23/2   (b) 15   (c) 55/4   (d) 37/3
21. (a) 2   (b) 4   (c) 0   (d) 300
23. -2 and -4

25. $\lim\limits_{x \to 2} \dfrac{x^2 - 3x + 2}{x - 2} = 1$

27. $\lim\limits_{x \to 1} \dfrac{3x^2 + 4x - 1}{5x^2 + 2x + 3} = \dfrac{3}{5}$

29. $\lim\limits_{x \to 3} \dfrac{2x^2 - 2x - 12}{x - 3} = 10$

Exercise Set 8.4, page 515

1. $f'(x) = 3$
3. $f'(x) = 3$
5. $f'(x) = 5/3$
7. $f'(x) = 6x, f'(0) = 0, f'(1) = 6, f'(-1) = -6$
9. $f'(x) = 7 + 2x, f'(0) = 7, f'(1) = 9, f'(-1) = 5$
11. $f'(x) = \dfrac{-4}{x^2}, f'(0)$ is not defined, $f'(1) = -4, f'(-1) = -4$
13. $f'(x) = \dfrac{\sqrt{3x}}{2x}, f'(0)$ is not defined, $f'(1) = \sqrt{3}/2, f'(-1)$ is not defined.

15. $f'(x) = \dfrac{-1}{2(x+1)^{3/2}}$, $f'(0) = -1/2$, $f'(1) = -\sqrt{2}/8$,
$f'(-1)$ is not defined.

17. $f'(x) = \dfrac{-4}{x^3}$, $f'(0)$ is not defined, $f'(1) = -4, f'(-1) = 4$

19. $f'(x) = 3x^2, f'(0) = 0, f'(1) = 3, f'(-1) = 3$

21. $y' = 2x, y'(1) = 2$

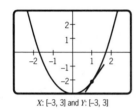

X: [-3, 3] and Y: [-3, 3]

23. $m = 6, y = 6x - 2$

25. (a) $f'(x) = 2x$ (b) $f'(2) = 4$ (c) $y = 4x - 3$
(d)

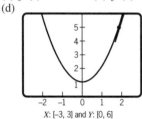

X: [-3, 3] and Y: [0, 6]

27. (a) $C'(x) = 300 + 2x$ (b) $C'(2) = 304$
(c) $C'(3) = 306$

29. (a) $p'(x) = -2x$ (b) $p'(3) = -6$
(c) $p'(5) = -10$

31. (a) $P'(x) = 10 - 10x$ (b) $P'(2) = -10$
(c) $P'(3) = -20$

33. (a) $N(2) = 2000$ (b) $N'(2) = -10$
(c) $N(4) = 1960$ (d) $N'(4) = -30$

35. $\displaystyle\lim_{x \to -2/3} \dfrac{3x^2 + 4x - 1}{3x + 2}$ is not defined.

Exercise Set 8.5, page 527

1. $f'(x) = 0$

3. $\dfrac{dy}{dx} = -1$

5. $D_x[3x^2 + 5] = 6x$

7. $y' = 15x^2 - 6x$

9. $\dfrac{d}{dx}(2x^{-4}) = -8x^{-5}$

11. $f'(x) = -x^{-3/2} = \dfrac{-1}{x^{3/2}}$

13. $f'(x) = 24x^3 - 3, f'(0) = -3,$
$f'(2) = 189, f'(-3) = -651$

15. $f'(x) = \dfrac{-3}{x^2} + 4$, $f'(0)$ is not defined, $f'(2) = \dfrac{13}{4}$,
$f'(-3) = \dfrac{11}{3}$

17. $f'(x) = 8x + \dfrac{4}{x^2}$, $f'(0)$ is not defined, $f'(2) = 17$,
$f'(-3) = \dfrac{-212}{9}$

19. $f'(x) = 2x^{-1/2}, f'(0)$ is not defined, $f'(2) = \sqrt{2}, f'(-3)$ is
not defined.

21. $f'(x) = 4x^{-1/3} + 2x^{-1/2}$, $f'(0)$ is not defined, $f'(2) =$
$2^{5/3} + 2^{1/2} = 4.58902, f'(-3)$ is not defined.

23. $f'(x) = 6x + 2x^{-3/2}$, $f'(0)$ is not defined, $f'(2) = 12 +$
$1/\sqrt{2} = 12.7071, f'(-3)$ is not defined.

25. $f'(x) = (7/2)x^{-1/2} + 3x^{-2/3}$, $f'(0)$ is not defined, $f'(2) =$
$\dfrac{7}{2^{3/2}} + \dfrac{3}{2^{2/3}} = 4.36476, f'(-3)$ is not defined.

27. $(0, 0)$ 29. $(1/8, -1/16)$

31. $y = 6x - 1$ 33. $y = 0.16x + 0.02$

35. (c) $8 + 0.02x$

37. (a) $C'(x) = 400 - 2x$ (b) $C'(1) = 398$
(c) $C'(2) = 396$ (d) $\overline{C}(2) = 798$
(e) $C'(1) = 398, C'(2) = 396, \overline{C}(2) = 798$

39. (a) $R'(x) = 20 - \dfrac{x}{250}$ (b) $x = 5000$

(c) $R(5000) = 50,000$ (d) $\overline{R}(x) = 20 - \dfrac{x}{500}$

41. (a) $C'(4) = 6; C'(10) = 6$ (b) Always 6
(c) $R'(5) = 39.9$ (d) $P'(4) = 33.92$
(e) $P(3) = 1.91$

(f) $\overline{P}(x) = 34 - 0.01x - \dfrac{100}{x}$

43. (a) $P'(x) = -0.4x^{-3}$ (b) $P'(1) = -0.4$
(c) $P'(2) = -0.05$

45. (a) $f'(x) = 24x^{-2/3}$ (b) $f'(1) = 24$ (c) $f'(8) = 6$
(d) $f(8) = 144$ (e) $f'(1) = 24, f'(8) = 6, f(8) = 144$

47. $\displaystyle\lim_{x \to 2} \dfrac{x^2 + x + 4}{x - 2}$ is not defined.

49. 19.4   51. $y' = 8x + 3$

Exercise Set 8.6, page 538

1. $\dfrac{dy}{dx} = 6x - 7$ 3. $\dfrac{dy}{dx} = 4x^3$

5. $\dfrac{dy}{dx} = 3x^2$ 7. $\dfrac{dy}{dx} = \dfrac{-15}{(2x - 3)^2}$

9. $\dfrac{dy}{dx} = \dfrac{-13}{(2x - 3)^2}$

11. $\dfrac{dy}{dx} = 36x^2 + 14x - 6, \left.\dfrac{dy}{dx}\right|_{x=1} = 44$

13. $\dfrac{dy}{dx} = \dfrac{-10x}{(x^2 + 1)^2}, \left.\dfrac{dy}{dx}\right|_{x=1} = -\dfrac{5}{2}$

15. $\dfrac{dy}{dx} = \dfrac{-3x^2 + 4x - 9}{(x^2 - 3)^2}$, $\dfrac{dy}{dx}\Big|_{x=1} = -2$

17. $\dfrac{dy}{dx} = \dfrac{-9x^2 - 12}{2\sqrt{x}(x^2 - 4)^2}$, $\dfrac{dy}{dx}\Big|_{x=1} = \dfrac{-21}{18}$

19. $y' = \dfrac{20x^2 + 40x - 40}{(x + 1)^2}$

21. $y' = \dfrac{6x^3 - 6x^2 - 6}{(3x - 2)^2}$

23. $y' = -6(2x + 5)^{-2} + (9/2)x^{1/2} - (5/2)x^{-1/2}$

$= \dfrac{-6}{(2x + 5)^2} + \dfrac{9\sqrt{x}}{2} - \dfrac{5}{2\sqrt{x}}$

25. $y' = \dfrac{-2x^4 + 3x^2 + 4x}{(x^3 + 1)^2} - \dfrac{1}{2\sqrt{x^3}}$

27. $y = \dfrac{15x}{32} + \dfrac{25}{16}$

29. (a) $p'(x) = \dfrac{4000(x - 50)}{x^3}$ (b) $p'(2) = -24,000$

(c) $\bar{p}'(x) = \dfrac{4000[2x - 75]}{x^4}$

31. (a) $R'(x) = 1000(2x) = 2000x$

(b) $\bar{R}'(x) = \dfrac{1000x^2 + 3000}{x^2}$

33. (a) $L'(t) = \dfrac{6800}{(90t + 85)^2}$ (b) $L'(1) = \dfrac{272}{1225}$

(c) $L'(8) = \dfrac{272}{25,921}$

35. $\dfrac{dy}{dx} = 14x - 3$

37. $\dfrac{dy}{dx} = 6x - 3x^2 + 8x^3$

39. $\dfrac{dy}{dx} = 15x^2 + 2x^{-2}$

41. $\dfrac{dy}{dx} = 24x^3 + 10x^{-3}$

43. 12.3

45. $y' = \lim\limits_{h \to 0} \dfrac{(6xh + 3h^2)}{h} = 6x$

47. $f'(x) = x^{-1/2} + 6x = \dfrac{1}{\sqrt{x}} + 6x$

49. $f'(x) = x^{-1/2} + 2x^{-3/2} = \dfrac{2 + x}{x^{3/2}}$

51. $f'(x) = -2x^{-3/2} + 4x^{-5/3} = \dfrac{4}{x^{5/3}} - \dfrac{2}{x^{3/2}}$

Exercise Set 8.7, page 549

1. $f[g(x)] = x^2 - 2x + 4$, $g[f(x)] = x^2 + 2$
3. $f[g(x)] = \sqrt{x^3 - 7}$, $g[f(x)] = (x - 7)^{3/2}$

5. $y = x^4 + 2x^2 + 1$ 7. $y = \dfrac{1}{\sqrt{x^4 + 3}}$

9. $y = u^5$, $u = 3x + 4$ 11. $y = 1/u$, $u = x + 8$

13. $\dfrac{dy}{dx} = 9$ 15. $\dfrac{dy}{dx} = -24x$

17. $\dfrac{dy}{dx} = 72x + 120$

19. $\dfrac{dy}{dx} = 18x + 24$

21. $\dfrac{dy}{dx} = -12x(6 - 2x^2)^2$

23. $\dfrac{dy}{dx} = \dfrac{-2}{(2x + 4)^2}$

25. $\dfrac{dy}{dx} = \dfrac{2x^2}{(2x^3 - 1)^{2/3}}$

27. $\dfrac{dy}{dx} = (3x^2 - 2x + 1)(12x - 4)$

29. $\dfrac{dy}{dx} = -30(2x + 1)^2$

31. $\dfrac{dy}{dx} = \dfrac{(x - 1)^2(7x - 1)}{2\sqrt{x}}$

33. $\dfrac{dy}{dx} = \dfrac{-2x}{(2x^2 - 1)^{3/2}}$

35. $f'(x) = 36x^2 - 32x - 33$

37. $f'(x) = \dfrac{8x + 2}{\sqrt{2x^2 + x - 1}}$

39. $f'(x) = \dfrac{(3x - 1)(30x - 2)}{\sqrt{x}}$

41. $f'(x) = \dfrac{-24x}{(2x - 3)^3}$

43. $f'(x) = 6x(x + 3)^3(x + 1)$

45. $f'(x) = \dfrac{10(x - 3)}{(x + 2)^3}$

47. (a) $C'(x) = 18x - 60$
(b) $C'(3) = -6$
(c) $C'(4) = 12$
(d) $\bar{C}'(x) = \dfrac{9x^2 - 124}{x^2}$

49. (a) $L'(1) = 72$
(b) $L'(14) = 72$

51. $y' = \dfrac{-19}{(3x - 2)^2}$

53. $y' = 36x^2 + 16x - 6$

55. $y' = 18(2x + 5)^2$

Chapter 8 Test, page 551

1. $f'(x) = \lim\limits_{h \to 0} \dfrac{f(x + h) - f(x)}{h}$

2. $f'(x) = 6x$ 3. 14

4. $f'(0) = 4$; $f'(-1) = 10$ 5. $y' = 0$

6. $\dfrac{dy}{dx} = -16x^3$ 7. $\bar{C}'(2) = -249$

8. -2

9. $\lim\limits_{x \to 1} \dfrac{x^2 - 3x + 3}{x - 1}$ does not exist.

10. $f'(x) = \dfrac{-3}{x^2} + \dfrac{2}{\sqrt{x}}$

11. $p'(x) > 0$ for $x > -4$
$p'(x) < 0$ for $x < -4$

12. $y' = 6x\sqrt{x^2 - 3}$; $y'(4) = 24\sqrt{13}$; $y'(4) = 86.5332$

13. $y'(x) = \dfrac{18x - 1}{2\sqrt{3x + 2}}$, $y'(1) = \dfrac{17}{2\sqrt{5}}$

14. $y' = \dfrac{-5x - 2}{(x^2 - 1)^{3/2}}$; $y'(2) = \dfrac{-12}{3^{3/2}}$; $y'(2) = -2.3094$

15. (a) $p'(x) = \dfrac{6000(x - 40)}{x^3}$

(b) $\bar{p}'(x) = \dfrac{12,000x - 360,000}{x^4}$

Exercise Set 9.1, page 562

1. $y' = 0$; no critical point; increasing, none
3. $f'(x) = 3$; no critical point; increasing, $(-\infty, \infty)$
5. $f'(x) = 2x$; critical point at $(0, 0)$; increasing, $(0, \infty)$
7. $y' = 2x + 2$; critical point at $(-1, -1)$; increasing, $(-1, \infty)$
9. $y' = 8x + 3$; critical point at $\left(-\dfrac{3}{8}, -\dfrac{41}{16}\right)$; increasing, $\left(-\dfrac{3}{8}, \infty\right)$
11. $y' = 6x + 12$; critical point at $(-2, -17)$; increasing, $(-2, \infty)$
13. $y' = 8x^{-1/3} + 1$; critical points at $(0, 0)$, $(-512,256)$; increasing, $(-\infty, -512)$, $(0, \infty)$
15. Increasing, $(2, \infty)$; decreasing, $(-\infty, 2)$
17. Increasing, $(-\infty, 4)$, $(10, \infty)$; decreasing, $(4, 10)$
19. Answers will vary, but should contain the points given and have a relative minimum at the point $(1, 1)$, the only extremum.

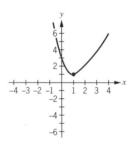

21. Answers will vary, but should contain the points given and have a relative maximum at the point $(3, 11)$, the only extremum.

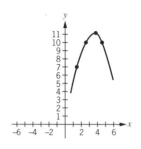

23. (a) $(a, c), (e, f), (i, k)$ (b) $(c, e), (f, i)$
(c) c, e, f, i
25. Critical point, $(0, 0)$; increasing, $(0, \infty)$; decreasing, $(-\infty, 0)$
27. Critical point, $(0, 3)$; increasing, $(-\infty, 0)$; decreasing, $(0, \infty)$
29. Critical point, $\left(-\dfrac{1}{2}, \dfrac{5}{4}\right)$; increasing, $\left(-\dfrac{1}{2}, \infty\right)$; decreasing, $\left(-\infty, -\dfrac{1}{2}\right)$
31. Critical point, $(0, 0)$; increasing, $(-\infty, \infty)$
33. Critical point, none; increasing, $(-\infty, \infty)$
35. Critical point, $(0, 0)$, $(-4, 32)$; increasing, $(-\infty, -4)$, $(0, \infty)$; decreasing, $(-4, 0)$
37. Critical point, $(0, 0)$, $(2, -8)$; increasing, $(-\infty, 0)$, $(2, \infty)$; decreasing, $(0, 2)$
39. Critical point, $(3, 0)$, $(1, 4)$; increasing, $(-\infty, 1)$, $(3, \infty)$; decreasing, $(1, 3)$
41. Critical point, $(0, 0)$; increasing, $(-\infty, \infty)$
43. Critical point, $(0, 0)$; increasing, $(0, \infty)$
45. Critical point, $(1, -4)$; increasing, $(1, 3)$, $(3, \infty)$; decreasing, $(-\infty, 1)$
47. Up to \$20 the profit increases; after \$20 it decreases.
49. Decreasing for x in the interval $(1, 3)$ and increasing for $x > 3$.
51. Since the critical value $x = c$ means that $P'(c) = 0$ and also, $P'(x) = R'(x) - C'(x)$; therefore, $R'(c) - C'(c) = 0$, so $R'(c) = C'(c)$.
53. Increasing on $(3, 8)$ and decreasing on $(0, 3)$

Exercise Set 9.2, page 572

1. Relative minimum of -2 at $x = -3$ and -2 at $x = 1$. Relative maximum of 3 at $x = -1$.
3. (a) Relative maximum   (b) Neither
(c) Relative minimum   (d) Neither
(e) Relative minimum   (f) Neither
(g) Relative maximum   (h) Neither
5. $x = e, x = i$

In Exercises 7–27 (a) contains critical points, (b) relative extrema, (c) where the function is increasing, and (d) where the function is decreasing.

7. (a) $(0, 0)$ (b) Min. at $(0, 0)$ (c) $(0, \infty)$
   (d) $(-\infty, 0)$

9. (a) $(0, 3)$ (b) Max. at $(0, 3)$ (c) $(-\infty, 0)$
   (d) $(0, \infty)$

11. (a) $(-\frac{1}{2}, \frac{5}{4})$ (b) Min. at $(-\frac{1}{2}, \frac{5}{4})$ (c) $(-\frac{1}{2}, \infty)$
    (d) $(-\infty, -\frac{1}{2})$

13. (a) $(0, 0)$ (b) None   (c) $(-\infty, \infty)$ (d) None

15. (a) None   (b) None   (c) $(-\infty, \infty)$ (d) None

17. (a) $(0, 0)$, $(4, -32)$
    (b) Max. at $(0, 0)$; min. at $(4, -32)$
    (c) $(-\infty, 0)$, $(4, \infty)$ (d) $(0, 4)$

19. (a) $\left(\dfrac{2}{3}, \dfrac{2\sqrt{3}}{9}\right)$ (b) Max. at $\left(\dfrac{2}{3}, \dfrac{2\sqrt{3}}{9}\right)$
    (c) $(-\infty, \frac{2}{3})$ (d) $(\frac{2}{3}, 1)$

21. (a) $(-2, 0)$ (b) Min. at $(-2, 0)$ (c) $(-2, \infty)$
    (d) $(-\infty, -2)$

23. (a) $\left(\dfrac{2}{3}, -\dfrac{16\sqrt{6}}{9}\right)$
    (b) Min. at $\left(\dfrac{2}{3}, -\dfrac{16\sqrt{6}}{9}\right)$
    (c) $(\frac{2}{3}, \infty)$ (d) $(0, \frac{2}{3})$

25. (a) None   (b) None   (c) None
    (d) $(-\infty, 1)$, $(1, \infty)$

27. (a) $(2, 4)$, $(-2, -4)$
    (b) Min. at $(2, 4)$; max. at $(-2, -4)$
    (c) $(-\infty, -2)$, $(2, \infty)$ (d) $(-2, 0)$, $(0, 2)$

29. (d)

31. Relative minimum at $(1, 0)$

33. Minimum cost of 182 at $x = 3$.

35. A price of \$20 yields \$100,000 maximum profit.

37. Minimum population of 20,000 at $t = 3$

39. Critical point, $(1, 2)$; decreasing on $(-\infty, 1)$; increasing on $(1, \infty)$

41. Critical point, $(1, 1)$; decreasing on $(-\infty, 1)$; increasing on $(1, \infty)$

Exercise Set 9.3, page 582

$\dfrac{dx}{dy}$	$\dfrac{d^2y}{dx^2}$	$\dfrac{d^3y}{dx^3}$
1. 0	0	0
3. 3	0	0
5. $70x - 27$	70	0
7. $6x^2 + 6x - 1$	$12x + 6$	12
9. $12x^5 + 9x^2 - 2$	$60x^4 + 18x$	$240x^3 + 18$

11. $2x^3 + x^2$ $6x^2 + 2x$ $12x + 2$

13. $x^{-2/3} = \dfrac{1}{x^{2/3}}$ $\dfrac{-2}{3}x^{-5/3} = \dfrac{-2}{3x^{5/3}}$ $\dfrac{10}{9}x^{-8/3} = \dfrac{10}{9x^{8/3}}$

15. $\dfrac{4}{3}x^{1/3}$ $\dfrac{4}{9}x^{-2/3} = \dfrac{4}{9x^{2/3}}$ $\dfrac{-8}{27}x^{-5/3} = \dfrac{-8}{27x^{5/3}}$

17. $-3(x - 1)^{-2}$ $6(x - 1)^{-3}$ $-18(x - 1)^{-4}$

19. (a) x_2, x_5, x_7, x_9 (b) x_3, x_6, x_8
    (c) x_4, x_6 (d) x_2, x_7 (e) x_5, x_9

21. (a) Positive, zero, negative
    (b) Negative, zero, positive
    (c) Zero, positive, zero   (d) Positive, positive, positive
    (e) Positive, negative, zero   (f) Positive
    (g) Positive   (h) Positive   (i) Positive
    (j) (x_2, x_5), (x_7, x_9) (k) (x_1, x_3), (x_6, x_8)

23. f is increasing and concave down.

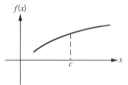

25. f is decreasing and concave down.

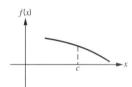

27. f is increasing for $x < c$ and decreasing for $x > c$ and concave down.

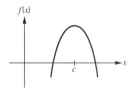

29.

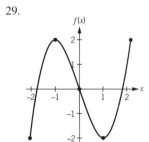

31.

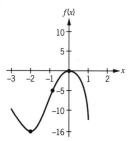

33. $f''(x) = 6x + 10$

35. $f''(x) = \dfrac{2x - 12}{(x + 3)^4}$

37. $f''(x) = \dfrac{2x^2 - 4}{(4 + x^2)^{5/2}}$

Inflection Points	Concave Up	Concave Down	Relative Extrema
39. None	None	$(-\infty, \infty)$	Max. at $(0, 0)$
41. None	None	$(-\infty, \infty)$	Max. at $(0, 0)$
43. None	None	$(-\infty, \infty)$	Max. at $(\frac{1}{3}, \frac{1}{3})$
45. $(0, 0)$	$(-\infty, 0)$	$(0, \infty)$	None
47. $(-\frac{4}{3}, \frac{128}{27})$	$(-\frac{4}{3}, \infty)$	$(-\infty, -\frac{4}{3})$	Min. at $(0, 0)$; max. at $(-\frac{8}{3}, \frac{256}{27})$
49. None	$(-\infty, \infty)$	None	Min. at $(0, 0)$
51. $(1, 6)$	$(1, \infty)$	$(-\infty, 1)$	Max. at $(-2, 4.76)$; min. at $(0,0)$

53.

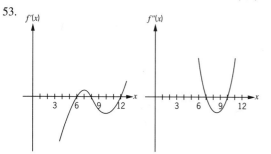

55.

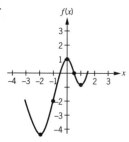

57. (a) $v(t) = t + 3t^2$, $a(t) = 1 + 6t$
    (b) $v(8) = 200$, $a(8) = 49$

59. (a) $P'(7) = 8000 > 0$; profit is increasing.
    (b) $(0, 15)$ (c) $S = 15$ (d) $112,500
61. (a) $x = 2664$ (b) $783,936
63. 324.15 million
65. Critical point, $(-1.5, -2.25)$; increasing, $(-1.50, \infty)$; decreasing, $(-\infty, -1.50)$

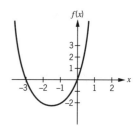

67. Critical points, $(1, 0)$, $(-1, 4)$; increasing, $(-\infty, -1)$, $(1, \infty)$; decreasing, $(-1, 1)$

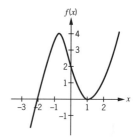

69. Critical point, $(1, -2)$; increasing $(1, \infty)$; decreasing, $(-\infty, 1)$

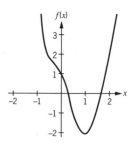

Exercise Set 9.4, page 592

1. Absolute max. at x_2; absolute min. at x_3
3. Absolute max. at x_4; absolute min. at x_1
5. (a) Absolute max. 0; Absolute min. -0.25
   (b) Absolute max. 6; Absolute min. -0.25
   (c) Absolute max. 6; Absolute min. -0.25

7. (a) Absolute max. 2; Absolute min. -2
   (b) Absolute max. 16; Absolute min. -16
   (c) Absolute max. 54; Absolute min. -54

9. (a) Absolute max. 3; Absolute min. 0
   (b) Absolute max. 24; Absolute min. 0
   (c) Absolute max. 99; Absolute min. 0

11. $f(-1) = 2$, relative max.; $f(0) = 0$ is relative min.

13. Test fails, $f(1) = -1$, relative min.

15. $f(-5) = 12.5$, relative max.; $f(-\frac{4}{3}) = -12.148$, relative min.

Critical Point(s)	Relative Extrema	Absolute Extrema	Inflection Point(s)
17. $(\frac{1}{8}, -\frac{1}{16})$	Min. $-\frac{1}{16}$	Min. $-\frac{1}{16}$	None
19. $(0, 0)$	None	None	$(0, 0)$
21. $(0, 0)$, $(2, -4)$	Max. 0, min -4	None	$(1, -2)$
23. $(0, 0)$	Min. 0	Min. 0	None
25. $(1/\sqrt{2}, 2\sqrt{2})$, $(-1/\sqrt{2}, -2\sqrt{2})$	Min. $2\sqrt{2}$, max. $-2\sqrt{2}$	None None	None None

27. Increasing, $(0, \infty)$; concave down, $(0, \infty)$
    No points of inflection

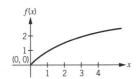

29. Increasing, $\left(-\infty, \frac{1}{3}\right)$, $(1, \infty)$; decreasing, $\left(\frac{1}{3}, 1\right)$; concave up, $\left(\frac{2}{3}, \infty\right)$; concave down, $\left(-\infty, \frac{2}{3}\right)$; max. at $\left(\frac{1}{3}, \frac{4}{27}\right)$; min at $(1, 0)$; inflection at $\left(\frac{2}{3}, \frac{2}{27}\right)$

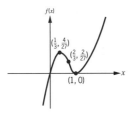

31. Increasing, $(-1, \infty)$; decreasing, $(-\infty, -1)$; concave up, $(-\infty, 0)$, $(2, \infty)$; concave down, $(0, 2)$; min. at $(-1, -3)$; inflection at $(0, 0)$, $(2, 6\sqrt[3]{2})$

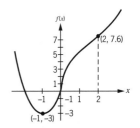

33. (c)

35. (d)

37. 20

39. $x = 5$; $\overline{P}(5) = \$0$

41. Profit is between $x = 3.28$ and $x = 9.5$. Loss is from $x = 0$ to $x = 3.28$ and for $x > 9.5$. Maximum profit is 73 when $x \approx 7$.

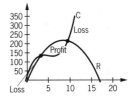

43. Critical point, $\left(-2, -\frac{1}{4}\right)$; increasing, $(-2, 0)$; decreasing, $(-\infty, -2)$, $(0, \infty)$; min. at $\left(-2, -\frac{1}{4}\right)$, inflection point, $\left(-3, \frac{-2}{9}\right)$; concave up, $(-3, 0)$, $(0, \infty)$; concave down, $(-\infty, -3)$

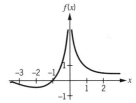

45. Critical points, $(-3, 3)$, $(0, 0)$; increasing, $(-3, -2)$, $(-2, \infty)$; decreasing, $(-\infty, -3)$; relative min. at $(-3, 3)$; inflection point, $(0, 0)$; concave up, $(-\infty, -2)$, $(0, \infty)$; concave down, $(-2, 0)$

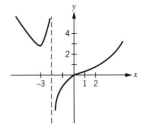

47. Critical points $(-2, 4.76)$, $(0, 0)$; increasing, $(-\infty, -2)$, $(0, \infty)$; decreasing, $(-2, 0)$; min. at $(0, 0)$; max. at $(-2, 4.76)$; inflection point, $(1, 6)$; concave up, $(1, \infty)$; concave down, $(-\infty, 1)$

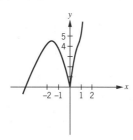

Exercise Set 9.5, page 607

1. 0.333, 0.039, 0.00399, 0.0003999; $\lim_{x \to \infty} f(x) = 0$

3. 8.33, 98.039, 998.0, 9998; $\lim_{x \to \infty} f(x) \to \infty$

5. -0.5, $\quad -0.0408$, $\quad -0.004008$, $\quad -0.0004000$; $\lim_{x \to -\infty} f(x) = 0$

7. $-12.5, -102.04, -1002.0, -10002.0$; $\lim_{x \to -\infty} f(x) \to -\infty$

9. $-20, -200, -2000, -20,000$; $\lim_{x \to 1^-} f(x) \to -\infty$

11. 10, 100, 1000, 10,000; $\lim_{x \to 1^-} f(x) \to \infty$

13. 20, 200, 2000, 20,000; $\lim_{x \to 1^+} f(x) \to \infty$

15. 10, 100, 1000, 10,000; $\lim_{x \to 1^+} f(x) \to \infty$

17. (a) $x = -2$ \quad (b) $y = 0$ \quad (c) 0 \quad (d) 0

19. 0 \hspace{3cm} 21. 7

23. 1 \hspace{3cm} 25. $\dfrac{3}{4}$

27. $\to \infty$ \hspace{2.7cm} 29. $\to \infty$

31. 0

33. Horizontal asymptote, $y = 1$; vertical asymptote, $x = -2$; no relative extrema; no inflection points

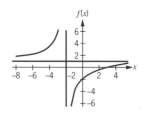

35. Horizontal asymptote, $y = 2$; vertical asymptote, $x = -1$; no relative extrema; no inflection points

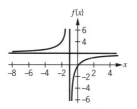

37. Horizontal asymptote, $y = 2$; vertical asymptote, $x = 1$; no relative extrema; no inflection points

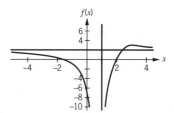

39. Horizontal asymptote, $y = 2$; vertical asymptote, $x = 1$; relative max. at $(3, 3)$; inflection point, $(4, 26/9)$

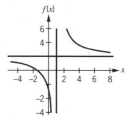

41. Horizontal asymptote, $y = 0$; vertical asymptote, $x = 0$; relative min. at $(12, -1/24)$; inflection point, $(18, -1/27)$

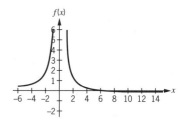

43. Horizontal asymptote, $y = 0$; vertical asymptote, $x = 0$; relative min. at $(1, -1)$; inflection point, $(3/2, -8/9)$

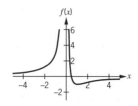

45. Vertical asymptotes, $x = 1$, $x = -1$
47. Horizontal asymptote, $y = 2$
49. $\rightarrow \infty$ 51. 3.50
53. $\rightarrow \infty$ 55. $\dfrac{1}{3}$

57. (a) At $x = a$, the rate of increase begins to go down.
    (b) This is the point of inflection where rate of increase changes from positive to negative.
    (c) The horizontal asymptote.
59. Absolute max. $= 3$; absolute min. $= -75$

Exercise Set 9.6, page 624

1. 26 and 26   3. 7 and 7
5. 1000 cubic inches   7. 24 by 24
9. 30 by 45
11. 10 centimeters; 16,000 cubic centimeters
13. 300 by 250
15. (a) $C'(x) = 10 + 0.02x$

    (b) $\overline{C}(x) = 10 + \dfrac{30}{x} + 0.01x$

    (c) $\overline{C}'(x) = \dfrac{-30}{x^2} + 0.01$

    (d) Both at $x = 10\sqrt{30}$
17. 7 units
19. $\dfrac{50}{3}$ or 17 tables; profit $= \$233$
21. Maximum profit $= \$516{,}750$ at $x = 4050$
23. 110 people   25. ≈ 100 items
27. 18 orders per year of 22 cars
29. 6 weeks   31. 30 trees
33. 40 people   35. 5 more trees
37. (a) absolute minimum of 1.376 at $x = 6.4$
    (b) $x = 5$, max. at $N(5) = 10$
39. (a) $N'(t) = 3 - 3t^2$ (b) $t = 1$
41. Vertical asymptote, $x = 2$; Horizontal asymptote, $y = 0$; no critical point; no extrema; decreasing, $(-\infty, 2)$, $(2, \infty)$; concave down, $(-\infty, 2)$; concave up, $(2, \infty)$

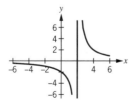

43. Relative max. at $(2\sqrt{2}, 4)$; relative min. at $(-2\sqrt{2}, -4)$; inflection point, $(0, 0)$; concave up, $(-4, 0)$; concave down, $(0, 4)$; increasing, $(-2\sqrt{2}, 2\sqrt{2})$; decreasing, $(-4, -2\sqrt{2})$, $(2\sqrt{2}, 4)$; No asymptotes

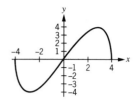

Exercise Set 9.7, page 635

1. (a) 1   (b) 3   (c) 3

3. $\left|E(1)\right| = \dfrac{2}{3}$; inelastic

5. $\left|E(3)\right| = 1$; unit elasticity

7. $\left|E(1)\right| = 2$; elastic

9. (a) $E(p) = \dfrac{p}{40 - p}$

    (b) $\left|E(10)\right| = \dfrac{1}{3}$

    (c) Inelastic
    (d) Price increase of 5% at $p = 10$ would result in $\dfrac{1}{3} \cdot 0.05 = 1.67\%$ change in demand.

    (e) $\left|E(30)\right| = 3$
    (f) Elastic
    (g) Price increase of 5% at $p = 30$ would result in $3 \cdot 0.05 = 15\%$ change in demand.

11. Elastic   13. Inelastic
15. Inelastic for $0 \leq p < 20$, elastic for $20 < p \leq 40$
17. Inelastic for $0 \leq p < 33\frac{1}{3}$, elastic for $33\frac{1}{3} < p \leq 100$
19. At $x = 4$, $\left|E(4)\right| = 0.17 < 1$, so the elasticity of cost is inelastic and a 10% increase in production produces a $(10\%)(0.17) = 1.7\%$ change in cost. At $x = 10$, $\left|E(10)\right| = 0.52 < 1$, so the elasticity of cost is inelastic and a 10% increase in production produces a $(10\%)(0.52) = 5.2\%$ change in cost. At $x = 20$, $\left|E(20)\right| = 1$, so there is unit elasticity and a 10% increase in production produces a 10% change in cost.
21. $R = p[100(100 - p)] = 10{,}000p - 100p^2$; increases and is inelastic over $(0, 50)$; decreases and is elastic over $(50, 100)$

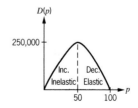

23. $R = 1,000,000p - 20,000p^2 + 100p^3$; increases and is inelastic over $\left(0, \dfrac{100}{3}\right)$; decreases and is elastic over $\left(\dfrac{100}{3}, 100\right)$

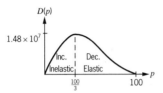

25. Critical points, $(0, 0)$, $(3, -9)$; relative minimum, $(3, -9)$; inflection points, $(0, 0)$, $\left(2, -\dfrac{16}{3}\right)$; decreasing, $(-\infty, 0)$, $(0, 3)$; increasing, $(3, \infty)$; concave up, $(-\infty, 0)$, $(2, \infty)$; concave down, $(0, 2)$; no horizontal or vertical asymptotes

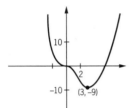

27. Critical points, $\left(\dfrac{2}{3}, \dfrac{64}{27}\right)$, $(2, 0)$; relative minimum, $(2, 0)$; relative maximum, $\left(\dfrac{2}{3}, \dfrac{64}{27}\right)$; inflection point, $\left(\dfrac{4}{3}, \dfrac{32}{27}\right)$; decreasing, $\left(\dfrac{2}{3}, 2\right)$; increasing, $\left(-\infty, \dfrac{2}{3}\right)$, $(2, \infty)$; concave up, $\left(\dfrac{4}{3}, \infty\right)$; concave down, $\left(-\infty, \dfrac{4}{3}\right)$; no horizontal or vertical asymptotes

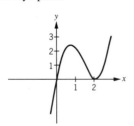

Chapter 9 Test, page 637

1. (a) (b, c), (d, e)
   (b) (a, b), (c, d), (e, f)
   (c) $x = c$, $x = e$

2. (a) $x = b$, $x = d$
   (b) $x = b$
   (c) $x = a$

3. (a) $x = g$, $x = h$
   (b) (a, g), (h, e), (e, f)
   (c) (g, h)

4. (a) (a, g), (h, f)
   (b) (g, h)
   (c) $x = e$

5. (a) Increasing, $(-\infty, 1)$; decreasing $(1, \infty)$
   (b) Relative max. at $(1, 2)$
   (c)

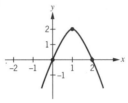

6. (a) Concave up, $(-\infty, 0)$, $\left(\frac{1}{2}, \infty\right)$; concave down, $\left(0, \frac{1}{2}\right)$
   (b) Inflection points, $(0, 0)$, $\left(\frac{1}{2}, -\frac{1}{16}\right)$
   (c) Relative min. at $\left(\frac{3}{4}, -\frac{27}{256}\right)$
   (d)

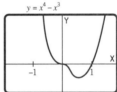 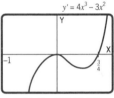

$X: [-2, 2]$ and $Y: [-.25, .50]$ $X: [-1, 1]$ and $Y: [-0.5, 0.5]$

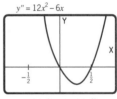

$X: [-1, 1]$ and $Y: [-1, 2]$

7. (a) Increasing, $(-\infty, -1)$, $(1, \infty)$; decreasing $(-1, 0)$, $(0, 1)$
   (b) Concave up, $(0, \infty)$; concave down, $(-\infty, 0)$
   (c) Relative min. at $(1, 2)$; rel. max. at $(-1, -2)$
   (d) Vertical asymptote, $x = 0$
   (e)

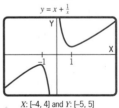

$X: [-4, 4]$ and $Y: [-5, 5]$ $X: [-3, 3]$ and $Y: [-2, 2]$

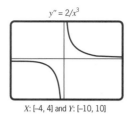

X: [-4, 4] and Y: [-10, 10]

$y'' = 2/x^3$

8. Order 100 radios, 50 times per year.

9. 300 cars

10. $|E(2)| = \frac{1}{2} < 1$; therefore, the demand is inelastic and a 5% price increase will cause a 2.5% change in demand.

Exercise Set 10.1, page 646

1. $\dfrac{dy}{dx} = 4e^{4x}$

3. $\dfrac{dy}{dx} = 2e^{2x}$

5. $\dfrac{dy}{dx} = 4 + 3e^{-3x}$

7. $\dfrac{dy}{dx} = 4 - 3e^{3x}$

9. $\dfrac{dy}{dx} = 2xe^{x^2}$

11. $\dfrac{dy}{dx} = \dfrac{e^x}{2\sqrt{e^x + 4}}$

13. $\dfrac{dy}{dx} = 3e^{-x} + 2e^x$

15. $\dfrac{dy}{dx} = 12x^2e^{4x} + 6xe^{4x}$

17. $\dfrac{dy}{dx} = 3e^{3x} - e^x$

19. $\dfrac{dy}{dx} = \dfrac{2e^{2x} + e^x}{2\sqrt{e^{2x} + e^x}}$

21. $f''(x) = e^{x-1}(x + 2)$

23. $f''(x) = e^{x^2+x}(4x^2 + 4x + 3)$

25. $y = 3x - 1$

27. $\dfrac{dy}{dx} = 3e^{3x} + 5e^x$

29. $\dfrac{dy}{dx} = 2xe^{2x} + e^{2x} + 4xe^{2x^2}$

31. $\dfrac{dy}{dx} = 3e^{3x} + 2xe^{x^2}$

33. (a) $\dfrac{dV}{dt} = -36{,}193.50$ (b) $\dfrac{dV}{dt} = -29{,}632.73$

(c) $\dfrac{dV}{dt} = -14{,}715.18$

35. (a) $\dfrac{dS}{dt} = 7.2$ (b) $\dfrac{dS}{dt} = 2.2$ (c) $\dfrac{dS}{dt} = 0.3$

37.

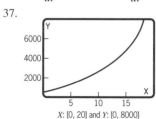

X: [0, 20] and Y: [0, 8000]

39.

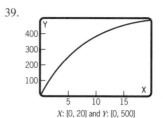

X: [0, 20] and Y: [0, 500]

41.

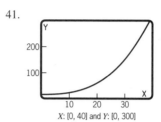

X: [0, 40] and Y: [0, 300]

43.

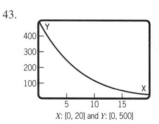

X: [0, 20] and Y: [0, 500]

45. (a) $N'(10) = 2.2$ (b) $N'(20) = 1.6$

47. $N'(t) = 240e^{-0.2t}(1 - 0.4e^{-0.2t})^2$

Exercise Set 10.2, page 654

1. (a) $\log_3 81 = 4$ (b) $\log_2 128 = 7$
   (c) $\log_3 \dfrac{1}{9} = -2$

3. (a) $7^2 = 49$ (b) $6^2 = 36$ (c) $3^{-2} = \dfrac{1}{9}$

5. (a) $\log_2 16 = 4$ (b) $\log_2 \dfrac{1}{8} = -3$
   (c) $\log_2 \dfrac{1}{8} = 3$ (d) $\log_3 27 = 3$

7. (a) $b = 4$ (b) $b = \dfrac{1}{2}$

9. (a) $x = 2$ (b) $x = 1000$

11. $2 \ln x + 4 \ln z - \ln y$

13. $\dfrac{2}{3} \log_b x - \dfrac{1}{3} \log_b y$ 15. $\ln x + \dfrac{1}{2} \ln y - \dfrac{1}{2} \ln z$

17. 0.73916   19. 0.74055

21. 5.98348   23. $x = 0$

25. $x = 1.125$

27. $x = e$ and $-e$

29. $x = -400$

31. $x = 1$

33. $x = 1.9769$

35. $x = 0.6442$

37. $x = -0.1567$

39. 2002

41. 8.7 years

43. (a) $RS = 6$ (b) $I = 50,118,723 I_0$

45. $\dfrac{dy}{dx} = 15x^2 e^{5x^3}$

Exercise Set. 10.3, page 659

1. $\dfrac{dy}{dx} = \dfrac{3}{x}$

2. $\dfrac{dy}{dx} = \dfrac{2}{2x + 3}$

5. $\dfrac{dy}{dx} = \dfrac{12}{3x + 2}$

7. $\dfrac{dy}{dx} = \dfrac{12x + 4}{3x^2 + 2x + 5}$

9. $\dfrac{dy}{dx} = 6x + 2 + \dfrac{2}{2x + 5}$

11. $\dfrac{dy}{dx} = \dfrac{9x^2}{3x + 2} + 6x \ln (3x + 2)$

13. $\dfrac{dy}{dx} = \dfrac{3}{x} (\ln x)^2$

15. $\dfrac{dy}{dx} = \dfrac{60x^3}{x^2 + 1} + 60x \ln (x^2 + 1)$

17. $\dfrac{dy}{dx} = \dfrac{14x + 1}{4x^2 + 2x}$

19. $\dfrac{dy}{dx} = \dfrac{(12x - 4) \ln (3x^2 - 2x + 5)}{3x^2 - 2x + 5}$

21. 15 items

23. 5000 pounds

25. $\dfrac{dR}{dt} = \dfrac{0.67}{E \ln 10}$

27. $pH' = \dfrac{-1}{H^+ \ln 10}$. As H^+ increases, $1/H^+$ decreases as does $\ln (1/H^+)$, thus, pH decreases.

29. $x = 5^3 = 125$

31. $\dfrac{dy}{dx} = \dfrac{x + 3}{x e^{3/x}}$

33. $\dfrac{dy}{dx} = \dfrac{12e^{-3x} - 12e^{2x} - 10e^{-x}}{(3 + e^{-3x})^2}$

Exercise Set 10.4, page 664

1.

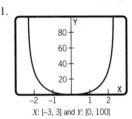

X: [-3, 3] and Y: [0, 100]

3.

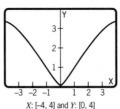

X: [-4, 4] and Y: [0, 4]

5. (a) x-intercept: none; y-intercept: $y = 1/e^4$
   (b) Domain: all real x
   (c) Always increasing
   (d) No relative extrema
   (e) Always concave up
   (f) As $x \to \infty$, $y \to \infty$; as $x \to -\infty$, $y \to 0$

7. (a) x-intercept: $x = 5$; y-intercept: none
   (b) Domain: $x > 4$
   (c) Always increasing
   (d) No relative extrema
   (e) Concave down on $(4, \infty)$
   (f) As $x \to \infty$, $y \to \infty$; as $x \to 4^+$, $y \to -\infty$

9. (a) x-intercept: none; y-intercept: $y = 1/e^4$
   (b) Always increasing; no relative extrema
   (c) Always concave up
   (d) $x \to \infty$, $y \to \infty$; $x \to -\infty$, $y \to 0$
   (e)

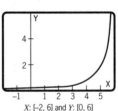

X: [-2, 6] and Y: [0, 6]

11. (a) x-intercept: $x = 5$; y-intercept: none
    (b) Increasing $(4, \infty)$; no relative extrema
    (c) Concave down $(4, \infty)$
    (d) $x \to \infty$, $y \to \infty$; $x \to -\infty$, the function is not defined
    (e)

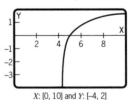

X: [0, 10] and Y: [-4, 2]

13. (a) x-intercept: $x = 0$; y-intercept: $y = 0$
    (b) Increasing for $x < 1$; decreasing for $x > 1$; at $x = 1$ is a relative maximum
    (c) Concave down for $x < 2$; concave up for $x > 2$
    (d) As $x \to \infty$, $y \to 0$; as $x \to -\infty$, $y \to -\infty$

(e)

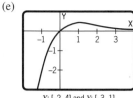

X: [–2, 4] and Y: [–3, 1]

15. (a) x-intercept: $x = 1$; y-intercept: none
(b) Increasing for $x > 1/e^{1/2}$; decreasing for $x < 1/e^{1/2}$; there is a relative minimum at $x = 1/e^{1/2}$.
(c) Concave down on $(0, 1/e^{3/2})$; concave up for $x > 1/e^{3/2}$
(d) As $x \to \infty$, $y \to \infty$; as $x \to 0$, $y \to 0$
(e)

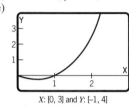

X: [0, 3] and Y: [–1, 4]

17. Relative minimum at $x = 0$; relative maxima at $x = \pm 1$

19.

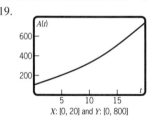

X: [0, 20] and Y: [0, 800]

21. (a) Relative maximum at $x = 0$; there are inflection points at $x = \pm 1$.
(b)

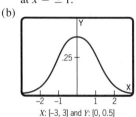

X: [–3, 3] and Y: [0, 0.5]

23. (a) Always increasing, $t > 0$.
(b) Always concave down, $t > 0$ (c) $N \to 1000$
(d)

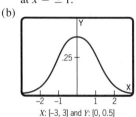

X: [0, 10] and Y: [0, 900]

25. $\dfrac{dy}{dx} = 6x^2 e^{3x^2} + e^{3x^2}$

27. $\dfrac{dy}{dx} = x + 2x \ln x$

29. $\dfrac{dy}{dx} = \dfrac{2x}{1 + x^2} - \dfrac{1}{x}$

Exercise Set 10.5, page 673

1. $\dfrac{dy}{dx} = 8x - 3$

3. $\dfrac{dy}{dx} = \dfrac{-1 - 6x}{2}$

5. $\dfrac{dy}{dx} = \dfrac{3 - 8x}{7}$

7. $\dfrac{dy}{dx} = \dfrac{3 - 2x}{5}$

9. $\dfrac{dy}{dx} = \dfrac{1 + 2x - 9x^2}{2}$

11. $\dfrac{dy}{dx} = \dfrac{8x - 3y}{3x}$

13. $\dfrac{dy}{dx} = \dfrac{3 - 4xy}{2x^2}$

15. $\dfrac{dy}{dx} = \dfrac{-3y}{1 + 3x}$

17. $\dfrac{dy}{dx} = \dfrac{-3 - 4xy}{1 + 2x^2}$

19. $\dfrac{dy}{dx} = \dfrac{2 + 6xy}{1 - 3x^2}$

21. $\dfrac{dy}{dx} = \dfrac{3}{2y + 1}$

23. $\dfrac{dy}{dx} = \dfrac{-8x}{6y - 1}$

25. $\dfrac{dy}{dx} = \dfrac{4 - y}{2y + x}$

27. $\dfrac{dy}{dx} = \dfrac{3y^2}{2 - 6xy}$

29. $\dfrac{dy}{dx} = \dfrac{3y - 2y^2 - 3x^2}{4xy - 3x}$

31. $y = -\dfrac{2}{3}x + \dfrac{5}{3}$

33. $y = -\dfrac{1}{7}x + \dfrac{9}{7}$

35. $y = \dfrac{3}{2}x + 5$

37. $\dfrac{dy}{dx} = \dfrac{-1}{2y - 5}\Big|_{(2, 3)} = -1$

39. $\dfrac{dy}{dx} = \dfrac{-1 - 4x - 2y}{2x + y}\Big|_{(-1, 3)} = -3$

41. $\dfrac{dy}{dx} = -\dfrac{y}{x}\Big|_{(3, 2)} = -\dfrac{2}{3}$

43. $\dfrac{dy}{dx} = (10^{3x+4})(3 \ln 10)$

45. $\dfrac{dy}{dx} = \left(\dfrac{\sqrt{x + 1}}{x^2}\right)\left(\dfrac{1}{2x + 2} - \dfrac{2}{x}\right)$

47. $\dfrac{dy}{dx} = \left(\dfrac{3x^2 (2x + 3)^{1/3}}{(4x^3 - 3x + 1)^2}\right)\left(\dfrac{2}{x} + \dfrac{2}{6x + 9} - \dfrac{24x^2 - 6}{4x^3 - 3x + 1}\right)$

49. $y'' = \dfrac{2y' - 6x - 6y(y')^2}{3y^2 - x}$

51. $y = \pm\sqrt{\dfrac{5x^2 - 4}{4}};\ y' = \dfrac{5x}{4y},\ y'(2, 2) = \dfrac{5}{4},$

$y'(-2, 2) = -\dfrac{5}{4},\ y'(2, -2) = -\dfrac{5}{4},\ y'(-2, -2) = \dfrac{5}{4};$

yes, there is agreement.

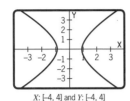

X: [-4, 4] and Y: [-4, 4]

53. (c)

55. $\dfrac{dp}{dx} = \dfrac{-x^2}{p^2}$

57. $p' = -\dfrac{1200}{1331}$

59. $\dfrac{dy}{dx} = \dfrac{4x}{x^2 + 3} + \dfrac{5}{10x + 2}$

Exercise Set 10.6, page 679

1. $\dfrac{dy}{dt} = -3$

3. $\dfrac{dy}{dt} = \dfrac{9}{2}$

5. $\dfrac{dy}{dt} = \dfrac{3}{\sqrt{2}}$

7. $\dfrac{dy}{dt} = \dfrac{9}{2\sqrt{2}}$

9. $\dfrac{dy}{dt} = 12$

11. $\dfrac{dy}{dt} = -3$

13. $\dfrac{dy}{dt} = -\dfrac{9}{2}$

15. $\dfrac{dy}{dt} = -\dfrac{5}{3}$

17. $\dfrac{dy}{dt} = 15$

19. $\dfrac{dy}{dt} = -\dfrac{14}{5}$

21. (a) $y'(2) = \sqrt{2}$ cm/sec   (b) $y'(9) = 2/3$ cm/sec

23. (a) $A'(1) = 16$ cm²/sec   (b) $A'(2) = 32$ cm²/sec

25. $\dfrac{dA}{dt} = 72$ cm²/sec

27. $\dfrac{dp}{dt} = -\dfrac{9}{5}$

29. $\dfrac{dR}{dt} = -\dfrac{5}{98}$

31. $\dfrac{dy}{dx} = -3$

33. $\dfrac{dy}{dx} = -\dfrac{2}{3}$

35. $\dfrac{dy}{dx} = \dfrac{-2y^2}{4xy + 1}$

37. $\dfrac{dy}{dx} = \dfrac{ye^x + 1}{1 - e^x}$

39. $\dfrac{dy}{dx} = \dfrac{2x}{x^2 + 2}$

Chapter 10 Test, page 680

1. $\dfrac{dy}{dx} = \dfrac{8x - 3y}{3x}$

2. $\dfrac{dy}{dt} = -\dfrac{16}{3}$

3. $\dfrac{dw}{dz} = 8ze^{z^2 + 5}$

4. $\dfrac{dr}{ds} = \dfrac{12s + 24}{s^2 + 4s}$

5. $\dfrac{dy}{dx} = 2 + \ln x^2$

6. (a) Maximum: approximately 1/3   (b) Maximum: 1/e

7. $y' = \dfrac{e^x}{1 + e^x}$

8. (a) (i)   Increasing for $x < 1$; decreasing for $x > 1$
       (ii) Concave up for $x > 2$; concave down for $x < 2$
       (iii)

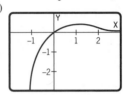

X: [-2, 3] and Y: [-3, 1]

(b) (i)   Increasing for $x < 1$; decreasing for $x > 1$
    (ii) Concave up for $x > 2$; concave down for $x < 2$
    (iii) Same as in part (a)

9. $y' = \dfrac{x + 2x \ln x + e^y}{1 - xe^y}$ 10. 1193.46

11. $y' = \dfrac{[1/(x + 1)] - 2xe^y}{x^2e^y + (1/y)}$

12.

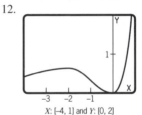

X: [-4, 1] and Y: [0, 2]

13. Relative maximum, $(-2, 0.54)$; relative minimum, $(0, 0)$

14. Increasing for $x > 0$ and $x < -2$; decreasing for $-2 < x < 0$

15. Concave up for $x < -2 - \sqrt{2}$ and $x > -2 + \sqrt{2}$; concave down for $-2 - \sqrt{2} < x < -2 + \sqrt{2}$

Exercise Set 11.1, page 693

1. $5x + C$

3. $\sqrt{3}x + C$

5. $-x^2 + C$

7. $\dfrac{2\sqrt{2}}{3}x^{3/2} + C$

9. $2x^3 + C$

11. $\dfrac{3}{4}x^4 + C$

13. $-\dfrac{6}{x} + C$

15. $8u^2 + C$

17. $\dfrac{8}{3}t^{3/2} + C$

19. $\dfrac{2}{3}x^{3/2} + C$

21. $-\dfrac{1}{2x^2} + C$

23. $-\dfrac{3}{2x^2} + C$

25. $-\dfrac{1}{x} + C$

27. $-\dfrac{1}{x} - x^2 + C$

29. $8\sqrt{u} - \dfrac{2}{3}u^{3/2} + C$

31. $\dfrac{-4x^{-3}}{3} + x + \dfrac{x^2}{2} + C$

33. $f(x) = 2x + \dfrac{3}{2}x^2 + 10$

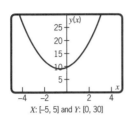

X: [-5, 5] and Y: [0, 30]

35. $y(u) = \dfrac{-10}{u} + 4$

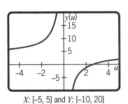

X: [-5, 5] and Y: [-10, 20]

37. $C(x) = 0.1x^2 + \dfrac{4}{3}x^3 + 10$

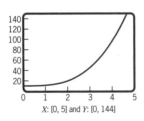

X: [0, 5] and Y: [0, 144]

39. $-\dfrac{2}{3}x^{3/2} + \dfrac{4}{5}x^{5/2} + \dfrac{2}{7}x^{7/2} + C$

41. $\dfrac{1}{x} + 3x - \dfrac{2}{3}x^3 + x^4 + C$

43. $-3x + \dfrac{8}{3}x^{3/2} + C$

45. $\dfrac{3}{x} + x - \dfrac{7}{2}x^2 + \dfrac{x^3}{3} + C$

47. $-\dfrac{9}{5}x^{5/3} + \dfrac{6}{11}x^{11/3} + C$

49. $\dfrac{9}{2}x^{2/3} - \dfrac{6}{5}x^{5/3} + \dfrac{3}{8}x^{8/3} + C$

51. $\dfrac{4}{3}\sqrt{x} - 5x^{3/5} + C$

53. 28

55. $P(x) = 500x - 2x^2$

57. $N(t) = -6t^2 + 60t + 1200$, $N(5) = 1350$, $N(8) = 1296$, $N(t) = 0$ when $t = 20$

59. $L(30) = 21.6$ and $L(40) = 35.2$

Exercise Set 11.2, page 703

1. $6\ln|x| + C$

3. $4e^x + C$

5. $6e^x + x^2 + C$

7. $e^x + 2\ln|x| + C$

9. $3\ln|x| + C$

11. $\dfrac{1}{2}e^{2t} + \dfrac{1}{t} + C$

13. $\dfrac{1}{10e^{10x}} + 2\ln|x| + C$

15. $\dfrac{-150}{e^{0.02t}} + C$

17. $\dfrac{-1}{8e^{2x}} - \dfrac{\ln|x|}{2} + C$

19. $-2e^x + \dfrac{e^{2x}}{2} + C$

21. $\dfrac{x^2}{2} + \ln|x| + C$

23. $\dfrac{-10}{3e^{0.3x}} - \dfrac{10}{e^{0.1x}} + 2\ln|x| + C$

25. $\dfrac{1}{2}e^{2x} + x + \dfrac{11}{2}$

27. $x + 3\ln|x| + 3$

29. 600,000 at $x = 1000$; $R(2000) = 800,000$ is maximum revenue

31. $D(x) = 6000e^{-0.05x} + 6360.82$

33. $R(x) = 4000x - \dfrac{5}{2}x^2$

35. (a) $P'(x) = 1000 - x$

(b) $P(x) = 1000x - \dfrac{x^2}{2} - 5000$, $P(100) = 90,000$

(c) $P(10) = 4950$

37. $\dfrac{10}{3}x^{3/2} + C$

39. $2\sqrt{x} - \dfrac{2}{5}x^{5/2} + C$

Exercise Set 11.3, page 710

1. $dy = 0.2$, $\Delta y = 0.2$

3. $dy = 0.4$, $\Delta y = 0.41$

5. $dy = 0.075$, $\Delta y = 0.0761905$

7. $dy = 0.02$, $\Delta y = 0.02$

9. $dy = 0.04$, $\Delta y = 0.0401$

11. $dy = 0.0075$, $\Delta y = 0.00751244$

13. $du = (2t + 4)\, dt$

15. $du = 0.3e^{0.3x}\, dx$

17. $dv = -\dfrac{1}{(x-1)^2}\, dx$

19. $dp = \dfrac{2t}{(1+t)^2}\, dt$

21. $du = 0.6xe^{0.3x^2 + 6}\, dx$

23. $dv = -\dfrac{x}{(x^2-1)^{3/2}}\, dx$

25. $dp = \dfrac{t}{1 + t^2}\, dt$

27. $9x - 3x^2 + \dfrac{1}{3}x^3 + C$ or $\dfrac{(x - 3)^3}{3} + C$

29. $9x - 6x^2 + \dfrac{4}{3}x^3 + C$ or $\dfrac{(2x - 3)^3}{6} + C$

31. $4x + 6x^2 + 3x^3 + C$ or $\dfrac{(3x + 2)^3}{9} + C$

33. $\dfrac{1}{5}e^{5x} + C$ 35. $\dfrac{1}{3}\ln|3x + 2| + C$

37. $-7x + \dfrac{5}{2}x^2 + x^3 + C$

Exercise Set 11.4, page 718

1. $4(x + 3)^{3/2} + C$ 3. $-\dfrac{4}{3}(4 - 3x)^{3/2} + C$

5. $\dfrac{1}{3}e^{3x^2} + C$ 7. $\dfrac{-3}{10e^{5x^2}} + C$

9. $\dfrac{3}{28}(3 + 7x)^{4/3} + C$ 11. $\dfrac{(2x^2 - 7)^{7/4}}{7} + C$

13. $\dfrac{(x^2 + 4)^{3/2}}{3} + C$ 15. $\dfrac{2(1 + x^3)^{3/2}}{9} + C$

17. $x + 2\ln|x - 1| + C$ 19. $2x + 6\ln|3 - x| + C$

21. $\dfrac{(x + 1)^6}{6} - \dfrac{(x + 1)^5}{5} + C$

23. $\dfrac{4}{5}(x - 5)^{5/2} + \dfrac{20}{3}(x - 5)^{3/2} + C$

25. $\dfrac{1}{2}\ln(1 + e^{2x}) + C$ 27. $-\dfrac{x}{3} - \dfrac{\ln|1 - 3x|}{9} + C$

29. $3(x + 1)^{4/3} - \dfrac{3}{7}(x + 1)^{7/3} + C$

31. $2e^{\sqrt{x-1}} + C$ 33. $\dfrac{1}{3}(\sqrt{x} + 3)^6 + C$

35. Because $1/(3x + 4)^2$ can be written as $(3x + 4)^{-2}$, the power rule can be used if $n \neq -1$; however, for $1/(3x + 4)$, $n = -1$ and the power rule does not apply.

37. $C(10) = 1900$

39. $R(x) = (x^2 + 16)^{1/2} - 4; R(3) = 1$

41. $N(t) = -80e^{0.02t} + 160$

43. $-\dfrac{1}{4}e^{-4x} + C$ 45. $\ln|x + 1| + C$

Exercise Set 11.5, page 729

1. 2   3. 7

5. 12   7. 30

9. $\dfrac{e^8 - 1}{2}$ 11. 31.5

13. $\dfrac{8}{3}$ 15. 3.159

17. 3.810   19. 1.39

21. 0.25   23. 0.5728

25. 0.6095   27. 4.20

29. 218.8   31. 10

33. 36   35. -18

37. 192.7 hours   39. 5500

41. $2000; $4000; $2000   43. (a) -0.9 (b) -3

45. 60   47. $\dfrac{3}{8}(2x + 3)^{4/3} + C$

49. $\dfrac{2(x - 1)^{5/2}}{5} + \dfrac{2(x - 1)^{3/2}}{3} + C$

Exercise Set 11.6, page 742

1. 15   3. 12

5. 4   7. $\dfrac{14}{3}$

9. $\dfrac{76}{3}$ 11. 26.7991

13. 0.202733   15. $\dfrac{31}{2}$

17. $\dfrac{4}{3}$ 19. 18

21. $166\frac{2}{3}$ 23. 4.25

25. $9\frac{1}{3}$ 27. $\dfrac{4}{3}$

29. 4.5   31. 37/12

33. $21\frac{1}{3}$ 35. $49\frac{1}{3}$

37. $22\frac{2}{3}$ 39. 241.54

41. $\dfrac{-1}{3e^{x^3}} + C$ 43. $\dfrac{1}{10}$

Exercise Set 11.7, page 752

1. 780   3. 11.5

5. $\Delta x = \dfrac{1 - 0}{10} = .1;\ c_1 = .05,\ c_2 = .15,\ c_3 = .25 \ldots$

$c_{10} = .95; 0.3325$

7. 3   9. 16

11. 10   13. 5.1875

15. 53.044

17. Answers will vary. Select $x = 1, 2, 3, 4, 5$

$F(1) = .667$, $F(2) = 1.886$, $F(3) = 3.464$, $F(4) = 5.333$,
$F(5) = 7.453$

Graph is same as graph of $f(x) = \dfrac{2}{3}x^{3/2}$

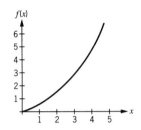

19. Answers will vary. Select $x = 1, 2, 3, 4, 5$

$F(1) = -.75$, $F(2) = 0$, $F(3) = 11.25$, $F(4) = 48$,
$F(5) = 131.25$

Graph is same as graph of $f(x) = \dfrac{x^4}{4} - x^2$ for $x \geq 0$

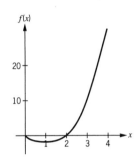

21. (c)

23. 75,875.80

25. $-6\sqrt{x} + \dfrac{2}{3}x^{3/2} + \dfrac{4}{5}x^{5/2} + C$

27. $\ln(1 + e^x) + C$

29. $\dfrac{10}{3}(x + 2)^{3/2} - 20(x + 2)^{1/2} + C$

31. -3.93147

33. 93

Exercise Set 11.8, page 764

1. (a) $\dfrac{1}{3}$ (b) $\dfrac{1}{2}$

3. 15,401.86

5. 16.229

7. 4.51188

9. (a)–(c)

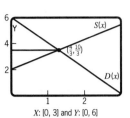

X: [0, 3] and Y: [0, 6]

(d) $CS = \dfrac{16}{9}$, $PS = \dfrac{8}{9}$

11. (a)–(c)

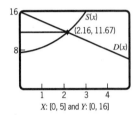

X: [0, 5] and Y: [0, 16]

(d) $CS = 4.70$, $PS = 6.73$

13. (a)–(c)

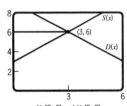

X: [0, 6] and Y: [0, 8]

(d) $CS = \dfrac{9}{2}$, $PS = \dfrac{9}{2}$

15. (a)–(c)

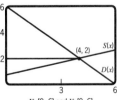

X: [0, 6] and Y: [0, 6]

(d) $CS = 8$, $PS = 2$

17. (a)–(c)

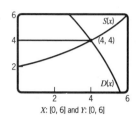

X: [0, 6] and Y: [0, 6]

(d) $CS = 10\frac{2}{3}$, $PS = 5\frac{1}{3}$

19. (a) 7.2%   (b) 67.2%   (c) .267

21. $CS = 49.78$

23. (a) (45, 11.4)   (b) $CS = \$211.20$
    (c) $PS = \$103.25$
    (d)

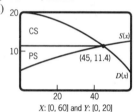

X: [0, 60] and Y: [0, 20]

25. $A = \$76,596.31$, $P = \$34,416.94$

27. (a)

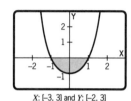

(b) Bottom 60% receive only 37.2% of income.
    (c) Coefficient = 0.317

29. Coefficient = 0.67

31. (a) $x = 125$ (b) 25,000   (c) 38,020.83
    (d) 13,020.83   (e) Yes

33. $4\frac{2}{3}$

Chapter 11 Test, page 766

1. $\frac{1}{2}x + \frac{4}{3}x^3 + \frac{1}{4}x^4 + C$ 2. $\frac{3}{2}(1 + x)^{2/3} + C$

3. $\frac{4}{x} + \frac{7}{2x^2} + \frac{-2}{x^3} + C$ 4. $-\frac{1}{3}(1 - x^2)^{3/2} + C$

5. $\frac{\ln|4x + 1|}{2} + C$ 6. $5e^{4x} - \frac{2}{x^2} + C$

7. $2e^{3x^2} + C$ 8. $2x + 3\ln|x - 1| + C$

9. $\frac{1}{3}(x^2 - 3)^{3/2} + C$

10. $\frac{(x + 3)^6}{6} - \frac{3(x + 3)^5}{5} + C$

11. 105   12. 6.22

13. $du = 3x(3x^2 - 1)^{-1/2}\,dx$

14. $\frac{4}{3}$

X: [-3, 3] and Y: [-2, 3]

15. $\$8409.64$ 16. 426.67

17. 0.3

Exercise Set 12.1, page 780

1. $(3x - 3)e^x + C$ 3. $\frac{-2}{e^x} - \frac{2x}{e^x} + C$

5. $e^x(x - 2) + C$ 7. $\frac{-5}{9}e^{3x} + \frac{2}{3}xe^{3x} + C$

9. $\frac{-5}{4e^{2x}} - \frac{3x}{2e^{2x}} + C$ 11. $\frac{-5x^2}{4} + \frac{5x^2\ln|x|}{2} + C$

13. $\frac{-x^3}{9} + \frac{x^3\ln|x|}{3} + C$ 15. $\frac{-x^4}{16} + \frac{x^4\ln|x|}{4} + C$

17. $-\frac{4}{5}(x + 1)^{5/2} + 2x(x + 1)^{3/2} + C$

19. $(2x - 10)(2 + x)^{1/2} + C$

21. $-\frac{x}{4(x + 4)^4} - \frac{1}{12(x + 4)^3} + C$

23. $\frac{-1}{2(\ln|x|)^2} + C$ 25. $\frac{e^{2x}}{4(1 + 2x)} + C$

27. 111,883.7   29. 0.5

31. 0.647918   33. 0.718282

35. 0.386294   37. 0.0721318

39. $\frac{e^{3x^2}}{6} + C$ 41. $\frac{-2}{x} - \frac{2\ln|x|}{x} + C$

43. 1.3419   45. 2.12818

47. 12,002.04   49. $345.779 \approx 346$

51. $\$821.04$ 53. 7.68399

Exercise Set 12.2, page 788

1. $2 \ln|3x^2 + x| + C$

3. $-2x + \dfrac{x^2}{2} + 4 \ln|x + 2| + C$

5. $\dfrac{1}{8} e^{8x} + C$

7. $\dfrac{3}{2(x + 3)^2} - \dfrac{1}{x + 3} + C$

9. $\sqrt{2x} - 6 \ln|6 + \sqrt{2x}| + C$

11. $2\sqrt{x} - 2 \ln(1 + \sqrt{x}) + C$

13. $\dfrac{1}{27} (x^3 + 1)^9 + C$

15. $\dfrac{1}{2} \ln|x^2 - 2x + 5| + C$

17. $\dfrac{e^{x^2 + x + 1}}{2} + C$

19. $\dfrac{2}{3} (\ln|x|)^{3/2} + C$

21. $\dfrac{-1 - x}{e^x} + C$

23. 37.2222

25. 6.67

27. 2.65827

29. 2.667

31. 16

33. 3.66384

35. $-1 + \dfrac{e}{2} \approx 0.359$

37. $D(x) = 2000\sqrt{25 - x^2} + 5000$

39. $\displaystyle\int_0^8 \dfrac{t}{\sqrt{t^2 + 1}}\, dt = 7.06226$

41. $-\dfrac{4}{9} x^3 + \dfrac{4}{3} x^3 \ln|x| + C$

Exercise Set 12.3, page 801

1. $\dfrac{1}{2}$

3. 1

5. $\sqrt{2}$

7. Divergent

9. 1

11. $\dfrac{1}{2}$

13. Divergent

15. $\dfrac{1}{2}$

17. $\dfrac{1}{10}$

19. $\dfrac{1}{3}$

21. 0

23. Divergent

25. 1

27. $-\dfrac{1}{2}$

29. Divergent

31. \$33,333.33

33. \$187,500

35. \$16,666.67

37. $\displaystyle\int_0^\infty Re^{-rt}\, dt = \lim_{b \to \infty} \int_0^b Re^{-rt}\, dt = \lim_{b \to \infty} -\dfrac{R}{r} e^{-rt} \Big|_0^b =$

$0 + \dfrac{R}{r} = \dfrac{R}{r}$

39. $\displaystyle\int_{-\infty}^0 0\, dx + \int_0^\infty \dfrac{1}{a} e^{-x/a}\, dx = \lim_{b \to \infty} \int_0^b \dfrac{1}{a} e^{-x/a}\, dx =$

$\lim_{b \to \infty} - e^{-x/a} \Big|_0^b = 0 + 1 = 1$

41. $\sigma^2 = \displaystyle\int_a^b x^2 \left(\dfrac{1}{b - a}\right) dx - \left(\dfrac{a + b}{2}\right)^2 = \dfrac{(b - a)^2}{12}$ and

$\sigma = \dfrac{b - a}{\sqrt{12}}$

43. $\sigma^2 = \displaystyle\int_a^b x^2 \left(\dfrac{1}{a} e^{-x/a}\right) dx - a^2 = a^2$ and $\sigma = a$

45. 0.135

47. 0.63212

49. $\dfrac{1}{2}$

51. (a) $\dfrac{1}{2}$ (b) 25%   (c) 220

53. (a) 0.393   (b) 0.123

55. (a) $x = 1$ (b) $x = 1.628$

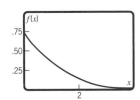

57. $\dfrac{15}{2} (e^x + 2)^{2/3} + C$

59. $(x - 2)e^{2x} + C$

61. $\dfrac{-x^5}{25} + \dfrac{x^5}{5} \ln|x| + C$

Exercise Set 12.4, page 810

1. $\dfrac{1}{10} \ln \left|\dfrac{x - 5}{x + 5}\right| + C$

3. $\dfrac{1}{12} \ln \left|\dfrac{6 + x}{6 - x}\right| + C$

5. $\ln|x + \sqrt{x^2 + 36}| + C$

7. $-\dfrac{1}{5} \ln \left|\dfrac{5 + \sqrt{25 + x^2}}{x}\right| + C$

9. $-\dfrac{\sqrt{49 + x^2}}{49x} + C$

11. $\dfrac{\sqrt{x^2 - 49}}{49x} + C$

13. $\dfrac{1}{8} [x(2x^2 + 25)\sqrt{x^2 + 25} - 625 \ln|x + \sqrt{x^2 + 25}|] + C$

15. $\sqrt{25 + x^2} - 5 \ln \left|\dfrac{5 + \sqrt{25 + x^2}}{x}\right| + C$

17. $\dfrac{1}{2} [x\sqrt{x^2 + 49} - 49 \ln|x + \sqrt{x^2 + 49}|] + C$

19. $\dfrac{1}{5} \ln \left| \dfrac{x-2}{x+3} \right| + C$

21. $2x^2 e^{x/2} - 8x e^{x/2} + 16 e^{x/2} + C$

23. $\dfrac{1}{3}(x-3)\sqrt{2x+3} + C$ 25. 0.52159

27. $\dfrac{1}{5} \ln \left| 5x + \sqrt{25x^2 + 16} \right| + C$

29. $-\dfrac{8}{5} \ln \left| \dfrac{5 + \sqrt{25 - 4x^2}}{2x} \right| + C$

31. $\dfrac{1}{72} [3x(18x^2 + 16)\sqrt{9x^2 + 16}$
    $- 256 \ln|3x + \sqrt{9x^2 + 16}|] + C$

33. $-\dfrac{1}{3} \dfrac{\sqrt{25x^2 + 16}}{x} + \dfrac{5}{3} \ln|5x + \sqrt{25x^2 + 16}| + C$

35. $\dfrac{1}{13} \ln \left| \dfrac{2x-3}{3x+2} \right| + C$

37. $PS = 135.21$

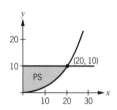

39. (a) \$2051.63   (b) \$2385.09   (c) \$10,239.50
41. 150.85   43. 4.2426
45. 1.5

Exercise Set 12.5, page 816

1. Trapezoidal rule: 3.34375; numeric: 3.333
3. Trapezoidal rule: 1.21819; numeric: 1.21895
5. Simpson's rule: 3.33333; numeric: 3.333
7. Simpson's rule: 1.21895; numeric: 1.21895
9. 0.001302   11. 8700
13. 933.08
15. (a) Divergent   (b) Divergent

Chapter 12 Test, page 817

1. Divergent

2. $\dfrac{2}{15}(3x-4)(x+2)^{3/2} + C$

3. $\dfrac{1}{60} \ln \left| \dfrac{5x-6}{5x+6} \right| + C$

4. $\dfrac{7e^8 + 1}{16} = 1304.23$

5. Simpson's rule: 4; numeric: 4

6. 21.5037
7. Trapezoidal rule: 22; numeric: 21.333

8. $e^x(2 - 2x + x^2) + C$ 9. $\dfrac{38}{3} = 12.66667$

10. $\text{pdf} = \dfrac{1}{7} e^{-x/7}, 0.681$

Exercise Set 13.1, page 826

1. 1   3. 3
5. 9   7. 7
9. 7   11. 6
13.

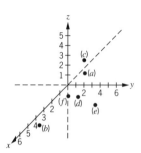

15. $3x + 7y = 14$, $(0, 2, 0)$, $\left(\dfrac{14}{3}, 0, 0 \right)$; $3x + 2z = 14$,
    $(0, 0, 7)$, $\left(\dfrac{14}{3}, 0, 0 \right)$; $7y + 2z = 14$, $(0, 2, 0)$, $(0, 0, 7)$

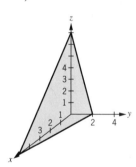

17. $5x + 2y = 15$, $(3, 0, 0)$, $\left(0, \dfrac{15}{2}, 0 \right)$; $5x + 3z = 15$,
    $(3, 0, 0)$, $(0, 0, 5)$; $2y + 3z = 15$, $\left(0, \dfrac{15}{2}, 0 \right)$, $(0, 0, 5)$

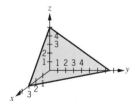

19. $2x - 3y = 12$, $(6, 0, 0)$, $(0, -4, 0)$; $2x - 4z = 12$, $(6, 0, 0)$, $(0, 0, -3)$; $-3y - 4z = 12$, $(0, -4, 0)$, $(0, 0, -3)$

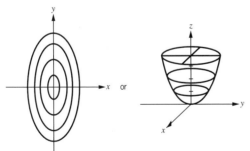

21.

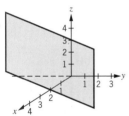

23.

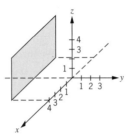

25.

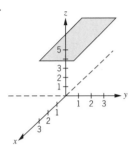

27.

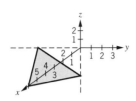

29.

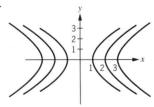

31. (a) $P(3, 2) = 44$ (b) $P(4, 1) = 53$
    (c) $P(3, 4) = 78$

33. $R(110, 10) = 11$. Ratio is 11 to 1.

35. 1600

37. (a) $\dfrac{6755}{13} = 519.615$ (b) $\dfrac{525}{16} = 32.8125$

Exercise Set 13.2, page 832

1. $f_x = 2$, $f_y = 0$, $f_x(3, 2) = 2$, $f_y(3, 2) = 0$
3. $f_x = 3$, $f_y = -2$, $f_x(3, 2) = 3$, $f_y(3, 2) = -2$
5. $f_x = -6x - 2y$, $f_y = -2x + 3y^2$,
   $f_x(3, 2) = -22$, $f_y(3, 2) = 6$
7. $f_x = 3e^x$, $f_y = -2e^y$, $f_x(3, 2) = 60.257$,
   $f_y(3, 2) = -14.778$
9. $f_x = \dfrac{3y + 6x}{xy + x^2}$, $f_y = \dfrac{3x}{xy + x^2}$, $f_x(3, 2) = \dfrac{8}{5}$, $f_y(3, 2) = \dfrac{3}{5}$
11. $z_x = ywe^{xyw}$, $z_y = xwe^{xyw}$, $z_w = xye^{xyw}$
13. $z_x = \dfrac{2x + y}{x^2 + xy + w^2}$, $z_y = \dfrac{x}{x^2 + xy + w^2}$,
    $z_w = \dfrac{2w}{x^2 + xy + w^2}$
15. $f_x(1, 1, 2) = 3$, $f_y(1, 1, 2) = 2$
17. $f_x\left(2, \dfrac{1}{2}, 1\right) = \dfrac{1}{2}$, $f_y\left(2, \dfrac{1}{2}, 1\right) = 2$
19. $A_i = nP(1 + i)^{n-1}$
21. (a) 28   (b) 22   (c) 6   (d) 26
23. $P_I(400, 10{,}000) = 2$, $P_C(400, 10{,}000) = 0.08$
25. (a) 130   (b) 76.923   (c) 10
    (d) The rate of change of I with respect to M is increasing

27.

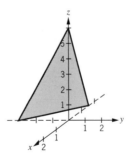

Exercise Set 13.3, page 839

1. (a) $C_x = 30$. Each additional item A costs $30 to produce.
   (b) $C_x = 20$. Each additional item B costs $20 to produce.
   (c) $R_x = 50$. Each additional item A yields $50 revenue.
   (d) $R_y = 30$. Each additional item B yields $30 revenue.
   (e) $P_x = 20$. Each additional item A produces $20 profit.
   (f) $P_y = 10$. Each additional item B produces $10 profit.
3. $R_x(8, 10) = 99.8$ is additional revenue by increasing production of Product A at (8, 10). $R_y(8, 10) = 59.84$ is addi-

tional revenue by increasing production of Product B at (8, 10).

5. $P_x = 4x - 2y$; additional profit from selling one additional unit of item 1 for given values of x and y. $P_y = 4y - 2x$; additional profit from selling one additional unit of item 2 for given values of x and y.

7. Neither   9. Complementary

11. Complementary

13. (a) $f_l(l, c) = 45 \left(\dfrac{c^{1/4}}{l^{1/4}} \right)$ (b) $f_c(l, c) = 15 \left(\dfrac{l^{3/4}}{c^{3/4}} \right)$

15. $f_x = 8xy - 3y^3$, $f_y = 4x^2 - 9xy^2$, $f_x(1, 2) = -8$, $f_y = (1, 2) = -32$

17. $f_x = 4y$, $f_y = 4x - 5e^{-y}$, $f_x(1, 2) = 8$, $f_y(1, 2) = 3.3233$

Exercise Set 13.4, page 843

1. $f_{xx} = 0$, $f_{yy} = 0$, $f_{xy}(1, 3) = 0$

3. $f_{xx} = 20y^3$, $f_{yy} = 60x^2y$, $f_{xy}(1, 3) = 540$

5. $f_{xx} = 10y$, $f_{yy} = -18xy$, $f_{xy}(1, 3) = -71$

7. $f_{xx} = 3e^x$, $f_{yy} = -2e^y$, $f_{xy}(1, 3) = 0$

9. $f_{xx} = 3y^2e^{xy}$, $f_{yy} = 3x^2e^{xy}$, $f_{xy}(1, 3) = 241.0264431$

11. $f_{xx} = \dfrac{18x(x^2 + 2y^2)}{(3x^2 + 2y^2)^2}$, $f_{yy} = \dfrac{4x(3x^2 - 2y^2)}{(3x^2 + 2y^2)^2}$,
$f_{xy}(1, 3) = 0.4081632655$

13. $f_{xx} = e^x \ln(3x + 2y) + \dfrac{6e^x(3x + 2y) - 9e^x}{(3x + 2y)^2}$,
$f_{yy} = \dfrac{-4e^x}{(3x + 2y)^2}$, $f_{xy}(1, 3) = 0.4027084189$

15. $f_{xx} = \dfrac{-1}{(2x + 3y)^{3/2}}$, $f_{yy} = \dfrac{-9}{4(2x + 3y)^{3/2}}$,
$f_{xy}(1, 3) = -0.04111518336$

17. $f_{xx} = 0$, $f_{yy} = \dfrac{2x}{y^3}$, $f_{xy}(1, 3) = \dfrac{-1}{9}$

19. $f_{xx} = e^{x-y}y^2$, $f_{yy} = (e^{x-y})(y^2 - 4y + 2)$,
$f_{xy}(1, 3) = -0.4060058498$

21. $f_{xxx} = 6y$, $f_{yyy} = 6x$, $f_{xyx} = 6x$, $f_{yxy} = 6y$

23. $f_{xxx} = e^{x+y}$, $f_{yyy} = e^{x+y}$, $f_{xyx} = e^{x+y}$, $f_{yxy} = e^{x+y}$

25. $f_{xxx} = \dfrac{2}{(x + y)^3}$, $f_{yyy} = \dfrac{2}{(x + y)^3}$, $f_{xyx} = \dfrac{2}{(x + y)^3}$,
$f_{yxy} = \dfrac{2}{(x + y)^3}$

27. $f_{ll} = \dfrac{-800C^{2/3}}{9l^{5/3}}$, $f_{CC} = \dfrac{-800l^{1/3}}{9C^{4/3}}$, $f_{lC} = \dfrac{800}{9l^{2/3}C^{1/3}}$

29. $f_x = 3y + \dfrac{2x}{x^2 + 2y^2}$, $f_x(1, 2) = \dfrac{56}{9}$, $f_y = 3x + \dfrac{4y}{x^2 + 2y^2}$,
$f_y(1, 2) = \dfrac{35}{9}$

Exercise Set 13.5, page 850

1. $(0, 1)$

3. $(1, 0)$, $(1, 1)$, $(1, -1)$

5. $(-3, -\frac{3}{2})$

7. $(\frac{1}{2}, -1)$ is a relative minimum.

9. $(-1, 0)$ is a relative minimum.

11. $(0, 0)$, relative minimum   13. $(0, 0)$, relative maximum

15. $(0, 0)$, relative maximum   17. $(0, 0)$, saddle point

19. $(2, 0)$, relative maximum

21. $(-2, 3)$, saddle point

23. $(0, 0)$, saddle point

25. $(2, -1)$, relative minimum

27. $(3, 2)$, relative maximum

29. $(0, 0)$, saddle point; $(0, 3)$, saddle point; $(3, 0)$, saddle point; $(1, 1)$, relative maximum

31. $(0, -1)$, relative maximum; $(0, 3)$, saddle point; $(4, -1)$, saddle point; $(4, 3)$, relative minimum

33. If the line segment is divided into three equal parts, the sum of the squares of their length is a minimum

35. $C(2, 3) = 138$; $\$138,000$

37. Critical point is $(\frac{91}{38}, \frac{103}{19})$; $\$211.68$

39. $f_{xx} = \dfrac{-2x^2 + 6y^2}{(x^2 + 3y^2)^2}$, $f_{xy} = \dfrac{-12xy}{(x^2 + 3y^2)^2}$

Exercise Set 13.6, page 859

1. $f(5, 5) = 75$ 3. $f\left(3, \dfrac{3}{2} \right) = 18$

5. $f\left(\dfrac{2}{3}, 2 \right) = 6$ 7. $f\left(\dfrac{68}{3}, \dfrac{34}{3} \right) = \dfrac{157,216}{27}$

9. $f(\frac{5}{6}, \frac{10}{9}) = \frac{500}{243}$

11. $f(47, 47) = 2209$

13. $f\left(\dfrac{35}{17}, \dfrac{15}{17} \right) = \dfrac{50}{17}$

15. $f\left(\dfrac{146}{3}, \dfrac{146}{3}, \dfrac{146}{3} \right) = \dfrac{3,112,136}{27}$

17. $x = 33,670$, $y = 67,330$; $p = 2,137,700$

19. $x = \dfrac{412}{41}$, $y = \dfrac{510}{41}$, $U(x, y) = \dfrac{5484}{41}$

21. 132.64 by 132.64 by 56.84

23. 75 by 75

25. 14.14 by 14.14 by 7.07

27. Relative minimum: $f(0, 2) = -4$

29. Relative minimum: $f\left(\dfrac{10}{3}, -\dfrac{2}{3} \right) = -\dfrac{28}{3}$

Chapter 13 Test, page 860

1. 10

2.

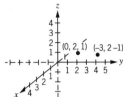

3.

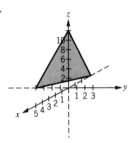

4. $f_{yy} = 8x$

5. $f_{xx} = \dfrac{6}{x^2 + 3y^2} - \dfrac{12x^2}{(x^2 + 3y^2)^2}$

6. $f_y = xe^{xy}$

7. $f_{xy} = 2x + 2y$

8.

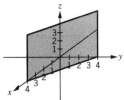

9. The critical point is $(2, -3)$

10. Relative minimum

11.

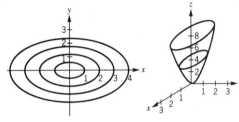

12. $f\left(\dfrac{4}{3}, 2, \dfrac{8}{3}\right) = 64$

Index